LANDOLT-BÖRNSTEIN

Numerical Data and Functional Relationships in Science and Technology

New Series

Editor in Chief: O. Madelung

Group II: Atomic and Molecular Physics

Volume 17

Supplement and Extension to Volume 9

Magnetic Properties of Free Radicals

Subvolume h

Organic Cation Radicals, Bi- and Polyradicals
Index of Substances for Volumes II/1, II/9, and II/17

H.C. Chandra · A.R. Forrester · K. Ishizu · G. Kothe
P. Meier · S.F. Nelsen · H. Ohya-Nishiguchi
M.C.R. Symons · K. Tajima · A. Terahara

Editor: H. Fischer

Springer-Verlag Berlin Heidelberg New York
London Paris Tokyo Hong Kong Barcelona

LANDOLT-BÖRNSTEIN

Zahlenwerte und Funktionen aus Naturwissenschaften und Technik

Neue Serie

Gesamtherausgabe: O. Madelung

Gruppe II: Atom- und Molekularphysik

Band 17

Ergänzung und Erweiterung zu Band 9

Magnetische Eigenschaften freier Radikale

Teilband h

Organische Kation-Radikale, Bi- und Polyradikale
Substanzenverzeichnis für die Bände II/1, II/9 und II/17

H.C. Chandra · A.R. Forrester · K. Ishizu · G. Kothe
P. Meier · S.F. Nelsen · H. Ohya-Nishiguchi
M.C.R. Symons · K. Tajima · A. Terahara

Herausgeber: H. Fischer

Springer-Verlag Berlin Heidelberg New York
London Paris Tokyo Hong Kong Barcelona

ISBN 3-540-51167-9 Springer-Verlag Berlin Heidelberg New York
ISBN 0-387-51167-9 Springer-Verlag New York Berlin Heidelberg

CIP-Titelaufnahme der Deutschen Bibliothek

Zahlenwerte und Funktionen aus Naturwissenschaften und Technik / Landolt-Börnstein. – Berlin; Heidelberg; New York; London; Paris; Tokyo; Hong Kong: Springer. Parallelt.: Numerical data and functional relationships in science and technology. – N.S. teilw. Gesamthrsg.: K.-H. Hellwege. – N.S. teilw. Gesamthrsg.: K.-H. Hellwege; O. Madelung N.S. Gesamthrsg.: O. Madelung. Gruppe 2, Atom- und Molekularphysik.
NE: Landolt, Hans [Begr.]; Hellwege, Karl-Heinz [Hrsg.]; Madelung, Otfried [Hrsg,]; PT Bd. 17. Magnetische Eigenschaften freier Radikale: Erg. u. Erw. zu Bd. 9 / Hrsg.: H. Fischer. Teilbd. h. Organische Kation-Radikale, Bi- und Polyradikale; Substanzenverzeichnis für die Bände II/1, II/9 und II/17 / H.C. Chandra ... – 1990

ISBN 3-540-51167-9 (Berlin ...)
ISBN 0-387-51167-9 (New York ...)

NE: Fischer, Hanns [Hrsg.]; Chandra, H. C. [Mitverf.]

Printed in Germany

Typesetting: Universitätsdruckerei H. Stürtz AG, Würzburg
Printing: Druckhaus Langenscheidt KG, Berlin
Bookbinding: Lüderitz & Bauer-GmbH, Berlin

2163/3020-543210 – Printed on acid-free paper

Preface

Volume II/17h is the eighth and last subvolume of a supplement to the previous compilations II/1 and II/9a–d2 on magnetic properties of free radicals published in 1965 and 1977–1980 which covered the literature up to 1975. Due to the still very rapid development of the field and the inclusion of new subjects the supplement covering the period 1975–1985 had to be split into subvolumes which appeared in fast succession. Together with the volumes published earlier II/17 offers a rather up-to-date and comprehensive survey and collection of data on important chemical intermediates.

The radical species have been grouped according the chemical aspects. The contents of the subvolumes are indicated on the inside of the front cover. Each group of substances has been compiled by experts in the fields. Some small overlap between chapters is intentional to allow a maximum of coherence and conprehensiveness of the display. A comprehensive index of substances is found in this subvolume for the reader's convenience.

We wish to thank the authors for their careful and experienced work and the most agreeable cooperation, the Landolt-Börnstein office, especially Frau H. Weise for the untiringly alert and careful checking of manuscripts and galley proofs, and Springer-Verlag for their cutomary care in the preparation of the volume which ist published without external financial support. It is planned to supplement the volumes in due time.

Zürich, May 1990 **The Editor**

Vorwort

Der vorliegende Band II/17h ist der achte und letzte Teilband eines Ergänzungswerks zu den 1965 bzw. 1977 bis 1980 erschienenen Bänden II/1 und II/9a–d2 über magnetische Eigenschaften freier Radikale, in denen die Literatur bis 1975 berücksichtigt war. Infolge der weiterhin hohen Publikationsrate besteht das Ergänzungswerk, das die Literatur der Periode 1975–1985 umfaßt, erneut aus mehreren Teilbänden, die in rascher Folge erschienen. Weiterhin wird neueren Entwicklungen durch die Aufnahme neuer Substanzklassen Rechnung getragen. Zusammen mit den früher publizierten Bänden wird durch II/17 eine aktuelle und weitgehend umfassende Zusammenstellung aller Meßergebnisse eines aktiven Forschungsgebiets angeboten.

Die radikalischen Spezies sind nach chemischen Kriterien in Gruppen zusammengefaßt, die aus der Übersicht auf der Innenseite des vorderen Einbanddeckels jedes Teilbandes ersichtlich sind. Die einzelnen Gruppen wurden jeweils von Spezialisten bearbeitet. Kleinere Überlappungen der Gebiete wurden bewußt in Kauf genommen, um eine möglichst kohärente und vollständige Darstellung der Fachgebiete zu erreichen. Ein Gesamtverzeichnis der Substanzen in diesem Teilband wird dem Benutzer die Suche erleichtern.

Den Autoren ist für ihre sorgfältige und sachkundige Arbeit und die angenehme Kooperation herzlich zu danken. Dank gebührt weiter der Landolt-Börnstein-Redaktion und hier vor allem Frau H. Weise für die unermüdlich aufmerksame und gründliche Bearbeitung der Originalmanuskripte und Fahnen. Ferner danken wir dem Springer-Verlag für die sorgfältige Fertigstellung des Bandes, der, wie alle Landolt-Börnstein-Bände, ohne externe finanzielle Unterstützung publiziert wird. Wir planen die Herausgabe weiterer Ergänzungen in angemessenen Zeitabständen.

Zürich, im Mai 1990 **Der Herausgeber**

Contents

Magnetic properties of free radicals
Subvolume h: Organic cation radicals, bi- and polyradicals

General introduction

H. Fischer, Physikalisch-Chemisches Institut der Universität Zürich, Switzerland

A Definition and substances . . . 1
B Magnetic properties . . . 1
C Arrangement of the tables . . . 3
D Monographs . . . 3

18 Carbon-oxygen and halogen-centred organic cation radicals

M.C.R. Symons, H.C. Chandra, Department of Chemistry, The University, Leicester, U.K.

18.1 Introduction . . . 5
18.1.1 Radicals included in this chapter . . . 5
18.1.2 Arrangement within sections . . . 5
18.1.3 Method of preparation . . . 5
18.1.4 Review articles . . . 5

18.2 Tables . . . 6
18.2.1 Alkane cations . . . 6
18.2.1.1 Linear alkane cations . . . 6
18.2.1.2 Methyl substituted alkane cations . . . 12
18.2.1.3 Cyclic alkane cations . . . 14
18.2.2 Alkene cations . . . 15
18.2.2.1 Unsubstituted alkene cations . . . 17
18.2.2.2 Monosubstituted alkene cations . . . 19
18.2.2.3 Disubstituted alkene cations . . . 21
18.2.2.4 Tetrasubstituted alkene cations . . . 23
18.2.3 Cyclic alkene cations . . . 25
18.2.3.1 Unsubstituted alkene cations . . . 25
18.2.3.2 Substituted alkene cations . . . 27
18.2.4 Diene cations . . . 28
18.2.4.1 Acyclic diene cations . . . 28
18.2.4.2 Cyclic diene cations . . . 28
18.2.5 Ether cations . . . 36
18.2.5.1 Mono- and di-ether cations . . . 36
18.2.5.2 Cyclic polyether cations . . . 38
18.2.6 Cations of carbonyl derivatives . . . 40
18.2.6.1 Aldehydes . . . 40
18.2.6.2 Acyclic ketones . . . 42
18.2.6.3 Cyclic ketone cations . . . 43
18.2.6.4 Ketene cations . . . 43
18.2.6.5 Dicarbonyl derivatives . . . 44
18.2.6.6 Carbocyclic acid and ester cations . . . 45
18.2.6.7 Diesters . . . 47
18.2.6.8 Cyclic esters (lactones) . . . 48
18.2.6.9 Carbonates . . . 49
18.2.6.10 Amides . . . 49

18.2.7 Dialkylperoxide cations 50
18.2.8 Nitroalkane cations 52
18.2.9 Halogenalkane cations 53
18.2.9.1 Chloroalkanes 53
18.2.9.2 Bromoalkanes 55
18.2.9.3 Iodoalkanes 56

18.3 References for 18.2 57

19 Cation radicals of aromatic hydrocarbons and their derivatives and S-heterocycles

H. Ohya-Nishiguchi, A. Terahara, K. Tajima, K. Ishizu, Department of Chemistry, Faculty of Science, Kyoto University, Kyoto, Japan

19.0 Introduction 59

19.1 Cation radicals of aromatic hydrocarbons and their derivatives 60
19.1.1 Benzenes, excluding $(OR)_n$ and $(SR)_n$ substituents 60
19.1.2 Naphthalenes, excluding $(OR)_n$ and $(SR)_n$ substituents 70
19.1.3 Biphenyls, excluding $(OR)_n$ and $(SR)_n$ substituents 75
19.1.4 Anthracenes, excluding $(OR)_n$ and $(SR)_n$ substituents 76
19.1.5 Further condensed aromatics 77
19.1.6 $(OR)_n$ substituted aromatics 79
19.1.6.1 Benzenes 79
19.1.6.2 Other aromatics 88
19.1.6.3 $(OR)_n$ *and* $(SR)_n$ substituted benzenes 91
19.1.7 $(SR)_n$ substituted aromatics 93
19.1.8 Non-benzeneoid aromatics 98
19.1.9 Annulenes 100
19.1.10 Bridged systems 102
19.1.11 Cyclophanes 106

19.2 *S*-Heterocycles 108

19.3 Miscellaneous compounds 119

19.4 References for 19 120

20 Cation radicals from nitrogen containing compounds

S.F. Nelsen, Department of Chemistry, University of Wisconsin, Madison, Wisconsin, U.S.A.

20.1 Introduction 122

20.2 Non-conjugated amine cation radicals 126
20.2.1 Monoamines and localized polyamine cation radicals 126
20.2.2 Delocalized cation radicals from non-conjugated amines 129

20.3 Heteroatom substituted amines: 3e-π-based 131
20.3.1 Haloamine cation radical 132
20.3.2 Hydroxylamine cation radicals 132
20.3.3 Thiohydroxylamine cation radicals (and other cation radicals with NS bonds 136
20.3.4 Hydrazine cation radicals 142

20.4 1-Azaallyl cation radical systems (and relatives) 154
20.4.1 (Enamine), hydrazone, N-nitrosamine cation radicals 154
20.4.2 Amide cation radicals (and R_2N-π-acceptor relatives) 156

20.5 Amino substituted unsaturated cation radicals 159
20.5.1 "Amino-terminated" systems 159
20.5.1.1 1,4-Enediamines and 2,3-aza analogues 159
20.5.1.2 Anilines (aminobenzenes) 161
20.5.1.3 Phenylenediamines (diaminobenzenes) 164

20.5.1.4 Benzidines (diaminobiphenyls) . . . 168
20.5.1.5 Other amino-terminated systems . . . 172
20.5.2 Hydrazoaromatic cation radicals . . . 176
20.5.2.1 1,4-Dihydropyridazines and viologens . . . 176
20.5.2.1.1 1,4-Dihydropyridazines . . . 176
20.5.2.1.2 Viologens: diazadihyrodipyridines . . . 186
20.5.2.2 Phenoxazines, phenothiazines, dihydro-thioalloxazines . . . 191
20.5.2.3 Biomolecules, including dihydroalloxazines . . . 204
20.5.2.4 Miscellaneous hydrazoaromatics . . . 219

20.6 Imino and more highly oxidized N functional group cation radicals . . . 224
20.6.1 "sp_2" hybridized N: imino compounds . . . 224
20.6.1.1 Derivatives of pyridine . . . 224
20.6.1.2 Two or more imino N atoms . . . 227
20.6.1.3 $-N=O$ compounds . . . 231
20.6.2 "sp" hybridized: N_2, diazo, nitriles and relatives . . . 232
20.6.3 Nitro compounds . . . 237

20.7 References for 20 . . . 240

21 Organic C-, O- and N-centered bi- and polyradicals

G. Kothe, Institut für Physikalische Chemie der Universität Stuttgart, FRG,
P. Meier, Laboratorium für Physikalische Chemie der Eidgenössischen Technischen Hochschule Zürich, Switzerland, present address: Varian GmbH, Darmstadt, FRG

21.0 Introduction . . . 246
21.0.1 Definition . . . 246
21.0.2 Magnetic properties . . . 246
21.0.3 Substances . . . 247
21.0.4 Arrangement of the tables . . . 247
21.0.5 Symbols and abbreviations . . . 248
21.0.6 References for 21.0 . . . 248

21.1 Carbon-centered polyradicals . . . 250
21.1.1 Trimethylenemethane biradicals . . . 250
21.1.2 *m*-Quinodimethane biradicals . . . 251
21.1.3 *p*-Quinodimethane biradicals . . . 252
21.1.4 Naphthoquinodimethane biradicals . . . 253
21.1.5 Cyclopentadienyl biradicals . . . 255

21.2 Oxygen-centered polyradicals . . . 257
21.2.1 Aroxyl biradicals . . . 257
21.2.2 Aroxyl triradicals . . . 260
21.2.3 Oxygen- and carbon-centred biradicals . . . 261
21.2.4 Oxygen- and carbon-centered polyradicals . . . 261

21.3 Nitrogen-centered polyradicals . . . 261
21.3.1 Hydrazyl biradicals . . . 261
21.3.2 Verdazyl biradicals . . . 262
21.3.3 Trisiminomethane biradicals . . . 264
21.3.4 Nitrogen- and carbon-centered biradicals . . . 264
21.3.5 Nitrogen- and oxygen-centered biradicals . . . 265

21.4 References for 21.1···21.3 . . . 268

22 Organic bis- and polynitroxides

A.R. Forrester, Department of Chemistry, University of Aberdeen, U.K.

22.0 Introduction . . . 270
22.0.1 General remarks . . . 270
22.0.2 Nomenclature and classification . . . 270
22.0.3 Reviews . . . 270

22.1 Bisnitroxides 271
22.1.1 Bis(dialkyl nitroxides) 271
22.1.2 Monocyclic bis(dialkyl nitroxides) 271
22.1.3 Bicyclic bis(dialkyl nitroxides) 271
22.1.4 Bis(pyrrolidinyl-1-oxyls) 272
22.1.5 Bis(pyrrol-3-inyl-1-oxyls) 275
22.1.6 Pyrrol-3-inyl-1-oxyls, galvinoxyls 277
22.1.7 Bis(pyrrolid-3-enyl-1-oxyls) 280
22.1.8 Bis(oxazolidinyl-1-oxyls) 281
22.1.9 Bis(imidazolinyl-1-oxyl-3-oxides) 284
22.1.10 Bis(imidazolinyl-1-oxyls) 286
22.1.11 (Pyrrolinyl-1-oxyl)-(pyrrolinyl-1-oxyl-3-oxide) 297

22.2 Bis(piperidinyl-1-oxyls), ·O—N⟨ring⟩—X—⟨ring⟩N—O· 298
22.2.1 X=0 atom 298
22.2.2 X=1 atom 299
22.2.3 X=2 atoms 303
22.2.4 X=3 atoms 304
22.2.5 X=4 atoms 308
22.2.6 X=5 atoms 311
22.2.7 X=6 atoms 311
22.2.8 X=7 atoms 313
22.2.9 X=8 atoms 314
22.2.10 X=9 atoms 315
22.2.11 X=10 atoms 316
22.2.12 X=12 atoms 317
22.2.13 X=14 atoms 318
22.2.14 X=16 atoms 320
22.2.15 X=17 atoms 320

22.3 (Pyrrolinyl-1-oxyl)-(piperidinyl-1-oxyl), ·O—N⟨ring⟩—X—⟨ring⟩N—O· 321

22.4 (Piperidinyl-1-oxyl)-(phenoxyl) 322

22.5 Trisnitroxides 324

22.6 Tetrakisnitroxides 326

22.7 Pentakis- and other poly-nitroxides 327

22.8 Conjugated bisnitroxides 328

22.9 References for 22 329

General symbols and abbreviations 332

Index of substances 334

General introduction

A Definition and substances

In the context of these tables the term free radical means a chemically stable or transient paramagnetic atomic or molecular species which derives its paramagnetism from a single, unpaired valence shell electron. Following this definition the tables cover

a) Atoms and atomic ions in ground and excited ^2S and ^2P states,
b) Diatomic and linear polyatomic molecules in $^2\Sigma$ and $^2\Pi$ states,
c) Polyatomic molecules and molecular ions which arise or may be thought to arise from the break of a single bond of a diamagnetic molecule or molecular ion,
d) Mono-(tri-, penta-, etc.) – negative or – positive ions of neutral organic or inorganic compounds.

Not classified as free radicals are atoms or molecules in ground or excited electronic states with multiplicities larger than two (e.g. O, ^3P; O_2, $^3\Sigma$; N, ^4S; molecules in excited triplet states), transition metal ions and their complexes deriving their paramagnetism exclusively or mainly from *d*- and *f*-electrons and charge transfer complexes. However, a number of polyatomic molecular species are included which do not fulfill the above definition because their properties closely resemble those of structurally very similar free radicals. These are

e) Polyradicals with electron exchange or dipolar couplings not greatly exceeding the Zeeman or hyperfine interactions.
f) Selected transition metal complexes deriving their paramagnetism from free radical ligands and electrons of the center atom.

Within, the individual chapters further deviations from the definition occur and are explained in the appropriate places.

Only compounds with unambiguously verified or at least very plausible structures are included. Papers which only state the presence of free radicals in a sample and do not give detailed structures nor magnetic properties of the radicals have not been reviewed. Also not reviewed are papers which deal exclusively with other topics than experimental determinations of magnetic properties of free radicals. Such papers may however be mentioned in footnotes or as further references at the appropriate entries.

The ordering of the substances classified as free radicals into subclasses is to be seen in the general table of contents. The ordering within the subclasses is explained, where not selfexplanatory, in introductory sections to the individual tables.

The literature was considered for the period of 1975 to 1985 with minor deviations. The literature published before 1964 and in the period of 1964 to 1975 was covered in: Magnetic Properties of Free Radicals, Landolt-Börnstein, New Series, Group II, Vol. 1. Berlin: Springer **1965**. Magnetic Properties of Free Radicals, Landolt-Börnstein, New Series, Group II, Vols. 9a–9d2. Berlin: Springer **1977–80**.

Further information on free radicals can also been found in: Radical Reaction Rates in Liquids, Landolt-Börnstein, New Series, Group II, Vols. 13a–13e. Berlin: Springer **1984–85**.

B Magnetic properties

The magnetic properties of most free radicals can conveniently be represented by parameters describing their interaction with an external magnetic field and the intra-molecular hyperfine interactions, i.e. the parameters $\boldsymbol{g}$ and $\boldsymbol{a}_\lambda$ of the Spin-Hamiltonian

$$\mathscr{H}=\mu_B \boldsymbol{B}_0 \cdot \boldsymbol{g} \cdot \boldsymbol{S} - \sum_\lambda \mu_N g_{N\lambda} \boldsymbol{B}_0 \cdot \boldsymbol{I}_\lambda + \sum_\lambda \boldsymbol{S} \cdot \boldsymbol{a}_\lambda \cdot \boldsymbol{I}_\lambda$$

where μ_B, μ_N, $\boldsymbol{B}_0$, $\boldsymbol{g}$, $\boldsymbol{S}$, $g_{N\lambda}$, $\boldsymbol{a}_\lambda$, $\boldsymbol{I}_\lambda$ are the Bohr magneton, the nuclear magneton, the magnetic induction, the *g*-tensor of the radical, the electron spin operator, the nuclear *g*-factor of nucleus λ, the hyperfine coupling tensor of nucleus λ, and the spin operator of nucleus λ, respectively.

$\boldsymbol{g}$ is symmetric and the mean value of its diagonal elements

$$g=\frac{1}{3}\sum_{i=1}^{3} g_{ii}$$

is called the isotropic *g*-factor. For many radicals g deviates only slightly from the *g*-factor of the free electron

$$g_e=2.0023193134(70)$$

$\boldsymbol{a}_\lambda$, the hyperfine coupling tensor, describes the dipolar and contact interaction between the electron spin momentum and the nuclear spin momentum of nucleus λ of the radical. $\boldsymbol{a}_\lambda$ is most often also symmetric and the mean value

$$a_\lambda = \frac{1}{3} \sum_{i=1}^{3} a_{ii,\lambda}$$

is called the isotropic hyperfine coupling constant or splitting parameter. If a radical contains several nuclei which interact there are several tensors $\boldsymbol{a}_\lambda$. In general their principal axes do not coincide, nor do they with the principal axes of $\boldsymbol{g}$.

For polyatomic radicals in the gas phase the above Spin-Hamiltonian does not apply and four magnetic hyperfine coupling constants a, b, c, d are needed to describe the interaction between a nuclear and the electron spin. These are defined and explained in the introduction to the tables on inorganic radicals. Polyradicals and certain radicals on transition metal complexes have N unpaired electrons located on different molecular segments k. Their Spin-Hamiltonian is

$$\mathscr{H} = \mu_B \sum_k \boldsymbol{B}_0 \cdot \boldsymbol{g}^k \cdot \boldsymbol{S}^k + J \sum_{l>k=1}^{N} \boldsymbol{S}^k \cdot \boldsymbol{S}^l + \boldsymbol{S} \cdot \boldsymbol{D} \cdot \boldsymbol{S} + \sum_{k=1}^{N} \sum_\lambda \boldsymbol{S}^k \cdot \boldsymbol{a}_\lambda^k \cdot \boldsymbol{I}_\lambda^k$$

where the nuclear Zeeman terms are omitted and

$$\boldsymbol{S} = \sum_k \boldsymbol{S}^k .$$

J is the electron exchange parameter and $\boldsymbol{D}$ the zero-field splitting tensor. $\boldsymbol{D}$ is symmetric and traceless, i.e.

$$\sum_{i=1}^{3} D_{ii} = 0$$

and consequently the two zero-field splitting parameters

$$D = \frac{3}{2} D_{33}$$

$$E = \frac{1}{2} (D_{11} - D_{22})$$

completely determine the tensor. J determines the energy separation of different spin states of the N-spin system. For N = 2

$$J = E_{\text{triplet}} - E_{\text{singlet}}$$

and for N = 3

$$\frac{3}{2} J = E_{\text{quartet}} - E_{\text{doublet}} .$$

Further information on the description of N-electron spin systems are found in the introductions of the appropriate chapters.

There are many experimental techniques for the determination of the Spin-Hamiltonian parameters $\boldsymbol{g}$, $\boldsymbol{a}_\lambda$, J, D, E. Often applied are Electron Paramagnetic or Spin Resonance (EPR, ESR), Electron Nuclear Double Resonance (ENDOR) or Triple Resonance, Electron-Electron Double Resonance (ELDOR), Nuclear Magnetic Resonance (NMR), occasionally utilizing effects of Chemically Induced Dynamic Nuclear Polarization (CIDNP), Optical Detections of Magnetic Resonance (ODMR) or Microwave Optical Double Resonance (MODR), Laser Magnetic Resonance (LMR), Atomic Beam Spectroscopy, and Muon Spin Rotation (μSR). The extraction of data from the spectra varies with the methods, the system studied and the physical state of the sample (gas, liquid, unordered or ordered solid). For these procedures the reader is referred to the monographs (D). Further, effective magnetic moments μ_{eff} of free radicals are often obtained from static susceptibilities. In recent years such determinations are rare, but they are mentioned in the tables. A list of references covering the more abundant literature up to 1964 is found in:
Magnetic Properties of Free Radicals, Landolt-Börnstein, New Series, Group II, Vol. 1, Berlin: Springer **1965** and Vols. 9a–9d2, Berlin: Springer **1977–80**.

C Arrangements of the tables

For the display of the data these tables on magnetic properties are devided into chapters each dealing with a specific class of compounds and prepared by authors who are experts in the fields. Each chapter is headed by an introduction which specifies the coverage, the ordering of substances, details of the arrangement, the special general literature and special abbreviations, if necessary. The tables are followed by the references belonging to the chapter. Grossly, the overall arrangement is equal to that of previous volumes on the same topics. A small overlap between chapters has been allowed for reasons of comprehensiveness and consistency of the chapters. An index of all substances covered appears at the end of the last subvolume of the series.

Within the individual chapters the data are arranged in columns in a manner, which, as far as possible, holds for all chapters:

The *first* column describes the structure of the species. It contains the gross formula including charge and, where appropriate, information on the electronic state. Whenever possible a structural formula is also given or a letter or number referring to a structural formula which is displayed elsewhere.

The *second* column briefly describes the method of radical generation and specifies the matrix or solvent in which the radical was studied.

The *third* column states the experimental technique applied to obtain the magnetic properties and the temperature for which the data are valid in Kelvin. 300 normally means an unspecified room temperature.

The *fourth* column refers to the g-tensor. If only one value is given it is the isotropic g-factor. If four values are listed the first three are the principal elements of the diagonal form of g, the fourth denoted by is: is the mean value. For axially symmetric g occasionally only the two principal elements and the isotropic g are listed. Errors are quoted in parentheses after the values in units of the last digit quoted for the value.

In most of the tables the *fifth* column contains the information on the hyperfine interactions. It states the nuclei by their chemical symbols, a left upper index specifying the isotope, if necessary. Numbers preceeding the chemical symbols note the number of equivalent nuclei, i.e. 3H means three equivalent ^1H nuclei. Right hand indices of the symbols or information given in parentheses following the chemical symbols point to positions of the nuclei in the structural formulae. The data are displayed following the symbols. If only one value is given it is the isotropic part of the coupling tensor. If four values are listed the first three are the principal values of the diagonalized form of $\boldsymbol{a}$, the fourth is the isotropic part. Signs are given whereever known, and errors are quoted in parentheses. In the tables on polyradicals the *fifth* column also gives the available information on the exchange and zero-field parameters J, D and E. Further, in some tables where liquid-crystal data are reported column five may give besides the isotropic coupling constant a the shift Δa caused by the partial alignment. It is related to the elements of $\boldsymbol{a}$ by

$$\Delta a = \frac{2}{3} \sum_{i,j} O_{ij} a_{ji}$$

where O_{ij} are the elements of the traceless ordering matrix. In these cases, appropriate entries may also occur in column four. For the extraction of the parameters from the spectra the original literature and the introduction to the individual chapters should be consulted. Finally, for radicals observed in the gas phase the fifth column lists the hyperfine coupling constants a, b, c, d. The general unit of column five is milli-Tesla [mT] with the occasional and well founded exception of Mc/s (MHz) for a few cases. The original literature often quotes coupling constants in Gauss and the conversion is

$$1\ \text{mT} \mathrel{\hat{=}} 10\ \text{Gauss} \mathrel{\hat{=}} 28.0247\ (g/g_e)\ \text{Mc/s}\,.$$

In some footnotes the unit cm^{-1} may be used for some interaction energy terms such as J, D and E with $1\ cm^{-1} \mathrel{\hat{=}} c_0^{-1} \cdot 1$ c/s where c_0 is the vacuum light velocity.

The *sixth* column lists the reference from which the data are taken. This reference is followed by additional and secondary references to the same subject. All references belonging to one chapter are collected in a bibliography at the end of the chapter, the respective pages are referred to at the top of each page.

Throughout the chapters footnotes give additional informations or explanations. A list of general symbols and abbreviations is given at the end of this volume.

D Monographs

Atkins, P. W., Symons, M. C. R.: The Structure of Inorganic Radicals. Amsterdam: Elsevier **1967**.
Ayscough, P. B.: Electron Spin Resonance in Chemistry. London: Methuen **1967**.
Carrington, A., McLauchlan, A. D.: Introduction to Magnetic Resonance. Harper International **1967**.
Gerson, F.: Hochauflösende ESR-Spektroskopie. Weinheim: Verlag Chemie **1967**.

Poole C. P., Jr.: Electron Spin Resonance. New York: Interscience **1967**.
Alger, R. S.: Electron Paramagnetic Resonance. New York: Interscience **1968**.
Kaiser, E. T., Kevan, L.: Radical Ions. New York: Interscience **1968**.
Scheffler, K., Stegmann, H. B.: Elektronenspinresonanz. Berlin, Heidelberg, New York: Springer **1970**.
Geschwind, S., (Editor): Electron Paramagnetic Resonance. New York: Plenum Press **1972**.
Muus, L. T., Atkins, P. W., (Editors): Electron Spin Relaxation in Liquids. New York: Plenum Press **1972**.
Swartz, H. M., Bolton, J. R., Borg, D. C.: Biological Applications of Electron Spin Resonance. New York: Wiley **1972**.
Wertz, J. E., Bolton, J. R.: Electron Spin Resonance. New York: McGraw-Hill **1972**.
Atherton, N. M.: Electron Spin Resonance, Theory and Applications. New York: Halsted **1973**.
Buchachenko, A. L., Wassermann, A. L.: Stable Radicals. Weinheim: Verlag Chemie **1973**.
Kochi, J. K., (Editor): Free Radicals. New York: Wiley **1973**.
Norman, R. O. C., (Editor): Electron Spin Resonance. London: The Chemical Society **1973ff.**
Carrington, A.: Microwave Spectroscopy of Free Radicals. London: Academic Press **1974**.
Ayscough, P. B., (Editor): Electron Spin Resonance. London: The Chemical Society **1977ff.**
Box, H. C.: Radiation Effects, ESR and ENDOR Analysis. New York: Academic Press **1977**.
Muus, L. T., Atkins, P. W., McLauchlan, K. A., Pedersen, J. B., (Editors): Chemically Induced Magnetic Polarization. Dordrecht: Reidel **1977**.
Rånby, B., Rabek, J. F.: ESR Spectroscopy in Polymer Research. Berlin: Springer **1977**.
Slichter, C. P.: Principles of Magnetic Resonance. Berlin: Springer **1978**.
Harriman, J. E.: Theoretical Foundations of Electron Spin Resonance. New York: Academic Press **1978**.
Symons, M. C. R.: Chemical and Biochemical Aspects of Electron Spin Resonance Spectroscopy. New York: van Nostrand-Reinhold **1978**.
Dorio, M. M., Freed, J. H., (Editors): Multiple Electron Resonance Spectroscopy. New York: Plenum Press **1979**.
Kevan, L., Schwartz, R.: Time Domain Electron Spin Resonance. New York: Wiley **1979**.
Shulman, R. G., (Editor): Biological Applications of Magnetic Resonance. New York: Academic Press **1979**.
Bertini, I., Drago, R. S.: ESR and NMR of Paramagnetic Species in Biological and Related Systems. Hingham: Kluver Boston **1980**.
Gordy, W.: Theory and Applications of Electron Spin Resonance. New York: Wiley **1980**.
Carrington, A., Hudson, A., McLauchlan, A. D.: Introduction to Magnetic Resonance, 2nd ed. New York: Chapman and Hall, **1983**.
Weltner, W., Jr.: Magnetic Atoms and Molecules. New York: van Nostrand-Reinhold **1983**.
Poole, C. P.: Electron Spin Resonance, 2nd ed. New York: Wiley **1983**.
Walker, D. C.: Muon and muonium Chemistry. Cambridge: Cambridge University Press **1983**.
Salikhov, K. M., Molin, Yu. N., Sagdeev, R. Z., Buchachenko, A. L.: Spin Polarization and Magnetic Effects in Radical Reactions. Amsterdam: Elsevier **1984**.

18 Carbon-oxygen and halogen centred organic cation radicals

18.1 Introduction

18.1.1 Radicals included in this chapter

The major section (18.2.1) is concerned with alkane cations. Since these had not been studied by ESR spectroscopy at the time of the previous report, we have tried to make a complete survey. Since the first cation was studied $[(CH_3)_3C-C(CH_3)_3]^{\cdot +}$ there has been a spate of studies, dominated by the fine work of the late Machio Iwasaki and his coworkers.

Other carbon-centred radical-cations include those of alkenes, etc., but exclude aromatic cations [see chapter 19].

These are followed by oxygen-centred radical-cations (18.2.2), starting with ether cations, and continuing with carbonyl cations and peroxyl cations. Finally, halogen-centred cations are listed (18.2.3).

18.1.2 Arrangement within sections

In general, we have tried to work from the simple to the complex, treating linear systems before branched-chain systems and cyclic systems.

18.1.3 Methods of preparation

The method of preparation for the vast majority of the radical-cations listed here is exposure of the parent neutral molecules as very dilute solutions in rigid halogenated solvents to ionizing radiation. Solvents include CCl_4, CCl_2F-CF_2Cl, C_2F_6, SF_6, etc., but by far the most favoured is $CFCl_3$. This is in part because the radical anions, always derived from the solvent rather than the solute give only a broad single feature at low temperatures, which does not interfere seriously with the normally narrow lines from solute radical-cations. Thus for many studies, spectroscopic interpretation has been relatively simple. Nevertheless, most spectra remain anisotropic up to the melting point of the solvents, and hence interpretation suffers all the usual disadvantages associated with solidstate "powder" spectra.

Various complications have been encountered in these studies. These include the appearance of extra hyperfine splitting associated with ^{19}F or $^{35/37}Cl$ nuclei, and unimolecular rearrangements or decomposition of the parent radical-cations. These changes are indicated in the tables.

18.1.4 Review articles

1 M.C.R. Symons: Radical Cations in Condensed Phases. Chem. Soc. Rev **13** (1984) 393.
2 T. Shida, E. Haselbach and T. Bally: Organic Radical Ions in Rigid Systems. Acc. Chem. Res. **17** (1984) 180.
3 L.B. Knight, Jr.: ESR Investigations of Molecular Cation Radicals in Neon Matrices at 4 K: Generation, Trapping, and Ion-Neutral Reactions. Acc. Chem. Res. **19** (1986) 313.
4 J.L. Courtneidge and A.G. Davies: Hydrocarbon Radical Cations. Acc. Chem. Res. **20** (1987) 90.
5 M. Shiotani: Maqn. Res. Rev. **12** (1987) 333.

18.2 Tables

18.2.1 Alkane cations

18.2.1.1 Linear alkane cations

Substance	Generation/ Matrix or Solvent	Method/ T [K]	g-Factor	a-Value [mT]	Ref./ add. Ref.
$[CH_4]^{\cdot +}$ 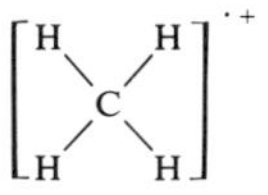	Open-tube neon discharge photoionization; electron bombardment; high energy neutral-atom bombardment of Ne matrix/ Ne	EPR/ 4	2.0039	4H, is: 5.48	84Kni1
$[CH_2D_2]^{\cdot +}$ 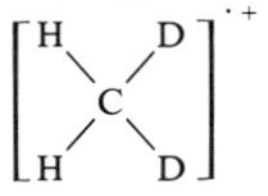	Open-tube neon discharge photoionization; electron bombardment; high energy neutral-atom bombardment of Ne matrix/ Ne	EPR/ 4	2.0029	2H, is: 12.17 2D, is: 0.222	84Kni
$[C_2H_6]^{\cdot +}$	γ-irr. of C_2H_6/	EPR/			82Tor1, 81Iwa1
	SF_6	4	–	2H, is: 15.25	
		77	–	6H, is: 5.03	
	C_2F_6	4	–	2H, is: 15.31	
$[C_2H_5D]^{\cdot +}$	γ-irr. of C_2H_5D/	EPR/			84Iwa1
	SF_6	4.2	–	2H, is: 15.23	
		77	–	2H, is: 14.80	
$[C_2H_4D_2]^{\cdot +}$	γ-irr. of $C_2H_4D_2$/	EPR/			84Iwa1
	SF_6	4.2	–	2H, is: 15.22	
		77	–	2H, is: 14.80	

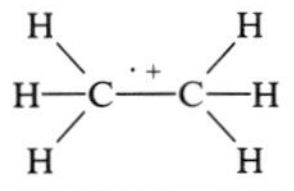

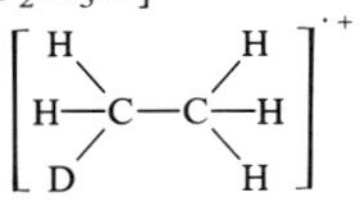

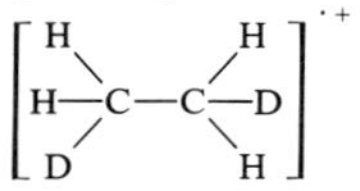

Substance	Generation/ Matrix or Solvent	Method/ T [K]	g-Factor	a-Value [mT]	Ref./ add. Ref.
$[C_2H_3D_3]^{\cdot +}$ 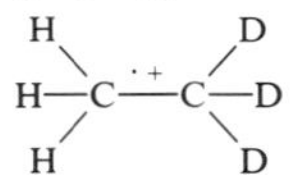	γ-irr. of $C_2H_3D_3$/ SF_6	EPR/ 4	–	1H, is: 14.05 2H, is: 0.9 1D, is: 2.19	82Tor1
$[C_2H_2D_4]^{\cdot +}$ 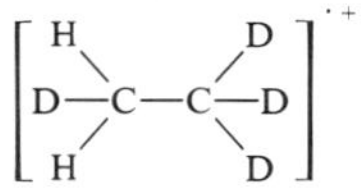	γ-irr. of $C_2H_2D_4$/ SF_6	EPR/ 4.2 77	 – –	 1H, is: 14.10 1D, is: 2.20 1H, is: 13.90 1D, is: 2.00	84Iwa1
$[C_2D_6]^{\cdot +}$ 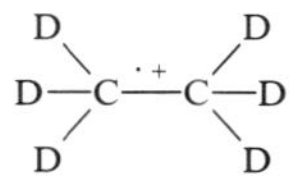	γ-irr. of C_2D_6/ SF_6	EPR/ 4	–	2D, is: 2.3	82Tor1
$[C_3H_8]^{\cdot +}$ 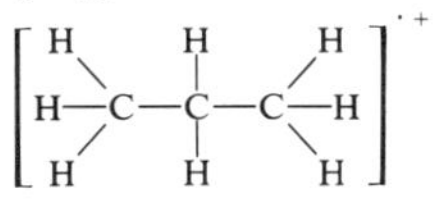	γ-irr. of C_3H_8/ SF_6 $CFCl_2CF_2Cl$ $CFCl_3$	EPR/ 4 77 4 77	 – – – –	 2H, is: 9.8 2H, is: 9.5 2H, is: 10.55 4H, is: 5.25 2H, is: 10.00 4H, is: 5.2	82Tor1, 84Iwa1
$[C_3H_6D_2]^{\cdot +}$ 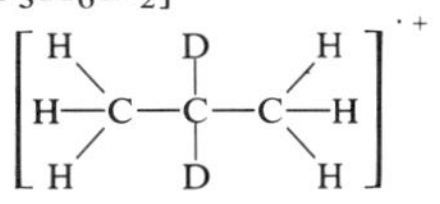	γ-irr. of $C_3H_6D_2$/ SF_6	EPR/ 4	–	2H, is: 9.88	82Tor1
$[C_3H_2D_6]^{\cdot +}$ 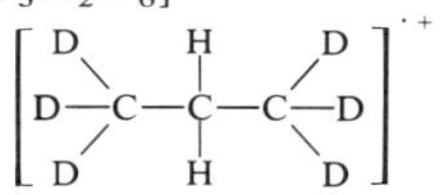	γ-irr. of $C_3H_2D_6$/ SF_6	EPR/ 77	–	2D, is: 1.48	82Tor1
$[C_4H_{10}]^{\cdot +}$ $[CH_3—CH_2—CH_2—CH_3]^{\cdot +}$ *(continued)*	γ-irr. of n-C_4H_{10}/ $CFCl_3$	EPR/ 150	2.0037	2H, is: 7.56 4H, is: 0.56	81Wan1

Substance	Generation/ Matrix or Solvent	Method/ T [K]	g-Factor	a-Value [mT]	Ref./ add. Ref.
$[C_4H_{10}]^{\cdot+}$ *(continued)*	γ-irr. of n-C_4H_{10}/	EPR/			82Tor1
	$CFCl_2CF_2Cl$	77	–	2H, is: 6.13	
	n-C_4F_{10}	77	–	2H, is: 6.3	
				4H, is: 0.8	
$[C_4H_6D_4]^{\cdot+}$	X-irr./	EPR/			87Lin1
$[\overset{1}{C}D_2H\overset{2}{C}H_2\overset{3}{C}H_2\overset{4}{C}D_2H]^{\cdot+}$	CF_3CCl_3 (i) [1]	4.2	2.0037	2H(1, 4) is: 6.10	
		77	–	2H(1, 4) is: 6.12	
		122	–	2H(1, 4) is: 5.97	
	(ii) [1]	4.2	–	H, D(1, 4): 6.08	
		77	–	H, D(1, 4): 6.12	
		122	–	H, D(1, 4): 5.88	
$[C_5H_{12}]^{\cdot+}$	γ-irr. of n-C_5H_{12}/	EPR/			82Tor1
$[\underset{1}{H_3C}—\underset{2}{CH_2}—\underset{3}{CH_2}—\underset{4}{CH_2}—\underset{5}{CH_3}]^{\cdot+}$	$CFCl_2CF_2Cl$ (i) [2]	77	–	$2H(2)_2$ is: 5.8	
				$2H(2)_w$ is: 9.0	
	(ii) [2]	–	–	2H(1) is: 5.7	
	X-irr./	EPR/			85Dol1
	CF_3CCl_3 (i) [3]	77	–	2H(1, 5) is: 5.7	
				6H(2, 3, 4) is: 0.8	
		135	–	2H(2, 4) is: 8.5	
				6H(1, 5) is: 1.7	
	$CF_2ClCFCl_2$ (ii)	77	–	2H(1, 5) is: 5.7	
$[C_5H_{10}D_2]^{\cdot+}$	X-irr./	EPR/			85Dol1
$[\underset{1}{C}H_3\underset{2}{C}H_2\underset{3}{C}D_2\underset{4}{C}H_2\underset{5}{C}H_3]^{\cdot+}$	CF_3CCl_3 (i) [3]	77	–	2H(1, 5) is: 5.7	
				4H(2, 4) is: 0.8	
		135	–	2H(2, 4) is: 8.5	
				6H(1, 5) is: 1.7	
	$CF_2ClCFCl_2$ (ii)	77	–	2H(1, 5) is: 5.7	

[1]) (i) and (ii) are H—H and H—D rotational isomers, respectively.
[2]) (i) and (ii) are set of constants for *gauche* and extended conformers, respectively.
[3]) At 77 K and 135 K two different conformers were identified.

Substance	Generation/ Matrix or Solvent	Method/ T [K]	g-Factor	a-Value [mT]	Ref./ add. Ref.
$[C_5H_8D_4]^{\cdot+}$	X-irr./	EPR/			85Dol1
$[CH_3CD_2CH_2CD_2CH_3]^{\cdot+}$ (1 2 3 4 5)	CF_3CCl_3 (i) [3]	77	–	H(5) is: 4.5	
		135	–	2D(2, 4) is: 1.4	
				6H(1, 5) is: 1.7	
	$CF_2ClCFCl_2$ (ii)	77	–	2H(1, 5) is: 5.7	
$[C_5H_6D_6]^{\cdot+}$	X-irr./	EPR/			85Dol1
$[CD_3CH_2CH_2CH_2CD_3]^{\cdot+}$ (1 2 3 4 5)	CF_3CCl_3 (i) [3]	77	–	H(2) is: 7.4	
				H(4) is: 2.6	
		135	–	2H(2, 4) is: 8.5	
	$CF_2ClCFCl_2$ (ii)	77	–	2D(1, 5) is: 0.8	
$[C_6H_{14}]^{\cdot+}$	γ-irr. of n-C_6H_{14}/	EPR/	2.0031	2H, is: 4.36	81Wan1
$[CH_3CH_2CH_2CH_2CH_2CH_3]^{\cdot+}$	$CFCl_3$	100		2H, is: 0.39	
	γ-irr. of n-C_6H_{14}/	EPR/			82Tor1
	$CFCl_2CF_2Cl$	77	–	2H, is: 4.10	
	$CFCl_3$	77	–	2H, is: 4.40	
				8H, is: 0.41	
	X-irr. of C_6H_{14}/	EPR/			85Dol2
	CF_3CCl_3 (i) [4]	77	–	2H, is: 4.1	
				8H, is: 0.4	
	(ii) [4]			2H, is: 10.0	
				4H, is: 2.2	
	$CF_2ClCFCl_2$ (i)	77	–	2H, is: 4.1	
				8H, is: 0.4	
	(ii)			1H, is: 7.4	
				1H, is: 5.3	
	$CFCl_3$ (i)	77	–	2H, is: 4.4	
				8H, is: 0.4	

[3]) At 77 K and 135 K two different conformers were identified.
[4]) (i) and (ii) are different conformers.

Substance	Generation/ Matrix or Solvent	Method/ T [K]	g-Factor	a-Value [mT]	Ref./ add. Ref.
$[C_6H_{10}D_4]^{\cdot +}$	X-irr. of $C_6H_{10}D_4$/	EPR/			85Dol2
$[CH_3CH_2CD_2CD_2CH_2CH_3]^{\cdot +}$	CF_3CCl_3	77	–	2H, is: 4.1	
				4H, is: 0.3	
	$CF_2ClCFCl_2$ (i) [4])	77	–	2H, is: 4.1	
	(ii) [4])			1H, is: 7.4	
				1H, is: 5.3	
	$CFCl_3$	77	–	2H, is: 4.1	
				4H, is: 0.4	
$[C_6H_8D_6]^{\cdot +}$	X-irr. of $C_6H_8D_6$/	EPR/			85Dol2
$[CD_3CH_2CH_2CH_2CH_2CD_3]^{\cdot +}$	CF_3CCl_3 (i) [5])	77	–	2D, is: 0.65	
				4H, is: 0.65	
	(ii) a			2H, is: 8.5	
				1H, is: 5.5	
				1H, is: 1.8	
	b			2H, is: 11.0	
				1H, is: 1.8	
	c			2H, is: 6.0	
				1H, is: 1.8	
	$CF_2ClCFCl_2$ (i)	77	–	2D, is: 0.65	
				4H, is: 0.65	
	(ii)			1H, is: 7.8	
	$CFCl_3$	77	–	2D, is: 0.65	
				4H, is: 0.65	
$[C_7H_{16}]^{\cdot +}$	γ-irr. of n-C_7H_{16}/	EPR/	–	2H(1) is: 3.0	82Tor1
$[H_3CCH_2CH_2CH_2CH_2CH_2CH_2CH_3]^{\cdot +}$ [6])	$CFCl_2CF_2Cl$	77			
	X-irr. of C_7H_{16}/	EPR/			85Dol2
	CF_3CCl_3	77	–	2H, is: 3.0	
				10H, is: 0.4	
	$CF_2ClCFCl_2$	77	–	2H, is: 2.9	
	$CFCl_3$	77	–	2H, is: 3.0	

[4]) (i) and (ii) are different conformers.
[5]) (i) and (ii) are different conformers and a, b, c are alternative assignments.
[6]) Only extended conformer was analyzed.

Substance	Generation/ Matrix or Solvent	Method/ T [K]	g-Factor	a-Value [mT]	Ref./ add. Ref.
$[C_7H_{10}D_6]^{\cdot+}$	X-irr. of $C_7H_{10}D_6$/	EPR/			85Dol2
$[CD_3CH_2CH_2CH_2CH_2CH_2CD_3]^{\cdot+}$	CF_3CCl_3	77	–	2D, is: 0.4	
				10H, is: 0.4	
	$CF_2ClCFCl_2$	77	–	2D, is: 0.46	
				10H, is: 0.4	
	$CFCl_3$	77	–	2D, is: 0.4	
				10H, is: 0.4	
$[C_8H_{18}]^{\cdot+}$	γ-irr. of C_8H_{18}/	EPR/			82Tor1
$[H_3CCH_2CH_2CH_2CH_2CH_2CH_2CH_3]^{\cdot+}$	$CF_2ClCFCl_2$ (i) [7])	77	–	2H(2): 2.2	
				$2H(2)_w$: 4.1	
	(ii) [7])	–	–	2H(1): 2.2	
	X-irr. of C_8H_{18}/	EPR/			85Dol2
	CF_3CCl_3	77	–	2H(1, 8) is: 2.4	
	$CF_2ClCFCl_2$	77	–	2H(1, 8) is: 2.4	
	$CFCl_3$ (i) [8])	77	–	2H(1, 8) is: 2.2	
	(ii) [8])			1H(2) is: 4.1	
				1H(8) is: 2.2	
$[C_8H_{12}D_6]^{\cdot+}$	X-irr. of $C_8H_{12}D_6$/	EPR/			85Dol2
$[CD_3CH_2CH_2CH_2CH_2CH_2CH_2CD_3]^{\cdot+}$	CF_3CCl_3 (i) [8])	77	–	unresolved	
	(ii)			1H, is: 4.1	
	$CF_2ClCFCl_2$	77	–	unresolved	
	$CFCl_3$ (i)	77	–	unresolved	
	(ii)			1H, is: 4.1	
$[C_9H_{20}]^{\cdot+}$	γ-irr. of n-C_9H_{20}/	EPR/	–	2H, is: 1.7	82Tor1
$[H_3CCH_2CH_2CH_2CH_2CH_2CH_2CH_2CH_3]^{\cdot+}$ [9])	$CF_2ClCFCl_2$	77			

[7]) (i) and (ii) are constants for *gauche* and extended conformers, respectively.
[8]) (i) and (ii) are different conformers.
[9]) Only extended conformer was analyzed.

Substance	Generation/ Matrix or Solvent	Method/ T [K]	g-Factor	a-Value [mT]	Ref./ add. Ref.
18.2.1.2 Methyl substituted alkane cations					
$[C_4H_{10}]^{\cdot+}$ $[(H_3C)_3C{-}H]^{\cdot+}$	γ-irr. of C_4H_{10}/ SF_6 $CFCl_2CF_2Cl$ $CFCl_3$	EPR/ 4 4 77 4 77	 – – – – –	 2H, is: 5.80 2H, is: 5.25 1H, is: 25.00 3H, is: 4.75 1H, is: 25.12 2H, is: 5.50 3H, is: 4.88	82Tor1
$[C_4H_9D]^{\cdot+}$ $[(H_3C)_3C{-}D]^{\cdot+}$	γ-irr. of C_4H_9D/ SF_6 $CFCl_3$	EPR/ 4 4	 – –	 2H, is: 5.78 2H, is: 5.42	82Tor1
$[C_4HD_9]^{\cdot+}$ $[(D_3C)_3C{-}H]^{\cdot+}$	γ-irr. of C_4HD_9/ SF_6 $CFCl_3$	EPR/ 4 4	 – –	 H: unresolved single line 2D, is: 0.25 1H, is: 1.8	82Tor1
$[C_5H_{12}]^{\cdot+}$ $[H_3C{-}C(CH_3)_2{-}CH_3]^{\cdot+}$	γ-irr. of C_5H_{12} (neopentane)/ SF_6 $CFCl_2CF_2Cl$ $CFCl_3$	EPR/ 4 4 4	 – – –	 3H, is: 4.0 3H, is: 3.98 3H, is: 4.07	82Tor1
$[C_6H_{14}]^{\cdot+}$ $[H_3C{-}CH(CH_3){-}CH(CH_3){-}CH_3]^{\cdot+}$	γ-irr. of C_6H_{14}/ $CFCl_3$	EPR/ 145	2.0033	4H, is: 4.10	81Wan1
$[C_6H_{14}]^{\cdot+}$ $[\overset{1}{C}H_3\overset{2}{C}H_2\overset{3}{C}H(\overset{3'}{C}H_3)\overset{4}{C}H_2\overset{5}{C}H_3]^{\cdot+}$	γ-irr./ $CFCl_2CF_2Cl$	EPR/ 77	–	3H(2, 5, 3′) is: 4.5	86Tor1
	γ-irr./ $CF_2ClCFCl_2$	EPR/ 77	–	3H(2, 5, 3′) is: ≃4.73	87Oht1

Substance	Generation/ Matrix or Solvent	Method/ T [K]	g-Factor	a-Value [mT]	Ref./ add. Ref.
$[C_6H_{13}D]^{\cdot +}$ $[CH_3CH_2CD(CH_3)CH_2CH_3]^{\cdot +}$ 1 2 3 3′ 4 5	γ-irr./ $CF_2ClCFCl_2$	EPR/ 77	–	3H(2, 5, 3′) is: 4.73	87Oht1
$[C_6H_{11}D_3]^{\cdot +}$ $[CH_3CH_2CH(CD_3)CH_2CH_3]^{\cdot +}$ 1 2 3 3′ 4 5	γ-irr./ $CF_2ClCFCl_2$	EPR/ 77	–	3D(3′) is: 5.13 H(2, 5) is: 5.13	87Oht1
$[C_6H_{10}D_4]^{\cdot +}$ $[CH_3CD_2CH(CH_3)CD_2CH_3]^{\cdot +}$ 1 2 3 3′ 4 5	γ-irr./ $CF_2ClCFCl_2$	EPR/ 77	–	D, is: ≈4.45	87Oht1
$[C_6H_8D_6]^{\cdot +}$ $[CD_3CH_2CH(CH_3)CH_2CD_3]^{\cdot +}$ 1 2 3 3′ 4 5	γ-irr./ $CF_2ClCFCl_2$	EPR/ 77	–	D, is: ≈4.6	87Oht1
$[C_7H_{16}]^{\cdot +}$ $[CH_3CH_2CH(CH_3)CH_2CH_2CH_3]^{\cdot +}$ 1 2 3 3′ 4 5 6	γ-irr./ $CF_2ClCFCl_2$	EPR/ 77	–	H(2) is: ≃5.71 H(3′) is: ≃3.91 H(5) is: ≃6.19	87Oht1
$[C_7H_{15}D]^{\cdot +}$ $[CH_3CH_2CD(CH_3)CH_2CH_2CH_3]^{\cdot +}$ 1 2 3 3′ 4 5 6	γ-irr./ $CF_2ClCFCl_2$	EPR/ 77	–	H(2) is: ≃5.29 H(3′) is: ≃3.93 H(5) is: ≃4.99	87Oht1
$[C_7H_{13}D_3]^{\cdot +}$ $[CH_3CH_2CH(CD_3)CH_2CH_2CH_3]^{\cdot +}$ 1 2 3 3′ 4 5 6	γ-irr./ $CF_2ClCFCl_2$	EPR/ 77	–	H(2) is: ≃5.71 D(3′) is: ≃3.91 H(5) is: ≃6.19	87Oht1
$[C_7H_{11}D_5]^{\cdot +}$ $[CH_3CD_2CD(CH_3)CD_2CH_2CH_3]^{\cdot +}$ 1 2 3 3′ 4 5 6	γ-irr./ $CF_2ClCFCl_2$	EPR/ 77	–	D(2) is: ≃5.71 H(3′) is: ≃3.91 H(5) is: ≃6.19	87Oht1
$[C_7H_{10}D_6]^{\cdot +}$ $[CD_3CH_2CH(CH_3)CH_2CH_2CD_3]^{\cdot +}$ 1 2 3 3′ 4 5 6	γ-irr. of $C_7H_{10}D_6$/ $CF_2ClCFCl_2$	EPR/ 77	–	H(2) is: 5.71 H(3′) is: 3.91 H(5) is: 6.19	87Oht1
$[C_7H_{10}D_6]^{\cdot +}$ $[CH_3CD_2CH(CH_3)CD_2CD_2CH_3]^{\cdot +}$ 1 2 3 3′ 4 5 6	γ-irr. of $C_7H_{10}D_6$/ $CF_2ClCFCl_2$	EPR/ 77	–	H(1) is: 5.29 H(3′) is : 3.91 D(4) is: −0.9	87Oht1

Substance	Generation/ Matrix or Solvent	Method/ T [K]	g-Factor	a-Value [mT]	Ref./ add. Ref.
$[C_7H_9D_7]^{\cdot +}$ $[\underset{1}{CH_3}\underset{2}{CD_2}\underset{3}{CD}(\underset{3'}{CH_3})\underset{4}{CD_2}\underset{5}{CD_2}\underset{6}{CH_3}]^{\cdot +}$	γ-irr. of $C_7H_9D_7$/ $CF_2ClCFCl_2$	EPR/ 77	–	H(1) is: ≃5.29 H(3′) is: ≃3.91 D(4) is: ≃−0.9	87Oht1
$[C_8H_{18}]^{\cdot +}$ $[H_3C—C(CH_3)_2—C(CH_3)_2—CH_3]^{\cdot +}$	γ-irr. of C_8H_{18}/ CCl_4 CBr_4	EPR/ 77 77	 – –	 6H, is: 3.20 12H, is: 0.45	79Sym1
	γ-irr. of C_8H_{18}/ $CFCl_3$	EPR/ 77	2.0031	6H, is: 2.9	80Wan1
	γ-irr. of C_8H_{18}/ $CFCl_3$	EPR/ 77	–	6H, is: 2.90 12H, is: 0.42	81Shi1
	γ-irr. of C_8H_{18}/ $CFCl_2CF_2Cl$ $CFCl_3$	EPR/ 77 77	 – –	 6H, is: 2.9 6H, is: 2.88 12H, is: 0.38	82Tor1
$[C_{10}H_{22}]^{\cdot +}$ $[CH_3CH_2CH_2CH(CH_3)CH_2CH_2CH_2CH_2CH_3]^{\cdot +}$	γ-irr. of $C_{10}H_{22}$/ $CF_2ClCFCl_2$	EPR/ 77	–	ΔH: 4.8	87Ich1

18.2.1.3 Cyclic alkane cations

Substance	Generation/ Matrix or Solvent	Method/ T [K]	g-Factor	a-Value [mT]	Ref./ add. Ref.
$[C_3H_6]^{\cdot +}$ $[cyclo\text{-}(CH_2)_3]^{\cdot +}$	γ-irr. of C_3H_6/ $CFCl_3$	EPR/ 77	2.0035	6H, is: ≈0.5	83Oht1
	γ-irr. of C_3H_6/ SF_6 $CFCl_2CF_2Cl$ $CFCl_3$	EPR/ 4.2 4.2 4.2	 – 2.004 2.0039; 2.0060; 2.0023; is: 2.0040	 singlet 4H(α) is: −1.25 2H(β) is: 2.10 4H(α): −0.28; −1.23; −1.79; is: −1.10 2H(β): 2.72; 2.30; 2.18; is: 2.40	83Iwa1

(continued)

Substance	Generation/ Matrix or Solvent	Method/ T [K]	g-Factor	a-Value [mT]	Ref./ add. Ref.
$[C_3H_6]^{\cdot+}$ *(continued)*	γ-irr. of C_3H_6/ SF_6 $CFCl_2CF_2Cl$ $CFCl_3$	EPR/ 77 77 77 140	 — — — —	 singlet singlet singlet singlet	83Iwa1
$[C_3H_6]^{\cdot+}$ $H_2C—CH_2—CH_2^{+}$	γ-irr. of C_3H_6/ $CFCl_2CF_2Cl$	EPR/ 108	2.0028	2H(β) is: 3.02 2H(α) is: 2.24	84Qin1
$[C_3D_6]^{\cdot+}$ $D_2C—CD_2—CD_2^{+}$	γ-irr. of C_3D_6/ $CFCl_2CF_2Cl$	EPR/ 110	2.0028	2D(β) is: 0.464 2D(α) is: 0.344	84Qin1
$[C_5H_{10}]^{\cdot+}$ $[H_3CHC^{1}—C^{3}H_2—C^{2}HCH_3]^{\cdot+}$ (cyclopropane ring)	γ-irr. of C_5H_{10}/ $CFCl_3$ (i) [10] (ii) [10]	EPR/ 150 150	 2.0041 2.0040	 2H(1, 2) is: 1.04 2H(3) is: 2.05 6H(CH_3) is: 2.05 2H(1, 2) is: 1.19 2H(3) is: 2.18 6H(CH_3) is: 2.18	86Qin1
$[C_6H_{12}]^{\cdot+}$ $[H_3CHC^{1}—C^{3}H_2—C^{2}(CH_3)_2]^{\cdot+}$ (cyclopropane ring)	γ-irr. of C_6H_{12}/ $CFCl_3$	EPR/ 150	2.0040	H(1) is: 0.98 2H(3) is: 1.79 6H(2) is: 2.06 3H(1) is: 1.45	86Qin1
$[C_7H_{14}]^{\cdot+}$ $[(CH_3)_2C^{1}—C^{3}H_2—C^{2}(CH_3)_2]^{\cdot+}$ (cyclopropane ring)	γ-irr. of C_7H_{14}/ $CFCl_3$ CF_2ClCCl_3	EPR/ 145 155	 2.0033 2.0033	 12H(1, 2) is: 1.50 2H(3) is: 1.87 12H(1, 2) is: 1.50 2H(3) is: 1.87	86Qin1, 84Qin2
$[C_7H_{14}]^{\cdot+}$ $H_3C(CH_3)C\cdot^{+}—CH_2—C(CH_3)_2$ (structure with H, H; CH_3, CH_3, CH_3)	γ-irr. of 1,1,2,2-tetra- methylcyclopropane/ $CF_2ClCFCl_2$	EPR/ 109 117	 2.0032 2.0029	 2H, is: 1.87 12H, is: 1.49 2H, is: 1.17 6H, is: 2.33	84Qin2

(continued)

[10]) (i) and (ii) are *cis* and *trans* forms, respectively.

Substance	Generation/ Matrix or Solvent	Method/ T [K]	g-Factor	a-Value [mT]	Ref./ add. Ref.
$[C_7H_{14}]^{\cdot +}$ *(continued)*	γ-irr. of C_7H_{14}/	EPR/			84Qin2
	$CF_2ClCFCl_2$	109	2.0032	2H(3) is: 1.87 12H(1, 2) is: 1.49	
	CD_3CCl_3	145	2.0034	2H(3) is: 1.81 12H(1, 2) is: 1.47	
	$CFCl_3$	145	2.0033	2H(3) is: 1.87 12H(1, 2) is: 1.50	
$[C_4H_8]^{\cdot +}$	γ-irr. of C_4H_8/ $CFCl_3$	EPR/ 77	—	8H, is: 1.6	83Oht1
$[C_5H_{10}]^{\cdot +}$	γ-irr. of C_5H_{10}/ CF_3CCl_3	EPR/			83Tab1
		6	2.0037	2H(3) is: 2.24	
		113	2.0037	10H, is: 0.63	
	γ-irr. of C_5H_{10}/ $CFCl_3$	EPR/ 77	2.0029	H(2) is: 0.77	83Oht1
$[C_6H_{12}]^{\cdot +}$	X-irr of c-C_6H_{12}/ CF_3CCl_3	EPR/ 11	—	1H, is: 8.4 2H, is: 3.9 2H, is: 1.3	85Lun1
	γ-irr. of C_6H_{12}/ CF_3CCl_3	EPR/ 141	2.0069	6H, is: 4.3	83Tab1
$[C_7H_{12}]^{\cdot +}$	X-irr. of C_7H_{12} (norbornane)/	EPR/			83Tor1
	$CFCl_2CF_2Cl$	4.2	—	4H, is: 6.51	
	$CFCl_3$	4.2	—	4H, is: 6.68	

Substance	Generation/ Matrix or Solvent	Method/ T [K]	g-Factor	a-Value [mT]	Ref./ add. Ref.
$[C_{11}H_{18}]^{\cdot +}$ 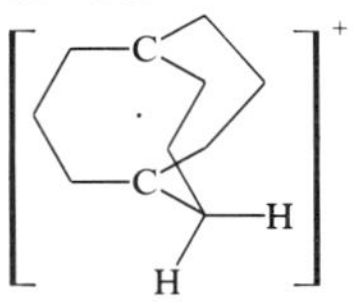	γ-irr. of $C_{11}H_{18}$/ $CFCl_3:CCl_4:CBr_4$	EPR/ 77	–	6H, is: 1.7	81Ald1

18.2.2 Alkene cations

18.2.2.1 Unsubstituted alkene cations

Substance	Generation/ Matrix or Solvent	Method/ T [K]	g-Factor	a-Value [mT]	Ref./ add. Ref.
$[C_2H_4]^{\cdot +}$ 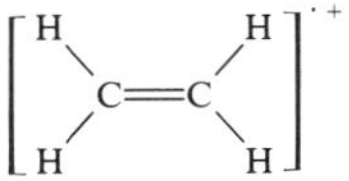 [1]	γ-irr. of C_2H_4/ CCl_2FCClF_2	EPR/ 105	2.0030	4H, is: 2.33	84Shi1
$[C_2D_4]^{\cdot +}$ 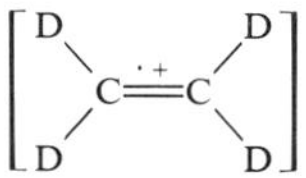 [1]	γ-irr. of C_2D_4/ CCl_2FCClF_2	EPR/ 105	2.0030	4D, is: 0.36	84Shi1
$[C_2H_4]^{\cdot +}$ 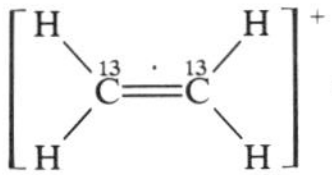 [1]	γ-irr. of $^{13}C_2H_4$/ CCl_2FCClF_2	EPR/ 105	–	2 ^{13}C: 3.25; 0.8; 0.8; is: 1.61 4H, is: 2.33	84Shi1
$[C_3H_6]^{\cdot +}$ $[CH_3CH{=}CH_2]^{\cdot +}$ *(continued)*	Warming up to 100···140 K, the cation of $CH_3CH{=}CH_2$ in $CFCl_3$, γ-irr. of $CH_3CH{=}CH_2$/ $CFCl_3$	EPR/ 77	–	1H(CH_3) is: 4.7 2H(CH_3) is: 2.35 1H(CH_2) is: 1.1 1H(CH_2) is: 0.6 H(CH) is: 0	82Tor1

[1]) Assignment in doubt.

Substance	Generation/ Matrix or Solvent	Method/ T [K]	g-Factor	a-Value [mT]	Ref./ add. Ref.
$[C_3H_6]^{\cdot+}$ *(continued)*	γ-irr. of C_3H_6/ $CFCl_3$	EPR/ 77	2.0033	$3H(CH_3)$ is: 2.4 1H(CH) is: 0.7 $2H(CH_2)$ is: 2.3, 1.2	84Shi2
		130	2.0033	$3H(CH_3)$ is: 2.4 1H(CH) is: 0.9 $2H(CH_2)$ is: 1.6, 0.9	
	SF_6	77	2.0033	$3H(CH_3)$ is: 2.7 1H(CH) is: 0.5 $2H(CH_2)$ is: 1.6, 1.4	
$[C_3H_4D_2]^{\cdot+}$ $[CH_3CH{=}CD_2]^{\cdot+}$	γ-irr. of $C_3H_4D_2$/ $CFCl_3$	EPR/ 77	2.0033	$3H(CH_3)$ is: 2.4 1H(CH) is: 0.7 $2D(CD_2)$ is: 0.35, 0.18	84Shi2
		130	2.0033	$3H(CH_3)$ is: 2.4 1H(CH) is: 0.9 $2D(CD_2)$ is: 0.25, 0.14	
	SF_6	77	2.0033	$3H(CH_3)$ is: 2.7 1H(CH) is: 0.5 $2D(CD_2)$ is: 0.25, 0.22	
$[C_4H_8]^{\cdot+}$ $[CH_3{-}CH{=}CH{-}CH_3]^{\cdot+}$	γ-irr. of C_4H_8/ $CFCl_3$ *cis*	EPR/ 77	–	$6H(CH_3)$ is: 2.21 2H(CH) is: −0.9	80Shi1
	trans	77	–	$6H(CH_3)$ is: 2.34 2H(CH) is: −0.88	
$[C_4H_8]^{\cdot+}$ $[(H_3C)_2C{=}CH_2]^{\cdot+}$	γ-irr. of C_4H_8/ $CFCl_3$	EPR/ 77	–	$6H(CH_3)$ is: 1.65 $2H(CH_2)$ is: 1.40	80Shi1
	γ-irr. of C_4H_8/ $CFCl_3$	EPR/ 160	2.0033	$8H(CH_3, CH_2)$ is: 1.62 1F(1): 0.52 from $CFCl_3$	84Shi2
	$CClF_2CClF_2$	77	2.0033	$8H(CH_3, CH_2)$ is: 1.61	
(continued)	SF_6	77	2.0033	$8H(CH_3, CH_2)$ is: 1.7	

Substance	Generation/ Matrix or Solvent	Method/ T [K]	g-Factor	a-Value [mT]	Ref./ add. Ref.
$[C_4H_8]^{\cdot +}$ *(continued)*	γ-irr. of $(CH_3)_4C$/ $CFCl_3$	EPR/ 77	–	$8H(CH_3, CH_2)$ is: 1.57	82Tor1
	Warming up to 100···140 K, the cation of $(CH_3)_3CH^+$ in $CFCl_3$, γ-irr. of $(CH_3)_3CH$/ $CFCl_3$	EPR/ 155	–	$6H(CH_3)$ is: 1.68 $2H(CH_2)$ is: 1.2	82Tor1
$[C_5H_{10}]^{\cdot +}$ $[(CH_3)_2C{=}CH(CH_3)]^{\cdot +}$	γ-irr. of C_5H_{10}/ $CFCl_2CF_2Cl$	EPR/ 77	–	1H(CH) is: 6.2 (2.6) [2] $3H(CHCH_3)$ is: 2.1 (2.18) [2] $6H((CH_3)_2)$ is: 1.67(1.68) [2]	84Tor1
$[C_6H_{12}]^{\cdot +}$ 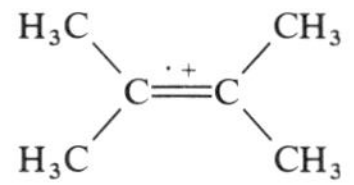	γ-irr. of C_6H_{12}/ $CFCl_3$	EPR/ 77	–	12H, is: 1.72	80Shi1
$[C_{10}H_{18}]^{\cdot +}$ 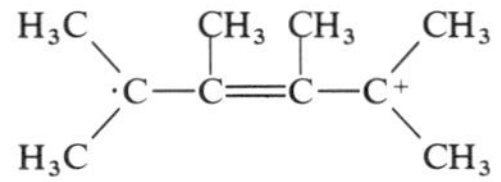	Oxidation with $AlCl_3:CH_2Cl_2$/ CH_2Cl_2	EPR/ 173	2.0028	6H, is: 0.42 6H, is: 1.05 6H, is: 1.07	84Cou1
$[C_{12}H_{24}]^{\cdot +}$ 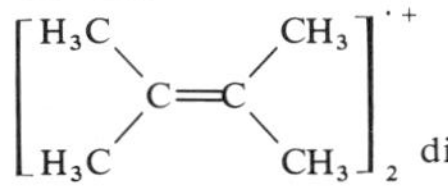 dimer	γ-irr. of $C_{12}H_{24}$/ $C_2H_5CH(CH_3)C_2H_5$	EPR/ 77	–	24H, is: 0.78	73Ich1

18.2.2.2 Monosubstituted alkene cations

Substance	Generation/ Matrix or Solvent	Method/ T [K]	g-Factor	a-Value [mT]	Ref./ add. Ref.
$[C_2H_3Br]^{\cdot +}$ 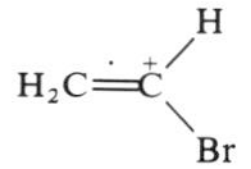	γ-irr. of C_2H_3Br/ $CFCl_3$	EPR/ 77	–	^{81}Br: 24.0 1H, is: ≃1.5	84Eas1

[2]) The value in the parenthesis is the alternative coupling constant.

Substance	Generation/ Matrix or Solvent	Method/ T [K]	g-Factor	a-Value [mT]	Ref./ add. Ref.
$[C_3H_3N]^{\cdot+}$ $H_2C\dot{=}\overset{+}{C}HCN$	γ-irr. of C_3H_3N/ $CFCl_3$	EPR/ 77	–	H(CH) is: 1.2 N: 2.0; 0; 0; is: ≃0.7	84Eas1
$[C_3H_5Cl]^{\cdot+}$ $H_2C\dot{=}\overset{+}{C}HCH_2Cl$	γ-irr. of C_3H_5Cl/ $CFCl_3$	EPR/ 77	–	^{35}Cl: 4.5; ≃1.0; ≃1.0; is: ≃2.2 $2H(H_2C{=})$ is: ≃1.6	84Eas1
$[C_3H_5Br]^{\cdot+}$ $H_2C\dot{=}\overset{+}{C}HCH_2Br$	γ-irr. of C_3H_5Br/ $CFCl_3$	EPR/ 77	–	^{81}Br: 28.7; 7.0; 7.0; is: 14.2 $2H(H_2C{=})$ is: ≃1.6	84Eas1
$[C_3H_6O]^{\cdot+}$ $H_2C\dot{=}\overset{+}{C}(H)OCH_3$	γ-irr. of C_3H_6O/ $CFCl_3$	EPR/ 77	–	$2H(CH_2)$ is: 1.9	84Eas1
$[C_4H_6O_2]^{\cdot+}$ $H_2C\dot{=}\overset{+}{C}(H)OCOCH_3$	γ-irr. of $C_4H_6O_2$/ $CFCl_3$	EPR/ 77	–	$2H(CH_2)$ is: 1.8	84Eas1
$[C_4H_8O]^{\cdot+}$ $H_2\dot{C}{=}C(H){-}\overset{+}{O}{-}C_2H_5$	γ-irr. of C_4H_8O/ $CFCl_3$	EPR/ 77	–	2H(α) is: 1.94 2H(ethyl) is: 0.35	84Sym1
$[C_5H_{12}Si]^{\cdot+}$ $H_2C\dot{=}\overset{+}{C}(H)Si(CH_3)_3$	γ-irr. of $C_5H_{12}Si$/ $CFCl_3$	EPR/ 77	–	$2H(CH_2)$ is: 4.0 1H(CH) is: 1.8	84Eas1
$[C_6H_9NO]^{\cdot+}$ $H_2C\dot{=}\overset{+}{C}(H){-}N$ (ring: $N{-}CH_2{-}CH_2{-}CH_2{-}C({=}O){-}N$)	γ-irr. of C_6H_9ON/ $CFCl_3$	EPR/ 77	–	$2H(CH_2)$ is: 1.3, 1.0 N, is: 3.2	84Eas1
$[C_6H_{12}O]^{\cdot+}$ $H_2C\dot{=}\overset{+}{C}(H){-}O{-}CH_2CH_2CH_2CH_3$	γ-irr. of $C_6H_{12}O$/ $CFCl_3$	EPR/ 77	–	$2H({-}OCH_2{-})$ is: 0.35 $2H(H_2C{=})$ is: 1.9	84Eas1

Substance	Generation/ Matrix or Solvent	Method/ T [K]	g-Factor	a-Value [mT]	Ref./ add. Ref.
$[C_6H_{14}N_2]^{\cdot +}$ 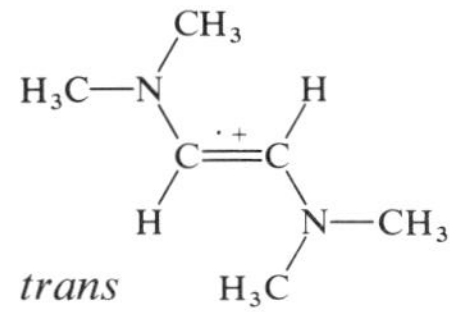 *trans*	$AlCl_3$:CH_2Cl_2/ CH_2Cl_2	EPR/ 293	–	2N, is: 0.813 6H(CH_3) is: 0.813 6H(CH_3) is: 0.695 2H(CH_2) is: 0.425 2N, is: 0.695 [3]) 12H(CH_3) is: 0.813 [3]) 2H(CH_2) is: 0.425 [3])	84Boc1, 70Gil1
$[C_7H_7N]^{\cdot +}$ 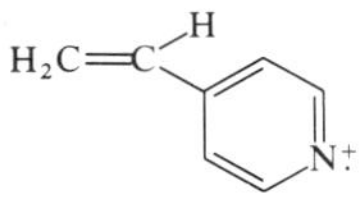	γ-irr. of C_7H_7N/ $CFCl_3$	EPR/ 77	–	N: 5.6; 2.9; 2.9; is: 3.8 2H(2, 6) is: $\simeq$2.8 2H(3, 5) is: $\simeq$1.0	84Eas1
$[C_7H_7N]^{\cdot +}$ 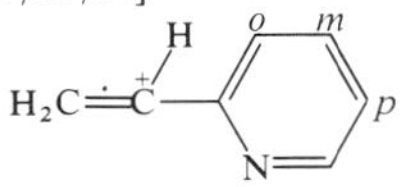	γ-irr. of C_7H_7N/ $CFCl_3$	EPR/ 77	–	H(p) is: 1.0 2H(CH_2) is: 1.0	84Eas1
$[C_{10}H_{22}N_2]^{\cdot +}$ 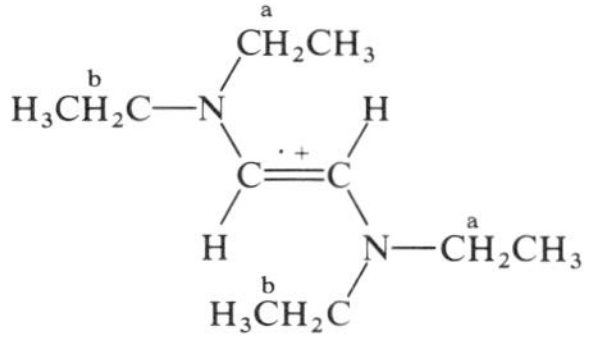 *trans*	$AlCl_3$:CH_2Cl_2/ CH_2Cl_2	EPR/ 293	–	2N, is: 0.800 2H(CH) is: 0.400 4H(CH_2, a) is: 0.5300 4H(CH_2, b) is: 0.400	84Boc1

18.2.2.3 Disubstituted alkene cations

Substance	Generation/ Matrix or Solvent	Method/ T [K]	g-Factor	a-Value [mT]	Ref./ add. Ref.
$[C_4H_6O_2]^{\cdot +}$ 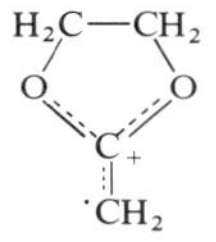	γ-irr. of $C_4H_6O_2$/ H_2O	EPR/ 276	is: 2.003	2H(α) is: 2.057 2H(OCH_2) is: 0.245 2H(OCH_2) is: 0.245	80Beh1

[3]) From [70Gil1].

Substance	Generation/ Matrix or Solvent	Method/ T [K]	g-Factor	a-Value [mT]	Ref./ add. Ref.
$[C_4H_8O_2]^{\cdot +}$ 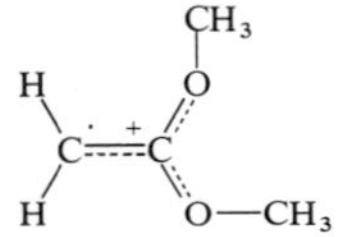	γ-irr. of $C_4H_8O_2$:$(CH_3O)_2\dot{C}CH_2Cl$/ H_2O	EPR/ 276	–	2H(α) is: 2.055 3H(CH_3) is: 0.31 3H(CH_3) is: 0.066	80Beh1
$[C_5H_8O_2]^{\cdot +}$ 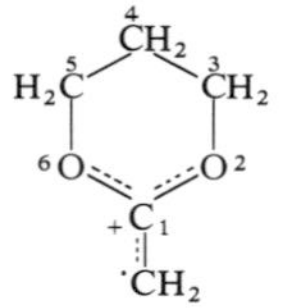	γ-irr. of $C_5H_8O_2$/ H_2O	EPR/ 276	2.003	2H(α) is: 2.076 2H(3) is: 0.337 2H(4) is: 0.008 2H(5) is: 0.337	80Beh1
$[C_5H_8O_2]^{\cdot +}$ 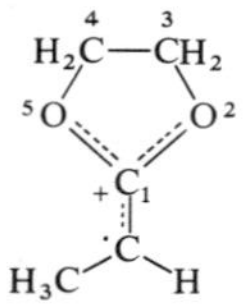	γ-irr. of $C_5H_8O_2$/ H_2O	EPR/ 276	2.003	3H, is: 2.511 1H, is: 1.87 2H(4) is: 0.22 2H(3) is: 0.258	80Beh1
$[C_5H_{10}O_2]^{\cdot +}$ 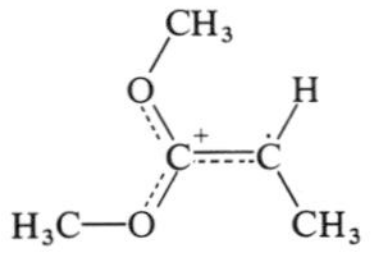	γ-irr. of $C_5H_{10}O_2$/ H_2O	EPR/ 276	is: 2.003	1H(α) is: 1.88 3H(β) is: 2.45 3H(OCH_2) is: 0.284 3H(OCH_3) is: 0.073	80Beh1
	γ-irr. of $C_5H_{10}O_2$/ $CFCl_3$	EPR/ 77	–	1H(α): 1.9 3H(β): 2.4 3H(OCH_3): 0.15	84Sym1
$[C_5H_{10}O_2]^{\cdot +}$ 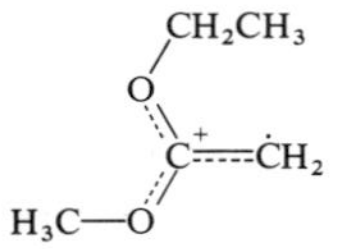	γ-irr. of $C_5H_{10}O_2$/ H_2O	EPR/ 276	is: 2.003	2H(α) is: 2.06 3H(OCH_3) is: 0.316 2H(OCH_2) is: 0.067 3H(CH_2CH_3) is: 0.007	80Beh1

Substance	Generation/ Matrix or Solvent	Method/ T [K]	g-Factor	a-Value [mT]	Ref./ add. Ref.
$[C_5H_{10}O_2]^{\cdot +}$ 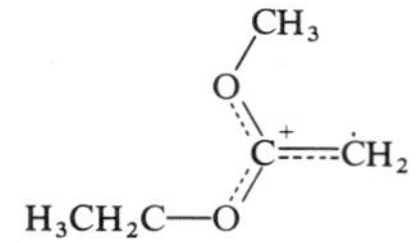	γ-irr. of $C_5H_{10}O_2$/ H_2O	EPR/ 276	is: 2.003	2H(α) is: 2.06 2H(OCH_2) is: 0.345 3H(CH_2CH_3) is: 0.065 3H(OCH_3) is: 0.003	80Beh1
$[C_6H_{12}O_2]^{\cdot +}$ 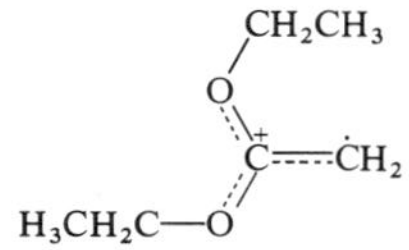	γ-irr. of $C_6H_{12}O_2$/ H_2O	EPR/ 276	2.003	2H(α) is: 2.06 2H(OCH_2) is: 0.336 2H(OCH_2) is: 0.06 3H(CH_2CH_3) is: 0.007 3H(CH_2CH_3) is: 0.0003	80Beh1
$[C_7H_{14}O_2]^{\cdot +}$ 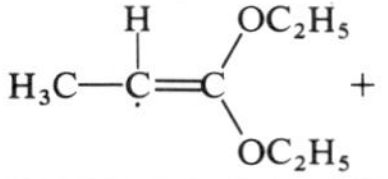	γ-irr. of $C_7H_{14}O_2$/ $CFCl_3$	EPR/ 77	–	1H(α) is: 1.9 3H(β) is: 2.4	84Sym1

18.2.2.4 Tetrasubstituted alkene cations

Substance	Generation/ Matrix or Solvent	Method/ T [K]	g-Factor	a-Value [mT]	Ref./ add. Ref.
$[C_2F_4]^{\cdot +}$ 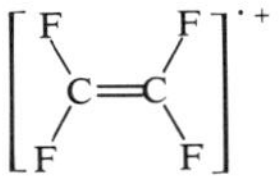	γ-irr. of C_2F_4/ $CFCl_3$	EPR/ 77	1.9905; 1.9903; 1.9903; is: 1.9903	4F: 14.6; 0; 0; is: 4.86 1F [4]): 1.39; ≤1.39; ≤1.39; is: ≃1.39	83Has1
$[C_6H_4S_4]^{\cdot +}$ 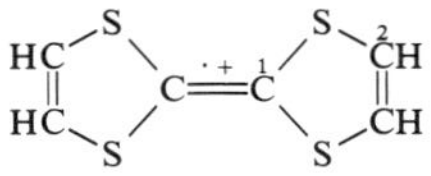	$AlCl_3$:CH_3Cl_2/ CH_2Cl_2	EPR/ 200	is: 2.0081	^{33}S, is: 0.42 4H, is: 0.124	80Boc1
$[C_6H_8S_4]^{\cdot +}$ 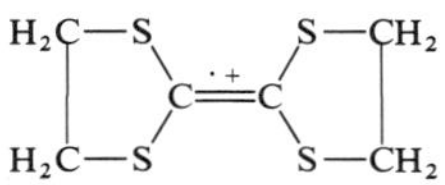	$AlCl_3$:CH_2Cl_2/ CH_2Cl_2	EPR/ 210	2.0089	8H, is: 0.238	80Boc1

[4]) From one F nucleus of $CFCl_3$.

Substance	Generation/ Matrix or Solvent	Method/ T [K]	g-Factor	a-Value [mT]	Ref./ add. Ref.
$[C_8H_{12}S_4]^{\cdot+}$ 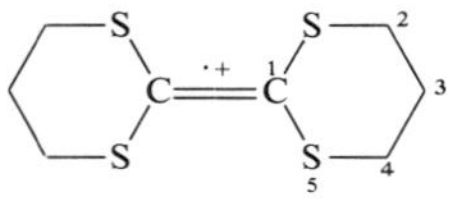	$AlCl_3:CH_2Cl_2/$ CH_2Cl_2	EPR/ 200	2.0102	8H(2, 4) is: 0.298 4H(3) is: 0.028	80Boc1
$[C_{10}H_{24}Si_2]^{\cdot+}$ 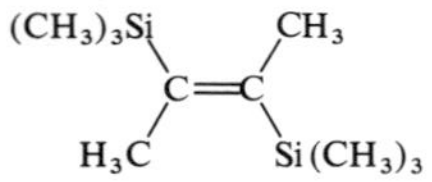	Solution in $AlCl_3:CH_2Cl_2/$ CH_2Cl_2	EPR/ 190		4H(1, 4): 1.072 (2H) 0.762 (2H) 6H(2, 3): 1.072 18H(5′, 6′): 0.046 2Si(5, 6): 1.4	80Boc2
$[C_{12}H_{28}Si_2]^{\cdot+}$ 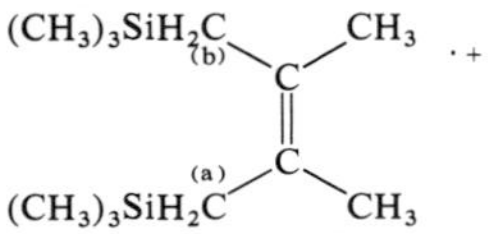	$AlCl_3:CH_2Cl_2/$ CH_2Cl_2	EPR/ 300	—	9H(Si(CH$_3$)$_3$) is: 0.046 1 ^{29}Si, is: 1.4 2H(CH$_2$, a) is: 1.07 2H(CH$_2$, b) is: 0.76 6H(CH$_3$) is: 1.07	77Boc1
$[C_{18}H_{44}Si_4]^{\cdot+}$ 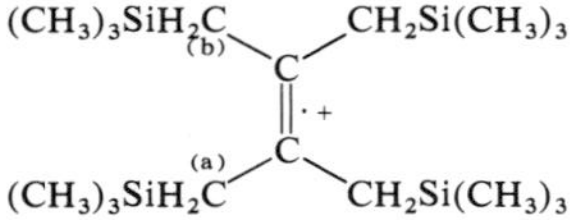	$AlCl_3:CH_2Cl_2/$ CH_2Cl_2	EPR/ 300	—	36H(Si(CH$_3$)$_3$) is: 0.031 2 ^{29}Si, is: 1.27 2H(CH$_2$, a) is: 0.855 2H(CH$_2$, b) is: 0.730	77Boc1
	Solution in $AlCl_3:CH_2Cl_2/$ CH_2Cl_2	EPR/ 243		4H(CH$_2$, a): 0.855 4H(CH$_2$, b): 0.72 36H(Si(CH$_3$)$_3$): 0.031 4Si: 1.25	80Boc2
$[C_{30}H_{68}Si_4]^{\cdot+}$ 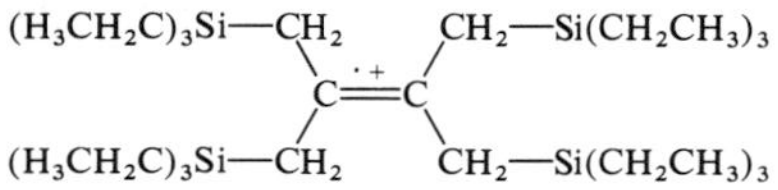	$AlCl_3:CH_2Cl_2/$ CH_2Cl_2	EPR/ 190···310	—	4H(CH$_2$) is: 0.83 4H(CH$_2$) is: 0.71 4 ^{29}Si, is: 1.0	80Boc2

Substance	Generation/ Matrix or Solvent	Method/ T [K]	g-Factor	a-Value [mT]	Ref./ add. Ref.

18.2.3 Cyclic alkene cations

18.2.3.1 Unsubstituted alkene cations

Substance	Generation/ Matrix or Solvent	Method/ T [K]	g-Factor	a-Value [mT]	Ref./ add. Ref.
$[C_5H_8]^{\cdot+}$ 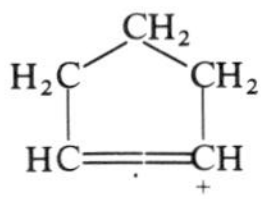	γ-irr. of C_5H_8 (cyclopentene)/ $CFCl_3$	EPR/ 77	–	2H(γ, ax) is: 1.25 2H(γ, eq) is: 0.55 2H(β, ax) is: 5.3 2H(β, eq) is: 4.85 2H(CH) is: 0.86	80Shi1
		130	–	4H(β) is: 4.93 2H(γ) is: 0.7 2H(CH) is: 0.98	
	γ-irr. of C_5H_8/ CF_3CCl_3	EPR/ 110	2.0039	4H(β) is: 4.96 2H(CH) is: 1.05 2H(γ) is: 0.7	83Tab1
$[C_6H_8]^{\cdot+}$ 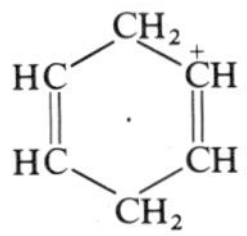	γ-irr. of C_6H_8/ $CFCl_3$	EPR/ 77	–	$4H(CH_2)$ is: 6.8 4H(CH) is: 0.48	80Shi1
$[C_6H_{10}]^{\cdot+}$ 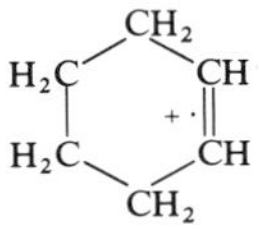	γ-irr. of C_6H_{10}/ CF_3CCl_3	EPR/ 140	2.0037	2H(β) is: 5.5 2H(β) is: 2.2 2H(CH) is: 0.9	83Tab1
	γ-irr. of C_6H_{10} (cyclohexene)/ $CFCl_3$	EPR/ 77	–	2H(γ, eq) is: 0.6 2H(γ, ax) is: 0.8 2H(β, ax) is: 5.4 2H(β, eq) is: 2.25 2H(CH) is: 0.88	80Shi1

Substance	Generation/ Matrix or Solvent	Method/ T [K]	g-Factor	a-Value [mT]	Ref./ add. Ref.
$[C_{16}H_{28}]^{\cdot +}$ 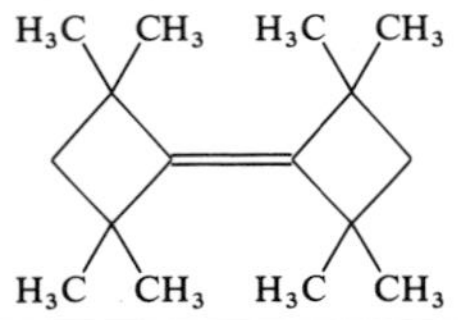	Oxidation by CF_3COOH/ $CF_3COOH:(CF_3CO)_2O:$ CH_2Cl_2	ESR/ 293	2.0027	24H: 0.123 4H: 0.049	81Ger1
$[C_{17}H_{30}]^{\cdot +}$ 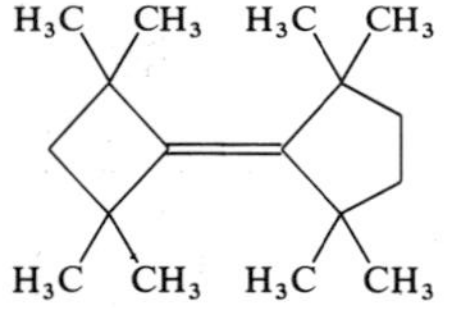	Oxidation by CF_3COOH/ $CF_3COOH:(CF_3CO)_2O:$ CH_2Cl_2	ESR/ —	2.00265		81Ger1
$[C_{18}H_{28}]^{\cdot +}$ 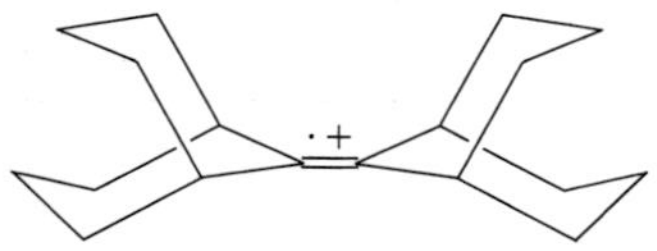	Solution in CF_3COOH, $(CF_3CO)_2O:CH_2Cl_2$/ CH_2Cl_2	EPR/ 193	2.0033	4H(1, eq): 0.105 8H(2, eq): 0.370 8H(2, ax): 0.055 4H(3, eq): 0.320 4H(3, ax): <0.020	81Ger2
$[C_{18}H_{32}]^{\cdot +}$ 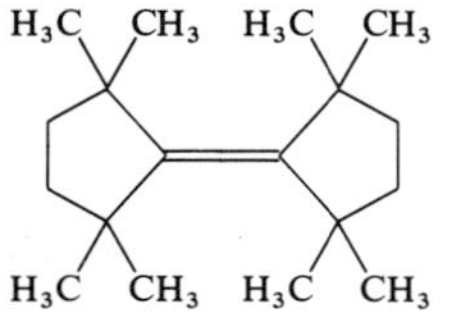	Oxidation by CF_3COOH/ $CF_3COOH:(CF_3CO)_2O:$ CH_2Cl_2	ESR/ 313	2.0026	12H: 0.203 4H: 0.028	81Ger1
$[C_{20}H_{26}D_2]^{\cdot +}$ 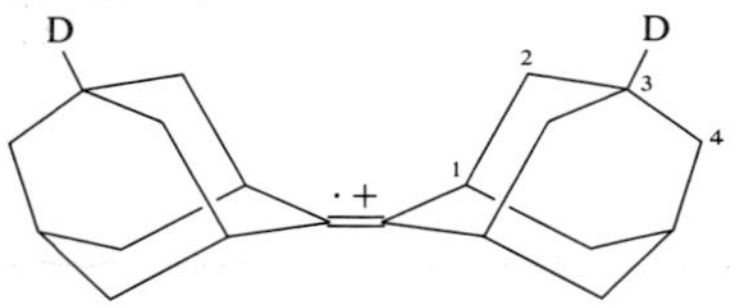	Solution in CF_3COOH, $(CF_3CO)_2O:CH_2Cl_2$/ CH_2Cl_2	EPR/ 193	2.0032	4H(1, eq): 0.057 8H(2, eq): 0.325 8H(2, ax): 0.046 2H(3): 0.602 4H(4): 0.012 2D(3): 0.095	81Ger2

Substance	Generation/ Matrix or Solvent	Method/ T [K]	g-Factor	a-Value [mT]	Ref./ add. Ref.
$[C_{20}H_{27}D]^{\cdot+}$ 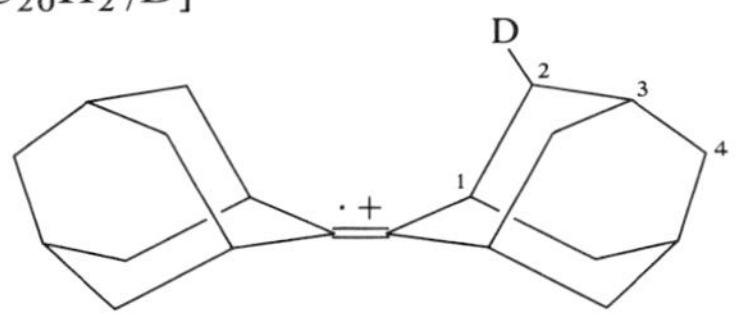	Solution in CF_3COOH, $(CF_3CO)_2O:CH_2Cl_2$/ CH_2Cl_2	EPR/ 193	2.0032	4H(1, eq): 0.056 7H(2, eq): 0.329 8H(2, ax): 0.046 4H(3, eq): 0.609 4H(4): 0.012 1D(2, eq): 0.049	81Ger1
$[C_{20}H_{28}]^{\cdot+}$ 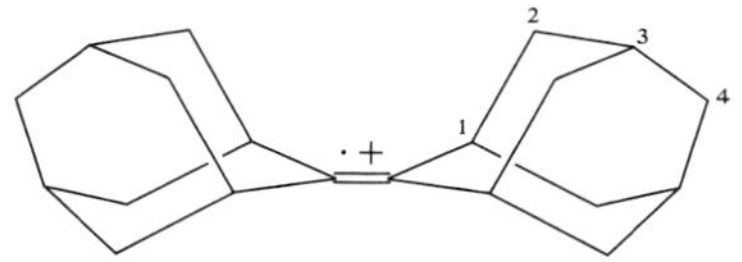	Solution in CF_3COOH, $(CF_3CO)_2O:CH_2Cl_2$/ CH_2Cl_2	EPR/ 193	2.0032	4H(1, eq): 0.058 8H(2, eq): 0.327 8H(2, ax): 0.047 4H(3, eq): 0.605 4H(4): 0.012	81Ger2
18.2.3.2 Substituted alkene cations					
$[C_{12}H_{28}Si_4]^{\cdot+}$ 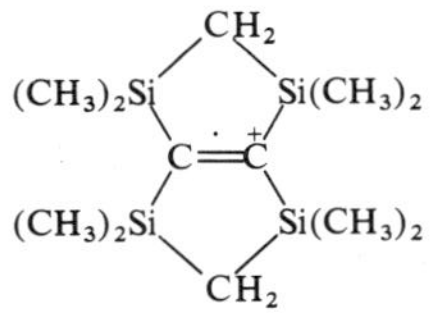	$AlCl_3:CH_2Cl_2$/ CH_2Cl_2	EPR/ 183	–	$24H(CH_3)$ is: 0.062 $4H(CH_2)$ is: 0.248 $4\,^{29}Si$, is: 2.271	78Fri1, 76Boc1
$[C_{22}H_{48}Si_4]^{\cdot+}$ 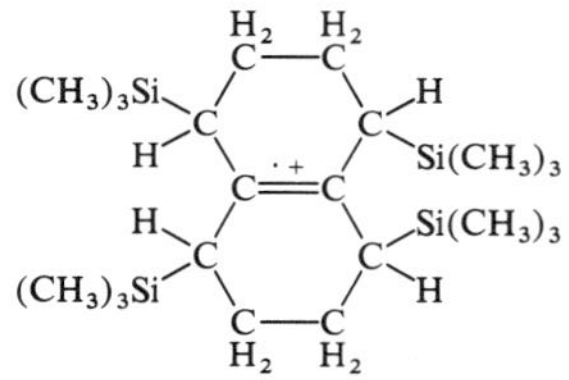	$AlCl_3:CH_2Cl_2$/ CH_2Cl_2	EPR/ 190···310		$4H(CH_2)$ is: 1.38 4H(ax) is: 0.08 4H(eq) is: 0.013 $4\,^{29}Si$, is: 1.3 $36H(CH_3)$ is: 0.027	80Boc2

18.2.4 Diene cations

18.2.4.1 Acyclic diene cations

Substance	Generation/ Matrix or Solvent	Method/ T [K]	g-Factor	a-Value [mT]	Ref./ add. Ref.
$[C_4H_6]^{\cdot +}$ $H_2C{=}CH{-}CH{=}\overset{\cdot +}{C}H_2$	γ-irr. of C_4H_6/ $CFCl_3$	EPR/ 77	–	$4H(CH_2)$ is: 1.22 $2H(CH)$ is: 0.2	80Shi1
$[C_4F_6]^{\cdot +}$ $[F_2C{=}CF{-}CF{=}CF_2]^{\cdot +}$	γ-irr. of C_4F_6/ $CClF_2CClF_2$ (i) [1]	EPR/ 77	2.0026; 2.0056; 2.0056; is: 2.0046	2F: 4.22 4F: 10.64; $\simeq 0$; $\simeq 0$; is: $\simeq 3.55$	84Shi3
	(ii) [1]	77	2.0028; 2.0056; 2.0056; is: 2.0042	2F: 10.62 2F: 9.7 2F: 6.07	
$[C_{16}H_{36}Si_4]^{\cdot +}$ $[((CH_3)_3Si)_2C{=}C{=}C{=}C(Si(CH_3)_3)_2]^{\cdot +}$	$AlCl_3:CH_2Cl_2$/ CH_2Cl_2	EPR/ 300	–	$36H(CH_3)$: 0.033 $4\,^{29}Si$: 1.210	79Kai1
$[C_{22}H_{54}Si_6]^{\cdot +}$ $[((CH_3)_3Si)_3C{-}C{\equiv}C{-}C(Si(CH_3)_3)_3]^{\cdot +}$	$AlCl_3:CH_2Cl_2$/ CH_2Cl_2	EPR/ 200	–	Doublet $\simeq 0.738$	79Kai1

18.2.4.2 Cyclic diene cations

Substance	Generation/ Matrix or Solvent	Method/ T [K]	g-Factor	a-Value [mT]	Ref./ add. Ref.
$[C_5H_6]^{\cdot +}$ (cyclopentadiene: CH_2, HC=CH, HC=CH ring)	γ-irr. of C_5H_6/ $CFCl_3$	EPR/ 77	–	$2H(CH_2)$: <0.2 2H(terminal): 1.16 2H(inner): 0.35	80Shi1

[1]) (i) and (ii) are *cis* and *trans* conformers of hexafluorobutadiene with C_{2v} and C_{2h} symmetry, respectively.

Substance	Generation/ Matrix or Solvent	Method/ T [K]	g-Factor	a-Value [mT]	Ref./ add. Ref.
$[C_6H_8]^{\cdot +}$ 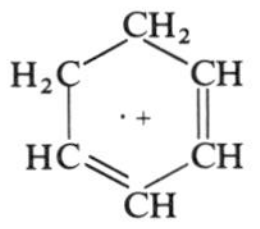	γ-irr. of C_6H_8 (1,3-cyclohexadiene)/ CF_3CCl_3	EPR/ 141	2.0037	4H(β): 3.02 2H(CH): 0.84 2H(CH): 0.42	83Tab1
	γ-irr. of C_6H_8 (1,3-cyclohexadiene)/ $CFCl_3$	EPR/ 77	–	4H(β, ax) is: 4.24 4H(β, eq) is: 1.34 1H(terminal) is: 0.84 1H(inner) is: 0.42	80Shi1
		130	–	4H(β, ax) is: 3.10 4H(β, eq) is: 2.80 1H(terminal) is: 0.84 1H(inner) is: 0.42	
$[C_6H_8]^{\cdot +}$ 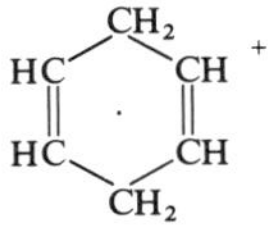	γ-irr. of C_6H_8 (1,4-cyclohexadiene)/ $CFCl_3$	EPR/ 77	–	$4H(CH_2)$ is: 6.8 4H(CH) is: 0.48	80Shi1
	γ-irr. of C_6H_8/ CF_3CCl_3	EPR/ 141	2.0038	$4H(CH_2)$: 6.71 4H(CH): 0.4	83Tab1
$[C_7H_8]^{\cdot +}$ 	X-irr. of C_7H_8 (norbornadiene)/ $CFCl_2CF_2Cl$ or $CFCl_3$	EPR/ 4.2	–	4H, is: 0.8 2H, is: 0.33	83Tor1
$[C_8H_{12}]^{\cdot +}$ 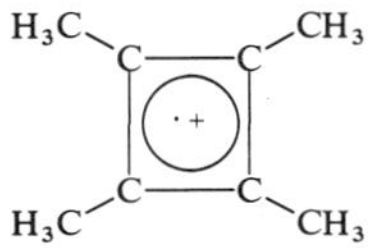	hν/ CH_2Cl_2	EPR/ 173	–	12H, is: 0.87	83Cou1
	UV-irr. of $CH_3{-}C{\equiv}C{-}CH_3{:}AlCl_3$/ CH_2Cl_2	EPR/ 150	2.0030	12H(1, 2, 3, 4): 0.870	85Cou1
	Oxidation by $AlCl_3$/ CH_2Cl_2	ESR, ENDOR/ 173	2.0030	12H: 0.870 ^{13}C: 0.404	84Cou/ 81Bro1
		253		12H: 0.863	

Substance	Generation/ Matrix or Solvent	Method/ T [K]	g-Factor	a-Value [mT]	Ref./ add. Ref.
$[C_{10}H_{14}]^{\cdot +}$ 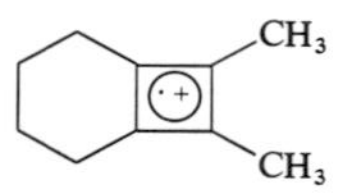	γ-irr. of $C_{10}H_{16}$/ CH_2Cl_2	EPR/ 183	2.0030	$6H(CH_3)$: 0.875 $2H(CH_2)$: 0.895 $2H(CH_2)$: 0.860 $2H(CH_2)$: 0.034 $2H(CH_2)$: 0.036	85Cou1
$[C_{10}H_{16}]^{\cdot +}$ 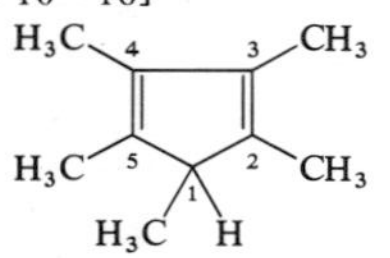	UV-irr./ CF_3COOH	ESR/ 263	–	6H(2, 5): 1.50 6H(3, 6): 0.40 1H(1): 0.16 3H(1): 0.08	84Cou3
$[C_{10}H_{16}]^{\cdot +}$ 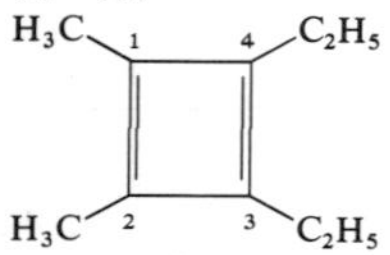 *cis* 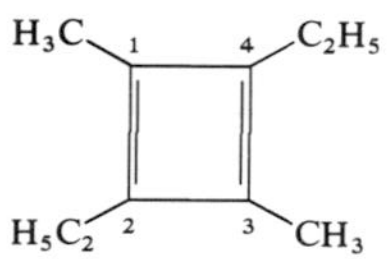 *trans*	Oxidation with $AlCl_3$/ CH_2Cl_2	ESR, ENDOR/ 182		*cis:* 6H: 0.840 4H: 0.937 *trans:* 6H: 0.915 4H: 0.900	84Cou1
$[C_{11}H_{18}]^{\cdot +}$ 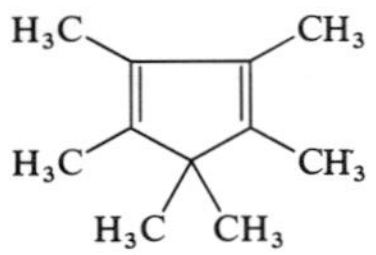	UV-irr./ CF_3COOH	ESR/ 223		6H: 1.44 6H: 0.40 6H: 0.13	84Cha1
$[C_{12}H_{20}]^{\cdot +}$ 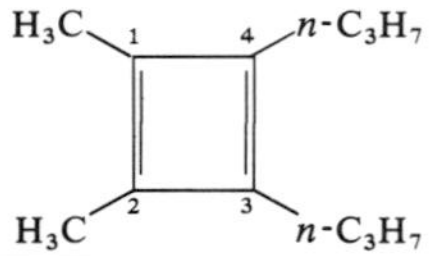	Oxidation with $AlCl_3$/ CH_2Cl_2	ESR, ENDOR/ 193	2.0029	6H: 0.838 4H: 0.762	84Cou2
$[C_{12}H_{20}]^{\cdot +}$ 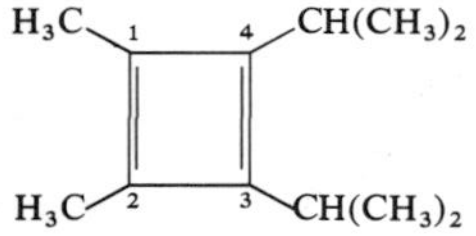	Oxidation with $AlCl_3$/ CH_2Cl_2	ESR, ENDOR/ 173 258	 – –	 6H: 0.894 2H: 0.630 6H: 0.908 2H: 0.588	84Cou2

Substance	Generation/ Matrix or Solvent	Method/ T [K]	g-Factor	a-Value [mT]	Ref./ add. Ref.
$[C_{12}H_{20}]^{\cdot+}$ 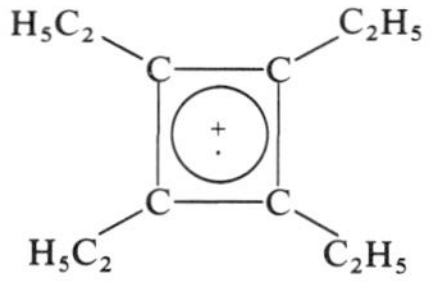	$h\nu$/ CH_2Cl_2	EPR/ 173	–	$8H(CH_2)$: 0.799	83Cou1
	Oxidation with $AlCl_3$/ CH_2Cl_2	ESR, ENDOR/ 173	2.0029	8H: 0.799 ^{13}C: 0.372 ^{13}C: 0.433 ^{13}C: 0.500	84Cou2/ 83Bro1
		253		8H: 0.772	
$[C_{12}H_{24}Si_2]^{\cdot+}$ 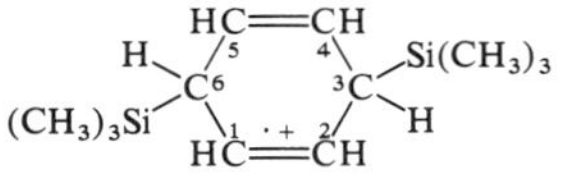	$AlCl_3$:CH_2Cl_2/ CH_2Cl_2	EPR/ 190···310	–	4H(1, 2, 4, 5) is: 0.32 2H(3, 6) is: 1.7 $18H(CH_3)$ is: 0.055	80Boc1
$[C_{16}H_{32}Si_2]^{\cdot+}$ 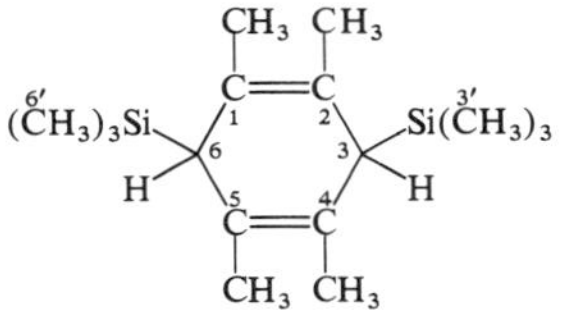	Solution in $AlCl_3$:CH_2Cl_2/ CH_2Cl_2	EPR/ 190		12H(1, 2, 4, 5): 0.314 1H(3, 6): 1.98 1H(3, 6): 0.314 18H(3′, 6′): 0.047	80Boc1
$[C_{14}H_{24}]^{\cdot+}$ *cis* *trans*	Oxidation with $AlCl_3$/ CH_2Cl_2	ESR, ENDOR/ 193	2.0028	*cis* 6H: 0.800 18H: 0.024 ^{13}C: 0.40	84Cou2/ 83Cou2, 83Cou1
		243	2.0028	6H: 0.803 18H: 0.024	
		193	2.0028	*trans* 6H: 0.900 18H: 0.020 ^{13}C: 0.40	
		243		6H: 0.917 18H: 0.20	

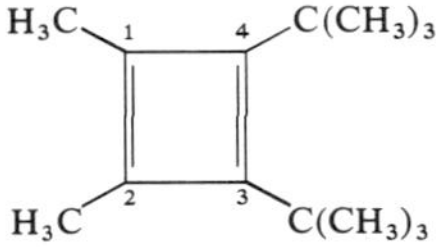

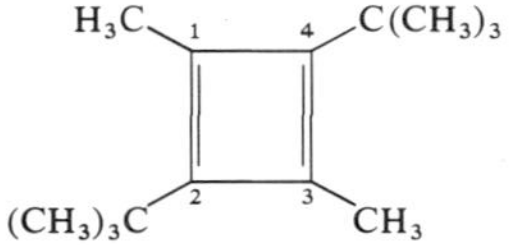

Substance	Generation/ Matrix or Solvent	Method/ T [K]	g-Factor	a-Value [mT]	Ref./ add. Ref.
$[C_{14}H_{24}]^{\cdot +}$	Oxidation with $AlCl_3$/ CH_2Cl_2	ESR, ENDOR/ 193 253	 2.0028 2.0028	 4H, ethyl: 0.812 4H, propyl: 0.692 4H, ethyl: 0.779 4H, propyl: 0.689	84Cou2
$[C_{14}H_{24}]^{\cdot +}$	Oxidation with $AlCl_3$/ CH_2Cl_2	ESR, ENDOR/ 183 204	 — —	 4H: 0.716 2H: 0.461 4H: 0.693 2H: 0.459	84Cou1
$[C_{14}H_{24}O]^{\cdot +}$ 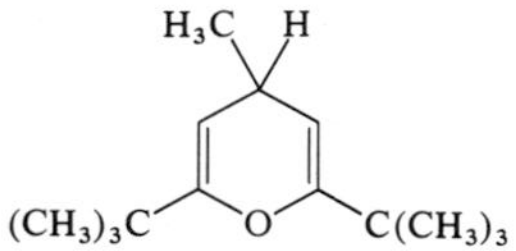	Oxidation with $AlCl_3$/ CH_3NO_2	ESR/ 243	—	1H: 0.53 3H: 1.12 2H: 0.08	79Okh1
$[C_{16}H_{28}]^{\cdot +}$ 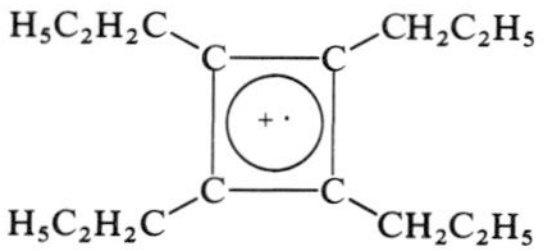	$h\nu$/ CH_2Cl_2	EPR/ 173	—	8H, is: 0.715	83Cou1
$[C_{16}H_{28}]^{\cdot +}$	Oxidation with $AlCl_3$/ CH_2Cl_2	ESR, ENDOR/ 173 253	 2.0028 2.0028	 4H(ethyl): 0.844 4H(butyl): 0.695 4H(ethyl): 0.800 4H(butyl): 0.693	84Cou2

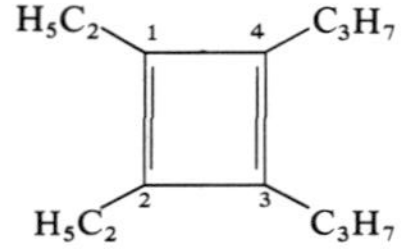

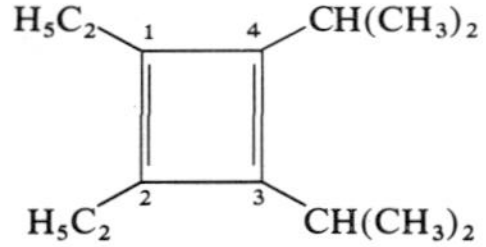

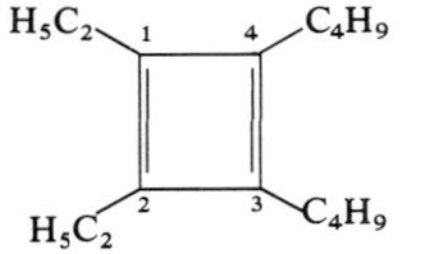

Substance	Generation/ Matrix or Solvent	Method/ T [K]	g-Factor	a-Value [mT]	Ref./ add. Ref.
$[C_{16}H_{28}]^{\cdot +}$ 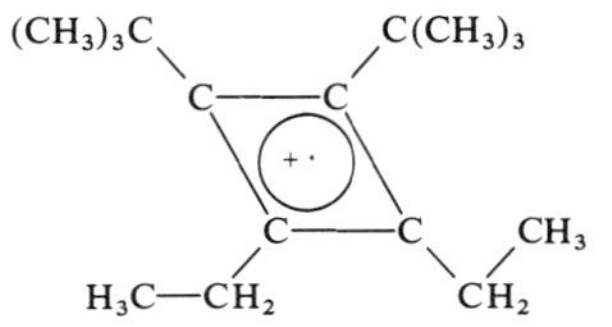	$h\nu$/ CH_2Cl_2 [2] [2]	EPR/ 203 203	 — —	 $4H(CH_2)$ is: 0.445 $18H(C(CH_3)_3)$ is: 0.02 $4H(CH_2)$ is: 0.431 $18H(C(CH_3)_3)$ is: 0.02	83Cou1
	Oxidation with $AlCl_3$/ CH_2Cl_2	ESR, ENDOR/ 183 220	 2.0028 2.0028	 4H: 0.400 4H: 0.448 18H: 0.022	84Cou2/ 83Cou1
$[C_{16}H_{28}]^{\cdot +}$ 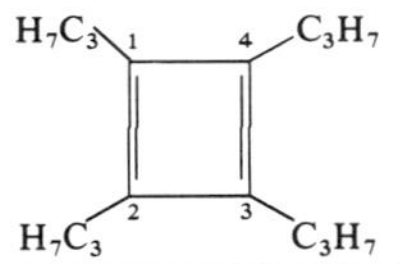	Oxidation with $AlCl_3$/ CH_2Cl_2	ESR, ENDOR/ 173 253	 2.0027	 8H: 0.715 8H: 0.701	84Cou2
$[C_{16}H_{28}]^{\cdot +}$ 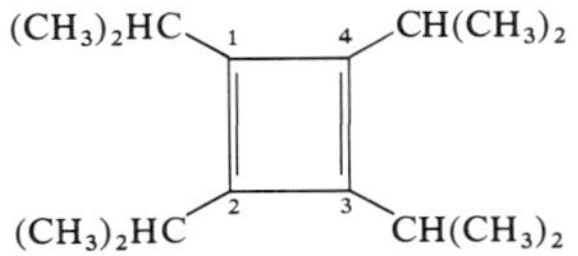	Oxidation with $AlCl_3$/ CH_2Cl_2	ESR, ENDOR/ 173 253	 2.0027	 4H: 0.276 24H: 0.028 4H: 0.274 24H: 0.028	84Cou2
$[C_{16}H_{32}Si_2]^{\cdot +}$ 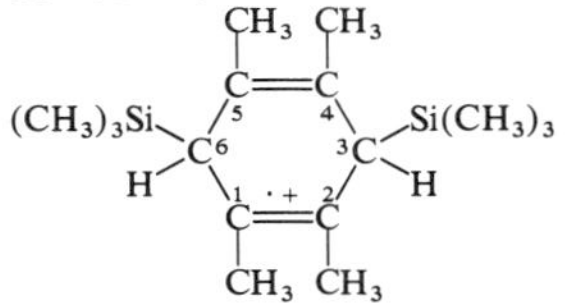	$AlCl_3$:CH_2Cl_2/ CH_2Cl_2	EPR/ 200	—	12H(1, 2, 4, 5) is: 0.314 2H(3, 6) is: 1.98 $18H(3, 6, CH_3)$ is: 0.047	80Boc2
$[C_{17}H_{30}O]^{\cdot +}$ 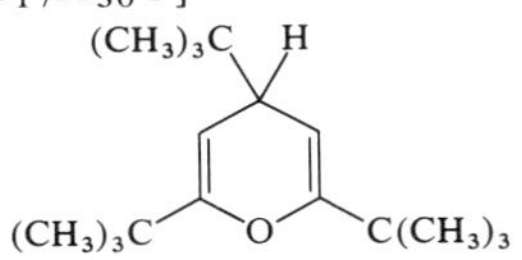	Oxidation with $AlCl_3$/ CH_3NO_2	ESR/ 243	—	1H: 0.43 2H: 0.08	79Okh1

[2]) These are *cis* and *trans* isomers.

Substance	Generation/ Matrix or Solvent	Method/ T [K]	g-Factor	a-Value [mT]	Ref./ add. Ref.
$[C_{18}H_{32}]^{\cdot +}$	Oxidation with $AlCl_3$/ CH_2Cl_2	ESR, ENDOR/			84Cou2
		183	2.0027	2H: 0.205 2H: 0.195	
		253	2.0027	2H: 0.230 2H: 0.210	
$[C_{18}H_{40}Si_4]^{\cdot +}$ 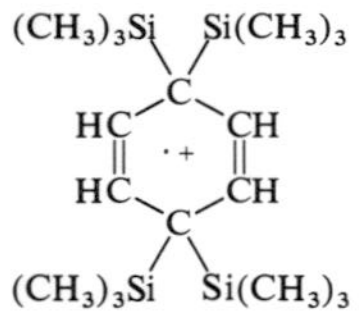	$AlCl_3:CH_2Cl_2$/ CH_2Cl_2	EPR/ 300	—	36H(CH_3) is: 0.018 4 ^{29}Si, is: 2.09 4H(CH) is: 0.303	77Boc1
$[C_{20}H_{28}O]^{\cdot +}$ 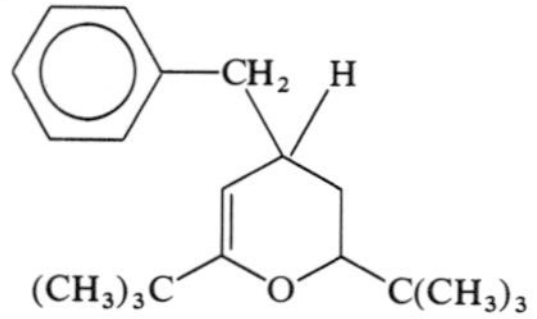	Oxidation with $AlCl_3$/ CH_3NO_2	ESR/ 243	—	1H: 0.72 2H: 1.44 2H: 0.08	79Ohk1
$[C_{20}H_{36}]^{\cdot +}$	Oxidation with $AlCl_3$/ CH_2Cl_2	ESR, ENDOR/			84Cou2
cis		173	2.0027	*cis* 4H: 0.455	
		273	2.0027	4H(β): 0.487	
trans		173	2.0027	*trans* 4H: 0.550	
		275	2.0027	4H: 0.577	

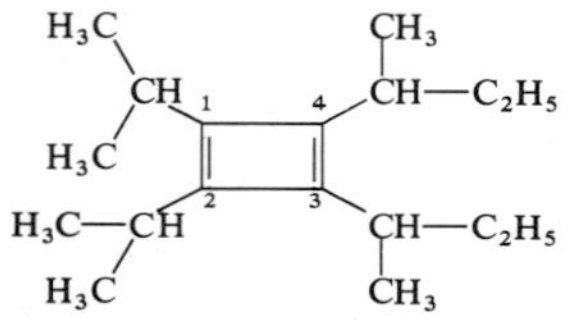

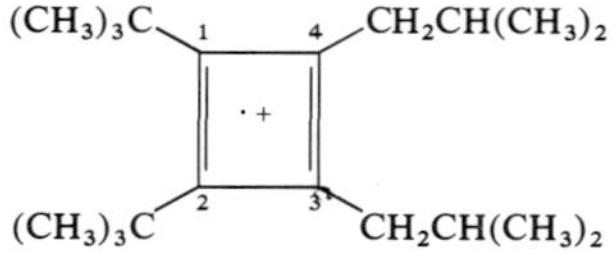

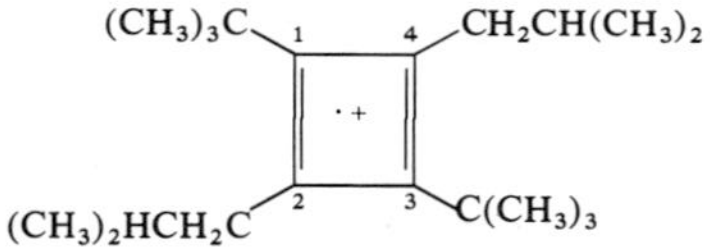

Substance	Generation/ Matrix or Solvent	Method/ T [K]	g-Factor	a-Value [mT]	Ref./ add. Ref.
$[C_{20}H_{36}]^{\cdot +}$ (tetrakis($CH_2CH(CH_3)_2$)cyclobutadiene cation)	$h\nu$/ CH_2Cl_2	EPR/ 173	–	8H(CH_2) is: 0.65	83Con1
	Oxidation with $AlCl_3$/ CH_2Cl_2	ESR, ENDOR/ 200	2.0027	8H: 0.650	84Con2
$[C_{20}H_{36}]^{\cdot +}$ (tetrakis($C(CH_3)_3$)cyclobutadiene cation)	$AlCl_3:CH_2Cl_2$/ CH_2Cl_2	EPR/ 200	2.0024	36H, is: 0.027	80Boc3
	Oxidation with $AlCl_3$/ CH_2Cl_2	ESR, ENDOR/ 200	2.0023	36H: 0.027	84Cou2/ 80Boc3, 84Boc2
$[C_{22}H_{40}]^{\cdot +}$ (1,2-bis($(C_2H_5)(CH_3)_2C$)-3,4-bis($C(CH_3)_3$)cyclobutadiene cation)	Oxidation with $AlCl_3$/ CH_2Cl_2	ESR, ENDOR/ 254	2.0022	≥18H: 0.029	84Cou2
$[C_{24}H_{44}]^{\cdot +}$ (tetrakis($CH_2C_4H_9$)cyclobutadiene cation)	$h\nu$/ CH_2Cl	EPR/ 191	–	8H: 0.71	83Bro1
$[C_{26}H_{36}]^{\cdot +}$ (1,2-bis(H_3C)-3,4-bis(Ad)cyclobutadiene cation; Ad = (adamantyl))	UV-irr./ CF_3COOH	ESR/ 223		6H(CH_3): 0.79	84Cha1

Substance	Generation/ Matrix or Solvent	Method/ T [K]	g-Factor	a-Value [mT]	Ref./ add. Ref.
$[C_{44}H_{60}]^{\cdot +}$ (tetra-Ad-cyclobutadiene; Ad = 1-adamantyl, positions 1′, 2′, 3′, 4′)	UV-irr. of (Ad—C≡)$_2$ + $AlCl_3$ CFCl$_3$, CF_3COOH	—/ 205	2.0029	C(1): 1.076 C(1′): 0.3 C(2′): 0.3 C(3′): 0.3	85Cha1

18.2.5 Ether cations

18.2.5.1 Mono- and di-ether cations

Substance	Generation/ Matrix or Solvent	Method/ T [K]	g-Factor	a-Value [mT]	Ref./ add. Ref.
$[C_2H_4O]^{\cdot +}$ (H_2C—CH_2 with bridging O^+) [1]	γ-irr. of C_2H_4O/ $CFCl_3$	EPR/ 77	—	4H: 1.63	84Sym1, 83Sym1
$[C_2H_4O]^{\cdot +}$ $[H_2C—O—CH_2]^{\cdot +}$ [2]	γ-irr. of C_2H_4O/ $CFCl_3$	EPR/ 110	2.0024	4H: 1.62 $2^{13}C$: 5.72; 0.9; 0.9; is: 25.1	85Qin1
$[C_2H_6O]^{\cdot +}$ (H_3C—$\dot{O}^+$—CH_3)	γ-irr. of C_2H_6O/ $CFCl_3$	EPR/ 77	—	6H: 4.23	81Kub1
$[C_3H_6O]^{\cdot +}$ (oxetane ring: CH_2, H_2C, CH_2, $\dot{O}^+$)	γ-irr. of C_3H_6O/ $CFCl_3$	EPR/ 77	—	4H: 6.4 2H: 1.1	84Sym1
$[C_3H_6O]^{\cdot +}$ $[H_2C—O—CHCH_3]^{\cdot +}$	γ-irr. of C_3H_6O/ $CFCl_3$	EPR/ 77	2.0024	4H: 1.2 2H: 2.1	86Rid1
$[C_3H_8O_2]^{\cdot +}$ $[CH_3—O—CH_2—O—CH_3]^{\cdot +}$	γ-irr. of $C_3H_8O_2$/ $CFCl_3$	EPR/ 133	2.0072	2H(β, CH_2): 13.61 2H(β): 3.13 4H(β): 0.6	82Sno1

[1] May have ring-opened structure, $[H_2C—O—CH_2]^{\cdot +}$
[2] May have ring-closed structure.

Substance	Generation/ Matrix or Solvent	Method/ T [K]	g-Factor	a-Value [mT]	Ref./ add. Ref.
$[C_4H_8O]^{\cdot +}$ $[H_3CHC—O—CHCH_3]^{\cdot +}$	γ-irr. of C_4H_8O/ $CFCl_3$	EPR/ 77	2.0024	8H: 1.65	84Sym1
$[C_4H_8O]^{\cdot +}$ $H_2C{=}\overset{+}{C}(H)OC_2H_5$	γ-irr. of C_4H_8O/ $CFCl_3$	EPR/ 77	–	2H, is: 1.94	83Rao1, 84Eas1
$[C_4H_8O]^{\cdot +}$ (ring: $H_2C—CH_2$, H_2C, CH_2, $\dot{O}^+$)	γ-irr. of C_4H_8O/ D_2SO_4 $CFCl_3$	EPR/ 77 77	 – –	 2H: 5.9 2H: 2.2 2H: 8.9 2H: 4.0	84Sym1
	γ-irr. of C_4H_8O/ $CFCl_3$	EPR/ 77 155	 – –	 2H(eq): 4.0 2H(ax): 8.9 4H: 6.5	81Kub1
$[C_4H_{10}O]^{\cdot +}$ $H_5C_2—\dot{O}^+—C_2H_5$	γ-irr. of $C_4H_{10}O$/ $CFCl_3$	EPR/ 77	–	4H: 6.87	84Sym1
$[C_4H_{10}O_2]^{\cdot +}$ $[CH_3—O—CH_2—CH_2—O—CH_3]^{\cdot +}$	γ-irr. of $C_4H_{10}O_2$/ $CFCl_3$	EPR/ 77	–	10H: 1.0	84Sym1
$[C_5H_8O]^{\cdot +}$ (bicyclic structure with O)$^{\cdot +}$	γ-irr. of C_5H_8O/ $CFCl_3$	EPR/ 77	–	2H: 0.9 2H: 1.53 2H: 3.1	86Rid1
$[C_5H_{10}O]^{\cdot +}$ (ring: $H_2C—CH_2$, H^a, H_{eq}, H^{eq}, CH_3, $\overset{+}{O}$)	γ-irr. of $C_5H_{10}O$/ $CFCl_3$ [3]) [3])	EPR/ 77 146	 – –	 2H(eq): 4.2 1H(ax): 8.3 2H(ax): 4.9 1H(eq): 7.8	81Kub1

[3]) These are two possible conformers at 77 K and 146 K, respectively.

Substance	Generation/ Matrix or Solvent	Method/ T [K]	g-Factor	a-Value [mT]	Ref./ add. Ref.
$[C_5H_{10}O]^{\cdot+}$ 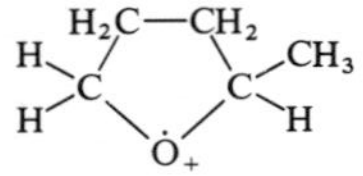	γ-irr. of $C_5H_{10}O$/ $CFCl_3$	EPR/ 77	—	1H: 8.3 2H: 4.2	84Sym1
$[C_5H_{10}O]^{\cdot+}$ 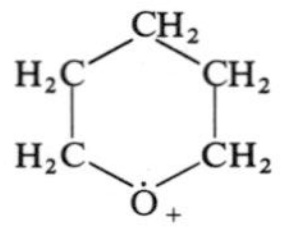	γ-irr. of $C_5H_{10}O$/ $CFCl_3$	EPR/ 77	—	2H(β): 3.45 2H(β): 1.4 2H(γ): 1.1 2H(γ): 0.3	84Sym1
$[C_6H_{12}O]^{\cdot+}$ 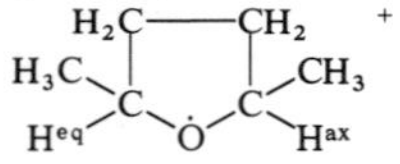 (*cis*) or 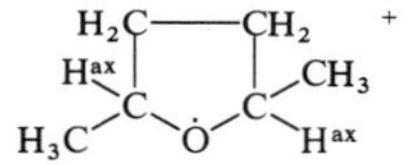 (*trans*)	γ-irr. of $C_6H_{12}O$/ $CFCl_3$ (i) [4] (ii) [4]	EPR/ 77···130 77···130	— —	H(ax): 9.5 H(eq): 3.5 2H: 9.7	81Kub1
$[C_6H_{12}O]^{\cdot+}$ $[(CH_3)_2C$—O—$C(CH_3)_2]^{\cdot+}$	γ-irr. of $C_6H_{12}O$/ $CFCl_3$	EPR/ 110	2.0024	12H: 15.2	85Qin2, 86Rid1
$[C_6H_{14}O]^{\cdot+}$ 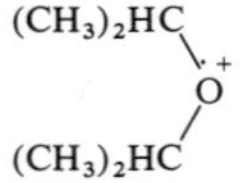	γ-irr. of $C_6H_{14}O$/ $CFCl_3$	EPR/ 77	—	2H: $\simeq$4.5	84Sym1

18.2.5.2 Cyclic polyether cations

Substance	Generation/ Matrix or Solvent	Method/ T [K]	g-Factor	a-Value [mT]	Ref./ add. Ref.
$[C_3H_6O_2]^{\cdot+}$ 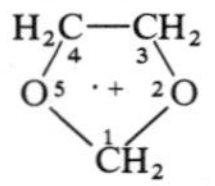	γ-irr. of $C_3H_6O_2$/ $CFCl_3$	EPR/ 77	—	2H(1): 15.3 4H(3, 4): 1.1	84Sym1
	γ-irr. of $C_3H_6O_2$/ $CFCl_3$	EPR/ 96	2.0070	2H(1): 15.32 4H(3, 4): 1.12	82Sno1

[4]) (i) and (ii) are constants for *cis* and *trans* isomers, respectively.

Substance	Generation/ Matrix or Solvent	Method/ T [K]	g-Factor	a-Value [mT]	Ref./ add. Ref.
$[C_3H_6O_3]^{\cdot+}$ 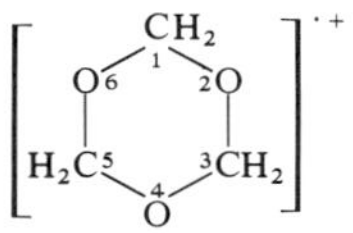	γ-irr. of $C_3H_6O_3$/ $CFCl_3$	EPR/ 89	is: 2.0065	2H: 16.02	82Sno1
	γ-irr. of $C_3H_6O_3$/ $CFCl_3$	EPR/ 77	–	2H(1) is: 16.2 H(β) is: 2.3 H(β) is: 1.1	84Sym1
$[C_4H_8O_3]^{\cdot+}$ 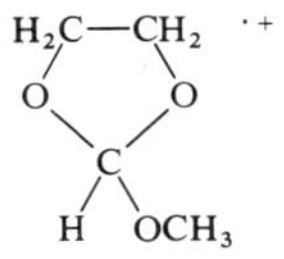	γ-irr. of $C_4H_8O_3$/ $CFCl_3$	EPR/ 77	–	1H, is: 15.0	84Sym1
$[C_4H_8O_2]^{\cdot+}$ 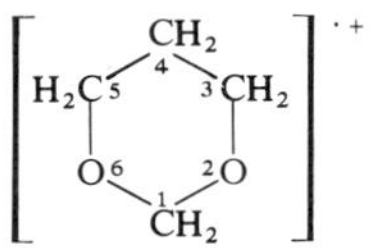	γ-irr. of $C_4H_8O_2$/ $CFCl_3$	EPR/ 83	is: 2.0086	2H(1): 14.06 2H: 2.63 2H: 1.24	82Sno1
	γ-irr. of $C_4H_8O_2$/ $CFCl_3$	EPR/ 77	–	2H(1) is: 14.0 2H(β) is: 2.6 2H(β) is: 1.25	84Sym1
$[C_4H_8O_2]^{\cdot+}$ 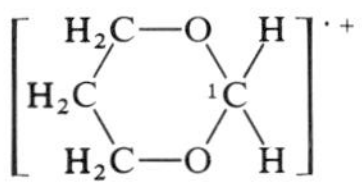	γ-irr. of $C_4H_8O_2$/ $CFCl_3$	EPR/ 77	–	2H(1) is: 14.0	82Rao1
$[C_4H_8O_2]^{\cdot+}$ 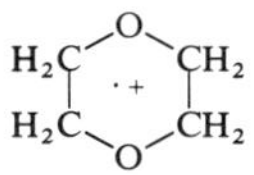	γ-irr. of $C_4H_8O_2$/ $CFCl_3$	EPR/ 77	–	4H(β) is: 1.1 4H, is: 0.8	84Sym1
$[C_6H_{12}O_3]^{\cdot+}$ 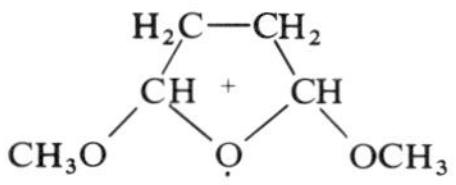	γ-irr. of $C_6H_{12}O_3$/ $CFCl_3$	EPR/ 125	2.0066	2H(β, CH) is: 13.5 2H, is: 3.2	82Sno1

Substance	Generation/ Matrix or Solvent	Method/ T [K]	g-Factor	a-Value [mT]	Ref./ add. Ref.
$[C_{11}H_{20}O_2]^{\cdot +}$ 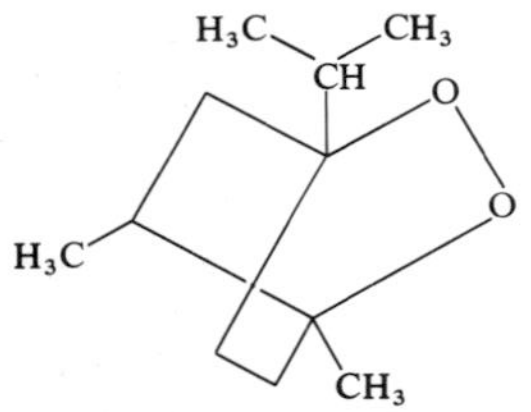	Electrolytic/ CH_2Cl_2 : CF_3COOH : $(CF_3CO)_2O$	EPR/ RT	is: 2.0091	4H, is: 0.47	84Nel1

18.2.6 Cations of carbonyl derivatives

18.2.6.1 Aldehydes

Substance	Generation/ Matrix or Solvent	Method/ T [K]	g-Factor	a-Value [mT]	Ref./ add. Ref.
$[CH_2O]^{\cdot +}$ 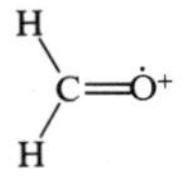	γ-irr. of CH_2O/ Ne	EPR/ 4	2.0069; 2.0015; 2.0025; is: 2.0036	2H: 12.86; 13.36; 12.61; is: 12.89	84Kni2
$[C_2H_4O]^{\cdot +}$	γ-irr. of C_2H_4O/	EPR/			83Sno1, 83Sno2
	CF_3CCl_3	130	2.0056	1H(β) is: 13.62	
	$CFCl_3$	130	2.0054	1H(β) is: 13.62	
$[C_2H_4O]^{\cdot +}$ 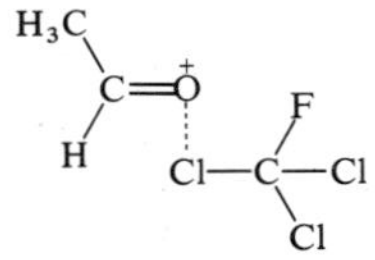	γ-irr. of C_2H_4O/ $CFCl_3$	EPR/ 77	2.0035	1H(β): 13.6; ^{35}Cl: 1.9; 0.6; 0.6; is: 10.3	84Boo1
$[C_2H_3DO]^{\cdot +}$ 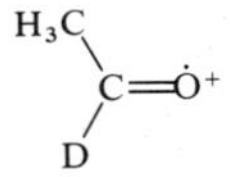	γ-irr. of C_2H_3DO/ $CFCl_3$	EPR/ 140	2.0054	D: 2.07	83Sno1

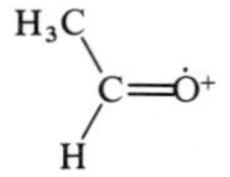

Substance	Generation/ Matrix or Solvent	Method/ T [K]	g-Factor	a-Value [mT]	Ref./ add. Ref.
$[C_2HD_3O]^{\cdot +}$ $D_3C(H)C{=}\dot{O}^{+}$	γ-irr. of C_2HD_3O/ $CFCl_3$	EPR/ 158	2.0048	1H: 13.65; 0; 0; is: 4.55	83Sno1
	$CFCl_3$	50	2.003; 2.008; 2.008; is: 2.006	^{35}Cl: 2.09; 0.6; 0.6; is: 1.1	
$[C_2D_4O]^{\cdot +}$ $D_3C(D)C{=}\dot{O}^{+}$	γ-irr. of C_2D_4O/ $CFCl_3$	EPR/ 150	2.0058	1D, is: 2.08	83Sno1
$[C_3H_6O]^{\cdot +}$ $H_3C{-}CH_2(H)C{=}\dot{O}^{+}$	γ-irr. of C_3H_6O/ $CFCl_3$	EPR/ 120	2.0048	1H(CH) is: 13.51 1H(CH_3) is: 1.25	83Sno2
	γ-irr. of C_3H_6O/ $CFCl_3$	EPR/ 120	is: 2.0035	1H(CH) is: 13.5 1H(CH_3) is: 1.25	84Boo1
$[C_3H_4D_2O]^{\cdot +}$ $H_3C{-}CD_2(H)C{=}\dot{O}^{+}$	γ-irr. of $C_3H_4D_2O$/ $CFCl_3$	EPR/ 120	2.0048	1H(CH) is: 13.53 1H(CH_3) is: 1.25	83Sno2
$[C_4H_8O]^{\cdot +}$ $(CH_3)_2CH(H)C{=}\dot{O}^{+}$	γ-irr. of C_4H_8O/ $CFCl_3$	EPR/ 77	is: 2.0035	1H(β, CH): 13.8 2H(CH_3): 2.0	84Boo1
	γ-irr. of C_4H_8O/ $CFCl_3$	EPR/ 120	2.0044	1H(β, CH) is: 12.03 2H(CH_3) is: 2.04	83Sno2

Substance	Generation/ Matrix or Solvent	Method/ T [K]	g-Factor	a-Value [mT]	Ref./ add. Ref.
18.2.6.2 Acyclic ketones					
$[C_3H_6O]^{\cdot+}$ $(H_3C)_2C{=}\dot{O}^{+}$	γ-irr. of C_3H_6O/ $CFCl_3$	EPR, ENDOR/ 77	–	2H: 0.15 4H: 0.03	84Boo2
	γ-irr. of C_3H_6O/ $CFCl_3$	EPR/ 77	2.008; 2.0033; 2.0071; is: 2.0037		84Ush1
	γ-irr. of C_3H_6O/ $CFCl_3$	EPR/ 88	is: 2.003	–	83Sno2
	γ-irr. of $(^{13}CH_3)_2CO$/ CCl_4	EPR/ 77	2.0061; 2.0035; 2.0023; is: 2.0039	$2\,^{13}C$: 1.20; 1.64; 1.84; is: 1.56 2H: 0.15 4H: 0.03	84Boo2
	$CFCl_3$	77	2.0077; 2.0017; 2.0030; is: 2.0041	$2\,^{13}C$: 2.8; 0.9; 0.9; is: 1.53	
	γ-irr. of $C_2H_6{}^{13}CO$/ $CFCl_3$	EPR/ 77	2.0008; 2.0033; 2.0071; is: 2.0037	^{13}C: 2.4; 3.5; 2.6; is: 2.83	84Ush1
$[C_5H_{10}O]^{\cdot+}$ $(H_3CH_2C)_2C{=}\dot{O}^{+}$	γ-irr. of $C_5H_{10}O$/ $CFCl_3$	EPR/ 77	2.0035	$2H(CH_3)$: 1.1	84Boo1
$[C_7H_{14}O]^{\cdot+}$ $((CH_3)_2HC)_2C{=}\dot{O}^{+}$	γ-irr. of $C_7H_{14}O$/ $CFCl_3$	EPR/ 77	is: 2.0035	$4H(CH_3)$: 1.5	84Boo1
	γ-irr. of $C_7H_{14}O$/ $CFCl_3$	EPR/ 88	is: 2.0033	$4H(CH_3)$: 1.52	83Sno2

Substance	Generation/ Matrix or Solvent	Method/ T [K]	g-Factor	a-Value [mT]	Ref./ add. Ref.
$[C_9H_{18}O]^{\cdot +}$ 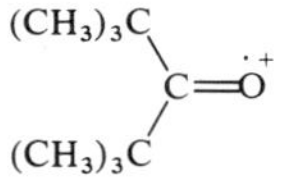	γ-irr. of $C_9H_{18}O$/ $CFCl_3$	EPR/ 77	is: 2.0035	4H(CH_3): 1.5	84Boo1

18.2.6.3 Cyclic ketone cations

Substance	Generation/ Matrix or Solvent	Method/ T [K]	g-Factor	a-Value [mT]	Ref./ add. Ref.
$[C_5H_8O]^{\cdot +}$ 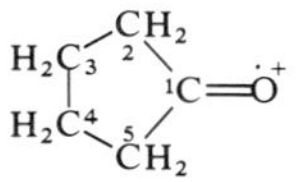	γ-irr. of C_5H_8O/ $CFCl_3$	EPR/ 77	is: 2.0035	2H(3, 4): 1.3	84Boo1
$[C_5H_8O_2]^{\cdot +}$ 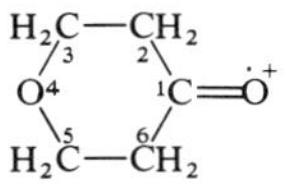	γ-irr. of $C_5H_8O_2$/ $CFCl_3$	EPR/ 77	is: 2.0035	2H(3, 5): 1.95	84Boo1
$[C_6H_{10}O]^{\cdot +}$ 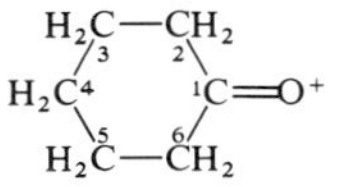	γ-irr. of $C_6H_{10}O$/ $CFCl_3$	EPR/ 77	is: 2.0035	2H(3, 5): 2.75	84Boo1
	γ-irr. of $C_6H_{10}O$/ $CFCl_3$	EPR/ 88	2.0032	2H: 2.75	83Sno2
$[C_6H_6D_4O]^{\cdot +}$ 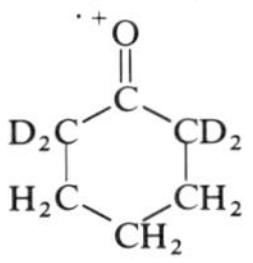	γ-irr. of $C_6H_6D_4O$/ $CFCl_3$	EPR/ 88	2.0032	2H: 2.75	83Sno2

18.2.6.4 Ketene cations

Substance	Generation/ Matrix or Solvent	Method/ T [K]	g-Factor	a-Value [mT]	Ref./ add. Ref.
$[C_2H_2O]^{\cdot +}$ $H_2C{=}C{=}O^{\cdot +}$	γ-irr. of C_2H_2O/	EPR/			83Shi1
	$CFCl_3$	103	is: 2.0032	2H, is: 2.04	
	SF_6	77	is: 2.0031	2H, is: 2.07	
	CCl_2FCClF_2	77	is: 2.0028	2H, is: 2.11	

Substance	Generation/ Matrix or Solvent	Method/ T [K]	g-Factor	a-Value [mT]	Ref./ add. Ref.
$[C_3H_4O]^{\cdot +}$ $H_3C—CH=C=\dot{O}^{\cdot +}$	γ-irr. of C_3H_4O/ SF_6	EPR/ 77	is: 2.0034	3H, is: 3.00 1H, is: 1.76	83Shi1
	$CFCl_3$	103	is: 2.0032	3H, is: 2.9 1H, is: 1.91	
	CCl_2FCClF_2	77	is: 2.0030	3H, is: 2.86 1H, is: 1.97	
	CCl_2FCClF_2	77	is: 2.0033	3H, is: 2.83 1H, is: 2.12	
$[C_4H_6O]^{\cdot +}$ $H_5C_2—CH=C=\dot{O}^{\cdot +}$	γ-irr. of C_4H_6O/ CCl_2FCCl_2F	EPR/ 103	is: 2.0029	2H, is: 4.33 1H, is: 1.81	83Shi1
	CCl_2FCClF_2	77	is: 2.0029	2H, is: 4.43 1H, is: 1.96	
$[C_4H_6O]^{\cdot +}$ $(H_3C)_2C=C=\dot{O}^{\cdot +}$	γ-irr. of C_4H_6O/ $CFCl_3$	EPR/ 77	is: 2.0034	6H, is: 2.34	83Shi1
	CCl_2FCCl_2F	77	is: 2.0034	6H, is: 2.30	
	CCl_2FCClF_2	77	is: 2.0034	6H, is: 2.31	
	γ-irr. of C_4H_6O/ CCl_4	EPR/ 153	2.0028; 2.0037; 2.0037; is: 2.0034	6H: 2.275; 2.362; 2.362; is: 2.333	85Fuj1
	CCl_2FCClF_2	77	is: 2.0034	6H, is: 2.31	
	CCl_2FCCl_2F	77	is: 2.0034	6H, is: 2.30	

18.2.6.5 Dicarbonyl derivatives

Substance	Generation/ Matrix or Solvent	Method/ T [K]	g-Factor	a-Value [mT]	Ref./ add. Ref.
$[C_2H_2O_2]^{\cdot +}$ $[O=CH—CH=O]^{\cdot +}$ or $O=CH—CH=O$ with $^{+\cdot}$ between the O atoms	γ-irr. of $C_2H_2O_2$ (glyoxal)/ $CFCl_3$	EPR/ 77	2.0029	2H, is: 8.42	86Por1

Substance	Generation/ Matrix or Solvent	Method/ T [K]	g-Factor	a-Value [mT]	Ref./ add. Ref.
$[C_3H_4O_2]^{\cdot+}$ 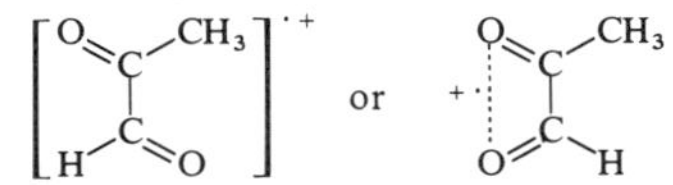	γ-irr. of $C_3H_4O_2$ (methyl glyoxal)/ $CFCl_3$	EPR/ 77	2.0061; 2.0025; 1.9989; is: 2.0025	H(1): 8.1; 8.2; 8.62; is: 8.31	86Por1
$[C_4H_6O_2]^{\cdot+}$ 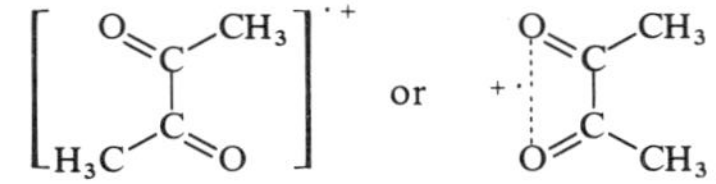	γ-irr. of $C_4H_6O_2$ (dimethyl glyoxal)/ $CFCl_3$	EPR/ 77	2.0019	singlet	86Por1

18.2.6.6 Carboxylic acid and ester cations

Substance	Generation/ Matrix or Solvent	Method/ T [K]	g-Factor	a-Value [mT]	Ref./ add. Ref.
$[CH_2O_2]^{\cdot+}$ 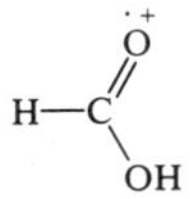	γ-irr. of CH_2O_2/ $CFCl_3$	EPR/ 77	2.016; 2.007; 2.003; is: 2.0087	H(1): 8.8; 8.7; 9.2; is: 8.9	86Cla1
$[C_2H_4O_2]^{\cdot+}$ 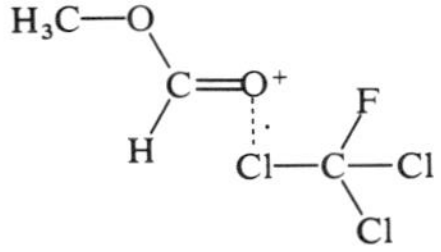	γ-irr. of $C_2H_4O_2$/ $CFCl_3$	EPR/ 77	–	H(1): 1.7; ^{35}Cl: 8.45	84Rao1
$[C_2H_4O_2]^{\cdot+}$ 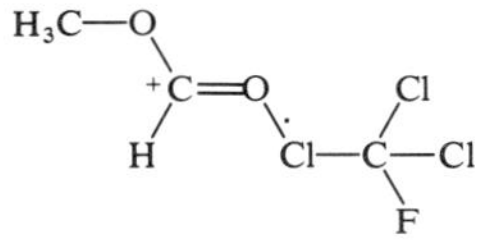	γ-irr. of $C_2H_4O_2$/ $CFCl_3$	EPR/ 77	2.0032	^{35}Cl: 8.44; ^{37}Cl: 7.03; H: 1.7	83Bec1
$[C_2H_4O_2]^{\cdot+}$ 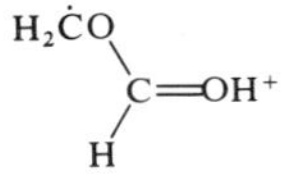	γ-irr. of $C_2H_4O_2$/ $CFCl_3$	EPR/ 140	is: 2.0026	H(OH): 0.55; 2H(α): 2.3; H(β): 0.4	84Rao1

Substance	Generation/ Matrix or Solvent	Method/ T [K]	g-Factor	a-Value [mT]	Ref./ add. Ref.
$[C_2H_3DO_2]^{\cdot +}$ 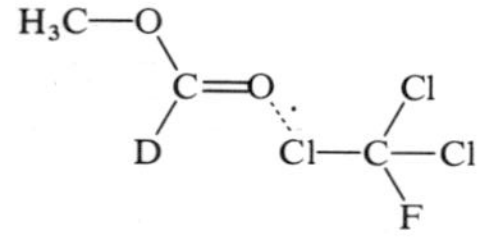 $H_3C{-}O{-}C(D){=}O\cdots ClCCl_2F$	γ-irr. of $C_2H_3DO_2$/ $CFCl_3$	EPR/ 77	2.0032	^{35}Cl: 8.4 ^{37}Cl: 7.06	83Bec1
$[C_3H_6O_2]^{\cdot +}$ 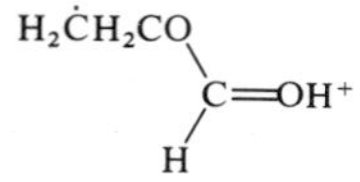 $H_2\dot{C}H_2CO{-}C(H){=}OH^+$ [1]	γ-irr. of $C_3H_6O_2$/ $CFCl_3$	EPR/ 77	2.0026	1H(OH): ≤0.4 2H(α): 2.2 2H(β): 1.6	84Rao1
	–	160	–	2H(β): 1.0	
$[C_3H_6O_2]^{\cdot +}$ 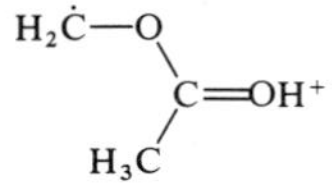 $H_2\dot{C}{-}O{-}C(H_3C){=}OH^+$ [1]	γ-irr. of $C_3H_6O_2$/ $CFCl_3$	EPR/ 77	is: 2.0026	3H(OH): 0.5 2H(α): 2.2	84Rao1
$[C_3H_4D_2O_2]^{\cdot +}$ 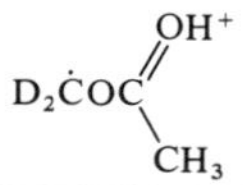 $D_2\dot{C}OC(=OH^+)CH_3$	γ-irr. of $C_3H_4D_2O_2$/ $CFCl_3$	EPR/ 77	–	broad singlet	86Rid2 [2]
$[C_4H_8O_2]^{\cdot +}$ 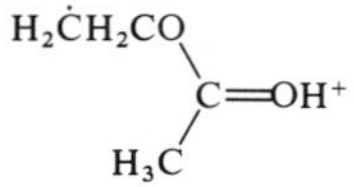 $H_2\dot{C}H_2CO{-}C(H_3C){=}OH^+$ [1]	γ-irr. of $C_4H_8O_2$/ $CFCl_3$	EPR/ 77	is: 2.0026	1H(α): 2.2 1H(α): 3.1 1H(β): 0.8 1H(β): 1.7	84Rao1
$[C_5H_{10}O_2]^{\cdot +}$ 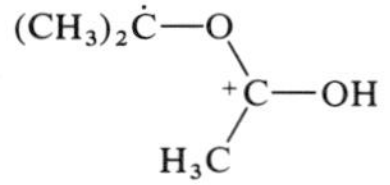 $(CH_3)_2\dot{C}{-}O{-}{}^{+}C(H_3C){-}OH$	γ-irr. of $C_5H_{10}O_2$/ $CFCl_3$	EPR/ 77	–	6H: 2.2	84Rao1

[1]) Probably the rearranged cation.
[2]) Establishes rearrangement.

18.2.6.7 Diesters

Substance	Generation/ Matrix or Solvent	Method/ T [K]	g-Factor	a-Value [mT]	Ref./ add. Ref.
$[C_4H_6O_4]^{\cdot+}$ 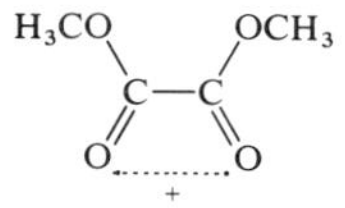 [3]	γ-irr. of $C_4H_6O_4$/ $CFCl_3$	EPR/ 77	2.0090; 2.0046; 2.0023; is: 2.0053	singlet	86Rid3
$[C_5H_8O_4]^{\cdot+}$ 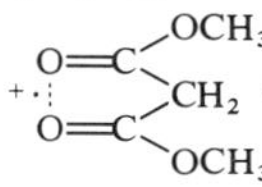 [3]	γ-irr. of $C_5H_8O_4$/ $CFCl_3$	EPR/ 77	2.0045; 2.0026; 2.0022; is: 2.0031	singlet	86Rid3
$[C_6H_8O_4]^{\cdot+}$ 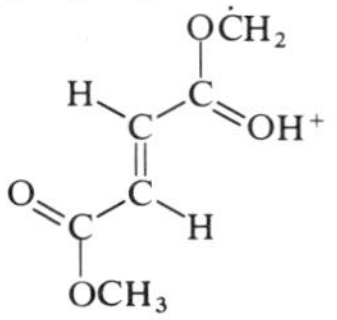 [4]	γ-irr. of $C_6H_8O_4$/ $CFCl_3$	EPR/ 77	is: 2.0030	2H, is: 1.9	86Rid3
$[C_6H_8O_4]^{\cdot+}$ 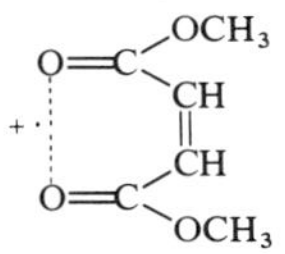	γ-irr. of $C_6H_8O_4$/ $CFCl_3$	EPR/ —	2.006, 2.003	2H, is: 0.7	86Rid3
$[C_6H_{10}O_4]^{\cdot+}$ 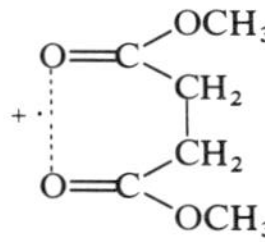 [3]	γ-irr. of $C_6H_{10}O_4$/ $CFCl_3$	EPR/ 88	2.0043; 2.0025; 2.0022; is: 2.0030	2H, is: 0.55	86Rid3

[3]) May be *trans-* rather than *cis.*
[4]) Rearranged cation.

Substance	Generation/ Matrix or Solvent	Method/ T [K]	g-Factor	a-Value [mT]	Ref./ add. Ref.
18.2.6.8 Cyclic esters (lactones)					
$[C_3H_4O_2]^{\cdot +}$ $H_2\dot{C}CH_2OCO^+$	γ-irr. of $C_3H_4O_2$ (propiolactone)/ $CFCl_3$	EPR/ 77 148	 — —	 3H, is: 2.2 2H, is: 2.4 2H, is: 1.05	85Rid1
$[C_3H_2D_2O_2]^{\cdot +}$ $H_2\dot{C}CD_2OCO^+$	γ-irr. of $C_3H_2D_2O_2$ (propiolactone-d_2)/ $CFCl_3$	EPR/ 77	—	2H: 2.2 1D: 0.3	85Rid1
$[C_4H_6O_2]^{\cdot +}$ 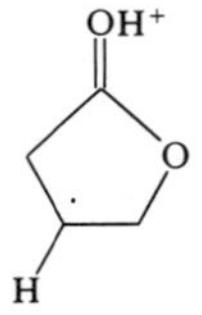	γ-irr. of $C_4H_6O_2$ (γ-butyrolactone)/ $CFCl_3$	EPR/ 145	—	2H, is: 4.2 1H, is: 2.2	85Rid1
$[C_5H_8O_2]^{\cdot +}$ 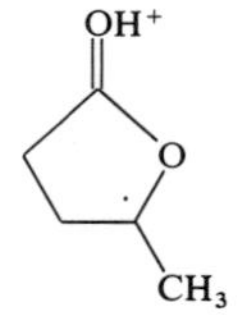	γ-irr. of $C_5H_8O_2$ (γ-valerolactone)/ $CFCl_3$	EPR/ 145	—	4H, is: 2.1 1H, is: 4.0	85Rid1
$[C_6H_{10}O_2]^{\cdot +}$ 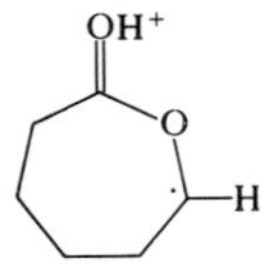	γ-irr. of $C_6H_{10}O_2$ (ε-caprolactone)/ $CFCl_3$	EPR/ 145	—	1H, is: 4.4 2H, is: 2.1	85Rid1

Substance	Generation/ Matrix or Solvent	Method/ T [K]	g-Factor	a-Value [mT]	Ref./ add. Ref.
18.2.6.9 Carbonates					
$[C_3H_4O_3]^{\cdot +}$ (cyclic: $\dot{C}H$–O–C(=OH^+)–O–CH_2)	γ-irr. of $C_3H_4O_3$/ $CFCl_3$	EPR/ 77	–	1H, is: 1.6 2H, is: 3.4	86Gan1
$[C_3H_6O_3]^{\cdot +}$ $H_2\dot{C}OC(=OH^+)OCH_3$	γ-irr. of $C_3H_6O_3$/ $CFCl_3$	EPR/ 77	–	2H: 3.1; 1.6; 1.6; is: 2.1	86Gan1
$[C_4H_7O_3]^{\cdot +}$ (cyclic: $\dot{C}(CH_3)$–O–C(=OH^+)–O–CH_2)	γ-irr. of $C_4H_7O_3$/ $CFCl_3$	EPR/ 77	–	3H, is: 2.1 1H, is: 3.0 1H, is: 3.6	86Gan1
$[C_5H_{10}O_3]^{\cdot +}$ $H_2\dot{C}CH_2COC(=OH^+)OCH_2CH_3$	γ-irr. of $C_5H_{10}O_3$/ $CFCl_3$	EPR/ 77	–	2H, is: 2.2 1H, is: 1.7 1H, is: 0.4	86Gan1
		130	–	2H, is: 2.2 2H, is: 1.1	
18.2.6.10 Amides					
$[C_3H_7NO]^{\cdot +}$ $(CH_3)_2\dot{N}^{+}$–C(H)=O	γ-irr. of C_3H_7NO/ $CFCl_3$	EPR/ 77	2.002; 2.007; 2.007; is: 2.005	N: 3.8 (1); ≈0.4; ≈0.4 6H: 3.2 (1); 3.3 (1); 3.3 (1); is: 3.3	86Eas1

Substance	Generation/ Matrix or Solvent	Method/ T [K]	g-Factor	a-Value [mT]	Ref./ add. Ref.
$[C_4H_9NO]^{\cdot+}$ $(CH_3)_2\dot{N}^{+}–C(CH_3)=O$	γ-irr. of C_4H_9NO/ $CFCl_3$	EPR/ 77	2.002; 2.007; 2.007; is: 2.005	N: 3.8 (1); ≈0.4; ≈0.4 6H: 3.2 (1); 3.3 (1); 3.3 (1); is: 3.3	86Eas1
$[C_4D_{10}NO]^{\cdot+}$ $(CD_3)_3\dot{N}^{+}–C(D)=O$	γ-irr. of $C_4H_{10}NO$/ $CFCl_3$	EPR/ 77	2.002; 2.007; 2.007; is: 2.005	N: 3.9 (1); ≈0.2; ≈0.2 D: 0.45; 0.5; ≈0.5; is: 0.48	86Eas1
$[C_6H_{11}NO]^{\cdot+}$ H–C(=O)–$\dot{N}^{+}$(H_2C–CH_2–CH_2–CH_2–CH_2)	γ-irr. of $C_6H_{11}NO$/ $CFCl_3$	EPR/ 77	2.002; 2.007; 2.007; is: 2.005	N: 3.6 (1); ≈0.4; ≈0.4 4H: 2.5 (3); 2.5 (3); 2.5 (3); is: 2.5 (3)	86Eas1
$[C_5H_9NO]^{\cdot+}$ $CH_3\dot{N}^{+}$(CH_2–CH_2–CH_2–C=O)	γ-irr. of C_5H_9NO/ $CFCl_3$	EPR/ 77	2.002; 2.007; 2.007; is: 2.005	N: 3.8 (1); ≈0.4; ≈0.4 3H, is: 3.3 (3) H, is: 6.2 (3) H, is: 3.6 (3)	86Eas1

18.2.7 Dialkylperoxide cations

Substance	Generation/ Matrix or Solvent	Method/ T [K]	g-Factor	a-Value [mT]	Ref./ add. Ref.
$[C_5H_8O_2]^{\cdot+}$ [bicyclic peroxide: C^7H_2, H_2C^5, C^4H, O^3, O^2, C^1H, H_2C^6]$^{\cdot+}$	$CF_3COOH:(CF_3CO)_2O$/ CH_2Cl_2	EPR/ 190	2.0094	2H(5, 6, *exo*) is: 0.99 H(7, *anti*) is: 0.22 H(7, *syn*) is: 0.11 2H, is: 0.12 2H, is: 0.03	85Nel1

Substance	Generation/ Matrix or Solvent	Method/ T [K]	g-Factor	a-Value [mT]	Ref./ add. Ref.
$[C_6H_{10}O_2]^{\cdot +}$ 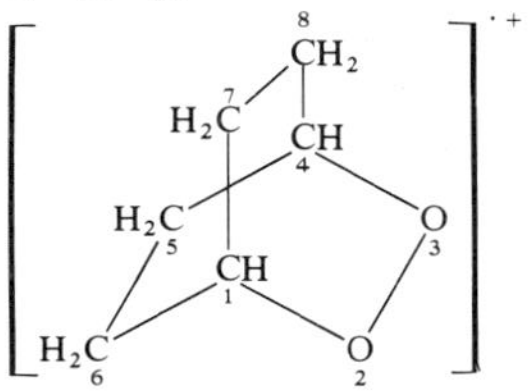	$CF_3COOH:(CF_3CO)_2O/$ CH_2Cl_2	EPR/ 190	2.0094	4H(5⋯8, *exo*) is: 0.47 2H(1, 4) is: 0.11 4H(5⋯8, *endo*) is: 0.02	85Nel1
$[C_7H_{12}O_2]^{\cdot +}$ 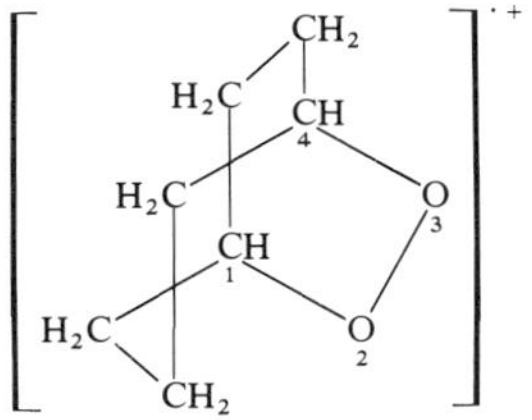	$CF_3COOH:(CF_3CO)_2O/$ CH_2Cl_2	EPR/ 190	2.0093	2H, is: 0.32 2H, is: 0.22 2H, is: 0.1	85Nel1
$[C_8H_{14}O_2]^{\cdot +}$ 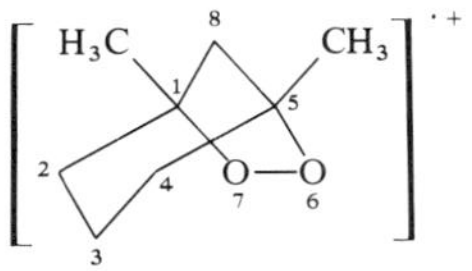	Electrolytic/ $CH_2Cl_2:CF_3COOH:$ $(CF_3CO)_2O$	EPR/ 195	is: 2.0084	4H(2, 4, ax) is: 0.48 4H(2, 4, eq) is: 0.07 1H(8, ax) is: 0.16 1H, is: 0.04 1H, is: 0.02 1H, is: 0.01 $6H(CH_3)$ is: <0.005	84Nel1
$[C_8H_{18}O_2]^{\cdot +}$ $[(CH_3)_3C—O—O—C(CH_3)_3]^{\cdot +}$	γ-irr. of $C_8H_{18}O_2/$ $CFCl_3$	EPR/ 77	2.015; 2.008; 2.002; is: 2.0084	—	83Cha1

Substance	Generation/ Matrix or Solvent	Method/ T [K]	g-Factor	a-Value [mT]	Ref./ add. Ref.
$[C_{11}H_{20}O_2]^{\cdot +}$ 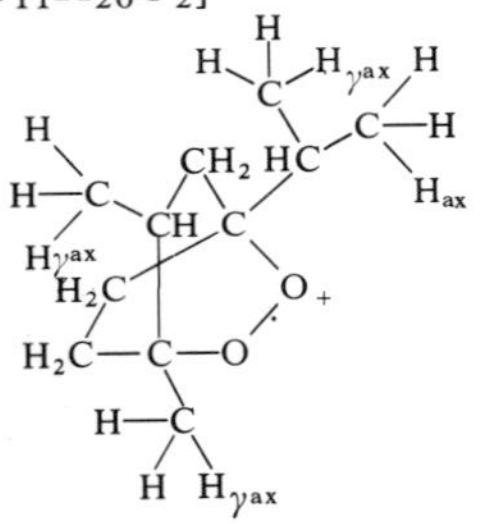	Electrolytic/ $CH_2Cl_2:CF_3COOH:(CF_3CO)_2O$	EPR/ 195	2.0091	4H(γ, ax) is: 0.47	84Nel1
18.2.8 Nitroalkane cations					
$[CN_4O_8]^{\cdot +}$ 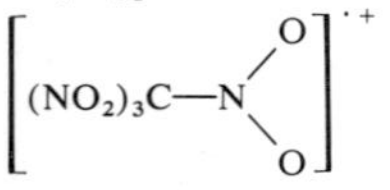	γ-irr. of CN_4O_8/ $CFCl_3$	EPR/ 77	2.004; 1.992; 2.002; is: 1.999	$^{14}N(1)$: 5.0; 4.8; 6.7; is: 5.5	85Rao1
$[CH_3NO_2]^{\cdot +}$ 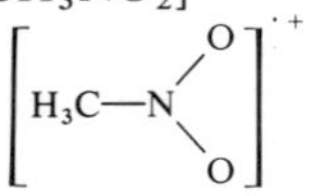	γ-irr. of CH_3NO_2/ $CFCl_3$	EPR/ 77	2.0045; 1.994; 2.002; is: 2.000	$^{14}N(1)$: 5.2; 4.7; 6.65; is: 5.52 $^{13}C(1)$: 12.0; 12.0; 15.5; is: 13.2	85Rao1
$[C_2H_5NO_2]^{\cdot +}$ 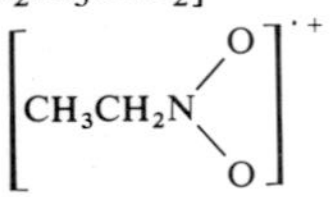	γ-irr. of $C_2H_5NO_2$/ $CFCl_3$	EPR/ 77	2.005; 1.994; 2.002; is: 2.000	$^{14}N(1)$: 5.3; 4.9; 7.0; is: 5.73	85Rao1
$[C_3H_6ClNO_2]^{\cdot +}$ 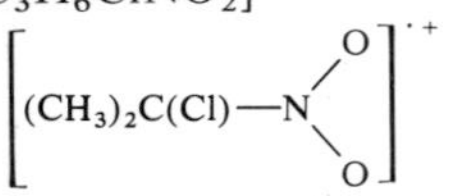	γ-irr. of $C_3H_6NO_2Cl$/ $CFCl_3$	EPR/ 77	2.002; 1.995; 2.002; is: 2.000	$^{14}N(1)$: 5.25; 5.0; 6.7; is: 5.65 $^{35}Cl(1)$: 0; 0; 0.5; is: 0.17	85Rao1
$[C_3H_6BrNO_2]^{\cdot +}$ 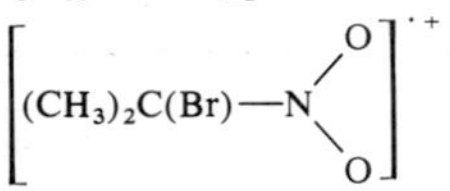	γ-irr. of $C_3H_6NOP_2Br$/ $CFCl_3$	EPR/ 77	2.004; 1.995; 2.002; is: 2.000	$^{14}N(1)$: 5.2; 4.9; 6.6; is: 5.56 ^{81}Br: 0.4; 0.4; 0.75; is: 0.52	85Rao1

Substance	Generation/ Matrix or Solvent	Method/ T [K]	g-Factor	a-Value [mT]	Ref./ add. Ref.
$[C_3H_7NO_2]^{\cdot+}$ $[CH_3CH_2CH_2N(O)O]^{\cdot+}$	γ-irr. of $C_3H_7NO_2$/ $CFCl_3$	EPR/ 77	2.0045; 1.994; 2.002; is: 2.000	$^{14}N(1)$: 5.2; 4.9; 6.7; is: 5.6	85Rao1
$[C_3H_7NO_2]^{\cdot+}$ $[(CH_3)_2CHN(O)O]^{\cdot+}$	γ-irr. of $C_3H_7NO_2$/ $CFCl_3$	EPR/ 77	2.005; 1.994; 2.002; is: 2.000	$^{14}N(1)$: 5.2; 4.8; 6.7; is: 5.57	85Rao1
$[C_4H_9NO_2]^{\cdot+}$ $[(CH_3)_3C{-}N(O)O]^{\cdot+}$	γ-irr. of $C_4H_9NO_2$/ $CFCl_3$	EPR/ 77	1.999; 1.996; 2.002; is: 1.999	$^{14}N(1)$: 5.2; 5.2; 6.3; is: 5.57	85Rao1
$[C_6H_{12}N_2O_4]^{\cdot+}$ $[(CH_3)_2C(NO_2){-}C(CH_3)_2N(O)O]^{\cdot+}$	γ-irr. of $C_6H_{12}N_2O_4$/ $CFCl_3$	EPR/ 77	2.005; 1.993; 2.002; is: 2.000	$^{14}N(1)$: 5.3; 4.9; 6.8; is: 5.67	85Rao1
$[C_7H_{11}NO_2]^{\cdot+}$ $[(CH_3)_2C(CO_2C_2H_5)N(O)O]^{\cdot+}$	γ-irr. of $C_7H_{11}NO_2$/ $CFCl_3$	EPR/ 77	2.005; 1.994; 2.002; is: 2.000	$^{14}N(1)$: 5.2; 4.9; 6.7; is: 5.6	85Rao1

18.2.9 Halogenalkane cations

18.2.9.1 Chloroalkanes

Substance	Generation/ Matrix or Solvent	Method/ T [K]	g-Factor	a-Value [mT]	Ref./ add. Ref.
$[CCl_4]^{\cdot+}$ $[Cl{-}C(Cl)(Cl){-}Cl]^{\cdot+}$	γ-irr. of CCl_4/ CCl_4	EPR/ 77	2.005; 2.043; 2.043; is: 2.0303	$3\,^{35}Cl$: 5.94; 4.0; 4.0; is: 4.65	82Sym1

Substance	Generation/ Matrix or Solvent	Method/ T [K]	g-Factor	a-Value [mT]	Ref./ add. Ref.
$[C_2H_5Cl]^{\cdot+}$ $[C_2H_5Cl\cdots\!\therefore\!\cdots ClCCl_3]^+$	γ-irr. of C_2H_5Cl/ CCl_4 (I) [1] (II) [1]	EPR/ 96	2.003; 2.000; 2.000; is: 2.001	^{35}Cl: 10.5; 0.35; 0.35; is: 3.73 ^{37}Cl: 8.7; 3.0; 3.0; is: 4.9 ^{35}Cl: 8.8; 3.5; 3.5 is: 5.26 ^{37}Cl: 7.3; 3.0; 3.0; is: 4.43 H: 0.7; 0.7; 0.7; is: 0.7	84Eas2
$[C_2H_5Cl]^{\cdot+}$ $[C_2H_5Cl\cdots\!\therefore\!\cdots ClCFCl_2]^+$	γ-irr. of C_2H_5Cl/ $CFCl_3$ (I) [1] (II) [1]	EPR/ 77	2.006; 2.000; 2.000 is: 2.002	^{35}Cl: 11.0; 3.5; 3.5; is: 6.00 ^{37}Cl: 9.2; 3.0; 3.0; is: 5.06 ^{35}Cl: 8.5; 3.5; 3.5; is: 13.16 ^{37}Cl: 7.1; 3.0; 3.0; is: 4.36 H: 1.0; 1.0; 1.0; is: 1.0	84Eas2
$[C_2Cl_6F_2]^{\cdot+}$	γ-irr. of $CFCl_3$/ $CFCl_3$	EPR/ 77	2.0049; 2.0110; 2.0396; is: 2.0185	$3^{35}Cl$: 9.4; 0.5; 0.5; is: 3.47	86Sym1
$[C_4H_{10}Cl_2]^{\cdot+}$ $[C_2H_5Cl\dot{-}ClC_2H_5]^+$	γ-irr. of $C_4H_{10}Cl_2$/ $CFCl_3$	EPR/ 77	2.006; 2.000; 2.000; is: 2.002	^{35}Cl: 9.9; 3.5; 3.5; is: 5.63 ^{37}Cl: 8.2; 3.0; 3.0; is: 4.73	84Eas2

[1]) (I) and (II) are different possible conformers.

Substance	Generation/ Matrix or Solvent	Method/ T [K]	g-Factor	a-Value [mT]	Ref./ add. Ref.
18.2.9.2 Bromoalkanes					
$[CBr_4]^{\cdot +}$ Br_2C(Br)(Br) $^{\cdot +}$	γ-irr. of CBr_4/ $CFCl_3$	EPR/ 77	2.052; 2.079; [2])	^{81}Br: 32.0; 23.5; [2])	85Sym1
$[CHBr_3]^{\cdot +}$ HC(Br)(Br)(Br) $^{\cdot +}$	γ-irr. of $CHBr_3$/ $CFCl_3$	EPR/ 77	2.050; 2.078; [2])	^{81}Br: 31.2; 23.0; [2])	85Sym1
$[CH_2Br_2]^{\cdot +}$ H_2C(Br)(Br) $^{\cdot +}$	γ-irr. of CH_2Br_2/ $CFCl_3$	EPR/ 77	2.051; 2.078; [2])	^{81}Br: 31.5; 23.0; [2])	85Sym1
$[C_2H_5Br]^{\cdot +}$ H_3CH_2C—Br $\cdots^{\cdot +}$ Cl—$CFCl_3$	γ-irr. of C_2H_5Br/ $CFCl_3$	EPR/ 4.2	2.490; 2.100; 1.902; is: 2.163	^{81}Br: 49.0; 20.0; 20.0; is: 29.7 ^{35}Cl: 5.3; 1.8; 1.8; is: 2.97	86Mut1
$[C_2H_5Br]^{\cdot +}$ $[C_2H_5Br\cdots\cdots ClCCl_3]^+$	γ-irr. of C_2H_5Br/ CCl_4	EPR/ 77	1.955; 2.329; 2.329; is: 2.2043	^{79}Br: 48.0; 11.9; 11.9; is: 23.93 ^{81}Br: 51.7; 12.9; 12.9; is: 25.83 ^{35}Cl: 6.8; 1.7; 1.7; is: 3.40	84Eas2
$[C_2BrD_5]^{\cdot +}$ $[C_2D_5Br\cdots\cdots ClCFCl_2]^+$	γ-irr. of C_2D_5Br/ $CFCl_3$	EPR/ 77	1.922; 2.392; 2.392; is: 2.2353	^{79}Br: 46.4; 15.9; 15.9; is: 26.06 ^{81}Br: 50.0; 17.1; 17.1; is: 28.06 ^{35}Cl: 5.7; 1.7; 1.7; is: 3.03	84Eas2

[2]) g_z and A_z not defined.

Substance	Generation/ Matrix or Solvent	Method/ T [K]	g-Factor	a-Value [mT]	Ref./ add. Ref.
$[C_4H_{10}Br_2]^{\cdot +}$ $[C_2H_5Br \cdots BrC_2H_5]^+$	γ-irr. of $C_4H_{10}Br_2$/ $CFCl_3$	EPR/ 77	1.988; 2.000; 2.000; is: 1.996	^{79}Br: 42.1; 12.0; 12.0; is: 22.03 ^{81}Br: 45.4; 13.0; 13.0; is: 23.80	84Eas2

18.2.9.3 Iodoalkanes

Substance	Generation/ Matrix or Solvent	Method/ T [K]	g-Factor	a-Value [mT]	Ref./ add. Ref.
$[C_2H_5I]^{\cdot +}$ $[C_2H_5I \cdots IC_2H_5]^+$	γ-irr. of C_2H_5I/ $CFCl_3$	EPR/ 77	1.877; 2.100; 2.100; is: 2.0256	^{127}I: 46.0; 12.0; 12.0; is: 23.33	84Eas2

18.3 References for 18.2

70Gil1 Gilbert, B.C., Schlossel, R.H., Gulick jr., W.M.: J. Am. Chem. Soc. **92** (1970) 2974.
73Ich1 Ichikawa, T., Ohta, N.: J. Am. Chem. Soc. **95** (1973) 8175.
76Boc1 Bock, H., Brähler, G., Fritz, G., Matern, E.: Angew. Chem. **88** (1976) 765.
77Boc1 Bock, H., Kaim, W.: Tetrahedron Lett. **1977**, 2343.
78Fri1 Fritz, G., Matern, E., Bock, H., Brähler, G.: Z. Anorg. Allg. Chem. **439** (1978) 173.
79Kai1 Kaim, W., Bock, H.: J. Organomet. Chem. **164** (1979) 281.
79Okh1 Okhlobystin, O.Yu., Samarskii, V.A., Nekhoroshev, M.V., Pokhodenko, V.D.: Zh. Org. Khim. **15** (1979) 1110.
79Sym1 Symons, M.C.R., Smith, I.G.: J. Chem. Res. (S) **1979**, 382.

80Beh1 Behrens, G., Bothe, E., Koltzenburg, G., Schulte-Frohlinde, D.: J. Chem. Soc., Perkin Trans. II **1980**, 883.
80Boc1 Bock, H., Brähler, G., Henkel, U., Schlecker, R., Seebach, D.: Chem. Ber. **113** (1980) 289.
80Boc2 Bock, H., Kaim, W.: J. Am. Chem. Soc. **102** (1980) 4429.
80Boc3 Bock, H., Roth, B., Maier, G.: Angew. Chem. Int. Ed. Engl. **19** (1980) 209.
80Shi1 Shida, T., Egawa, Y., Kubodera, H., Kato, T.: J. Chem. Phys. **73** (1980) 5963.
80Wan1 Wang, J.T., Williams, F.: J. Phys. Chem. **84** (1980) 3156.

81Ald1 Alder, R.W., Sessions, R.B., Symons, M.C.R.: J. Chem. Res. (S) **1981**, 82.
81Bro1 Broxterman, Q.B., Hogeveen, H., Kok, D.M.: Tetrahedron Lett. **22** (1981) 173.
81Ger1 Gerson, F., Lopez, J., Krebs, A., Ruger, W.: Angew. Chem. **93** (1981) 106.
81Ger2 Gerson, F., Lopez, J., Akaba, R., Nelsen, S.F.: J. Am. Chem. Soc. **103** (1981) 6716.
81Iwa1 Iwasaki, M., Toriyama, K., Nunome, K.: J. Am. Chem. Soc. **103** (1981) 3591.
81Kub1 Kubodera, H., Shida, T., Shimokoshi, K.: J. Phys. Chem. **85** (1981) 2583.
81Shi1 Shida, T., Kubodera, H., Egawa, Y.: Chem. Phys. Lett. **79** (1981) 179.
81Wan1 Wang, J.T., Williams, F.: Chem. Phys. Lett. **82** (1981) 177.

82Rao1 Rao, D.N.R., Symons, M.C.R., Wren, B.W.: Tetrahedron Lett. **23** (1982) 4739.
82Sno1 Snow, L.D., Wang, J.T., Williams, F.: J. Am. Chem. Soc. **104** (1982) 2062.
82Sym1 Symons, M.C.R., Albano, E., Slater, F.T., Tomasi, A.: J. Chem. Soc., Faraday Trans. I **78** (1982) 2205.
82Tor1 Toriyama, K., Nunome, K., Iwasaki, M.: J. Chem. Phys. **77** (1982) 5891

83Bec1 Becker, D., Plante, K., Sevilla, M.D.: J. Phys. Chem. **87** (1983) 1648.
83Bro1 Broxterman, Q.B., Hogeveen, H.: Tetrahedron Lett. **23** (1983) 639.
83Cha1 Chandra, H., Rao, D.N.R., Symons, M.C.R.: J. Chem. Res. (S) **1983**, 68.
83Cou1 Courtneidge, J.L., Davies, A.G., Parkin, J.E.: J. Chem. Soc., Chem. Commun. **1983**, 1262.
83Cou2 Courtneidge, J.L., Davies, A.G., Lusztyk, J.: J. Chem. Soc., Chem. Commun. **1983**, 893.
83Has1 Hasegawa, A., Symons, M.C.R.: J. Chem. Soc., Faraday Trans. I **79** (1983) 93.
83Iwa1 Iwasaki, M., Toriyama, K., Nunome, K.: J. Chem. Soc., Chem. Commun. **1983**, 202.
83Oht1 Ohta, K., Nakatsuji, H., Kubodera, H., Shida, T.: Chem. Phys. **76** (1983) 271.
83Rao1 Rao, D.N.R., Symons, M.C.R.: Tetrahedron Lett. **24** (1983) 1293.
83Shi1 Shimokoshi, K., Fujisawa, J., Nakamura, K., Sato, S., Shida, T.: Chem. Phys. Lett. **99** (1983) 483.
83Sno1 Snow, L.D., Williams, F.: Chem. Phys. Lett. **100** (1983) 198.
83Sno2 Snow, L.D., Williams, F.: J. Chem. Soc., Chem. Commun. **1983**, 1090.
83Sym1 Symons, M.C.R., Wren, B.W.: Tetrahedron Lett. **24** (1983) 2315.
83Tab1 Tabata, M., Lund, A.: Chem. Phys. **75** (1983) 379.
83Tor1 Toriyama, K., Nunome, K., Iwasaki, M.: J. Chem. Soc., Chem. Commun. **1983**, 1346.

84Boc1 Bock, H., Kaim, W., Kira, M., Rene, L., Viehe, H.-G.: Z. Naturforsch. **39b** (1984) 763.
84Boc2 Bock, H., Roth, B., Maier, G.: Chem. Ber. **117** (1984) 172.
84Boo1 Boon, P.J., Symons, M.C.R., Ushida, K., Shida, T.: J. Chem. Soc., Perkin Trans. II **1984**, 1213.
84Boo2 Boon, P.J., Harris, L., Olm, M.T., Wyatt, J.L., Symons, M.C.R.: Chem. Phys. Lett. **106** (1984) 408.

84Cha1 Chan, W., Courtneidge, J.L., Davies, A.G., Djap, W.H., Gregory, P.S., Yadzi, S.N.: J. Chem. Soc., Chem. Commun. **1984**, 1541.

84Cou1 Courtneidge, J.L., Davies, A.G.: J. Chem. Soc., Chem. Commun. **1984**, 136.

84Cou2 Courtneidge, J.L., Davies, A.G., Lusztyk, E., Lusztyk, J.: J. Chem. Soc., Perkin Trans. II **1984**, 155.

84Cou3 Courtneidge, J.L., Davies, A.G., Yadzi, S.N.: J. Chem. Soc., Chem. Commun **1984**, 570.

84Eas1 Eastland, G.W., Kurita, Y., Symons, M.C.R.: J. Chem. Soc., Perkin Trans. II **1984**, 1843.

84Eas2 Eastland, G.W., Maj, S.P., Symons, M.C.R., Hasegawa, A., Glidewell, C., Hayashi, M., Wakabayashi, T.: J. Chem. Soc., Perkin Trans. II **1984**, 1439.

84Ger1 Gerson, F., Huber, W., Lopez, J.: J. Am. Chem. Soc. **106** (1984) 5808.

84Iwa1 Iwasaki, M., Toriyama, K., Nunome, K.: Chem. Phys. Lett. **111** (1984) 309.

84Kni1 Knight, jr., L.B., Steadman, J., Feller, D., Davidson, E.R.: J. Am. Chem. Soc. **106** (1984) 3700.

84Kni2 Knight, L.B., Steadman, J.: J. Chem. Phys. **80** (1984) 1018.

84Nel1 Nelson, S.F., Teasley, M.F., Kapp, D.L., Wilson, R.M.: J. Org. Chem. **49** (1984) 1843.

84Qin1 Qin, X.Z., Williams, F.: Chem. Phys. Lett. **112** (1984) 79.

84Qin2 Qin, X.Z., Snow, L.D., Williams, F.: J. Am. Chem. Soc. **106** (1984) 7640.

84Rao1 Rao, D.N.R., Rideout, J., Symons, M.C.R.: J. Chem. Soc., Perkin Trans. II **1984**, 1221.

84Shi1 Shiotani, M., Nagata, Y., Sohma, J.: J. Am. Chem. Soc. **106** (1984) 4640.

84Shi2 Shiotani, M., Nagata, Y., Sohma, J.: J. Phys. Chem. **88** (1984) 4078.

84Shi3 Shiotani, M., Kawazoe, H., Sohma, J.: Chem. Phys. Lett. **111** (1984) 254.

84Sym1 Symons, M.C.R., Wren, B.W.: J. Chem. Soc., Perkin Trans. II **1984**, 511.

84Tor1 Toriyama, K., Nunome, K., Iwasaki, M.: Chem. Phys. Lett. **107** (1984) 86.

84Ush1 Ushida, K., Shida, T.: Chem. Phys. Lett. **108** (1984) 200.

85Cha1 Chan, W., Courtneidge, J.L., Davies, A.G., Evance, J.C., Neville, A.G., Rowlands, C.C.: Tetrahedron Lett. **26** (1985) 4212.

85Cou1 Courtneidge, J.L., Davies, A.G., Tollerfield, S.M., Rideout, J., Symons, M.C.R.: J. Chem. Soc., Chem. Commun. **1985**, 1092.

85Dol1 Dolivo, G., Lund, A.: J. Phys. Chem. **89** (1985) 3977.

85Dol2 Dolivo, G., Lund, A.: Z. Naturforsch. **40a** (1985) 52.

85Fuj1 Fujisawa, J., Sato, S., Shimokoshi, K., Shida, T.: Bull. Chem. Soc. Jpn. **58** (1985) 1267.

85Lun1 Lunell, S. Huang, M.B., Claesson, O., Lund, A.: J. Chem. Phys. **82** (1985) 5121.

85Nel1 Nelson, S.F., Teasley, M.F., Bloodworth, A.J., Eggelte, H.J.: J. Org. Chem. **50** (1985) 3299.

85Qin1 Qin, X.Z., Snow, L.D., Williams, F.: J. Am. Chem. Soc. **107** (1985) 3366.

85Qin2 Qin, X.Z., Snow, L.D., Williams, F.: J. Phys. Chem. **89** (1985) 3602.

85Rao1 Rao, D.N.R., Symons, M.C.R.: J. Chem. Soc., Faraday Trans. I **81** (1985) 565.

85Rid1 Rideout, J., Symons, M.C.R., Swarts, S., Besler, B., Sevilla, M.D.: J. Phys. Chem. **89** (1985) 5251.

85Sul1 Sullivan, P.D., Bannoura, F., Daub, G.H.: J. Am. Chem. Soc. **107** (1985) 32.

85Sym1 Symons, M.C.R.: J. Chem. Res. (S) **1985**, 256.

86Cla1 Claxton, T.A., Rideout, J., Symons, M.C.R.: J. Chem. Res. (S) **1986**, 72.

86Eas1 Eastland, G.W., Rao, D.N.R., Symons, M.C.R.: J. Chem. Soc., Faraday Trans. I **82** (1986) 2833.

86Gan1 Ganghi, N.S., Rao, D.N.R., Symons, M.C.R.: J. Chem. Soc., Faraday Trans. I **82** (1986) 2367.

86Mut1 Muto, H., Toriyama, K., Nunome, K., Iwasaki, M.: J. Chem. Soc., Perkin Trans. II **1986**, 2011.

86Por1 Portwood, L., Rhodes, C.J., Symons, M.C.R.: J. Chem. Soc., Chem. Commun. **1986**, 1535.

86Qin1 Qin, X.Z., Williams, F.: Tetrahedron **42** (1986) 6301.

86Rid1 Rideout, J., Symons, M.C.R., Wren, B.W.: J. Chem. Soc., Faraday Trans. I **82** (1986) 167.

86Rid2 Rideout, J., Symons, M.C.R.: J. Chem. Soc., Perkin Trans. II **1986**, 625.

86Rid3 Rideout, J., Symons, M.C.R.: J. Chem. Soc., Perkin Trans. II **1986**, 969.

86Sym1 Symons, M.C.R., Wren, B.W., Muto, H., Triyama, K., Iwasaki, M.: Chem. Phys. Lett. **127** (1986) 424.

86Tor1 Toriyama, K., Nunome, K., Iwasaki, M.: Chem. Phys. Lett. **132** (1986) 456.

87Ich1 Ichikawa, T., Ohta, N.: J. Phys. Chem. **91** (1987) 3244.

87Lin1 Lindgren, M., Lund, A.: J. Chem. Soc., Faraday Trans. I **83** (1987) 1851.

87Oht1 Ohta, N., Ichikawa, T.: J. Phys. Chem. **91** (1987) 3736.

19 Cation radicals of aromatic hydrocarbons and their derivatives and S-heterocycles

19.0 Introduction

In the last decade, a number of the cation radicals of aromatic hydrocarbons and O and S containing heteroaromatics have been investigated by using ESR and ENDOR spectroscopies. In this chapter their data are rather arbitrarily divided into conventional catagories as follows:

19.1 Cation radicals of aromatic hydrocarbons and their derivatives
- 19.1.1 Benzenes, excluding $(OR)_n$ and $(SR)_n$ substituents
- 19.1.2 Naphthalenes, excluding $(OR)_n$ and $(SR)_n$ substituents
- 19.1.3 Biphenyls, excluding $(OR)_n$ and $(SR)_n$ substituents
- 19.1.4 Anthracenes, excluding $(OR)_n$ and $(SR)_n$ substituents
- 19.1.5 Further condensed aromatics
- 19.1.6 $(OR)_n$ substituted aromatics
  - 19.1.6.1 Benzenes
  - 19.1.6.2 Other aromatics
  - 19.1.6.3 $(OR)_n$ *and* $(SR)_n$ substituted benzenes
- 19.1.7 $(SR)_n$ substituted aromatics
- 19.1.8 Nonbenzenoid aromatics
- 19.1.9 Annulenes
- 19.1.10 Bridged systems
- 19.1.11 Cyclophanes

19.2 S-Heterocycles

19.3 Miscellaneous compounds

In each category, the derivatives are ordered according to increasing number of substituents, and then according to the gross formula. The data of the dimer cation radicals follow directly those of the respective monomer cation radicals for comparison. The S-heterocycles are arranged in the order of increasing number of π-electrons in the main cycles. Derivatives of each heterocycle are ordered according to the gross formula.

A number of ways of generating cation radicals for ESR measurements have been devised by many workers over the last 30 years. Although any method of preparation is specific to the radicals to be generated, the representative methods of generating methods are reviewed in [77Mc1] and [86Hi1]. Recently, some new methods combining acidic solvents and oxidizing reagents have been developed and have led to the ESR observation of the radicals which were undetectable so far [79Ru2]. It has been clarified that the UV-irradiation [84Co1] or electrolysis [86Te1] is effective to generate the cation radicals in the combination of strong acids and oxidizing reagents.

The literature included in this chapter were collected based on the result of a search of CA SEARCH database using the DIALOG Information Retrieval Service, the keywords being ESR, EPR, radical, and cation.

Ohya-Nishiguchi, Terahara, Tajima, Ishizu

19.1 Cation radicals of aromatic hydrocarbons and their derivatives

19.1.1 Benzenes, excluding $(OR)_n$ and $(SR)_n$ substituents

Substance	Generation/ Matrix or Solvent	Method/ T [K]	g-Factor	a-Value [mT]	Ref./ add. Ref.
$[C_{14}H_{26}Si_2]^{\cdot+}$ (1,4-bis($CH_2Si(CH_3)_3$)benzene cation)	Oxidation by $AlCl_3$/ CH_2Cl_2	ESR/ 190		4H(CH_2): 0.84 18H(CH_3): 0.029 4H(ring): 0.177	78Bo1/ 78Bo2, 77Bo1
$[C_{20}H_{42}Si_4]^{\cdot+}$ ($((CH_3)_3Si)_2$—HC—$C_6H_4^{+}$—CH—$(Si(CH_3)_3)_2$)	Oxidation by $AlCl_3$/ CH_2Cl_2	ESR/ 225		2H(CH): 0.154 [1] 4H(ring): 0.175 36H(CH_3): 0.021 4Si: 0.97	77Bo1
	Oxidation by $AlCl_3$/ CH_2Cl_2	ESR, ENDOR/ 300		4H(ring): 0.175 2H(CH): 0.138 36H(CH_3): 0.021 4Si: 0.962	78Bo2
		180		4H(ring): 0.175 2H(CH): 0.162 36H(CH_3): 0.021 4Si: 0.985	
$[C_{26}H_{58}Si_6]^{\cdot+}$ ($((CH_3)_3Si)_3$—C—$C_6H_4^{+}$—C—$(Si(CH_3)_3)_3$)	Oxidation by $AlCl_3$/ CH_2Cl_2	ESR, ENDOR [2]/ 250		4H(ring): 0.171 54H(CH_3): 0.013 6Si: 0.63	77Bo1, 78Bo2
$[C_8H_9NO_2]^{\cdot+}$ (1-NO_2, 2-CH_3, 5-H_3C benzene cation)	Oxidation by $HFSO_3$:PbO_2/ HSO_3F	EPR/ 198		3H(2, CH_3): 1.855 H(3): 0.270 H(4): 0.090 3H(5, CH_3): 1.715 H(6): 0.415	79Ru2

[1]) Temperature dependent.
[2]) ENDOR measurement from [78Bo2].

Substance	Generation/ Matrix or Solvent	Method/ T [K]	g-Factor	a-Value [mT]	Ref./ add. Ref.
$[C_8H_9NO_2]^{\cdot +}$ 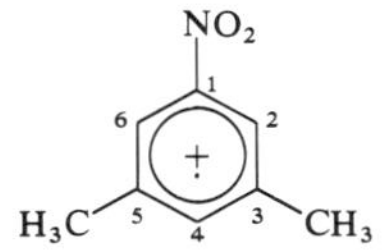	Oxidation by $HSO_3F:PbO_2/$ HSO_3F	EPR/ 198		2H(2, 6): 0.905 6H(3, 5, CH_3): 1.155	82De1
$[C_9H_9N]^{\cdot +}$ 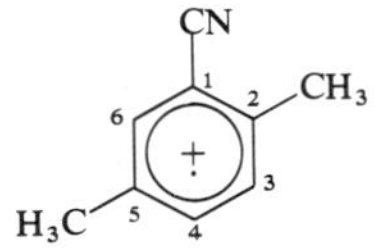	Oxidation by $HSO_3F:PbO_2/$ HSO_3F	EPR/ 198		3H(2, CH_3): 1.865 H(3): 0.245 H(4): 0.110 3H(5, CH_3): 1.715 H(6): 0.385	83Ru1
$[C_9H_9N]^{\cdot +}$ 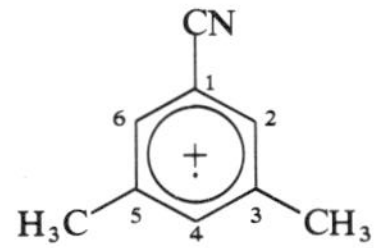	Oxidation by $HSO_3F:PbO_2/$ HSO_3F	EPR/ 198		2H(2, 6): 0.915 6H(3, 5, CH_3): 1.145	79Ru2
$[C_9H_9ClO]^{\cdot +}$ 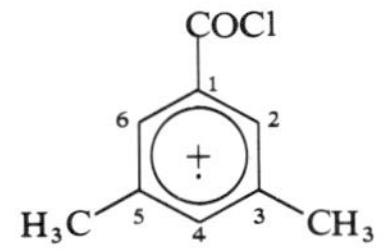	Oxidation by $HSO_3F:PbO_2/$ HSO_3F	EPR/ 198		2H(2, 6): 0.905 6H(3, 5, CH_3): 1.155	79Ru2
$[C_9H_{10}O_2]^{\cdot +}$ 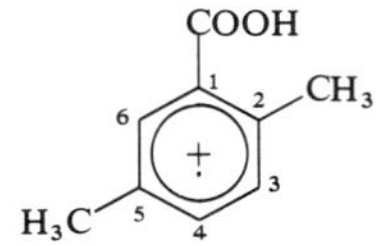	Oxidation by $HSO_3F:PbO_2/$ HSO_3F	EPR/ 198		3H(2, CH_3): 1.815 H(3): 0.330 H(4): 0.050 3H(5, CH_3): 1.700 H(6): 0.470	79Ru2

Ohya-Nishiguchi, Terahara, Tajima, Ishizu

Substance	Generation/ Matrix or Solvent	Method/ T [K]	g-Factor	a-Value [mT]	Ref./ add. Ref.
$[C_9H_{10}O_2]^{\cdot +}$ 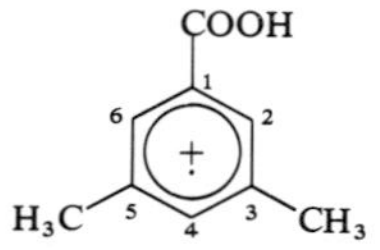	Oxidation by $HSO_3F:PbO_2/$ HSO_3F	EPR/ 198		2H(2, 6): 0.905 6H(CH_3): 1.155 H(4): 0.040	79Ru2
$[C_9H_{10}O_4]^{\cdot +}$ 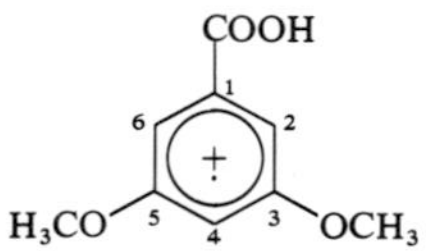	Oxidation by $HSO_3F:PbO_2/$ HSO_3F	EPR/ 198		2H(2, 6): 1.110 6H(CH_3): 0.330	79Ru2
$[C_{10}H_{11}N]^{\cdot +}$ 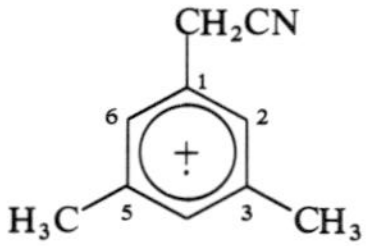	Solution in $HSO_3F/$ HSO_3F	EPR/ 198		2H(CH_2): 0.095 2H(2, 6): 0.85 6H(CH_3): 1.110	81Ru1
$[C_9H_{11}NO]^{\cdot +}$ 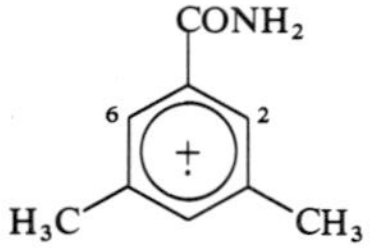	Oxidation by $HSO_3F:PbO_2/$ HSO_3F	EPR/ 198		2H(2, 6): 0.905 6H(CH_3): 1.155	79Ru2
$[C_{10}H_{12}O]^{\cdot +}$ 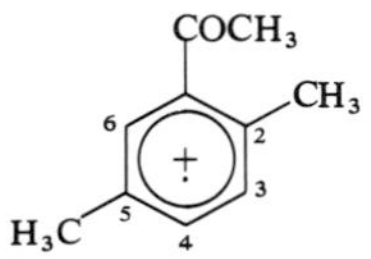	Oxidation by $HSO_3F:PbO_2/$ HSO_3F	EPR/ 198		3H(2, CH_3): 1.820 H(3): 0.345 3H(5, CH_3): 1.700 H(6): 0.495	79Ru2

Substance	Generation/ Matrix or Solvent	Method/ T [K]	g-Factor	a-Value [mT]	Ref./ add. Ref.
$[C_{10}H_{13}NO]^{\cdot+}$	Oxidation by $HSO_3F:PbO_2$/ HSO_3F	EPR/ 198		3H(2, CH_3): 1.846 H(3): 0.231 H(4): 0.127 3H(5, CH_3): 1.721 H(6): 0.415	79Ru2
$[C_{10}H_{11}N]^{\cdot+}$	Oxidation by $HSO_3F:PbO_2$/ HSO_3F	EPR/ 198		3H(2, CH_3): 0.900 3H(3, CH_3): 1.615 H(5): 0.460 3H(6, CH_3): 1.440	83Ru1
$[C_{10}H_{11}N]^{\cdot+}$	Oxidation by $HSO_3F:PbO_2$/ HSO_3F	EPR/ 198		3H(2, CH_3): 1.440 H(3): 0.037 3H(4, CH_3): 0.455 3H(5, CH_3): 1.877 H(6): 0.175 N: 0.037	83Ru1
$[C_{10}H_{11}N]^{\cdot+}$	Oxidation by $HSO_3F:PbO_2$/ HSO_3F	EPR/ 198		3H(2, CH_3): 1.580 3H(3, CH_3): 0.780 H(4): 0.040 3H(5, CH_3): 1.370 H(6): 0.640	83Ru1
$[C_{10}H_{11}N]^{\cdot+}$	Oxidation by $HSO_3F:PbO_2$/ HSO_3F	EPR/ 198		6H(2, 6, CH_3): 1.170 2H(3, 5): 0.765	79Ru2

Substance	Generation/ Matrix or Solvent	Method/ T [K]	g-Factor	a-Value [mT]	Ref./ add. Ref.
$[C_{10}H_{12}O_2]^{\cdot +}$ 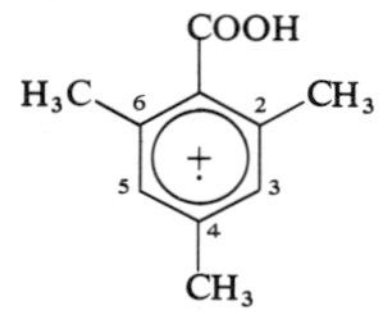	Oxidation by $HSO_3F:PbO_2$/ HSO_3F	EPR/ 198		6H(2, 6, CH_3): 1.225 2H(3, 5): 0.775 3H(4, CH_3): 0.087	79Ru2
$[C_{11}H_{14}O]^{\cdot +}$ 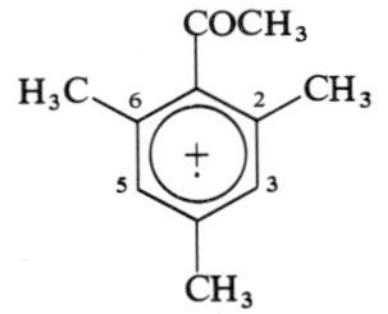	Oxidation by $HSO_3F:PbO_2$/ HSO_3F	EPR/ 198		6H(2, 6, CH_3): 1.186 2H(3, 5): 0.755	79Ru2
$[C_{13}H_{17}N]^{\cdot +}$ 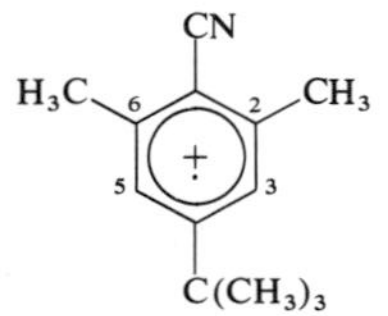	Oxidation by $HSO_3F:PbO_2$/ HSO_3F	EPR/ 198		6H(2, 6, CH_3): 1.190 H(3, 5): 0.762	83Ru1
$[C_{13}H_{18}O_2]^{\cdot +}$ 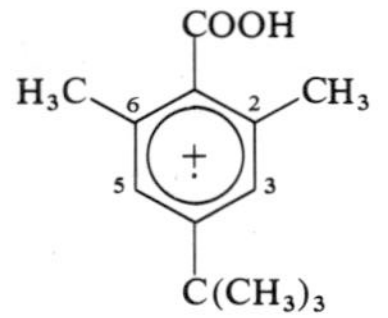	Oxidation by $HSO_3F:PbO_2$/ HSO_3F	EPR/ 198		6H(2, 6, CH_3): 1.230 H(3, 5): 0.768	79Ru2
$[C_{16}H_{30}Si_2]^{\cdot +}$ 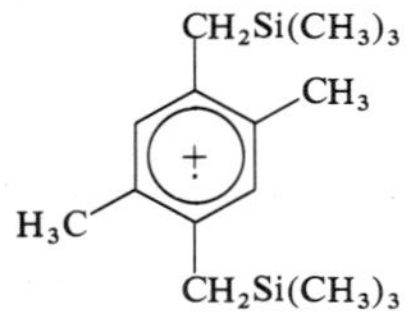	Oxidation by $AlCl_3$/ CH_2Cl_2	ESR/ 200		H(CH_2): 1.00 H(CH_2): 0.65 18H($SiCH_3$): 0.029 6H(CH_3): 0.345 2H(ring): 0.095	78Bo1

Substance	Generation/ Matrix or Solvent	Method/ T [K]	g-Factor	a-Value [mT]	Ref./ add. Ref.
$[C_{22}H_{46}Si_4]^{\cdot +}$ 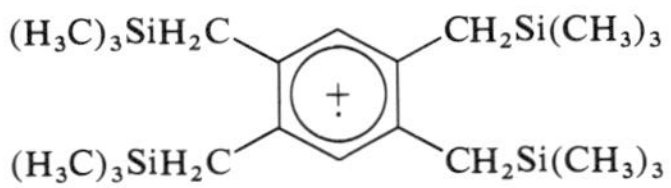	Oxidation by $AlCl_3$/ CH_2Cl_2	ESR/ 250		8H(CH_2): 0.579 36H(CH_3): 0.018 2H(ring): 0.061 4Si: 0.825	78Bo1/ 77Bo1 [3])
$[C_{11}H_{13}N]^{\cdot +}$ 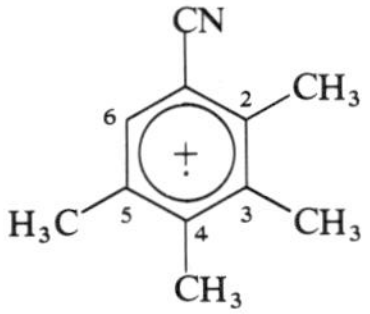	Oxidation by $HSO_3F:PbO_2$/ HSO_3F	EPR/ 198		3H(2, CH_3): 1.540 3H(3, CH_3): 0.580 3H(4, CH_3): 0.037 3H(5, CH_3): 1.480 H(6): 0.560	83Ru1
$[C_{11}H_{13}N]^{\cdot +}$ 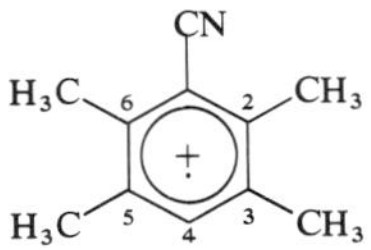	Oxidation by $HSO_3F:PbO_2$/ HSO_3F	EPR/ 198		6H(2, 6, CH_3): 1.200 6H(3, 5, CH_3): 1.060 H(4): 0.080	79Ru2
$[C_{11}H_{14}S_2]^{\cdot +}$ 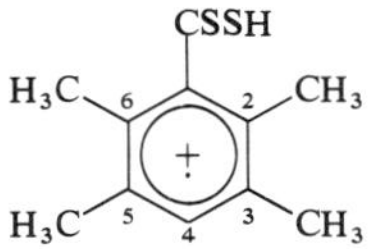	Oxidation by $HSO_3F:PbO_2$/ HSO_3F	EPR/ 198		6H(2, 6, CH_3): 1.180 6H(3, 5, CH_3): 1.050 H(4): 0.080	79Ru2
$[C_{11}H_{14}O_2]^{\cdot +}$ 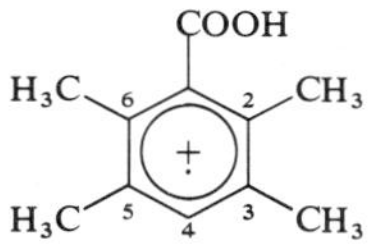	Oxidation by $HSO_3F:PbO_2$/ HSO_3F	EPR/ 198		6H(2, 6, CH_3): 1.185 6H(3, 5, CH_3): 1.045 H(4): 0.075	79Ru2

[3]) Similar hfcc at 275 K.

Substance	Generation/ Matrix or Solvent	Method/ T [K]	g-Factor	a-Value [mT]	Ref./ add. Ref.
$[C_{12}H_{16}O]^{\cdot +}$	Oxidation by $HSO_3F:PbO_2$/ HSO_3F	EPR/ 198		6H(2, 6, CH_3): 1.165 6H(3, 5, CH_3): 1.060 H(4): 0.075	79Ru2
$[C_{10}H_{12}N_2O_4]^{\cdot +}$	Oxidation by $HFSO_3:PbO_2$/ HSO_3F	EPR/ 198		12H(CH_3): 1.145	79Ru2
$[C_{11}H_{13}N]^{\cdot +}$	Oxidation by $HSO_3F:PbO_2$/ HSO_3F	EPR/ 198		3H(2, CH_3): 0.668 3H(3, CH_3): 1.765 3H(4, CH_3): 0.127 H(5): 0.254 3H(6, CH_3): 1.405	83Ru1
$[C_{11}H_{15}NO_2]^{\cdot +}$	Oxidation by $HSO_3F:PbO_2$/ HSO_3F	EPR/ 198		6H(2, 6, CH_3): 1.125 6H(3, 5, CH_3): 1.020 3H(4, CH_3): 0.055 N: 0.01	82De1
$[C_{12}H_{15}N]^{\cdot +}$	Oxidation by $HSO_3F:PbO_2$/ HSO_3F	EPR/ 198		3H(2, CH_3): 1.160 6H(3, 5, CH_3): 1.040 6H(4, 6, CH_3): 0.097	83Ru1

Substance	Generation/ Matrix or Solvent	Method/ T [K]	g-Factor	a-Value [mT]	Ref./ add. Ref.
$[C_{12}H_{16}F_2O_6S_2]^{\cdot +}$ 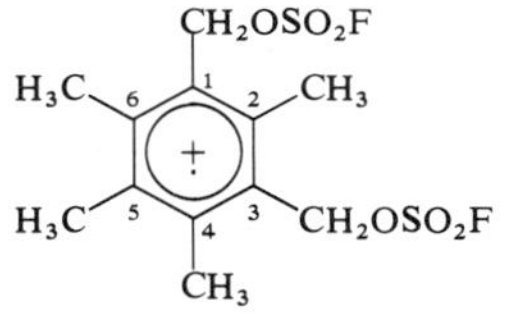	Oxidation by $HSO_3F:PbO_2$/ HSO_3F	EPR/ 198		$4H(CH_2)$: 0.070 $6H(4, 6, CH_3)$: 0.375 $3H(2, CH_3)$: 1.34 $3H(5, CH_3)$: 1.71	81Ru2, 81Ru1
$[C_{12}H_{17}FO_3S]^{\cdot +}$ 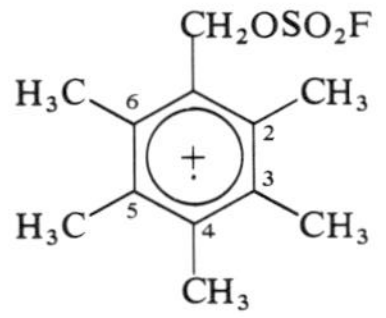	Oxidation by $HSO_3F:PbO_2$/ HSO_3F	EPR/ 198		$2H(CH_2)$: 0.070 $6H(2, 6, CH_3)$: 1.065 $6H(3, 5, CH_3)$: 1.005 $3H(4, CH_3)$: 0.070	81Ru1, 81Ru2
$[C_{12}H_{17}FO_3S]^{\cdot +}$ 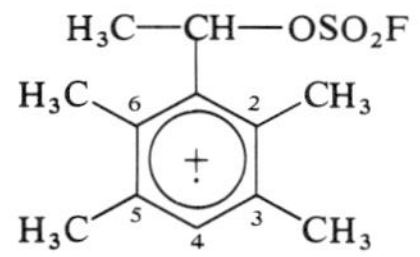	Solution in HSO_3F/ HSO_3F	EPR/ 198		$6H(2, 6, CH_3)$: 1.130 $6H(3, 5, CH_3)$: 1.000 H(4): 0.060	81Ru1
$[C_{12}H_{18}]^{\cdot +}$ 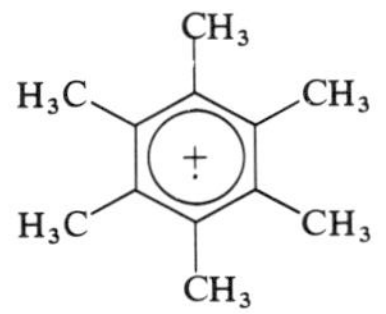	UV-irr. with $Hg(OOCF_3)_2$/ CH_2Cl_2	EPR/ 173	2.0023	$18H(CH_3)$: 0.65	86La1
$[C_{24}H_{36}]^{\cdot +}$ 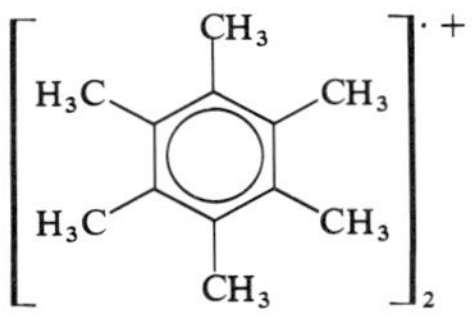	UV-irr. with $Hg(OOCF_3)_2$/ CH_2Cl_2	ESR/ 173		$36H(CH_3)$: 0.031	86La1

Ohya-Nishiguchi, Terahara, Tajima, Ishizu

Substance	Generation/ Matrix or Solvent	Method/ T[K]	g-Factor	a-Value [mT]	Ref./ add. Ref.
$[C_{13}H_{17}N]^{\cdot+}$	Solution in HSO_3F/ HSO_3F	EPR/ 198		2H(CH_2): 0.068 6H(2, 6, CH_3): 1.070 6H(3, 5, CH_3): 1.010 3H(4, CH_3): 0.068	81Ru1
$[C_{13}H_{18}O_2]^{\cdot+}$	Solution in HSO_3F/ HSO_3F	EPR/ 198		2H(CH_2): 0.060 6H(2, 6, CH_3): 1.060 6H(3, 5, CH_3): 1.005 3H(4, CH_3): 0.060	81Ru1
$[C_{14}H_{16}N_2]^{\cdot+}$	Solution in HSO_3F/ HSO_3F	EPR/ 198		4H(CH_2): 0.080 6H(3, 6, CH_3): 1.475 6H(4, 5, CH_3): 0.275	81Ru1
$[C_{14}H_{16}N_2]^{\cdot+}$	Solution in HSO_3F/ HSO_3F	EPR/ 198		4H(CH_2): 0.070 3H(2, CH_3): 1.340 6H(4, 6, CH_3): 0.375 3H(5, CH_3): 1.710	81Ru1
$[C_{14}H_{16}N_2]^{\cdot+}$	Solution in HSO_3F/ HSO_3F	EPR/ 198		4H(CH_2): 0.080 12H(CH_3): 1.090	81Ru1

Substance	Generation/ Matrix or Solvent	Method/ T [K]	g-Factor	a-Value [mT]	Ref./ add. Ref.
$[C_{14}H_{22}O_2]^{\cdot +}$ 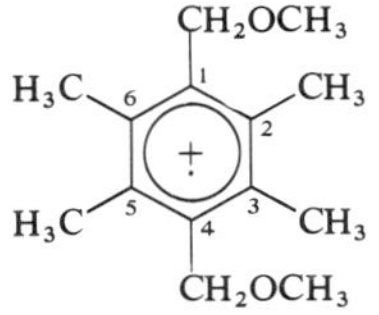	Solution in HSO_3F/ HSO_3F	EPR/ 198		$4H(CH_2)$: 0.060 $12H(2, 3, 5, 6, CH_3)$: 1.083	81Ru1
$[C_{16}H_{22}O_6]^{\cdot +}$ 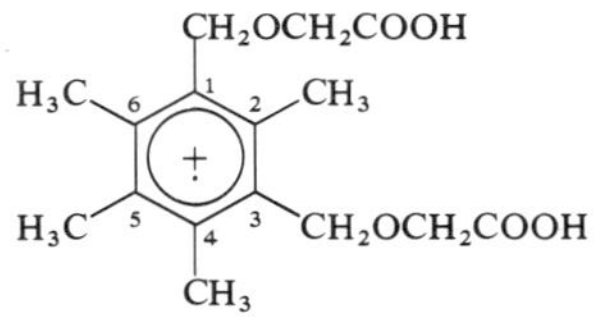	Solution in HSO_3F/ HSO_3F	EPR/ 198		$4H(CH_2)$: 0.070 $3H(2, CH_3)$: 1.340 $6H(4, 6, CH_3)$: 0.375 $3H(5, CH_3)$: 1.710	81Ru1
$[C_{16}H_{22}O_6]^{\cdot +}$ 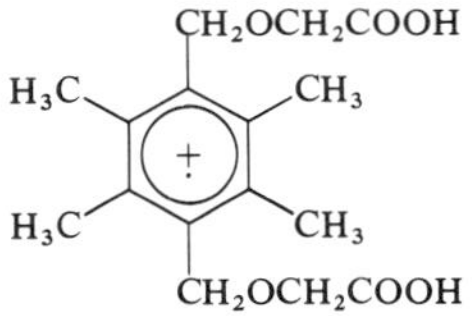	Solution in HSO_3F/ HSO_3F	EPR/ 198		$4H(CH_2)$: 0.060 $12H(CH_3)$: 1.083	81Ru1
$[C_{18}H_{34}Si_2]^{\cdot +}$ 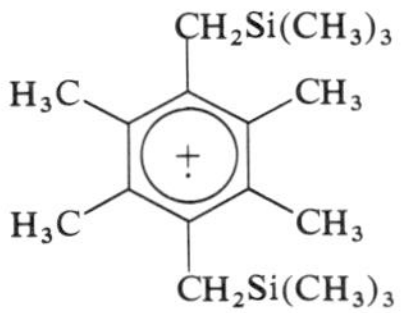	Oxidation by $AlCl_3$/ CH_2Cl_2	ESR, ENDOR/ 200		$4H(CH_2)$: 0.858 $18H(SiCH_3)$: 0.028 $12H(ring, CH_3)$: 0.188	78Bo1
$[C_{21}H_{42}Si_3]^{\cdot +}$ 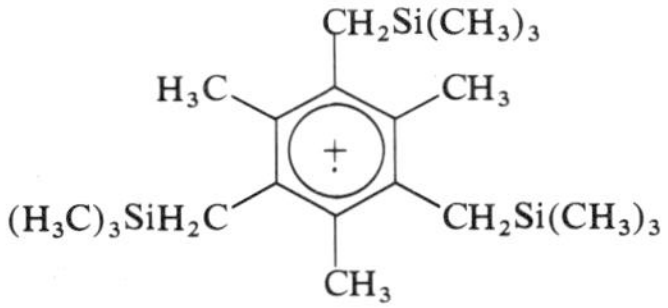	Oxidation by $AlCl_3$/ CH_2Cl_2	ESR/ –		$6H(CH_2)$: 0.48 $27H(SiCH_3)$: 0.020	78Bo1

Ohya-Nishiguchi, Terahara, Tajima, Ishizu

Substance	Generation/ Matrix or Solvent	Method/ T [K]	g-Factor	a-Value [mT]	Ref./ add. Ref.
$[C_{30}H_{66}Si_6]^{\cdot +}$	Oxidation by $AlCl_3$/ CH_2Cl_2	ESR/ 300		$54H(CH_3)$: 0.013 6Si: 0.54 $12H(CH_2)$: 0.353	77Bo1, 78Bo1 [4])

19.1.2 Naphthalenes, excluding $(OR)_n$ and $(SR)_n$ substituents

Substance	Generation/ Matrix or Solvent	Method/ T [K]	g-Factor	a-Value [mT]	Ref./ add. Ref.
$[C_{20}H_{16}]^{\cdot +}$	Electrochem. oxidation/ CH_2Cl_2 $+CF_3COOH:(CF_3CO)_2O$	EPR/ 170		4H(1, 4, 5, 8): 0.276 4H(2, 3, 6, 7): 0.103	86Te2
$[C_{22}H_{20}]^{\cdot +}$	Electrochem. oxidation/ CH_2Cl_2 $+CF_3COOH:(CF_3CO)_2O$	EPR/ 190		$6H(CH_3)$: 0.589 2H(2): 0.093 2H(3): 0.031 2H(5): 0.186 2H(8): 0.124	86Te2
$[C_{24}H_{24}]^{\cdot +}$	Electrochem. oxidation/ CH_2Cl_2 $+CF_3COOH:(CF_3CO)_2O$	EPR/ 170		$6H(1, CH_3)$: 0.580 $6H(2, CH_3)$: 0.106 2H(3): 0.023 2H(4): 0.342 2H(5): 0.212 2H(6): 0.083 2H(7): 0.023 2H(8): 0.212	86Te2
$[C_{12}H_{12}]^{\cdot +}$	Electrochem. oxidation/ CH_2Cl_2 $+CF_3COOH:(CF_3CO)_2O$	EPR/ 170		$6H(1, 4, CH_3)$: 0.961 2H(5, 8): 0.359 2H(2, 3): 0.217 2H(6, 7): 0.132	86Te2

[4]) Same hfcc at 225 K.

Substance	Generation/ Matrix or Solvent	Method/ T [K]	g-Factor	a-Value [mT]	Ref./ add. Ref.
$[C_{24}H_{24}]^{\cdot +}$ 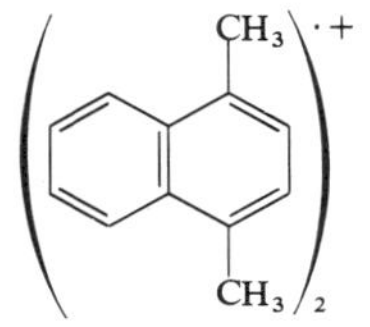	Electrochem. oxidation/ CH_2Cl_2 + $F_3COOH:(CF_3CO)_2O$	EPR/ 170		12H(1, 4, CH_3): 0.328 4H(5, 8): 0.212 4H(2, 3): 0.119 4H(6, 7): 0.090	86Te2
$[C_{12}H_{12}]^{\cdot +}$ 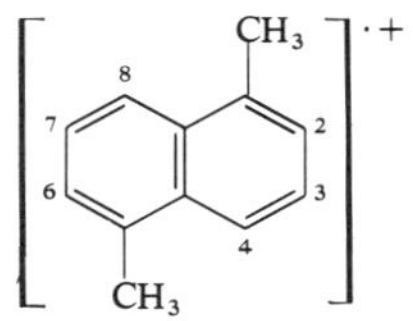	Electrochem. oxidation/ CH_2Cl_2 + $CF_3COOH:(CF_3O)_2O$	EPR/ —		6H(1, 5, CH_3): 0.707 2H(4, 8): 0.535 2H(2, 7): 0.169 2H(3, 6): 0.169	86Te2
$[C_{24}H_{24}]^{\cdot +}$ 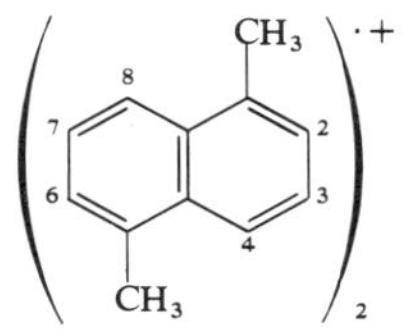	Electrochem. oxidation/ CH_2Cl_2 + $CF_3COOH:(CF_3CO)_2O$	EPR/ 200		6H(1, 5): 0.405 6H(1, 5): 0.270 2H(4, 8): 0.199 2H(4, 8): 0.109	86Te2
$[C_{12}H_{10}]^{\cdot +}$ 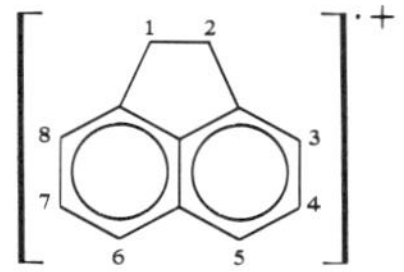	Oxidation by molten $SbCl_3$: 8mol% $AlCl_3$/ $SbCl_3$: 8mol% $AlCl_3$	EPR/ 353		4H(1, 2): 1.318 2H(3, 8): 0.313 2H(4, 7): 0.059 2H(5, 6): 0.659	80Bu1

Ohya-Nishiguchi, Terahara, Tajima, Ishizu

Substance	Generation/ Matrix or Solvent	Method/ T [K]	g-Factor	a-Value [mT]	Ref./ add. Ref.
$[C_{24}H_{20}]^{\cdot +}$	Electrochem. oxidation/ CH_2Cl_2 $+CF_3COOH:(CF_3CO)_2O$	EPR/ —		2H(1, 8): 0.962 4H(1, 8): 0.882 2H(1, 8): 0.400 2H(2, 7): 0.080 2H(2, 7): 0.064 2H(3, 6): 0.064 2H(3, 6): 0.016 2H(4, 5): 0.567 2H(4, 5): 0.080	86Te2
$[C_{12}H_{12}]^{\cdot +}$	Electrochem. oxidation/ CH_2Cl_2 $+CF_3COOH:(CF_3CO)_2O$	EPR/ —		6H(CH_3): 0.825 2H(2, 7): 0.245 2H(3, 6): 0.116 2H(4, 5): 0.573	86Te2
$[C_{24}H_{24}]^{\cdot +}$	Electrochem. oxidation/ CH_2Cl_2 $+CF_3COOH:(CF_3CO)_2O$	ESR/ —		6H(CH_3): 0.568 6H(CH_3): 0.284 2H(2, 7): 0.364 2H(2, 7): 0.284 2H(3, 6): 0.080 2H(3, 6): 0.019 2H(4, 5): 0.364 2H(4, 5): 0.284	86Te2
$[C_{14}H_{14}]^{\cdot +}$	Electrochem. oxidation/ CH_2Cl_2 $+CF_3COOH:(CF_3CO)_2O$	ESR/ 273		4H(CH_2): 1.289 6H(CH_3): 0.859 2H(2, 7): 0.313 2H(3, 6): 0.114	86Te2

Substance	Generation/ Matrix or Solvent	Method/ T [K]	g-Factor	a-Value [mT]	Ref./ add. Ref.
$[C_{14}H_{12}]^{\cdot +}$ 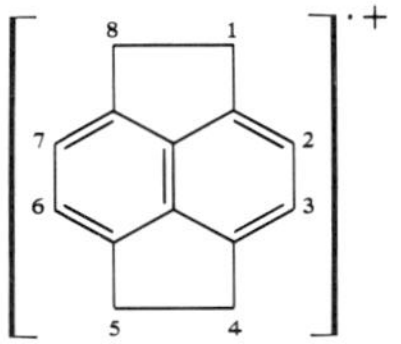	Electrochem. oxidation/ CH_2Cl_2 $+CF_3COOH:(CF_3CO)_2O$	EPR/ 273		8H(CH_2): 1.318 4H(2, 3, 6, 7): 0.203	86Te2
$[C_{28}H_{24}]^{\cdot +}$ 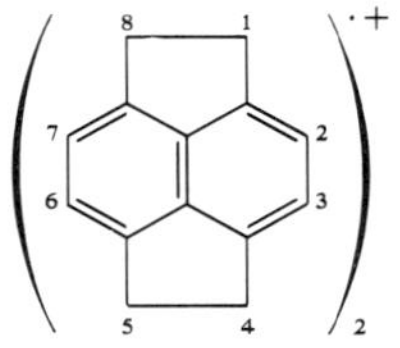	Electrochem. oxidation/ CH_2Cl_2 $+CF_3COOH:(CF_3CO)_2O$	ESR/ 273		8H(1, 4, 5, 8): 0.562 8H(1, 4, 5, 8): 0.548 8H(2, 3, 6, 7): 0.115	86Te2
$[C_{14}H_{16}]^{\cdot +}$ 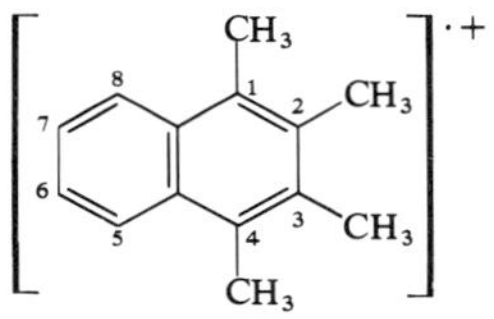	Electrochem. oxidation/ CH_2Cl_3 $+CF_3COOH:(CF_3CO)_2O$	EPR/ 170		6H(1, 4, CH_3): 0.995 2H(5, 8): 0.337 6H(2, 3, CH_3): 0.236 2H(6, 7): 0.092	86Te2
$[C_{28}H_{32}]^{\cdot +}$ 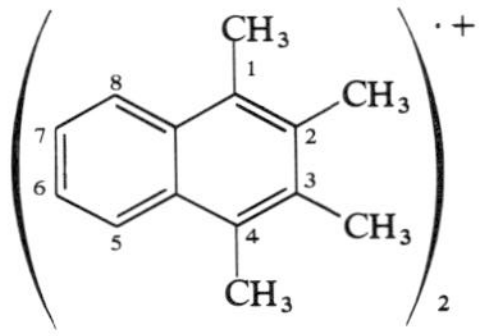	Electrochem. oxidation/ CH_2Cl_2 $+CF_3COOH:(CF_3O)_2O$	EPR/ 170		12H(1, 4, CH_3),: 0.349 4H(6, 7): 0.092 12H(2, 3, CH_3): 0.137 4H(5, 8): 0.222	86Te2

Ohya-Nishiguchi, Terahara, Tajima, Ishizu

Substance	Generation/ Matrix or Solvent	Method/ T [K]	g-Factor	a-Value [mT]	Ref./ add. Ref.
$[C_{14}H_{16}]^{\cdot +}$ 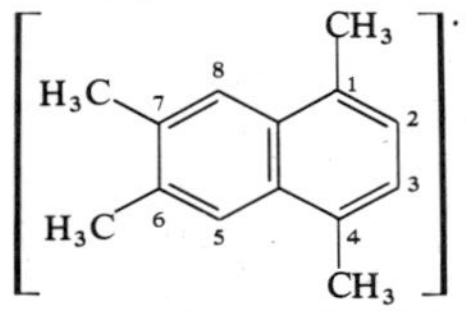	Electrochem. oxidation/ CH_2Cl_2 $+CF_3COOH:(CF_3CO)_2O$	EPR/ —		6H(1, 4, CH_3): 0.850 6H(6, 7, CH_3): 0.205 2H(5, 8): 0.205 2H(2, 3): 0.439	86Te2
$[C_{28}H_{32}]^{\cdot +}$ 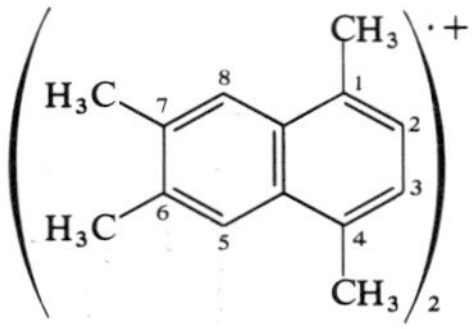	Electrochem. oxidation/ CH_2Cl_2 $+CF_3COOH:(CF_3CO)_2O$	EPR/ 170		12H(1, 4, CH_3): 0.314 4H(5, 8): 0.243 4H(2, 3): 0.128 12H(6, 7, CH_3): 0.120	86Te2
$[C_{16}H_{16}]^{\cdot +}$ 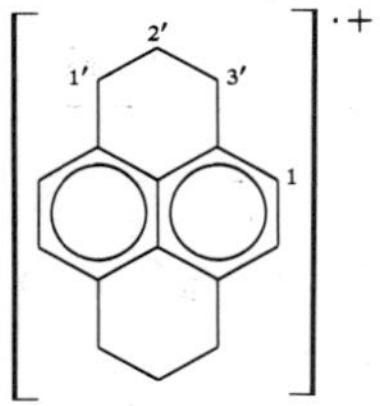	Solution in $HSO_3F:SO_2ClF$/ $HSO_3F:SO_2ClF$	EPR, ENDOR, TRIPLE/ 168		ESR measurement Conformer I: H(1′, ax): 1.485 H(1′, eq): 0.362 2H(2′, ax, eq): 0.042 H(1): 0.190 Confomer II: H(1′, ax): 1.478 H(1′, eq): 0.388 2H(2′, ax, eq): 0.041 H(1): 0.190 TRIPLE measurement Conformer I: H(1′, ax): 1.475 H(1′, eq): 0.359 2H(2′, ax, eq): −0.042 H(1): −0.190 Conformer II: H(1′, ax): 1.468 H(1′, eq): 0.385 H(2′, ax): −0.041 H(2′, eq): −0.042 H(1): −0.190	85Ma1

Substance	Generation/ Matrix or Solvent	Method/ T [K]	g-Factor	a-Value [mT]	Ref./ add. Ref.
$[C_{18}H_{24}]^{\cdot +}$	Oxidation by $AlCl_3$/ $CH_2Cl_2 : CHCl_2F$	EPR/ 115		12H(2, 3, 6, 7): 0.180 4H(1, 4, 5, 8): 1.245 4H(1, 4, 5, 8): 0.448 4H(1, 4, 5, 8): 0.271	86Ba1
		241		12H(2, 3, 6, 7): 0.185 12H(1, 4, 5, 8): 0.654	

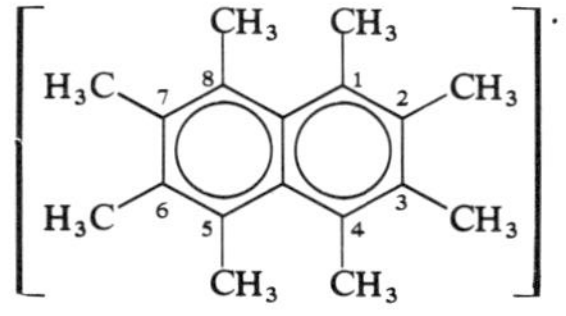

19.1.3 Biphenyls, excluding $(OR)_n$ and $(SR)_n$ substituents

Substance	Generation/ Matrix or Solvent	Method/ T [K]	g-Factor	a-Value [mT]	Ref./ add. Ref.
$[C_{12}H_{10}]^{\cdot +}$ 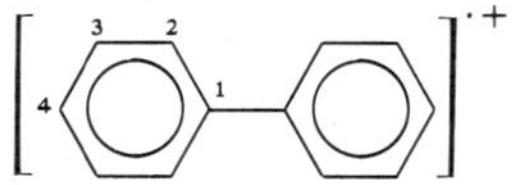	UV-irr. with $Hg(CF_3CO_2)_2$/ CF_3COOH	ESR/ 273	2.0027	H(2): 0.315 H(3): 0.051 H(4): 0.630	84Co1
$[C_{12}H_8F_2]^{\cdot +}$ 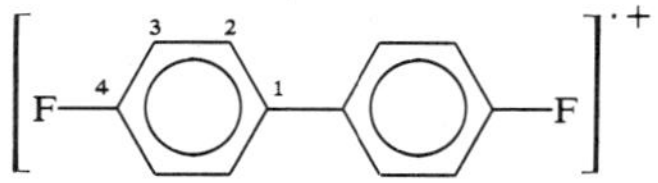	Oxidation with $SbCl_5$/ CH_2Cl_2	ESR/ 193 [5]	–	H(2): 0.276 F(4): 1.920 $^{13}C(1)$: −0.078 $^{13}C(2)$: 0.190 $^{13}C(3)$: −0.357 $^{13}C(4)$: 0.692	77Ne1
$[C_{20}H_{30}Si_2]^{\cdot +}$ 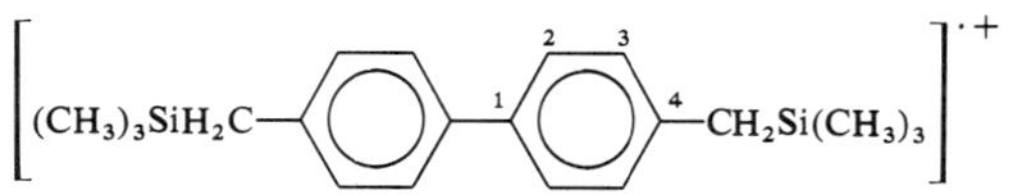 $(CH_3)_3SiH_2C$–biphenyl–$CH_2Si(CH_3)_3$	Oxidation by $AlCl_3$/ CH_2Cl_2	ESR/ 200		18H(CH_3): 0.012 4H(CH_2): 0.554 H(2): 0.201 H(3): 0.053	77Bo1/ 78Ka1

[5]) Temperature dependent linewidth.

Ohya-Nishiguchi, Terahara, Tajima, Ishizu

19.1.4 Anthracenes, excluding $(OR)_n$ and $(SR)_n$ substituents

Substance	Generation/ Matrix or Solvent	Method/ T [K]	g-Factor	a-Value [mT]	Ref./ add. Ref.
$[C_{14}H_{10}]^{\cdot +}$ 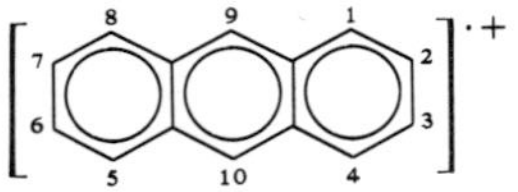	Oxidation by molten $SbCl_3$: 8mol% $AlCl_3$/ $SbCl_3$: 8mol% $AlCl_3$	EPR/ 253		4H(1, 4, 5, 8): 0.308 4H(2, 3, 6, 7): 0.136 2H(9, 10): 0.651	80Bu1
$[C_{16}H_{14}]^{\cdot +}$ 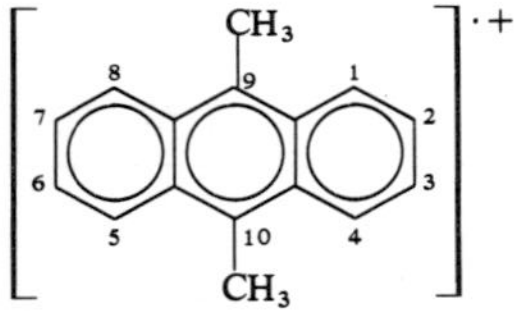	Oxidation by molten $SbCl_3$: 8mol% $AlCl_3$/ $SbCl_3$: 8mol% $AlCl_3$	EPR/ 353		4H(1, 4, 5, 8): 0.248 4H(2, 3, 6, 7): 0.122 6H(CH_3): 0.789	80Bu1
$[C_{22}H_{30}Si_2]^{\cdot +}$ 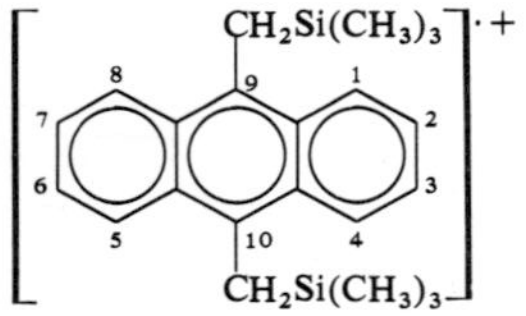	Oxidation by $AlCl_3$/ CH_2Cl_2	ESR/ 200		4H(1, 4, 5, 8): 0.183 4H(2, 3, 6, 7): 0.105 4H(CH_2): 0.610 18H(CH_3): 0.022	78Ka1
$[C_{26}H_{18}]^{\cdot +}$ 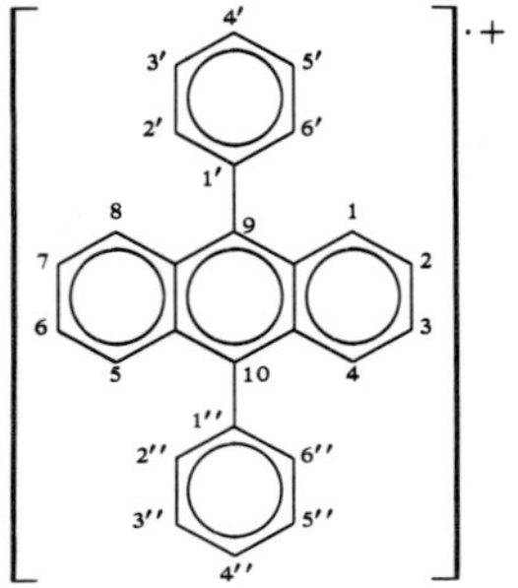	Oxidation by molten $SbCl_3$: 8mol% $AlCl_3$/ $SBCl_3$: 8mol% $AlCl_3$	EPR/ 353		4H(1, 4, 5, 8): 0.265 4H(2, 3, 6, 7): 0.125 10H(C_6H_5): 0.044	80Bu1

Substance	Generation/ Matrix or Solvent	Method/ T [K]	g-Factor	a-Value [mT]	Ref./ add. Ref.
19.1.5 Further condensed aromatics					
$[C_{16}H_{10}]^{\cdot+}$	Oxidation by molten $SbCl_3$:8mol% $AlCl_3$/ $SbCl_3$:8mol% $AlCl_3$	EPR/ 253		4H(1, 3, 6, 8): 0.540 2H(2, 7): 0.120 4H(4, 5, 9, 10): 0.210	80Bu1
$[C_{18}H_{12}]^{\cdot+}$	Oxidation by molten $SbCl_3$:8mol% $AlCl_3$/ $SbCl_3$:8mol% $AlCl_3$	EPR/ 353		4H(1, 4, 7, 10): 0.169 4H(2, 3, 8, 9): 0.101 4H(5, 6, 11, 12): 0.495	80Bu1
$[C_{20}H_{12}]^{\cdot+}$	Oxidation by molten $SbCl_3$:8mol% $AlCl_3$/ $SbCl_3$:8mol% $AlCl_3$	EPR/ 353		H(1): 0.402 H(2): 0.044 H(3): 0.304	80Bu1
	Oxidation by molten $SbCl_3$:8mol% $AlCl_3$, (naphthalene → perylene)/ $SbCl_3$:8mol% $AlCl_3$	EPR/ 353		H(1): 0.405 H(2): 0.043 H(3): 0.303	80Bu1

Substance	Generation/ Matrix or Solvent	Method/ T [K]	g-Factor	a-Value [mT]	Ref./ add. Ref.
$[C_{20}H_{12}]^{\cdot +}$	Solution in H_2SO_4/ H_2SO_4	EPR/ 298		H(1): 0.457 H(2): 0.054 H(3): 0.377 H(4): 0.037 H(5): 0.211 H(6): 0.663 H(7): 0.223 H(8): 0.019 H(9): 0.295 H(10): 0.194 H(11): 0.0824 H(12): 0.275 C(1): 0.601 C(2): 0.455 C(3): 0.444 C(4): 0.093 C(5): 0.132 C(6): 0.830 C(7): 0.374 C(8): 0.291 C(9): 0.301 C(10): 0.129 C(11): 0.196 C(12): 0.360	85Su1/ 83Su1, 86Su1
$[C_{21}H_{14}]^{\cdot +}$	Oxidation by CF_3COOH: $(CF_3CO)_2O$ $(Tl(III)(CF_3COOH)_3)$/ CF_3COOH	ESR, ENDOR/ 238		3H(6): 0.738 H(1): 0.445 H(3): 0.363 H(9): 0.306 H(12): 0.247 H(7): 0.221 H(5): 0.181 H(10): 0.170 H(11): 0.079 H(2): 0.044 H(4): 0.034	86Su1

Substance	Generation/ Matrix or Solvent	Method/ T [K]	g-Factor	a-Value [mT]	Ref./ add. Ref.
$[C_{22}H_{11}F_3O_2]^{\cdot+}$	Oxidation by CF_3COOH: $(CF_3CO)_2O$ $(Tl(III)(CF_3COO)_3)$/ CF_3COOH	ENDOR/ 238		3F(6): 0.023 H(1): 0.468 H(3): 0.373 H(9): 0.293 H(12): 0.267 H(7): 0.234 H(5): 0.180 H(10): 0.145 H(11): 0.127 H(2): 0.070 H(4): 0.043 H(8): 0.023	86Su1

19.1.6 $(OR)_n$ substituted aromatics

19.1.6.1 Benzenes

Substance	Generation/ Matrix or Solvent	Method/ T [K]	g-Factor	a-Value [mT]	Ref./ add. Ref.
$[C_7H_6O_2]^{\cdot+}$	Oxidation by Ce(IV)SO_4/ $H_2SO_4 \cdot 4H_2O$ (flow)	ESR/ —	2.0039	2H(CH_2): 2.19 2H(3, 6): 0.04 2H(4, 5): 0.49	76Di1
$[C_8H_8O_2]^{\cdot+}$	Oxidation by Ce(IV)SO_4/ $H_2SO_4 \cdot 4H_2O$ (flow)	ESR/ —	2.0034	4H(CH_2): 0.25 2H(3, 6): 0.04 2H(4, 5): 0.465	76Di1
$[C_8H_{10}O_2]^{\cdot+}$	Oxidation by $AlCl_3$/ $C_6H_5NO_2$	EPR/ 283		6H(CH_3): 0.3289 4H(ring): 0.2635	86Qu1

Ohya-Nishiguchi, Terahara, Tajima, Ishizu

Substance	Generation/ Matrix or Solvent	Method/ T [K]	g-Factor	a-Value [mT]	Ref./ add. Ref.
$[C_{10}H_{14}O_2]^{\cdot +}$ 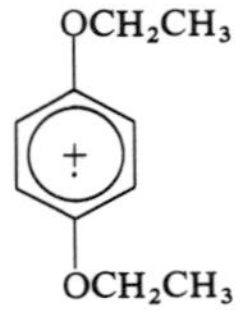	$AlCl_3$/ $C_6H_5NO_2$	EPR/ 283		$4H(CH_2)$: 0.3802 $6H(CH_3)$: 0.0152 4H(ring): 0.1880	86Qu1
$[C_8H_8O_2]^{\cdot +}$ 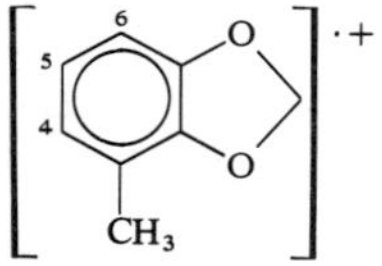	Oxidation by Ce(IV)SO_4/ $H_2SO_4 \cdot 4H_2O$ (flow)	ESR/ —	2.0039	$2H(CH_2)$: 2.14 $3H(CH_3)$: 0.04 H(4): 0.42 H(5): 0.53 H(6): 0.06	76Di1
$[C_8H_8O_2]^{\cdot +}$ 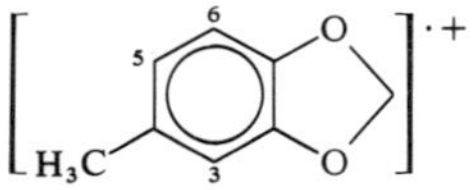	Oxidation by Ce(IV)SO_4/ $H_2SO_4 \cdot 4H_2O$ (flow)	ESR/ —	2.0038	$2H(CH_2)$: 1.94 H(3): 0.075 $3H(CH_3)$: 0.795 H(5): 0.515 H(6): 0.025	76Di1
$[C_8H_8O_3]^{\cdot +}$ 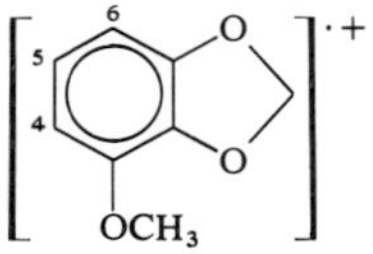	Oxidation by Ce(IV)SO_4/ $H_2SO_4 \cdot 4H_2O$ (flow)	ESR/ —	2.0040	$2H(CH_2)$: 1.925 $3H(CH_3)$: 0.15 2H(4, 6): 0.085 H(5): 0.605	76Di1
$[C_8H_8O_3]^{\cdot +}$ 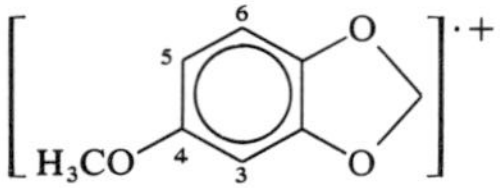	Oxidation by Ce(IV)SO_4/ $H_2SO_4 \cdot 4H_2O$ (flow)	ESR/ —	2.0040	$2H(CH_2)$: 1.445 2H(3, 6): 0.045 $3H(CH_3)$: 0.265 H(5): 0.625	76Di1

Substance	Generation/ Matrix or Solvent	Method/ T [K]	g-Factor	a-Value [mT]	Ref./ add. Ref.
$[C_9H_{12}O_3]^{\cdot +}$	Oxidation by Ce(IV)SO_4/ $H_2SO_4 \cdot 4H_2O$ (flow)	ESR/ —	2.0038	6H(1, 3): 0.205 3H(2): 0.525 2H(4, 6): 0.045 H(5): 0.625	79Ho1/ 76Di1
$[C_{10}H_{12}O_2]^{\cdot +}$	Oxidation by Ce(IV)SO_4/ $H_2SO_4 \cdot 4H_2O$ (flow)	ESR/ —	2.0039	2H(CH_2): 2.145 H(4): 0.415 H(5): 0.52 H(6): 0.06	76Di1
$[C_7H_6O_3]^{\cdot +}$	Oxidation by Ce(IV)SO_4/ $H_2SO_4 \cdot 4H_2O$ (flow)	ESR/ —	2.00395	2H(CH_2): 1.35 H(5): 0.625 H(6): 0.04	76Di1
$[C_7H_7NO_3]^{\cdot +}$	Oxidation by $HFSO_3$:PbO_2/ HSO_3F	EPR/ 198		H(OH): 0.390 H(3): 0.485 H(4): 0.060 3H(CH_3): 1.680 H(6): 0.235 N: 0.060	79Ru2
$[C_8H_5NO_2]^{\cdot +}$	Oxidation by Ce(IV)SO_4/ $H_2SO_4 \cdot 4H_2O$ (flow)	ESR/ —	2.00405	2H(CH_2): 2.42 H(5): 0.40	76Di1
$[C_8H_6O_3]^{\cdot +}$	Oxidation by Ce(IV)SO_4/ $H_2SO_4 \cdot 4H_2O$ (flow)	ESR/ —	2.00435	2H(CH_2): 2.28 H(3): 0.025 H(5): 0.405	76Di1

Ohya-Nishiguchi, Terahara, Tajima, Ishizu

Substance	Generation/ Matrix or Solvent	Method/ T [K]	g-Factor	a-Value [mT]	Ref./ add. Ref.
$[C_8H_6O_4]^{\cdot+}$ 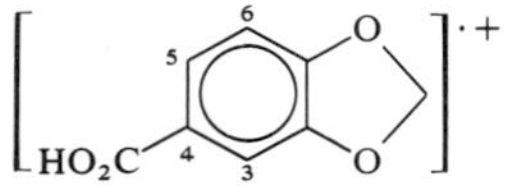	Oxidation by Ce(IV)SO_4/ $H_2SO_4 \cdot 4H_2O$ (flow)	ESR/ —	2.00415	$2H(CH_2)$: 2.325 H(3): 0.04 H(5): 0.425	76Di1
$[C_8H_8O_3]^{\cdot+}$ 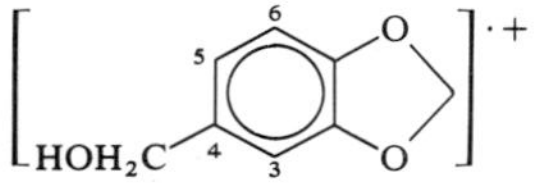	Oxidation by Ce(IV)SO_4/ $H_2SO_4 \cdot 4H_2O$ (flow)	ESR/ —	2.0038	$2H(CH_2)$: 2.00 H(3): 0.075 2H(4): 1.05 H(5): 0.50 H(6): 0.025	76Di1
$[C_9H_8O_4]^{\cdot+}$ 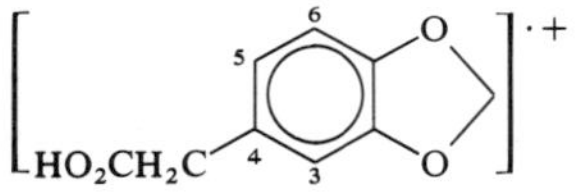	Oxidation by Ce(IV)SO_4/ $H_2SO_4 \cdot 4H_2O$ (flow)	ESR/ —	2.0039	$2H(CH_2)$: 2.04 H(3): 0.06 2H(4): 0.465 H(5): 0.49 H(6): 0.025	76Di1
$[C_{10}H_8O_4]^{\cdot+}$ 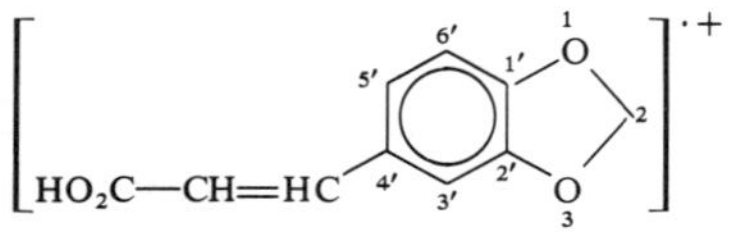	Oxidation by Ce(IV)SO_4/ $H_2SO_4 \cdot 4H_2O$ (flow)	ESR/ —	2.0038	$2H(CH_2)$: 1.85 H(3′): 0.11 1H(4′): 0.40 1H(4′): 0.19 H(5′): 0.40 H(6′): 0.04	76Di1
$[C_{10}H_{10}O_2]^{\cdot+}$ 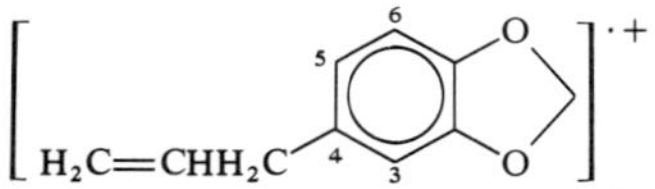	Oxidation by Ce(IV)SO_4/ $H_2SO_4 \cdot 4H_2O$ (flow)	ESR/ —	2.0038	$2H(CH_2)$: 1.99 H(3): 0.075 $2H(4, CH_2)$: 0.725 H(5): 0.525 H(6): 0.025	76Di1
$[C_{10}H_{10}O_4]^{\cdot+}$ 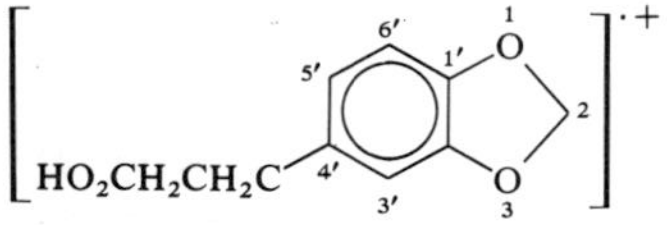	Oxidation by Ce(IV)SO_4/ $H_2SO_4 \cdot 4H_2O$ (flow)	ESR/ —	0.0038	$2H(CH_2)$: 2.025 H(3′): 0.07 $2H(4', CH_2)$: 0.675 H(5′): 0.51 H(6′): 0.025	76Di1

Substance	Generation/ Matrix or Solvent	Method/ T [K]	g-Factor	a-Value [mT]	Ref./ add. Ref.
$[C_8H_6Cl_2O_2]^{\cdot +}$ 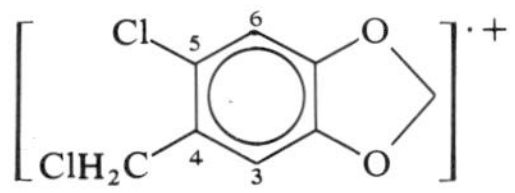	Oxidation by Ce(IV)SO_4/ $H_2SO_4 \cdot 4H_2O$ (flow)	ESR/ —	2.00495	2H(CH_2): 1.95 H(3): 0.075 2H(4, CH_2): 0.89 H(6): 0.025 Cl(5): 0.125	76Di1
$[C_8H_6O_4]^{\cdot +}$ 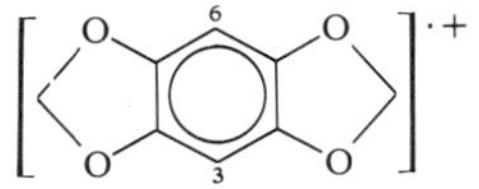	Oxidation by Ce(IV)SO_4/ $H_2SO_4 \cdot 4H_2O$	ESR/ —	2.00435	4H(CH_2): 1.175 2H(3, 6): 0.1025	79Ho1
$[C_9H_8O_3]^{\cdot +}$ 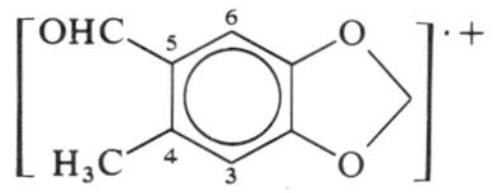	Oxidation by Ce(IV)SO_4/ $H_2SO_4 \cdot 4H_2O$ (flow)	ESR/ —	2.0042	2H(CH_2): 2.125 H(3): 0.065 3H(CH_3): 0.685 2H(5, 6): 0.025	79Ho1
$[C_{10}H_{12}O_4]^{\cdot +}$ 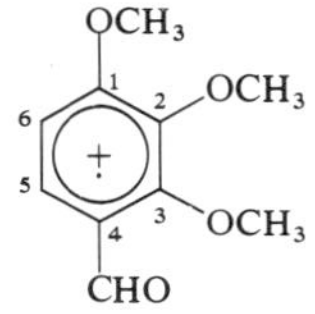	Oxidation by Ce(IV)SO_4/ $H_2SO_4 \cdot 4H_2O$ (flow)	ESR/ —	2.0040	3H(1, CH_3): 0.175 3H(2, CH_3): 0.570 3H(3, CH_3): 0.225 H(5): 0.555 H(6): 0.045	79Ho1
$[C_{10}H_{12}O_4]^{\cdot +}$ 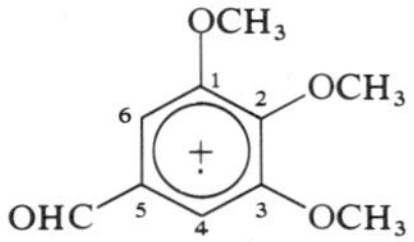	Oxidation by Ce(IV)SO_4/ $H_2SO_4 \cdot 4H_2O$ (flow)	ESR/ —	2.0043	6H(1, 3, CH_3): 0.230 3H(2, CH_3): 0.530 2H(4, 6): 0.080 H(OH): 0.010	79Ho1
$[C_{10}H_{12}O_5]^{\cdot +}$ 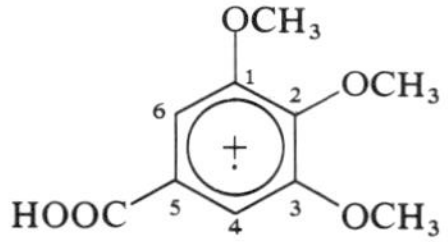	Oxidation by Ce(IV)SO_4/ $H_2SO_4 \cdot 4H_2O$ (flow)	ESR/ —	2.0040	6H(1, 3, CH_3): 0.225 3H(2, CH_3): 0.323 2H(4, 6): 0.075	79Ho1

Substance	Generation/ Matrix or Solvent	Method/ T [K]	g-Factor	a-Value [mT]	Ref./ add. Ref.
$[C_{10}H_{12}O_5]^{\cdot +}$ 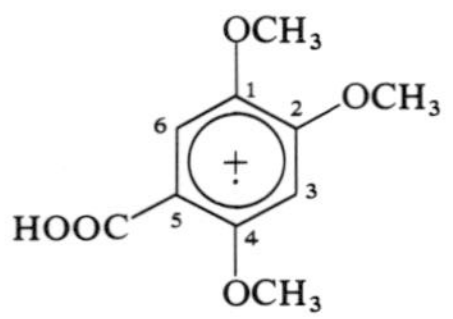	Oxidation by $Ce(IV)SO_4$/ $H_2SO_4 \cdot 4H_2O$ (flow)	ESR/ —	2.00385	$3H(1, CH_3)$: 0.340 $3H(2, CH_3)$: 0.185 2H(3, 6): 0.035 $3H(4, CH_3)$: 0.265	79Ho1
$[C_{10}H_{14}O_3]^{\cdot +}$ 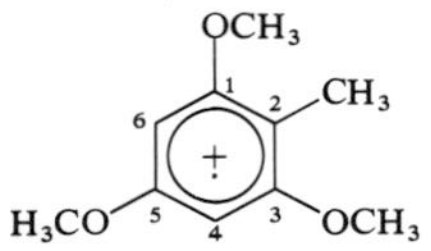	Oxidation by $Ce(IV)SO_4$/ $H_2SO_4 \cdot 4H_2O$ (flow)	ESR/ —	2.00315	$6H(1, 3, CH_3)$: 0.050 $3H(2, CH_3)$: 1.560 2H(4, 6): 0.250 $3H(5, CH_3)$: 0.450	79Ho1
$[C_{10}H_{14}O_3]^{\cdot +}$ 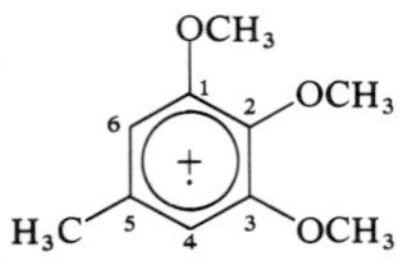	Oxidation by $Ce(IV)SO_4$/ $H_2SO_4 \cdot 4H_2O$ (flow)	ESR/ —	2.0037	$6H(1, 3, CH_3)$: 0.180 $3H(2, CH_3)$: 0.50 2H(4, 6): 0.005 $3H(5, CH_3)$: 0.895	79Ho1
$[C_{10}H_{14}O_4]^{\cdot +}$ 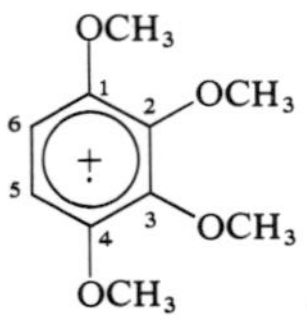	Oxidation by $Ce(IV)SO_4$/ $H_2SO_4 \cdot 4H_2O$ (flow)	ESR/ —	2.0039	$6H(1, 4, CH_3)$: 0.280 $6H(2, 3, CH_3)$: 0.055 2H(5, 6): 0.300	79Ho1
$[C_{10}H_{14}O_4]^{\cdot +}$ 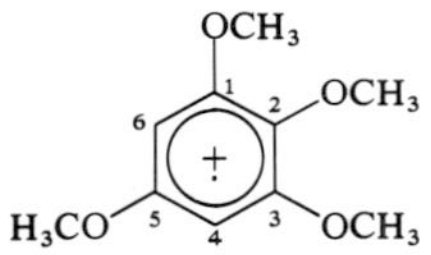	Oxidation by $Ce(IV)SO_4$/ $H_2SO_4 \cdot 4H_2O$ (flow)	ESR/ —	2.00385	$6H(1, 3, CH_3)$: 0.220 $3H(2, CH_3)$: 0.470 2H(4, 6): 0.020 $3H(5, CH_3)$: 0.260	79Ho1

Ohya-Nishiguchi, Terahara, Tajima, Ishizu

Substance	Generation/ Matrix or Solvent	Method/ T [K]	g-Factor	a-Value [mT]	Ref./ add. Ref.
$[C_{10}H_{14}O_4]^{\cdot +}$ 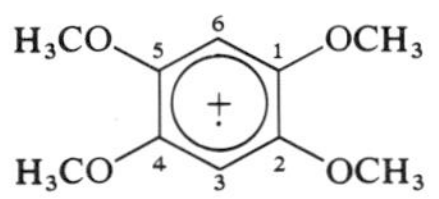	Oxidation by Ce(IV)SO_4/ $H_2SO_4 \cdot 4H_2O$ (flow)	ESR/ —	2.0039	12H(CH_3): 0.235 2H(3, 6): 0.095	79Ho1
$[C_{11}H_{12}O_3]^{\cdot +}$ 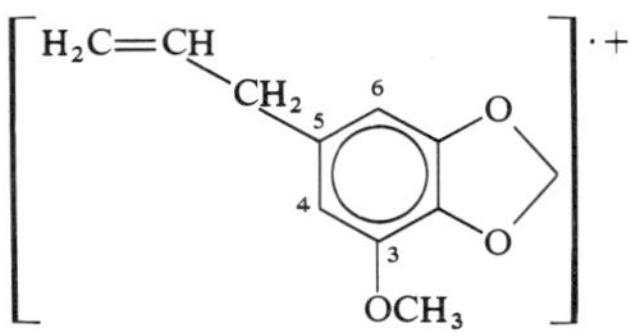	Oxidation by Ce(IV)SO_4/ $H_2SO_4 \cdot 4H_2O$ (flow)	ESR/ —	2.0044	2H(CH_2): 1.890 3H(OCH_3): 0.135 H(4): 0.025 2H(5, CH_2): 0.765 H(6): 0.065	79Ho1
$[C_{11}H_{14}O_2]^{\cdot +}$ 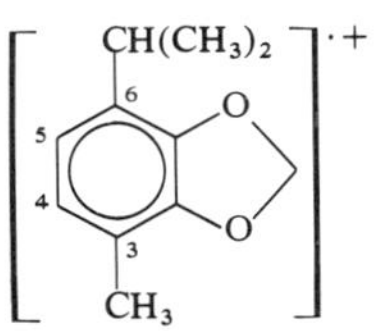	Oxidation by Ce(IV)SO_4/ $H_2SO_4 \cdot 4H_2O$ (flow)	ESR/ —	2.0039	2H(CH_2): 2.123 3H(3, CH_3): 0.067 2H(4, 5): 0.461 H(6, CH): 0.027	79Ho1
$[C_{11}H_{14}O_5]^{\cdot +}$ 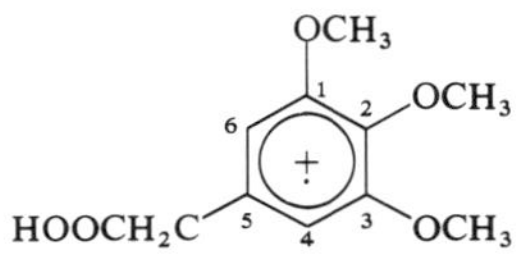	Oxidation by Ce(IV)SO_4/ $H_2SO_4 \cdot 4H_2O$ (flow)	ESR/ —	2.0037	6H(1, 3, CH_3): 0.20 5H(2, CH_3, 5, CH_2): 0.520 2H(4, 6): 0.033	79Ho1
$[C_{11}H_{16}NO_2]^{\cdot 2+}$ 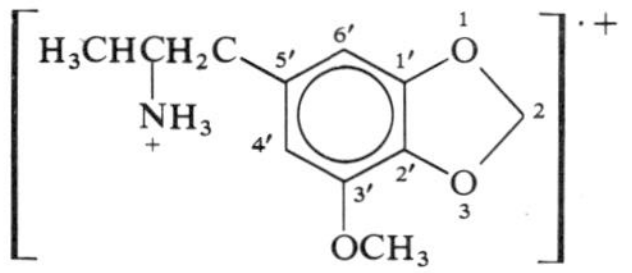	Oxidation by Ce(IV)SO_4/ $H_2SO_4 \cdot 4H_2O$ (flow)	ESR/ —	2.0040	2H(CH_2): 1.85 3H(OCH_3): 0.15 H(4′): 0.08 1H(5′, CH_2): 0.56 1H(5′, CH_2): 0.49 H(6′): 0.03	76Di1

Ohya-Nishiguchi, Terahara, Tajima, Ishizu

Substance	Generation/ Matrix or Solvent	Method/ T [K]	g-Factor	a-Value [mT]	Ref./ add. Ref.
$[C_{11}H_{18}NO_3]^{\cdot 2+}$ 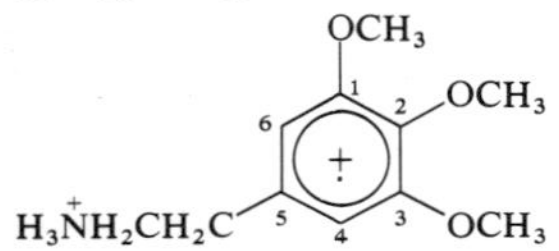	Oxidation by $Ce(IV)SO_4$/ $H_2SO_4 \cdot 4H_2O$ (flow)	ESR/ —	2.0038	6H(1, 3, CH_3): 0.20 3H(2, CH_3): 0.515 2H(4, 6): 0.030 2H(5, CH_2): 0.565	79Ho1
$[C_{16}H_{22}O_2]^{\cdot +}$ 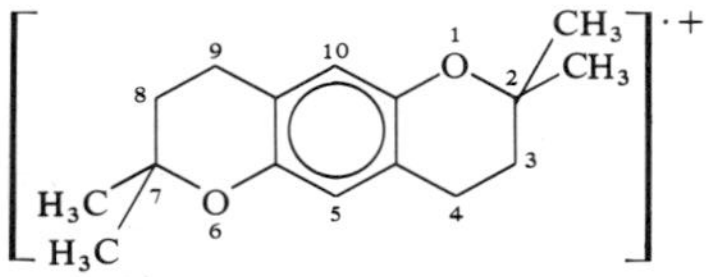	Oxidation by CF_3COOH/ Benzonitrile:ACN (1:1)	ESR, ENDOR, TRIPLE/ —		2H(5, 10): 0.294 2H(4, 9, ax): 0.242 2H(4, 9, eq): 0.036 2H(3, 8, ax): 0.036 2H(3, 9, eq): 0.014 6H(2, 7, ax): 0.005	82Fa1
$[C_{10}H_{12}O_5]^{\cdot +}$ 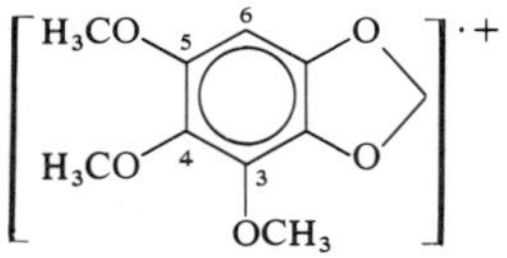	Oxidation by $Ce(IV)SO_4$/ $H_2SO_4 \cdot 4H_2O$ (flow)	ESR/ —	2.0040	2H(CH_2): 1.125 3H(3, CH_3): 0.005 6H(4, 5, CH_3): 0.215 H(6): 0.075	79Ho1
$[C_{10}H_{12}O_5]^{\cdot +}$ 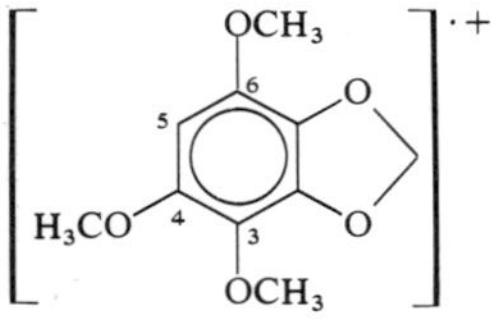	Oxidation by $Ce(IV)SO_4$/ $H_2SO_4 \cdot 4H_2O$ (flow)	ESR/ —	2.0040	2H(CH_2): 0.435 3H(3, CH_3): 0.145 6H(4, 6, CH_3): 0.230 H(5): 0.020	79Ho1
$[C_{11}H_{16}O_5]^{\cdot +}$ 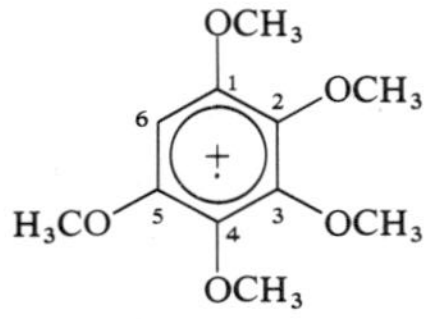	Oxidation by $Ce(IV)SO_4$/ $H_2SO_4 \cdot 4H_2O$ (flow)	ESR/ —	2.0039	6H(1, 5, CH_3): 0.225 6H(2, 4, CH_3): 0.205 3H(3, CH_3): 0.025 H(6): 0.060	79Ho1

Substance	Generation/ Matrix or Solvent	Method/ T [K]	g-Factor	a-Value [mT]	Ref./ add. Ref.
$[C_{12}H_{14}O_4]^{\cdot +}$	Oxidation by $Ce(IV)SO_4$/ $H_2SO_4 \cdot 4H_2O$ (flow)	ESR/ —	2.0038	2H(CH_2): 1.215 3H(4, CH_3): 0.30 2H(5, CH_2): 0.610	79Ho1
$[C_{12}H_{14}O_4]^{\cdot +}$	Oxidation by $Ce(IV)SO_4$/ $H_2SO_4 \cdot 4H_2O$ (flow)	ESR/ —	2.0038	2H(CH_2): 1.742 6H(3, 6, CH_3): 0.132 2H(4, CH_2): 0.770 H(5): 0.060	79Ho1
$[C_{16}H_{22}O_2]^{\cdot +}$	Oxidation by $CHCl_2COOH$/ $CHCl_2COOH:CH_2Cl_2$ (1:3 V/V)	ESR, ENDOR, TRIPLE/ 310	—	6H(4, 8, CH_3): 0.140 4H(3, 7, CH_2): 0.618	82Fa1
$[C_{18}H_{26}O_2]^{\cdot +}$	Oxidation by CF_3COOH/ Propionitrile	ESR, ENDOR, TRIPLE/	—	6H(9, 10, CH_3): 0.1557 2H(4, 5, ax): 0.623 2H(3, 6, ax): 0.0257 2H(3, 6, eq): 0.0257	82Fa1

19.1.6.2 Other aromatics

Substance	Generation/ Matrix or Solvent	Method/ T [K]	g-Factor	a-Value [mT]	Ref./ add. Ref.
$[C_{12}H_{10}O_2]^{\cdot +}$	Solution in $AlCl_3:CH_3NO_2$/ CH_3NO_2	EPR/ 216		2H(2, 2′): 0.1874 (50) 2H(3, 3′): 0.0752 (50) 2H(5, 5′): 0.0752 (50) 2H(6, 6′): 0.2029 (50) 2H(4, 4′, OH): 0.1722 (50)	77Su1
		257		2H(2, 2′): 0.1951 (50) 2H(3, 3′): 0.0752 (50) 2H(5, 5′): 0.0752 (50) 2H(6, 6′): 0.1951 (50) 2H(4, 4′, OH): 0.1682 (50)	
$[C_{14}H_{14}O_2]^{\cdot +}$	Solution in CF_3COOH $+(CF_3CO)_2O:CH_3NO_2$/ CH_3NO_2	EPR/ 216		2H(2, 2′): 0.23008 (290) 2H(3, 3′): 0.0729 (36) 6H(4, 4′, CH_3): 0.17768 (220) 2H(5, 5′): 0.0838 (40) 2H(6, 6′): 0.1568 (30)	77Su1
		328		4H(2, 2′, 6 ,6′): 0.19136 (100) 4H(3, 3′, 5, 5′): 0.07759 (160) 6H(4, 4′, CH_3): 0.17629 (100)	
$[C_{14}H_{14}O_2]^{\cdot +}$	Solution in $AlCl_3:CH_3NO_2$/ CH_3NO_2	EPR/ 265	—	2H(2, 2′): 0.1454 6H(3, 3′, CH_3): 0.0820 2H(5, 5′): 0.0566 2H(6, 6′): 0.2342 2H(4, 4′, OH): 0.1641	77Su1

Substance	Generation/ Matrix or Solvent	Method/ T [K]	g-Factor	a-Value [mT]	Ref./ add. Ref.
$[C_{17}H_{18}O_4]^{\cdot +}$	Oxidation by $AlCl_3$/ CH_3NO_2	ESR/ 263	2.00336	4H(CH_2, 3, 3′): 0.0141 6H(4, 4′, CH_3): 0.1601 6H(5, 5′, CH_3): 0.0802 2H(6, 6′): 0.0375	78Su1
$[C_{18}H_{20}O_4]^{\cdot +}$	Oxidation by $AlCl_3$/ CH_3NO_2	ESR/ —	2.00332	4H(2, 2′, CH_2): 0.5204 6H(4, 4′, CH_3): 0.1505 6H(5, 5′, CH_3): 0.0848 2H(6, 6′): 0.0425	78Su1
$[C_{16}H_{16}O_4]^{\cdot +}$	Oxidation by $AlCl_3$/ CH_3NO_2	ESR/ 233	2.00343	4H: 0.0502 12H(CH_3): 0.1260	78Su1
$[C_{16}H_9F_3O_2]^{\cdot +}$	Anthracene in $Tl(CF_3CO_2)_3$:CF_3COOH/ CF_3COOH	ESR, ENDOR, TRIPLE/ 255	—	3F: 0.026 2H(1, 8): 0.285 2H(4, 5): 0.330 2H(2, 7): 0.158 2H(3, 6): 0.100 H(10): 0.641	85El1
$[C_{18}H_8F_6O_4]^{\cdot +}$	Anthracene in $Tl(CF_3CO_2)_3$:CF_3COOH/ CF_3COOH	ESR, ENDOR, TRIPLE/ 255	—	6F: 0.0262 4H(1. 4. 5, 8): 0.309 4H(2, 3, 6, 7): 0.122	85El1

Substance	Generation/ Matrix or Solvent	Method/ T [K]	g-Factor	a-Value [mT]	Ref./ add. Ref.
$[C_{22}H_{18}O_2]^{\cdot +}$ [a])	Oxidation by $AlCl_3$/ $C_6H_5NO_2$	EPR/ 224	—	6H(1, 4): 0.322 2H(2, 3): 0.264 2H(9, 10): 0.008 8H(others): 0.011	86Qu1
$[C_{24}H_{22}O_2]^{\cdot +}$ [b])	Oxidation by $AlCl_3$/ $C_6H_5NO_2$	EPR/ 224	—	4H(1, 4): 0.345 6H(1, 4): 0.0138 2H(2, 3): 0.258 2H(9, 10): 0.0105	86Qu1
$[C_{24}H_{22}O_4]^{\cdot +}$ [c])	Oxidation by $AlCl_3$/ $C_6H_5NO_2$	EPR/ 224	—	6H(1, 4): 0.324 2H(2, 3): 0.262 2H(9, 10): 0.0071 6H(others): 0.011	86Qu1
$[C_{24}H_{22}O_4]^{\cdot +}$ 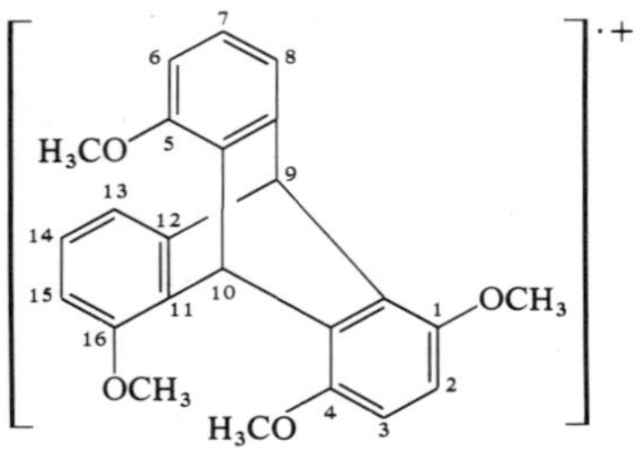	Oxidation by $AlCl_3$/ $C_6H_5NO_2$	EPR/ 224	—	6H(1, 4): 0.320 2H(2, 3): 0.263 2H(9, 10): 0.0073 6H(others): 0.010	86Qu1

a) b) 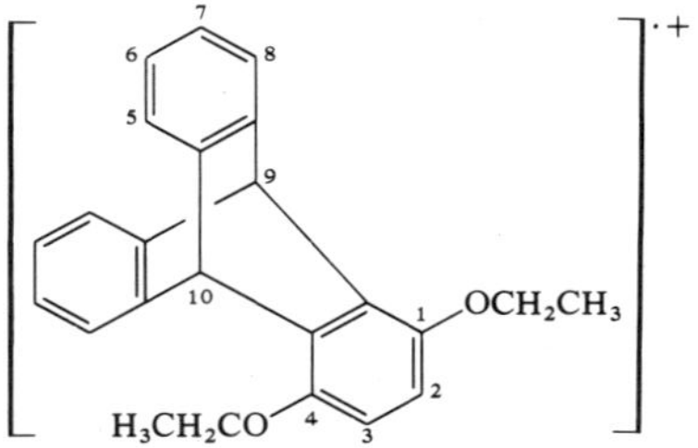 c)

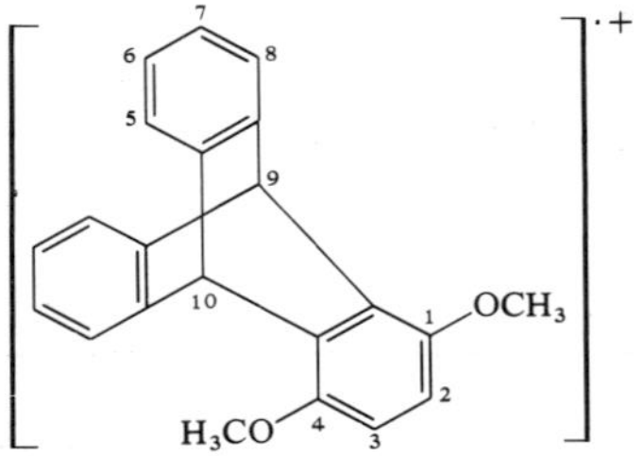

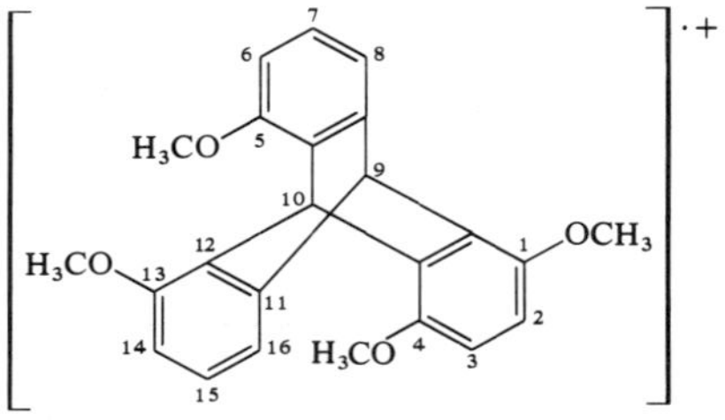

Substance	Generation/ Matrix or Solvent	Method/ T [K]	g-Factor	a-Value [mT]	Ref./ add. Ref.
19.1.6.3 $(OR)_n$ *and* $(SR)_n$ substituted benzenes					
$[C_7H_7FO_3S]^{\cdot+}$ 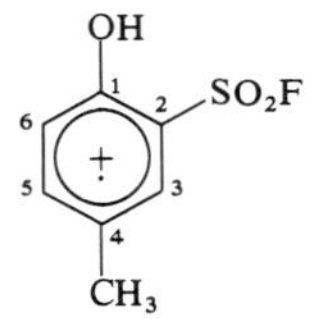	Oxidation by HSO_3:PbO_2/ HSO_3F	EPR/ 198	—	F: 0.085 H(OH): 0.380 H(3): 0.215 $3H(CH_3)$: 1.720 H(5): 0.025 H(6): 0.470	79Ru1
$[C_7H_8O_4S]^{\cdot+}$ 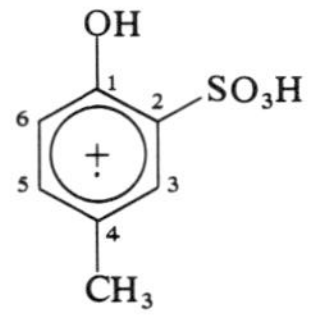	Oxidation by HSO_3:PbO_2/ HSO_3F	EPR/ 198	—	H(OH): 0.380 H(3): 0.180 $3H(CH_3)$: 1.700 H(6): 0.450	79Ru1, 79Ru2
$[C_8H_9FO_3S]^{\cdot+}$ 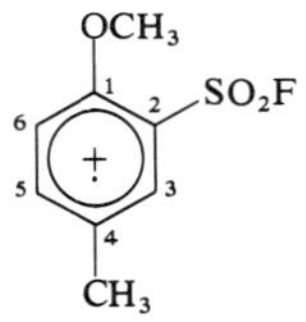	Oxidation by HSO_3:PbO_2/ HSO_3F	EPR/ 198, 203	—	F: 0.085 $3H(OCH_3)$: 0.425 H(3): 0.155 $3H(CH_3)$: 1.690 H(5): 0.025 H(6): 0.470	79Ru1, 77Ru1
$[C_8H_{10}O_4S]^{\cdot+}$ 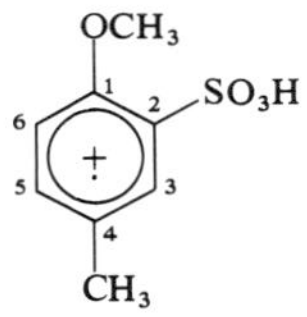	Oxidation by HSO_3:PbO_2/ HSO_3F	EPR/ 198	—	$3H(OCH_3)$: 0.410 H(3): 0.120 $3H(CH_3)$: 1.680 H(6): 0.440	79Ru1

Substance	Generation/ Matrix or Solvent	Method/ T [K]	g-Factor	a-Value [mT]	Ref./ add. Ref.
$[C_8H_{10}O_5S]^{\cdot +}$ 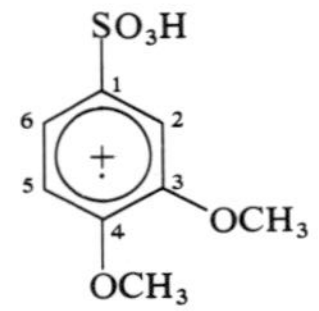	Oxidation by HSO_3F:PbO_2/ HSO_3F	EPR/ 198	—	H(2): 0.075 3H(3, CH_3): 0.420 3H(4, CH_3): 0.365 H(6): 0.410	79Ru2
$[C_9H_{11}FO_3S]^{\cdot +}$ 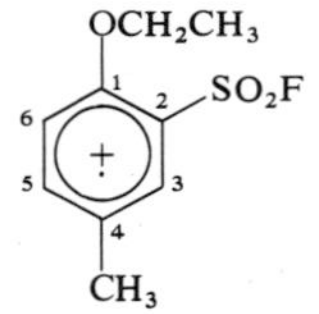	Oxidation by HSO_3:PbO_2/ HSO_3F	EPR/ 198	—	F: 0.085 3H(1, CH_2, 6): 0.475 H(3): 0.150 3H(4, CH_3): 1.670 H(5): 0.025	79Ru1
$[C_9H_{12}O_4S]^{\cdot +}$ 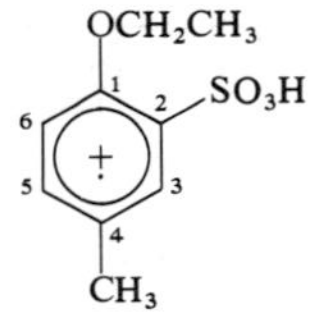	Oxidation by HSO_3:PbO_2/ HSO_3F	EPR/ 198	—	2H(CH_2): 0.455 H(3): 0.120 3H(4, CH_3): 1.665 H(6): 0.455	79Ru1
$[C_{11}H_{15}FO_3S]^{\cdot +}$ 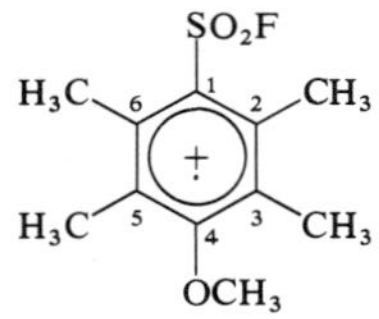	Oxidation by $HFSO_3$:PbO_2/ HSO_3F	EPR/ 198	—	F: 0.205 6H(2, 6, CH_3): 0.069 6H(3, 5, CH_3): 0.712 3H(4, CH_3): 0.512	79Ru2

Substance	Generation/ Matrix or Solvent	Method/ T [K]	g-Factor	a-Value [mT]	Ref./ add. Ref.
19.1.7 $(SR)_n$-substituted aromatics					
$[C_8H_9FO_2S]^{\cdot +}$ 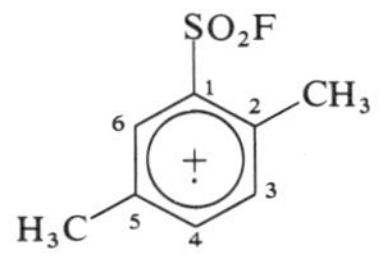	Oxidation by $HSO_3F:PbO_2/$ HSO_3F	EPR/ 198	–	F: 0.067 $3H(2, CH_3)$: 1.860 H(3): 0.300 2H(4, 6): 0.445 $3H(5, CH_3)$: 1.700	79Ru3
$[C_8H_{10}O_3S]^{\cdot +}$ 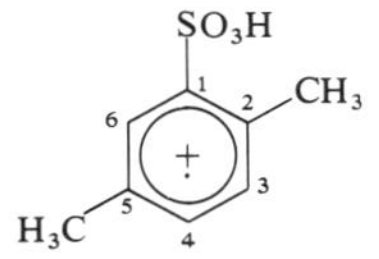	Oxidation by $HSO_3F:PbO_2/$ HSO_3F	EPR/ 198	–	$3H(2, CH_3)$: 1.850 H(3): 0.085 H(4): 0.400 $3H(5, CH_3)$: 1.710 H(6): 0.285	79Ru3
$[C_9H_{11}FO_2S]^{\cdot +}$ 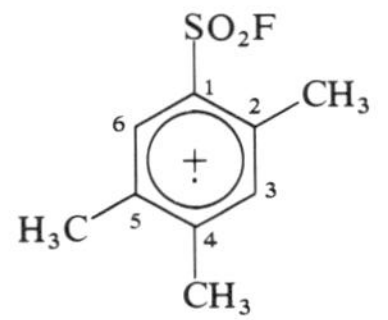	Oxidation by $HSO_3F:PbO_2/$ HSO_3F	EPR/ 198	–	F: 0.085 $3H(2, CH_3)$: 1.505 H(3): 0.085 $3H(4, CH_3)$: 1.950 $3H(5, CH_3)$: 0.315 H(6): 0.290	79Ru3
$[C_9H_{11}FO_2S]^{\cdot +}$ 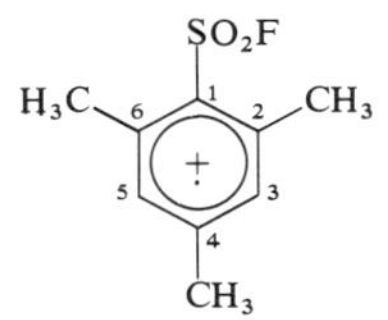	Oxidation by $HSO_3F:PbO_2/$ HSO_3F	EPR/ 203	–	F: 0.10 $6H(2, 6, CH_3)$: 1.260 H(3, 5): 0.770 $3H(4, CH_3)$: 0.10	77Ru1
$[C_9H_{12}O_3S]^{\cdot +}$ 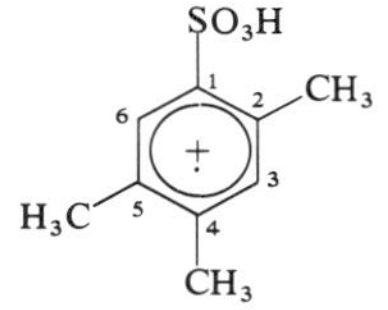	Oxidation by $HSO_3F:PbO_2/$ HSO_3F	EPR/ 198	–	$3H(2, CH_3)$: 0.230 H(3): 0.075 $3H(4, CH_3)$: 1.917 $3H(5, CH_3)$: 0.365 H(6): 1.475	79Ru3

Substance	Generation/ Matrix or Solvent	Method/ T [K]	g-Factor	a-Value [mT]	Ref./ add. Ref.
$[C_9H_{12}O_3S]^{\cdot+}$ 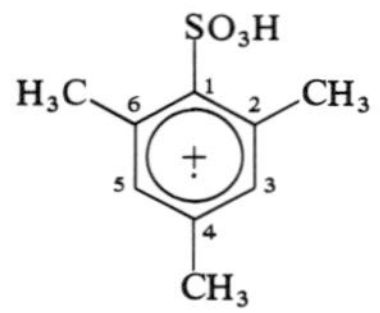	Oxidation by $HSO_3F:PbO_2/HSO_3F$	EPR/ 203	—	$6H(2, 6, CH_3)$: 1.240 $2H(3, 5)$: 0.775 $3H(4, CH_3)$: 0.080	77Ru1
$[C_{10}H_{13}FO_2S]^{\cdot+}$ 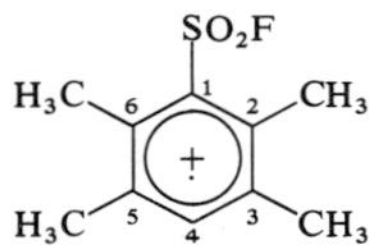	Oxidation by $HSO_3F:PbO_2/HSO_3F$	EPR/ 203	—	F: 0.10 $6H(2, 6, CH_3)$: 1.180 $6H(3, 5, CH_3)$: 1.026 H(4): 0.078	77Ru1
$[C_{10}H_{14}O_3S]^{\cdot+}$ 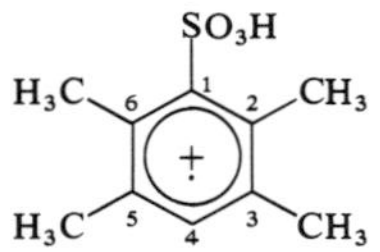	Oxidation by $HSO_3F:PbO_2/HSO_3F$	EPR/ 203	—	$6H(2, 6, CH_3)$: 1.173 $6H(3, 5, CH_3)$: 1.031 H(4): 0.078	77Ru1
$[C_{10}H_{14}O_4S]^{\cdot+}$ 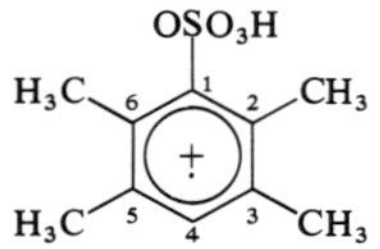	Oxidation by $HSO_3F:PbO_2/HSO_3F$	EPR/ 198	—	$6H(2, 6, CH_3)$: 1.140 $6H(3, 5, CH_3)$: 1.045 H(4): 0.060	79Ru2
$[C_{10}H_{14}O_3S]^{\cdot+}$ 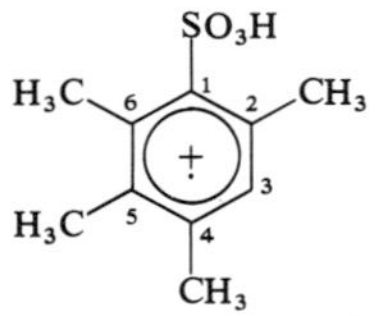	Oxidation by $HSO_3F:PbO_2/HSO_3F$	EPR/ 198	—	$6H(2, 6, CH_3)$: 1.375 H(3): 0.320 $3H(4, CH_3)$: 1.666 $3H(5, CH_3)$: 0.815	79Ru3

Substance	Generation/ Matrix or Solvent	Method/ T [K]	g-Factor	a-Value [mT]	Ref./ add. Ref.
$[C_{10}H_{13}FO_2S]^{\cdot +}$ 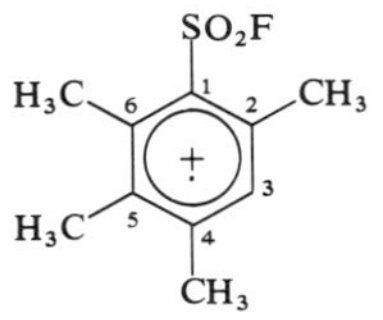	Oxidation by $HSO_3F:PbO_2$/ HSO_3F	EPR/ 198	–	F: 0.080 6H(2, 6, CH_3): 1.380 H(3): 0.330 3H(4, CH_3): 1.668 3H(5, CH_3): 0.840	79Ru3
$[C_{10}H_{14}O_3S]^{\cdot +}$ 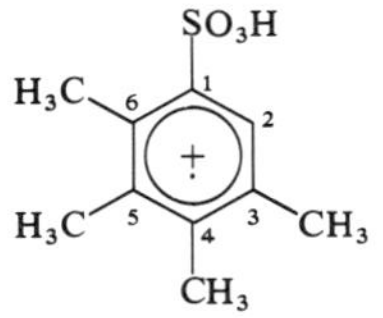	Oxidation by $HSO_3F:PbO_2$/ HSO_3F	EPR/ 198	–	H(2): 0.540 6H(3, 5, CH_3): 0.635 3H(4, CH_3): 1.545 3H(6, CH_3): 1.455	79Ru3
$[C_{11}H_{15}FO_2S]^{\cdot +}$ 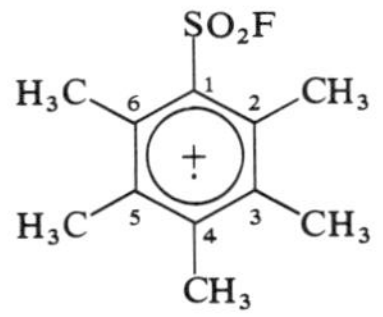	Oxidation by $HSO_3F:PbO_2$/ HSO_3F	EPR/ 203	–	F: 0.100 6H(2, 6, CH_3): 1.150 6H(3, 5, CH_3): 1.000 3H(4, CH_3): 0.100	77Ru1
$[C_{11}H_{16}O_4S]^{\cdot +}$ 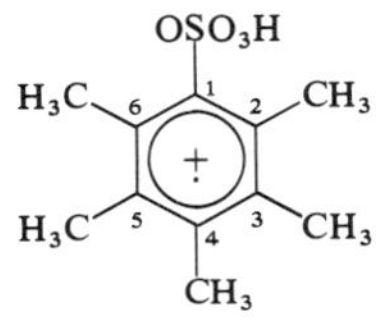	Oxidation by $HSO_3F:PbO_2$/ HSO_3F	EPR/ 198	–	H(2, 6, CH_3): 1.060 H(3, 5, CH_3): 1.000	79Ru2

Substance	Generation/ Matrix or Solvent	Method/ T [K]	g-Factor	a-Value [mT]	Ref./ add. Ref.
$[C_{12}H_{18}S_2]^{\cdot +}$ 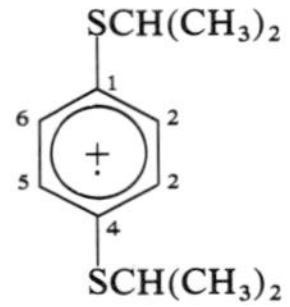	Oxidation by $AlCl_3:SbCl_3$ (3:1)/ CH_2Cl_2	ESR/ 228	2.0087	*cis* form: 2H: 0.336 2H: 0.109 2H: 0.062 12H: 0.011 *trans* form: 2H: 0.348 2H: 0.179 2H: 0.098 12H: 0.011	84Al1
$[C_9H_{12}S_3]^{\cdot +}$ 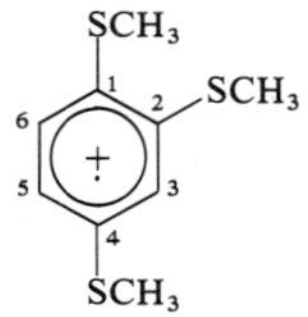	Oxidation by $AlCl_3:SbCl_3$ (3:1)/ CH_2Cl_2	ESR/ 233	2.0082	6H: 0.44 1H: 0.29 3H: 0.16 2H: 0.02	84Al1
$[C_{15}H_{24}S_3]^{\cdot +}$ 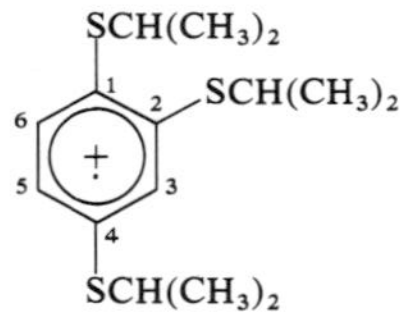	Oxidation by $AlCl_3$: $SbCl_3$ (3:1)/ CH_2Cl_2	ESR/ 233	2.0082	1H: 0.30 2H: 0.30 1H: 0.11 2H: <0.2	84Al1
$[C_{10}H_{14}S_4]^{\cdot +}$ 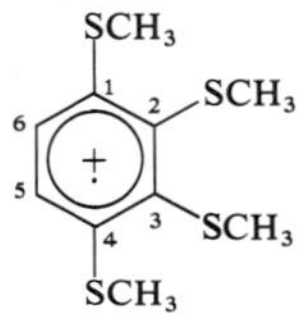	Oxidation by $AlCl_3:SbCl_3$ (3:1)/ CH_2Cl_2	ESR/ 263	2.0089	2H: 0.147 6H: 0.501	84Al1

Substance	Generation/ Matrix or Solvent	Method/ T [K]	g-Factor	a-Value [mT]	Ref./ add. Ref.
$[C_{10}H_{14}S_4]^{\cdot +}$	Oxidation by $AlCl_3:SbCl_3$ (3:1)/ CH_2Cl_2	ESR/ 298	2.0077	2H(ring): 0.073 12H(CH_3): 0.259	84Al1
$[C_{18}H_{30}S_4]^{\cdot +}$	Oxidation by $AlCl_3:SbCl_3$ (3:1)/ CH_2Cl_2	ESR/ 298	–	2H: 0.142 2H: 0.328	84Al1
$[C_{18}H_{30}S_4]^{\cdot +}$	Oxidation by $AlCl_3:SbCl_3$ (3:1)/ CH_2Cl_2	ESR/ 213	2.0077	2H(ring): 0.069 4H(CH): 0.193 [6]	84Al1
$[C_{12}H_{18}S_6]^{\cdot +}$	Oxidation by $AlCl_3:SbCl_3$ (3:1)/ CH_2Cl_2	ESR/ 298	2.0070	18H(CH_3): 0.255	84Al1
$[C_{24}H_{42}S_6]^{\cdot +}$	Oxidation by $AlCl_3:SbCl_3$ (3:1)/ CH_2Cl_2	ESR/ 213	2.0076	6H(CH): 0.379 [6]	84Al1

[6]) Temperature dependent hfcc.

Substance	Generation/ Matrix or Solvent	Method/ T[K]	g-Factor	a-Value [mT]	Ref./ add. Ref.
$[C_{12}H_{12}S_2]^{\cdot +}$ 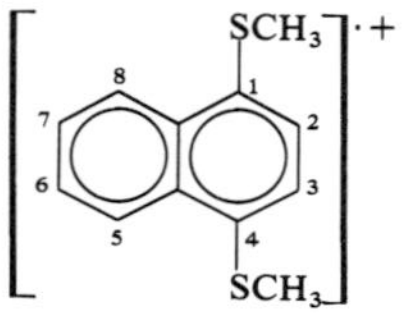	Oxidation by $AlCl_3$/ CH_3NO_2	ESR/ —	2.0075	2H(2): 0.215 2H(5): 0.090 2H(6): 0.040 6H(CH_3): 0.365	77Bo2
$[C_{12}H_{12}S_2]^{\cdot +}$ 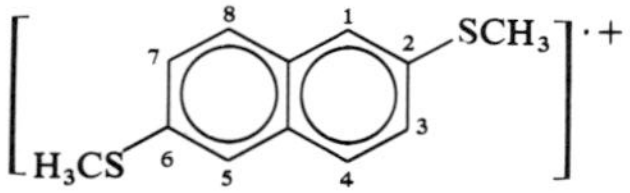	Oxidation by $AlCl_3$/ CH_3NO_2	ESR/ —	2.0068	2H(1): 0.371 2H(3): <0.04 2H(4): 0.059 6H(CH_3): 0.220	77Bo2
$[C_{16}H_{14}S_2]^{\cdot +}$ 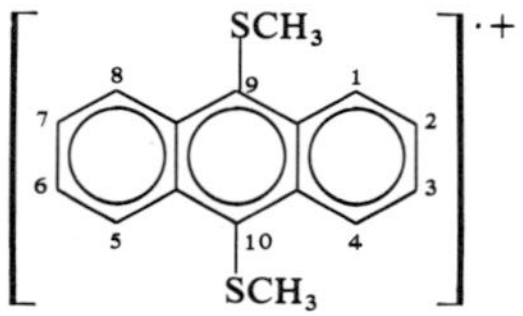	Oxidation by $AlCl_3$/ CH_3NO_2	ESR/ —	2.0058	4H(1): 0.131 4H(2): 0.075 6H(CH_3): 0.220	77Bo2

19.1.8 Non-benzenoid aromatics

Substance	Generation/ Matrix or Solvent	Method/ T[K]	g-Factor	a-Value [mT]	Ref./ add. Ref.
$[C_{28}H_{20}]^{\cdot +}$ Oxidation with $AgClO_4$/ 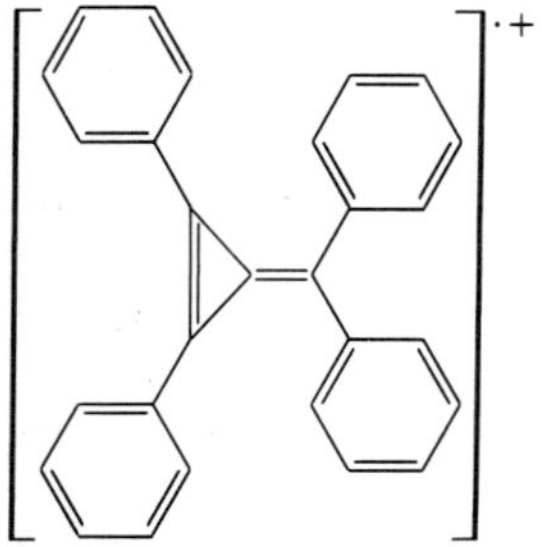	CH_2Cl_2	ESR/ —	2.002616	[7]	81Ko1

[7]) About 160 lines, not analyzed.

Substance	Generation/ Matrix or Solvent	Method/ T [K]	g-Factor	a-Value [mT]	Ref./ add. Ref.
$[C_{20}H_{30}]^{\cdot +}$ 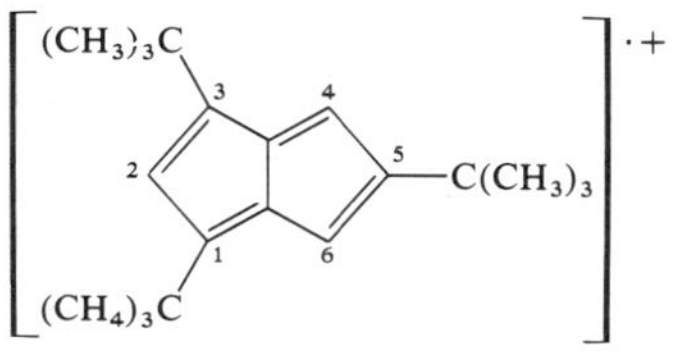	Oxidation with $AlCl_3$/ CH_2Cl_2	ESR, ENDOR/ 283	–	18H(1, 3, $C(CH_3)_3$): 0.006 H(2): 0.918 2H(4, 6): 0.040 9H(5, $C(CH_3)_3$): 0.045	78Fu1
	CH_3NO_2	283	–	18H(1, 3, $C(CH_3)_3$): not resolved H(2): 0.919 2H(4, 6): 0.044 9H(5, $C(CH_3)_3$): 0.044	
	Electrolytic oxidation; oxidation with CF_3COOH/ CH_2Cl_2	ESR, ENDOR/ 213	–	18H(1, 3, $C(CH_3)_3$): not resolved H(2): 0.920 2H(4, 6): 0.046 9H(5, $C(CH_3)_3$): 0.046	78Fu1
	CH_2Cl_2 : CF_3COOH : $(CF_3CO)_2O$	293	–	18H(1, 3, $C(CH_3)_3$): 0.007 H(2): 0.920 2H(4, 6): 0.039 9H(5, $C(CH_3)_3$): 0.044	
$[C_{28}H_{20}]^{\cdot +}$ 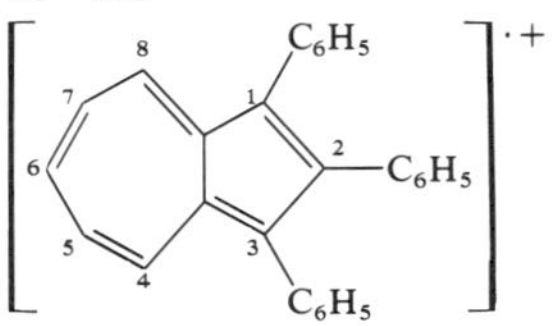	UV-irr. to 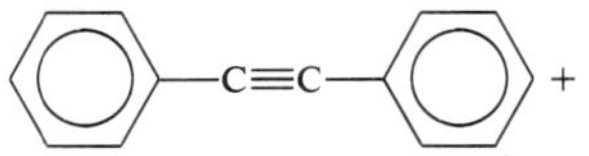 $[C_6H_5{-}C{\equiv}C{-}C_6H_5]^{+}$ $(CF_3CO_2)_2Hg$/ CF_3COOH	EPR/ 208	–	2H(4, 8): 0.067 2H(5, 7): 0.413 4H(2′, 6′, 2″, 6″): 0.059 2H(4′, 4″): 0.124 H(6): 0.007	86Co1
$[C_{16}H_{10}]^{\cdot +}$ 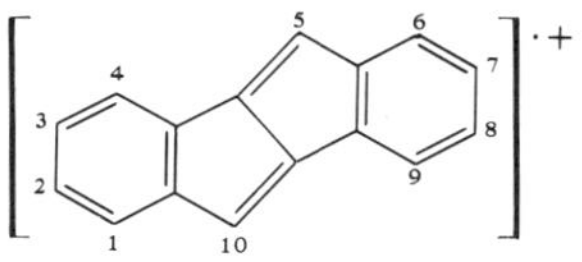	Oxidation with $AlCl_3$/ CH_2Cl_2	ESR/ 273	–	2H(1, 6): 0.058 2H(2, 7): 0.353 2H(3, 8): 0.005 [8] 2H(4, 9): 0.147 2H(5, 10): 0.033	78Fu2

[8]) Not resolved, estimated from linewidth of simulation.

Substance	Generation/ Matrix or Solvent	Method/ T [K]	g-Factor	a-Value [mT]	Ref./ add. Ref.
$[C_{18}H_{14}]^{\cdot +}$ 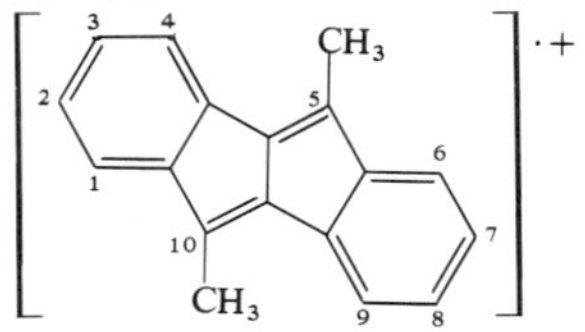	Oxidation with $AlCl_3$/ CH_2Cl_2	ESR/ 243	–	2H(1, 6): 0.051 2H(2, 7): 0.361 2H(3, 8): 0.017 2H(4, 9): 0.119 6H(5, 10, CH_3): 0.068	78Fu2
19.1.9 Annulenes					
$[C_{16}H_{12}]^{\cdot +}$ 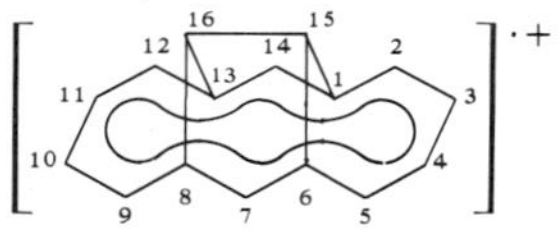	Solution in $AlCl_3$:CH_2Cl_2/ CH_2Cl_2	EPR/ 183···253	2.0027	4H(2, 5, 9, 12): 0.014 4H(3, 4, 10, 11): 0.248 2H(7, 14): 0.092 4H(15, 16): 2.815	84Ge1
$[C_{17}H_{14}]^{\cdot +}$ 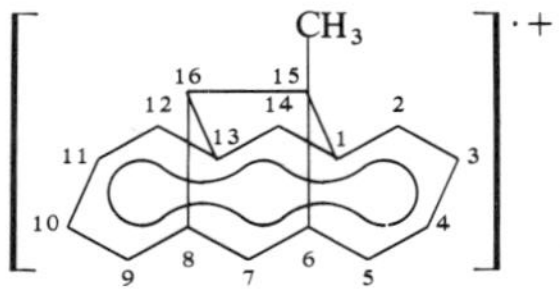	Solution in $AlCl_3$:CH_2Cl_2/ CH_2Cl_2	EPR/ 183···253	2.0027	2H(2, 5, 9, 12): 0.036 2H(2, 5, 9, 12): 0.013 2H(3, 4, 10, 11): 0.257 2H(3, 4, 10, 11): 0.254 2H(7, 14): 0.091 H(16): 2.747 3H(15, CH_3): 0.040	84Ge1
$[C_{18}H_{16}]^{\cdot +}$ 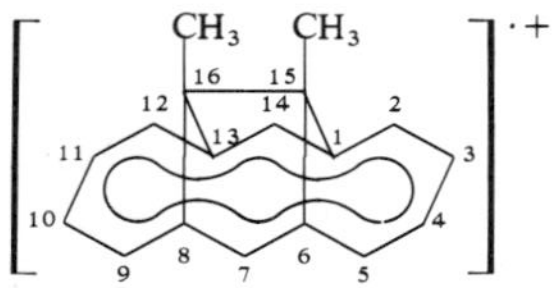	Solution in $AlCl_3$:CH_2Cl_2/ CH_2Cl_2	EPR/ 183···253	2.0027	4H(2, 5, 9, 12): 0.032 4H(3, 4, 10, 11): 0.257 2H(7, 14): 0.081 6H(15, 16, CH_3): 0.038	84Ge1
$[C_{17}H_{14}]^{\cdot +}$ 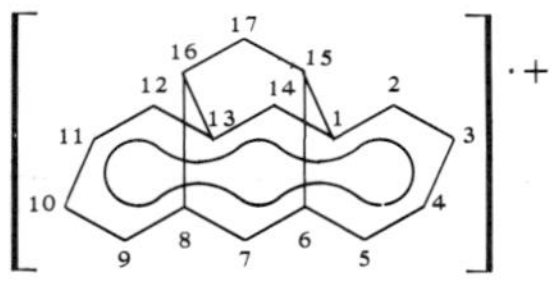	Solution in $AlCl_3$:CH_2Cl_2/ CH_2Cl_2	EPR/ 183···253	2.0022	4H(2, 5, 9, 12): 0.325 4H(3, 4, 10, 11): 0.058 2H(7, 14): 0.457 2H(15, 16): 0.140 2H(17): 0.140	84Ge1

Substance	Generation/ Matrix or Solvent	Method/ T [K]	g-Factor	a-Value [mT]	Ref./ add. Ref.
$[C_{18}H_{16}]^{\cdot +}$ 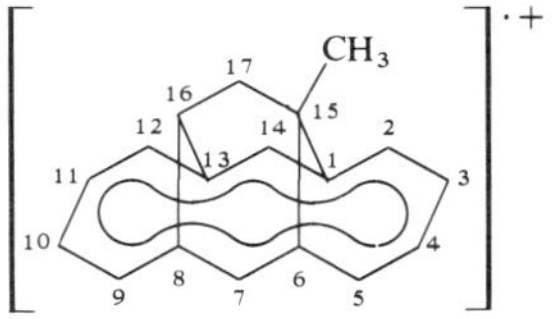	Solution in $AlCl_3:CH_2Cl_2/CH_2Cl_2$	EPR/ 183···253	2.0022	2H(2, 5, 9, 12): 0.325 2H(2, 5, 9, 12): 0.295 2H(3, 4, 10, 11): 0.056 2H(3, 4, 10, 11): 0.056 2H(7, 14): 0.4000 H(16): 0.145 3H(15, CH_3): 0.019 2H(17): 0.145	84Ge1
$[C_{19}H_{18}]^{\cdot +}$ 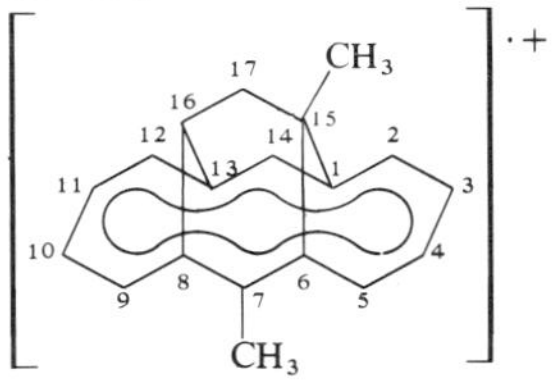	Solution in $AlCl_3:CH_2Cl_2/CH_2Cl_2$	EPR/ 183···253	2.0024	4H(2, 5, 9, 12): 0.266 4H(3, 4, 10, 11): 0.058 6H(7, 14, CH_3): 0.536 2H(15, 16): 0.137 H(17): 0.137	84Ge1
$[C_{17}H_{12}O]^{\cdot +}$ 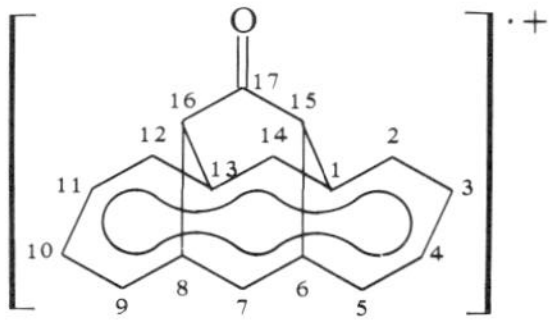	Solution in $AlCl_3:CH_2Cl_2/CH_2Cl_2$	EPR/ 183···253	2.0020	4H(2, 5, 9, 12): 0.318 4H(3, 4, 10, 11): 0.038 2H(7, 14): 0.413 2H(15, 16): 0.136	84Ge1
$[C_{19}H_{16}O]^{\cdot +}$ 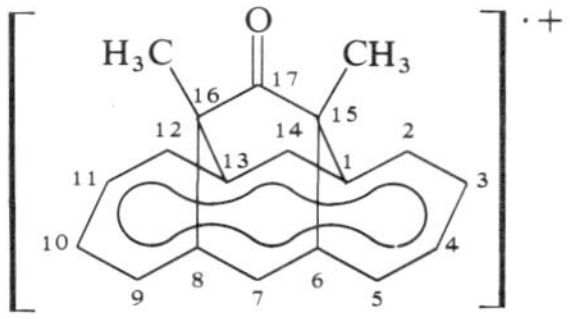	Solution in $AlCl_3:CH_2Cl_2/CH_2Cl_2$	EPR/ 183···253	2.0020	4H(2, 5, 9, 12): 0.301 4H(3, 4, 10, 11): 0.080 2H(7, 14): 0.322 6H(15, 16, CH_3): 0.021	84Ge1

Substance	Generation/ Matrix or Solvent	Method/ T [K]	g-Factor	a-Value [mT]	Ref./ add. Ref.
$[C_{30}H_{42}]^{\cdot +}$ 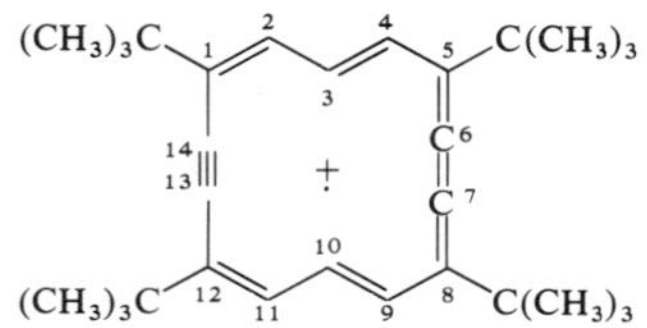	Oxidation by $AlCl_3$/ CH_2Cl_2	EPR/ 300	—	36H(1, 5, 8, 12): 0.019 4H(2, 4, 9, 11): 0.106 2H(3, 10): 0.462	85Hu1
$[C_{34}H_{46}]^{\cdot +}$ 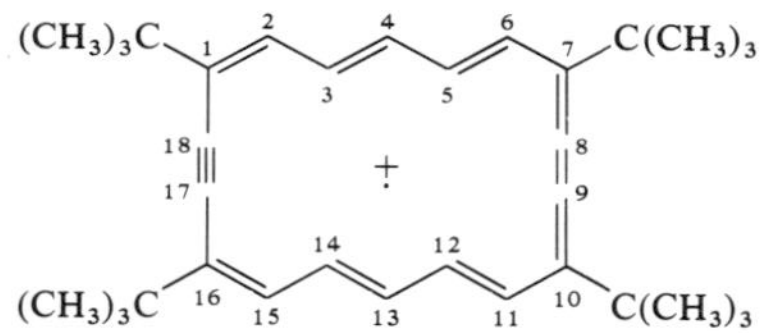	Oxidation by $AlCl_3$/ CH_2Cl_2	EPR/ 300	—	36H(1, 7, 10, 16): 0.018 4H(2, 6, 11, 15): 0.086 4H(3, 5, 12, 14): 0.394 2H(4, 13): 0.129	85Hu1
$[C_{38}H_{50}]^{\cdot +}$ 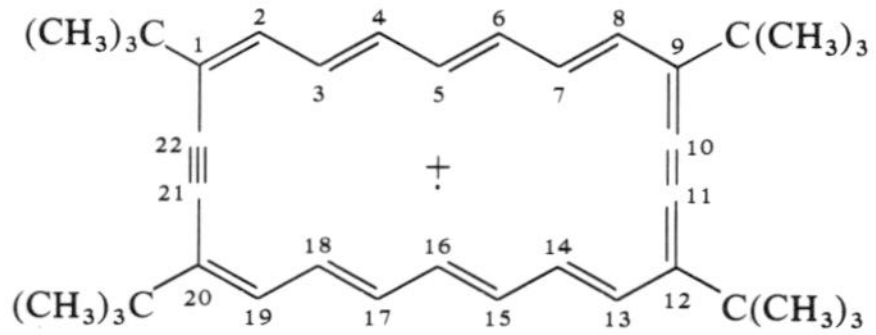	Oxidation of $AlCl_3$/ CH_2Cl_2	EPR/ 300	—	36H(1, 9, 12, 20): 0.016 4H(2, 8, 13, 19): 0.069 4H(3, 7, 14, 18): 0.328 4H(4, 6, 15, 17): 0.123 2H(5, 16): 0.364	85Hu1

19.1.10 Bridged systems

Substance	Generation/ Matrix or Solvent	Method/ T [K]	g-Factor	a-Value [mT]	Ref./ add. Ref.
$[C_{16}H_{18}]^{\cdot +}$ 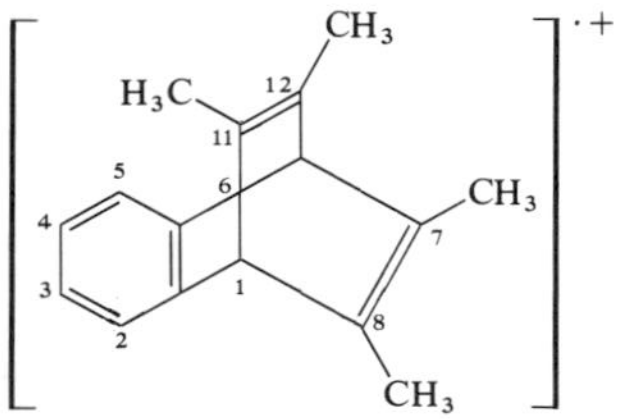	UV-irr. with $AlCl_3$/ CH_2Cl_2	EPR/ 195	2.0022	12H(7, 8, 11, 12, CH_3): 0.858 2H(3, 4): 0.068	86Cl1

Substance	Generation/ Matrix or Solvent	Method/ T [K]	g-Factor	a-Value [mT]	Ref./ add. Ref.
$[C_{18}H_{22}]^{\cdot +}$ 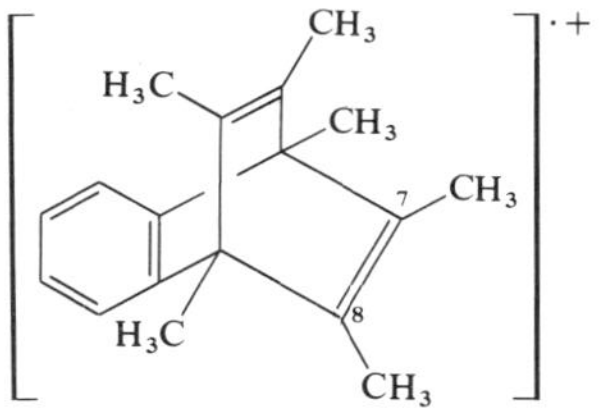	UV-irr. with $AlCl_3$/ CH_2Cl_2	EPR/ 195	2.0022	12H(7, 8, 11, 12, CH_3): 0.856 2H(3, 4): 0.063	86Cl1
$[C_{20}H_{28}Si_2]^{\cdot +}$ 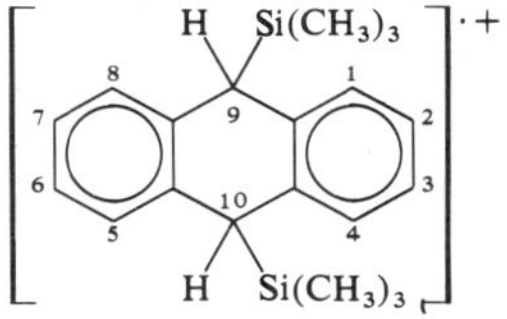	Oxidation by $AlCl_3$/ CH_2Cl_2	ESR/ —	—	4H(1, 4, 5, 8): 0.075 4H(2, 3, 6, 7): 0.319 2H(9, 10): 1.055 18H(CH_3): 0.026	78Ka1
$[C_{18}H_{14}Cl_2O_2]^{\cdot +}$ 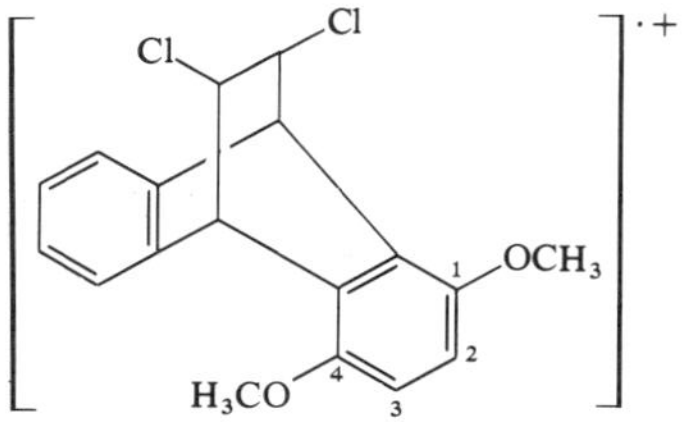	Oxidation by $C_6H_5NO_2:AlCl_3$/ conc. H_2SO_4	EPR/ 235	—	6H(1, 4, CH_3): 0.329 2H(2, 3): 0.258	86Qu1
$[C_{18}H_{18}O_2]^{\cdot +}$ 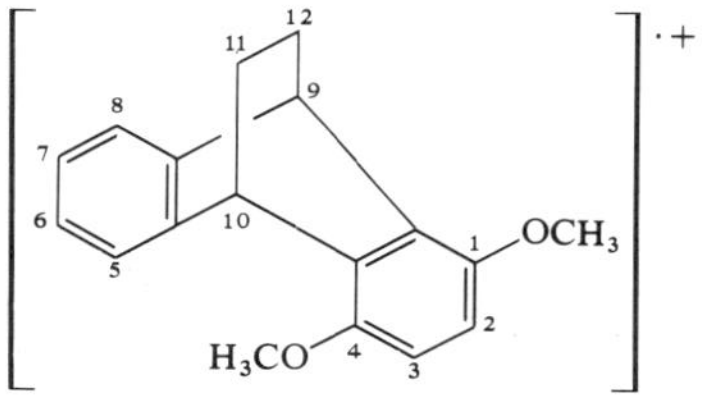	Oxidation by $C_6H_5NO_2:AlCl_3$/ conc. H_2SO_4	EPR/ 235	—	6H(1, 4, CH_3): 0.312 2H(2, 3): 0.272 4H(5, 6, 7, 8): 0.008 2H(11, 12; *anti*): 0.069 2H(11, 12; *syn*): 0.013	86Qu1

Ohya-Nishiguchi, Terahara, Tajima, Ishizu

Substance	Generation/ Matrix or Solvent	Method/ T [K]	g-Factor	a-Value [mT]	Ref./ add. Ref.
$[C_{18}H_{16}O_3]^{\cdot +}$ [a])	Oxidation by $C_6H_5NO_2:AlCl_3$/ conc. H_2SO_4	EPR/ 235	—	6H(1, 4, CH_3): 0.339 2H(2, 3): 0.259	86Qu1
$[C_{20}H_{22}O_4]^{\cdot +}$ [b])	Oxidation by $H_3CNO_2:AlCl_3$/ H_3CNO_2	EPR/ 248	—	6H(1, 4, CH_3): 0.320 2H(2, 3): 0.260 2H(11, 12, *anti*): 0.067	86Qu1
$[C_{20}H_{20}Cl_2O_4]^{\cdot +}$ [c])	Oxidation by $H_3CNO_2:AlCl_3$/ H_3CNO_2	EPR/ 248	—	12H(1, 4, 5, 8, CH_3): 0.164 4H(2, 3, 6, 7): 0.127 2H(11, 12): 0.037	86Qu1
$[C_{20}H_{16}O_5]^{\cdot +}$	Oxidation by $C_6H_5NO_2:AlCl_3$/ conc. H_2SO_4	EPR/ 235	—	6H(1, 4, CH_3): 0.335 2H(2, 3): 0.257 4H(5, 6, 7, 8): 0.010 2H(11, 12, *endo*): 0.059	86Qu1

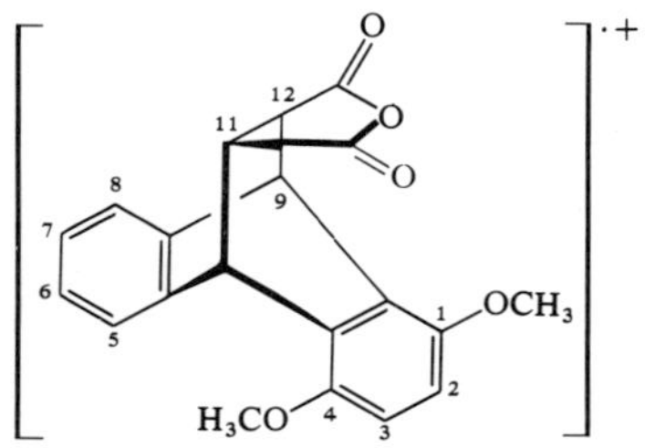

endo configuration

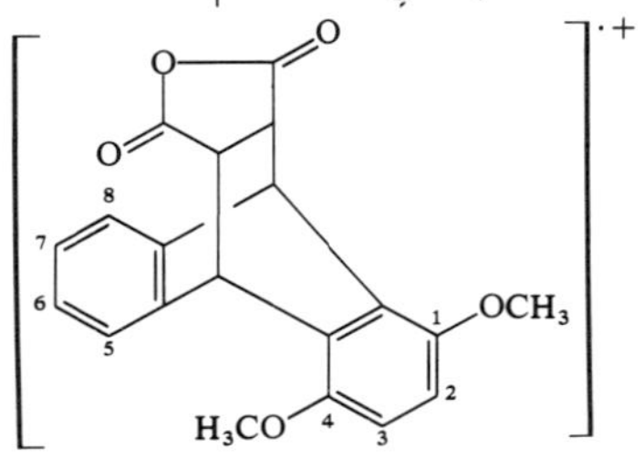

exo configuration

a)

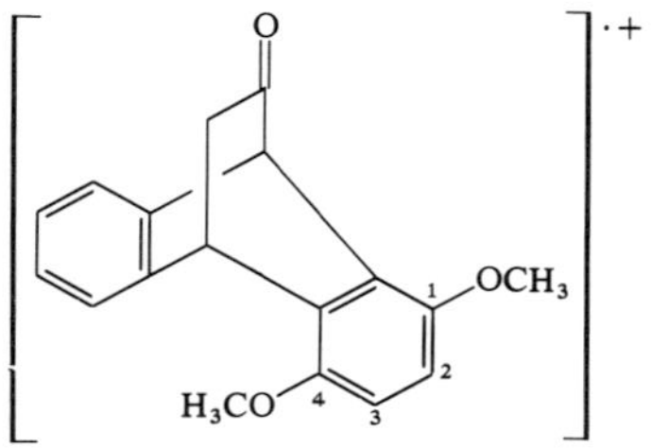

b)

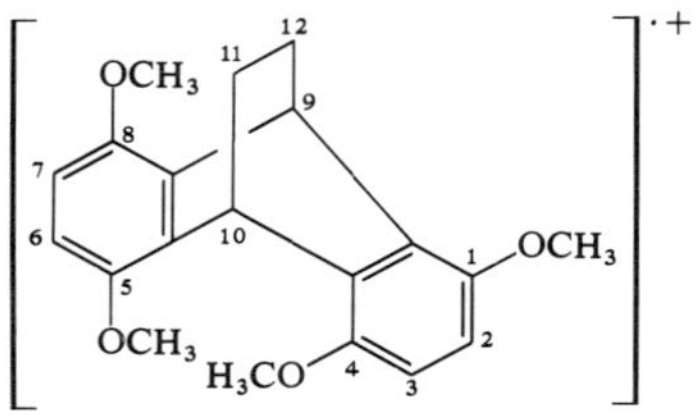

c)

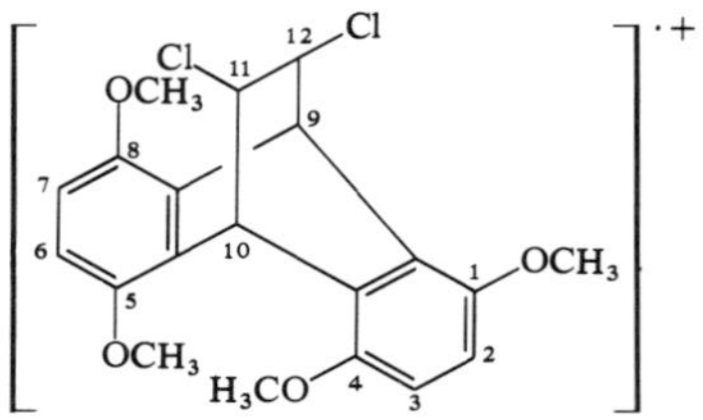

Substance	Generation/ Matrix or Solvent	Method/ T [K]	g-Factor	a-Value [mT]	Ref./ add. Ref.
$[C_{22}H_{20}O_7]^{\cdot+}$ d)	Oxidation by $H_3CNO_2:AlCl_3/H_3CNO_2$	EPR/ 248		6H(1, 4, CH_3): 0.329 2H(2, 3): 0.259	86Qu1
$[C_{29}H_{20}]^{\cdot+}$ e)	Electrochem. oxidation/ CH_2Cl_2	ESR/ 230	—	4H(1, 4, 5, 8): 0.260 4H(2, 3, 6, 7): 0.120 H(10): 0.785 1H(CH_2): 0.912 1H(CH_2): 0.120	86Te1
$[C_{30}H_{22}]^{\cdot+}$ f)	Electrochem. oxidation/ CH_2Cl_2	ESR/ 230		4H(1, 4, 5, 8): 0.262 4H(2, 3, 6, 7): 0.138 H(10): 0.780 H(CH_2): 1.350 H(CH_2): 0.370	86Te1
$[C_{31}H_{24}]^{\cdot+}$ g)	Electrochem. oxidation/ CH_2Cl_2	ESR/ 230		8H(1, 4, 5, 8): 0.143 8H(2, 3, 6, 7): 0.101 2H(10): 0.396 2H(CH_2): 0.241 2H(CH_2): 0.041	86Te1
$[C_{32}H_{26}]^{\cdot+}$ h)	Electrochem. oxidation/ CH_2Cl_2	ESR/ 230		4H(1, 4, 5, 8): 0.276 4H(2, 3, 6, 7): 0.138 H(10): 0.720 H(CH_2): 0.404	86Te1

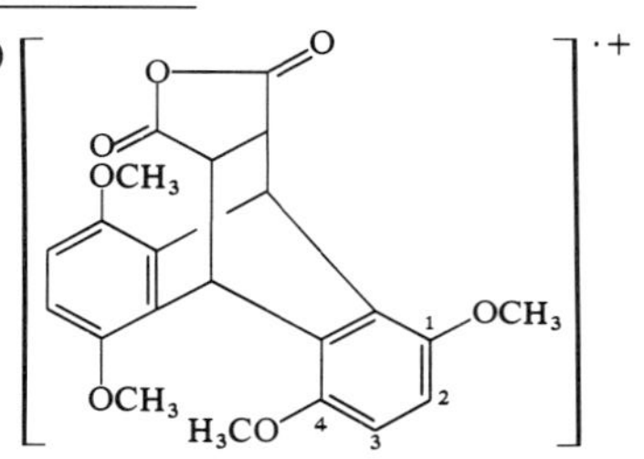

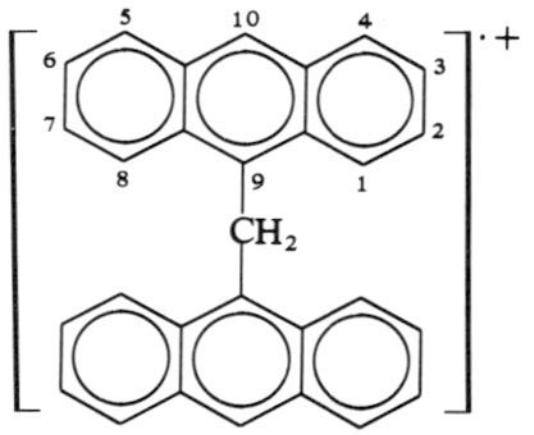

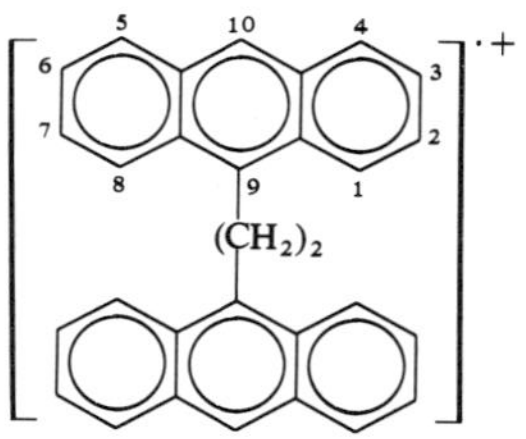

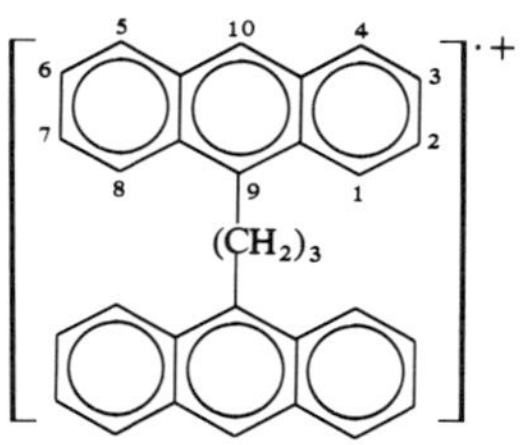

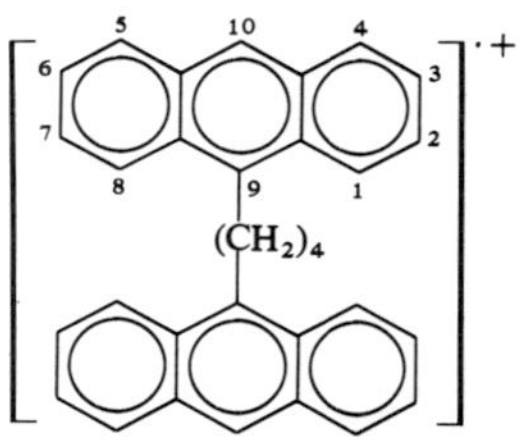

Ohya-Nishiguchi, Terahara, Tajima, Ishizu

Substance	Generation/ Matrix or Solvent	Method/ T [K]	g-Factor	a-Value [mT]	Ref./ add. Ref.
		19.1.11 Cyclophanes			
$[C_{20}H_{24}]^{\cdot +}$ [a])	Electrolytic oxidation/ CH_2Cl_2	ESR/ 188	—	12H(4, 7, 12, 15, CH_3): 0.6486 8H(1, 2, 9, 10, CH_2): 0.0486 4H(5, 8, 13, 16): 0.0334	82Oh1
$[C_{40}H_{48}]^{\cdot +}$ [b])	Electrolytic oxidation/ CH_2Cl_2	ESR/ 161	—	12H: 0.3074 12H: 0.0768 8H: 0.0177 [9]) 4H: 0.0138 [9])	82Oh1
$[C_{20}H_{24}]^{\cdot +}$ [c])	Electrolytic oxidation/ CH_2Cl_2	ESR/ 203	—	12H(4, 7, 13, 16, CH_3): 0.6365 8H(1, 2, 9, 10, CH_2): 0.041 4H(5, 8, 12, 15): 0.043	82Oh1
$[C_{22}H_{28}]^{\cdot +}$ [d])	Electrolytic oxidation/ CH_2Cl_2:CF_3COOH: $(CF_3CO)_2O$	ESR/ 193	2.0026	H(7, CH_3): 0.542 H(5, CH_3): 0.111 H(4, CH_3): 0.660 H(8): <0.03 H(1, CH_2): <0.03	82Ge1
$[C_{24}H_{32}]^{\cdot +}$ [e])	Electrolytic oxidation/ CH_2Cl_2:CF_3COOH: $(CF_3CO)_2O$	ESR/ 193	2.0026	24H(CH_3): 0.435 C(4, ^{13}C): 0.323 C(3, ^{13}C): 0.182	82Ge1
$[C_{26}H_{26}]^{\cdot +}$ [f])	Electrolytic oxidation/ CH_2Cl_2	ESR/ 217	—	4H: 0.3279 4H: 0.3039 2H: 0.0789 4H: 0.0320 4H: 0.0238 8H: 0.0091	82Oh1
$[C_{36}H_{36}]^{\cdot +}$ [g])	Electrolytic oxidation/ CH_2Cl_2	ESR/ 188	—	[10])	82Oh1
$[C_{40}H_{44}]^{\cdot +}$ [h])	Electrolytic oxidation/ CH_2Cl_2	ESR/ 188	—	[11])	82Oh1

[9]) Not resolved, assumed for simulation. [10]) Gaussian-type line with a linewidth of 0.56 mT. [11]) Gaussian-type line with a linewidth of 0.45 mT.

Substance	Generation/ Matrix or Solvent	Method/ T [K]	g-Factor	a-Value [mT]	Ref./ add. Ref.
$[C_{36}H_{24}]^{\cdot +}$ i)	Electrolytic oxidation or oxidation with CF_3COOH/CH_2Cl_2	ESR, ENDOR/ 188	—	8H: 0.2342 8H: 0.1107 8H: 0.0316	82Te1

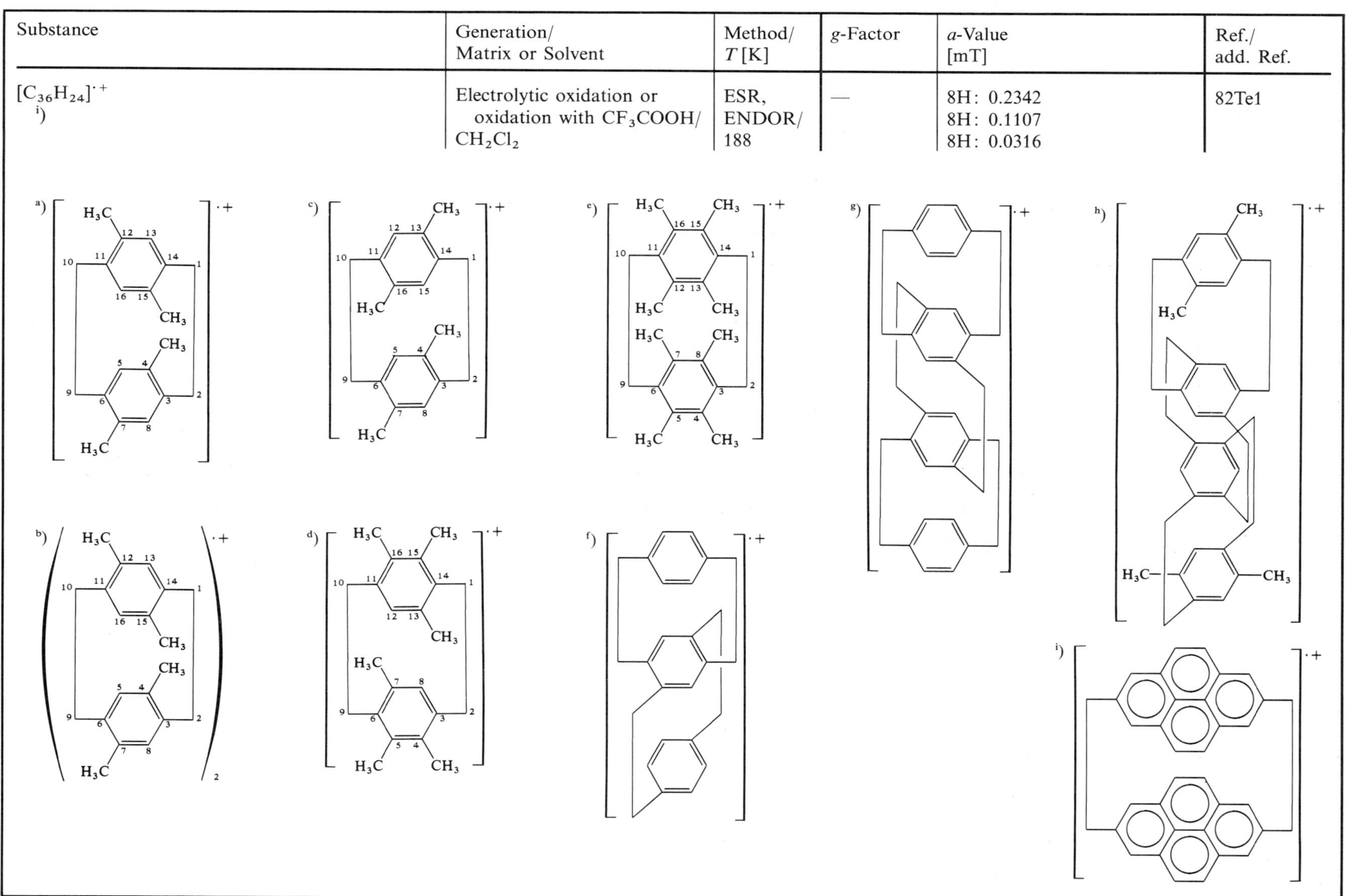

Ohya-Nishiguchi, Terahara, Tajima, Ishizu

Substance	Generation/ Matrix or Solvent	Method/ T [K]	g-Factor	a-Value [mT]	Ref./ add. Ref.
$[C_{36}H_{24}]^{\cdot+}$ 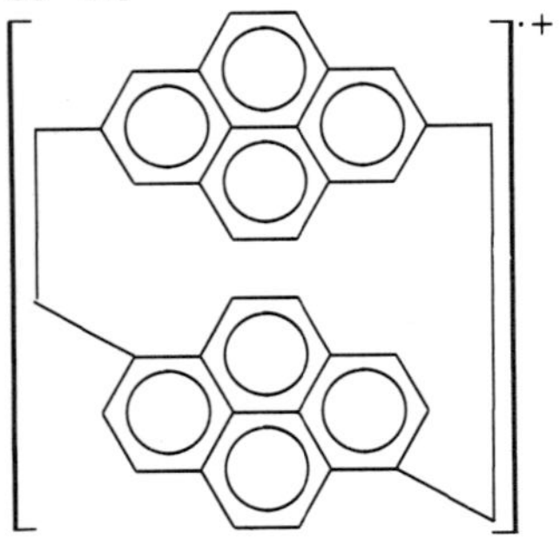	Electrolytic oxidation or oxidation with CF_3COOH/ CH_2Cl_2	ESR, ENDOR/ 188	—	2H: 0.3702 2H: 0.2390 2H: 0.2090 2H: 0.2090 2H: 0.1653 2H: 0.1290 2H: 0.1155 2H: 0.0840 2H: 0.0720 2H: 0.0450 2H: 0.0130	82Te1
$[C_{36}H_{24}]^{\cdot+}$ 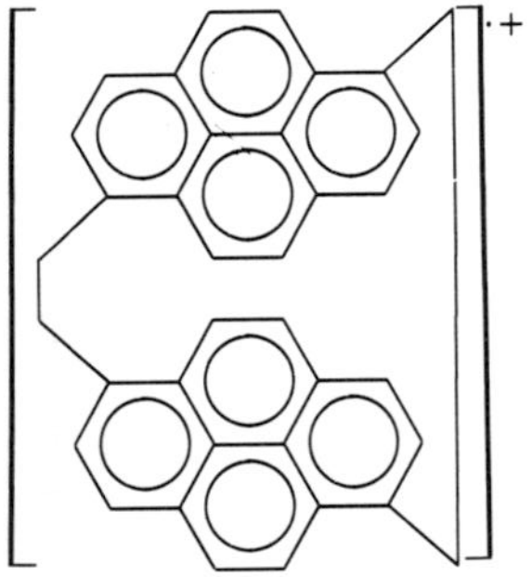	Electrolytic oxidation or oxidation with CF_3COOH/ CH_2Cl_2	ESR, ENDOR/ 188	—	4H: 0.3080 4H: 0.1144 4H: 0.0704 4H: 0.0625 4H: 0.0616 4H: 0.0440	82Te1

19.2 *S*-heterocycles

Substance	Generation/ Matrix or Solvent	Method/ T [K]	g-Factor	a-Value [mT]	Ref./ add. Ref.
$[C_2H_2S_2]^{\cdot+}$ 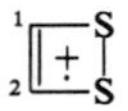	Oxidation by H_2SO_4/ CH_2Cl_2	EPR/ 163	2.0155	2H(1, 2): 0.28	86Ru1
$[C_4H_6S_2]^{\cdot+}$ 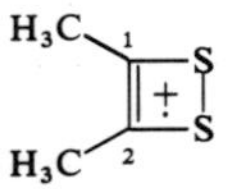	Oxidation by H_2SO_4/ $H_2SO_4:CH_2Cl_2$	EPR/ 163	2.0155	6H(CH_3): 0.22	86Ru1

Substance	Generation/ Matrix or Solvent	Method/ *T* [K]	*g*-Factor	*a*-Value [mT]	Ref./ add. Ref.
$[C_4H_6S_2]^{\cdot +}$ 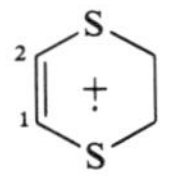	Oxidation by H_2SO_4/ CH_2Cl_2	EPR/ 163	2.0080	2H(1, 2): 0.34	86Ru1
$[C_6H_{10}S_2]^{\cdot +}$ 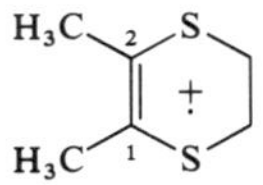	Oxidation by H_2SO_4/ CH_2Cl_2	EPR/ 163	2.0080	$6H(CH_3)$: 0.57	86Ru1
$[C_7H_{12}S_2]^{\cdot +}$ 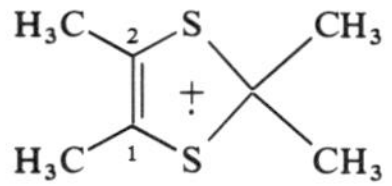	Oxidation by H_2SO_4/ CH_2Cl_2	EPR/ 163	2.0095	H(1, 2): 0.62	86Ru1
$[C_7H_{12}S_2]^{\cdot +}$ 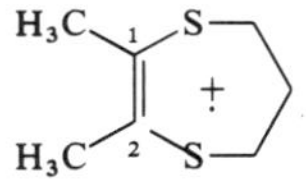	Oxidation by H_2SO_4/ CH_2Cl_2	EPR/ 163	2.0093	$6H(CH_3)$: 0.53	86Ru1
$[C_8H_{12}S_2]^{\cdot +}$ 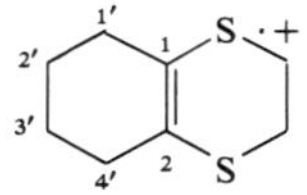	Oxidation by H_2SO_4/ CH_2Cl_2	EPR/ 163	2.008	4H(1′, 2′, 3′, 4′): 0.734 2H(other): 0.685 2H(other): 0.226 2H(other): 0.073	86Ru1
$[C_{12}H_{18}S_2]^{\cdot +}$ 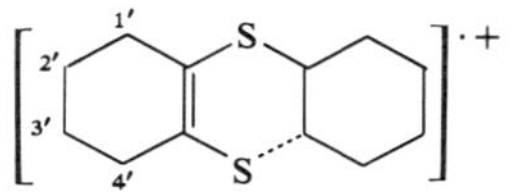	Oxidation by H_2SO_4/ CH_2Cl_2	EPR/ 168	2.0082	4H(1′, 2′, 3′, 4′): 0.79 2H(other): 0.61	86Ru1
$[C_{12}H_{18}S_2]^{\cdot +}$ 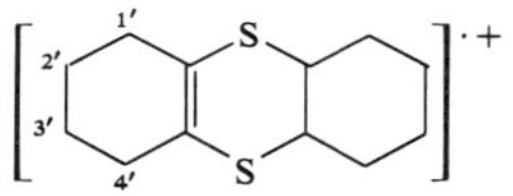	Oxidation by H_2SO_4/ CH_2Cl_2	EPR/ 168	2.0082	4H(1′, 2′, 3′, 4′): 0.76 1H(other): 0.76 2H(other): 0.16	86Ru1

Substance	Generation/ Matrix or Solvent	Method/ T [K]	g-Factor	a-Value [mT]	Ref./ add. Ref.
$[C_4H_4S_2]^{\cdot +}$ 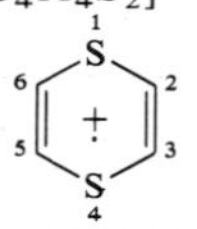	Oxidation by $AlCl_3$/ CH_2Cl_2	EPR/ —	2.0080	4H(2, 3, 5, 6): 0.282 $2\,^{33}S(1, 4)$: 0.984	84Bo1
	Oxidation by H_2SO_4/ CH_2Cl_2	EPR/ 163	2.0089	4H(2, 3, 5, 6): 0.28	86Ru1
$[C_8H_{10}S_2]^{\cdot +}$ 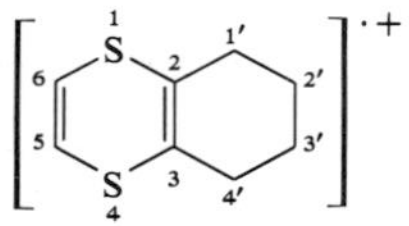	Oxidation by H_2SO_4/ CH_2Cl_2	EPR/ 168	2.0088	4H(1′, 2′, 3′, 4′): 0.32 2H(other): 0.26	86Ru1
$[C_8H_{12}S_2]^{\cdot +}$ 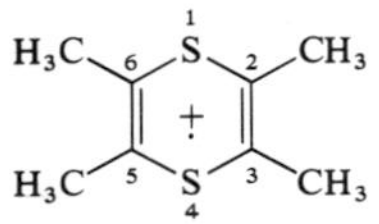	Oxidation by H_2SO_4/ CH_2Cl_2	EPR/ 163	2.0089	12H(2, 3, 5, 6): 0.21	86Ru1
$[C_{12}H_{16}S_2]^{\cdot +}$ 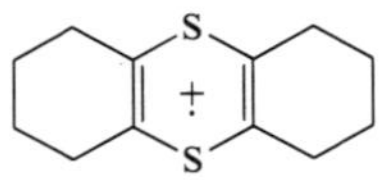	Oxidation by H_2SO_4/ CH_2Cl_2	EPR/ 168	2.0092	8H: 0.288	86Ru1
$[C_6H_8S_4]^{\cdot +}$ 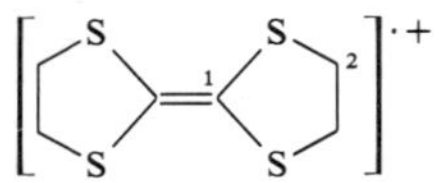	Oxidation by $AlCl_3$/ CH_2Cl_2	ESR/ 210	2.0089	$8H(CH_2)$: 0.238	80Bo2
$[C_8H_{12}S_4]^{\cdot +}$ 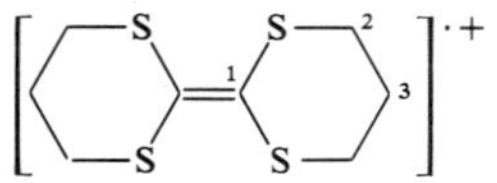	Oxidation by $AlCl_3$/ CH_2Cl_2	ESR/ 200	2.0102	$8H(2, CH_2)$: 0.298 $4H(3, CH_2)$: 0.028	80Bo2

Substance	Generation/ Matrix or Solvent	Method/ T [K]	g-Factor	a-Value [mT]	Ref./ add. Ref.
$[C_{12}H_{12}S_2]^{\cdot +}$ 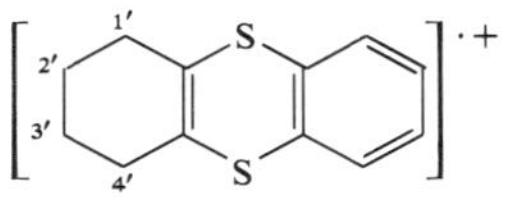	Oxidation by H_2SO_4/ CH_2Cl_2	EPR/ 168	2.0082	4H(1′, 2′, 3′, 4′): 0.41 2H(other): 0.095	86Ru1
$[C_6H_4S_4]^{\cdot +}$ 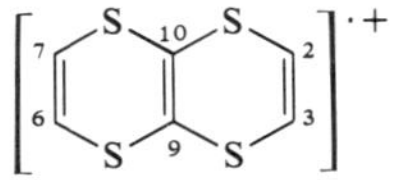	Oxidation by $AlCl_3$/ CH_2Cl_2	EPR/ —	2.0082	4H(2, 3, 6, 7): 0.03 $2^{13}C$(9, 10): 1.251 $4^{33}S$(1, 4, 5, 8): 0.417	84Bo1
$[C_6H_4S_4]^{\cdot +}$ 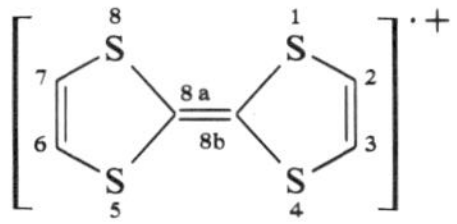	Oxidation by $CF_3COOH:(CF_3CO)_2O$/ CH_2Cl_2	EPR/ 233	2.0081	4H(2, 3, 6, 7): 0.125 $4^{33}S$(1, 4, 5, 8): 0.425 $2^{13}C$(8a, 8b): 0.285	86Ca1
	Electrolytic oxidation/ CH_2Cl_2	ESR/ 293	2.0084	4H(2, 3, 6, 7): 0.126 $4^{33}S$(1, 4, 5, 8): 0.427	84Te1
	Oxidation by $AlCl_3$/ CH_2Cl_2	EPR/ —	2.0081	4H(2, 3, 6, 7): 0.124 4S(1, 4, 5, 8): 0.42	84Bo1
$[C_8H_8S_4]^{\cdot +}$ 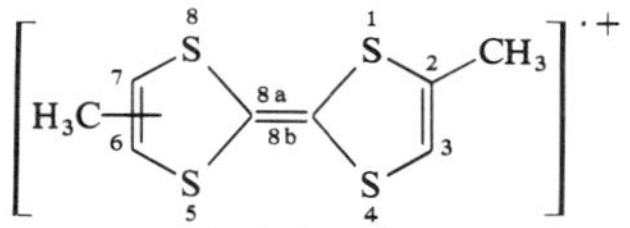 (mixture)	Oxidation by $CF_3COOH:(CF_3CO)_2O$/ CH_2Cl_2	EPR/ 233	2.0079	2H(3, 6 or 7): 0.132 6H(2, 7 or 6, CH_3): 0.066 $2^{33}S$(1, 5 or 8): 0.435 $2^{33}S$(4, 8 or 5): 0.390	86Ca1
$[C_{10}N_4S_4]^{\cdot +}$ 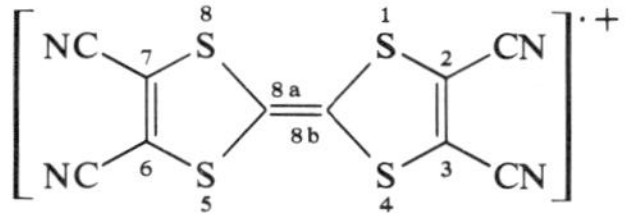	Oxidation by $CF_3COOH:(CF_3CO)_2O$/ CH_2Cl_2	EPR/ 253	2.0080	$4^{14}N$(2, 3, 6, 7): 0.012 $4^{33}S$(1, 4, 5, 8): 0.470 $2^{13}C$(8a, 8b): 0.360	86Ca1

Substance	Generation/ Matrix or Solvent	Method/ T [K]	g-Factor	a-Value [mT]	Ref./ add. Ref.
$[C_{10}F_{12}S_4]^{\cdot +}$ 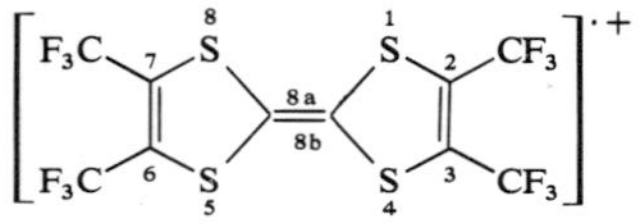	Oxidation by $CF_3COOH:(CF_3CO)_2O$/ CH_2Cl_2	EPR/ 233	2.0079	$12\ ^{19}F(2, 3, 6, 7)$: 0.067 $4\ ^{33}S(1, 4, 5, 8)$: 0.465	86Ca1
$[C_{10}H_{12}S_4]^{\cdot +}$ 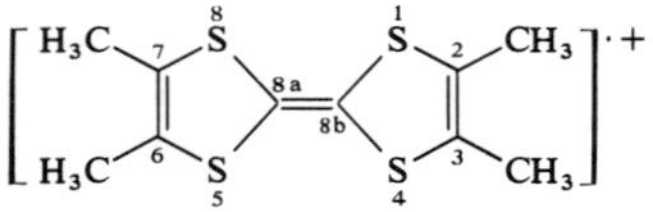	Electrolytic oxidation/ CH_2Cl_2	ESR/ 293	2.0081	12H(2, 3, 6, 7, CH_3): 0.078 $4\ ^{33}S(1, 4, 5, 8)$: 0.399	84Te1
	Oxidation by $CF_3COOH:(CF_3CO)_2O$/ CH_2Cl_2	EPR/ 233	2.0078	12H(2, 3, 7, 6, CH_3): 0.074 $4\ ^{33}S(1, 4, 5, 8)$: 0.395	86Ca1
$[C_{10}H_8O_4S_6]^{\cdot +}$ 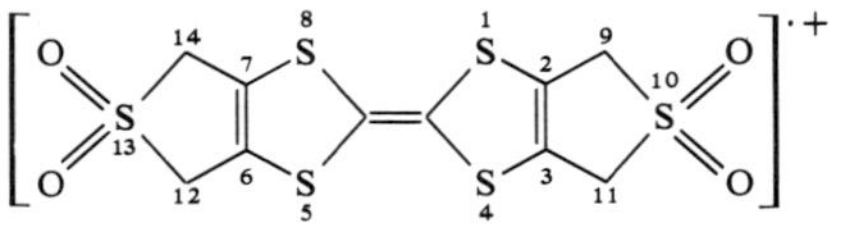	Oxidation by $CF_3COOH:(CF_3CO)_2O$/ CH_2Cl_2	EPR/ 233	2.0077	8H(9, 11, 12, 14, CH_2): 0.158 $4\ ^{33}S(1, 4, 5, 8)$: 0.425	86Ca1
$[C_{12}H_{12}S_4]^{\cdot +}$ 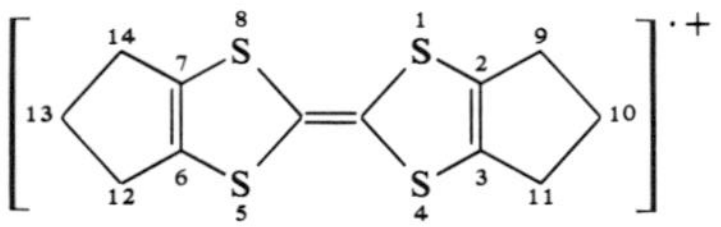	Oxidation by $CF_3COOH:(CF_3CO)_2O$/ CH_2Cl_2	EPR/ 233	2.0078	8H(9, 11, 12, 14, CH_2): 0.211 4H(10, 13): 0.010 $4\ ^{33}S(1, 4, 5, 8)$: 0.420	86Ca1
$[C_{14}H_{12}O_8S_4]^{\cdot +}$ 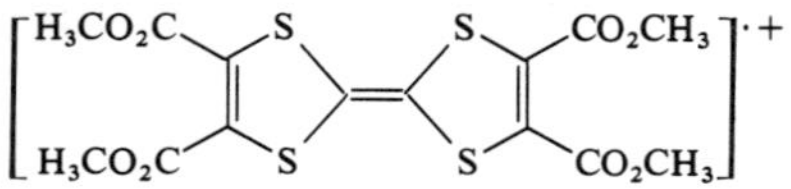	Electrolytic oxidation/ CH_2Cl_2	ESR/ 293	2.0081	$4\ ^{33}S$: 0.452	84Te1

Substance	Generation/ Matrix or Solvent	Method/ T [K]	g-Factor	a-Value [mT]	Ref./ add. Ref.
$[C_{26}H_{40}S_2]^{\cdot +}$ 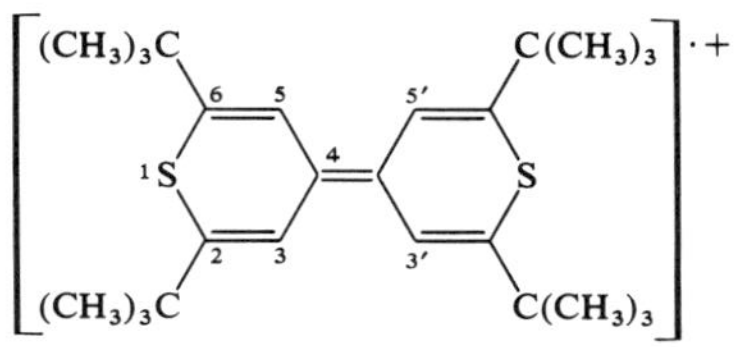	Electrochem. generation/ CH_2Cl_2	EPR/ 298	2.009	4H(3.3′, 5,5′): 0.0532	85De1
$[C_{26}H_{40}STe]^{\cdot +}$ 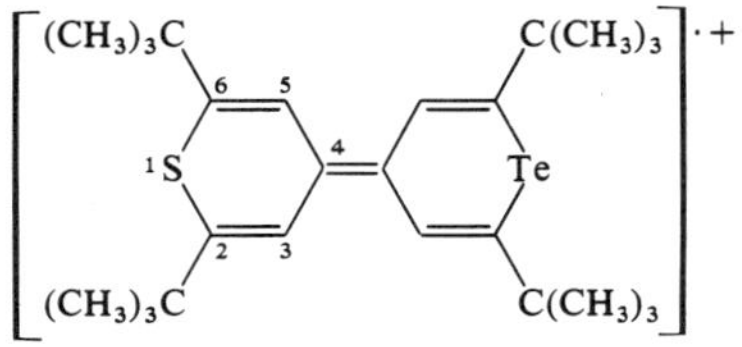	Electrochem. generation/ CH_2Cl_2	EPR/ 298	2.0027	[12])	85De1
$[C_{10}H_6S_2]^{\cdot +}$ 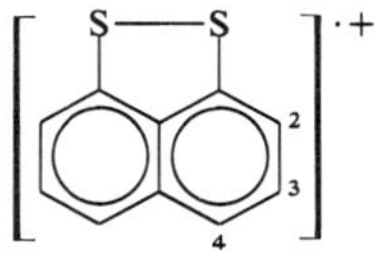	Oxidation by $AlCl_3$/ CH_2Cl_2	ESR/ 200	2.0086	2H(2): 0.432 2H(3): 0.092 2H(4): 0.525 $2^{33}S$: 0.733	81Bo1
$[C_{10}H_6SSe]^{\cdot +}$ 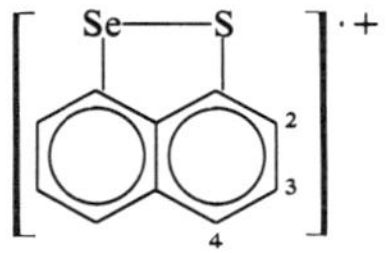	Oxidation by $AlCl_3$/ CH_2Cl_2	ESR/ 200	2.0209	2H(2): 0.44 2H(3): 0.08 [13]) 2H(4): 0.48	81Bo1
$[C_{10}H_6STe]^{\cdot +}$ 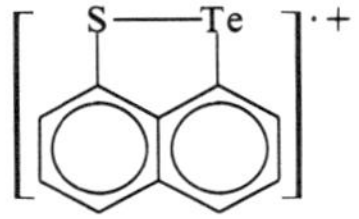	Oxidation by $AlCl_3$/ CH_2Cl_2	ESR/ 200	2.0318	[14])	81Bo1

[12]) Linewidth 0.202 mT.
[13]) Not detectable due to large linewidth.
[14]) Single broad line.

Substance	Generation/ Matrix or Solvent	Method/ T [K]	g-Factor	a-Value [mT]	Ref./ add. Ref.
$[C_{12}H_8S]^{\cdot +}$ 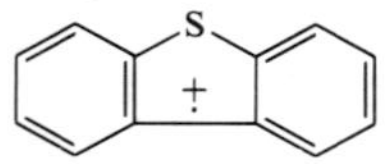	Oxidation of C_6H_5SH with AsF_5/ $AsF_5:AsF_3$	ESR/ —	2.0076	H: 0.12	85Mu1
$[C_{12}H_{10}S]^{\cdot +}$ 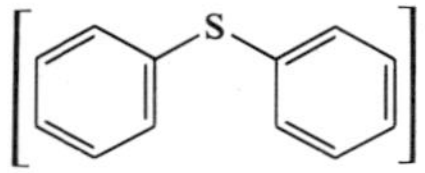	Oxidation of C_6H_5SH with AsF_5/ $AsF_5:AsF_3$	EPR/ —	2.0079	H: 0.12	85Mu1
$[C_{12}H_{12}S_4]^{\cdot +}$ 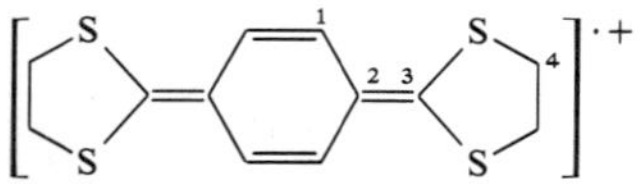	Oxidation by $AlCl_3$/ CH_2Cl_2	ESR/ 190	2.0063	H(1): 0.111 2H(4, CH_2): 0.136	80Bo2
$[C_8O_4S_4]^{\cdot +}$ 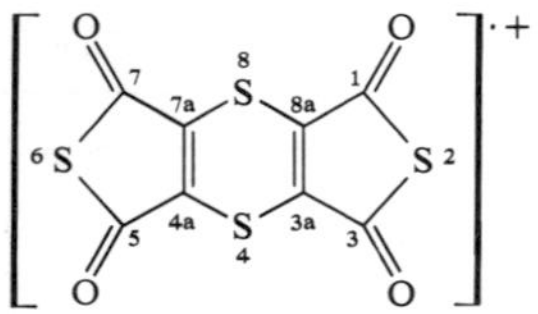	Solution in $AlCl_3:CH_3NO_2$/ CH_3NO_2	EPR/ 243	—	$4\,^{13}C$(1, 3, 5, 7): −0.18 $4\,^{13}C$(3a, 4a, 7a, 8a): −0.05 $2\,^{33}S$(2, 6): 0.05 $2\,^{33}S$(4, 8): 0.90 $4\,^{17}O$(1, 3, 5, 7): −0.06	77Ge1
$[C_{12}H_{12}S_4]^{\cdot +}$ 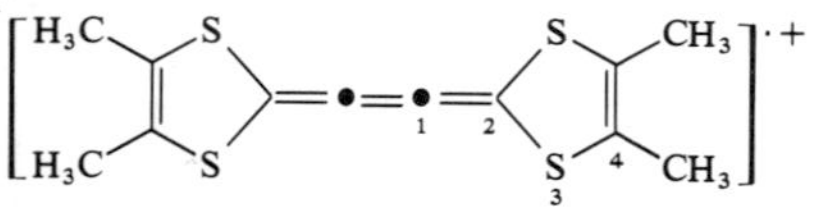	Electrolytic red. of di-perchlorate salt/ CH_2Cl_2	ESR/ 293	2.0087	3H(4, CH_3): 0.069 ^{33}S(3): 0.393	84Te1
$[C_{32}H_{20}S_4]^{\cdot +}$ 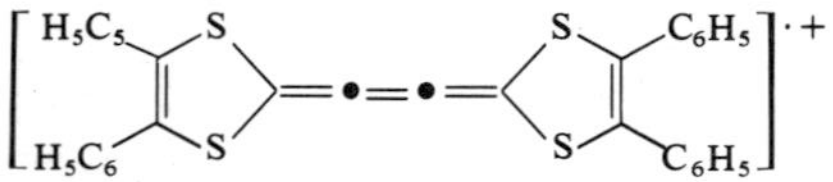	Electrolytic red. of di-perchlorate salt/ CH_2Cl_2	ESR/ 293	2.0085	—	84Te1

Substance	Generation/ Matrix or Solvent	Method/ T [K]	g-Factor	a-Value [mT]	Ref./ add. Ref.
$[C_8H_6S_4]^{\cdot +}$ 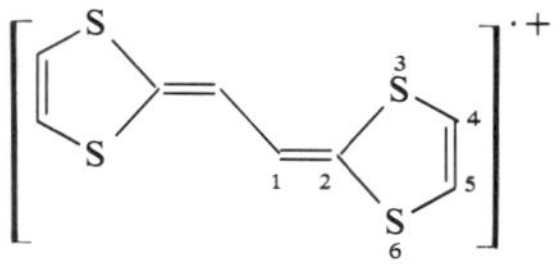	Electrolytic oxidation/ CH_2Cl_2	ESR/ 293	2.0081	H(1): 0.313 H(4): 0.123 H(5): 0.082 $^{33}S(6)$: 0.398	84Te1
$[C_{12}H_{14}S_4]^{\cdot +}$ 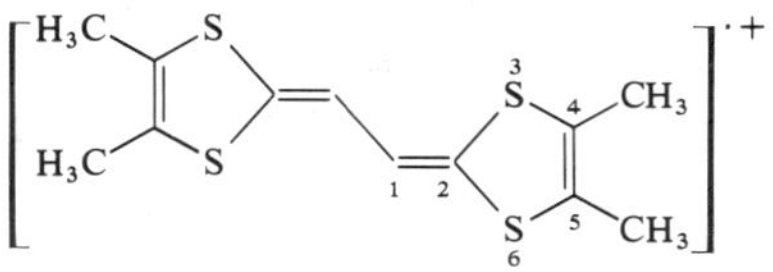	Electrolytic oxidation/ CH_2Cl_2	ESR/ 293	2.0072	H(1): 0.290 3H(4, CH_3): 0.084 3H(5, CH_3): 0.038	84Te1
$[C_{16}H_{14}O_8S_4]^{\cdot +}$ 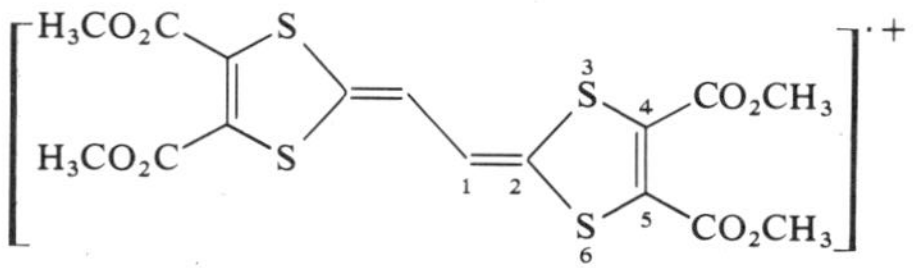	Electrolytic oxidation/ CH_2Cl_2	ESR/ 293	2.0080	H(1): 0.293 ^{33}S: 0.489 ^{33}S: 0.426	84Te1
$[C_{22}H_{16}S_4]^{\cdot +}$ 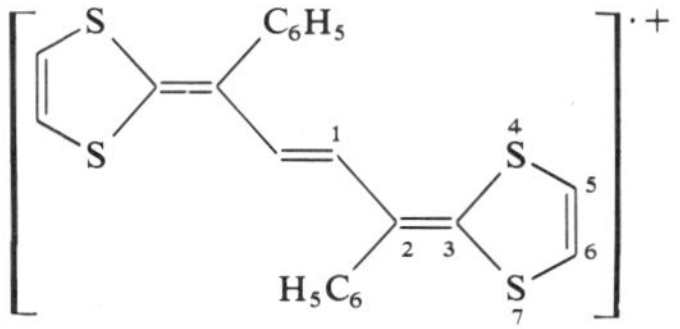	Electrolytic oxidation/ CH_2Cl_2	ESR/ 293	2.0072	H(1): 0.191 2H(5, 6): 0.092 $^{33}S(7)$: 0.266	84Te1
$[C_{14}H_{12}S_4]^{\cdot +}$ 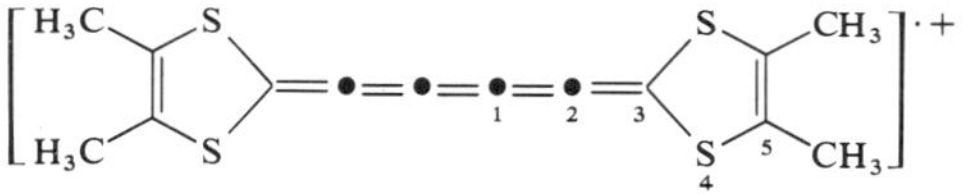	Electrolytic red. of di-perchlorate salt/ CH_2Cl_2	ESR/ 293	2.0066	3H(5, CH_3): 0.05	84Te1

Ohya-Nishiguchi, Terahara, Tajima, Ishizu

Substance	Generation/ Matrix or Solvent	Method/ T [K]	g-Factor	a-Value [mT]	Ref./ add. Ref.
$[C_{34}H_{20}S_4]^{\cdot +}$ 	Electrolytic red. of di-perchlorate salt/ CH_2Cl_2	ESR/ 293	2.0053	^{33}S: 0.369	84Te1
$[C_{10}H_8S_4]^{\cdot +}$ 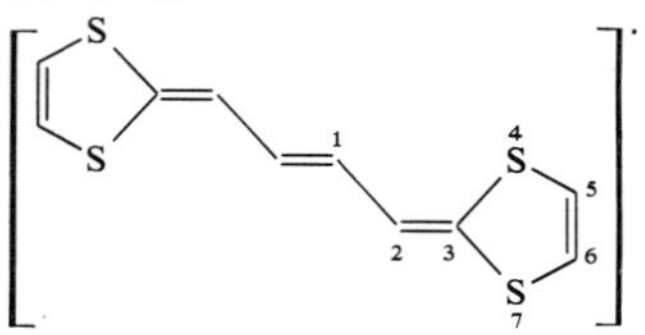	Electrolytic oxidation/ CH_2Cl_2	ESR/ 293	2.0072	2H(1, 2): 0.214 H(5): 0.099 H(6): 0.063	84Te1
$[C_{14}H_{16}S_4]^{\cdot +}$ 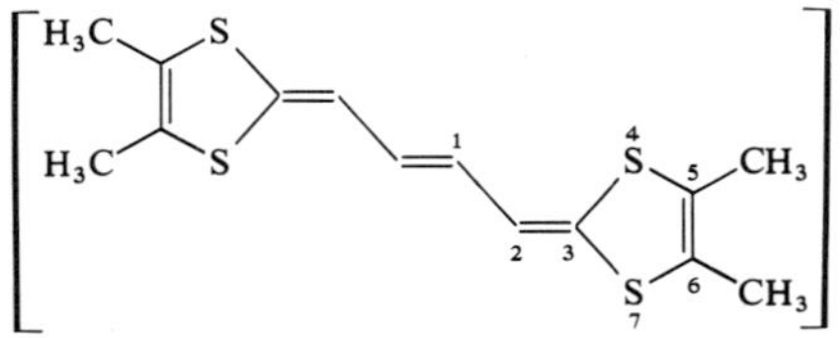	Electrolytic oxidation/ CH_2Cl_2	ESR/ 293	2.0070	H(2): 0.232 H(1): 0.199 3H(5, CH_3): 0.099 3H(6, CH_3): 0.033	84Te1
$[C_{18}H_{16}O_8S_4]^{\cdot +}$ 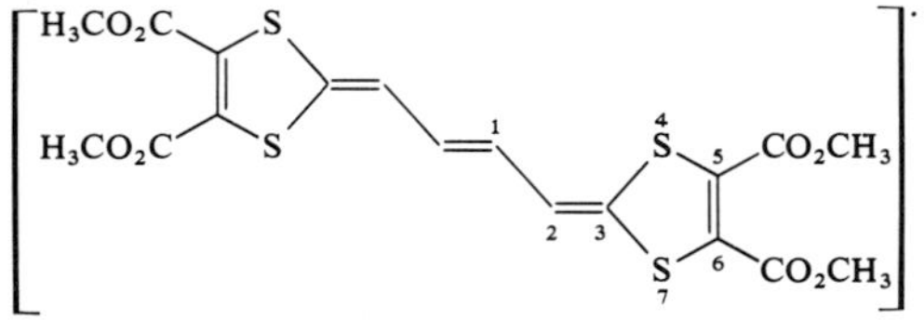	Electrolytic oxidation/ CH_2Cl_2	ESR/ 293	2.0072	H(1): 0.221 ^{33}S: 0.270	84Te1
$[C_{12}H_{10}S_4]^{\cdot +}$ 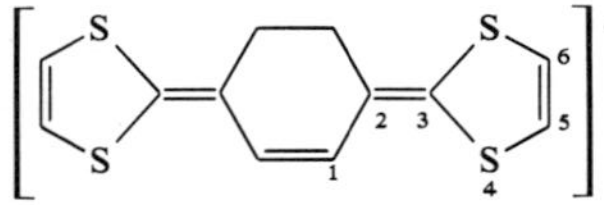	Electrolytic oxidation/ CH_2Cl_2	ESR/ 293	2.0068	H(1): 0.189 2H(CH_2): 0.189 2H(5, 6): 0.086	84Te1

Substance	Generation/ Matrix or Solvent	Method/ T [K]	g-Factor	a-Value [mT]	Ref./ add. Ref.
$[C_{20}H_{18}O_8S_4]^{\cdot +}$ 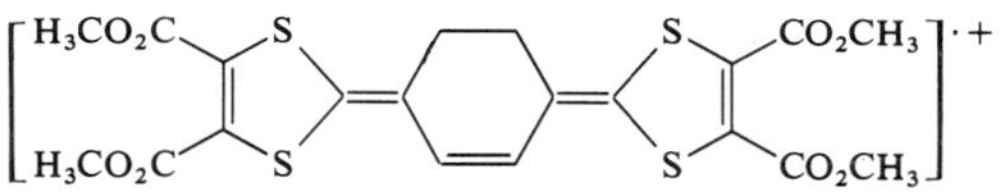	Electrolytic oxidation/ CH_2Cl_2	ESR/ 293	2.0066	H(1): 0.203 $2H(CH_2)$: 0.167 ^{33}S: 0.289 ^{33}S: 0.252	84Te1
$[C_{10}H_8S_8]^{\cdot +}$ 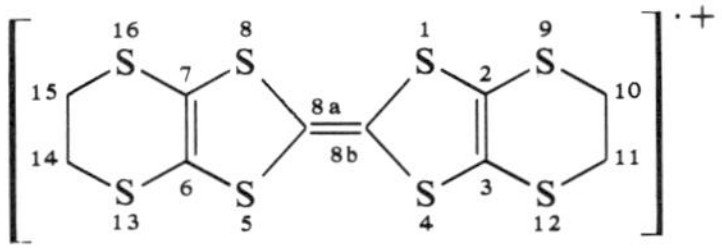	Oxidation by $CF_3COOH:(CF_3CO)_2O$/ CH_2Cl_2	EPR/ 233	2.0074	8H(10, 11, 14, 15): 0.005 $4^{33}S$(1, 4, 5, 8): 0.370 $4^{33}S$(9, 12, 13, 16): 0.080 $2^{13}C$(8a, 8b): 0.255	86Ca1
$[C_{10}H_4S_6]^{\cdot +}$ 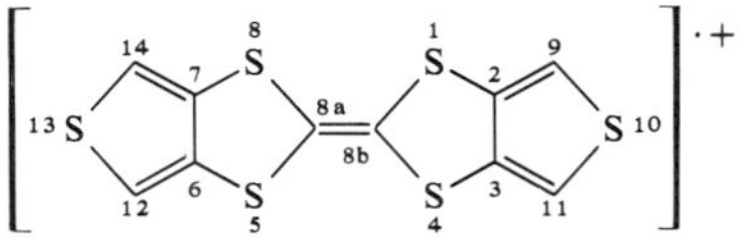	Oxidation by $CF_3COOH:(CF_3CO)_2O$/ CH_2Cl_2	EPR/ 233	2.0077	4H(9, 11, 12, 14): 0.020 $4^{33}S$(1, 4, 5, 8): 0.405	86Ca1
$[C_{14}H_{12}S_6]^{\cdot +}$ 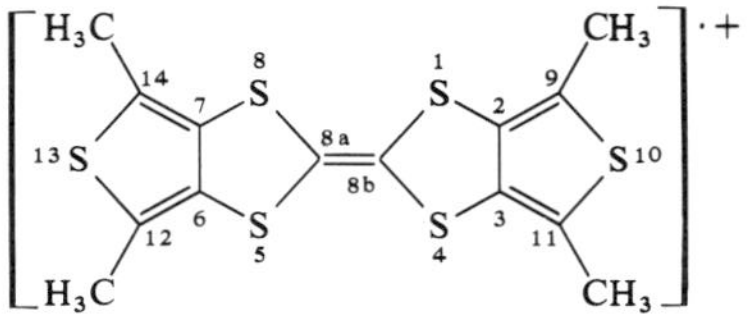	Oxidation by $CF_3COOH:(CF_3CO)_2O$/ CH_2Cl_2	EPR/ 233	2.0076	12H(9, 11, 12, 14): 0.005 $4^{33}S$(1, 4, 5, 8): 0.390 $2^{13}C$(8a, 8b): 0.355	86Ca1
$[C_{14}H_8S_4]^{\cdot +}$ 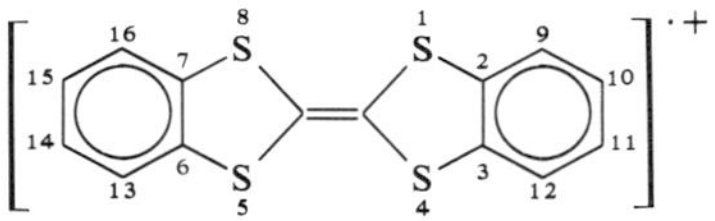	Oxidation with $CF_3COOH:(CF_3CO)_2O$/ CH_2Cl_2	EPR/ 233	2.0077	4H(9, 12, 13, 16): 0.049 or 0.015 4H(10, 11, 14, 15): 0.015 or 0.049	86Ca1

Substance	Generation/ Matrix or Solvent	Method/ T [K]	g-Factor	a-Value [mT]	Ref./ add. Ref.
$[C_{16}H_{10}S_4]^{\cdot +}$ 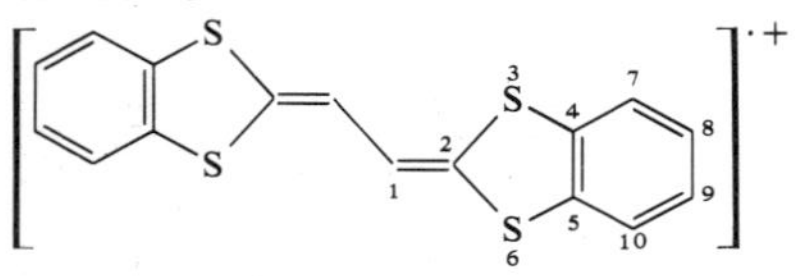	Electrolytic oxidation/ CH_2Cl_2	ESR/ 293	2.0076	H(1): 0.292 H(8): 0.050 H(9): 0.028 2H(7, 10): 0.024 $^{33}S(6)$: 0.407 $^{33}S(3)$: 0.303	84Te1
$[C_{18}H_{12}S_4]^{\cdot +}$ 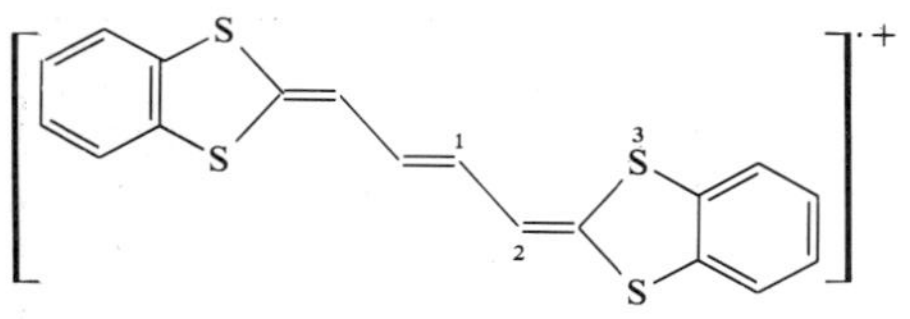	Electrolytic oxidation/ CH_2Cl_2	ESR/ 293	2.0067	2H(1, 2): 0.210	84Te1
$[C_{30}H_{24}O_8S_4]^{\cdot +}$ 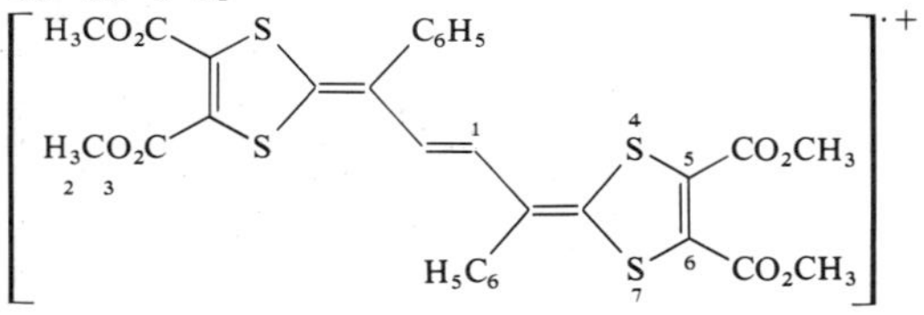	Electrolytic oxidation/ CH_2Cl_2	ESR/ 293	2.0070	H(1): 0.204 $^{33}S(7)$: 0.296 $^{33}S(4)$: 0.259	84Te1
$[C_{26}H_{12}S_4]^{\cdot +}$ 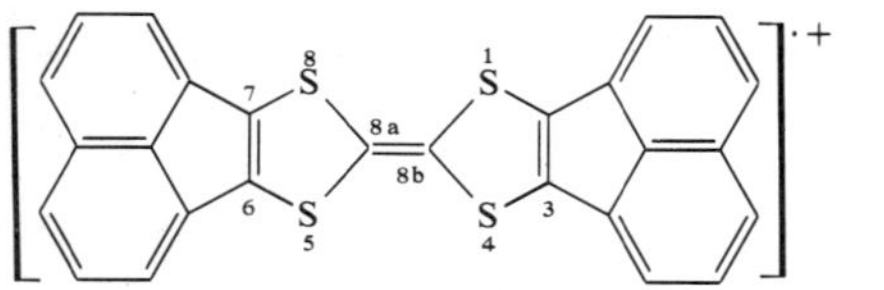	Oxidation by $CF_3COOH:(CF_3CO)_2O$/ CH_2Cl_2	EPR/ 233	2.0073	4H: 0.017 $4\,^{33}S(1, 4, 5, 8)$: 0.390	86Ca1
$[C_{30}H_{16}S_4]^{\cdot +}$ 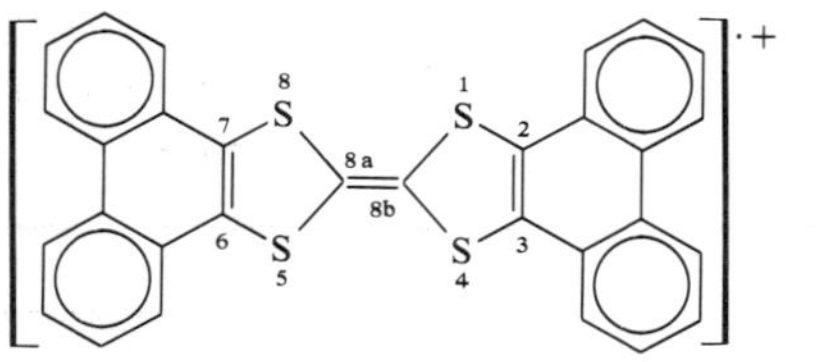	Oxidation by $CF_3COOH:(CF_3CO)_2O$/ CH_2Cl_2	EPR/ 233	2.0074	4H: 0.028 $4\,^{33}S(1, 4, 5, 8)$: 0.385	86Ca1

Substance	Generation/ Matrix or Solvent	Method/ T [K]	g-Factor	a-Value [mT]	Ref./ add. Ref.

19.3 Miscellaneous compounds

Substance	Generation/ Matrix or Solvent	Method/ T [K]	g-Factor	a-Value [mT]	Ref./ add. Ref.
$[C_{26}H_{40}O_2]^{\cdot +}$ a)	Electrochem. generation/ CH_2Cl_2	EPR/ 298	2.008	4H(3, 5, 3′, 5′): 0.0438	85De1
$[C_{26}H_{40}Se_2]^{\cdot +}$ b)	Electrochem. generation/ CH_2Cl_2	EPR/ 298	2.024	[15]	85De1
$[C_{26}H_{40}Te_2]^{\cdot +}$ c)	Electrochem. generation/ CH_2Cl_2	EPR/ 298	2.0041	[16]	85De1
$[C_{10}H_6Se_2]^{\cdot +}$ 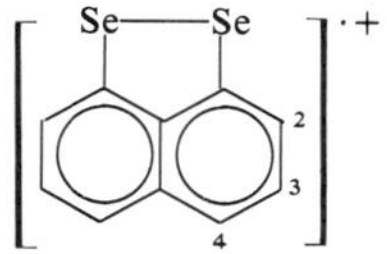	Oxidation by $AlCl_3$/ CH_2Cl_2	ESR/ 200	2.00397	H(2): 0.37 H(3): 0.06 [17] H(4): 0.43	81Bo1
$[C_{10}H_6SeTe]^{\cdot +}$ 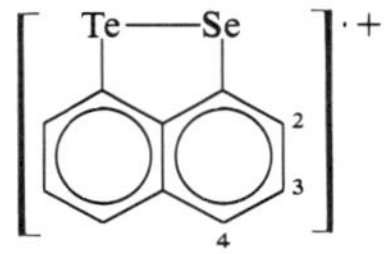	Oxidation by $AlCl_3$/ CH_2Cl_2	ESR/ 200	2.0409	[18]	81Bo1

[15] Single broad line with linewidth of 0.125 mT.
[16] Linewidth: 0.406 mT.
[17] Not detectable due to large linewidth.
[18] Single broad line.

a)

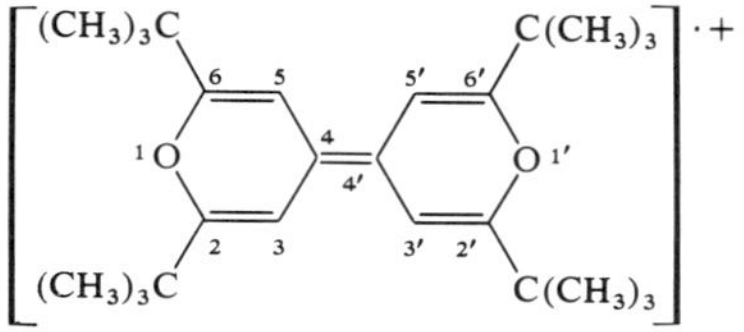

b)

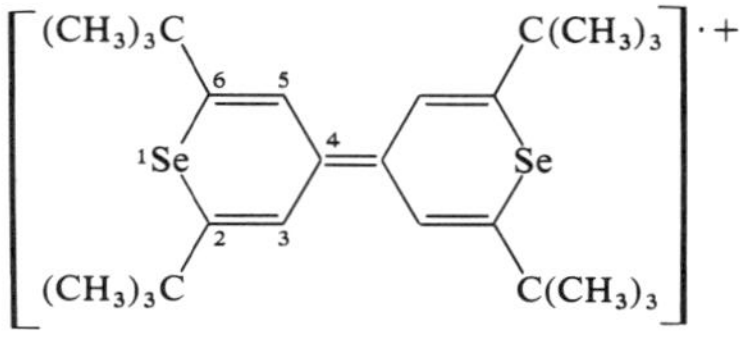

c)

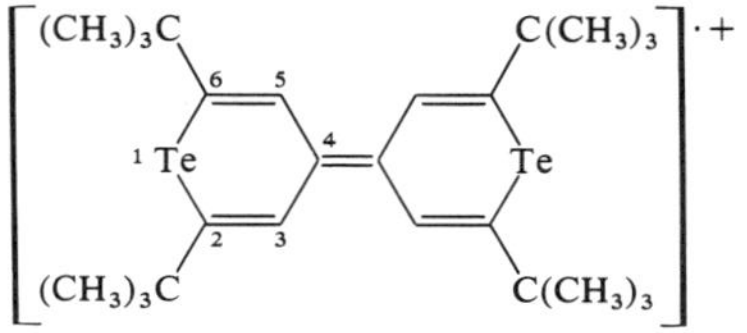

19.4 References for 19

76Bo1 Bock, H., Brähler, G., Fritz, G., Matern, E.: Angew. Chem. **88** (1976) 765.
76Di1 Dixon, W.T., Murphy, D.: J. Chem. Soc., Perkin Trans. II **1976**, 1823.
76Su1 Sullivan, P.D., Norman, L.J.: J. Magn. Reson. **23** (1976) 395.

77Bo1 Bock, H., Kaim, W., Rohwer, H.E.: J. Organomet. Chem. **135** (1977) C14.
77Bo2 Bock, H., Brähler, G.: Angew. Chem. **89** (1977) 893.
77Ge1 Gerson, F., Wydler, C., Kluge, F.: J. Magn. Reson. **26** (1977) 271.
77Mc1 McKinny, T.M.: Electroanal. Chem. **10** (1977) 97.
77Ne1 Nemoto, F., Ishizu, K., Shimoda, F.: Bull. Chem. Soc. Jpn. **50** (1977) 570.
77Ru1 Rudenko, A.P., Kutnevich, A.M., Arapov, O.V., Zarubin, M.Ya., Reishakhrit, L.S.: Dokl. Akad. Nauk SSSR **233** (1977) 882.
77Su1 Sullivan, P.D., Fong, J.Y.: J. Phys. Chem. **81** (1977) 71.

78Bo1 Bock, H., Kaim, W.: Chem. Ber. **111** (1978) 3552.
78Bo2 Bock, H., Kaim, W., Rohwer, H.E.: Chem. Ber. **111** (1978) 3585.
78Fu1 Fuerderer, P., Gerson, F., Hafner, K.: Helv. Chim. Acta **61** (1978) 2974.
78Fu2 Fuerderer, P., Gerson, F., Rabinovitz, M., Willner, I.: Helv. Chim. Acta **61** (1978) 2981.
78Ka1 Kaim, W., Bock, H.: Chem. Ber. **111** (1978) 3585.
78Su1 Sullivan, P.D., Fong, J.Y., Williams, M.L., Parker, V.D.: J. Phys. Chem. **82** (1978) 1181.

79Ho1 Holton, D.M., Murphy, D.: J. Chem. Soc., Faraday Trans. II **75** (1979) 1185.
79Ru1 Rudenko, A.P., Zarubin, M.Ya., Kutnevich, A.M.: Zh. Obshch. Khim. **49** (1979) 954.
79Ru2 Rudenko, A.P., Zarubin, M.Ya., Aver'yanov, S.F., Bersheva, N.S.: Dokl. Akad. Nauk SSSR **249** (1979) 117.
79Ru3 Rudenko, A.P., Arapov, O.V., Zarubin, M.Ya.: Zh. Org. Khim. **15** (1979) 653.

80Bo1 Bock, H., Roth, B., Maier, G.: Angew. Chem. **92** (1980) 213.
80Bo2 Bock, H., Brähler, G., Henkel, U., Schlecker, R., Seebach, D.: Chem. Ber. **113** (1980) 289.
80Bu1 Buchanan III, A.C., Livingston, R., Dworkin, A.S., Smith, G.P.: J. Phys. Chem. **84** (1980) 423.

81Bo1 Bock, H., Brähler, G., Dauplaise, D., Meinwald, J.: Chem. Ber. **114** (1981) 289.
81Ge1 Gerson, F., Lopez, J., Krebs, A., Ruger, W.: Angew. Chem. **93** (1981) 106.
81Ko1 Komatsu, K., Moriyama, T., Nishiyama, T., Okamoto, K.: Tetrahedron **37** (1982) 721.
81Ru1 Rudenko, A.P., Zarubin, M.Ya., Bubenchikova, E.M.: Dokl. Akad. Nauk SSSR **260** (1981) 642.
81Ru2 Rudenko, A.P., Zarubin, M.Ya., Bubenchikova, E.M.: Zh. Org. Khim. **17** (1981) 1336.

82De1 Detsina, A.N., Rudenko, A.P., Cheremisin, A.A., Starichenko, V.F., Zarubin, M.Ya.: Zh. Org. Khim. **18** (1982) 105.
82Fa1 Fairhurst, S.A., Sutcliffe, L.H., Taylor, S.M.: J. Chem. Soc., Faraday Trans. I **78** (1982) 2743.
82Ger1 Gerson, F., Lopez, J., Hopf, H.: Helv. Chim. Acta **65** (1982) 1389.
82Oh1 Ohya-Nishiguchi, H., Terahara, T., Hirota, H., Sakata, Y., Misumi, S.: Bull. Chem. Soc. Jpn. **55** (1982) 1782.
82Te1 Terahara, T., Ohya-Nishiguchi, H., Hirota, H., Sakata, Y., Misumi, S., Ishizu, K.: Bull Chem. Soc. Jpn. **55** (1982) 3896.

83Ru1 Rudenko, A.P., Cheremisin, A.A., Shchegoleva, L.N., Detsina, A.N., Zarubin, M.Ya.: Zh. Org. Khim. **19** (1983) 1910.
83Su1 Sullivan, P.D.: J. Magn. Reson. **54** (1983) 314.

84Al1 Alberti, A., Pedulli, G.F., Tiecco, M., Testaferri, L., Tingoli, M.: J. Chem. Soc., Perkin Trans. II **1984**, 875.
84Bo1 Bock, H., Roth, B., Maier, G.: Chem. Ber. **117** (1984) 172.
84Co1 Courtneidge, J.L., Davies, A.G., Clark, T., Wilhelm, D.: J. Chem. Soc., Perkin Trans. II **1984**, 1197.

84Ge1 Gerson, F., Huber, W., Lopez, J.: J. Am. Chem. Soc. **106** (1984) 5808.

84Te1 Terahara, A., Ohya-Nishiguchi, H., Hirota, H., Awaji, H., Kawase, T., Yoneda, S., Sugimoto, T., Yoshida, Z.: Bull. Chem. Soc. Jpn. **57** (1984) 1760.

85De1 Detty, R.M., Hassett, W.J., Murray, J.B., Reynolds, A.G.: Tetrahedron **41** (1985) 4835.

85El1 Eloranta, J., Kasa, S.: Acta Chem. Scand., Ser. A **39** (1985) 63.

85Hu1 Huber, W.: Helv. Chim. Acta **68** (1985) 1140.

85Ma1 Maekela, R., Vuolle, M.: Magn. Reson. Chem. **23** (1985) 666.

85Mu1 Murray, D.P., Kispert, L.D., Frommer, J.E.: J. Chem. Phys. **83** (1985) 3681.

85Su1 Sullivan, P.D., Bannoura, F., Daub, G.H.: J. Am. Chem. Soc. **107** (1985) 32.

86Ba1 Bakker, M.G., Claridge, R.F.C.: J. Magn. Reson. **67** (1986) 438.

86Ca1 Cavara, L., Gerson, F., Cowan, D.O., Lerstrup, K.: Helv. Chim. Acta **69** (1986) 141.

86Cl1 Clark, T., Courtneidge, J.L., Davies, A.G., Schoety, K.: J. Chem. Soc., Chem. Commun. **1986**, 547.

86Co1 Cooksey, C.J., Courtneidge, J.L., Davies, A.G., Evance, J.C., Gregory, P.S., Rowlands, C.C.: J. Chem. Soc., Chem. Commun. **1986**, 549.

86Hi1 Hirota, N., Ohya-Nishiguchi, H., in: Investigation of Rates and Mechanisms of Reactions, (C.F. Bernasconi, ed.) New York: John Wiley & Sons **1986**, Part II Chapter 11, p. 605.

86La1 Lau, W., Kochi, J.K.: J. Org. Chem. **51** (1986) 1901.

86Qu1 Quast, H., Fuchsbauer, H.-L.: Chem. Ber. **119** (1986) 2414.

86Ru1 Russell, G.A., Law, W.C.: Heterocycles **1986**, 321.

86Su1 Sullivan, P.D., Ocasio, I.J., Chen, X., Bonnoura, F.: J. Am. Chem. Soc. **108** (1986) 257.

86Te1 Terahara, A., Ohya-Nishiguchi, H., Hirota, N., Higuchi, H., Misumi, S.: J. Phys. Chem. **90** (1986) 4958.

86Te2 Terahara, A., Ohya-Nishiguchi, H., Hirota, N., Oku, A.: J. Phys. Chem. **90** (1986) 1564.

20 Cation radicals from nitrogen containing compounds

20.1 Introduction

Nitrogen-centered cation radicals (CR) are a very heterogeneous group of compounds. CR from neutral compounds with amino N atoms (three formally single bonded substituents at N) will be found in Sections 20.2···20.5, which approximately correspond to 1, 2, 3, and more conjugated centers available for π delocalization. Although only compounds with saturated directly bonded substituents (H and formally sp^3 hybridized C and Si) appear in Section 20.2, 20.3 includes compounds with unsaturated substituents such as phenyl; α-heteroatoms are assigned precedence over unsaturated substituents. Compounds with only imino nitrogen (-N=X) or higher nitrogen oxidation levels appear in Section 20.6.

20.2 [one center] Non-conjugated amine CR
- 20.2.1 Monoamines and localized polyamine CR
- 20.2.2 Delocalized CR from non-conjugated amines

20.3 [2 center] Heteroatom substituted amines: 3e-π-based
- 20.3.1 Haloamine CR
- 20.3.2 Hydroxylamine CR
- 20.3.3 Thiohydroxylamine CR (and other CR with NS bonds)
- 20.3.4 Hydrazine CR

20.4 [3 center] 1-Azaallyl CR systems (and relatives)
- 20.4.1 [Enamine], hydrazone, N-nitrosamine CR
- 20.4.2 Amide CR (and R_2N-π-acceptor relatives)

20.5 [4 and more center] Amino substituted unsaturated CR
- 20.5.1 "Amino-terminated" systems:
  - 20.5.1.1 1,4-Enediamines and 2,3-aza analogues
  - 20.5.1.2 Anilines (aminobenzenes)
  - 20.5.1.3 Phenylenediamines (diaminobenzenes)
  - 20.5.1.4 Benzidines (diaminobiphenyls)
  - 20.5.1.5 Other amino-terminated systems
- 20.5.2 Hydrazaaromatic CR
  - 20.5.2.1 1,4-dihydropyridazines and viologens
    - 20.5.2.1.1 1,4-Dihydropyridazines
    - 20.5.2.1.2 Viologens: diazadihydrodipyridines
  - 20.5.2.2 Phenoxazines, phenothiazenes, dihydrothioalloxazines
  - 20.5.2.3 Biomolecules, including dihydroalloxazines
  - 20.5.2.4 Miscellaneous hydrazaaromatics

20.6 Imino and more highly oxidized N functional group CR
- 20.6.1 "sp^2" hybridized N: imino compounds
  - 20.6.1.1 Derivatives of pyridine
  - 20.6.1.2 Two or more imino N atoms
  - 20.6.1.3 -N=O compounds
- 20.6.2 "sp" hybridized: N_2, diazo, nitriles and relatives
- 20.6.3 Nitro compounds

N is used in this chapter to mean ^{14}N ($I=1$, natural abundance 99.635%) unless ^{15}N ($I=1/2$) is explicitly stated. The ratio of ^{15}N to ^{14}N gyromagnetic ratios is 1.402, which should be the ratio of nitrogen splitting constants a(N) observed for ^{15}N labelled radicals to those of their ^{14}N analogues. Although the tables use milli-Tesla (mT) as the unit for a(N) values, the use of Gauss (1 G=0.1 mT) appears to be more common, particularly in the U.S. literature, and care must be taken to be sure which units were used. Much ENDOR and TRIPLE work now reports the splittings in the MHz units in which they are experimentally determined, and numbers so reported have not been converted. The conversion equation is:

$$a(\text{mT}) = 0.03568\ R\ a(\text{MHz})$$

where R is 2.002319/g, experimentally 1 for most data.

Spin-bearing orbital hybridization ratios may be calculated from solid state ESR data using the free atom comparison method (FACM). The fraction of s character is taken as the ratio of the observed $A_{iso} = (A_{\parallel} + 2A_{\perp})/3$ value to the free atom A^0_{iso} value, and the fraction of p character as the ratio of the observed $A_{dip} = (A_{\parallel} - A_{\perp})/3$ value to the free atom A^0_{dip} value. The most recent and detailed study [87Kn1] uses Morton and Preston's calculations (see [78Mo1], which does not actually contain A^0 values), quoted for ^{15}N atoms as $A^0_{iso} = 2541$, $A^0_{dip} = 78$ MHz. For ^{14}N, $R=1$, these values correspond to $A^0_{iso} = 65.6$, $A^0_{dip} =$

1.99 mT in the units of this compilation. Knight and coworkers note that the Morton and Preston values are somewhat larger than the more recent values of Koh and Miller [85Ko2], for which they quote ^{15}N atoms: $A^0_{iso}=2158$, $A^0_{dip}=67$ MHz, but are closer to the Koh and Miller values for N^+, and recommend them as suitable ones to use for N centered CR. Use of the Morton and Preston values for N_4 CR produces the sum of % *s* and % *p* character at all atoms of the molecule as 122% [87Kn1]; this crucial check on internal consistency is not usually available. Whether the error found is more likely to indicate inaccuracy in the method or inaccurate A^0 values is not discussed.

Despite often being called "non bonding" electrons, removal of one lone pair electron from a nitrogen atom causes significant changes in bonding, and in many cases the equilibrium geometry of the CR is very different from that of the neutral compound. N-containing CR have traditionally been classified as "π radicals" when the spin at nitrogen is formally in a pure $N(2p)$ orbital (as in Sections 20.2.1 and 20.5), and as "σ radicals" when there is a significant $N(2s)$ component to the spin-bearing orbital at N (as in 20.6.1 and many 20.6.2 examples).

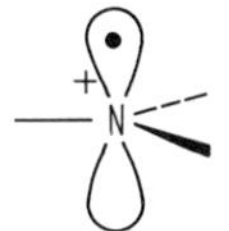
Amine CR : π

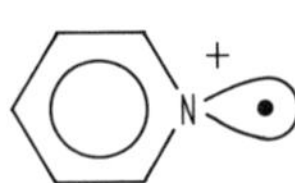
Pyridine CR : σ

A simple trialkylamine is pyramidal at N and has significant *s*-hybridization for its lone pair electrons, but electron removal gives a CR whose equilibrium geometry is planar at N, clearly a π-radical. The presence of most unsaturated substituents at amino nitrogen atoms results in delocalized π CR being produced upon electron removal. The ESR splitting constant a(N) for such π radicals is approximately 2.0···2.5 mT times the $2p$ spin density at N, ϱ (N^π). Although included in Section 20.2.1 because the starting materials are saturated amines, neither aziridine nor cyclopropylamine has been observed to produce the amine cation radical upon electron loss, even in a matrix at low temperature. In both cases, only the product of $C_\alpha-C_\beta$ bond cleavage (2-azaallyl CR [86Qi1] and 4-aza-3-buten-1-yl CR [86Qi2], respectively) were observed.

Because the isotropic a(N) value is proportional to the *s*-orbital spin density at N (which arises from spin polarization in π radicals), when significant spin density is present in N-centered orbitals having substantial *s* hybridization, higher a(N) values are observed than would be seen for π radicals. Such CR are frequently called σ radicals, and pyridine CR is a good example. The electron removed clearly comes not from the 6 electron aromatic π system, but from the nitrogen lone pair perpendicular to it. Significant *s*-hybridization at N is present even in the cation radical, for which $a(N)=4.2$ mT is observed. Some delocalization through the σ framework also occurs, and rather small hydrogen splittings, largest at the *ortho* positions, are observed [79Sh1, 84Ra1]. This σ vs. π classification roughly corresponds to spin density at amino and imino nitrogen, but this implicitly assumes that amino nitrogen atom RC will be planar at N. This expectation has many exceptions, especially for α-heteroatom systems, which appear in Section 20.3.

Three fundamentally different types of spin and charge delocalization have been found for CR of compounds with two equivalent formally saturated N atoms.

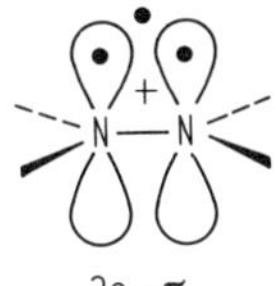
3e – π

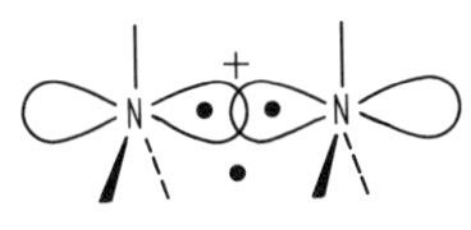
3e – σ

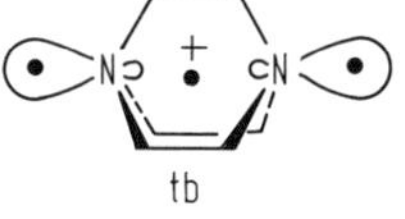
tb

CR from hydrazines with saturated substituents, which have their charge and spin formally delocalized over both nitrogens, will be used for illustration of the most common type, two center, 3 e-π radicals. Neutral hydrazines have strongly pyramidal nitrogens which electronically prefer to have perpendicular lone pair orbital axes, but electron removal causes flattening at both nitrogens and N—N bond rotation to make the spin-bearing lone pair orbital axes coplanar; the cations may be qualitatively described as two center 3e-π bonded systems. However, pyramidalization at N of hydrazine CR is particularly facile, and many examples of R_4N_2 CR have equilibrium geometries which are significantly pyramidalized. Differences in symmetry for pyramidalizing the nitrogens *syn* and *anti* cause large differences in a(N) for a given amount of pyramidalization (for discussion, see [86Ne1]). Even N_2H_4 CR in an $N_2H_6^{2+}$ oxalate single crystal appears from recent ENDOR studies to have one N slightly pyramidal at equilibrium [86Sa1]. The nitrogens being slightly non-equivalent is presumably a result of crystal packing forces. Observed ESR splittings are a time-average over the energy surface of the radical. The energy surface for pyramidalization

at nitrogen can be especially shallow, in which case the equilibrium geometry is not a very complete description of the radical being observed.

Examples of the second bonding type, 3e-σ, were first demonstrated by Alder and coworkers for CR produced by adding an electron to certain hexaalkylhydrazine dications (for a review, see [82Al1]; for X-ray structural proof, [85Al1]). The σ here has nothing to do with the fraction of *s*-orbital character in the spin-bearing orbitals, but specifies the geometry of interaction of the orbitals at N. 3e-π radicals have the π^* orbital as the singly occupied MO. Their *p*-rich orbital axes are roughly perpendicular to the NN σ bond. 3e-σ radicals have the σ^* orbital as the singly occupied MO. Their *p*-rich orbital axes are colinear with the NN σ bond. Alder's examples of 3e-σ NN bonds were all intramolecular, as are some other examples appearing in Section 20.2.2. The initial species from $N_2H_6^{2+}$ SO_4^{2-} has been assigned as $H_3N \cdots NH_3^+$ [86Ga1]. Quinuclidine (1-azabicyclo[2.2.2]octane) CR has recently been shown to give an intermolecular example of a 3e-σ bond [88Di1], as does N_2 CR [87Kn1]. Presumably N_2CO CR [86Kn1] represents an unsymmetrical 3e-σ bonded species. The dimer RC of 4-methoxy-dimethylaniline has been assigned as an NO 3e-σ bonded species [85Cl1], although this seems unlikely to us on thermodynamic grounds. Bis-hydrazines do not form even intramolecular 3e-σ bonded RC [86Ne2], and the strength of such bonds falls off rapidly when the atoms have different electronegativities [88Cl1]. a(2N) values for 3e-σ bonds currently range from the 1.47 mT of 1,5-diazabicyclo-[3.3.3]undecane CR [82Al1] to the 10.0 mT of N_4 CR; the latter is the largest reported N splitting, and corresponds to $sp^{1.9}$ hybridization at N.

1,4-Diazabicyclo[2.2.2]octane CR exhibits the third type of delocalization of spin and charge over two nitrogen atoms mentioned above, through bond (tb). The nitrogen-centered orbitals do not exhibit significant direct overlap with each other, but combine with the σ orbitals of the hydrocarbon framework to form an effectively delocalized system. For detailed discussion of how ring size affects spin delocalization in bridgehead diazabicylic CR, see [81Al1].

Section 20.3 contains CR from amines which are substituted at nitrogen with a heteroatom. Although most give 3e-π bonded species, little if any 3e-π stabilization appears to be present for haloamines. Both σ electron withdrawal and π electron donation favor pyramidal N structures, and NF_3 CR [78Go1] is both strongly pyramidal and thermodynamically difficult to form; it is a σ radical in the traditional sense. Monohaloamine CR are probably planar or nearly planar at N, although definite experimental verification appears not to exist. The 3e-π bond formulation of CR with NN, NS, and NO systems present is qualitatively valid even when π delocalizing substituents such as aryl groups are present, and replacement of saturated alkyl groups by phenyl makes the compounds harder to oxidize, not easier, despite the increased π delocalization possible in the phenyl substituted compounds. Even with strongly electron-withdrawing π groups on the N atoms of 3e-π bonded CR, little spin delocalization onto the substituent occurs; for N,N'-diacylhydrazine CR, see [87Ne1], and for N-acylhydroxylamine CR, [87Ne2]. The strong resonance interaction between the NR_2 group and the C=O group present in the neutral form is mostly relinquished in the CR, for which 3e-π delocalization is present. Feller, Davidson and Borden pointed out that allylic resonance becomes weaker when the ends of the allylic system become more electronegative [83Fe1]. Some multiple π center S and N containing cation radicals which are not related to simple sulfenamides by any structural feature besides having NS bonds were also included in Section 20.3.3 for convenience.

Because two center CR containing N rather clearly form a fundamental unit for spin delocalization (except when the second heteroatom is halogen), we have grouped $R_2N{-}X{=}Y$ compounds together in Section 20.4, to look for similarities and differences in bonding in three center 1-azaallyl CR. Enamine (X=Y is $CR{=}CR_2$), hydrazone ($N{=}CR_2$), N-nitrosamine (N=O) and amide (C(=O)R) cation radicals are examples of such heteroallyl CR systems. Ureas and other examples having amino nitrogens substituted by strongly electron withdrawing unsaturated substituents such as CN, NO_2, C=S, and P→O also appear here. N-lone pair, X=Y resonance interaction in the neutral form is well known to increase as X=Y becomes more electron withdrawing; N-nitrosamine and simple amide nitrogens are planar, and have $R_2N{-}X{=}Y$ rotational barriers on the order of 20 kcal/mol. Enamines and hydrazones have less pyramidal nitrogens than saturated amines, and prefer geometries with N lone pair, X=Y overlap, but have less resonance stabilization. We have found no recent work on simple enamine CR, but they are certainly expected to be conjugated, 1-azaallyl "π radicals" with substantial $R_2N{-}C$ rotational barriers like their aniline analogues. An N-nitrosamine CR is isoelectronic with a neutral iminoxy radical, and the only example we have found [82Ma1] also has its spin in the NO 3e-π system perpendicular to the allyl π system. Because of rather large spin density at an *s*-rich N orbital, this is a σ radical in the traditional sense (a(N) is 4.5 mT). Berndt and coworkers have shown that in contrast to their neutral forms, hydrazone cation radicals assume conformations with the NR_2 lone pair axis perpendicular to the $N{=}CR_2$ π bond, unless they are constrained not to do so [81Be1]. This allows a 3e-π interaction between the *p*-rich NR_2 orbital and the "sp^2" $N{=}CR_2$ lone pair, which must therefore be more stabilizing than the allylic resonance (plus

steric interaction) of the neutral compound geometry. There is still considerable discussion about the geometry of amide cation radicals, as will be seen in the references in Section 20.4 but we believe that it is unwise to expect significant allylic resonance in $R_2NC(=O)R'$ RC to hold them in the planar geometry of the neutral compound. Molecular orbital calculations predict a great sensitivity of the shape of the $NH_2—X=Y$ CR energy surface to the presence of heteroatoms in X=Y. $NH_2—CH=CH_2$ CR is predicted to be substantially more stable when the N lone pair is parallel to the $CH=CH_2$ π system to allow allylic resonance. The species with the NH_2^+ p orbital twisted 90° from the vinyl π system is calculated to be 25 kcal/mol higher in energy by AM1-UHF, but twisting the $CH_2^{\cdot}$ group 90° from the $^+NH_2=CH$ group is calculated only to raise the energy 5.3 kcal/mol, so $NH_2^+=CH—CH_2^{\cdot}$ is clearly the best single resonance structure. There has obviously been a substantial change in the bonding between the neutral and the CR forms, although NH_2, $CH=CH_2$ conjugation is favorable for both species. In contrast, the energy increase for 90° rotation of the NH_2^+ group out of conjugation with the X=Y group is calculated to drop to only 0.3 kcal/mol for formamide CR, $H_2N—CH=O$ [87Ne1], and the parent hydrazone CR $H_2N—N=CH_2$ is calculated to be 3.8 kcal/mol more stable in the 90° twisted form, in qualitative agreement with Berndt's experimental data. These calculations suggest that 1-azaallyl resonance is not very large when X=Y is electron withdrawing.

Section 20.5 contains amino-stabilized unsaturated CR, which have been arbitrarily divided into the "amino terminated" categories of 20.5.1 and "azahydro aromatics" of 20.5.2, and further subdivided to try to organize the large amount of data available. Section 20.5.1 includes unsaturated substituents, such as di- and triarylamines, but compounds with non-aromatic unsaturated substituents such as systems containing indole and isoindole systems have been arbitrarily placed in Section 20.5.2.4. The biologically important phenothiazines and their analogues, and other compounds of biological significance have been separated in Sections 20.5.2.2 and 20.5.2.3, hopefully for convenience in finding systems for which empirical formulas are cumbersome at best. Kurreck's group especially has done a thorough investigation of flavin model compounds and even flavoenzymes using ENDOR and TRIPLE [84Ku1]. Somewhat illogically, the isothioalloxazine derivatives appear in 20.5.2.2, while the dihydroalloxazines are in 20.5.2.3. Kaim originally formulated [81Ka1-3] the radicals produced by reaction of azaaromatics (1,4-diazine and its benzo derivatives and 4,4′ and 2,2′ bipyridine) with AlR_3, BR_3, GaR_3, $BePh_2$, MgR_2, $ZnPh_2$, and related organometallics as cation radicals such as **A**, but has more recently shown [84Ka1] that they are anion radicals such as **B**; these species will be found in this work formulated as the anion radicals in Chapter 13.1 (Subvolume II(17f):

A $[Me_2Al—N(C_4H_4)N—AlMe_2]^{+\cdot}$ $AlMe_4^-$ B $[Me_3Al—N(C_4H_4)N—AlMe_3]^{-\cdot}$ $AlMe_2^+$

Section 20.6 contains CR of compounds with their nitrogens more highly oxidized than the amino level. 20.6.1 contains compounds for which the nitrogen losing an electron in the CR is an imino nitrogen. An azoalkane CR which is long-lived at 110 K and gives a nearly isotropic spectrum (that from 2,3-diazabicyclo[2.2.2]oct-2-ene) has been recently reported [88Ge1, 88Wi1]. Section 20.6.2 includes data for formally *sp* hybridized N compounds. Knight's use of neon as the matrix at 4 K has made even N_2 CR available [83Kn1]. Its N splitting of 3.72 mT only corresponds to 5.7% *s* character, far lower than might have been predicted. An interesting difference in a(N) values for aryldiazomethane CR depending on the rotation angle of the aryl group has been found by Sawaki, Iwamura and coworkers [87Is1]. Both a(N) values for aryl, *t*-butyl compounds are under 0.5 mT, as might be expected for linear CNN structures, but diaryl and aryl, methyl compounds show both a(N) values above 1.0 mT, which the authors interpret as requiring a non-linear CNN fragment. Only the N_2-derived species, diazo compounds, and acrylonitrile CR of Section 20.6.2 can reasonably be considered nitrogen centered, because the aromatic nitriles show no nitrogen splittings. Finally, section 20.6.3 contains data measured for CR derived from nitro compounds. The aliphatic cases obviously give nitrogen-centered radicals, but how their structures should be formulated is still under discussion in the literature. We will only refer to data on highly methylated protonated aniline dication radicals and diprotonated *p*-phenylenediamine trication radicals here [85Ru2]; these species are clearly not reasonably considered nitrogen centered.

Despite the availability of computer searching, we have probably missed important papers on N-centered CR, for which we apologize. Our computer search file was last updated to CA **107** issue 24. None of the extensive work on porphyrins and their relatives, which is increasingly focusing on the transition metal complexed natural product analogues, or on solid "organic metal" complexes, which do not show ESR hyperfine splittings, has been included in this chapter.

20.2 Non-conjugated amine cation radicals

20.2.1 Monoamines and localized polyamine cation radicals

Substance	Generation/ Matrix or Solvent	Method/ T [K]	g-Factor	a-Value [mT]	Ref./ add. Ref.
$[H_3N]^{\cdot+}$ $(NH_3)^{\cdot+}$	Photoionization/ Neon, $^{14}NH_3$	ESR/ 4	 2.0032 (2)	[1] 3H: 2.74 N: 1.95	82Kn1
	$^{15}NH_3$	4	2.0031 (2)	3H: 2.75 N: 2.75	
$[CH_5N]^{\cdot+}$ $CH_3NH_2^{\cdot+}$	γ-irr./ $CFCl_3$	ESR/ 138		2H: is: 0.225 N: 4.9; 0.55; 0.55; is: 2	85Be3/ 83Kh2
$[C_2H_5N]^{\cdot+}$ $CH_2{=}N^+H\dot{C}H_2$	γ-irr. of parent compound [2])/ $CFCl_3$	ESR/ 150	 2.0026	 4H: 1.61 [3]) H(NH): 0.43 N: 0.77	86Qi1
	$CFCl_3$	150	2.0024	4H: 1.61 [4]) D(ND): 0.066 N: 0.77	
$[C_2H_7N]^{\cdot+}$ $((CH_3)_2NH)^{\cdot+}$	γ-irr./ $CFCl_3$	ESR/ 77⋯148		6H: is: 3.4 1H: 3.6; 1.9; 1.9; is: 2.47 1N: 4.9; ≦0.4; ≦0.4; is: 1.6	85Be3
$[C_3H_7N]^{\cdot+}$ $N^+H_2{=}CHCH_2\dot{C}H_2$	γ-irr. of parent compound [5])/ CF_3CCl_3	ESR/ 80⋯140	 2.0026	 2H(α): 2.26 2H(β): 1.70	87Qi2
	$CFCl_3$	80⋯160	2.0027	2H(α): 2.25 2H(β): 1.70	

[1]) Isotropic spectra even at 3 K.

[2]) (aziridine ring)NH. Three-membered ring opens to the 2-azaallyl cation radical.

[3]) 2H: 1.72 and 1.40 mT also reported.

[4]) N-deuterated compound.

[5]) (cyclopropyl)—NH_2. Ring-closed cyclopropylamine cation radical not observed. Compare with neutral ($C{=}N^+H_2$ deprotonated) system, where H transfer from N to CH_2 occurs; it is not seen here.

Substance	Generation/ Matrix or Solvent	Method/ T [K]	g-Factor	a-Value [mT]	Ref./ add. Ref.
$[C_3H_7N]^{\cdot+}$ NH	γ-irr./ $CFCl_3$	ESR/ 90	2.0037	4H: 5.42 H(NH): 2.26 N: 4.14; 0.08; 0.08; is: 1.91	86Qi1
		140	2.0038	4H: 5.41 H(NH): 2.27 N: 3.82; 0.95; 0.95; is: 1.91	
$[C_3H_9N]^{\cdot+}$ $((CH_3)_3N)^{\cdot+}$	3 MeV e-pulse/ n-Hexane	FDMR/ 295		N: 2.06 [6]) 9H: 2.89	84Le1
	γ-irr./ $CFCl_3$	ESR/ 77		9H: 2.82 N: 4.7; 0.0; 0.0; is: 1.6	84Ea1
	γ-irr./ $CFCl_3$	ESR/ 77	2.0040	9H: 2.83 N: 4.7; <0.4; <0.4; is: 1.57	85Be3
$[C_5H_{11}N]^{\cdot+}$ N—CH_3	γ-irr./ $CFCl_3$	ESR/ 120		3H: 2.8 2H: 2.8 2H: 5.7 N: 4.8; 0.0; 0.0; is: 1.6	84Ea1
$[C_5H_{12}N_2]^{\cdot+}$ CH_3, N, N, CH_3	γ-irr./ $CFCl_3$	ESR/ 77		4H: 2.75 1H: 3.6 1H: 5.6 1H: 6.4 N: 4.4; 0.0; 0.0; [7]) is: 1.5	84Ea1

[6]) Used to analyze spectrum.
[7]) Spin localized on one N.

Substance	Generation/ Matrix or Solvent	Method/ T [K]	g-Factor	a-Value [mT]	Ref./ add. Ref.
$[C_5H_{14}N_2]^{\cdot +}$ $((CH_3)_2NCH_2N(CH_3)_2)^{\cdot +}$	γ-irr./ $CFCl_3$	ESR/ 77		6H: 2.7 2H: 1.8 N: ≈0.0 [8]	84Ea1
$[C_6H_{14}N_2]^{\cdot +}$ 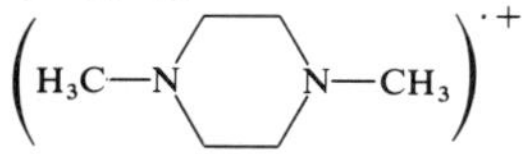	γ-irr./ $CFCl_3$	ESR/ 77		≈10H: is: 0.9 N: ≈0.0 [8]	84Ea1
$[C_6H_{15}N]^{\cdot +}$ $((CH_3CH_2)_3N)^{\cdot +}$	3 MeV e-pulse/ n-Hexane	FDMR [9]/ 295		N=6H(β): 2.08 (2) [10]	84Le1 [11]
	γ-irr./ $CFCl_3$	ESR/ 77 140		3H: 3.8 N: 4.8; 0.0; 0.0; is: 1.6 6H: is: 1.9	84Ea1
$[C_6H_{13}N]^{\cdot +}$ 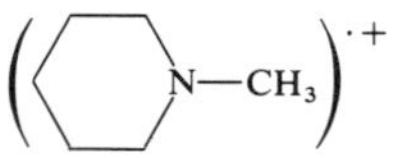	γ-irr./ $CFCl_3$	ESR/ 77		3H: is: 2.9 2H: is: 3.8 N: 4.8; 0.0; 0.0; is: 1.6	84Ea1
$[C_9H_{21}N]^{\cdot +}$ $((CH_3CH_2CH_2)_3N)^{\cdot +}$	3 MeV e-pulse/ —	FDMR/ 295		N: ≈1.93 (2) 6H: ≈1.93 (2)	84Le1
$[C_{12}H_{27}N]^{\cdot +}$ $((CH_3CH_2CH_2CH_2)_3N)^{\cdot +}$	3 MeV e-pulse/ —	FDMR/ 295		N: ≈1.88 (1) 6H: ≈1.88 (1)	84Le1
$[C_{11}H_{26}N_2Si]^{\cdot +}$ 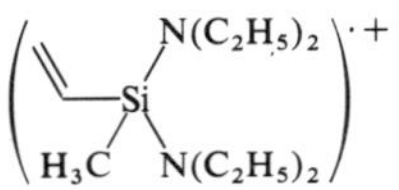	Electrochem. ox., $CH_3CN:HClO_4$/ $CH_3CN:HClO_4$	ESR/ RT		2N: 0.83 8H: 0.415	79Ko1

[8]) N ≦ 0.4 mT; $N_{\parallel}$ not observed. Spin localized on one N.
[9]) 2,5-Diphenylloxazole or anthracene as probe.
[10]) Deuterated compounds also studied.
[11]) Discussion of hole trapping rates.

Substance	Generation/ Matrix or Solvent	Method/ T [K]	g-Factor	a-Value [mT]	Ref./ add. Ref.
$[C_{11}H_{22}N_2]^{\cdot+}$ 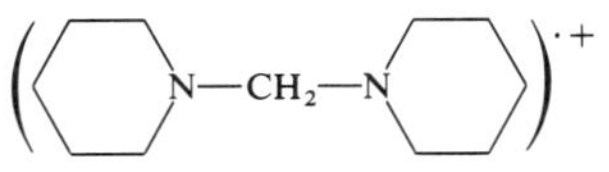	γ-irr./ $CFCl_3$	ESR/ 77		2H: 2.0 2H: 4.0 N: ≈0.0 [8]	84Ea1
$[C_{12}H_{33}NSi_3]^{\cdot+}$ $(((CH_3)_3SiCH_2)_3N)^{\cdot+}$	Ox. with $AlCl_3$/ CH_2Cl_2	ESR/ 180 [12]		N: 1.6 2H: 1.0 2H: 1.6 2H: 2.8 ^{29}Si: 1.5	79Bo1
$[C_{13}H_{25}N]^{\cdot+}$ 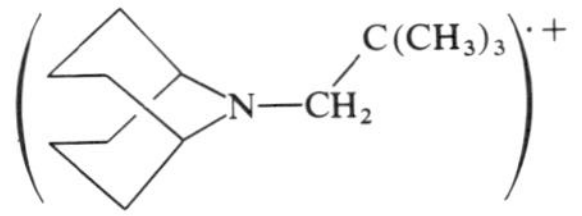	Electrochem. ox./ n-C_3H_7CN	ESR/ 178	2.0043 (1)	Not given	85Ne4

20.2.2 Delocalized cation radicals from non-conjugated amines

Substance	Generation/ Matrix or Solvent	Method/ T [K]	g-Factor	a-Value [mT]	Ref./ add. Ref.
$[H_6N_2]^{\cdot+}$ $(N_2H_6)^{\cdot+}$	γ-irr. of $N_2H_6^{2+}SO_4^{2-}$/ —	ESR/ 77		2N: 5.6; 3.0; 3.0; [13] is: ≈3.9	86Ga1
$[C_6H_{12}N_2]^{\cdot+}$ 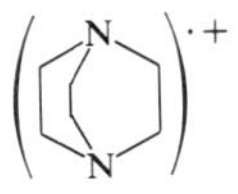	γ-irr./ $CFCl_3$	ESR/ 77		12H: is: 0.73 2N: 2.4; 1.35; 1.35; is: 1.7	84Ea1
	Laser flash phot. of xanthone + DABCO/ $H_2O:CH_3CN$ (1:1)	ESR [14]/ RT	2.00477	12H: 0.749 2N: 1.691	87Ka1

[8]) $N \leq 0.4$ mT; $N_{\parallel}$ not observed. Spin localized on one N.
[12]) 310 K: $N \approx 1.8$, $6H \approx 1.8$ mT.
[13]) 3e-σ-bonded species. Decomposes to $(NH_3)^{\cdot+}$ upon annealing. Deuterated compound also studied.
[14]) Time-resolved, 0.1···1.1 µs after flash, both xanthone$^{\cdot-}$ and DABCO$^{\cdot}$ in emission.

Substance	Generation/ Matrix or Solvent	Method/ T [K]	g-Factor	a-Value [mT]	Ref./ add. Ref.
$[C_{12}H_{24}N_2]^{\cdot+}$ [15]	Isolated as BF_4^- salt/	ENDOR, ESR/			85Ki1
	CF_3CO_2H:toluene	230	2.00290	$2^{14}N$: +3.596 [16] 6H: +1.765 1H: +0.086 1H: −0.028 1H: −0.01 $2^{15}N$: −5.036 [17]	
	CH_2Cl_2	160		6H: +1.763 1H: +0.079 1H: −0.028 1H: −0.010	
$[C_{14}H_{26}N_2]^{\cdot+}$	Ox. of quinuclidine with O_2+SbF_6/ $CHClF_2$	ESR/ 133		2N: 3.869 2H(δ): 0.406 12H(β): 0.337 [18] 12H(γ): 0.066	88Di1
$[C_{17}H_{16}N_2]^{\cdot+}$ [19]	Ox. with $(p\text{-}BrC_6H_4)_3N^+SbCl_6^-$/ CH_2Cl_2	ESR/ 298	2.0036 (1)	2N: 1.685 (15) 4H(NCH_2): 1.108 (10) [20] 8H: 0.180 (5) [21]	86Ge1

[15]) 3e-σ-bonded species.
[16]) Second-order ENDOR splittings resolved.
[17]) ^{15}N species.
[18]) Proven by deuteration.
[19]) 3e-σ-bonded species. X-ray structure of ClO_4 salt reported. NN distance 2.160 Å.
[20]) Estimated E_a for $(CH_2)_3$ ring inversion 6.8(9) kcal mol^{-1}. Only sum of the splittings observed.
[21]) Assigned to 6 H in unsaturated ring, plus 2 γ-H.

Substance	Generation/ Matrix or Solvent	Method/ T [K]	g-Factor	a-Value [mT]	Ref./ add. Ref.

20.3 Heteroatom substituted amines: 3e-π-based

20.3.1 Haloamine cation radicals

Substance	Generation/ Matrix or Solvent	Method/ T [K]	g-Factor	a-Value [mT]	Ref./ add. Ref.
$[F_3N]^{\cdot +}$ $(NF_3)^{\cdot +}$ [1])	γ-irr./ $^{14}NF_4^+AsF_6^-$	ESR/ 26	2.003 (2); 2.006 (2); 2.006 (2); is: 2.005 (2)	F: 30.8; (−)3.3; (−)3.3; is: 8.1 ^{14}N: 11.49 (20); 7.8 (5); 7.8 (5); is: 9.0	78Go1
		240	2.0073 (10); 2.0040 (7); 2.0040 (7); is: 2.005(7)	F: 2.0025; 12.48 (5); 12.48 (5); is: 8.99 (14) ^{14}N: 8.70 (25); 9.36 (5); 9.36 (5); is: 9.14 (14)	
	$^{15}NF_4^+AsF_6^-$	24	2.003 (2); 2.005 (2); 2.005 (2); is: 2.005 (2)	F: 30.6; (−)3.8; (−)3.8; is: 7.7 N: −16.08 (20); −10.6 (10); −10.6 (10); is: −12.4	
		240	2.0079 (12); 2.0047 (7); 2.0047 (7); is: 2.0058 (9)	F: 2.40 (25); 11.97 (5); 11.97 (5); is: 8.78 (14) N: −12.20 (25); −13.3 (5); −13.3 (5); is: −12.75 (14)	
$[C_9H_{18}BrN]^{\cdot +}$ 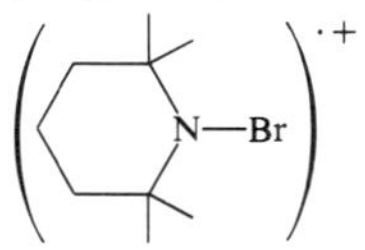	Electrochem. ox./ CH_3CN	ESR/ RT(?)	2.0027	N: 2.07 Br: 2.07	82Re1

1) N pyramidal, but less than $\dot{C}F_3$.

Substance	Generation/ Matrix or Solvent	Method/ T [K]	g-Factor	a-Value [mT]	Ref./ add. Ref.
$[C_9H_{18}ClN]^{\cdot+}$ 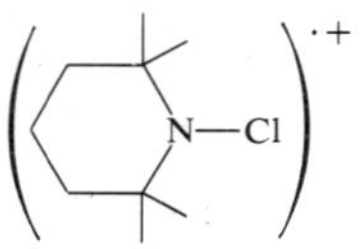	UV-irr./ H_2SO_4: CH_3COOH	ESR/ RT(?)	2.0098	N: 2.093 Cl: 0.628	82Re1
$[C_9H_{18}FN]^{\cdot+}$ 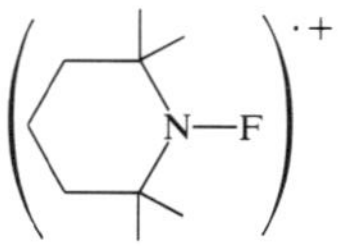	Electrochem. ox./ CH_3CN	ESR/ RT(?)	2.0051	N: 2.461 F: 6.796	82Re1

20.3.2 Hydroxylamine cation radicals

Substance	Generation/ Matrix or Solvent	Method/ T [K]	g-Factor	a-Value [mT]	Ref./ add. Ref.
$[C_7H_{10}ClNO_2]^{\cdot+}$ 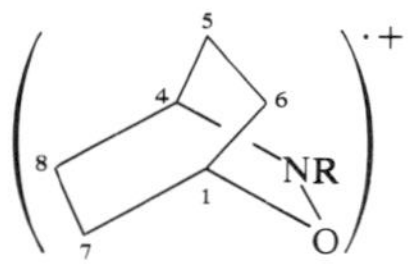 **A**, R = C(=O)Cl	Electrochem. ox./ CH_2Cl_2	ESR/ 209	2.0056	N: 1.375 2H(5, 8, *exo*): 0.407 2H(6, 7, *exo*): 0.275	87Ne2
$[C_7H_{11}NO_2]^{\cdot+}$ **A**, R = CH(=O)	Electrochem. ox./ CH_2Cl_2	ESR/ 218		N: 1.343 2H(5, 8, *exo*): 0.411 2H(6, 7, *exo*): 0.263 2H(5, 8, *endo*): 0.072 2H(1, CHO): 0.056	87Ne1
$[C_7H_{13}NO]^{\cdot+}$ **A**, R = CH_3	Electrochem. ox./ CH_2Cl_2	ESR/ 210	2.0047	N: 1.99 3H(CH_3): 1.99 2H(5, 8, *exo*): 0.430 2H(6, 7, *exo*): 0.165 2H(5, 8, *endo*): 0.103 H(4): 0.090	87Ne2

Substance	Generation/ Matrix or Solvent	Method/ T [K]	g-Factor	a-Value [mT]	Ref./ add. Ref.
$[C_8H_{10}F_3NO_2]^{\cdot+}$ **A**, R = C(=O)CF_3	Electrochem. ox./ CH_2Cl_2	ESR/ 195	2.0055	N: 1.28 2H(5, 8, *exo*): 0.41 2H(6, 7, *exo*): 0.31	87Ne2
$[C_8H_{13}NO_2]^{\cdot+}$ 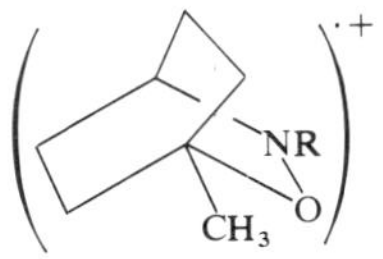 **B**, R = C=O)H	Electrochem. ox./ CH_2Cl_2	ESR/ 205	2.0058	N: 1.336 2H(5, 8, *exo*): 0.380 2H(6, 7, *exo*): 0.290	87Ne2
$[C_8H_{15}NO]^{\cdot+}$ **A**, R = CH_2CH_3	Electrochem. ox./ CH_2Cl_2	ESR/ 187		N: 1.994 2H(NCH_2): 1.543 2H(5, 8, *exo*): 0.440 2H(6, 7, *exo*): 0.170 2H(5, 8, *endo*): 0.099 H(1): 0.074	87Ne2
$[C_8H_{13}NO_2]^{\cdot+}$ **A**, R = C(=O)CH_3	Electrochem. ox./ CH_2Cl_2	ESR/ 249	2.0051	N: 1.435 [2] 2H(5, 8, *exo*): 0.414 2H(6, 7, *exo*): 0.235 2H(5, 8, *endo*): 0.071 H(4): 0.061 3H(CH_3): 0.069 [3]	87Ne2
$[C_8H_{13}NO_3]^{\cdot+}$ **A**, R = C(=O)OCH_3	Electrochem. ox./ CH_2Cl_2	ESR/ 234		N: 1.466 2H(5. 8, *exo*): 0.435 2H(6, 7, *exo*): 0.251 2H(5, 8, *endo*): 0.067 4H(1, CH_3): 0.065	87Ne1
$[C_8H_{15}NO]^{\cdot+}$ **B**, R = CH_3	Electrochem. ox./ CH_2Cl_2	ESR/ 214		N: 1.99 3H(CH_3): 1.99 2H(5, 8, *exo*): 0.43 2H(6, 7, *exo*): 0.165 2H(5, 8, *endo*): 0.103	87Ne2

[2]) Temperature dependence also studied.
[3]) Proven by deuteration.

Substance	Generation/ Matrix or Solvent	Method/ T [K]	g-Factor	a-Value [mT]	Ref./ add. Ref.
$[C_9H_{15}NO_2]^{\cdot+}$ **B**, R = C(=O)CH$_3$	Electrochem. ox./ CH_2Cl_2	ESR/ 228	2.0054	N: 1.414 2H(5, 8, *exo*): 0.395 2H(6, 7, *exo*): 0.256 2H(5, 8, *endo*): 0.035 3H(CH_3): 0.072	87Ne2
$[C_8H_{19}NO]^{\cdot+}$	Protonation of the nitroxide/ CH_2Cl_2 [3a]), $TiBr_4:H_2O$ $BF_3:H_2O$ $TiCl_4:H_2O$ $SbCl_5:H_2O$	ESR/ 253 (?)	 2.00466 2.00437 2.00422 2.00405	 N: 2.17(1) [3b]) H: −0.41(1) [3b])	74Co1 [3c])
$[C_9H_{19}NO]^{\cdot+}$	Protonation of the nitroxide/ CH_2Cl_2 [3a]), $BF_3:H_2O$	ESR/ 253		N: 2.15 ^{17}O (labelled compound): 1.38 H: −0.38	74Co1 [3c])

3a) Solvent not mentioned, but stated as CH_2Cl_2 in [69Ho1].
3b) Independent of Lewis Acid.
3c) For Lewis Acid complexes, see [74Co2, 71Ea1].

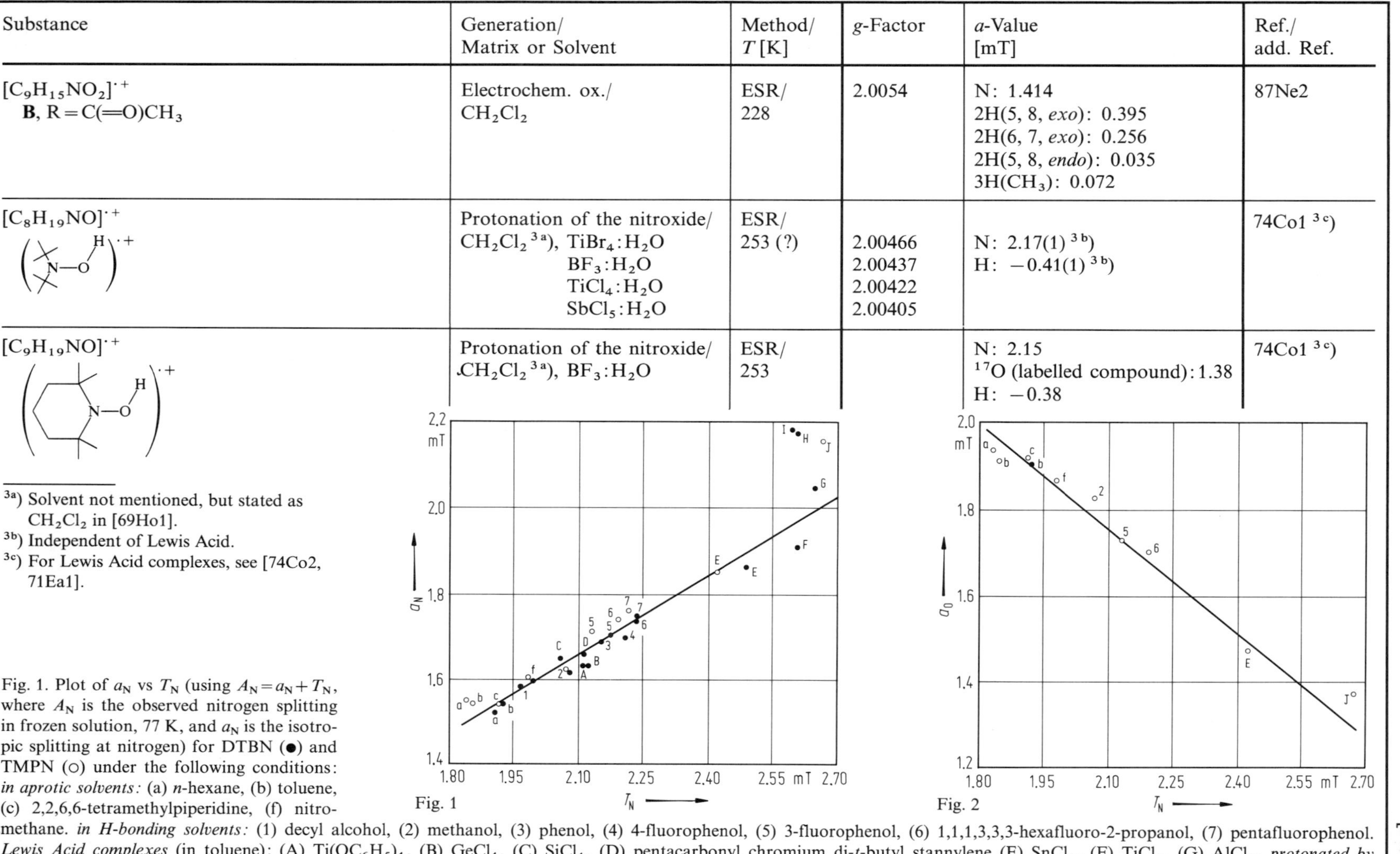

Fig. 1. Plot of a_N vs T_N (using $A_N = a_N + T_N$, where A_N is the observed nitrogen splitting in frozen solution, 77 K, and a_N is the isotropic splitting at nitrogen) for DTBN (●) and TMPN (○) under the following conditions: *in aprotic solvents:* (a) *n*-hexane, (b) toluene, (c) 2,2,6,6-tetramethylpiperidine, (f) nitromethane. *in H-bonding solvents:* (1) decyl alcohol, (2) methanol, (3) phenol, (4) 4-fluorophenol, (5) 3-fluorophenol, (6) 1,1,1,3,3,3-hexafluoro-2-propanol, (7) pentafluorophenol. *Lewis Acid complexes* (in toluene): (A) $Ti(OC_6H_5)_4$, (B) $GeCl_4$, (C) $SiCl_4$, (D) pentacarbonyl chromium di-*t*-butyl stannylene (E) $SnCl_4$, (F) $TiCl_4$, (G) $AlCl_3$. *protonated by* (in CH_2Cl_2) (H) $H_2O/TiCl_4$, (I) $H_2O:TiBr_4$, (J) $H_2O:BF_3$.

Fig. 2. Plot of a_O vs. T_N, where a_O is the isotropic ^{17}O splitting for labelled DTBN (●) or TMPN (○), and the points are labelled as in Fig. 1.

Substance	Generation/ Matrix or Solvent	Method/ T [K]	g-Factor	a-Value [mT]	Ref./ add. Ref.
$[C_9H_{16}N_2O_2]^{\cdot+}$ **A**, R = C(=O)N$(CH_3)_2$	Electrochem. ox./ CH_2Cl_2	ESR/ 208	2.0050	N: 1.70	87Ne2
$[C_{11}H_{19}NO_2]^{\cdot+}$ **A**, R = C(=O)C$(CH_3)_3$	Electrochem. ox./ CH_2Cl_2	ESR/ 209	2.0053	N: 1.462 2H(5, 8, *exo*): 0.451 2H(6, 7, *exo*): 0.231 2H(5, 8, *endo*): 0.053 H(1): 0.055	87Ne2
$[C_{11}H_{19}NO_3]^{\cdot+}$ **A**, R = $CO_2C(CH_3)_3$	Electrochem. ox./ CH_2Cl_2	ESR/ 212	2.0056	N: 1.488 2H(5, 8, *exo*): 0.435 2H(6, 7, *exo*): 0.255 2H(5, 8, *endo*): 0.068 H(1): 0.065	87Ne2
$[C_{11}H_{21}NO]^{\cdot+}$ **A**, R = $CH_2C(CH_3)_3$	Ox. with (p-Br$C_6H_4)_3N^+SbCl_6^-$/ CH_2Cl_2	ENDOR/ 220	2.00479	N: +1.865 2H(NCH_2): +1.148 2H: +0.448 1H: +0.167 [4] 1H: +0.029 [4] 1H: −0.093 [4]	87Ne2
$[C_{11}H_{21}NO]^{\cdot+}$ 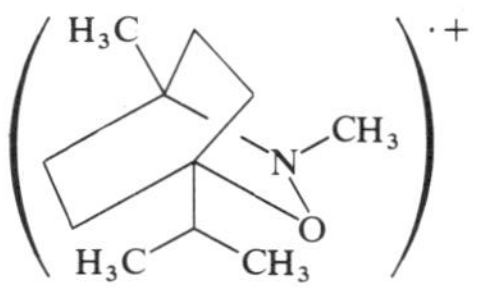	Electrochem. ox./ CH_2Cl_2	ESR/ 207		N: 1.97 3H(NCH_3): 1.97 2H(5, 8, *exo*): 0.44 2H(6, 7, *exo*): 0.16	87Ne2
$[C_{13}H_{20}N_2O_3]^{\cdot+}$ 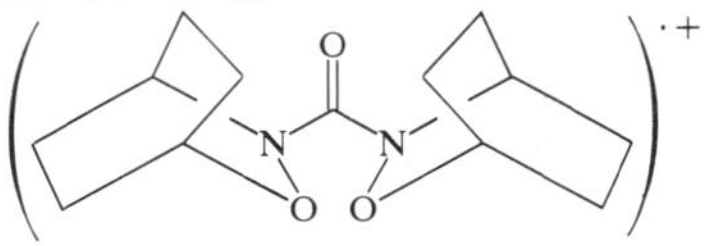	Ox. with (4-Br$C_6H_4)_3NSbCl_6$/ CH_2Cl_2	ESR/ 321 [5]		2N: 0.920 4H: 0.250 4H: 0.120	86To1

[4]) Proton not assigned.

[5]) At low temperature, alternating linewidth effects due to showing electron transfer. At 190, 208, 226, 244 K $k(\mathrm{et}) = 3.8\cdot10^8$, $8.8\cdot10^8$, $1.8\cdot10^9$, $4.4\cdot10^9\ \mathrm{s}^{-1}$, respectively.

Substance	Generation/ Matrix or Solvent	Method/ T [K]	g-Factor	a-Value [mT]	Ref./ add. Ref.
$[C_{13}H_{21}NO_3]^{\cdot+}$ 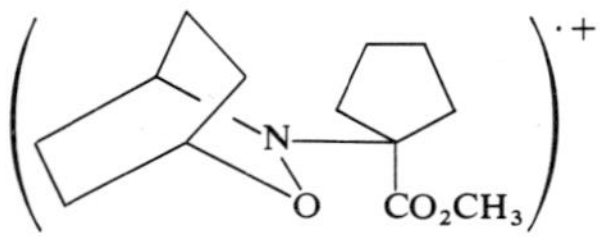	Ox. with $(4\text{-}BrC_6H_4)_3NSbCl_6$/ CH_2Cl_3	ESR/ 188 209 230 249 254		N: 1.837 N: 1.839 N: 1.842 N: 1.845 N: 1.855	87Ne2
$[C_{14}H_{20}N_2O_4]^{\cdot+}$ 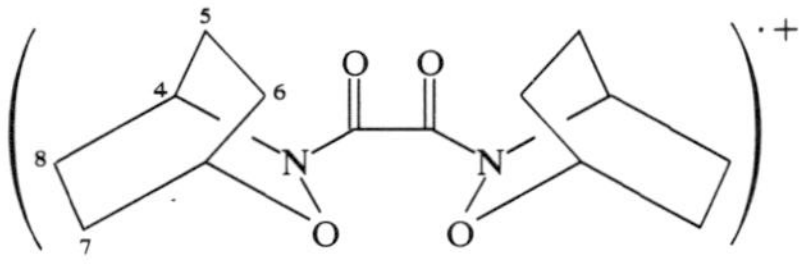	Electrochem. ox./ CH_2Cl_2	ESR/ 186		1N: 1.325 2H(5, 8, *exo*): 0.400 2H(6, 7, *exo*): 0.275	87Ne2

20.3.3 Thiohydroxylamine cation radicals (and other cation radicals with NS bonds)

Substance	Generation/ Matrix or Solvent	Method/ T [K]	g-Factor	a-Value [mT]	Ref./ add. Ref.
$[N_2S_3]^{\cdot+}$ 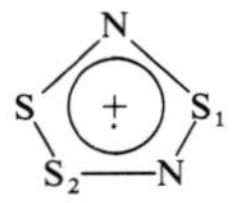	Dissolved NS_2^+ in THF/ THF	ESR/ 295	2.01085 (9)	$1^{14}N$: 0.319 (1) $1^{33}S$: 0.872 (7)	86Fa1/ 84Fa1
	Dissolved dimer dichloride $(S_3N_2)_2Cl_2$ in D_2SO_4/ D_2SO_4	ESR/ 113	2.0013; 2.0062; 2.0250; is: 2.0108	$1^{14}N$: 0.918; 0; 0; is: 0.306 $1^{15}N$: 1.285; 0; 0; is: 0.428 $1^{33}S(2)$: 3.784; 0; 0.882; is: 1.555 $1^{33}S(1)$: 0.893; 0; 0; is: 0.298	86Fa1/ 84Fa1
$[N_4S_4]^{\cdot+}$ $(S_4N_4)^{\cdot+}$	γ-irr./ $CFCl_3$	ESR/ 77	2.004	No N splitting detected	87Ch2

Substance	Generation/ Matrix or Solvent	Method/ T [K]	g-Factor	a-Value [mT]	Ref./ add. Ref.
$[C_6H_2N_2S_4]^{\cdot+}$, $[C_6D_2N_2S_4]^{\cdot+}$ R = H or D	Ox. of diradical/ H_2O	ESR/ RT	 2.0068 (2)	 2N: 0.624 (3) 2H: 0.220 (3)	87Do1/ 86Wo1
		100	2.0026 (7) [6]	2N: 1.63 (1) [6]	
		RT		2N: 0.627 (2) [7] 2D: 0.035 (2) [7]	
$[C_{10}H_6N_2S]^{\cdot+}$ 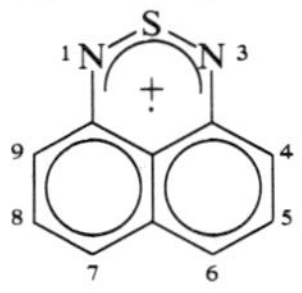	Ox. with AlCl/ CH_2Cl_2	ESR/ 253	2.0027	2N(1, 3): +0.196 [8] 2H(4, 9): −0.508 2H(5, 8): +0.077 2H(6, 7): −0.674 $2^{13}C(4, 9)$: +0.75 $2^{13}C(5, 8)$: −0.59 $2^{13}C(6, 7)$: +0.93	80Ge1
$[C_{12}H_8N_2S]^{\cdot+}$ 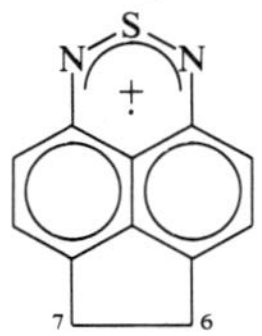	Ox. with $AlCl_3$/ CH_2Cl_2	ESR/ 213	2.0027	N(1, 3): +0.186 H(4, 9): −0.416 H(5, 8): +0.021 4H(6, 7): +1.264	80Ge1
$[C_3H_9NS]^{\cdot+}$ 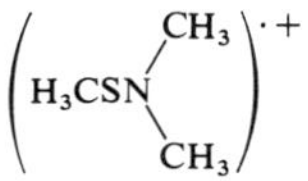 $(H_3CSN(CH_3)_2)^{\cdot+}$	Ox. with $AlCl_3$/ CH_3NO_2	ESR/ 295	2.0069	N: 1.21 3H(SCH_3): 0.87 3H(NCH_3): 1.43 3H(NCH_3): 1.44	81Iz1
$[C_4H_{12}N_2S]^{\cdot+}$ 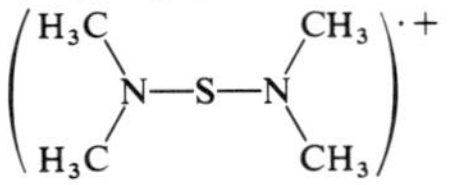 $((H_3C)_2N{-}S{-}N(CH_3)_2)^{\cdot+}$	Ox. with $AlCl_3$/ CH_3NO_2	ESR/ 295	2.0050	2N: 0.75 12H: 0.75	81Iz1
	Electrochem. ox./ CH_2Cl_2	ESR/ 218	 2.0053	 2N: 0.76 12H: 0.76	79Ga1
		188	2.0052		
		295	2.0055		

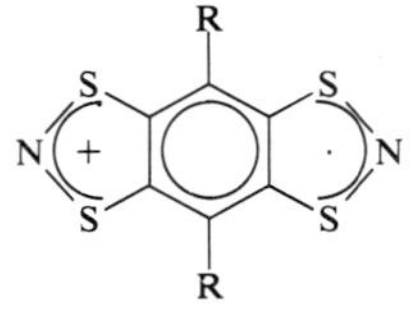

[6]) One principal value.
[7]) Deuterated species.
[8]) Signs from line broadening measurements.

Substance	Generation/ Matrix or Solvent	Method/ T[K]	g-Factor	a-Value [mT]	Ref./ add. Ref.
$[C_4H_{12}N_2S_2]^{\cdot+}$ $((CH_3)_2NSSN(CH_3)_2)^{\cdot+}$	Ox. with $AlCl_3$/ CH_2Cl_2	ESR/ 210		2N: 0.75 12H: 0.75	81Bo2
$[C_4H_{12}N_2OS]^{\cdot+}$ 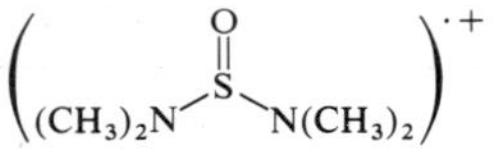	Electrochem. ox./ n-C_3H_7CN	ESR/ 268	2.0037	2N: 1.30 12H: 1.24	79Ga1
$[C_5H_{11}NS]^{\cdot+}$ 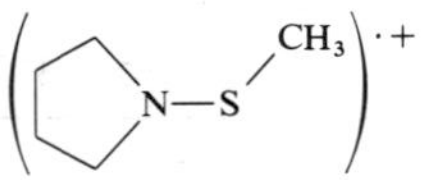	Ox. with $(BrC_6H_4)_3N^+SbCl_6^-$/ CH_2Cl_2	ESR/ 295		N: 1.38 [9] 3H: 0.85 2H: 2.17 2H: 1.82	82Ne2
$[C_5H_{13}NS]^{\cdot+}$ 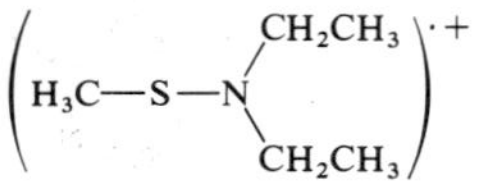	Ox. with $AlCl_3$/ CH_3NO_2	ESR/ 273	2.0071	N: 1.43 $3H(SCH_3)$: 0.86 2H: 0.99 2H: 0.86	81Iz1
$[C_8H_{20}N_2S]^{\cdot+}$ 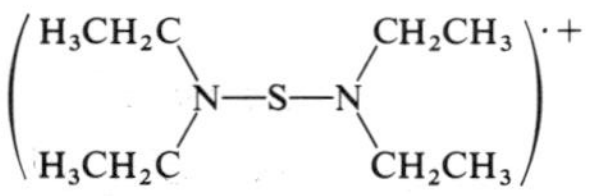	Ox. with $AlCl_3$/ CH_3NO_2	ESR/ 295	2.0056	2N: 0.75 8H: 0.50	81Iz1
	Electrochem. ox./ CH_2Cl_2	ESR/ 208	2.0058	2N: 0.73 8H: 0.50	79Ga1
$[C_8H_{16}N_2S]^{\cdot+}$ 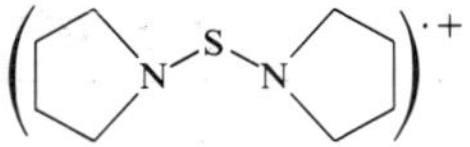	Electrochem. ox./ CH_2Cl_2	ESR/ 253	2.0055	2N: 0.76 8H: 1.03	79Ga1
$[C_9H_{17}NS]^{\cdot+}$ 	Ox. with $(BrC_6H_4)_3N^+SbCl_6^-$/ CH_2Cl_2	ESR/ 295		N: 1.41 $3H(CH_3)$: 0.83 [10]	82Ne2

[9]) γ-splitting unresolved, linewidth 0.15 mT.
[10]) Further unresolved splittings, linewidth 0.4 mT.

Substance	Generation/ Matrix or Solvent	Method/ T [K]	g-Factor	a-Value [mT]	Ref./ add. Ref.
$[C_{10}H_{12}N_2O_2S]^{\cdot +}$	Electrochem. ox./ CH_3CN	ESR/ 298	2.0061	N: 1.43 2H(NCH_2): 2.00 2H(NCH_2): 1.90	85Sa1/ 85Sa2
	n-C_3H_7CN	183	2.0061	N: 1.40 1H(NCH_2): 2.15 1H(NH_2): 2.03 1H(NH_2): 1.97 1H(NH_2): 1.47	
$[C_{10}H_{12}N_2O_3S]^{\cdot +}$	Electrochem. ox./ CH_3CN	ESR/ 298	2.0061	N: 1.42 1H: 2.16 1H: 1.52	85Sa1/ 85Sa2
		233	2.0061	N: 1.34 1H: 2.17 1H: 1.46 1H: 0.28 1H: 0.13	
$[C_{10}H_{12}N_2O_2S_2]^{\cdot +}$	Electrochem. ox./ CH_3CN	ESR/ 298	2.0061	N: 1.37 1H: 1.97 1H: 1.34	85Sa1
		233	2.0062	N: 1.31 1H: 2.05 1H: 1.39	
$[C_{11}H_{14}N_2O_2S]^{\cdot +}$ 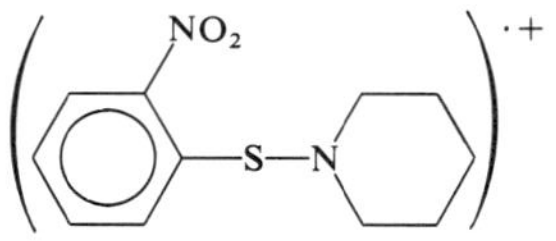	Electrochem. ox./ CH_3CN	ESR/ 243	2.0061	N: 1.40 1H: 2.27 1H: 1.61 1H: 0.18 8H: 0.095 [11])	85Sa1

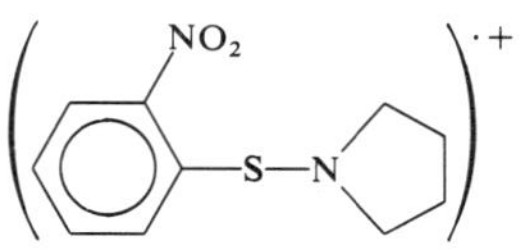

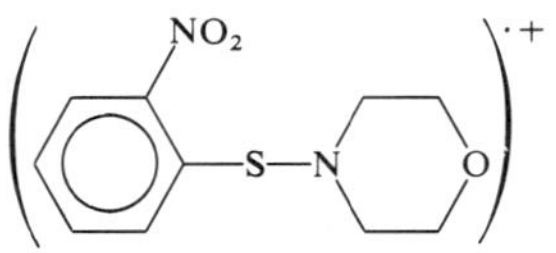

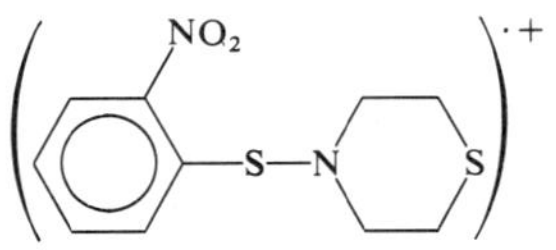

[11]) Mixture of conformations, must include aryl-H.

Substance	Generation/ Matrix or Solvent	Method/ T [K]	g-Factor	a-Value [mT]	Ref./ add. Ref.
$[C_{12}H_{16}N_2O_2S]^{\cdot+}$ 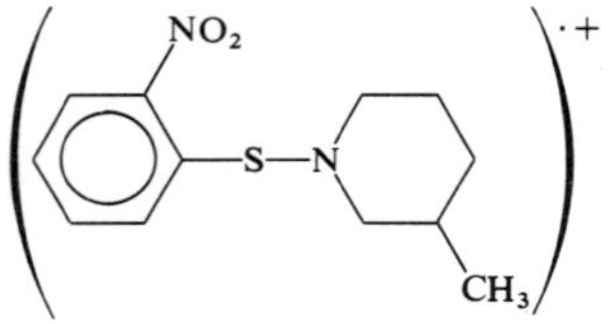	Electrochem. ox./ CH_3CN	ESR/ 298	2.0061	N: 1.43 1H: 2.21 1H: 1.64	85Sa1
$[C_{12}H_{16}N_2O_2S]^{\cdot+}$ 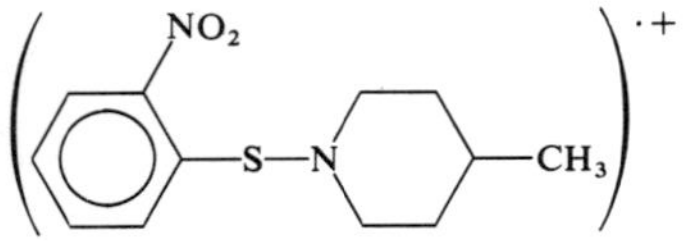	Electrochem. ox./ CH_3CN	ESR/ 298	2.0061	N: 1.40 1H: 2.25 1H: 1.73 1H: 0.18 7H: 0.95	85Sa1
$[C_{12}H_{16}N_2O_2S]^{\cdot+}$ 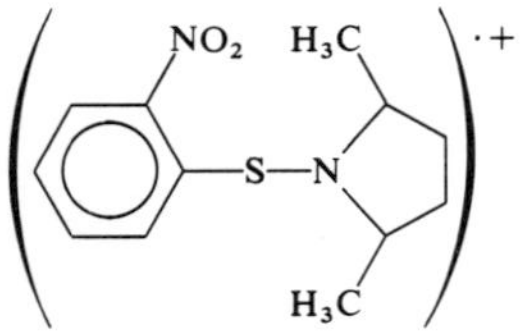	Electrochem. ox./ CH_3CN	ESR/ 298	2.006	N: 1.36 (2) 1H: 2.15 (2) [12] 1H: 1.48 (2) [12]	85Sa1
$[C_{12}H_{16}N_2O_2S]^{\cdot+}$ 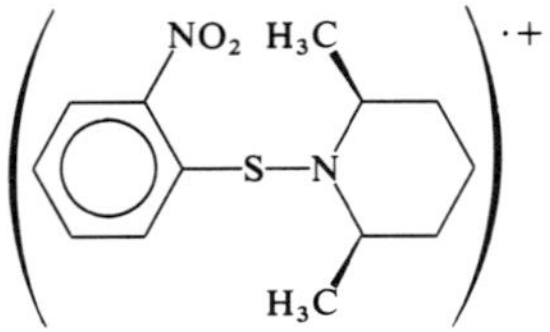	Electrochem. ox./ CH_3CN	ESR/ 243	2.0062	A: N: 1.46 [13] 1H: 0.55 1H: 0.27 B: N: 1.46 [13] 1H: 0.24 1H: 0.19	85Sa1

[12]) Mixture of *cis* and *trans* isomers.
[13]) A, B two species in ratio A:B = 3:1. Authors propose N pyramidal, with S-aryl occupying equatorial and axial positions.

Substance	Generation/ Matrix or Solvent	Method/ T [K]	g-Factor	a-Value [mT]	Ref./ add. Ref.
$[C_{13}H_{18}N_2O_2S]^{\cdot +}$	Electrochem. ox./ CH_3CN	ESR/ 298	2.0061	N: 1.41 [12] 1H: 2.19 1H: 1.16	85Sa1
$[C_{14}H_{14}N_2O_2S]^{\cdot +}$ [14]	Electrochem. ox./ CH_3CN	ESR/ 298	2.0031 [15]	N: 0.97 6H: 0.94 2H: 0.45 2H: 0.10	86Sa2
$[C_{15}H_{22}N_2O_2S]^{\cdot +}$	Electrochem. ox./ CH_3CN	ESR/ 298	2.0067	N: 1.27 [16] 2H: 0.20 SH: 0.10	85Sa1
$[C_{16}H_{16}N_2O_4S]^{\cdot +}$	Electrochem. ox./ CH_3CN	ESR/ 233	2.0042	N: 1.02 3H: 0.96 2H: 0.415 2H: 0.10	86Sa2

12) Mixture of *cis* and *trans* isomers.
14) N-phenyl analogues (which lack the *p*-substituent on the N-aromatic ring) give cation radicals formulated as *p*-azaquinodiphenyls or benzidines. See 20.5.1.4
15) Note low *g*-factor, but assigned as given.
16) Line broadening at 233 K; N: 1.24, 1H: 0.28 mT resolved.

Substance	Generation/ Matrix or Solvent	Method/ T [K]	g-Factor	a-Value [mT]	Ref./ add. Ref.
$[C_{17}H_{20}N_2O_2S]^{\cdot +}$ 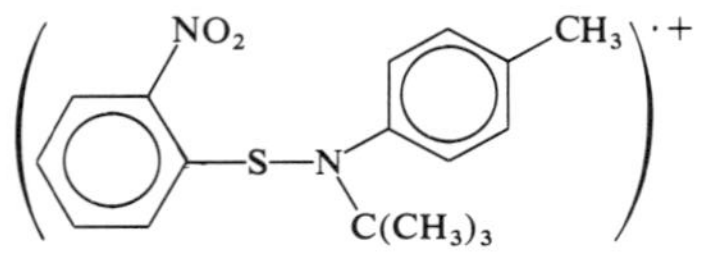	Electrochem. ox./ CH_3CN	ESR/ 298	2.0030 [15]	N: 0.95 3H: 0.92 2H: 0.44 2H: 0.10	86Sa2
$[C_{20}H_{18}N_2O_2S]^{\cdot +}$ 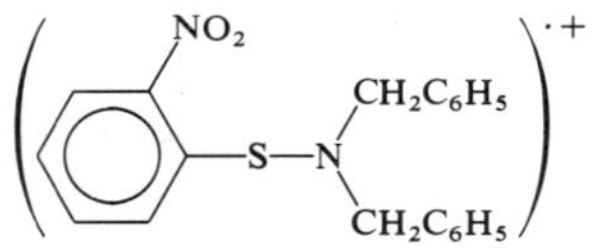	Electrochem. ox./ CH_3CN	ESR/ 253	2.0063	N: 1.34 2H: 1.02 2H: 0.99 2H: 0.10 1H: 0.09	86Sa2

20.3.4 Hydrazine cation radicals

Substance	Generation/ Matrix or Solvent	Method/ T [K]	g-Factor	a-Value [mT]	Ref./ add. Ref.
$[H_4N_2]^{\cdot +}$ $(H_2NNH_2)^{\cdot +}$ [17]	γ-irr./ $N_2H_6SO_4$	ESR/ 295	2.006 (1)	2N: 2.95; 0.67; 0.70; is: 1.44	81Ga1
	γ-irr./ $N_2H_6SO_4$, single crystal	ESR/ 295	2.0034	2N: 1.15 4H: 1.10	85Ko1
[18]	X-irr. of hydrazinium oxalate/ Single crystal	ENDOR, ESR/ 295		1H [19]: −2.183 (6); −1.402 (4); +0.072 (9); is: −1.171 1H [19]: −2.119 (5); −1.34 (3); +0.110 (9); is: −1.116 H [20]: +0.144; −0.092; −0.105; is: −0.018 H [20]: +0.121; −0.038; −0.063; is: +007	86Sa1/ 68Ed1

[15]) Note low g-factor, but assigned as given.
[17]) Deuterated compound also measured.
[18]) One N planar, the other slightly bent, H bonding present.
[19]) Strongly coupled H.
[20]) Weakly coupled H.

Substance	Generation/ Matrix or Solvent	Method/ T [K]	g-Factor	a-Value [mT]	Ref./ add. Ref.
$[C_4H_{12}N_2]^{\cdot +}$ $((CH_3)_2NN(CH_3)_2)^{\cdot +}$	Ox. with horseradish peroxidase + $H_2O_2 + H_2O$/ H_2O	ESR/ RT		2N: 1.375 12H: 1.27	82Ka1
$[C_6H_{14}N_2]^{\cdot +}$ 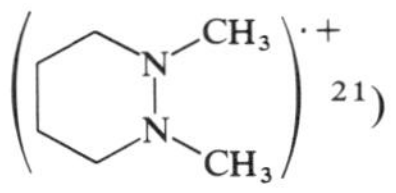 [21]	Ox. with $(4\text{-}BrC_6H_4)_3N^+SbCl_6^-$/ $n\text{-}C_3H_7CN$	ESR/ 173		2N: ≅1.28 6H: ≅1.28 2H: 2.76	85Ne2
$[C_6H_8N_2O_2]^{\cdot +}$ 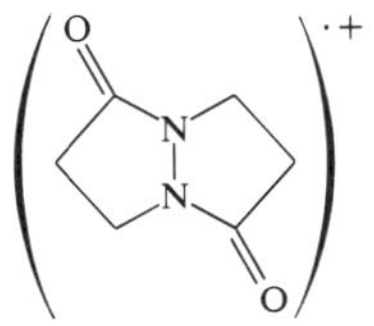	Electrochem. ox./ CH_2Cl_2	ESR/ 213		2N: 0.98 4H: 1.48	87Ne1
$[C_8H_{12}N_2O_2]^{\cdot +}$ 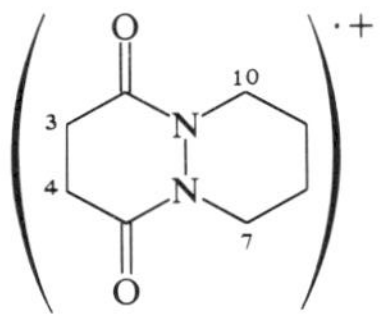	Electrochem. ox./ CH_2Cl_2	ESR/ 295		2N: 1.02 4H (7, 10): 1.65 4H (3, 4): 0.215	87Ne1
$[C_8H_{16}N_2]^{\cdot +}$ 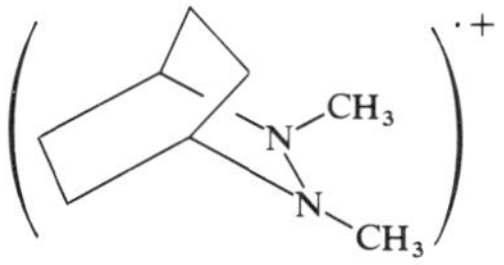	Ox. with $(4\text{-}BrC_6H_4)_3NSbCl_6$/ CH_2Cl_2	ESR/ 199 219 237 258		 2N: 1.370 2N: 1.370 2N: 1.375 2N: 1.378	87Ne2

[21]) Half-chair conformation.

Substance	Generation/ Matrix or Solvent	Method/ T [K]	g-Factor	a-Value [mT]	Ref./ add. Ref.
$[C_8H_{18}N_2]^{\cdot+}$	Ox. with $(4\text{-}BrC_6H_4)_3NSbCl_6$/ n-C_3H_7CN	ESR/ 333		2N: 1.365 6H(NCH_3): 1.262 2H(3, 6): ≈1.365	85Ne2
		173		2N: ≈1.29 6H(NCH_3): ≈1.29 1H(3): 2.75 (pseudoaxial)	
$[C_8H_{18}N_2]^{\cdot+}$ 22)	Ox. with $(4\text{-}BrC_6H_4)_3NSbCl_6$/ n-C_3H_7CN	ESR/ 358		2N≈1.30 6H(NCH_3): ≈1.30 2H(3, 6): 0.82	85Ne2
		268		2N: 1.325 6H(NCH_3): 1.325 2H(3, 6): 0.36	
$[C_8H_{24}N_4Si_2]^{\cdot+}$ 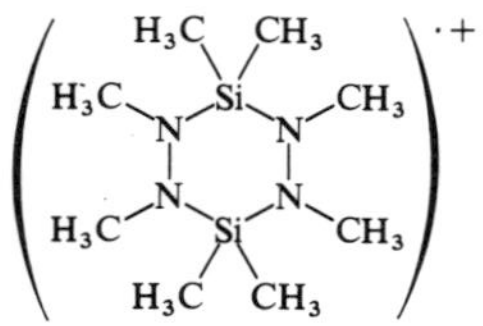	Ox. with $AlCl_3$/ CH_2Cl_2	ESR/ 300		2N: 1.050 6H(NCH_3): 1.375 ^{29}Si: 0.5 6H($SiCH_3$): 0.03	80Bo1, 78Bo1
$[C_{10}H_{13}N_3O]^{\cdot+}$ 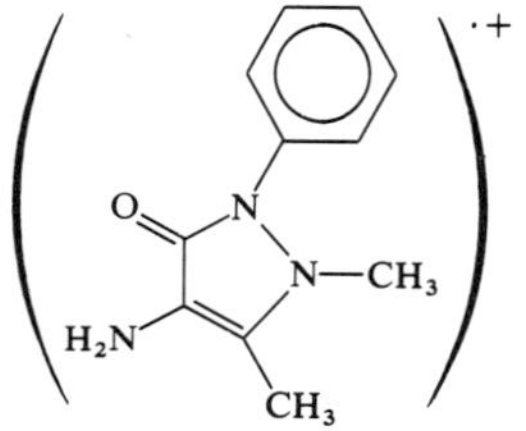	Ox. with horseradish peroxidase, metmyoglobin and protohemin $+H_2O_2$/ H_2O	ESR/ RT	2.003	hfs not analyzed	78Gr1

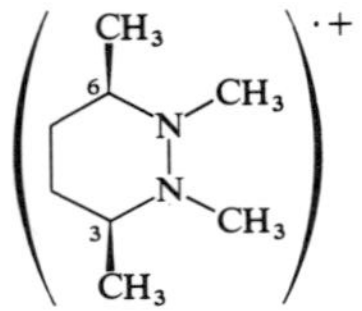

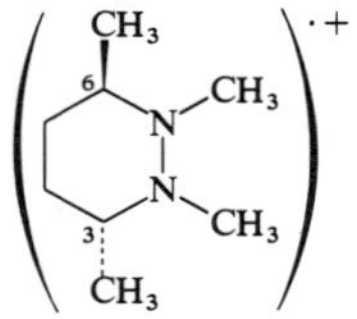

22) Diaxial 3,6-CH_3 conformation estimated to be ≈1.0 kcal mol^{-1} more stable than the diequatorial conformation.

Substance	Generation/ Matrix or Solvent	Method/ T [K]	g-Factor	a-Value [mT]	Ref./ add. Ref.
$[C_{10}H_{14}N_2]^{\cdot +}$	Ox. with $AgNO_3$/ n-C_3H_7CN	ESR/ 193		2N: 2.187 2H: 0.50	88Fr1
$[C_{10}H_{14}N_2O_2]^{\cdot +}$	Ox. with $NOPF_6$/ CH_2Cl_2	ESR/ 295	2.0044	2N: 0.96 4H(*anti*): 0.29 [23]	83Ne3/ 87Ne1
$[C_{10}H_{16}N_2]^{\cdot +}$	Ox. with $AgNO_3$/ n-C_3H_7CN	ESR/ 295 171		 2N: 2.06 4H: 0.41 2N: 2.11	88Fr1
$[C_{10}H_{15}D_2N_3]^{\cdot +}$ [24]	Electrolytic ox./ n-C_3H_7CN	ESR/ 233		1N: 1.72 1N: 0.77 No further splittings determined	83Ne1
$[C_{10}H_{16}N_2]^{\cdot +}$	NO_3^- salt/ n-C_3H_7CN	ESR/ 296 167 172 182 189 196 213 239 247 295		 2N: 2.08 [25] 2N: 2.108 [26] 2H(exo, 1): 0.685 [26] [27] 2H(exo, 2): 0.140 [26] [27] 2N: 2.106 [26] 2N: 2.105 [26] 2N: 2.104 [26] 2N: 2.103 [26] 2N: 2.096 [26] 2N: 2.095 [26] 2N: 2.093 [26] 2N: 2.075 [26] 4H(*exo*) 0.42 [26]	86Ne3

[23]) Other splittings observed.
[24]) d_0-species also studied.
[25]) Fully protiated species.
[26]) Partially deuterated species.
[27]) $\Delta G^{\ddagger} = 4.6$ kcal mol^{-1} for double N inversion.

Substance	Generation/ Matrix or Solvent	Method/ T [K]	g-Factor	a-Value [mT]	Ref./ add. Ref.
$[C_{10}H_{20}N_2]^{\cdot+}$ 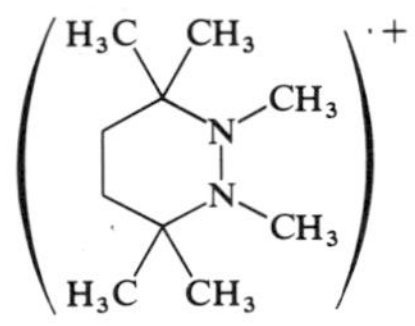	Ox. with $(4\text{-}BrC_6H_4)_3NSbCl_6$/ $n\text{-}C_3H_7CN$	ESR/ 333 213		2N, 6H: 1.30 2N, 6H: 1.28	85Ne2
$[C_{10}H_{20}N_2]^{\cdot+}$ 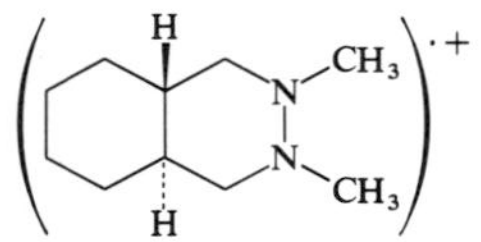	Ox. with $(4\text{-}BrC_6H_4)_3NSbCl_6$/ $n\text{-}C_3H_7CN$	ESR/ 168		2N, 6H: ≈1.27 2H (pseudoaxial): 2.32 2H (pseudoequatorial): 0.42	85Ne2
$[C_{10}H_{28}N_2Si_4]^{\cdot+}$ 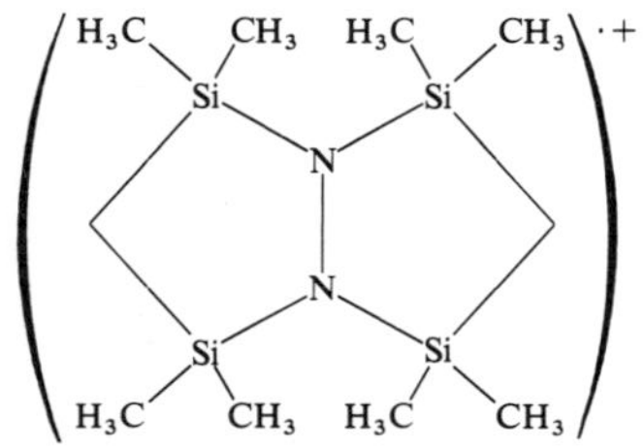	Ox. with $AlCl_3$/ CH_2Cl_2	ESR/ 300 170 180		2N: 0.868 4H(γ): 0.038 24H(CH_3): 0.030 ^{29}Si: 0.485 4H(γ): 0.043 4 ^{29}Si: 0.515	80Bo1
$[C_{11}H_{16}N_2]^{\cdot+}$ 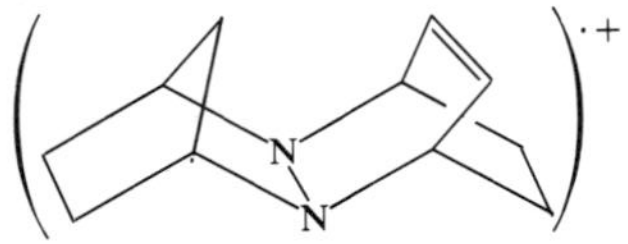	NO_3^- salt/ CH_2Cl_2	ESR/ 193		2N: 1.72 [28])	86Ne3
$[C_{11}H_{18}N_2]^{\cdot+}$ 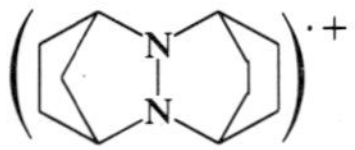	Ox. with $AgNO_3$/ CH_2Cl_2	ESR/ 198		2N: 1.61 [29]) 2H: 0.55 2H: 0.38 2H: 0.12	88Fr1/ 86Ne3 [29])

[28]) At least 4 H near 0.3 mT.

[29]) 2N: 1.606 mT at 215 K, 1.614 mT at 200 K, 1.611 mT at 193 K.

Substance	Generation/ Matrix or Solvent	Method/ T [K]	g-Factor	a-Value [mT]	Ref./ add. Ref.
$[C_{12}H_{18}N_2]^{\cdot +}$ 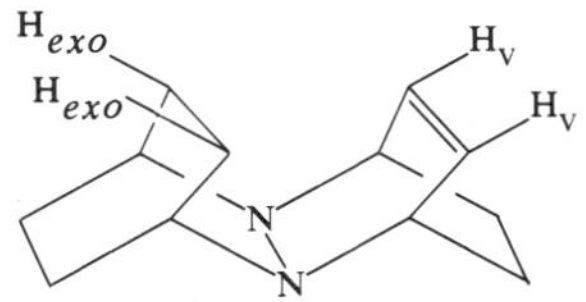	NO_3^- salt/ CH_3CN	ESR/ 297	2.0038	2N: 1.92 2H(*exo*): 0.54 2H (vinyl): 0.54	84Ne2/ 86Ne3
$[C_{12}H_{18}N_2]^{\cdot +}$ 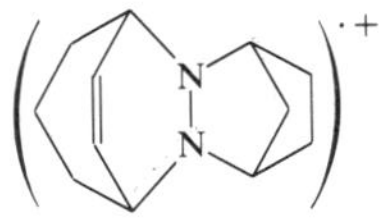	Ox. with $AgNO_3$/ CH_3CN	ESR/ 297		2N: 1.52	88Fr1
$[C_{12}H_{20}N_2]^{\cdot +}$ 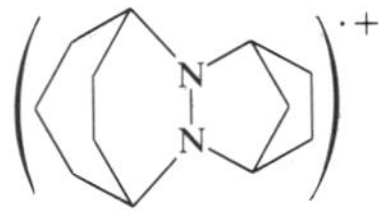	Ox. with $AgNO_3$/ n-C_3H_7CN	ESR/ 297		2N: 1.44	88Fr1
$[C_{12}H_{20}N_2]^{\cdot +}$	NO_3^- salt/ H_3CN	ESR/ 297	2.0036	2N: 1.524 8H(*exo*): 0.287 8H(*endo*): 0.058 4H(bridge): <0.02	84Ne2/ 86Ne3
	n-C_3H_7CN [30]	295 238 213 188		2N: 1.525 2N: 1.5288 2N: 1.5363 2N: 1.5525	
$[C_{12}H_{20}N_2O_4]^{\cdot +}$ 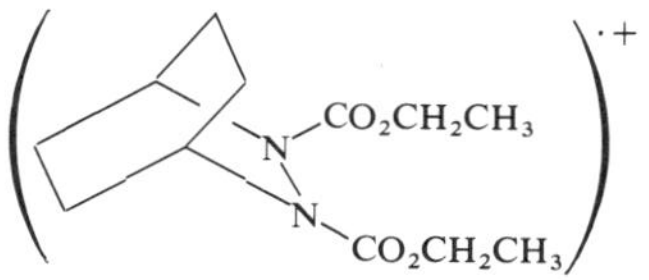	Ox. with $(4\text{-}BrC_6H_4)_3NSbCl_6$/ CH_2Cl_2	ESR/ RT		2N: 1.19	85To1

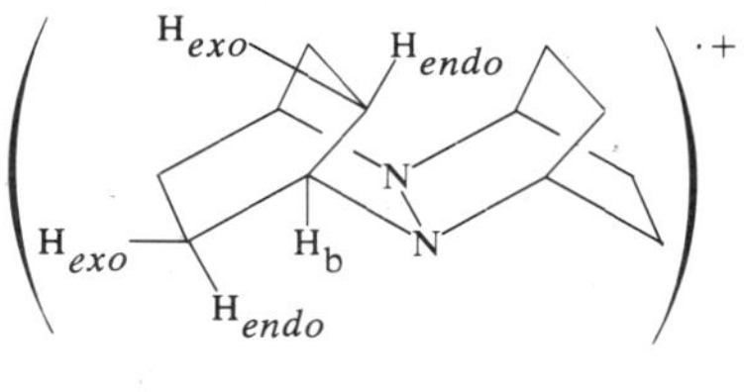

[30]) From [86Ne3].

Substance	Generation/ Matrix or Solvent	Method/ T [K]	g-Factor	a-Value [mT]	Ref./ add. Ref.
$[C_{12}H_{36}N_2Si_4]^{\cdot +}$ 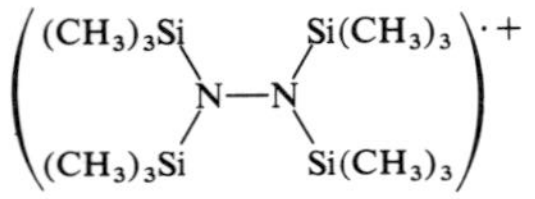	Ox. with $AlCl_3$/ CH_2Cl_2	ESR/ 300		2N: 0.752 36H: 0.013 ^{29}Si: 0.16	81Bo1
$[C_{13}H_{18}N_2]^{\cdot +}$ 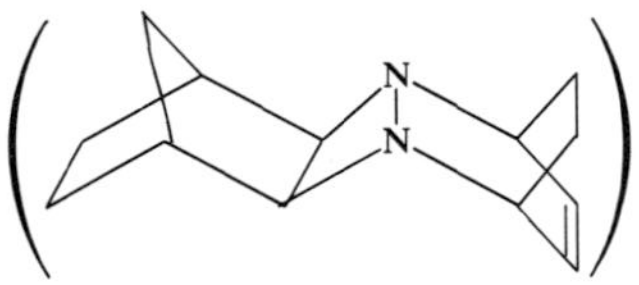	Ox. with $AgNO_3$/ n-C_3H_7CN	ESR/ 295 269 223 199		2N: 2.15 2H: 1.65 2H: 0.72 2N: 2.175 2N: 2.188 2N: 2.193	88Fr1
$[C_{13}H_{20}N_2]^{\cdot +}$ 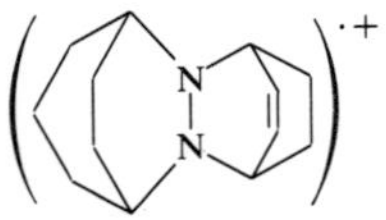	Ox. with $AgNO_3$/ CH_3CN	ESR/ 297 255		2N: 1.601 2H: 0.42 2H: 0.40 2H: 0.14 2N: 1.63	88Fr1
$[C_{13}H_{22}N_2]^{\cdot +}$ 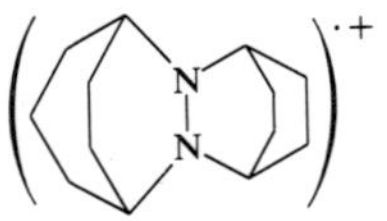	Ox. with $AgNO_3$/ CH_3CN	ESR/ 297		2N: 1.394 6H: 0.28 8H: 0.14 2H: 0.07	88Fr1
$[C_{13}H_{24}D_2N_2]^{\cdot +}$ 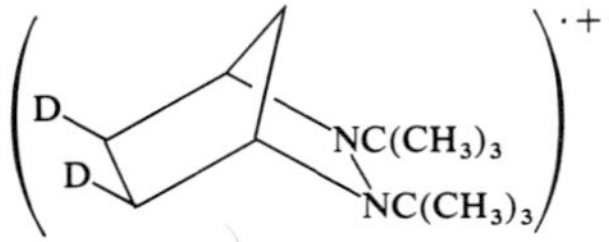	Ox. with $(4\text{-}BrC_6H_4)_3NSbCl_6$/ CH_3CN	ESR/ 233		2N: 1.35	81Ne2

Substance	Generation/ Matrix or Solvent	Method/ T [K]	g-Factor	a-Value [mT]	Ref./ add. Ref.
$[C_{14}H_{22}N_2]^{\cdot +}$ 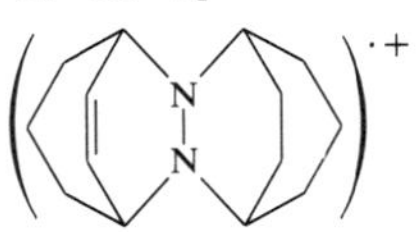	Ox. with $AgNO_3$/ C_3H_7CN	ESR/ 297		2N: 1.35	88Fr1
$[C_{14}H_{24}N_2]^{\cdot +}$ 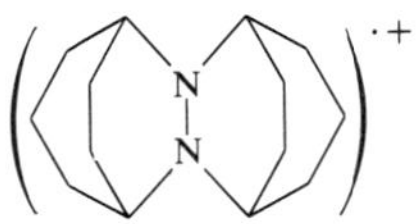	Ox. with $AgNO_3$/ C_3H_7CN	ESR/ 297		2N: 1.30 4H: 0.32 4H: 0.16 4H: 0.08 4H: 0.04	88Fr1
$[C_{14}H_{14}N_2]^{\cdot +}$ 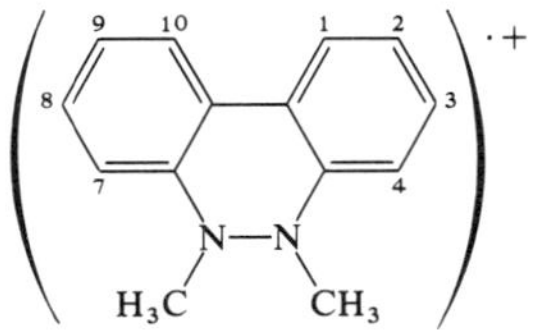	Ox. of the parent hydrazine with lead tetraacetate or benzoyl peroxide/ Toluene: CF_3CO_2H (9:1)	ENDOR/ 260		2N: 0.878 2H(2, 9): 0.236 H(4, 7): 0.173 2 or 4H: 0.044 6H(CH_3): 0.822	86Neu1
$[C_{14}H_{20}N_2O_2]^{\cdot +}$ 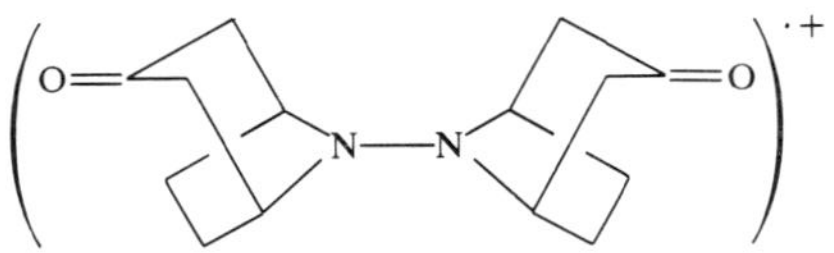	PF_6^- salt/ n-C_3H_7CN	ESR/ 233		2N: 1.36 4H(β): 0.414 8H(γ): 0.138	85Neu3
$[C_{14}H_{24}N_2]^{\cdot +}$ 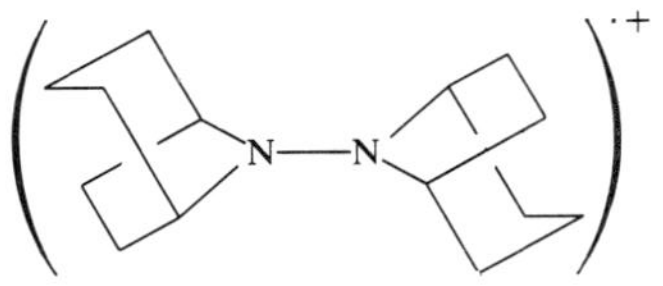	PF_6^- salt/ CH_2Cl_2	ENDOR, ESR/ 240		2N: 1.373 4H(β): 0.449 4H(γ, eq): 0.103 4H(γ, ax): 0.024 2H(δ, eq): 0.343	85Ne3 [32])/ 83Ne2
	CH_3CN [31])	295	2.0033	2N: 1.37	

[31]) From [83Ne2].
[32]) Discussion of structure.

Substance	Generation/ Matrix or Solvent	Method/ T [K]	g-Factor	a-Value [mT]	Ref./ add. Ref.
$[C_{15}H_{34}N_4]^{\cdot+}$ 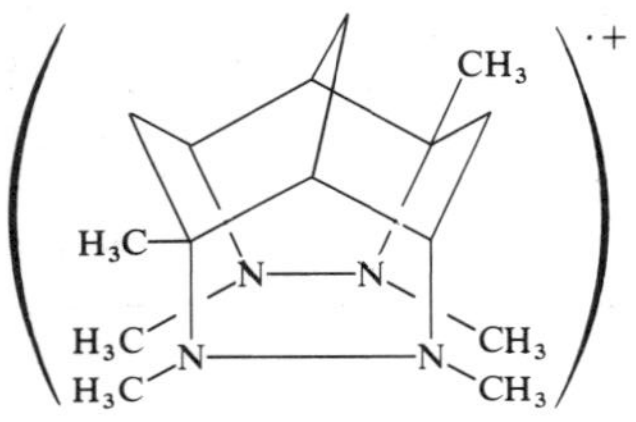	Ox. with $(4\text{-}BrC_6H_4)_3NSbCl_6$/ CH_2Cl_2	ESR/ 195⋯298		2N: 1.5 [33] 6H(CH_3): 1.28 1H: 1.4	86Ne2
$[C_{15}H_{45}N_3Si_5]^{\cdot+}$ 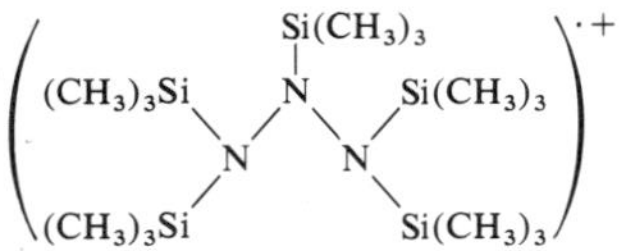	Ox. of parent compound [34] with $AlCl_3$/ CH_2Cl_2	ESR/ –		2N: 1.09 1N: 0.95	78Bo2
$[C_{16}H_{18}N_2]^{\cdot+}$ 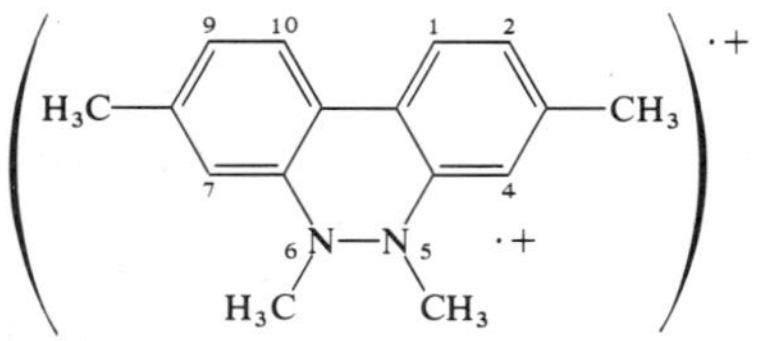	Ox. of the hydrazine with lead tetraacetate or benzoyl peroxide/ Toluene: CF_3CO_2H	EPR, ENDOR/ 260		2N: 0.840 2H(2, 9): −0.248 2H(4, 7): −0.149 6 or 8H: +0.041 6H(5, 6, CH_3): +0.800	86Ne4
$[C_{16}H_{18}N_2]^{\cdot+}$ 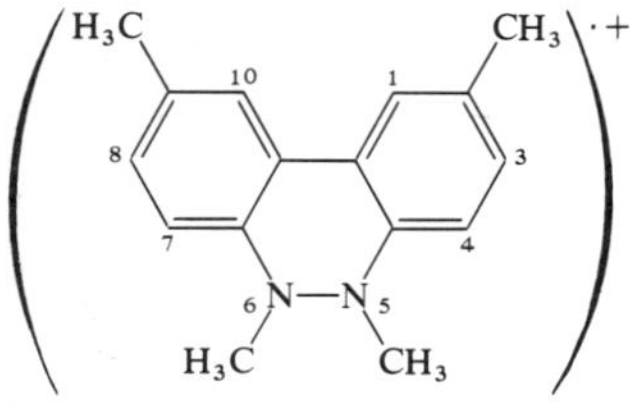	Ox. of the hydrazine with lead tetraacetate or benzoyl peroxide/ CH_2Cl_2:CF_3CO_2H (4:1)	EPR/ 295	2.0033	2N: 0.880 6H(2, 9, CH_3): 0.260 2H(4, 7): 0.162 2 or 4H: 0.060 6H(5, 6, CH_3): 0.781	86Ne4
	Toluene:CF_3CO_2H (9:1)	ENDOR/ 260		6H(2, 9, CH_3): 0.260 2H(4, 7): 0.173 2 or 4H: 0.051 6H(5, 6, CH_3): 0.816	

[33]) Spin localized on one R_4N_2 unit.

[34]) $(CH_3)_3Si$—N(N=N)(Si(CH_3)(CH_3))N—N($Si(CH_3)_3$)$_2$

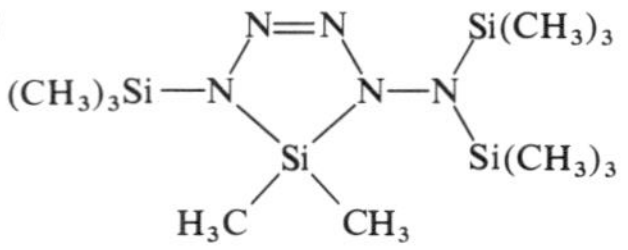

Substance	Generation/ Matrix or Solvent	Method/ T [K]	g-Factor	a-Value [mT]	Ref./ add. Ref.
$[C_{16}H_{28}N_2]^{\cdot+}$ 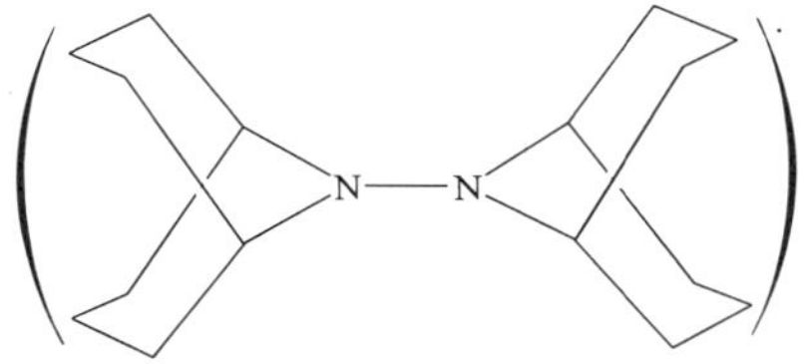	PF_6^- salt/ CH_2Cl_2	ENDOR, ESR/ 243	2.0038	2N: 1.33 4H(β): 0.099 8H(γ, eq): 0.128 8H(γ, ax): 0.059 4H(δ, eq): 0.160 4H(δ, ax): 0.027	81Ge1
$[C_{18}H_{30}N_4]^{\cdot+}$ 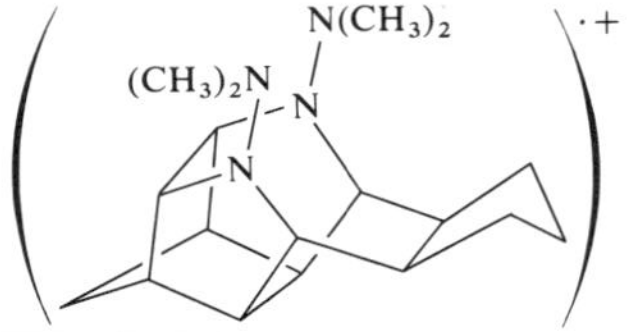	Ox. with $(4\text{-}BrC_6H_4)_3NSbCl_6$/ CH_2Cl_2	ESR/ 195…298		1N: 1.3 [35] 1N: 1.2 6H(CH_3): 1.2	86Ne2
$[C_{18}H_{32}N_2]^{\cdot+}$ 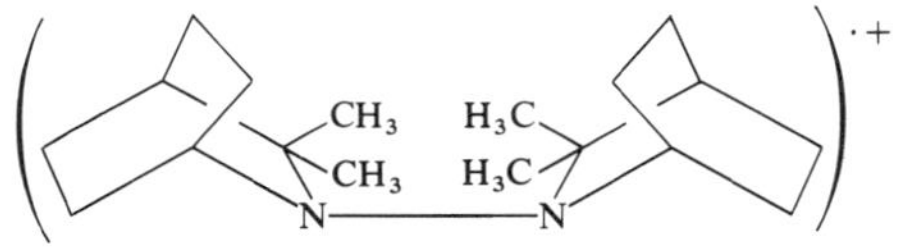	Ox. with $NOPF_6$ or $(BrC_6H_4)_3N^+SbCl_6^-$/ n-C_3H_7CN	ESR/ 295		2N: 1.35 4H(*exo*): 0.25	82Ne1
$[C_{24}H_{16}N_2]^{\cdot+}$ 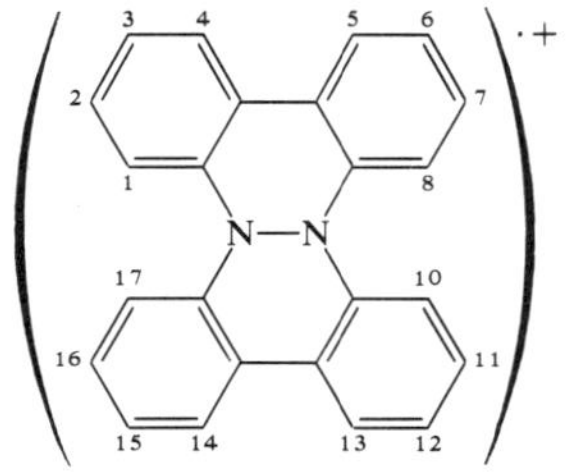	Ox. of the hydrazine with lead tetraacetate or benzoyl peroxide/ Toluene: CF_3CO_2H (9:1)	EPR/ 260	2.0032	2N: 0.720	86Ne4
	Ox. of the hydrazine with lead tetraacetate or benzoyl peroxide/ Toluene: CF_3CO_2H (9:1)	ENDOR/ 260	2.0032	2N: 0.717	86Ne4
(continued)					

[35]) Spin localized on one N_2 unit. Similar hfs for ethylated compound.

Substance	Generation/ Matrix or Solvent	Method/ T [K]	g-Factor	a-Value [mT]	Ref./ add. Ref.
$[C_{24}H_{16}N_2]^{\cdot+}$ *(continued)*	Ox. of the hydrazine with lead tetraacetate or benzoyl peroxide/ Toluene: CF_3CO_2H (9:1)	EPR, ENDOR/ 260		4H(3, 6, 12, 15): 0.144 4H(1, 8, 10, 17): 0.116 4H(2, 7, 11, 16): 0.063 or 0.040 4H(4, 5, 13, 14): 0.040 or 0.063	86Ne4
$[C_{24}H_{18}N_2]^{\cdot+}$ 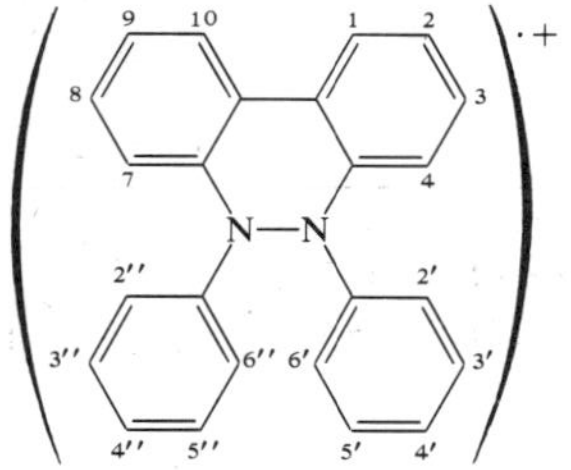	Ox. of the hydrazine with lead tetraacetate or benzoyl peroxide/ Toluene: CF_3CO_2H (9:1)	EPR/ 260	2.0032	2N: 0.890 2H(2, 9): 0.260 2H(4, 7): 0.178 2H: 0.055 8H: 0.040	86Ne4
	Ox. of the hydrazine with lead tetraacetate or benzoyl peroxide/ Toluene: CF_3CO_2H (9:1)	ENDOR/ 260	2.0032	2N: 0.895 2H(2, 9): −0.260 2H(4, 7): −0.180 2H(1, 10): +0.055 or +0.048 2H(3, 8): +0.048 or +0.055 6H(2′, 4′, 6′, 2″, 4″, 6″): −0.040	86Ne4
$[C_{24}H_{20}N_2]^{\cdot+}$ 	Ox. of the parent hydrazine with lead tetraacetate or benzoyl peroxide/ Toluene: CF_3CO_2H (3:2)	ENDOR/ 260		2N: 0.740 [36] 4H(2): −0.167 or −0.138 4H(6): −0.138 or −0.167 8H(3, 5): +0.050 4H(4): −0.191	86Ne4
$[C_{24}H_{20}N_2O_6S_2]^{\cdot+}$ 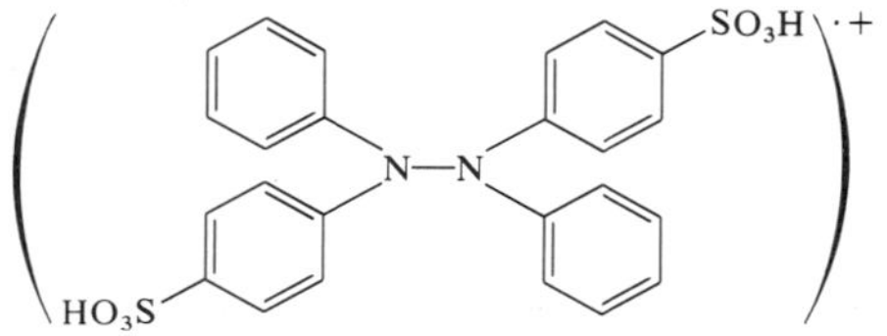	Ox. with 4-$HO_3SC_6H_4NHC_6H_5$ with chromic acid/ 8N H_2SO_4	ESR/ 293		2N: 0.84 [37]	84Pa1

[36]) 4-deuterated species also studied.
[37]) Additional H-splittings 0.17 mT observed.

Substance	Generation/ Matrix or Solvent	Method/ T [K]	g-Factor	a-Value [mT]	Ref./ add. Ref.
$[C_{32}H_{26}N_4]^{\cdot +}$ 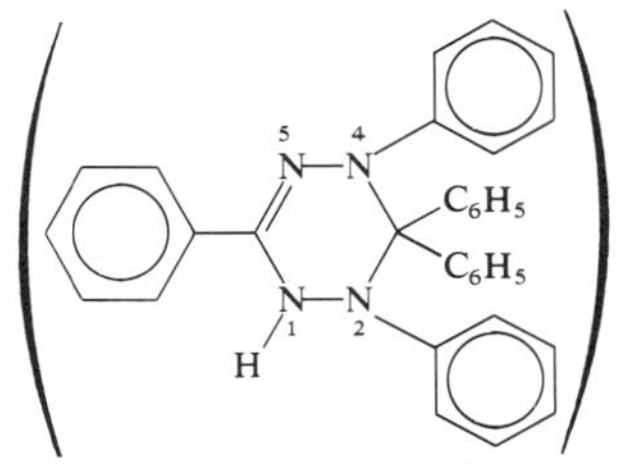	Ox. with $BF_3 \cdot CH_3CH_2O$/ Dioxane	ESR/ 295		N(1): 0.592 N(2): 0.395 N(4): 0.318 N(5): 0.310 H(2, C_6H_5): (+)0.158 H(4, C_6H_5): (−)0.152 H(NH): (−)1.387	83Ne4/ 82Ne3 [38]
$[C_{46}H_{44}N_2O_2]^{\cdot +}$ 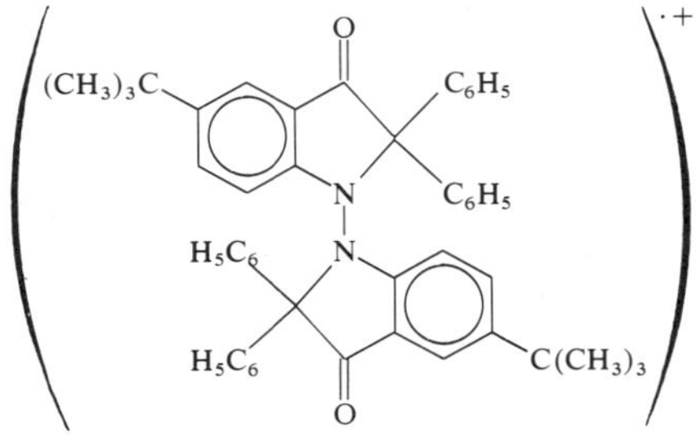	Electrochem. ox./ CH_3CN	ESR/ RT		2H: 0.055 2H: 0.225 2N: 0.585	85Be1
$[C_{56}H_{84}N_2]^{\cdot +}$ 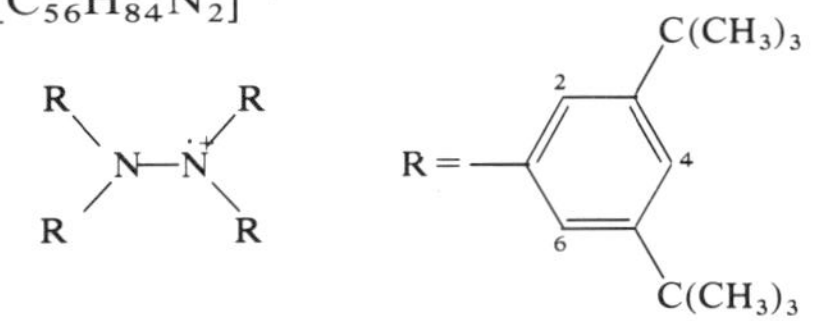	Ox. of the hydrazine with lead tetraacetate or benzoyl peroxide/ Toluene:CF_3CO_2H (3:2)	ENDOR/ 260		2N: 0.724 4H(2): 0.159 or 0.141 4H(6): 0.141 or 0.159 4H(4): 0.194	84Ne1

[38]) Structure of parent compound calculated.

20.4 1-Azaallyl cation radical systems (and relatives)

20.4.1 (Enamine), hydrazone, N-nitrosamine cation radicals

Substance	Generation/ Matrix or Solvent	Method/ T [K]	g-Factor	a-Value [mT]	Ref./ add. Ref.
$[C_6H_{14}N_2O]^{\cdot+}$ 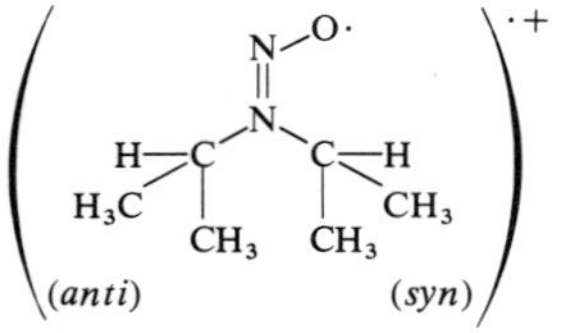	Electrochem. ox./ n-C_3H_7CN	ESR/ 213	2.0026	N(2): 4.55 N(1): 0.838 1H (*syn*): 0.510 6H (*syn*): 0.252 7H (*anti*): 0.076	83Ma2
$[C_8H_{18}N_2]^{\cdot+}$ 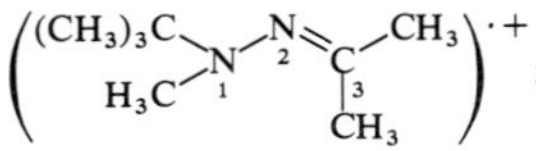 [39]	Ox. with $(p\text{-}BrC_6H_4)_3N^+SbCl_6^-$/ CH_2Cl_2	ESR/ RT(?)		N(1): 1.88 N(2): 1.50 3H(1, CH_3): 1.74 3H(3, CH_3): 0.09	81Be1
$[C_9H_{16}N_2O_2]^{\cdot+}$ 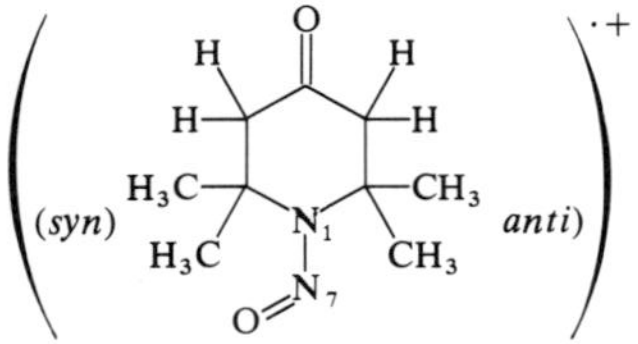	Electrochem. ox./ CH_3CN	ESR/ 243	2.0027	1N(7): 4.60 1N(1): 0.75 2H (*syn*): 0.25 6H(CH_3, *syn*): 0.23 6H(CH_3, *anti*): 0.088 2H (*anti*): 0.15	86Oh1
$[C_9H_{18}N_2O]^{\cdot+}$	Electrochem. ox./ CH_3CN	ESR/ 243	2.0026	N(7): 0.450 N(1): 0.658 8H: 0.282 [40] 6H(CH_3): 0.094	83Ma2/ 82Ma1
		243	2.0025	^{15}N(7): 0.620 [41] N(1): 0.658 8H: 0.282 6H(CH_3): 0.094	

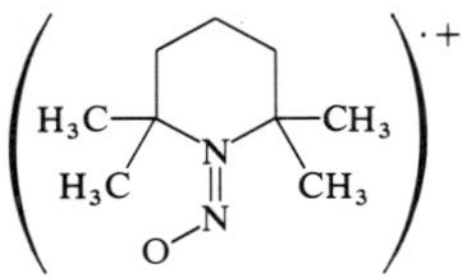

[39]) Twisted structure.
[40]) Rapid ring inversion at 243 K.
[41]) ^{15}N-labelled compound.

Substance	Generation/ Matrix or Solvent	Method/ T [K]	g-Factor	a-Value [mT]	Ref./ add. Ref.
$[C_{10}H_{14}N_2]^{\cdot +}$ [39]	Ox. with $(p\text{-}BrC_6H_4)_3N^+SbCl_6^-$/ CH_2Cl_2	ESR/ RT(?)		N(1): 1.37 N(2): 1.18 3H(3, CH_3): 0.39 6H(1, CH_3): 1.50	81Be1
$[C_{10}H_{20}N_2]^{\cdot +}$ [42]	Ox. with $(p\text{-}BrC_6H_4)_3N^+SBCl_6^-$/ CH_3Cl_2	ESR/ RT(?)		N(1): 1.56 N(2): <0.08 2H(4): 1.62 3H(1, CH_3): 2.00	81Be1
$[C_{12}H_{22}N_2O]^{\cdot +}$ [43]	Electrochem. ox./ $n\text{-}C_3H_7CN$	ESR/ 213	2.0028	N(2): 4.56 N(1): 0.77 1H(bridge): 0.65 2H(ax): 0.29	83Ma2
$[C_{10}H_{22}N_2]^{\cdot +}$ [39]	Ox. with $(p\text{-}BrC_6H_4)_3N^+SbCl_6^-$/ CH_2Cl_2	ESR/ RT(?)		N(1): 1.62 N(2): 1.02 H(3): 0.25 3H(1, CH_3): 1.82	81Be1
$[C_{15}H_{16}N_2]^{\cdot +}$ [39]	Ox. with $(p\text{-}BrC_6H_4)_3N^+SbCl_6^-$/ CH_2Cl_2	ESR/ RT(?)		N(1): 1.38 N(2): 0.75 6H(1, CH_3): 1.38	81Be1

[39]) Twisted structure.
[42]) Planar structure.
[43]) Slow ring reversal. Assignment made by analogy with iminoxy radicals.

Nelsen

Substance	Generation/ Matrix or Solvent	Method/ T [K]	g-Factor	a-Value [mT]	Ref./ add. Ref.
$[C_{15}H_{28}N_2]^{\cdot +}$ [39]	Ox. with $(p\text{-}BrC_6H_4)_3N^+SbCl_6^-$/ CH_2Cl_2	ESR/ RT(?)		N(1): 1.74 N(2): 1.57 H(3): 1.28	81Be1
$[C_{17}H_{26}N_2]^{\cdot +}$ [39]	Ox. with $(p\text{-}BrC_6H_4)_3N^+SbCl_6^-$/ CH_2Cl_2	ESR/ RT(?)		N(1): 1.95 N(2): 1.48 3H(3, CH_3): <0.1	81Be1
$[C_{20}H_{34}N_2]^{\cdot +}$ [42]	Ox. with $(p\text{-}BrC_6H_4)_3N^+SbCl_6^-$/ CH_2Cl_2	ESR/ —		N(1): 1.16 N(2): <0.04 H(3): 0.62 1H(p): 0.42 2H(o): 0.28 3H(1, CH_3): 1.10	81Be1

20.4.2 Amide cation radicals (and R_2N-π-acceptor relatives)

Substance	Generation/ Matrix or Solvent	Method/ T [K]	g-Factor	a-Value [mT]	Ref./ add. Ref.
$[CH_4N_2S]^{\cdot +}$	γ-irr./ Single crystal	ESR/ 77		[44]	84Ve1
$[C_2H_6N_2O_2]^{\cdot +}$ $((CH_3)_2N{-}NO_2)^{\cdot +}$ [45]	UV-phot. at 203 K/ Single crystal	ESR/ 77		1N: 3.385; 0.580; 0.560; is: 1.510 6H(CH_3): 1.185; 1.355; 1.450; is: 1.330	81Mi1

[39]) Twisted structure.
[42]) Planar structure.
[44]) Several species present.
[45]) Previously assigned to $(CH_3)_2N^{\cdot}$.

Substance	Generation/ Matrix or Solvent	Method/ T [K]	g-Factor	a-Value [mT]	Ref./ add. Ref.
$[C_3H_6N_2]^{\cdot+}$ $((CH_3)_2N{-}C{\equiv}N)^{\cdot+}$	γ-irr./ $CFCl_3$	ESR/ 110	2.002 [46]	$6H(CH_3)$: is: 2.2 1N: 2.2; 0.8; 0.8; is: 1.3	87Ch1
$[C_3H_7NO]^{\cdot+}$ $(H{-}C({=}O){-}N(CH_3)_2)^{\cdot+}$	γ-irr./ $CFCl_2CF_2Cl$	ESR/ 82	2.002; 2.0043; 2.0043; is: 2.0035	$6H(CH_3)$: 3.22 N: 3.96 [46]	87Qi1
	$CFCl_3$	81	2.002; 2.0045; 2.0045; is: 2.0037	$6H(CH_3)$: 3.18 N: 3.96 [46]	
	γ-irr./ $CFCl_3$	ESR/ 77	2.002; 2.007; 2.007; is: 2.005	$6H(CH_3)$: 3.2; 3.3; 3.3; is: 3.7 N: 3.8 [46]	82Ra1, 86Ea1
		77	2.002; 2.007; 2.007; is: 2.005	6D: 0.45; 0.50; 0.50; [47] is: 0.48 N: 3.9; 0(2); 0(2); is: 1.3	
$[C_4H_9NO]^{\cdot+}$ $(H_3C{-}C({=}O){-}N(CH_3)_2)^{\cdot+}$	γ-irr./ $CFCl_2CF_2Cl$	ESR/ 82	2.002; 2.0042; 2.0042; is: 2.0035	$6H(CH_3)$: 3.25 N: 3.97 [46]	87Qi1
	$CFCl_3$	81	2.002; 2.0044; 2.0044; is: 2.0036	$6H(CH_3)$: 3.24 N: 4.00 [46]	
	γ-irr./ $CFCl_3$	ESR/ 77	2.002; 2.007; 2.007; is: 2.005	$6H(CH_3)$: 3.2; 3.3; 3.3; is: 3.7 N: 3.8 [46]	82Ra1, 86Ea1

[46]) One principal value.
[47]) d_7-species [86Ea1].

Nelsen

Substance	Generation/ Matrix or Solvent	Method/ T [K]	g-Factor	a-Value [mT]	Ref./ add. Ref.
$[C_5H_9NO]^{\cdot +}$	γ-irr./ $CFCl_3$	ESR/ 77	2.002; 2.007; 2.007; is: 2.005	N: 3.8 (1); 0.0 (4); 0.0 (4); is: 1.3 3H: 3.3 (3) 1H: 6.2 (3) 1H: 3.6 (3)	86Ea1
$[C_5H_9NO_2]^{\cdot +}$ 48)	γ-irr./ $CFCl_3$	ESR/ 77	2.0030; 2.0073; 2.0085; is: 2.0063	N: +0.1 (1); −3.5 (5); −3.5 (5); is: −2.3	86Ea1
$[C_5H_{10}N_2O]^{\cdot +}$	γ-irr./ $CFCl_2CF_2Cl$	ESR/ 84	2.0044	$6H(CH_3)$: 1.84 4H: 3.22	87Qi1/ 87Sy1
	CF_3CCl_3	120	2.0037	$6H(CH_3)$: 1.83 4H: 3.23 2N: 2.49 [49]	
	$CFCl_2CFCl_2$	135	2.0036	$6H(CH_3)$: 1.84 4H: 3.25	
$[C_5H_{12}N_2O]^{\cdot +}$ $((CH_3)_2NC(O)N(CH_3)_2)^{\cdot +}$ 52)	γ-irr./ $CFCl_2CF_2Cl$	ESR/ 84	2.002; 2.0042; 2.0042; is: 2.0035	$6H(CH_3)$: 2.70 N: 4.16 [50]	87Qi1 51)
	$CFCl_3$	81	2.003; 2.0043; 2.0043; is: 2.0039	$6H(CH_3)$: 2.61 N: 4.18 [50]	
	F_3CCCl_3	81	2.003; 2.0043; 2.0043; is: 2.0039	$6H(CH_3)$: 2.83 N: 4.2 [50]	
	$CFCl_3 : CF_2BrCF_2Br$ (1:1)	77	2.003; 2.0040; 2.0040; is: 2.0037	$6H(CH_3)$: 2.80 N: 4.26 [50]	

(continued)

48) C=O O(p_z) combination orbital in *syn* structure postulated; to give some stabilization via in-plane O, O σ-bonding.
49) Tentative, one principal value.
50) One principal value.
51) Discussion of radical structure.
52) Alternative assignment $(CH_3)_2N^{\cdot}$; see also [87Sy1] and [87Qi1].

Substance	Generation/ Matrix or Solvent	Method/ T [K]	g-Factor	a-Value [mT]	Ref./ add. Ref.
$[C_5H_{12}N_2O]^{\cdot +}$ *(continued)*	γ-irr./ $CFCl_3$	ESR/ 77	2.002; 2.005; 2.005; is: 2.004	N: 4.1; ≈0; ≈0; is: 1.37; $6H(CH_3)$: 2.7	86Sy1/ 87Sy1
$[C_6H_9NO]^{\cdot +}$ 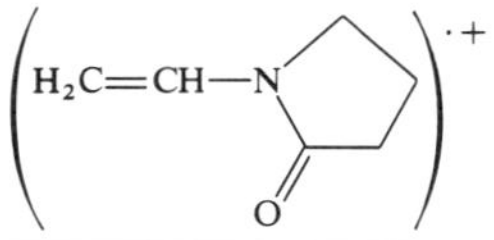	γ-irr./ $CFCl_3$	ESR/ 77		N: 1.7; 0; 0; is: 0.57; 2H: is: 2.8; 1H: 1.7, 1.3 [53]	84Ea2
$[C_6H_{11}NO]^{\cdot +}$ 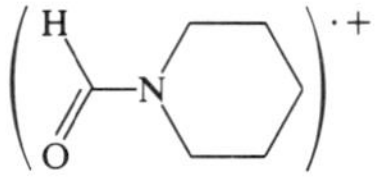	γ-irr./ $CFCl_3$	ESR/ 77	2.002; 2.007; 2.007; is: 2.0053	N: 3.6 (1); 0.0 (4); 0.0 (4); is: 1.2; 4H: 2.5 (3)	86Ea1
$[C_6H_{18}N_3OP]^{\cdot +}$ $([(CH_3)_2N]_3PO)^{\cdot +}$	γ-irr./ $CFCl_3$	ESR/ 77	2.007··· 2.002	$6H(CH_3)$: 2.75; P: 1.6; 1N: 3.6; 0.0 (4); 0.0 (4); is: 1.2	87Sy2 [54]

20.5 Amino substituted unsaturated cation radicals

20.5.1 "Amino terminated" systems

20.5.1.1 1,4-Enediamines and 2,3-aza analogues

Substance	Generation/ Matrix or Solvent	Method/ T [K]	g-Factor	a-Value [mT]	Ref./ add. Ref.
$[C_4H_{12}N_4]^{\cdot +}$ 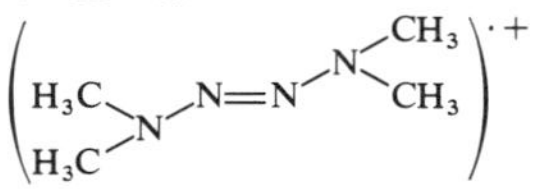	–/ CF_3CO_2H: CH_3CN	ESR/ RT	2.0040(5)	2N(1, 4): 1.085; 2N(2, 3): 0.115; $6H(CH_3)$: 1.040; $6H(CH_3)$: 1.080	78Ma1
$[C_6H_{14}N_2]^{\cdot +}$ 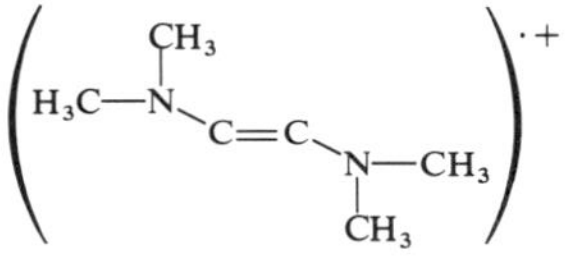	Ox. with $AlCl_3$/ CH_2Cl_2	–/ 293		2N: 0.813; 2H: 0.425; $6H(CH_3$, *anti*): 0.813; $6H(CH_3$, *syn*): 0.695	84Bo1

[53]) Principal values. Author's interpretation. Original table lists 3.2, 1.7, 1.3 mT.

[54]) Complex annealing behaviour. Change localized at one $(CH_3)_2N$ group. Perdeuterated compound also studied.

Substance	Generation/ Matrix or Solvent	Method/ T [K]	g-Factor	a-Value [mT]	Ref./ add. Ref.
$[C_{10}H_{22}N_2]^{\cdot+}$ $((H_5C_2)_2N{-}C{=}C{-}N(C_2H_5)_2)^{\cdot+}$ 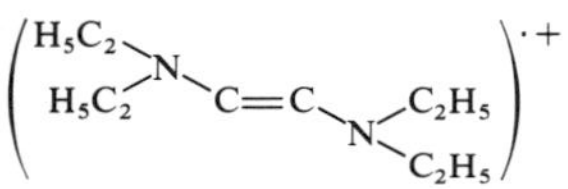	Ox. with $AlCl_3$/ CH_2Cl_2	–/ 293		2N: 0.800 2H: 0.400 4H(CH_2, *anti*): 0.530 4H(CH_2, *syn*): 0.400	84Bo1
$[C_{10}H_{22}N_2]^{\cdot+}$	Phot. of DTBO + CF_3CO_2H (trace) + parent compound [1])/	ESR/			83Al2
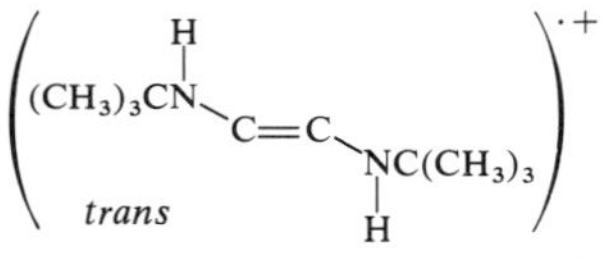 $((CH_3)_3CNH{-}C{=}C{-}NHC(CH_3)_3)^{\cdot+}$ *trans*	*trans* isomer	293	2.0032	N(1, 4): 0.735 H(2, 3): 0.485 H(NH): 0.754	
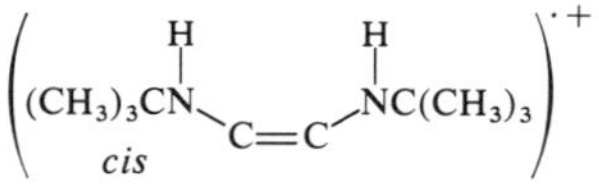 $((CH_3)_3CNH{-}C{=}C{-}NHC(CH_3)_3)^{\cdot+}$ *cis*	*cis* isomer	293	2.0033	N(1, 4): 0.657 H(2, 3): 0.569 H(NH): 0.694	
$[C_{12}H_{36}N_4Si_4]^{\cdot+}$ $([(CH_3)_3Si]_2\overset{1}{N}{-}\overset{2}{N}{=}\overset{3}{N}{-}\overset{4}{N}[Si(CH_3)_3]_2)^{\cdot+}$	Ox. with $AlCl_3$/ CH_2Cl_2	–/ 295		2N(1, 4): 0.69 2N(2, 3): 0.24 ^{29}Si: 0.385 36H(CH_3): 0.015	78Bo2
$[C_{16}H_{28}N_4]^{\cdot+}$ 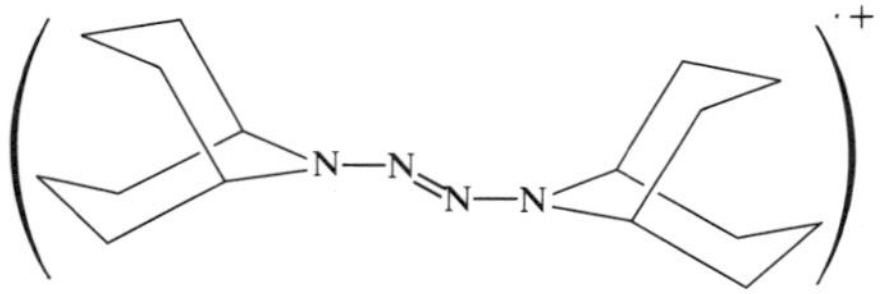 (N–N=N–N)$^{\cdot+}$	Ox. with (4-$BrC_6H_4)_3NSbCl_6$/ CH_3CN	ESR/ 295		2N: 1.11	80Ne1

[1]) $(CH_3)_3CN{=}CH{-}CH{=}NC(CH_3)_3$

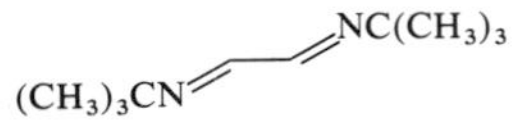

Substance	Generation/ Matrix or Solvent	Method/ T [K]	g-Factor	a-Value [mT]	Ref./ add. Ref.

20.5.1.2 Anilines (aminobenzenes)

For p-phenylenediamines, see 20.5.1.3; for other hydrazoaromatics, see 20.5.2.4

Substance	Generation/ Matrix or Solvent	Method/ T [K]	g-Factor	a-Value [mT]	Ref./ add. Ref.
$[C_5H_6N_2]^{\cdot+}$ 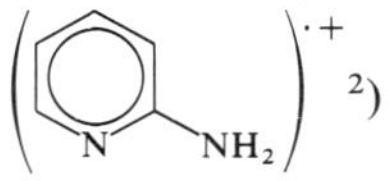 [2]	γ-irr./ $CFCl_3$	ESR/ 77	2.002 (1)	N: 2.4; ≈0.0; ≈0.0; is: 0.8 2H: ≈0.8	84Ra1
$[C_8H_8NO_3]^{\cdot+}$ 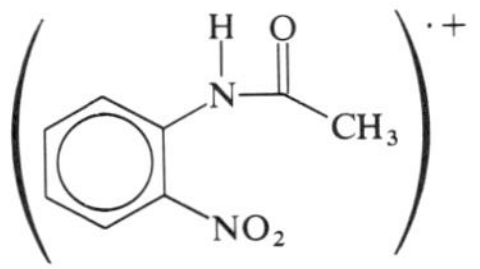	γ-irr./ $CFCl_3$	ESR/ 77		2H: 0.8 1H: 0.6 N: 1.8; ≈0.0; ≈0.0; is: 0.6	85Ra1
$[C_8H_9NO]^{\cdot+}$ 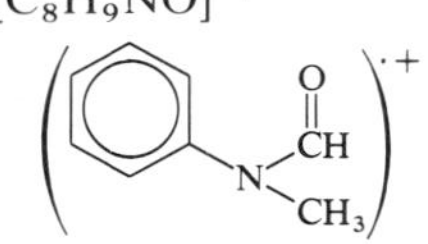	γ-irr./ $CFCl_3$	ESR/ 77		1H(ring): 0.8 $3H(CH_3)$: 0.8 N: 1.8; ≈0; ≈0; is: 0.6	85Ra1
$[C_8H_{11}N]^{\cdot+}$ 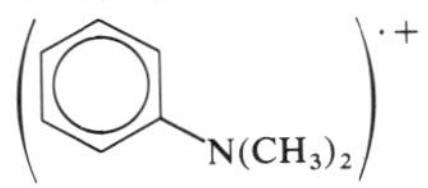	γ-irr./ $CFCl_3$	ESR/ 77		3H(ring): 0.5 $6H(CH_3)$: 1.2 N: 3.0; ≈0; ≈0; is: 1.0	85Ra1
$[C_9H_{13}NO]^{\cdot+}$ 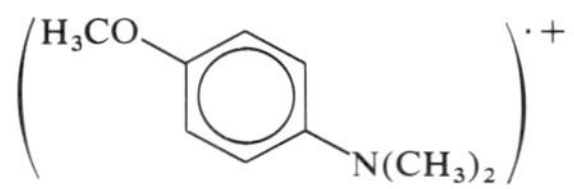	Ox. with $(BrC_6H_4)_3NSbCl_6$/ CH_2Cl_2	ESR/ RT		N: 1.00 H: 0.425 $3H(OCH_3)$: 0.185	79Wo1/ 85Sa3 [3], 86Sa4 [4]

[2]) Alternative assignment

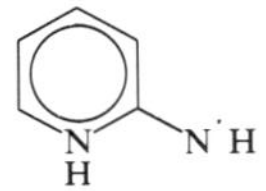

[3]) Detected in catalase catalyzed enzymatic deamination of the amine, as was the 4-methyl analogue, starting from the 4-methylamine.

[4]) Detected in the myeloperoxidase/Cl^- enzymatic system, but not in absence of Cl^-.

Substance	Generation/ Matrix or Solvent	Method/ T [K]	g-Factor	a-Value [mT]	Ref./ add. Ref.
$[C_{14}H_{23}N]^{\cdot+}$ 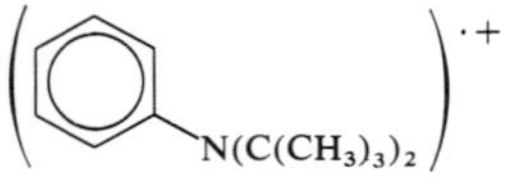 $(C_6H_5N(C(CH_3)_3)_2)^{\cdot+}$	Ox. with $(4\text{-}BrC_6H_4)_3NSbCl_6$/ CH_2Cl_2	ESR/ RT		N: 1.96 [5]	79Wo1
$[C_{15}H_{25}NO]^{\cdot+}$ 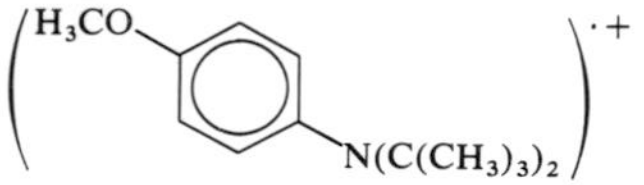 $(4\text{-}H_3CO\text{-}C_6H_4N(C(CH_3)_3)_2)^{\cdot+}$	Ox. with $(4\text{-}BrC_6H_4)_3NSbCl_6$/ CH_2Cl_2	ESR/ RT		N: 1.27 H: 0.295 $3H(OCH_3)$: 0.155	79Wo1
$[C_{16}H_{25}NO]^{\cdot+}$ 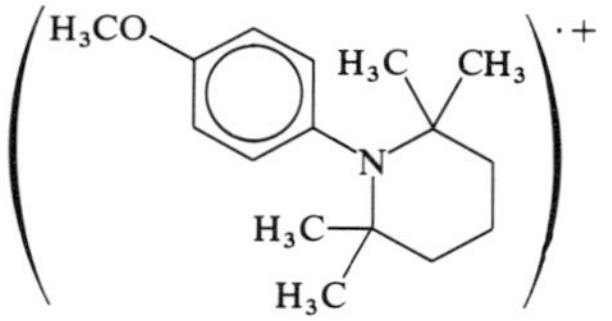 (1-(4-methoxyphenyl)-2,2,6,6-tetramethylpiperidine; H_3CO, H_3C, CH_3, N, H_3C, H_3C)$^{\cdot+}$	Ox. with $(4\text{-}BrC_6H_4)_3NSbCl_6$/ CH_2Cl_2	ESR/ RT		N: 1.14 H: 0.36 $3H(OCH_3)$: 0.17	79Wo1
$[C_{16}H_{27}N]^{\cdot+}$ 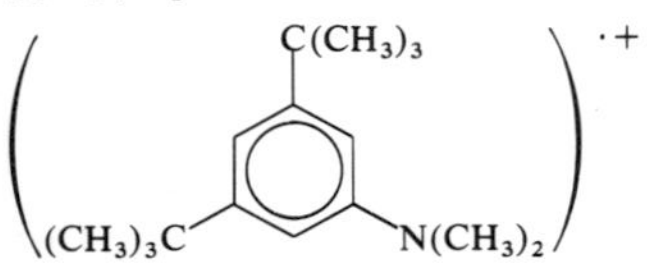 $(3,5\text{-}((CH_3)_3C)_2C_6H_3N(CH_3)_2)^{\cdot+}$	Ox. with $(4\text{-}BrC_6H_4)_3NSbCl_6$/ CH_2Cl_2	ESR/ RT		N: 1.06 H: 0.53 H: 0.85	79Wo1
$[C_{18}H_{26}N_2O_2]^{\cdot+}$ $(4\text{-}H_3CO\text{-}C_6H_4N(CH_3)_2)_2^{\cdot+}$ 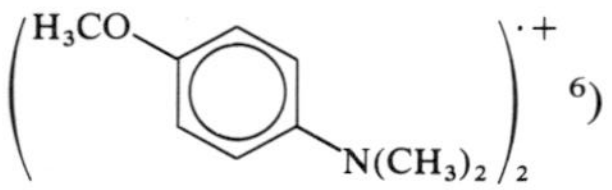 [6]	Ox. with $Tl(OOCCH_3)_4$/ $CHCl_3$ or CH_3NO_2	ESR/ RT		1N: 0.679 9H: 0.679 4H: 0.20	85Cl1
$[C_{20}H_{27}N]^{\cdot+}$ 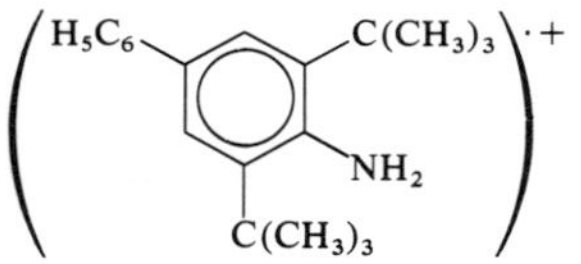 (H_5C_6, $C(CH_3)_3$, NH_2, $C(CH_3)_3$ substituted benzene)$^{\cdot+}$	Electrochem. ox. or ox. with $AgClO_4:I_2$/ CH_3CN	ESR/ RT	2.00268	hfs not analyzed	83Sp1

[5]) Additional ring proton splittings.

[6]) Dimer cation radical. Blue species with $\lambda_{max} \approx 608$ nm. Monomer (red species) has $\lambda_{max} \approx 482$ nm. Authors concluded from splittings an N, O 3e-σ-bonded species.

Substance	Generation/ Matrix or Solvent	Method/ T [K]	g-Factor	a-Value [mT]	Ref./ add. Ref.
$[C_{20}H_{29}NO]^{\cdot +}$	Electrochem. ox. or ox. with $AgClO_4 : I_2$/ CH_3CN	ESR/ RT	2.00277	hfs not analyzed	83Sp1
$[C_{20}H_{35}N]^{\cdot +}$	–/ CH_2Cl_2	ESR/ RT		N: 1.11 H: 0.48 H: 0.83	79Wo1
$[C_{21}H_{37}N]^{\cdot +}$	–/ CH_2Cl_2	ESR/ RT		N: 1.33 H: 0.43 H: 0.66	79Wo1
$[C_{22}H_{39}N]^{\cdot +}$	Ox. with $(4\text{-}BrC_6H_4)_3NSbCl_6$/ CH_2Cl_2	ESR/ RT		N: 1.845 H: 0.240 H: 0.205	79Wo1
$[C_{23}H_{39}N]^{\cdot +}$	Ox. with $(4\text{-}BrC_6H_4)_3NSbCl_6$/ CH_2Cl_2	ESR/ RT		N: 1.42 H: 0.39 H: 0.57	79Wo1

Substance	Generation/ Matrix or Solvent	Method/ T [K]	g-Factor	a-Value [mT]	Ref./ add. Ref.
$[C_{24}H_{25}NO]^{\cdot+}$ 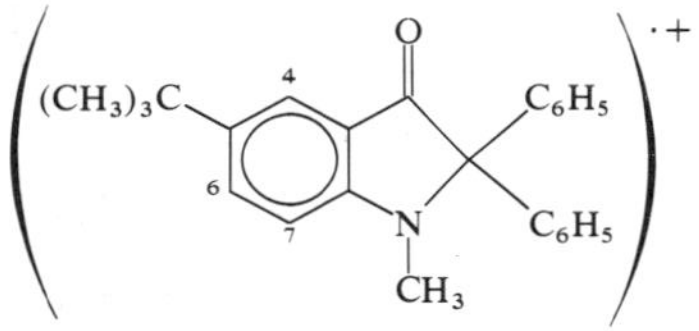	Electrochem. ox./ CH_3CN	ESR/ RT		$3H(NCH_3)$: 0.58 2H(4, 6): 0.05 H(7): 0.237 N: 0.482	85Be1
$[C_{29}H_{25}NO_6]^{\cdot+}$ 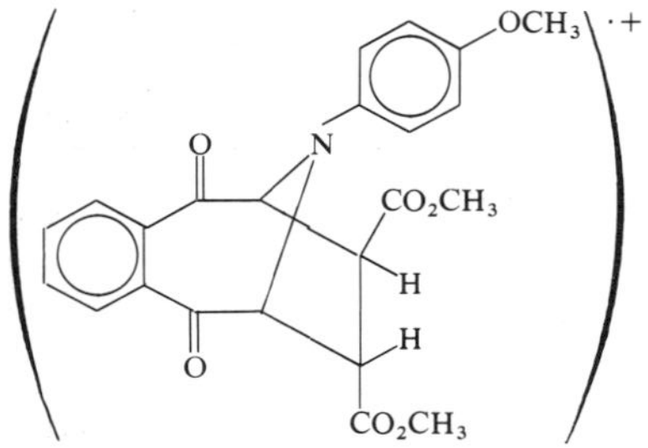	Phot. in presence of O_2/ CF_3CO_2H	ESR/ —	2.0032 [7])	hfs not analyzed	83Ma1
$[C_{30}H_{37}N]^{\cdot+}$ 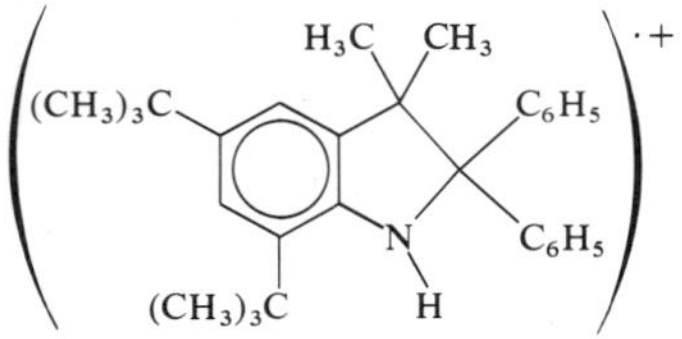	Electrochem. ox./ CH_3CN	ESR/ RT		H(NH): 0.825 N: 0.775	85Be1

20.5.1.3 Phenylenediamines (diaminobenzenes)

Substance	Generation/ Matrix or Solvent	Method/ T [K]	g-Factor	a-Value [mT]	Ref./ add. Ref.
$[C_6H_8N_2]^{\cdot+}$ 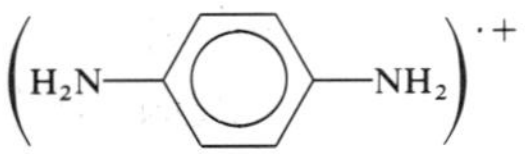	ClO_4^- salt/ CH_3OH	ESR/ 293		2N: 0.529 4H(ring): 0.213 $4H(NH_2)$: 0.588	84Gr1

[7]) 5 analogues gave similar spectra, all $g = 2.0032$.

Substance	Generation/ Matrix or Solvent	Method/ T [K]	g-Factor	a-Value [mT]	Ref./ add. Ref.
$[C_6H_6Cl_2N_2]^{\cdot +}$	Electrochem. ox./ DMF	ESR/ 293		2N: 0.552 4H(NH_2): 0.643 2H(ring): 0.190	84Gr1
$[C_6H_4Cl_4N_2]^{\cdot +}$	Electrochem. ox./ CH_3CN	ESR/ 293		2N: 0.515 4H(NH_2): 0.576 [8])	84Gr1
$[C_6H_{12}N_6]^{\cdot +}$	Ox. with $Pb(OOCCH_3)_4$/ DMSO	ESR/ 293	2.0046	4N: 0.117 2N: 0.809 12H: 0.875	86Fi1
$[C_8H_{12}N_2]^{\cdot +}$	ClO_4^- salt/ CH_3OH	ESR/ 293		N(7): 0.762 N(8): 0.478 H(2, 6): 0.265 H(3, 5): 0.146 6H(CH_3): 0.775 2H(NH_2): 0.561	67La1
	ClO_4^- salt/ CH_3OH	ESR/ 293		N(7): 0.778 N(8): 0.512 H(2, 6): 0.273 H(3, 5): 0.205 6H(CH_3): 0.749 2H(NH_2): 0.478	84Gr1

(continued)

[8]) Deuterated species 4D(ND_2): 0.088 mT.

Substance	Generation/ Matrix or Solvent	Method/ T [K]	g-Factor	a-Value [mT]	Ref./ add. Ref.
$[C_8H_{12}N_2]^{\cdot+}$ (*continued*)	Electrochem. ox./ CH_3CN	ESR/ RT(?)		N(7): 0.719 N(8): 0.453 H(2, 6): 0.167 H(3, 5): 0.111 6H(CH_3): 0.739 2H(NH_2): 0.469	74Ya1
$[C_8H_{10}D_2N_2]^{\cdot+}$ 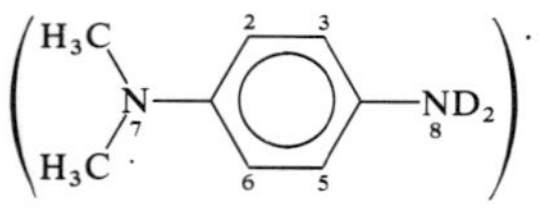	–/ CH_3OD	ESR/ 293		N(7): 0.778 N(8): 0.512 6H(CH_3): 0.749 2H(2, 6): 0.273 2H(3, 5): 0.205	84Gr1
$[C_8H_6D_6N_2]^{\cdot+}$ 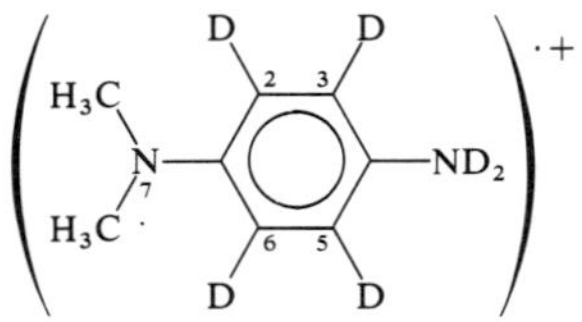	–/ CH_3OD	ESR/ 293		N(7): 0.778 N(8): 0.512 6H(CH_3): 0.749 2D(2, 6): 0.042 2D(3, 5): 0.032	84Gr1
$[C_{10}H_{16}N_2]^{\cdot+}$ 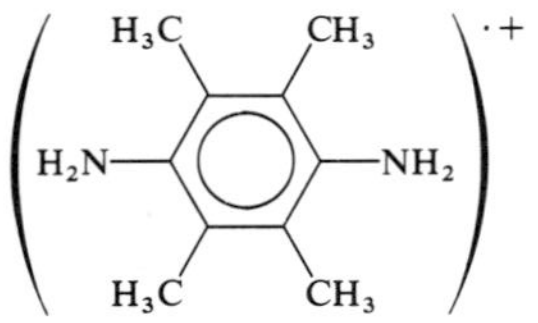	ClO_4^- salt/ CH_3OH	ESR/ 293		2N: 0.465 4H(NH_2): 0.509 12H(CH_3): 0.155	84Gr1
$[C_{10}H_{16}N_2]^{\cdot+}$ 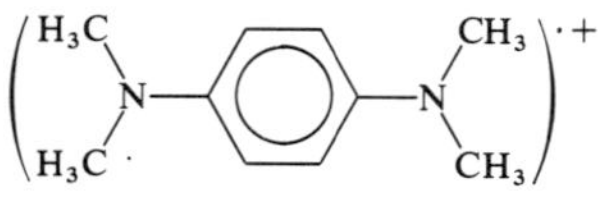	ClO_4^- salt/ CH_3OH	ESR/ 293		2N: 0.705 12H(CH_3): 0.677 4H(CH): 0.199	84Gr1/ 86Ka2 [9])
(*continued*)	PF_6^- salt/ CH_3OH	ESR/ 323		N: 0.702 H(CH_3): 0.668 H(ring): 0.197	83Fu1 [10])

[9]) Line broadening study, electrochem. ox.
[10]) da/dT also given.

Substance	Generation/ Matrix or Solvent	Method/ T [K]	g-Factor	a-Value [mT]	Ref./ add. Ref.
$[C_{10}H_{16}N_2]^{\cdot +}$ *(continued)*	AsF_6^- salt/ CH_3OH	ESR/ 313		N: 0.703 $H(CH_3)$: 0.668 H(ring): 0.197	83Fu1 [10])
	SbF_6^- salt/ CH_3OH	ESR/ 313		N: 0.708 $H(CH_3)$: 0.674 H(ring): 0.196	83Fu1 [10])
	BF_4^- salt/ CH_3OH	ESR/ 313		N: 0.707 $N(CH_3)$: 0.671 H(ring): 0.198	83Fu1 [10])
	ClO_4^- salt/ CH_3OH	ESR/ 313		N: 0.700 $H(CH_3)$: 0.666 H(ring): 0.197	83Fu1 [10])
$[C_{10}H_{16}N_2]^{\cdot +}$ $(H_2N_8-C_6H_4-N_7(CH_2CH_3)_2)^{\cdot +}$	ClO_4^- salt/ CH_3OH	ESR/ 293		N(7): 0.760 N(8): 0.420 H(2, 6): 0.274 H(3, 5): 0.208 $4H(NCH_2)$: 0.450 $2H(NH_2)$: 0.476	84Gr1
$[C_{14}H_{12}N_2]^{\cdot +}$ 9,10-diaminoanthracene (NH_2, NH_2)	Electrochem. ox./ CH_3CN	ESR/ 293		2N: 0.392 $4H(NH_2)$: 0.392 8H: 0.083	84Gr1
$[C_{14}H_{24}N_2]^{\cdot +}$ $((H_3CH_2C)_2N-C_6H_4-N(CH_2CH_3)_2)^{\cdot +}$	ClO_4^- salt/ CH_3OH	ESR/ 293		2N: 0.688 $8H(NH_2)$: 0.369 4H(CH): 0.185	84Gr1

[10]) da/dT also given.

Substance	Generation/ Matrix or Solvent	Method/ T [K]	g-Factor	a-Value [mT]	Ref./ add. Ref.
$[C_{48}H_{44}N_2O_4]^{\cdot+}$	Electrochem. CH_3CN	ESR/ RT		H(NH): 0.675 N: 0.637 N: 0.600	85Be1
For further radicals of similar structure, see [83Ga1].					
20.5.1.4 Benzidines (diaminobiphenyls)					
$[C_{12}H_{12}N_2]^{\cdot+}$	Electrochem. ox./ CH_3CN	ESR/ 293		2N: 0.339 4H(2): 0.144 4H(3): 0.110 4H(NH_2): 0.352	84Gr1/ 65Sm1
$[C_{14}H_{14}N_2]^{\cdot+}$ [11]	Electrochem. ox. of parent compound [12])/ CH_3CN	ESR/ 298	2.0025	2N: 0.375 [13]) 6H(CH_3): 0.375 4H: 0.25 4H: 0.125	86Sa2
$[C_{16}H_{18}N_2]^{\cdot+}$ [11]	Electrochem. ox. of parent compound [14])/ CH_3CN	ESR/ 298	2.0026	2N: 0.385 4H: 0.255 2H: 0.25 2H: 0.185 4H: 0.135	86Sa2
$[C_{16}H_{10}D_8N_2]^{\cdot+}$ [11]	Electrochem. ox. of parent compound [15])/ CH_3CN	ESR/ 298	2.0025	2N: 0.39 2H: 0.245 2H: 0.195	86Sa2

[11]) Author's assignment.

[12]) 2-nitrophenyl–SN(CH_3)(C_6H_5)

[13]) Deuterated compound also studied.

[14]) 2-nitrophenyl–SN(C_2H_5)(C_6H_5)

[15]) 2-nitrophenyl–SN(C_2H_5)(C_6D_5)

Substance	Generation/ Matrix or Solvent	Method/ T [K]	g-Factor	a-Value [mT]	Ref./ add. Ref.
$[C_{16}H_{20}N_2]^{\cdot +}$ 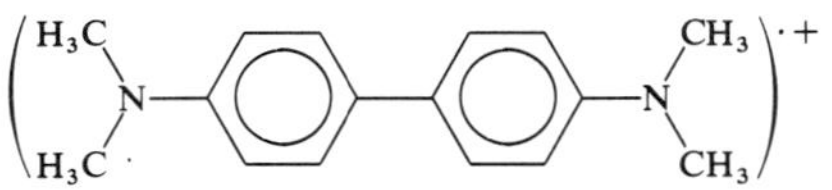	Electrochem. ox./ CH_3CN	ESR/ 293		2N: 0.488 12H(CH_3): 0.470 4H(2): 0.165 4H(3): 0.076	84Gr1/ 85Sa3 [16])
	Photochem. ox./ $(CH_3)_4N^+$ and Na^+ dodecyl- sulfate micelles: H_2O	ESR/ 77 295	 2.0020	 2N: ≈0.48 12H(CH_3): ≈0.48 8H: ≈0.085	84Sz1, 81Na1/ 82Na1 [17])
	Photochem. ox./ $(n\text{-}C_{18}H_{37})_2N(CH_3)_2^+Cl^-$ vesicles: H_2O dihexadecylphosphate vesicles: H_2O	ESR/ 295 77 77	 2.004 2.004 2.004	 ΔHpp: ≈2.8 ΔHpp: ≈2.8 ΔHpp: ≈2.8	83Li1
	$(CH_3)_2NC_6H_5$ + benzoyl peroxide in acetic acid/ CH_3CO_2H	ESR/ 295		2N: 0.489 12H: 0.476 4H: 0.163 4H: 0.079	85Ti1
$[C_{20}H_{28}N_2]^{\cdot +}$ 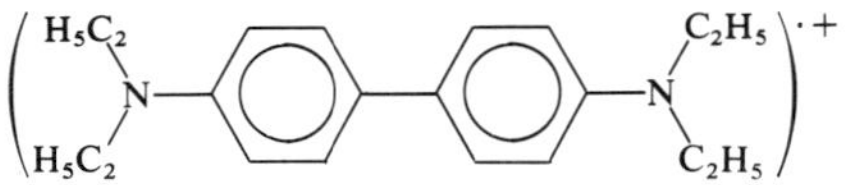	$(C_2H_5)_2NC_6H_5$ + benzoyl peroxide/ CH_3CO_2H	ESR/ 295		2N: 0.792 H(NCH_2): 0.76 4H: 0.088	85Ti1
$[C_{24}H_{36}N_2]^{\cdot +}$ 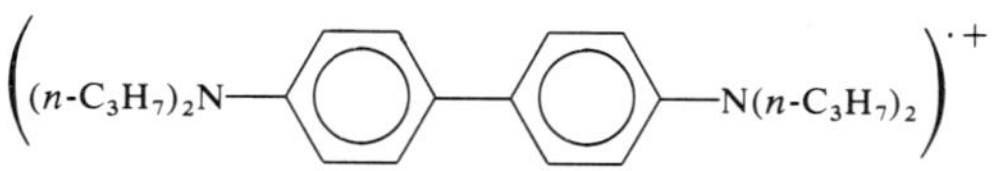	$(n\text{-}C_3H_7)_2NC_6H_5$ + benzoyl peroxide/ CH_3CO_2H	ESR/ 295		2N: 1.320 H(NCH_2): 0.220	85Ti1

[16]) Detected in catalase catalyzed enzymatic demethylation of N,N-dimethylaniline.
[17]) Triton X-100/H_2O and dodecyltrimethylammonium chloride/H_2O micelles.

Substance	Generation/ Matrix or Solvent	Method/ T [K]	g-Factor	a-Value [mT]	Ref./ add. Ref.
$[C_{22}H_{32}N_2]^{\cdot+}$ 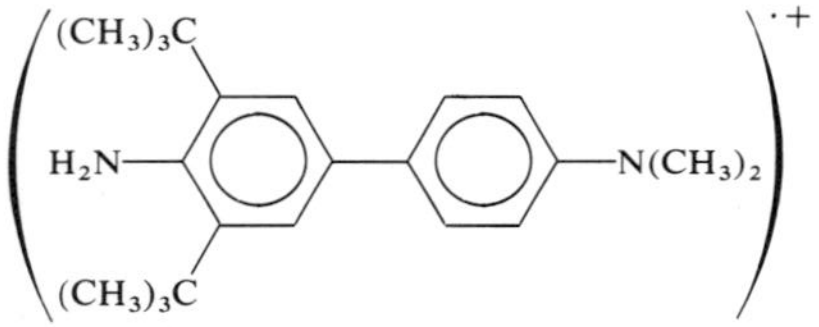	Electrochem. ox./ CH_3CN	ESR/ RT	2.00277	hfs not analyzed	83Sp1
$[C_{26}H_{22}N_2]^{\cdot+}$ 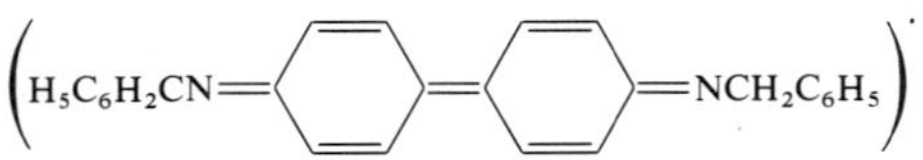 [18]	Electrochem. ox. of parent compound [19])/ CH_3CN	ESR/ 298	2.0027	2N: 0.37 6H: 0.25 2H: 0.195 4H: 0.125	86Sa2
$[C_{26}H_{24}N_2O_6S_2]^{\cdot+}$ 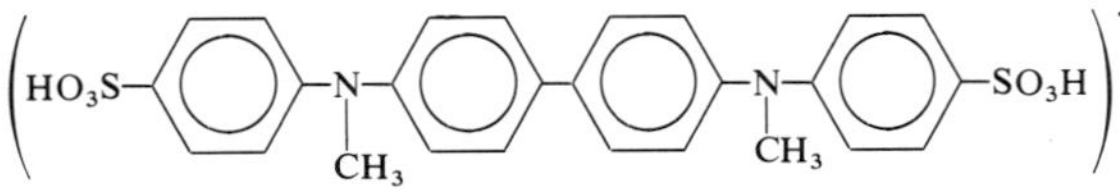 +	Ox. of 4-$HO_3SC_6H_4N(CH_3)C_6H_5$/ 8M H_2SO_4	ESR/ 293		2N: ≈0.4 [20]) 6H: ≈0.4	84Pa1
$[C_{30}H_{26}N_4O_4S_2]^{\cdot+}$ 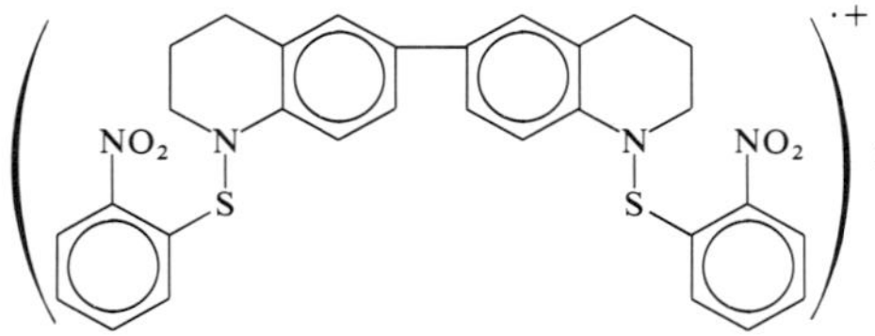 [21]	Electrochem. ox. of parent compound [22])/ CH_3CN	ESR/ 298	2.0026	2N: 0.50 2H: 0.50 4H: 0.26 4H: 0.13	86Sa3

[18]) Author's assignment.

[19])

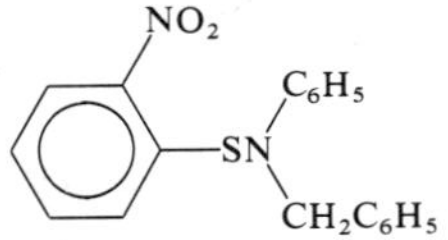

[20]) Seven lines assigned as central point at 2N, 6H multiplet of species shown.

[21]) 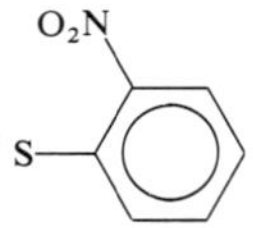 group not lost as in other cases because the cation radical crystallizes out.

[22])

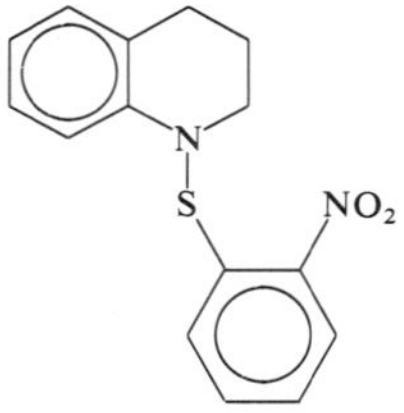

Substance	Generation/ Matrix or Solvent	Method/ T [K]	g-Factor	a-Value [mT]	Ref./ add. Ref.
$[C_{30}H_{24}N_2O_2]^{\cdot +}$ 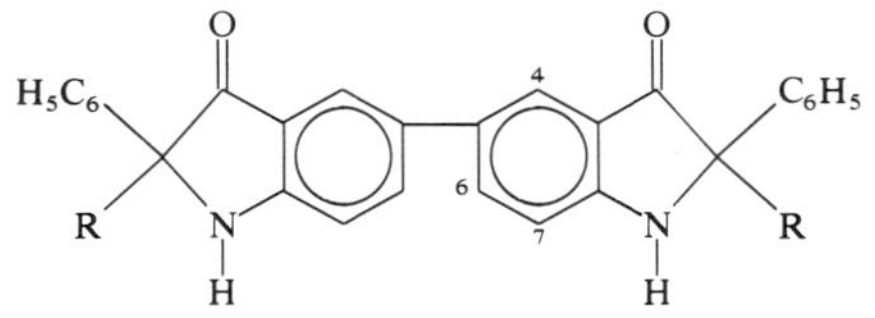 **D**, $R = CH_3$	Electrochem. ox./ CH_3CN	ESR/ RT		2H(NH): 0.400 4H(4, 6): 0.040 2H(7): 0.162 2N: 0.400 [23])	85Be1
$[C_{36}H_{36}N_2O_2]^{\cdot +}$ **D**, $R = C(CH_3)_3$	Electrochem. ox./ CH_3CN	ESR/ RT		2H(NH): 0.398 4H(4, 6): 0.045 2H(7): 0.153 2N: 0.393	85Be1
$[C_{40}H_{28}N_2O_2]^{\cdot +}$ **D**, $R = C_6H_5$	Electrochem. ox./ CH_3CN	ESR/ RT		2H(NH): 0.390 4H(4, 6): 0.045 2H(7): 0.160 2N: 0.393	85Be1
$[C_{42}H_{36}N_2O_2]^{\cdot +}$ 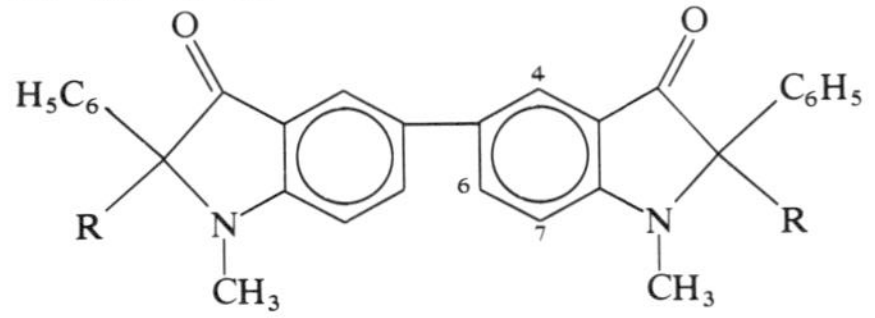 **E**, $R = CH_2C_6H_5$	Electrochem. ox./ CH_3CN	ESR/ RT		6H(NCH_3): 0.500 2H(7): 0.100 2N: 0.475	85Be1
$[C_{40}H_{32}N_2O_2]^{\cdot +}$ **E**, $R = C_6H_5$	Electrochem. ox./ CH_3CN	ESR/ RT		6H(NCH_3): 0.457 4H(4, 6): 0.037 2H(7): 0.128 2N: 0.457	85Be1
$[C_{42}H_{32}N_2O_2]^{\cdot +}$ **D**, $R = CH_2C_6H_5$	Electrochem. ox./ CH_3CN	ESR/ RT		2H(NH): 0.374 4H(4, 6): 0.045 2H(7): 0.140 2N: 0.374	85Be1

[23]) Same splittings reported for the ethylated compound.

Substance	Generation/ Matrix or Solvent	Method/ T [K]	g-Factor	a-Value [mT]	Ref./ add. Ref.
$[C_{48}H_{44}N_2O_4]^{\cdot +}$	Electrochem. ox./ CH_3CN	ESR/ RT		2H(NH): 0.355 4H(4, 6): 0.032 2N: 0.360	85Be1
$[C_{60}H_{48}N_4O_2]^{\cdot +}$	Electrochem. ox./ CH_3CN	ESR/ RT		$6H(NCH_3)$: 0.482 2H(7): 0.125 2N: 0.462	85Be1
20.5.1.5 Other amino-terminated systems					
$[C_{12}H_{11}NO_3S]^{\cdot +}$	Ox. of the amine with 3N H_2SO_4, $K_2Cr_2O_7$/ 3N H_2SO_4	ESR/ 293		1H: 0.74 1N: 0.74	84Pa1 [24])
$[C_{14}H_{15}NO_2]^{\cdot +}$	Ox. with $NOBF_4$/ CH_3CN	ESR/ 295		N: 0.81 H(NH): 0.96 4H(o): 0.27 $6H(OCH_3)$: 0.10 4H(m): 0.04	83Ko1/ 71Ne1

[24]) Same species observed from $4\text{-}HO_3SC_6H_4N(CH_3)C_6H_5$; demethylation occurs under reaction conditions.

Substance	Generation/ Matrix or Solvent	Method/ T [K]	g-Factor	a-Value [mT]	Ref./ add. Ref.
$[C_{14}H_{24}N_4O_2]^{\cdot +}$	Electrochem. ox./ DMF: 0.1 M $(n\text{-}C_4H_9)_4NClO_4$	ESR/ RT	2.0032	4N: 0.332 24H(CH_3): 0.332	85Bo1
$[C_{16}H_{14}N_4S_2]^{\cdot +}$, $[C_{16}H_8D_6N_4S_2]^{\cdot +}$ **BTA**, R = CH_3	Ox. with $Pb(OOCCH_3)_4$/ Acetone	ESR/ 295		2N(1): 0.409 2N(10): 0.409 2H(3): 0.112 2H(4): 0.0335 2H(5): 0.0865 2H(6): 0.0335 6H(CH_3): 0.297	84Lo1/ 83Ok1 [25])
	Ox. with $Pb(OOCCH_3)_4$/ CH_2Cl_2	ENDOR/ 223		2H(3): 0.114 2H(4): 0.029 2H(5): 0.074 2H(6): 0.029 6H(CH_3): 0.269	84Lot1
R = CD_3	Ox. with $Pb(OOCCH_3)_4$/ Acetone	ESR/ 295		2N(1), 2N(10): 0.391 2H(3): 0.134 2H(4): 0.0435 2H(5): 0.087 2H(6): 0.0435 6D(CD_3): 0.435	84Lot1/ 83Ok1 [25])
	Ox. with $Pb(OOCCH_3)_4$/ CH_2Cl_2	ENDOR/ 223		2H(3): 0.114 2H(3): 0.026 2H(5): 0.074 2H(6): 0.026	74Lot1

[25]) Slightly different splittings reported.

Substance	Generation/ Matrix or Solvent	Method/ T [K]	g-Factor	a-Value [mT]	Ref./ add. Ref.
$[C_{18}H_{18}N_4S_2]^{\cdot+}$ **BTA**, R = CH_2CH_3	Ox. with $Pb(OOCCH_3)_4$/ Acetone	ESR/ 295		2N(1): 0.409 2N(10): 0.409 2H(3): 0.119 2H(4): 0.040 2H(5): 0.080 2H(6): 0.040 4H(NCH_2): 0.158	84Lo1/ 83Ok1 [25])
	Ox. with $Pb(OOCCH_3)_4$/ CH_2Cl_2	ENDOR/ 223		2H(3): 0.119 2H(4): 0.024 2H(5): 0.074 2H(6): 0.024 4H(NCH_2): 0.154	84Lo1
$[C_{18}H_{24}N_6]^{\cdot+}$	Ox. with I_2, Br_2 or $NOSbF_6$/ CH_2Cl	ESR/ 295		[26])	84Br1
$[C_{20}H_{22}N_2]^{\cdot+}$	–/ CH_2Cl_2	ESR/ 210		4 or 2H: 0.362 2 or 4H: 0.300 2N: 0.245 12H: 0.097 2H: 0.045 2H: 0.008	84Bo2
	–/ CH_2Cl_2	ENDOR/ 200		4 or 2H: 0.366 2 or 4H: 0.299 2N: 0.238 12H: 0.089 2H: 0.041 2H: 0.009	84Bo2

[25]) Slightly different splittings reported.
[26]) 23···25 lines, dication is a triplet.

Substance	Generation/ Matrix or Solvent	Method/ T [K]	g-Factor	a-Value [mT]	Ref./ add. Ref.
$[C_{28}H_{21}N_3]^{\cdot+}$, $[C_{32}H_{29}N_3]^{\cdot+}$ 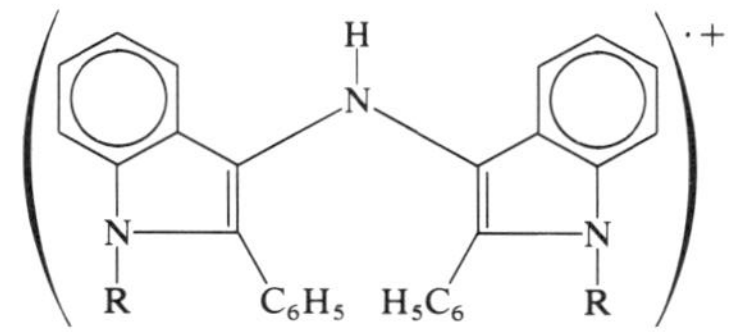 R = H	Ox. of the amino indole with $AgClO_4$/ —	ESR/ RT		N: ≅0.7 H(NH): ≅0.7	79Br1
R = CH_2CH_3	Ox. of the amino indole with $AgClO_4$ —	ESR/ RT		N: ≅0.75 H(NH): ≅0.75	79Br1
$[C_{36}H_{24}N_2]^{\cdot+}$ 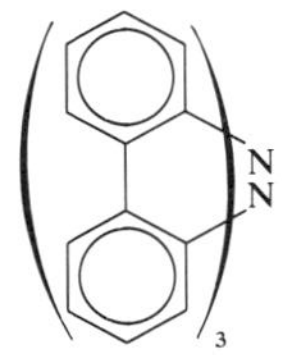	Ox. with benzoyl peroxide/ Toluene: CF_3CO_2H (9:1)	ESR, ENDOR/ 300 190		2N: 0.699 2N: 0.699 8H: −0.132 [27] 4H: +0.067 [28] 4H: +0.037 [28]	86Ne5/ 85Ne1
$[C_{44}H_{39}N]^{\cdot+}$ 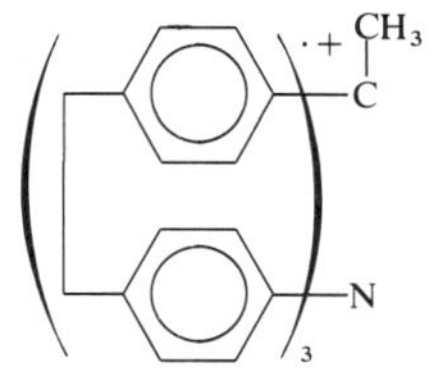	Ox. with $AgClO_4$:I_2/ $(CH_3CH_2)_2O$	ESR/ —	2.0027	N: 1.07	79Vö1

[27]) *o*- and *p*-hydrogens at two of the biphenyl bridging rings.
[28]) *m*-hydrogens of two of the biphenyl bridging rings.

20.5.2 Hydrazoaromatic cation radicals

20.5.2.1 1,4-Dihydropyridazines and viologens

20.5.2.1.1 1,4-Dihydropyridazines

(For alloxazine derivatives, see 20.5.2.3)

Substance	Generation/ Matrix or Solvent	Method/ T [K]	g-Factor	a-Value [mT]	Ref./ add. Ref.
$[C_2H_4N_4]^{\cdot +}$ 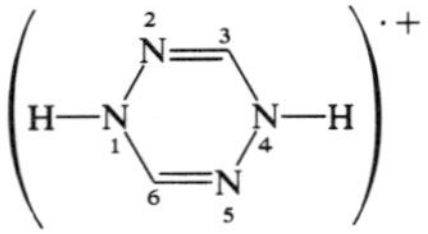	Ox. with $Pb(OOCCH_3)_4$/ CF_3CO_2H	ESR/ RT	2.0039	2N(1, 4): 0.729 2N(2, 5): 0.455 2H(NH): 0.910	83Ne5/ 84Ne1 [29]
$[C_4H_6N_2]^{\cdot +}$ 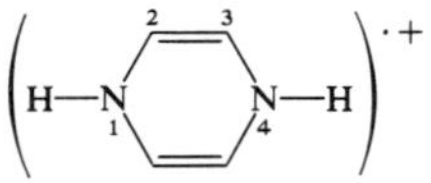	Electrochem. red. of the dication/ H_2O:1M $HClO_4$	ESR/ 295	2.0034 (1)	2N(1, 4): 0.745 2H(NH): 0.805 4H(2, 3, 5, 6): 0.316	84Ka
	Ox. with $AlCl_3$/ CH_2Cl_2: $(CH_3)_3SiCl$	–/ 295	2.0034	2N: 0.745 4H(2): 0.316 2H(NH): 0.805 $^{13}C(2)$: 0.12	83Ka1
	Photochem. red. of the dication/ $2\text{-}C_3H_7ON$:acetone:1M HCl (85:10:5)	ESR/ 297	2.00324	2N(1, 4): 0.746 2H(NH): 0.788 4H: 0.315	79Ra1
$[C_4H_8N_4O_2]^{\cdot +}$ 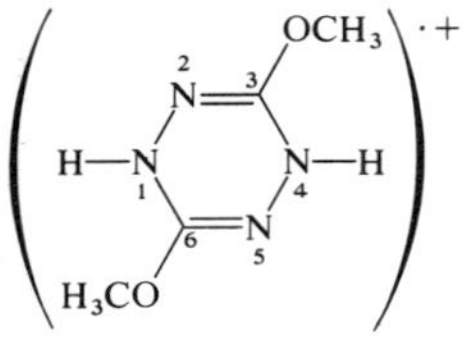	Ox. with $Pb(OOCCH_3)_4$/ CF_3CO_2H	ESR/ RT	2.0038	2N(1, 4): 0.666 2N(2, 5): 0.522 2H(NH): 0.825 6H(CH_3): 0.019	83Ne5

[29]) Calculations on verdazyls and their cations.

Substance	Generation/ Matrix or Solvent	Method/ T [K]	g-Factor	a-Value [mT]	Ref./ add. Ref.
$[C_4H_8N_4S_2]^{\cdot +}$ 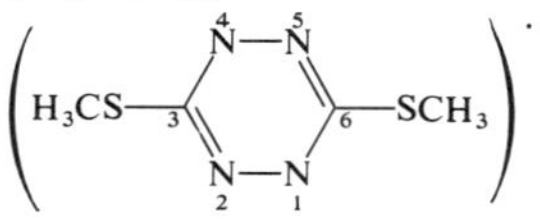	Ox. with $Pb(OOCCH_3)_4$ CF_3CO_2H	ESR/ RT	2.0038	2N(1, 4): 0.675 2N(2, 5): 0.520 2H(NH): 0.830	83Ne5
$[C_5H_8N_2]^{\cdot +}$ 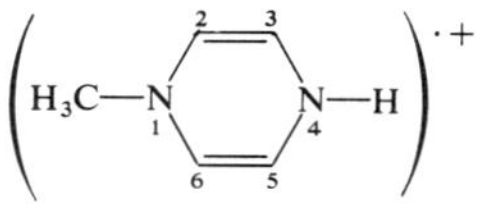	Red. with sodium dithionite/ Methanol	ESR/ RT	2.0032	N(1): 0.714 H(2, 6): 0.305 H(3, 5): 0.295 N(4): 0.762 $3H(CH_3)$: 0.853 1H(NH): 0.906	85Ea1
	Electrochem. red. of the dication/	ESR			84Ka2
	H_2O: 1M $HClO_4$	295	2.0034 (1)	N(1): 0.891 N(4): 0.714 $3H(CH_3)$: 0.855 H(4): 0.773 2H(2, 6): 0.290 2H(3, 5): 0.318	
	DMF:1M $HClO_4$	295		N(1): 0.891 N(4): 0.705 $3H(CH_3)$: 0.852 H(4): 0.756 2H(2, 6): 0.292 2H(3, 5): 0.310	
$[C_5H_7DN_2]^{\cdot +}$ 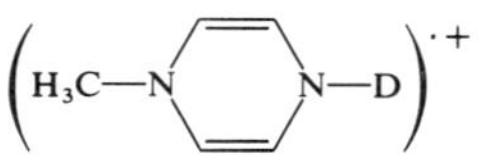	Electrochem. red. of the dication/ DMF:1 M $DClO_4$	ESR/ 295	2.0034 (1)	N(1): 0.890 N(4): 0.710 $3H(CH_3)$: 0.850 D(4): 0.121 2H(2, 6): 0.292 2H(3, 5): 0.315	84Ka2

Substance	Generation/ Matrix or Solvent	Method/ T [K]	g-Factor	a-Value [mT]	Ref./ add. Ref.
$[C_5H_8N_2]^{\cdot +}$ 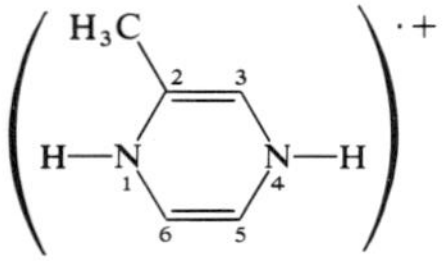	Photochem. red. of the dication/ 2-C_3H_7ON: acetone: 1 M HCl (85:10:5)	ESR/ 297	2.00312	N(1): 0.696 N(4): 0.751 1H(NH, 1): 0.727 3H(CH_3): 0.251 H(3): 0.379 H(NH, 4): 0.791 H(5): 0.248 H(6): 0.351	79Ra1
$[C_6H_{10}N_2]^{\cdot +}$ 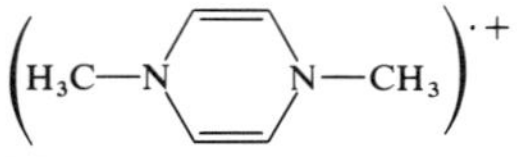	Electrochem. red. of the dication/ H_2O: 1 M $HClO_4$	ESR/ 295	2.0034 (1)	2N: 0.840 6H(CH_3): 0.800 4H: 0.285	84Ka2
$[C_6H_{10}N_2]^{\cdot +}$ 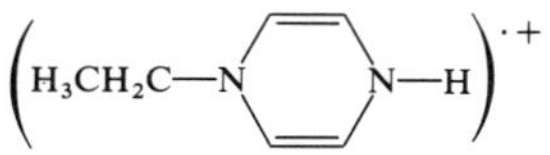	Electrochem. red. of the dication/ H_2O: 1 M $HClO_4$	ESR/ 295	2.0034 (1)	N(1) + N(4): 1.600 2H(CH_2): 0.534 3H(CH_3): 0.023 H(NH): 0.771 2H(2, 6): 0.292 2H(3, 5): 0.315	84Ka2
$[C_6H_{12}N_4S_2]^{\cdot +}$ 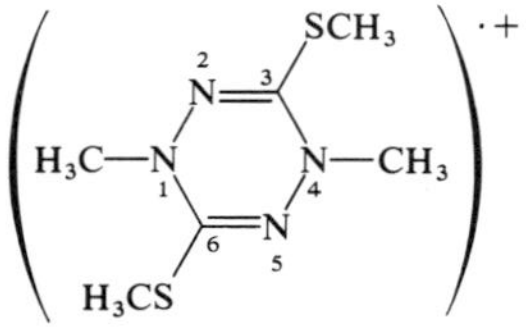	Ox. with $Pb(OOCCH_3)_4$/ CF_3CO_2N	ESR/ RT	2.0038	2N(1, 4): 0.760 2N(2, 5): 0.495 6H(CH_3): 0.780 6H(SCH_3): 0.020	83Ne5
$[C_7H_{12}N_2]^{\cdot +}$ 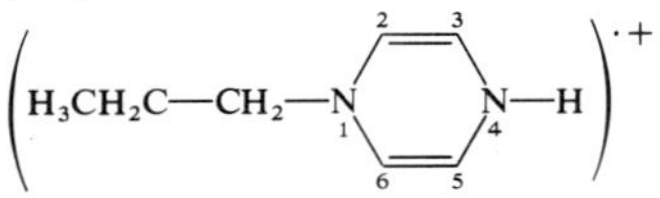	Electrochem. red. of the dication/ H_2O: 1 M $HClO_4$	ESR/ 295	2.0034 (1)	N(1): 0.890 N(4): 0.712 H(NH): 0.775 2H(NH_2): 0.540 2H(2, 6): 0.290 2H(3, 5): 0.312	84Ka2

Substance	Generation/ Matrix or Solvent	Method/ T [K]	g-Factor	a-Value [mT]	Ref./ add. Ref.
$[C_7H_{12}N_2]^{\cdot +}$ 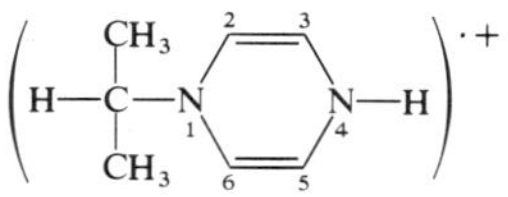	Electrochem. red. of the dication/ H_2O: 1 M $HClO_4$	ESR/ 295	2.0034 (1)	N(1): 0.890 N(4): 0.713 1H(NCH): 0.270 6H(CH_3): 0.015 H(NH): 0.773 2H(2, 6): 0.287 2H(3, 5): 0.318	84Ka2
$[C_7H_{12}N_2]^{\cdot +}$ 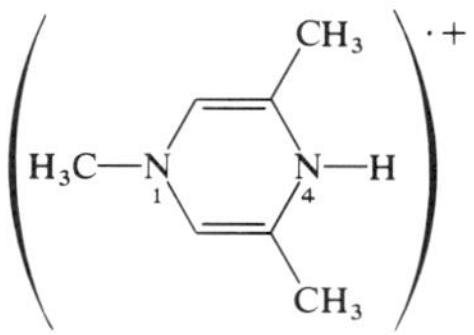	Electrochem. red. of the dication/ H_2O: 1 M $HClO_4$	ESR/ 295		N(1): 0.890 N(4): 0.615 3H(1, CH_3): 0.635 H(NH): 0.645 2H(2, 6): 0.300 6H(3, 5, CH_3): 0.280	84Ka2
$[C_7H_{12}N_2]^{\cdot +}$ 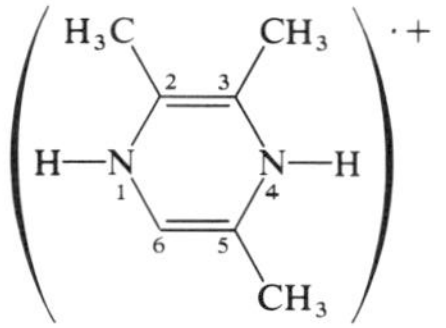	Photochem. red. of the dication/ 2-C_3H_7ON: acetone: 1 M HCl (85:10:5)	ESR/ 297	2.00317	N(1): 0.653 N(2): 0.652 H(NH, 1): 0.803 3H(CH_3, 2): 0.223 3H(CH_3, 3): 0.313 H(NH, 4): 0.779 3H(CH_3, 5): 0.214 H(6): 0.312	79Ra1
$[C_8H_8N_2]^{\cdot +}$ 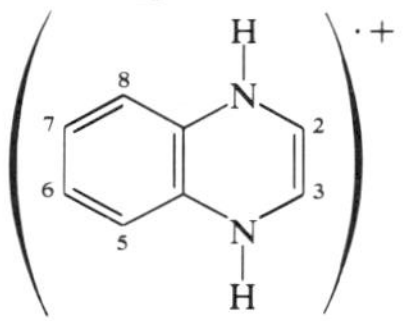	Electrochem. red. of the dication/ DMF: 1 M $HClO_4$	ESR/ 295		2N: 0.665 2H(NH): 0.717 2H(2, 3): 0.399 2H(5, 8): 0.078 2H(6, 7): 0.138	84Ka1
$[C_8H_{10}N_2O_2]^{\cdot +}$ 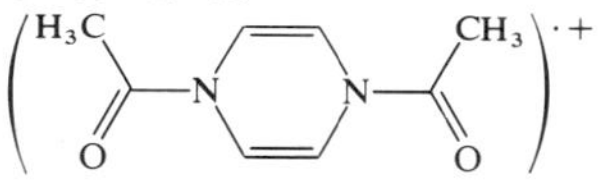	Ox. with $AlCl_3$/ CH_2Cl_2	ESR/ RT		2N: 0.58 [30]) 4H: 0.29 [30])	85Be2

[30]) Average for *cis*-, *trans*-isomer mixture.

Substance	Generation/ Matrix or Solvent	Method/ T [K]	g-Factor	a-Value [mT]	Ref./ add. Ref.
$[C_8H_{14}N_2]^{\cdot +}$ 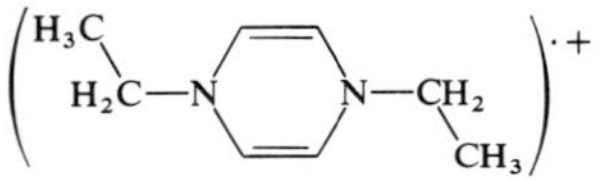	Electrochem. red. of dication/ H_2O: 0.1 M KCl	ESR/ RT		2N: 0.890 4H: 0.540 4H: 0.290 6H(CH_3): 0.023	85Ka2
$[C_8H_{14}N_2]^{\cdot +}$ 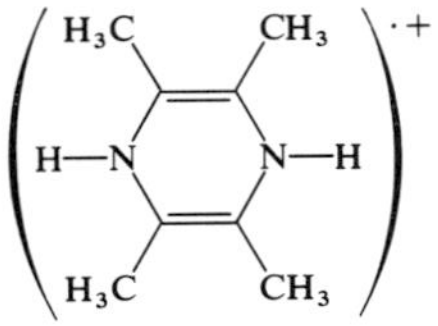	Photochem. red. of the dication/ 2-C_3H_7ON:acetone:1 M HCl (85:10:5)	ESR/ 297	2.00310	2N(1, 4): 0.671 2H(NH): 0.692 12H(CH_3): 0.253	79Ra1
$[C_8H_{16}N_2P_2S_2]^{\cdot +}$ 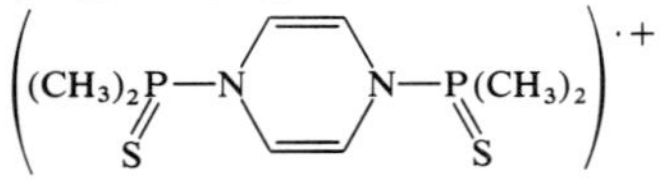	Electrochem. ox./ CH_3CN	ESR/ RT	2.0031	2N: 0.68 4H: 0.34 2P: 0.78	85Be2
$[C_8H_{18}N_2Si_2]^{\cdot +}$ 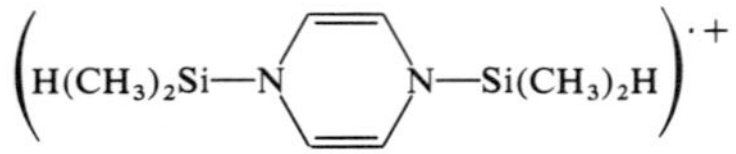	Ox. with $AlCl_3$/ CH_2Cl_2 : $(CH_3)_3SiCl$	ESR/ 295	2.0033	2N: 0.668 4H(ring): 0.313 2H(SiH): 0.103	83Ka1
$[C_{10}H_7FeN_7]^{\cdot 2-}$ 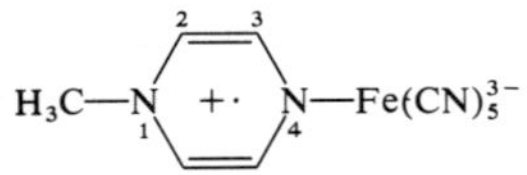	Red. of the dication with sodium dithionite/ H_2O:CH_3OH:CH_2Cl_2 (1:1:1)	ESR/ RT	2.0032	N(1): 0.821 2H(2): 0.477 2H(3): 0.102 N(4): 0.718 3H(CH_3): 0.718	85Ea1
$[C_{10}H_8FeN_7]^{\cdot 2-}$ 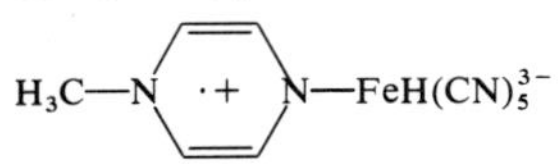	Red. of the dication with sodium dithionite/ H_2O:CH_3OH:CH_2Cl_2 (1:1:1)	ESR/ RT	2.0032	N(1): 0.968 2H(2): 0.428 2H(3): 0.090 N(4): 0.690 3H(CH_3): 0.690	85Ea1

Substance	Generation/ Matrix or Solvent	Method/ T [K]	g-Factor	a-Value [mT]	Ref./ add. Ref.
$[C_{10}H_{12}N_2]^{\cdot+}$ 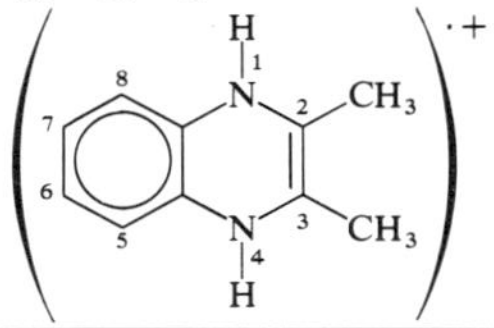	Electrochem. red. of the dication/ DMF: 1 M $HClO_4$	ESR/ 295	2.0031 (1)	2N(1, 4): 0.612 2H(NH): 0.660 6H(CH_3): 0.397 2H(5, 8): 0.078 2H(6, 7): 0.128	84Ka1
$[C_{10}H_{12}N_2]^{\cdot+}$ 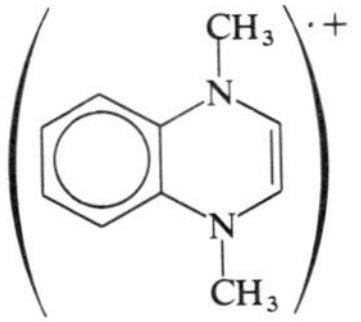	Electrochem. ref. of the dication/ H_2O: 0.1 M KCl	ESR/ RT		2N: 0.742 6H(CH_3): 0.690 2H: 0.370 2H: 0.141 2H: 0.092	85Ka2
$[C_{10}H_{12}N_2]^{\cdot+}$ 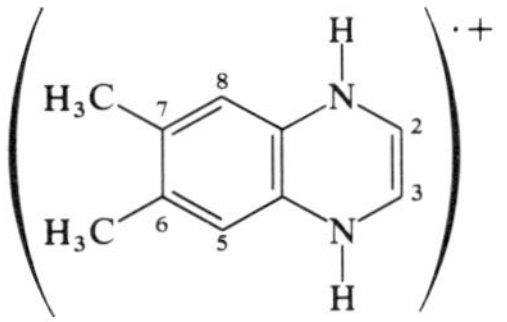	Electrochem. red. of the dication/ DMF: 1 M $HClO_4$	ESR/ 295	2.0031 (1)	2N(1, 4): 0.612 2H(NH): 0.712 2H(2, 3): 0.375 2H(5, 8): 0.053 6H(6, 7, CH_3): 0.165	84Ka1
$[C_{10}H_{18}N_2]^{\cdot+}$ 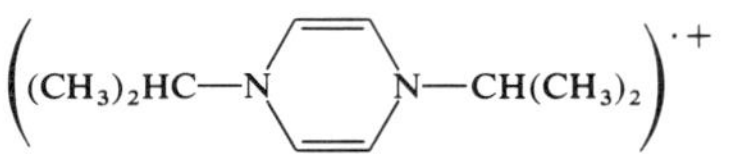	Ox. of parent compound [31] with $SbCl_3$/ THF	ESR/ ≈343	2.0034	2N: 0.840 4H: 0.256 2H(NCH): 0.290	83Al1
$[C_{10}H_{22}N_2Si_2]^{\cdot+}$ 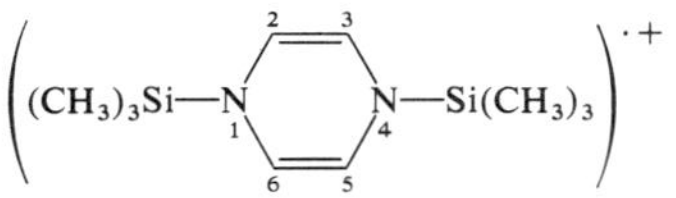	Ox. with $AlCl_3$/ CH_2Cl_2 : $(CH_3)_3SiCl$	ESR/ 295	2.0033	2N: 0.664 4H(ring): 0.313 ^{13}C(ring): 0.11 ^{29}Si: 0.279	83Ka1/ 80Ka1
	Ox. with TCNE or TCNQ/ CH_2Cl_2	ESR/ 295	2.0033		84Ka3

[31] $(CH_3)_2CHN$, H, $NCH(CH_3)_2$, H

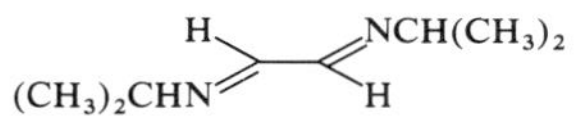

Substance	Generation/ Matrix or Solvent	Method/ T [K]	g-Factor	a-Value [mT]	Ref./ add. Ref.
$[C_{12}H_{16}N_2]^{\cdot +}$	Electrochem. red. of the dication/ H_2O: 0.1 M KCl	ESR/ RT		2N: 0.763 4H: 0.407 2H: 0.360 2H: 0.136 2H: 0.099	85Ka2
$[C_{12}H_{16}N_2]^{\cdot +}$	Electrochem. red. of the dication/ DMF: 1 M $HClO_4$	ESR/ 295	2.0031 (1)	2N: 0.611 2H(1, 4): 0.662 6H(2, 3, CH_3): 0.374 2H(5, 8): 0.052 6H(6, 7, CH_3): 0.148	84Ka1
$[C_{12}H_{10}N_2]^{\cdot +}$	Electrochem. red. of oxidized form in acid/	ESR/			80El1
	CH_3CN	233		4H(1, 4, 6, 9): 0.060 4H(2, 3, 7, 8): 0.177 2N(9, 10): 0.610 2H(9, 10): 0.670	
	$CH_3CH_2CO_2H$	258		4H(1, 4, 6, 9): 0.064 4H(2, 3, 7, 8): 0.171 2N(9, 10): 0.615 2H(9, 10): 0.660	
	CH_3CO_2H	288		4H(1, 4, 6, 9): 0.062 4H(2, 3, 7, 8): 0.171 2N(2, 9): 0.610 2H(9, 10): 0.667	
	HCO_2H	265		4H(1, 4, 6, 9): 0.058 4H(2, 3, 7, 8): 0.171 2N(9, 10): 0.615 2H(9, 10): 0.622	
(*continued*)					

Substance	Generation/ Matrix or Solvent	Method/ T [K]	g-Factor	a-Value [mT]	Ref./ add. Ref.
$[C_{12}H_{10}N_2]^{\cdot+}$ *(continued)*	Electrochem. red. of oxidized form in acid/	ESR/			80El1
	CCl_3CO_2H	353		4H(1, 4, 6, 9): 0.056 4H(2, 3, 7, 8): 0.172 2N(9, 10): 0.612 2H(9, 10): 0.675	
	CF_3CO_2D	263		4H(1, 4, 6, 9): 0.056 4H(2, 3, 7, 8): 0.177 2N(9, 10): 0.605 2H(9, 10): –	
	CF_3CO_2H	263		4H(1, 4, 6, 9): 0.053 4H(2, 3, 7, 8): 0.173 2N(9, 10): 0.620 2H(9, 10): 0.675	
	Electrochem. red. of oxidized form in acid/ CF_3CO_2H	ENDOR/ 263		4H(1, 4, 6, 9): 0.0532 4H(2, 3, 7, 8): 0.1751	80El1
$[C_{12}H_{22}N_2]^{\cdot+}$ $((CH_3)_3C$—N⟨ ⟩N—$(CH_3)_3)^{\cdot+}$	Ox. of parent compound [32] with $SbCl_3$/ THF	ESR/ 343.5	2.0031	2N: 0.862 4H: 0.274 18H(CH_3): 0.011	83Al1
$[C_{13}H_{12}N_2]^{\cdot+}$ (structure: H on N10, CH_3 on N5; positions 1, 2, 3, 4)$^{\cdot+}$	Electrochem. red./ H_2O: 1 M $HClO_4$	–/ 295		N(5): 0.693 N(10): 0.600 3H(5, CH_3): 0.635 H(NH): 0.645	80Ch2/ 84Ka2
	Phot./ H_2O	ESR/ ≈295	2.0035 (2)	N(10): 0.600 H(NH): 0.645 [33] N(5): 0.693 3H(CH_3): 0.635 2H(2 or 3): 0.203 2H(3 or 2): 0.139 2H(1 or 4): 0.094 2H(4 or 1): 0.045	80Ch1/ 80Ch2

[32]) $(CH_3)_3CN$=CH–CH=$NC(CH_3)_3$

[33]) Proven by deuteration.

Substance	Generation/ Matrix or Solvent	Method/ T [K]	g-Factor	a-Value [mT]	Ref./ add. Ref.
$[C_{14}H_{12}N_4]^{\cdot +}$ 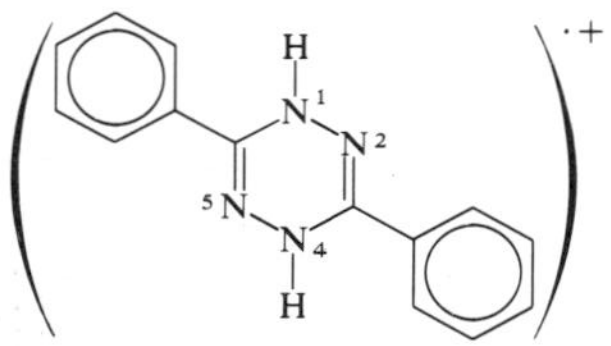	Ox. of $Pb(OOCCH_3)_4$/ CF_3CO_2H	ESR/ RT	2.0039	2N(1, 4): 0.670 2N(2, 5): 0.500 2H(NH): 0.840	83Ne5
$[C_{14}H_{24}N_4]^{\cdot +}$ 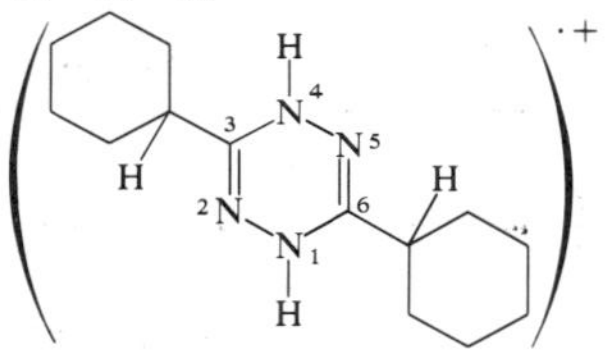	Ox. with $Pb(OOCCH_3)_4$/ CF_3CO_2H	ESR/ RT	2.0039	2N(1, 4): 0.675 2N(2, 5): 0.49 2H(NH): 0.88 2H: 0.08	83Ne5
$[C_{16}H_{16}N_4]^{\cdot +}$ 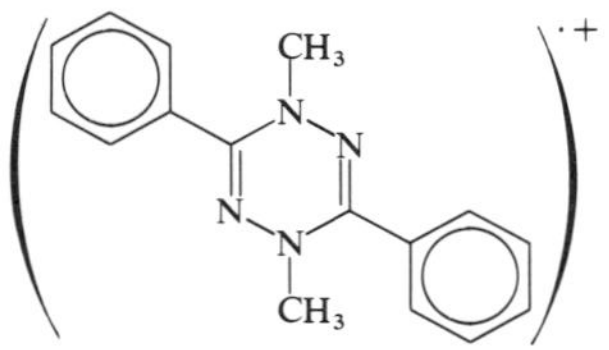	Ox. with $Pb(OOCCH_3)_4$/ CF_3CO_2H	ESR/ RT	2.0038	2N(1, 4): 0.782 2N(2, 5): 0.454 $6H(CH_3)$: 0.782	83Ne5
$[C_{16}H_{28}N_2]^{\cdot +}$ 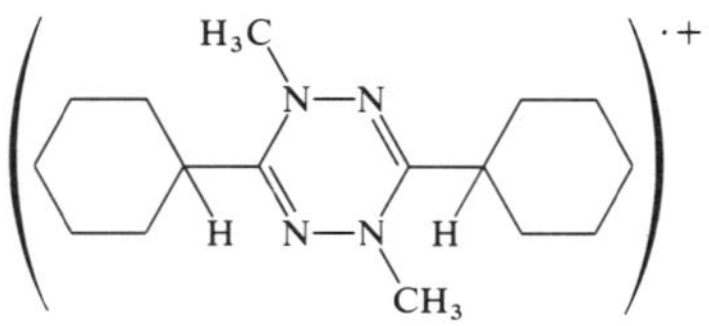	Ox. with $Pb(COOCCH_3)_4$/ CF_3CO_2H	ESR/ RT	2.0039	2N(1, 4): 0.785 2N(2, 5): 0.435 $6H(CH_3)$: 0.785 2H: 0.080	83Ne5

Substance		Generation/ Matrix or Solvent	Method/ T [K]	g-Factor	a-Value [mT]	Ref./ add. Ref.
$[C_{17}H_{19}N_3]^{\cdot+}$ (+labelled analogues)	$R=C_6H_5$	Protonation of neutral radical/ CF_3CO_2H	ESR/ RT	2.0032	N(1): 0.980 N(2): 0.240 N(4): 0.485 H(NH): 0.490	81Ne1
		(6, 8)-d_2			N(1): 0.990 N(2): 0.250 N(4): 0.485 H(NH): 0.485	
	$R=C_6D_5$	(5, 6, 7, 8)-d_4			N(1): 0.990 ^{15}N(2): 0.330 N(4): 0.480 H(NH): 0.480	
		(5, 6, 7, 8)-d_4			N(1): 0.990 ^{15}N(2): 0.245 N(4): 0.480 H(NH): 0.485	
		(5, 7)-d_2			N(1): ≈0.995 N(2): 0.245 N(4): ≈0.470 H(NH): ≈0.520 other: 0.275, 0.05	
		(6, 8)-d_2			N(1): 0.990 N(2): 0.250 N(4): 0.485 H(NH): 0.490	
		(6, 8)-d_2			^{15}N(1): 1.410 N(2): 0.245 N(4): 0.485 H(NH): 0.485	
		(6, 8)-d_2			N(1): 0.990 N(2): 0.250 ^{15}N(4): 0.660 H(NH): 0.510	

Substance	Generation/ Matrix or Solvent	Method/ T [K]	g-Factor	a-Value [mT]	Ref./ add. Ref.
$[C_{19}H_{10}D_5N_2]^{\cdot+}$, $[C_{19}H_8D_7N_2]^{\cdot+}$ 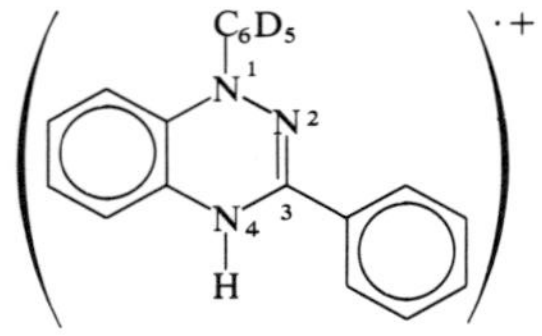	Protonation of neutral radical/ CF_3CO_2H	ESR/ RT	2.0032	N(1): 0.990 N(2): 0.245 N(4): 0.480 H(NH): 0.485 H(6, 8): 0.270, 0.150	81Ne1
	(6, 8)-d_2		2.0032	N(1): 0.995 N(2): 0.245 N(4): 0.480 H(NH): 0.485	

20.5.2.1.2 Viologens: Diazadihydrodipyridines

Substance	Generation/ Matrix or Solvent	Method/ T [K]	g-Factor	a-Value [mT]	Ref./ add. Ref.
$C_{12}H_{12}N_2]^{\cdot+}$ 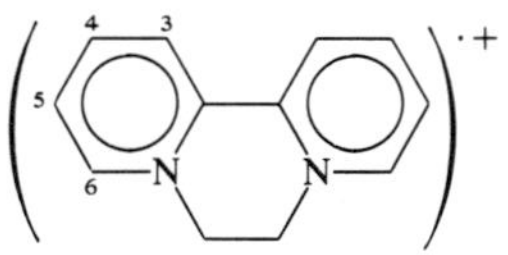 (diquat)	Red. of dication with Zn(Hg)/ CH_3OH	ESR/ 200		2N: 0.402 2H(3): 0.060 2H(4): 0.247 2H(5): 0.308 2H(6): 0.030 4H(CH_2): 0.353	84Ri1 [34])
		295		2N: 0.410 2H(3): 0.0572 2H(4): 0.248 2H(5): 0.300 2H(6): 0.026 4H(CH_2): 0.346	
$[C_{12}H_{14}N_2]^{\cdot+}$ 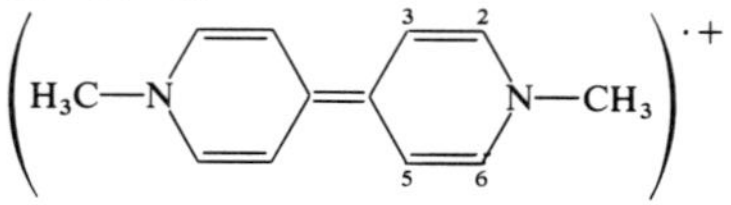 (methylviologen)	Red. of dication with Zn(Hg)/ CH_3OH	ESR/ 233		2N: 0.4219 4H(2, 6):0.1354 4H(3, 5): 0.1638 6H(CH_3): 0.4058	84Ri1 [35])/ 85Yu1, 85Ka1, 85Ga1, 87Co1
		295		2N: 0.4247 4H(2, 6): 0.1343 4H(3, 5): 0.1591 6H(CH_3): 0.4012	

[34]) Electron exchange broadening studied. Cation non-planar: exchange rate of non-equivalent H in CH_2 groups $\approx 10^7\ s^{-1}$.
[35]) Electron exchange broadening studied.

Nelsen

Substance	Generation/ Matrix or Solvent	Method/ T [K]	g-Factor	a-Value [mT]	Ref./ add. Ref.
$[C_{22}H_{18}N_2]^{\cdot+}$ 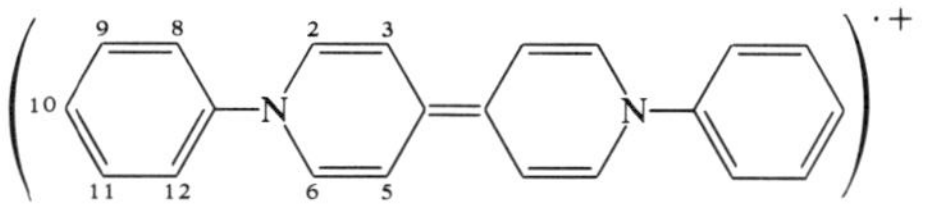	Red. of the dication with Zn/ CH_3OH	ENDOR, TRIPLE/ 203		2N(1): 0.013 4H(2, 6): 0.0057 4H(3, 5): 0.005 4H(8, 12): 0.002 4H(9, 11): 0.001 2H(10): 0.0018	85Cl1/ 83Cl1
$[C_{22}H_{16}F_2N_2]^{\cdot+}$ 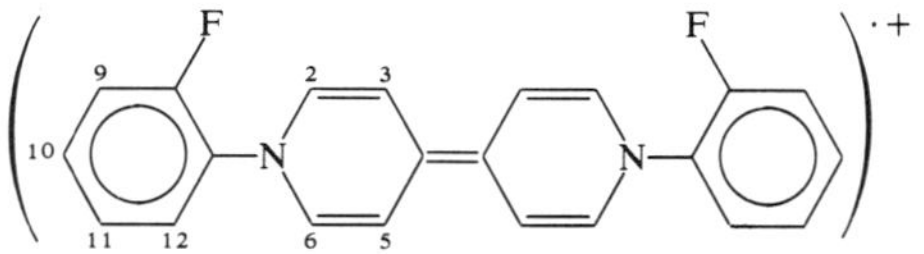	Red. of the dication with Zn/ CH_3OH	ENDOR, TRIPLE/ 203		2N(1): +0.013 4H(2, 6): −0.006 4H(3, 5): −0.005 2F(8): −0.0003 2H(9): +0.0008 2H(10): −0.0013 2H(11): +0.0011 2H(12): −0.0017	85Cl1/ 83Cl1
$[C_{22}H_{16}F_2N_3]^{\cdot+}$ 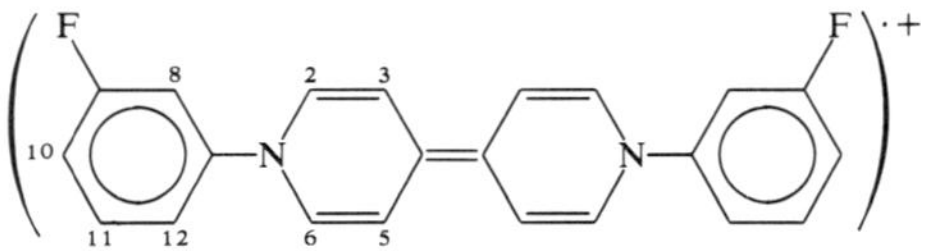	Red. of the dication with Zn/ CH_3OH	ENDOR, TRIPLE/ 203		2N(1): +0.013 4H(2, 6): −0.006 4H(3, 5): −0.005 2H(8): −0.002 2F(9): +0.0002 2H(10): −0.002 2H(11): +0.001 2H(12): −0.002	85Cl1/ 83Cl1
$[C_{22}H_{16}F_2N_2]^{\cdot+}$ 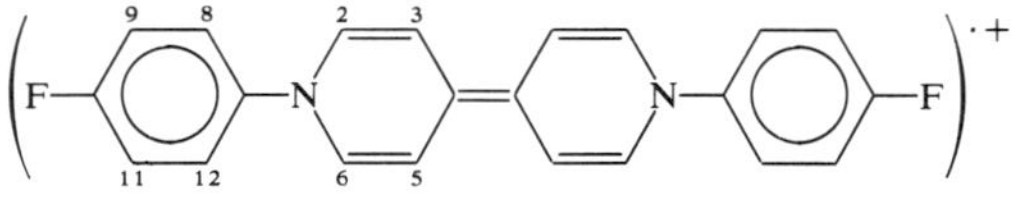	Red. of the dication with Zn/ CH_3OH	ENDOR, TRIPLE/ 203		2N(1): +0.013 4H(2, 6): −0.006 4H(3, 5): −0.005 2H(8): −0.001 2H(9): +0.001 2F(10): +0.004 2H(11): +0.001 2H(12): −0.001	85Cl1/ 83Cl1

Substance	Generation/ Matrix or Solvent	Method/ T [K]	g-Factor	a-Value [mT]	Ref./ add. Ref.
$[C_{22}H_{14}F_4N_2]^{\cdot +}$ 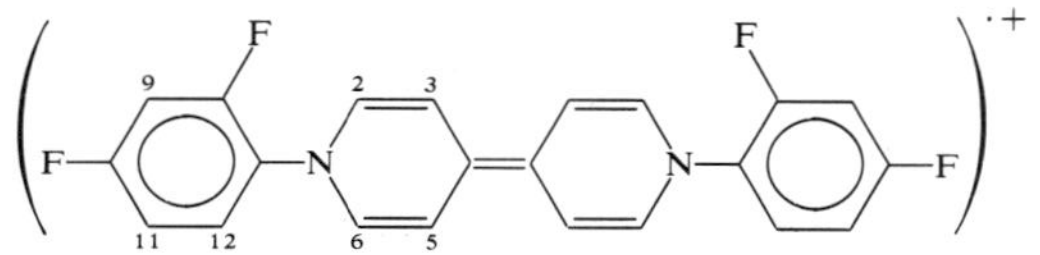	Red. of the dication with Zn/ CH_3OH	ENDOR, TRIPLE/ 203		2N(1): +0.013 4H(2, 6): −0.006 4H(3, 5): −0.005 2F(8): −0.0004 2H(9): +0.0007 2H(11): +0.0009 2H(12): −0.0016 2F(10): +0.0027	85Cl1/ 85Ev1
$[C_{22}H_{14}F_4N_2]^{\cdot +}$ 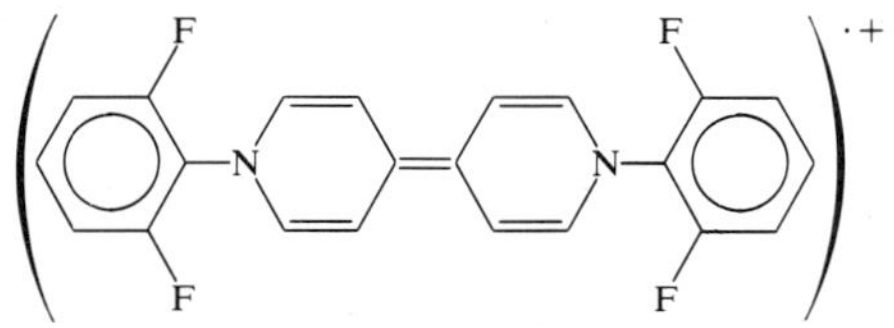	Red. of the dication with Zn/ CH_3OH	ENDOR, TRIPLE/ 203		2N(1): +0.012 4H(2, 6): −0.0065 4H(3, 5): −0.005 2F(8): −0.0005 2H(9): +0.0008 2H(10): −0.0009 2H(11): +0.0008 2F(12): −0.0005	85Cl1/ 85Ev1
$[C_{22}H_{12}F_6N_2]^{\cdot +}$ 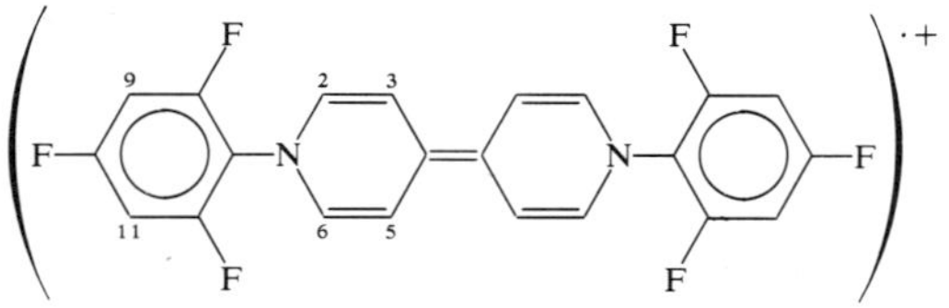	Red. of the dication with Zn/ CH_3OH	ENDOR, TRIPLE/ 203		2N(1): +0.013 4H(2, 6): −0.0066 4H(3, 5): −0.005 2F(8): −0.0009 2H(9): +0.0007 2F(10): +0.0019 2H(11): +0.0007 2F(12): −0.0004	85Cl1/ 85Ev1
$[C_{22}H_{10}F_8N_2]^{\cdot +}$ 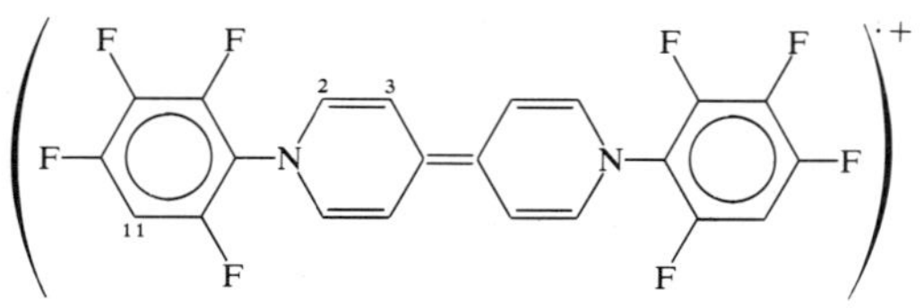	Red. of the dication with Zn/ CH_3OH	ENDOR, TRIPLE/ 203		2N(1): +0.012 4H(2, 6): −0.007 4H(3, 5): −0.005 2F(8): −0.0005 2F(9): +0.0008 2F(10): +0.0023 2H(11): +0.0006 2F(12): −0.0005	85Cl1

Substance	Generation/ Matrix or Solvent	Method/ T [K]	g-Factor	a-Value [mT]	Ref./ add. Ref.
$[C_{22}H_{10}F_8N_2]^{\cdot+}$ 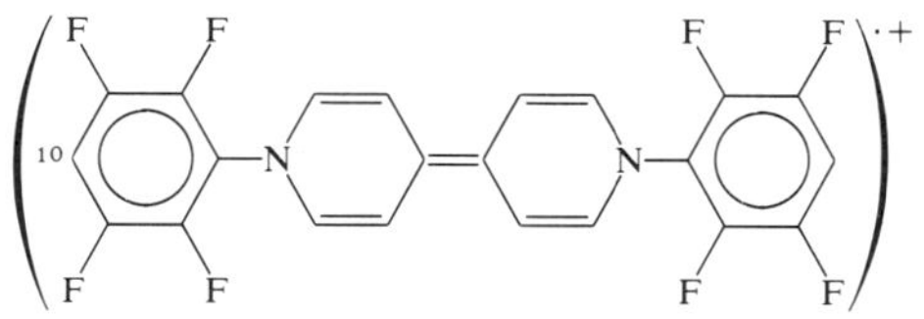	Red. of the dication with Zn/ CH_3OH	ENDOR, TRIPLE/ 203		2N(1): +0.012 4H(2, 6): −0.007 4H(3, 5): −0.0046 2F(8): −0.0001 2F(9): +0.0001 2H(10): −0.0010 2F(11): +0.0001 2F(12): −0.0001	85Cl1
$[C_{22}H_8F_{10}N_2]^{\cdot+}$ 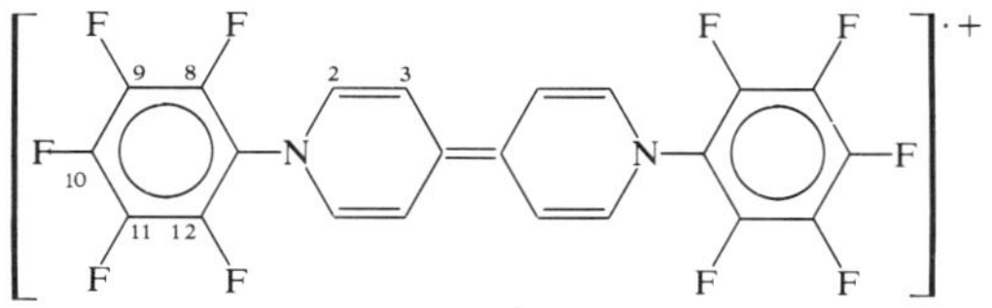	Red. of the dication with Zn/ CH_3OH	ENDOR, TRIPLE/ 203		2N(1): +0.012 4H(2, 6): −0.007 4H(3, 5): −0.0046 2F(8): −0.0003 2F(9): +0.0003 2F(10): +0.0025 2F(11): +0.0003 2F(12): −0.0003	85Cl1
$[C_{24}H_{22}N_2]^{\cdot+}$ 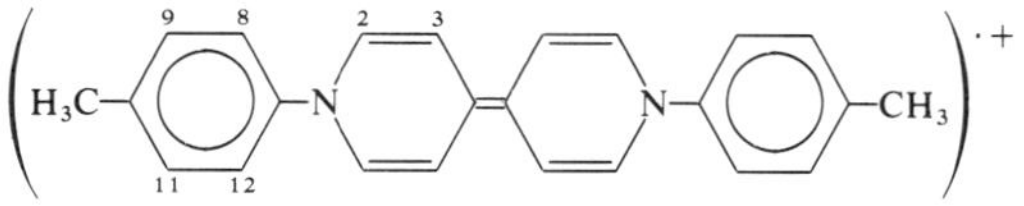	Red. of the dication with Zn/ CH_3OH	ENDOR, TRIPLE/ 233		2N(1): +0.0136 4H(2, 6): −0.0056 4H(3, 5): −0.005 2H(8): −0.0020 2H(9): +0.0107 6H(10, CH_3): +0.0020 2H(11): +0.0011 2H(12): −0.0020	85Ev1
$[C_{24}H_{22}N_2]^{\cdot+}$ 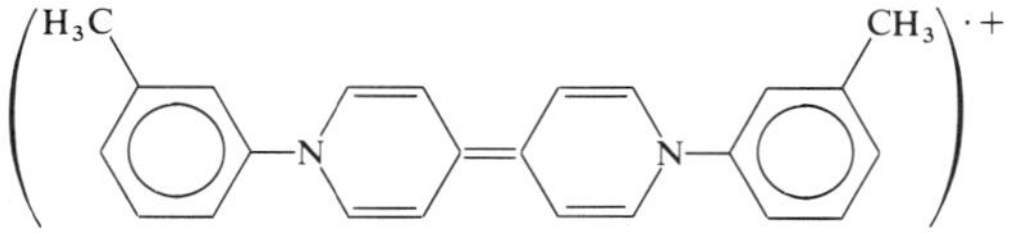	Red. of the dication with Zn/ CH_3OH	ENDOR, TRIPLE/ 233		2N(1): +0.0136 4H(2, 6): −0.0056 4H(3, 5): −0.005 2H(8): −0.0020 6H(9, CH_3): −0.0006 2H(10): −0.0020 2H(11): +0.001 2H(12): −0.0020	85Ev1

Substance	Generation/ Matrix or Solvent	Method/ T [K]	g-Factor	a-Value [mT]	Ref./ add. Ref.
$[C_{24}H_{22}N_2]^{\cdot +}$ 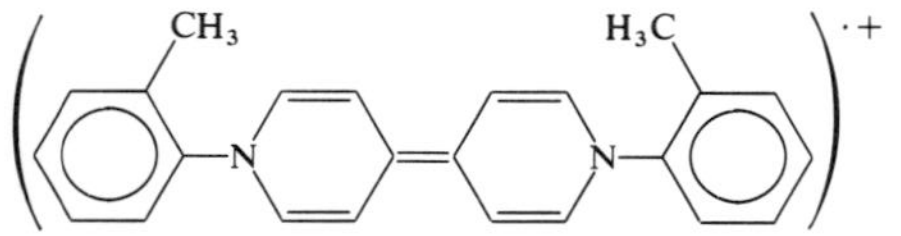	Red. of the dication with Zn/ CH_3OH	ENDOR, TRIPLE/ 233		2N(1): +0.0143 4H(2, 6): −0.0057 4H(3, 5): −0.0054 6H(8, CH_3): +0.0011 2H(9): +0.0004 2H(10): −0.0008 2H(11): +0.0007 2H(12): −0.0013	85Ev1
$[C_{24}H_{22}N_2]^{\cdot +}$ 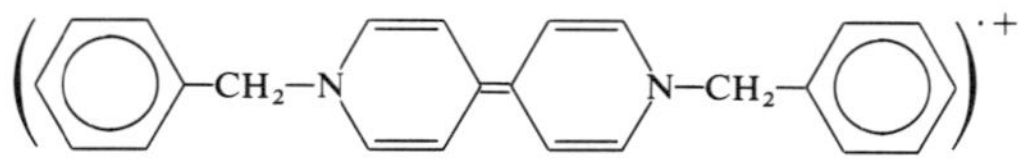	Red. of the dication with Zn(Hg)/ CH_3OH	ESR/ 240 295		2N: 0.408 8H(2, 3, 5, 6): 0.156 4H(CH_2): 0.262 2N: 0.414 8H(2, 3, 5, 6): 0.156 4H(CH_2): 0.262	84Ri1 [35]
$[C_{25}H_{40}N_2]^{\cdot +}$ 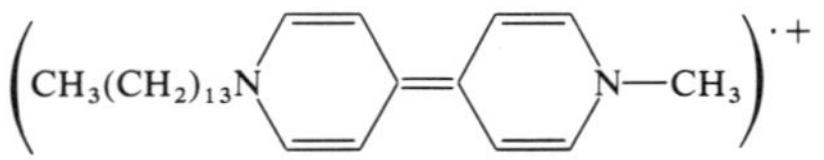	Electrochem. red. of the dication [36]/ Film on electrode	ESR/ RT		ΔHpp: 0.53···0.25	86Cr1
$[C_{28}H_{30}N_2]^{\cdot +}$ 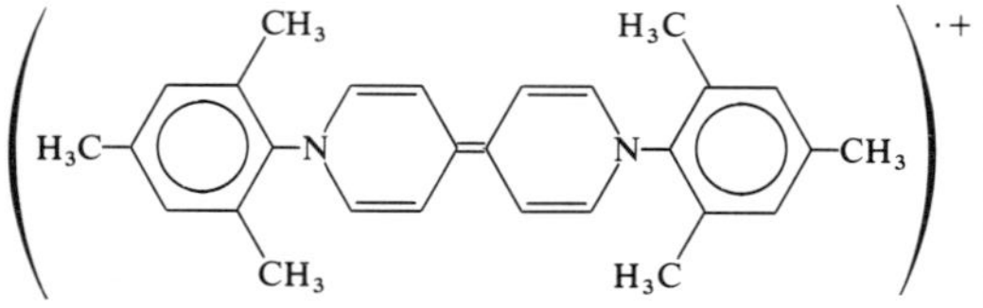	Red. of the dication with Zn/ CH_3OH	ENDOR, TRIPLE/ 233		2N(1): 0.0014 4H(2 6): −0.0056 4H(3, 5): −0.0056 24H(8, 12, CH_3): +0.0008 4H(9, 11): +0.00003 6H(10, CH_3): +0.0008	85Ev1
$[C_{32}H_{32}N_4]^{\cdot 3+}$ 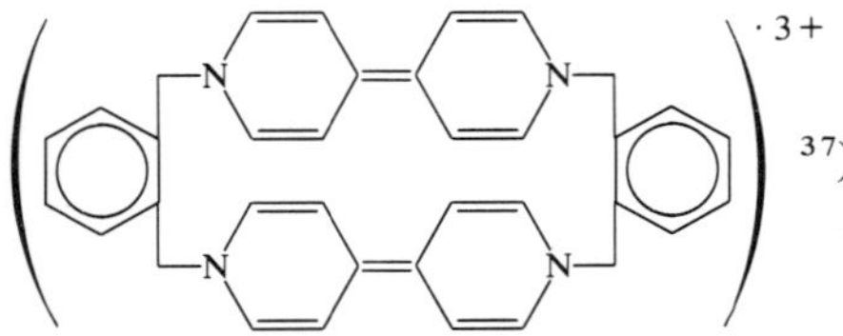 [37]	Irr. in presence of TiO_2 particles/ H_2O: TiO_2 particles	ESR/ RT	2.005	[38]	86Cr1

[35]) Electron exchange broadening studied.
[36]) Also by irr. in presence of TiO_2 particles.
[37]) Mixture of blue (3+) and red (2+) species.
[38]) >61 lines. Separation ≈ 0.03 mT.

Substance	Generation/ Matrix or Solvent	Method/ T [K]	g-Factor	a-Value [mT]	Ref./ add. Ref.
20.5.2.2 Phenoxazines, phenothiazines, dihydrothioalloxazines					
$[C_{11}H_9N_3O_2S]^{\cdot +}$ 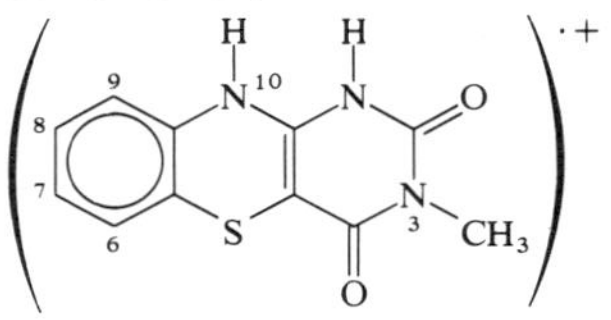	Ox. with dibenzyl peroxide/ CF_3CO_2H: toluene	ENDOR, TRIPLE/ 260···310		N(10): 0.3893 3H(3, CH_3): −0.044 H(6): −0.0139 H(7): −0.0971 H(8): −0.1378 H(10): −0.4211	84Ku1
$[C_{12}H_9NO]^{\cdot +}$ 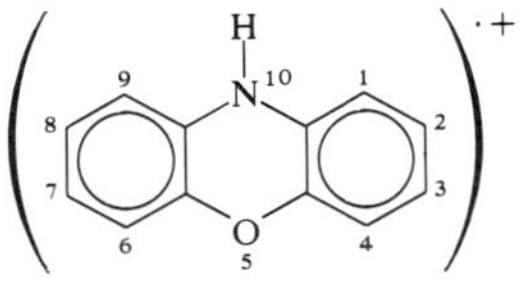 (phenoxazine)	Ox. with H_2SO_4/ CH_3CN	ESR/ 295		N: 0.775 2H(2: 8): 0.065 2H(3, 7): 0.325	81He1
	Phot./ Benzophenone	ESR/ RT		N: 0.78 H(NH): 0.78 2H(3, 7): 0.39 2H(1, 9): 0.087 H(4): 0.074 or 0.086 H(6): 0.086 or 0.074	86Li1
	Ox. with $AlCl_3$/ CH_3NO_2:$C_6H_5CH_3$ (4:1)	ESR, ENDOR/ 230···238	2.0037	N: 0.755 H(NH): 0.860 2H(1, 9): 0.160 2H(2, 8): 0.065 2H(3, 7): 0.328 2H(4, 6): 0.037	86Ka1
	−/ $SbCl_3$ [39])	ESR/ 403		H(NH): 0.845 N: 0.769 H(3, 7): 0.314 H(1, 9): 0.141 H(5, 6): 0.064 H(2, 8): 0.064 ΔHpp: 0.0050	86Ch1

[39]) Also reported at 403 and 363 K in $SbCl_3$, $AlCl_3$, N-1-butyl-pyridimium $AlCl_4^-$ mixtures.

Substance	Generation/ Matrix or Solvent	Method/ T [K]	g-Factor	a-Value [mT]	Ref./ add. Ref.
$[C_{12}H_9NS]^{\cdot+}$ (phenothiazine)	ClO_4^- salt/ CH_3CN	ESR/ 295		N: 0.650 H(2): 0.050 H(3): 0.257	81He1
	Photochem. ox./ CF_3CO_2H	ESR, CIDEP [40])/ 296	2.0051 (10)	N: 0.634 H(NH): 0.729 2H(1, 9): 0.113 4H: 0.050 2H(3, 7): 0.249	83De1
	Ox. with O_2/ Dilute H_2SO_4	ESR/ RT		N: 0.650 H(NH): 0.736 2H(1, 9): 0.127 2H(2, 8): 0.052 2H(3, 7): 0.263 2H(4, 6): 0.039	85Mo1/ 66Gi1, 65Lh1
	Isolated cation/ H_2SO_4	ESR/ 253	2.0052	N: 0.652 H(NH): 0.736 2H(3, 7): 0.258 2H(2, 8): 0.048 2H(1, 9): 0.126 2H(4, 6): 0.046	84Ru1/ 86Ru1
	–/ $SbCl_3$ [39])	ESR/ 403		H(NH): 0.689 N: 0.622 2H(3, 7): 0.242 2H(1, 9): 0.102 2H(4, 6): 0.065 2H(2, 8): 0.045 ΔHpp: 0.075	86Ch1
(continued)	Ox. with $AlCl_3$/ $CH_3NO_2:C_6H_5CH_3$ (4:1)	ESR, ENDOR/ 230···238	2.0055	N: 0.615 H(NH): 0.717 2H(1, 9): 0.119 2H(2, 8): 0.047 2H(3, 7): 0.251 2H(4, 6): 0.047	86Ka1

[39]) Also reported at 403 and 363 K in $SbCl_3$, $AlCl_3$, N-1-butyl-pyridimium $AlCl_4^-$ mixtures.
[40]) T_1 for CIDEP 1.8(5) µs at 296 K.

Substance	Generation/ Matrix or Solvent	Method/ T [K]	g-Factor	a-Value [mT]	Ref./ add. Ref.
$[C_{12}H_9NS]^{\cdot +}$ (*continued*)	Isolated cation/ 25 Vol.-% H_2SO_4	ESR/ 253	2.0052	N: 0.650 H(NH): 0.736 2H(1, 9): 0.126 2H(2, 8): 0.048 2H(3, 7): 0.258 2H(4, 6): 0.046 ΔHpp: 0.022	84Ru1/ 64Me1
	Powdered perchlorate salt	77	2.0072; 2.0060; 2.0024; is: 2.0052	H(NH): 0.85 [41] N: 0.13; 0.13; 1.70; is: 0.65	
		293	2.0056; 2.0023; 2.0023; is: 2.0045		
$[C_{12}H_8ClNS]^{\cdot +}$ 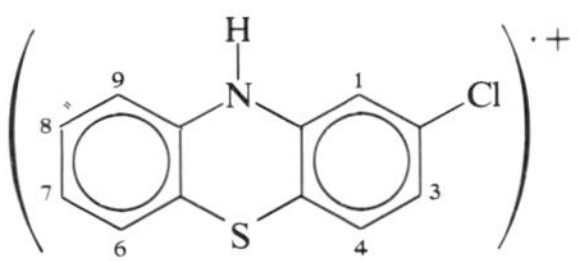	Air ox./ Dilute H_2SO_4	–/ RT		N: 0.632 H(NH): 0.699 2H(1, 9): 0.110, 0.119 H(8): 0.050 H(3, 7): 0.253, 0.260 H(4): 0.040 or 0.050 H(6): 0.050 or 0.040	85Mo1
$[C_{12}H_{11}N_3O_2S]^{\cdot +}$	Ox. with dibenzyl peroxide/ CF_3CO_2H: toluene	ENDOR, TRIPLE/ 260···310		N(10): 0.467 3H(3, CH_3): +0.035 H(7): −0.065 H(8): −0.154 3H(10, CH_3): +0.505	84Ku1
	CF_3CO_2H	110···150		3H(10, CH_3): 0.596; 0.479; 0.518; is: 0.531	

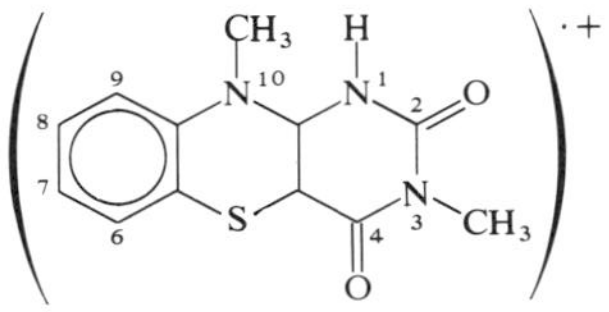

[41]) One principal value.

Substance	Generation/ Matrix or Solvent	Method/ T [K]	g-Factor	a-Value [mT]	Ref./ add. Ref.
$[C_{13}H_{11}NS]^{\cdot+}$ 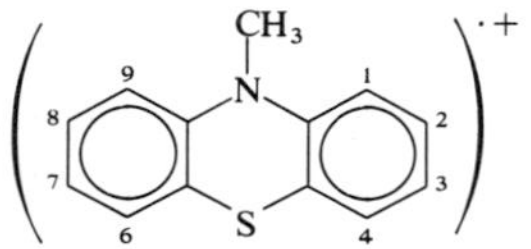	Ox. with $AlCl_3$/ $CH_3NO_2 : C_6H_5CH_3$	ESR, ENDOR/ 230⋯238	2.0055	N: 0.750 3H(CH_3): 0.725 2H(1, 9): 0.099 2H(2, 8): 0.070 2H(3, 7): 0.213 2H(4, 6): 0.023	86Ka1/ 82Ho1
	ClO_4^- salt/ CH_3CN	ESR/ 295		N: 0.755 H(2): 0.072 H(3): 0.213	81He1
	ClO_4^- salt/ CH_3NO_2	ESR/ 213	2.0050	2H(1, 9): 0.101 2H(2, 8): 0.077 2H(3, 7): 0.224 2H(4, 6): 0.026 N: 0.765 3H(CH_3): 0.742 2H(1, 9): 0.076 [42] 2H(2, 8): 0.070 2H(3, 7): 0.217 2H(4, 6): 0.037 N: 0.764 3D(CD_3): 0.112	80Fu1
	Ox. with $Pb(OOCCH_3)_4$/ CCl_4	ESR/ 243	2.0050		80Fu1
$[C_{13}H_{13}N_3O_2S]^{\cdot+}$ 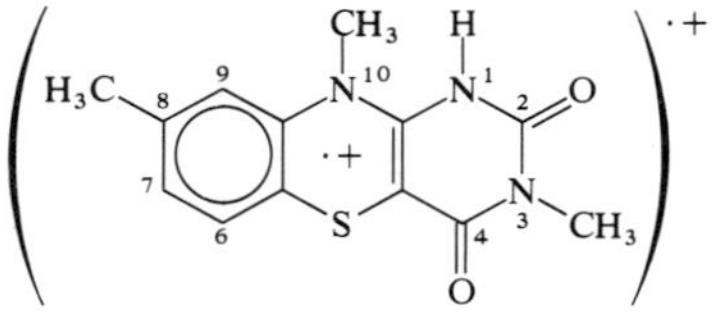	Ox. with dibenzyl peroxide/ CF_3CO_2H: toluene	ENDOR, TRIPLE/ 260⋯310		N(10): 0.342 3H(3, CH_3): 0.02 H(7): 0.069 3H(8, CH_3): 0.121 3H(10, CH_3): 0.397	84Ku1

(continued)

[42]) Deuterated species.

Substance	Generation/ Matrix or Solvent	Method/ T [K]	g-Factor	a-Value [mT]	Ref./ add. Ref.
$[C_{13}H_{13}N_3O_2S]^{\cdot+}$ *(continued)*	Ox. with dibenzyl peroxide/ CF_3CO_2H	ENDOR, TRIPLE/ 110···150		3H(10, CH_3): 0.579; 0.479; 0.479; is: 0.511 3H(8, CH_3): 0.271; 0.225; 0.225; is: 0.239	84Ku1
$[C_{13}H_{13}N_3O_2S]^{\cdot+}$ 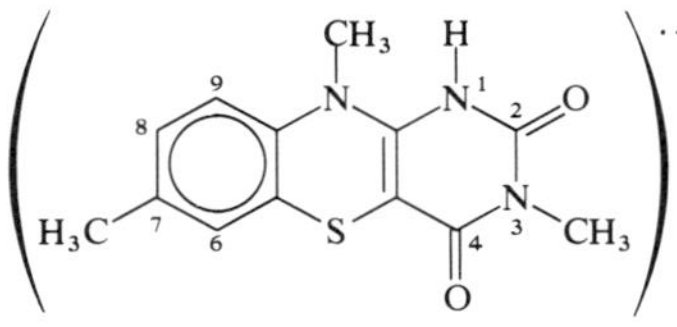	Ox. with dibenzyl peroxide/ CF_3CO_2H: toluene	ENDOR, TRIPLE/ 260···310		N(10): 0.476 3H(3, CH_3): +0.0379 3H(7, CH_3): +0.1 H(8): −0.174 3H(10, CH_3): +0.511	84Ku1
$[C_{14}H_{10}N_2S]^{\cdot+}$ 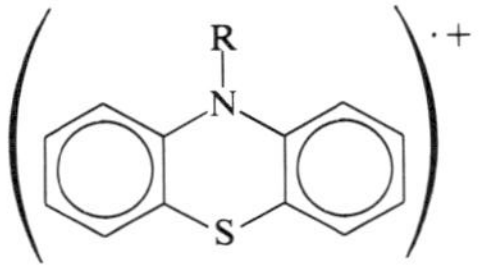 **PT**, $R=CH_2CN$	Ox. with $Tl(OOCCH_3)_4$/ CH_3NO_2	ESR/ 295		N: 0.648	81Ha1 [43])
$[C_{14}H_{11}NOS]^{\cdot+}$ **PT**, $R=CH_2C({=}O)H$	Ox. with $Tl(OOCCH_3)_4$/ CH_3NO_2	ESR/ 295		N: 0.685	81Ha1 [43])
$[C_{14}H_{12}ClNS]^{\cdot+}$ **PT**, $R=CH_2CH_2Cl$	Ox. with $Tl(OOCCH_3)_4$/ CH_3NO_2	ESR/ 295		N: 0.700	81Ha1 [43])

[43]) Correlations of a(N) with σ_T.

Substance	Generation/ Matrix or Solvent	Method/ T [K]	g-Factor	a-Value [mT]	Ref./ add. Ref.
$[C_{14}H_{13}NS]^{\cdot +}$ 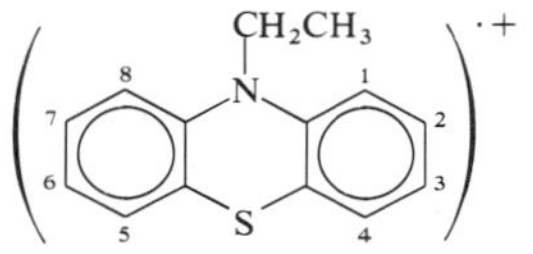	ClO_4^- salt/ CH_3NO_2	ESR/ 243	2.0051	H(1): 0.099 [44] H(2): 0.080 H(3): 0.214 H(4): 0.023 N: 0.753 2H(CH_2): 0.374 3H(CH_3): 0.018	80Fu1, 78Cl1/ 81Ha1
	CH_3NO_2	295		N: 0.751	
	Ox. with $AlCl_3$/ CH_3NO_2 : $C_6H_5CH_3$ (4:1)	ESR, ENDOR/ 230···238	2.0055	N: 0.720 2H(CH_2): 0.374 2H(1, 9): 0.098 2H(2, 8): 0.076 2H(3, 7): 0.214	86Ka1
$[C_{14}H_{15}N_3O_2S]^{\cdot +}$ 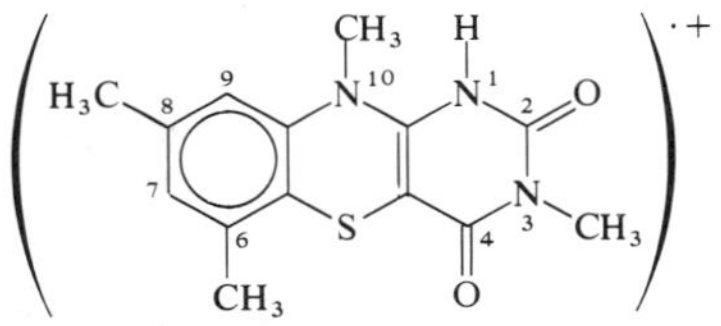	Ox. with dibenzyl peroxide/ CF_3CO_2H : toluene	ENDOR, TRIPLE/ 260···310		N(10): 0.457 3H(3, CH_3): +0.035 3H(6, CH_3): −0.013 H(7): −0.099 3H(8, CH_3): +0.220 3H(10, CH_3): +0.496	84Ku1
	CF_3CO_2H	110···150		3H(10, CH_3): 0.579; 0.475; 0.475; is: 0.511 3H(8, CH_3): 0.261; 0.214; 0.214; is: 0.229	

[44]) $N-CD_2CD_3$ species: N: 0.74 mT, $g = 2.0051$.

Substance	Generation/ Matrix or Solvent	Method/ T [K]	g-Factor	a-Value [mT]	Ref./ add. Ref.
$[C_{14}H_{15}N_3O_2S]^{\cdot +}$	Ox. with dibenzyl peroxide/ CF_3CO_2H: toluene	ENDOR, TRIPLE/ 260···310 (265)		N(10): 0.467 3H(3, CH_3): +0.038 H(6): 0.011 3H(7, CH_3): +0.114 3H(8, CH_3): +0.225 3H(10, CH_3): +0.502	84Ku1/ 81Bo1
		110···150		3H(10, CH_3): 0.586; 0.468; 0.468; is: 0.507 3H(8, CH_3): 0.286; 0.225; 0.225; is: 0.246 3H(7, CH_3): 0.161; 0.104; 0.104; is: 0.121	
$[C_{14}H_{13}D_2N_3O_2S]^{\cdot +}$ 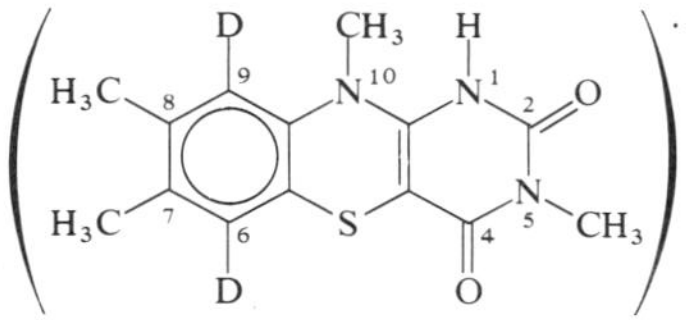	Ox. with dibenzyl peroxide/ CF_3CO_2D: toluene-d_8	ENDOR/ 260···310		N(10): 0.468 3H(3, CH_3): 0.036 3H(7, CH_3): 0.111 3H(8, CH_3): 0.218 3H(10, CH_3): 0.505	84Ku1
$[C_{15}H_{12}N_2S]^{\cdot +}$ 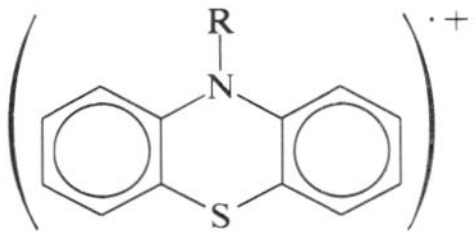 **PT**, R = CH_2CH_2CN	Ox. with $Tl(OOCCH_3)_3$/ CH_3NO_2	ESR/ 295		N: 0.692	78Cl1/ 81Ha1
$[C_{15}H_{13}NO_2S]^{\cdot +}$ **PT**, R = $CO_2C_2H_5$	Ox. with $Tl(OOCCH_3)_3$/ CH_3NO_2	ESR/ 295		N: 0.632	78Cl1/ 81Ha1

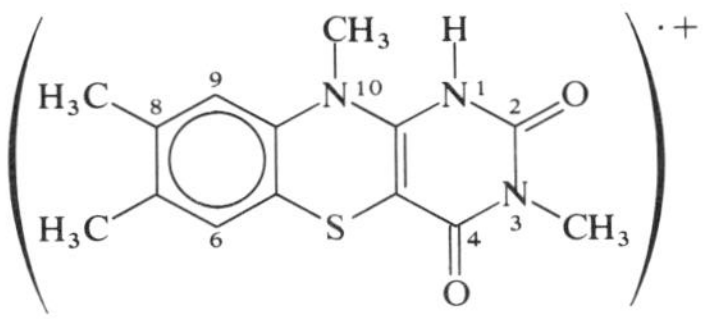

Substance	Generation/ Matrix or Solvent	Method/ T [K]	g-Factor	a-Value [mT]	Ref./ add. Ref.
$[C_{15}H_{14}ClNS]^{\cdot +}$ **PT**, $R=(CH_2)_3Cl$ [45]	Ox. with $Tl(OOCCH_3)_3$/ CH_3NO_2	ESR/ 295		N: 0.732	81Ha1
$[C_{15}H_{15}NS]^{\cdot +}$ **PT**, $R=CH_2CH_2CH_3$ [46]	Ox. with $Tl(OOCCH_3)_3$/ CH_3NO_2	ESR/ 295		N: 0.751	81Ha1
$[C_{16}H_{14}N_2S]^{\cdot +}$ **PT**, $R=(CH_2)_3CN$	Ox. with $Tl(OOCCH_3)_3$/ CH_3NO_2	ESR/ 295		N: 0.730	81Ha1
$[C_{16}H_{15}NO_2S]^{\cdot +}$ **PT**, $R=CH_2CO_2C_2H_5$	Ox. with $Tl(OOCCH_3)_3$/ CH_3NO_2	ESR/ 295		N: 0.685	78Cl1/ 81Ha1
$[C_{16}H_{21}N_3S]^{\cdot 3+}$ [47]	Rad. of methylene blue/ H_2O: H_2SO_4	ESR/ 295	2.0059	1N: 0.66	83Co1
$[C_{17}H_{16}N_2S]^{\cdot +}$ **PT**, $R=(CH_2)_4CN$	Ox. with $Tl(OOCCH_3)_3$/ CH_3NO_2	ESR/ 295		N: 0.745	81Ha1
$[C_{17}H_{17}NO_2S]^{\cdot +}$ **PT**, $R=CH_2CH_2CO_2C_2H_5$	Ox. with $Tl(OOCCH_3)_3$/ CH_3NO_2	ESR/ 295		N: 0.725	81Ha1
$[C_{17}H_{20}N_2S]^{\cdot +}$ **PT**, $R=(CH_2)_3N(CH_3)_2$ (promazine) *(continued)*	Ox. with horseradish peroxidase + H_2O_2/ H_2O (pH = 3.7)	ESR/ –		N(10): 0.726 2H(10, CH_2): 0.361 2H(10, CH_2CH_2): 0.017 2H(1, 9): 0.092 2H(2, 8): 0.084 2H(3, 7): 0.204 2H(4, 6): 0.042	85Mo1

[45]) $N-(CH_2)_4Cl$ and $N(CH_2)_5Cl$ species also measured.
[46]) NC_4H_9 species also measured.
[47]) Identification as trication radical is based on pH study; maximal concentration at 22.2 N H_2SO_4.

Substance	Generation/ Matrix or Solvent	Method/ T [K]	g-Factor	a-Value [mT]	Ref./ add. Ref.
$[C_{17}H_{20}N_2S]^{\cdot +}$ *(continued)*	ClO_4^- salt/ H_2SO_4	ESR/ 295	2.0053	N(10): 0.708 2H(10, CH_2): 0.352 2H(3, 7): 0.196 2H(2, 8): 0.092 2H(1, 9): 0.080 2H(4, 6): 0.040	83Ru1/ 86Ru1
$[C_{17}H_{20}N_2S]^{\cdot +}$ **PT**, R = $CH_2CH(CH_3)N(CH_3)_2$ (promethazine)	ClO_4^- salt/ H_2SO_4	ESR/ 273	2.0052	N(10): 0.640 1H(N(10)CH_2): 0.262 H(3, 7): 0.130 1H(N(10)CH_2): 0.730 H(2, 8): 0.124 H(1, 9): 0.048 H(4, 6): 0.046	86Ru1
	H_2SO_4	77	2.0072; 2.0060; 2.0024; is: 2.0052	N(10): 13; 0.13; 1.66; is: 0.04	
	Ox. with O_2/ 5N DCl	–/ –		N(10): 0.682 1H(N(10)CH_2): 0.036 1H(N(10)CH_2): 0.739 1H(NCH_2CH): 0.015 2H(1, 9): 0.113 2H(2, 8): 0.068 2H(3, 7): 0.173 2H(4, 6): 0.015	85Mo1
$[C_{17}H_{19}ClN_2S]^{\cdot +}$ (10-$(CH_2)_3N(CH_3)_2$-2-Cl-phenothiazine)$^{\cdot +}$ (chlorpromazine) *(continued)*	ClO_4^- salt/ H_2SO_4	ESR/ 295	2.0056	N: 0.681 2H(NCH_2): 0.339 2H(3, 7): 0.192 2H(1, 9): 0.063 H(8): 0.111 2H(4, 6): 0.033	85Ru3

Substance	Generation/ Matrix or Solvent	Method/ T [K]	g-Factor	a-Value [mT]	Ref./ add. Ref.
$[C_{17}H_{19}ClN_2S]^{\cdot +}$ *(continued)*	ClO_4^- salt/ H_2SO_4	ESR/ 77	2.0076; 2.0069; 2.0022; is: 2.0056	N: 1.68; 0.18; 0.18; is: 0.68 2H(NCH_2): 0.339	85Ru3
	H_2SO_4 powder	un-specified	2.0073; 2.0058; 2.0022; is: 2.0051		
	H_2SO_4 single crystal	un-specified	2.0074; 2.0060; 2.0023; is: 2.0052		
	Ox. with O_2/ 20% DCl in D_2O [48]	/		N: 0.684 2H(NCH_2): 0.342 1H(NCH_2CH): 0.042 2H(1, 9): 0.067, 0.106 H(8): 0.042 2H(3, 7): 0.195 2H(4, 6): 0.042	85Mo1
$[C_{17}H_{21}N_2S]^{\cdot 2+}$ **PT**, R = $CH_2CH_2N(CH_3)_3^+$	Ox. with $Tl(OOCCH_3)_3$/ CH_3NO_2	ESR/ 295		N: 0.667	81Ha1
$[C_{18}H_{13}NS]^{\cdot +}$ (structure: N-phenylphenothiazine, C_6H_5 on N; $^{\cdot +}$)	–/ 80% aqueous H_2SO_4	ESR/ 295	2.0050	[49]	84Jo1
$[C_{18}H_{18}N_2S]^{\cdot +}$ **PT**, R = $(CH_2)_5CN$	Ox. with $Tl(OOCCH_3)_3$/ CH_3NO_2	ESR/ 295		N: 0.750	81Ha1

[48]) Also in dilute H_2SO_4.
[49]) Unresolved. Spin on charge in phenothiazine ring.

Substance	Generation/ Matrix or Solvent	Method/ T [K]	g-Factor	a-Value [mT]	Ref./ add. Ref.
$[C_{18}H_{22}N_2S]^{\cdot +}$ **PT**, R $=(CH_2)_2N(CH_2CH_3)_2$	—/ H_2SO_4	ESR/ 273	2.0054	N: 0.670 2H(NCH_2): 0.342 H(3, 7): 0.180 H(2, 8): 0.108 H(1, 9): 0.062 H(4, 6): 0.038	86Ru1
	H_2SO_4	77	2.0074; 2.0061; 2.0022; is: 2.0052	N: 0.16; 0.16; 1.69; is: 0.67	
$[C_{18}H_{22}N_2S]^{\cdot +}$ **PT**, R $=CH_2CH(CH_3)CH_2N(CH_3)_2$ (alimemazine, trimeprazine)	Ox. with horseradish peroxidase + H_2O_2/ H_2O (pH = 3.5)	EPR/ —		N(10): 0.752 1H(N(10)CH_2): 0.273 1H(N(10)CH_2): 0.425 1H(NCH_2CH): <0.03 2H(1, 9): 0.094 2H(2, 8): 0.092 2H(3, 7): 0.205 H(4, 6): <0.03	85Mo1
	ClO_4^- salt/ H_2SO_4	ESR/ 77	2.0075; 2.0059; 2.0023; is: 2.0052	N(10): 36; 0.20; 1.56; is: 0.71 H: 0.18, 0.56 [50]	82Ru1
(continued)	pure		2.0078; 2.0061; 2.0022; is: 2.0053		

[50]) Two principal values.

Substance	Generation/ Matrix or Solvent	Method/ T [K]	g-Factor	a-Value [mT]	Ref./ add. Ref.
$[C_{18}H_{22}N_2S]^{\cdot +}$ *(continued)*	ClO_4^- salt/ 25% $H_2SO_4:H_2O$	ESR/ 293	2.0053	N(10): 0.708 1H(β_1, NCH_2): 0.196 [51] 1H(β_2, NCH_2): 0.548 2H(3, 7): 0.196 2H(1, 9): 0.104 2H(2, 8): 0.096 2H(4, 6): 0.028 ΔHpp [52]: 0.060 ΔHpp [53]: 0.076	82Ru2
		273		N: 0.708 1H(β_1, NCH_2): 0.192 1H(β_2, NCH_2): 0.548 2H(3, 7): 0.196 2H(1, 9): 0.104 2H(2, 8): 0.096 2H(4, 6): 0.028 ΔHpp [52]: 0.080 ΔHpp [53]: 0.096	
		253		N: 0.708 1H(β_1, NCH_2): 0.188 1H(β_2, NCH_2): 0.552 2H(3, 7): 0.196 2H(1, 9): 0.104 2H(2, 8): 0.096 2H(4, 6): 0.028 ΔHpp [52]: 0.100 ΔHpp [53]: 0.108	
		233		N: 0.708 1H(β_1, NCH_2): 0.176 1H(β_2, NCH_2): 0.564 2H(3, 7): 0.196 2H(1, 9): 0.104 2H(2, 8): 0.096 2H(4, 6): 0.028 ΔHpp [52]: 0.140 ΔHpp [53]: 0.140	

[51]) *Gauche* conformation.
[52]) Linewidth of lateral groups of lines of the H(β) quartets.
[53]) Linewidth of central groups of lines of the H(β) quartets.

Substance	Generation/ Matrix or Solvent	Method/ T [K]	g-Factor	a-Value [mT]	Ref./ add. Ref.
$[C_{18}H_{23}N_2S]^{\cdot 2+}$ **PT**, $R=(CH_2)_3N(CH_3)_3^+$	Ox. with $Tl(OOCCH_3)_3$/ CH_3NO_2	ESR/ 295		N: 0.715	81Ha1
$[C_{19}H_{15}NS]^{\cdot +}$ **PT**, $R=CH_2C_6H_5$	Ox. with $Tl(OOCCH_3)_3$/ CH_3NO_2	ESR/ 295		N: 0.710	18Cl1/ 81Ha1
$[C_{19}H_{24}N_2S]^{\cdot +}$ **PT**, $R=CH_2CH(CH_3)N(CH_2CH_3)_2$ (isothiazine)	–/ H_2SO_4	ESR/ 253	2.0052	N(10): 0.646 1H(N(10)CH_2): 0.264 1H(N(10)CH_2): 0.736 2H(3, 7): 0.130 2H(2, 8): 0.124 2H(1, 9): 0.048 2H(4, 6): 0.046	86Ru1
		77	2.0072; 2.0080; 2.0024; is: 2.0059	N(10): 0.17; 0.17; 1.60; is: 0.646	
$[C_{19}H_{25}N_2S]^{\cdot 2+}$ **PT**, $R=(CH_2)_4N(CH_3)_3^+$	Ox. with $Tl(OOCCH_3)_3$/ CH_3NO_2	ESR/ 295		N: 0.735	81Ha1
$[C_{20}H_{20}N_2S]^{\cdot +}$ **PT**, $R=C_6H_5N(CH_3)_2$	–/ 80% aqueous H_2SO_4	ESR/ 295	2.0030	[54])	84Jo1
$[C_{20}H_{17}NS]^{\cdot +}$ **PT**, $R=CH_2CH_2C_6H_5$	Ox. with $Tl(OOCCH_3)_3$/ CH_3NO_2	ESR/ 295		N: 0.737	78Cl1/ 81Ha1
$[C_{20}H_{25}N_3S]^{\cdot +}$ **PT**, $R=(CH_2)_3$—N(piperazine ring)NCH$_3$ (perazine)	ClO_4^- salt/ H_2SO_4	ESR/ 295	2.0054	N(10): 0.706 2H(N(10)CH_2): 0.352 2H(3, 7): 0.200 2H(2, 8): 0.096 2H(1, 9): 0.080 2H(4, 6): 0.040	85Ru1/ 86Ru1
$[C_{20}H_{27}N_2S]^{\cdot 2+}$ **PT**, $R=(CH_2)_5N(CH_3)_3^+$	Ox. with $Tl(OOCCH_3)_3$/ CH_3NO_2	ESR/ 295		N: 0.748	81Ha1

[54]) Unresolved. Spin and charge in diaminobenzene ring.

Substance	Generation/ Matrix or Solvent	Method/ T [K]	g-Factor	a-Value [mT]	Ref./ add. Ref.
$[C_{21}H_{19}NS]^{\cdot +}$ **PT**, $R=(CH_2)_3C_6H_5$	Ox. with $Tl(OOCCH_3)_3$/ CH_3NO_2	ESR/ 295		N: 0.750	81Ha1
$[C_{28}H_{44}NO_2]^{\cdot +}$ 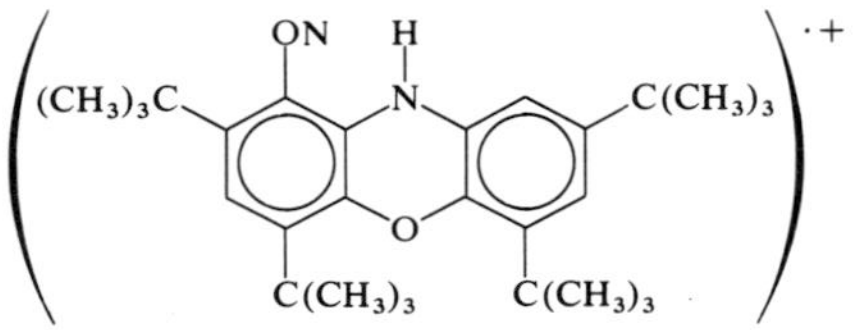	ClO_4^- salt or neutral radical $+CH_3CO_2H$ (in toluene)/ Toluene(?)	ESR/ RT		N: 0.78 H(NH): 0.78 [55]) 2H: 0.35 1H: 0.16	86Iv1

20.5.2.3 Biomolecules, including dihydroalloxazines

Substance	Generation/ Matrix or Solvent	Method/ T [K]	g-Factor	a-Value [mT]	Ref./ add. Ref.
$[C_3H_5N_3O]^{\cdot +}$ 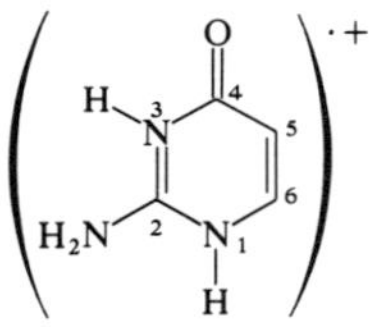 (isocytosine)	γ-irr./ Single crystal	ESR/ 77	2.0022; 2.0047; 2.0037; is: 2.0035	H(5): -1.78; -0.73; -2.37; [56]) is: -1.17 N(1): 0.79; ≈ 0; ≈ 0; is: ≈ 0.26 N(3): 1.39; ≈ 0; ≈ 0; is: ≈ 0.46	79He1
$[C_5H_5BrN_2O_2]^{\cdot +}$ 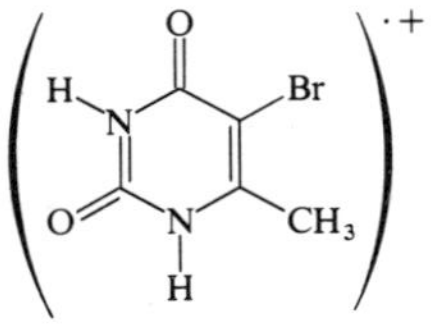	X-irr./ 5.3 M H_2SO_4	ESR/ 77	1.996; 2.033; 2.030; is: 2.020	Br: 13.4; -2.6; -4.0; is: -1.75 N: 1.3; 0; 0; is: 0.43	82Ri1
$[C_6H_5BrN_2O_2]^{\cdot +}$ 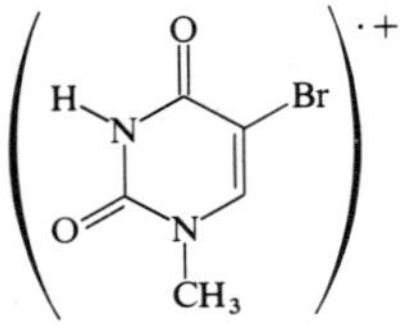	X-irr./ 5.3 M H_2SO_4	ESR/ 77	1.996; 2.033; 2.030; is: 2.020	Br: 15.6; -3.0; -4.5; is: 2.7 N: 2.030; 0; 0; is: 0.43	82Ri1

[55]) ND compound and once and twice deprotonated species also studied.
[56]) ND species also studied.

Substance	Generation/ Matrix or Solvent	Method/ T [K]	g-Factor	a-Value [mT]	Ref./ add. Ref.
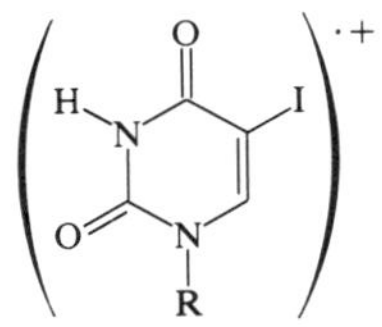 R = ribose	X-irr./ 5.3 M H_2SO_4	ESR/ 77	1.94; 2.05; 2.01; is: 2	I: ≈10.0 [57] N: ≈1.3; 0; 0; is: ≈0.43	82Ri1
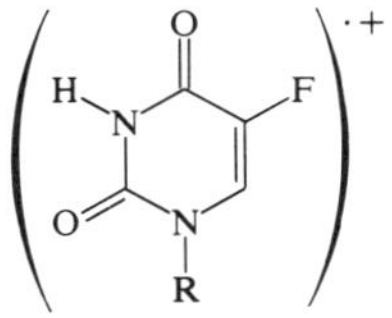 R = deoxyribose	X-irr./ 5.3 M H_2SO_4	ESR/ 77	2.0000; 2.0075; 2.0050; [57a] is: 2.0032	F: 1.70; −1.8; −1.8 is: −0.63 N: 1.5; 0; 0; is: 0.5 H: 0.5	82Ri1
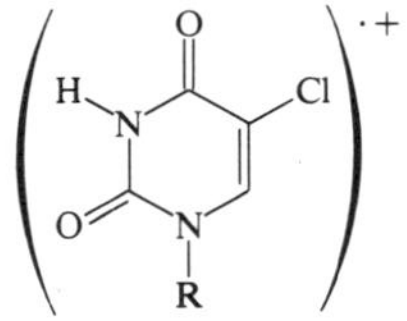 R = deoxyribose	X-irr./ 5.3 M H_2SO_4	ESR/ 77	2.0000; 2.0070; 2.0080; is: 2.005	Cl: 2.2; −0.5; −0.6; is: 0.37 N: 1.3; 0; 0; is: 0.43 H: ≈0.5	82Ri1
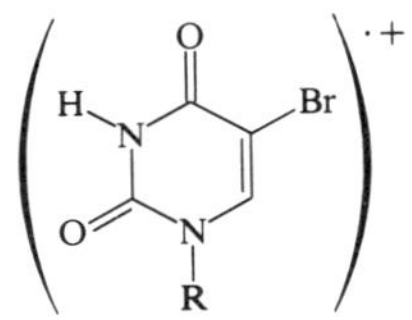 R = deoxyribose	X-irr./ 5.3 M H_2SO_4	ESR/ 77	1.996; 2.033; 2.030; is: 2.020	Br: 5.5; −3.0; −4.5; is: 2.7 N: ≈1.3; 0; 0; is: ≈0.43 H: ≈0.5 [58]	82Ri1

[57]) One principal value.
[57a]) $z \| C–F$ bond.
[58]) H = 0.5 mT for R = ribose, unobserved for R = H; other spectral parameters as above.

Substance	Generation/ Matrix or Solvent	Method/ T [K]	g-Factor	a-Value [mT]	Ref./ add. Ref.
$[C_5H_5N_5O]^{\cdot +}$ 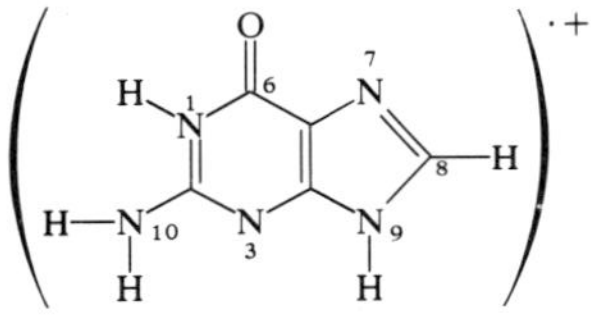 (guanine)	X-irr. of guanine $+HCl+H_2O$/ Single crystal	ESR/ 15	2.0052; 2.0040; 2.0025; is: 2.0039	[59]) N(3): 1.54; 0.27; 0.00; is: 0.60 N(10): 0.92; 0.17; 0.00; is: 0.36 H(8): 0.75; 0.50; 0.30; is: 0.52 H(10, 10′): 0.79; 0.50; 0.05; is: 0.43	85Cl2
$[C_8H_{10}N_4O_2]^{\cdot +}$ 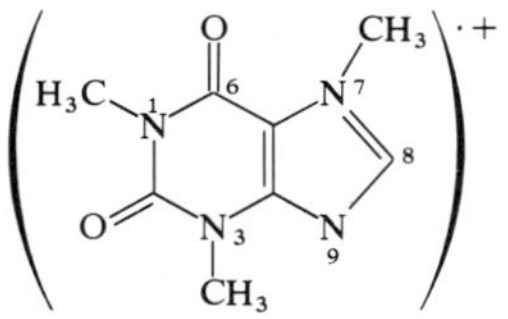	X-irr. of caffeine hydrochloride dihydrate/ Single crystal	ENDOR/ –		3H(7, CH_3): 0.653; 0.614; 0.803; is: 0.69 3H(3, CH_3): 0.167; 0.152; 0.253; is: 0.191	84Le2
$[C_8H_{12}N_4O_2]^{\cdot +}$ 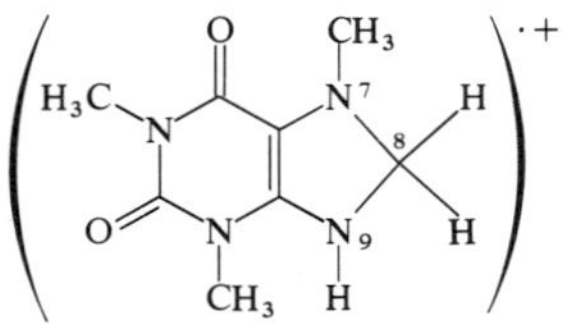 (caffeine)	X-irr. of caffeine hydrochloride dihydrate/ Single crystal	ESR, ENDOR/ –		H(9): −0.225; −0.066; 0.021; is: −0.311 3H(7, CH_3): 1.047; 1.015; 1.25; is: 1.104 ^{14}N [60]): 0.405; 0.366; 2.540; is: 1.104 1H(8): 3.516; 3.567; 3.877; is: 3.653 1H(8′): 3.745; 3.805; 4.102; is: 3.884	84Le2
$[C_{10}H_{17}N_5O]^{\cdot +}$ 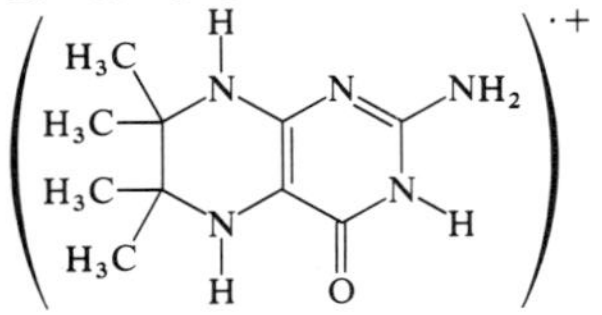	Mixture of reduced and oxidized forms/ CF_3CO_2H	ESR/ 295	2.0008 (15)		84Eb1

[59]) Data for $DCl \cdot D_2O$ spectrum used in assignment.
[60]) Quadrupole coupling tensor also given.

Substance	Generation/ Matrix or Solvent	Method/ T [K]	g-Factor	a-Value [mT]	Ref./ add. Ref.
$[C_7H_8N_4O_2]^{\cdot +}$ 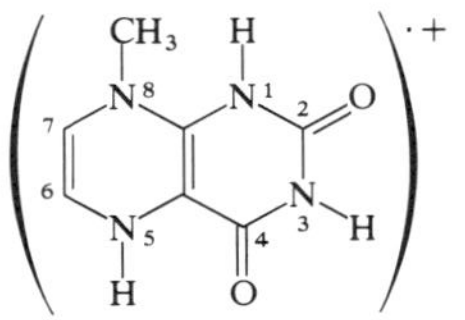	–/ CF_3CO_2H	ESR/ –		N(8): 0.586 3N(NH): 0.739 3H(8, CH_3): 0.586 3H(NH): 0.816 H(7): 0.586 H(6): 0.127	84Ka2
$[C_{10}H_8N_4O_2]^{\cdot +}$ 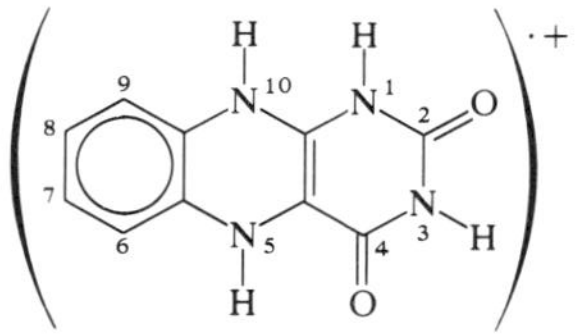 (dihydroalloxazine)	Red. of the alloxazine with Na_2SiO_4 or Zn in acidic solution/ CF_3CO_2H:toluene	ENDOR/ 260···310	un-reported	N(5): 0.766 N(10): 0.401 H(NH, 5): 0.8 H(6): 0.14844 H(7): 0.045 H(NH, 10): 0.431	84Ku1
$[C_{12}H_{12}N_4O_2]^{\cdot +}$ 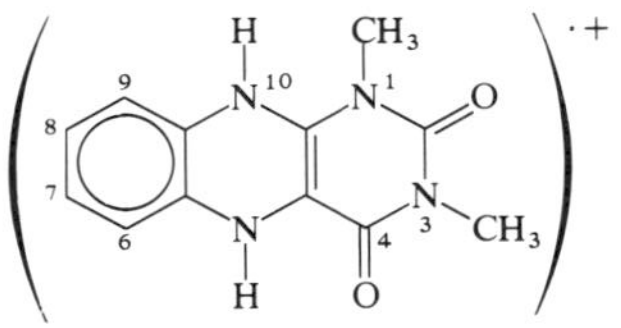	Red. of the alloxazine with $TiCl_3$ in acidic solution/ 6N HCl:CH_3CN (1:1)	ESR/ 294		N(5): 0.762 N(10): 0.402 H(5): 0.816 H(6): 0.152 H(7): 0.040 H(8): 0.251 H(10): 0.408 linewidth: 0.056	81Mü1
$[C_{12}H_{12}N_4O_2]^{\cdot +}$ 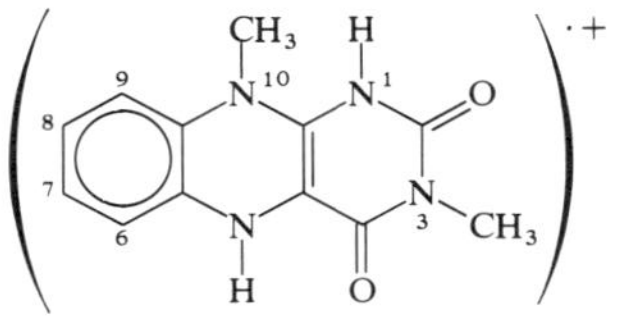	Red. of the isoalloxazine with $TiCl_3$ in acidic solution/ 6M HCl:CH_3CN (1:1)	ESR/ 294		N(5): 0.736 N(10): 0.484 H(5): 0.775 H(6): 0.175 H(7): 0.040 H(8): 0.270 H(9): 0.049 3H(10, CH_3): 0.503	81Mü1

Substance	Generation/ Matrix or Solvent	Method/ T [K]	g-Factor	a-Value [mT]	Ref./ add. Ref.
$[C_{12}H_8D_4N_4O_2]^{\cdot +}$	Red. of the isoalloxazine with $TiCl_3$ in acidic solution/ 6M HCl:CH_3CN (1:1)	ESR/ 294		N(5): 0.734 N(10): 0.481 H(5): 0.784 D(6): 0.035 D(8): 0.040 3H(10, CH_3): 0.502	81Mü1
$[C_{12}H_8D_4N_4O_2]^{\cdot +}$	Red. of the alloxazine with $TiCl_3$ in acidic solution/ 6N HCl:CH_3CN (1:1)	ESR/ 294		N(5): 0.760 N(10): 0.400 H(5): 0.807 D(6): 0.028 D(8): 0.040 H(10): 0.400	81Mü1
$[C_{12}H_6D_6N_4O_2]^{\cdot +}$	Red. of the alloxazine with $TiCl_3$ in acidic solution/ 6N DCl:CH_3CN (1:1)	ESR/ 294		N(5): 0.751 N(10): 0.394 D(5): 0.126 D(6): 0.029 D(8): 0.040 D(10): 0.062	81Mü1
$[C_{12}H_{12}N_4O_2]^{\cdot +}$	Red. of the alloxazine with $Na_2S_2O_4$ or Zn in acidic solution/	ENDOR/			84Ku1
	CF_3CO_2H:toluene	260···310		N(5): 0.750 N(10): 0.4075 H(NH, 5): 0.8125 H(6): 0.118 3H(7, CH_3): 0.067 H(8): 0.273 H(NH, 10): 0.427	
	HCl	110···150		H(6): is: 0.192	

Substance	Generation/ Matrix or Solvent	Method/ T [K]	g-Factor	a-Value [mT]	Ref./ add. Ref.
$[C_{12}H_{12}N_4O_2]^{\cdot +}$ 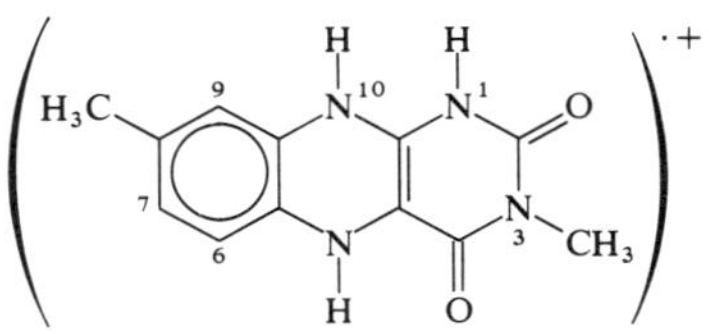	Red. of the alloxazine with $Na_2S_2O_4$ or Zn in acidic solution/	ENDOR, TRIPLE/			84Ku1
	CF_3CO_2H:toluene	260···310		N(5): 0.766 N(10): 0.395 3H(3, CH_3): 0.044 H(NH, 5): −0.837 H(6): −0.146 H(7): 0.075 3H(8, CH_3): 0.319 H(NH, 10): −0.426	
	DCl	110···150		3H(8, CH_3): 0.354; 0.296; 0.296; is: 0.314 H(6): is: 0.204	
$[C_{12}H_{12}N_4O_2]^{\cdot +}$ 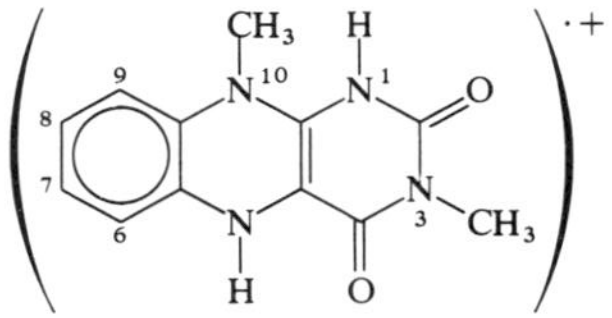	Red. of the isoalloxazine with $Na_2S_2O_4$ or Zn in acidic solution/	ENDOR, TRIPLE/			84Ku1
	CF_3CO_2H:toluene	260···310		N(5): 0.738 N(10): 0.481 H(NH, 5): −0.824 H(6): −0.172 H(7): −0.032 H(8): −0.278 3H(10, CH_3): 0.503	
	DCl	110···150		3H(10, CH_3): 0.593; 0.471; 0.471; is: 0.511 H(6): is: 0.218	

Substance	Generation/ Matrix or Solvent	Method/ T [K]	g-Factor	a-Value [mT]	Ref./ add. Ref.
$[C_{12}H_{12}N_4O_2]^{\cdot +}$	Red. of the alloxazine with $Na_2S_2O_4$ or Zn in acidic solution/	ENDOR, TRIPLE/			84Ku1
	CF_3CO_2H:toluene	260···310		N(5): 0.758 N(10): 0.395 H(3): −0.045 H(NH, 5): −0.828 H(6): −0.118 3H(7, CH_3): 0.084 3H(8, CH_3): 0.334 H(NH, 10): −0.426	
	HCl	110···150		3H(8, CH_3): 0.35; 0.307; 0.307; is: 0.321 3H(7, CH_3): is: 0.062 H(6): is: 0.171	
$[C_{13}H_{14}N_4O_2]^{\cdot +}$	Red. of the alloxazine with $Na_2S_2O_4$ or Zn in acidic solution/	ENDOR, TRIPLE/			84Ku1
	CF_3CO_2H:toluene	260···310		N(5): 0.762 N(10): 0.400 3H(3, CH_3): 0.042 H(NH, 5): 0.815 3H(6, CH_3): 0.066 H(7): −0.076 3H(8, CH_3): 0.3125 H(NH, 10): −0.4375	
	DCl	110···150		3H(8, CH_3): 0.336; 0.289; 0.289; is: 0.304 H(6): is: 0.196	

Substance	Generation/ Matrix or Solvent	Method/ T [K]	g-Factor	a-Value [mT]	Ref./ add. Ref.
$[C_{13}H_{10}D_4N_4O_2]^{\cdot +}$	Red. of the alloxazine with $Na_2S_2O_4$ or Zn in acidic solution/ CF_3CO_2D:toluene-d_8	ENDOR/ 260···310		N(5): 0.750 N(10): 0.395 3H(3, CH_3): 0.045 D(ND, 5): 0.132 H(6): 0.118 3H(7, CH_3): 0.083 3H(8, CH_3): 0.329 D(ND, 10): 0.071	84Ku1
$[C_{13}H_9D_5N_4O_2]^{\cdot +}$	Red. of the alloxazine with $Na_2S_2O_4$ or Zn in acidic solution/	ENDOR/			84Ku1
	CF_3CO_2D:toluene-d_8	260···320		N(10): 0.395 D(ND, 5): 0.132 3H(7, CH_3): 0.084 3H(8, CH_3): 0.325 D(ND, 10): 0.071	
	DCl	110···150		3H(8, CH_3): 0.354; 0.307; 0.307; is: 0.321 3H(7, CH_3): is: 0.064	
$[C_{13}H_{14}N_4O_2]^{\cdot +}$	Red. of the isoalloxazine with $TiCl_3$ in acidic solution/ 6N HCl:CH_3CN (1:1)	ESR/ 294		N(5): 0.723 N(10): 0.489 H(NH, 5): 0.771 H(6): 0.156 3H(7, CH_3): 0.042 H(8): 0.287 H(9): 0.031	81Mü1

(continued)

Substance	Generation/ Matrix or Solvent	Method/ T [K]	g-Factor	a-Value [mT]	Ref./ add. Ref.
$[C_{13}H_{14}N_4O_2]^{\cdot+}$ *(continued)*	Red. of the isoalloxazine with $Na_2S_2O_4$ or Zn in acidic solution/ CF_3CO_2H:toluene	ENDOR, TRIPLE/ 260···310		N(5): 0.726 N(10): 0.492 3H(3, CH_3): 0.035 H(NH, 5): 0.811 H(6): −0.1475 3H(7, CH_3): 0.052 H(8): −0.309 3H(10, CH_3): 0.506	84Ku1
	HCl	110···150		3H(10, CH_3): 0.596; 0.464; 0.464; is: 0.507 3H(7, CH_3): is: 0.204	
$[C_{13}H_{14}N_4O_2]^{\cdot+}$ 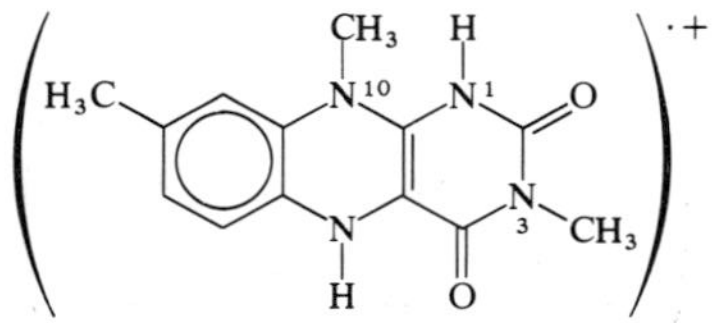	Red. of the isoalloxazine with $TiCl_3$ in acidic solution/ 6N HCl:CH_3CN (1:1)	ESR/ 294		N(5): 0.746 N(10): 0.477 H(NH, 5): 0.782 H(6): 0.176 H(7): 0.047 3H(8, CH_3): 0.318 H(9): 0.060	81Mü1
(continued)	Red. of the isoalloxazine with $Na_2S_2O_4$ or Zn in acidic solution/ CF_3CO_2H:toluene	ENDOR, TRIPLE/ 260···310		N(5): 0.75 N(10): 0.468 3H(3, CH_3): 0.031 H(NH, 5): −0.831 H(6): −0.170 H(7): −0.051 3H(8, CH_3): 0.341 3H(10, CH_3): 0.496	84Ku1

Substance	Generation/ Matrix or Solvent	Method/ T [K]	g-Factor	a-Value [mT]	Ref./ add. Ref.
$[C_{13}H_{14}N_4O_2]^{\cdot+}$ *(continued)*	Red. of the isoalloxazine with $Na_2S_2O_4$ or Zn in acidic solution/ HCl	ENDOR, TRIPLE/ 110···150		3H(10, CH_3): 0.579; 0.45; 0.45; is: 0.493 3H(8, CH_3): 0.361; 0.3; 0.3; is: 0.321 H(6): is: 0.229	84Ku1
$[C_{13}H_{14}N_4O_2]^{\cdot+}$ (lumiflavin)	Red. of the isoalloxazine with $Na_2S_2O_4$ or Zn in acidic solution/ CF_3CO_2H:toluene	ESR, ENDOR, TRIPLE/ 260···310		N(5): 0.740 N(10): 0.471 H(3): 0.032 H(NH, 5): −0.811 H(6): −0.144 3H(7, CH_3): 0.055 3H(8, CH_3): 0.355 3H(10, CH_3): 0.494	84Ka1, 81Bo1, 84Ka2
		265 [60a]		3H(7, CH_3): 0.031 H(10, CH_3): 0.495 H(9): 0.055 all other a-values are the same as from [84Ka1]	
$[C_{14}H_{16}N_4O_2]^{\cdot+}$ *(continued)*	Red. of the isoalloxazine with $TiCl_3$ in acidic solution/ 6M HCl:CH_3CN (1:1)	ESR/ 294		N(5): 0.732 N(10): 0.475 H(NH, 5): 0.779 H(6): 0.154 3H(7, CH_3): 0.050 3H(8, CH_3): 0.322 H(9): 0.040 3H(10, CH_3): 0.497	81Mü1

[60a]) From [81Bo1].

Substance	Generation/ Matrix or Solvent	Method/ T [K]	g-Factor	a-Value [mT]	Ref./ add. Ref.
$[C_{14}H_{16}N_4O_2]^{\cdot +}$ *(continued)*	Red. of the isoalloxazine with $Na_2S_2O_4$ or Zn in acidic solution/	ENDOR/			84Ku1
	HCl	110⋯150		$3H(10, CH_3)$: 0.593; 0.464; 0.464; is: 0.507 $3H(8, CH_3)$: 0.382; 0.321; 0.321; is: 0.343 $3H(7, CH_3)$: is: 0.068 H(6): is: 0.196	
	CF_3CO_2H:toluene	110⋯150		$3H(10, CH_3)$: 0.586; 0.464; 0.464; is: 0.504 $3H(8, CH_3)$: 0.407; 0.35; 0.35; is: 0.368 H(6): is: 0.193	
	CCl_3CO_2H:toluene			$3H(10, CH_3)$: 0.589; 0.471; 0.471; is: 0.511 $3H(8, CH_3)$: 0.386; 0.336; 0.336; is: 0.354 H(6): is: 0.193	
$[C_{14}H_{15}DN_4O_2]^{\cdot +}$ 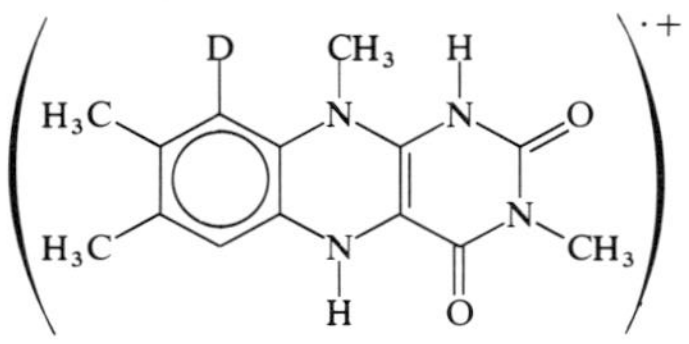	Red. of the isoalloxazine with $TiCl_3$ in acidic solution/ 6M $HCl:CH_3CN$ (1:1)	ESR/ 294		N(5): 0.731 N(10): 0.468 H(NH, 5): 0.779 H(6): 0.170 $3H(7, CH_3)$: 0.054 $3H(8, CH_3)$: 0.332	81Mü1

Substance	Generation/ Matrix or Solvent	Method/ T [K]	g-Factor	a-Value [mT]	Ref./ add. Ref.
$[C_{14}H_{14}D_2N_4O_2]^{\cdot +}$ 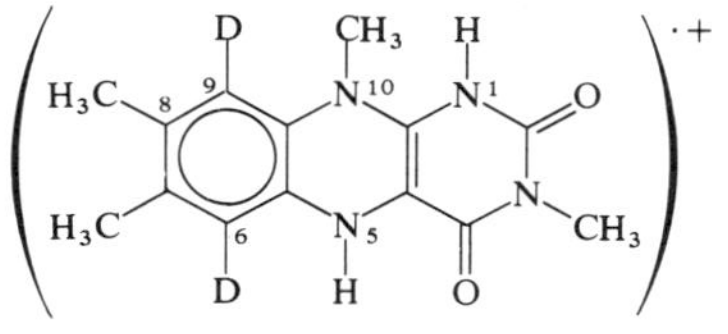	Red. of the isoalloxazine with $TiCl_3$ in acidic solution/ 6N HCl:CH_3CN (1:1)	ESR/ 294		N(5): 0.734 N(10): 0.476 H(5): 0.782 D(6): 0.025 3H(7, CH_3): 0.051 3H(8, CH_3): 0.318 3H(10, CH_3): 0.501	81Mü1
$[C_{14}H_{13}D_3N_4O_2]^{\cdot +}$ 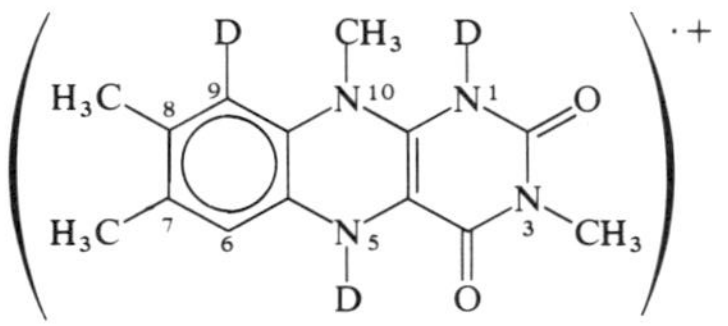	Red. of the isoalloxazine with $Na_2S_2O_4$ or Zn in acidic solution/ CF_3CO_2D:toluene-d_8	ENDOR/ 260···310		N(10): 0.476 3H(3, CH_3): 0.032 D(ND, 5): 0.131 H(6): 0.142 3H(7, CH_3): 0.059 3H(8, CH_3): 0.354 3H(10, CH_3): 0.5025	84Ku1
$[C_{14}H_{11}D_5N_4O_2]^{\cdot +}$	Red. of the isoalloxazine with $Na_2S_2O_4$ or Zn in acidic solution/	ENDOR/			84Ku1
	CF_3CO_2D:toluene-d_8	260···310		N(5): 0.806 N(10): 0.471 3H(3, CH_3): 0.031 D(ND, 5): 0.127 H(6): 0.138 3H(7, CH_3): 0.056 3H(8, CH_3): 0.353 3D(10, CD_3): 0.071	
	DCl	110···150		3D(10, CD_3): 0.089; 0.075; 0.075; is: 0.079 3H(8, CH_3): 0.375; 0.311; 0.311; is: 0.332 H(6): 0.193	

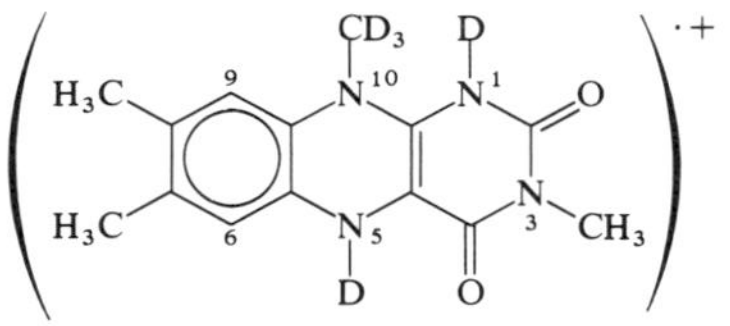

Nelsen

Substance	Generation/ Matrix or Solvent	Method/ T [K]	g-Factor	a-Value [mT]	Ref./ add. Ref.
$[C_{14}H_{11}D_5N_4O_2]^{\cdot+}$	Red. of the isoalloxazine with $Na_2S_2O_4$ or Zn in acidic solution/ CF_3CO_2H:toluene	ENDOR/ 260···310		N(10): 0.479 D(ND, 5): 0.133 H(6): 0.141 3H(7, CH_3): 0.057 3H(8, CH_3): 0.353 3H(10, CH_3): 0.501	84Ku1
$[C_{14}H_{12}D_4N_4O_2]^{\cdot+}$	Red. of the isoalloxazine with $Na_2S_2O_4$ or Zn in acidic solution/ CF_3CO_2D:toluene-d_8	ENDOR/ 260···310		N(10): 0.476 3H(3, CH_3): 0.032 D(ND, 5): 0.130 3H(7, CH_3): 0.059 3H(8, CH_3): 0.351 3H(10, CH_3): 0.501	84Ku1
	DCl	110···150		3H(10, CH_3): 0.589; 0.468; 0.468; is: 0.507 3H(8, CH_3): 0.375; 0.318; 0.318; is: 0.336 3H(7, CH_3): is: 0.068	
$[C_{14}H_{16}N_4O_2]^{\cdot+}$	Red. of the isoalloxazine with $Na_2S_2O_4$ or Zn in acidic solution/ CF_3CO_2H:toluene	ENDOR, TRIPLE/ 260···310		N(5): 0.738 N(10): 0.479 3H(3, CH_3): 0.030 H(NH, 5): −0.788 3H(6, CH_3): 0.089 H(7): −0.047 3H(8, CH_3): 0.317 3H(10, CH_3): 0.508	84Ku1

(continued)

Substance	Generation/ Matrix or Solvent	Method/ T [K]	g-Factor	a-Value [mT]	Ref./ add. Ref.
$[C_{14}H_{16}N_4O_2]^{\cdot +}$ *(continued)*	Red. of the isoalloxazine with $Na_2S_2O_4$ or Zn in acidic solution/ HCl	ENDOR, TRIPLE/ 110···150		$3H(10, CH_3)$: 0.568; 0.457; 0.457; is: 0.493 $3H(8, CH_3)$: 0.364; 0.290; 0.290; is: 0.311	84Ku1
$[C_{15}H_{19}N_4O_2]^{\cdot}$ 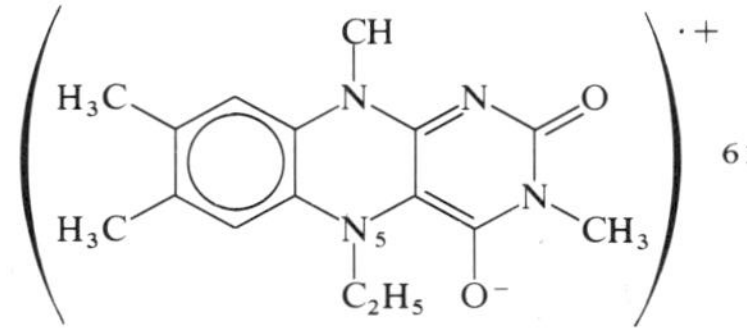 [61]	Electrolytic red./ DMF (0.1 M $(C_2H_5)_4NCl_4^-$)	–/ 295		N: ≈0.3 H: ≈0.17	81Na1
$[C_{12}H_{11}N_4O_2R]^{\cdot +}$ 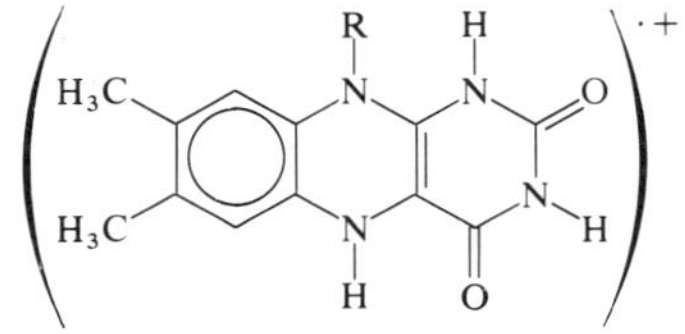 R = D-ribityl (riboflavin)	Red. of the isoalloxazine with $Na_2S_2O_4$ or Zn in acidic solution/ CF_3CO_2H:toluene	ENDOR, TRIPLE/ 265		N(5): 0.756 N(10): 0.464 H(3): −0.029 H(NH, 5): −0.832 H(6): −0.145 $3H(7, CH_3)$: 0.055 $3H(8, CH_3)$: 0.370 2H(10): 0.262	84Ku1/ 81Bo1
	HCl	110···150		$3H(8, CH_3)$: 0.389; 0.329; 0.329; is: 0.35 $3H(7, CH_3)$: is: 0.068 H(6): is: 0.2	

[61] Formulated as the N(1) deprotonated species, a neutral radical.

Substance	Generation/ Matrix or Solvent	Method/ T [K]	g-Factor	a-Value [mT]	Ref./ add. Ref.
$[C_{12}H_{11}N_4O_2R]^{\cdot +}$	Red. of the isoalloxazine with $Na_2S_2O_4$ or Zn in acidic solution/	ENDOR/			84Ku1
	CF_3CO_2H:toluene	260···310		N(5): 0.806 N(10): 0.099 H(3): 0.031 H(NH, 5): 0.794 H(6): 0.141 3H(7, CH_3): 0.0575 3H(8, CH_3): 0.365 2H(10): 0.266	
(FMN) R = D-ribitylphosphate	HCl	110···150		3H(8, CH_3): 0.382; 0.321; 0.321; is: 3.429 3H(7, CH_3): is: 0.068 H(6): is: 0.196	
$[C_{12}H_{11}N_4O_2R]^{\cdot +}$	Red. of the isoalloxazine with $Na_2S_2O_4$ or Zn in acidic solution/	ENDOR/			84Ku1
	CF_3CO_2H:toluene	260···310		N(5): 0.753 N(10): 0.46 H(NH, 3): 0.028 H(NH, 5): 0.816 H(6): 0.141 3H(7, CH_3): 0.0575 3H(8, CH_3): 0.367 2H(10): 0.2675; 0.247	
R = adenosyl-D-ribitylphosphate		110···150		3H(8, CH_3): 0.418; 0.364; 0.364; is: 0.382 3H(7, CH_3): is: 0.054 H(6): is: 0.196	

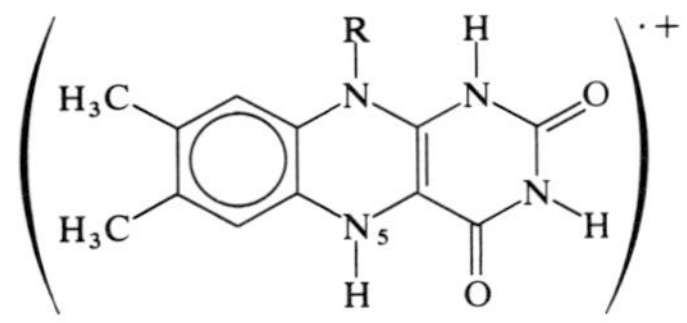

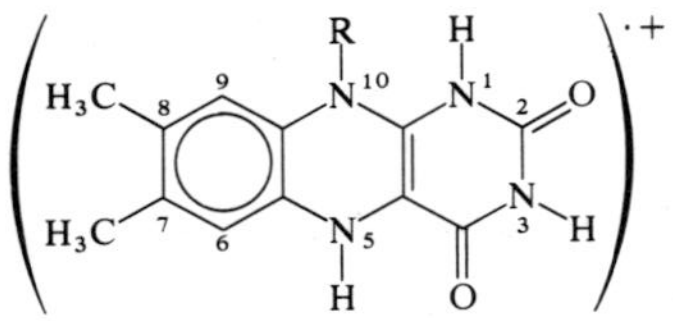

Substance	Generation/ Matrix or Solvent	Method/ T [K]	g-Factor	a-Value [mT]	Ref./ add. Ref.
$[C_{28}H_{37}N_5O_{11}]^{\cdot+}$ 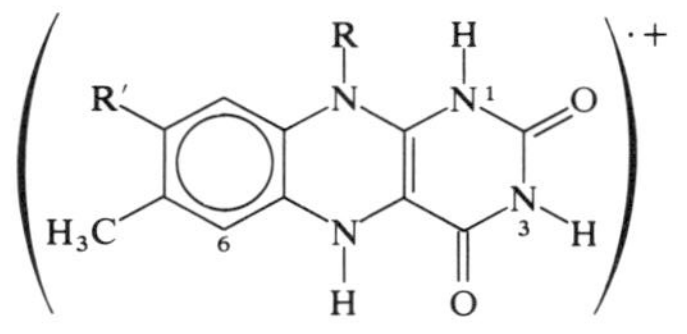 $R = CH_2(CHOOCCH_3)_3CH_2OOCCH_3$ 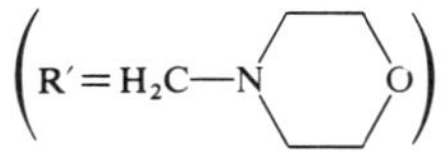 ($R' = H_2C$—N morpholino)	Red. of the isoalloxazine with $Na_2S_2O_4$ or Zn in acidic solution/ HCl	ENDOR/ 110···150		H(6): 0.236	84Ku1
20.5.2.4 Miscellaneous hydrazoaromatics					
$[C_4H_5N]^{\cdot+}$ 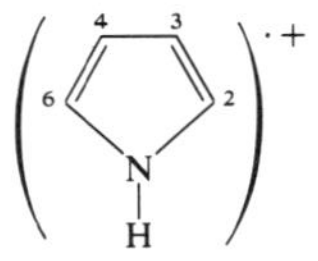	γ-irr./ $CFCl_3$	ESR/ 77		N: ≈0.3 2H(2, 5): 1.6 2H(3, 4): 0.3	83Ra1
	γ-irr./ $CFCl_3$	ESR/ 150	2.003	2H(2, 5): ≈1.80 H(1): ≈0.35 2H(3, 4): ≈0.20	83Sh1
$[C_4H_6N_2]^{\cdot+}$ 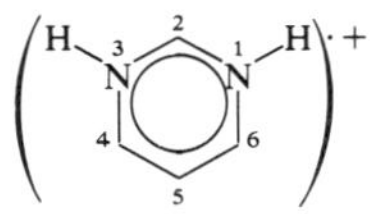	Phot./ 2-C_3H_7OH:acetone:HCl (85:10:5)	ESR/ 297	2.00290	N(1, 3): 0.141 2H(NH): 0.249 H(2): 0.037 2H(4, 6): 1.298 H(5): 0.237	79Ra1
$[C_5H_7N]^{\cdot+}$ 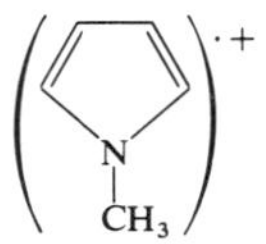	γ-irr./ $CFCl_3$	ESR/ 77		N: ≈0.35 2H(2, 5): 1.55 2H(3, 4): 0.36	83Ra1

Substance	Generation/ Matrix or Solvent	Method/ T [K]	g-Factor	a-Value [mT]	Ref./ add. Ref.
$[C_5H_8N_2]^{\cdot+}$ 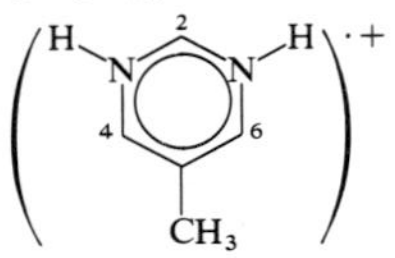	Phot./ 2-C_3H_7OH : acetone : HCl (85:10:5)	ESR/ 297	2.00283	N(1, 3): 0.146 2H(NH): 0.247 H(2): 0.03 H(4, 6): 1.270 3H(CH_3): 0.247	79Ra1
$[C_5H_8N_2]^{\cdot+}$ 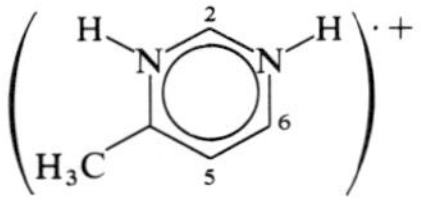	Phot./ 2-C_3H_7OH : acetone : HCl (85:10:5)	ESR/ 297	2.00286	N(1): 0.055 N(3): 0.170 H(NH, 1): 0.229 H(NH, 3): 0.254 3H(4, CH_3): 1.221 H(5): 0.229 H(6): 1.365	79Ra1
$[C_6H_9N]^{\cdot+}$ 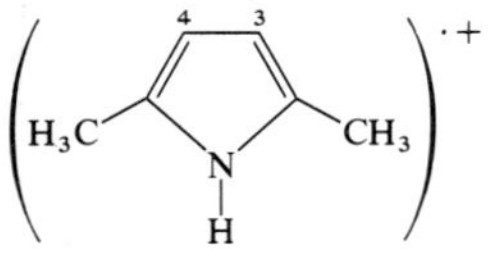	γ-irr./ $CFCl_3$	ESR/ 77		N: $\approx$0.3 6H(2, 5, CH_3): 1.75 2H(3, 4): 0.35	83Ra1
$[C_6H_{10}N_2]^{\cdot+}$ 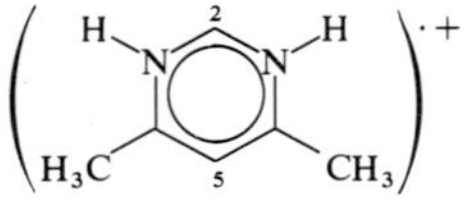	Phot./ 2-C_3H_7OH : acetone : HCl (85:10:5)	ESR/ 297	2.00285	N(1, 3): 0.074 2H(NH): 0.250 6H(4, 6, CH_3): 1.322 H(5): 0.180	79Ra1
$[C_{11}H_{13}N_3]^{\cdot+}$ 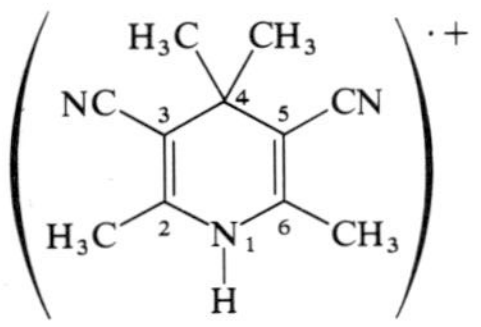	Electrochem. ox./ CH_3CN	ESR/ —		N(1): 0.645 H(NH): 0.770 2N(CN): 0.113 6H(2, 6, CH_3): 0.155	82Lu1

Substance	Generation/ Matrix or Solvent	Method/ T [K]	g-Factor	a-Value [mT]	Ref./ add. Ref.
$[C_{13}H_{15}N_3]^{\cdot +}$ 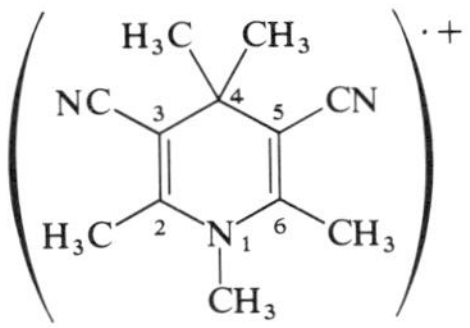	Electrochem. ox./ CH_3CN	ESR/ —		N(1): 0.805 3H(NCH_3): 1.035 2N(CN): 0.115 6H(2, 6, CH_3): 0.130	82Lu1
$[C_{14}H_9N]^{\cdot +}$ 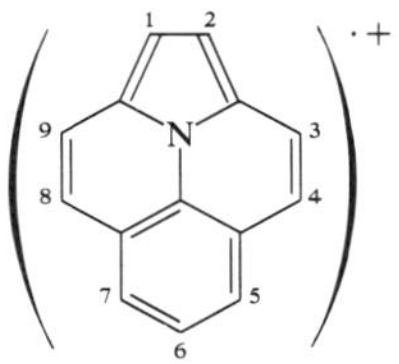	Ox. with $AlCl_3$/ CH_2Cl_2	ESR, ENDOR/ 273	2.0027	2H(1, 2): 0.181 2H(3, 9): 0.171 2H(4, 8): 0.528 2H(5, 7): 0.595 H(6): 0.158 N: 0.208	84Ge1
$[C_{16}H_{13}N]^{\cdot +}$ 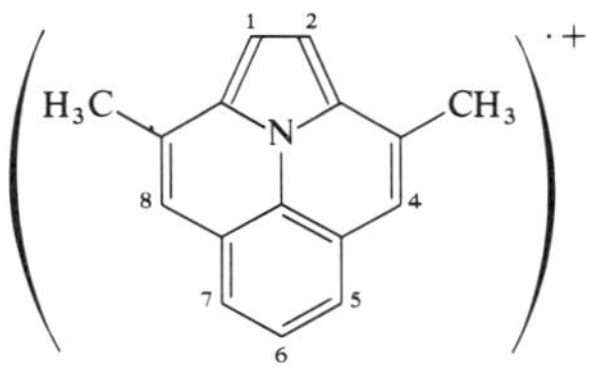	Ox. with $AlCl_3$/ CH_2Cl_2	ESR, ENDOR/ 273	2.0027	2H(1, 2): 0.180 6H(3, 9, CH_3): 0.070 2H(4, 8): 0.537 2H(5, 7): 0.626 H(6): 0.165 N: 0.195	84Ge1
$[C_{16}H_{15}NS]^{\cdot +}$ 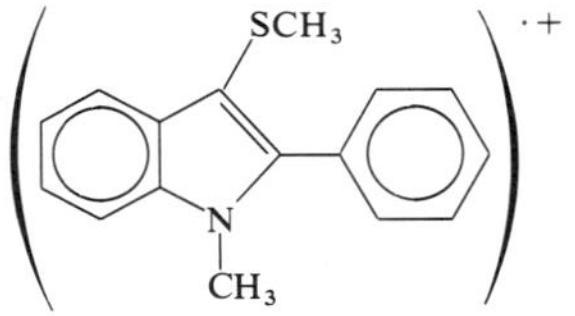	Ox. with $C_6H_5N_2^+$ or mixture of neutral with dication radicals or electrochem. ox./ CH_3CN(?)	ESR/ 295(?)		N: 0.412 3H: 0.845 3H: 0.187 2H: 0.062	82Co1

Substance	Generation/ Matrix or Solvent	Method/ T [K]	g-Factor	a-Value [mT]	Ref./ add. Ref.
$[C_{18}H_{11}N]^{\cdot +}$	$AgClO_4$/ CH_2Cl_2	ESR/ RT		H(1): 0.415 2H(8, 12): 0.334 2H(6, 14): 0.318 2H(5, 15): 0.268 2H(7, 13): 0.087 2H(2, 18): 0.069 N: unobserved	85Ma1
$[C_{18}H_{19}N]^{\cdot +}$, $[C_{19}H_{21}N]^{\cdot +}$	Electrochem. ox./ —	ESR/ 243	2.00288	N: 0.398 2H: 0.142 6H(1, 3, CH_3): 1.253 6H(4, 7, CH_3): 0.322 [62]	80Ku1
For further radical cations of similar structure, see [83Ba1] and [84Og1].					
$[C_{21}H_{17}N_3]^{\cdot +}$	Electrochem. ox./ CH_3CN	ESR/ RT		N(1): 0.645 H(NH): 0.725 2N(CN): 0.110	82Lu1
$[C_{22}H_{19}N_3]^{\cdot +}$	Electrochem. ox./ CH_3CN	ESR/ RT		N(1): 0.826 3H(NCH_3): 1.019 2N(CN): 0.107	82Lu1

[62] p-CH_3 species same data.

Substance	Generation/ Matrix or Solvent	Method/ T [K]	g-Factor	a-Value [mT]	Ref./ add. Ref.
$[C_{22}H_{25}N]^{\cdot +}$ 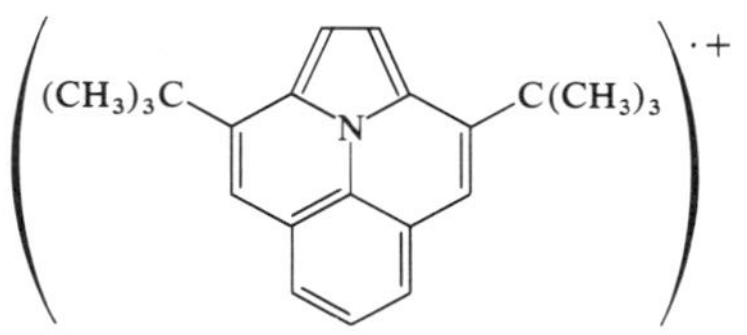	Ox. with $AlCl_3$/ CH_2Cl_2	ESR, ENDOR/ 273	2.0027	2H(1, 2): 0.179 2H(4, 8): 0.518 2H(5, 7): 0.604 H(6): 0.158 N: 0.201 18H(3, 9, $C(CH_3)_3$): 0.003	84Ge1
$[C_{25}H_{15}NO]^{\cdot +}$ 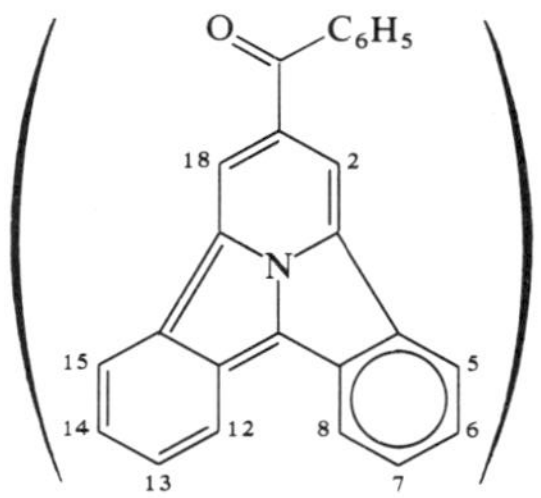	Ox. with $AgClO_4$/ CH_2Cl_2	ESR/ RT		2H(8, 12): 0.309 2H(6, 14): 0.288 2H(5, 15): 0.232 2H(7, 13): 0.077 2H(2, 18): 0.069 N: 0.018	85Ma1
$[C_{26}H_{17}N]^{\cdot +}$ 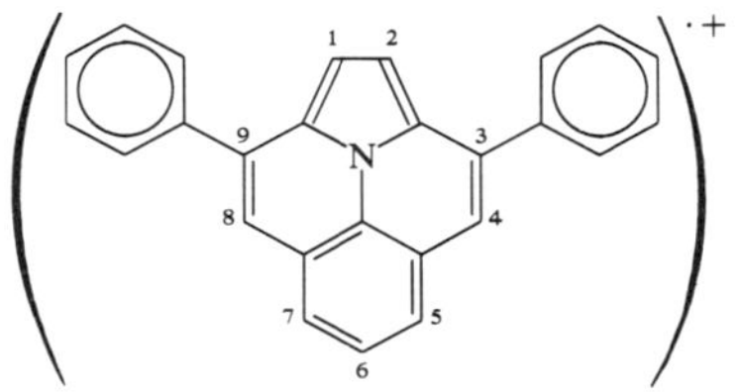	Ox. with $AlCl_3$/ CH_2Cl_2	ESR, ENDOR/ 273	2.0027	2H(1, 2): 0.165 2H(4, 8): 0.525 2H(5, 7): 0.621 H(6): 0.165 N: 0.189 10H(3, 5, C_6H_5): <0.01	84Ge1

20.6 Imino and more highly oxidized N functional group cation radicals

20.6.1 "sp²" hybridized N: imino compounds

20.6.1.1 Derivatives of pyridine

Substance	Generation/ Matrix or Solvent	Method/ T [K]	g-Factor	a-Value [mT]	Ref./ add. Ref.
$[C_5H_5N]^{\cdot+}$	γ-irr./ $CFCl_3$	ESR/ 77	2.002 (1)	N: 6.2; 3.1; 3.1; is: 4.2 2H(2, 6): 2.9 (3) 2H(3, 5): 0.87 (4) H(6): 0.87 (2)	84Ra1 [1]
	γ-irr./ $CFCl_3$	ESR/ 77		N: 6.34; 3.1; 3.1; is: 4.1 2H(2, 6): 2.93 2H(3, 5): 0.86 H(6): 1.12	79Sh1
$[C_5H_4ClN]^{\cdot+}$	γ-irr./ $CFCl_3$	ESR/ 77	2.002 (1)	^{35}Cl: 2.8 1H: 1.0 N: not reported	84Ra1 [2]
$[C_5H_4BrN]^{\cdot+}$	γ-irr./ $CFCl_3$	ESR/ 77		^{81}Br: 19.5 N, H: not reported	84Ra1
$[C_5H_4FN]^{\cdot+}$	γ-irr./ $CFCl_3$	ESR/ 77		F(2): 12.174; 0; 0; is: 4.058 3H(3, 5, 6): 0.955	84Sh1

[1]) Perdeuterated species also studied.
[2]) *m*-Cl species also studied.

Substance	Generation/ Matrix or Solvent	Method/ T [K]	g-Factor	a-Value [mT]	Ref./ add. Ref.
$[C_5H_3F_2N]^{\cdot+}$ 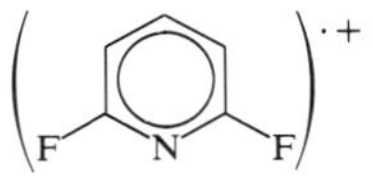	γ-irr./ $CFCl_3$	ESR/ 77	2.0035; 2.003; 2.003; is: 2.0032	2F(2, 6): 10.18; 0; 0; is: 3.39 2H(3, 5): 1.05; 1.40; 1.40; is: 1.28	84Sh1
	cyclo-C_6F_{12}	77	2.0019; 2.0046; 2.0046; is: 2.0037	2F(2, 6): 9.288; 0; 0; is: 3.10	
$[C_5F_5N]^{\cdot+}$ 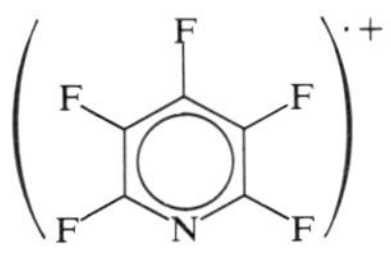	γ-irr./ *cyclo*-C_6F_{12}	ESR/ 77	2.002; 2.0068; 2.0068; is: 2.0052	4F(2, 3, 5, 6): 14.02; 0; 0; is: 4.67 F(4): 2.27; 0; 0; is: 0.756	84Sh1
$[C_6H_5NO]^{\cdot+}$ 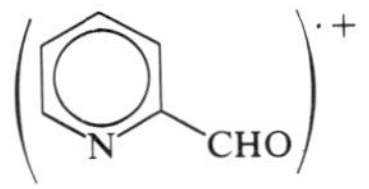	γ-irr./ $CFCl_3$	ESR/ 77	2.002 (1)	N: 5.8; 3.0; 3.0; is: 3.93 1H: 2.8 (3) 2H: 0.9 (2)	84Ra1 [3]
$[C_6H_7N]^{\cdot+}$ 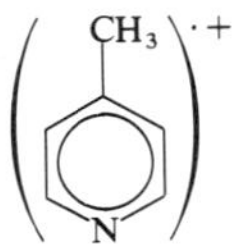	γ-irr./ $CFCl_3$	ESR/ 77	2.002 (1)	N: 5.9; 3.0; 3.0; is: 3.97 2H(2, 6): 2.9 (3) 2H(3, 5): 0.85 (20)	84Ra1
$[C_6H_7N]^{\cdot+}$ 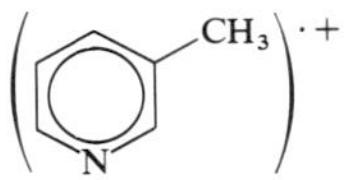	γ-irr./ $CFCl_3$	ESR/ 77	2.002 (1)	N: 6.0; 3.0; 3.0; is: 4.0 H(2, 6): 2.6 (4)	84Ra1

[3]) *m*-CHO species also studied.

Substance	Generation/ Matrix or Solvent	Method/ T [K]	g-Factor	a-Value [mT]	Ref./ add. Ref.
$[C_7H_7N]^{\cdot +}$ [4]	γ-irr./ $CFCl_3$	ESR/ 77		$2H(CH_2)$: 1.0 H(p): 1.0	84Ea2
$[C_7H_7N]^{\cdot +}$ [5]	γ-irr./ $CFCl_3$	ESR/ 77		N: 5.6; 2.9; 2.9; is: 3.8 H(2, 6): 2.8 H(3, 5): 1.0	84Ea2
$[C_7H_9N]^{\cdot +}$	γ-irr./ $CFCl_3$	ESR/ 77	2.002 (1)	N: 6.0; 3.0; 3.0; is: 4.0 H(2, 6): 3.0 (3)	84Ra1
$[C_7H_9N]^{\cdot +}$	γ-irr./ $CFCl_3$	ESR/ 77	2.002 (1)	N: not reported $3H(CH_3)$: 2.1 or 2.3 H: 3.5 or 2.8	84Ra1
$[C_7H_9N]^{\cdot +}$	γ-irr./ $CFCl_3$	ESR/ 77	2.002 (1)	N: 6.0 or 5.5; 3.0; 3.0; is: 4.0 or 3.83 2H(3, 5): 0.9 H(4): 0.9 or 1.7	84Ra1
$[C_9H_7N]^{\cdot +}$	γ-irr./ $CCl_3F:CF_2BrCF_2Br$ (1:1)	ESR/ 77		ΔHpp: 1.4···1.7	79Ka1

[4]) π-radical.
[5]) σ-radical at N.

Substance	Generation/ Matrix or Solvent	Method/ T [K]	g-Factor	a-Value [mT]	Ref./ add. Ref.
$[C_9H_7N]^{\cdot +}$ 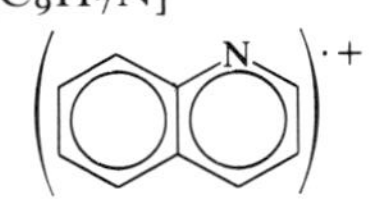	γ-irr./ $CCl_3F:CF_2BrCF_2Br$ (1:1)	ESR/ 77		ΔHpp: 1.4···1.7	79Ka1 [6])
20.6.1.2 Two or more imino N atoms					
$[C_2H_2N_4]^{\cdot +}$ 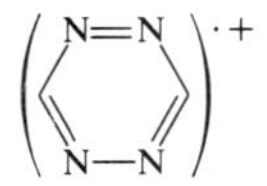	γ-irr./ $CFCl_3$	ESR/ 77		1N: 0.30; 0.19; 0.19; is: 0.227	86Fi1
$[C_4H_6N_4O_2]^{\cdot +}$ 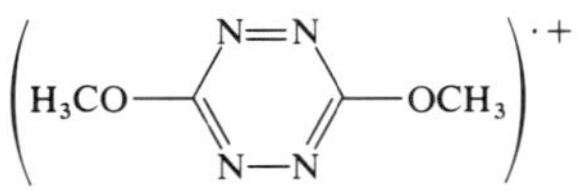	γ-irr./ $CFCl_3$	ESR/ 77		1N: 0.30; 0.185; 0.185; is: 0.223	86Fi1
$[C_4H_6N_4S_2]^{\cdot +}$ 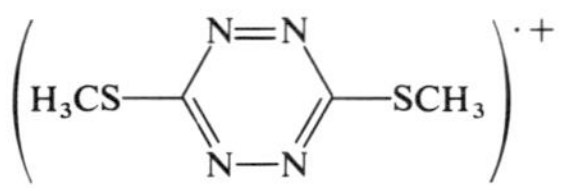	γ-irr./ $CFCl_3$	ESR/ 77		1N: 0.32; 0.20; 0.20; is: 0.24	86Fi1
For 3,6-(bis-dimethylamino) *syn*-tetrazine, see 20.5.1.3					
$[C_4H_4N_2]^{\cdot +}$ 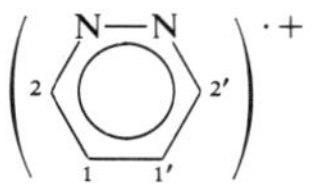 (*continued*)	γ-irr./ $CFCl_3$	ESR/ 77		2N: 5.4; 6.8; 5.4; is: 5.87 2H: 1.25; 1.35; 1.25; is: 1.28 2H: 0.35; 0.30; 0.35; is: 0.33	85Ra3

[6]) , , also studied.

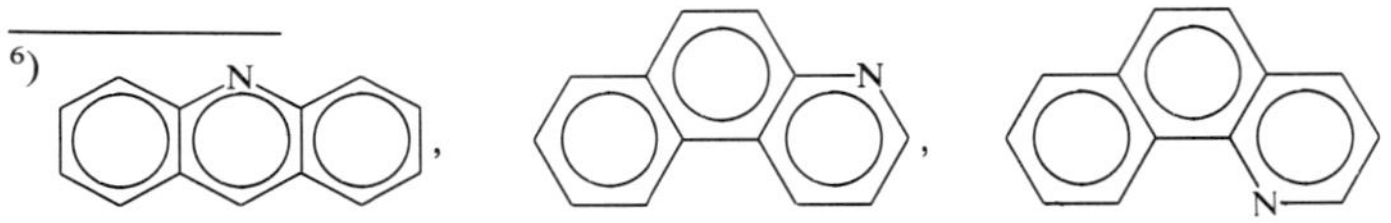

Substance	Generation/ Matrix or Solvent	Method/ T [K]	g-Factor	a-Value [mT]	Ref./ add. Ref.
$[C_4H_4N_2]^{\cdot+}$ *(continued)*	γ-irr./ $CCl_3F:CF_2BrCF_2Br$ (1:1)	ESR/ 77	2.00236; 2.00150; 2.00235; is: 2.00207	2N: 5.04; 6.86; 5.14; is: 5.68 2H(1): 1.34; 1.23; 1.21; is: 1.26 2H(2): 0.76; 0.13; 0.03; is: 0.31	79Ka1
$[C_4H_4N_2]^{\cdot+}$ 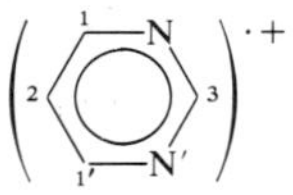	γ-irr./ $CFCl_3$	ESR/ 77		N: 2.8; 6.0; 2.8; is: 3.87 2H: ≈2.8 1H: ≈1.2 1H: ≈0.8	85Ra3
	γ-irr./ $CCl_3F:CF_2BrCF_2Br$ (1:1)	ESR/ 77	2.00348; 2.00221; 2.00463; is: 2.00344	N,N′: 0.61; 2.58; 0.7; is: 1.30 2H(1): 2.97; 2.66; 2.62; is: 2.75 H(2): 0.86; 0.49; 0.44; is: 0.60 H(3): 0.51; 0; −0.15; is: 0.12	79Ka1
$[C_4H_4N_2]^{\cdot+}$ 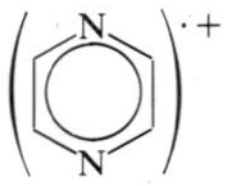	γ-irr./ $CCl_3F:CF_2BrCF_2Br$ (1:1)	ESR/ 77	2.00239; 2.00342; 2.00231; is: 2.00271	2N: 1.84; 3.27; 1.84; is: 2.32 4H: 3.40; 2.92; 2.86; is: 3.06	79Ka1
	γ-irr./ $CFCl_3$	ESR/ 77		2N: 1.82; 3.27; 1.82; is: 2.30 4H: 3.2; 2.9; 3.0; is: 3.0	85Ra3

Substance	Generation/ Matrix or Solvent	Method/ T [K]	g-Factor	a-Value [mT]	Ref./ add. Ref.
$[C_6H_{10}N_2]^{\cdot +}$ [6a]	γ-irr./ $CF_2ClCFCl_2$ [6b]	ESR/ 110		2N: 3.10 4H(*exo*): 1.55	88Wi1
	γ-irr./ $CF_2ClCFCl_2$	ESR, ENDOR, TRIPLE/ 115	2.0022 (5)	2N: +3.14 4H(*exo*): +1.51 4H(*endo*): +0.135 2H(*bridge*): −0.336	88Ge1
$[C_8H_6N_2]^{\cdot +}$	γ-irr./ $CCl_3F:CF_2BrCF_2Br$ (1:1)	ESR/ 77	2.00239; 2.00341; 2.00232; is: 2.00271	2N: 2.99; 1.49; 1.55; is: 2.01 2H(1): 3.35; 2.93; 2.86; is: 3.05 2H(2): 2.43; 2.17; 2.15; is: 1.53 2H(3): 0.65; 0.58; 0.57; is: 0.6	79Ka1
$[C_8H_6N_2]^{\cdot +}$	γ-irr./ $CCl_3F:CF_2BrCF_2Br$ (1:1)	ESR/ 77	2.00236; 2.00150; 2.00235; is: 2.00207	1N: 6.75; 4.55; 4.84; is: 5.38 1N: 5.95; 4.85; 4.81; is: 5.20 H(1): 0.69; 0.25; 0.20; is: 0.38 H(2): 1.39; 1.42; 1.36; is: 1.39 H(3): 0.45; 0.42; 0.39; is: 0.42 H(4): 0.09; 0.00; 0.02; is: 0.04 H(5): 0.05; 0.02; −0.01; is: 0.02 H(6): 0.66; 0.24; 0.27; is: 0.39	79Ka1

[6a]) Nearly isotropic. $H_{exo}-d_2$ species not isotropic, but spectral width proved assignment.
[6b]) See [87Bl1] for unanalyzed anisotropic spectra in other matrices.

Substance	Generation/ Matrix or Solvent	Method/ T [K]	g-Factor	a-Value [mT]	Ref./ add. Ref.
$[C_8H_{16}N_2]^{\cdot +}$ 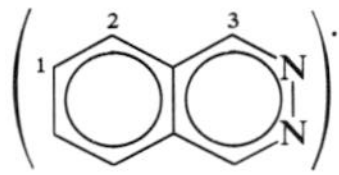	γ-irr./ $CCl_3F:CF_2BrCF_2Br$ (1:1)	ESR/ 77	2.00236; 2.00150; 2.00235; is: 2.00207	2N: 5.23; 7.01; 5.31; is: 5.85 2H(1): 0.36; 0.33; 0.33; is: 0.34 2H(2): 0.13; −0.01; −0.03; is: 0.03 2H(3): 0.64; 0.11; 0.0; is: 0.25	79Ka1
$[C_{12}H_8N_2]^{\cdot +}$ 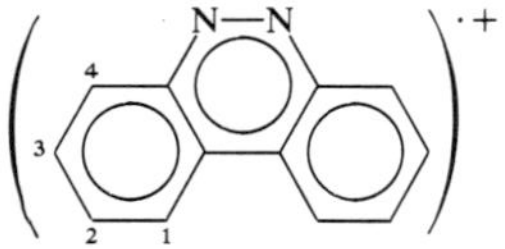	γ-irr./ $CCl_3F:CF_2BrCF_2Br$ (1:1)	ESR/ 77	2.00236; 2.00150; 2.00235; is: 2.00207	2N: 4.43; 6.28; 4.53; is: 5.08 2H(1): 0.45; 0.32; 0.31; is: 0.36 2H(2): 0.59; 0.33; 0.33; is: 0.42 2H(3): 0.14; 0.10; 0.09; is: 0.11 2H(4): 0.06; −0.01; −0.02; is: 0.01	79Ka1 [7])
$[C_{26}H_{20}N_2]^{\cdot +}$ 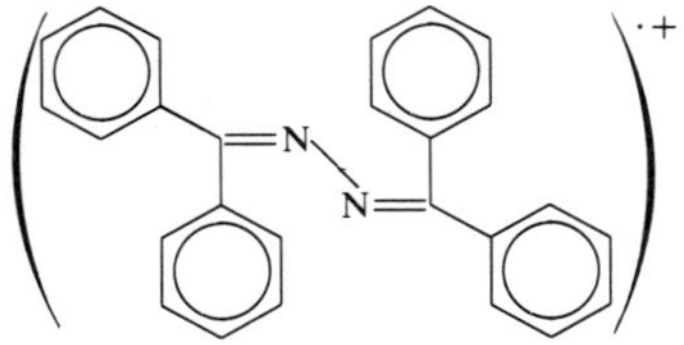	Ox. with $(C_6H_5)_2CN_2$ + $Cu(II)(ClO_4)_2$/ $C_7H_{15}CN$	ESR/ 293···313		2N: 0.256 4H(p): 0.128 8H(o): 0.063	79Be1

[7]) also studied. All gave a broad singlet assign as π radicals.

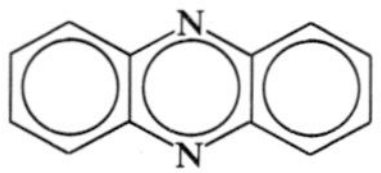

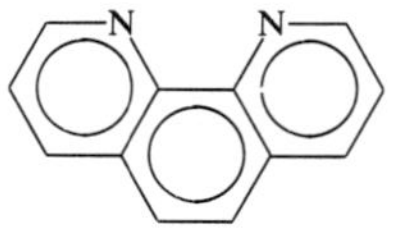

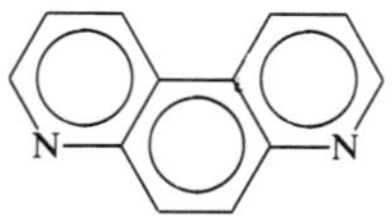

Substance	Generation/ Matrix or Solvent	Method/ T [K]	g-Factor	a-Value [mT]	Ref./ add. Ref.
20.6.1.3 —N=O compounds					
$[N_2O_3]^{\cdot+}$ (O=N—O—N=O)$^{\cdot+}$ [8])	Reaction of N_2O_4 with $(CH_3)_2N$—N=N—$N(CH_3)_2$/ Absorbed on $MgCl_2$	ESR/ 223		2N: 1.28	83Kh1
$[C_9H_{11}NO]^{\cdot+}$	Electrochem. ox./ CH_3CN	ESR/ 243		N: 3.353 [9])	82So1
$[C_{10}H_{13}NO]^{\cdot+}$	Electrochem. ox./ CH_3CN	ESR/ 243		N: 3.386 [9])	82So1
$[C_{11}H_{15}NO]^{\cdot+}$	Reaction of FSO_3H with (pentamethylnitrobenzene, NO_2)/ FSO_3N	ESR/ 233···300		N: 3.4	82De1

[8]) Identification speculative.
[9]) Proton hfs observed but not reported.

Substance	Generation/ Matrix or Solvent	Method/ T [K]	g-Factor	a-Value [mT]	Ref./ add. Ref.
$[C_{18}H_{29}NO]^{\cdot+}$	γ-irr./ CCl_4	ESR/ 77	1.9978; 2.0061; 2.0031; is: 2.0023	N: 0.28; 0.30; 0.51; is: 3.63	79Mu1
	CF_3CO_2H	77	2.0020	N: 3.59 1H(*m*): 0.35	
	Electrochem. ox./ CH_3CN	ESR/ 243		N: 3.390	82So1

20.6.2 "sp" hybridized: N_2, diazo, nitriles and relatives

Substance	Generation/ Matrix or Solvent	Method/ T [K]	g-Factor	a-Value [mT]	Ref./ add. Ref.
$[N_2]^{\cdot+}$ $(N{\equiv}N)^{\cdot+}$	e^--irr./ Neon	ESR/ 4	2.0004 (2)	2 ^{14}N: 3.72 (2) 2 ^{15}N: 5.22 (2)	83Kn1
$[N_4]^{\cdot+}$ $N_1{\equiv}N_2{\cdot}^{+}{\cdot}N_3{\equiv}N_4$ [10]	e-irr./ Ne	ESR/ 4	2.0016 (4); 1.9998 (2); 1.9998 (2); is: 2.0004	^{14}N(2, 3): 11.10; 9.43; 9.43; is: 10.0 ^{14}N(1, 4): 0.37; −0.73; −0.73; is: −0.36	87Kn1
			2.0016 (4); 1.9996 (2); 1.9996 (2); is: 2.0003	^{15}N(2, 3): 15.57; 13.21; 13.21; is: 14 ^{15}N(1, 4): 0.5; −1.02; −1.02; is: −0.51	
$[CN_2O]^{\cdot+}$ $(N{\equiv}N{\cdots}C{=}O)^{\cdot+}$	Photoionization/ Ne	ESR/ 4		Not analyzed	86Kn1

[10]) Linear, $^2\Sigma\mu$ ground electronic state. Calculation of spin densities.

Substance	Generation/ Matrix or Solvent	Method/ T [K]	g-Factor	a-Value [mT]	Ref./ add. Ref.
$[C_3H_3N]^{\cdot+}$ ($H_2C{=}CH{-}CN$)$^{\cdot+}$	γ-irr./ $CFCl_3$	ESR/ 77		^{14}N: 2.0; 0; 0; is: 0.7 $2H(CH_2)$: 2.4 H(CH): 1.2	84Ea2
$[C_4H_6N_2S_2]^{\cdot+}$ $(CH_3CNS)_2^{\cdot+}$ [11]	γ-irr./ $CFCl_3$	ESR/ 77	2.021; 2.012; 2.002; is: 2.012	2N: 0.3	84Sy1
$[C_7H_5N]^{\cdot+}$ (C_6H_5CN)$^{\cdot+}$	γ-irr./ $CFCl_3$	ESR/ 77		1H: 1.0	85Ra1
$[C_8H_8N_2]^{\cdot+}$ ($C_6H_5C(CH_3){=}N^+{=}N^-$)$^{\cdot+}$	Electrochem. ox./ CH_2Cl_2 :0.1 M $(n\text{-}C_4H_9)_4NBF_4$	ESR/ 183⋯203	2.0015	1N: 1.49 [12] 1N: 1.01 [12]	87Is1
$[C_9H_{10}N]^{2+}$ (3,5-$(CH_3)_2C_6H_3C{\equiv}NH$)$^{\cdot 2+}$ [13]	Ox. with PbO_2/ FSO_3H	ESR/ 198		$6H(3,5,CH_3)$: 1.155 2H(2, 6): 0.905 H(4): 0.040	79Ru1, 83Ru2
$[C_9H_{10}N]^{2+}$ (2,5-$(CH_3)_2C_6H_3C{\equiv}NH$)$^{\cdot 2+}$ [13]	Ox. with PbO_2/ FSO_3H	ESR/ 198		$3H(2,CH_3)$: 1.865 H(3): 0.245 H(4): 0.110 $3H(5,CH_3)$: 1.715 H(6): 0.385	83Ru2

[11]) Planar structure for dimer cation suggested.
[12]) Suggests bent at N, σ-radical structure.
[13]) Assignment not certain.

Substance	Generation/ Matrix or Solvent	Method/ T [K]	g-Factor	a-Value [mT]	Ref./ add. Ref.
$[C_{10}H_{12}N]^{\cdot 2+}$ 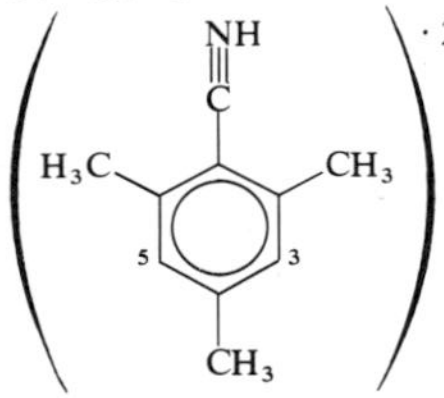 [13]	Ox. with PbO_2/ FSO_3H	ESR/ 198		6H(2, 6, CH_3): 1.170 2H(3, 5): 0.265	79Ru1, 83Ru2
$[C_{10}H_{12}N]^{\cdot 2+}$ 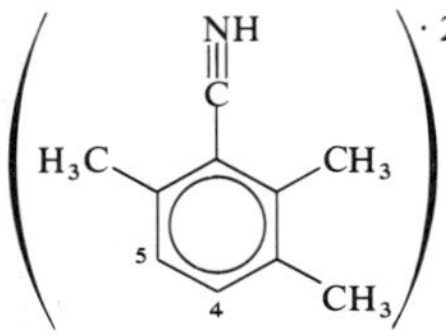 [13]	Ox. with PbO_2/ FSO_3H	ESR/ 198		3H(2, CH_3): 0.900 3H(3, CH_3): 1.615 3H(6, CH_3): 1.440 H(5): 0.460	83Ru2
$[C_{10}H_{12}N]^{\cdot 2+}$ 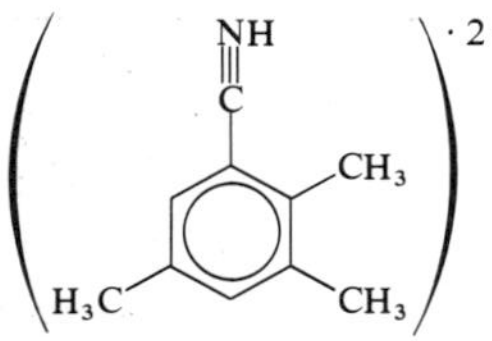	Ox. with PbO_2/ FSO_3H	ESR/ 198		3H(2, CH_3): 1.580 3H(3, CH_3): 0.780 3H(5, CH_3): 1.370 H(4): 0.040 H(6): 0.640	83Ru2
$[C_{10}H_{12}N]^{\cdot 2+}$ 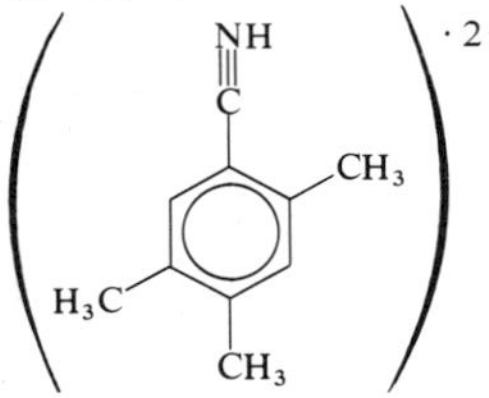	Ox. with PbO_2/ FSO_3H	ESR/ 198		3H(2, CH_3): 1.440 3H(4, CH_3): 0.455 3H(5, CH_3): 1.877 H(3): 0.037 H(6): 0.175 N: 0.037	83Ru2

[13]) Assignment not certain.

Substance	Generation/ Matrix or Solvent	Method/ T [K]	g-Factor	a-Value [mT]	Ref./ add. Ref.
$[C_{11}H_{14}N]^{\cdot 2+}$ 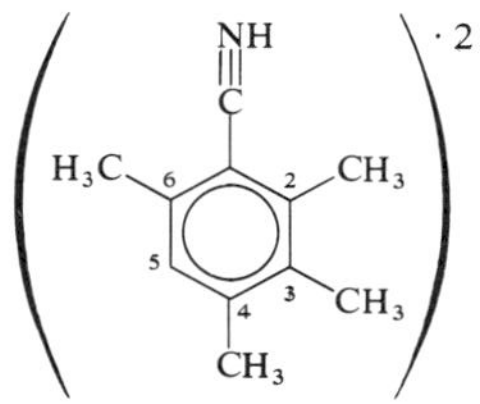 [13]	Ox. with PbO_2/ FSO_3H	ESR/ 198		3H(2, CH_3): 0.668 3H(3, CH_3): 1.765 3H(4, CH_3): 0.127 3H(6, CH_3): 1.405 H(5): 0.254	83Ru2
$[C_{11}H_{14}N]^{\cdot 2+}$ 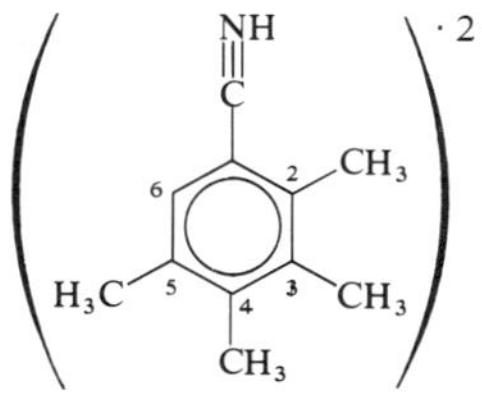 [13]	Ox. with PbO_2/ FSO_3H	ESR/ 198		3H(2, CH_3): 1.540 3H(3, CH_3): 0.580 3H(4, CH_3): 0.037 3H(5, CH_3): 1.480 H(6): 0.560	83Ru2
$[C_{12}H_{16}N]^{\cdot 2+}$ 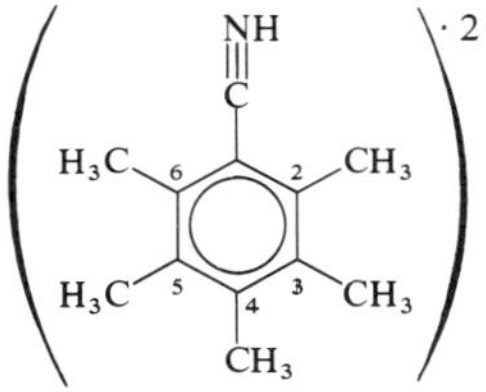 [13]	Ox. with PbO_2/ FSO_3H	ESR/ 198		6H(2, 6, CH_3): 1.160 6H(3, 5, CH_3): 1.040 3H(4, CH_3): 0.097	83Ru2
$[C_{11}H_{14}N]^{\cdot 2+}$ 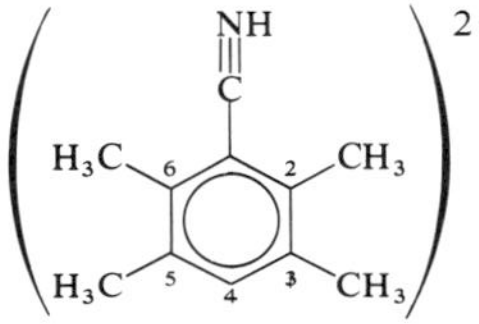 [13]	Ox. with PbO_2/ FSO_3H	ESR 198		6H(2, 6, CH_3): 1.200 6H(3, 5, CH_3): 1.060 H(4): 0.080	79Ru1, 83Ru2

[13]) Assignment not certain.

Substance	Generation/ Matrix or Solvent	Method/ T [K]	g-Factor	a-Value [mT]	Ref./ add. Ref.
$[C_{13}H_{16}N]^{\cdot 2+}$ 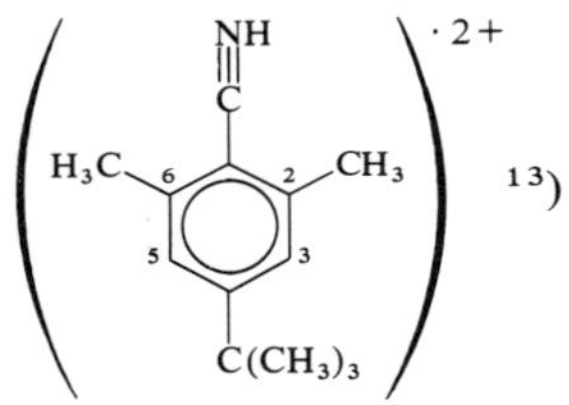 [13]	Ox. with PbO_2/ FSO_3H	ESR/ 198		6H(2, 6, CH_3): 1.190 2H(3, 5): 0.762	83Ru2
$[C_{12}H_{16}N_2]^{\cdot +}$ 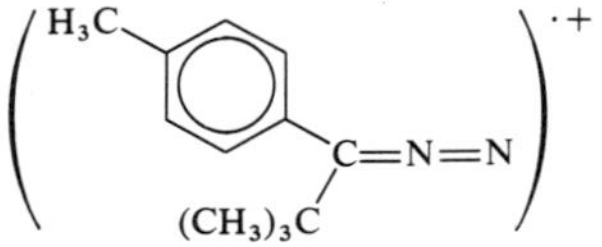	Electrochem. ox./ CH_2Cl_2 : 0.1 M $(n\text{-}C_4H_9)_4NBF_4$	ESR/ 183···203	2.0017	N: 0.47, 0.35 [14] 2H(o): 0.35 2H(m): 0.08 3H(CH_3): 0.67	87Is1
$[C_{12}H_{16}N_2O]^{\cdot +}$ 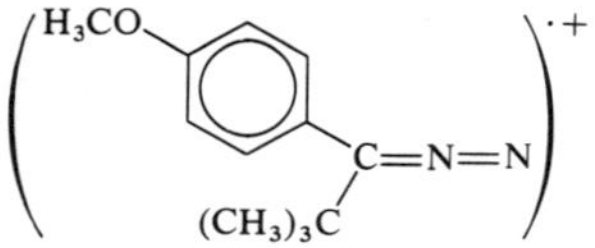	Electrochem. ox./ CH_2Cl_2 : 0.1 M $(n\text{-}C_4H_9)_4NBF_4$	ESR/ 183···203	2.0021	N: 0.39, 0.34 [14] 2H(o): 0.33 2H(m): 0.26 3H(OCH_3): 0.14	87Is1
$[C_{13}H_{10}N_2]^{\cdot +}$ 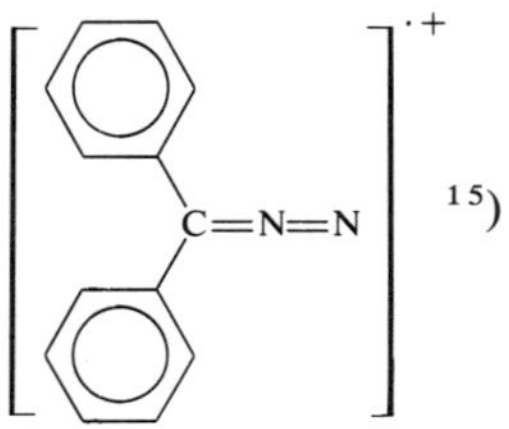 [15]	Ox. with $(C_6H_5)_2CN_2 + CuCClO_4$/ $C_7H_{15}CN$	ESR/ 266		2N: 0.323 2H(p): 0.115 4H(o): 0.057	79Be1
	Electrochem. ox./ CH_2Cl_2 : 0.1 M $(n\text{-}C_4H_9)_4NBF_4$	ESR/ 183···203	2.0009	1N: 1.72 [16] 1N: 1.01	87Is1

[13]) Assignment not certain.
[14]) Suggests linear CNN group, π electronic state.
[15]) Only tentative assignment.
[16]) Suggests bent at N, σ radical structure.

Substance	Generation/ Matrix or Solvent	Method/ T [K]	g-Factor	a-Value [mT]	Ref./ add. Ref.
$[C_{15}H_{14}N_2]^{\cdot+}$ $((H_3CO-C_6H_4-)_2C=N^+=N^-)^{\cdot+}$	Electrochem. ox./ CH_2Cl_2 :0.1 M $(n\text{-}C_4H_9)_4NBF_4$	ESR/ 183···203	2.0012	1N: 1.65 [16] 1N: 1.02	87Is1

20.6.3 Nitro compounds

Substance	Generation/ Matrix or Solvent	Method/ T [K]	g-Factor	a-Value [mT]	Ref./ add. Ref.
$[CN_4O_8]^{\cdot+}$ $(C(NO_2)_4)^{\cdot+}$	γ-irr./ $CFCl_3$	ESR/ 77	2.004; 1.992; 2.002; is: 1.999	N: 5.0; 4.8; 6.7; is: 5.5	85Ra2
$[CH_3NO_2]^{\cdot+}$ $(CH_3NO_2)^{\cdot+}$	γ-irr./ $CFCl_3$	ESR/ 77	2.0045; 1.994; 2.002; is: 2.000	N: 5.2; 4.7; 6.65; is: 5.52 [17] ^{13}C: 12.0; 12.0; 15.5; is: 13.2 [18]	85Ra2/ 83Ra2
$[CH_3NO_2]^{\cdot+}$ $(CH_3ONO)^{\cdot+}$	γ-irr. of the nitro compound/ $CFCl_3$	ESR/ 77	2.005; 1.994; 2.002; is: 2.000	N: 5.25; 4.49; 6.62; is: 5.45	83Ra2
$[CH_3NO_3]^{\cdot+}$ $(H_3CO-\overset{+}{N}(=O)-O^-)^{\cdot+}$ [19]	γ-irr./ $CFCl_3$	ESR/ 77	2.0045; 1.992; 2.0020; is: 2.000	N: 5.2; 4.9; 6.7; is: 5.6	86Ga2
$[C_2H_5NO_2]^{\cdot+}$ $(CH_3CH_2NO_2)^{\cdot+}$	γ-irr./ $CFCl_3$	ESR/ 77	2.005; 1.994; 2.002; is: 2.000	N: 5.3; 4.9; 7.0; is: 5.73	85Ra2
$[C_2H_5NO_2]^{\cdot+}$ $(CH_3CH_2ONO)^{\cdot+}$	γ-irr. of nitro compound/ $CFCl_3$	ESR/ 77	2.0050; 1.9935; 2.0020; is: 2.000	N: 5.20; 4.47; 6.68; is: 5.45	83Ra2

[16] Suggests bent at N, σ radical structure.
[17] σ-structure suggested.
[18] After annealing ^{13}C features 1 st, but N features little changed, this may be rearranged for $O-\dot{N}-OR^+$, originally suggested at 77K.
[19] Assignment tentative.

Nelsen

Substance	Generation/ Matrix or Solvent	Method/ T [K]	g-Factor	a-Value [mT]	Ref./ add. Ref.
$[C_3H_7NO_2]^{\cdot+}$ $(CH_3CH_2CH_2NO_2)^{\cdot+}$	γ-irr./ $CFCl_3$	ESR/ 77	2.0049; 1.994; 2.002; is: 2.0003	N: 5.2; 4.9; 6.7; is: 5.6	85Ra2
$[C_3H_7NO_2]^{\cdot+}$ $((CH_3)_2CHNO_2)^{\cdot+}$	γ-irr./ $CFCl_3$	ESR/ 77	2.005; 1.994; 2.002; is: 2.0003	N: 5.2; 4.9; 6.7; is: 5.57	85Ra2
$[C_3H_6BrNO_2]^{\cdot+}$ $((CH_3)_2C(Br)NO_2)^{\cdot+}$	γ-irr./ $CFCl_3$	ESR/ 77	2.004; 1.995; 2.002; is: 2.0003	N: 5.2; 4.9; 6.6; is: 5.57 ^{87}Br: 0.4; 0.4; 0.75; is: 0.52	85Ra2
$[C_3H_6ClNO_2]^{\cdot+}$ $((CH_3)_2C(Cl)NO_2)^{\cdot+}$	γ-irr./ $CFCl_3$	ESR/ 77	2.003; 1.995; 2.002; is: 2	N: 5.25; 5.0; 6.7; is: 5.65 ^{35}Cl: 0.0; 0.0; 0.5; is: 0.17	85Ra2
$[C_3H_6N_2O_4]^{\cdot+}$ $((CH_3)_2C(NO_2)_2)^{\cdot+}$	γ-irr./ $CFCl_3$	ESR/ 77	2.005; 1.993; 2.002; is: 2	1N: 5.3; 4.9; 6.8; is: 5.67	85Ra2
$[C_4H_9NO_2]^{\cdot+}$ $((CH_3)_3CNO_2)^{\cdot+}$	γ-irr./ $CFCl_3$	ESR/ 77	1.999; 1.996; 2.002; is: 1.999	N: 5.2; 5.2; 6.3; is: 5.57	85Ra2

Substance	Generation/ Matrix or Solvent	Method/ T [K]	g-Factor	a-Value [mT]	Ref./ add. Ref.
$[C_6H_5NO_2]^{\cdot+}$ $(C_6H_5{-}NO_2)^{\cdot+}$ [20]	γ-irr./ $CFCl_3$	ESR/ 77		4H: 0.6	85Ra1/ 83Ra2
$[C_6H_{11}NO_4]^{\cdot+}$ $((CH_3)_2C(CO_2C_2H_5)(NO_2))^{\cdot+}$	γ-irr./ $CFCl_3$	ESR/ 77	2.005; 1.994; 2.002; is: 2.0003	N: 5.2; 4.9; 6.7; is: 5.6	85Ra2
$[C_6H_{12}N_2O_4]^{\cdot+}$ $((CH_3)_2C(O_2N){-}C(CH_3)_2(NO_2))^{\cdot+}$	γ-irr./ $CFCl_3$	ESR/ 77	2.0045; 1.994; 2.002; is: 2.002	N: 5.1; 4.8; 6.7; is: 5.53	85Ra2
$[C_9H_{10}N_2O_4]^{\cdot+}$ $(O_2N{-}C_6H_4{-}C(CH_3)_2(NO_2))^{\cdot+}$	γ-irr./ $CFCl_3$	ESR/ 77	2.005; 1.993; 2.002; is: 2	N: 5.3; 5.0; 6.8; is: 5.7	85Ra2

[20]) o-, m-, p-tolyl derivatives also reported.

20.7 References for 20

64Me1 Merkle, F.H., Disher, C.A.: J. Pharm. Soc. **53** (1964) 955.
64Me2 Merkle, F.H., Disher, C.A.: J. Pharm. Soc. **53** (1964) 965.

65Lh1 Lhoste, J.M., Tonnard, F.: J. Chim. Phys. **63** (1965) 678.
65Sm1 Smejtek, P., Honzl, J., Metalova, V.: Coll. Czech. Chem. Commun. **30** (1956) 3875.

66Gi1 Gilbert, B.C., Hanson, P., Norman, R.O.C., Sutcliffe, B.T.: J. Chem. Soc., Chem. Commun. **1966**, 161.

67La1 Latta, B.M., Taft, R.W.: J. Am. Chem. Soc. **89** (1967) 5172.

68Ed1 Edlund, O., Lund, A., Nilsson, A.: J. Chem. Phys. **49** (1968) 749.

69Ho1 Hoffman, B.M., Eames, T.B.: J. Am. Chem. Soc. **91** (1969) 2169.

71Ea1 Eames, T.B., Hoffman, B.M.: J. Am. Chem. Soc. **93** (1971) 3141.
71Ne1 Neugebauer, F.A., Barbega, S.: Angew. Chem. **83** (1971) 48.

74Co1 Cohen, A.H., Hoffman, B.M.: J. Phys. Chem. **78** (1974) 1313.
74Co2 Cohen, A.H., Hoffman, B.M.: Inorg. Chem. **13** (1974) 1484.
74Ya1 Yao, T., Muska, S., Munemori, M.: Chem. Lett. **1974**, 939.

78Bo1 Bock, H., Kaim, A., Semkow, A., Nöth, H.: Angew. Chem. **90** (1978) 308; Angew. Chem. Int. Ed. Engl. **17** (1978) 286.
78Bo2 Bock, H., Kaim, W., Wiberg, N., Ziegleder, G.: Chem. Ber. **111** (1978) 3150.
78Cl1 Clarke, D., Gilbert, B.C., Hanson, P., Kirk, C.M.: J. Chem. Soc., Perkin Trans. II **1978**, 1103.
78Go1 Goldberg, I.B., Crowe, H.R., Christe, K.O.: Inorg. Chem. **17** (1978) 3189.
78Gr1 Griffin, B.W., Ting, P.L.: Biochem. **17** (1978) 2206.
78Ma1 Madjzinski, L.J., Pillay, K.S., Richard, H., Chow, Y.L.: Can. J. Chem. **56** (1978) 1657.
78Mo1 Morton, J.R., Preston, K.F.: J. Magn. Reson. **30** (1978) 577.

79Be1 Bethell, D., Hando, K.L., Fairhurst, S.A., Sutcliffe, L.H.: J. Chem. Soc., Perkin Trans. II **1979**, 707.
79Bo1 Bock, H., Kaim, W., Kira, M., Osawa, H., Sakurai, H.: J. Organomet. Chem. **164** (1979) 295.
79Br1 Bruni, P., Cardellini, L., Fava, G., Tosi, G.: J. Heterocycl. Chem. **16** (1979) 779.
79Ga1 Gara, W.B., Giles, J.R.M., Roberts, B.P.: J. Chem. Soc., Perkin Trans. II **1979**, 1444.
79He1 Herak, J.N., Rakvin, B., Bytyci, M.: J. Magn. Reson. **33** (1979) 319.
79Ka1 Kato, T., Shida, T.: J. Am. Chem. Soc. **101** (1979) 6869.
79Ko1 Kolodyazhnyi, Yu.V., Sadimenko, A.P., Sadimenko, L.P., Tkachenko, K.I., Prokof'ev, A.I., Morgunova, M.M., Osipov, O.A.: Teor. Eskp. Khim. **14** (1978) 230; Theor. Exper. Chem. (English Transl.) **14** (1978) 181.
79Mu1 Murabayashi, S., Shiotani, M., Sohma, J.: J. Phys. Chem. **83** (1979) 844.
79Ra1 Rakowsky, T., Dohrmann, J.K.: Ber. Bunsenges. Phys. Chem. **83** (1979) 495.
79Ri1 Richardson, P.F., Chang, C.K., Spaulding, L.D., Fajer, J.: J. Am. Chem. Soc. **101** (1979) 7736.
79Ru1 Rudenko, A.P., Zarubin, M.Ya., Aver'yanov, S.F., Barsheva, N.S.: Dokl. Akad. Nauk SSSR **249** (1979) 117.
79Sh1 Shida, T., Kato, T.: Chem. Phys. Lett. **68** (1979) 106.
79Vo1 Vögtle, F., Winkel, J.: Tetrahron Lett. **1979**, 1561.
79Wo1 Woynar, H., Schafer, H., Berndt, A., Thiel, W., Schweig, W.: Z. Naturforsch. **34b** (1979) 1339.

80Bo1 Bock, H., Kaim, W., Nöth, H., Semkov, A.: J. Am. Chem. Soc. **102** (1980) 4421.
80Ch1 Chen, V.S.F., Bolton, J.R.: J. Phys. Chem. **84** (1980) 1903.
80Ch2 Chen, V.S.F., Bolton, J.R.: J. Magn. Reson. **37** (1980) 231.
80El1 Eloranta, J., Salo, E., Mäkinen, S.: Acta Chem. Scand., Ser. A **34** (1980) 427.
80Fu1 Fujita, H., Yamauchi, J.: J. Heterocycl. Chem. **17** (1980) 1053.

80Ge1 Gerson, F., Plattner, G., Bartetzko, R., Gleiter, R.: Helv. Chim. Acta **63** (1980) 2144.
80Ka1 Kaim, W.: Angew. Chem. **94** (1980) 940; Angew. Chem. Int. Ed. Engl. **19** (1980) 911.
80Ku1 Kubacek, P.: Coll. Czech. Chem. Commun. **45** (1980) 1669.
80Ne1 Nelsen, S.F., Kessel, C.R., Grezzo, L.A., Steffek, D.J.: J. Am. Chem. Soc. **102** (1980) 5482.
80St1 Stolzenberg, A.M., Spreer, L.O., Holm, R.H.: J. Am. Chem. Soc. **102** (1980) 364.

81Al1 Alder, R.W., Arrowsmith, R.J., Casson, A., Sessions, R.B., Heilbronner, E., Kovac, B., Huber, H., Taagepera, M.: J. Am. Chem. Soc. **103** (1981) 6137.
81Be1 Berndt, A., Bolze, R., Schnaut, R., Woynar, H.: Angew. Chem. **93** (1981) 400.
81Bo1 Bock, M., Lubitz, W., Kurreck, H., Fenner, H., Grauert, R.: J. Am. Chem. Soc. **103** (1981) 5567.
81Bo2 Bock, H., Schulz, W., Stein, V.: Chem. Ber. **114** (1981) 2632.
81Ga1 Gade, S., Johnson, B., Hendrich, K., Knispel, R., Harrell Jr., J.W.: J. Chem. Phys. **75** (1981) 4742.
81Ge1 Gerson, F., Lopez, J., Akaba, R., Nelsen, S.F.: J. Am. Chem. Soc. **103** (1981) 6716.
81Ha1 Hanson, P., Isham, W.J., Lewis, R.J., Stockburn, W.A.: J. Chem. Soc., Perkin Trans. II **1981**, 1492.
81He1 Hester, R.E., Williams, K.P.J.: J. Chem. Soc., Perkin Trans. II **1981**, 852.
81Iz1 Izuoka, A., Kobayoshi, M.: Chem. Lett. **1981**, 1603.
81Ka1 Kaim, W.: J. Organometal. Chem. **215** (1981) 325.
81Ka2 Kaim, W.: J. Organometal. Chem. **215** (1981) 337.
81Ka3 Kaim, W.: Z. Naturforsch. **36b** (1981) 1110.
81Mi1 Michaut, N.P., Roncin, J.: J. Chem. Phys. **74** (1981) 68.
81Mü1 Müller, F., Graude, H.J., Harding, L.J., Dunham, W.R., Visser, J.W.G., Reindars, J.H., Hemmerich, P., Ehrenberg, A.: Eur. J. Biochem. **116** (1981) 17.
81Na1 Narayana, P.A., Li, A.S.W., Kevan, L.: J. Am. Chem. Soc. **103** (1981) 3603.
81Na2 Nanni Jr., E.J., Sawyer, D.T., Ball, S.S., Bruice, T.C.: J. Am. Chem. Soc. **103** (1981) 2797.
81Ne1 Neugebauer, F.A., Umminger, I.: Chem. Ber. **114** (1981) 2423.
81Ne2 Nelsen, S.F., Parmelee, W.P.: J. Org. Chem. **46** (1981) 3453.
81Ya1 Yasukouchi, K., Taniguchi, T., Yanaguchi, H., Arakawa, K.: J. Electroanal. Chem. **121** (1981) 231.

82Al1 Alder, R.W., Sessions, R.B.: "The Chemistry of Amino, Nitroso, and Nitro Compounds and Their Derivatives"; Patai, S. (Ed.); New York: Wiley **1982**; Part 2, Chap. 18.
82Co1 Colonna, M., Greci, L., Polini, M.: J. Chem. Soc., Perkin Trans. II **1982** 455.
82De1 Detsina, A.N., Efremova, N.V., Starichenko, V.F.: Zh. Org. Khim. **18** (1982) 1120; J. Org. Chem. (USSR) (English Transl.) **18** (1982) 970.
82Gr1 Grigoryants, V.M., Anisimov, O.A., Molin, Yu.N.: J. Struct. Chem. (USSR) (English Transl.) **23** (1982) 4.
82Ho1 Hovey, M.C.: J. Am. Chem. Soc. **104** (1982) 4196.
82Ka1 Kalyanaraman, B., Mason, R.P., Sivarajah, K.: Biochem. Biophys. Res. Commun. **105** (1982) 217.
82Kn1 Knight, L.B., Steadman, J.: J. Chem. Phys. **77** (1982) 1750.
82Lu1 Ludvik, J., Klima, J., Volke, J., Kurfürst, A., Kuthan, J.: J. Electroanal. Chem. Interfacial Electrochem. **138** (1982) 131.
82Ma1 Masuri, M., Nose, K., Ohmori, H., Sayo, H.: J. Chem. Soc., Chem. Commun. **1982**, 879.
82Na1 Narayana, P.A., Li, A.S.W., Kevan, L.: J. Am. Chem. Soc. **104** (1982) 6502.
82Ne1 Nelsen, S.F., Gannett, P.M.: J. Am. Chem. Soc. **104** (1982) 5292.
82Ne2 Nelsen, S.F., Steffek, D.J., Cunkle, G.T., Gannett, P.M.: J. Am. Chem. Soc. **104** (1982) 6641.
82Ne3 Nesterenko, A.M., Polumbrik, O.M., Markovskii, L.N.: Dokl. Akad. Nauk Ukr. SSR, Ser. B **1982**, 40.
82Ra1 Rao, D.N.R., Symons, M.C.R.: Chem. Phys. Lett. **93** (1982) 495.
82Re1 Reddoch, A.H.: private communication.
82Ri1 Riederer, H., Hüttermann, J.: J. Phys. Chem. **86** (1982) 3454.
82Ru1 Ruperez, F.L., Conesa, J.C., Soria, J.: J. Org. Magn. Reson. **20** (1982) 162.
82Ru2 Ruperez, F.L., Conesa, J.C., Soria, R.: J. Chem. Soc., Perkin Trans. II **1982**, 1517.
82So1 Sosonkin, I.M., Belevskii, V.N., Strogov, G.N., Domarev, A.N., Yarkov, S.P.: Zh. Org. Khim. **18** (1982) 1504; J. Org. Chem. (USSR) (English Transl.) **18** (1982) 1313.

83Al1 Alberti, A., Hudson, A.: J. Organometallic Chem. **248** (1983) 199.
83Al2 Alberti, A., Hudson, A.: J. Organometallic Chem. **241** (1983) 313.
83Ba1 Baumans, L., Ogle, J., Gavor, R., Stradins, J., Duburs, G.: Khim. Getertsikl. Soedin **1983**, 1422.
83Cl1 Clack, D.W., Evans, J.C., Obaid, A.Y., Rowlands, C.C.: Tetrahedron **39** (1983) 3615.
83Co1 Contineanu, M., Iliescu, C., Ciureanu, M.: Radiochem. Radioanal. Lett. **57** (1983) 9.
83De1 Depew, M.C., Zhongli, L., Wan, J.K.S.: J. Am. Chem. Soc. **105** (1983) 2480.
83Fe1 Feller, D., Davidson, E.R., Borden, W.T.: J. Am. Chem. Soc. **105** (1983) 3347.
83Fu1 Fujita, H.: J. Chem. Soc., Faraday Trans. I **79** (1983) 2077.
83Ga1 Gascoyne, P.R.C., Symons, M.C.R., McLaughlin, J.A.: Int. J. Quantum Chem.: Quantum Biol. Syn. **10** (1983) 123.
83Ge1 Gerson, F., Metzger, A.: Helv. Chim. Acta **66** (1983) 2031.
83Ka1 Kaim, W.: J. Am. Chem. Soc. **105** (1983) 707.
83Ka2 Kaim, W.: J. Organomet. Chem. **241** (1983) 157.
83Kh1 Khusidman, M.B., Grigor'eva, N.V., Vyatkin, V.P., Dobychin, S.L.: Zh. Obshch. Kim. **53** (1983) 1568; J. Gen. Chem. (USSR) (English Transl.) **53** (1983) 1415.
83Kh2 Khusidman, M.B., Vyatkin, V.P., Grigor'eva, N.V., Dobychin, S.L.: Zh. Prikl. Khim. (Leningrad) **56** (1983) 222; CA **98**, 160106r.
83Kn1 Knight, L.B., Bostick, J.M., Woodward, R.W., Steadman, J.: J. Chem. Phys. **78** (1983) 6415.
83Ko1 Koshechko, V.G., Inozemtsev, A.N., Pokhodenko, V.D.: Zh. Org. Khim. **19** (1983) 751; J. Org. Chem. (USSR) (English Transl.) **19** (1983) 662.
83Li1 Li, A.S.W., Kevan, L.: J. Am. Chem. Soc. **105** (1983) 5752.
83Ma1 Maruyama, K., Ogawa, T.: Bull. Chem. Soc. Jpn. **56** (1983) 3525.
83Ma2 Masui, M., Ohmori, H., Veda, C., Nose, K., Sayo, H.: Chem. Pharm. Bull. **31** (1983) 3385.
83Ne1 Nelsen, S.F., Parmelee, W.P., Rumack, D.T.: J. Org. Chem. **48** (1983) 4219.
83Ne2 Nelsen, S.F., Cunkle, G.T., Evans, D.H.: J. Am. Chem. Soc. **105** (1983) 5928.
83Ne3 Nelsen, S.F., Blackstock, S.C., Rumack, D.T.: J. Am. Chem. Soc. **105** (1983) 3115.
83Ne4 Nesterenko, A.M., Polumbrik, O.M., Markovskii, L.N.: Zh. Org. Khim. **19** (1983) 1961; J. Org. Chem. (USSR) (English Transl.) **19** (1983) 1722.
83Ne5 Neugebauer, F.A., Krieger, C., Fischer, F., Siegel, R.: Chem. Ber. **116** (1983) 2261.
83Ok1 Okada, K., Yamauchi, J., Deguchi, Y.: J. Heterocycl. Chem. **20** (1983) 723.
83Ra1 Rao, D.N.R., Symons, M.C.R.: J. Chem. Soc. Perkin Trans. II **1983**, 135.
83Ra2 Rao, D.N.R., Symons, M.C.R.: Tetrahedron Lett. **24** (1983) 1293.
83Ru1 Ruperez, F.L., Conesa, J.C., Soria, J.: J. Mol. Struct. **98** (1983) 165.
83Ru2 Rudenko, A.P., Cheremisin, A.A., Shchegoleva, L.N., Detsina, A.A., Zarubin, M.Ya.: Zh. Org. Khim. **19** (1983) 1910; J. Org. Chem. (USSR) (English Transl.) **19** (1983) 1675.
83Sh1 Shiotani, M., Nagata, Y., Tasaki, M., Sohma, J., Shida, T.: J. Phys. Chem. **87** (1983) 1170.
83Sp1 Speiser, B., Rieker, A., Pons, S.: J. Electroanal. Chem., Interfacial Electrochem. **147** (1983) 205.

84Bo1 Bock, H., Kaim, W., Kira, M., Rene, L., Viche, H.-G.: Z. Naturforsch. **39b** (1984) 763.
84Bo2 Bock, H., Roth, B., Daub, J.: Z. Naturforsch. **39b** (1984) 771.
84Br1 Breslow, R., Maslak, P., Thomaides, J.S.: J. Am. Chem. Soc. **106** (1984) 6453.
84Ea1 Eastland, G.W., Rao, D.N.R., Symons, M.C.R.: J. Chem. Soc., Perkin Trans. II **1984**, 1551.
84Ea2 Eastland, G.W., Kurita, Y., Symons, M.C.R.: J. Chem. Soc., Perkin Trans. II **1984**, 1843.
84Eb1 Eberlein, G., Bruice, T.C., Lazarus, R.A., Henrie, R., Benkovic, S.J.: J. Am. Chem. Soc. **106** (1984) 7916.
84Fa1 Fairhurst, S.A., Preston, K.F., Sutchiffe, L.H.: Can. J. Chem. **63** (1984) 1124.
84Gr1 Grampp, G., Stiegler, G.: Z. Phys. Chem., N.F. **141** (1984) 185.
84Jo1 Jovanovic, M.V., Biehl, E.R., de Meester, P., Chu, S.S.C.: J. Heterocycl. Chem. **21** (1984) 1425.
84Ka1 Kaim, W.: J. Chem. Soc., Perkin Trans. II **1984**, 1767.
84Ka2 Kaim, W.: J. Chem. Soc., Perkin Trans. II **1984**, 1357.
84Ka3 Kaim, W.: Angew. Chem. **96** (1984) 609; Angew. Chem. Int. Ed. Engl. **23** (1984) 613.
84Ka4 Kaim, W.: J. Am. Chem. Soc. **106** (1984) 1712.
84Ku1 Kurreck, H., Bock, M., Bretz, N., Elsner, M., Kraus, H., Lubitz, W., Müller, R., Geissler, J., Kroneck, P.M.H.: J. Am. Chem. Soc. **106** (1984) 737.
84Le1 Lefkowitz, S.M., Trifunac, A.D.: J. Phys. Chem. **88** (1984) 77.
84Le2 Lenard, D.R., McDowell, C.A.: J. Mol. Strukt. **118** (1984) 21.

84Lo1 Lopez, J., Yamauchi, J., Okada, K., Deguchi, Y.: Bull. Chem. Soc. Jpn. **57** (1984) 673.

84Ne1 Nesterenko, A.M., Polumbrik, O.M., Ryabokon', I.G., Markovskii, L.M.: Zh. Org. Khim **20** (1984) 1465.

84Ne2 Nelsen, S.F., Blackstock, S.C., Frigo, T.B.: J. Am. Chem. Soc. **106** (1984) 3366.

84Og1 Ogle, J., Baumane, L., Gavars, R., Kodis, V., Stradins, J., Lusis, V., Muceniece, D., Duburs, G.: Khim. Geterosikl. Soedin **1984**, 651; CA **101**, 90153h.

84Pa1 Pankratov, A.N., Morozov, V.L., Mushtakova, S.P., Il'yasov, A.V.: Izv. Akad. Nauk SSSR, Ser. Khim. **1984**, 1483; Bull. Acad. Sci. USSR, Div. Chem. Sci. **33** (1984) 1363.

84Ra1 Rao, D.N.R., Eastland, G.W., Symons, M.C.R.: J. Chem. Soc., Faraday Trans. I **80** (1984) 2803.

84Ri1 Rieger, A.L., Rieger, P.H.: J. Phys. Chem. **88** (1984) 5845.

84Ru1 Ruperez, F.L., Conesa, J.C., Soria, J.: Spectrochim. Acta **40A** (1984) 1021.

84Sh1 Shiotani, M., Kawazoe, H., Sohma, J.: J. Phys. Chem. **88** (1984) 2220.

84Sy1 Symons, M.C.R., Trousson, P.M.R.: Radiation Phys. Chem. **23** (1984) 127.

84Sz1 Szajdzinska-Pietek, E., Maldonado, R., Kevan, L., Jones, R.R.M.: J. Am. Chem. Soc. **106** (1984) 4675.

84Ve1 Veda, Y., Okazaki, T., Ichikawa, T.: Kenkyo Hokoku-Hiroshima Daigaku Kogakubu **32** (1984) 145.

85Al1 Alder, R.W., Orpen, A.G., White, J.M.: J. Chem. Soc., Chem. Commun. **1985**, 949.

85Be1 Berti, C., Greci, L., Andruzzi, R., Trazza, A.: J. Org. Chem. **50** (1985) 368.

85Be2 Bessenbacher, C., Gross, R., Kaim, W.: J. Chem. Soc., Chem. Commun. **1985**, 1369.

85Be3 Belevskii, V.N., Khvan, O. In., Belopushkin, S.I., Fel'dman, V.I.: Dokl. Akad. Nauk SSSR **281** (1985) 869; Dokl. Phys. Chem. (English Transl.) **281** (1985) 338.

85Bo1 Bock, H., Haenel, P., Kaim, W., Lechner-Knoblauch, U.: Tetrahedron Lett. **26** (1985) 5115.

85Ci1 Ciminale, F., Curci, R., Troisi, L.: Tetrahedron Lett. **26** (1985) 6369.

85Cl1 Clack, D.W., Evans, J.C., Obaid, A.Y., Rowlands, C.C.: J. Chem. Soc., Perkin Trans. II **1985**, 1653.

85Cl2 Close, D.M., Sagstuen, E., Nelson, W.H.: J. Chem. Phys. **82** (1985) 4368.

85Ea1 Eaton, D.R., Watkins, J.M., Buist, R.J.: J. Am. Chem. Soc. **107** (1985) 5604.

85Ev1 Evans, J.C., Evans, A.G., Nouri-Sorkhobi, N.H., Obaid, A.Y., Rowlands, C.C.: J. Chem. Soc., Perkin Trans. II **1985**, 315.

85Ga1 Gaudiello, J.G., Ghosh, P.K., Bard, A.J.: J. Am. Chem. Soc. **107** (1985) 3027.

85Ka1 Kaifer, A.E., Bard, A.J.: J. Phys. Chem. **89** (1985) 4876.

85Ka2 Kaim, W.: Heterocycles **23** (1985) 1363.

85Ki1 Kirste, B., Kurreck, H.: private communication.

85Ki2 Kirste, B., Alder, R.W., Sessions, R.B., Bock, M., Kurreck, H., Nelsen, S.F.: J. Am. Chem. Soc. **107** (1985) 2635.

85Ko1 Koksal, F., Cakir, O., Gumrukcu, I., Birey, M.: Z. Naturforsch. **40a** (1985) 903.

85Ko2 Koh, A.K., Miller, D.J.: St. Data Nucl. Data Tables **33** (1985) 235.

85Ma1 Matsumoto, K., Yamauchi, J., Uchida, T.: Heterocycles **23** (1985) 2773.

85Mo1 Motten, A.G., Chignell, C.F.: Magn. Reson. Chem. **23** (1985) 834.

85Ne1 Neugebauer, F.A., Kuhnhäuser, S.: Angew. Chem. Int. Ed. Engl. **24** (1985) 596.

85Ne2 Nelsen, S.F., Yumibe, N.P.: J. Org. Chem. **50** (1985) 4749.

85Ne3 Nelsen, S.F., Cunkle, G.T., Evans, D.H., Haller, K.J., Kaftory, M., Kirste, B., Kurreck, H., Clark, T.: J. Am. Chem. Soc. **107** (1985) 3825.

85Ne4 Nelsen, S.F., Cunkle, G.T.: J. Org. Chem. **50** (1985) 3701.

85Ra1 Rao, D.N.R., Symons, M.C.R.: J. Chem. Soc., Perkin Trans. II **1985**, 991.

85Ra2 Rao, D.N.R., Symons, M.C.R.: J. Chem. Soc., Faraday Trans. I **81** (1985) 565.

85Ra3 Rao, D.N.R., Eastland, G.W., Symons, M.C.R.: J. Chem. Soc., Faraday Trans. I **81** (1985) 727.

85Ru1 Ruperez, F.L., Conesa, J.C., Soria, J.: An. Fis., Ser. B **81** (1985) 127.

85Ru2 Rudenko, A.P., Aver'yanov, S.F., Zarubin, M.J.: Dokl. Akad. Nauk SSSR **281** (1985) 1122; CA **103**, 177802j.

85Ru3 Ruperez, F.L., Conesa, J.C., Soria, J., Apreda, M.C., Cano, F.H., Foces-Foces, C.: J. Phys. Chem. **89** (1985) 1178.

85Sa1 Sayo, H., Michida, T.: Chem. Pharm. Bull. **33** (1985) 3271.

85Sa2 Sayo, H., Michida, T.: Chem. Pharm. Bull. **33** (1985) 2541.

85Sa3 Sayo, H., Hosokawa, M.: Chem. Pharm. Bull. **33** (1985) 4471.

85Ti1 Titov, E.V., Alaev, Yu.N., Luk'yaneko, L.V.: Teor. Eksp. Khim. **21** (1985) 623; Theor. Exp. Chem. (USSR) (English Transl.) **21** (1985) 596.

85Yu1 Yu, F.R., Wang, Y.Y., Wan, C.C.: Electrochem. Acta **30** (1985) 1693.

86Ch1 Chapman, D.M., Buchanan', III. A.C., Smith, G.P., Mamantov, G.: J. Am. Chem. Soc. **108** (1986) 654.

86Cr1 Crovigneau, P., Enea, O., Lamy, C.: Nouv. J. Chim. **10** (1986) 539.

86Ea1 Eastland, G.W., Rao, D.N.R., Symons, M.C.R.: J. Chem. Soc., Faraday Trans. I **82** (1986) 2388.

86Fa1 Fairhurst, S.A., Johnson, K.M., Sutcliffe, L.H., Preston, K.F., Banistery, A.N., Hauptman, Z.V., Passmore, J.: J. Chem. Soc., Dalton Trans. **1986**, 1465.

86Fi1 Fischer, H., Umminger, I., Neugebauer, F.A., Chandra, H., Symons, M.C.R.: J. Chem. Soc., Chem. Commun. **1986**, 837.

86Ga1 Ganghi, N.S., Wyatt, J.L., Symons, M.C.R.: J. Chem. Soc., Chem. Commun. **1986**, 1424.

86Ga2 Ganghi, N.S., Rao, D.N.R., Symons, M.C.R.: J. Chem. Soc., Faraday Trans. I **82** (1986) 2367.

86Ge1 Gerson, F., Knöbel, J., Buser, U., Vogel, E., Zehnder, M.: J. Am. Chem. Soc. **108** (1986) 3781.

86Iv1 Ivakhnenko, E.P., Karsanov, I.V., Khandrakova, V.S., Rubeshov, A.Z., Okhlobystin, O.Yu., Minkin, V.I., Prokof'ev, A.I., Kabachnik, M.I.: Izv. Akad. Nauk SSSR, Ser. Khim. **1986**, 2755; Bull. Acad. Sci. USSR, Div. Chem. Sci. (English Transl.) **1986**, 2755.

86Ka1 Kadirov, M.K., Il'yasov, A.V., Kamalov, R.M., Yarmukhametova, D.Kh., Vafina, A.A.: Izv. Akad. Nauk SSSR, Ser. Khim. **1986**, 1094; Bull. Acad. Sci. USSR, Div. Chem. Sci. (English Transl.) **1986**, 991.

86Ka2 Kaifer, A.E., Bard, A.J.: J. Chem. Phys. **90** (1986) 868.

86Kn1 Knight, L.B.: Acc. Chem. Res. **19** (1986) 313.

86Li1 Lin, T.-S., Retsky, J.: J. Phys. Chem. **90** (1986) 2687.

86Ne1 Nelsen, S.F.: Mol. Structure and Energetics. V.3: Deerfield Beach, FL: VCH Publ. **1986**, 1.

86Ne2 Nelsen, S.F., Willi, M.R., Mellor, J.M., Smith, N.M.: J. Org. Chem. **51** (1986) 2081.

86Ne3 Nelsen, S.F., Frigo, T.B., Kim, Y., Thompson-Colon, J.A., Blackstock, S.C.: J. Am. Chem. Soc. **108** (1986) 7926.

86Ne4 Neugebauer, F.A., Bock, M., Kuhnhäuser, S., Kurreck, H.: Chem. Ber. **119** (1986) 980.

86Ne5 Neugebauer, F.A., Kuhnhäuser, S.: unpublished.

86Oh1 Ohmori, N., Maeda, H., Ueda, C., Sayo, H., Masui, M.: Chem. Pharm. Bull. **34** (1986) 3079.

86Ru1 Ruperez, F.L., Conesa, J.C., Soria, J.: J. Chem. Soc., Perkin Trans. II **1986**, 391.

86Qi1 Qin, X.Z., Williams, F.: J. Phys. Chem. **90** (1986) 2292.

86Sa1 Sagstuen, E., Awadelkarim, O., Lund, A., Masiakowski, J.: J. Chem. Phys. **85** (1986) 3222.

86Sa2 Sayo, H., Michida, T., Hatsumura, H.: Chem. Pharm. Bull. **34** (1986) 558; CA **106**, 66654e.

86Sa3 Sayo, H., Hiromi, H., Michida, T.: Chem. Pharm. Bull. **34** (1986) 4139.

86Sa4 Sayo, H., Mikio, H., Lee, E., Kariya, K., Kono, M.: Biochim. Biophys. Acta **874** (1986) 187.

86Sy1 Symons, M.C.R.: J. Chem. Soc., Chem. Commun. **1986**, 11.

86Th1 Thompson-Colon, J.A.: Ph.D. Thesis, University of Wisconsin-Madison **1986**.

86Wo1 Wolmershaeuser, G., Schauber, M., Wilhelm, T., Sutcliffe, L.H.: Synth. Met. **14** (1986) 239; CA **105**, 33595n.

87Bl1 Blackstock, S.C., Kochi, J.K.: J. Am. Chem. Soc. **109** (1987) 2484.

87Ch1 Chandra, H., Mishra, S.P., Symons, M.C.R.: Chem. Phys. Lett. **134** (1987) 307.

87Ch2 Chandra, H., Rao, D.N.R., Symons, M.C.R.: J. Chem. Soc., Dalton Trans. **1987**, 726; CA **106**, 206702w.

87Co1 Colameri, M.J., Kevan, L., Thompson, D.H.P., Hurst, J.K.: J. Phys. Chem. **91** (1987) 4072.

87Do1 Dormann, E., Nowak, M.J., Williams, K.A., Augus jr., R.O., Wudl, F.: J. Am. Chem. Soc. **109** (1987) 2594.

87Is1 Ishiguro, K., Sawaki, Y., Izuoka, A., Sugawara, T., Iwamura, H.: J. Am. Chem. Soc. **109** (1987) 2530.

87Ka1 Kaise, M., Someno, K.: Chem. Lett. **1987**, 1295.

87Kn1 Knight, Jr., L.B., Johannessen, K.D., Cobranchi, D.C., Earl, E.A., Feller, D., Davidson, E.R.: J. Chem. Phys. **87** (1987) 885.

87Ne1 Nelsen, S.F., Blackstock, S.C., Petillo, P.A., Agmon, I., Kaftory, M.: J. Am. Chem. Soc. **109** (1987) 5724.

87Ne2 Nelsen, S.F., Thompson-Colon, J.A., Kirste, B., Rosenhouse, A., Kaftory, M.: J. Am. Chem. Soc. **109** (1987) 7128.

87Qi1 Qin, X.-Z., Pentecost, T.C., Wang, J.T., Williams, F.: J. Chem. Soc., Chem. Commun. **1987**, 451.

87Qi2 Qin, X.-Z., Williams, F.: J. Am. Chem. Soc. **109** (1987) 595.

87Sy1 Symons, M.C.R.: J. Chem. Soc., Chem. Commun. **1987**, 866.

87Sy2 Symons, M.C.R., Janes, R.: J. Chem. Soc., Faraday Trans. I **83** (1987) 383.

88Cl1 Clark, T.: J. Am. Chem. Soc. **110** (1988) 1672.

88Di1 Dinnocenzo, J.P., Banach, T.E.: J. Am. Chem. Soc. **110** (1988) 971.

88Fr1 Frigo, T.B.: Ph.D. Thesis, University of Wisconsin-Madison **1988**.

88Ge1 Gerson, F., Qin, X.-Z.: Helv. Chim. Acta **71** (1988) 1498.

88Wi1 Williams, F., Guo, Q.-X., Petillo, P.A., Nelsen, S.F.: J. Am. Chem. Soc. **110** (1988) 7887.

21 Organic C-, O-, and N-centered bi- and polyradicals

21.0 Introduction

21.0.1 Definition

Organic polyradicals are paramagnetic molecules, having N unpaired electrons, which can be described by the spin Hamiltonian (1)

$$\mathscr{H} = \mu_B \sum_{i=1}^{N} \boldsymbol{B} \cdot \mathbf{g}^{(i)} \cdot \boldsymbol{S}^{(i)} + J \sum_{i>j=1}^{N} \boldsymbol{S}^{(i)} \cdot \boldsymbol{S}^{(j)} + \boldsymbol{S} \cdot \mathbf{D} \cdot \boldsymbol{S} + \sum_{i=1}^{N} \sum_{k=1}^{n} \boldsymbol{S}^{(i)} \cdot \mathbf{A}^{(ik)} \cdot \boldsymbol{I}^{(ik)}, \qquad \boldsymbol{S} = \sum_{i=1}^{N} \boldsymbol{S}^{(i)} \tag{1}$$

where μ_B, $\boldsymbol{B}$, $\mathbf{g}^{(i)}$, J, $\boldsymbol{S}^{(i)}$, $\mathbf{D}$, $\mathbf{A}^{(ik)}$, $\boldsymbol{I}^{(ik)}$, n are the Bohr magneton, the external magnetic field, the g-tensor of segment i, the electron exchange parameter, the electron spin operator of segment i, the zero-field splitting tensor, the hyperfine tensor of nucleus k in segment i, the nuclear spin operator of nucleus k in segment i, and the number of nuclei per segment, respectively.

The spin Hamiltonian (1), which is based on a particular model of the electronic structure of polyradicals, has been derived by Reitz and Weissman [1]. The authors assume, that a polyradical with N unpaired electrons can be divided into N segments. To each segment proper spatial and spin functions are assigned. The electron spin operator $\boldsymbol{S}^{(i)}$ refers to segment i. It is assumed, that the corresponding spatial function vanishes at segment $j \neq i$ and that overlap can be neglected.

We will not justify this Hamiltonian, but should note that it reflects specific assumptions about the electronic structure of a polyradical. While there is a whole range of systems for which it is difficult to draw a distinction between a polyradical and a multiplet state (triplet, quartet, etc.) it is worth while to recognize that both terms are useful for describing limiting behaviour.

21.0.2 Magnetic properties

As magnetic properties of polyradicals we consider the quantities $\mathbf{g}$, $\mathbf{A}$, $\mathbf{D}$ and J of the spin Hamiltonian (1). They are determined by a variety of magnetic resonance methods, such as Electron Paramagnetic or Spin Resonance (EPR, ESR), Nuclear Magnetic Resonance (NMR), Electron Nuclear Double Resonance (ENDOR) and Electron Electron Double Resonance (ELDOR).

$\mathbf{g}$ is the spectroscopic splitting tensor (g-tensor). It is symmetric. The mean value of its diagonal elements (principal values)

$$g = \tfrac{1}{3} \sum_{l=1}^{3} g_{ll} \tag{2}$$

is called isotropic g-factor.

$\mathbf{A}^{(k)}$, the hyperfine tensor, describes the magnetic interactions between an electron and the nucleus k. This tensor is symmetric, too. Its isotropic part

$$a^{(k)} = \tfrac{1}{3} \sum_{l=1}^{3} A_{ll}^{(k)} \tag{3}$$

is called hyperfine coupling constant.

The zero-field splitting tensor $\mathbf{D}$ describes the dipolar magnetic interactions of the unpaired electrons. $\mathbf{D}$ is symmetric and traceless, i.e.

$$\sum_{l=1}^{3} D_{ll} = 0. \tag{4}$$

Consequently two parameters

$$D = \tfrac{3}{2} D_{33}, \tag{5}$$

$$E = \tfrac{1}{2}(D_{11} - D_{22}) \tag{6}$$

completely determine this tensor. The definitions (5) and (6) refer to a standard system, proposed for the choice of the tensor elements.

The parameter J describes the electron exchange interactions and determines the energy separation of the different spin states of a polyradical. In the case of biradicals this separation is found to be:

$$E_{\text{triplet}} - E_{\text{singlet}} = J. \tag{7}$$

For triradicals one obtains:

$$E_{\text{quartet}} - E_{\text{doublet}} = 3/2\,J. \tag{8}$$

21.0.3 Substances

The following tables deal with all paramagnetic molecules, that fulfill the definition 21.0.1. However, only compounds with established or at least plausible structures are included. Papers which only state the presence of polyradicals in a sample and do not give detailed structures nor magnetic properties have not been reviewed.

For the display of the data the substances are subdivided into several classes and subclasses. Their arrangement is to be seen from the table of contents. Within the individual subclasses structural related radicals are collected in groups. Their ordering is determined by the gross formula of the typical representative heading the group. It is followed by the derivatives, arranged according to the gross formula.

Not classified as polyradicals are the following systems:

1. Paramagnetic molecules with dominant electron exchange interactions (for instance triplet carbenes). They cannot be described by the spin Hamiltonian (1).
2. Complexes or aggregates of monoradicals.
3. Polymers, which are spin-labelled by monoradicals.

21.0.4 Arrangement of the tables

In the tables 21.1 to 21.3 the spin Hamiltonian parameters of polyradicals are given.

The first column (Substance) describes the structure of the radical and contains the gross formula. Radicals with related structures have been collected in groups. The typical representative, heading the group, is followed by derivatives, arranged according to the gross formula.

The second column (Generation/Matrix or Solvent) briefly describes the generation of the free radical. This information, however, is restricted to radicals, which have not been isolated as pure compounds prior to the measurement. For a detailed description of the radical generation we refer to the original papers. The second column also specifies the matrix or solvent in which the radical was measured.

The third column (Physical Method/T) specifies the method, used to determine the magnetic parameters, and states the sample temperature during the measurement (in units K).

The fourth column (g-Factor) lists the elements of the g-tensor $\mathbf{g}$. If for one compound four values are given, the first three are the principal values of $\mathbf{g}$, the fourth is the isotropic part g. If only one value is given, this is always the isotropic part. Principal axes of $\mathbf{g}$ are not listed.

The fifth column (a-Value) contains the absolute values of the elements of the hyperfine tensor $\mathbf{A}$. Where possible the signs of these elements are given in parentheses. If for one nuclei four values are given, the first three are the principal values of $\mathbf{A}$, the fourth is its isotropic part a. If only one value is given, this is always the isotropic part. Principal axes of $\mathbf{A}$ are not listed. The hyperfine parameters are labelled as described in the General Introduction.

The unit for the elements of $\mathbf{A}$, including its isotropic part, is milli-Tesla (mT).

The same fifth column contains D, E-values, the elements of the zero-field splitting tensor $\mathbf{D}$. Where possible the signs of D and E are given in parentheses.

Furthermore, the fifth column contains the absolute value of the electron exchange parameter J. If known, the sign for J is given in parentheses.

The unit for D and E parameters is milli-Tesla (mT). For parameter J Tesla(T) is applied.

The last column (Ref./add. Ref.) lists the reference from which the data are taken. This reference is followed by additional references to the same subject. All references belonging to the tables 21.1 to 21.3 are found in the bibliography 21.4. The literature is considered for the period from September 1976 to March 1987.

21.0.5 Symbols and abbreviations

$\mathbf{A}^{(ik)}$	hyperfine tensor of nucleus k in segment i
$A_{ll}^{(k)}$	principal value of the hyperfine tensor of nucleus k
$a^{(k)}$	isotropic hyperfine coupling constant of nucleus k
$\boldsymbol{B}$	external magnetic field
μ_B	Bohr magneton
$\mathbf{D}$	zero-field splitting tensor
D_{ll}	principal value of the zero-field splitting tensor
D	zero-field splitting parameter
E	zero-field splitting parameter
$\mathbf{g}^{(i)}$	g-tensor of segment i
g_{ll}	principal value of the g-tensor
g	isotropic g-tensor
$\mathcal{H}$	spin Hamiltonian
$\boldsymbol{I}^{(ik)}$	nuclear spin operator of nucleus k in segment i
J	electron exchange parameter
N	number of unpaired electrons
n	number of nuclei per segment
$\boldsymbol{S}^{(i)}$	electron spin operator of segment i
$\boldsymbol{S}$	total electron spin operator
T	temperature
ν	frequency

For abbreviations, see the list at the end of the volume

21.0.6 References for 21.0

Theoretical background

[1]	Reitz, D.C., Weissman, S.I.: J. Chem. Phys. **33** (1960) 700.	Biradicals, spin Hamiltonian
[2]	Briere, R., Dupeyre, R.M. Lemaire, H., Morat, C., Rassat, A., Rey, P.: Bull. Soc. Chim. France **1965**, 3290.	Biradicals, analysis of ESR spectra
[3]	Falle, H.R., Luckhurst, G.R., Lemaire, H., Marechal, Y., Rassat, A., Rey, P.: Mol. Phys. **11** (1966) 49.	Biradicals, ESR in liquid crystals
[4]	Luckhurst, G.R.: Mol. Phys. **10** (1966) 543.	Biradicals, alternating linewidths
[5]	Glarum, S.H., Marshall, J.H.: J. Chem. Phys. **47** (1967) 1374.	Biradicals, spin exchange
[6]	Lemaire, H.: J. Chim. Phys. **64** (1967) 559.	Biradicals, spin exchange
[7]	Hudson, A., Luckhurst, G.R.: Mol. Phys. **13** (1967) 409.	Triradicals, alternating linewidths
[8]	Luckhurst, G.R., Pedulli, G.F.: Mol. Phys. **20** (1971) 1043.	Biradicals, ESR line shape theory
[9]	Brickmann, J., Kothe, G.: J. Chem. Phys. **59** (1973) 2807.	Triradicals, rigid limit ESR spectra
[10]	Parmon, V.N., Zhidomirov, G.M.: Mol. Phys. **27** (1974) 367.	Dynamic biradicals, ESR line shape theory
[11]	Luckhurst, G.R., Poupko, R., Zannoni, C.: Mol. Phys. **30** (1975) 499.	Biradicals, spin relaxation in liquid crystals
[12]	van Willigen, H., Plato, M., Möbius, K., Dinse, K.-P., Kurreck, H., Reusch, J.: Mol. Phys. **30** (1975) 1359.	Biradicals, analysis of ENDOR spectra
[13]	Weissman, S.I., Kothe, G.: J. Am. Chem. Soc. **97** (1975) 2537.	Triradicals, $\Delta m = 3$ transitions
[14]	Vollmann, W.: Mol. Phys. **32** (1976) 395.	Biradicals, theory of ENDOR spectra
[15]	Kothe, G., Naujok, A., Ohmes, E.: Mol. Phys. **32** (1976) 1215.	Triradicals, theory of fast rotational ESR spectra
[16]	Parmon, V.N., Zhidomirov, G.M.: Mol. Phys. **32** (1976) 613.	Dynamic biradicals, ESR line shape theory
[17]	Luckhurst, G.R., Zannoni, C.: J. Magn. Reson. **23** (1976) 275.	Triradicals, spin relaxation in liquid crystals

[18]	Luckhurst, G.R., in: Berliner, L.J., (ed.), Spin Labeling, New York: Academic Press, **1976**, p. 133.	Biradicals as spin probes
[19]	Kothe, G.: Mol. Phys. **33** (1977) 147.	Triradicals, theory of slow-motional ESR spectra
[20]	Luckhurst, G.R., Setaka, M.: J. Magn. Reson. **25** (1977) 539.	Biradicals, ESR line shape theory
[21]	Fukui, K., Tanaka, K.: Bull. Chem. Soc. Jpn. **50** (1977) 1391.	Biradicals, MO theory
[22]	Kothe, G., Wassmer, K.-H., Naujok, A., Ohmes, E., Rieser, J., Wallenfels, K.: J. Magn. Reson. **30** (1979) 425.	Triradicals, ESR line shapes of $\Delta m = 1, 2, 3$ transitions
[23]	Döhnert, D., Koutecky, J.: J. Am. Chem. Soc. **102** (1980) 1789.	Biradicals, MO theory
[24]	Kothe, G., Berthold, T., Ohmes, E.: Mol. Phys. **40** (1980) 1441.	Triradicals, angular dependent linewidths
[25]	Meier, P., Blume, A., Ohmes, E., Neugebauer, F.A., Kothe, G.: Biochemistry **21** (1982) 526.	Biradicals, theory of slow-motional ESR spectra
[26]	Coffmann, R.E., Pezesh, K.A.: J. Magn. Reson. **65** (1985) 62.	Triradicals, rigid limit ESR spectra

Reviews and monographs

[27]	Zhidomirov, G.M., Buchachenko, A.L.: Zh Strukt. Khim. **8** (1967) 1110; J. Struct. Chem. (USSR) (English Transl.) **8** (1967) 987.	Polyradicals, review article
[28]	Morozova, I.D., Dyatkina, M.E.: Usp. Khim. **37** (1968) 865; Russ. Chem. Rev. (English Transl.) **37** (1968) 376.	Biradicals, review article
[29]	Kreilick, R.W.: Adv. Magn. Reson. **6** (1973) 171.	Polyradicals, discussion of NMR spectra
[30]	Kothe, G., Wilker, W., in: Fischer, H., Hellwege, K.-H., (eds.), Magnetic Properties of Free Radicals, Landolt-Börnstein, New Series, Vol. II/9, Part d2, Berlin: Springer-Verlag, **1980**, p. 148.	Polyradicals, data tables
[31]	Bordon, W.T.: Diradicals, New York: Wiley, **1982**.	Biradicals, monograph on C-centered species
[32]	Closs, G.L., Miller, R.J., Redwine, O.D.: Accounts Chem. Res. **18** (1985) 196.	Biradicals, review on Chemically Induced Dynamic Nuclear Polarization
[33]	Bordon, W.T., in: Maitland Jr., J., Moss, R.A. (eds.), Reactive Intermediates, Vol. 3, New York: Wiley, **1985**, p. 151.	Biradicals, review on C-centered species
[34]	Torres, M., Safarik, I., Murai, H., Strausz, O.P.: Rev. Chem. Interm. **7** (1986) 243.	Biradicals, review article including carbenes

21.1 Carbon-centered polyradicals

21.1.1 Trimethylenemethane biradicals

Substance	Generation/ Matrix or Solvent	Method T [K]	g-Factor	a-Value [mT]; D,E [mT]; J [T]	Ref./ add. Ref.
$[C_4H_6]^{2\cdot}$ 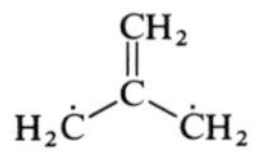	Irr. of 4-methylene-Δ-pyrazoline/ Hexafluorobenzene	EPR/ 77		D: 26.8	66Dow1/ 72Dow1
	Irr. of methylenecyclo-propane/ Single crystal	EPR/ 4.2···106		H: (−)1.00 [1]; (−)2.71 [1]; (−)1.86 [1]; is: (−)1.86 [1]) D: (+)26.5 E: ≤0.3	80Cla1
	Irr. of methylene-cyclobutanone/ Methylcyclohexane	20···80		$(-)9350 < J < (-)6730$	76Bas1/ 82Dow1
$[C_6H_8]^{2\cdot}$ 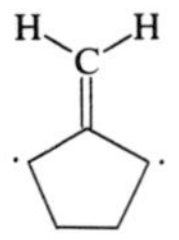	Irr. of the corresp. azo-compound/ MTHF	EPR/ <100		D: 28.3 E: 5.9	76Pla1
$[C_{12}H_{12}]^{2\cdot}$ 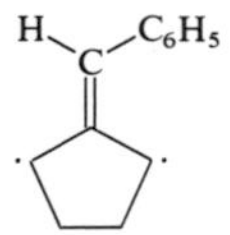	Irr. of the corresp. azo-compound/ MTHF	EPR/ <100		D: 21.0 E: 4.3	76Pla1
$[C_8H_{12}]^{2\cdot}$ 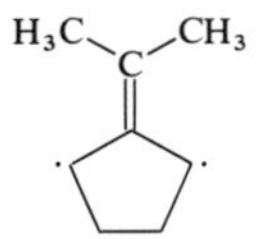	Irr. of the corresp. azo-compound/ MTHF	EPR/ <100		D: (+)27.4 E: 3.6	76Pla1/ 84Chi1 [2])

[1]) Analysis based on an extended (triplet) Hamiltonian.
[2]) Transient EPR and magnetophotoselection investigations; determination of the sign of D.

Substance	Generation/ Matrix or Solvent	Method T [K]	g-Factor	a-Value [mT]; D,E [mT]; J [T]	Ref./ add. Ref.
$[C_{18}H_{16}]^{2\cdot}$ 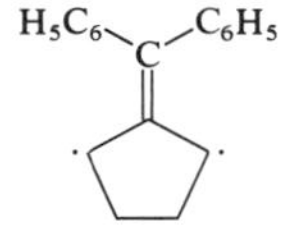	Irr. of the corresp. azo-compound/ MTHF	EPR/ <100		D: 19.3 $1.4 < E < 3.4$	76Pla1
$[C_6H_7Cl]^{2\cdot}$ 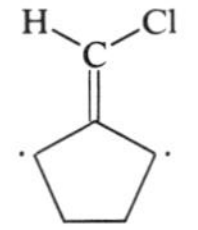	Irr. of the corresp. azo-compound/ MTHF	EPR/ <100		D: 28.5 E: 5.5	76Pla1
$[C_8H_{12}O_2]^{2\cdot}$ 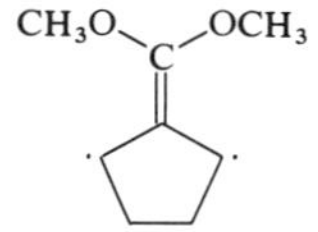	Irr. of the corresp. azo-compound/ MTHF	EPR/ <100		D: 29.7 E: <1.4	76Pla1
$[C_7H_{10}O]^{2\cdot}$ 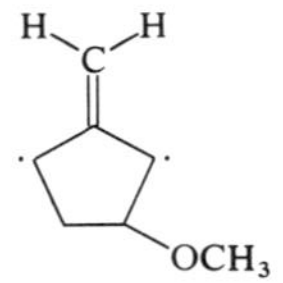	Irr. of the corresp. azo-compound/ MTHF	EPR/ 77		D: 28.8 E: 6.1	82Laz1
$[C_6H_6]^{2\cdot}$ 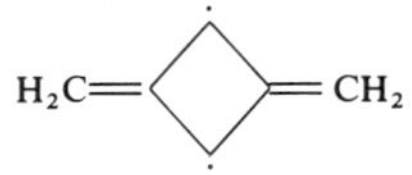	Irr. of the corresp. azo-compound/ MTHF	EPR/ 77		$0.6 < H < 0.7$ [3]) D: 21.9 E: 3.0	85Sny1/ 86Dow1

21.1.2 *m*-Quinodimethane biradicals

Substance	Generation/ Matrix or Solvent	Method T [K]	g-Factor	a-Value [mT]; D,E [mT]; J [T]	Ref./ add. Ref.
$[C_8H_8]^{2\cdot}$ 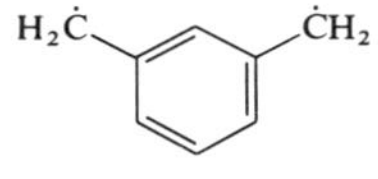	Irr. of the corresp. azo-compound/ Ethanol-d_6	EPR/ 77		D: 11.8 E: ≤1.1	82Wri1/ 85Goo1

[3]) Determined from the $\Delta m = 2$ transition; analysis based on an extended (triplet) Hamiltonian.

Substance	Generation/ Matrix or Solvent	Method T [K]	g-Factor	a-Value [mT]; D,E [mT]; J [T]	Ref./ add. Ref.
21.1.3 *p*-Quinodimethane biradicals					
$[C_{16}H_{10}O_2]^{2\cdot}$ 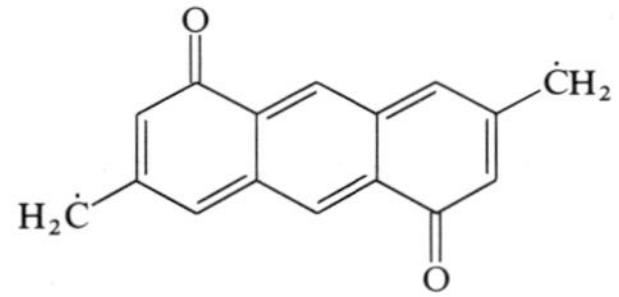	Irr. of a pentacyclic diketone/ MTHF	EPR/ 15···90		D: 21.4 E: 7.5 J: ≈9	83See1
$[C_{16}H_6D_4O_2]^{2\cdot}$ 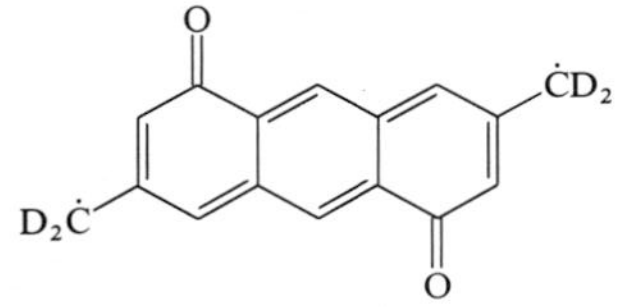	Irr./ MTHF	EPR/ 15···90		D: 26.7 E: 1.0 J: ≈9	83See1
$[C_{38}H_{28}]^{2\cdot}$ 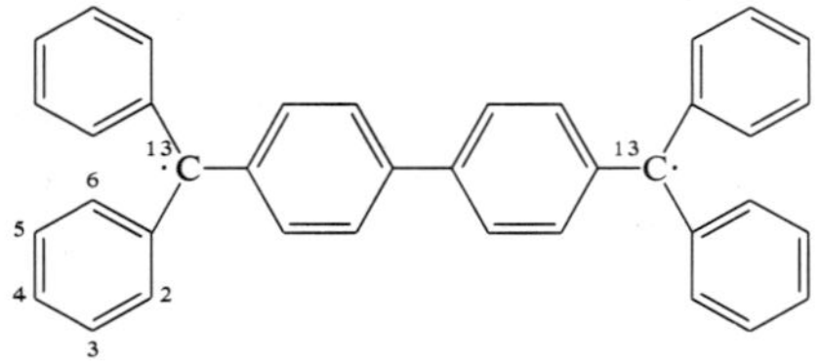	–/ Toluene	EPR/ 294		^{13}C: 2.30 H(2, 4, 6): 0.26 [4]) H(3, 5): 0.11 [4]) J: <(+)0.0025	78Pop1
$[C_{54}H_{60}]^{2\cdot}$ 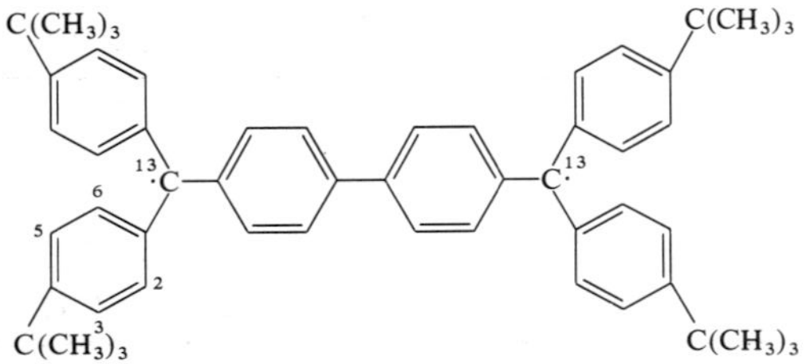	–/ Toluene	EPR/ 294		^{13}C: 2.27 H(2, 6): 0.26 [4]) H(3, 5): 0.11 [4]) J: <(+)0.0025	78Pop1

[4]) Determined by simulation of the experimental spectrum; assignment to the biradical not unambiguous.

Substance	Generation/ Matrix or Solvent	Method T [K]	g-Factor	a-Value [mT]; D,E [mT]; J [T]	Ref./ add. Ref.
$[C_{55}H_{60}]^{2\cdot}$ 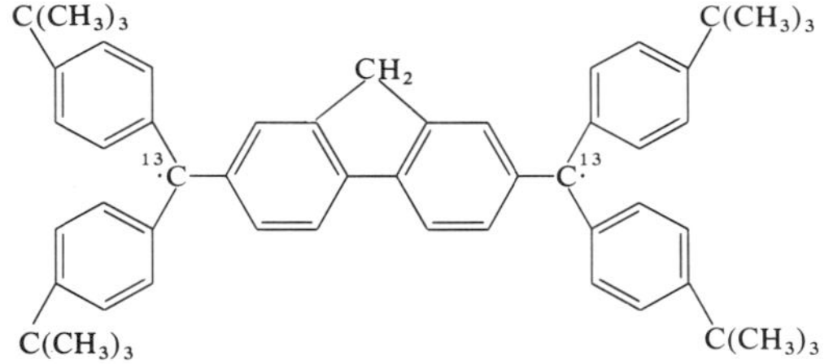 $C(CH_3)_3$, CH_2, ^{13}C	–/ Toluene	EPR/ 293		^{13}C: 2.25 J: $<(+)0.0025$	78Pop1
21.1.4 Naphthoquinodimethane biradicals					
$[C_{12}H_{10}]^{2\cdot}$ $\dot{C}H_3$ $\dot{C}H_2$ 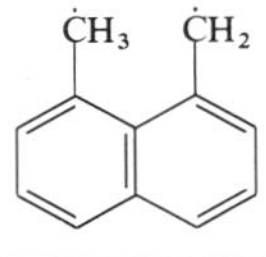	Irr. of the corresp. diazepine/ Hexfluorobenzene	EPR/ 77		D: 23.3 E: 2.2 J: $(-)16.8$	77Pag1/ 79Pla1, 79Gis1
$[C_{13}H_{10}]^{2\cdot}$ 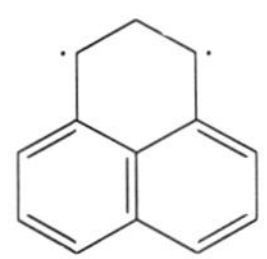	Irr. of the corresp. hydrocarbon/ 3-Methylpentane	EPR/ 77		H: 5.2 [5]) D: 27.8 E: <2.1 J: 239	78Mul1
$[C_{15}H_{14}]^{2\cdot}$ 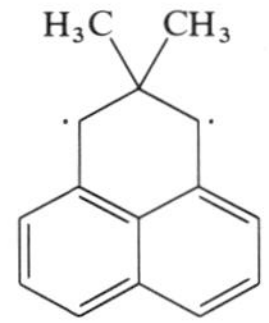 H_3C CH_3	Irr. of the corresp. hydrocarbon/ Diethylether: isopentane: ethanol (5:5:2)	EPR/ 3.1		D: 31.0 E: <2.5 J: $(-)2.1$	85Has1

[5]) Determined from the $\Delta m = 2$ transition.

Substance	Generation/ Matrix or Solvent	Method T [K]	g-Factor	a-Value [mT]; D,E [mT]; J [T]	Ref./ add. Ref.
$[C_{14}H_{12}]^{2\cdot}$ 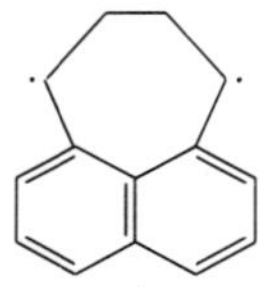	Irr. of the corresp. azo-compound/ Hexafluorobenzene	EPR/ 77		D: 19.3 E: <3.2 J: 74.8	74Pag1/ 76Wat1, 80Chi1 [6])
$[C_{12}H_9Cl]^{2\cdot}$ 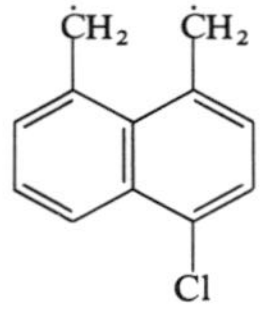	Irr. of the corresp. diazomethane compound/ MTHF Hexafluorobenzene	EPR/ 77		D: 23.0 E: 0.9	85Fri1
$[C_{12}H_9Br]^{2\cdot}$ 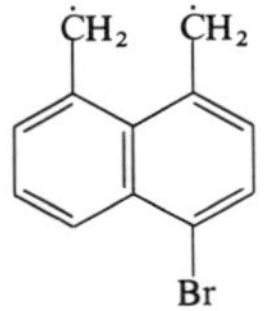	Irr. of the corresp. diazomethane compound/ MTHF	EPR/ 77		D: 28.8 E: 1.6	85Fri1
$[C_{18}H_{14}S]^{2\cdot}$ 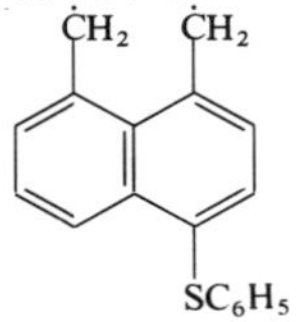	Irr. of the corresp. diazomethane compound/ MTHF	EPR/ 77		D: 23.0 E: 0.7	85Fri1
$[C_{18}H_{14}O_2S]^{2\cdot}$ 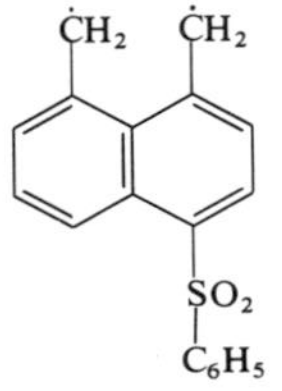	Irr. of the corresp. diazomethane compound/ MTHF	EPR/ 77		D: 23.0 E: 0.6	85Fri1

[6]) Time-resolved EPR studies.

Substance	Generation/ Matrix or Solvent	Method T [K]	g-Factor	a-Value [mT]; D,E [mT]; J [T]	Ref./ add. Ref.
21.1.5 Cyclopentadienyl biradicals					
$[C_{72}H_{52}]^{2\cdot}$ 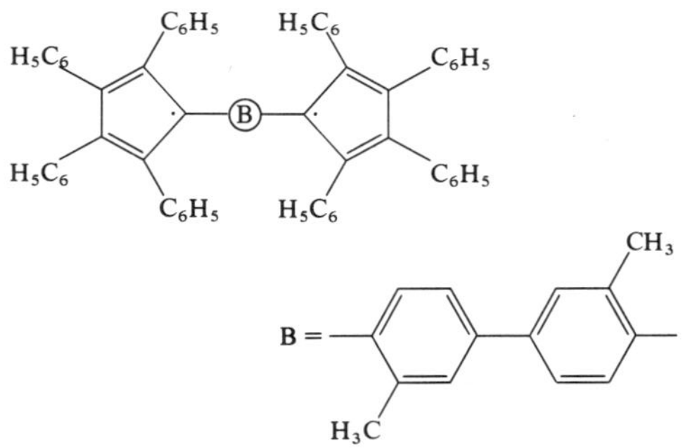	Dehalog. of the corresp. dibromide with Ag/ Toluene	EPR/ 160		D: 1.4	76Bro1
$[C_{70}H_{48}]^{2\cdot}$ 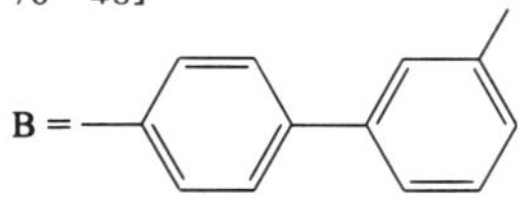	Dehalog. of the corresp. dibromide with Ag/ Toluene	EPR/ 160		D: 2.1	76Bro1
$[C_{72}H_{52}]^{2\cdot}$ 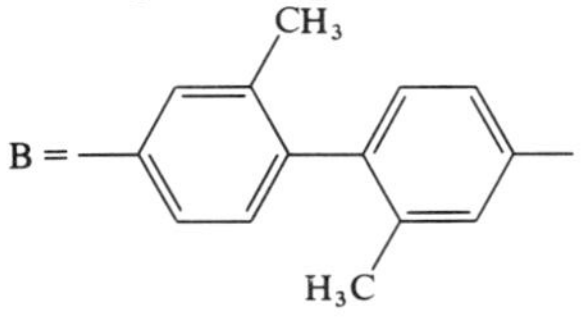	Dehalog. of the corresp. dibromide with Ag/ Toluene	EPR/ 160		D: 1.8	76Bro1
$[C_{74}H_{56}]^{2\cdot}$ 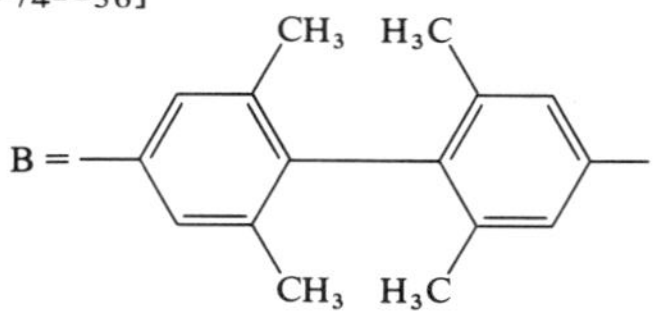	Dehalog. of the corresp. dibromide with Ag/ Toluene	EPR/ 160		D: 1.5	76Bro1

Substance	Generation/ Matrix or Solvent	Method T [K]	g-Factor	a-Value [mT]; D,E [mT]; J [T]	Ref./ add. Ref.
$[C_{76}H_{52}]^{2\cdot}$ B = 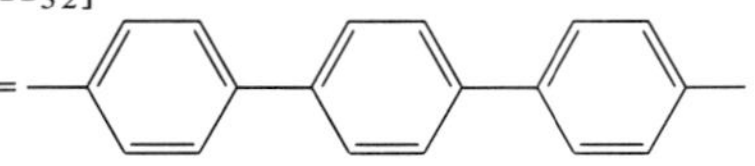	Dehalog. of the corresp. dibromide with Ag/ Toluene	EPR/ 160		D: 1.1	76Bro1
$[C_{76}H_{52}]^{2\cdot}$ B = 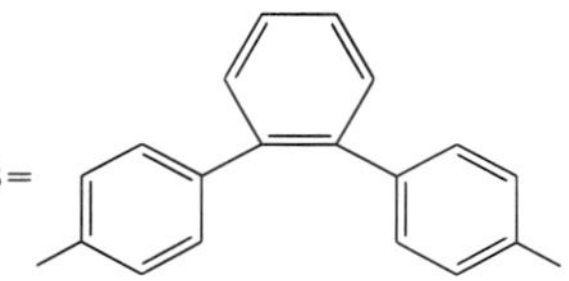	Dehalog. of the corresp. dibromide with Ag/ Toluene	EPR/ 160		D: 3.6	76Bro1
$[C_{72}H_{60}]^{2\cdot}$ 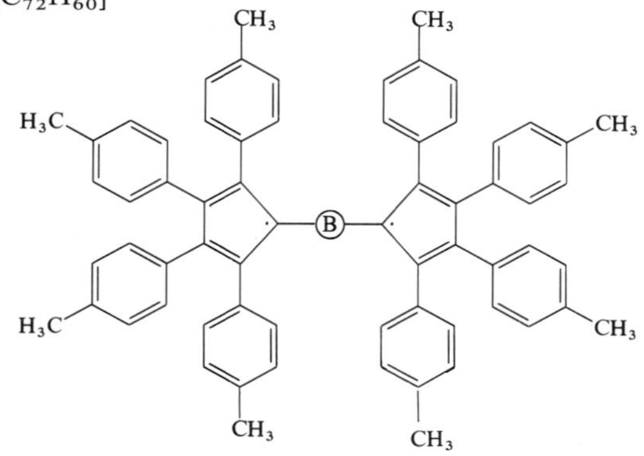 CH_3 CH_3 H_3C CH_3 (B) H_3C CH_3 CH_3 CH_3 B 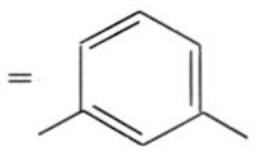 =	Dehalog. of the corresp. dibromide with Ag/ Toluene	EPR/ 160		D: 5.2	76Bro1

21.2 Oxygen-centered polyradicals

21.2.1 Aroxyl biradicals

Substance	Generation/ Matrix or Solvent	Method T [K]	g-Factor	a-Value [mT]; D,E [mT]; J [T]	Ref./ add. Ref.
$[C_{37}H_{48}O_3]^{2\cdot}$	Dehydr. of the corresp. bisphenol with PbO_2/ Toluene	EPR/ 293 77	 2.0046; 2.0041; 2.0035; is: 2.00457	H(3, 5): 0.172 H(CH_3): 0.516 D: 2.33 E: 0.14	79Muk1
$[C_{40}H_{54}O_3]^{2\cdot}$	Dehydr. of the corresp. bisphenol with PbO_2/ Toluene	EPR/ 293 77	 2.0052; 2.0055; 2.0030; is: 2.00451	H(3, 5): 0.172 H(CH_3): 0.518 D: 3.42 E: 0.22	79Muk1
$[C_{43}H_{60}O_3]^{2\cdot}$	Dehydr. of the corresp. bisphenol with $K_3[Fe(CN)_6]$/ Toluene	EPR, ENDOR/ 320 150	2.0048	H: (+)0.176 ^{13}C: (−)0.65; (−)0.65; (−)1.292; [7]) is: (−)0.864 [7]) D: 3.21 $(-)107 < J < 32.1$	78Kir1/ 82Wil1

[7]) Analysis based on an extended Hamiltonian.

Substance	Generation/ Matrix or Solvent	Method T [K]	g-Factor	a-Value [mT]; D,E [mT]; J [T]	Ref./ add. Ref.
$[C_{64}H_{80}D_4O_4]^{2\cdot}$ 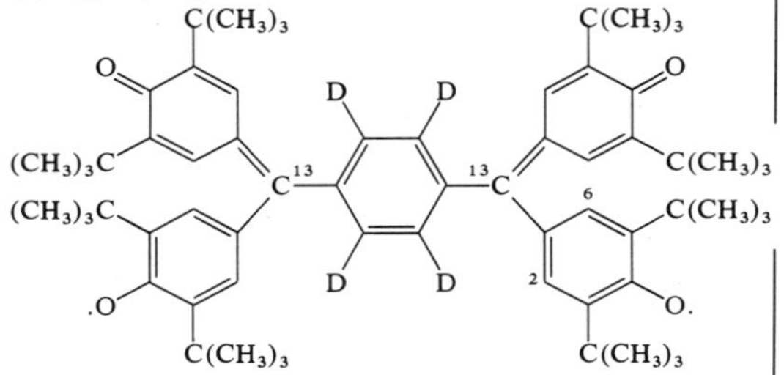	Dehydr. of the corresp. bisphenol with PbO_2/ Toluene	EPR, ENDOR/ 330		H(2, 6): 0.137 ^{13}C: 0.986 J: (−)0.3209	78Kie2/ 82Wil1
$[C_{43}H_{59}DO_3]^{2\cdot}$ 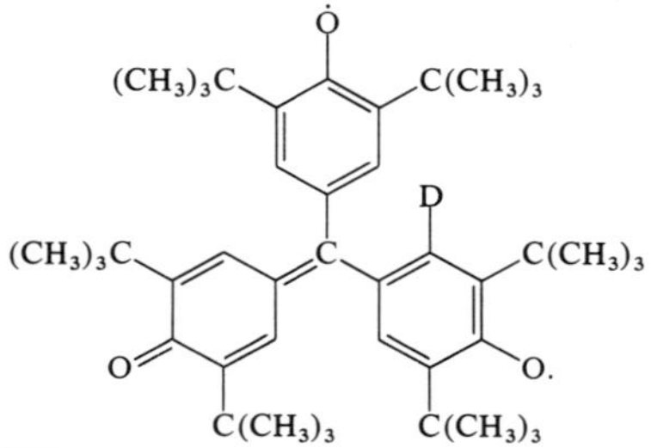	Dehydr. of the corresp. bisphenol with PbO_2/ Toluene	EPR/ 293 77	 2.0055; 2.0054; 2.0025; is: 2.00449	H: 0.180 D: 3.42 E: 0.23	77Muk1
$[C_{43}H_{58}D_2O_3]^{2\cdot}$ 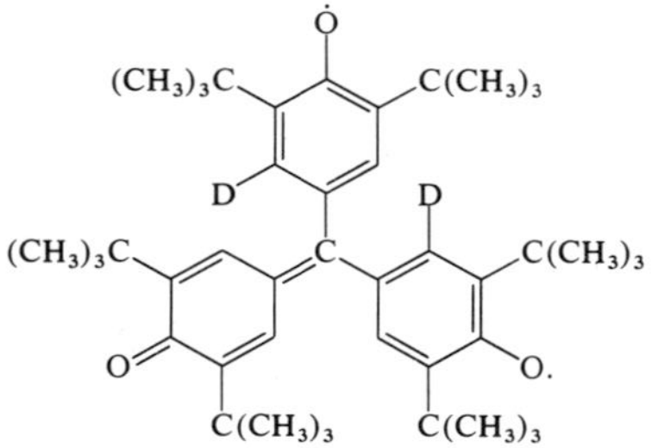	Dehydr. of the corresp. bisphenol with PbO_2/ Toluene	EPR/ 293 77	 2.0054; 2.0053; 2.0025; is: 2.00449	H: 0.180 D: 3.42 E: 0.24	77Muk1

Substance	Generation/ Matrix or Solvent	Method T [K]	g-Factor	a-Value [mT]; D,E [mT]; J [T]	Ref./ add. Ref.
$[C_{43}H_{54}D_6O_3]^{2\cdot}$	–/ Toluene	EPR, ENDOR/ 100		D: 3.4 E: 0.2	85Kir1
$[C_{45}H_{64}O_3]^{2\cdot}$	–/ Toluene	EPR/ 293 77	 2.0050; 2.0053; 2.0025; is: 2.00447	 H(3, 5): 0.172 D: 3.33 E: 0.27	77Muk2
$[C_{34}H_{44}O_2]^{2\cdot}$	Dehydr. of the corresp. bisphenol with PbO_2/ Toluene	EPR/ 293 77	 2.0051; 2.0035; 2.0024; is: 2.00414	 H(2, 4, 5): 0.172 [8]) H(1, 3, 6): 0.172 [8]) D: 3.40 E: 0.30	79Muk2

[8]) Determined by simulation of the experimental spectrum; analysis based on an extended Hamiltonian.

Substance	Generation/ Matrix or Solvent	Method T [K]	g-Factor	a-Value [mT]; D,E [mT]; J [T]	Ref./ add. Ref.
$[C_{40}H_{48}O_2]^{2\cdot}$ 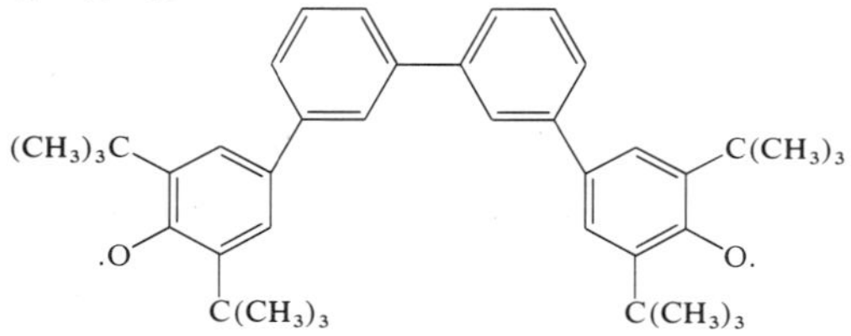	Dehydr. of the corresp. bisphenol with PbO_2/ Toluene	EPR/ 77	2.0045; 2.0051; 2.0024; is: 2.0040	D: 3.54 E: 0.26	80Muk1
$[C_{140}H_{178}O_8Si]^{2\cdot}$ [a]), R = $-C(CH_3)_3$	–/ Toluene	EPR/ 290		D: ≈0.9	81Kir1

21.2.2 Aroxyl triradicals

Substance	Generation/ Matrix or Solvent	Method T [K]	g-Factor	a-Value [mT]; D,E [mT]; J [T]	Ref./ add. Ref.
$[C_{93}H_{123}O_6]^{3\cdot}$ [b])	Dehydr. of the corresp. tris-phenol with $K_3[Fe(CN)_6]$/ Toluene	EPR, ENDOR/ 250		H(2, 6): (+)0.129 ^{13}C: (−)1.007	83Sch1
		240		D: 0.78	

a)

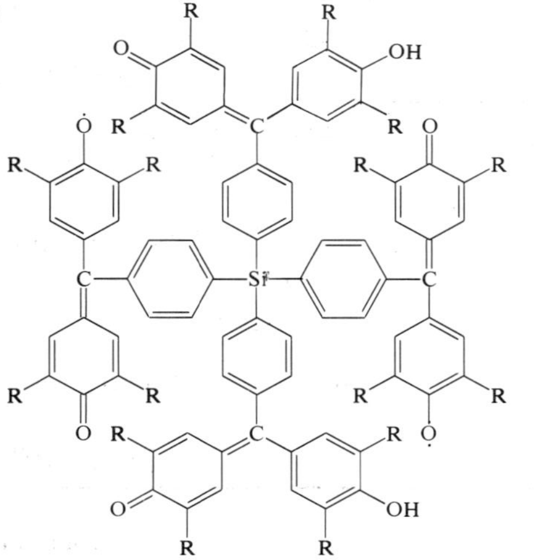

b)

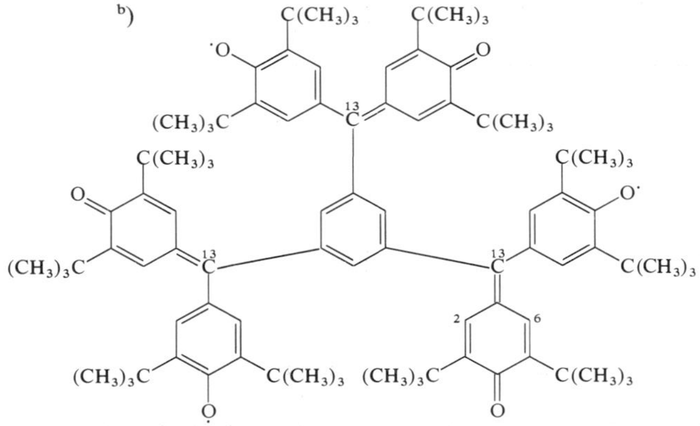

21.2.3 Oxygen- and carbon-centered biradicals

Substance	Generation/ Matrix or Solvent	Method T [K]	g-Factor	a-Value [mT]; D,E [mT]; J [T]	Ref./ add. Ref.
$[C_7H_6O]^{2\cdot}$ 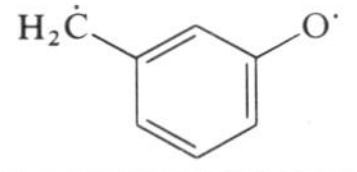	Irr. of the corresp. dienone/ MTHF	EPR/ 11		D: ≈28.9 E: ≈8.6	79Rul1/ 82Rul1
$[C_{11}H_8O]^{2\cdot}$ 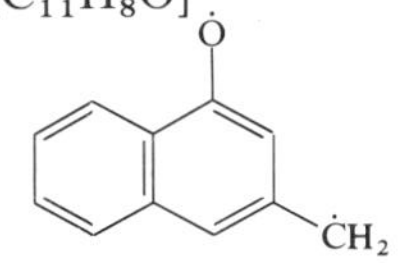	Irr. of the corresp. dienone/ MTHF	EPR/ 77		D: 21.8 E: 5.6	81See1

21.2.4 Oxygen- and carbon-centered polyradicals

Substance	Generation/ Matrix or Solvent	Method T [K]	g-Factor	a-Value [mT]; D,E [mT]; J [T]	Ref./ add. Ref.
$[C_{16}H_{10}O_2]^{4\cdot}$ 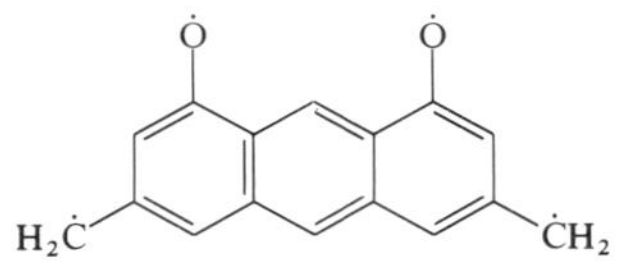	Irr. of the corresp. dienone/ MTHF	EPR/ 77		D: 12.6 E: 3.4	83See2

21.3 Nitrogen-centered polyradicals

21.3.1 Hydrazyl biradicals

Substance	Generation/ Matrix or Solvent	Method T [K]	g-Factor	a-Value [mT]; D,E [mT]; J [T]	Ref./ add. Ref.
$[C_{36}H_{20}N_8O_8]^{2\cdot}$ 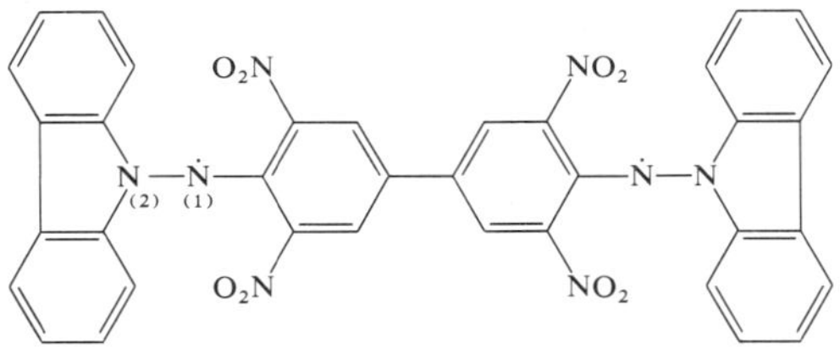	–/ –	EPR/ 293		N(1): 0.563 [9]) N(2): 1.091 [9]) J: >0.0150 [9])	75Kor1

[9]) Determined by simulation of the experimental spectrum.

Substance	Generation/ Matrix or Solvent	Method T [K]	g-Factor	a-Value [mT]; D,E [mT]; J [T]	Ref./ add. Ref.
$[C_{38}H_{28}N_8O_{10}]^{2\cdot}$ 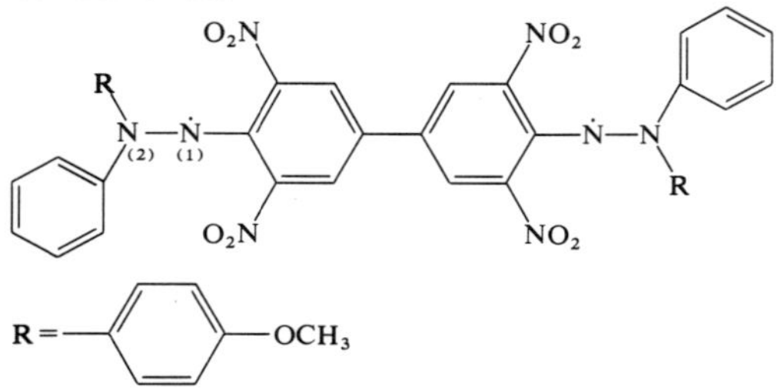	–/ –	EPR/ 293		N(1): 0.801 [9] N(2): 0.805 [9] J: >0.0150 [9]	75Kor1
$[C_{48}H_{28}N_8O_8S_2]^{2\cdot}$ 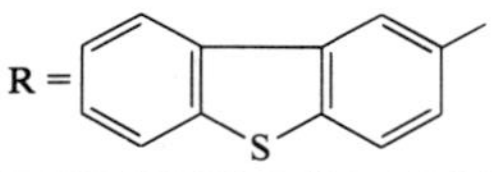	–/ –	EPR/ 293		N(1): 0.879 [9] N(2): 0.701 [9] J: 0.0003 [9]	75Kor1

21.3.2 Verdazyl biradicals

Substance	Generation/ Matrix or Solvent	Method T [K]	g-Factor	a-Value [mT]; D,E [mT]; J [T]	Ref./ add. Ref.
$[C_{28}H_{24}N_8]^{2\cdot}$ 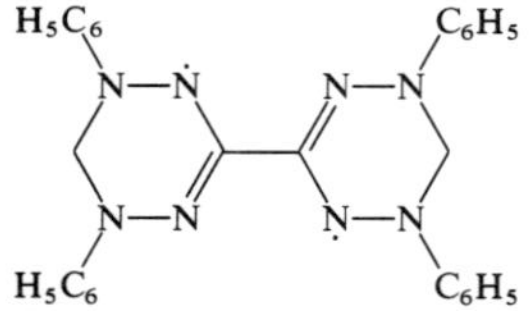	–/ Toluene	EPR/ 88		D: 30.5 [9]	80Neu1
$[C_8H_{12}N_8O_2]^{2\cdot}$ 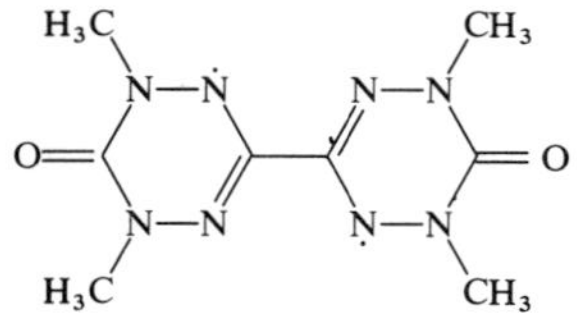	–/ –	EPR/ 88		D: 32.0	80Neu2

[9]) Determined by simulation of the experimental spectrum.

Substance	Generation/ Matrix or Solvent	Method T [K]	g-Factor	a-Value [mT]; D,E [mT]; J [T]	Ref./ add. Ref.
$[C_{40}H_{36}D_{10}N_8]^{2\cdot}$ 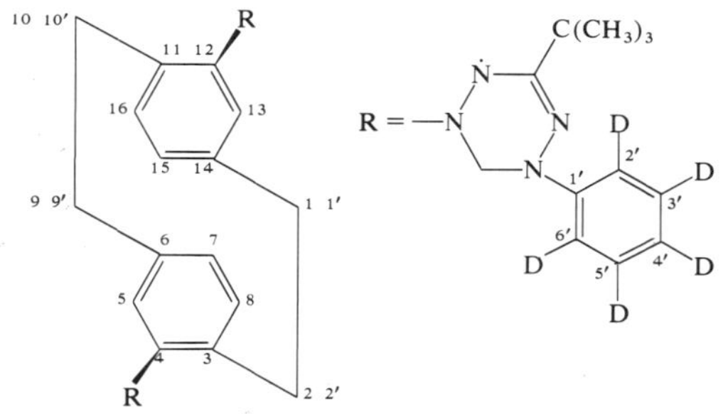	–/ DBNO	NMR/ 300		H(2, 10): (+)0.071 H(2′, 10′): (+)0.022 H(5, 13): (−)0.099 H(7, 15): (−)0.104 H(8, 16): (+)0.058 $H(C(CH_3)_3)$: (+)0.011 D(2′, 6′): (−)0.0163 D(3′, 5′): (+)0.0060 D(4′): 0.0174	81Neu1
	–/ Benzene nematic liquid crystal (phase V)	EPR/ 293 240		N: 0.59 D: 5.62	81Neu1
$[C_{40}H_{36}D_{10}N_8]^{2\cdot}$ 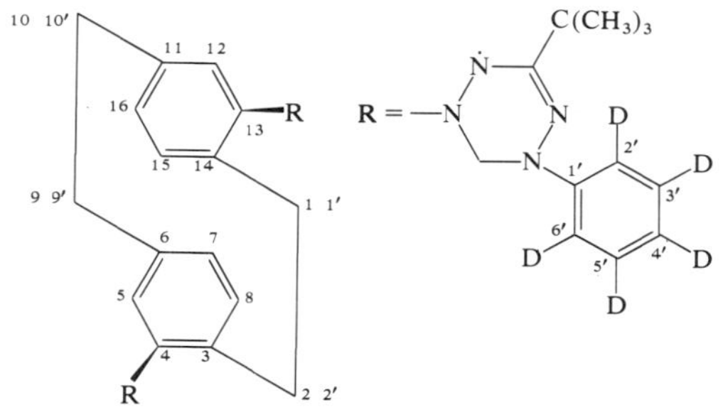	–/ DBNO	NMR/ 300		H(1, 2): (+)0.082 H(1′, 2′): (+)0.019 H(5, 7, 12, 16): (−)0.102 H(8, 15): (+)0.055 H(9, 10): (−)0.011 $H(C(CH_3)_3)$: (+)0.011 D(2′, 6′): (−)0.0161 D(3′, 5′): (+)0.0060 D(4′): (−)0.0171	81Neu1
	–/ Benzene nematic liquid crystal (phase V)	EPR/ 293 240		N: 0.59 D: 6.82	81Neu1

Substance	Generation/ Matrix or Solvent	Method T [K]	g-Factor	a-Value [mT]; D,E [mT]; J [T]	Ref./ add. Ref.
$[C_{40}H_{36}D_{10}N_8]^{2\cdot}$ 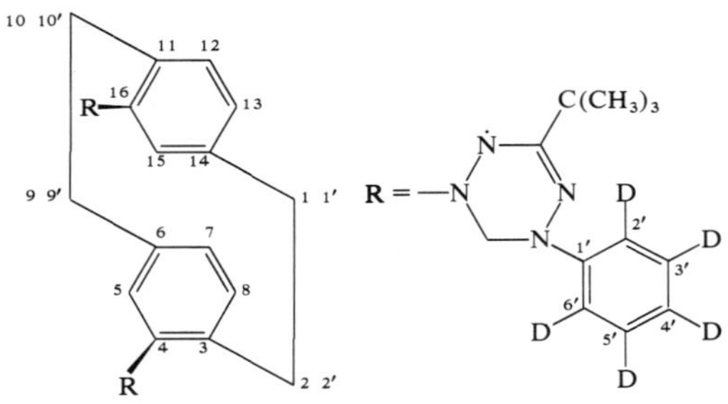	–/ DBNO	NMR/ 300		H(1, 9): (–)0.011 H(1′, 9′): (–)0.006 H(2, 10): (+)0.096 H(2′, 10′): (+)0.018 H(5, 15): (–)0.075 H(7, 13): (+)0.091 H(8, 12): (+)0.050 H(C(CH$_3$)$_3$): (+)0.012 D(2′, 6′): (–)0.0158 D(3′, 5′): (+)0.0059 D(4′): (–)0.0173	81Neu1
	–/ Benzene nematic liquid crystal (phase V)	EPR/ 293 240		N: 0.59 D: 11.18	81Neu1

21.3.3 Trisiminomethane biradicals

Substance	Generation/ Matrix or Solvent	Method T [K]	g-Factor	a-Value [mT]; D,E [mT]; J [T]	Ref./ add. Ref.
$[C_4H_9N_3]^{2\cdot}$ 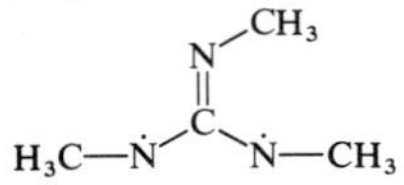	Irr. of the corresp. diaziridine/ Butyronitrile	EPR/ 78		H: 2.50 [10]) N: 1.14 D: 35.3	78Qua1

21.3.4 Nitrogen- and carbon-centered biradicals

Substance	Generation/ Matrix or Solvent	Method T [K]	g-Factor	a-Value [mT]; D,E [mT]; J [T]	Ref./ add. Ref.
$[C_{11}H_9N]^{2\cdot}$ 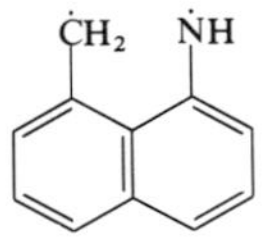	Irr. of the corresp. azide/ MTHF	EPR/ 77		D: 27.3 E: 0.9	79Pla2

[10]) Determined from the $\Delta m=2$ transition by simulation; analysis based on an extended (triplet) Hamiltonian.

21.3.5 Nitrogen- and oxygen-centered biradicals

Substance	Generation/ Matrix or Solvent	Method T [K]	g-Factor	a-Value [mT]; D,E [mT]; J [T]	Ref./ add. Ref.
$[C_{24}H_{37}NO_4]^{2\cdot}$ 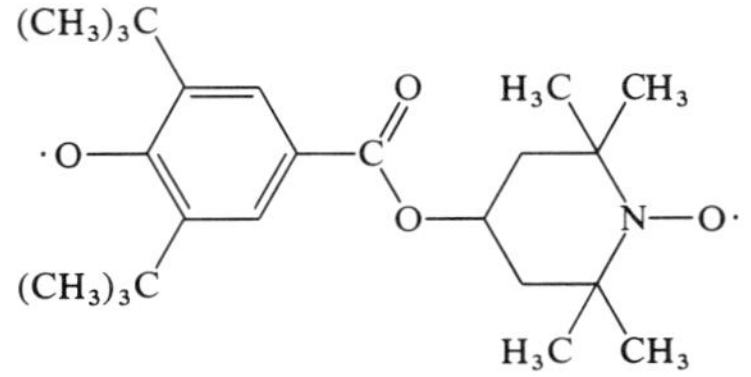	–/ Toluene	EPR/ 293	2.0053	N: 1.52	81Muk1
$[C_{37}H_{53}NO_3]^{2\cdot}$ 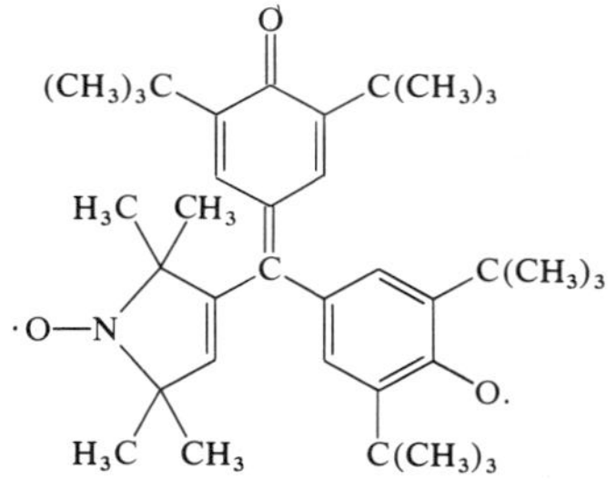	–/ Toluene	EPR, ENDOR/ 290 160		N: (+)1.46 D: 7.00	82Kir1
$[C_{38}H_{57}NO_3]^{2\cdot}$ 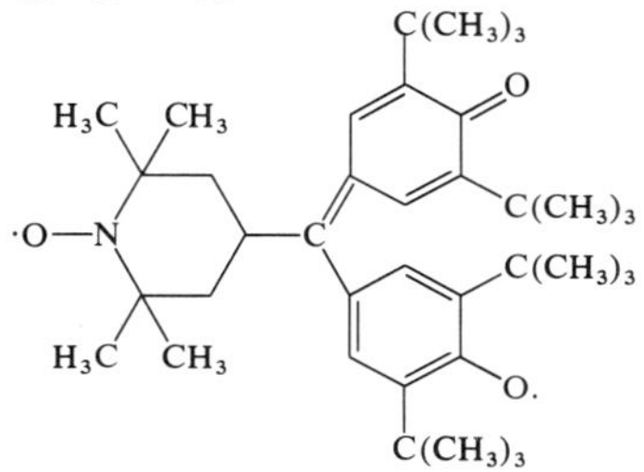	–/ Toluene	ENDOR, EPR/ 290 160		N: (+)1.54 D: 7.14	82Kir1

Substance	Generation/ Matrix or Solvent	Method T [K]	g-Factor	a-Value [mT]; D,E [mT]; J [T]	Ref./ add. Ref.
$[C_{44}H_{57}NO_5]^{2\cdot}$ 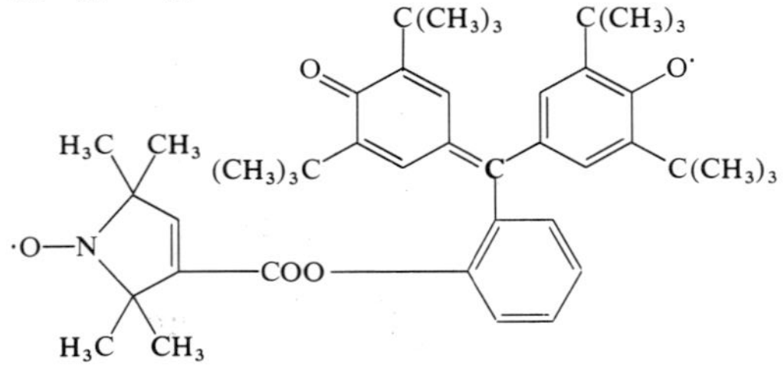	–/ Toluene	EPR, ENDOR/ 290 160	≈2.0054	N: (+)1.42 D: 5.82	82Kir1
$[C_{44}H_{57}NO_5]^{2\cdot}$ 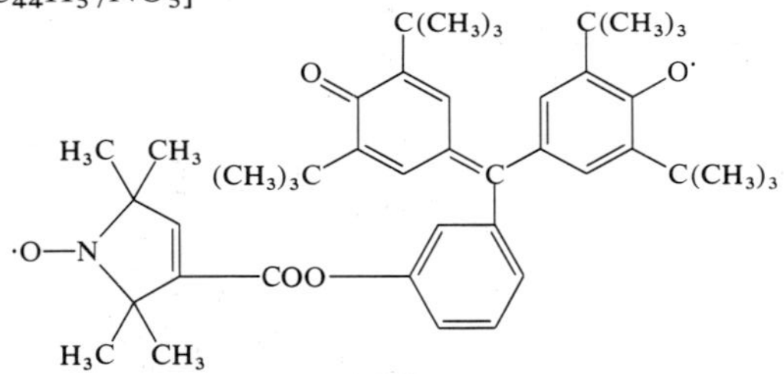	–/ Toluene	EPR, ENDOR/ 290 160	2.00529	J: 0.00582 D: 3.04	82Kir1
$[C_{44}H_{57}NO_5]^{2\cdot}$ 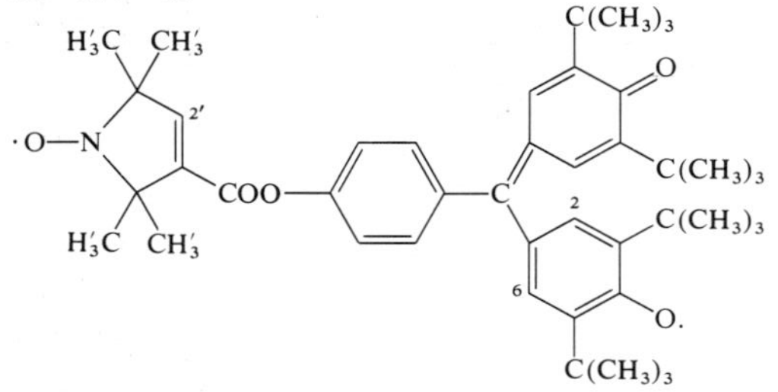	–/ Toluene	EPR, ENDOR/ 290 225 160	2.00530	N: (+)1.42 J: (–)0.00596 H(2, 6): (+)0.1286 H(2′, CH_3'): (–)0.0229 D: 2.07	82Kir1

Substance	Generation/ Matrix or Solvent	Method T [K]	g-Factor	a-Value [mT]; D,E [mT]; J [T]	Ref./ add. Ref.
$[C_{45}H_{61}NO_5]^{2\cdot}$	–/ Toluene	EPR, ENDOR/			82Kir1
		290	2.00542	N: (+)1.55 J: (–)0.00518	
		225		H(2, 6): (+)0.1314 H(2′, 6′): (–)0.0475	
		160		D: 2.07	
$[C_{50}H_{61}NO_5]^{2\cdot}$	–/ Toluene	EPR, ENDOR/ 290	2.00531	J: 0.00114	82Kir1 [11])
$[C_{51}H_{65}NO_5]^{2\cdot}$	–/ Toluene	EPR, ENDOR/			82Kir1
		290		N: 1.56 J: (–)0.00089	
		225		H(2, 6): 0.1307 H(2′, 6′): (–)0.0482	

[11]) Zero-field splitting smaller than ^{14}N hyperfine anisotropy.

21.4 References for 21.1···21.3

66Dow1 Dowd, P.: J. Am. Chem. Soc. **88** (1966) 2587.

72Dow1 Dowd, P.: Accounts Chem. Res. **5** (1972) 242.

74Pag1 Pagni, R.M., Watson, Jr., C.R., Bloor, J.E., Dodd, J.R.: J. Am. Chem. Soc. **96** (1974) 4064.

75Kor1 Koryakov, V.I., Maksimov, A.A., Gubanov, V.A.: Tr. Inst. Khim., Ural. Nauchn. Tsentr., Akad. Nauk SSSR **34** (1975) 106.

76Bas1 Baseman, R.J., Pratt, D.W., Chow, M., Dowd, P.: J. Am. Chem. Soc. **98** (1976) 5726.
76Bro1 Broser, W., Janzen, D., Kurreck, H., Braasch, D., Oestreich, S., Plato, M.: Tetrahedron **32** (1976) 1819.
76Pla1 Platz, M.S., McBride, J.M., Little, R.D., Harrison, J.J., Shaw, A., Potter, S.E., Berson, J.A.: J. Am. Chem. Soc. **98** (1976) 5725.
76Wat1 Watson, Jr., C.R., Pagni, R.M., Dodd, J.R., Bloor, J.E.: J. Am. Chem. Soc. **98** (1976) 2551.

77Muk1 Mukai, K., Mishima, T., Ishizu, K.: J. Chem. Phys. **66** (1977) 1680.
77Muk2 Mukai, K., Yorimitsu, K., Mishima, T.: Bull. Chem. Soc. Jpn. **50** (1977) 2471.
77Pag1 Pagni, R.M., Burnett, M.N., Dodd, J.R.: J. Am. Chem. Soc. **99** (1977) 1972.

78Kir1 Kirste, B., Kurreck, H., Lubitz, W., Schubert, K.: J. Am. Chem. Soc. **100** (1978) 2292.
78Kir2 Kirste, B., Kurreck, H., Schubert, K.: Tetrahedron Lett. **9** (1978) 777.
78Mul1 Muller, J.-P., Muller, D., Dewey, H.J., Michl., J.: J. Am. Chem. Soc. **100** (1978) 1629.
78Pop1 Popp, F., Bickelhaupt, F., McLean, C.: Chem. Phys. Lett. **55** (1978) 327.
78Qua1 Quast, H., Bieber, L., Danen, W.C.: J. Am. Chem. Soc. **100** (1978) 1306.

79Gis1 Gisin, M., Rommel, E., Wirz, J. Burnett, M.N., Pagni, R.M.: J. Am. Chem. Soc. **101** (1979) 2216.
79Muk1 Mukai, K.: Bull. Chem. Soc. Jpn. **52** (1979) 1911.
79Muk2 Mukai, K., Hara, T., Ishizu, K.: Bull. Chem. Soc. Jpn. **52** (1979) 1853.
79Pla1 Platz, M.S.: J. Am. Chem. Soc. **101** (1979) 3398.
79Pla2 Platz, M.S., Burns, J.R.: J. Am. Chem. Soc. **101** (1979) 4425.
79Rul1 Rule, M., Matlin, A.R., Hilinksi, E.F., Dougherty, D.A., Berson, J.A.: J. Am. Chem. Soc. **101** (1979) 5098.

80Chi1 Chisholm, W.P., Weissman, S.I., Burnett, M.N., Pagni, R.M.: J. Am. Chem. Soc. **102** (1980) 7104.
80Cla1 Claesson, O., Lund, A., Gillbro, T., Ichikawa, T., Edlund, O., Yoshida, H.: J. Chem. Phys. **72** (1980) 1463.
80Muk1 Mukai, K., Inagaki, N.: Bull. Chem. Soc. Jpn. **53** (1980) 2695.
80Neu1 Neugebauer, F.A., Fischer, H., Meier P.: Chem. Ber. **113** (1980) 2049.
80Neu2 Neugebauer, F.A., Fischer, H.: Angew. Chem. **92** (1980) 766.

81Kir1 Kirste, B., Harrer, W., Kurreck, H.: Angew. Chem. **93** (1981) 912.
81Muk1 Mukai, K., Yano, H., Ishizu, K.: Tetrahedron Lett. **22** (1981) 1903.
81Neu1 Neugebauer, F.A., Fischer, H.: J. Chem. Soc., Perkin Trans II **1981**, 896.
81See1 Seeger, D.E., Hilinski, E.F., Berson, J.A.: J. Am. Chem. Soc. **103** (1981) 720.

82Dow1 Dowd, P., Chow, M.: Tetrahedron **38** (1982) 799.
82Kir1 Kirste, B., Krüger, A., Kurreck, H.: J. Am. Chem. Soc. **104** (1982) 3850.
82Laz1 Lazzara, M.G., Harrison, J.J., Rule, M., Hilinski, E.F., Berson, J.A.: J. Am. Chem. Soc. **104** (1982) 2233.
82Rul1 Rule, M., Matlin, A.R., Seeger, D.E., Hilinski, E.F., Dougherty, D.A., Berson, J.A.: Tetrahedron **38** (1982) 787.
82Wil1 van Willigen H., Kirste, B., Kurreck, H., Plato, M.: Tetrahedron **38** (1982) 759.
82Wri1 Wright, B.B., Platz, M.S.: J. Am. Chem. Soc. **105** (1983) 628.

83Sch1 Schubert, K., Kirste, B., Kurreck, H.: Angew. Chem. **95** (1983) 149.
83See1 Seeger, D.E., Berson, J.A.: J. Am. Chem. Soc. **105** (1983) 5146.
83See2 Seeger, D.E., Berson, J.A.: J. Am. Chem. Soc. **105** (1983) 5144.

84Chi1 Chisholm, W.P., Yu, H.L., Murugesan, R., Weissman, S.I., Hilinski, E.F., Berson, J.A.: J. Am. Chem. Soc. **106** (1984) 4419.

85Fri1 Fritz, M.J., Ramos, E.L., Platz, M.S.: J. Org. Chem. **50** (1985) 3522.
85Goo1 Goodman, J.L., Berson, J.A.: J. Am. Chem. Soc. **107** (1985) 5409.
85Has1 Hasler, E., Gassmann, E., Wirz, J.: Helv. Chim. Acta **68** (1985) 777.
85Kir1 Kirste, B., Kurreck, H., Sordo, M.: Chem. Ber. **118** (1985) 1782.
85Sny1 Snyder, G.J., Dougherty, D.A.: J. Am. Chem. Soc. **107** (1985) 1774.

86Dow1 Dowd, P., Paik, Y.H.: J. Am. Chem. Soc. **108** (1986) 2788.

22 Organic bis- and polynitroxides

22.0 Introduction

22.0.1 General remarks

The underlying theory of spin exchange mechanisms is now largely understood. Methods for the evaluation of J (spin exchange interaction) are reasonably well described and much of the current interest is centred on exchange between nitroxide groups and paramagnetic metal ions (usually transition metals). This chapter does not attempt to cover the many studies of, for example, copper and cobalt-ion nitroxide interaction because the spectra observed are more characteristic of the metal ion than of the nitroxide. This subject has been covered in several other reviews. However, many bis- and tris-nitroxides linked by complexation through other metal ions, notably Pt and Pd are included and D, E and J values are reported for them. In such cases the metal atom merely has the same function as an organic linking group. A high proportion of the recent work has been done in the USSR and published in Russian Journals which are either not readily available or not readily available in translation. Accordingly, full details of such work is not always reported. The literature has been covered through Chemical Abstracts 1989, Vol. 110.

22.0.2 Nomenclature and classification

The vast majority of bisnitroxides listed are cyclic and they have been arranged in sections according to the basic heterocycle, the molecular formula and/or the number of atoms in the bridge linking the nitroxide groups. Coupling constants, D, E and J values are given where possible but in a large number of cases only the number of lines observed and the approximate relationship between a_N and J is given, e.g. $a_N \gg J$, $J \gg a_N$ od $A_N \approx J$.

22.0.3 Reviews

Gulin, V.I., Dikanov, S.A., Marin, V.V., Grigorev, I.A., Volodarskii, L.B.: Izv. Sib. Otd. Akad. Nauk SSSR, Ser. Khim. Nauk **1988**, 99; C.A. **110** (1989) 38518.

Dobryakov, S.N., Dmitriev, P.I., Shapiro, A.B.: Dokl. Akad. Nauk SSSR **229** (1988) 1153; C.A. **110** (1989) 212968.

22.1 Bisnitroxides

22.1.1 Bis(dialkyl nitroxides)

Substance	Generation/ Matrix or Solvent	Method/ T[K]	g-Factor	a-Value, D, E, J [mT]	Ref./ add. Ref.
$[C_{18}H_{38}N_4O_4]^{2\cdot}$	$(CH_3)_3CN(O\cdot)C(CH_3)_2C(=NOH)CH_3$ + PbO_2/ C_6H_{14}	EPR/ 213		5-line spectrum $J \gg a_N$	81Tka1

22.1.2 Monocyclic bis(dialkyl nitroxides)

Substance	Generation/ Matrix or Solvent	Method/ T[K]	g-Factor	a-Value, D, E, J [mT]	Ref./ add. Ref.
$[C_{12}H_{22}N_2O_2]^{2\cdot}$	Not given/ –	EPR/ –		D: 80.3 [1])	80Mic2

22.1.3 Bicyclic bis(dialkyl nitroxides)

Substance	Generation/ Matrix or Solvent	Method/ T[K]	g-Factor	a-Value, D, E, J [mT]	Ref./ add. Ref.
$[C_8H_{12}N_2O_2]^{2\cdot}$	Not given/ DMF	Calc./ –		D: 25.0 [2]) N: 1.9	80Mic2

[1]) D calculated as 88 mT.
[2]) D calculated as 28.0 mT.

22.1.4 Bis(pyrrolidinyl-1-oxyls)

Substance	Generation/ Matrix or Solvent	Method/ T[K]	g-Factor	a-Value, D, E, J [mT]	Ref./ add. Ref.
$[C_{16}H_{34}Cl_2N_4O_2Pt]^{2\cdot}$ 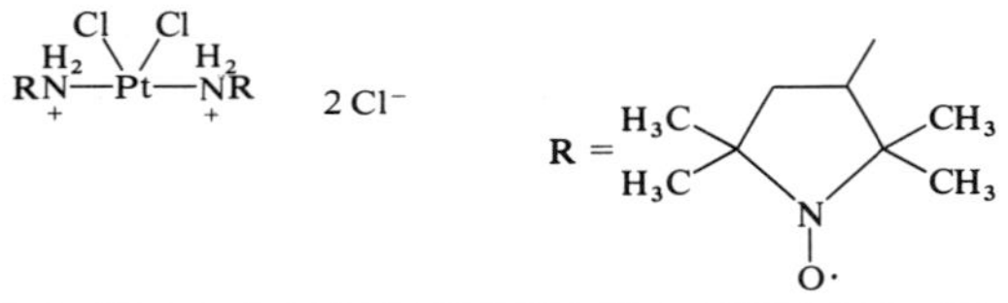	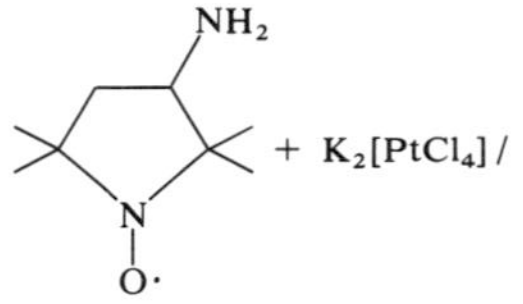 + $K_2[PtCl_4]$ / H_2O	EPR/ 300	2.0054		79Mat1
$[C_{18}H_{28}N_2O_4]^{2\cdot}$	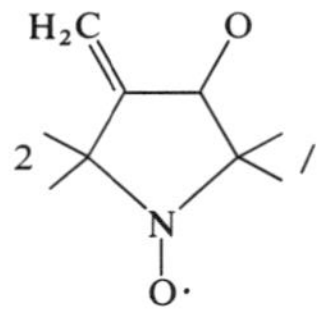 2 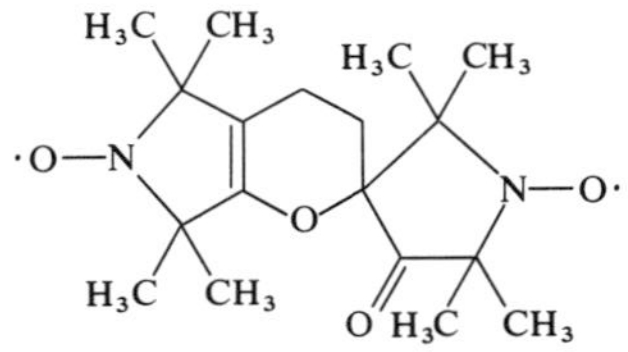 … / C_2H_5OH	EPR/ 300		5-line spectrum $J \gg a_N$	82Vla1 [3]), 80Kok2 [4])
$[C_{18}H_{32}N_4O_2]^{\cdot}$ RN=CHCH=NR 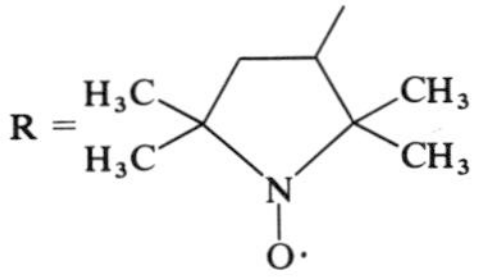	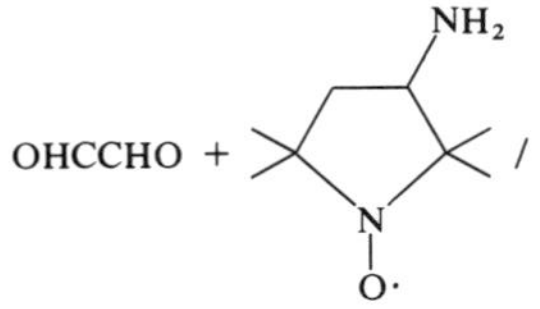 OHCCHO + … / $C_6H_5CH_3$	EPR/ 257…328		$\bar{J}/a = 9.2$ [5])	81Kok2 [6])
$[C_{20}H_{36}N_4O_4]^{2\cdot}$ $RC(O)NHCH_2CH_2NHC(O)R$	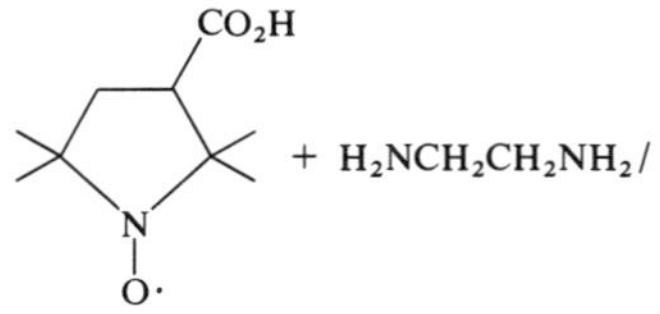 + $H_2NCH_2CH_2NH_2$ 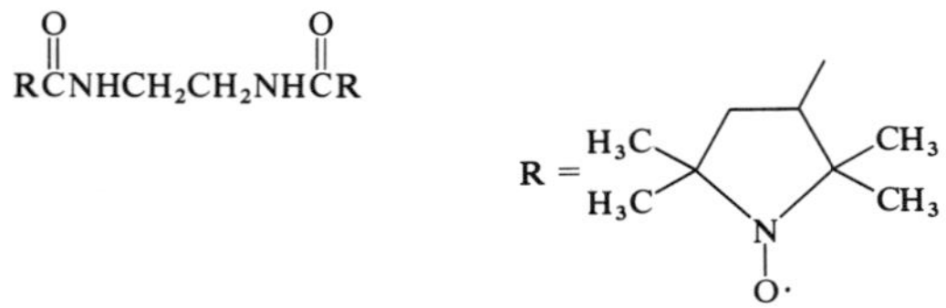 / Not given	EPR/ 300		5-line spectrum complex $J \approx a_N$	86Ehm1

³) Spectra of biradical when attached to photoreceptor membrane reported.
⁴) Measurements also in the presence of a photoreceptor membrane.
⁵) $\bar{J}$ = mean exchange integral, $a = 2.7 \cdot 10^8$ rad s^{-1}.
⁶) Variable temperature study.

Substance	Generation/ Matrix or Solvent	Method/ T [K]	g-Factor	a-Value, D, E, J [mT]	Ref./ add. Ref.
$[C_{22}H_{34}N_4O_2S]^{2\cdot}$ RN=HC–C_6H_4–CH=NR, R = 3,3,5,5-tetramethylpyrrolidine-1-oxyl (H₃C, CH₃, N–O·)	OHC–C_6H_4–CHO + 3-amino-2,2,5,5-tetramethylpyrrolidine-1-oxyl (NH_2, N–O·) / $C_6H_5CH_3$	EPR/ 248 ··· 347		$\bar{J}/a = 3.0$ [5]	81Kok2 [6]
$[C_{22}H_{36}N_4O_2]^{2\cdot}$ RN=CHCH=CH–CH=CHCH=NR, R = (H₃C, CH₃, N–O·)	OHCCH=CHCH=CHCHO + (NH_2, N–O·) / $C_6H_5CH_3$	EPR/ 285 ··· 346		$\bar{J}/a = 3.1$ [5]	81Kok2 [6]
$[C_{24}H_{36}N_4O_2]^{2\cdot}$ RN=CH–(thiophene-2,5-diyl, S)–CH=NR, R = (H₃C, CH₃, N–O·)	OHC–(thiophene-2,5-diyl, S)–CHO + (NH_2, N–O·) / $C_6H_5CH_3$	EPR/ 249 ··· 347		$\bar{J}/a = 1.74$ [5]	81Kok2 [6]
$[C_{24}H_{38}N_4O_2]^{2\cdot}$ RN=CH(CH=CH)$_3$CH=NR, R = (H₃C, CH₃, N–O·)	OHC(HC=CH)$_3$CHO + (NH_2, N–O·) / $C_6H_5CH_3$	EPR/ 284 ··· 346		$\bar{J}/a = 1.8$ [5]	81Kok2 [6]

[5]) $\bar{J}$ = mean exchange integral, $a = 2.7 \cdot 10^8$ rad s^{-1}.
[6]) Variable temperature study.

Substance	Generation/ Matrix or Solvent	Method/ T[K]	g-Factor	a-Value, D, E, J [mT]	Ref./ add. Ref.
$[C_{24}H_{38}N_4O_2S_3]^{2\cdot}$ 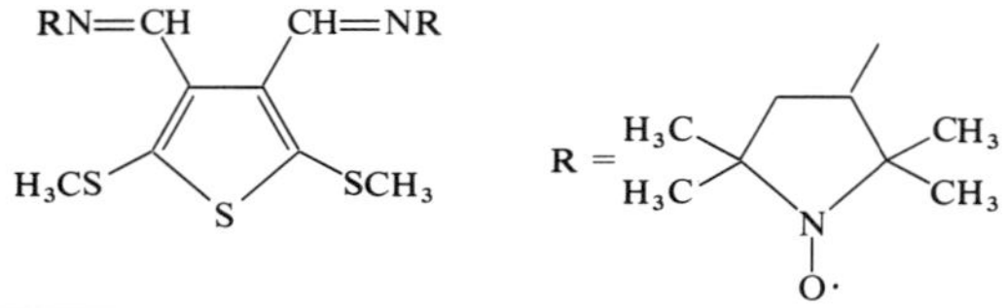 RN=CH, CH=NR, H_3CS, S, SCH_3; R = H_3C, H_3C, CH_3, CH_3, N, O·	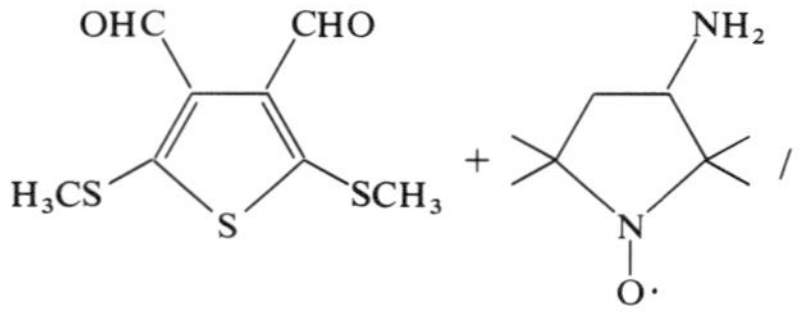 OHC, CHO, H_3CS, S, SCH_3 + NH_2, N, O· / $C_6H_5CH_3$	EPR/ 247⋯335		$\bar{J}/a=12.4$[5]	81Kok2[6]
$[C_{25}H_{38}N_4O_3]^{2\cdot}$ 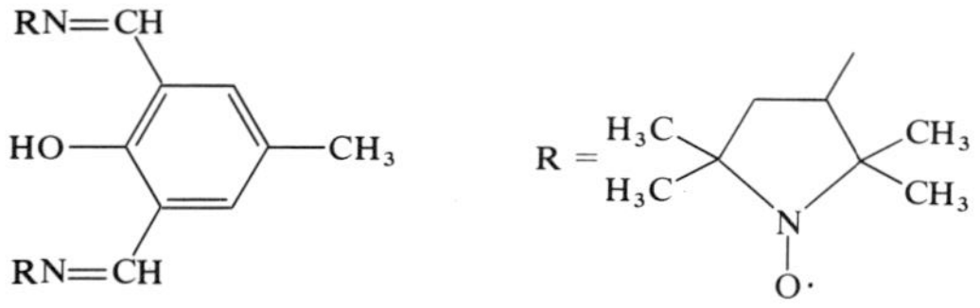 RN=CH, HO, CH_3, RN=CH; R = H_3C, H_3C, CH_3, CH_3, N, O·	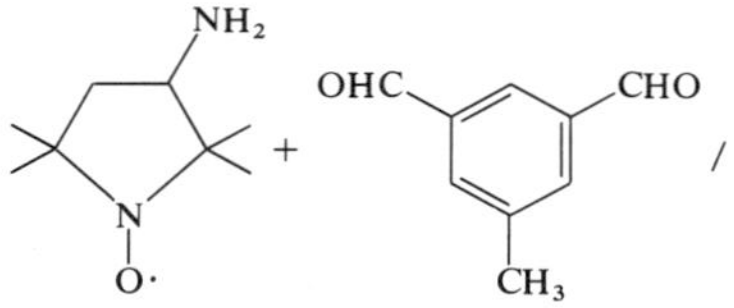 NH_2, N, O· + OHC, CHO, CH_3 / $C_6H_5CH_3$	EPR/ 298		$J/a_N \approx 0.2$	85Med1
$[C_{26}H_{42}N_4O_2]^{2\cdot}$ 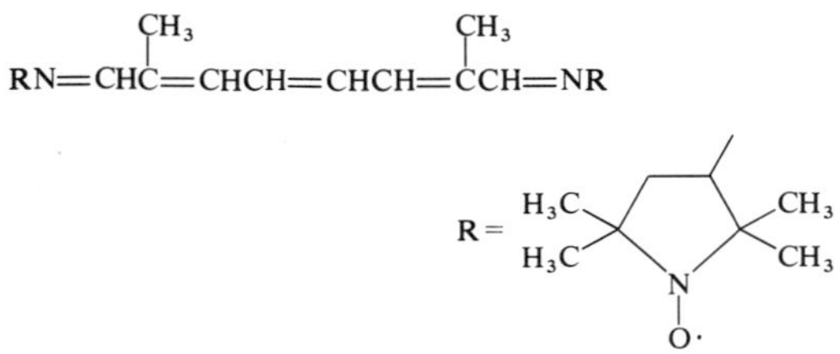 RN=CHC(CH₃)=CHCH=CHCH=C(CH₃)CH=NR; R = H_3C, H_3C, CH_3, CH_3, N, O·	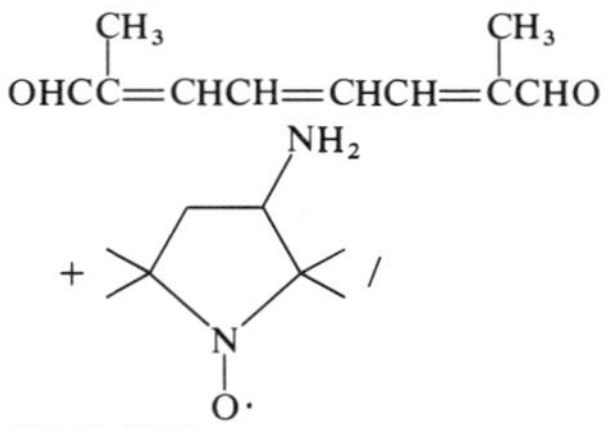 OHCC(CH₃)=CHCH=CHCH=C(CH₃)CHO + NH_2, N, O· / $C_6H_5CH_3$	EPR/ 247⋯336		$\bar{J}/a=1.82$[5] $\bar{J}/a=1.57$[5]	81Kok2[6] [7]
$[C_{28}H_{40}N_4O_2]^{2\cdot}$ 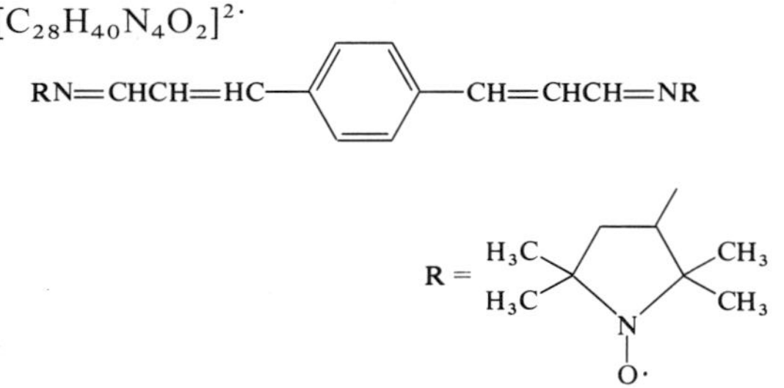 RN=CHCH=HC–C₆H₄–CH=CHCH=NR; R = H_3C, H_3C, CH_3, CH_3, N, O·	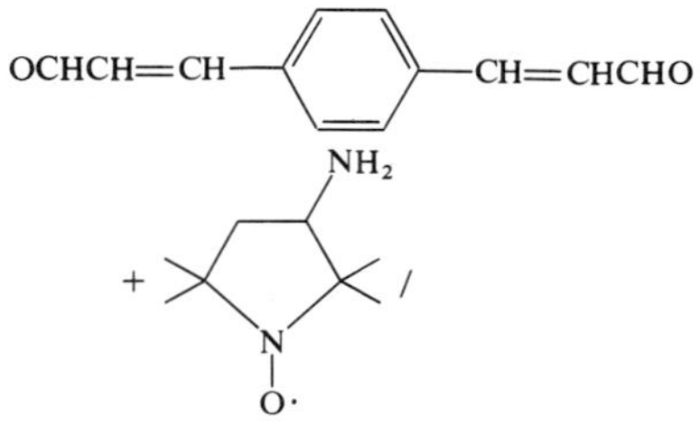 OCHCH=CH–C₆H₄–CH=CHCHO + NH_2, N, O· / $C_6H_5CH_3$	EPR/ 283⋯343		$\bar{J}/a=0.57$[5] $\bar{J}/a=0.32$[5]	81Kok2[6] [7]

[5]) $\bar{J}$ = mean exchange integral, $a = 2.7 \cdot 10^8$ rad s^{-1}.
[6]) Variable temperature study.
[7]) Biradical has two different exchange conformations.

Substance	Generation/ Matrix or Solvent	Method/ T [K]	g-Factor	a-Value, D, E, J [mT]	Ref./ add. Ref.
22.1.5 Bis(pyrrol-3-inyl-1-oxyls)					
$[C_{16}H_{26}HgN_2O_2]^{2\cdot}$	+ NH_3/ $C_2H_5OH:H_2O$	EPR/ 300		5-line spectrum $J \gg a_N$ a_{Hg}: 0.59	81Dmi1
$[C_{18}H_{24}Br_2N_2O_5]^{2\cdot}$	+ $SOCl_2$/ $C_6H_5CH_3$	EPR/ 298		7-line spectrum $J \approx a_N$	83Chu1
$[C_{18}H_{24}N_2O_6Pd]^{2\cdot}$	+ $PdCl_2$/ CH_2Cl_2	EPR/ 293		$\bar{J}$: 4.8 [8])	81Esp1

[8]) $L = \bar{J} + a_N + (J^2 + a_N^2)^{1/2}$.

Forrester

Substance	Generation/ Matrix or Solvent	Method/ T[K]	g-Factor	a-Value, D, E, J [mT]	Ref./ add. Ref.
$[C_{18}H_{31}N_3O_2]^{2\cdot}$ 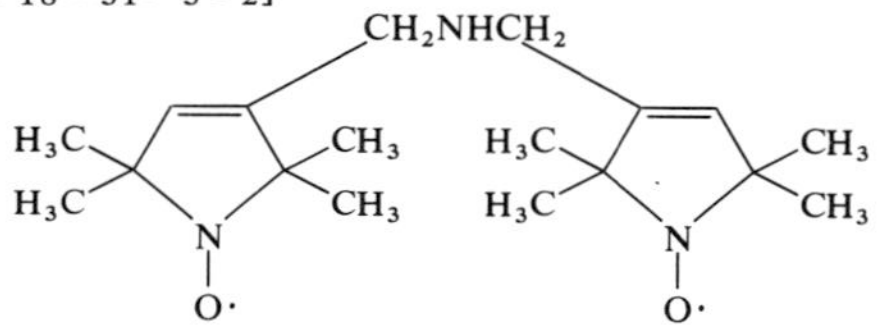 CH₂NHCH₂-linked bis(2,2,5,5-tetramethyl-3-pyrrolin-1-oxyl)	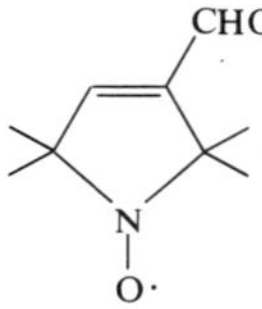 CHO-substituted tetramethylpyrrolinyloxyl + $NaB(CN)H_3$ + NH_4OCOCH_3/ $CHCl_3$	EPR/ 300		N: 1.45 5-line spectrum $J \gg a_N$	80Hid1
$[C_{19}H_{32}N_4O_2S]^{2\cdot}$ 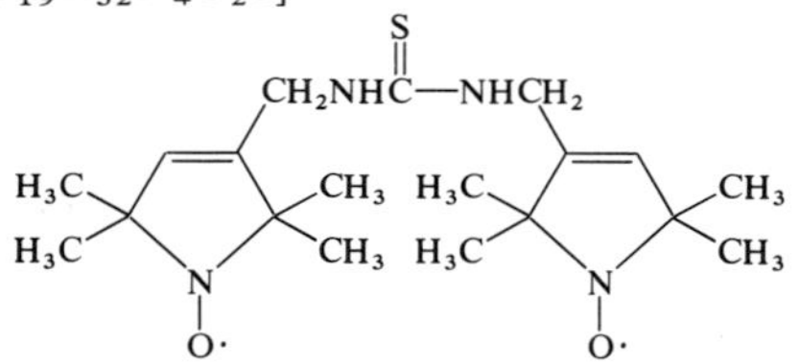 $CH_2NHC(=S)—NHCH_2$-linked bis(tetramethylpyrrolinyloxyl)	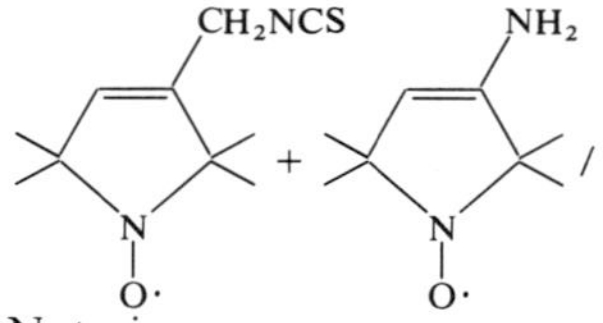 CH_2NCS-substituted + NH_2-substituted tetramethylpyrrolinyloxyl/ Not given	EPR/ 300		N: 1.49 5-line spectrum $J \gg a_N$	81Han1 [9])
$[C_{19}H_{30}N_4O_2]^{2\cdot}$ 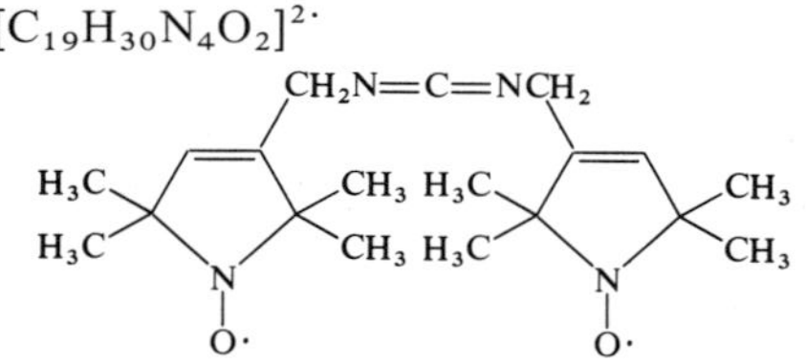 $CH_2N{=}C{=}NCH_2$-linked bis(tetramethylpyrrolinyloxyl)	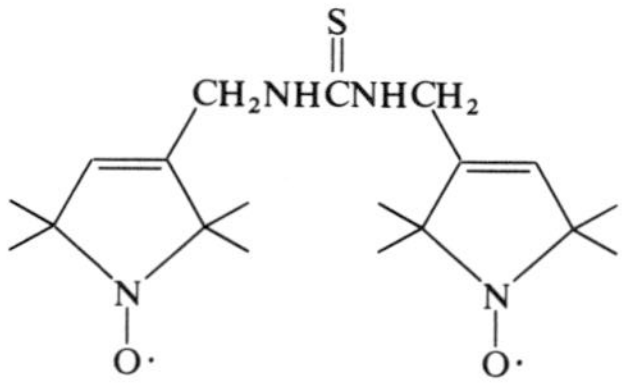 $CH_2NHC(=S)NHCH_2$-linked bis(tetramethylpyrrolinyloxyl) + HgO/ Not given	EPR/ 300		N: 1.49 5-line spectrum $J \gg a_N$	81Han1 [9])

[9]) Corresponding pyrrolidinyl-1-oxyl also reported.

Substance	Generation/ Matrix or Solvent	Method/ T [K]	g-Factor	a-Value, D, E, J [mT]	Ref./ add. Ref.
$[C_{20}H_{32}N_8O_2PdS_2]^{2\cdot}$ 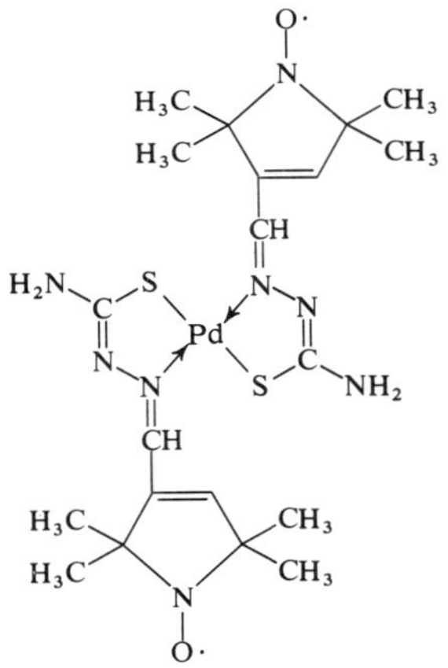	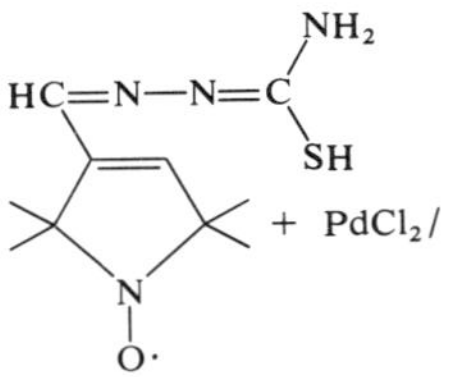 + $PdCl_2$/ Not given	EPR/ –		N: 1.40 5-line spectrum $J \gg a_N$	81Ovc1 [10]

22.1.6 Pyrrol-3-inyl-1-oxyls, galvinoxyls

Substance	Generation/ Matrix or Solvent	Method/ T [K]	g-Factor	a-Value, D, E, J [mT]	Ref./ add. Ref.
$[C_{37}H_{53}NO_3]^{2\cdot}$ 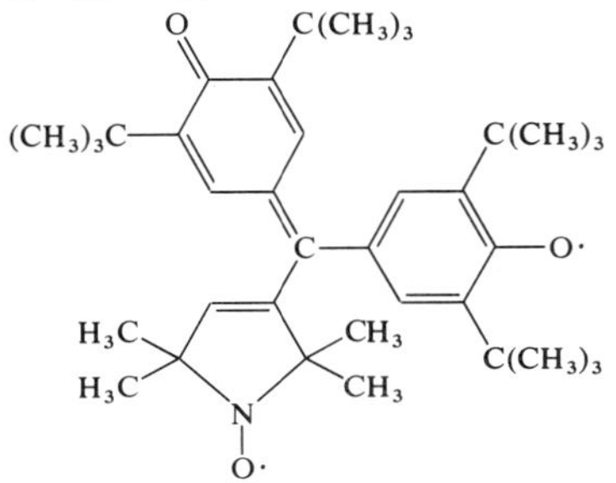	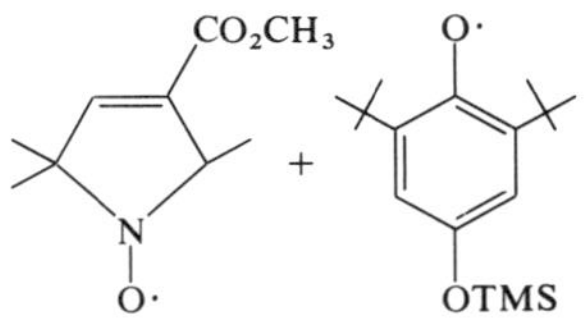 + / $C_6H_5CH_3$	EPR, ENDOR/ 290 160		 N: +1.457 $J \gg a_N$ D: 7.0	82Kir1

[10]) Zn(II), Cd(II) and Pt(II) complexes give similar spectra.

Substance	Generation/ Matrix or Solvent	Method/ T [K]	g-Factor	a-Value, D, E, J [mT]	Ref./ add. Ref.
$[C_{44}H_{57}NO_5]^{2\cdot}$ 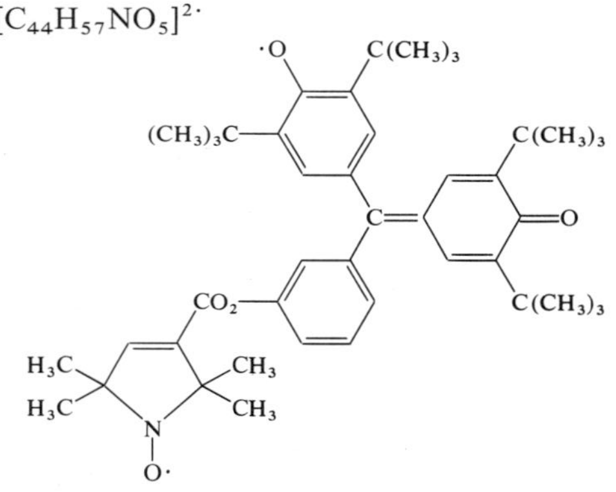	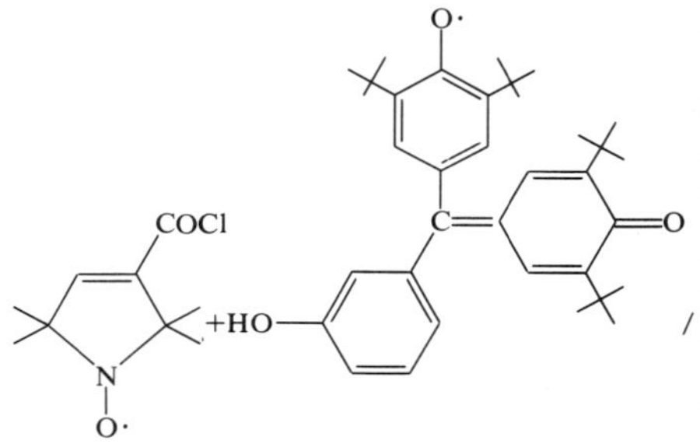 COCl + HO / $C_6H_5CH_3$	EPR, ENDOR/ 290 160	2.00529	J: 5.82 D: 3.03	82Kir1
$[C_{44}H_{57}NO_5]^{2\cdot}$ 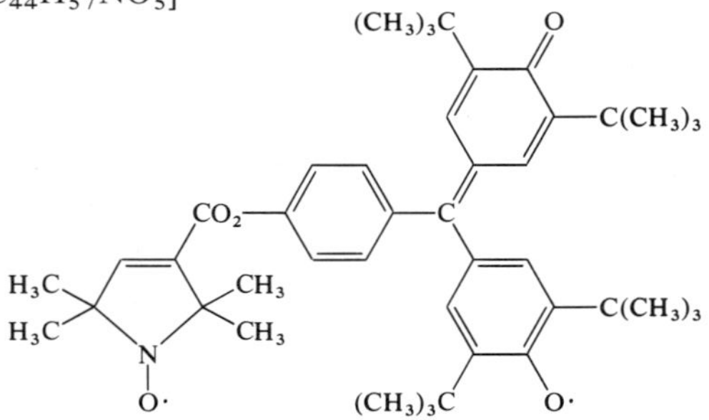	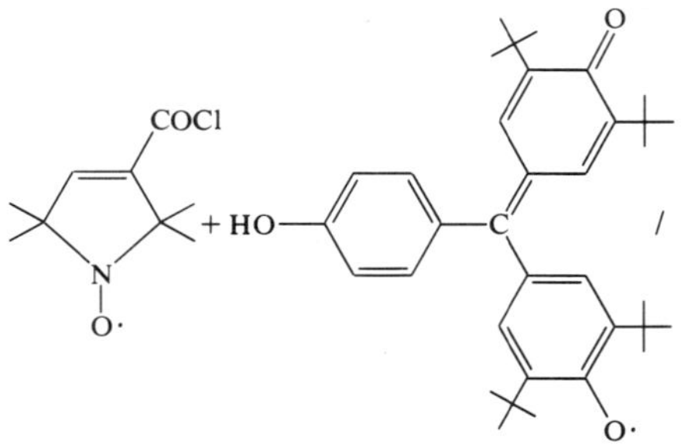 COCl + HO / $C_6H_5CH_3$	EPR, ENDOR/ 290 225 160	2.00530	N: 1.418 J: −5.96 H (phenoxyl): +0.128 H (nitroxide): −0.228 D: 2.07	82Kir1

Substance	Generation/ Matrix or Solvent	Method/ T[K]	g-Factor	a-Value, D, E, J [mT]	Ref./ add. Ref.
$[C_{44}H_{57}NO_5]^{2\cdot}$	+ / $C_6H_5CH_3$	EPR, ENDOR/ 290	2.0054	N: 1.42	82Kir1
		160		$J \gg a_N$ D: 5.82	
$[C_{50}H_{61}NO_5]^{2\cdot}$ 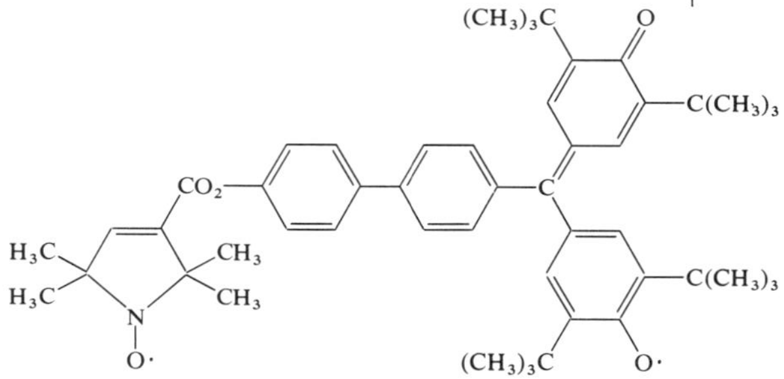	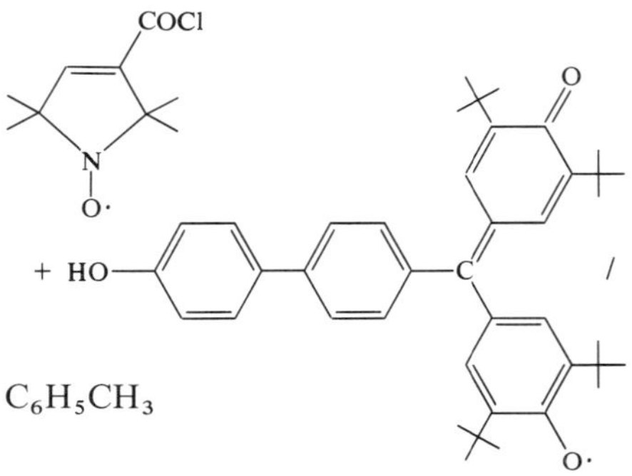 + HO / $C_6H_5CH_3$	EPR, ENDOR/ 290	2.00531	J: 1.14 D: $< a_N$	82Kir1

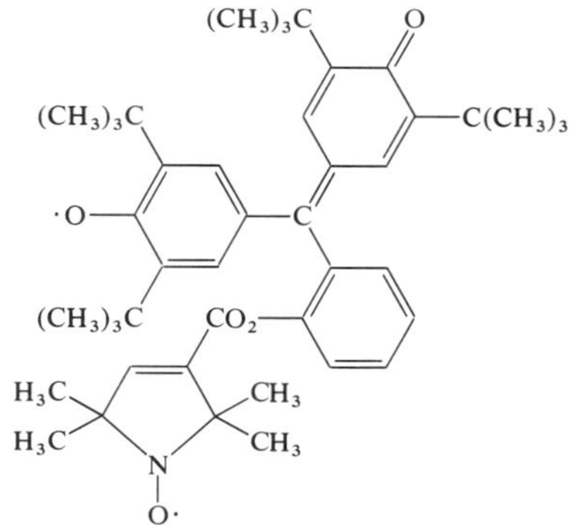

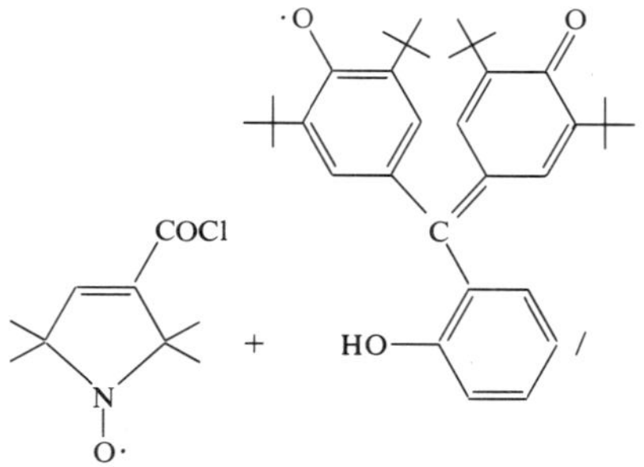

Substance	Generation/ Matrix or Solvent	Method/ T[K]	g-Factor	a-Value, D, E, J [mT]	Ref./ add. Ref.
22.1.7 Bis(pyrrolid-3-enyl-1-oxyls)					
$[C_{16}H_{28}N_4O_2]^{2\cdot}$	Not given/ DMF	Calc./ 300		N: 1.45 D: 4.4 [11]	80Mic2
$[C_{22}H_{32}N_6O_2S]^{2\cdot}$ R=N—N=CH—(thiophene-2,5-diyl)—CH=N—N=R	O=CH—(thiophene-2,5-diyl)—CH=O + (3-hydrazono-2,2,5,5-tetramethylpyrrolidin-1-oxyl) / $C_6H_5CH_3$	EPR/ 285···348		$\bar{J}/a = 1.84$ [12]	81Kok2 [13]
$[C_{24}H_{34}N_6O_2]^{2\cdot}$ R=N—N=CH—(1,4-phenylene)—CH=N—N=R	OHC—(1,4-phenylene)—CHO + (3-hydrazono-2,2,5,5-tetramethylpyrrolidin-1-oxyl) / $C_6H_5CH_3$	EPR/ 245···335		$\bar{J}/a = 1.06$ [12]	81Kok2 [13]
$[C_{24}H_{36}N_6O_2]^{2\cdot}$ R=NN=CH(CH=CH)$_3$CH=NN=R	OHC(HC=CH)$_3$CHO + (3-hydrazono-2,2,5,5-tetramethylpyrrolidin-1-oxyl) / $C_6H_5CH_3$	EPR/ 263···335		$\bar{J}/a = 1.15$ [12]	81Kok2 [13]

R = 2,2,5,5-tetramethylpyrrolidin-3-ylidene-1-oxyl (H_3C, H_3C, CH_3, CH_3, N—O·)

[11]) D calculated as 4.4 mT.
[12]) $\bar{J}$ = mean exchange integral, $a = 2.7 \cdot 10^8$ rad s^{-1}.
[13]) Variable temperature study.

Substance	Generation/ Matrix or Solvent	Method/ T[K]	g-Factor	a-Value, D, E, J [mT]	Ref./ add. Ref.
$[C_{28}H_{38}N_6O_2]^{2\cdot}$ R=N–N=CHCH=CH–C₆H₄–CH=CHCH=N–N=R	OCHCH=CH–C₆H₄–CH=CHCHO + (3-hydrazono-2,2,5,5-tetramethylpyrrolidin-1-oxyl) / $C_6H_5CH_3$	EPR/ 245 ··· 335		[14] $\bar{J}/a = 0.36$[12] $\bar{J}/a = 0.22$[12]	81Kok2[13]

22.1.8 Bis(oxazolidinyl-1-oxyls)

Substance	Generation/ Matrix or Solvent	Method/ T[K]	g-Factor	a-Value, D, E, J [mT]	Ref./ add. Ref.
$[C_{12}H_{20}N_2O_3]^{2\cdot}$	Not given/ Not given	Calc./ –		D: 12.1[15]	80Mic2
$[C_{13}H_{24}N_2O_3]^{2\cdot}$	Not given/ $C_6H_5CH_3$: THF	EPR/ 93		D: 22.25[16]	83Eat1

[12] $\bar{J}$ = mean exchange integral, $a = 2.7 \cdot 10^8$ rad s^{-1}.
[13] Variable temperature study.
[14] Biradical has two different exchange integrals.
[15] D calculated as 11.6 mT.
[16] $r = 4.94$ Å measured from intensity of half-field transition. $r = 5.02$ Å measured using point-dipole approximation.

Substance	Generation/ Matrix or Solvent	Method/ T [K]	g-Factor	a-Value, D, E, J [mT]	Ref./ add. Ref.
$[C_{14}H_{24}N_2O_4]^{2\cdot}$ 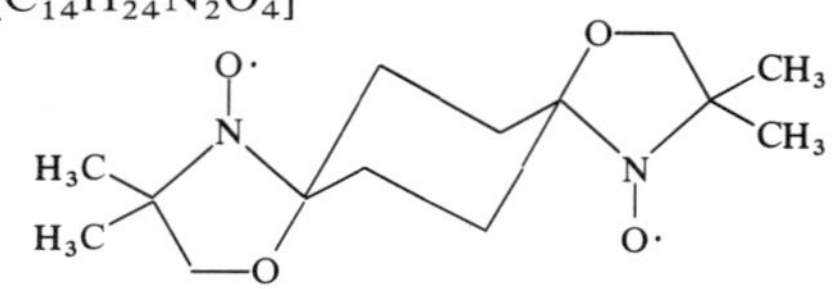	Oxidation of the corresp. diamine/	EPR/			80Chi1 [19]), 80Mic2
	C_2H_5OH	123		D: 12.1	
	cyclo-C_6H_{12}	263		D: 19.0 } two D: 8.7 } conformers D: 14.6 } two D: 6.2 } conformers	
	$CHCl_3$ [17])	300		N: 1.5 D: 12.1 [18])	
$[C_{16}H_{28}N_2O_3]^{2\cdot}$ 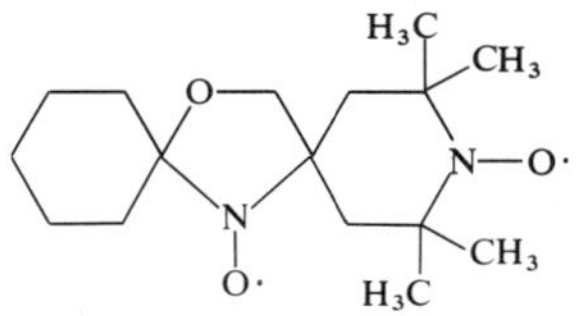	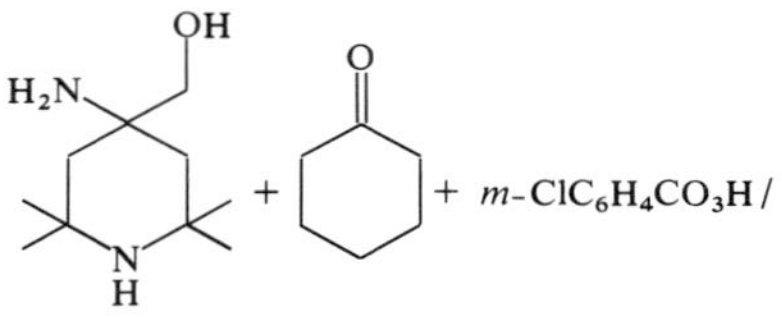 + + m-$ClC_6H_4CO_3H$/	EPR, NMR/			80Mic1 [20])
	C_2H_5OH	123	g_{xx}: 2.0059; g_{yy}: 2.0053 or 2.0048; g_{zz}: 2.0073	$2D$: 45.0 (2) E: 0.75	
	CD_2Cl_2	300		$2D$: 50.0 (5)	
$[C_{18}H_{28}N_2O_4]^{2\cdot}$ 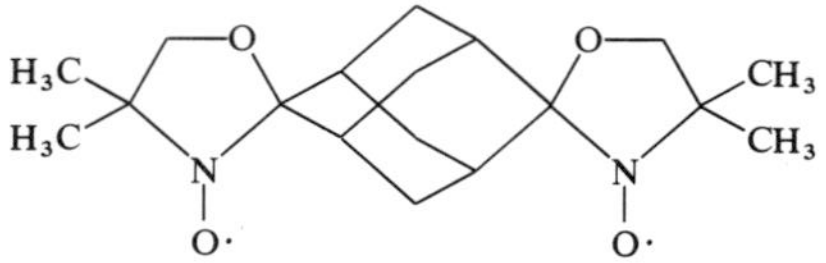	Ox. of the corresp. diamine with m-$ClC_6H_4CO_3H$/ C_6H_6	EPR/ 300		5-line-spectrum $J \gg a_N$ N: 1.425	79Mor1

[17]) From [80Mic2].
[18]) D calculated as 11.6 mT.
[19]) Second spectrum observed after sample had been rotated 90° about its vertical axis.
[20]) Zero-point approximation gives electron-electron separation of 4.98 (1) Å and 4.81 (2) Å for two isomers. $J = 3 \cdot 10^6$ MHz $= 0.3$ kcal mol^{-1}: NMR hfcc expressed as Hz.

Substance	Generation/ Matrix or Solvent	Method/ T [K]	g-Factor	a-Value, D, E, J [mT]	Ref./ add. Ref.
$[C_{18}H_{30}N_2O_4]^{2\cdot}$ 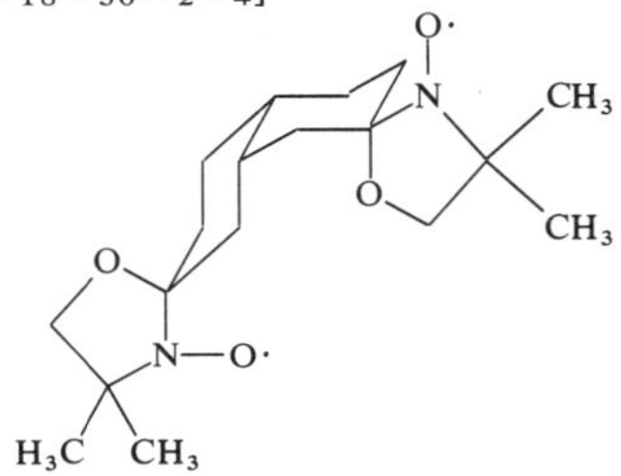	Ox. of the corresp. diamine with $ClC_6H_4CO_3H$/ $CDCl_3$	EPR, NMR/ 293		5-line spectrum $J \gg a_N$ N: 1.47	85Mic1 [21])
	C_4H_9OH	163	g_1: 2.0066; g_2: 2.0068; g_3: 2.0038; is: 2.0057	D: 7.6	
	$CDCl_3$	293		H (CH_3): −0.010 (2) H (CH_2): 0.0092 (2) H (eq, 1, 3, 6, 8): −0.063 (2) H (ax, 1, 3, 6, 8): −0.063 (2) H (eq, 4, 5): +0.110 (2) H (9): +0.110 (2) H (ax, 4, 5): +0.023 and +0.017 (2)	
	Not given/ $CHCl_3$	Calc./ 300		N: 1.5 D: 7.6 [22])	80Mic2
$[C_{32}H_{40}N_2O_4]^{2\cdot}$ 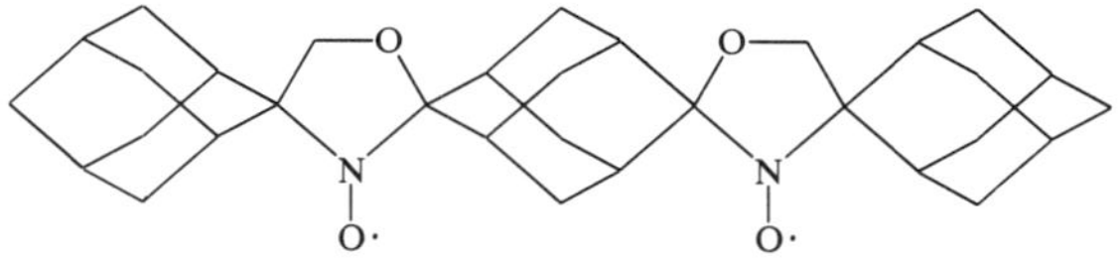	Ox. of the corresp. diamine with m-$ClC_6H_4CO_3H$/ C_6H_6	EPR/ 300		5-line spectrum $J \gg a_N$ N: 1.440	79Mor1

[21]) $r = 7.15$ (5) Å; J calculated as $= 0 \pm 20$ K.
[22]) D calculated as 8.0 mT.

Substance	Generation/ Matrix or Solvent	Method/ T[K]	g-Factor	a-Value, D, E, J [mT]	Ref./ add. Ref.
$[C_{45}H_{86}N_3O_{12}P]^{2\cdot}$ 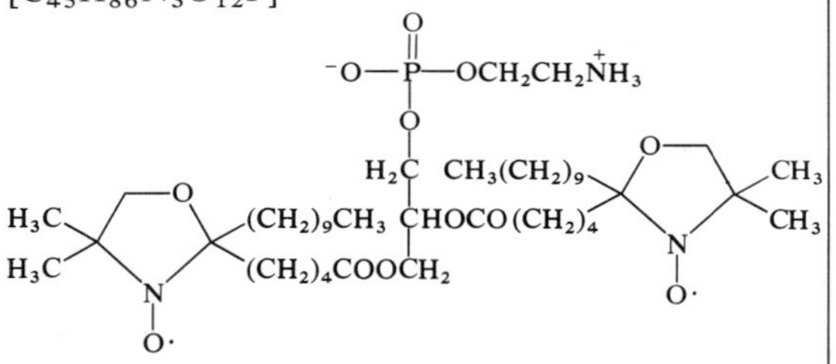	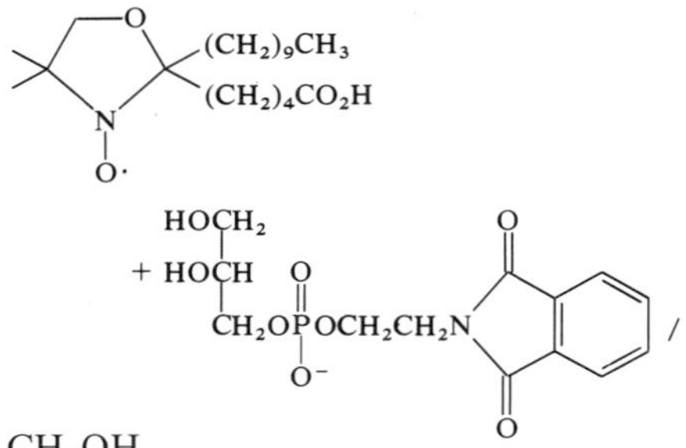 + / CH_3OH	EPR/ 300		3-line spectrum	79Zhd1 [23])

22.1.9 Bis(imidazolinyl-1-oxyl 3-oxides

Substance	Generation/ Matrix or Solvent	Method/ T[K]	g-Factor	a-Value, D, E, J [mT]	Ref./ add. Ref.
$[C_{16}H_{26}Br_2N_4O_4]^{2\cdot}$, $[C_{16}H_{26}Cl_2N_4O_4]^{2\cdot}$ 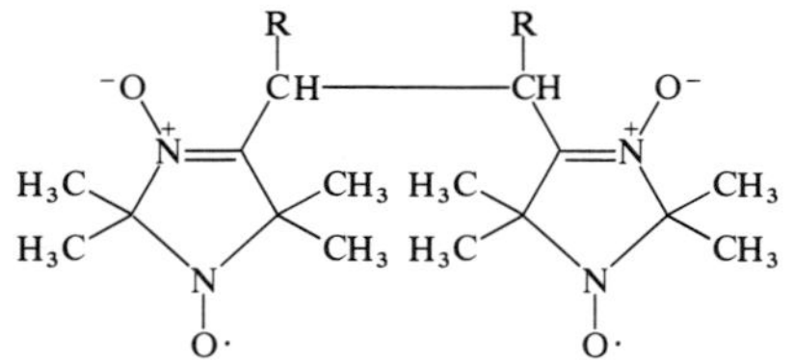 R = Br or Cl	Not translated/ $C_6H_5CH_3$ (R = Br) (R = Cl)	EPR/ 294		J/a_N: 7.2 (7) J/a_N: 6.40(25)	88Gul1 [23a])
$[C_{16}H_{26}N_6O_4]^{2\cdot}$ 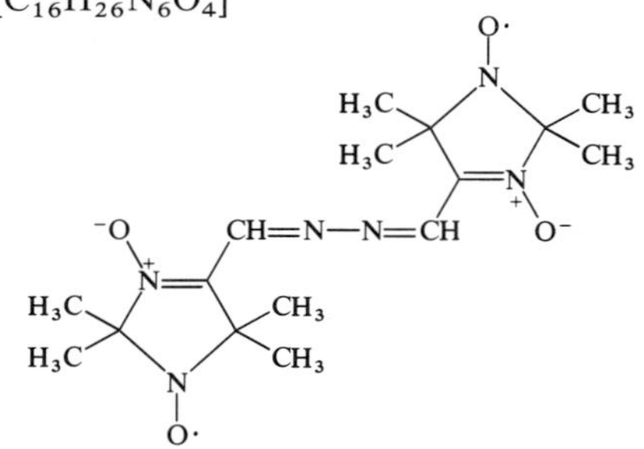	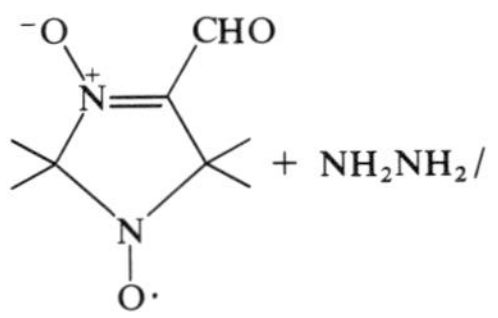 + NH_2NH_2/ $C_6H_5CH_3$	EPR/ 300		$J/a \geqslant 30$	83Gri1

[23]) Measurements of bisnitroxide when associated with liver microsomal membranes.
[23a]) Variable temperature study; measurements in two other solvents.

Substance	Generation/ Matrix or Solvent	Method/ T [K]	g-Factor	a-Value, D, E, J [mT]	Ref./ add. Ref.
$[C_{16}H_{26}N_6O_4]^{2\cdot}$ 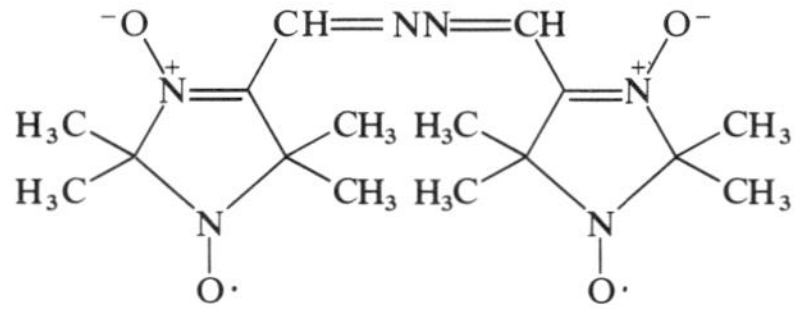	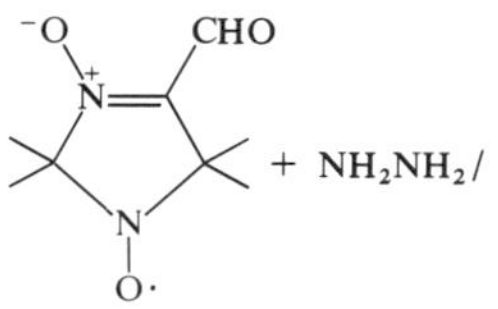 + NH_2NH_2 / $C_6H_5CH_3$	EPR/ 300		$J/a > 25 \cdots 30$	82Gri1
$[C_{18}H_{30}N_6O_4]^{2\cdot}$	+ $NH_2CH_2CH_2NH_2$ / $C_6H_5CH_3$	EPR/ 233		5-line spectrum $J \approx a_N$	85Dik1 [24])
		273		5-line spectrum $J \gg a_N$ N: 1.42	
$[C_{24}H_{32}N_4O_4]^{2\cdot}$ 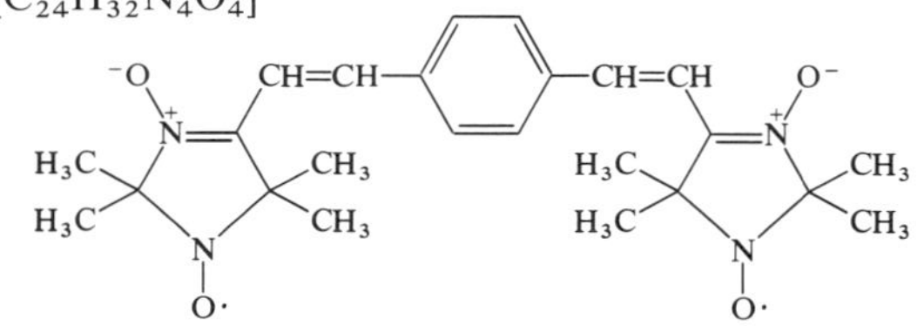	Not given/ $C_6H_5CH_3$	EPR/ 300		$J/a > 25 \cdots 30$	82Gri1
$[C_{24}H_{32}N_8O_4]^{2\cdot}$ 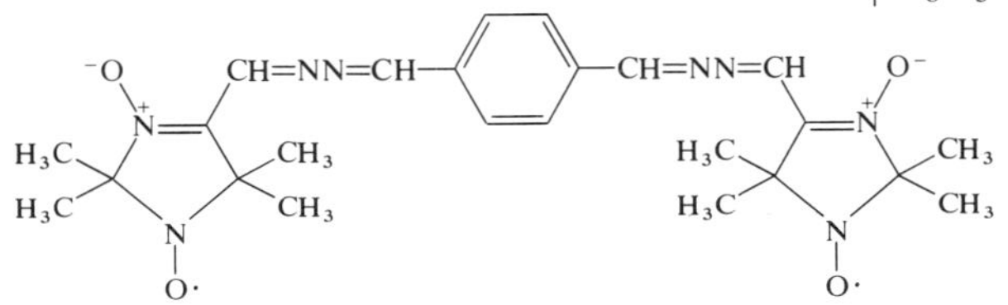	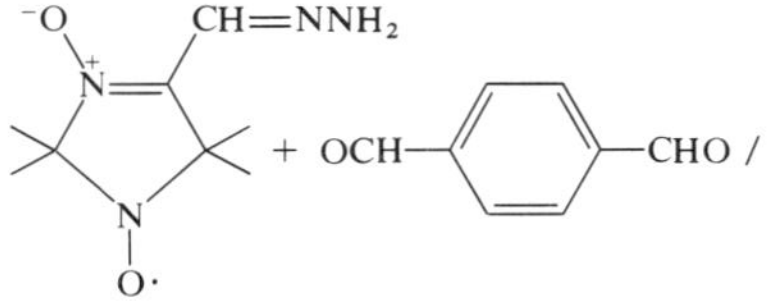 + OCH–C_6H_4–CHO / $C_6H_5CH_3$	EPR/ 300		J/a: 8.75 (50) 7-line spectrum	82Gri1

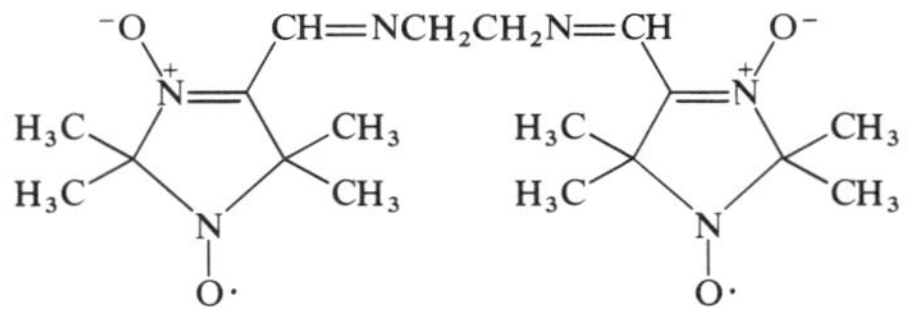

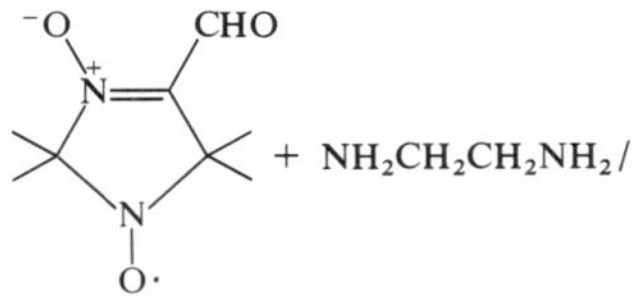

[24]) Variable temperature study, several homologous biradicals also measured.

Substance	Generation/ Matrix or Solvent	Method/ T[K]	g-Factor	a-Value, D, E, J [mT]	Ref./ add. Ref.
$[C_{24}H_{42}N_6O_4]^{2\cdot}$ 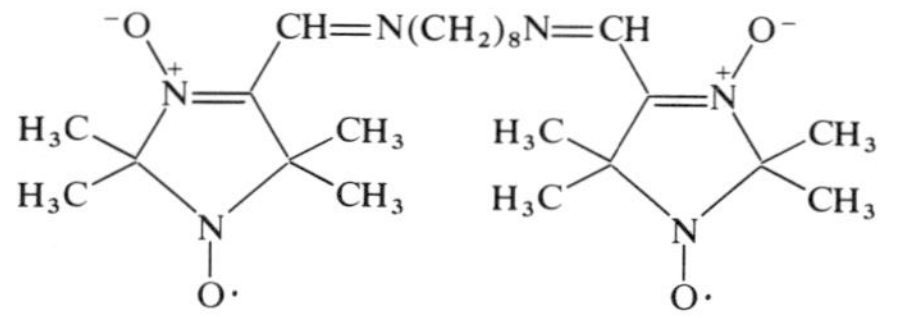	Not given/ $C_6H_5CH_3$ glass	ELDOR, ESE/ 77			84Mil1 [25]
	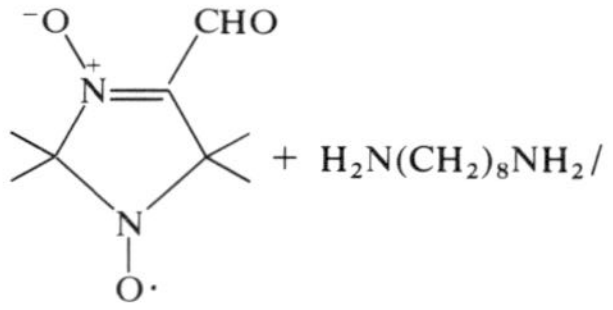 + $H_2N(CH_2)_8NH_2$/ $C_6H_5CH_3$	EPR/ 293		3-line spectrum $a_N \gg J$	85Dik1 [26]
		323		5-line spectrum $J \approx a_N$ N: 1.42	

22.1.10 Bis(imidazolinyl-1-oxyls)

Substance	Generation/ Matrix or Solvent	Method/ T[K]	g-Factor	a-Value, D, E, J [mT]	Ref./ add. Ref.
$[C_{14}H_{24}AgN_4O_2S_2]^{2\cdot}$ 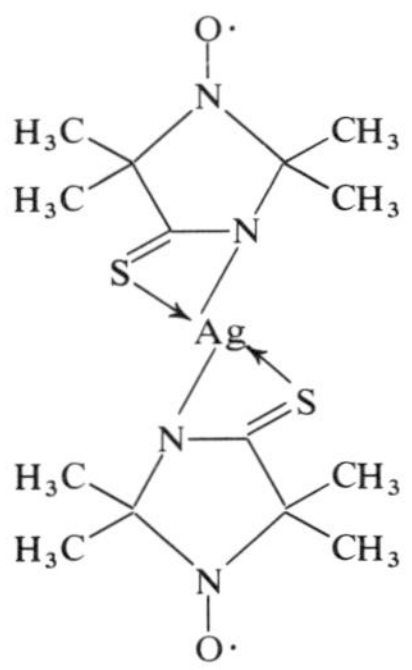	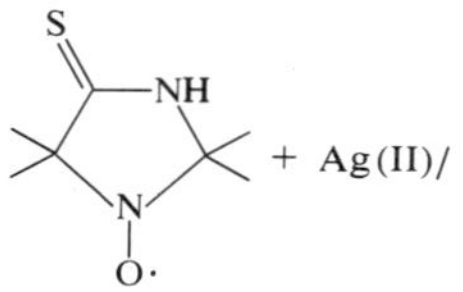 + Ag(II)/ $CHCl_3$	EPR/ 300		11-line spectrum $J \approx a_N$	88Van1

[25]) Measurements on radical pairs.
[26]) Variable temperature study, several homologous biradicals also measured.

Substance	Generation/ Matrix or Solvent	Method/ T[K]	g-Factor	a-Value, D, E, J [mT]	Ref./ add. Ref.
$[C_{14}H_{24}N_4O_2PtS_2]^{2\cdot}$ 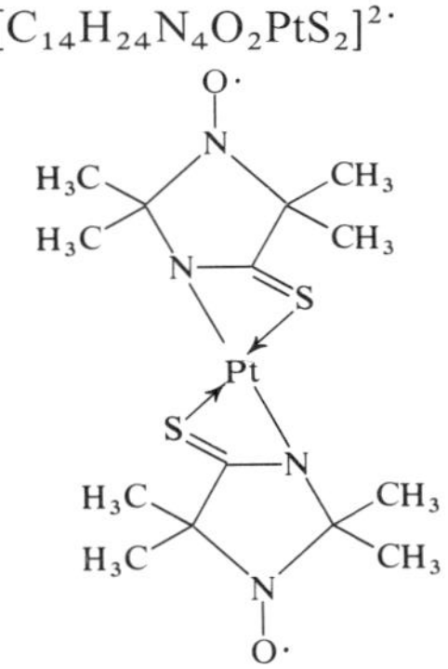	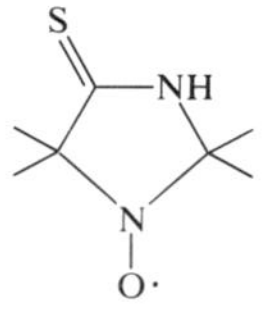 + $PtCl_2$/ $CHCl_3$	EPR/ 300		7-line spectrum $J \approx a_N$	88Van1 [27])
$[C_{14}H_{24}N_4O_2PdS_2]^{2\cdot}$ 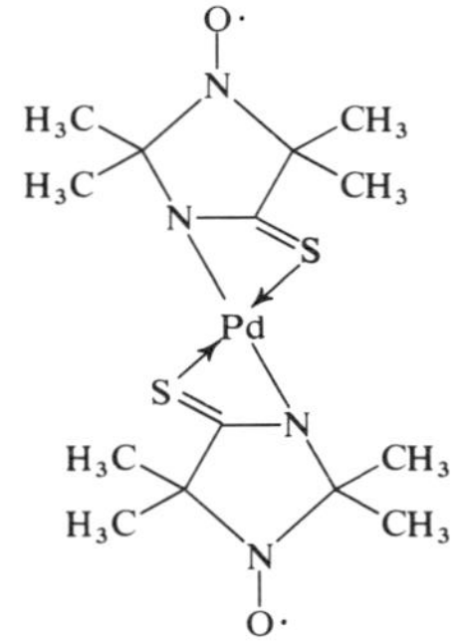	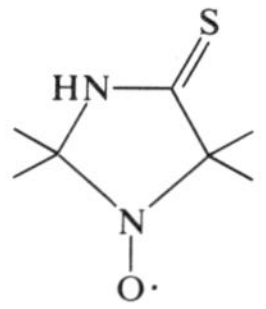 + $PdCl_2$/ $CHCl_3$	EPR/ 300		9-line spectrum $J \approx a_N$	88Van1
$[C_{14}H_{26}Br_2N_4O_2PdS_2]^{2\cdot}$ 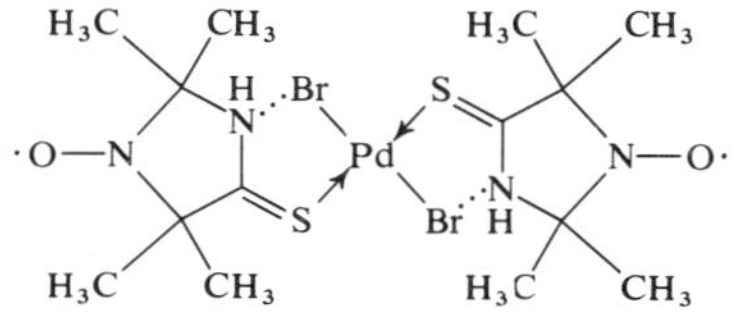	$[PdBr_4]^{2-}$ + 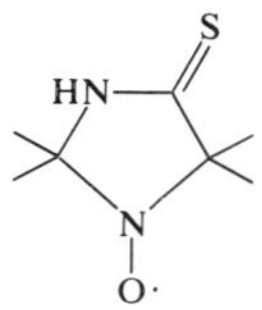 / $C_6H_5CH_3$	EPR/ 293		$J/a_N = 1.37$ (2)	88Evs1 [29])

[27]) Hg and Au complexes give similar spectra.

[29]) r calculated as 11.3 (3) Å. Measurement of J/a at several temperatures.

Substance	Generation/ Matrix or Solvent	Method/ T [K]	g-Factor	a-Value, D, E, J [mT]	Ref./ add. Ref.
$[C_{14}H_{26}Cl_2N_4O_2PdS_2]^{2\cdot}$	$[PdCl_4]^{2-}$ + / $CHCl_3$	EPR/ 293		9-line spectrum $J \approx a_N$	88Evs1 [28])
$[C_{16}H_{24}N_4O_6Pd]^{2\cdot}$	+ PdX_2/	EPR/			81Kok1 [30])
	$CHCl_3$	293		N: 1.422 J_1/a: 1.56 (18) J_2/a: −1.03 (8)	
	C_2H_5OH	293		N: 1.466 J_1/a: 1.05 (24) J_2/a: −0.78 (17)	
	$C_6H_5CH_3:CHCl_3$	77	g_{xx}: 2.0089; g_{yy}: 2.0061; g_{zz}: 2.0027; is: 2.0059	$A_{xx}=A_{yy}$: 0.6; A_{zz}: 3.2 D: 3.2	
	+ PdX_2/ $CHCl_3$	EPR/ 250···363		N: 1.43 J/a: 1.0	81Kok3

[28]) Temperature dependence of J in toluene, chloroform, toluene+pentane, bromoform. $r = 10.6$ (4) Å: point-dipole approximation; measurements of spectra in toluene glass 77 K. Values also calculated using other methods, $r = 11.2$ (4) Å.

[30]) Variable temperature study of J values: spectrum simulated.

Substance	Generation/ Matrix or Solvent	Method/ T[K]	g-Factor	a-Value, D, E, J [mT]	Ref./ add. Ref.
$[C_{16}H_{26}N_6O_2]^{2\cdot}$	+ NH_2NH_2/ $C_6H_5CH_3$	EPR/ 300		9-line spectrum $J/a = 0.35$	82Gri1
$[C_{16}H_{26}N_6O_2]^{2\cdot}$	+ NH_2NH_2/ $CHCl_3$	EPR/ 300		J/a_N: 0.35	83Gri1
$[C_{16}H_{26}N_6O_4Pd]^{2\cdot}$	+ PdX_2/ $CHCl_3$	EPR/ 243···328		N: 1.43 J/a: 6.65	81Kok3

Substance	Generation/ Matrix or Solvent	Method/ T[K]	g-Factor	a-Value, D, E, J [mT]	Ref./ add. Ref.
$[C_{17}H_{28}N_6O_2]^{2\cdot}$ 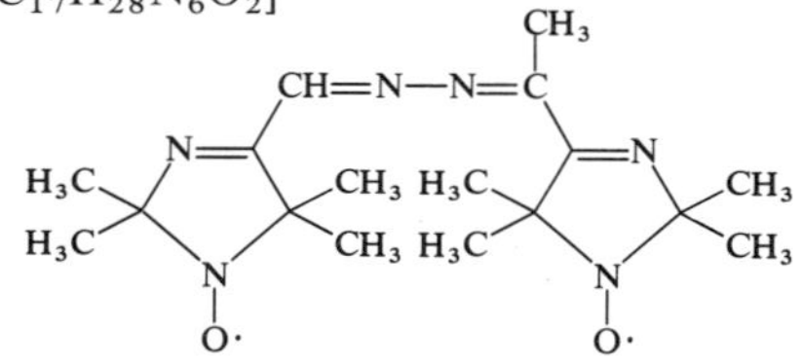	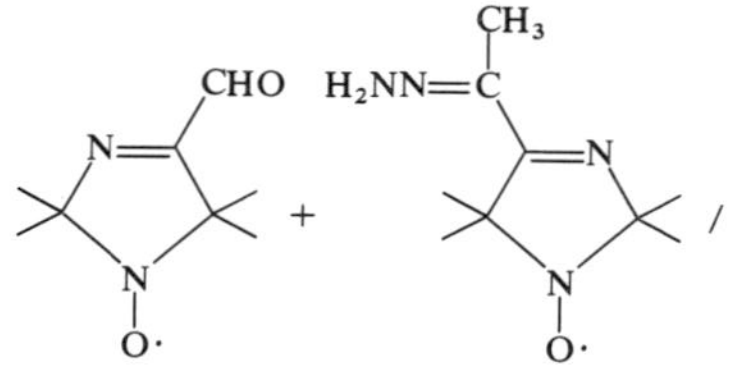 + / $C_6H_5CH_3$	EPR/ 300		J/a: 0.15	82Gri1
$[C_{18}H_{30}N_6O_2]^{2\cdot}$	+ $NH_2CH_2CH_2NH_2$/ $C_6H_5CH_3$	EPR/ 293 233		5-line spectrum $J \gg a$ 5-line spectrum $J \approx a_N$	85Dik1 [31])
$[C_{18}H_{30}N_6O_2]^{2\cdot}$	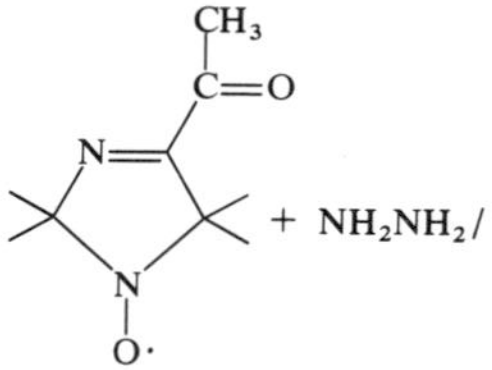 + NH_2NH_2/ 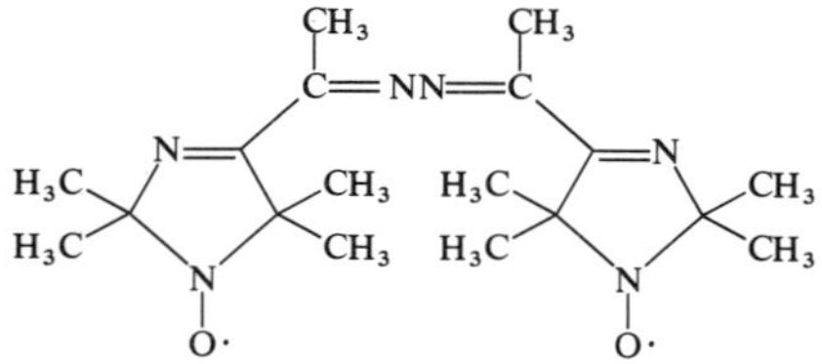 $C_6H_5CH_3$	EPR/ 300		11-line spectrum J/a: 0.65	82Gri1
$[C_{18}H_{30}Br_2N_6O_4Pd_2]^{2\cdot}$	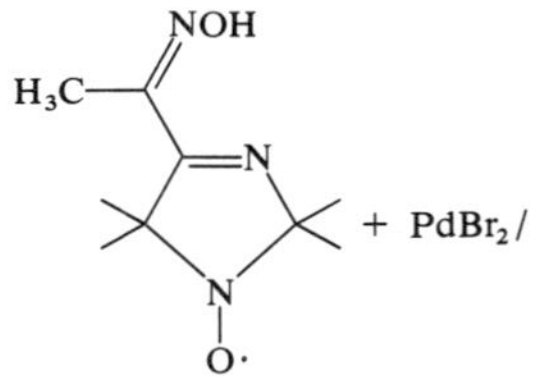 + $PdBr_2$/ 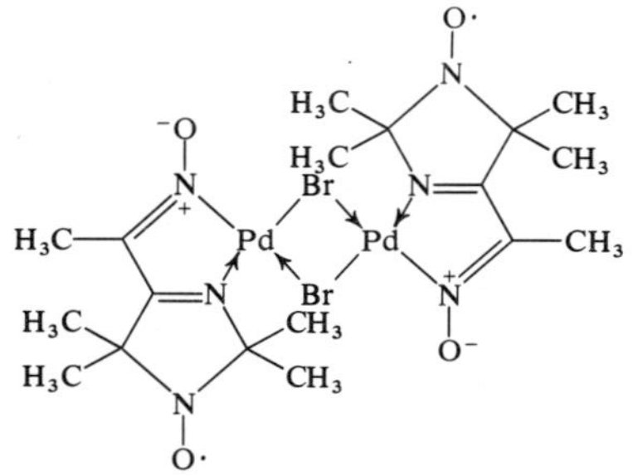 $CHCl_3$	EPR/ 273 ⋯ 343		N: 1.426 J/a: 2.95 J/a: 0.56	81Kok3

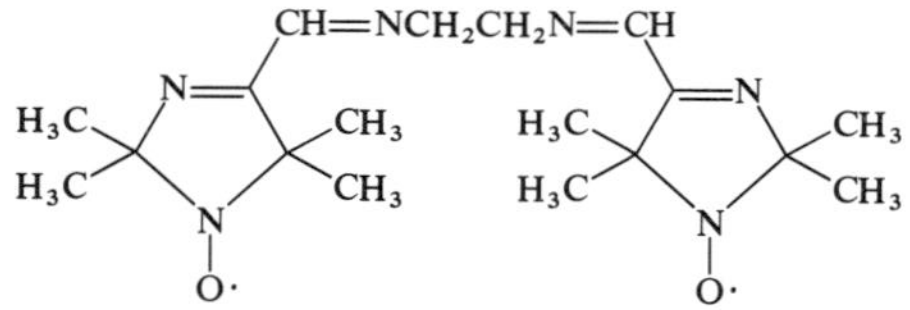

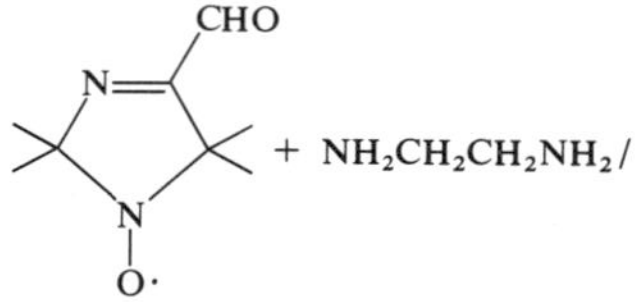

[31]) Variable temperature study, several homologous biradicals also measured.

Substance	Generation/ Matrix or Solvent	Method/ T[K]	g-Factor	a-Value, D, E, J [mT]	Ref./ add. Ref.
$[C_{18}H_{30}Cl_2N_6O_4Pd_2]^{2\cdot}$	+ $PdCl_2$/ $CHCl_3$	EPR/ 284 ··· 343		N: 1.43 J/a: 2.3	81Kok3
$[C_{18}H_{30}N_6O_4Pd]^{2\cdot}$	+ $PdCl_2$/ $CHCl_3$	EPR/ 286 ··· 346		N: 1.438 J/a: 10.7	81Kok3
$[C_{18}H_{30}Cl_4N_6O_2Pd_2]^{2\cdot}$	+ $PdCl_2$/ $C_6H_6:CH_3COCH_3$ (1:1)	EPR/ 248 ··· 333		$J \approx 2a_N$	88Avd1

Substance	Generation/ Matrix or Solvent	Method/ T[K]	g-Factor	a-Value, D, E, J [mT]	Ref./ add. Ref.
$[C_{18}H_{34}Cl_4N_6O_4Pd_2]^{2\cdot}$ 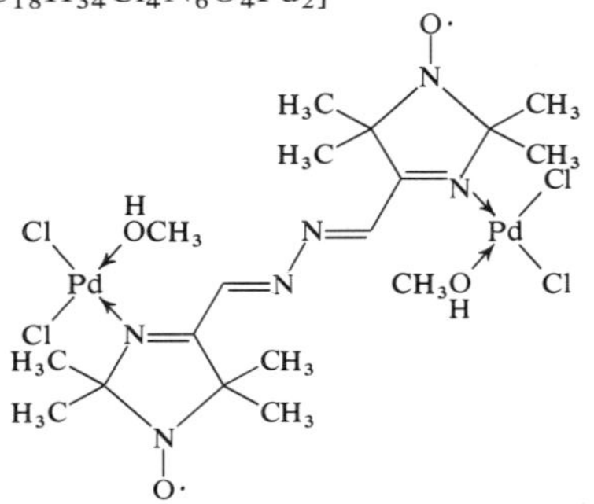	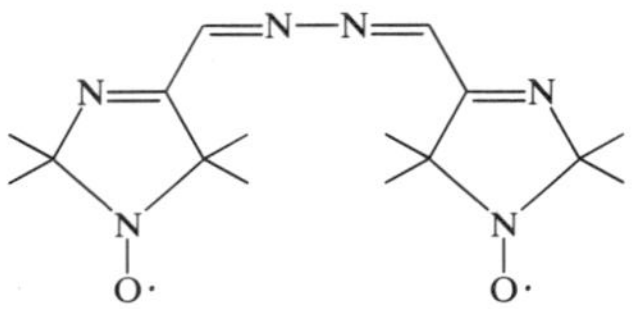 + $PdCl_2$/ $CHCl_3$	EPR/ 203 ··· 323		3-line-spectrum N: 1.49 $a_N \gg J$ 32) $a_N \approx 2.5 \cdots 5.0\ J$ 32)	88Avd1
$[C_{20}H_{34}N_6O_2]^{2\cdot}$	+ $H_2N(CH_2)_4NH_2$/ $C_6H_5CH_3$	EPR/ 293 323		3-line spectrum $a_N \gg J$ 5-line spectrum $J \approx a_N$	85Dik1 33)
$[C_{24}H_{32}N_8O_2]^{2\cdot}$	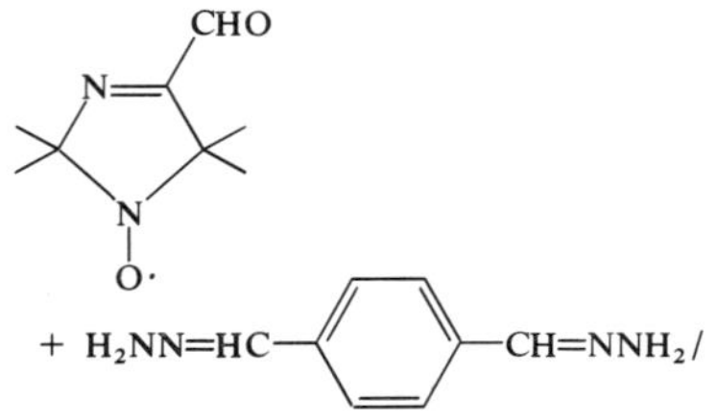 + $H_2NN{=}HC{-}C_6H_4{-}CH{=}NNH_2$/ 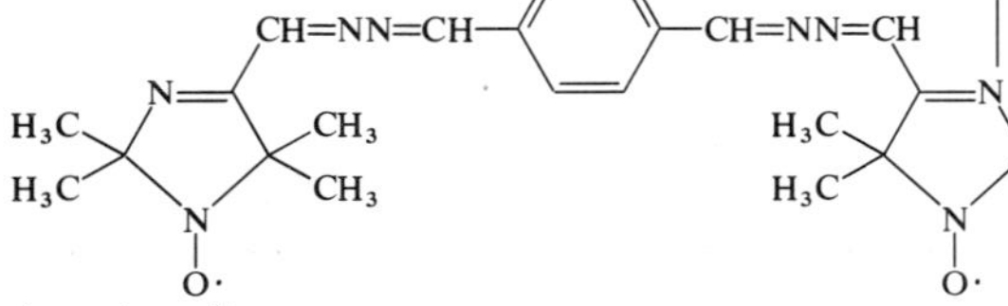 $C_6H_5CH_3$	EPR/ 300		9-line spectrum $J/a < 0.05$	82Gri1

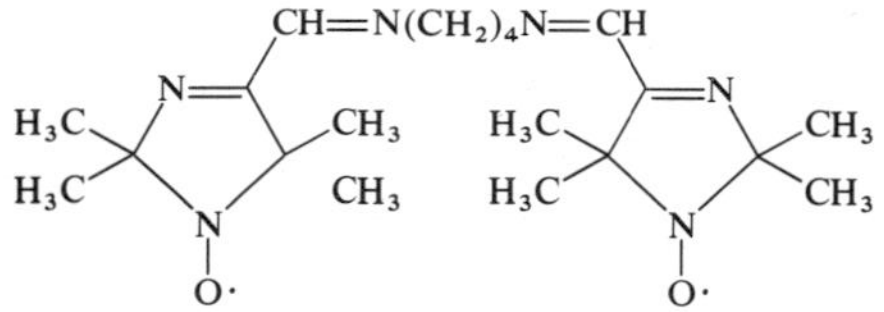

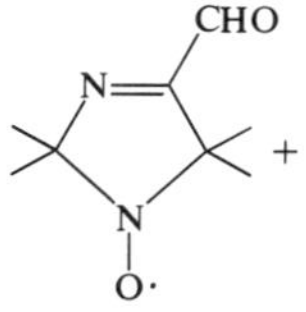

(continued)

32) Exchanging conformations present A, B, C. $J_A \ll a_N$; $J_B \ll a_N$; $J_C \approx a_N$. Exchange between B and C is rapid but transition A→B and A→C is slow. Conformation C is energetically favoured.

33) Variable temperature study. Several homologous biradicals also measured.

Substance	Generation/ Matrix or Solvent	Method/ T[K]	g-Factor	a-Value, D, E, J [mT]	Ref./ add. Ref.
$[C_{24}H_{32}N_8O_2]^{2\cdot}$ *(continued)*	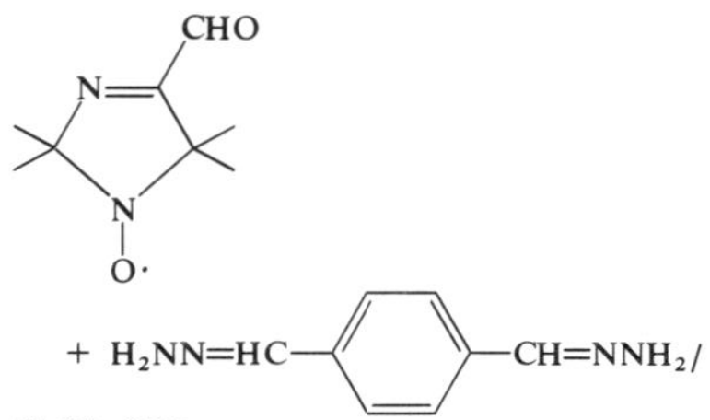 + $H_2NN{=}HC{-}C_6H_4{-}CH{=}NNH_2$/ $C_6H_5CH_3$	[35]/ 77		$J<0.2$	84Mil1 [36]
$[C_{24}H_{40}N_4O_4Pd]^{2\cdot}$ 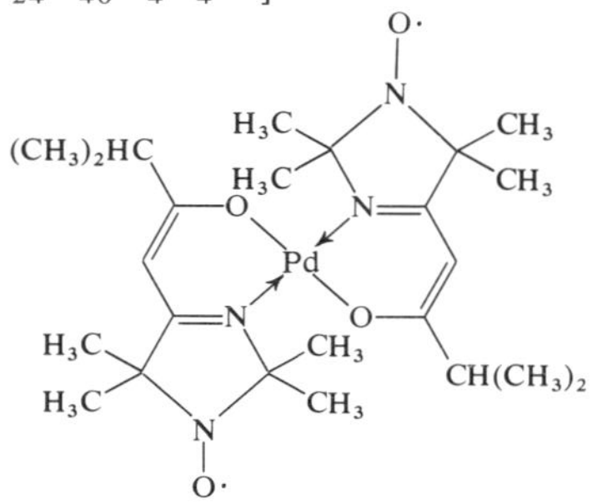	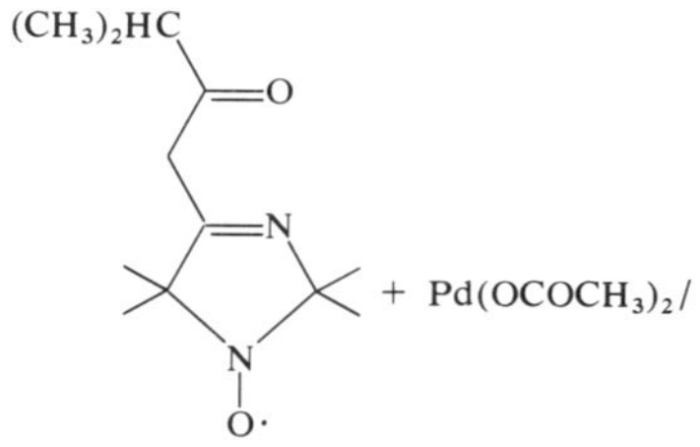 + $Pd(OCOCH_3)_2$/ $CHCl_3$	EPR/ 300	2.0052	N: 1.46 5-line spectrum $J \gg a_N$	88Lar1 [34]
$[C_{26}H_{32}Br_2N_4O_2Pd_2]^{2\cdot}$ 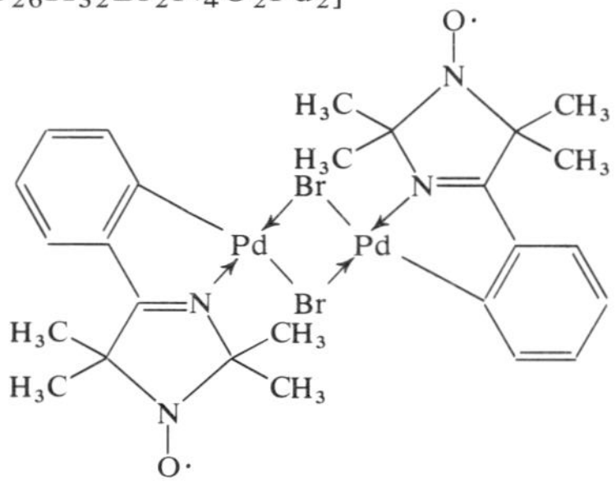	Not given/ $CHCl_3$	EPR/ 298		N: 1.465 J/a: 9.53	81Kok3

[34]) Zn complex gives similar spectrum.
[35]) Double electron-electron resonance.
[36]) r estimated as 19 (1.6)$\cdot 10^{-8}$ cm; dimethyl homologue gives similar results.

Substance	Generation/ Matrix or Solvent	Method/ T[K]	g-Factor	a-Value, D, E, J [mT]	Ref./ add. Ref.
$[C_{26}H_{32}Cl_2N_4O_2Pd_2]^{2\cdot}$	Not given/ $CHCl_3$	EPR/ 273···334		N: 1.463 J/a: 4.7	81Kok3
$[C_{26}H_{32}Cl_2N_4O_2Pd_2]^{2\cdot}$	K_2PdCl_4 + $CHCl_3$	EPR/ 300		5-line spectrum $J \gg a_N$	79Ovc1 [37])
$[C_{26}H_{32}I_2N_4O_2Pd_2]^{2\cdot}$	Not given/ $CHCl_3$ $C_6H_5CH_3$	EPR/ 283···315 298		N: 1.458 J/a: 9.53 N: 1.45 J/a: 9.7	81Kok3

[37]) Br and I complexes give similar spectra.

Substance	Generation/ Matrix or Solvent	Method/ T[K]	g-Factor	a-Value, D, E, J [mT]	Ref./ add. Ref.
$[C_{26}H_{32}N_4O_2TlX]^{2\cdot}$ 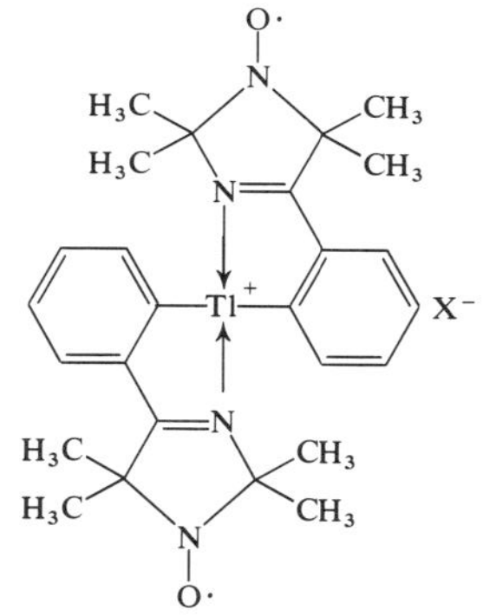	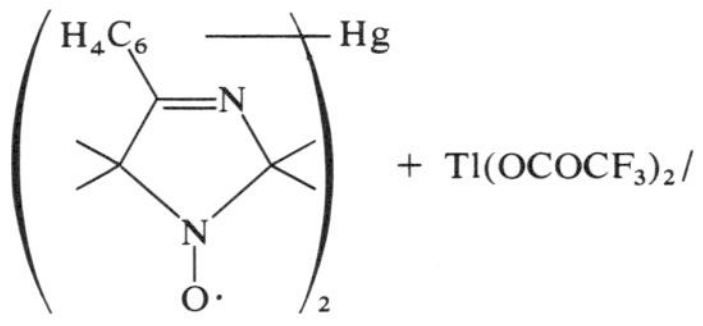 + $Tl(OCOCF_3)_2$/	EPR/			80Sha2 [38]
	$C_6H_5CH_3$	300		5-line spectrum $J \gg a_N$	
	H_2O	300		7-line spectrum $J \approx a_N$	
$[C_{26}H_{36}N_8O_2]^{2\cdot}$ 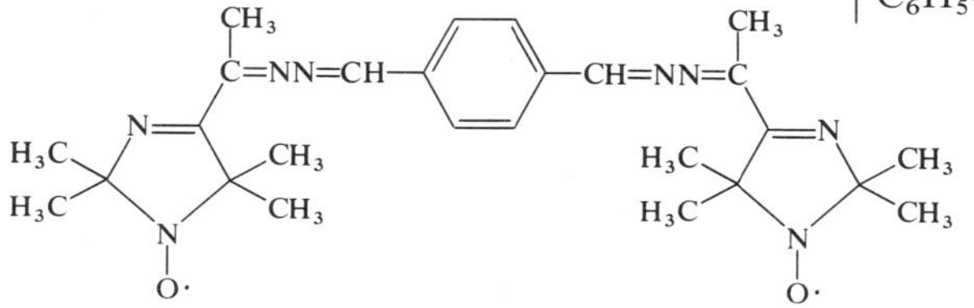	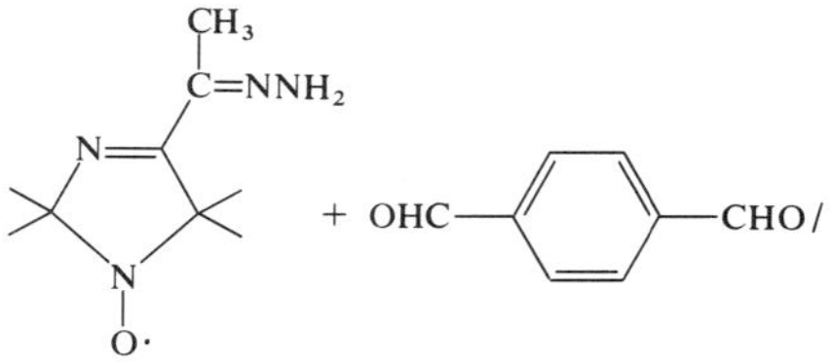 + OHC–C_6H_4–CHO/ $C_6H_5CH_3$	EPR/ 300		$J/a < 0.05$	82Gri1

[38]) Biradical bound to albumin has $2A_{\parallel}$: 0.65 mT.

Substance	Generation/ Matrix or Solvent	Method/ T[K]	g-Factor	a-Value, D, E, J [mT]	Ref./ add. Ref.
$[C_{28}H_{40}N_8O_4Pd]^{2\cdot}$ 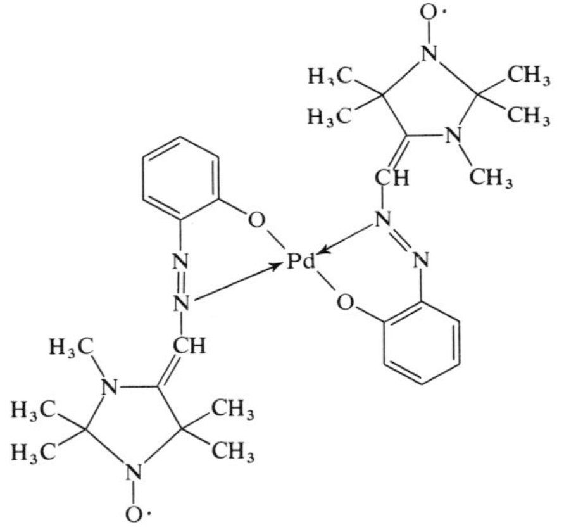	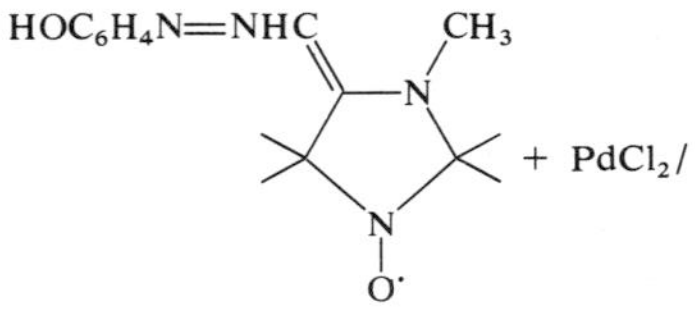 $HOC_6H_4N{=}NHC$ … + $PdCl_2$/ $CHCl_3$	EPR/ 300		8-line spectrum $J \approx a_N$	82Zol1
$[C_{30}H_{36}N_4O_4Pd]^{2\cdot}$ 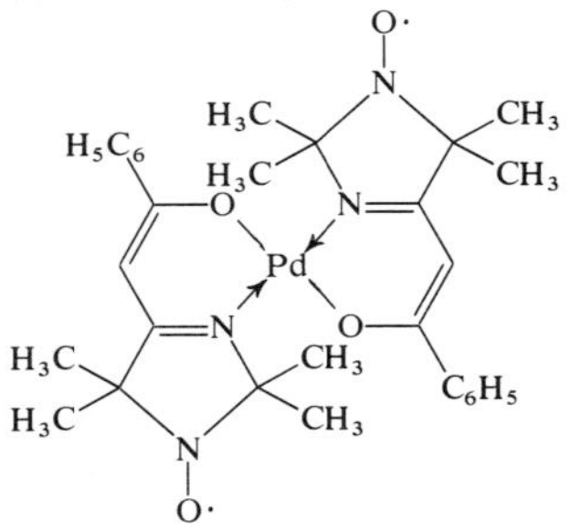	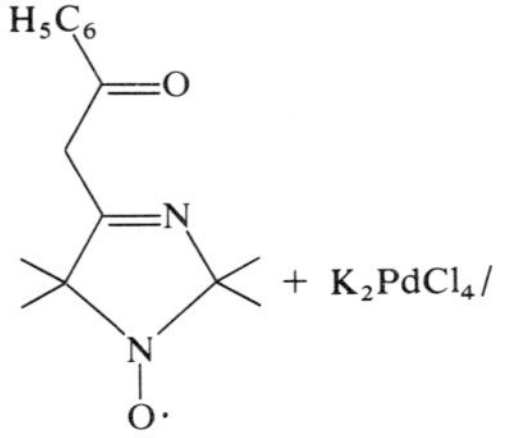 H_5C_6 … + K_2PdCl_4/ Not given	EPR/ not given	2.0058	N: 1.42 5-line-spectrum $J \gg a_N$	82Lar1 [39])

[39]) Zn complex gives similar spectrum.

22.1.11 (Pyrrolinyl-1-oxyl)-(pyrrolinyl-1-oxyl-3-oxide)

Substance	Generation/ Matrix or Solvent	Method/ T[K]	g-Factor	a-Value, D, E, J [mT]	Ref./ add. Ref.
$[C_{16}H_{26}N_6O_3]^{2\cdot}$ 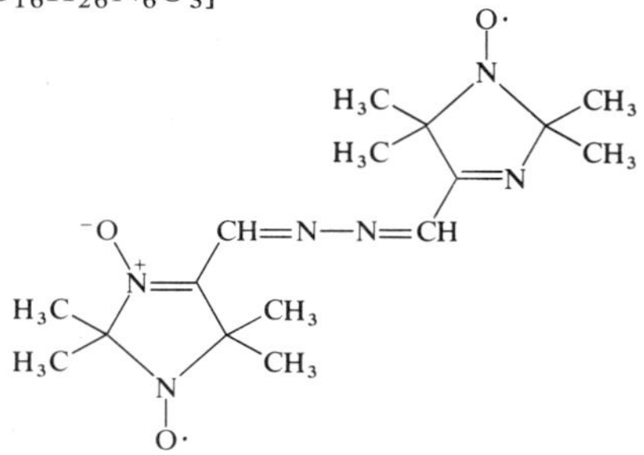	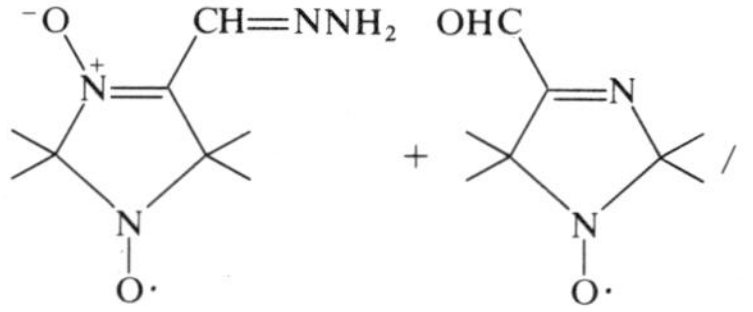 $C_6H_5CH_3$	EPR/ 300		J/a: ≈3.5	83Gri1
$[C_{16}H_{26}N_6O_3]^{2\cdot}$ 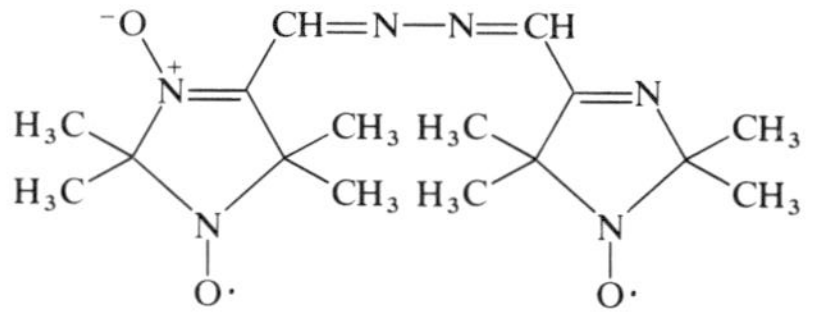	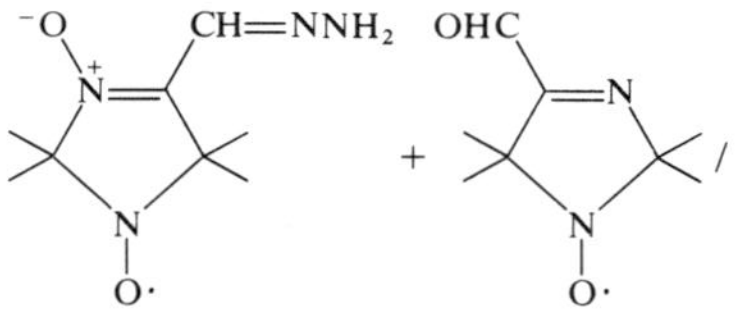 $C_6H_5CH_3$	EPR/ 300		J/a: 3.4 (1)	82Gri1
$[C_{17}H_{28}N_6O_3]^{2\cdot}$ 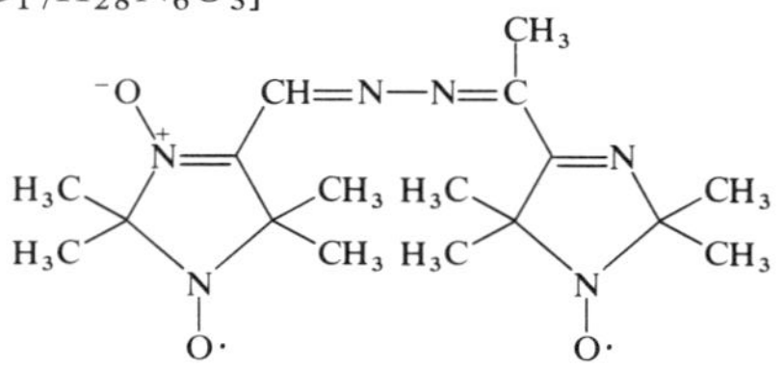	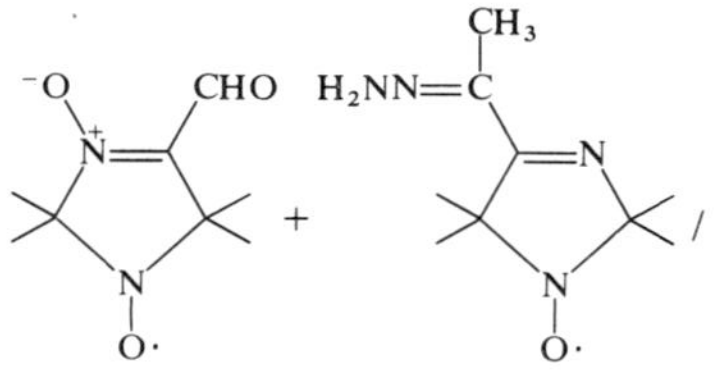 $C_6H_5CH_3$	EPR/ 300		J/a: 5.6 (2)	82Gri1

22.2 Bis(piperidinyl-1-oxyls), ·O–N X N–O·

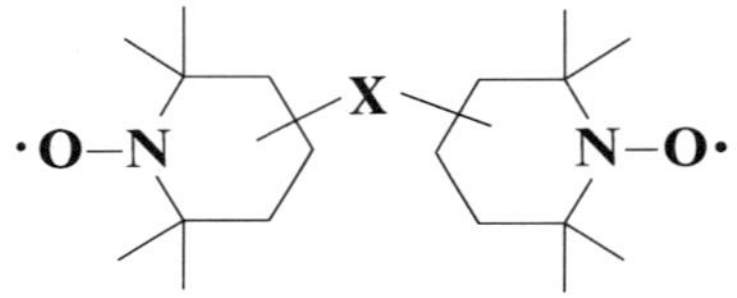

22.2.1 X=0 atom

Substance	Generation/ Matrix or Solvent	Method/ T [K]	g-Factor	a-Value, D, E, J [mT]	Ref./ add. Ref.
$[C_{18}H_{32}N_4O_6]^{2\cdot}$ 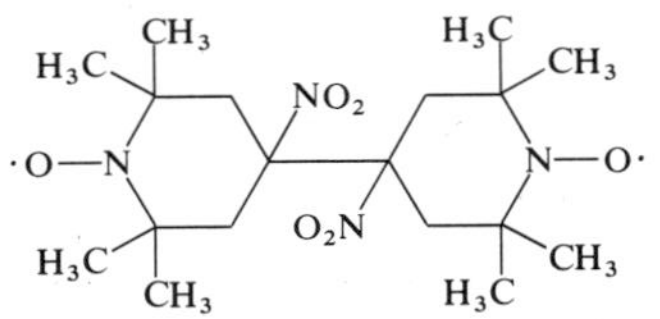	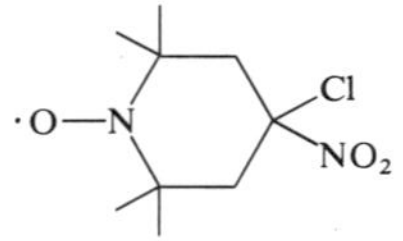 + $NaBH_4$/	EPR/			80Mys1 [1])
	–	–			
$[C_{18}H_{34}N_2O_4]^{2\cdot}$	Ox. of the corresp. diamine with pertungstic acid/	EPR/		5-line spectrum	86Smi1
	Not given	300		$J \gg a_N$	
$[C_{16}H_{20}N_3O_7]^{2\cdot}$ 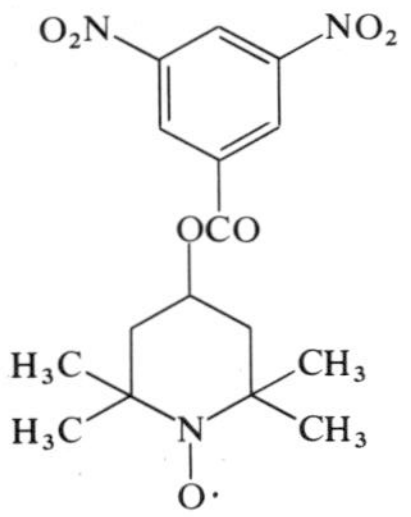	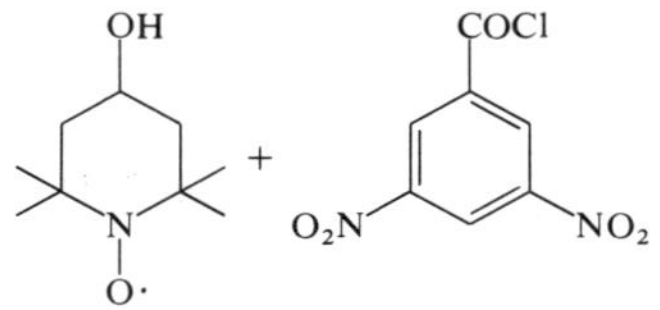 +	EPR/			81Ras1 [4])
	C_2H_5OH	153		$2D$: 34.0 [2])	
	C_2H_5OH 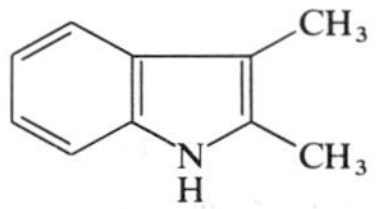 +	153		$2D$: 25.0 [3])	

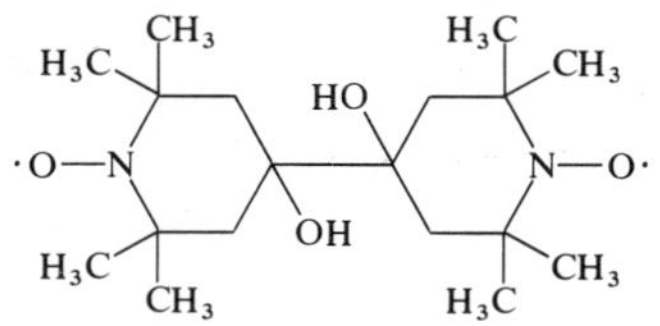

[1]) Details in Russian, not translated.
[2]) $r = 5.6$ (1) Å (dipole approximation).
[3]) $r = 6.1$ (1) Å (dipole approximation).
[4]) Measurement of radical-radical interaction in solid.

Substance	Generation/ Matrix or Solvent	Method/ T[K]	g-Factor	a-Value, D, E, J [mT]	Ref./ add. Ref.
22.2.2 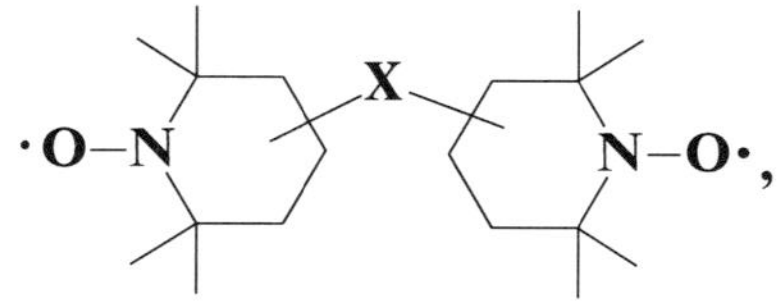 ·O–N ... X ... N–O·, X=1 atom					
$[C_{18}H_{30}HgN_2O_2]^{2\cdot}$ 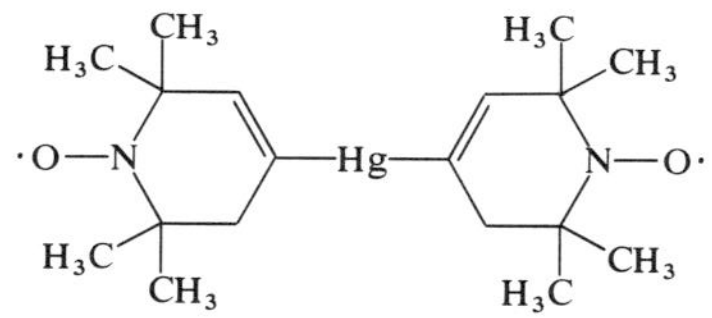	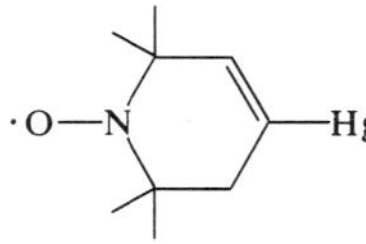 ·O–N ... –HgCl + NH_3 + Na_2SnO_2 $H_2O:C_2H_5OH$ $C_6H_5CH_3$	EPR/ 293		Hg: 0.30 N: 1.62 J/a: 4.92	81Sha1
$[C_{18}H_{30}N_2O_2Tl]^{2\cdot}$ 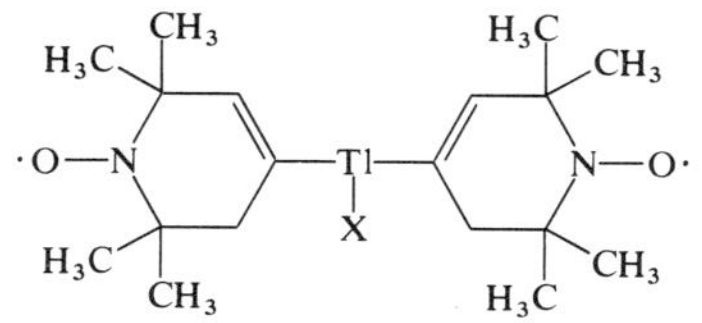	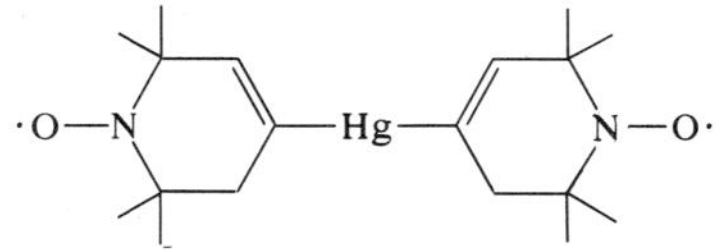 ·O–N ... –Hg– ... N–O· + TlX_3/ $H_2O:C_2H_5OH$	EPR/ 293		Tl: 1.32 N: 1.66 J/a: 12.0	81Sha1
$[C_{18}H_{35}N_3O_2]^{2\cdot}$ 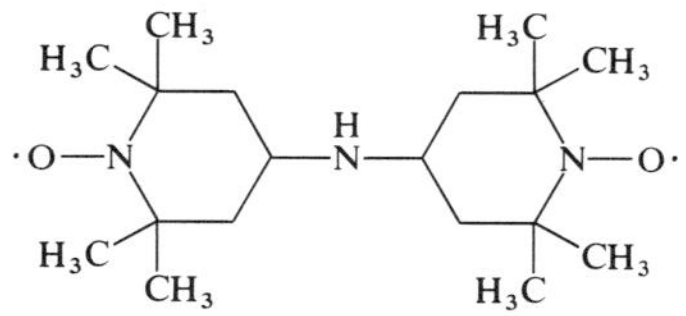	Not given/ H_2O	EPR/ 300		5-line spectrum $J \gg a_N$	84Vor1 [5])
$[C_{19}H_{32}N_2O_4]^{2\cdot}$ 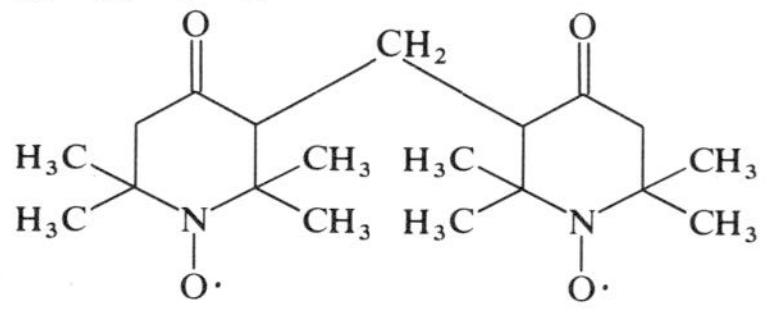	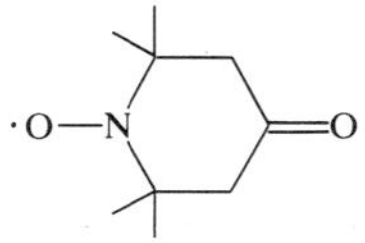 ·O–N ... =O + CH_2O/ $CHCl_3$	EPR/ 300		5-line spectrum $J \gg a_N$	86Sha1

[5]) Study of the oxidation of the bisnitroxide.

Substance	Generation/ Matrix or Solvent	Method/ T [K]	g-Factor	a-Value, D, E, J [mT]	Ref./ add. Ref.
$[C_{20}H_{32}N_2O_4]^{2\cdot}$ 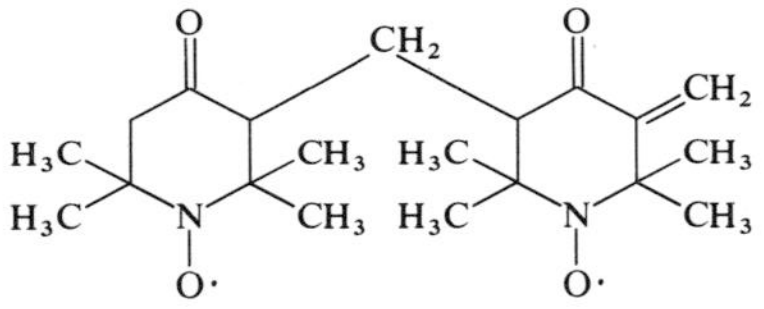	Dimerization of 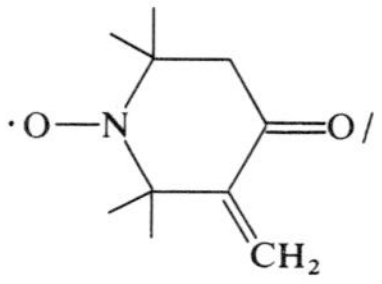 / $CHCl_3$	EPR/ 300		5-line spectrum $J \gg a_N$	86Sha1
$[C_{21}H_{32}N_2O_4]^{2\cdot}$ 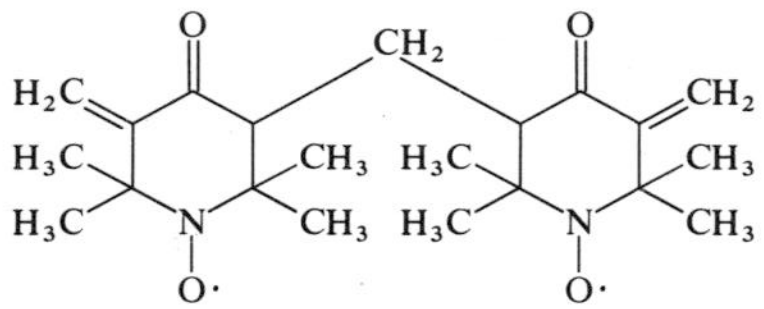	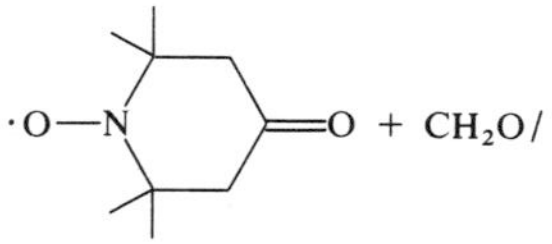 + CH_2O/ $CHCl_3$	EPR/ 300		5-line spectrum $J \gg a_N$	86Sha1 [6])
$[C_{22}H_{32}N_2O_4]^{2\cdot}$ 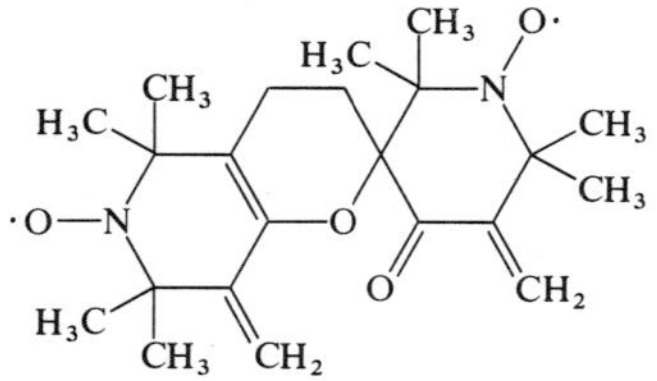	Not given/ C_2H_5OH	EPR/ 300		5-line spectrum $J \gg a_N$ [7])	80Kok2, 86Sha1
	Not given/ –	EPR/ –			79Vla1 [8]), 82Vla1
$[C_{22}H_{32}N_2O_5]^{2\cdot}$ 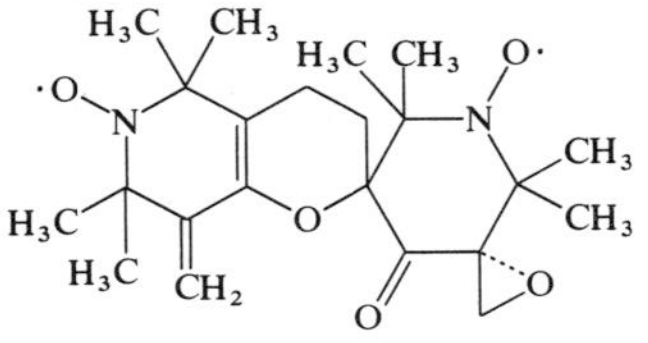	Ox. of the corresp. diamine with $Na_2WO_4 + H_2O_2$/ C_6H_6	EPR/ 300		5-line spectrum $J \gg a_N$	85Sha1 [9])

[6]) Stereoisomer also reported gives identical spectrum.
[7]) Measurements also in the presence of a photoreceptor membrane.
[8]) Measurement in frog rod outer segments.
[9]) Stereoisomer (oxiran ring) gives identical spectrum.

Substance	Generation/ Matrix or Solvent	Method/ T[K]	g-Factor	a-Value, D, E, J [mT]	Ref./ add. Ref.
$[C_{22}H_{35}BrN_2O_4]^{2\cdot}$ 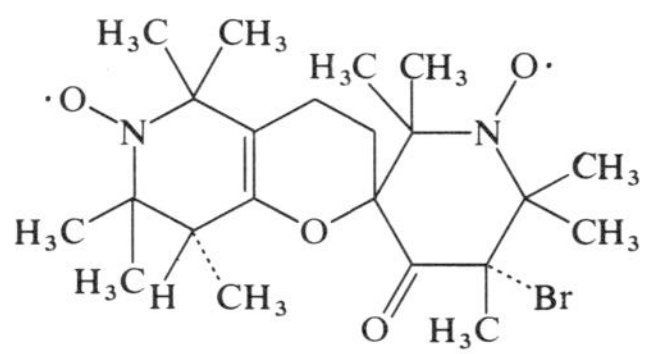	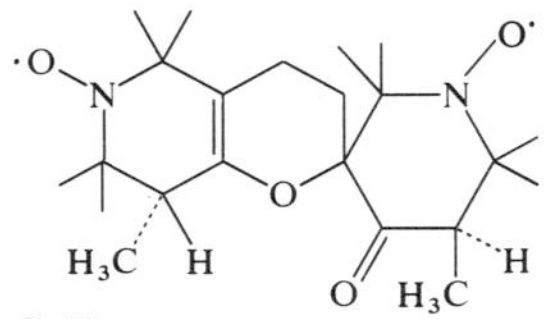 + NaOBr/ C_6H_6	EPR/ 300		5-line spectrum $J \gg a_N$	85Sha1
$[C_{22}H_{36}N_2O_4]^{2\cdot}$ 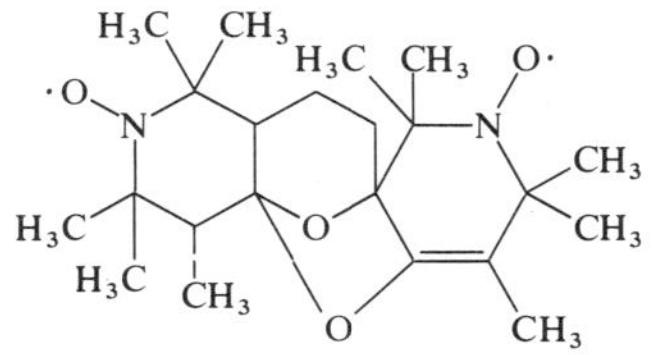	Ox. of the corresp. diamine with $Na_2WO_4 + H_2O_2$/	EPR/			82Sha1 11), 85Sha1
	C_2H_5OH	300		5-line spectrum $J \gg a_N$	
	C_6H_6 10)	300		5-line spectrum 10) $J \gg a_N$	
$[C_{22}H_{36}N_2O_4]^{2\cdot}$ 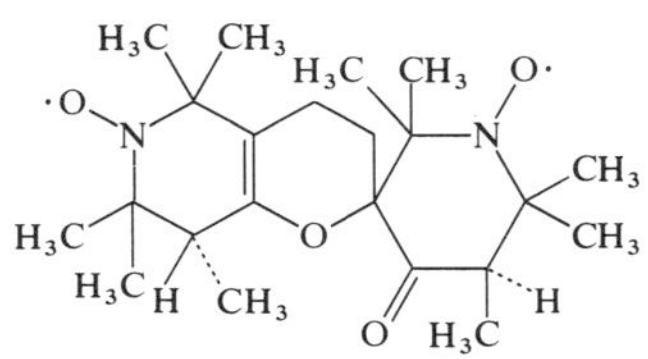	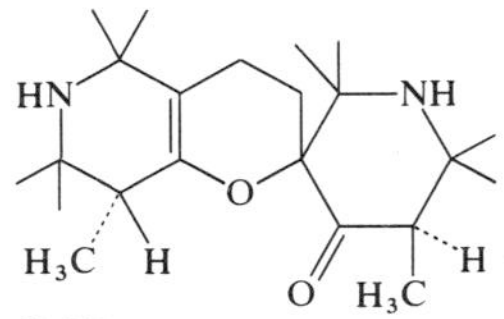 + H_2O_2 + Na_2WO_4/ C_6H_6	EPR/ 300		5-line spectrum $J \gg a_N$	85Sha1
$[C_{22}H_{36}N_2O_4]^{2\cdot}$ 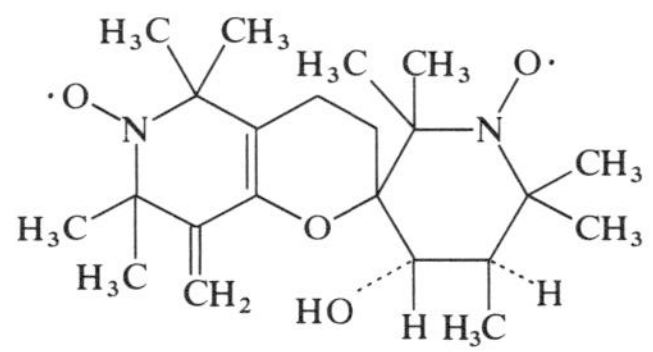	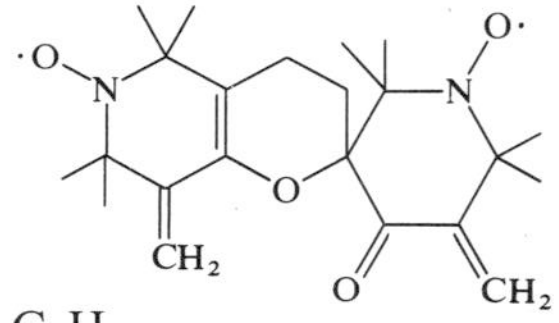 + $NaBH_4$/ C_6H_6	EPR/ 300		5-line spectrum $J \gg a_N$	85Sha1

10) From [85Sha1].
11) Crystal structure determined.

Substance	Generation/ Matrix or Solvent	Method/ T [K]	g-Factor	a-Value, D, E, J [mT]	Ref./ add. Ref.
$[C_{22}H_{36}N_2O_4]^{2\cdot}$	+ NaCN/ C_6H_6	EPR/ 300		5-line spectrum $J \gg a_N$	85Sha1
$[C_{22}H_{36}N_2O_4]^{2\cdot}$	Not given/ C_2H_5OH	EPR/ 300		5-line spectrum $J \gg a_N$	82Vla1 [12]), 80Kok2
$[C_{22}H_{38}N_2O_4]^{2\cdot}$	+ H_2O_2 + Na_2WO_4/ C_6H_6	EPR/ 300		5-line spectrum $J \gg a_N$	85Sha1
$[C_{23}H_{33}N_3O_4]^{2\cdot}$	+ NaCN + NH_4Cl/ C_6H_6	EPR/ 300		5-line spectrum $J \gg a_N$	85Sha1

[12]) Spectra of biradical when attached to photoreceptor membrane reported.

Substance	Generation/ Matrix or Solvent	Method/ T[K]	g-Factor	a-Value, D, E, J [mT]	Ref./ add. Ref.
22.2.3 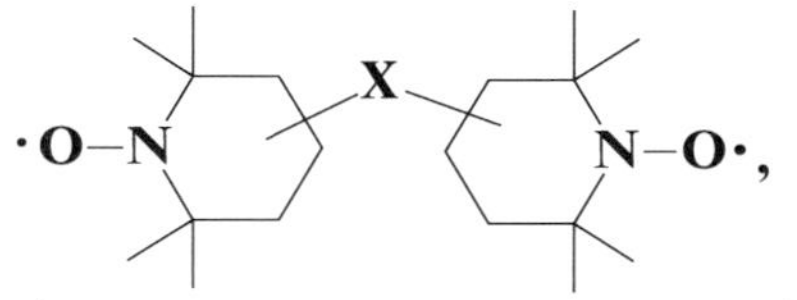 ·O–N(...)–X–(...)N–O·, X=2 atoms					
$[C_{18}H_{32}N_4O_2]^{2\cdot}$ 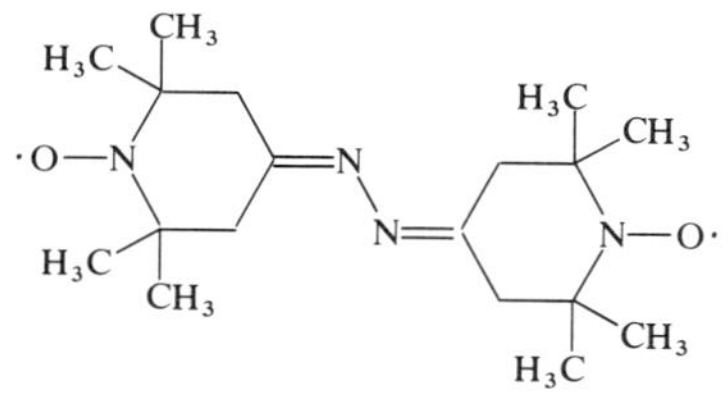	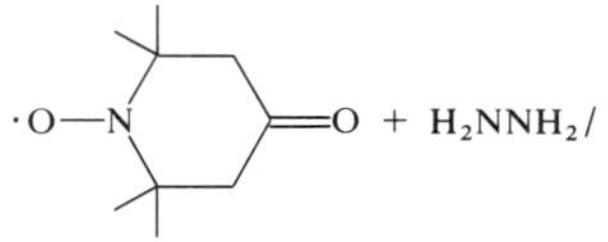 ·O–N(...)=O + H_2NNH_2/ $C_6H_5CH_3$: THF	EPR/ 93		$J\approx 50.0$	83Eat1 [13]), 82Eat1 [14])
	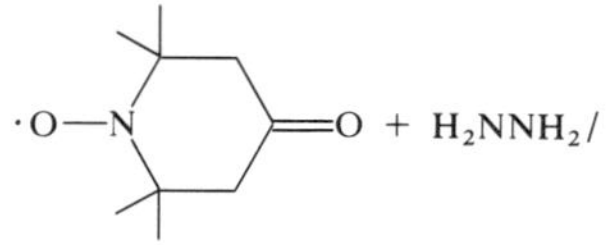 ·O–N(...)=O + H_2NNH_2/ DMF	Calc./ 300		N: 1.4 D: 3.65 [15])	80Mic2
$[C_{20}H_{30}N_2O_2]^{2\cdot}$ 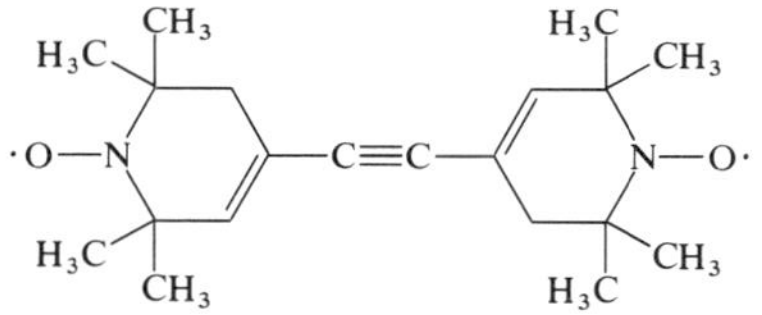	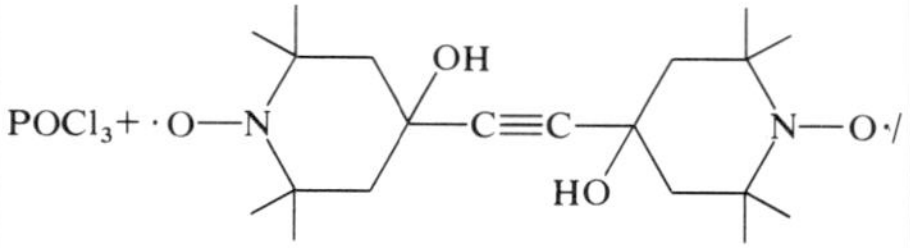 $POCl_3$+ ·O–N(...)(OH)–C≡C–(HO)(...)N–O·/ $C_6H_5CH_3$	EPR/ 305		5-line spectrum $J \gg a_N$ J/a: 35 (4)	80Pav3, 80Kok1
$[C_{20}H_{30}Cl_3GaN_2O_2]^{\cdot 2}$ 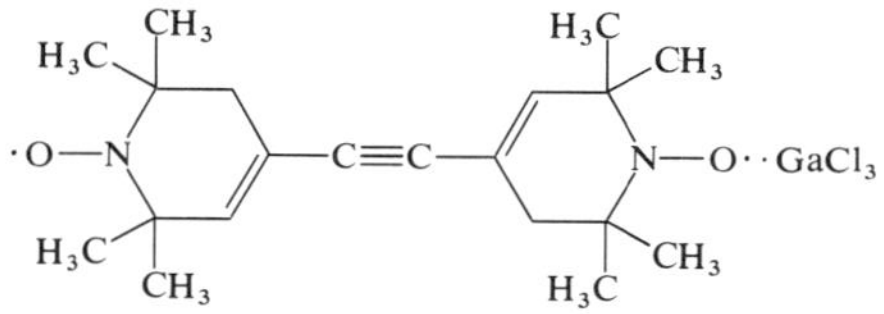	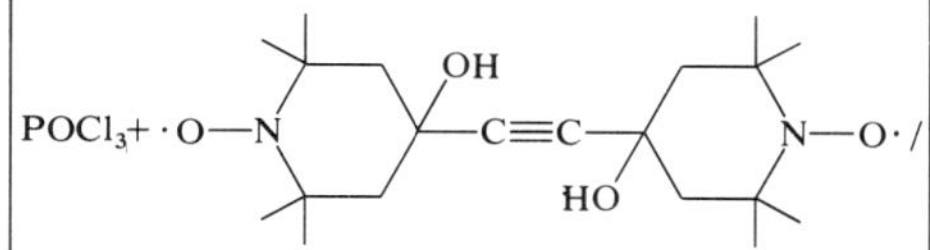 $POCl_3$+ ·O–N(...)(OH)–C≡C–(HO)(...)N–O·/ $GaCl_3$ complex (1:1)	EPR/ 300		N: 1.90 ^{69}Ga: 3.76 ^{71}Ga: 4.74	82Sol2

[13]) $r = 8.6$ Å measured from intensity of half-field transition; $r = 9.0$ Å measured using point-dipole approximation.
[14]) r evaluated using intensity of $\Delta M_s = \pm 2 = I$; $I \propto 1/r^{-6}$.
[15]) D calculated as 3.60 mT.

Substance	Generation/ Matrix or Solvent	Method/ T[K]	g-Factor	a-Value, D, E, J [mT]	Ref./ add. Ref.
$[C_{20}H_{34}N_2O_4]^{2\cdot}$	·O—N(piperidinone)=O + HC≡CH + 2 metal/ $C_6H_5CH_3$	EPR/ 305		5-line spectrum $J \approx a_N$	80Pav3, 80Kok1

22.2.4 ·O–N ... X ... N–O·, X = 3 atoms

Substance	Generation/ Matrix or Solvent	Method/ T[K]	g-Factor	a-Value, D, E, J [mT]	Ref./ add. Ref.
$[C_{18}H_{38}Cl_2N_4O_2Pt]^{2\cdot}$	·O—N(piperidine)—NH_2 + $K_2[PtCl_4]$/ H_2O	EPR/ 300	2.0054		79Mat1, 86Cla1 [16]
$[C_{18}H_{38}Cl_2N_4O_2Pd]^{2\cdot}$	·O—N(piperidine)—NH_2 + $PdCl_2$/ $CHCl_3$	EPR/ 273 373		3-line spectrum 5-line spectrum N: 1.50	85Zak1 [17]
$[C_{19}H_{34}N_2O_5]^{2\cdot}$	·O—N(piperidine)—OH + ClC(O)Cl/ $C_6H_5CH_3$: THF	EPR/ 93		$J \approx -2.5$ [18]	83Eat1

[16]) Toxicity measurements.

[17]) Variable temperature study; dibromide gives similar spectrum.

[18]) $r = 11.3$ Å measured from intensity of half-field transition; $r = 11.0$ Å measured from point-dipole approximation.

Substance	Generation/ Matrix or Solvent	Method/ T[K]	g-Factor	a-Value, D, E, J [mT]	Ref./ add. Ref.
$[C_{19}H_{34}N_4O_2]^{2\cdot}$	Thermolysis of (·O—N piperidinyl —NP(O)(OC₂H₅)₂Na⁺, CO₂⁻) or (·O—N piperidinyl —NHC(S)HN— piperidinyl N—O·) + HgO [19])/	EPR/			84Sen1, 81Han1
	C_6H_6	293	2.00590	11-line spectrum N: 1.548 $a_N \approx J$	
	not given [19])	300		5-line spectrum [19]) N: 1.48 $J > a_N$	
$[C_{19}H_{35}N_3O_4]^{2\cdot}$	·O—N piperidinyl —OH + ·O—N piperidinyl —NHC(O)N((O)P(OC₂H₅)₂)— piperidinyl N—O·/ C_6H_6	EPR/ 293		13-line spectrum $J \approx a_N$ N: 1.54 (1)	87Tka1
$[C_{19}H_{36}N_4O_2S]^{2\cdot}$	·O—N piperidinyl —NH₂ + ·O—N piperidinyl —NCS/ Not given	EPR/ 300		5-line spectrum N: 1.48 $J \gg a_N$	81Han1

[19]) From [81Han1].

Substance	Generation/ Matrix or Solvent	Method/ T[K]	g-Factor	a-Value, D, E, J [mT]	Ref./ add. Ref.
$[C_{20}H_{30}N_2O_6Pd]^{2\cdot}$ 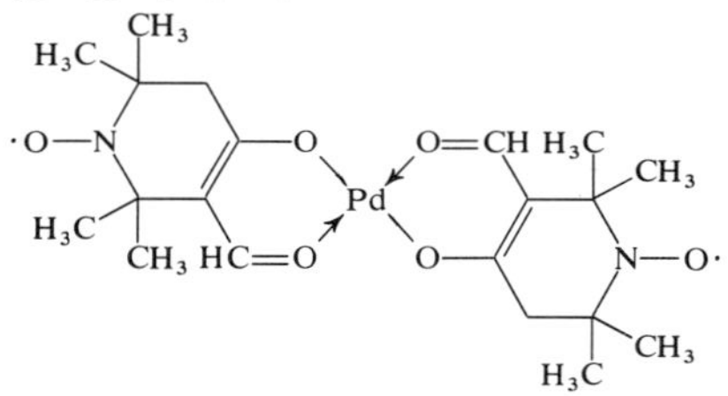	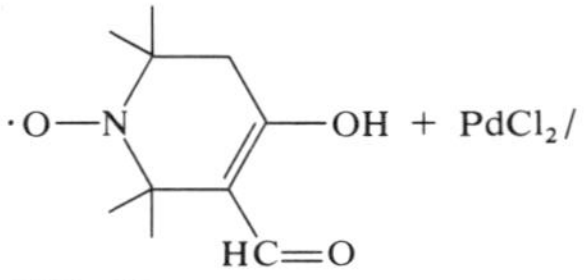 CH_2Cl_2	EPR/ 293		$\bar{J}$: 0.6[20])	81Esp1
$[C_{20}H_{40}N_2O_4Si]^{2\cdot}$ 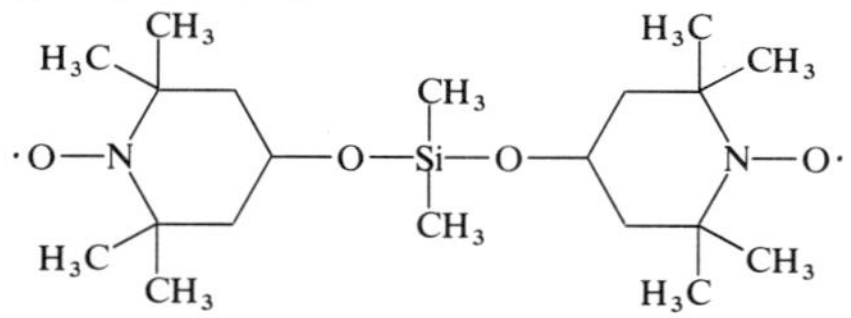	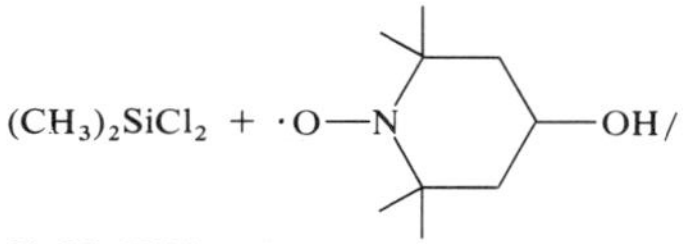 C_2H_5OH	EPR/ 300		5-line spectrum $J \gg a_N$	80Kok2[21]), 82Vla1
$[C_{22}H_{42}N_3O_5P]^{2\cdot}$ 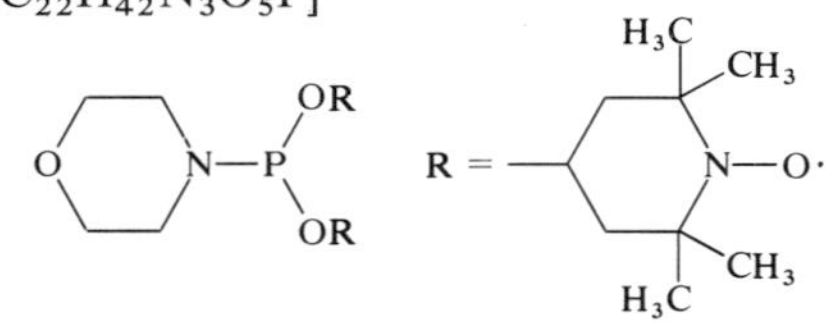	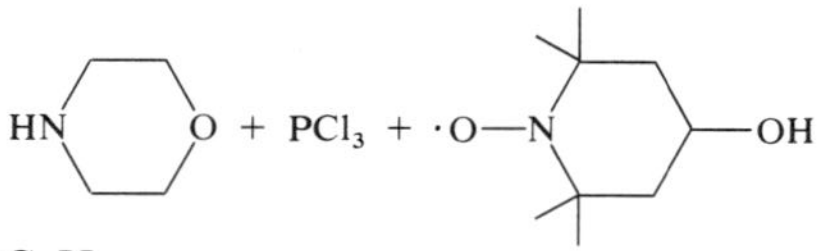 C_6H_6	EPR/ 300		5-line spectrum N: 16.6 $J \gg a_N$	78Kon1
$[C_{22}H_{43}N_4O_4PS]^{2\cdot}$ 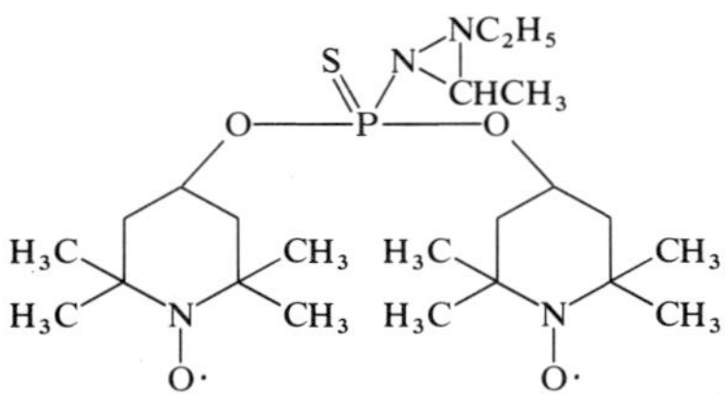	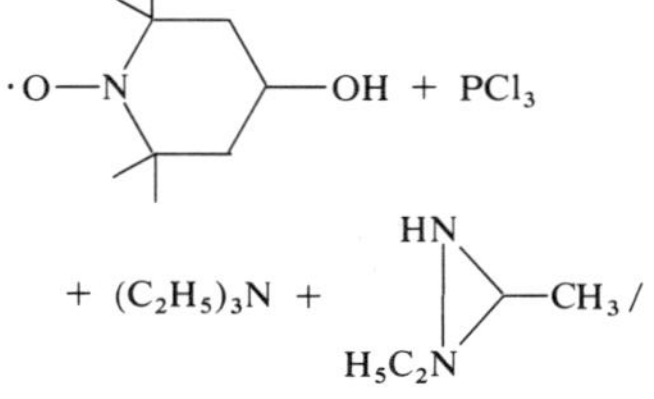 C_6H_6	EPR/ 300		N: 1.50 $J \gg a_N$	83Sos1

[20]) $L = \bar{J} + a_N + (\bar{J}^2 + a_N^2)^{1/2}$.
[21]) Measurements also in the presence of a photoreceptor membrane.

Substance	Generation/ Matrix or Solvent	Method/ T[K]	g-Factor	a-Value, D, E, J [mT]	Ref./ add. Ref.
$[C_{23}H_{44}N_3O_4P]^{2\cdot}$ 	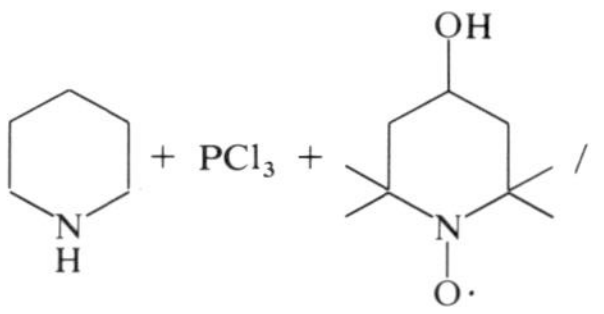 + PCl_3 + / C_6H_6	EPR/ 300		5-line spectrum N: 1.68	78Kon1
$[C_{23}H_{45}N_4O_6P]^{2\cdot}$ 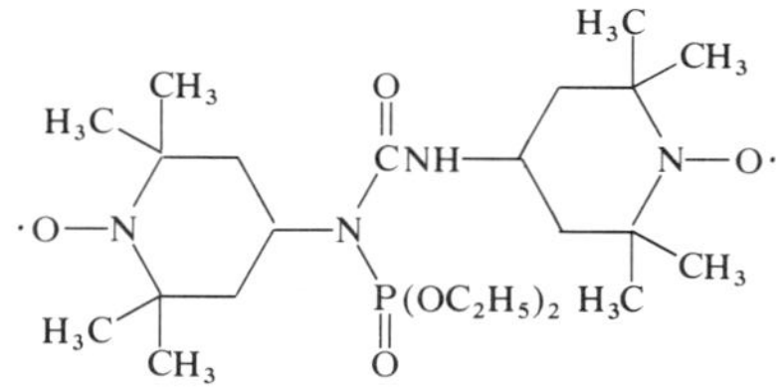	Thermolysis of 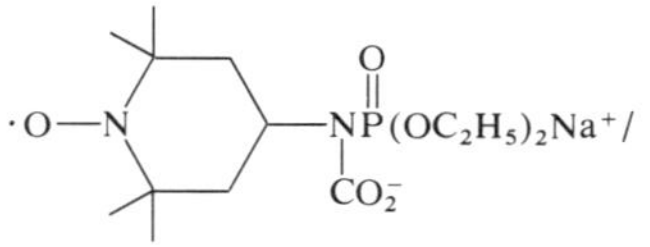 $NP(OC_2H_5)_2Na^+$/ C_6H_6	EPR/ 298	2.00587	13-line spectrum N: 1.551 $a_N \approx J$	84Sen1
$[C_{26}H_{34}F_{12}N_4O_2]^{2\cdot}$ 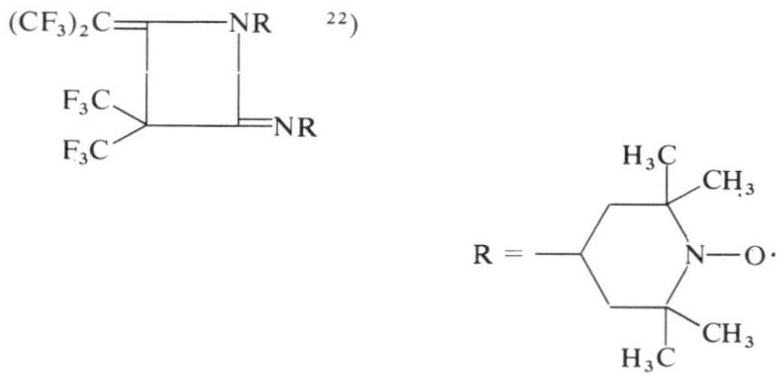 [22]	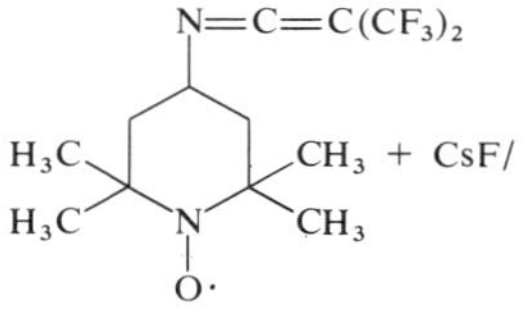 $N{=}C{=}C(CF_3)_2$ + CsF/	EPR/			80Sha1
	Not given	298		7-line spectrum $J \approx a_N$	
	Not given	353		5-line spectrum $J > a_N$	

[22]) 3-Conformations detected. One has NO˙–NO˙ separation 9.2 Å; point-dipole approximation.

Substance	Generation/ Matrix or Solvent	Method/ T[K]	g-Factor	a-Value, D, E, J [mT]	Ref./ add. Ref.
22.2.5 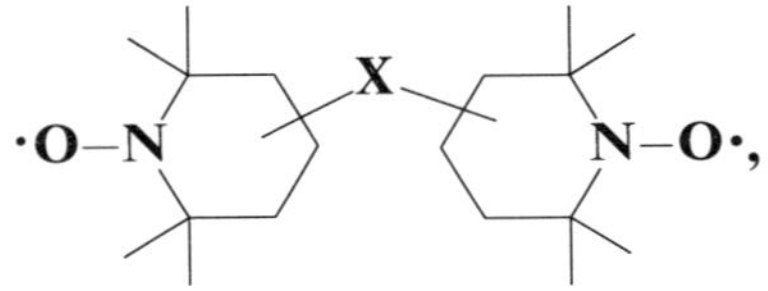 ·O–N(piperidine)–X–(piperidine)N–O·, X=4 atoms					
$[C_{19}H_{35}N_3O_3S_2]^{2\cdot}$ 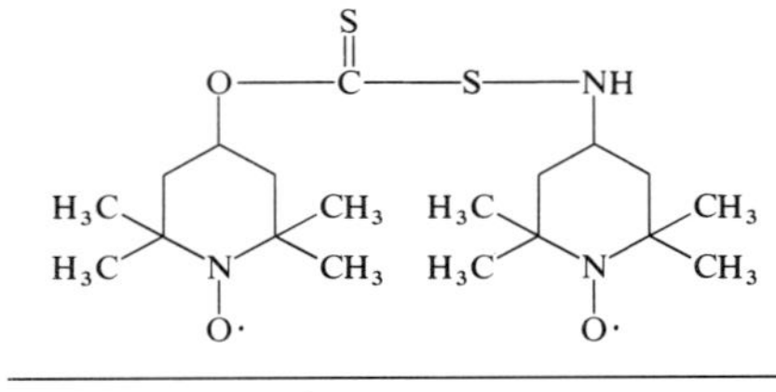	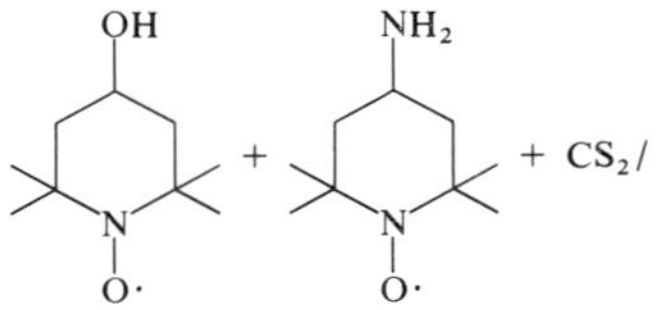 4-hydroxy-TEMPO + 4-amino-TEMPO (NH_2) + CS_2/ Several solvents	EPR/ 300		7-line spectrum $J \approx a_N$	86Iva1
$[C_{20}H_{36}N_4O_4]^{2\cdot}$ 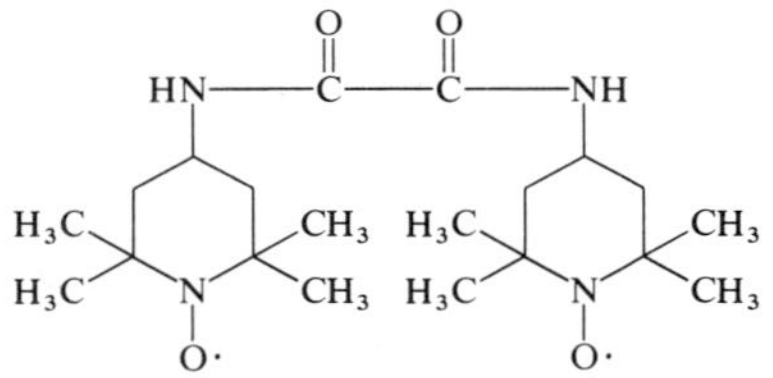	Ox. of corresp. diamine with pertungstic acid/ Not given	EPR/ 300		3-line spectrum N: 1.5 $a_N \gg J$	86Ehm1
$[C_{20}H_{37}N_3O_3S_2]^{2\cdot}$ 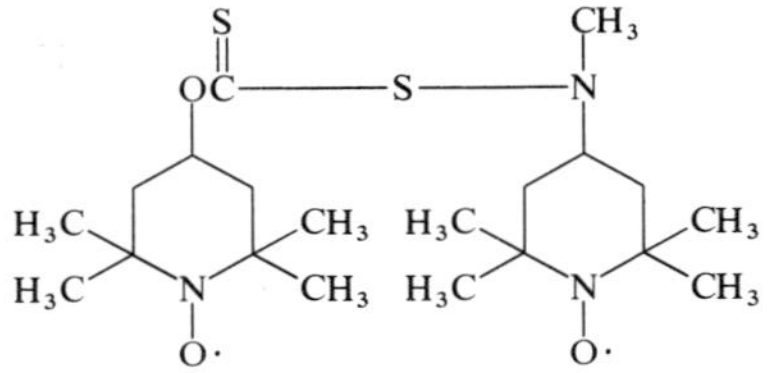	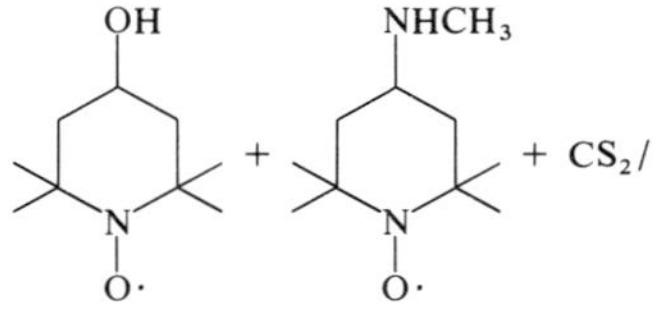 4-hydroxy-TEMPO + 4-($NHCH_3$)-TEMPO + CS_2/ Several solvents	EPR/ 300		9-line spectrum $J \approx a_N$	86Iva1

Substance	Generation/ Matrix or Solvent	Method/ T[K]	g-Factor	a-Value, D, E, J [mT]	Ref./ add. Ref.
$[C_{20}H_{40}N_4O_2]^{2\cdot}$ 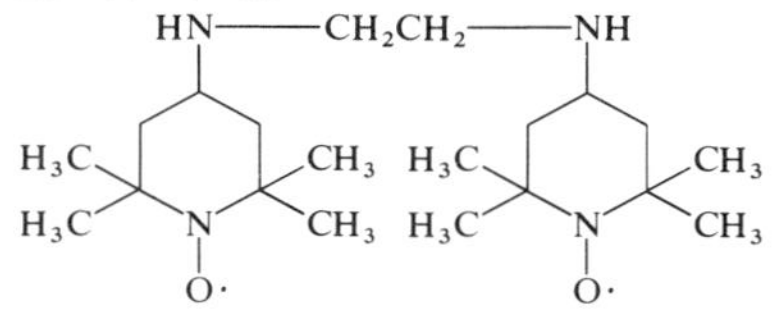	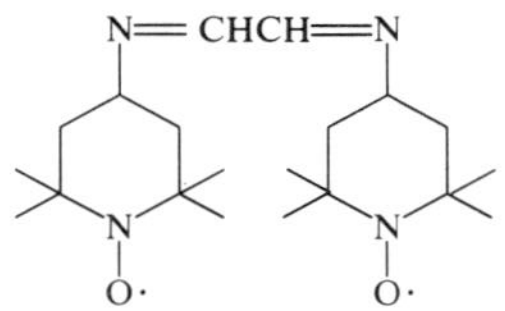 N═CHCH═N (bis-piperidinyl-1-oxyl) + $NaBH_4$/	EPR/			86Abs1[23]), 86Ehm1
	H_2O, pH = 2.47	294		J: 0.42	
	H_2O, pH = 11.1	294		J: 2.22	
	H_2O, pH = 2.47	343		J: 5.8 (3)	
	H_2O, pH = 11.1	349		J: 6.6 (3)	
	DMSO	294		J: 2.0 (2) N: 1.58	
	C_6H_6	294		J: 2.6 (2) N: 1.55	
	CS_2	294		J: 2.6 (5) N: 1.55	
	$CHCl_3$	294		J: 3.0 (2) N: 1.63	
	DMF	294		J: 3.1 (2) N: 1.58	
	1-butanol	294		J: 1.5 (2) N: 1.50	
$[C_{21}H_{40}N_4O_2]^{2\cdot}$ 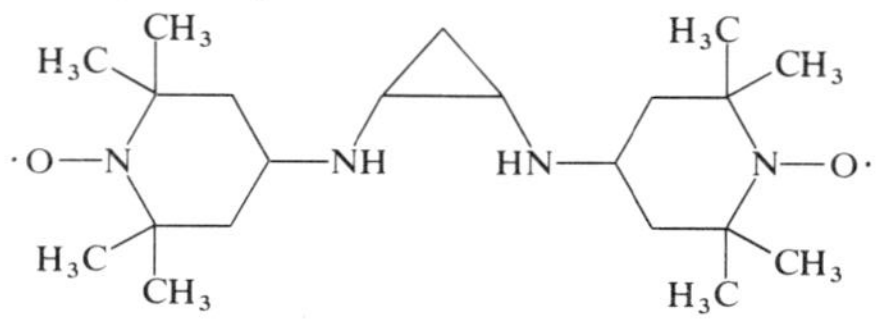	Ox. of corresp. amine/ H_2O	EPR/ 294		3-line spectrum $a_N \gg J$ N: 1.71 [24])	85Mil1
$[C_{22}H_{30}N_2O_2]^{2\cdot}$ 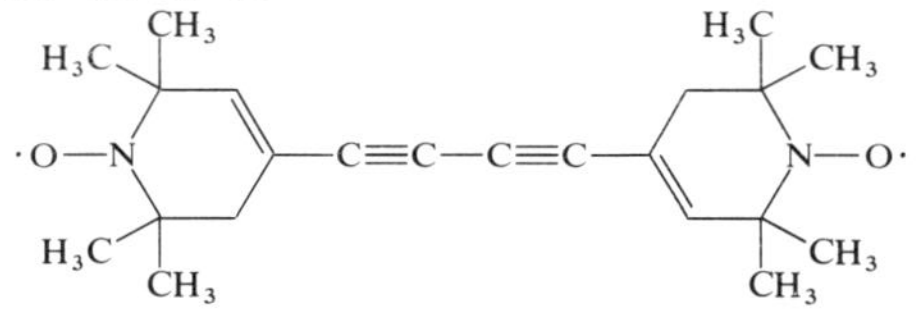	RC≡CH + $Cu(CH_3CO_2)_2$ · pyridine/ $C_6H_5CH_3$	EPR/ 305		5-line spectrum $J \gg a_N$ J/a: 14.5 (2)	80Pav1, 80Pav2, 80Kok1, 81Yan1

[23]) Study of conformational analysis of bisnitroxide.

[24]) *Trans* isomer also shows $a_N \gg J$ and a_N 1.71 mT.

Substance	Generation/ Matrix or Solvent	Method/ T[K]	g-Factor	a-Value, D, E, J [mT]	Ref./ add. Ref.
$[C_{22}H_{30}AlCl_3N_2O_2]^{2\cdot}$ 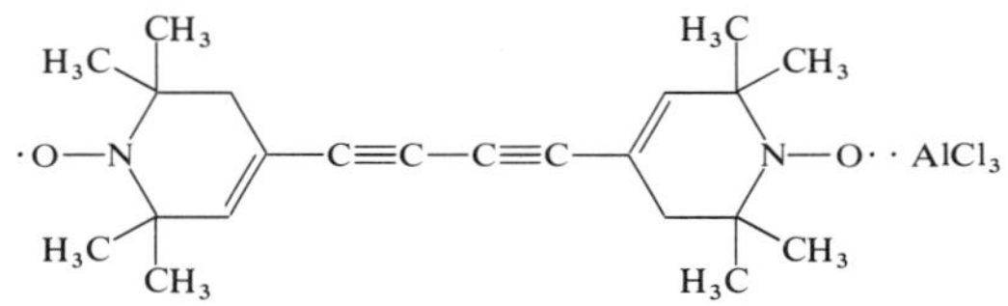	$RC{\equiv}CH + Cu(CH_3CO_2)_2$: C_5H_5N/ $AlCl_3$ complex (1:1) $AlCl_3$ complex (2:1)	EPR/ 300		N: 2.02 Al: 1.00 N: 2.02 Al: 0.50	82Sol2
$[C_{22}H_{34}N_2O_2]^{2\cdot}$ 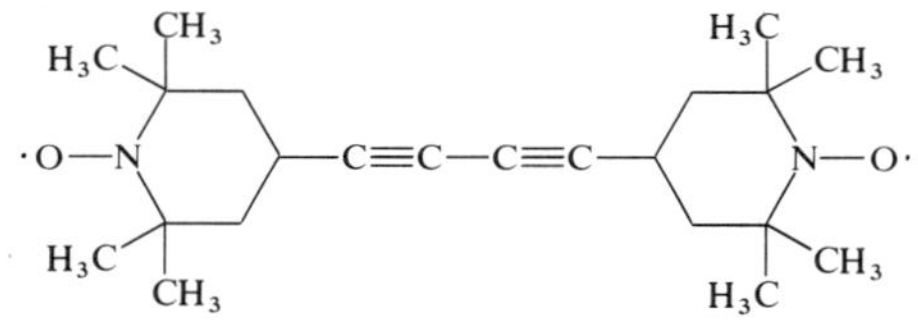	Not given/ –	EPR/ –			88Mil2 [25]), 87Ovc1
$[C_{22}H_{34}N_2O_4]^{2\cdot}$ 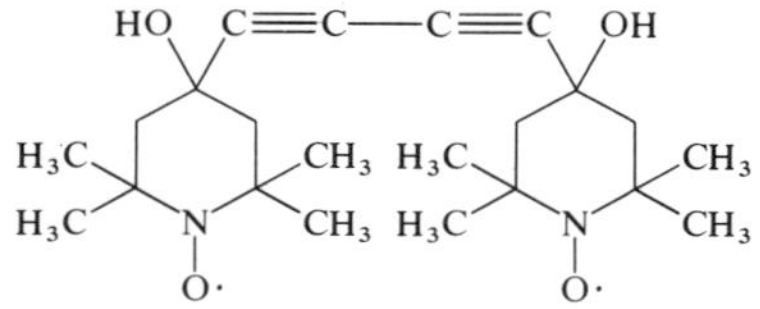	Not given/ –	EPR/ –			87Kor1 [26]), 86Kor1 [27]), 87Buc1 [28]), 88Mil2 [29])

[25]) Magnetic susceptibility measurements. $\mu_{eff} = 2.47\ \mu_B$ per molecule $= 1.79\ \mu_B$ per NO group.
[26]) EPR spectra of polymers derived from this monomer measured ($J_{\parallel}$: 1 eV); magnetic properties also reported.
[27]) Measurement of polymers derived therefrom.
[28]) Crystal structure determined.
[29]) Magnetic measurements.

Substance	Generation/ Matrix or Solvent	Method/ T [K]	g-Factor	a-Value, D, E, J [mT]	Ref./ add. Ref.
22.2.6 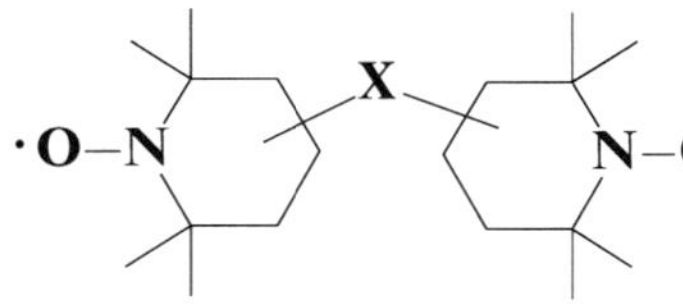 ·O–N(…)–X–(…)N–O·, X=5 atoms					
$[C_{21}H_{36}N_2O_6]^{2\cdot}$	Ox. of corresp. diamine with m-$ClC_6H_4CO_3H$/ DMF	EPR/ 293		3-line spectrum $a_N \gg J$ N: 1.525	80Bri1
		418		5-line spectrum $J \gg a_N$ N: 1.52	
$[C_{22}H_{30}HgN_2O_2]^{2\cdot}$ 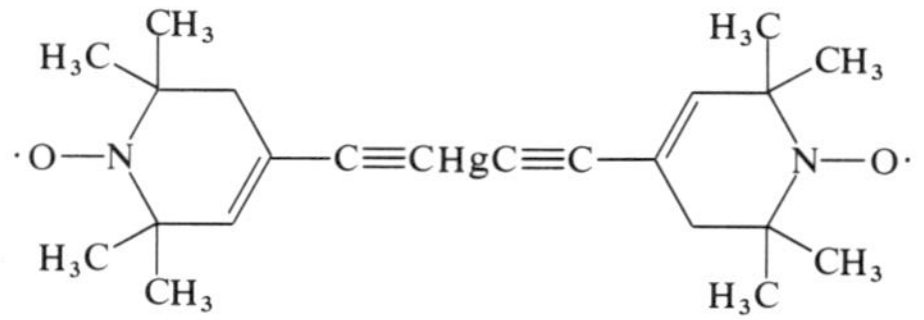	$RC{\equiv}CH + HgCl_2 + KI$, R = 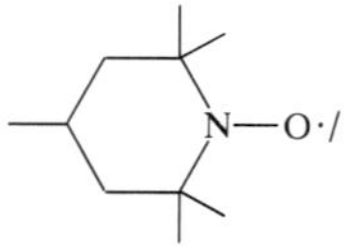 / $C_6H_5CH_3$	EPR/ 293		11-line spectrum J/a: 0.60 (3)	80Pav1, 80Kok1
22.2.7 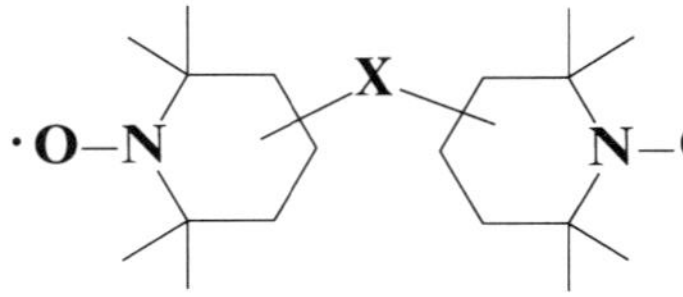 ·O–N(…)–X–(…)N–O·, X=6 atoms					
$[C_{20}H_{34}N_2O_4S_4]^{2\cdot}$ 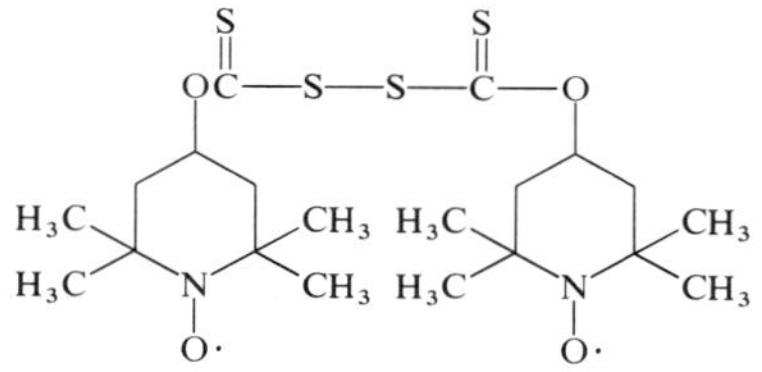 (continued)	Dimerization of 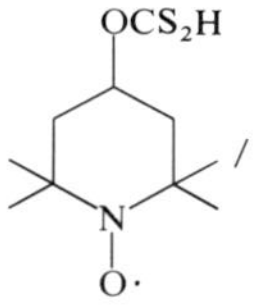 (4-OCS_2H-piperidinyl-1-oxyl)/ Several solvents	EPR/ 300		5-line spectrum $J \approx a_N$	86Iva1

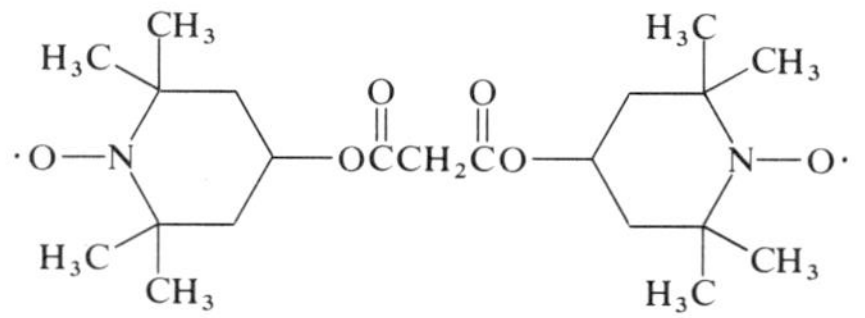

Substance	Generation/ Matrix or Solvent	Method/ T [K]	g-Factor	a-Value, D, E, J [mT]	Ref./ add. Ref.
$[C_{20}H_{34}N_2O_4S_4]^{2\cdot}$ *(continued)*	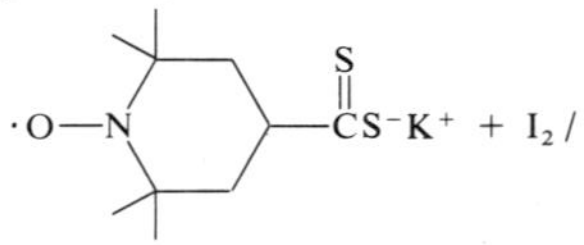 $\cdot$O—N ... —$\overset{S}{\overset{\parallel}{C}}S^-K^+$ + I_2 /	EPR/			80Sol1
	Organic solvent	300		5-line spectrum $J \gg a_N$ N: 1.56	
	$C_6H_5CH_3$ two $AlCl_3$ complexes	300		(a) N: 1.96 (6)[29a] Al: 0.86 (3) (b) N: 1.96[29a] Al: 1.10 (3)	
	two $GaCl_3$ complexes	300		(a) N: 1.81 (6)[29a] ^{69}Ga: 3.44 (7) ^{71}Ga: 4.36 (8) (b) N: 1.82 (6)[29a] ^{69}Ga: 3.64 (7) ^{71}Ga: 4.61 (3)	
$[C_{22}H_{38}N_2O_6]^{2\cdot}$ 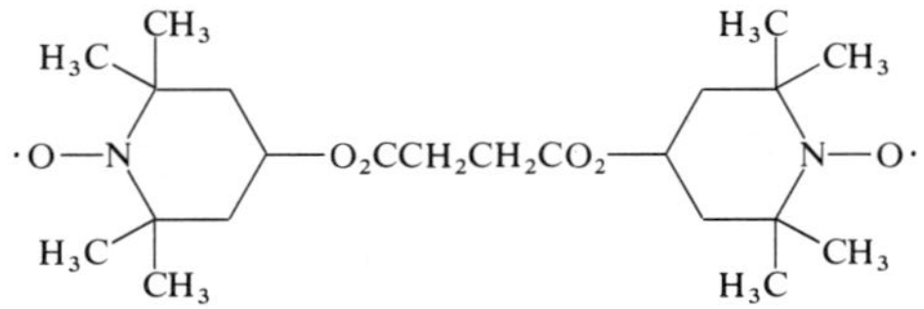 $\cdot$O—N ... —$O_2CCH_2CH_2CO_2$— ... N—O$\cdot$ (H_3C, CH_3 substituents)	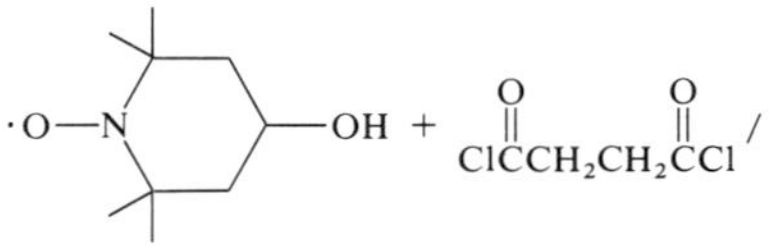 $\cdot$O—N ... —OH + $Cl\overset{O}{\overset{\parallel}{C}}CH_2CH_2\overset{O}{\overset{\parallel}{C}}Cl$ /	EPR/			80Bar1[30]
	Methyl cellosolve-water	353		5-line spectrum $J \gg a_N$	
	C_2H_5OH	90		–	
$[C_{22}H_{40}N_4O_2S_4]^{2\cdot}$ $H_3CN—\overset{S}{\overset{\parallel}{C}}—S—S—\overset{S}{\overset{\parallel}{C}}—NCH_3$ (piperidine rings with H_3C, CH_3 substituents, N—O$\cdot$) 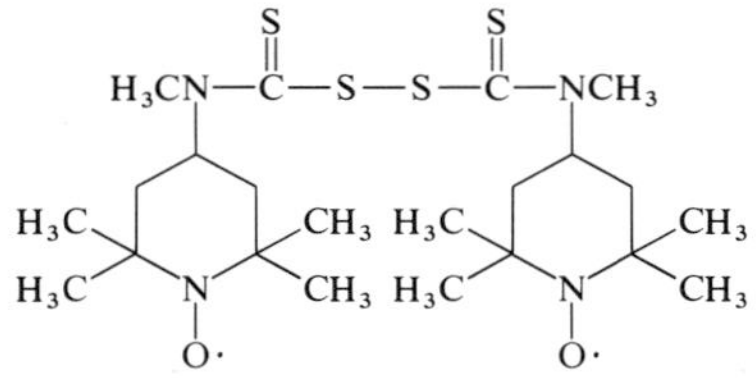	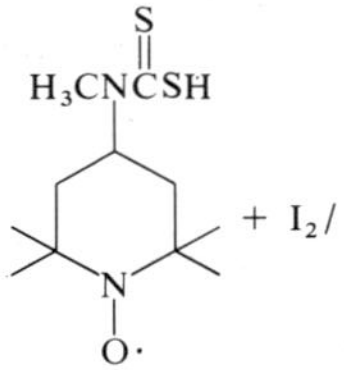 $H_3CN\overset{S}{\overset{\parallel}{C}}SH$ ... N—O$\cdot$ + I_2 / Several solvent	EPR/ 300		3-line spectrum $J \ll a_N$	86Iva1, 84Sol1

[29a]) (a) and (b) refer to two different complexes of aluminum with the bisnitroxide present together in solution.

[30]) Measurement of NO$^\cdot$ – NO$^\cdot$ separation in C_2H_5OH: 12 (1) Å; $C_6H_5C_8H_{17}$: 13.5 Å; DMF: 12.5 (1) Å; methyl cellosolve-H_2O: 14.5 Å. Measurements in liquid crystal.

Substance	Generation/ Matrix or Solvent	Method/ T[K]	g-Factor	a-Value, D, E, J [mT]	Ref./ add. Ref.
$[C_{32}H_{42}N_4O_6]^{2\cdot}$	Ox. of corresp. diamine with *m*-chloroperbenzoic acid/ H_2O	EPR/ 300		3-line spectrum $a_N \gg J$	84Mat1

22.2.8 ·O–N(piperidine)–X–(piperidine)N–O·, X=7 atoms

Substance	Generation/ Matrix or Solvent	Method/ T[K]	g-Factor	a-Value, D, E, J [mT]	Ref./ add. Ref.
$[C_{22}H_{40}N_4O_2S_4Zn]^{2\cdot}$	4-(NHCH$_3$)-piperidinyl-1-oxyl + CS_2 + NaOH + $ZnCl_2$/ $C_6H_5CH_3$	EPR/ 300		3-line spectrum $a_N \gg J$	82Sol1
$[C_{22}H_{40}HgN_4O_2S_4]^{2\cdot}$	Treatment of corresp. Fe(II) complex with $HgCl_2$/ $CHCl_2$	EPR/ 300		3-line spectrum $a_N \gg J$	88Sol1 [31a]

[31a]) Pb^{2+} complex gives similar spectrum.

Substance	Generation/ Matrix or Solvent	Method/ T [K]	g-Factor	a-Value, D, E, J [mT]	Ref./ add. Ref.
$[C_{25}H_{44}N_2O_6]^{2\cdot}$ 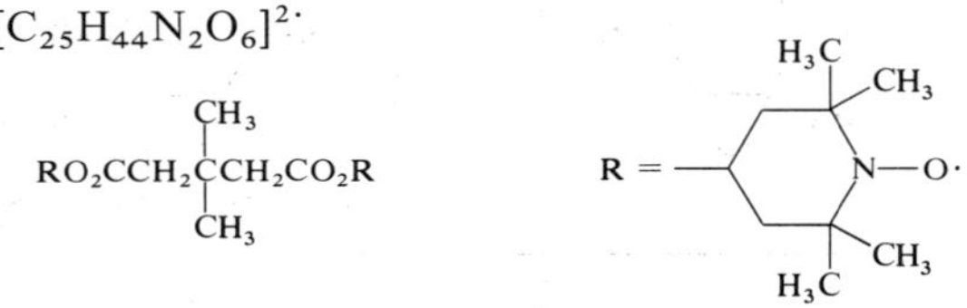 $RO_2CCH_2C(CH_3)_2CH_2CO_2R$	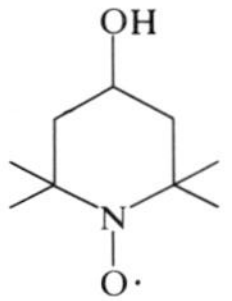 + $ClCOCH_2C(CH_3)_2CH_2COCl$/ Methyl cellosolve-water	EPR/ –		No details	80Bar1 [31])
$[C_{27}H_{42}N_4O_2]^{2\cdot}$ 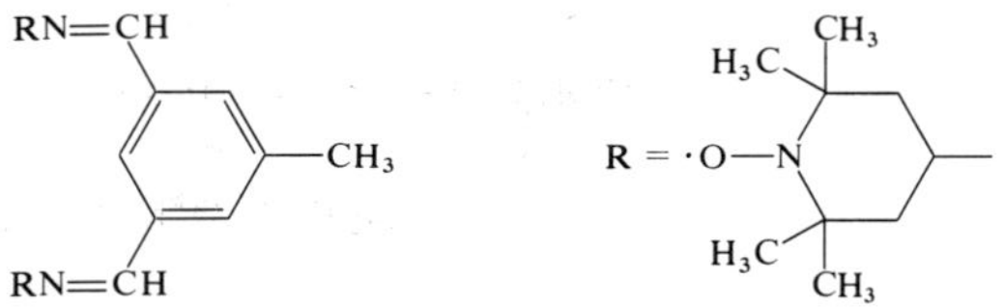	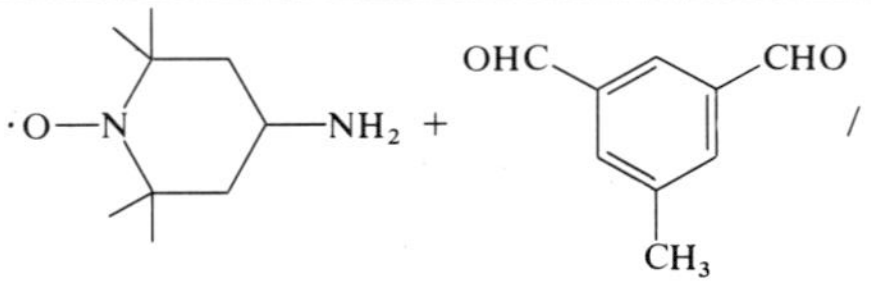 $-NH_2$ + / $C_6H_5CH_3$	EPR/ 298		3-line spectrum $J \ll a_N$	85Med1

22.2.9 ·O–N ... X ... N–O·, X=8 atoms

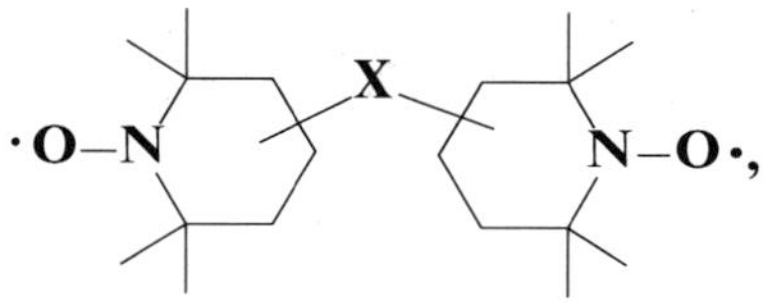

Substance	Generation/ Matrix or Solvent	Method/ T [K]	g-Factor	a-Value, D, E, J [mT]	Ref./ add. Ref.
$[C_{24}H_{42}N_2O_6]^{2\cdot}$ 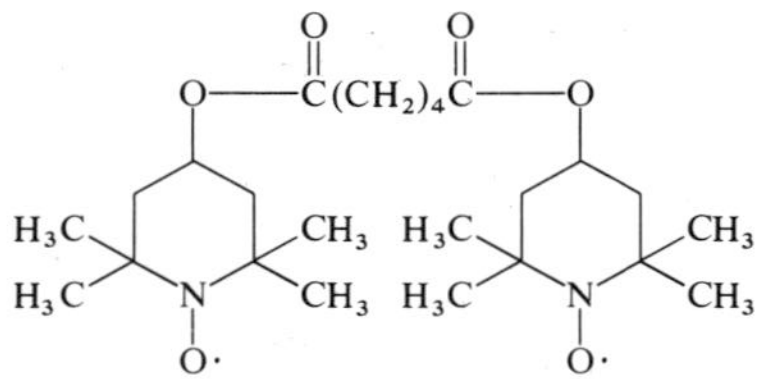	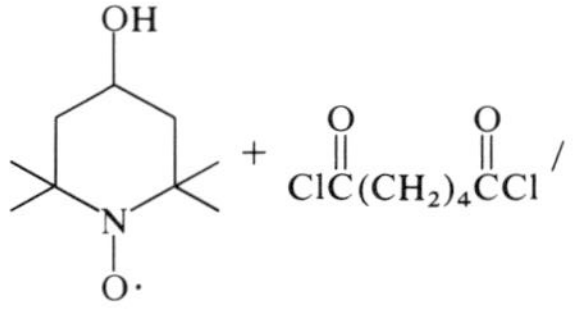 + $ClC(O)(CH_2)_4C(O)Cl$/	EPR/			86Har1 [32])
	$H_2O:CH_3CN$	300		5-line spectrum $J \gg a_N$	
	$H_2O:CH_3CN$: β-cyclodextrin	300		3-line spectrum $a_N \gg J$	

[31]) Measurement of NO˙–NO˙ separation in C_2H_5OH: 12 (1) Å; $C_6H_5C_8H_{17}$: 13.5 Å; DMF: 12.5 (1) Å; methyl cellosolve-H_2O: 14.5 Å. Measurements also in liquid crystals.
[32]) Complexation with amylose, methylated amylose and methylated β-cyclodextrin leads to 3-line spectra with exchange interaction decreasing.

Substance	Generation/ Matrix or Solvent	Method/ T[K]	g-Factor	a-Value, D, E, J [mT]	Ref./ add. Ref.
$[C_{28}H_{34}N_2O_2]^{2\cdot}$ 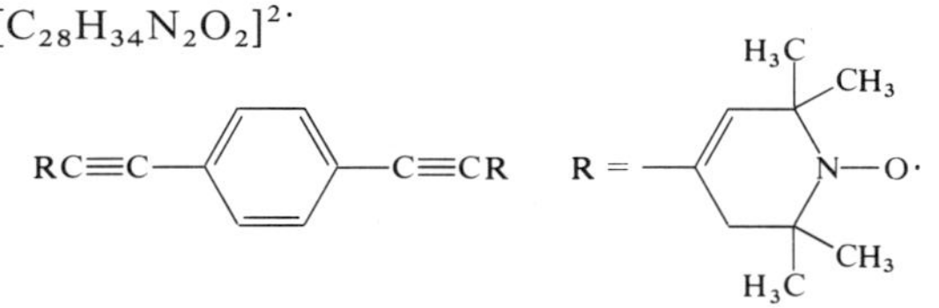 RC≡C–C₆H₄–C≡CR, R = 2,2,6,6-tetramethyl-1-oxyl-1,2,3,6-tetrahydropyridin-4-yl	Ox. of corresp. diamine with pertungstate/ $C_6H_5CH_3$	EPR/ 293		7-line spectrum J/a_N: 2.25	80Pav1, 80Kok1
$[C_{28}H_{46}N_2O_2]^{2\cdot}$ 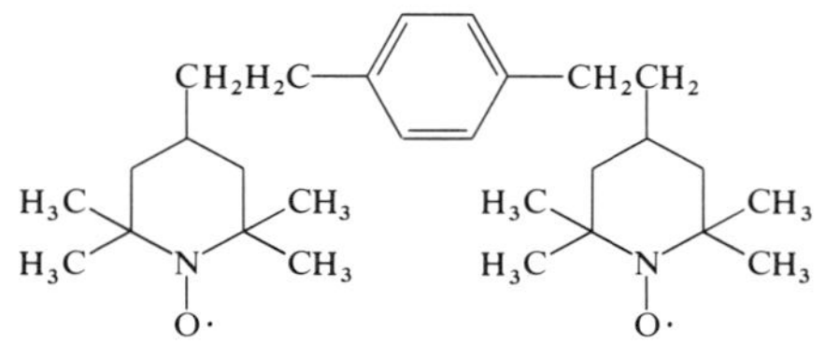	Not given/ $C_6H_5CH_3$	ELDOR, ESE/ 71			84Mil1 [33]

22.2.10 ·O–N⟨ ⟩–X–⟨ ⟩N–O·, X=9 atoms

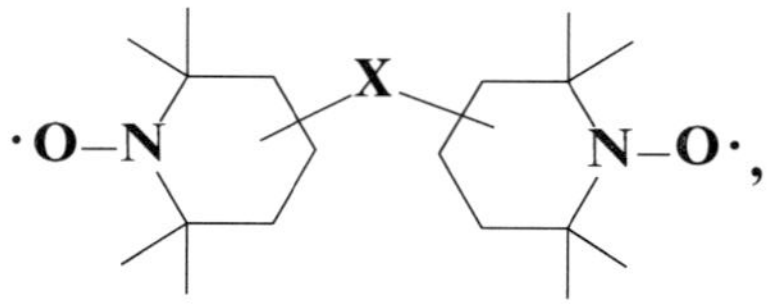

Substance	Generation/ Matrix or Solvent	Method/ T[K]	g-Factor	a-Value, D, E, J [mT]	Ref./ add. Ref.
$[C_{30}H_{40}FeN_2O_6]^{2\cdot}$ $Fe(C_5H_4CO_2R)_2$, R = 2,2,6,6-tetramethyl-1-oxylpiperidin-4-yl 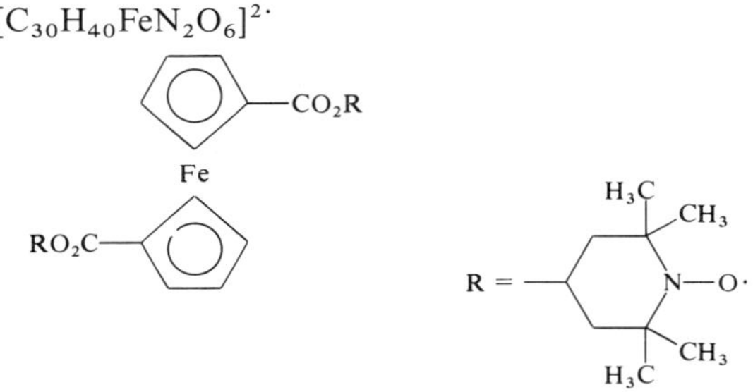	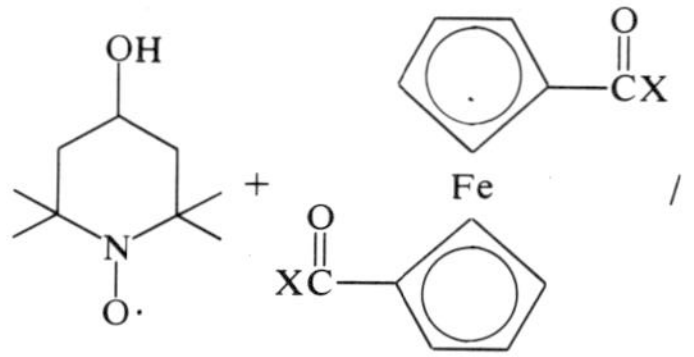 4-hydroxy-2,2,6,6-tetramethylpiperidin-1-oxyl + $Fe(C_5H_4COX)_2$ / $C_6H_5CH_3$ glass	EPR/ –		D: 5.0 [34] D: 16.85 [34]	80Nic1 [35]

[33]) Measurements on radical pairs.
[34]) Two conformations.
[35]) Measurements in decalin, cyclohexane, methylcyclohexane, trimethylbenzene, DMSO, benzaldehyde, propan-2-ol, ethanol, DMF.

22.2.11 ·O–N(piperidinyl)–X–(piperidinyl)N–O·, X = 10 atoms

Substance	Generation/ Matrix or Solvent	Method/ T[K]	g-Factor	a-Value, D, E, J [mT]	Ref./ add. Ref.
$[C_{26}H_{38}HgN_2O_5S_2]^{2\cdot}$ 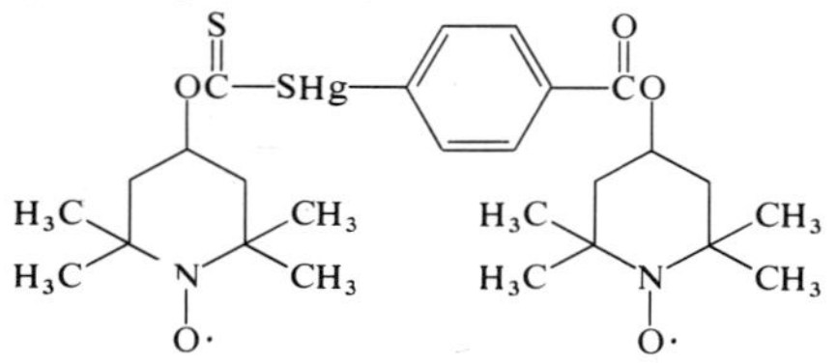	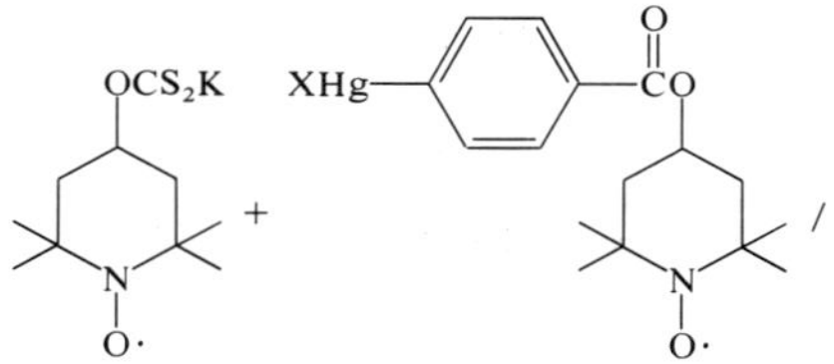 4-(OCS_2K)-2,2,6,6-tetramethylpiperidin-1-oxyl + XHg–C_6H_4–CO–O-(2,2,6,6-tetramethylpiperidin-1-oxyl-4-yl) /	EPR/			86Iva1, 80Sol1
	Several solvents	300		3-line spectrum $a_N \gg J$ N: 1.55	
	Al complex [36]	300		N: 2.00 (4) Al: 1.12 (3)	
	$C_6H_5CH_3$ [36]			N: 2.00 (6) Al: 0.85 (3)	
$[C_{26}H_{46}N_2O_6]^{2\cdot}$ 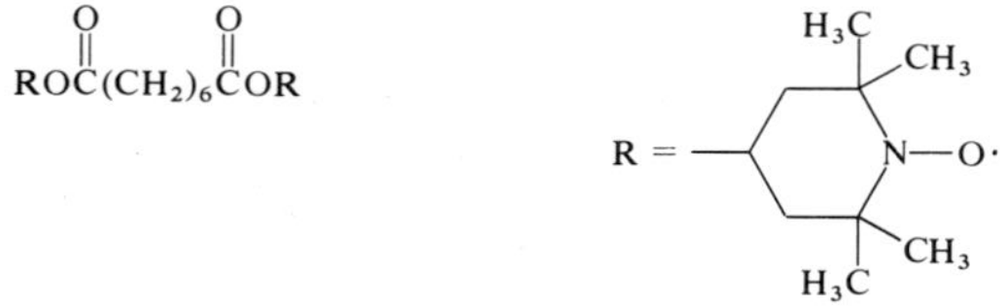 $ROC(O)(CH_2)_6C(O)OR$, R = 2,2,6,6-tetramethylpiperidin-1-oxyl-4-yl	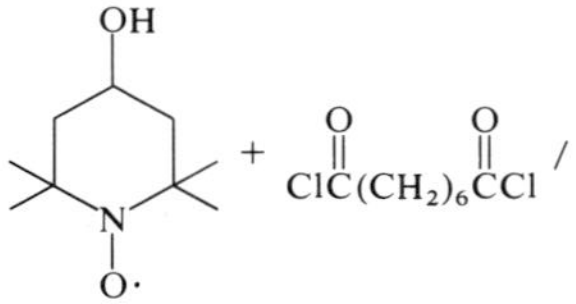 4-OH-2,2,6,6-tetramethylpiperidin-1-oxyl + $ClC(O)(CH_2)_6C(O)Cl$ / –	EPR/ –		No details	79Bro1 [37] 83Bro1 [37]
$[C_{28}H_{50}N_6O_8]^{2\cdot}$ 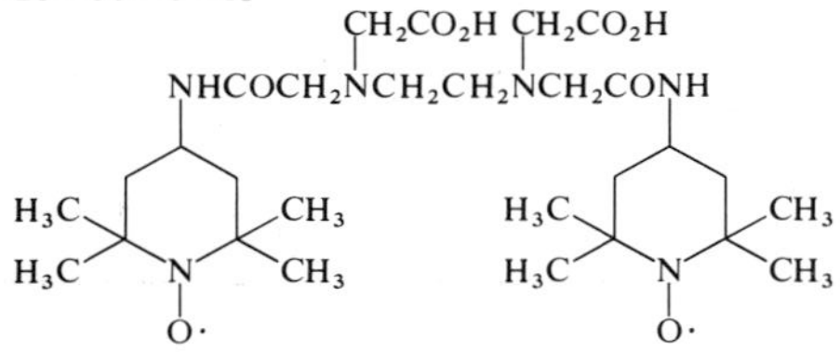 (2,2,6,6-tetramethylpiperidin-1-oxyl-4-yl)$NHCOCH_2N(CH_2CO_2H)CH_2CH_2N(CH_2CO_2H)CH_2CONH$(2,2,6,6-tetramethylpiperidin-1-oxyl-4-yl)	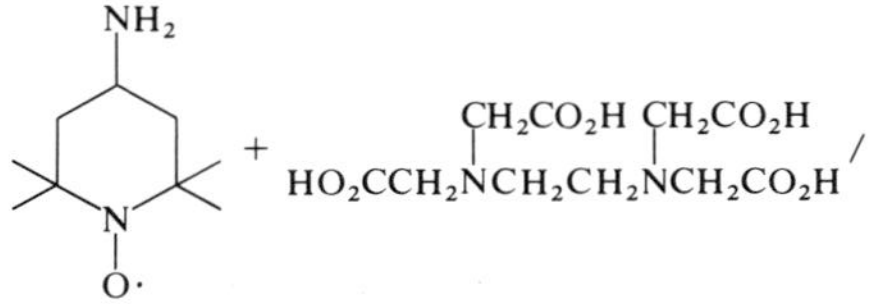 4-NH_2-2,2,6,6-tetramethylpiperidin-1-oxyl + $HO_2CCH_2N(CH_2CO_2H)CH_2CH_2N(CH_2CO_2H)CH_2CO_2H$ / Not given	EPR/ 300		$J \approx a_N$ complex spectrum	86Ehm1

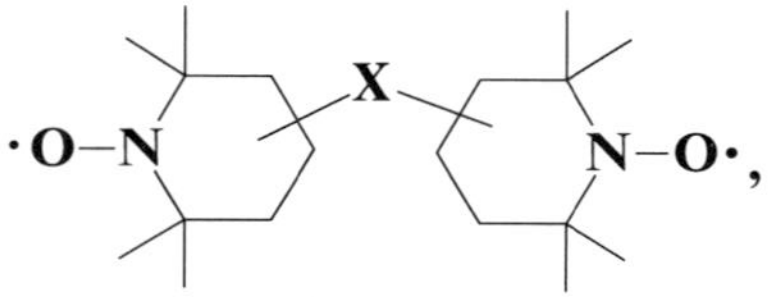

[36]) From [80Sol1].
[37]) Determination of spin densities.

Substance	Generation/ Matrix or Solvent	Method/ T[K]	g-Factor	a-Value, D, E, J [mT]	Ref./ add. Ref.
$[C_{30}H_{54}N_6O_8]^{2\cdot}$	4-(methylamino)-2,2,6,6-tetramethylpiperidin-1-oxyl + $HO_2CCH_2N(CH_2CO_2H)CH_2CH_2N(CH_2CO_2H)CH_2CO_2H$ / Not given	EPR/ 300		3-line spectrum $a_N \gg J$	86Ehm1
$[C_{34}H_{62}N_6O_6S_2]^{2\cdot}$	4-amino-2,2,6,6-tetramethylpiperidin-1-oxyl + DCC + $HOOC-CH(NHCOOC(CH_3)_3)-CH_2-S-S-CH_2-CH(NHCOOC(CH_3)_3)-COOH$ / H_2O, pH = 7.0	EPR/ 300		3-line spectrum $a_N \gg J$	86DiG1

22.2.12 ·O–N(piperidine)–X–(piperidine)N–O·, X = 12 atoms

Substance	Generation/ Matrix or Solvent	Method/ T[K]	g-Factor	a-Value, D, E, J [mT]	Ref./ add. Ref.
$[C_{26}H_{38}N_2O_{10}]^{2\cdot}$	·O–N(tetramethyltetrahydropyridine)–CH_2CO_2H + $ClCOOCH_2CH_2OCOCl$ / C_2H_5OH or C_6H_6	EPR/ 300		5-line spectrum $J \gg a_N$	83Sen1

Substance	Generation/ Matrix or Solvent	Method/ T [K]	g-Factor	a-Value, D, E, J [mT]	Ref./ add. Ref.
$[C_{26}H_{42}N_2O_{10}]^{2\cdot}$ 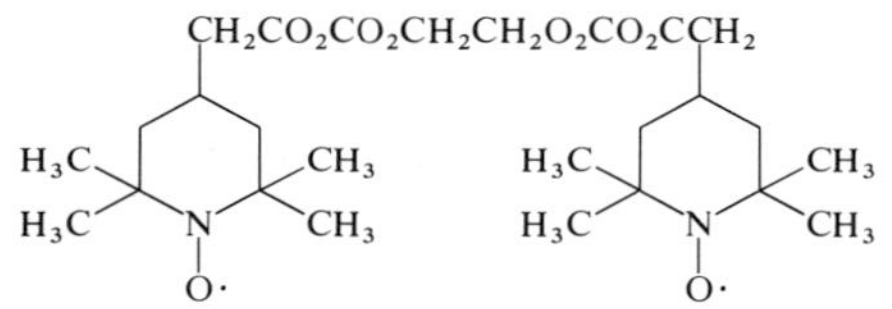	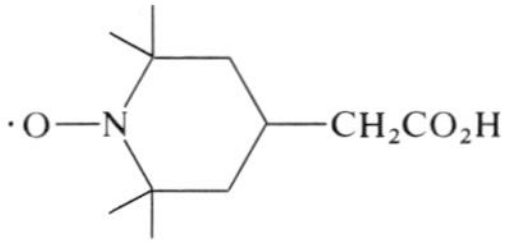 + $ClCO_2CH_2CH_2O_2CCl$/ C_2H_5OH or C_6H_6	EPR/ 300		5-line spectrum $J \gg a_N$	83Sen1

22.2.13 ·O–N…X…N–O·, X=14 atoms

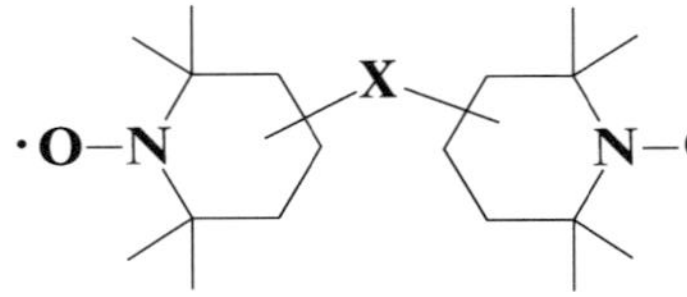

Substance	Generation/ Matrix or Solvent	Method/ T [K]	g-Factor	a-Value, D, E, J [mT]	Ref./ add. Ref.
$[C_{34}H_{58}N_4O_{14}]^{2\cdot}$ 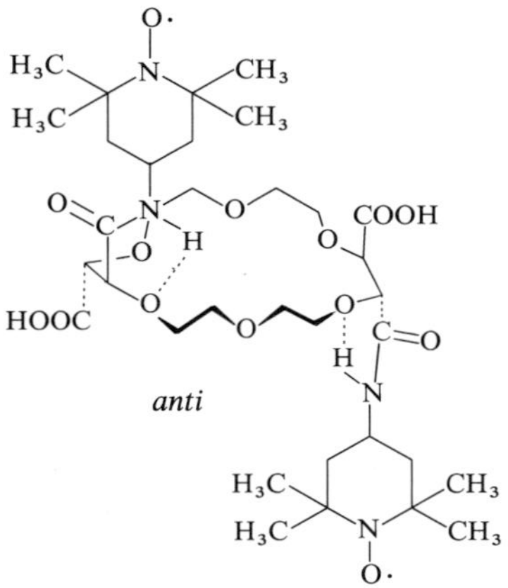 *anti*	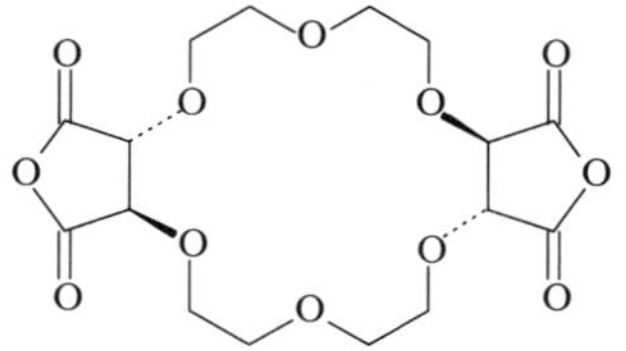 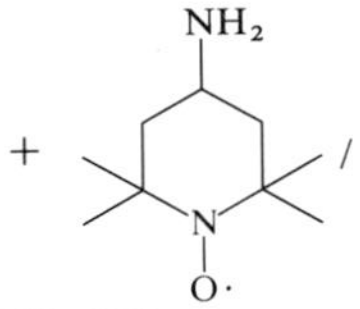 + (4-amino-TEMPO)/	EPR/			82Dug1, 84Dug1
	C_2H_5OH	293		N: 1.62	
	C_2H_5OH: KSCN	293		3-line spectrum $a_N \gg J$	
	H_3O 38)	300		3-line spectrum $a_N \gg J$	

38) From [84Dug1].

Substance	Generation/ Matrix or Solvent	Method/ T[K]	g-Factor	a-Value, D, E, J [mT]	Ref./ add. Ref.
$[C_{34}H_{58}N_4O_{14}]^{2\cdot}$ 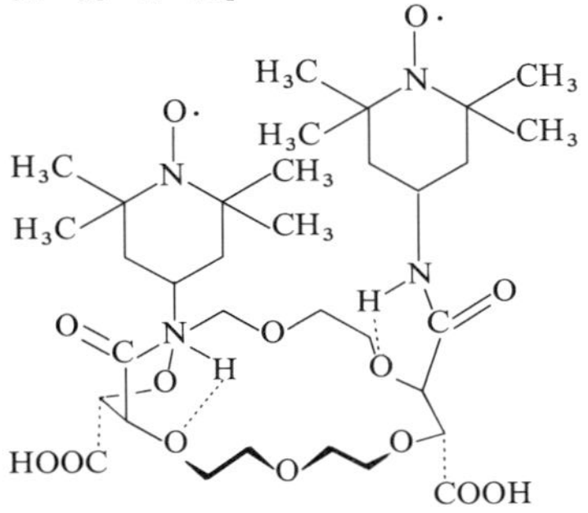 *syn*	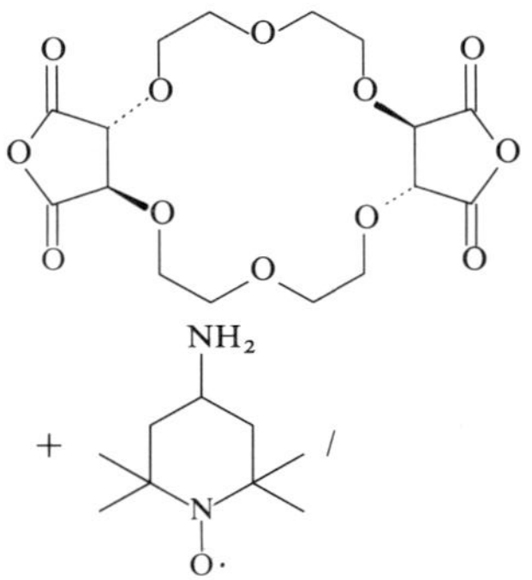 + /	EPR/			82Zhd1, 84Dug1[41])
	C_2H_5OH	293		N: 1.62 $a_N \gg J$	
	C_2H_5OH: KSCN	293		$J \geqq a_N$	
	C_2H_5OH	110		$2D$: 12.1 (4)[39])	
	C_2H_5OH:KSCN	110		$2D$: 15.3 (5)[40])	
	H_2O[38])	293		5-line spectrum $J \approx 2\,a_N$	
$[C_{36}H_{62}N_4O_{14}]^{2\cdot}$ 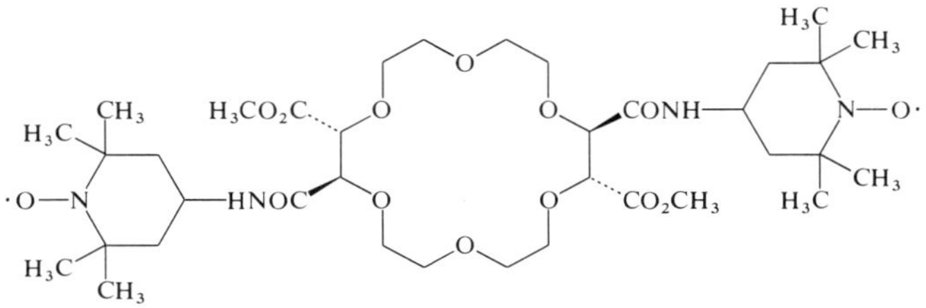	Corresp. acid + CH_2N_2/ H_2O	EPR/ –		5-line spectrum $J \approx a_N$	84Dug1

38) From [84Dug1].
39) $r = 7.4$ (2) Å point-dipole approximation.
40) $r = 6.8$ (2) Å point-dipole approximation.
41) Measurements also in C_2H_5OH, $CHCl_3$, DMSO. Variable temperature measurements and pH variable measurements.

Substance	Generation/ Matrix or Solvent	Method/ T[K]	g-Factor	a-Value, D, E, J [mT]	Ref./ add. Ref.
22.2.14 ·O–N(piperidine)–X–(piperidine)N–O·, X=16 atoms					
$[C_{38}H_{38}N_2O_2]^{2\cdot}$ RC≡C–C_6H_4–C≡C–C≡C–C_6H_4–C≡CR R = 2,2,6,6-tetramethyl-1-oxyl-tetrahydropyridin-4-yl (H_3C, CH_3, N–O·, CH_3, H_3C)	Ox. of corresp. diamine with pertungstate/ $C_6H_5CH_3$	EPR/ 293		3-line spectrum $a_N \gg J$ a_N/J: $\ll 0.1$	80Pac1, 80Kok1
22.2.15 ·O–N(piperidine)–X–(piperidine)N–O·, X=17 atoms					
$[C_{38}H_{38}HgN_2O_2]^{2\cdot}$ RC≡C–C_6H_4–C≡CHgC≡C–C_6H_4–C≡CR R = 2,2,6,6-tetramethyl-1-oxyl-tetrahydropyridin-4-yl (H_3C, CH_3, N–O·, CH_3, H_3C)	RC≡C–C_6H_4–C≡CH + $HgCl_2$ + KI/ $C_6H_5CH_3$	EPR/ 323		3-line spectrum J/a_N: $\ll 1$	80Pav1, 80Kok1

22.3 (Pyrrolinyl-1-oxyl)-(piperidinyl-1-oxyl), ·O–N X N–O·

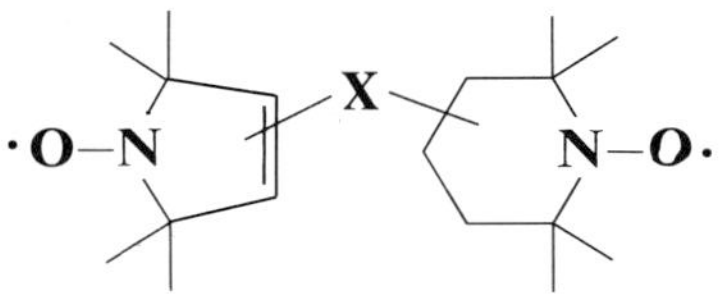

Substance	Generation/ Matrix or Solvent	Method/ T [K]	g-Factor	a-Value, D, E, J [mT]	Ref./ add. Ref.
$[C_{18}H_{26}Br_2N_2O_4]^{2\cdot}$	+ $(C_2H_5)_3N$/	EPR/			86Bla1 [1])
	CH_3COCH_3	300		9-line spectrum J: 2.6 J/a_N: 1.7	
	$CHCl_3$	100		D: 2.8	
$[C_{18}H_{31}N_3O_2]^{2\cdot}$ 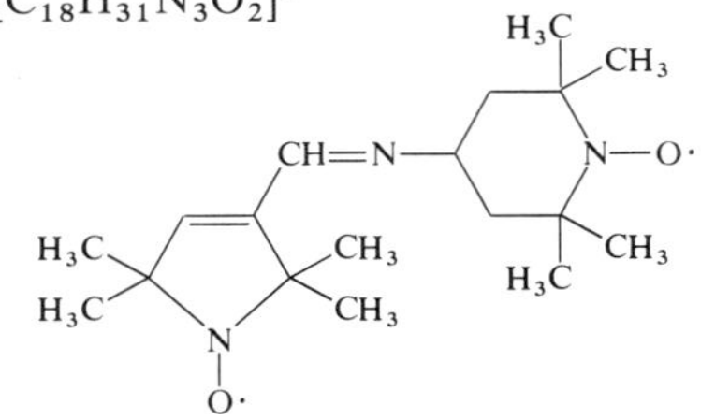	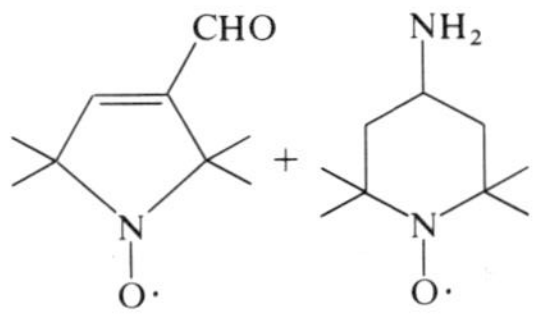 + /	EPR/ 300		5-line spectrum N: 1.50 $J \gg a_N$	80Hid1
	$CHCl_3$				
$[C_{18}H_{31}N_3O_3]^{2\cdot}$	+ + DCC/	EPR/			86Bla1
	CH_3COCH_3	300	2.006	5-line spectrum $J > a_N$ N: 1.46 $J/a_N = 3.6$ [2])	
	CH_3OH	143			

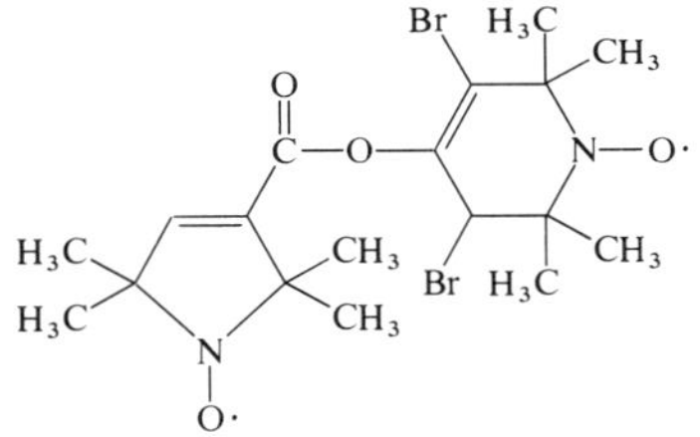

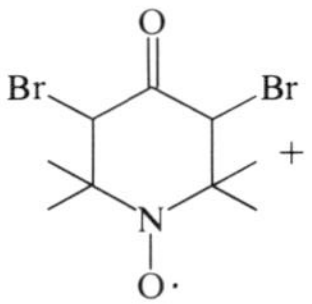

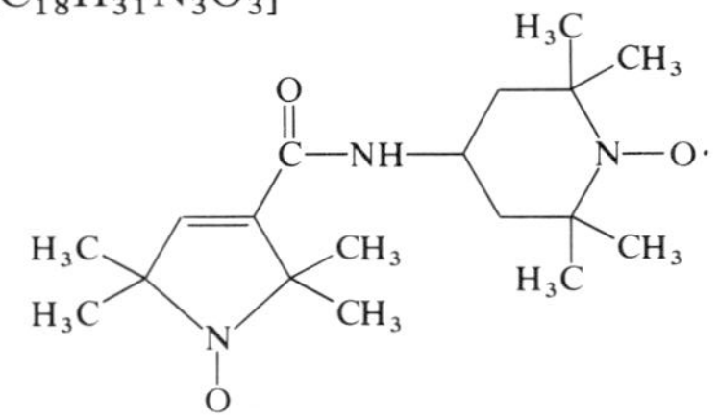

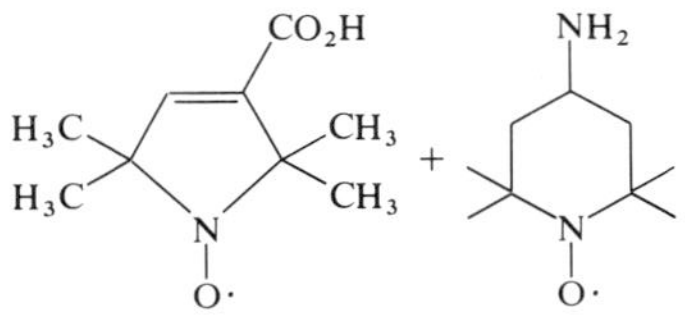

[1]) r calculated using point-dipole approximation 10.0 (1) Å.
[2]) r calculated using point-dipole approximation.

Substance	Generation/ Matrix or Solvent	Method/ T[K]	g-Factor	a-Value, D, E, J [mT]	Ref./ add. Ref.
$[C_{23}H_{35}N_3O_4]^{2\cdot}$ 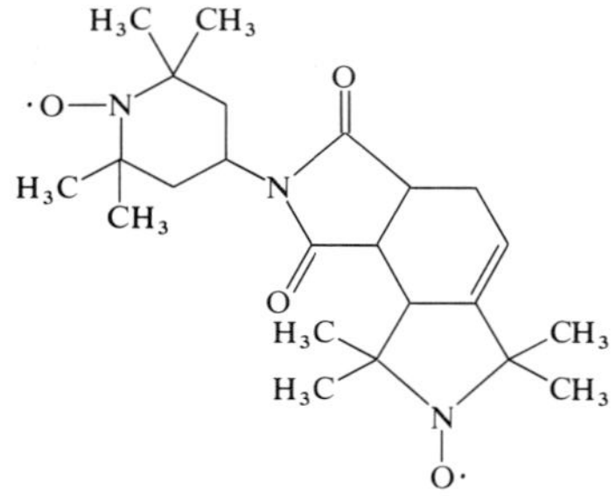	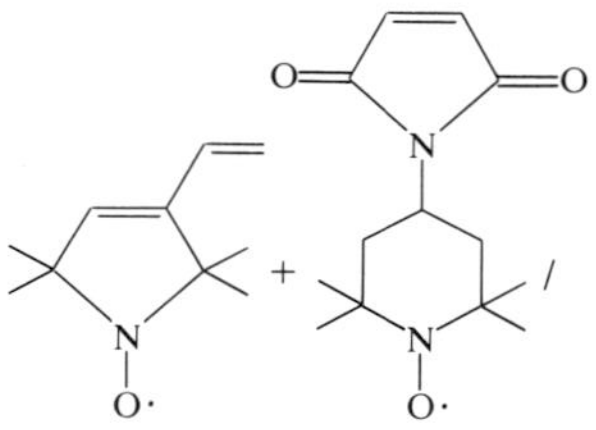 + / Not given	EPR/ 300		5-line spectrum N: 1.742 $J \gg a_N$	86Hid1

22.4 (Piperidinyl-1-oxyl)-(phenoxyl)

Substance	Generation/ Matrix or Solvent	Method/ T[K]	g-Factor	a-Value, D, E, J [mT]	Ref./ add. Ref.
$[C_{24}H_{37}NO_4]^{2\cdot}$ 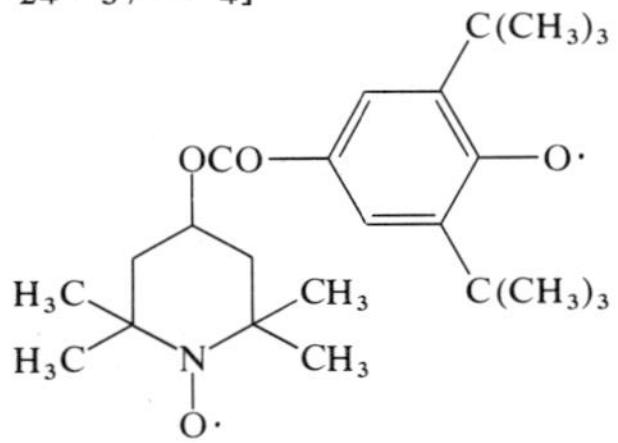	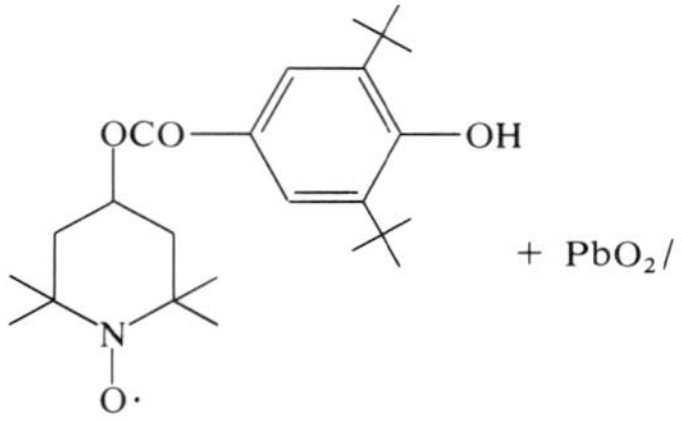 + PbO_2/ $C_6H_5CH_3$	EPR/ 300	2.0053	N: 1.52 $J \gg a_N$	81Ji1
$[C_{37}H_{57}NO_3]^{2\cdot}$ 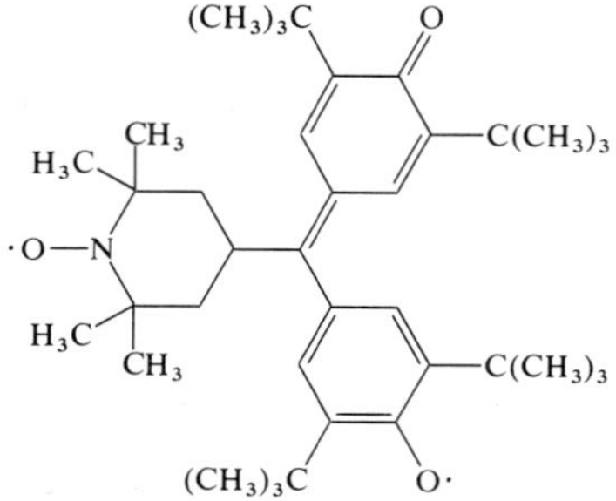	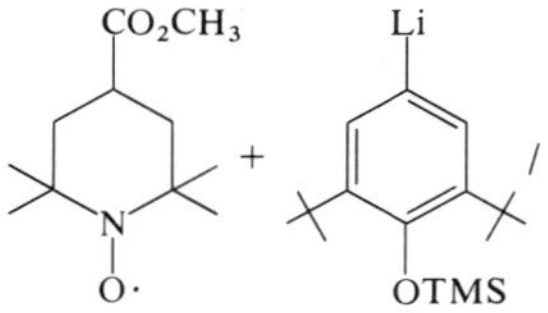 + / $C_6H_5CH_3$	EPR, ENDOR/ 290 160		$J \gg a_N$ N: +1.54 (2) D: 7.14	82Kir1

Substance	Generation/ Matrix or Solvent	Method/ T[K]	g-Factor	a-Value, D, E, J [mT]	Ref./ add. Ref.
$[C_{45}H_{61}NO_5]^{2\cdot}$	$C_6H_5CH_3$	EPR, ENDOR/ 225 225 290 290 160	 2.00542	 H (phenoxy): +0.13 H (nitroxide): −0.0475 N: +1.549 J: −5.17 D: 2.07	82Kir1
$[C_{50}H_{65}NO_5]^{2\cdot}$	$C_6H_5CH_3$	EPR, ENDOR/ 290 225 160		 N: +1.55 J: 0.89 H (phenoxy): +0.13 H (nitroxide): −0.0475 D: < a_N	82Kir1

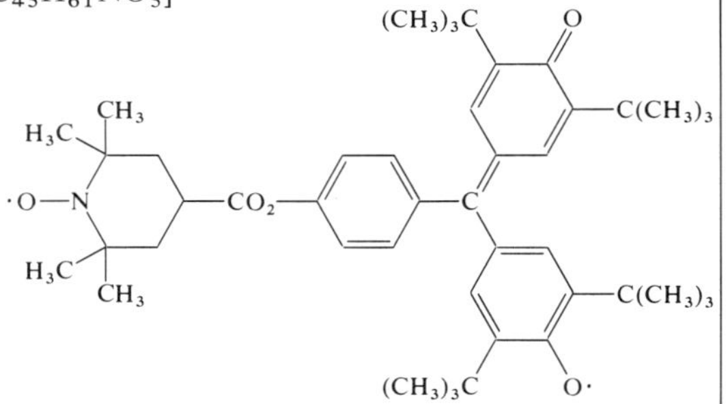

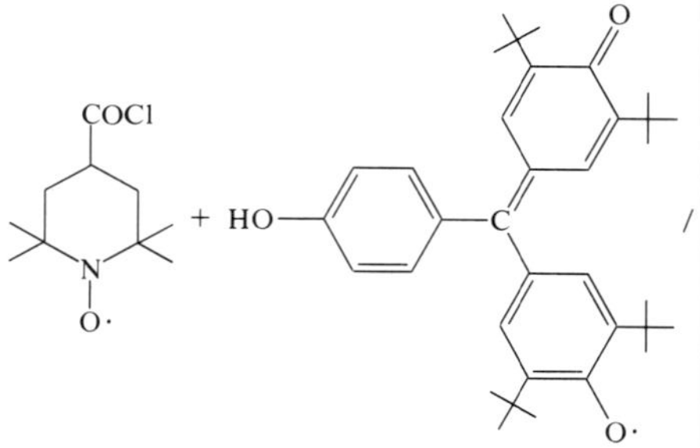

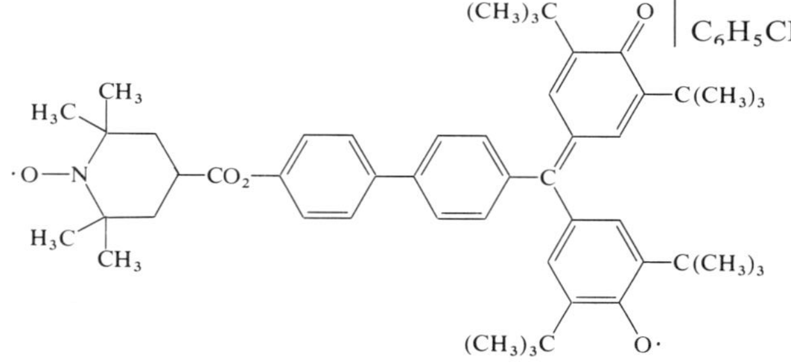

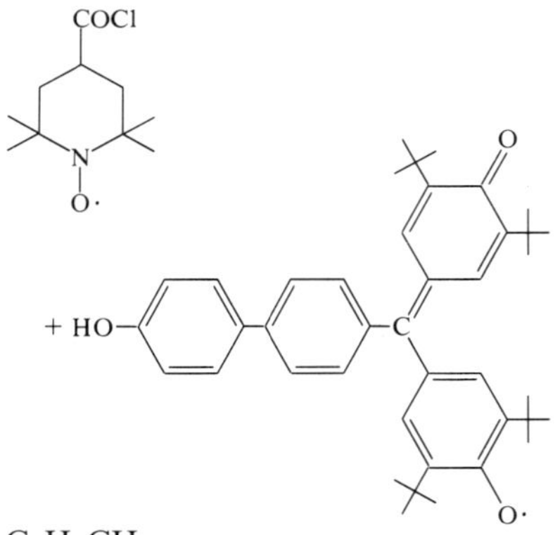

22.5 Trisnitroxides

Substance	Generation/ Matrix or Solvent	Method/ T [K]	g-Factor	a-Value, D, E, J [mT]	Ref./ add. Ref.
$[C_{21}H_{39}Cl_2N_6O_3PdS_3]^{3\cdot}$ 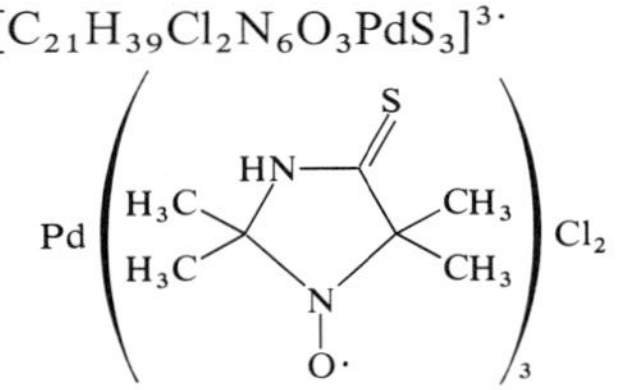	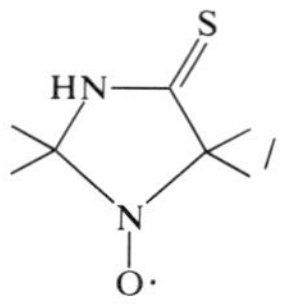 $[PdCl_4]^{2-}$ + (thioxo-imidazolidine nitroxide)/	EPR/ 293		7-line spectrum $J \gg a_N$	88Evs1 [3])
	$CHCl_3$				
$[C_{30}H_{57}N_6O_3]^{3\cdot}$ 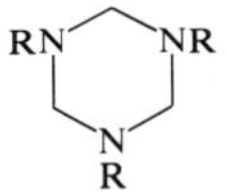 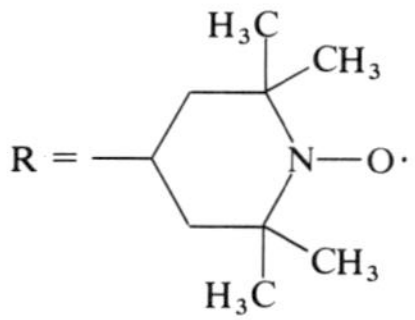 (RN–NR–N–R triazinane; R = 2,2,6,6-tetramethylpiperidin-4-yl-N–O·)	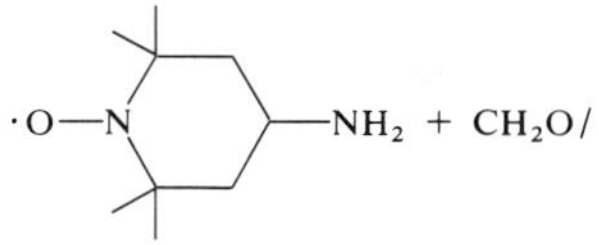 ·O–N(piperidine)–NH_2 + CH_2O/	EPR/			81Gol1 [4])
	$CHCl_3$	298		3-line spectrum $a_N \gg J$	
	C_7H_{16}	298		9-line spectrum $J \gg a_N$	
	–/ H_2O	EPR/ 300		3-line spectrum $a_N \gg J$	87Yud1 [5])
	C_2H_5OH	77		d_1/a: 0.58 (r = 16 Å)	
$[C_{33}H_{63}N_6O_3]^{3\cdot}$ 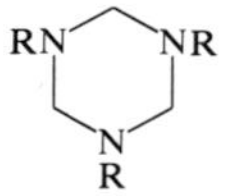 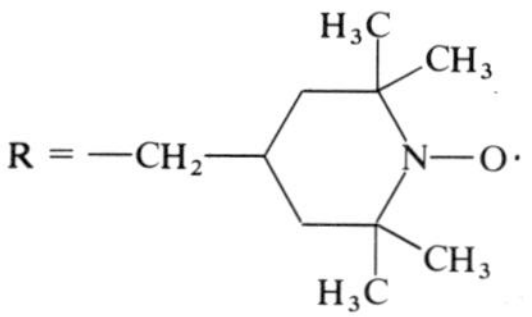 (RN–NR–N–R triazinane; R = —CH_2—(2,2,6,6-tetramethylpiperidin-4-yl-N–O·))	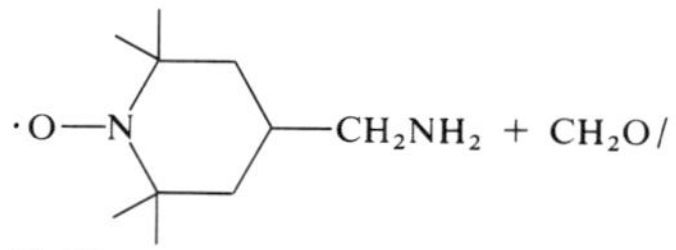 ·O–N(piperidine)–CH_2NH_2 + CH_2O/ C_7H_{16}	EPR/ 298		7-line spectrum $J \gg a_N$ and $J \approx a_N$	82Gol1 [4])

[3]) r calculated as 10.1 (4) Å from low temperature. Glass spectrum probably trigonal bipyramid structure with Cl at apices.
[4]) Mixture of conformers with weak and strong exchange interaction.
[5]) Measurement of spin exchange with $K_3Fe(CN)_6$, $CuSO_4$, $NiCl_2$, $CoCl_2$, ferrihaemoglobin.

Substance	Generation/ Matrix or Solvent	Method/ T [K]	g-Factor	a-Value, D, E, J [mT]	Ref./ add. Ref.
$[C_{36}H_{54}N_3O_9]^{3\cdot}$	$C_6H_5CH_3$	EPR/			81Sho1 [6])
		213		3-line spectrum $a_N \gg J$	
		428		7-line spectrum $J \approx a_N$	
$[C_{36}H_{54}N_3O_9]^{3\cdot}$ 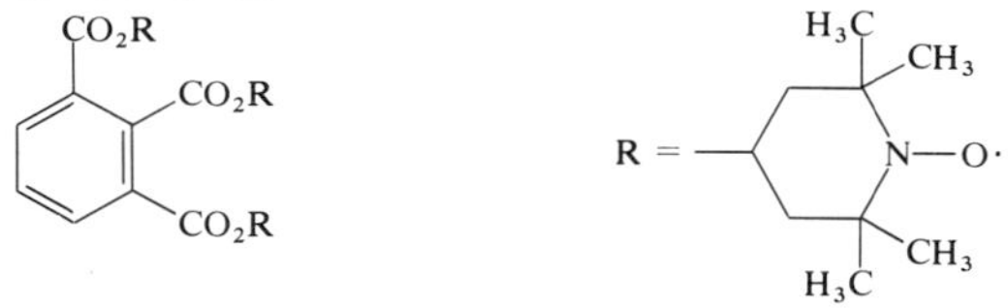	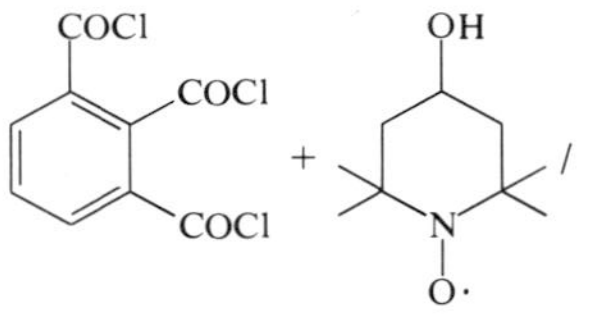 $C_6H_5CH_3$	EPR/ 213…453		7-line spectrum $J \gg a_N$	81Sho1
$[C_{39}H_{60}N_9O_3]^{3\cdot}$ 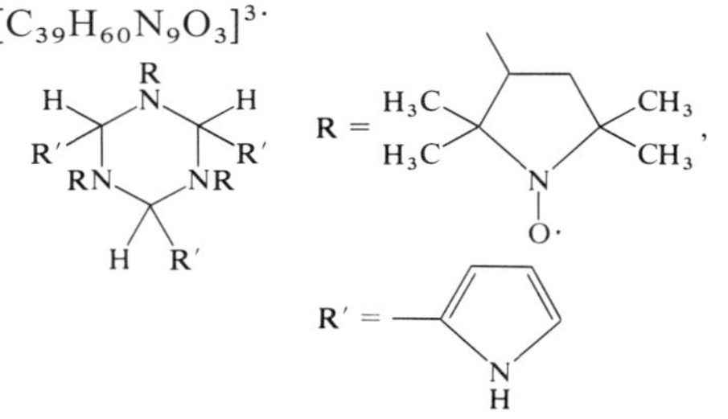	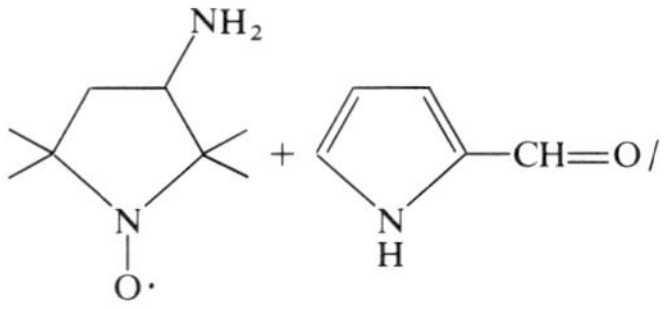 $CHCl_3$	EPR/ 220…320		$J/a \approx 0.2$	84Mam1

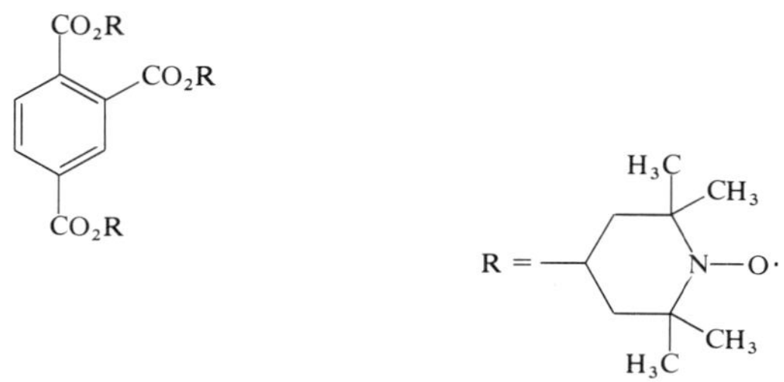

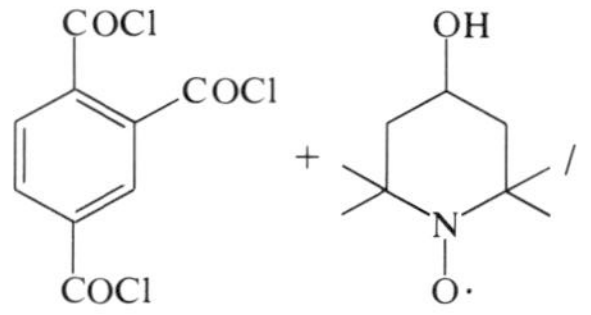

[6]) Variable temperature study.

22.6 Tetrakisnitroxides

Substance	Generation/ Matrix or Solvent	Method/ T [K]	g-Factor	a-Value, D, E, J [mT]	Ref./ add. Ref.
$[C_{28}H_{48}N_8O_4Pd_2S_4]^{4\cdot}$ [7)	$[PdCl_4]^{2-}$ +	EPR/			88Evs1
	$CHCl_3$	293		9-line spectrum $J \gg a_N$	
	$C_6H_5CH_3$	77		$J/a > 300$	
$[C_{46}H_{70}N_4O_{12}]^{4\cdot}$	$C_6H_5CH_3$	EPR/ 253		5-line spectrum $J \approx a_N$ 9-line spectrum $J \gg a_N$	81Sho1 [8)
$[C_{46}H_{70}N_4O_{12}]^{4\cdot}$	$C_6H_5CH_3$	EPR/ 213		5-line spectrum $J \approx a_N$	81Sho1 [8)
		428		9-line spectrum $J \gg a_N$	

[7) Spectra measured at several temperatures. Arrangement of groups around Pd atoms:

[8) Variable temperature study.

Substance	Generation/ Matrix or Solvent	Method/ T[K]	g-Factor	a-Value, D, E, J [mT]	Ref./ add. Ref.
$[C_{46}H_{70}N_4O_{12}]^{4\cdot}$ 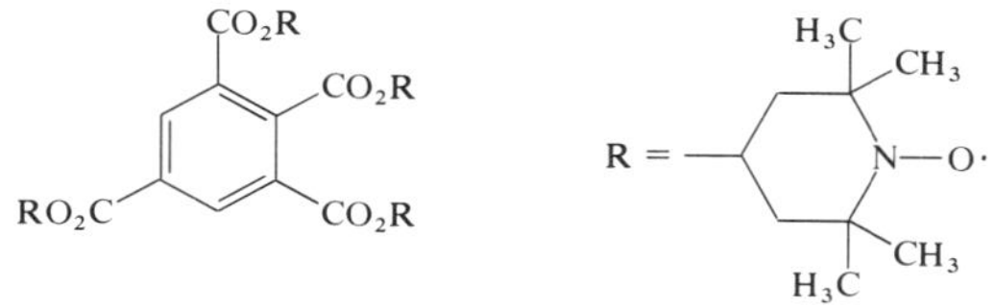	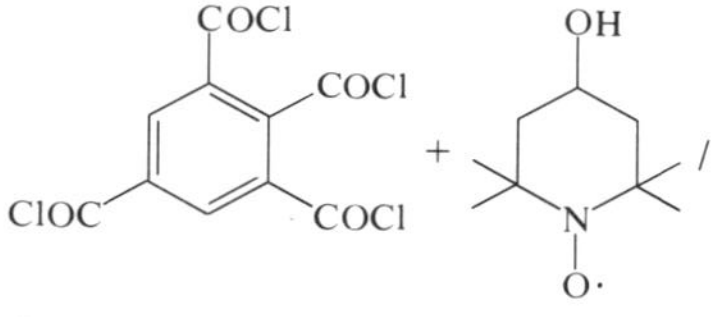 $C_6H_5CH_3$	EPR/ 253 423		7-line spectrum $J \approx a_N$ 9-line spectrum $J \gg a_N$	81Sho1[9])
$[C_{52}H_{92}N_8O_{14}]^{4\cdot}$ 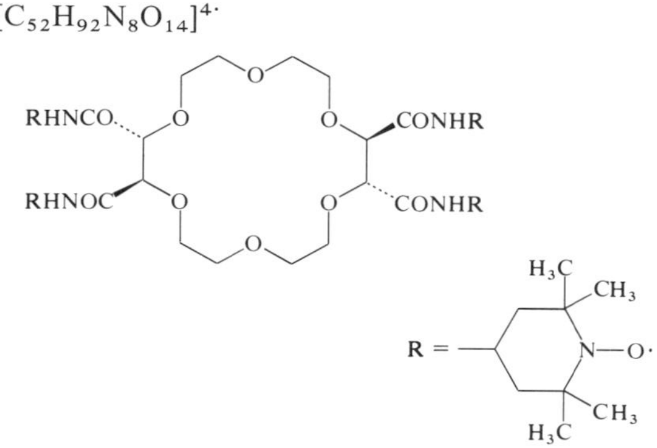	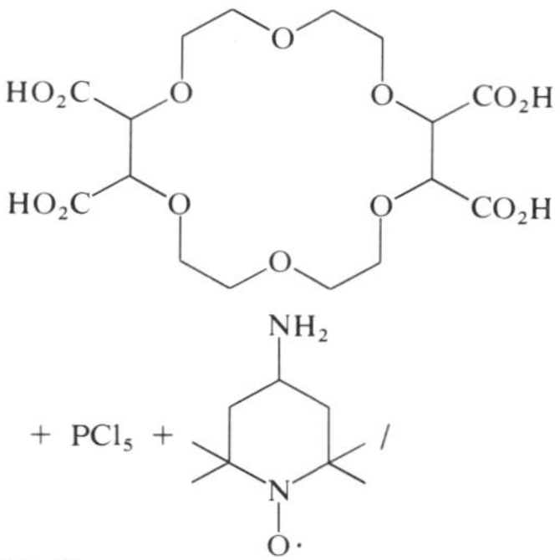 + PCl_5 + H_2O	EPR/ 293		13-line spectrum $J \approx a_N$	84Dug1

22.7 Pentakis- and other poly-nitroxides

Substance	Generation/ Matrix or Solvent	Method/ T[K]	g-Factor	a-Value, D, E, J [mT]	Ref./ add. Ref.
$[C_{56}H_{86}N_5O_{15}]^{5\cdot}$ 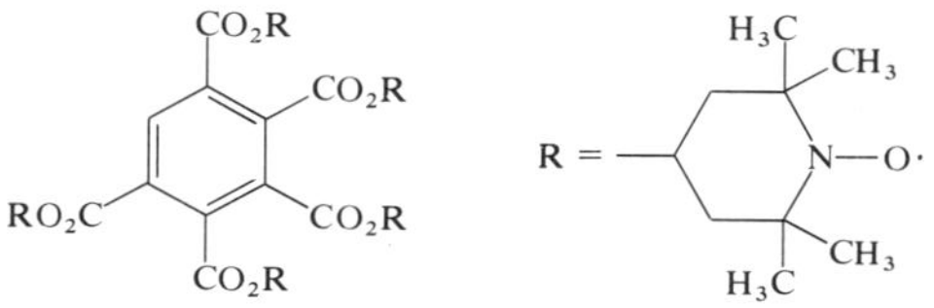	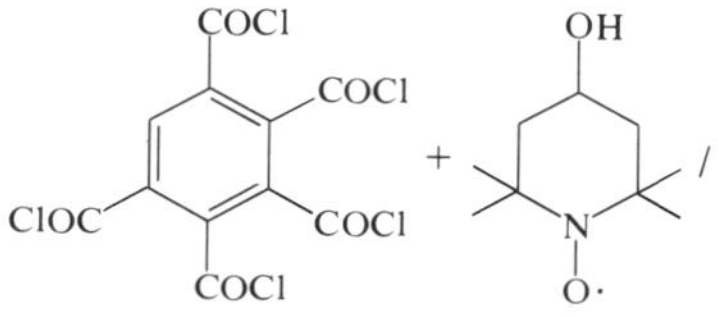 $C_6H_5CH_3$	EPR/ 300 428		broad singlet 11-line spectrum $J \gg a_N$	81Sho1[9])

[9]) Variable temperature study.

Substance	Generation/ Matrix or Solvent	Method/ T[K]	g-Factor	a-Value, D, E, J [mT]	Ref./ add. Ref.
$[C_{66}H_{102}N_6O_{18}]^{6\cdot}$ 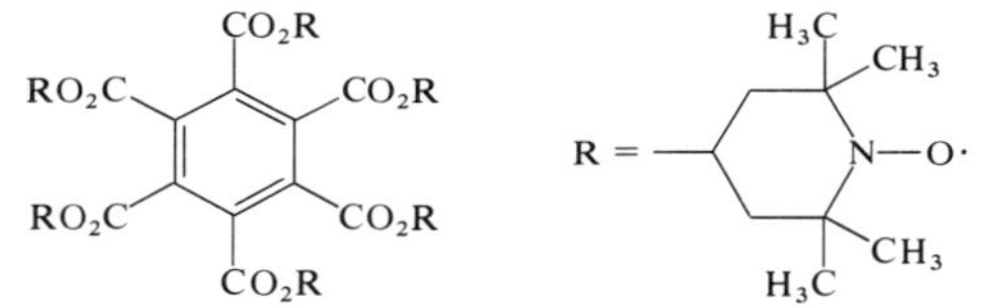	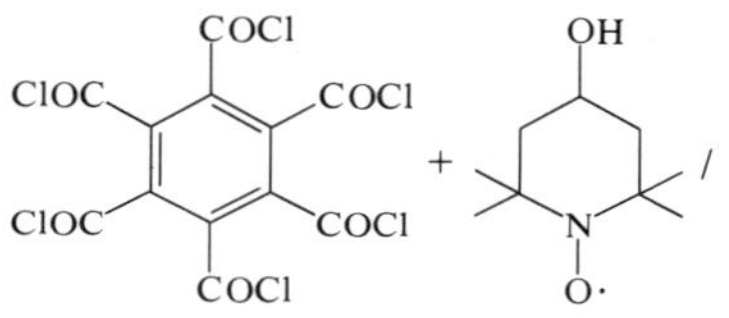 $C_6H_5CH_3$	EPR/ 293 453		broad singlet 13-line spectrum $J \gg a_N$	81Sho1

22.8 Conjugated bisnitroxides

Substance	Generation/ Matrix or Solvent	Method/ T[K]	g-Factor	a-Value, D, E, J [mT]	Ref./ add. Ref.
$[C_{18}H_{26}N_2O_4]^{2\cdot}$ 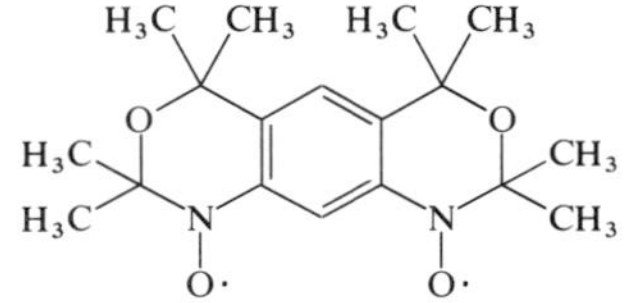	Not given/ C_2H_5OH	Calc. 300		N: 1.1 D: 14.0[10])	80Mic2
$[C_{31}H_{28}N_2O_6]^{2\cdot}$ 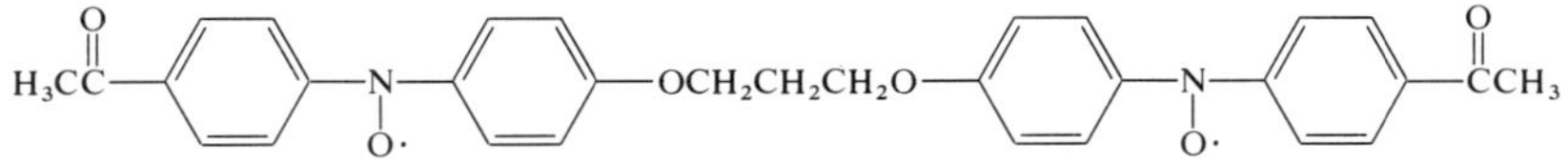	Not given/ –	EPR/ –			81Kot1[11])

[10]) D calculated as 14.4 and 13.8 mT.
[11]) Measurement of exchange interaction from IR band intensities.

22.9 References for 22

76Vla1 Vlasova, S.A., Surnakova, N.E.: Mekh. Sens. Retseptsii (Mater. Vses. Simp. 3rd, **1976**, p. 31; C.A. **91** (1979) 34213.

78Kon1 Konieczny, M., Sosnovsky, G.: Z. Naturforsch. **33b** (1978) 1040.

79Bro1 Brown, P.J., Capiomont, A., Gillon, B., Schweizer, A.: J. Magn. Magn. Mater. **14** (1979) 289.

79Mat1 Mathew, A., Bergquist, B., Zimbrick, J.: J. Chem. Soc., Chem. Commun. **1979**, 222.

79Mor1 Morat, C., Rassat, A.: Tetrahedron Lett. **1979**, 4561.

79Ovc1 Ovcharenko, V.I., Larionov, S.V., Mironova, G.N., Volodarskii, L.B.: Bull. Acad. Sci. USSR (English Transl.) **28** (1979) 599.

79Zhd1 Zhdanov, R.I., Koltover, V.K., Shveta, V.A.: Dokl. Akad. Nauk SSSR **245** (1979) 242.

80Bar1 Barbarin, F., Fabre, C., Germain, J.P., Michon, J., Rassat, A., Tchapla, A.: Nouv. J. Chim. **4** (1980) 437.

80Bri1 Briere, R., Giroud, A.-M., Rassat, A., Rey, P.: Bull. Soc. Chim. France II **1980**, 147.

80Chi1 Chiarelli, R., Jeunet, A., Michon, J., Michon, P., Morat, C., Rassat, A., Sieveking, H.U.: Org. Magn. Reson. **13** (1980) 216.

80Hid1 Hideg, K., Hankovszki, H.O., Lex, L., Kulcsar, G.: Synthesis **1980**, 911.

80Kok1 Kokorin, A.I., Pavlikov, V.V., Shapiro, A.B.: Dokl. Akad. Nauk SSSR **253** (1980) 147.

80Kok2 Kokorin, A.I., Vlasova, S.A.: Biophys. **25** (1980) 92.

80Lar1 Larionov, S.V., Mironova, G.N., Ovcharenko, V.I., Volodarskii, L.B.: Bull. Acad. Sci. USSR (English Transl.) **29** (1980) 686.

80Mic1 Michon, J., Rassat, A.: Tetrahedron **36** (1980) 871.

80Mic2 Michon, J., Michon, P., Rassat, A.: Nouv. J. Chim. **4** (1980) 523.

80Mys1 Myshkina, L.A., Rozynov, B.V., Rozantsev, E.D.: Izv. Akad. Nauk SSSR, Ser. Khim. **1980**, 1416.

80Nic1 Nickel, B., Rassat, A.: Tetrahedron Lett. **21** (1980) 2409.

80Par1 Parmon, V.N., Kokorin, A.I., Zhidomirov, G.M.: Stable Biradicals; Moscow: Nauka **1980**.

80Pav1 Pavlikov, V.V., Shapiro, A.B., Rozantsev, E.G.: Bull. Acad. Sci. USSR (English Transl.) **29** (1980) 113.

80Pav2 Pavlikov, V.V., Rozantsev, E.G., Shapiro, A.B., Sholle, V.D.: Izv. Akad. Nauk SSSR **1980**, 197.

80Pav3 Pavlikov, V.V., Murav'ev, V.V., Shapiro, A.B., Taits, S.Z., Rozantsev, E.G.: Izv. Akad. Nauk SSSR **1980**, 1200.

80Sha1 Shapiro, A.B., Deltsova, D.P., Kokorin, A.I., Rozantsev, E.G.: Izv. Akad. Nauk SSSR **1980**, 413.

80Sha2 Shapiro, A.B., Volodarskii, L.B., Krasochka, O.N., Atovmyan, L.O.: Doklady Chem. (English Transl.) **254** (1980) 473.

80Sol1 Solozhenkin, P.M., Ivanov, A.V., Shvengler, F.A.: Doklady Chem. (English Transl.) **254** (1980) 432.

81Dmi1 Dmitriev, P.I., Shapiro, A.B.: Bull. Acad. Sci. USSR (English Transl.) **30** (1981) 2175.

81Esp1 Espi, J.-C., Ramasseul, R., Rassat, A., Rey, P.: Bull. Soc. Chim. France II **1981**, 33.

81Han1 Hankovsky, H.O., Hideg, K., Lex, L.: Synthesis **1981**, 147.

81Ji1 Ji, A.X., Li, C.X., Mukai, K., Yano, H., Ishizu, K.: Tetrahedron Lett. **22** (1981) 1903.

81Kok1 Kokorin, A.I., Shubin, A.A., Parmon, V.N.: Bull. Acad. Sci. USSR (English Transl.) **30** (1981) 1660.

81Kok2 Kokorin, A.I., Novozhilova, G.A., Shapiro, A.B.: Bull. Acad. Sci. USSR (English Transl.) **30** (1981) 215.

81Kok3 Kokorin, A.I., Parmon, V.N., Ovcharenko, V.I., Larionov, S.V., Volodarskii, L.B.: Bull. Acad. Sci. USSR (English Transl.) **30** (1981) 1433.

81Kot1 Kotorlenko, L.A., Aleksandrova, V.S.: Teor. Eksp. Khim. **17** (1981) 273.

81Muk1 Mukai, K., Yano, H., Ishizu, H.: Tetrahedron Lett. **22** (1981) 1903.

81Ovc1 Ovcharenko, V.I., Larionov, S.V.: Russ. J. Inorg. Chem. (English Transl.) **26** (1981) 1477.

81Ras1 Rassat, A., Terech, P.: Bull. Soc. Chim. France II **1981**, 468.

81Sch1 Schwarzhans, K.I., Stuffer, A.: Naturforsch. **36b** (1981) 195.

81Sha1 Shapiro, A.B., Dmitriev, P.I.: Doklady Chem. (English Transl.) **257** (1981) 138.

81Sho1 Sholle, V.D., Prokofev, A.I., Sartova, K.A., Sarymaakov, Sh.S., Rozantsev, E.G.: Bull. Acad. Sci. USSR (English Transl.) **30** (1981) 2138.
81Tka1 Tkacheva, O.P., Marin, V.V., Volodarskii, L.B., Buchachenko, A.L.: Bull. Acad. Sci. USSR (English Transl.) **30** (1981) 1153.
81Yan1 Yang, J.S., Chae, H.J., Yo, C.H.: Taehan Hwahakhoe Chi **25** (1981) 13; C.A. **95** (1981) 42853x.

82Dug1 Dugas, H., Ptak, M.: J. Chem. Soc., Chem. Commun. **1982**, 710.
82Eat1 Eaton, S.S., Eaton, G.R.: J. Am. Chem. Soc. **104** (1982) 5002.
82Gol1 Golubev, V.A., Rashba, Y.E.: Bull. Acad. Sci. USSR (English Transl.) **31** (1982) 2445.
82Gri1 Grigorev, I.A., Dikanov, S.A., Shchukin, G.I., Volodarskii, L.B., Tsvetkov, Yu.D.: J. Struct. Chem. (English Transl.) **23** (1982) 870.
82Kir1 Kirste, B., Krüger, A., Kurreck, H.: J. Am. Chem. Soc. **104** (1982) 3850.
82Lar1 Larionov, S.V., Ovcharenko, V.I., Kirichenko, V.N, Mokhoseva, V.K., Volodarskii, L.B.: Bull. Acad. Sci. USSR (English Transl.) **31** (1982) 7.
82Lar2 Larionov, S.V.: J. Struct. Chem. USSR (English Transl.) **23** (1982) 594.
82Sha1 Shapiro, A.B., Lobkovskaya, R.M., Malkin, Y.N., Shibaeva, R.P.: Doklady Chem. (English Transl.) **264** (1982) 157.
82Sol1 Solozhenkin, P.M., Shvengler, F.A. Kopitsya, N.I., Semikopnyi, A.I.: Doklady Chem. (English Transl.) **262** (1982) 48.
82Sol2 Solozhenkin, P.M., Shapiro, A.B., Shvengler, F.A.: Dokl. Akad. Nauk SSSR **265** (1982) 137.
82Vla1 Vlasov, S.A., Kokorin, A.I., Ostrovsky, M.A.: Stud. Biophys. **89** (1982) 161.
82Zol1 Zolotov, Yu.A., Bodnya, V.A., Kelareva, M.P., Morosanova, E.I., Volodarskii, L.B., Reznikov, V.A.: Zh. Anal. Khim. **37** (1982) 981.

83Bro1 Brown, P.J., Capiomont, A., Gillon, B., Schweizer, J.: Mol. Phys. **48** (1983) 753.
83Chu1 Chudinov, A.V., Rozantsev, E.G.: Bull. Acad. Sci. USSR (English Transl.) **32** (1983) 361.
83Dor1 Dorfer, A., Schwarzhans, K.E.: Z. Naturforsch. **38b** (1983) 265.
83Eat1 Eaton, S.S., More, K.M., Sawant, B.M., Eaton, G.R.: J. Am. Chem. Soc. **105** (1983) 6560.
83Gri1 Grigorev, I.A., Shchukin, G.I., Volodarskii, L.B.: Bull. Acad. Sci. USSR (English Transl.) **32** (1983) 1030.
83Sen1 Sen, V.D., Kapustina, E.V., Golubev, V.A.: Bull. Acad. Sci. USSR (English Transl.) **32** (1983) 1659.
83Sos1 Sosnovsky, G., Lukszo, J.: Z. Naturforsch. **38b** (1983) 884.

84Dug1 Dugas, H., Keroak, P., Ptak, M.: Can. J. Chem. **62** (1984) 489.
84Mam1 Mamedov, I.M., Manafova, R.A.: Azerb. Khim. Zh. **1984**, 75.
84Mat1 Mathew, A.E., Cheng, C.C.: Tetrahedron Lett. **25** (1984) 269.
84Mil1 Milov, A.D., Ponomarev, A.B., Tsvetkov, Yu.D.: J. Struct. Chem. (English Transl.) **25** (1984) 710.
84Sen1 Sen, V.D., Kapustina, E.V., Golubev, V.A.: Bull. Acad. Sci. USSR (English Transl.) **33** (1984) 1906.
84Sol1 Solozhenkin, P.M., Semikopnyi, A.I., Rakitina, E.V., Burichenko, V.K.: Doklady Chem. (English Transl.) **274** (1984) 4.
84Vor1 Vorobyeva, T.P., Kozlov, Y.M., Kokorin, A.I., Petrov, A.N.: Sov. J. Chem. Phys. (English Transl.) **1** (1984) 2563.

85Dik1 Dikanov, S.A., Shchukin, G.I., Grigorev, A.I., Rukin, S.I., Volodarskii, L.B.: Bull. Acad. Sci. USSR (English Transl.) **34** (1985) 515.
85Evs1 Evstiferov, M.V., Petrukhin, O.M., Kokorin, A.I., Volodarskii, L.B., Zolotov, Yu.A.: Bull. Acad. Sci. USSR (English Transl.) **34** (1985) 59.
85Med1 Medzhidov, A.A., Manofova, R.A., Mamedov, I.M.: J. Gen. Chem. USSR (English Transl.) **55** (1985) 708.
85Mic1 Michon, P., Rassat, A.: Nouv. J. Chim. **9** (1985) 431.
85Mil1 Millar, B.C., Jenkins, T.C., Smithen, C.E., Jinks, S.: Radiat. Res. **101** (1985) 111.
85Sha1 Shapiro, A.B., Suskina, V.I.: Bull. Acad. Sci. USSR (English Transl.) **34** (1985) 378.
85Zak1 Zakharova, I.A., Kurbakova, A.P., Ivanova, N.A., Lokshin, B.V., Prokofev, A.I.: Bull. Acad. Sci. USSR (English Transl.) **34** (1985) 1963.

86Als1 Alster, E., Silver, B.L.: Mol. Phys. **58** (1986) 977.
86Bla1 Blaquiere, C.: J. Mol. Struct. **144** (1986) 377.
86Cla1 Claycamp, H.G., Shaw, E.I., Zimbrick, J.D.:Radiat. Res. **106** (1986) 141.
86DiG1 DiGleria, K., Hill, H.A.O., Page, D.J., Tew, D.G.: J. Chem. Soc., Chem. Commun. **1986,** 460.
86Ehm1 Ehman, R.L., Brasch, R.C., McNamara, M.T., Erikkson, U., Sosnovsky, G., Lukszo, J., Li, S.W.: Invest. Radiat. **21** (1986) 125.
86Har1 Harten, B., Darcy, R., Lyons, R.: Nouv. J. Chim. **10** (1986) 569.
86Hid1 Hideg, K., Cseko, J., Hankovsky, H.O., Sohar, P.: Can. J. Chem. **64** (1986) 1482.
86Iva1 Ivanov, A.V., Solozhenkin, P.M., Shvengler, F.A., Kopitsya, N.I., Semikopnyi, A.I.: Dokl. Akad. Nauk SSSR **289** (1986) 909.
86Kor1 Korshak, Y.V., Ovchinnikov, A.A., Shapiro, A.M., Medvedeva, T.V., Spektor, V.N.: Pisma Zh. Eksp. Teor. Fiz. **43** (1986) 309.
86Sha1 Shapiro, A.B., Suskina, V.I.: Bull. Acad. Sci. USSR (English Transl.) **35** (1986) 1728.
86Smi1 Smirnov, V.A., Kagan, E.Sh., Kondrashov, S.V., Smushkevich, Yu.I.: Doklady Chem. (English Transl.) **288** (1986) 203.

87Buc1 Buchachenko, A.L., Shivaeva, R.P., Rozenberg, L.P., Dadali, A.A.: Khim. Fiz. **6** (1987) 773.
87Des1 Desroches, J., Dugas, H., Bouchard, M., Fyles, T.M., Robertson, G.D.: Can. J. Chem. **65** (1987) 1513.
87Kor1 Korshak, Yu.V., Medvedeva, T.V., Ovchinnikov, A.A., Spektor, V.N.: Nature (London) **326** (1987) 370.
87Mam1 Mamedova, Y.G., Chudinov, A.V., Tuarsheva, Z.O., Rozantsev, E.G.: Zh. Org. Khim. **23** (1987) 563.
87Ovc1 Ovchinnikov, A.A., Cheranovskii, V.O.: Fiz. Tverd. Tela (Leningrad) **29** (1987) 3100.
87Tka1 Tkachev, V.V., Sen, V.D., Atovmyan, L.O.: Bull. Acad. Sci. USSR (English Transl.) **36** (1987) 1829.
87Yud1 Yudanova, E.I., Kulikov, A.V.: Russ. J. Phys. Chem. (English Transl.) **61** (1987) 1084.

88Avd1 Avdeeva, M.I., Ovcharenko, V.I., Larionov, S.V., Volodarskii, L.B.: Bull. Acad. Sci. USSR (English Transl.) **37** (1988) 2328.
88Evs1 Evstiferov, M.U., Vanifatova, N.G., Petrukhin, O.M., Kokorin, A.I., Volodarskii, L.B.: Bull. Acad. Sci. USSR (English Transl.) **37** (1988) 2241.
88Gul1 Gulin, Y.I., Dikanov, S.A., Marin, V.V., Grigorev, I.A., Volodarskii, L.B.: Izv. Sib. Otd. Akad. Nauk SSSR, Ser. Khim. Nauk **1988**, 99.
88Lar1 Larionov, L.A., Patrina, L.A., Dolgoruk, S.N., Boguslavskii, E.G., Kovacik, I., Durasov, V.B.: Bull. Acad. Sci. USSR (English Transl.) **37** (1988) 997.
88Mil1 Miller, J.S., Calabrese, J.C., Glatzhofer, D.T., Epstein, A.J.: J. Appl. Phys. **63** (1988) 2949.
88Mil2 Miller, J.S., Glatzhofer, D.T., Calabrese, J.C., Epstein, A.J.: J. Chem. Soc., Chem. Commun. **1988**, 322.
88Sol1 Solozhenkin, P.M., Semenev, E.V., Semikopnyi, A.I., Shvengler, F.A., Kopitsya, N.I., Frolov, E.N.: Doklady Chem. (English Transl.) **297** (1988) 1141.
88Van1 Vanifatova, N.G., Evstiferov, M.V., Marin, V.V., Petrukhin, O.M., Volodarskii, L.B., Zolotov, Yu.A.: J. Anal. Chem. USSR (English Transl.) **43** (1988) 341.

General symbols and abbreviations

Symbols

a, b, c, d	hyperfine coupling constants for polyatomic radicals in the gas phase. Unit MHz=Mc/s
$\boldsymbol{a}$	hyperfine coupling tensor with elements a_{ij}. Unit milli-Tesla [mT]
a	isotropic coupling constant. Unit milli-Tesla [mT]
Δa	shift of a in liquid crystals, i.e. observed average $\bar{a}=a+\Delta a$. Unit [mT]
$\boldsymbol{B}$	magnetic induction. Unit Tesla [T]
D, E	zero-field splitting parameters. Units [mT] or [cm^{-1}]
$\boldsymbol{g}$	g-tensor with elements g_{ij}
g	isotropic part of g, i.e. mean value of principal elements
g_N	nuclear g-factor
$\mathscr{H}$	spin Hamiltonian operator
$\boldsymbol{I}$	nuclear spin operator
J	exchange coupling parameter. Units [mT] or [cm^{-1}]
L	separation between extreme lines in the spectrum
μ_B	Bohr magneton
μ_{eff}	effective magnetic moment in units μ_B
μ_N	nuclear magneton
r	average distance between the unpaired electrons. Units Å
$\boldsymbol{S}$	electron spin operator
S	total electron spin quantum number
T	temperature in Kelvin [K]

Abbreviations

General

ax	axial
CIDEP	chemical induced dynamic electron polarization
CIDNP	chemical induced dynamic nuclear polarization
corresp.	corresponding
e	electron
ELDOR	electron electron double resonance
ENDOR	electron nuclear double resonance
EPR	electron paramagnetic resonance
ESE	electron spin echo
ESR	electron spin resonance
eV	electron Volt
FDMR	fluorescence detected magnetic resonance
hfcc	hyperfine coupling constant
hfs	hyperfine splitting
INDO	intermediate neglect of differential overlap
irr.	irradiation
is	isotropic
mol.	molecular
MO	molecular orbital
NMR	nuclear magnetic resonance
ox.	oxidation
pH	pH-value
phot.	photolysis
red.	reduction
RT	room temperature
theor.	theoretical
Tris-buffer	tris (hydroxymethyl) aminomethane
UHF	unrestricted Hartree-Fock
UV	ultraviolet
X	X-ray
γ	γ-irradiation
α, β, γ ...	notation of position of proton in radical structure

Substances or part of substances

ACN	acetonitrile
DABCO	diazabicyclooctane
DBNO	di-*t*-butyl nitric oxide
DMF	dimethyl formamide
DTBN	di-*t*-butyl nitroxide
DTBO	di-*t*-butyl oxide
DTBP	di-*t*-butyl peroxide
EDTA	ethylene diamine tetracetic acid
FMN	flavin mononucleotide (riboflavin-5'-phosphate)
HMPA	hexamethyl phosphoric acid triamide
MTHF	2-methyltetrahydrofuran
TCNE	tetracyanoethylene
TCNQ	tetracyanoquinodimethane
THF	tetrahydrofuran
TMS	tetramethylsilane

Index of substances

Preliminary remarks

The index of substances shows the gross formulae and the corresponding volume and page numbers for all substances treated in volumes II/1, II/9 and II/17.

Organic compounds are arranged according to the Hill System, i.e. according to increasing C number, then to increasing H number, and finally alphabetically according to the other element symbols.

Inorganic substances and complex compounds containing C are listed under C atom following consequently the Hill system, while those without any C are listed alphabetically according to the alphabetically ordered gross formula (e.g. $Ag(CO)_3$ is to be found under C_3AgO_3, and $SbFH_3$ under FH_3Sb).

Radicals derived from high-polymers and biological compounds are listed in a separate index after the low-molecular species.

The italic numbers, *1, 9a, 9b, 9c1, 9c2, 9d1, 9d2,* and *17a, 17b, 17c, 17d1, 17d2, 17e, 17f, 17g, 17h* indicate the localization of the radical in the respective subvolumes of II/1, II/9 and II/17. They are followed by the page numbers. The *1, 9a ⋯ 9d2* and *17a ⋯ 17h* will also help to identify an individual compound if several substances have the same gross formula. For the reader's convenience, shortened tables of contents of II/1, II/9a ⋯ d2 and II/17a ⋯ h are preceding the index of substances.

Shortened tables of contents of II/1, II/9 and II/17

II/1: Magnetic properties of free radicals

2.1 Inorganic free radicals and radical ions . . . *1* 9
2.2 Uncharged organic free radicals . . . *1* 20
2.2.1 Carbon radicals . . . *1* 20
2.2.2 Nitrogen radicals . . . *1* 45
2.2.3 Nitric oxides radicals . . . *1* 54
2.2.4 Oxy radicals . . . *1* 57
2.2.5 Thio radicals . . . *1* 76
2.3 Organic negative radical ions . . . *1* 77
2.3.1 Acyclic compounds . . . *1* 77
2.3.2 Isocyclic compounds . . . *1* 79
2.3.3 Heterocyclic compounds . . . *1* 113
2.3.4 Organic compounds of silicon, phosphorus, and boron . . . *1* 120
2.4 Organic positive radicals ions . . . *1* 121
2.4.1 Acyclic compounds . . . *1* 121
2.4.2 Isocyclic compounds . . . *1* 121
2.4.3 Heterocyclic compounds . . . *1* 125
2.5 Organic biradicals and organic donor-acceptor-complexes . . . *1* 128

II/9a: Atoms, inorganic radicals, and radicals in metal complexes

1 Inorganic radicals . . . *9a* 5

2 Radicals in metal complexes . . . *9a* 290

II/9b: Organic C-centered radicals

3 Nonconjugated carbon radicals
3.1 Alkyl radicals . . . *9b* 7
3.2 σ-electronic carbon radicals . . . *9b* 303

4 Carbon radicals with conjugated π-electrons
4.2 Radicals with 3 conjugated π-electrons . . . *9b* 345
4.3 Radicals with 5 conjugated π-electrons . . . *9b* 435
4.4 Radicals with 7 conjugated π-electrons . . . *9b* 535
4.5 Radicals with 9 conjugated π-electrons . . . *9b* 616
4.6 Radicals with 11 conjugated π-electrons . . . *9b* 626
4.7 Radicals with 13 conjugated π-electrons . . . *9b* 627
4.8 Radicals with 19 conjugated π-electrons . . . *9b* 718

II/9c1: Organic N-centered radicals und nitroxide radicals

5 Organic N-centered radicals

5.1 Acyclic aminyl radicals 9c1 6
5.2 Cyclic aminyl radicals 9c1 19
5.3 Acyclic hydrazyl radicals 9c1 40
5.4 Cyclic hydrazyl radicals 9c1 68
5.5 Hydrazonyl radicals 9c1 92
5.6 Acyclic oxyaminyl radicals 9c1 96
5.7 Cyclic oxyaminyl radicals 9c1 107
5.8 Thioaminyl radicals 9c1 107
5.9 Iminyl radicals 9c1 115
5.10 Triazenyl radicals 9c1 121
5.11 Acyclic iminoxyl radicals 9c1 123
5.12 Cyclic iminoxyl radicals 9c1 150
5.13 (Arylazo)oxyls 9c1 178

6 Organic nitroxide radicals

6.1 Dihydro and monoalkyl nitroxides 9c1 195
6.2 Dialkyl nitroxides 9c1 208
6.3 Cyclic dialkyl nitroxides 9c1 385
6.4 Monoaryl nitroxides 9c1 538
6.5 Diaryl nitroxides 9c1 556
6.6 Conjugated cyclic nitroxides 9c1 583
6.7 Alkyl-aryl nitroxides 9c1 610
6.8 Cycloalkyl-aryl nitroxides 9c1 729
6.9 Alkyl and alkoxy-vinyl nitroxides 9c1 752
6.10 Aryl-vinyl and cyclic vinyl nitroxides 9c1 757
6.11 Mono-acyl nitroxides 9c1 762
6.12 Alkyl-acyl nitroxides 9c1 770
6.13 Aryl-acyl nitroxides 9c1 790
6.14 Cycloaryl-acyl nitroxides 9c1 798
6.15 Cycloaryl-acyl nitroxides 9c1 800
6.15A Other conjugated cycloacyl nitroxides 9c1 805
6.16 Diacyl nitroxides 9c1 805
6.17 Alkyl-imidoyl nitroxides 9c1 807
6.18 Aryl-imidoyl nitroxides 9c1 826
6.19 Alkyl-imidoyl cyclonitroxides 9c1 835
6.20 Benz- and conjugated -imidoyl cyclonitroxides 9c1 858
6.21 Nitronyl nitroxides 9c1 862
6.22 Imidazolyl-3-oxyl 1-oxides 9c1 891
6.23 Cyano nitroxides 9c1 895
6.24 Alkyl amino nitroxides 9c1 895
6.25 Cyclic amino nitroxides 9c1 901
6.26 Aryl-amino and -imino nitroxides 9c1 902
6.27 Acyl-amino nitroxides 9c1 908
6.28 Nitroso-aryl nitroxides 9c1 909
6.29 Azo and azoxy nitroxides 9c1 910
6.30 Alkyl-hydroxy nitroxides 9c1 912
6.31 Alkyl alkoxy nitroxides 9c1 914
6.32 Aryl hydroxy nitroxides 9c1 945
6.33 Aryl-oxy nitroxides 9c1 947
6.34 Cyclic oxy nitroxides 9c1 982
6.35 Thio nitroxides 9c1 986
6.36 Alkyl-sulfo, -sufino, and -sulfonyl nitroxides 9c1 989
6.37 Aryl-sulfino, -sulfo, and -sulfonyl nitroxides 9c1 999
6.38 Disulfo, disulfino, and disulfonyl nitroxides 9c1 1003
6.39 Silyl nitroxides 9c1 1005
6.40 Germyl nitroxides 9c1 1006
6.41 Phospho nitroxides 9c1 1006
6.42 Cobalto nitroxides 9c1 1011
6.43 Aryl-cobalto nitroxides 9c1 1024
6.44 Ferro-, manganeso-, rhenio-, molybdeno-nitroxides 9c1 1037
6.45 Thionitroxides 9c1 1042

II/9c2: Organic O-, P-, S-, Se-, Si-, Ge-, Sn-, Pb-, As-, Sb-centered radicals

7 Oxy- and peroxy-alkyl radicals

7.1 Oxyalkyl radicals . . . *9c2* 5
7.2 Peroxyalkyl radicals from hydrocarbon compounds . . . *9c2* 8
7.3 Peroxyalkyl radicals from hetero-substituted hydrocarbon compounds . . . *9c2* 18
7.4 Other peroxy radicals . . . *9c2* 22
7.5 Trioxy radicals . . . *9c2* 26

8 Aroxyl radicals

8.1 Carbocycles . . . *9c2* 29
8.2 Heterocycles . . . *9c2* 203

9 P-centered organic radicals

9.1 Phosphinyl radicals of type $X_2P^{\cdot}$ . . . *9c2* 215
9.2 Phosphonyl radicals of type $X_2P^{\cdot}{=}O$ . . . *9c2* 217
9.3 Phosphoranyl radicals of type $X_4P^{\cdot}$, including the phosphate radical anions and related species . . . *9c2* 221
9.4 Phosphonium cation radicals of type $X_3P^{\cdot+}$, phosphine dimer cation radicals of type $X_3P^{\cdot}P^{+}X_3$, and related species . . . *9c2* 263

10 S- and Se-centered organic radicals

10.1 S-centered radicals . . . *9c2* 271
10.2 Se-centered radicals . . . *9c2* 303

11 Organic radicals centered on heteroatoms

11.1 Si-centered radicals . . . *9c2* 307
11.2 Ge-centered radicals . . . *9c2* 310
11.3 Zn-centered radicals . . . *9c2* 312
11.4 Pb-centered radicals . . . *9c2* 316
11.5 As-centered radicals . . . *9c2* 316
11.6 Sb-centered radicals . . . *9c2* 318

II/9d1: Organic anion radicals

12 Semidiones and related species

12.1 Semidiones . . . *9d1* 6
12.2 Thiosemidiones . . . *9d1* 73
12.3 Semitriones . . . *9d1* 79
12.4 Anion radicals from aromatic polycarbonyl and -thiocarbonyl compounds and polycarboxylic acids . . *9d1* 83

13 Semiquinones and related species

13.1 Semiquinone anion radicals . . . *9d1* 113
13.2 Semiquinone anions from heterocyclic compounds . . . *9d1* 334
13.3 Semiquinone cation radicals . . . *9d1* 351
13.4 Charge-transfer complexes of quinones . . . *9d1* 361

14 Anion radicals from nitro compounds

14.1 Nitro compounds without double bonds conjugated with the NO_2 group . . . *9d1* 377
14.2 Nitro compounds with double bonds conjugated with the NO_2 group . . . *9d1* 414
14.3 Aromatic mononitro compounds . . . *9d1* 430
14.4 Aromatic dinitro compounds . . . *9d1* 609
14.5 Aromatic trinitro compounds . . . *9d1* 661
14.6 Aromatic tetranitro compounds . . . *9d1* 668

15 Anion radicals from nitroso compounds

15.1 Nitroso compounds without double bonds conjugated with the N=O group . . . *9d1* 680
15.2 Nitroso compounds with double bonds conjugated with the N=O group . . . *9d1* 687
15.3 Aromatic nitroso compounds . . . *9d1* 690

15A Anion radicals from B compounds

15A.1 Anion radicals containing one boron atom . . . *9d1* 699
15A.2 Anion radicals containing two boron atoms . . . *9d1* 705

16 Hydrocarbon anion radicals

16.1 Cyclic hydrocarbon anion radicals . . . *9d1* 708
16.2 Non-aromatic hydrocarbon anion radicals . . . *9d1* 844

17 Anion radicals from heterocyclic compounds

17.1 N containing anion radicals . . . *9d1* 858
17.2 Anion radicals of heterocycles containing O, S, and Se . . . *9d1* 883
17.3 Anion radicals of heterocycles containing P and Si . . . *9d1* 895
17.4 Anion radicals of heterocycles containing two or more different heteroatoms . . . *9d1* 898

II/9d2: Organic cation radicals, and polyradicals

18 Aromatic hydrocarbon cation radicals

18.1 Cation radicals from aromatic compounds and their alkyl derivatives . . . *9d2* 6
18.2 Cation radicals from halogenated aromatic hydrocarbons . . . *9d2* 19

19 Cation radicals from nitrogen-containing compounds

19.1 Alkyl and H-substituted amines . . . *9d2* 25
19.2 Aromatic amines . . . *9d2* 28
19.3 Unsaturated amines . . . *9d2* 60
19.4 [N—N] compounds . . . *9d2* 93
19.5 [N—O] compounds . . . *9d2* 116
19.6 Miscellaneous/uncertain structure . . . *9d2* 119

20 Aromatic cation radicals containing O and S atoms

20.1 Radicals containing O and/or S atoms . . . *9d2* 125
20.2 Heterocyclic compounds containing O and/or S atoms as ring members . . . *9d2* 142

21 Organic C-, O-, and N-centered polyradicals

21.1 C-centered polyradicals . . . *9d2* 151
21.2 O-centered polyradicals . . . *9d2* 160
21.3 N-centered polyradicals . . . *9d2* 172

22 Organic bis- and polyradicals

22.1 Bisnitroxides . . . *9d2* 191
22.2 Trisnitroxides . . . *9d2* 293
22.3 Tetrakisnitroxides . . . *9d2* 299
22.4 Pentakis(piperidinyl-1-oxyls) . . . *9d2* 303
22.5 Hexakis(piperidinyl-1-oxyls) . . . *9d2* 305
22.6 Phenoxyl-nitroxide . . . *9d2* 308

II/17a: Inorganic radicals, radical ions and radicals in metal complexes

1 Inorganic radicals and radical ions . . . *17a* 5

2 Radicals in metal complexes . . . *17a* 199

II/17b: Nonconjugated carbon radicals

3 Nonconjugated carbon radicals

3.1 Alkyl radicals . . . *17b* 8
3.2 σ-electronic carbon radicals . . . *17b* 492

II/17c: Conjugated carbon-centered and nitrogen radicals

4 Carbon radicals with conjugated π-systems

4.2 Radicals with 3 conjugated π-electrons . . . *17c* 8
4.3 Radicals with 5 conjugated π-electrons . . . *17c* 58
4.4 Radicals with 7 conjugated π-electrons . . . *17c* 165
4.5 Radicals with 9 conjugated π-electrons . . . *17c* 242
4.6 Radicals with 11 conjugated π-electrons . . . *17c* 251
4.7 Radicals with 13 conjugated π-electrons . . . *17c* 251
4.8 Radicals with 15 conjugated π-electrons . . . *17c* 347
4.9 Radicals with 17 conjugated π-electrons . . . *17c* 348
4.10 Radicals with 19 conjugated π-electrons . . . *17c* 349

5 Nitrogen-centered radicals

5.1 Aminyl radicals: $R—N^{\cdot}—R$ ($R=H$, C) . . . *17c* 388
5.2 Borylaminyl radicals: $—N^{\cdot}—B$ . . . *17c* 438
5.3 Germylaminyl radicals: $—Ge—N^{\cdot}—Ge—$ . . . *17c* 439
5.4 Hydrazyl and hydrazonyl radicals: $—N^{\cdot}—N$, $—N^{\cdot}—N=$ . . . *17c* 439
5.5 Oxyaminyl radicals: $—N^{\cdot}—O—$ . . . *17c* 481
5.6 Phosphorylaminyl radicals: $R—N^{\cdot}—P(O)R_2$ . . . *17c* 498
5.7 Thioaminyl radicals: $—N^{\cdot}—S—$, $—N^{\cdot}—\overset{O}{S}—$, $—N^{\cdot}—\overset{O}{\underset{NR}{S}}—$, $—N^{\cdot}—SO_2$ . . . *17c* 499
5.8 Silylaminyl radicals: $R—N^{\cdot}—SiR_3$ . . . *17c* 560
5.9 Iminyl radicals: $C=N^{\cdot}$ . . . *17c* 563
5.10 Diazenyl radicals: $R—N^{\cdot+}=N^{-}$ . . . *17c* 575

5.11 Triazenyl radicals: $R{-}N^{-}{-}N^{\cdot +}{=}N{-}R$ . . . *17c* 580
5.12 Iminoxyl radicals: $C{=}N{-}O^{\cdot}$ . . . *17c* 584
5.13 (Azo)oxyl radicals of type $R_2N^{+}{=}N{-}O^{\cdot}$ . . . *17c* 633
5.14 Iminothiyl radicals: $C{=}N{-}S^{\cdot}$ . . . *17c* 634

II/17d1: Nitroxide radicals I

6 Ntroxides
6.2 Nitroxides formed by trapping with nitrosoalkanes . . . *17d1* 8
6.3 Nitroxides formed by trapping with nitrosoarenes, nitrosoheterocycles and nitroarenes . . . *17d1* 212

II/17d2: Nitroxide radicals II

6 Nitroxides
6.4 Nitroxides formed by trapping with acyclic nitrones . . . *17d2* 8
6.5 Nitroxides formed by trapping with cyclic nitrones . . . *17d2* 126
6.6 Dialkyl nitroxides . . . *17d2* 199
6.7 Cyclic dialkyl nitroxides . . . *17d2* 222
6.8 Alkyl, aryl and vinyl nitroxides . . . *17d2* 346
6.9 Aryl, diaryl and divinylnitroxides . . . *17d2* 367
6.10 Alkyl imidoyl and nitronyl nitroxides . . . *17d2* 374
6.11 Acyl nitroxides . . . *17d2* 391
6.12 Amino nitroxides . . . *17d2* 404
6.13 Oxy nitroxides . . . *17d2* 408
6.14 Thio and sulfonyl nitroxides . . . *17d2* 413
6.15 Phospho nitroxides . . . *17d2* 418
6.16 Other nitroxides . . . *17d2* 420

II/17e: Radicals centered on heteroatoms with Z > 7 and selected anion radicals I

7 Oxy- and peroxy-alkyl radicals
7.1 Alkoxyls . . . *17e* 5
7.2 Alkylperoxyls . . . *17e* 11
7.3 Other peroxyls . . . *17e* 16

8 Aroxyl radicals
8.1 Carbocycles . . . *17e* 36
8.2 Heterocycles . . . *17e* 211
8.3 Cations . . . *17e* 234

9 Phosphorus-centered organic radicals and related radical ions
9.1 Phosphinyl radicals of type $X_2P^{\cdot}$ . . . *17e* 255
9.2 Phosphonyl radicals of type $X_2P^{\cdot}{=}O$ . . . *17e* 262
9.3 Phosphoranyl radicals of type $X_4P^{\cdot}$ including phosphate anions and related species . . . *17e* 267
9.4 Cation and anion radicals . . . *17e* 307

10 Sulfur-, selenium- amd tellurium-centered organic radicals
10.1 Neutral radicals . . . *17e* 316
10.2 Cation radicals . . . *17e* 361
10.3 Anion radicals of type $R{-}S{-}S^{\cdot -}{-}R$, $R{-}N{-}S^{\cdot -}{-}N{-}R$ . . . *17e* 385

II/17f: Radicals centered on heteroatoms with Z > 7 and selected anion radicals II

11 Organic radicals centered on other heteroatoms and related radical ions
11.1 Silicon-centered radicals and related radical ions . . . *17f* 6
11.2 Germanium-centered radicals and related radical ions . . . *17f* 38
11.3 Tin-centered radicals and related radical ions . . . *17f* 46
11.4 Lead-centered radicals and related radical ions . . . *17f* 53
11.5 Arsenic-centered radicals and related radical ions . . . *17f* 55
11.6 Antimony-centered radicals . . . *17f* 59
11.7 Boron-centered radicals $X_2B^{\cdot}$ . . . *17f* 60
11.8 Aluminum-centered radicals and related radical ions . . . *17f* 62
11.9 Mercury-centered radicals and related radical ions . . . *17f* 64

12 Hydrocarbon and substituted hydrocarbon anion radicals
12.1 Hydrocarbon anion radicals . . . *17f* 69
12.2 Substituted hydrocarbon anion radicals . . . *17f* 181

13 Anion radicals from heterocyclic compounds
13.1 Nitrogen containing anion radicals . . . *17f* 196
13.2 Anion radicals of heterocycles containing O and S . . . *17f* 211
13.3 Anion radicals of heterocycles containing two different heteroatoms . . . *17f* 217

14 Anion radicals from organo-boron compounds
14.2 Anion radicals containing one boron atom . . . *17f* 221
14.3 Anion radicals containing two boron atoms . . . *17f* 222
14.4 Anion radicals containing four boron atoms . . . *17f* 227

15 Anion radicals from nitro and nitroso compounds
15.2 Nitro compounds without double bonds conjugated with the NO_2 group . . . *17f* 230
15.3 Aromatic mononitro compounds . . . *17f* 238
15.4 Aromatic dinitro compounds . . . *17f* 294
15.5 Aromatic trinitro compounds . . . *17f* 303
15.6 Nitroso compounds . . . *17f* 305

II/17g: Semidiones and semiquinones, and related species

16 Semidiones and related species
16.1 Semidiones [R—CO^-(=C—C)$_n$—CO$^{\cdot}$—R′; n = 0, 1, 2, 3] . . . *17g* 6
16.2 Thiosemidiones and phosphasemidiones . . . *17g* 44
16.3 Semitriones (R—CO^-=CO$^{\cdot}$—R′) . . . *17g* 56
16.4 Anion radicals from aromatic polycarbonyl and -thiocarbonyl compounds and polycarboxylic acids . . *17g* 61

17 Semiquinones and related species
17.1 Semiquoinone anion radicals from carbocyclic compounds . . . *17g* 70
17.2 Semiquinone anion radicals from heterocyclic compounds . . . *17g* 347
17.3 Related compounds . . . *17g* 378
17.4 Semiquinone cation radicals . . . *17g* 383

II/17h: Cation radicals, bi- and polyradicals

18 Carbon-oxygen and halogen-centred organic cation radicals
18.2.1 Alkane cations . . . *17h* 6
18.2.2 Alkene cations . . . *17h* 17
18.2.3 Cyclic alkene cations . . . *17h* 25
18.2.4 Diene cations . . . *17h* 28
18.2.5 Ether cations . . . *17h* 36
18.2.6 Cations of carbonyl derivatives . . . *17h* 40
18.2.7 Dialkylperoxide cations . . . *17h* 50
18.2.8 Nitroalkane cations . . . *17h* 52
18.2.9 Halogenalkane cations . . . *17h* 53

19 Cation radicals of aromatic hydrocarbons and their derivatives and S-heterocycles
19.1 Cation radicals of aromatic hydrocarbons and their derivatives . . . *17h* 60
19.2 S-heterocycles . . . *17h* 108
19.3 Miscellaneous compounds . . . *17h* 120

20 Cation radicals from nitrogen containing compounds
20.2 Non-conjugated amine cation radicals . . . *17h* 126
20.3 Heteroatom substituted amines: 3e-π-based . . . *17h* 131
20.4 1-Azaallyl cation radical systems (and relatives) . . . *17h* 154
20.5 Amino substituted unsaturated cation radicals . . . *17h* 159
20.6 Imino and more highly oxidized N functional group cation radicals . . . *17h* 224

21 Organic C-, O- and N-centered bi- and polyradicals
21.1 Carbon-centered polyradicals . . . *17h* 250
21.2 Oxygen-centered polyradicals . . . *17h* 257
21.3 Nitrogen-centered polyradicals . . . *17h* 261

22 Organic bis- and polynitroxides
22.1 Bisnitroxides . . . *17h* 271
22.2 Bis(piperidinyl-1-oxyls) . . . *17h* 298
22.3 (Pyrrolinyl-1-oxyl)-(piperidinyl-1-oxyl) . . . *17h* 321
22.4 (Piperidinyl-1-oxyl)-(phenoxyl) . . . *17h* 322
22.5 Trisnitroxides . . . *17h* 324
22.6 Tetrakisnitroxides . . . *17h* 326
22.7 Pentakis- and other poly-nitroxides . . . *17h* 327
22.8 Conjugated bisnitroxides . . . *17h* 328

Organic and inorganic substances

(for polymers and compounds of biological interest, see separate index, page 569)

Ag
9a 233ff, *17a* 151ff
$AgAlO_4$
17a 81
AgBa
17a 171
AgCa
17a 154
AgCd
9a 243, *17a* 162
$AgCd^{2+}$
9a 243
$AgCu_2$
17a 157
$AgGe^{4+}$
17a 155
AgH^+
9a 239, *17a* 154
AgHg
9a 261, *17a* 177
$AgHg^{2+}$
9a 260
AgMg
17a 154
$AgNO^+$
9a 64
$AgNO_3^-$
17a 46
AgO_2
17a 70, *17e* 33
AgO_2Si
17a 82
AgO_4
17e 33
$AgO_8S_2^{4-}$
9a 240
AgS
9a 239
AgSr
17a 155
AgZn
9a 239, *17a* 154
Ag_2^+
9a 239ff, *17a* 150, 156
Ag_2^{3+}
17a 156
Ag_2NO^{2+}
9a 76
Ag_3
17a 157
Ag_3^{2+}
9a 240, *17a* 157
Ag_3Cu_2
17a 158
Ag_4^{3+}
9a 241, *17a* 157ff
Ag_4Cu
17a 158
Ag_5
17a 158
Ag_6^{n+}
17a 158
Ag_8^{x+}
17a 159
Al
9a 135, *17a* 79ff
Al^{2+}
9a 134ff
AlH^+
17a 80
AlH_3^-
9a 137, *17a* 80
AlO
9a 136
Al_2^{5+}
9a 136
Al_2^{n+}
9a 136
Al_2O^-
9a 120
ArK
9a 200
As
17a 136
$AsCl_3^-$
9a 210
$AsFH_3$
9a 211
AsF_3^-
9a 210
AsF_4
9a 213
AsF_4O^{2-}
17a 139
AsF_5^-
9a 213, *17a* 139
AsF_6^{2-}
17a 139
$AsHO_3^-$
9a 210
AsH_2O_3
17a 137
$AsH_2O_4^{2-}$
9a 212
AsH_4O_4
9a 213
AsO_2^{2-}
9a 208, *17a* 136
AsO_3^{2-}
9a 209ff, *17a* 137
AsO_4^{4-}
9a 211ff, *17a* 138
AsO_4^{2-}
9a 211, *17a* 137
$AsO_4 \cdots PO_4$
17a 138
Au
9a 258, *17a* 174
AuCd
17a 175
AuFe
17a 175
AuHg
17a 177
AuMg
17a 174
AuO_2
17a 70, *17e* 33
AuZn
17a 175
Au_3
17a 175

B
9a 16
B^{2+}
9a 16

BCl_2
17a 15
BF_2
9a 17
BF_3^-
9a 18
BF_4
17a 17
BHO^-
9a 17
BH_2O
9a 17
BH_2O_3
9a 18
BH_3^-
9a 17, *17a* 15
BH_4
9a 19, *17a* 16
BH_5N
17a 47
BO
9a 95, *17a* 57
BO_2
9a 17
BO_3^{4-}
9a 18
BO_3^{2-}
9a 17ff, *17a* 15
BO_3^{n-}
17a 16
BO_4^{4-}
17a 16
BO_4^{2-}
17a 16
BO_4
9a 19
BO_n
9a 19
BS
9a 158
B_2H_5
9a 19
$B_2H_6^-$
9a 19, *17a* 17
$B_2O_4^{x-}$
17a 17
$B_3H_7^-$
17a 17
$B_3N_3^{+-}$
9a 19
Ba
9a 257
Ba^+
9a 256, *17a* 170
BaCl
17a 171
BaF
9a 257, *17a* 170
BaH
9a 257
Be^+
9a 15, *17a* 14
BeCl
17a 108
BeF
17a 74
BeH
9a 16
BeHO
9a 109, *17a* 70
Bi
9a 267
BiF_6^{2-}
17a 184
Br^{2-}
9a 222
Br
9a 222, *17a* 143
BrCa
17a 143ff
$BrCl^-$
9a 223, *17a* 143
BrD_3Si^-
17a 83
BrF^-
9a 223
BrF_3Si^-
17a 85
BrF_6
9a 227, *17a* 146
$BrGeH_3^-$
17a 135
$BrHO^-$
9a 111
$BrH_2O_4^{2-}$
17a 145
BrH_3Si^-
17a 83
BrO
9a 222
BrOS
9a 165
BrO_2
9a 225ff, *17a* 144
BrO_3^{2-}
9a 226, *17a* 145
BrO_3
9a 226, *17a* 144ff
BrO_4^{2-}
9a 227, *17a* 145
BrO_4
9a 226, *17a* 69
Br_2^-
9a 224ff, *17a* 144
$Br_2O_8^-$
9a 227
Br_2Sc
17a 118
$Br_6N_3P_3^-$
17a 95
Br_6Ru^{3-}
17a 149

CAgO
17a 154
CAuO
17a 174
$CBrCl_2$
17b 322
$CBrF_3^-$
17a 29, *17f* 181
$CBrN^-$
9a 31, *17a* 22
$CBrN_2S_2$
17c 532
CBr_2N
9a 48
CBr_4^-
9a 48
CBr_4^+
17h 55
CCdN
9a 31
$CClF_2$
9a 47, *17b* 322
$CClF_3^-$
17a 29
$CClN^-$
9a 30

$CClN_2S_2$
17c 532
$CClO$
9a 31
CCl_2F
9a 47, *9b* 251, *17a* 28, *17b* 322
$CCl_2F_3N_2O$
9c1 895
CCl_3
9a 47ff, *9b* 250ff, *17b* 322
CCl_3NO^+
9a 55
CCl_3O_2
9a 112, *9c2* 21
CCl_4^+
17h 53
$CCuO$
17a 130
CDO
9b 318, *17b* 512
CD_3
1 22, *9b* 10, *17a* 23, *17b* 11ff
$CD_3 \cdots CD_3OS^-$
9a 34
$CD_3(As(C_6H_5)_3)$
17b 11
$CD_3(Br^-)$
17b 11
$CD_3(CN^-)$
17b 11
$CD_3 \cdots I^-$
17a 23, *17b* 12
CD_3O
9c2 5
CD_3OS
17e 328
CF
9a 121, *17a* 74
CFN^-
9a 24
CFN_2S_2
17c 532
CFO
9a 30, *9b* 322
$CF_2N_2O_4^-$
9a 48, *9d1* 388
CF_2OR
17b 324
CF_3
9a 45ff, *9b* 251ff, *17a* 26ff, *17b* 322ff
CF_3I^-
17a 29, *17f* 181
CF_3NO^-
9d1 680
$CF_3NO_2^-$
9d1 386
CF_3O_2
9a 112, *9c2* 21
CF_3O_2S
17e 334
CF_3O_3
9a 116, *9c2* 26
CF_3O_3S
9a 170
CF_4O_2P
9a 151, *9c2* 221
CF_4R
17b 323
CF_6OP
9a 151, *9c2* 221
CF_7OS
9a 175, *9c2* 295
$CF_{11}O_3S_3$
9a 167
$CF_{15}O_4S_4$
9a 167
$CFeN_3O_4$
17a 123
$CHAgN$
17a 18, 155, 373, *17c* 566
$CHAuN$
17a 375, *17c* 566
$CHBrCl$
17b 235
$CHBrN$
9a 47
$CHBr_3^+$
17h 55
$CHCl_2$
9a 46ff, *9b* 175, *17b* 235
$CHCl_3OP$
17b 237
$CHCuN$
17a 130, 355, *17c* 566, 567
CHD_2
9b 10
CHD_2O_2
17b 235
CHF_2
9a 41, *9b* 175
$CHF_2NO_2^-$
9a 70, *9d1* 386
CHF_3NO
17d1 206
CHF_3NO_4S
9c1 989
CHN^-
9a 21, *17a* 18
$CHNO^-$
9a 40, 53, *9c1* 121
$CHNO_2^-$
9a 54, *9c1* 18
$CHNS^-$
9a 53
$CHN_2O_2^{2-}$
9d1 688
$CHN_2O_3^{2-}$
9d1 689
$CHN_2O_4^{2-}$
9a 40, *9d1* 421
CHO
1 21, *9a* 22, *9b* 318, *17a* 19, *17b* 512
CHO_2^{2-}
9a 40, *9b* 176, *17b* 235
CHO_2
9a 30
CHO_3^{2-}
9a 45
CHO_3
9a 44
CHO_3S^{2-}
1 19
$CHO_6S_2^{4-}$
9b 178
CH_2^-
17a 17, *17b* 8
$CH_2AsNO_4^-$
9d1 414
$CH_2AsO_3^{2-}$
9a 39, *9b* 55
CH_2BN^-
17a 15, *17f* 222
CH_2Br
9b 55

CH_2BrHg
9a 39, *9b* 57
$CH_2Br_2^+$
17h 55
CH_2Cl
9b 56
CH_2ClHg
9a 39
CH_2ClMg
9b 57
CH_2ClO_2S
17e 334
CH_2Cl_2OP
17b 89
$CH_2Cl_2O_3P$
17b 324
CH_2Cl_3Si
17b 94
CH_2D
9b 10
$CH_2D_2^+$
17h 6
CH_2F
9a 38, *9b* 56
CH_2HgI
9b 57
CH_2N
1 15, *9a* 36ff, *9c1* 115, *17a* 23, *17c* 563
CH_2NO
9a 23, 40, 54, *9b* 323, *9c1* 121, 123, *17b* 518
$CH_2NO_2^{2-}$
9d1 378
CH_2NO_2
9a 37, *9b* 62, *9c1* 762, *17d2* 391
$CH_2NO_2S^{2-}$
9d1 402
$CH_2NO_3^-$
9c1 769, *9d1* 390
$CH_2NO_5P^{3-}$
9a 70, *9d1* 411
$CH_2NO_5S^{2-}$
9a 70, *9d1* 403
$CH_2NO_5S_2^{2-}$
9d1 403
$CH_2NO_6P^{3-}$
9a 69, *9d1* 392
$CH_2NO_6S^{2-}$
9a 69, *9d1* 392
$CH_2NO_9P_2^{4-}$
9a 69, *9d1* 392
$CH_2N_4O_2^-$
9d1 407
CH_2O^-
9a 37, *9b* 63
CH_2O^+
9a 37, *17a* 23, *17h* 40
$CH_2OSi\equiv$
9a 38
CH_2OSiX
(X = silica gel matrix)
9b 71
$CH_2O_2^-$
9a 40, *9b* 176
$CH_2O_2^+$
17h 45
$CH_2O_3S^-$
17a 23, *17b* 93
$CH_2O_4P^{2-}$
9b 71
$CH_2O_4S^-$
9a 38
CH_3
1 21ff, *9a* 32ff, *9b* 7ff, *17b* 8ff
$CH_3(As(C_6H_5)_3)$
17b 10
$CH_3(Br^-)$
9b 9, *17b* 10
$CH_3 \cdots Br$
9a 34
$CH_3((CH_3)_2P(O)^-)$
9a 34, *17b* 11
$CH_3(Cl^-)$
17b 10
$CH_3 \cdots H^+$
9a 34, *17b* 10
$CH_3(I^-)$
9b 9, *17b* 11
$CH_3 \cdots I^-$
17a 22
$CH_3 \cdots I$
9a 35
$CH_3(NH_3)$
9b 10
$CH_3 \cdots SO_2CH_3^-$
9a 34, *17b* 11
CH_3AgO^+
9a 239
$CH_3AsO_3^{3-}$
9a 211, *17f* 55
$CH_3Cl_2O_2P^-$
9a 153, *9c2* 222
CH_3Cl_2Si
9a 139, *9c2* 307
CH_3Cl_2Sn
9c2 312
CH_3F_4OS
9a 175, *9c2* 294
$CH_3NO_2^-$
9a 69, *9d1* 377, 684
$CH_3NO_2^+$
17h 52, 237
$CH_3NO_2S^-$
9d1 402
$CH_3NO_3^-$
9a 69, *9d1* 390
$CH_3NO_3^+$
17h 237
$CH_3NO_4P^{2-}$
9a 70, *9d1* 410
CH_3N_2
17c 565
CH_3N_2O
9a 54, *9c1* 18
$CH_3N_2O_2$
9a 67, *9c1* 762, 769, *17a* 24, *17d2* 391
$CH_3N_2O_3$
9c1 769
CH_3O
1 24, *9a* 38, *9b* 63ff, *9c2* 5, *17a* 28, *17b* 79ff
CH_3OS
9a 160, *9c2* 285, *17e* 328
CH_3O_2
9b 176, *9c2* 8, *17b* 235, *17e* 11
CH_3O_2S
9a 167, *9c2* 288, *17e* 333ff
CH_3O_3
9c2 18
$CH_3O_3P^-$
9a 146, *9c2* 217
CH_3O_3S
17b 93

CH_3Pb^{2+}
9a 48
CH_3S
9a 48, *9b* 73, *9c2* 271ff, *17b* 89, *17e* 316
CH_3S_2
9c2 295
CH_4^+
17a 28, *17h* 6
CH_4N
9a 37, 54, *9b* 57, *9c1* 6, *17b* 75, *17c* 388
CH_4NO
9a 67, *9b* 70, *9c1* 96, 196, *17d2* 200
CH_4NO_2
9c1 196, 912
CH_4NO_2S
17a 101, *17e* 343
CH_4NO_3
9c1 912
$CH_4NO_3P^-$
9c1 196
$CH_4N_2O_2^-$
9a 69, *9d1* 406
$CH_4N_2S^+$
9a 46, *17h* 156
CH_4N_3O
9a 54, *9c1* 63
CH_4N_3S
9c1 18
CH_4O^+
9b 64
CH_4O_3P
17b 89
CH_4S^+
17a 98, *17e* 361
$CH_5As_2O_5$
17f 55
$CH_5As_2O_6$
17b 235
CH_5BrGe^-
17a 135, *17f* 43
CH_5BrSi^-
17a 83, *17f* 9
CH_5GeI^-
17a 135, *17f* 43
CH_5ISi^-
17a 84, *17f* 9
CH_5N^+
9b 58, *17a* 44, *17b* 75, *17h* 126
CH_5N_2
9c1 40
CH_5N_4O
9a 55, *9c1* 63
$CH_5O_5P_2$
9a 143
$CH_5O_6P_2$
17b 238
CH_5P^+
9a 143
CH_5Si
9a 137, *9c2* 307
$CH_6As_2O_6^-$
17f 55, 58
$CH_6N_2^+$
9d2 94
$CH_6N_2O^+$
9a 67, *9c1* 196
CH_6Sn^-
17f 50
CH_6Sn^+
17a 165, *17f* 52
$CH_7As_2O_7^{2-}$
17a 138
$CH_7O_6P_2$
9a 149
$CHgN$
9a 31
CIN_2S_2
17c 532, 540
CKO_2
17a 212
$CLiO_2$
17a 200
CN
1 15, *9a* 50, *17a* 34
CNO^{2-}
9a 23, *17a* 36
CNO
9a 23, *17a* 19
CNO_2^{2-}
17a 25, 37
CNO_4^{2-}
9a 68, *9d1* 421
CNS^{2-}
9a 30
CNS
9a 30
CN_2O^-
9a 41
CN_2O^+
17h 232
$CN_3O_5^{2-}$
9d1 690
CN_3O_5
17c 621
$CN_3O_6^{2-}$
9a 41, *9d1* 425
CN_3O_6
9a 41
$CN_4O_8^+$
17h 52, 237
$CNaO_2$
17a 204
CO^-
9a 95
CO^+
9a 95, *17a* 57
$CORuX_3^{3+}$
17a 149
COS^-
9a 30, *9c2* 301
CO_2^-
1 17, *9a* 24ff, *17a* 20ff
$CO_2^- \cdots H^+$
1 17
$CO_2^- \cdots Li$
17a 20
CO_2S^{3-}
9a 46
CO_3^{3-}
9a 43ff, *17a* 26
CO_3^-
1 18, *9a* 42ff, 111, *17a* 25
$CO_3RuX_3^{3+}$
17a 149
CO_4^-
9c2 21, *17a* 26, *17e* 11
CRh
9a 233
CS_2^-
9a 31, *9c2* 301
CSi_n
9a 47
C_2^-
9a 20

C_2AgD_3N
17a 18, 28, 155, *17c* 571
$C_2AgN_2^{2-}$
9a 240, *17a* 155
C_2AgN_2
9a 240, *17a* 155
$C_2AgN_2X_2^{2+}$
17a 150
C_2AlO_2
17a 80
C_2AuD_3N
17a 28, 175, *17c* 572
C_2B
9a 22
$C_2BrD_5^+$
17h 55
$C_2BrF_3^-$
17a 32
C_2Br_2N
17b 318
C_2ClD_2FNO
9b 244
$C_2ClF_3^-$
17a 32
C_2ClF_4
9b 244
$C_2ClF_4NO_2^-$
9d1 387
$C_2Cl_2F_3$
9b 243, *17a* 27, *17b* 318, 319
$C_2Cl_2O_2^-$
9a 39, *9b* 243, *17b* 318
$C_2Cl_3N_2S_2$
17c 532
$C_2Cl_4F_2NO$
9c1 384
$C_2Cl_4NO_2$
9c1 789
$C_2Cl_6F_2^+$
17h 54
C_2Cl_6NO
9a 68, *9c1* 379
$C_2CoN_2O_4$
17a 126
C_2D_2F
9b 304, *17b* 494
C_2D_2N
9b 55
$C_2D_3O_2$
17b 70, 521
$C_2D_4^+$
17h 17
C_2D_4Mu
17b 15
$C_2D_4O^+$
17h 41
C_2D_5
1 22
C_2D_5O
9c2 5
$C_2D_6^+$
17h 7
$C_2D_6NO_3S$
17d2 414
$C_2D_6OS^+$
17e 385
$C_2D_6S^+$
17e 362
$C_2F_2O_2^-$
9a 45, *9b* 247
$C_2F_3N_2S_2$
17c 532, 540
C_2F_3O
9a 23, *9b* 320
$(C_2F_3)_nO_2$
17e 15
$C_2F_4^-$
17a 32, *17f* 181ff
$C_2F_4^+$
17a 31, *17h* 23
$C_2F_4NO_2^{2-}$
9d1 388
$C_2F_4N_2^-$
9a 79
$C_2F_4N_2O_3^-$
9d1 683
$C_2F_4N_2O_4^{2-}$
9d1 388
$C_2F_4N_2O_4^-$
9d1 685
C_2F_4OR
17b 319
$C_2F_4O_2R$
17b 324
C_2F_5
9b 244ff
$C_2F_5NO_2^-$
9d1 387
C_2F_5O
9a 45
C_2F_5Se
17b 327
C_2F_6N
17a 36, *17c* 395
C_2F_6NO
9a 68, *9c1* 383, *17a* 44, *17d1* 210
$C_2F_6NO_2$
17d1 211
$C_2F_6NS_2$
17c 517
$C_2F_7O_2S$
9a 171, *9c2* 285
$C_2FeNO_4^-$
17a 123
C_2GaO_2
17a 134
C_2H
1 21, *9a* 20ff, *17a* 18
C_2HBrFO_2
17a 27, *17b* 318
C_2HBrN
17b 225
C_2HClF_3
9b 164
C_2HClN
17b 225
$C_2HClO_2^-$
9b 164, *17b* 226
$C_2HCl_2O_2$
9b 244, *17b* 318
$C_2HCl_3NO_2$
9c1 762
C_2HDNOR_2
17b 227
$C_2HD_2NO_2^-$
9b 167
C_2HD_3Mu
17b 15
$C_2HD_3O^+$
17h 41
C_2HD_4
1 22, *9b* 12
C_2HFNaO_2
9b 165
$C_2HFO_2^-$
9b 165

C_2HF_2O
17b 299
$C_2HF_2O_2$
9b 247
C_2HF_3NO
9a 54, *9c1* 17
$C_2HF_3NO_2$
9c1 790
$C_2HF_3O_2^-$
9b 248
C_2HF_4
9b 165
C_2HNO^-
9b 174
$(C_2HNO_3R)_n$
17e 14
$C_2HN_4O_2^{2-}$
17f 288
$C_2HO_2S^{2-}$
9b 174
$C_2HO_3^{2-}$
9b 173, *17b* 231
C_2HO_3
9b 322
$C_2HO_4^{2-}$
9b 177, *17b* 237
$C_2H_2^-$
17b 492
C_2H_2Ag
17a 31, 373, *17b* 492
C_2H_2Al
17a 30, 393, *17b* 493
C_2H_2Au
17a 31, 175, 375, 376, *17b* 493, 494
$C_2H_2BrF_2NO^-$
17a 33
$C_2H_2BrO_2$
9b 164
C_2H_2ClFNO
17a 27, *17b* 319
$C_2H_2ClF_2$
9b 225
C_2H_2ClO
9c2 6, *17a* 29
$C_2H_2ClO_2$
9b 165, *17b* 226
$C_2H_2Cl_2NO$
9b 243
$C_2H_2Cl_2NO_2$
9c1 763
$C_2H_2Cl_3$
9b 225
$C_2H_2Cl_3NO_2^-$
9d1 389
C_2H_2Cu
17a 130, 355, *17b* 493
$C_2H_2DO_2$
9b 53
$C_2H_2D_2NO_2^{2-}$
9b 228
$C_2H_2D_2O^+$
17c 29
$C_2H_2D_2O_2^+$
17b 40
$C_2H_2D_4^+$
17h 7
C_2H_2F
17a 30, *17b* 493ff
$C_2H_2F_2^-$
17a 31, *17f* 181
$C_2H_2F_2NO$
1 39, *9b* 246
$C_2H_2F_2NO_3$
9c2 20
$C_2H_2F_2N_2O_3^-$
9d1 387
$C_2H_2F_3$
9b 46, *17b* 64
$C_2H_2F_3NO^-$
17a 25, *17b* 320
$C_2H_2F_3O$
9b 172, *17b* 230
$C_2H_2F_3O_2S$
17e 334
$C_2H_2F_3S$
17b 90
$C_2H_2KO_3$
1 31
$C_2H_2LiO_3$
1 31
C_2H_2N
9a 35, *9b* 54ff, *17b* 65, 79
C_2H_2NO
9a 40, *9b* 62, 74, *17b* 79
$C_2H_2NO_2$
9b 304, 322
$C_2H_2NO_2S_2^-$
9d1 404
$C_2H_2NO_4^{2-}$
9d1 396
C_2H_2NS
9b 62, *17b* 79
$C_2H_2NS_2$
17c 533
$C_2H_2N_2O_2^-$
9a 69, *9d1* 408
$C_2H_2N_2O_2S^-$
9d1 403
$C_2H_2N_2O_3^-$
9d1 46
$C_2H_2N_3O_5^{2-}$
9d1 425
$C_2H_2N_4^-$
1 116
$C_2H_2N_4^+$
17h 227
C_2H_2Na
17a 23, 204
$C_2H_2Na_2O_5P$
17b 222
$C_2H_2O^+$
17b 65, *17h* 43
$C_2H_2O_2^-$
1 23, *9a* 35ff, *9b* 50ff, *9d1* 6, *17b* 69ff, *17g* 6
$C_2H_2O_2^+$
17h 44
$C_2H_2O_2S^-$
9c2 273
$C_2H_2O_2SiX$
(X = Vycor glass)
9b 54
$C_2H_2O_2Tl$
9b 54
$C_2H_2O_3^-$
9b 173, *9d1* 49, *17a* 24, *17b* 231
$C_2H_2O_3S^-$
9c2 286
$C_2H_2O_4^-$
17b 320
$C_2H_2O_4S^-$
9c2 290
$C_2H_2O_6^-$
9a 49
$C_2H_2S_2^+$
17e 368ff, *17h* 108

C_2H_3
1 20, *9b* 303ff, *17b* 492
$C_2H_3BD_3N^-$
17a 24, *17c* 572
$C_2H_3BD_5N_2$
17c 573
$C_2H_3Br^+$
17h 19
C_2H_3BrNO
9b 164
$C_2H_3BrO_2^-$
9b 228, *17a* 23, *17b* 293
$C_2H_3ClNO_2$
9c1 763
$C_2H_3ClO_2^-$
17b 293
$C_2H_3ClO_3S^-$
17b 181
$C_2H_3Cl_2$
9b 213
$C_2H_3Cl_2F_2N_2O$
9c1 895
$C_2H_3Cl_2O$
9b 225, *17b* 237, 323
$C_2H_3Cl_2O_2$
9b 233
C_2H_3CuN
17a 28, *17c* 574
C_2H_3DMu
17b 15
$C_2H_3DO^+$
17h 40
$C_2H_3DO_2^-$
17b 273
$C_2H_3DO_2^+$
17h 46
$C_2H_3D_2NO^-$
9b 213, *17b* 272
$C_2H_3D_3^+$
17h 7
$C_2H_3D_3NO$
17b 272
$C_2H_3D_4NO^+$
9b 131
C_2H_3FNO
1 30, *9b* 165
$C_2H_3F_2$
9b 42, *9b* 213, *17a* 26, *17b* 272
$C_2H_3F_2O$
17b 290
$C_2H_3F_3NO$
17d1 206
$C_2H_3F_3NO_2$
17d1 206
C_2H_3INO
9b 165ff
$C_2H_3IO_2^-$
17b 293
$C_2H_3NO^-$
17a 36, *17b* 76, 508
C_2H_3NO
9c1 121
C_2H_3NOX
(X = various alkyl substituents)
9b 61, *17b* 69
$C_2H_3NO_2^-$
9b 166ff, *9c1* 18, *17c* 389
$C_2H_3NS^-$
17b 511
$C_2H_3N_2$
17b 226, *17c* 479
$C_2H_3N_2O_2^{2-}$
9d1 688
$C_2H_3N_2O_2$
9c1 138
$C_2H_3N_2O_3^{2-}$
9d1 689
$C_2H_3N_2O_4^{2-}$
9d1 421ff
$C_2H_3N_2O_5^{2-}$
9d1 422
$C_2H_3N_2S_2$
17c 530ff, 539
$C_2H_3O^-$
17b 492
C_2H_3O
1 21, *9b* 47, 254ff, 319, *17a* 19, *17b* 56ff, 220, 512
C_2H_3OS
17b 529
$C_2H_3O_2^{2-}$
9a 41, *9b* 214, *17a* 25, *17b* 272, 273
$C_2H_3O_2$
1 23, *9b* 52ff, 69, 324, *17a* 24, *17b* 70, 86, 220, 520
$C_2H_3O_2S$
9c2 273, *17e* 316, 336
$(C_2H_3)_nO_2$
17e 14
$C_2H_3O_3$
1 31, *9b* 173ff, *17b* 231
$C_2H_3O_3S$
9c2 286
$C_2H_3O_4$
9b 248, *17b* 320
$C_2H_3O_6S_2^{2-}$
17b 200
$C_2H_3O_7S_2^{2-}$
17b 199
$C_2H_4^-$
17a 18, *17b* 131
$C_2H_4^+$
17h 17
C_2H_4Al
9a 137, *17a* 80, 393, 394
$C_2H_4AlCl_3$
9b 95
C_2H_4Au
17a 175, 376
C_2H_4Br
9b 28, 95
$C_2H_4BrNO_2^-$
17f 230
$C_2H_4Br_2NO_3S$
9c1 989
C_2H_4Cl
9b 28, *17b* 37
C_2H_4ClHg
9b 95
C_2H_4ClMg
9b 96
$C_2H_4ClNO^-$
9d1 681
$C_2H_4ClNO_2^-$
9d1 388
C_2H_4ClO
9b 129, 136, *17b* 82, 235
$C_2H_4ClO_2S$
17e 335
$C_2H_4Cl_3Si$
9b 36, *17b* 46
C_2H_4Cu
17a 131, 355
C_2H_4D
17b 14

$C_2H_4DN^+$
17c 29
$C_2H_4D_2^+$
17h 6
$C_2H_4D_2N$
9b 96
C_2H_4F
9b 28, 95, *17b* 37ff
C_2H_4FMg
9b 96
C_2H_4FO
9b 129, 213
C_2H_4I
9b 29
$C_2H_4INO_2^-$
17f 230
$C_2H_4LiO_4$
1 31
C_2H_4Mu
17b 14
C_2H_4N
9c1 19ff, 116, *17c* 396, 406, 563
C_2H_4NO
9b 49, 60, 323, *9c1* 117, 121, 123, *17b* 68, 77, 518
$C_2H_4NO_2$
1 33, *9a* 40, *9b* 69, 98, 167, 168, *9c1* 7, 763, *17b* 226ff, *17d2* 391, 398
$C_2H_4NO_2S^{2-}$
9d1 404
$C_2H_4NO_2S$
17c 528
$C_2H_4NO_3$
9c1 197, 764, 770
$C_2H_4NO_3S$
17d2 415
$C_2H_4NO_5^{2-}$
9d1 392
$C_2H_4NO_5P^{3-}$
9d1 412
$C_2H_4NO_5S^{2-}$
9d1 405
$C_2H_4NO_5S_2^{2-}$
9d1 405
$C_2H_4N_2O_3^-$
9d1 408
$C_2H_4N_2O_4^-$
9d1 408
$C_2H_4N_2S_2^-$
9d1 75
$C_2H_4N_3$
9c1 69
$C_2H_4N_4^+$
17h 176
$C_2H_4N_5$
9b 79
$C_2H_4N_6^-$
9d1 880
$C_2H_4O^-$
9b 30, 99, *17b* 39
$C_2H_4O^+$
17a 23, *17b* 80, *17c* 29, *17h* 36, 40
C_2H_4OR
17b 186
$C_2H_4O_2^-$
9b 176, 214ff, *17b* 235
$C_2H_4O_2^+$
17b 83, *17h* 45
$C_2H_4O_3S^-$
9b 35, 104, *17b* 46, 149
$C_2H_4O_4P^{2-}$
9c2 218
$C_2H_4O_4S^-$
9b 102, *17b* 44, 199
$C_2H_4Pd^+$
17a 150
$C_2H_4S^+$
17e 363
$C_2H_4S_2^+$
17e 367, 377
C_2H_4X
(X = Al surface site)
9b 104
C_2H_5
1 22, *9b* 10ff, *17b* 12ff
$C_2H_5(Br^-)$
17b 14
$C_2H_5(I^-)$
9b 12, *17b* 14
$C_2H_5Br^+$
17h 55
$C_2H_5BrSi^-$
17f 9
$C_2H_5Cl^+$
17h 54
$C_2H_5ClNO_3S$
9c1 989
$C_2H_5Cl_2O_2P^-$
9c2 222
$C_2H_5Cl_2Si$
17b 94
$C_2H_5D^+$
17h 6
$C_2H_5F_3OP$
9c2 222
$C_2H_5F_4OS$
9c2 294
C_2H_5Hg
9b 57, *17a* 389, *17f* 64
$C_2H_5I^+$
17h 56
$C_2H_5ISi^-$
17f 9
C_2H_5Mg
9b 57
$C_2H_5N^+$
17c 29, *17h* 126
$C_2H_5NOS^-$
9c1 197
$C_2H_5NO_2^-$
1 77, *9b* 228, *9d1* 378, 685
$C_2H_5NO_2^+$
17h 52, 237
$C_2H_5NO_3^-$
9d1 391, 393, 399
$C_2H_5NO_4P^{2-}$
9c1 197, *9d1* 411
$C_2H_5N_2O$
1 24
$C_2H_5N_2O_2$
9c1 764
$C_2H_5N_4$
9c1 121, *17c* 566
$C_2H_5N_4O_3$
9c1 809
C_2H_5O
1 31, *9b* 30ff, 64ff, 99ff, *9c2* 5, *17b* 39, 143
$C_2H_5OPS_2^-$
17e 262
C_2H_5OS
9b 74, *9c2* 275, 285, *17b* 92, *17e* 328
$C_2H_5O_2$
9b 42, 137ff, 176, 215, *9c2* 9, *17b* 187

$C_2H_5O_2PS^-$
17e 263
$C_2H_5O_2S$
9b 75, *9c2* 286, 288, *17b* 92, *17e* 329, 332, 334
$C_2H_5O_2S_2$
9c2 290
$C_2H_5O_3$
9c2 19
$C_2H_5O_3P^-$
9c2 218
$C_2H_5O_3S$
9c2 290, 294, *17e* 336
C_2H_5S
9b 73, *9c2* 272, *17b* 89, *17e* 316
$C_2H_5S_2$
9c2 296, *17b* 93
$C_2H_6^+$
17a 28, *17h* 6
$C_2H_6AlBr^-$
17a 81, *17f* 62
$C_2H_6AlCl^-$
17a 81, *17f* 63
$C_2H_6AlI^-$
17a 81, *17f* 63
C_2H_6As
9c2 316
$C_2H_6AsO_2^{2-}$
17f 55
$C_2H_6BD_2N_2$
17c 573
$C_2H_6BrO_2PS^-$
9c2 222
$C_2H_6BrPS^-$
17e 267
C_2H_6BrS
9c2 284, *17e* 349
$C_2H_6ClNO_2^-$
9b 228
$C_2H_6ClNS^-$
9b 131
C_2H_6ClS
17e 349
C_2H_6ClSi
9c2 307
C_2H_6ClSn
9c2 312ff
$C_2H_6DN_2O$
9c1 895
C_2H_6FSi
9c2 308
$C_2H_6F_2NSi$
9c2 308
$C_2H_6Hg^-$
17f 64
C_2H_6IS
17e 349
C_2H_6N
9b 58, 96, *9c1* 6, 7, *17b* 38, 140, *17c* 388, 389
C_2H_6NO
9c1 97, 197, 208, *17d2* 205
C_2H_6NOS
17e 332
$C_2H_6NO_2$
9c1 198, 209, 912
$C_2H_6NO_2S$
17a 101, *17c* 526, *17e* 343
$C_2H_6NO_3$
9c1 213, 912, 914, 917
$C_2H_6NO_3P^-$
9c1 198
$C_2H_6NO_3S$
9b 132, 145, *9c1* 994, *17d2* 414
$C_2H_6NO_4P^-$
9b 143
C_2H_6NS
9c2 274, *17e* 316
$C_2H_6NS_2$
17e 357
$C_2H_6N_2O^-$
9d1 685
$C_2H_6N_2O^+$
17c 633
$C_2H_6N_2O_2^-$
9d1 407, 409
$C_2H_6N_2O_2^+$
17h 156
$C_2H_6N_3$
17c 580
$C_2H_6O^+$
17h 36
C_2H_6OP
9c2 218
$C_2H_6OS^+$
17e 384ff
$C_2H_6O_2P$
9c2 215
$C_2H_6O_2S$
17e 357
$C_2H_6O_3P$
9b 70, *9c2* 218ff
$C_2H_6O_4P^{2-}$
9c2 223
$C_2H_6O_4P$
9b 70
$C_2H_6PS_2$
9c2 215
$C_2H_6S^+$
17e 361ff
$C_2H_6S_2^-$
9c2 302
$C_2H_6S_2^+$
9c2 299, *17a* 106ff, *17e* 365
$C_2H_6Se^+$
17e 362
$C_2H_7Al^-$
17b 37
$C_2H_7B^-$
17b 37
C_2H_7BN
17b 38
$C_2H_7BrGe^-$
17a 136, *17f* 43
$C_2H_7BrSi^-$
17a 84, *17f* 10
C_2H_7ClNS
9c2 274ff, *17b* 183
$C_2H_7ClNS_2$
9c2 296
C_2H_7Ge
9b 28
$C_2H_7GeI^-$
17a 136, *17f* 43
$C_2H_7ISi^-$
17a 84, *17f* 10
$C_2H_7N^+$
9b 29, 58, 96, *9d2* 25ff, *17h* 126
$C_2H_7NO^+$
9b 136, *17b* 182, 187
$C_2H_7NOS^+$
9c2 286
$C_2H_7NO_2S^+$
9c2 289
$C_2H_7NO_4P$
17b 194

$C_2H_7N_2$
9c1 40ff
$C_2H_7N_2O$
9c1 209
C_2H_7Si
9b 35, *9c2* 308
$C_2H_8N_2^+$
9d2 95
$C_2H_8N_2O^+$
9c1 198
C_2H_8OP
9c2 223
$C_2H_8O_7P_2^-$
9c2 223
$C_2H_8Sn^-$
17f 50
$C_2H_8Sn^+$
17f 52
C_2H_9BN
17c 390
$C_2H_9O_6P_2$
9c2 223
$C_2H_{10}N_3O$
9c1 213
$C_2N_2^-$
9a 48
$C_2N_2S_2^-$
9a 176
C_2N_2Sc
17a 118
C_2N_2Y
17a 148
$C_2N_3O_3^{2-}$
9d1 690
$C_2N_3O_4^{2-}$
9d1 425
$C_2N_4S_2^-$
9d1 901
$C_2NiO_2^+$
17a 128
$C_2O_2^-$
17a 30
$C_2O_2^+$
17a 29
$C_2O_4^-$
9a 49
C_3AgO_3
9a 240, *17a* 156
C_3BrClF_6NO
9c1 384
C_3BrF_6NOX
(X unknown)
17d1 210
$C_3Br_2F_6NO$
9c1 384
C_3ClF_4
9b 356
C_3ClF_6NOX
(X unknown)
17d1 210
C_3ClF_7NO
9c1 383
$C_3ClF_8N_2O_3$
9c1 382
$C_3Cl_2F_5NOX$
(X unknown)
17d1 211
$C_3Cl_2F_6NO$
9c1 384
C_3Cl_5O
17b 318
C_3CuO_3
9a 205, *17a* 131
$C_3D_2O_4^-$
17b 224
$C_3D_6^+$
17b 17, *17h* 15
$C_3D_6F_3S_2$
17e 344
C_3D_6NO
9c1 124
$C_3D_9O_3S$
17e 355
$C_3FO_4^{2-}$
9b 240
$C_3F_4NaO_2$
9b 239
C_3F_5
9b 355
$C_3F_5NO_3^-$
9d1 387
C_3F_5O
9b 321
$C_3F_5O_2R$
(R = polymer chain)
17e 15
C_3F_6INOX
(X unknown)
17d1 211
C_3F_6N
9c1 119
C_3F_6NO
9c1 124
$C_3F_6NO_2$
9c1 790
$C_3F_6O^-$
9a 41, *9b* 241, *17b* 310
C_3F_6OR
17b 306
$C_3F_6O_2R$
(R = polymer chain)
17e 15
C_3F_6R
17b 319
$C_3F_6S^-$
17b 313
C_3F_7
9b 238, 245, *17b* 306
$C_3F_7KNO_4S$
17d2 414
C_3F_7O
9b 247
$C_3F_7O_3S$
9a 167
$C_3F_9O_3S$
9a 170, *9c2* 285
C_3HAgF_3
17b 495
$C_3HBrN_3O_2$
17b 438, *17c* 481
$C_3HCl_9F_3Si_3$
9b 233
C_3HCuF_3
17b 496
$C_3HDN_3O_2$
17c 480
$C_3HD_2N_2O_2$
17b 346
$C_3HD_2N_4O$
17c 480
$C_3HD_6O_2$
17b 236
C_3HF_4
9b 355
$C_3HF_4O_2$
9b 239
C_3HF_5NO
9c1 17

C_3HF_6
9b 159, *17b* 223
C_3HF_6O
9a 41, *9b* 241, *17b* 310
$C_3HF_6S_2$
17b 239
$C_3HNO_2^-$
9b 163
C_3HN_2
9a 39, *9b* 164, *17b* 223
$C_3HN_2OS^{2-}$
17b 346
$C_3HN_2O_2^{2-}$
17b 345
$C_3HN_3O_5^{3-}$
9d1 607
C_3HNiO_3
17a 128
C_3HO_3R
17b 371
$C_3HO_4^{2-}$
9a 39, *9b* 162, *17b* 224
$C_3HO_5^{2-}$
9b 242
C_3H_2Br
9b 433
C_3H_2Cl
9b 432
$C_3H_2ClN_4O_8^{2-}$
9d1 423
$C_3H_2ClO_3$
17b 371
$C_3H_2Cl_2NO_4^{2-}$
9d1 389
$C_3H_2Cl_3$
9b 357
$C_3H_2CoO_3$
17a 126
$C_3H_2DO_2$
9b 304
$C_3H_2D_2O_2^+$
17h 48
$C_3H_2D_2O_4^-$
17b 215, 286
$C_3H_2D_3O_4$
9b 157, *17b* 215, 293
$C_3H_2D_4NO$
9b 61
$C_3H_2D_4O_2^+$
17b 41
$C_3H_2D_5NO_3^+$
17b 214
$C_3H_2D_6^+$
17h 7
C_3H_2F
9b 432
$C_3H_2F_3$
9b 354
$C_3H_2F_4NO$
9b 239
$C_3H_2F_4NO_3$
9c2 20
$C_3H_2F_5$
9b 46
$C_3H_2F_5O$
9b 225, *17b* 299
$C_3H_2F_5S$
17b 291
$C_3H_2KO_4$
1 28, *9b* 163
C_3H_2N
9b 305, *17b* 494
$C_3H_2NO_2$
9b 163
$C_3H_2N_2O$
9c1 24
$C_3H_2N_2OS^-$
17b 346
$C_3H_2N_2O_2^-$
17b 345
$C_3H_2N_2O_2S^-$
9d1 584
$C_3H_2N_3O_2^{2-}$
9d1 587, 588, *17f* 285, 286, 287
$C_3H_2O^-$
9b 433
$C_3H_2O_2^-$
9b 304, *17b* 494
$C_3H_2O_3^-$
9b 49, 161, *17b* 211
$C_3H_2O_4^-$
9b 162, *17b* 224
$C_3H_2S_3^-$
17c 45
C_3H_3
9b 46, 309, 404, 430ff, *17c* 32, 54
C_3H_3AgN
17c 571
C_3H_3AuN
17c 572
$C_3H_3BrNO_4^{2-}$
17f 231
$C_3H_3ClD_4NO_2S$
9c2 277
$C_3H_3ClF_3O$
9b 129
$C_3H_3ClNO_4^{2-}$
9d1 389
$C_3H_3ClO_2^-$
17b 180, 265
$C_3H_3Cl_2$
9b 357
$C_3H_3Cl_2F_2O$
9b 243
C_3H_3CuN
17c 574
$C_3H_3DNOR_2$
17b 266
$C_3H_3DO_2^-$
9b 92
$C_3H_3D_2$
9b 346, *17c* 10
$C_3H_3D_2NO_2^-$
9b 208
$C_3H_3D_2O_2R$
17b 201
$C_3H_3D_2O_3$
17b 267
$C_3H_3D_3O_2^+$
17b 41
$C_3H_3D_4NO_2^+$
17b 266
$C_3H_3D_5NO_3$
1 38, *9b* 236
$C_3H_3D_5N_2O^+$
17b 182
$C_3H_3F_2$
9b 354
$C_3H_3F_2NO_4^-$
9d1 387
$C_3H_3F_3NO_2$
9c1 790
$C_3H_3F_3NO_3^{2-}$
9d1 388
$C_3H_3F_4O$
9b 129
$C_3H_3F_4S$
17b 181

$C_3H_3KO_4^-$
9b 227
C_3H_3MuN
17b 134
$C_3H_3N^+$
17h 20, 233
$C_3H_3NNaO_3$
9c1 140
$C_3H_3NO^-$
9b 127, 309
$C_3H_3NO_2^-$
9b 228, *9c1* 117, *17b* 361
$C_3H_3NO_3^-$
9b 128, *17b* 519
$C_3H_3NO_3P^{2-}$
9b 128
$C_3H_3NO_3X$
(X = various alkyl substituents)
9b 171
$C_3H_3NS^-$
9b 128
$C_3H_3N_2$
9c1 23
$C_3H_3N_2O_2$
17b 345
$C_3H_3N_2O_2R_2$
17b 227
$C_3H_3N_2O_3$
9b 270, *17c* 497
$C_3H_3N_3O_2^-$
9d1 62, 587
$C_3H_3N_4O_2^{2-}$
17f 288
C_3H_3O
1 40, *9b* 433, *17b* 329
$C_3H_3OR_4$
17b 297
$C_3H_3O_2^{2-}$
9b 364
$C_3H_3O_2$
9b 258, 305, *17b* 515, 523
$C_3H_3O_2S^{2-}$
9b 122, 143
$C_3H_3O_3^{2-}$
9b 121, 210, *17b* 267
$C_3H_3O_3$
9b 161, 269, 306, 320, *17b* 222, 370, 514
$C_3H_3O_4$
1 28, *9a* 39, *9b* 162ff, *17b* 225
$C_3H_3O_5$
9b 242
$C_3H_3O_5P^{3-}$
9b 122
$C_3H_3O_5S^{2-}$
9b 122, *17b* 199
$C_3H_3O_6P^{3-}$
17b 180
$C_3H_3O_6S^{2-}$
9b 122, *17b* 180
$C_3H_3O_7P^{3-}$
9b 158
C_3H_4Ag
17a 373, *17b* 495, *17c* 24
C_3H_4Au
17a 376, *17b* 495, *17c* 25
C_3H_4Br
9b 357
$C_3H_4BrO_2$
9b 207
C_3H_4Cl
9b 304, 356, *17c* 12
$C_3H_4ClF_2O$
9b 164
C_3H_4ClO
9b 48, 362
$C_3H_4ClO_2$
9b 207, *17b* 231
$C_3H_4Cl_2NO_2$
17d2 398
$C_3H_4Cl_3$
9b 23, 225
C_3H_4Cu
17a 355, *17b* 496, *17c* 25
C_3H_4D
9b 346, *17c* 9
$C_3H_4DO_2$
9b 26ff, 93
$C_3H_4D_2^+$
17h 18
$C_3H_4D_2F$
17b 124
$C_3H_4D_2NO$
17b 35
$C_3H_4D_2NO_2^{2-}$
9b 227
$C_3H_4D_2NO_2S$
17c 540
$C_3H_4D_2O^+$
17h 41
$C_3H_4D_2O_2^+$
17b 87, *17h* 46
$C_3H_4D_4NO_2$
17b 299
$C_3H_4D_4N_2O^+$
9b 61
C_3H_4F
9b 354
$C_3H_4FN_2O_2$
9b 240
$C_3H_4F_3$
9b 23, *17b* 32
$C_3H_4F_3O$
9b 32, 209, *17b* 40
$C_3H_4F_3S$
9b 34ff, *17b* 45
$C_3H_4KO_3$
1 32
C_3H_4N
9b 27, 94, 387, *17b* 34, 134, *17c* 56
C_3H_4NO
1 31, *9b* 42, 98, 127ff, 136, 212, 305, *17b* 142, 167, 231
$C_3H_4NO_2$
9b 267, *9c1* 107, 117, *17b* 361, *17c* 496, 564
$C_3H_4NO_2P^-$
9b 128
$C_3H_4NO_2S^{2-}$
9b 222
$C_3H_4NO_4^{2-}$
9d1 397, 400
$C_3H_4NO_4S^{2-}$
9d1 402
$C_3H_4NO_6P^{3-}$
9b 222
C_3H_4NS
9b 98, 128, *17b* 234
$C_3H_4NS_2$
17c 533
$C_3H_4N_2^-$
9b 420, *9d1* 883
$C_3H_4N_3O$
9c1 63

$C_3H_4N_3O_4$
9c1 140
$C_3H_4O^-$
9b 358, 359
$C_3H_4O^+$
17b 134, 494, *17h* 44
$C_3H_4O_2^-$
9b 26, 91ff, 173, 364, *9d1* 6ff, *17b* 35, 135ff, *17g* 6
$C_3H_4O_2^+$
17b 40, *17h* 45, 48
$C_3H_4O_2S^-$
9b 74, 122
$C_3H_4O_3^-$
9b 42, 121, 210, *9d1* 7, 49, *17b* 180, *17g* 7
$C_3H_4O_3^+$
17h 49
$C_3H_4O_4^-$
17b 215
$C_3H_4O_4P^{2-}$
9b 122
$C_3H_4O_4S^-$
17b 496
$C_3H_4O_5P^{2-}$
9b 111
$C_3H_4O_5S^-$
9b 122, *17b* 181
$C_3H_4O_6S^-$
17b 180
$C_3H_4S_2^+$
17e 369
$C_3H_4Te_2^+$
17e 379
C_3H_5
1 21, 26, 39, *9b* 253, 305, 345, *17b* 328, *17c* 8ff
$C_3H_5BN_2^-$
17a 24, *17c* 572
$C_3H_5Br^+$
17h 20
$C_3H_5BrNO_5S^{2-}$
17f 231
$C_3H_5Br_2N^{2+}$ (?)
9d2 119
$C_3H_5Cl^+$
17h 20
C_3H_5ClNO
17c 398
$C_3H_5Cl_2$
9b 107, 224
$C_3H_5Cl_2N^{2+}$ (?)
9d2 119
$C_3H_5Cl_2O$
17b 81, 323
C_3H_5DNO
9b 60, 91
$C_3H_5D_2O_2$
17b 291
$C_3H_5D_3NO$
17b 291
$C_3H_5F_2O$
17b 291
$C_3H_5F_2S$
9b 225
$C_3H_5F_3NO$
9c1 248
$C_3H_5F_3NOS$
9c1 986
$C_3H_5F_3NO_2$
9c1 945, *17d1* 207
$C_3H_5NO^-$
9b 380, *17b* 77
C_3H_5NOX
(X = various alkyl substituents)
9b 98
$C_3H_5NO_2^-$
9b 131, 168, 207, *9d1* 382, *17b* 227
$C_3H_5NO_3^-$
9d1 395, 396, 398, 688
$C_3H_5NO_3$
9c2 7
$C_3H_5NO_6P^{2-}$
9b 157
$C_3H_5N_2$
9b 127, 420, *9c1* 22, *17c* 409, 479
$C_3H_5N_2O$
9b 266, 420, *9c1* 68, 118, *17c* 564
$C_3H_5N_2O_2$
9b 161, *17b* 519
$C_3H_5N_2O_3^{2-}$
9d1 689
$C_3H_5N_2O_3$
1 32
$C_3H_5N_2O_4^{2-}$
9d1 422
$C_3H_5N_3O^+$
17h 204
$C_3H_5N_4$
17c 475
C_3H_5O
1 39, *9b* 47ff, 77, 89, 255, 258, 319, 358, 359, *17b* 65, 85, 131
C_3H_5OS
9c2 286, *17b* 378
$C_3H_5O_2$
1 24, 27, *9b* 26, 32, 48, 54, 69, 92ff, 101, 110, 173, 267, 268, 320, 324, 366, *17b* 35, 40, 65, 71, 87, 136ff, 146, 164, 363, 367, 521
$(C_3H_5)_nO_2$
17e 14
$C_3H_5O_2S$
9c2 273, 278, *17b* 90, *17e* 317, 336, 337
$C_3H_5O_2S_2$
9c2 296, *17b* 93
$C_3H_5O_3$
1 27, *9b* 42, 66, 69, 121, 173, 174, 211, 293, *9c2* 19, *17b* 88, 232, 267, 471
$C_3H_5O_3R_2$
17b 297
$C_3H_5O_3S$
9b 270, 271, *9c2* 285, 287
$C_3H_5O_4$
9b 157, 248, *17b* 293
$C_3H_5O_4S$
9c2 289, *17e* 335
$C_3H_5O_5P^-$
9b 122
$C_3H_5R_2$
17b 150
C_3H_5S
9b 77
$C_3H_5S_2$
17b 380, 385
$C_3H_5S_3$
17b 474
$C_3H_6^-$
1 79, *9d1* 845

$C_3H_6^+$
17b 17, *17h* 14ff, 17ff
C_3H_6Br
9b 20
$C_3H_6BrNO^-$
9d1 681
$C_3H_6BrNO_2^-$
17f 230, 233
$C_3H_6BrNO_2^+$
17h 52, 238
$C_3H_6Br_2NO_3S$
9c1 989
C_3H_6Cl
9b 20, 84, 191, *17b* 124
$C_3H_6ClF_2N_2O$
9c1 895
C_3H_6ClMg
9b 191
$C_3H_6ClNO^-$
9d1 681
$C_3H_6ClNO_2^-$
9d1 389, *17f* 232ff
$C_3H_6ClNO_2^+$
17h 52, 238
$C_3H_6ClNO_2S^-$
9c2 276
C_3H_6ClO
9b 65, 135, *17b* 80, 157, 188, 236
C_3H_6ClOS
9c2 275, *17e* 316
$C_3H_6ClOS_2$
9c2 296
$C_3H_6ClO_2$
17b 212
C_3H_6DO
17b 185, 258
$C_3H_6D_2^+$
17h 7
C_3H_6F
9b 20, 85, 191, *17b* 124
C_3H_6FO
17b 181
C_3H_6FS
9b 129
$C_3H_6F_3S_2$
17a 104, *17e* 344
C_3H_6I
9b 191
C_3H_6Mu
17b 117
C_3H_6MuO
17b 258
C_3H_6N
9b 25, 30, 79, *9c1* 20, 116, 119, *17c* 406, 563, 567
C_3H_6NO
9b 26, 60, 62, 91, 323, *9c1* 118, 123, *17b* 20, 34, 77,135, 518, *17c* 399, 589
$C_3H_6NO_2$
9b 39, 40, 101, 114, 120, 192, 208, *9c1* 198, 764, 770, *17b* 54, 180, 257, *17d2* 391, 392
$C_3H_6NO_2^+$
9b 258
$C_3H_6NO_2S^{2-}$
9d1 406
$C_3H_6NO_2S$
9c2 275ff, *17c* 540, *17e* 316
$C_3H_6NO_3$
9b 157, *9c1* 198, 765, 772, *17b* 214, *17d2* 391, *17e* 5
$C_3H_6NO_3S$
17d2 415
$C_3H_6NO_5P^{3-}$
9d1 413
$C_3H_6NO_5S^{2-}$
9d1 405, 406, *17f* 232
$C_3H_6NO_6S^{2-}$
9d1 404
$C_3H_6N_2^+$
17b 167, *17h* 157
$C_3H_6N_2O_3^-$
9d1 683
$C_3H_6N_2O_4^-$
9d1 408, 409, *17f* 237
$C_3H_6N_2O_4^+$
17h 238
$C_3H_6N_2S_2^-$
9d1 75
$C_3H_6N_3$
17b 257
$C_3H_6N_4O_2^-$
17f 235
$C_3H_6N_5O_5$
9c1 512, *17d2* 336
$C_3H_6O^-$
9b 85, 192
$C_3H_6O^+$
17b 54, 82, *17h* 20, 36, 41, 42
$C_3H_6OSiX_3$
(X = silica gel matrix)
9b 195
$C_3H_6O_2^-$
9b 139, 215, 226, *17b* 273, 291
$C_3H_6O_2^+$
17b 41, 87, *17h* 38, 46
$C_3H_6O_2PS_2$
17e 327
$C_3H_6O_2R$
17b 189
$C_3H_6O_3^-$
17b 324
$C_3H_6O_3^+$
17h 39, 49
$C_3H_6O_4S^-$
9b 109, *17b* 126
$C_3H_6O_5P^{2-}$
17b 189, 194
$C_3H_6O_5P$
9b 111
$C_3H_6O_5S^-$
9b 109, *17b* 55, 158
$C_3H_6O_6P^{2-}$
9b 157
$C_3H_6S^+$
17e 363
$C_3H_6S_2^-$
9c2 303
$C_3H_6S_2^+$
17e 367ff, 378
C_3H_7
1 22, 25, *9b* 12ff, 80ff, *17b* 15ff, 116
$C_3H_7(Br^-)$
17b 116
$C_3H_7(I^-)$
17b 117
$C_3H_7AlCl_2$
9b 95
$C_3H_7AsO_2^-$
17f 56
$C_3H_7AsO_3^{3-}$
17f 56

$C_3H_7BN_2^-$
17a 24
$C_3H_7ClNO_2S$
1 76, *9c2* 276ff, *17e* 316
$C_3H_7ClNO_2S_2$
9c2 296ff
$C_3H_7ClNO_3$
9c1 917, 922
$C_3H_7ClNO_3S$
9c1 990
$C_3H_7Cl_2Si$
17b 240, 327
$C_3H_7F_4OS$
9c2 295
$C_3H_7N^-$
17b 181
$C_3H_7N^+$
17h 126, 127
$C_3H_7NO^+$
17b 513, *17h* 49, 157
$C_3H_7NOS^-$
9c1 201
$C_3H_7NO_2^-$
1 77, *9d1* 379, 381ff
$C_3H_7NO_2^+$
9b 168, *17b* 54, *17h* 53, 238
$C_3H_7NO_2S^+$
9c2 276
$C_3H_7NO_3^-$
9d1 391, 393, 394, 399, 400, 401
$C_3H_7NO_3^+$
17b 214
$C_3H_7NO_3S^-$
9d1 402
$C_3H_7NO_3S^+$
9c2 286
$C_3H_7NO_4^-$
9d1 395
$C_3H_7NO_4P^{2-}$
9c1 199, *9d1* 413
$C_3H_7NO_4S^+$
9c2 290
$C_3H_7NO_5S^-$
9b 236, *9d1* 404
$C_3H_7N_2O$
1 32, *9b* 61, 98
$C_3H_7N_2O_2$
9c1 62, 765
$C_3H_7N_2O_5$
9c1 922
$C_3H_7N_2O_5S$
9c1 990
C_3H_7O
1 24, 26, 38, *9b* 20, 31, 39, 65, 85, 100, 134, 193, *17b* 25, 39, 52, 124ff, 185, 257ff
$C_3H_7OPS_2^-$
17e 263
C_3H_7OS
9b 74, *9c2* 285, 287, *17e* 328, 329
$C_3H_7O_2$
1 24, 27, 31, *9b* 41, 42, 65, 66, 108, 135, 138, 139, 155, 176, 177, 204, *9c2* 9, 11, *17b* 55ff, 80, 82, 188, 236
$C_3H_7O_2S$
9c2 278, 288, 291, *17e* 335, 336
$C_3H_7O_3$
9b 155ff, 218, *9c2* 20, *17b* 214, 277
$C_3H_7O_3P^-$
9c2 219
$C_3H_7O_3S_2$
9b 75
$C_3H_7O_5P^-$
17b 189, 194
$C_3H_7O_6P^-$
9b 156
C_3H_7S
9b 34, 73, *17b* 45, 89
$C_3H_7S_2$
17b 238
$C_3H_8^+$
17h 7
$C_3H_8AsO_3$
9b 84
C_3H_8B
9b 55
$C_3H_8BD_2N_2$
17c 573
$C_3H_8BN^-$
17a 24, *17c* 573
$C_3H_8BO_3$
9b 69
$C_3H_8ClNO_2S^-$
17b 301
C_3H_8ClSi
9b 75, *17b* 94
C_3H_8ClSn
9b 76
C_3H_8FSi
9b 75
C_3H_8N
9b 59, 130, 191ff, *9c1* 7, *17b* 75ff, *17c* 390
C_3H_8NO
9b 40, 107, *9c1* 97, 199, *17d2* 205
C_3H_8NOS
17c 522
$C_3H_8NO_2$
9b 227, 235, *9c1* 98, 199, 201, 209, 913
$C_3H_8NO_2S$
17c 526
$C_3H_8NO_3$
9b 236, *9c1* 913, 914, 917
$C_3H_8NO_3P^-$
9c1 199
$C_3H_8NO_5S$
9b 236, 237
$C_3H_8N_2O^+$
9b 61, *17b* 77
$C_3H_8N_2O_2^-$
9d1 407, 409, 410
$C_3H_8O^+$
9b 20, 194
C_3H_8OP
9b 72
$C_3H_8O_2^+$
17h 36
$C_3H_8O_3P$
9b 70, 177, *17b* 88
$C_3H_8O_4P$
9b 70ff, 177
$C_3H_8O_5P$
17b 81, 190
$C_3H_9Al^-$
17f 62
$C_3H_9B^-$
17f 221
C_3H_9BN
17b 25, 38, 52, 124, *17c* 406

$C_3H_9ClO_3P$
17e 267
C_3H_9ClP
17e 267
$C_3H_9ClPb^-$
17f 54
$C_3H_9ClSn^-$
17f 51
C_3H_9DPS
17e 267
$C_3H_9F_4OSSi$
9c2 295
C_3H_9Ge
9c2 310, *17b* 75
$C_3H_9GeO_2$
9c2 23
$C_3H_9N^+$
1 20, *9b* 20, 39, 59, 85, 131, 192, *9d2* 26, *17h* 127
$C_3H_9NO^+$
9b 40, 136, 139, 203
$C_3H_9NS^+$
17h 137
$C_3H_9N_2$
9b 107, 131, *9c1* 42
$C_3H_9N_3^+$
17b 76
$C_3H_9OP^-$
17e 268
$C_3H_9OP^+$
9b 72
$C_3H_9O_2Pb$
9c2 24
$C_3H_9O_2Si$
9c2 22
$C_3H_9O_2Sn$
9c2 23
$C_3H_9O_3P^+$
9b 177, *17e* 307
$C_3H_9O_3S$
17e 355
$C_3H_9O_4P^-$
9c2 224
$C_3H_9P^+$
9c2 263, *17e* 307
$C_3H_9PS^-$
17e 268
$C_3H_9PS_3^+$
9c2 263
C_3H_9Pb
9c2 316
$C_3H_9S_2$
9c2 295
C_3H_9Si
9c2 308, *17b* 93, *17f* 6
C_3H_9Sn
9c2 313, *17f* 46
$C_3H_{10}N_2O^+$
9c1 201
$C_3H_{10}OP$
9c2 224
$C_3H_{10}O_3P$
9c2 224
$C_3H_{10}O_3PS$
17e 268
$C_3H_{10}P$
9c2 224
$C_3H_{10}PS$
17e 268
$C_3H_{10}Sn^-$
17f 51
$C_3H_{10}Sn^+$
17f 53
$C_3H_{11}BN$
9b 55, *17f* 60
$C_3H_{11}BO_3P$
17a 91, *17f* 60
C_3N
17a 30
C_3NX_9
17a 44
C_3N_3O
17c 598
$C_3NiO_3^+$
17a 128
$C_4CdN_4^{3-}$
9a 244
$C_4CdN_4S_4^{3-}$
9a 244
$C_4Cl_2F_8NO$
9a 68, *9c1* 382, *17d2* 220
$C_4Cl_6N_3S$
17c 541
$C_4CoO_4^{2-}$
9a 201
C_4CoO_4
17a 126
C_4CoO_6
9a 119
$C_4DO_4^{2-}$
17b 495
$C_4D_5O_3$
9b 319
C_4D_9
17b 242
$C_4D_{10}NO^+$
17h 50
$C_4F_3Na_2O_4$
1 39
$C_4F_3O_4^{2-}$
9b 240
$C_4F_6^+$
17h 28
$C_4F_6NS_2$
17c 536
$C_4F_6NaO_2$
9b 239
$C_4F_6O^-$
9b 257
$C_4F_6O_2^-$
9d1 17
$C_4F_6S_2^+$
17e 371
$C_4F_7NO_2^-$
9d1 388
$C_4F_7R_2$
17b 305
$C_4F_8^-$
9a 49
C_4F_8I
9b 246
C_4F_8NO
9c1 516
$C_4F_8N_3O_5$
9c1 382
C_4F_8OR
17b 324
C_4F_8R
17b 306
C_4F_9
9b 238, *17b* 305
$C_4F_9I^-$
17a 29
C_4F_9O
9b 242
$C_4F_9O_4S$
9a 167
$C_4F_9S_3$
17b 325

$C_4F_{10}NO$
17d2 220
$C_4F_{10}NO_5S_2$
9c1 381
$C_4F_{10}N_2^-$
9a 78
$C_4F_{10}N_2O_2^{2-}$
9d2 191
$C_4F_{12}OP$
9a 152, *9c2* 224
$C_4F_{13}OS$
9c2 295
$C_4FeO_4^-$
17a 123
C_4H
9a 23, *17a* 30
$C_4HBrDN_2O_2$
17b 429
$C_4HBrDN_3O^-$
17b 428, *17c* 413
$C_4HBrDO_2S_2$
17e 342, 343
$C_4HBr_2O_2S_2$
17e 342, 343
$C_4HClDN_2O_2$
17b 429
$C_4HCl_6F_6Si_2$
9b 233
$C_4HCoO_4^-$
17a 126
$C_4HDFN_2O_2$
17b 430
$C_4HDFN_3O^-$
17b 429, *17c* 413
$C_4HDIN_2O_2$
17b 430, *17c* 416
$C_4HD_2FN_3O$
17b 430, *17c* 413
$C_4HD_4O_6$
1 38
$C_4HD_5NO_3$
9b 170
$C_4HD_9^+$
17h 12
$C_4HF_6O_2$
9b 239, 246
C_4HF_8O
17b 319
$C_4HN_2O_2^{2-}$
17b 504
$C_4HN_2O_3^{2-}$
9b 288, 307, *9d1* 341, *17b* 425, *17c* 417
$C_4HN_3O_4^{3-}$
9d1 605
$C_4HN_4Ni^-$
17a 128
$C_4HN_4Pt^{2-}$
17a 173
C_4HO_3
9b 321
$C_4HO_3R_2$
17b 355
C_4HO_3X
17b 357
$C_4HO_4^{2-}$
9b 306, *17b* 495
$C_4HO_5^{2-}$
9b 162, *17b* 224, 499
$C_4H_2BrNO_2S^-$
17f 284
$C_4H_2BrNO_3^-$
9d1 574, *17f* 279
$C_4H_2BrN_2O_2$
17b 428, *17c* 416
$C_4H_2BrO_2S_2$
17e 342
$C_4H_2BrO_3$
9b 265
$C_4H_2ClNO_2S^-$
9d1 580
$C_4H_2ClNO_3^-$
9d1 574, *17f* 279
$C_4H_2ClN_2O_2$
17b 429, *17c* 416
$C_4H_2ClO_4^{2-}$
17b 206ff, 283
$C_4H_2DN_2O_2$
17b 423
$C_4H_2DN_3O^-$
17b 423, *17c* 412
$C_4H_2DO_5^{2-}$
17b 209
$C_4H_2D_2N_3O$
9c1 27
$C_4H_2D_2O_4^-$
9d1 64
$C_4H_2D_3KNaO_6$
17b 301
$C_4H_2D_4NO_4$
17b 283
$C_4H_2D_4O_6^-$
9b 237, *17b* 301
$C_4H_2D_5NO_3^-$
9b 229
$C_4H_2D_7N$-
17b 212
$C_4H_2FN_2O_2$
9c1 26, *17b* 430, *17c* 416
$C_4H_2F_3S_3$
17b 384
$C_4H_2F_6NO$
9b 239
$C_4H_2F_8NO$
17d2 220
$C_4H_2INO_2S^-$
17f 284
$C_4H_2INO_3^-$
9d1 574, *17f* 279
$C_4H_2KO_4$
1 57
$C_4H_2NO_2X$
17b 344
$C_4H_2N_2O_2^-$
17c 413
$C_4H_2N_2O_3^{3-}$
9b 530, *17b* 422
$C_4H_2N_2O_3^-$
9b 287, *9d1* 62, *17b* 410, 425, *17c* 417
$C_4H_2N_2O_4^-$
1 115, *9b* 287, *9d1* 80
$C_4H_2N_2O_4S^-$
9d1 657
$C_4H_2N_2O_4Zn^+$
17a 377
$C_4H_2N_2O_5^-$
17f 302
$C_4H_2N_3OS_3$
17c 542
$C_4H_2N_3O_3$
9c1 157
$C_4H_2N_3O_3S$
9c1 156
$C_4H_2N_3O_4^{2-}$
9d1 605
$C_4H_2N_3O_4$
9c1 155

$C_4H_2N_4S^-$
9d1 900
$C_4H_2O_2S^-$
17g 41
$C_4H_2O_3^-$
9d1 68, *17b* 514, *17g* 41ff
$C_4H_2O_3R$
17b 517
$C_4H_2O_4^{3-}$
9d1 64
$C_4H_2O_4^-$
9d1 80
$C_4H_2O_4R$
17b 499
$C_4H_2O_4S^{3-}$
9b 224
$C_4H_2O_5^{3-}$
9b 152, 223, *17b* 285
$C_4H_2O_6^{3-}$
9b 232, *9d1* 64, *17b* 298
$C_4H_2O_7P^{4-}$
9b 153ff
$C_4H_2O_8P^{4-}$
17b 209
$C_4H_2O_8S^{3-}$
9b 153, *17b* 210
C_4H_3
9b 309, 434
$C_4H_3BrN_2O_6S^-$
17b 418
$C_4H_3ClNO_2$
17b 343
$C_4H_3ClN_2O_2$
9b 424
$C_4H_3ClN_2O_6S^-$
17b 419
$C_4H_3DN_3O_2$
17b 438, *17c* 481
$C_4H_3D_2N_2O_2$
17b 406, 431, 438
$C_4H_3D_2O_4$
9b 120, *17b* 179
$C_4H_3D_3NO_3$
17b 177
$C_4H_3D_4NO_3^-$
9b 214
$C_4H_3D_4NO_4^-$
17b 300
$C_4H_3D_4N_2O_3$
17b 229
$C_4H_3D_4O$
17b 347
$C_4H_3D_5NO_4$
17b 300
$C_4H_3D_5N_2O_3^-$
17b 291, 300
$C_4H_3D_5N_2O_3^+$
17b 229
$C_4H_3D_6OS_2$
17e 346
$C_4H_3FN_2O_2^-$
9b 424
$C_4H_3KNaO_6$
1 38
$C_4H_3K_2O_6$
9b 232
$C_4H_3NO_2^-$
9b 206, *9d1* 69, *17b* 342, *17g* 43
$C_4H_3NO_2S^-$
9d1 578ff, 582, *17f* 284
$C_4H_3NO_2Se^-$
17f 285
$C_4H_3NO_3^-$
9d1 572ff, *17f* 277ff
$C_4H_3N_2$
9b 127, 316, *17b* 503, 504
$C_4H_3N_2OS$
17b 424, *17c* 418, *17e* 323
$C_4H_3N_2O_2^{2-}$
17b 425, *17c* 109, 417
$C_4H_3N_2O_2$
9c1 26, *17b* 423, *17c* 414, 463, *17e* 211
$C_4H_3N_2O_2S$
9b 289
$C_4H_3N_2O_3$
9b 287, 288, *9c1* 26, *17b* 410
$C_4H_3N_2O_4$
9b 288ff
$C_4H_3N_2O_5$
17b 519
$C_4H_3N_3O_2^-$
9d1 604
$C_4H_3N_3O_4^-$
9d1 605
$C_4H_3N_3O_5^-$
9d1 606
$C_4H_3N_4O_3^{2-}$
17f 289
$C_4H_3N_4O_3$
9c1 156
$C_4H_3OS_2$
17e 332
$C_4H_3O_2$
9b 416
$C_4H_3O_2S_2$
17e 342
$C_4H_3O_3^{2-}$
9b 369, 370
$C_4H_3O_3$
9b 265, 417, *17b* 356
$C_4H_3O_4^{2-}$
9b 118ff, *9d1* 7, 65, *17b* 177ff, *17g* 60
$C_4H_3O_4$
9b 306
$C_4H_3O_4S^{2-}$
9b 175
$C_4H_3O_4S_2^{2-}$
9b 175
$C_4H_3O_5^{2-}$
9b 153, 174, *17b* 208
$C_4H_3O_6^{2-}$
17b 298
$C_4H_3O_6P^{3-}$
9b 153
$C_4H_3O_8P^{3-}$
17b 210
$C_4H_4AgN_3O_3^+$
17a 374
$C_4H_4Al_2D_6N_2^-$
17a 394
$C_4H_4BrNO_2^-$
17c 407
$C_4H_4BrN_2O_2$
17b 417
$C_4H_4BrN_2O_3$
17b 418
$C_4H_4BrO_4$
9b 221, *17b* 206
$C_4H_4Br_2ClF_3NO$
9c1 329
C_4H_4Cl
9b 46, 433
$C_4H_4ClD_4N_2O_3$
9b 171

$C_4H_4ClF_3NO$
9c1 752
$C_4H_4ClNO_2^-$
17c 407
$C_4H_4ClN_2O_2$
17b 418ff
$C_4H_4ClN_2O_3$
17b 419
$C_4H_4ClN_3O^-$
17c 418
$C_4H_4ClO_4$
17b 207
C_4H_4Cu
17a 131, 356
$C_4H_4DNO_3^-$
17b 68
$C_4H_4D_2NO_3$
9b 170, *17b* 228
$C_4H_4D_2N_2O_2^-$
17b 437
$C_4H_4D_2O_4^-$
9b 227
$C_4H_4D_3NO_3^-$
17b 292
$C_4H_4D_3N_2O_3$
9b 170
$C_4H_4D_3O$
17b 348
$C_4H_4D_3O_4$
17b 292
$C_4H_4D_4NO_3$
17b 265
$C_4H_4D_4N_2O_3^+$
9b 171
$C_4H_4D_5NO_3^+$
17b 265
$C_4H_4D_5N_2O_3$
9b 226, 229
$C_4H_4FN_2O_2$
17b 420
$C_4H_4F_3S_3$
17b 382
$C_4H_4INO_2^-$
17c 408
$C_4H_4KO_5$
17b 232, 233
C_4H_4N
9b 380, 381, *9c1* 20, *17c* 21
C_4H_4NO
9b 163
$C_4H_4NO_2$
9b 69, 263, *17b* 342ff, *17c* 407
$C_4H_4NO_3^{2-}$
9b 171, *9d1* 66, *17b* 227
$C_4H_4NO_3$
9c1 806, *17b* 344
$C_4H_4NO_4^{2-}$
9b 169, 222, *17b* 207
$C_4H_4N_2^-$
1 115, *9d1* 870ff, *17f* 199, 200
$C_4H_4N_2^+$
17h 227ff
$C_4H_4N_2Na_2^+$
17a 205
$C_4H_4N_2O^-$
9d1 870, 871
$C_4H_4N_2OS^-$
9b 426, *17c* 48
$C_4H_4N_2O_2^-$
9b 422, *9d1* 871, *17b* 346, 437, *17c* 46, 157
$C_4H_4N_2O_2S^-$
9d1 584, 585
$C_4H_4N_2O_3S_2^-$
17e 318
$C_4H_4N_2O_6S^-$
17b 409
$C_4H_4N_3O_2^{2-}$
9d1 588, 589, *17f* 287
$C_4H_4N_3O_2$
9b 163, 404, *9c1* 73, *17b* 424, *17c* 418
$C_4H_4N_3O_2S_2$
17c 536
$C_4H_4N_3O_4$
9c1 157
$C_4H_4N_3O_5^{2-}$
9d1 424
$C_4H_4N_5O_2$
9c1 156
$C_4H_4O^-$
9b 304, 433
$C_4H_4O_2^-$
9b 369, 434, *9d1* 30, *17g* 20
$C_4H_4O_3^-$
17b 224
$C_4H_4O_4^{3-}$
9b 226
$C_4H_4O_4^-$
9b 119, *9d1* 65, *17g* 60
$C_4H_4O_4R$
1 29, *17b* 209
$C_4H_4O_7S^-$
9b 154, *17b* 211
$C_4H_4S_2^+$
1 127, *9d2* 143, *17e* 380, 382, *17h* 110
C_4H_5
9b 295, 387, 405, 431, *17c* 30, 54
$C_4H_5BrNO_4^{2-}$
17f 231
$C_4H_5Br_2F_3NO$
9c1 248
$C_4H_5ClNO_4^{2-}$
17f 231
$(C_4H_5Cl)_nO_2$
17e 14
$C_4H_5DNO_3$
9b 169
$C_4H_5DO_2^-$
17b 169
$C_4H_5D_2NO_3^-$
9b 229, *17b* 272, 293
$C_4H_5D_4NO_2^+$
17b 62
$C_4H_5F_3NO$
9c1 249
$C_4H_5N^-$
9b 418, *9d1* 883
$C_4H_5N^+$
17h 219
$C_4H_5NO^-$
17b 340
$C_4H_5NO_2^-$
17b 342
$C_4H_5NO_3^-$
9b 50, 169, *17b* 177, 227
$C_4H_5NO_3X$
(X = various alkyl substituents)
9b 209
$C_4H_5NO_4^-$
9b 152, *17b* 208
C_4H_5NR
17b 261
$C_4H_5NS^-$
9b 202

$C_4H_5N_2$
9b 530, *9c1* 23, 28, *17c* 419
$C_4H_5N_2OS$
17b 409, 433
$C_4H_5N_2O_2$
1 33, *9b* 278, 282, 422, *17b* 346, 404, 406, 431, 437, *17d2* 390
$C_4H_5N_2O_2R_2$
17b 266
$C_4H_5N_2O_3$
9b 161, 270, 279, 285, *17b* 405, 409, 434
$C_4H_5N_2O_5$
9c1 140, 141
$C_4H_5N_2O_6^{2-}$
9d1 425
$C_4H_5N_3O_2^-$
9d1 588, 589, 591, *17f* 286, 287, 288
C_4H_5O
9b 258, 321, 414, *17b* 331, 514, 521, *17c* 36
C_4H_5OS
17b 375, *17c* 41
$C_4H_5O_2^{2-}$
9b 365
$C_4H_5O_2$
9b 254, 305, 369, 375, 415, 434, *17b* 349, 515, 523, *17c* 37
$C_4H_5O_2R_2$
17b 262
$C_4H_5O_2S^{2-}$
9b 199
$C_4H_5O_3^{2-}$
9b 198, 223, *17b* 284
$C_4H_5O_3^-$
9b 45
$C_4H_5O_3$
9b 54, 79, 270, *17b* 224, 371, 372
$C_4H_5O_4$
1 28, 37, *9b* 119ff, 206, 371, *17b* 53, 178
$C_4H_5O_4S$
1 32, 76, *9c2* 279
$C_4H_5O_4S_2$
9b 175, *17b* 234
$C_4H_5O_5$
1 28, 32, *9b* 153, 223ff
$C_4H_5O_5P^{3-}$
9b 146, 199
$C_4H_5O_5S^{2-}$
9b 147
$C_4H_5O_5S$
9c2 287
$C_4H_5O_6^{2-}$
17b 301
$C_4H_5O_6$
9b 232
$C_4H_5O_6P^-$
9b 153
$C_4H_5O_6S^{2-}$
9b 88, 106, 147
$C_4H_5O_6S$
9c2 291
$C_4H_5O_7P^-$
9b 154
$C_4H_5O_8P^-$
17b 210
C_4H_5S
9b 417, *17c* 40
$C_4H_5S_3$
17b 384
$C_4H_6^-$
9d1 844
$C_4H_6^+$
17h 28
$C_4H_6^{2\cdot}$
17h 250
C_4H_6Al
17c 25
C_4H_6Cl
17c 12
$C_4H_6ClNO_3^-$
9d1 398
C_4H_6ClO
17b 349
$C_4H_6ClO_2$
17b 155, 367
$C_4H_6Cl_3$
9b 18, 224
$C_4H_6Cl_3NO_2^-$
9d1 389
$C_4H_6Cl_3O_2Sn$
17a 451
$C_4H_6Cl_4NO$
9c1 379
$C_4H_6CsO_2$
17a 223
$C_4H_6DO_2$
17b 169
$C_4H_6D_2NO$
17b 254
$C_4H_6D_3NO_2^-$
9b 248
$C_4H_6D_4^+$
17h 8
$C_4H_6FO_2$
17b 366, 367
$C_4H_6F_3$
9b 38
$C_4H_6F_3N_2O_3$
17d1 209
$C_4H_6F_3O$
9b 85
$C_4H_6F_3OS$
17b 239
$C_4H_6F_3OSe$
17b 274
$C_4H_6F_3S$
9b 86
$C_4H_6F_3S_2$
17e 346
$C_4H_6F_3S_3$
17b 325
$C_4H_6KO_2$
17a 212
C_4H_6Mu
17b 28
$C_4H_6MuO_2$
17b 146
$C_4H_6LiO_2$
17a 200
C_4H_6N
9b 20, 39, 84, 126, 190ff, 418, *9c1* 20, *17b* 24, 51, 123, 166, 253ff
C_4H_6NO
1 35, *9b* 40, 45, 50, 88, 106, 126, 133, 154, 192, 202, 254, 263, *9c1* 150, *17b* 129, 167, 202, 257, 341, 362, *17c* 44
$C_4H_6NO_2$
9b 69, 254, *9c1* 139
$C_4H_6NO_3^{2-}$
9b 214

$C_4H_6NO_3$
1 32, *9b* 169, 207, *9c1* 385, 805, *17b* 68, 177, 228, *17c* 588

$C_4H_6NO_4^{2-}$
9d1 397, 400

$C_4H_6NO_4$
9b 152, *17d2* 392

$C_4H_6NO_5^{2-}$
9d1 397

C_4H_6NS
9b 134, 192, 202, *17b* 378, *17c* 44

$C_4H_6N_2^-$
9b 420, *17f* 199

$C_4H_6N_2^+$
1 125, *9d2* 68, 89, 105, *17h* 176, 219

$C_4H_6N_2O_2^-$
17f 235

$C_4H_6N_2O_2^+$
17b 406

$C_4H_6N_2O_2S^-$
17f 236

$C_4H_6N_2O_3^-$
17b 229

$C_4H_6N_2O_5$
9c1 770

$C_4H_6N_2S_2^+$
17h 233

$C_4H_6N_3O$
1 33, *9b* 277, 282, 531, *17b* 436

$C_4H_6N_3O_2$
9b 278, 279, 285, *9c1* 118

$C_4H_6N_3O_5S$
9b 285

$C_4H_6N_4^-$
9d1 880

$C_4H_6N_4OS^-$
17c 473

$C_4H_6N_4O_2^-$
17c 471

$C_4H_6N_4O_2^+$
17h 227

$C_4H_6N_4S_2^-$
17c 474

$C_4H_6N_4S_2^+$
17h 227

$C_4H_6NaO_2$
17a 205

$C_4H_6O^-$
9b 360, *17c* 13

$C_4H_6O^+$
17b 167, 254, *17h* 44

$C_4H_6OS^-$
17g 47

$C_4H_6OS_3^-$
17g 44

$C_4H_6O_2^-$
9b 38, 114, 188ff, 364, 365, *9d1* 8, 10ff, *17b* 24, 123, 169, 254, *17g* 10ff

$C_4H_6O_2^+$
17b 54, 111, 355, *17h* 20, 21, 45, 48

$C_4H_6O_2R$
17b 193

$C_4H_6O_2S^-$
9b 199

$C_4H_6O_2S_2^-$
17g 44

$C_4H_6O_3^-$
9b 88, 106, 146, 199, *9d1* 50, *17b* 129, 156, 186, 208

$C_4H_6O_4^-$
9b 226, *9d1* 52

$C_4H_6O_4^+$
17h 47

$C_4H_6O_4P^{2-}$
9b 42, 113, 146, 199

$C_4H_6O_4S^-$
9b 229, *17b* 496, *17c* 19

$C_4H_6O_4S^+$
9c2 298

$C_4H_6O_4S_2^-$
9c2 302

$C_4H_6O_4S_2^+$
9c2 299

$C_4H_6O_5P^{2-}$
9b 142

$C_4H_6O_5S^-$
9b 142, 147, *17b* 63, 194, 211

$C_4H_6O_6P^{2-}$
17b 111

$C_4H_6O_6S^-$
9b 88, 106, *17b* 129, 158, 210, 262

$C_4H_6S^-$
17c 18

$C_4H_6S_2^+$
17e 370, 380, *17h* 108, 109

C_4H_7
1 23, 26, *9b* 23, 77, 256, 346, 347, *17b* 329, *17c* 10

$C_4H_7BN_2^-$
17c 573

$C_4H_7Ca_{1/2}N_3O_2$
1 33

C_4H_7ClNO
9c1 776

$C_4H_7ClNO_4S$
9c1 990

$C_4H_7Cl_2$
9b 164

$C_4H_7Cl_2NO_2^-$
9d1 389

$C_4H_7Cl_2O$
17b 83

$C_4H_7Cl_3O_4P$
9b 71

$C_4H_7D_2O_2$
17b 291

$C_4H_7D_3NO$
17b 299

$C_4H_7F_3NO$
9c1 329

$C_4H_7F_3NO_2$
17d1 207, 209

$C_4H_7NO^-$
17b 77, 141

$C_4H_7NO_2^-$
9b 45, *9d1* 414, *17b* 78, 185

$C_4H_7NO_2S^-$
9c1 201

$C_4H_7NO_3^-$
9d1 52, 391, 396, 398, *17b* 291

$C_4H_7NO_3PS$
17e 268

$C_4H_7NO_3S_2^-$
9d1 402

$C_4H_7NO_4^-$
9d1 393, 396, 399

$C_4H_7NO_5^-$
9d1 399
$C_4H_7NS^+$
17b 375, *17c* 42
$C_4H_7N_2$
9b 202, *17b* 128, 202
$C_4H_7N_2O_2$
9b 113, 206, *9c1* 138, *17b* 64, *17c* 596
$C_4H_7N_2O_3^{2-}$
9b 225
$C_4H_7N_2O_3$
1 32, *9b* 170, 171, *17b* 203, 229
$C_4H_7N_2O_3S$
9c1 990
$C_4H_7N_2O_4^{2-}$
9d1 422
$C_4H_7N_3O^+$
17b 421
$C_4H_7N_3O_2^+$
17b 422
$C_4H_7N_4O$
9b 278, *17c* 468
$C_4H_7N_4OS$
17c 474
$C_4H_7N_4O_2$
17c 471
$C_4H_7N_4S^{2+}$
17e 322
$C_4H_7N_4S$
17c 471ff
$C_4H_7N_4S_2$
17c 474
$C_4H_7N_4S_2^{2+}$
17e 358
C_4H_7O
9b 24, 89ff, 101, 187, 255, 257, 263, 264, 319, 320, 359, 360, 362, 367, *17b* 28, 29, 134ff, 145, 163, 164, 252, 347, 355, 499, 512, 514, *17c* 13, 14, 16
C_4H_7OS
9b 266, 293, *17b* 378, 379, 471, 472
$C_4H_7O_2$
1 30, 35, *9b* 20, 27, 32ff, 38, 54, 69, 84, 93, 101, 112, 114, 135, 138, 189, 210, 267, 269, 292, 293, 367, *9c2* 13, *17b* 25, 26, 36, 41, 51, 52, 55, 71, 123, 125, 137, 146, 168, 169, 186, 193, 202, 254, 258ff, 261, 266, 350, 355, 363, 364, 370, 372, 466, 470, 521, *17c* 14, 16
$(C_4H_7O_2)_n$
17e 11
$C_4H_7O_2S$
9b 266, *9c2* 275, 280, *17b* 234
$C_4H_7O_3$
1 27, 28, 36, *9b* 40, 41, 45, 54, 88, 106, 108, 117, 124, 142, 199,211, 217, 223, *17b* 62, 72ff, 81, 82, 84, 111, 129, 143, 156, 157, 168, 187, 193, 208, 232, 262, 263, 285, 366, 367
$C_4H_7O_3^+$
17h 49
$C_4H_7O_4$
9b 227, 248, 249, *17b* 292
$C_4H_7O_4P^-$
9b 199
$C_4H_7O_5$
9b 249
$C_4H_7O_5P^-$
9b 199
$C_4H_7R_2$
17b 260
C_4H_7S
9b 265ff, *17b* 374ff
$C_4H_7S_2$
9b 293, *17b* 380, 385, 472, 473
$C_4H_7S_3$
17b 381
$C_4H_8^+$
17h 16, 18ff
C_4H_8Ag
17a 156, 374
C_4H_8Al
17a 394
C_4H_8Au
17a 175, 376
$C_4H_8BD_2N_2$
17c 574
C_4H_8Br
9b 17, 105, 182ff, *17b* 246
C_4H_8Cl
9b 17, 87, 183
$C_4H_8ClNO_4^-$
17b 292, 300
$C_4H_8ClN_2O_3$
1 33
$C_4H_8ClN_3O$
1 34
C_4H_8ClO
17b 155
$C_4H_8ClO_2$
17b 84
$C_4H_8Cl_2O_4P$
9b 233ff
C_4H_8Cu
17a 132, 356
C_4H_8D
9b 37, 180, *17b* 242
C_4H_8DO
17b 186, 213
C_4H_8DSi
17f 6
$C_4H_8D_3N^+$
9b 155
C_4H_8F
9b 183, *17b* 246
$C_4H_8FN_3O$
1 34
C_4H_8FS
9b 130
C_4H_8I
9b 183
C_4H_8Mu
17b 242
C_4H_8MuO
17b 121, 128, 155
$C_4H_8MuO_2$
17b 156
C_4H_8N
9b 30, 77, 262, *9c1* 20, 116, *17c* 20, 564
C_4H_8NO
9b 38, 62, 188, *9c1* 124, 386ff, *17b* 24, 77, 123, 141, *17c* 399, *17d2* 222, 223

C_4H_8NOS
17b 183, *17e* 333
$C_4H_8NO_2$
9b 22, 45, 88, 107, 117, 197, 198, *9c1* 516, 765, 772, *17b* 79, 128, 187, 193, 207, *17d2* 337, 392
$C_4H_8NO_2S$
17b 90, 180, *17c* 529
$C_4H_8NO_3$
9b 42, 135, *9c1* 201, 202, 770, 772, 776, *17b* 265
$C_4H_8NO_3S$
17d2 415
$C_4H_8NO_5$
9b 224
C_4H_8NS
17e 325
$C_4H_8N_2O^-$
9d1 686
$C_4H_8N_2O_2^-$
9d1 52
$C_4H_8N_2O_2^+$
1 121
$C_4H_8N_2O_3^+$
9c1 19, *17b* 226, 229
$C_4H_8N_2O_4^-$
9d1 408, 409, 410
$C_4H_8N_2S_2^-$
9d1 75
$C_4H_8N_3O_2$
9b 208, *17b* 230
$C_4H_8N_3O_3$
9c1 512, 513
$C_4H_8N_4^+$
9d2 68, 69
$C_4H_8N_4O_2^+$
17h 176
$C_4H_8N_4O_4$
1 33
$C_4H_8N_4S_2^+$
17h 177
$C_4H_8N_7O_7$
9c1 523, *17d2* 339
$C_4H_8NaO_4S$
9b 197
$C_4H_8O^-$
9b 203, *17b* 213
$C_4H_8O^+$
9b 46, *17b* 54, 144, *17h* 20, 37, 41
C_4H_8OR
17b 191
$C_4H_8OSiX_3$
(X = silica gel matrix)
9b 44
$C_4H_8O_2^-$
9b 226, 234, *17b* 273, 291, 294, 299
$C_4H_8O_2^+$
17b 26, 71, 258, *17h* 22, 39, 46
$C_4H_8O_3^+$
17h 39
$C_4H_8O_3P^{2-}$
9b 184
$C_4H_8O_3P$
9c2 219
$C_4H_8O_4P^{2-}$
9b 140
$C_4H_8O_4P$
9c2 225, *17e* 269
$C_4H_8O_4S^-$
9b 141, *17b* 128, 249
$C_4H_8O_5S^-$
17b 129, 158, 260
$C_4H_8O_6P^{2-}$
17b 73
$C_4H_8Pd^+$
17a 150
$C_4H_8S^+$
17b 330, *17e* 363
$C_4H_8S_2^-$
9c2 303
$C_4H_8S_2$
9c2 272
$C_4H_8S_2^+$
17e 367, 368, 377, 379
C_4H_9
1 23, 25, 34, *9b* 13ff, 37, 81, 179ff, *17b* 16ff, 47ff, 117, 240ff
$C_4H_9(Br^-)$
17b 47, 241
$C_4H_9(Cl^-)$
17b 241
$C_4H_9(I^-)$
9b 180, *17b* 48, 241
$C_4H_9AsNO_3$
17d1 201
$C_4H_9AsNO_4M$
17d1 201
$C_4H_9BrNO_3S$
9c1 990, 994
C_4H_9ClNO
9c1 202
C_4H_9ClNOS
9c2 274
$C_4H_9ClNO_2S$
9c2 273, 277
$C_4H_9ClNO_3$
9c1 922, 933
$C_4H_9ClNO_3S$
9c1 994
$C_4H_9D^+$
17h 12
C_4H_9DNOS
17e 333
$C_4H_9FNO_3S$
9c1 994
$C_4H_9FO_3P$
9c2 225
$C_4H_9F_3OP$
9c2 225
$C_4H_9F_3O_3P$
9c2 226
$C_4H_9F_3O_3PS$
17e 269
$C_4H_9F_4OS$
9c2 295
C_4H_9Hg
17a 389, *17b* 38
C_4H_9Mg
9b 96
$C_4H_9N^-$
17b 181, 212
$C_4H_9N^+$
9b 262
$C_4H_9NNaO_3S$
9b 197
$C_4H_9NNaO_4S$
17d1 195
$C_4H_9NO^-$
9d1 680, *17f* 305
$C_4H_9NO^+$
9b 292, *17h* 50, 157

$C_4H_9NO_2^-$
1 77, *9b* 235, 248, *9d1* 382, 384ff, *17f* 232

$C_4H_9NO_2^+$
9b 22, 142, *17b* 155, *17h* 53, 238

$C_4H_9NO_3^-$
1 78, *9d1* 391, 394, 399, 400, 401

$C_4H_9NO_3^+$
9b 205, *17b* 265

$C_4H_9NO_3S^-$
9d1 404

$C_4H_9NO_3SM$
17d1 194

$C_4H_9NO_4^-$
1 78, *9d1* 401

$C_4H_9NO_4P$
17e 269

$C_4H_9NO_4S^-$
17f 235

$C_4H_9NO_4SM$
17d1 195

$C_4H_9NO_5^-$
9d1 401

$C_4H_9N_2$
9c1 41

$C_4H_9N_2O$
9b 192, *9c1* 513, *17b* 141

$C_4H_9N_2O_2$
9b 40, 108, *9c1* 202, *17b* 184. *17d2* 209

$C_4H_9N_2O_3$
17b 293

$C_4H_9N_2O_5$
9c1 933

$C_4H_9N_2O_5S$
9c1 990

$C_4H_9N_3^{2\cdot}$
17h 264

$C_4H_9N_3O_2^-$
17b 294

$C_4H_9N_3O_3$
1 33

$C_4H_9N_4$
17c 464

C_4H_9O
1 24, 31, 57, *9b* 17, 31, 38, 44, 85, 87, 100, 134, 155, 183, 203, *9c2* 7, *17b* 20ff, 48, 60, 128, 185ff, 213, 246, 263

C_4H_9OS
9b 103, *9c2* 286, 287, *17b* 90, 148, 149, 197, 238, *17e* 328, 329

$C_4H_9OS_2$
17e 346

$C_4H_9O_2$
9b 41, 42, 65, 66, 88, 135, 139ff, 140, 176, 177, 197, 205, 216, *9c2* 9, 11, 15ff, *17b* 56, 63, 80, 83, 129, 156, 162, 188ff, 191, 221, 260, 264, 273, *17e* 11

$C_4H_9O_2P^-$
9b 184

$C_4H_9O_2S$
9b 18, 104, 144, *9c2* 286, 289, 291, *17b* 55, 197, *17e* 335, 336

$C_4H_9O_2S_2$
9b 74, *9c2* 275, 278, 289, *17e* 317

$C_4H_9O_2S_3$
9c2 297

$C_4H_9O_3$
9b 142, 216, *9c2* 19, 26, *17b* 73, 162, 278, 294, 324

$C_4H_9O_3P^-$
9b 140, *184*, *9c2* 219

$C_4H_9O_3S$
9b 31, *9c2* 290, 294

$C_4H_9O_3S_2$
9b 74, 103

$C_4H_9O_4P^-$
9b 141, *17b* 43, 147

C_4H_9S
9b 73, 86, 103, *9c2* 272, 279, *17b* 45, 148

$C_4H_9S_2$
9b 104, *9c2* 272, *17b* 90, *17e* 317, 357, 358

$C_4H_9S_3$
17b 325

C_4H_9Si
17f 6

$C_4H_{10}^+$
17h 7ff, 12

$C_4H_{10}AlCl^-$
17f 63

$C_4H_{10}Al_2N_2^-$
17a 394, *17f* 200

$C_4H_{10}BD_2N_2$
17c 573

$C_4H_{10}Br_2^+$
17h 56

$C_4H_{10}ClNO^+$
9b 142

$C_4H_{10}ClO_3P^-$
9c2 226

$C_4H_{10}Cl_2^+$
17h 54

$C_4H_{10}Cl_2O_2P$
9c2 226

$C_4H_{10}Cl_2PS_2$
9c2 226

$C_4H_{10}F_2O_2P$
9c2 227

$C_4H_{10}Hg^+$
17f 64

$C_4H_{10}MuN_2$
17c 439

$C_4H_{10}N$
9b 59, 96ff, 130, 155, 202, *9c1* 6, 7, 8, *17b* 76, 212, *17c* 389, 390, 392

$C_4H_{10}NO$
1 54, *9b* 197, *9c1* 97, 98, 202ff, 213, *17b* 128, *17c* 391, *17d1* 110, *17d2* 206

$C_4H_{10}NOS$
17c 523, *17e* 333

$C_4H_{10}NOS_2$
9c1 213

$C_4H_{10}NO_2$
9c1 99, 203, 204, 210, 218, 913, *17d1* 184

$C_4H_{10}NO_2S$
17c 526

$C_4H_{10}NO_3$
9c1 204, 258, 913, 914, 917, 923, *17d2* 212

$C_4H_{10}NO_3S$
9c1 991

$C_4H_{10}NO_4$
9c1 915

$C_4H_{10}NS$
17e 325

$C_4H_{10}N_2^+$
9d2 95
$C_4H_{10}N_2O^-$
9d1 685
$C_4H_{10}N_2O^+$
9b 97, *9d2* 95
$C_4H_{10}N_2O_2^-$
9d1 407
$C_4H_{10}N_2O_2^+$
9c1 204
$C_4H_{10}N_2O_2P$
17e 270
$C_4H_{10}N_3O_5P$
1 33
$C_4H_{10}N_4^+$
9d2 60
$C_4H_{10}O^+$
17h 37
$C_4H_{10}O_2^+$
17b 143, *17h* 37
$C_4H_{10}O_2P$
9c2 216
$C_4H_{10}O_2PS_2$
17e 326
$C_4H_{10}O_2PS_3$
17e 359
$C_4H_{10}O_3P$
9b 33, 102, *9c2* 219ff
$C_4H_{10}O_3PS^{2-}$
17e 270
$C_4H_{10}O_3PS$
17e 326
$C_4H_{10}O_3PS_2$
17e 359
$C_4H_{10}O_4P^{2-}$
9c2 226, 227
$C_4H_{10}O_4P$
9b 102, 216, *17e* 270
$C_4H_{10}P$
17b 44
$C_4H_{10}PS$
17e 263
$C_4H_{10}PS_2^{2-}$
17e 271
$C_4H_{10}PS_2$
9c2 216
$C_4H_{10}S^+$
17e 361, 362
$C_4H_{11}Al$
9b 55
$C_4H_{11}BN$
17c 407
$C_4H_{11}Ge$
9b 56
$C_4H_{11}GeO_4$
9b 70
$C_4H_{11}Li_4$
9b 57
$C_4H_{11}N^+$
9b 17, 18, 29ff, 38, 60, 83, 87, 97, 183, *9d2* 26
$C_4H_{11}NO^+$
9b 88, 137, 140, 203, *17b* 191, *17d1* 111
$C_4H_{11}NO_2$
17d2 209
$C_4H_{11}NO_2^+$
9b 137
$C_4H_{11}NO_3^+$
9b 172
$C_4H_{11}NO_3P$
17b 79, *17e* 271
$C_4H_{11}NO_4P^{2-}$
9c1 1006
$C_4H_{11}NS^+$
17b 89, 196
$C_4H_{11}N_2$
9b 130, 196, *9c1* 41
$C_4H_{11}N_2O$
9c1 218, *17d2* 205
$C_4H_{11}N_2O_2$
17d2 205
$C_4H_{11}N_2O_3S$
17d1 195
$C_4H_{11}N_2O_6$
9b 232
$C_4H_{11}OSi$
9b 71
$C_4H_{11}O_2Si$
9b 71
$C_4H_{11}O_4P^+$
9b 178
$C_4H_{11}O_4Si$
9b 71
$C_4H_{11}P_2S_2$
9b 72
$C_4H_{11}Pb$
9b 72ff
$C_4H_{11}Si$
9b 75, *17b* 93
$C_4H_{11}Si_2$
9c2 308
$C_4H_{11}Sn$
9b 76
$C_4H_{12}AsS$
9c2 317
$C_4H_{12}BNOM$
17d1 198
$C_4H_{12}BN_3^-$
17c 581
$C_4H_{12}Ge^+$
17f 46
$C_4H_{12}GeNO_2$
17d2 408
$C_4H_{12}NOSi$
17c 492
$C_4H_{12}NO_2Si$
17d2 408
$C_4H_{12}N_2^+$
9d2 96, *17h* 143
$C_4H_{12}N_2OP$
9c2 220
$C_4H_{12}N_2OS^+$
17e 385, *17h* 138
$C_4H_{12}N_2P$
17e 255
$C_4H_{12}N_2S^+$
17e 364, *17h* 137
$C_4H_{12}N_2S_2^+$
17h 138
$C_4H_{12}N_3O$
9c1 214
$C_4H_{12}N_3OP$
9c1 1011
$C_4H_{12}N_3Si$
17c 581
$C_4H_{12}N_4^+$
9d2 60, *17h* 159
$C_4H_{12}OP$
9c2 227
$C_4H_{12}O_2S_2^+$
17e 385
$C_4H_{12}O_3P$
9c2 227
$C_4H_{12}O_3PS$
17e 271
$C_4H_{12}O_4P$
9c2 28, *17e* 271, 272
$C_4H_{12}PS$
17e 272

$C_4H_{12}P_2S_2^-$
9c2 263, *17e* 272, 307
$C_4H_{12}Pb^-$
17f 54
$C_4H_{12}Pb^+$
17a 493, *17f* 55
$C_4H_{12}S_2^+$
9c2 299, *17e* 374
$C_4H_{12}Se_2^+$
17e 374
$C_4H_{12}Si^+$
17f 32
$C_4H_{12}Sn^-$
17f 51
$C_4H_{12}Sn^+$
17a 451, *17f* 53
$C_4H_{13}BN$
17c 392
$C_4H_{13}Si_2$
9c2 308
$C_4H_{14}NOSi_2$
9c1 1005
$C_4H_{14}N_3O^{2+}$
9c1 258
$C_4H_{14}P_2^+$
9c2 263
$C_4HgN_4^{3-}$
9a 261
$C_4HgN_4S_4^{3-}$
9a 261
$C_4MnNO_5^-$
17a 122
C_4N_3
9a 40, *17b* 305
$C_4N_3O_4^{2-}$
17c 624
$C_4N_3O_5^{4-}$
9d1 607
$C_4N_4Zn^{3-}$
9a 206
$C_4N_6^-$
1 166
$C_4O_4^-$
9a 49
$C_5BD_8N^-$
17c 411
$C_5BrO_5Re^-$
17a 172
C_5Br_5
9b 439
$C_5ClMnO_5^-$
17a 122
$C_5Cl_3D_9NO$
9c1 379
$C_5Cl_3F_9NO$
9c1 384
C_5Cl_5
9b 439
$C_5CrIO_5^{2-}$
17a 121
$C_5D_9N_3O$
9c1 835ff
C_5F_5
17c 80ff
$C_5F_5N^+$
17h 225
C_5F_6N
9b 494
$C_5F_6NO_3$
9c1 143, *17c* 600
$C_5F_9N_2O$
17d1 209
$C_5F_9O_2$
9b 243
$C_5F_9R_2$
17b 306
$C_5F_{10}O^-$
9b 241
C_5F_{11}
17b 305, 319
$C_5FeKrO_5^+$
17a 124
$C_5FeO_5^-$
17a 124
$C_5FeO_5^+$
17a 124
$C_5HCl_2D_9NO$
9c1 327
$C_5HCl_2F_9NO$
9c1 328
$C_5HDN_2O_4^-$
17b 424, 428, *17c* 417
C_5HD_5N
17c 128
$C_5HF_{10}O$
9b 241, 242
$C_5HN_2O_5^{4-}$
9b 530, *17b* 423
$C_5HN_2O_5^{2-}$
9d1 62
$C_5H_2D_2N_4^-$
17f 210
$C_5H_2D_5O_3$
17b 513
$C_5H_2D_6N_3O_2$
9c1 836
$C_5H_2F_3NO_3^-$
17f 278
$C_5H_2F_3O_4^{2-}$
9b 151
$C_5H_2F_5O_4$
9b 240
$C_5H_2F_6O_2^-$
17c 13
$C_5H_2NO_3^{2-}$
9d1 336
$C_5H_2NO_4S^{2-}$
17f 284
$C_5H_2NO_4Se^{2-}$
17f 285
$C_5H_2NO_5^{2-}$
9d1 577, *17f* 280
$C_5H_2N_2O_2^-$
9b 316
$C_5H_2N_2O_2S$
9d1 582
$C_5H_2N_2O_3^-$
9d1 578, *17f* 283
$C_5H_2N_2O_4^-$
17b 427, *17c* 415
$C_5H_2N_2O_5^{3-}$
9b 280, 283, 494, *9c1* 23
$C_5H_2N_2O_5^-$
9b 287
$C_5H_2O_4^-$
9b 417, *17c* 40
$C_5H_2O_6^{3-}$
9b 151, *17b* 206
C_5H_3
17c 65
$C_5H_3ClN_2O_2^-$
9d1 591
$C_5H_3ClN_4O^-$
17c 426
$C_5H_3CsN_3O_4$
17a 223
$C_5H_3DN_4^-$
17f 210
$C_5H_3D_9NOSi$
17b 509

$C_5H_3F_2N^+$
17h 225
$C_5H_3F_3NO_3$
9c1 143, *17c* 600
$C_5H_3F_6O_2$
9b 242
$C_5H_3KN_3O_4$
17a 213
$C_5H_3LiN_3O_4$
17a 200ff
$C_5H_3NO_2^-$
9d1 334, *9d1* 335
$C_5H_3NO_2Se^-$
17f 285
$C_5H_3NO_3S^-$
9d1 581, *17f* 284
$C_5H_3NO_3Se^-$
17f 285
$C_5H_3NO_4^-$
9d1 576, *17f* 279
$C_5H_3NO_5^-$
9d1 576, *17f* 280
$C_5H_3NS^-$
9d1 891
$C_5H_3N_2$
9b 381
$C_5H_3N_2O_2^{2-}$
17b 504
$C_5H_3N_2O_2S$
17e 323
$C_5H_3N_2O_3^{2-}$
17b 427
$C_5H_3N_2O_3S$
9c1 27
$C_5H_3N_2O_4^{2-}$
9b 425, *9d1* 577
$C_5H_3N_2O_4$
9b 282, *9c1* 27, *17b* 428, *17c* 415
$C_5H_3N_2O_5^{2-}$
9b 283, 420, *9c1* 22
$C_5H_3N_2S_2$
17c 547
$C_5H_3N_3NaO_4$
17a 205ff
$C_5H_3N_3O_3^-$
17c 623
$C_5H_3N_3O_4^{3-}$
9d1 606
$C_5H_3N_3O_4^-$
9d1 657ff
$C_5H_3N_3O_6^-$
9d1 607
$C_5H_3N_4O_3$
9c1 805, *17c* 425, *17d2* 403, 404
$C_5H_3N_4S$
17e 323
$C_5H_3N_4S_2$
17e 358
C_5H_3OS
17b 524, 525
$C_5H_3O_2$
17b 524
$C_5H_3O_3$
9b 402
$C_5H_3O_4^{2-}$
17c 16
$C_5H_3O_4$
17c 40
$C_5H_3O_5^{2-}$
17c 16
$C_5H_4BrN^+$
17h 224
$C_5H_4BrN_2O_2$
17b 113
$C_5H_4ClD_5NO_4$
9b 118
$C_5H_4ClN^+$
17h 224
$C_5H_4ClN_5^-$
17c 425
$C_5H_4ClO_4^{2-}$
9b 148
$C_5H_4Cl_3Si$
17c 83
C_5H_4D
17c 71
$C_5H_4DN_2O_2$
17b 424, 426, *17c* 417
C_5H_4DO
17c 229
C_5H_4DS
17c 231, 233
$C_5H_4D_2N$
9c1 24
$C_5H_4D_3N_2O_3$
17b 412
$C_5H_4D_3O_5$
17b 277
$C_5H_4D_4NO_4$
17b 174, 283
$C_5H_4D_4O^-$
9b 260
$C_5H_4D_5NO_4^+$
17b 174
$C_5H_4D_5N_2O_4$
17b 284
$C_5H_4FN^+$
17h 224
$C_5H_4F_3O_2$
17b 516
$C_5H_4KN_2O_2$
17a 213ff
$C_5H_4LiN_2O_2$
17a 201
C_5H_4N
9b 309, 315, *9c1* 117
C_5H_4NO
9b 415, *17b* 523
$C_5H_4NO_2$
9b 416, *9c1* 130, 805, *9c2* 203, 204
$C_5H_4NO_3^{2-}$
9b 419, *17b* 342
$C_5H_4N_2NaO_2$
17a 206ff
$C_5H_4N_2O^-$
9d1 862
$C_5H_4N_2OS_3^-$
17f 218
$C_5H_4N_2O_2^-$
1 114, *9b* 403, *9d1* 589, 590, 594ff, *17c* 414, 416, *17f* 289
$C_5H_4N_2O_2S_2^-$
17f 218
$C_5H_4N_2O_3^-$
1 114, *9d1* 600, 601ff, *17b* 427, *17c* 415
$C_5H_4N_2O_4^-$
9b 425, *9d1* 577, *17f* 280
$C_5H_4N_2O_5^-$
9b 230, 281
$C_5H_4N_3O_4^{2-}$
9d1 606
$C_5H_4N_3O_4$
9c1 27, 28

$C_5H_4N_4^-$
9d1 882, *17f* 209ff
$C_5H_4N_5O_2$
17d2 390
$C_5H_4N_5S_2$
17e 359
$C_5H_4O^-$
9b 439, *17c* 81
$C_5H_4OS^-$
9b 496, 605, 607, *17c* 119
$C_5H_4O_2^-$
17c 229
$C_5H_4O_3^-$
9b 415, *9d1* 79, *17g* 43
$C_5H_4O_3S^-$
17b 376, 377, *17c* 42, 44
$C_5H_4O_4S^{3-}$
9b 217
$C_5H_4O_5^{3-}$
17b 285
$C_5H_4O_7P^{4-}$
17b 277
$C_5H_4O_7S^{3-}$
17b 277
$C_5H_4S_2^-$
9b 505, *17c* 119
$C_5H_4S_3^-$
17f 215
C_5H_5
1 40, *9b* 404, 437, 438, *17c* 64, 65, 70ff
$C_5H_5Al_2Cl_8Ti$
9a 302
$C_5H_5BD_3N^-$
17c 411
$C_5H_5BrN_2O_2^+$
17h 204
$C_5H_5BrN_2O_2X$
(X unknown)
17b 418
$C_5H_5ClD_5N_3^+$
17b 182
$C_5H_5Cl_2O_2$
17e 17
$C_5H_5Cl_2O_2Zr$
17e 32
C_5H_5DN
17c 128
$C_5H_5D_2O_4$
17b 173
$C_5H_5D_3NO_2$
9c1 762
$C_5H_5D_4NO_4^-$
17b 299
$C_5H_5D_4N_2O_3$
17b 229, 266
$C_5H_5D_5NO_4$
17b 299
$C_5H_5D_5N_2O_3^+$
17b 230, 266
$C_5H_5N^-$
9d1 858, *17f* 196
$C_5H_5N^+$
17h 224
$C_5H_5NO^-$
9b 492, 512, *9d1* 858
$C_5H_5NO_2^-$
9d1 69, 861, *17c* 106, *17g* 43
$C_5H_5NO_2S^-$
9d1 579ff, 583
$C_5H_5NO_2S_2^-$
9d1 580
$C_5H_5NO_3^-$
9b 418ff, *9d1* 80, 574, *17b* 342, *17f* 278
$C_5H_5NO_4R$
(R = polymer chain)
17e 14
$C_5H_5NO_4S_2^-$
9d1 581
$C_5H_5NO_6^{3-}$
9d1 398
$C_5H_5N_2$
9b 613
$C_5H_5N_2O$
9c1 555
$C_5H_5N_2O_2$
9b 80, 283, 403, *9c1* 26, *17b* 113, 424, 425, *17c* 32, 134, 414, 417, 463, *17e* 211
$C_5H_5N_2O_3$
9c1 147, *17b* 415, 427, *17c* 415
$C_5H_5N_2O_3S$
17b 405
$C_5H_5N_2O_4$
9b 281, 404, 424
$C_5H_5N_2O_5$
9b 280, 281, 286
$C_5H_5N_3O^-$
17b 425, *17c* 412
$C_5H_5N_3O_2^-$
9d1 592, 593, *17b* 426, *17c* 413
$C_5H_5N_3O_4^-$
9d1 606
$C_5H_5N_4O$
17c 423, 427
$C_5H_5N_4O_2$
9c1 31
$C_5H_5N_4S$
17c 425
$C_5H_5N_5O^+$
17c 429, *17h* 206
C_5H_5O
9b 601, *17c* 229
$C_5H_5O_2^{2-}$
9b 436
$C_5H_5O_2S$
9b 417, 418
$C_5H_5O_3^{2-}$
9b 370
$C_5H_5O_3$
1 40, *9b* 414, 415, *17b* 66, 357, 516
$C_5H_5O_3S$
17b 377, *17c* 41, 44
$C_5H_5O_4^{2-}$
9b 147, 371, *9d1* 8, 11, 66, *17b* 173, 204
$C_5H_5O_4$
1 40, *9b* 371, 416, *17b* 356, 497
$C_5H_5O_5^{2-}$
9b 148, 217, 218, *17b* 64, 205, 277
$C_5H_5O_5$
17b 497
$C_5H_5O_6$
9b 151
$C_5H_5P^-$
9d1 895
C_5H_5S
9b 603, 607, *17c* 231, 232
$C_5H_6^+$
17h 28
$C_5H_6BrN_2O_3$
17b 435

$C_5H_6DNO_3^-$
17b 69, 135
$C_5H_6D_2NO_3$
17b 266
$C_5H_6D_3N_2O_3$
17b 230
$C_5H_6D_3O_4$
17b 292
$C_5H_6D_5NO_2S^+$
17b 63
$C_5H_6D_6^+$
17h 9
$C_5H_6F_3O$
17b 166, 213
$C_5H_6F_5N_2O_3$
17d2 220
$C_5H_6F_7OSi$
17b 312
$C_5H_6KO_4$
17b 173
C_5H_6N
9b 492, 509ff, *17c* 127ff
C_5H_6NO
9b 206, *17c* 21ff, 106
$C_5H_6NOS_2$
17c 529
$C_5H_6NO_2$
9b 33, 163, *9c1* 151, *17c* 106
$C_5H_6NO_2S$
9b 128
$C_5H_6NO_2S_2$
17c 529
$C_5H_6NO_3^{2-}$
9d1 66
$C_5H_6NO_3$
9c1 142ff, 199, 806, *17c* 598ff
$C_5H_6NO_4^{2-}$
9b 148, 221
$C_5H_6NO_4$
9c1 144
$C_5H_6N_2^-$
9d1 870
$C_5H_6N_2^+$
17h 161
$C_5H_6N_2O_2^-$
9b 422, 423, *17c* 47, 157
$C_5H_6N_2O_2S^-$
9d1 585, *17g* 28
$C_5H_6N_2O_3^-$
9d1 46
$C_5H_6N_2O_3S_2^-$
17e 318
$C_5H_6N_2O_6S^-$
17b 412
$C_5H_6N_3$
17b 307
$C_5H_6N_3O$
9b 403
$C_5H_6N_3O_2$
17b 426
$C_5H_6N_3O_4$
9b 280, 426, *9c1* 158, 946
$C_5H_6N_5$
9c1 30, *17c* 422, 426
$C_5H_6N_5O$
9c1 30
$C_5H_6NaO_2$
1 37
$C_5H_6O^-$
9b 407, 435, *17c* 60
$C_5H_6O_2^-$
9d1 11, 30, 34, *17g* 21
$C_5H_6O_4^-$
9d1 82, *17b* 204, *17g* 60
$C_5H_6O_5^-$
9d1 82
$C_5H_6S^-$
17c 62
$C_5H_6S_2^+$
17e 372
C_5H_7
1 40, *9b* 387, 406, 431, 435, *17b* 96, 328, 340, 482, *17c* 30, 32, 33, 58ff
$C_5H_7Ag^+$
9a 328
$C_5H_7BeO_2$
9b 421
$C_5H_7ClNO_3$
1 38, *9b* 208
$C_5H_7ClN_5$
17b 479
$C_5H_7Cl_2$
17b 123
$C_5H_7DNO_3^{2-}$
17b 272
$C_5H_7DN_2^+$
17h 177
$C_5H_7DO_2^-$
17b 204
$C_5H_7D_2NO_2^-$
9b 231
$C_5H_7D_3O_2^+$
17b 51, 255
$C_5H_7D_4NO_2^+$
17b 132
$C_5H_7D_4NO_2S^+$
17b 89
$C_5H_7F_3NO$
9c1 249, *17c* 403
$C_5H_7F_3NO_3$
9c1 945, *17d1* 212
$C_5H_7MgO_2$
9b 421
$C_5H_7MgO_2^+$
17a 224
$C_5H_7N^+$
9b 428, 429, 492, *17h* 219
$C_5H_7NO^-$
17b 341
$C_5H_7NO_3^-$
9b 50, 91, *17b* 266
$C_5H_7NO_4^-$
9b 109, 118
$C_5H_7N_2$
17c 419
$C_5H_7N_2O_2$
9b 278, 279, 280, 283ff, 422, *17b* 404, 407, 409, 410, 432, 433, 435, *17c* 46, 412, 455
$C_5H_7N_2O_3$
9b 161, 206, 280, 285, 286, *17b* 412, 435
$C_5H_7N_2O_5^{2-}$
9d1 422, 689
$C_5H_7N_2O_6^{2-}$
9d1 423
$C_5H_7N_3O_2^-$
9d1 592, *17f* 286, 287, 288
$C_5H_7N_3O_5^-$
17f 234
$C_5H_7N_4O_8^{2-}$
9d1 423
$C_5H_7NiO_2$
17c 45

C_5H_7O
9b 261, 368, 407, *17b* 336, 521, *17c* 34, 36, 45, 60
C_5H_7OS
17b 375, 376, *17c* 39, 41, 44
$C_5H_7O_2^{2-}$
9b 365, 367
$C_5H_7O_2$
1 37, *9b* 161, 257, 304, 369, 372, 373, 414, *17b* 223, 332, 515, *17c* 15, 16, 18, 37, 40, *17e* 12
$C_5H_7O_3$
9b 161, 415, *17b* 65, 88, 164, 225
$C_5H_7O_4$
1 29, 35, 37, *9b* 116ff, 147, 198, 221, 291, *17b* 28, 62, 88, 127, 154, 173, 225, 298, 442, 457, 458
$C_5H_7O_4S$
9b 123
$C_5H_7O_4X$
(X = unknown)
1 37
$C_5H_7O_5$
1 36, *9b* 46, 149, 218, 223, *17b* 205, 298
$C_5H_7O_5P^{3-}$
9b 160
$C_5H_7S_2$
17c 43
$C_5H_8^-$
9d1 844
$C_5H_8^+$
17h 25
C_5H_8Br
17b 33, 94
C_5H_8BrO
9b 207
$C_5H_8ClN_4O$
9c1 836
C_5H_8ClO
17b 348
$C_5H_8ClO_2$
17b 155
$C_5H_8Cl_3$
9b 16, 182, 224
$C_5H_8DO_2$
9b 146, 190
$C_5H_8D_2O_2^+$
17b 126
$C_5H_8D_4^+$
17h 9
C_5H_8F
17b 330
$C_5H_8FO_2$
17b 365, 366, 369
$C_5H_8F_3O$
9b 183
$C_5H_8F_3OS$
17b 274
$C_5H_8F_3S$
9b 185
$C_5H_8F_3S_2$
17e 346
C_5H_8Mu
17b 127
$C_5H_8MuO_2$
17b 138, 255
C_5H_8N
9b 44, 201, *17b* 20, 202, *17c* 57, 570
C_5H_8NO
9b 133, 154, 201, 419, *9c1* 19, 151, *17b* 184, 341, 362
$C_5H_8NO_2$
9b 262, *9c1* 139, 772, *17c* 40, *17d2* 392, 402
$C_5H_8NO_3$
1 38, *9b* 208, 262, *9c1* 387, *17d2* 223, 400
$C_5H_8NO_3S$
9c2 278, *17e* 317
$C_5H_8NO_4^{2-}$
17f 231
$C_5H_8NO_4$
9c1 204, *17d2* 223, 392
$C_5H_8NO_5^{2-}$
9d1 397
C_5H_8NS
9b 134, *17b* 271
$C_5H_8N_2^+$
9d2 69, *17h* 177, 178, 220
$C_5H_8N_2O_2^-$
9b 287
$C_5H_8N_2O_2^+$
17b 407, 411
$C_5H_8N_2O_3^-$
9b 166
$C_5H_8N_3O$
9b 277, 282, *17b* 436
$C_5H_8N_3O_2$
9b 279, 285, *9c1* 837
$C_5H_8O^-$
9b 260, *17b* 339
$C_5H_8O^+$
17h 37, 43
$C_5H_8OS^-$
17g 47
$C_5H_8O_2^-$
9b 18, 44, 146, 365, 366, *9d1* 8, 11, *17b* 20, 120, 153, 170, 203, 246
$C_5H_8O_2^+$
17b 114, 133, 353, 441, *17h* 22, 43, 48, 50
$C_5H_8O_2S_2^-$
9d1 75, *17g* 44, 45
$C_5H_8O_3^-$
9b 159, 186, 212, *9d1* 50, *17b* 128, 132, 161, 162, 264
$C_5H_8O_3S^-$
17b 90, 196
$C_5H_8O_4^-$
9b 234
$C_5H_8O_4^+$
17h 47
$C_5H_8O_4P^{2-}$
9b 159, 186
$C_5H_8O_4S^-$
17b 496
$C_5H_8O_4X$
(X = purine)
9c2 6
$C_5H_8O_5^-$
17g 7
$C_5H_8O_5S^-$
17b 224
$C_5H_8O_6P^{2-}$
17b 36, 114, 133
$C_5H_8O_6S^-$
17b 158, 264
$C_5H_8S_2^+$
17e 368, 369, 370
C_5H_9
1 26, *9b* 19, 24, 25, 87,

259, 347, 348, *17b* 23, 49, 95, 122, 328, 335

C_5H_9AgN
17c 571

C_5H_9AuN
17c 572

$C_5H_9BrNO_2$
9c1 780

C_5H_9ClNO
17d1 177

$C_5H_9ClNO_2$
9c1 780ff, *17b* 341

$C_5H_9ClNO_3$
17b 345

$C_5H_9ClNO_4$
1 29, *9b* 118

$C_5H_9Cl_2F_3OP$
9c2 228

$C_5H_9Cl_3NO$
9c1 379, *17d1* 162

$C_5H_9Cl_3NO_3$
17d1 184

$C_5H_9Cl_3NS$
17c 502

C_5H_9CuN
17c 574

$C_5H_9D_3NO$
9c1 210

$C_5H_9D_3NO_2S$
17c 527

$C_5H_9D_6OSSi$
17e 350

C_5H_9FNO
9c1 781

$C_5H_9FNO_2$
17d1 177

$C_5H_9F_3NO$
9c1 381, *17d1* 162

$C_5H_9F_3NO_2$
17d1 207, 212

$C_5H_9F_3O_3P$
9c2 228

$C_5H_9NO^+$
17h 50, 158

$C_5H_9NOS_2^-$
17g 45

$C_5H_9NO_2^-$
9d1 383, 681

$C_5H_9NO_2^+$
17h 158

$C_5H_9NO_3^-$
17b 301, *17f* 233

$C_5H_9NO_3$
9d1 682

$C_5H_9NO_3M$
17d1 177

$C_5H_9NO_3S^+$
17e 317

$C_5H_9NO_4^-$
9d1 394, 398, 399

$C_5H_9NO_4^+$
9b 148, 152

$C_5H_9NO_4M$
17d2 398

$C_5H_9NO_4P$
9c2 228, *17b* 520

$C_5H_9NS^+$
17c 42

$C_5H_9NS_3^-$
17g 45

$C_5H_9N_2$
17c 480

$C_5H_9N_2O$
9c1 895, *17d1* 177

$C_5H_9N_2O_2$
9c1 853, *17b* 405, 406

$C_5H_9N_2O_3$
9b 121, 133, 170, 171, 208

$C_5H_9N_2O_3S$
9c1 991

$C_5H_9N_2O_4$
9c1 205

$C_5H_9N_2O_5$
9c1 908

$C_5H_9N_2O_5S$
9c1 991

$C_5H_9N_2S_2$
17c 531, 539

$C_5H_9N_4O$
9c1 837ff, *17c* 468

$C_5H_9N_4S$
17c 472

$C_5H_9N_4S_2$
17c 474

C_5H_9O
9b 19, 25, 45, 90, 111ff, 187ff, 255, 259, 261, 265, 290, 320, *17b* 29ff, 32, 48, 50, 55, 62, 110, 135, 164, 167ff, 214, 261, 340, 348, 351ff, 513, 514

C_5H_9OS
17b 45, 91, 351, 379, 380, 471, 472

$C_5H_9OS_2$
17e 358

$C_5H_9O_2$
1 35, *9b* 21, 33, 37, 38, 39, 44, 112, 115, 139, 182, 189ff, 194, 197, 203, 264, 267, 268, 292, 293, 324, 363, *9c2* 13, *17b* 20, 25, 27, 36, 48, 51, 52, 53, 60, 65, 72, 73, 120, 124, 125, 127, 138, 153, 164, 169, 170, 193, 204, 221, 246, 248, 255, 259, 261ff, 263, 348, 353, 364, 368, 371, 372, 439, 441, 466, 467, 468, 476, *17c* 15

$(C_5H_9)_nO_2$
17e 13

$C_5H_9O_2S$
9b 122, *17b* 91, 145, 210, 379

$C_5H_9O_2S_2$
17b 149

$C_5H_9O_3$
1 28, 36, *9b* 33, 41, 66, 101, 108, 125, 140, 159, 186, 198, 200, 204, 205, 212, *17b* 36, 42, 63, 72, 81, 114, 132, 138, 145, 146, 156, 157, 189, 194, 223, 232, 251, 264, 267, 268, 285, 468

$C_5H_9O_3S$
9c2 289

$C_5H_9O_4$
9b 205, 249, *17b* 291, 292, 299

$C_5H_9O_4P^-$
9b 160, 186

$C_5H_9O_5$
17b 214, 446, 450, 451, 458, 461, 462

$C_5H_9O_5P^-$
9b 160, 187

C_5H_9S
9b 291, 379, *17b* 375

$C_5H_9S_2$
17b 381, 385, 472, 473
C_5H_{10}
9b 77
$C_5H_{10}^+$
17h 15, 16, 19
$C_5H_{10}Br$
9b 16
$C_5H_{10}BrOS$
17e 350
$C_5H_{10}Cl$
9b 23, 185
$C_5H_{10}ClNO_2S$
9c2 280
$C_5H_{10}ClNO_3^-$
17b 302
$C_5H_{10}ClN_3^+$
17b 182
$C_5H_{10}ClO$
17b 166
$C_5H_{10}ClOS$
17e 350
$C_5H_{10}Cl_2NO$
9c1 327, *17d1* 151
$C_5H_{10}Cl_2O_4P$
9b 234
$C_5H_{10}D_2^+$
17h 8
$C_5H_{10}F_3S_2$
17e 345
$C_5H_{10}Mu$
17b 117, 150, 242
$C_5H_{10}MuO$
17b 125
$C_5H_{10}N$
9b 289, *9c1* 119, *17b* 339, 506, *17c* 567
$C_5H_{10}NO$
9b 30, 77, 98, 176, 324, *9c1* 123, 124, 443ff, *17b* 359, 519, *17c* 399, 400, 402, 404, *17d2* 281
$C_5H_{10}NO_2$
1 35, *9b* 38, 116, 142, 186, 200, 204, *9c1* 209, 259, 444, 771, 776, 781, *17b* 173, 250, *17d1* 178, *17d2* 392
$C_5H_{10}NO_2S$
9b 74, 143ff, 144, *9c1* 771, *9c2* 279, *17b* 89, 196, *17e* 317, 343
$C_5H_{10}NO_3$
9c1 106, 205, 772, 781, *9c1* 915, *17b* 174
$C_5H_{10}NO_3S$
17b 199
$C_5H_{10}NO_4$
9c1 205, 915
$C_5H_{10}NO_5S$
9c1 991
$C_5H_{10}NO_5S_2$
17c 621
$C_5H_{10}NS_2$
17e 324
$C_5H_{10}N_2O^+$
17h 158
$C_5H_{10}N_2OS^-$
9d1 75
$C_5H_{10}N_3^{2+}$
17c 32
$C_5H_{10}NaO_3S$
9b 196
$C_5H_{10}O^+$
17b 144, *17h* 37, 38, 42
$C_5H_{10}OR$
17b 192
$C_5H_{10}O_2^-$
9b 226, *9b* 248
$C_5H_{10}O_2^+$
17b 72, 126, 133, 258, 264, *17h* 22, 23, 46
$C_5H_{10}O_3^+$
17h 49
$C_5H_{10}O_3P$
9c2 220
$C_5H_{10}O_5P$
17e 273
$C_5H_{10}O_5S^-$
17b 157, 166
$C_5H_{10}S^+$
17b 331, *17e* 364
$C_5H_{10}S_2$
9c2 273
$C_5H_{10}S_2^+$
17e 368
C_5H_{11}
1 23, 25, 34, *9b* 14, 21ff, 43, 82, 104, 181, *17b* 17, 27, 57, 126, 242
$C_5H_{11}BN_4^-$
17c 581
$C_5H_{11}BrNO_2S$
17e 349
$C_5H_{11}BrNO_3S$
9c1 991
$C_5H_{11}ClNO$
9c1 249, *17d1* 131
$C_5H_{11}ClNO_2$
17b 173
$C_5H_{11}ClNO_2S$
9c2 277, 280, *17e* 349
$C_5H_{11}ClNO_2S_2$
9c2 297
$C_5H_{11}ClNO_3$
9c1 933, *17b* 174
$C_5H_{11}ClNO_3S$
9c1 991
$C_5H_{11}F_2Sn$
17b 291
$C_5H_{11}F_3OP$
9c2 229
$C_5H_{11}F_4OS$
9c2 295
$C_5H_{11}INO_2S$
17e 350
$C_5H_{11}N^+$
9b 259, 262, 289, *17h* 127
$C_5H_{11}NO^+$
9b 264, 292
$C_5H_{11}NO_2^-$
1 78, *9d1* 379
$C_5H_{11}NO_2^+$
9b 185, 205
$C_5H_{11}NO_2S^-$
9b 235, 402
$C_5H_{11}NO_2S^+$
9c2 279, 298
$C_5H_{11}NO_3^-$
9d1 393, 394, 395, 400, *17f* 230
$C_5H_{11}NO_3S^-$
9d1 402
$C_5H_{11}NO_4^-$
1 78

$C_5H_{11}NS^+$
17h 138
$C_5H_{11}N_2O$
9b 133, *9c1* 513, *17b* 141, *17d2* 374
$C_5H_{11}N_2O_2$
9c1 809, *17d1* 178
$C_5H_{11}N_2O_3S$
9c1 992
$C_5H_{11}N_2O_5S$
9c1 992
$C_5H_{11}N_4$
17c 464
$C_5H_{11}O$
9b 23, 37, 66, 100, 183, 194, 203, 217, *17b* 22, 84, 246
$C_5H_{11}OS$
9b 74, 144, 145, 196, *17b* 197, 259
$C_5H_{11}OS_2$
9b 158, *17e* 346
$C_5H_{11}O_2$
9b 41, 108, 177, *9c2* 10, 12, *17b* 144, 156, 166, 236, 251, 260
$C_5H_{11}O_2S$
9b 16
$C_5H_{11}O_3$
9b 41, 108
$C_5H_{11}O_3S_2$
9b 196
$C_5H_{11}O_4$
1 31, *9b* 172
$C_5H_{11}O_5$
9b 222, 231, *17b* 214, *17e* 5, 6
$C_5H_{11}S$
9b 73, 184, 196, *17b* 90
$C_5H_{11}S_2$
9b 74, 109, 143, *17b* 148, 158, 238
$C_5H_{12}^+$
17h 8, 12
$C_5H_{12}Cl_2OP$
9c2 229
$C_5H_{12}MuN_2$
17c 440
$C_5H_{12}N$
9b 59, 60, 97, 130, 166, 202, 217, *17b* 76, 226, *17c* 391
$C_5H_{12}NO$
9c1 97, 205, 210, 218, *17c* 391, *17d1* 111ff, *17d2* 72, 208
$C_5H_{12}NOS$
9c1 986, *17c* 524, *17d1* 190
$C_5H_{12}NOSi$
17b 509
$C_5H_{12}NO_2$
9c1 219, 241, 242, 259, 915, 934, *17d1* 130, 184, *17d2* 408
$C_5H_{12}NO_2S$
9c1 112, *17c* 527
$C_5H_{12}NO_3$
9c1 242, 913, 915ff, 917, 923
$C_5H_{12}NO_3S$
9c1 998, *17c* 495, 527, *17d1* 195
$C_5H_{12}NO_4$
9c1 916
$C_5H_{12}NO_5P^-$
9d1 411
$C_5H_{12}NS$
17c 499
$C_5H_{12}N_2^+$
9d2 96, *17h* 127
$C_5H_{12}N_2O^+$
9b 97, *9c1* 205, *17h* 158ff
$C_5H_{12}N_2O_2^-$
9d1 407
$C_5H_{12}OSi^-$
9b 216
$C_5H_{12}O_2P$
17b 73, *17e* 273
$C_5H_{12}O_3P$
9b 33, 102
$C_5H_{12}O_4P$
9b 102
$C_5H_{12}P$
17b 26
$C_5H_{12}Si^+$
17h 20
$C_5H_{13}BN$
17c 411, *17f* 60
$C_5H_{13}BrN$
9b 131
$C_5H_{13}ClNO_2$
9b 236
$C_5H_{13}ClO_2S$
9c2 26
$C_5H_{13}DO_3PSSi$
17e 272
$C_5H_{13}Ge$
9b 28
$C_5H_{13}N^+$
9d2 26
$C_5H_{13}NO^+$
9b 137, 203
$C_5H_{13}NO_3P$
17b 142, *17c* 498, *17e* 273
$C_5H_{13}NO_4P$
17c 493, *17d2* 418
$C_5H_{13}NO_5P$
17d2 419
$C_5H_{13}NS^+$
17h 138
$C_5H_{13}N_2$
9c1 43
$C_5H_{13}N_2O$
9c1 239, *17d1* 12, 127, 182
$C_5H_{13}N_2O_2$
17d2 208, 211
$C_5H_{13}N_3O_2$
9c1 205
$C_5H_{13}OSi$
9b 34, *17b* 147
$C_5H_{13}O_2S$
17e 353
$C_5H_{13}Si$
9b 35, *17b* 46, 260
$C_5H_{13}Sn$
9b 36, *17b* 46
$C_5H_{14}AsS$
9c2 317
$C_5H_{14}GeNO$
17a 431, *17d2* 205
$C_5H_{14}NOSi$
17d2 205
$C_5H_{14}NOX$
17d1 127
$C_5H_{14}NO_2SSi$
17c 523
$C_5H_{14}NO_2Si$
17d2 408

$C_5H_{14}N_2^+$
17h 128
$C_5H_{14}OP$
9c2 229
$C_5H_{14}O_3PSSi$
17e 273
$C_5H_{15}BN$
17c 392
$C_5H_{15}BSi^-$
17b 200
$C_5H_{15}OSSi$
17e 350
$C_5H_{15}Si_2$
9c2 308
$C_5H_{15}Sn$
9c2 313
$C_5H_{15}Sn_2$
9c2 313
$C_5IMoO_5^{2-}$
17a 148
$C_5IO_5Re^-$
17a 172
$C_5IO_5W^{2-}$
17a 171
C_5KrMnO_5
17a 122
C_5MnO_5
17a 121ff
C_5MnO_7
9a 119, *17a* 69, *17e* 17
$C_5N_2O_4^{3-}$
17b 504
$C_5N_2O_5^{3-}$
9d1 341
$C_5N_3O_6^{4-}$
9d1 607
$C_5O_5^{3-}$
17a 33
$C_5O_5^-$
9a 49
$C_5O_5Os^+$
17a 173
C_5O_5V
17a 120
$C_5O_6V^{2+}$
17a 120
$C_6Al(Br,I)_2Cl_4O_2$
9d1 233
$C_6AlBr_2Cl_4O_2$
9d1 154
$C_6AlBr_3Cl_4O_2$
9a 296
$C_6AlCl_4I_2O_2$
9d1 155
$C_6AlCl_6O_2$
9d1 154, *17a* 394, *17e* 193, *17g* 205
$C_6BrF_5^-$
17f 184
$C_6Br_2Cl_4InO_2$
9d1 155
$C_6Br_2F_4^-$
17f 184
$C_6Br_3Cl_4GaO_2$
9d1 155
$C_6Br_4O_2^-$
9d1 231, *17g* 243
$C_6Br_4O_2$ on ZnO, TiO_2, MgO surfaces
9d1 366
$C_6ClF_5^-$
17f 184
$C_6Cl_2O_4^{3-}$
9d1 274
$C_6Cl_4CsO_2$
9d1 154
$C_6Cl_4GaI_2O_2$
9d1 155
$C_6Cl_4I_2InO_2$
9d1 155
$C_6Cl_4KO_2$
9d1 154, 232ff
$C_6Cl_4LiO_2$
9d1 153
$C_6Cl_4NaO_2$
9d1 153, 233, *17g* 208
$C_6Cl_4O_2^-$
1 82, *9d1* 153, 231ff, *17g* 205, 244
$C_6Cl_4O_2 \overset{CT}{\leftarrow} C_{10}H_{16}N_2$
9d1 363
$C_6Cl_4O_2 \overset{CT}{\leftarrow}$ (D)
9d1 361, 363, 364
$C_6Cl_4O_2$ on ZnO, TiO_2, MgO surfaces
9d1 366
$C_6Cl_4O_2Rb$
9d1 154
$C_6Cl_4O_2Tl$
9d1 157
$C_6Cl_4O_5P^{2-}$
9c2 181
$C_6Cl_5NO_2^-$
9d1 488
$C_6Cl_6GaO_2$
9d1 155
$C_6Cl_6InO_2$
9d1 155
$C_6Cl_6LaO_2$
9d1 156
$C_6Cl_6LuO_2$
9d1 156
$C_6Cl_6O_2Sc$
9d1 154
$C_6Cl_6O_2Y$
9d1 156
C_6Cl_7
17c 104
$C_6Cl_7O_2Sn$
9d1 156
$C_6CoF_3N_6O$
9c1 1022
C_6CoN_7O
9c1 1023
$C_6D_2N_2S_4^+$
17h 137
$C_6D_4KO_2$
9d1 171
$C_6D_4LiO_2$
9d1 171
$C_6D_4NaO_2$
9d1 172
$C_6D_4O_2^-$
17g 213
$C_6D_5NO_2^-$
9d1 449
$C_6D_6^-$
9d1 710
C_6D_7
9b 453
$C_6F_2O_4^{3-}$
9d1 274
$C_6F_4I_2^-$
17f 185
$C_6F_4N_2Se^-$
9d1 900
$C_6F_4O_2^-$
1 81, *9d1* 234ff, *17g* 245
$C_6F_4O_2$ on Al_2O_3 surface
9d1 365

$C_6F_4O_2$ on ZnO, TiO_2, MgO surfaces
9d1 366
$C_6F_5I^-$
17f 185
$C_6F_5NO_2^-$
17f 245
C_6F_5O
9c2 179
$C_6F_5O_2$
9c2 22
$C_6F_5O_2S$
9c2 294
$C_6F_6^-$
9a 49, *17a* 33, *17f* 185
C_6F_7
9b 489
C_6F_{11}
9b 277
$C_6F_{11}O_2$
9c2 20
$C_6F_{14}N_2^-$
9a 79
$C_6F_{14}NO$
17d2 221
$C_6HBr_2O_3^{2-}$
9d1 248
$C_6HCl_3O_2^-$
1 81
$C_6HCl_4NO_2^-$
9d1 487
$C_6HCl_4O_2$
9c2 180, *17e* 189
$C_6HCoN_7O_3^-$
9c1 1011, 1012
$C_6HD_3O_2^-$
17g 213
C_6HD_6
9b 453
$C_6HD_{10}O$
9b 48
$C_6HF_4NO_2^-$
17f 242ff
C_6HF_4O
9c2 176
$C_6HF_4O_2$
9c2 179ff
$C_6HF_5^-$
17f 182
C_6HF_6
9b 464, *17c* 91
$C_6HF_9NO_3$
9c1 329
$C_6HF_{12}O$
9b 242
$C_6HO_5^{4-}$
17g 256
$C_6HO_8^{4-}$
9b 231
$C_6HO_9S_2^{4-}$
9d1 248
$C_6H_2Br_2K_2NO_7S_2$
17d1 261
$C_6H_2Br_2NNa_2O_6$
17d1 260
$C_6H_2Br_2NNa_2O_6S_2$
17d1 260
$C_6H_2Br_2O_2^-$
9d1 149
$C_6H_2ClD_2N_2$
17c 577
$C_6H_2ClN_2O_5^{2-}$
9d1 624
$C_6H_2Cl_2KO_2$
9d1 217
$C_6H_2Cl_2LiO_2$
9d1 200, 217
$C_6H_2Cl_2NaO_2$
9d1 149, 217
$C_6H_2Cl_2O_2^-$
1 81, *9d1* 200, 217
$C_6H_2Cl_3NO_2^-$
1 99, *9d1* 487
$C_6H_2Cl_3O$
17e 152
$C_6H_2Cl_3O_2S$
9c2 294, *17e* 341
$C_6H_2Cl_4NO_2$
9c1 946, *17d1* 371
$C_6H_2D_2FN_2$
17c 578
$C_6H_2D_{14}N_2^+$
9d2 61
$C_6H_2F_2O_2^-$
9d1 200
$C_6H_2F_3NO_2^-$
17f 242
$C_6H_2F_4^-$
17f 182
$C_6H_2F_5$
9b 455
$C_6H_2I_2O_2^-$
9d1 218
$C_6H_2KO_4^{2-}$
17g 256
$C_6H_2NO_4^{2-}$
9d1 336
$C_6H_2N_2O_7^{2-}$
17f 305
$C_6H_2N_2S_4^+$
17h 137
$C_6H_2N_3$
9b 382
$C_6H_2N_3O_2^{2-}$
17g 354, 357
$C_6H_2N_3O_7^{2-}$
9d1 665
$C_6H_2O_2S_2^-$
9d1 62
$C_6H_2O_4^{3-}$
9d1 273
$C_6H_2O_8S_2^{3-}$
9d1 143, *17g* 153
$C_6H_2O_9S_2^{3-}$
17g 255
$C_6H_3BD_5NNaO$
17d1 233
$C_6H_3BrNO_2$
9c1 161
$C_6H_3BrNS_2$
17c 545
$C_6H_3BrO_2^-$
9d1 188
$C_6H_3Br_2NO_2^-$
9d1 490
$C_6H_3Br_2O_2S$
17e 339
$C_6H_3ClNO_2$
9c1 160, 161
$C_6H_3ClNO_5S^{2-}$
9d1 551
$C_6H_3ClNS_2$
17c 544, 545
$C_6H_3ClN_2O_4^-$
9d1 623, 624
$C_6H_3ClO_2^-$
1 81, *9d1* 161, 188, *17g* 216
$C_6H_3ClO_6^{3-}$
9b 149

$C_6H_3Cl_2$
9b 313
$C_6H_3Cl_2NO_2^-$
1 98, *9d1* 486, 487, *17f* 246
$C_6H_3Cl_2O$
9c2 58, 61, *17e* 51,53
$C_6H_3Cl_2O_2S$
9c2 293, *17e* 339, 340
$C_6H_3Cl_3NO$
17d1 238
$C_6H_3Cl_3NO_2$
9c1 945
$C_6H_3Cl_3NO_3S$
9c1 1000
$C_6H_3D_2N_2$
17c 575
$C_6H_3D_3^-$
9d1 710
$C_6H_3D_4O_7$
17b 224
$C_6H_3FNO_2$
9c1 159, 160, 161
$C_6H_3FO_2^-$
9d1 188
$C_6H_3F_2NO_2^-$
9d1 475ff, *17f* 241
$C_6H_3F_4$
9b 455, 488
$C_6H_3NO_2^{\cdot-}$
9b 316
$C_6H_3NO_2S_2^-$
9d1 583
$C_6H_3NO_4^{3-}$
9d1 498
$C_6H_3NO_4^-$
9d1 160
$C_6H_3N_2O_4^{2-}$
9c1 28, 593
$C_6H_3N_2O_5^{2-}$
9d1 624, 625, *17f* 296
$C_6H_3N_3O_6^-$
1 102, *9d1* 661ff, *17f* 303ff
$C_6H_3N_3O_7^-$
17f 305
$C_6H_3N_4O_2^{2-}$
17f 292
$C_6H_3N_4O_3$
9c1 910, 911
$C_6H_3N_5^-$
9d1 882
$C_6H_3O_3^{2-}$
9d1 242ff, 249, 273, *17g* 249, 250
$C_6H_3O_3^-$
1 82
$C_6H_3O_3$
9c2 182
$C_6H_3O_3Zn$
17g 249
$C_6H_3O_5^{2-}$
9b 415, 416
$C_6H_3O_6^{2-}$
9b 265
$C_6H_3O_6R_2$
17b 373
$C_6H_4AlBr_3ClS$
17a 394, *17e* 320
$C_6H_4AlO_2^{2+}$
9d1 117, *17a* 395, *17g* 71
$C_6H_4Al_2Br_6ClS$
17a 395
$C_6H_4Al_2Br_6NO_2S$
17a 395, *17e* 320
$C_6H_4BaO_2^+$
9a 291, *9d1* 116ff, *17a* 236, *17g* 71
$C_6H_4BrNO_2^-$
1 96, *9d1* 488ff
$C_6H_4BrN_2$
17c 577
C_6H_4BrOS
17e 331
$C_6H_4BrO_2S$
9c2 292, *17e* 338
$C_6H_4CaO_2^+$
9a 291, *9d1* 116ff, *17a* 230, *17g* 71
$C_6H_4CdO_2^+$
9a 330, *9d1* 116ff, *17a* 386, *17g* 71
$C_6H_4ClNO^-$
9d1 692ff
$C_6H_4ClNO_2^-$
1 88, 89, 96, *9d1* 479ff, 481ff, 483ff, *17f* 246
$C_6H_4ClN_2$
17c 577
$C_6H_4ClN_2O_2$
9c1 909
C_6H_4ClO
9c2 32, 36, 42, *17e* 37, 39, 41
C_6H_4ClOS
17e 331
$C_6H_4ClO_2P^-$
17e 307
$C_6H_4ClO_2S$
9c2 292, 293, *17e* 338
$C_6H_4Cl_2O_2Sb$
17g 74
$C_6H_4Cl_4GaS$
17e 320
$C_6H_4Cl_4Ga_2S$
17a 406
$C_6H_4Cl_4N_2^+$
17h 165
$C_6H_4Cl_7Ga_2S$
17a 406
$C_6H_4CsO_2^-$
9d1 170
$C_6H_4D_2^-$
9d1 709, 710
$C_6H_4D_2NO$
17e 43
$C_6H_4D_3O_6$
17b 358
$C_6H_4D_4O_7^-$
17b 292
$C_6H_4D_5N_2O_5$
17b 284
$C_6H_4FNO_2^-$
1 89, 95ff, *9d1* 471ff, *17f* 240
$C_6H_4FNO_3^-$
9d1 494, 495
$C_6H_4FNO_4S^-$
9d1 549
C_6H_4FO
1 58, 60, *9c2* 32, 37, 42, *17e* 37, 39, 42
$C_6H_4FO_2S$
9c2 293, *17e* 338
$C_6H_4F_2^-$
17f 182
$C_6H_4F_2N$
9c1 10
$C_6H_4F_3$
9b 455

$C_6H_4F_6NO_2$
9c1 329
$C_6H_4F_9O$
17b 40
$C_6H_4GaO_2^{2+}$
17a 406, *17g* 72
C_6H_4I
17b 502
$C_6H_4INO_2^-$
1 96, *9d1* 490ff
$C_6H_4IN_2$
17c 578
$C_6H_4INa_2O_2$
9d1 173
$C_6H_4I_2^-$
17f 183
$C_6H_4InO_2^{2+}$
17a 408, *17g* 72
$C_6H_4KN_2$
17a 214
$C_6H_4KN_4$
17a 214
$C_6H_4KO_2$
9d1 114ff, 170ff
$C_6H_4LaO_2^{2+}$
9a 300, *9d1* 117, *17a* 240, *17g* 72
$C_6H_4LiN_2$
17a 202
$C_6H_4LiN_4$
17a 202
$C_6H_4LiO_2$
9d1 115
$C_6H_4MgO_2^+$
9a 291, *9d1* 116ff, *17a* 224, *17g* 72
$C_6H_4NO_2^{2-}$
9b 610, 612
$C_6H_4NO_2$
9c1 158, 160, 243
$C_6H_4NO_2S^{2-}$
9d1 505, *17f* 250
$C_6H_4NO_3^{2-}$
9d1 336, 495, 496, *17f* 247, *17g* 254
$C_6H_4NO_3$
1 59, *9c1* 159, *9c2* 33, 37, 44, *17e* 38, 39, 44
$C_6H_4NO_3S^{2-}$
9d1 547
$C_6H_4NO_3S$
17e 331
$C_6H_4NO_4^{2-}$
17c 107
$C_6H_4NO_4$
9c2 55, 203, 204
$C_6H_4NO_4S^{2-}$
9d1 549
$C_6H_4NO_4S$
9c2 293
$C_6H_4NO_5S^{2-}$
9d1 551, *17f* 269
$C_6H_4NS_2$
17c 543, 547
$C_6H_4N_2^-$
1 114, *9d1* 861
$C_6H_4N_2Na$
17a 207
$C_6H_4N_2O^-$
9d1 902
$C_6H_4N_2O_3^{2-}$
9d1 494
$C_6H_4N_2O_4^-$
1 100, 101, 102, *9d1* 609ff, 614ff, 634ff, *17f* 294, 295, 299
$C_6H_4N_2O_5^-$
1 101, *9d1* 578, 625, *17f* 295
$C_6H_4N_2Rb$
17a 222
$C_6H_4N_2S^-$
9d1 898ff
$C_6H_4N_2S_2^-$
9d1 901
$C_6H_4N_2Se^-$
9d1 899ff
$C_6H_4N_3O$
9c1 911
$C_6H_4N_3O_2$
9c1 911, *9d1* 553
$C_6H_4N_3O_3^{2-}$
9d1 553
$C_6H_4N_3O_3$
9c1 179
$C_6H_4N_3O_4$
9c1 909
$C_6H_4N_4^-$
9d1 881
$C_6H_4N_4Na$
17a 207
$C_6H_4N_4O_2^-$
9d1 62, *17f* 291
$C_6H_4N_4O_2^+$
9d2 69
$C_6H_4N_4O_6^-$
9d1 665
$C_6H_4N_6O_4^-$
17f 303
$C_6H_4NaO_2$
9d1 116, 172
$C_6H_4NaO_4S$
1 60
$C_6H_4Na_2O_2^+$
9d1 173
$C_6H_4Na_3O_7$
9b 161
$C_6H_4O^-$
9b 312
$C_6H_4O_2^-$
1 80, *9d1* 113, 159, 162ff, *17g* 70ff, *17g* 210, 211ff
$C_6H_4O_2^+$
17g 383
$C_6H_4O_2 \overset{CT}{\leftarrow}$ (D)
9d1 361ff
$C_6H_4O_2^-$ on metaloxide surface
9d1 365
$C_6H_4O_2$ on ZnO, TiO_2, MgO surfaces
9d1 366
$C_6H_4O_2P$
17e 255
$C_6H_4O_2Rb$
9d1 116, 173
$C_6H_4O_2S^-$
9d1 83, *17g* 61
$C_6H_4O_2Sc^{2+}$
17a 238, *17g* 74
$C_6H_4O_2Sr^+$
9a 291, *9d1* 116ff, *17a* 233, *17g* 75
$C_6H_4O_2Y^{2+}$
9a 300, *9d1* 117, *17a* 239, *17g* 77
$C_6H_4O_2Zn^+$
9a 330, *9d1* 116ff, *17a* 377, *17g* 77

$C_6H_4O_3^-$
9d1 46, *17g* 249, 251
$C_6H_4O_4^{3-}$
9d1 66
$C_6H_4O_4^-$
1 82
$C_6H_4O_5^{3-}$
9b 375, 376
$C_6H_4O_5P^{2-}$
9c2 34, 54
$C_6H_4O_6^{3-}$
9b 148, 217, *17b* 204, 276
$C_6H_4O_7^{3-}$
9b 160, *17b* 224, 278
$C_6H_4O_7P^{4-}$
17c 23
$C_6H_4O_8S^{3-}$
17c 20
$C_6H_4S_2^-$
9d1 891
$C_6H_4S_4^+$
9d2 146, *17h* 23, 111
C_6H_5
1 21, *9b* 310ff
$C_6H_5AlBr_3S$
17a 395, *17e* 319
$C_6H_5AlCl_3S$
17e 319
$C_6H_5Al_2Br_6S$
17a 395
$C_6H_5BBr_2NNa_2O_4S$
17d1 261
$C_6H_5Br^-$
17f 183
C_6H_5BrNO
9c1 541, 542
$C_6H_5BrNO_3S$
9c1 1000
$C_6H_5BrNO_4S$
9c1 1001
$C_6H_5BrN_2O_2^+$
17h 204
$C_6H_5BrO^-$
17c 104
$C_6H_5Br_3GaS$
17e 319
C_6H_5ClNO
9c1 542ff
$C_6H_5ClNO_3S$
9c1 1000
$C_6H_5ClNO_4S$
9c1 1001
$C_6H_5ClO^+$
17e 235, 238, 241
$C_6H_5Cl_2OP^-$
9c2 230
$C_6H_5Cl_3GaS$
17a 407, *17e* 319
$C_6H_5Cl_6Ga_2S$
17a 407
$C_6H_5D^-$
1 84, *9d1* 709
$C_6H_5D_5N_3O_2$
17b 300
$C_6H_5D_9NOSi$
17b 509
$C_6H_5F^-$
17f 183
C_6H_5FN
9c1 9, 10
C_6H_5FNO
9c1 543ff
$C_6H_5FNO_4S$
9c1 1001
$C_6H_5FO^+$
17e 235, 239, 241
$C_6H_5F_2$
9b 455, 488
$C_6H_5F_3NO_4$
9c1 145
C_6H_5INO
9c1 544
$C_6H_5IO^-$
17c 105, *17f* 183
$C_6H_5NO^-$
9d1 690ff, *17f* 305ff
$C_6H_5NO^+$
9d2 117, *17h* 225
C_6H_5NOM
17d1 213
$C_6H_5NO_2^-$
1 86ff, *9b* 513, 514, 515, 609, 611, *9c1* 545, *9d1* 335, 430ff, 861, *17c* 132, *17f* 238ff, *17g* 79, 110, 216
$C_6H_5NO_2^+$
17h 239
$C_6H_5NO_3^-$
1 88, 89, 94, *9b* 493, *9d1* 335, 493, 495ff, *17c* 106, 107, *17f* 247
$C_6H_5NO_3^+$
17e 236, 239, 242
$C_6H_5NO_3S^-$
9d1 581
$C_6H_5NO_4^-$
9d1 498, 576, *17c* 107, *17f* 280
$C_6H_5NO_4S^-$
9c2 58, *9d1* 581
$C_6H_5NO_5^-$
9d1 577
$C_6H_5NO_5P$
9c1 970
$C_6H_5NO_6^{3-}$
9b 171
$C_6H_5N_2$
9b 381, 520, 521, *17c* 23, 134, 575
$C_6H_5N_2O$
9c1 129, 130, 178
$C_6H_5N_2O_2$
9c1 909, *17d1* 230
$C_6H_5N_2O_2S$
9c1 107, *9c2* 287
$C_6H_5N_2O_3$
9c1 544, 545
$C_6H_5N_3O_4^-$
1 101, *9d1* 627
$C_6H_5N_3O_6S^-$
9d1 633
$C_6H_5N_4O_3$
9c1 805
$C_6H_5N_4S$
17b 92
C_6H_5O
1 58, *9b* 311, 312, *9c2* 29ff, *17e* 36
C_6H_5OS
17b 524, *17e* 329
C_6H_5OSe
17e 330
$C_6H_5O_2$
9c2 33, 37ff, 44ff, *17b* 524, *17e* 38, 40, 44ff
$C_6H_5O_2P^-$
9c2 220
$C_6H_5O_2S$
9c2 291, *17e* 337

$C_6H_5O_2Se$
17e 337

$C_6H_5O_3$
9c2 61, 205, *17e* 49, 54

$C_6H_5O_4^{2-}$
9b 375

$C_6H_5O_4$
9c2 205

$C_6H_5O_4S$
9c2 54

$C_6H_5O_5^{2-}$
9b 376, *17b* 206

$C_6H_5O_5$
9b 372, 376, 415, 416

$C_6H_5O_6$
9b 416

$C_6H_5O_7S_2$
9c2 59

$C_6H_5O_{10}S_3$
9c2 173

$C_6H_5O_{24}P_6^{12-}$
9b 276

C_6H_5S
1 76, *9c2* 280, *17e* 318

C_6H_5Se
9c2 303

$C_6H_6^-$
1 84, *9d1* 708ff, 845, *17f* 69

$C_6H_6^+$
9d2 6

$C_6H_6^{2\cdot}$
17h 251

C_6H_6Ag
17a 374

C_6H_6Al
17a 80

$C_6H_6AsNO_5^-$
9d1 519, *17f* 58

C_6H_6Au
17a 376

$C_6H_6BNO_4^-$
9d1 522

C_6H_6Br
9b 464, *17c* 91

$C_6H_6BrHgO_6$
17a 389

C_6H_6BrO
9b 465, *17c* 104

C_6H_6Cl
9b 455

$C_6H_6ClHgO_6$
17a 389

$C_6H_6ClO_2$
9b 466

$C_6H_6Cl_2N_2^+$
17h 165

C_6H_6Cu
17a 356

C_6H_6D
17c 48

$C_6H_6DN_2O_2$
17b 426, *17c* 415

$C_6H_6D_4NO$
9c1 151

$C_6H_6D_4O^+$
17h 43

$C_6H_6D_8N_4O_2^+$
17b 183

C_6H_6F
9b 454, 455, 463

C_6H_6FO
9b 465

$C_6H_6FO_2$
9b 465, 466

$C_6H_6F_5NO_4^-$
9d1 387

$C_6H_6F_7N_2O_3$
17d2 220

C_6H_6I
17c 92

C_6H_6N
9b 405, 609, 610, 611, *9c1* 9, *17c* 233, 234

C_6H_6NO
1 54, *9b* 512, *9c1* 524, 538ff, *9c2* 32ff, 37, 42, *17b* 523, *17c* 131, *17d1* 212, 213, *17e* 37, 42

$C_6H_6NOS_2$
17c 542

$C_6H_6NO_2$
1 55, *9b* 513, 514, *9c1* 545, 945, *9c2* 205, *17c* 132

$C_6H_6NO_3^{2-}$
9b 419

$C_6H_6NO_3$
9c1 154, 806, *17c* 487

$C_6H_6NO_3S$
9c1 999

$C_6H_6NO_4$
9b 475, *17d2* 398

$C_6H_6NO_4S$
9c1 1001

$C_6H_6NO_4S_2$
17c 530, 537ff

$C_6H_6NO_5P^-$
9d1 515

C_6H_6NS
17e 318

$C_6H_6N_2NaO_3$
17b 417

$C_6H_6N_2O_2^-$
1 88, 89, 97, *9d1* 201, 381, 508, 509, 510ff, *17b* 175, *17f* 251, 252, *17g* 112, 152, 198, 219, 223, 230

$C_6H_6N_2O_2^+$
9d2 28

$C_6H_6N_2O_3^-$
9d1 510, 591

$C_6H_6N_2O_4^-$
9d1 80, *17b* 414, *17f* 269

$C_6H_6N_2R^+$ (R = n-C_5H_{11}, n-C_6H_{13}, etc.)
9d2 57

$C_6H_6N_3O_4$
9c1 155ff

$C_6H_6N_4$
9b 79

$C_6H_6N_4O_4^-$
17f 281

$C_6H_6N_4O_6S^-$
9d1 633

$C_6H_6O^-$
9b 439, 456, *17c* 92, *17f* 183

$C_6H_6O^+$
17e 234

$C_6H_6OS^-$
9b 605, 606, *17c* 231

$C_6H_6O_2^-$
1 106, *9b* 456, 457, *9d1* 11, 34, 38, *9d1* 57, *17c* 230, *17g* 33

$C_6H_6O_2^+$
1 121, *9d1* 351, 352ff, *9d2* 125, *17e* 236, 239ff, 242, *17g* 384

$C_6H_6O_2P$
9c2 220
$C_6H_6O_3^-$
9d1 69, *17g* 44
$C_6H_6O_3^+$
9d1 357, *17e* 244
$C_6H_6O_3S^-$
17b 376, 377, *17c* 42
$C_6H_6O_4^+$
1 121, *9d1* 357, *9d2* 127
$C_6H_6O_6^-$
9d1 81, *17g* 58ff
$C_6H_6O_8S^-$
17c 19
$C_6H_6O_8U^+$
17a 241
C_6H_6V
17a 120
C_6H_7
1 40, *9b* 426, 452ff,
17c 71ff, 88ff
$C_6H_7AgBF_4$
9a 329
$C_6H_7AgClO_4$
9a 329
$C_6H_7Br_2F_3NO_3$
9c1 249
$C_6H_7Cl^{2\cdot}$
17h 251
$C_6H_7Cl_2F_3NO_3$
9c1 249
$C_6H_7D_2N_2O_3$
17b 413
$C_6H_7D_3N_3O_4$
9b 171
$C_6H_7D_4N_2O_3$
17b 266
$C_6H_7D_6$
17b 58
$C_6H_7F_3NO$
9c1 329
$C_6H_7F_3NO_3$
9c1 249, 752, *17d1* 207, 211
$C_6H_7N^-$
9d1 858, 859
$C_6H_7N^+$
9d2 29, *17h* 225
$C_6H_7NO^-$
9d1 859, 861, *17c* 136
$C_6H_7NO^+$
17e 236, 239, 241
$C_6H_7NOS^+$ *17e* 331
$C_6H_7NO_2^-$
9d1 69
$C_6H_7NO_2S^-$
9d1 580
$C_6H_7NO_3^-$
9b 419
$C_6H_7N_2$
9b 106, 126, *17c* 440
$C_6H_7N_2O$
9b 126, 127, 520
$C_6H_7N_2O_2$
9b 403, *17b* 175, *17c* 463,
17e 212
$C_6H_7N_2O_3^{2-}$
17b 414, *17c* 418
$C_6H_7N_2O_3$
9b 288, *17b* 416, *17c* 463,
17d1 365, *17e* 211
$C_6H_7N_2O_3S$
9c1 1000
$C_6H_7N_3O_2^-$
9d1 511, 592
$C_6H_7N_4O_3$
9c1 157
C_6H_7O
1 41, *9b* 456, 465, 602,
17b 523, *17c* 92, 229
$C_6H_7O_2$
9b 274, 456, 457, 465,
17c 38, 230
$C_{6...14}H_{7...15}O_2$
1 40
$C_6H_7O_2P$
9b 478
$C_6H_7O_2S$
17b 376, *17c* 41, 44
$C_6H_7O_3^{2-}$
9b 370
$C_6H_7O_3$
9b 466, *17b* 357, 516,
17c 37, 38
$C_6H_7O_3S$
17c 42
$C_6H_7O_4^{2-}$
9d1 12, 15, 67, *17b* 172, 204
$C_6H_7O_4$
9b 256, 306, 371, 495,
17b 497, *17c* 17
$C_6H_7O_4S$
9b 473
$C_6H_7O_5^{2-}$
9b 149, *17b* 205
$C_6H_7O_5$
9b 372, *17b* 497, *17c* 17
$C_6H_7O_6^{2-}$
9b 150 [1]), *17b* 205
$C_6H_7O_6$
17b 276, 450
$C_6H_7O_6S$
9b 154, *17b* 211
$C_6H_7O_7^{2-}$
17g 8
$C_6H_7O_7$
9b 160, *17b* 278
$C_6H_7O_7R_2$
17b 464
$C_6H_7O_7S_2$
9b 474
C_6H_7S
9b 461, 603, *17c* 231
$C_6H_7S^+$
1 128
$C_6H_8^-$
9d1 844, 846
$C_6H_8^+$
17h 25, 29
$C_6H_8^{2\cdot}$
17h 250
$C_6H_8BF_9O_3P$
17f 60
C_6H_8DS
9b 408
$C_6H_8D_2N_2O_2^-$
17b 437
$C_6H_8D_4S_2^+$
17e 379
$C_6H_8D_6^+$
17h 10, 13

[1]) In II/9b erroneously listed as $[C_6H_7O_7]^{\cdot 2-}$.

$C_6H_8F_3O_2$
17b 21, 61, 121, 157, 194, 249, 366
C_6H_8N
9b 260, 462, 510, 511, *17c* 128, 129, 135
C_6H_8NO
9b 260, 512, *9c1* 167, *17b* 339
$C_6H_8NO_2$
9b 201, 202, *9c1* 154, *17c* 622
$C_6H_8NO_2S_2$
17c 529
$C_6H_8NO_3^{2-}$
9d1 67
$C_6H_8NO_3$
9c1 140, *17b* 344
$C_6H_8NO_4$
9c1 145, *17c* 601
$C_6H_8N_2^-$
17f 201
$(C_6H_8N_2)_x$
17b 167
$C_6H_8N_2^+$
1 122, *9d2* 52ff [2]), *17h* 164
$C_6H_8N_2^+ \overset{CT}{\leftarrow} C_8Cl_2N_2O_2^-$
9d1 364
$C_6H_8N_2O_2^-$
9b 423
$C_6H_8N_2O_2^+$
17b 176, *17h* 143
$C_6H_8N_2O_2S^-$
9d1 585
$C_6H_8N_2O_4^-$
9d1 546
$C_6H_8N_2O_5^-$
17b 207
$C_6H_8N_5$
17c 422
$C_6H_8NaO_2$
1 37
C_6H_8O
9b 411
$C_6H_8OS^-$
17g 47
$C_6H_8O_2^-$
9d1 12, 34, 35, 38, *17b* 338, *17g* 8, 23
$C_6H_8O_2S_2^-$
17g 45
$C_6H_8O_4^-$
9b 45, *9d1* 67, 82
$C_6H_8O_4^+$
17h 47
$C_6H_8O_5R$
17b 446
$C_6H_8S_2^+$
17e 372
$C_6H_8S_4^+$
17h 23, 110
$C_6H_8Si^-$
17f 10
$C_6H_8Si^+$
17f 32
C_6H_9
1 40, *9b* 295, 407, 410, 432, *17b* 33, 100, 331, 482, 483, *17c* 10, 30, 33, 35, 59
$C_6H_9Ag^+$
9a 328, 329
$C_6H_9Al^-$
17c 98
$C_6H_9B^-$
17c 98
$C_6H_9ClF_4NO$
17d1 162
$C_6H_9ClF_6OP$
9c2 230
C_6H_9ClNO
9c1 776
$C_6H_9Cl_2F_3NO$
17d1 162, 163
$C_6H_9Cl_5NO$
17d1 163
$C_6H_9D_2NO_2^-$
9b 222
$C_6H_9D_3F_3S_2$
17e 344
$C_6H_9D_3NO_3$
9c1 916
$C_6H_9D_3O_2^+$
17b 60
$C_6H_9D_4$
17b 260
$C_6H_9D_4NO_2S^+$
17b 148
$C_6H_9F_3NO$
9c1 330, *17c* 404, *17d1* 208
$C_6H_9F_3NO_2$
9c1 781, *17d1* 203
$C_6H_9F_3NO_2S$
9c1 772
$C_6H_9F_3NS$
17b 511
$C_6H_9F_3O_5P$
9c 230
$C_6H_9F_5NO$
9c1 380, *17d1* 163
$C_6H_9F_6N_2$
9c1 43
$C_6H_9F_6OPb$
17a 493, *17b* 311
$C_6H_9F_6OSi$
17b 312
$C_6H_9F_6OSn$
17a 452, *17b* 312
$C_6H_9N^+$
17h 220
$C_6H_9NO^+$
17h 20, 159
$C_6H_9NO_2^-$
9d1 47
$C_6H_9NO_2S^+$
17c 43
$C_6H_9N_2O$
9c1 891
$C_6H_9N_2O_2$
17b 433, *17c* 589, *17d2* 389
$C_6H_9N_2O_3$
9c1 69
$C_6H_9N_2O_4^{2-}$
9b 214, *9d1* 423
$C_6H_9N_2O_4$
9c1 141
$C_6H_9N_2O_5$
9c1 141
$C_6H_9N_3O_2^-$
17b 300

[2]) In II/9d2 erroneously listed as $[C_8H_8N_2]^{\cdot}$.

$C_6H_9N_3O_3^-$
9d1 589
C_6H_9O
9b 272, 273, 363, 410, *17b* 387ff, 513, 517
$C_6H_9O_2$
9b 206, 260, 373, *9c2* 14, *17b* 131, 339, 390, 491, 513, 522
$C_6H_9O_3^{2-}$
17b 285
$C_6H_9O_3$
9b 94, 161, 277, *17b* 87, 138, 140, 146, 225, 349, 351, 491, *17c* 16, 17, 37
$C_6H_9O_4$
1 29, 30, 37, *9b* 83, 159, 204, 217, 221, 267, *17b* 22, 31, 49, 63, 73, 87, 121, 132, 154, 160, 172, 179, 194, 250, 481
$C_6H_9O_4S$
9b 147, 154, *17b* 197
$C_6H_9O_4S_2$
17b 271
$C_6H_9O_5$
9b 149ff, 273, 291, *17b* 209, 390, 445, 449, 450, 454, 457
$C_6H_9O_5PS^-$
17e 274
$C_6H_9O_5S$
17b 211
$C_6H_9O_6$
17b 358
$C_6H_9O_6Pb$
17f 53
$C_6H_9O_6R_2$
17b 219, 452, 456, 460, 463, 464
$C_6H_9O_7$
17b 358, 447, 452, 453, 454, 456, 460, 461, 463, 464, 465
C_6H_9S
9b 408, *17b* 376
$C_6H_{10}^-$
9d1 845
$C_6H_{10}^+$
17h 25
$C_6H_{10}Ag$
17b 495
$C_6H_{10}Au$
17b 496
$C_6H_{10}BrNO^-$
9d1 681
$C_6H_{10}Cl$
17b 337
$C_6H_{10}ClNO^-$
9d1 681
$C_6H_{10}Cl_2NO$
9c1 406
$C_6H_{10}Cl_2NO_2$
9c1 781
$C_6H_{10}Cl_2O_5P$
9b 234
$C_6H_{10}Cl_3Si$
9b 308
$C_6H_{10}Cl_4NO$
9c1 376, *17d1* 151, 163
$C_6H_{10}D_4^+$
17h 10, 13
$C_6H_{10}F$
17b 337
$C_6H_{10}FO_2$
17b 368
$C_6H_{10}F_2NO$
9c1 406
$C_6H_{10}F_3OS$
17b 239
$C_6H_{10}F_6OP$
9c2 230
$C_6H_{10}FeNO_3S_4$
17a 328
$C_6H_{10}K_2O_9P$
17b 219
$C_6H_{10}MnNO_3S_4$
17a 282
$C_6H_{10}Mu$
17b 49, 122, 245, 251
$C_6H_{10}MuO$
17b 21, 26, 39, 121, 125, 143, 392
$C_6H_{10}MuO_2$
17b 256
$C_6H_{10}N$
9b 201, *17b* 30, *17c* 395
$C_6H_{10}NO$
1 32, *9b* 294, *9c1* 151, *17b* 30, 106, 110, 233,362, 522, *17c* 20, 407, 410, 622, *17d2* 206, 340
$C_6H_{10}NO_2$
9c1 798, 799, *17d2* 401, 402
$C_6H_{10}NO_2S$
17b 360
$C_6H_{10}NO_3$
9c1 447, 805
$C_6H_{10}NO_3S$
1 76
$C_6H_{10}NO_3S_2$
9c2 298
$C_6H_{10}NO_4^{2-}$
9d1 397
$C_6H_{10}NO_4$
17b 207
$C_6H_{10}NO_5$
9c1 206, *17d2* 212
$C_6H_{10}NS$
9b 128, *17b* 234
$C_6H_{10}N_2^+$
9d2 69, 70, *17h* 178, 220, 229
$C_6H_{10}N_2O_2^-$
9d1 410
$C_6H_{10}N_2O_2^+$
17b 409, 411
$(C_6H_{10}N_2O_2)_x$
17b 168
$C_6H_{10}N_2O_4^-$
9d1 67
$C_6H_{10}N_3O_2$
9b 266, *17c* 410
$C_6H_{10}N_3O_3$
17d1 389
$C_6H_{10}N_3O_4$
9b 166
$C_6H_{10}O^-$
9b 274, 360
$C_6H_{10}O^+$
17h 43
$C_6H_{10}OS_3^-$
17g 45
$C_6H_{10}O_2^-$
9b 117, 146, 366, *9d1* 8, 12, 13, 16, *17b* 119, 170
$C_6H_{10}O_2^+$
17b 476, *17h* 48, 51

$C_6H_{10}O_2S^-$
17b 197
$C_6H_{10}O_2S_2^-$
9d1 52, 76, *17g* 45
$C_6H_{10}O_2S_3^-$
17g 51
$C_6H_{10}O_3^-$
17b 161, 162
$C_6H_{10}O_3S^-$
9d1 52, 76
$C_6H_{10}O_4^-$
9b 234, *9d1* 53
$C_6H_{10}O_4^+$
17h 47
$C_6H_{10}O_5^-$
17g 8
$C_6H_{10}O_6^-$
9d1 9, *17g* 8, 9
$C_6H_{10}O_6P^{2-}$
17b 37
$C_6H_{10}O_6S^-$
17b 161, 162
$C_6H_{10}S_2^+$
17e 367, 369, 370, 380, *17h* 109
C_6H_{11}
1 26, *9b* 16, 24, 78, 185, 260, 271, 348, 349, 350, *17b* 48, 59, 95, 96, 99, 120, 244, 336, 386
$C_6H_{11}AsNO_4$
17d2 418
$C_6H_{11}BD_3NOM$
17d2 171
$C_6H_{11}BrNO_3S$
9c1 992
$C_6H_{11}ClNO_3$
9c1 327, 923
$C_6H_{11}ClN_3O_2$
17b 300
$C_6H_{11}Cl_2$
17b 290
$C_6H_{11}Cl_2NO_4$
9c1 923
$C_6H_{11}Cl_3NO$
9c1 221
$C_6H_{11}Cl_6Si_2$
9b 145
$C_6H_{11}DNO$
17d2 127
$C_6H_{11}D_3^+$
17h 13
$C_6H_{11}F_2O$
17b 64
$C_6H_{11}F_3N$
17c 394
$C_6H_{11}F_3NO$
17d1 208
$C_6H_{11}F_3NO_2$
17d1 207
$C_6H_{11}GeO_2$
17b 516
$C_6H_{11}N^+$
9d2 26
$C_6H_{11}NO^+$
17h 50, 159
$C_6H_{11}NOS_2^-$
17g 46
$C_6H_{11}NO_2^-$
1 79, *9b* 116, *9d1* 50, 414
$C_6H_{11}NO_2S^-$
9d1 76
$C_6H_{11}NO_3^-$
9d1 398
$C_6H_{11}NO_3M$
17d1 8, *17d2* 158ff
$C_6H_{11}NO_3SM$
17d2 152
$C_6H_{11}NO_4^-$
17f 233, 234
$C_6H_{11}NO_4$
9b 149
$C_6H_{11}NO_4^+$
17h 239
$C_6H_{11}NO_4S^+$
17b 210
$C_6H_{11}NO_4SM$
17d2 152
$C_6H_{11}NO_6P$
17c 609
$C_6H_{11}NS_3^-$
17g 46
$C_6H_{11}N_2$
9b 382
$C_6H_{11}N_2O$
9c1 856, *17d1* 132, *17d2* 376
$C_6H_{11}N_2O_3$
9b 121, 133, 208, *17b* 77
$C_6H_{11}N_2O_5S$
9c1 992
$C_6H_{11}N_4O$
17d2 147
$C_6H_{11}O$
9b 16, 48, 90, 112, 188, 197, 256, 271, 274, *17b* 50, 114, 123, 127, 131, 192, 202ff, 335, 352, 392, 513, *17c* 14
$C_6H_{11}OS$
17b 91, 353, 379, 472
$C_6H_{11}OSi$
17c 57
$C_6H_{11}O_2$
9b 21, 44, 87, 115, 139, 140, 184, 190, 200, 203, 223, 267, 269, 275, 363, *9c2* 13, *17b* 21, 22, 28, 48, 53, 61, 73, 110, 119, 121, 127, 139, 151, 153, 157, 160, 162, 171, 193, 222, 243, 248, 255, 258, 262, 263, 276, 365, 368, 439, 441, 467, 469, 477
$C_6H_{11}O_2S$
9b 123, 146, *17b* 148
$C_6H_{11}O_2Si$
17b 516, 517
$C_6H_{11}O_2S_3$
17b 299
$C_6H_{11}O_3$
1 36, *9b* 33, 41, 100, 108, 109, 209, 293, *17b* 36, 82, 83, 128, 160, 161, 162, 221, 264, 285, 471
$C_6H_{11}O_4$
9b 205, 249
$C_6H_{11}O_4P^-$
9c2 231
$C_6H_{11}O_5$
9b 276, *17b* 443, 451, 454, 459, 462, *17e* 6
$C_6H_{11}O_6$
9b 276, *17b* 217, 219, 220, 393, 394, 446, 447, 451, 452, 453, 455, 456, 457, 459, 460, 461, 462, 464, *17e* 6, 7
$C_6H_{11}O_{24}P_6^{6-}$
9b 276

$C_6H_{11}S$
17b 90, 91, 332, 336
$C_6H_{11}S_2$
17b 473, 476, 477
$C_6H_{11}S_3Sn$
17b 384
$C_6H_{12}^+$
17h 15, 16, 19
$C_6H_{12}Br$
9b 15, *17b* 252
$C_6H_{12}Br_2NO$
9c1 373
$C_6H_{12}Cl$
9b 187
$C_6H_{12}ClNO_2$
1 30
$C_6H_{12}ClN_2O_2$
17d1 140
$C_6H_{12}ClO$
17b 154, 155
$C_6H_{12}ClO_2$
17b 144
$C_6H_{12}Cl_2NO$
9c1 328, 377, *17d1* 163
$C_6H_{12}Cl_2N_2^+$
9d2 60
$C_6H_{12}Cl_2O_4P$
9b 234
$C_6H_{12}Cl_3Si$
17b 200
$C_6H_{12}D$
17b 127
$C_6H_{12}FO$
17b 181
$C_6H_{12}FO_2$
17b 144, 294
$C_6H_{12}F_2NO$
17d1 163
$C_6H_{12}F_3OSSi$
17b 274
$C_6H_{12}F_3OSi$
17b 269
$C_6H_{12}F_3OSn$
17a 452, *17b* 270
$C_6H_{12}F_3O_2Si$
17b 320
$C_6H_{12}F_3S_2$
17e 344
$C_6H_{12}Mu$
17b 132, 250
$C_6H_{12}MuN_2$
17c 479
$C_6H_{12}MuO$
17b 25, 125
$C_6H_{12}N$
9b 59, 309, *17b* 30, *17c* 29
$C_6H_{12}NO$
1 30, *9c1* 19, 124, 200, 388, 444, 445, 448, 520, 752, *17b* 38, 102, 213, 361, 508, *17c* 399, 400, 401, 402, 404, *17d2* 126ff
$C_6H_{12}NOS$
17c 515
$C_6H_{12}NO_2$
9b 107, 181, 197, *9c1* 206, 209, 210ff, 394, 782, *17b* 227, 359, *17c* 403, 482, 487, *17d1* 126, 178, *17d2* 152ff, 393
$C_6H_{12}NO_2S$
9c1 771
$C_6H_{12}NO_2Si$
17b 272
$C_6H_{12}NO_3$
9c1 206, 234, 242, 395, 776, 782, 918, 934, *17d1* 178ff, *17d2* 160ff
$C_6H_{12}NO_3P^+$
17e 274
$C_6H_{12}NO_3S$
17d1 190
$C_6H_{12}NO_4$
9c1 918
$C_6H_{12}NO_6$
17d2 211
$C_6H_{12}NS$
17b 511
$C_6H_{12}N_2^+$
9d2 27, 96, 97, *17h* 129
$C_6H_{12}N_2O_2^-$
9d1 53
$C_6H_{12}N_2O_2S_2^-$
9d1 403
$C_6H_{12}N_2O_3^-$
17f 234
$C_6H_{12}N_2O_3M$
17d1 21, 74
$C_6H_{12}N_2O_3SM$
17d2 406
$C_6H_{12}N_2O_4^-$
9d1 410, *17f* 237
$C_6H_{12}N_2O_4^+$
17h 53, 239
$C_6H_{12}N_2O_4S_2^-$
17e 385
$C_6H_{12}N_2S_2^-$
9d1 76
$C_6H_{12}N_2S_2^+$
17e 365
$C_6H_{12}N_3O_5$
9c1 356
$C_6H_{12}N_4^-$
1 133
$C_6H_{12}N_4S_2^+$
17h 178
$C_6H_{12}N_6^+$
17h 165
$C_6H_{12}NaO_4S$
17b 119, 152
$C_6H_{12}O^-$
9b 209
$C_6H_{12}O^+$
17b 54, 258, *17c* 30, *17h* 20, 38
$C_6H_{12}OR$
17b 192
$C_6H_{12}O_2^-$
9b 215
$C_6H_{12}O_2^+$
17b 26, 61, 72, 126, 247, *17h* 23
$C_6H_{12}O_3^+$
17h 39
$C_6H_{12}O_5P$
9c2 231
$C_6H_{12}O_5S^-$
17b 157, 158, 160, 162
$C_6H_{12}S_2^+$
17e 379
C_6H_{13}
9b 14, 22, 37, 40, 82, 104, 181, 185, *17b* 17, 31, 58, 117, 126
$C_6H_{13}BrNO_2$
9b 181
$C_6H_{13}ClNO_2$
9b 182
$C_6H_{13}ClNO_3$
9c1 924

$C_6H_{13}ClNO_3S$
9c1 992
$C_6H_{13}ClO_3P$
9c2 231, 232
$C_6H_{13}D^+$
17h 13
$C_6H_{13}D_3O_3PSSi$
17e 274
$C_6H_{13}FO_3P$
9c2 232
$C_6H_{13}Mg$
9b 130, 191
$C_6H_{13}N^-$
17b 182, 212, 213
$C_6H_{13}N^+$
9b 271, 289, *17h* 128
$C_6H_{13}NO^+$
9b 292, *17b* 102, 110
$C_6H_{13}NO_2^-$
1 78
$C_6H_{13}NO_2^+$
9b 37, 87, 209
$C_6H_{13}NO_2S^-$
9d1 404
$C_6H_{13}NO_3^-$
9d1 391, 395
$C_6H_{13}NO_3^+$
9b 205
$C_6H_{13}NO_3PS$
17e 274
$C_6H_{13}NO_4P$
9c1 1007
$C_6H_{13}NO_4PSi$
17b 520, *17e* 275
$C_6H_{13}NO_5^+$
17b 217, 459, 462
$C_6H_{13}N_2$
9c1 42
$C_6H_{13}N_2O$
17d1 175, *17d2* 148, 374
$C_6H_{13}N_2OS$
17b 183
$C_6H_{13}N_2O_2$
9c1 809, 821, *17d1* 126, 127, 179, *17d2* 209, 212
$C_6H_{13}N_2O_3$
17d1 21ff, 112
$C_6H_{13}O$
9b 31, 39, 141, 194, 205, *17b* 40, 119, 145, 151, 253
$C_6H_{13}OS$
9b 145, *17b* 259
$C_6H_{13}OS_2Sn$
17b 530
$C_6H_{13}O_2$
1 24, *9b* 31, 42, 43, 100, 141, 177, 216, *9c2* 17, *17b* 42, 56, 128, 156, 160, 161, 162
$C_6H_{13}O_2S$
9b 21
$C_6H_{13}O_2Si$
17c 24
$C_6H_{13}O_3$
9b 66, 230, *17b* 187
$C_6H_{13}O_4$
9b 216
$C_6H_{13}O_4P^-$
17g 53
$C_6H_{13}O_4S$
17e 357
$C_6H_{13}O_5$
17e 6
$C_6H_{13}O_6$
17e 7
$C_6H_{13}O_6P^-$
17b 190
$C_6H_{13}S$
9b 34, 196, *17b* 45, 130, 197
$C_6H_{13}S_2$
17b 148, 274
$C_6H_{13}S_3Si$
17b 383
$C_6H_{13}S_3Sn$
17b 383
$C_6H_{13}Si$
9b 383, 384, *17b* 330
$C_6H_{13}Sn$
17b 385
$C_6H_{14}^+$
17h 9, 12
$C_6H_{14}As$
9b 27
$C_6H_{14}BNSi^-$
17b 200
$C_6H_{14}ClSi$
17b 149
$C_6H_{14}Cl_2$
9c2 233
$C_6H_{14}Cl_2N_2O_4S_2^-$
9c2 302, *17e* 386
$C_6H_{14}Cl_2N_2O_4S_2^+$
1 121, *9c2* 300, *17e* 365
$C_6H_{14}Cl_2O_2P$
9c2 233
$C_6H_{14}MuN_2$
17c 440
$C_6H_{14}N$
9b 97, 130, *9c1* 8, *17b* 140, *17c* 392
$C_6H_{14}NO$
9c1 98, 206, 214, 221, 250ff, *17c* 392, 393, *17d1* 112, *17d2* 206, 212
$C_6H_{14}NOS$
9c1 986, *17c* 522, *17d1* 153, 190ff
$C_6H_{14}NOSi$
17b 509
$C_6H_{14}NO_2$
9c1 206, 308, 934, *17d1* 113, 148, 184, *17d2* 72, 76
$C_6H_{14}NO_2S$
17c 522ff, *17d1* 191
$C_6H_{14}NO_3$
9c1 319, 360, 918, 924, 934, *17d1* 148, 184, *17d2* 212, 217
$C_6H_{14}NO_3S$
9c1 995
$C_6H_{14}NO_4$
9c1 935
$C_6H_{14}NO_5P^-$
9d1 412
$C_6H_{14}NS$
9c1 1042, *9c2* 284, *17c* 499, *17e* 325
$C_6H_{14}N_2^+$
9d2 60ff, 97, *17b* 101, *17h* 21, 128, 143, 159
$C_6H_{14}N_2O^-$
9d1 686
$C_6H_{14}N_2O^+$
9c1 207, *9d2* 97, *17c* 634, *17h* 154
$C_6H_{14}N_2O_2^-$
9b 235

$C_6H_{14}N_2O_4^+$
9c1 207
$C_6H_{14}N_3$
17c 580
$C_6H_{14}N_3O_2$
17d1 127
$C_6H_{14}O^+$
17b 40, *17h* 38
$C_6H_{14}OP$
17b 44
$C_6H_{14}OPS_2^{2-}$
17e 275
$C_6H_{14}O_2PS_2$
17e 326ff
$C_6H_{14}O_2PS_3$
17e 360
$C_6H_{14}O_3P$
9b 21, 39, 86, 102, 143, 195, *9c2* 220, *17b* 44ff, 74, 147, *17e* 276
$C_6H_{14}O_4P$
9b 34, 102, *9c2* 233
$C_6H_{14}O_4PS$
17c 276
$C_6H_{14}O_6P$
17b 190
$C_6H_{14}P$
9b 34, 103, *9c2* 216, *17b* 21, 249
$C_6H_{14}PS_2$
9c2 216
$C_6H_{14}S^+$
17e 362
$C_6H_{14}S_2^+$
17e 366
$C_6H_{14}Si^+$
17f 32
$C_6H_{15}Al^-$
17f 63
$C_6H_{15}AlK$
17f 62
$C_6H_{15}As^+$
9b 27, *17f* 59
$C_6H_{15}BN$
17b 110
$C_6H_{15}Cl_2N_2O_4S_2$
9c2 298
$C_6H_{15}FO_3P$
9c2 233
$C_6H_{15}Ge$
9c2 310
$C_6H_{15}N^+$
9b 30, *9d2* 27, *17h* 128
$C_6H_{15}NO^+$
9b 40, 141
$C_6H_{15}NOPS$
9c1 1007
$C_6H_{15}NO_2$
17d2 211
$C_6H_{15}NO_3^+$
9b 137
$C_6H_{15}NO_3P$
17b 184, 257, *17c* 498, *17e* 276
$C_6H_{15}NO_4P$
9c1 1007, *17d1* 200, *17d2* 418
$C_6H_{15}N_2$
9c1 41, 42
$C_6H_{15}N_2O$
9c1 301, 895, *17d1* 127, 182, *17d2* 75
$C_6H_{15}N_2O_2^+$
1 30
$C_6H_{15}N_2O_2S$
17c 527, *17d2* 406
$C_6H_{15}N_2O_3S$
17c 496, *17d1* 195
$C_6H_{15}OP^-$
17e 276
$C_6H_{15}OSSi$
17b 274
$C_6H_{15}OSSn$
17b 239
$C_6H_{15}OSeSn$
17b 274
$C_6H_{15}OSi$
9b 86, *17b* 259
$C_6H_{15}O_2P_2S_2$
17e 263
$C_6H_{15}O_2SSi$
17b 27
$C_6H_{15}O_2Si$
9b 178
$C_6H_{15}O_3P^+$
17e 308
$C_6H_{15}O_3S$
17e 355
$C_6H_{15}O_3Si$
9b 102
$C_6H_{15}O_4P^-$
9c2 234
$C_6H_{15}O_8$
17e 7
$C_6H_{15}P^+$
9b 103, *9c2* 263
$C_6H_{15}Pb$
9c2 316
$C_6H_{15}S_2$
9c2 296
$C_6H_{15}S_2Si$
17b 327
$C_6H_{15}S_3Si$
17b 326
$C_6H_{15}S_3Sn$
17b 327
$C_6H_{15}Si$
9c2 309, *17b* 126, *17f* 6
$C_6H_{15}Sn$
9c2 313, *17f* 46
$C_6H_{16}As$
17f 56
$C_6H_{16}BN_2O$
17d1 113, 198
$C_6H_{16}NOSi$
17d2 207
$C_6H_{16}NO_2$
9c1 782
$C_6H_{16}NO_2Si$
17d2 408
$C_6H_{16}NO_3$
17d2 207
$C_6H_{16}NO_4$
1 30
$C_6H_{16}N_2O^+$
9c1 222
$C_6H_{16}N_2OP$
17b 44, *17c* 450
$C_6H_{16}N_2OX$
17d1 15, 44, 113
$C_6H_{16}N_2O_2X$
17d1 44
$C_6H_{16}N_2O_3P$
17c 450
$C_6H_{16}N_3O^{2+}$
9c1 354
$C_6H_{16}N_4^+$
9d2 97

$C_6H_{16}OP$
9c2 234
$C_6H_{16}O_2S_2^+$
9c2 300
$C_6H_{16}O_3PSSi$
17e 277
$C_6H_{16}O_3PSi$
17e 276
$C_6H_{17}BN$
17c 395, *17f* 61
$C_6H_{17}BP$
17f 61
$C_6H_{17}GeSi$
17f 38
$C_6H_{17}Ge_2$
9b 57
$C_6H_{17}Ge_2O$
9b 57
$C_6H_{17}NOP$
9b 62
$C_6H_{17}N_4O$
9c1 214
$C_6H_{17}OSi_2$
9b 75
$C_6H_{17}O_2Si_2$
17b 88, 94
$C_6H_{17}O_{24}P_6$
9b 276
$C_6H_{17}Si_2$
9b 76, *9c2* 309
$C_6H_{17}Sn_2$
9b 76
$C_6H_{18}BNOP$
17d1 200
$C_6H_{18}B_2N_4^+$
9d2 97
$C_6H_{18}B_2O_6^-$
9d1 705
$C_6H_{18}Cl_3N_6P_3^-$
9c2 234
$C_6H_{18}GeSi^+$
17a 431, *17f* 32, 46
$C_6H_{18}GeSn^+$
17a 452, *17f* 46, 53
$C_6H_{18}Ge_2^+$
17a 431, *17f* 46
$C_6H_{18}Ge_2N$
17c 439
$C_6H_{18}Ge_2NO$
9c1 1006
$C_6H_{18}NOSi_2$
9c1 1005, *17c* 496, *17d2* 420
$C_6H_{18}NSi_2$
9b 75
$C_6H_{18}N_3OP^+$
17h 159
$C_6H_{18}N_3OX_2$
17d1 145
$C_6H_{18}N_3Si_2$
17c 583
$C_6H_{18}N_6P_2^+$
9d2 98, *17e* 308
$C_6H_{18}O_3PSi$
9c2 221
$C_6H_{18}O_6P_2^+$
9c2 263ff, *17e* 308
$C_6H_{18}P_2^+$
9c2 263
$C_6H_{18}P_2S_6^+$
9c2 264
$C_6H_{18}Si_2^+$
17f 37
$C_6H_{18}Sn_2^+$
17a 452, *17f* 53
$C_6H_{19}ClOP$
9c2 234
$C_6H_{20}AlN_2$
17f 62
$C_6H_{20}BN_3P$
17f 61
$C_6H_{20}Cl_2CoN_4$
9d1 137
$C_6H_{21}Cl_3N_3S_2$
9c2 296
$C_6I_4O_2^-$
9d1 235, *17g* 245
$C_6N_4^-$
1 78
$C_6N_4O_2^{3-}$
17g 358
$C_6Ni_2O_6^+$
17a 129
$C_6O_6^{5-}$
9d1 275
$C_6O_6^-$
9a 49
C_6O_6V
17a 120
C_7ClF_5NO
17c 603
$C_7ClF_9NS_3$
17c 409
C_7Cl_7
9b 565
C_7D_7
9b 538
$C_7D_{10}NS_2$
17c 552
C_7F_5O
9b 322
C_7F_7O
9c2 179
C_7F_8NO
9c1 727
C_7F_9O
17c 102
C_7F_{13}
9b 277
C_7F_{15}
9b 245, *17b* 306
$C_7F_{15}O$
9b 242
C_7HCl_6
9b 565
C_7HF_5NO
9c1 127, *17c* 587
$C_7HN_4^{2-}$
1 78
$C_7H_2Br_2D_3KNO_4S$
17d1 256
$C_7H_2ClD_3O^-$
17c 168
$C_7H_2F_3N_2O_5^{2-}$
9d1 624
$C_7H_2F_5$
9b 548
$C_7H_2F_5O$
9b 68
$C_7H_2N_2O_7^{3-}$
9d1 629
$C_7H_2N_3O_8^{2-}$
9d1 666
$C_7H_2O_5^{3-}$
9d1 245, 246
$C_7H_2O_6^{3-}$
9b 496, *17c* 110
$C_7H_2O_9P^{4-}$
17b 465, *17c* 110
$C_7H_3BrNO_5^{2-}$
9d1 575

$C_7H_3BrN_2O_2^-$
17f 273
$C_7H_3Br_2CoN_6O$
9c1 1015
$C_7H_3CdO_4$
17a 386, *17g* 108
$C_7H_3ClF_3NO_2^-$
9d1 481
$C_7H_3ClNO_5^{2-}$
9d1 575
$C_7H_3ClN_2O_2^-$
17f 273
$C_7H_3ClN_3O_2^{2-}$
17f 290
$C_7H_3Cl_2CoN_6O$
9c1 1015
$C_7H_3Cl_3NO$
17c 603
$C_7H_3Cl_5NO$
17d1 236
$C_7H_3CoN_7O_2^-$
9c1 1012
$C_7H_3CoN_7O_3^-$
9c1 1012
$C_7H_3D_3O^-$
9b 540
$C_7H_3F_3NO_3^{2-}$
9d1 497
$C_7H_3F_3NS_2$
17c 544
$C_7H_3F_3O_2^-$
9d1 188
$C_7H_3F_4NO_2^-$
17f 245
$C_7H_3F_4NO_3^-$
17f 249
$C_7H_3F_4O$
9c2 178
$C_7H_3F_6$
9b 489, *17c* 100
$C_7H_3F_6O$
17c 102
$C_7H_3IN_2O_2^-$
17f 274
$C_7H_3NO_2^-$
9d1 188
$C_7H_3NO_4^{3-}$
9d1 88
$C_7H_3NO_5^{3-}$
9b 519, *9d1* 535, *17c* 105
$C_7H_3NO_5$
17d2 412
$C_7H_3NO_6^{3-}$
17c 108
$C_7H_3NS_2^-$
9d1 892
$C_7H_3N_2O_2^{2-}$
9d1 339, 341
$C_7H_3N_2O_6^{2-}$
9d1 323, 613, 628, 629, 641, 652
$C_7H_3N_3O_4^-$
1 101, *9d1* 633ff, *17f* 297ff
$C_7H_3NaO_4^-$
17g 108
$C_7H_3O_4^{2-}$
9d1 120, 127, 160, 161, 162, 192, 244, *17g* 108, 210, 211
$C_7H_3O_4Zn$
17a 377, *17g* 109
$C_7H_3O_5^{4-}$
17g 255
$C_7H_3O_5$
9c2 182
$C_7H_3O_6^{2-}$
9b 495
$C_7H_3O_8^{4-}$
17b 278
$C_7H_4BrCoN_6O$
9c1 1012
C_7H_4ClFNO
17c 586, 587
$C_7H_4ClNO_3^-$
1 98
$C_7H_4ClNO_4^-$
9d1 535
$C_7H_4Cl_2FO$
17b 323
$C_7H_4Cl_2NO$
9c1 127, 128, *17c* 586, 603
C_7H_4FO
9b 321, 322
$C_7H_4FO_2^{2-}$
9b 573
$C_7H_4F_3$
9b 564
$C_7H_4F_3NO_2^-$
9d1 477ff, 479, *17f* 246
$C_7H_4F_3NO_2S^-$
9d1 508
$C_7H_4F_3NO_3^-$
9d1 497, 503, *17f* 279
$C_7H_4F_3NO_3S^-$
9d1 548
$C_7H_4F_3NO_4S^-$
9d1 551
$C_7H_4F_3O$
17e 36, 39, 41
$C_7H_4F_5$
9b 464, 489
$C_7H_4F_6NO_4$
9c1 753
C_7H_4NO
9c2 32, 36, 41
$C_7H_4NO_2^{2-}$
9b 518
$C_7H_4NO_2$
9c1 31
$C_7H_4NO_3^{2-}$
9d1 693
$C_7H_4NO_3$
9c1 983, *17d2* 412
$C_7H_4NO_4^{2-}$
9b 493, 515, 516, 517, *9c1* 24, *9d1* 322, 534, 535, 536ff, 570, *17c* 132ff, *17f* 265
$C_7H_4NO_5^{2-}$
9b 518, *9d1* 535, 575, *17c* 105, 107, 108, 109, 411
$C_7H_4NO_6^{2-}$
17c 108
$C_7H_4N_2OS_2^-$
9b 711
$C_7H_4N_2O_2^-$
1 89, 94, *9d1* 340, 554, 555ff, *17f* 273, 274ff
$C_7H_4N_2O_6^-$
9d1 613, 628, 629, 641
$C_7H_4N_3O_2^{2-}$
17f 289, *17g* 358
$C_7H_4N_3O_6^{2-}$
9d1 664
$C_7H_4Na_3O_4$
17g 256
$C_7H_4O_2^-$
9b 311, 312
$C_7H_4O_2S_2^-$
9d1 63, *17g* 38ff

$C_7H_4O_3^-$
9c2 31, 36, 41, *9d1* 44, 79, 322
$C_7H_4O_4^{3-}$
9d1 273, *17g* 256
$C_7H_4O_4^-$
9c2 55, 57, *9d1* 120, 127, 188, *17e* 47, 53, 54, *17g* 361, 364
$C_7H_4O_6^{3-}$
17b 206, *17c* 17
C_7H_5AgN
17a 374, *17c* 571
C_7H_5AuN
17c 572
$C_7H_5BD_3F_3^-$
17c 99
$C_7H_5BrMgO_2^-$
17a 225, *17c* 203
C_7H_5BrNO
9c1 128, *17c* 585
$C_7H_5BrO_3^-$
9b 467
$C_7H_5Br_2KNO_4S$
17d1 255
$C_7H_5Br_2NNaO_4S$
17d1 255
$C_7H_5Br_2NNaO_5S$
17d1 256
$C_7H_5Br_3NO$
17d1 251
$C_7H_5CaO_4$
9d1 270
C_7H_5ClNO
9c1 127, 128, *17c* 585, 586
$C_7H_5ClNO_2$
1 53
$C_7H_5ClN_2O_3^-$
9d1 542
$C_7H_5ClN_2O_4^-$
9d1 694ff
$C_7H_5ClN_2O_5^-$
9d1 626
$C_7H_5ClO^-$
17c 168, 169
$C_7H_5ClO_3^-$
9b 467
$C_7H_5Cl_2O$
17b 323
$C_7H_5Cl_2O_2S$
17e 340
$C_7H_5Cl_3NO$
9c1 610
$C_7H_5FNO^{2-}$
9b 581
C_7H_5FNO
9c1 127, 128
$C_7H_5FO^-$
9b 541, 553
$C_7H_5FO_3^-$
9b 467
$C_7H_5F_2$
9b 564
$C_7H_5F_2NO_2^-$
9d1 476ff
$C_7H_5F_2NO_2S^-$
9d1 507
$C_7H_5F_2NO_3^-$
9d1 503
$C_7H_5F_2NO_3S^-$
9d1 547
$C_7H_5F_2NO_4S^-$
9d1 551
$C_7H_5F_3NO$
9c1 545, 724
$C_7H_5F_3NOS$
9c1 986
$C_7H_5F_3N_2O_2^-$
9d1 511
$C_7H_5F_3O^+$
17e 234, 238, 240
$C_7H_5F_4$
9b 464
$C_7H_5FeO_4$
17e 18
$C_7H_5IN_2O_3^-$
17f 267
$C_7H_5KN_3O_8S$
9c1 105
$C_7H_5N^-$
1 103
$C_7H_5N^+$
17h 233
$C_7H_5NO^-$
9b 309
$C_7H_5NO_2^-$
9b 516, *9c1* 9, *9d1* 688
$C_7H_5NO_3^-$
1 87ff, 92ff, *9c2* 55, 57, 59, *9d1* 529ff, *17f* 258, 259, 261
$C_7H_5NO_4^-$
1 89, 92, *9b* 518, *9d1* 536, 575, *17f* 265, 282
$C_7H_5NO_5^-$
9b 475
$C_7H_5NO_7P$
9c1 971
$C_7H_5NS_2^-$
9d1 901
$C_7H_5N_2$
17c 480
$C_7H_5N_2^+$
9d2 29
$C_7H_5N_2O$
9c1 858, *17d2* 367
$C_7H_5N_2O_2$
9c1 892, *17d1* 382
$C_7H_5N_2O_3^{2-}$
9d1 543, 546, 689, *17f* 269
$C_7H_5N_2O_3$
9c1 126, 127, 128
$C_7H_5N_2O_4^{2-}$
9d1 424
$C_7H_5N_2O_4$
9c1 768
$C_7H_5N_2O_6^{2-}$
9d1 624
$C_7H_5N_2S$
17e 325
$C_7H_5N_2S_2$
17c 531, *17e* 358
$C_7H_5N_3O_5^-$
9d1 631, *17f* 296
$C_7H_5N_3O_6^-$
9d1 663ff
$C_7H_5N_3O_7^-$
9d1 665
$C_7H_5N_4O_2^{2-}$
17f 291, 292
$C_7H_5N_4O_7$
9c1 610, *17d1* 375
C_7H_5O
9b 321, *17b* 525
$C_7H_5OS^{2-}$
9b 577
$C_7H_5O_2^{2-}$
9b 571

$C_7H_5O_2$
1 59, *9c2* 31, 35, 40, 185, *17b* 480, *17e* 7
$C_7H_5O_3^{2-}$
9b 573, *9d1* 250, 260, *17c* 92, 93, *17g* 251
$C_7H_5O_3$
1 58, 59, *9c2* 31, 36, 41, *17e* 228, 229
$C_7H_5O_3S^{2-}$
9b 602
$C_7H_5O_3S$
17e 330
$C_7H_5O_4^{2-}$
9b 467, *9d1* 270
$C_7H_5O_4$
9c2 55, 63, 64, *17e* 229
$C_7H_5O_4S$
9c2 292
$C_7H_5O_5$
9c2 65, 165
$C_7H_5O_6$
9b 306, *17c* 110
$C_7H_5O_6S$
9c2 57
C_7H_6Br
9b 549
$C_7H_6BrNO_2^-$
1 91, *17f* 247
$C_7H_6BrN_2O^-$
9c1 807
C_7H_6BrO
9b 67
$C_7H_6BrO_2S$
17e 339
$C_7H_6Br_2NO$
17d1 369
C_7H_6Cl
9b 549, *17c* 174
$C_7H_6ClNO_2^-$
1 91, *9d1* 379, 482, 488, *17f* 247
$C_7H_6ClNO_4S^-$
9d1 548
$C_7H_6ClN_2$
17c 578
C_7H_6ClO
9b 67ff, *17b* 86, 237
$C_7H_6ClO_2$
9b 458, *9c2* 21
$C_7H_6ClO_2S$
17e 339
C_7H_6D
9b 538, *17c* 49
$C_7H_6D_5N_2O_5$
17b 284
C_7H_6F
9b 547ff, 548, 552, *17c* 173, 174, 182
$C_7H_6FNO^-$
9b 581
$C_7H_6FNO_2^-$
9d1 476
$C_7H_6FNO_4S^-$
9d1 551
C_7H_6FO
9b 68
$C_7H_6F_9O$
17b 125
$C_7H_6HgNO_6S$
17a 390
$C_7H_6KO_2$
9d1 175
$C_7H_6LiO_2$
9d1 175
C_7H_6N
9b 428, *9c1* 117, *17c* 49, 96, 565
$C_7H_6NO^{2-}$
9b 580
C_7H_6NO
9c1 126, 127, *17c* 584
$C_7H_6NOS_2$
17c 543, 545
$C_7H_6NO_2$
1 53, *9b* 551, *9c1* 158, 160, 161, 545, 767, *17d2* 367
$C_7H_6NO_2S$
17c 541
$C_7H_6NO_3^{2-}$
9d1 380, 494, *17g* 254
$C_7H_6NO_3$
9c1 159, 546, 547ff
$C_7H_6NO_3S$
17d2 415
$C_7H_6NO_4$
9b 516, 518
$C_7H_6NO_5^{2-}$
17g 59
$C_7H_6NO_6S^{2-}$
9d1 552
$C_7H_6NS^{2-}$
9b 584
$C_7H_6NS_2$
17c 544, 547
$C_7H_6N_2O_2S^-$
9d1 512, 554
$C_7H_6N_2O_3^-$
1 94, *9d1* 542, 545, *17f* 266, 267
$C_7H_6N_2O_4^-$
1 101, *9d1* 593, 611, 621
$C_7H_6N_2O_5^-$
9d1 625, 694
$C_7H_6N_3O$
9c1 147, *17c* 605
$C_7H_6N_3O_2$
9c1 911
$C_7H_6N_3O_3^-$
9c1 807
$C_7H_6N_3O_6$
17d1 282
$C_7H_6N_4O_2^-$
17f 292
$C_7H_6N_4O_5^-$
9d1 632
$C_7H_6N_4O_6^-$
17f 281
$C_7H_6NaO_2$
9d1 175, *17g* 81
$C_7H_6O^-$
1 103, *9b* 540, 550, 553, *17c* 167, 182
$C_7H_6O^{2\cdot}$
17h 261
$C_7H_6O_2^-$
1 80, *9b* 571, *9d1* 30, 35, 57, 118, 122, 160, 161, 174, *17g* 78, 80ff, 214ff
$C_7H_6O_2^+$
17h 79
$C_7H_6O_3^-$
9b 466, 467, 573, *9d1* 121, 132, 189, *17g* 79ff, 111, 216ff
$C_7H_6O_3^+$
17e 243, *17h* 81
$C_7H_6O_3S^-$
9b 551

$C_7H_6O_3Zn^+$
17g 111
$C_7H_6O_4^+$
17e 246
$C_7H_7^{2-}$
9d1 723
C_7H_7
1 41, *9b* 311, 312, 429, 438, 537ff, 543ff, *17b* 489, 501, *17c* 49, 165, 166, 171ff
$C_7H_7AlBr_3S$
17a 395, 396, *17e* 319
$C_7H_7Al_2Br_6S$
17a 396
$C_7H_7Cl_3GaS$
17a 407, *17e* 319
$C_7H_7Cl_6Ga_2S$
17a 407
$C_7H_7D^-$
9d1 711, 848
$C_7H_7D_8N_2O_4S_2^{2+}$
17b 238
$C_7H_7FO_3S^+$
17h 91
$C_7H_7F_8N_2O_3$
17d2 221
$C_7H_7HgO_2$
17g 72
$C_7H_7N^-$
9d1 860
$C_7H_7N^+$
17b 492, *17h* 21, 226
$C_7H_7NO^-$
1 114, *9b* 609, 610, 611
$C_7H_7NO_2^-$
1 87, 88, 90, *9b* 521, 523, 612, *9d1* 379, 449ff, 451ff, *17c* 139, *17f* 239ff, *17g* 98, 223
$C_7H_7NO_2S^-$
9d1 506, 507
$C_7H_7NO_3^-$
1 91, 95, *9b* 612, *9d1* 494, 497, 499, 500ff, *17f* 248ff, 250
$C_7H_7NO_3^+$
17h 81
$C_7H_7NO_3S^-$
9d1 546, 547
$C_7H_7NO_4^-$
17g 60
$C_7H_7NO_4P^{2-}$
9d1 411
$C_7H_7NO_4S^-$
9d1 548, 550
$C_7H_7NO_5^-$
17f 280
$C_7H_7NO_5S^-$
1 97, *17f* 269
$C_7H_7N_2$
9b 524, *17c* 144, 575, 576
$C_7H_7N_2O^-$
9c1 808
$C_7H_7N_2O$
9c1 179, *17c* 578, 579
$C_7H_7N_2O_2$
9c1 179, 766, 808, 910, *17c* 23
$C_7H_7N_2O_5$
9c1 949
$C_7H_7N_2S$
17c 580
$C_7H_7N_3^-$
9d1 880
$C_7H_7N_3O_4^-$
9d1 627, *17f* 280
$C_7H_7N_3O_6^-$
9d1 660
$C_7H_7N_5O^+$
9d2 70
C_7H_7O
1 58, *9b* 67, 312, 552, *9c2* 30, 34, 38, *17c* 49, 51, 166, 182, *17e* 36, 38, 40
C_7H_7OS
17e 329, 330ff
$C_7H_7O_2$
1 58, 59, *9b* 427, 457, *9c2* 11, 34, 38, 53ff, 55, 56, 64, *17c* 96, *17e* 38, 40, 45, 47, 49, 52, 53
$C_7H_7O_2S$
9c2 290, 92, *17e* 331, 337ff
$C_7H_7O_3$
9b 458, 466, *9c2* 19, *17e* 54, 55
$C_7H_7O_3S$
9b 551, *9c2* 293
$C_7H_7O_4^-$
17g 360
$C_7H_7O_5^{2-}$
17b 206
$C_7H_7O_6$
9b 376
$C_7H_7O_6S$
9b 474
$C_7H_7O_7$
9b 377
$C_7H_7O_7S$
9b 489
C_7H_7S
9b 74, *9c2* 272, *17b* 92
C_7H_7Se
9c2 303
$C_7H_8^-$
1 85, *9d1* 710ff, 847
$C_7H_8^+$
9d2 6, *17h* 29
C_7H_8Ag
17a 374
C_7H_8AgO
17a 374
$C_7H_8CdNO_2^{2+}$
17a 387, *17g* 99
C_7H_8D
9b 454
C_7H_8DO
17c 48, 49
$C_7H_8D_2NS_2$
17c 551, 555
C_7H_8N
9b 511
C_7H_8NO
9c1 97, 197, 548, 610, *17c* 131, *17d1* 213, *17e* 51
$C_7H_8NO_2$
9b 475, 519, *9c1* 169, 200, 548, 549, 623, 947, *17g* 98
$C_7H_8NO_2Zn^{2+}$
17a 377, *17g* 99
$C_7H_8NO_3$
9c1 947, *17d1* 230
$C_7H_8NO_3S$
17d1 232, *17d2* 414
$C_7H_8NO_4$
9c1 949
$C_7H_8NO_4S$
9c1 1002

$C_7H_8NO_6S$
17d1 383
C_7H_8NS
9c1 1042
$C_7H_8NS_2$
17c 554
$C_7H_8N_2O^-$
9d1 687
$C_7H_8N_2O_4^-$
9d1 591
$C_7H_8N_3$
17b 307
$C_7H_8N_3O$
17b 308
$C_7H_8N_3S$
17c 446, 447
$C_7H_8N_4O_2^+$
9d2 70, *17h* 207
$C_7H_8N_5$
17b 142
$C_7H_8O^-$
9b 439
$C_7H_8O^+$
9d2 129, *17e* 234, 237, 240
C_7H_8OP
9c2 216
$C_7H_8OS^-$
17g 47
$C_7H_8O_2^-$
9b 495, *9d1* 30, 35, 36, 38, 39, 56, 59, *17c* 109, *17g* 37
$C_7H_8O_2^+$
17e 237, 240, 242, 244, 245
$C_7H_8O_3^+$
17e 245, *17g* 387
$C_7H_8O_4S^+$
17h 91
$C_7H_8S_2^-$
9b 506, *17g* 48
$C_7H_8S_3^-$
17f 216
C_7H_9
9b 301, 429, 453, 454, *17b* 487, 488, *17c* 50, 51, 72, 73, 75, 91, 165
$C_7H_9ClF_4NO_2$
17d1 203
$C_7H_9Cl_7NO$
17d1 163
$C_7H_9D_2$
17b 483, 484
$C_7H_9D_3NO$
9c1 152
$C_7H_9D_4N_2O_3S$
17b 284
$C_7H_9D_5O_2^+$
17b 59, 248
$C_7H_9D_7^+$
17h 14
$C_7H_9F_3NO$
17d1 208
$C_7H_9F_5NO_2$
17d1 204
$C_7H_9F_6O$
17b 311
$C_7H_9F_7NO$
9c1 379, 380
$C_7H_9N^-$
1 114, *9d1* 859, 860
$C_7H_9N^+$
9b 611, *9d2* 29, *17h* 226
$C_7H_9NO^-$
9d1 860
$C_7H_9NO^+$
9d2 30
$C_7H_9NO_2^-$
9d1 48, 861
$C_7H_9NO_3$
9b 126
$C_7H_9N_2$
9c1 40
$C_7H_9N_2O_3$
17d1 365
$C_7H_9N_2O_3S$
9c1 980
$C_7H_9N_2O_5S$
17d1 384
$C_7H_9N_3O_4P$
9c2 234
C_7H_9O
9b 254, 296, 427, 476, *17c* 95
$C_7H_9O_2$
17c 109
$C_7H_9O_3^{2-}$
9b 370
$C_7H_9O_3$
17b 516, 522
$C_7H_9O_4^{2-}$
9d1 16, *17b* 204, 205
$C_7H_9O_4$
9b 371, *17b* 470, 497
$C_7H_9O_4S$
9b 473
$C_7H_9O_5^{2-}$
9b 150, *17b* 206
$C_7H_9O_5$
17b 498
$C_7H_9O_6S$
17b 210, 211
$C_7H_{10}Br$
17b 484
$C_7H_{10}BrO_4$
9b 238
$C_7H_{10}ClSi$
17c 83
$C_7H_{10}Cl_2NO_2^+$
17h 132
$C_7H_{10}DNO_3S^-$
17b 283
$C_7H_{10}D_6^+$
17h 11, 13
$C_7H_{10}FSi$
17c 82
$C_7H_{10}F_6NO$
9c1 260
$C_7H_{10}I$
17b 484
$C_7H_{10}N$
17c 129, 130, 135
$C_7H_{10}NO$
9c1 168, 169
$C_7H_{10}NO_2$
9c1 162, 524, *17b* 248, *17c* 629, *17d2* 340
$C_7H_{10}NO_3$
9c1 807, *17b* 343, *17c* 497
$C_7H_{10}NO_4$
9c1 145
$C_7H_{10}NO_5$
9c1 147, *17c* 601
$C_7H_{10}NS$
9c1 1043
$C_7H_{10}NS_2$
17c 550ff, 555
$C_7H_{10}N_2O_5^-$
17b 207

$C_7H_{10}N_3$
17b 307
$C_7H_{10}N_3O_3$
17b 41
$C_7H_{10}N_3O_3S$
17d2 407
$C_7H_{10}N_5$
9c1 30, *17c* 423, 426
$C_7H_{10}O^-$
9b 231, 411
$C_7H_{10}O^{2\cdot}$
17h 251
$C_7H_{10}OSi^-$
17c 164
$C_7H_{10}O_2^-$
9d1 13, 39, 44, *17b* 387, *17g* 24
$C_7H_{10}O_5^-$
9d1 83
$C_7H_{10}S_2^+$
17e 372, 373, 380
$C_7H_{10}Si^-$
17f 10
$C_7H_{10}Si^+$
17f 32
C_7H_{11}
9b 24, 145, 296, 301, 387, 402, 414, *17b* 23, 33, 113, 331, 483, 484, *17c* 31, 34, 59
$C_7H_{11}Ag^+$
9a 329
$C_7H_{11}Cl_3NO$
9c1 776, *17d2* 139
$C_7H_{11}Cl_3NO_3$
17d2 161
$C_7H_{11}Cl_5NO$
17d1 163, 164
$C_7H_{11}D_5^+$
17h 13
$C_7H_{11}F_3NO$
9c1 330, 388, *17d1* 208, 209
$C_7H_{11}F_5NO$
17d1 164
$C_7H_{11}F_6N_2$
9c1 43
$C_7H_{11}NO_2^+$
17h 53, 132
$C_7H_{11}NO_3^-$
9d1 418, 682
$C_7H_{11}NO_3M$
17d2 145
$C_7H_{11}NO_6^-$
9d1 401
$C_7H_{11}NO_6S^+$
17b 211
$C_7H_{11}N_2O$
9c1 207, 260, *17c* 564
$C_7H_{11}N_2OS$
9c1 822, *17d2* 163
$C_7H_{11}N_2O_3$
17b 230
$C_7H_{11}N_2O_3S$
9c1 993
$C_7H_{11}N_2O_4^{2-}$
9d1 424
$C_7H_{11}N_2O_4$
17c 452
$C_7H_{11}N_2O_5^{2-}$
9d1 424
$C_7H_{11}N_3O_2^-$
9d1 589
$C_7H_{11}N_3O_4^-$
9d1 586
$C_7H_{11}N_4O_3$
17c 470
$C_7H_{11}O$
9b 261, 272, 273, 298, 375, *17b* 388, 389, 391, 475
$C_7H_{11}O_2$
9b 274, 373, *17b* 62, 392, 516
$C_7H_{11}O_3$
9b 264, *17b* 87, 170, 349, 353, *17c* 38
$C_7H_{11}O_4$
1 29, 30, *9b* 83, 159, 200, 221, *17b* 23, 31, 41, 48, 63, 87, 119, 122, 132, 146, 152, 154, 161, 165, 172, 174, 201, 243
$C_7H_{11}O_4S$
9b 154
$C_7H_{11}O_5$
1 36, *9b* 117, 151, *17b* 311, 449
$C_7H_{11}O_5S_2$
9b 158
$C_7H_{11}S$
17c 57
$C_7H_{11}S_3$
17b 384
$C_7H_{11}S_4Sn$
17b 383
$C_7H_{11}Si$
17c 81
$C_7H_{12}^+$
17h 16
$C_7H_{12}BrN_2O$
9c1 838
$C_7H_{12}BrN_2O_2$
9c1 862
$C_7H_{12}ClNO_3^-$
9c1 222
$C_7H_{12}ClN_2O_2$
9c1 862, *17d2* 409
$C_7H_{12}Cl_4NO$
17d1 164
$C_7H_{12}D_2O_2^+$
17b 59, 248
$C_7H_{12}F_2NO_2$
17d1 204
$C_7H_{12}F_4NO$
17d1 164
$C_7H_{12}IN_2O$
9c1 838
$C_7H_{12}IN_2O_2$
9c1 862
$C_7H_{12}Mu$
17b 18, 119
$C_7H_{12}N$
17b 167, 491, *17c* 57
$C_7H_{12}NO$
9c1 152, 162, *17b* 24, 106, 113, 140, 362, 522, *17c* 399, 400, *17d2* 340
$C_7H_{12}NO_2$
9c1 219, 524, 525, 766, 799, *17b* 465, *17d2* 401, 402
$C_7H_{12}NO_3$
17b 250, 297, *17d2* 145ff
$C_7H_{12}NO_3S$
1 38, *9c2* 280
$C_7H_{12}NO_4$
9b 113, *9c1* 238, *17c* 421
$C_7H_{12}NO_5^{2-}$
9c1 299
$C_7H_{12}NO_5$
17d1 140

$C_7H_{12}N_2^+$
9d2 98, *17h* 178, 179

$C_7H_{12}N_2OSM$
17d2 378

$C_7H_{12}N_2O_3^-$
9d1 83

$C_7H_{12}N_2O_3SM$
17d2 407

$C_7H_{12}N_2O_4M$
17d1 179

$C_7H_{12}N_3O$
17d1 182

$C_7H_{12}N_3O_2$
17c 455

$C_7H_{12}N_3O_3$
17d1 140

$C_7H_{12}N_4O^+$
17b 184

$C_7H_{12}O^-$
9b 275

$C_7H_{12}OS_3^-$
17g 46

$C_7H_{12}O_2^-$
9d1 13, 14, 16, *17b* 118, 171

$C_7H_{12}O_2^+$
17h 51

$C_7H_{12}O_2R$
17b 175

$C_7H_{12}O_2S_2^-$
17g 46

$C_7H_{12}O_2S_2^+$
17e 368

$C_7H_{12}O_3PS_3$
17b 384

$C_7H_{12}O_4^-$
9b 234, *9d1* 53

$C_7H_{12}O_6^-$
17b 217, 457

$C_7H_{12}S_2^+$
17e 370, 380, 382, *17h* 109

C_7H_{13}
1 26, *9b* 182, 253, 274, 294, 348, 350, *17b* 95, 112, 243, 251, 475, *17c* 10

$C_7H_{13}ClNO_2$
9c1 782

$C_7H_{13}ClNO_3$
9c1 935

$C_7H_{13}ClNO_4$
9c1 935

$C_7H_{13}Cl_3NO$
17d1 151, 164

$C_7H_{13}D_3^+$
17h 13

$C_7H_{13}FNO$
9c1 406

$C_7H_{13}F_3NO$
9c1 381, *17d1* 208, 210

$C_7H_{13}F_3NO_2$
17d1 210

$C_7H_{13}F_3O_3P$
17b 196

$C_7H_{13}N^-$
17b 213

$C_7H_{13}N^+$
9d2 27

$C_7H_{13}NO^+$
17h 132

$C_7H_{13}NO_2^-$
9d1 414, *9d1* 415

$C_7H_{13}NO_3^-$
17b 272, 301

$C_7H_{13}NO_3M$
17d1 25

$C_7H_{13}NO_3S^+$
9c2 298

$C_7H_{13}NO_4^+$
9b 150

$C_7H_{13}NO_4M$
17d1 25

$C_7H_{13}NO_4P$
17b 520, *17c* 608, *17e* 277

$C_7H_{13}NO_5P$
17c 609

$C_7H_{13}NO_6P$
17c 609, 610

$C_7H_{13}NO_6PM_2$
17d1 140

$C_7H_{13}N_2O$
9b 110, *9c1* 22, 838, 857, *17b* 438, *17d1* 113, *17d2* 376

$C_7H_{13}N_2OS$
9c1 415, *17d2* 378

$C_7H_{13}N_2O_2$
9c1 389, 415, 419, 839, 853, 863, 890, *17d2* 401

$C_7H_{13}N_2O_3$
9c1 863, *17b* 49, 251, *17d2* 249, 261

$C_7H_{13}N_2O_4^{2-}$
9d1 423

$C_7H_{13}N_2O_4$
9c1 63, *17d1* 140

$C_7H_{13}N_4O_3$
17d1 80

$C_7H_{13}O$
9b 91, 110, 188, 197, 294, 360, 363, 367, 368, *17b* 53, 55, 203, 393, 513, *17c* 15

$C_7H_{13}OS$
17b 46, 149, 441, 529

$C_7H_{13}OS_3$
17b 382

$C_7H_{13}O_2$
9b 18, 94, 140, 190, 268, *9c2* 15, *17b* 22, 28, 118, 121, 127, 139, 150, 151, 154, 160, 171, 253, 256, 276, 363, 440

$C_7H_{13}O_2S$
9b 146

$C_7H_{13}O_2Si$
9b 385, *17b* 517, 518

$C_7H_{13}O_3$
9b 33, 41, 86, 109, *17b* 191, 267, 468, 469

$C_7H_{13}O_4$
17b 84

$C_7H_{13}O_6$
17b 85, 218, 219, 451, 452, 453, 455, 456, 459, 460, 461, 463

$C_7H_{13}S$
17b 91, 337, *17c* 18, 19

$C_7H_{13}S_2$
17b 473, 477

$C_7H_{13}S_3$
17b 381

$C_7H_{13}S_4Sn$
17b 381

$C_7H_{13}Si$
9c2 309, *17b* 112

$C_7H_{14}^+$
17b 244, *17h* 15ff

$C_7H_{14}F_3S_2$
17e 345

$C_7H_{14}N$
9b 310, *9c1* 119, *17b* 24, 506, *17c* 20, 393

$C_7H_{14}NO$
9c1 124, 125, 260, 389, 445, 446, 449, 523, *17b* 48, 102, 135, 359, *17c* 400, 401, 404, *17d1* 165, *17d2* 127ff, 189, 209
$C_7H_{14}NOS$
17c 515, 524, *17d2* 163
$C_7H_{14}NO_2$
9b 107, *9c1* 214, 234, 236, 259, 309, 386, 389, 395, 425, 782ff, 924, *17c* 405, *17d1* 113, 126, 179, *17d2* 74, 128, 155, 188, 270
$C_7H_{14}NO_2S$
9c1 773
$C_7H_{14}NO_2S_2$
17d1 191
$C_7H_{14}NO_3$
9c1 106, 290, 309, 783, 916, 919, 924, 935, *17d1* 113,126, 130, 140, *17d2* 72
$C_7H_{14}NO_3S$
17d1 191
$C_7H_{14}NO_4$
9c1 919, 924, *17d1* 114, 148
$C_7H_{14}NO_5$
17d1 148
$C_7H_{14}NO_5S$
9c1 993
$C_7H_{14}NS_2$
17e 324
$C_7H_{14}N_2^+$
9d2 98
$C_7H_{14}N_2O^-$
9d1 686
$C_7H_{14}N_2O_3M$
17d2 399
$C_7H_{14}N_3O$
9c1 839, *17d1* 145, *17d2* 255
$C_7H_{14}N_3O_3$
17d1 141
$C_7H_{14}NaO_4S$
17b 118, 152
$C_7H_{14}O^+$
17h 42
$C_7H_{14}O_2^-$
9b 215, 230
$C_7H_{14}O_2^+$
17b 133, 233, 247, *17h* 23
$C_7H_{14}O_3PS_3$
17b 382
$C_7H_{14}O_4P$
17e 278
C_7H_{15}
1 23, 34, *9b* 14, 37, 105, 187, 216, *17b* 22, 49, 252
$C_7H_{15}AlN$
17a 396
$C_7H_{15}ClNO$
9c1 374
$C_7H_{15}ClNO_2$
17d2 409
$C_7H_{15}ClNO_3$
9c1 935
$C_7H_{15}ClO_3P$
9c2 235
$C_7H_{15}D^+$
17h 13
$C_7H_{15}DO_3PS$
17e 277
$C_7H_{15}F_2OSi$
9b 252
$C_7H_{15}F_3O_3P$
9c2 235
$C_7H_{15}Ge$
9b 386
$C_7H_{15}N^-$
17b 213
$C_7H_{15}N^+$
9b 290
$C_7H_{15}NO$
9c1 222
$C_7H_{15}NO^+$
17b 102
$C_7H_{15}NO_2$
9c1 783
$C_7H_{15}NO_3^-$
9d1 395
$C_7H_{15}NO_3$
17d1 15
$C_7H_{15}NO_4P$
17c 602, 607, 608, *17e* 278
$C_7H_{15}NO_5P$
17c 602
$C_7H_{15}NO_7PM_2$
17d1 149, 159
$C_7H_{15}N_2O$
17c 405, *17d2* 148
$C_7H_{15}N_2O_2$
9c1 211, 240, *17d1* 13, 114, 128, 141, 145, 179, *17d2* 187, 215
$C_7H_{15}N_2O_2S$
17d2 407
$C_7H_{15}N_2O_3$
9c1 290, 356, *17d1* 25, 114
$C_7H_{15}N_2O_3S$
17d1 75
$C_7H_{15}N_2O_4$
17d1 63
$C_7H_{15}N_2O_5S$
17d1 195
$C_7H_{15}N_3O_2^-$
17f 234
$C_7H_{15}N_4O$
17d1 159
$C_7H_{15}N_4Si$
17c 475
$C_7H_{15}O$
9b 217, *17b* 32, 49, 54, 120, 132, 246, 247, 251, 276, 297
$C_7H_{15}OS$
17b 26, 81, 198
$C_7H_{15}OS_2$
9b 158, *17e* 347
$C_7H_{15}OS_3$
17b 326
$C_7H_{15}OSi$
9b 385, *17b* 33, 222
$C_7H_{15}O_2$
9b 204, 215, *9c2* 10, 12, *17b* 52, 81, 157, 189, 192, 236, 237, 258, 263
$C_7H_{15}O_2Pb$
17b 268
$C_7H_{15}O_2S$
17b 27
$C_7H_{15}O_2Sn$
17a 452, *17b* 270
$C_7H_{15}O_3$
9b 31, 101, 252
$C_7H_{15}O_3S$
17b 26
$C_7H_{15}O_4P^-$
17g 53ff

$C_7H_{15}S$
9b 86, 187
$C_7H_{15}S_2$
17b 149
$C_7H_{15}S_2Si$
17b 473
$C_7H_{15}S_3$
17b 148, 271, 325
$C_7H_{15}Si$
9b 385, *17b* 29, 329, 474, *17c* 24
$C_7H_{15}Sn$
17b 474
$C_7H_{16}^+$
17h 10, 13
$C_7H_{16}N$
17c 395
$C_7H_{16}NO$
9c1 219, 222, 261, *17c* 393, 394, *17d1* 132, *17d2* 72
$C_7H_{16}NOS$
9c1 986, 987, *17c* 523, *17d1* 191
$C_7H_{16}NOSi$
17b 509
$C_7H_{16}NO_2$
9c1 211, 222, 309, 358, 936, *17d1* 114, 149, 184ff, *17d2* 73, 76
$C_7H_{16}NO_2S$
9c1 112, *17c* 523
$C_7H_{16}NO_3$
9c1 223, 320, 360, 916, 919, 925, *17d1* 131, 149
$C_7H_{16}NO_4$
17d1 149, 185
$C_7H_{16}NO_4P^-$
9d1 684
$C_7H_{16}NO_4PS_2^-$
9d1 403
$C_7H_{16}NO_5P^-$
9d1 413, *17f* 236
$C_7H_{16}NO_5PSM$
17d2 414
$C_7H_{16}NO_6PSM$
17d2 414
$C_7H_{16}NS$
17c 499, 500
$C_7H_{16}N_2^+$
9d2 98, *17b* 101
$C_7H_{16}N_3O_2$
17d1 114, 128, 145
$C_7H_{16}N_3O_3S$
17d2 407
$C_7H_{16}O^+$
17b 125
$C_7H_{16}OSi^-$
9b 237
$C_7H_{16}O_2P$
17b 73, *17e* 278
$C_7H_{16}O_3P$
17b 195, 259
$C_7H_{16}O_3PS$
17e 278
$C_7H_{16}O_3PS_3$
17b 326
$C_7H_{16}O_4P$
9b 21, *9c2* 235
$C_7H_{16}O_4PS$
17b 239
$C_7H_{16}O_4PSe$
17b 239
$C_7H_{16}Si^+$
17f 33
$C_7H_{17}Ge$
17f 38
$C_7H_{17}NO_2^+$
9b 204
$C_7H_{17}NO_3P$
9c2 235, 236, *17b* 79, *17c* 498, *17e* 279
$C_7H_{17}NO_4P$
17c 493, *17d2* 418
$C_7H_{17}NO_5$
17d2 419
$C_7H_{17}N_2O$
9c1 355, *17d1* 128, *17d2* 75
$C_7H_{17}N_2O^+$
9c1 223
$C_7H_{17}N_3O_2X$
17d1 13, 44
$C_7H_{17}OSSn$
17b 274
$C_7H_{17}OSi$
9b 184, *17b* 249
$C_7H_{17}O_2SSi$
17b 21
$C_7H_{17}O_2Si$
17b 94
$C_7H_{17}Si$
17b 250
$C_7H_{17}Sn$
17b 94
$C_7H_{18}AsO_3$
9c2 25
$C_7H_{18}AsS$
9c2 317
$C_7H_{18}BN_2O$
17d1 114
$C_7H_{18}GeNO_2$
17d1 185
$C_7H_{18}NOSi$
17c 492, *17d1* 201
$C_7H_{18}NO_2Si$
17d1 185
$C_7H_{18}NO_2Sn$
17d1 185
$C_7H_{18}NSi$
17c 560, 561
$C_7H_{18}NSi_2$
17b 512
$C_7H_{18}N_2O$
9c1 355
$C_7H_{18}N_2OSX$
17d1 44
$C_7H_{18}N_2O_2X$
17d1 45
$C_7H_{18}N_3Si$
17c 582
$C_7H_{18}OP$
9c2 236
$C_7H_{18}O_3P$
17e 279
$C_7H_{18}O_3PS$
9c2 236, *17e* 279
$C_7H_{18}O_4P$
9c2 236
$C_7H_{18}O_5P$
9c2 8
$C_7H_{18}O_6P$
9c2 24
$C_7H_{18}Sn^+$
17a 452, *17f* 53
$C_7H_{19}OSSi$
17e 351
$C_7H_{19}Si_2$
9b 178
$C_7H_{20}BN_2O$
17d1 198

$C_7H_{20}N_3OX_2$
17d1 132, 146
$C_7H_{21}NO_3PSi$
17e 279
$C_7H_{21}Si$
9c2 309
$C_7Ni_2O_7^-$
17a 129
$C_8Cl_2F_{16}NO$
17d2 221
$C_8Cl_2N_2O_2^-$
9d1 233, *17g* 242ff
$C_8Cl_2N_2O_2^- \xrightarrow{CT} C_6H_8N_2^+$
9d1 364
$C_8Cl_2N_2O_2^- \xrightarrow{CT} C_{10}H_{16}N_2^+$
9d1 364
$C_8Cl_2N_2O_2 \xleftarrow{CT}$ (D)
9d1 365
C_8Cl_9
9b 565
$C_8Co_2O_8^-$
17a 127
$C_8D_{18}NO$
17d1 156
$C_8F_8^-$
17f 186
$C_8F_{10}NO$
9c1 728
$C_8F_{12}NS_4$
17c 409
$C_8Fe_2O_8^-$
17a 124ff
$C_8HCl_6N_2S$
17c 431
$C_8HCo_2O_8$
17a 127
$C_8HD_8NNaO_5S$
17d1 347
$C_8HF_{16}O$
9b 232
$C_8HFe_2O_8$
17a 125
$C_8H_2BrCl_3F_3NOX$
(X unknown)
17d1 249
$C_8H_2Br_3ClF_3X$
(X unknown)
17d1 254
$C_8H_2Br_3Cl_2F_2NOX$
(X unknown)
17d1 254
$C_8H_2Br_4F_3NOX$
(X unknown)
17d1 255
$C_8H_2Cl_3F_3INOX$
(X unknown)
17d1 249
$C_8H_2Cl_3F_5NO$
9c1 725
$C_8H_2Cl_4F_3NOX$
(X unknown)
17d1 249
$C_8H_2D_{15}N_2O$
17d2 224
$C_8H_2F_5O$
17b 67
$C_8H_2F_{16}NO$
17d2 221
$C_8H_2N_2O_2^-$
9d1 193, *17g* 219
$C_8H_2O_6^{3-}$
9d1 140, 215
$C_8H_3D_5NO$
17c 590, 591
$C_8H_3D_5N_3O$
17c 606
$C_8H_3D_7NO_2$
17c 143
$C_8H_3F_3N_3O_2^{2-}$
17f 290
$C_8H_3F_6O_2S$
17e 338
$C_8H_3NO_6^{3-}$
9d1 538, *17f* 265
$C_8H_3NO_7^{3-}$
17c 94
$C_8H_3N_2O_3^{2-}$
17g 360
$C_8H_3N_3O_2^-$
17f 275
$C_8H_3N_4O_2^{2-}$
17f 291
$C_8H_3O_4^{2-}$
9b 313
$C_8H_3O_5^{2-}$
9c2 63
$C_8H_3O_6^{2-}$
9c2 65
$C_8H_4DNO_3^+$
17g 353
$C_8H_4DNO_3Zn^+$
17a 377
$C_8H_4D_3O_2S$
17e 352
$C_8H_4D_6N_2O_2^-$
9d1 512
$C_8H_4KO_2S$
17a 215
$C_8H_4LiO_2S$
17a 202
$C_8H_4NO_2^{2-}$
9d1 100, 337, *17f* 272
$C_8H_4NO_2Zn^+$
17g 348
$C_8H_4NO_3^{2-}$
9d1 337
$C_8H_4NO_3$
9c1 800, 806
$C_8H_4NO_4$
9c1 553, 804
$C_8H_4NO_4S^{2-}$
9d1 48
$C_8H_4NO_5^{2-}$
17f 266
$C_8H_4NO_7^{2-}$
17c 93
$C_8H_4NS_3$
17c 548
$C_8H_4N_2^-$
1 104, 105
$C_8H_4N_2O_2^-$
17g 359
$C_8H_4N_2O_4^-$
9d1 543, 544
$C_8H_4N_2O_4S^-$
9d1 657
$C_8H_4N_2O_4S_2^-$
17f 303
$C_8H_4N_2O_6^-$
9d1 657, *17f* 303
$C_8H_4N_4O_2^-$
17g 357
$C_8H_4N_4S^-$
9d1 900
$C_8H_4NaO_2S$
17a 208
$C_8H_4O_2^-$
9d1 332

$C_8H_4O_2S^-$
9d1 91, 347
$C_8H_4O_2S_2^-$
9d1 84
$C_8H_4O_2S_3^-$
9d1 892, 893
$C_8H_4O_3^{2-}$
17g 381
$C_8H_4O_3^-$
9d1 91, 347, *17g* 65
$C_8H_4O_4^{3-}$
9d1 88, 89
$C_8H_4O_5^{3-}$
17c 93, 94, *17g* 254
$C_8H_4O_8^{3-}$
17b 209
$C_8H_4O_{12}P^{6-}$
17b 205
C_8H_5
9b 564
$C_8H_5^{3\cdot}$
17c 189
$C_8H_5AlCl_5O_2$
17e 192, *17g* 205
$C_8H_5BrF_3NOX$
17d1 221
$C_8H_5ClF_4NO$
17d1 221
$C_8H_5Cl_2CoN_6O$
9c1 1015
$C_8H_5Cl_3NO$
9c1 136
$C_8H_5CrO_3$
9a 303
C_8H_5DMu
17b 494
$C_8H_5DN_3O_2$
17c 456
$C_8H_5D_2O_2S$
17e 352
$C_8H_5D_3NO$
17c 590
$C_8H_5F_3NO$
9c1 135, 136
$C_8H_5F_3N_2O_5^-$
9d1 626
$C_8H_5F_3N_3O$
17c 606
$C_8H_5F_3O^-$
9b 571, *17c* 200
$C_8H_5F_3O_2S^-$
17c 29
$C_8H_5F_5NO$
9c1 724
$C_8H_5F_6$
17c 100
$C_8H_5F_6O$
17c 102
$C_8H_5KNO_2$
17a 215, 216
$C_8H_5LiNO_2$
17a 202
$C_8H_5NNaO_2$
17a 208ff
$C_8H_5NO^-$
1 106, *9b* 554
$C_8H_5NO_2^{2-}$
1 105
$C_8H_5NO_2^-$
9d1 100, 337, *17f* 269ff
$C_8H_5NO_2^+$
17h 81
$C_8H_5NO_3^-$
9d1 338
$C_8H_5NO_3Zn^+$
17a 378, *17g* 353
$C_8H_5N_2O_2^{2-}$
9d1 341
$C_8H_5N_2O_2$
9c1 163, *17c* 477, *17e* 214
$C_8H_5N_2O_3$
17c 497
$C_8H_5N_2O_6^{2-}$
9d1 696
$C_8H_5N_2O_7^{2-}$
9d1 630
$C_8H_5N_3O_3^-$
9d1 555
$C_8H_5N_3O_4^-$
17f 298, 299
$C_8H_5N_4$
17b 305
$C_8H_5O_4^{2-}$
9b 459, *9d1* 89, 123, 244, *17g* 98
$C_8H_5O_5^{2-}$
9b 468, *9d1* 140, 247, 272, *17g* 104
$C_8H_5O_5$
9c2 63
$C_8H_5O_6$
9b 469, *9c2* 65
$C_8H_5O_8^{4-}$
9b 149
$C_8H_5O_9^{4-}$
17b 205
$C_8H_5O_9$
9b 376
$C_8H_6^-$
9d1 729, *17f* 69
C_8H_6Ag
17a 375, *17b* 495
C_8H_6Au
17a 376, *17b* 496
$C_8H_6BrCoN_6O$
9c1 1013, 1015
$C_8H_6BrNO_2^-$
9d1 526, *17f* 256ff
$C_8H_6BrN_2O_3$
9c1 134
$C_8H_6ClCoN_6O$
9c1 1013
$C_8H_6ClNO_3^-$
9d1 530
$C_8H_6ClNO_4^-$
9d1 392
$C_8H_6ClN_2O_3$
9c1 133
C_8H_6ClO
17b 67
$C_8H_6Cl_2NO$
9c1 134
$C_8H_6Cl_2NOS$
17c 614
$C_8H_6Cl_2NO_2$
17c 593, *17d2* 394
$C_8H_6Cl_2NO_3$
9c1 796
$C_8H_6Cl_2O_2^+$
17h 83
$C_8H_6CoN_6O$
9a 324
$C_8H_6CoN_7O_3$
9c1 1015
C_8H_6Cu
17a 356, *17b* 496
$C_8H_6DO_2\dot{S}$
17e 352
$C_8H_6D_2N_2^+$
9d2 89

$C_8H_6D_4NO_2$
17c 143
$C_8H_6D_4O_2^+$
9d2 130
$C_8H_6D_6N_2^+$
17h 166
C_8H_6FO
17b 67
$C_8H_6F_3$
17c 174
$C_8H_6F_3NO_3^-$
9d1 499, 502
$C_8H_6F_3NO_5S^-$
9d1 549
$C_8H_6F_3O$
17c 200
$C_8H_6F_5O_2Si$
17e 192
$C_8H_6F_8NO_5$
9c1 382
$C_8H_6MnO_6$
17a 282
C_8H_6Mu
17b 494
C_8H_6N
9b 551, *17c* 50, 178, 183
C_8H_6NO
17b 86, *17e* 215
$C_8H_6NOS_2$
17c 544
$C_8H_6NO_2$
9c1 800, *17c* 587, *17d2* 351, 403
$C_8H_6NO_2S_2$
17c 543
$C_8H_6NO_3^-$
9c1 729
$C_8H_6NO_3$
9c1 587, 801, 804
$C_8H_6NO_4^{2-}$
9d1 380, 381, 505, 534
$C_8H_6NO_4$
17d2 413
$C_8H_6NO_4S_2$
17c 556
$C_8H_6NO_5^{2-}$
9d1 575
$C_8H_6NO_5$
9c1 553, 554
$C_8H_6NS_2$
17c 534
$C_8H_6N_2^-$
1 115, 116, *9d1* 873, 874ff, 876
$C_8H_6N_2^+$
17h 229
$C_8H_6N_2O_2^{2-}$
9d1 340
$C_8H_6N_2O_2^-$
1 91, *9d1* 341, 554, 555, 873, *17f* 273
$C_8H_6N_2O_6^-$
9d1 630
$C_8H_6N_3$
17c 144
$C_8H_6N_3O$
9c1 583
$C_8H_6N_3O_2^{2-}$
17f 289
$C_8H_6N_3O_2$
9c1 68, *17c* 456
$C_8H_6N_3O_3^{2-}$
17f 290
$C_8H_6N_4^-$
9d1 881, *17f* 208
$C_8H_6N_4O_5^-$
17f 281
$C_8H_6N_4O_6^-$
17f 297
$C_8H_6O_2^-$
1 104, 106, *9b* 550, 556, 615, *9d1* 9, 36, 39, 58, 84, *17c* 175, *17g* 9, 24
$C_8H_6O_2S_2^-$
9d1 63, *17g* 39
$C_8H_6O_3^-$
9b 67, *9d1* 70, 188
$C_8H_6O_3^+$
17h 81
$C_8H_6O_4^-$
9d1 89, 188, *17g* 215
$C_8H_6O_4^+$
17h 82, 83
$C_8H_6O_6Re$
17a 314
$C_8H_6S^-$
17f 212
$C_8H_6S_2^-$
9d1 891, *17f* 214
$C_8H_6S_2^+$
1 127, *9d2* 145, *17e* 369ff
$C_8H_6S_4^+$
17h 115
C_8H_7
9b 305, 563
$C_8H_7BrMgO_2^-$
17a 225, *17c* 204
C_8H_7BrNO
9c1 131, 132, 133, 135
$C_8H_7BrNO_2$
17d2 394
$C_8H_7BrO^-$
9b 567
$C_8H_7Br_2NNaO_5S$
17d1 256, 258
$C_8H_7Br_2O_2$
9b 68
$C_8H_7Br_3NO$
17d1 252
C_8H_7ClNO
9c1 131, 132, 133, *17c* 591
C_8H_7ClNOS
17c 614
$C_8H_7ClNO_2$
17c 592, 593, *17d2* 394
$C_8H_7ClNO_3$
9c1 793
$C_8H_7ClN_2O_5^-$
9d1 695
$C_8H_7ClNaO_2$
9d1 149
$C_8H_7ClS_2^-$
9b 579
$C_8H_7Cl_3NO$
17d1 238
$C_8H_7CoN_6O$
9c1 1013
$C_8H_7D^-$
9d1 724
$C_8H_7D_3NO_2$
9b 523
$C_8H_7F^-$
17f 185
C_8H_7FNO
9c1 130ff, 132, 133
$C_8H_7FNO_2$
17d2 394
$C_8H_7FO^-$
1 106

$C_8H_7F_2$
9b 549, *17c* 181
$C_8H_7F_6N_2O_2$
17d1 208
C_8H_7INO
9c1 132
$C_8H_7KN_3O_8S$
9c1 105
$C_8H_7N^-$
1 105, *9b* 620, *9d1* 883
$C_8H_7NO_2^-$
9d1 100, 465, 524ff, *17f* 255
$C_8H_7NO_2S_2^-$
9b 579, *9d1* 545
$C_8H_7NO_3^-$
1 89, 93, *9d1* 504, 530, 531ff, *17f* 262ff
$C_8H_7NO_4^-$
1 91, 92, *9d1* 392, 538, 539, 540ff, 570, *17f* 265, 266
$C_8H_7NO_5^-$
9d1 536, 576, *17f* 282
$C_8H_7N_2$
17c 479
$C_8H_7N_2O$
9c1 858
$C_8H_7N_2O_2$
9c1 892, *17c* 597
$C_8H_7N_2O_3$
9c1 130, 132, *17d1* 382
$C_8H_7N_3O_5^-$
9d1 631, *17f* 282
$C_8H_7N_3O_6^-$
9d1 632
C_8H_7O
9b 49, 320, *17b* 66, 514, 525, 526, 529, *17c* 51
$C_8H_7O_2^{2-}$
9b 572, *17c* 192
$C_8H_7O_2$
1 59, *9c2* 31, 35, 40, *17b* 525, *17c* 39, 182, *17e* 36, 38, 41
$C_8H_7O_2S$
17e 337, 351
$C_8H_7O_3^{2-}$
9b 574, *9d1* 250, 260, 271, 272, *17g* 255
$C_8H_7O_3$
1 60, *17c* 205, *17e* 7, 46
$C_8H_7O_4^{2-}$
9d1 271, *17g* 255
$C_8H_7O_4$
9b 459, *17e* 57
$C_8H_7O_5^{2-}$
9b 299, 300
$C_8H_7O_5$
17e 57
$C_8H_7O_8S$
9b 474
$C_8H_7S_2$
17c 43
$C_8H_8^-$
1 79, 85, *9d1* 711, 716ff, 724, *17f* 69ff
$C_8H_8^+$
9d2 19
$C_8H_8^{2-}\cdot$
17h 251
$C_8H_8AlO_4^{2+}$
9d1 149
$C_8H_8BaO_4^+$
9d1 149
$C_8H_8BrO_2S$
17e 340
$C_8H_8CaO_4^+$
9d1 149
$C_8H_8CdO_4^+$
9d1 149
C_8H_8Cl
17c 53
$C_8H_8ClNO_2^-$
9d1 383
$C_8H_8ClN_2O_6^{2-}$
9d1 428
$C_8H_8CsO_2$
9d1 193, 208
$C_8H_8D_2N_2O_2^-$
17g 197
C_8H_8F
17c 179, 180
$C_8H_8F_3O$
17b 487
$C_8H_8F_9O$
17b 247
$C_8H_8KO_2$
9d1 193, 195, 209
$C_8H_8LaO_4^+$
9d1 149
$C_8H_8LiO_2$
9d1 195, 210
$C_8H_8MgO_4^+$
9d1 149
$C_8H_8MoN_2O_4^-$
9a 309
C_8H_8N
9b 620, *17c* 568
$C_8H_8NO^{2-}$
9b 581
C_8H_8NO
9c1 130, 261, 730, *17b* 78, *17c* 585, 589
$C_8H_8NOS_2$
17c 544, 546
$C_8H_8NO_2$
9c1 127, 128, 135, 158, 159, 162, 549, 766, 771, 790, *17c* 488, 587, *17d1* 227, *17e* 43
$C_8H_8NO_3^{2-}$
17g 100, 252
$C_8H_8NO_3$
9c1 549ff, 768, 773, 791
$C_8H_8NO_3^+$
17h 161
$C_8H_8NO_3S$
9c1 149, *17d2* 415
$C_8H_8NO_5$
9c1 950, *17d1* 385
$C_8H_8NS_2$
17c 532
$C_8H_8NS_3$
17c 548
$C_8H_8N_2 \xrightarrow{CT} C_6Cl_4O_2$
9d1 363
$C_8H_8N_2^+$
9d2 71, 89, *17h* 179, 233
$C_8H_8N_2O_2S^-$
9d1 513, 544, *17f* 255
$C_8H_8N_2O_3^-$
17f 266
$C_8H_8N_2O_4^-$
1 102, *9d1* 611, 621, 639, 640, *17f* 295
$C_8H_8N_2O_5^-$
9d1 695
$C_8H_8N_2O_6^-$
9d1 626

$C_8H_8N_3O_8^{2-}$
9d1 429
$C_8H_8N_4O_2^-$
17f 292
$C_8H_8N_4O_2^+$
9d2 71
$C_8H_8N_4O_4^-$
17f 281
$C_8H_8NaO_2$
9d1 149, 195, 210ff
$C_8H_8NaO_4$
17g 230
$C_8H_8Na_2O_2^+$
9d1 193, 195, 211
$C_8H_8O^-$
1 103, *9b* 566
$C_8H_8OS^-$
9b 577, *17c* 207
$C_8H_8O_2^-$
1 81, *9b* 216, 461, 541, 542, 554, 574, *9d1* 31, 40, 58, 59, 123, 161, 175, 192, 194ff, 206ff , *17g* 23, 38, 220, 224
$C_8H_8O_2^+$
17e 235, 238, 241, *17g* 384, *17h* 79, 80
$C_8H_8O_2Rb$
9d1 211
$C_8H_8O_2S^-$
9d1 85
$C_8H_8O_2S_2^-$
9d1 205
$C_8H_8O_2S_2^+$
17g 384
$C_8H_8O_3^-$
9d1 70, *17c* 92, 93, *17g* 217
$C_8H_8O_3^+$
17e 243, *17h* 80, 82
$C_8H_8O_3S^-$
17b 199
$C_8H_8O_4^-$
9b 467, *9d1* 193 [3]), 203, 218, *17g* 113, 153, 199, 219, 224, 230
$C_8H_8O_4^+$
17e 245, *17g* 387
$C_8H_8O_4Sr^+$
9d1 149
$C_8H_8O_4Y^{2+}$
9d1 149
$C_8H_8O_4Zn^+$
9d1 149
$C_8H_8O_5^+$
17e 245
$C_8H_8O_6^{3-}$
9d1 17
$C_8H_8S_2^-$
9b 578
$C_8H_8S_4^+$
17h 111
$C_8H_9^{2-}$
9d1 723
C_8H_9
9b 25, 539, 543, 545, 551, *17c* 52, 172, 178
$C_8H_9CdNO_2^+$
17a 387, *17g* 87
$C_8H_9ClCrNO_5$
17a 253
$C_8H_9ClFeNO_5$
9c1 1037
$C_8H_9ClMoNO_5$
17a 267
$C_8H_9ClNO_3S$
9c1 993
$C_8H_9CoO^{3+}$
17e 53
$C_8H_9D_9NO_2$
17d1 117
$C_8H_9FO_2S^+$
17h 93
$C_8H_9FO_3S^+$
17h 91
$C_8H_9F_7NO_2$
17d1 204
$C_8H_9F_9NO$
17d1 164
$C_8H_9MnO_6$
17a 282
$C_8H_9NO^-$
9b 582
$C_8H_9NO^+$
17h 161
$C_8H_9NO_2^-$
1 90, 97, 98, 99, *9b* 613, *9d1* 383, 452, 453, 456, 457, *17f* 232, *17g* 86ff
$C_8H_9NO_2^+$
17h 60, 61
$C_8H_9NO_2S^-$
9d1 507, 508
$C_8H_9NO_2Y^{2+}$
17a 239, *17g* 88
$C_8H_9NO_2Zn^+$
17a 378, *17g* 88
$C_8H_9NO_3^-$
1 91, 95, *9b* 613, *9d1* 380, 499, 502, 503, 861, *17g* 99, 197
$C_8H_9NO_3Zn^+$
17a 378, *17g* 252
$C_8H_9NO_4P^{2-}$
9d1 412
$C_8H_9NO_4S^-$
9d1 550
$C_8H_9NO_5S^-$
9d1 549
$C_8H_9NO_7^-$
9d1 575
$C_8H_9N_2$
17c 576
$C_8H_9N_2O^-$
9c1 808
$C_8H_9N_2O$
9c1 132, 810, *17c* 579
$C_8H_9N_2O_2^-$
9c1 808
$C_8H_9N_2O_2$
9c1 766, 771, 810
$C_8H_9N_2O_6^{2-}$
9d1 428
$C_8H_9N_2S_2$
17c 553
$C_8H_9N_3O_3^-$
17b 69, 168
$C_8H_9N_3O_4^-$
9d1 627

[3]) In II/9d1 erroneously listed as $[C_8H_8O_4]^{\cdot +}$

C_8H_9O
1 59, 60, *9b* 32, 101, 550, 554ff, 566, *9c2* 56, 59, 63, *17c* 50, 89, 96, 174, *17e* 53
C_8H_9OS
17b 82, 239, *17c* 177
$C_8H_9O_2$
9b 68, *9c2* 10, 62, 64, 66, 69ff, *17e* 70, 76
$C_8H_9O_2S$
9c2 292, *17c* 177
$C_8H_9O_3$
1 61, *9b* 473, *9c2* 62, *17e* 54
$C_8H_9O_3S$
17c 177
$C_8H_9O_4$
17b 356
$C_8H_9O_5$
9b 376
$C_8H_9O_6$
9b 300, 372, 377
C_8H_9S
9b 103, *17c* 176
$C_8H_{10}^-$
1 85, 86, *9d1* 712, 715, 720, *17f* 70ff
$C_8H_{10}Al_2Br_6NS$
17a 396, *17e* 320
$C_8H_{10}CdNO_2^{2+}$
17a 387, *17g* 88
$C_8H_{10}D_2N_2^+$
17h 166
$C_8H_{10}FN^+$
9d2 30
$C_8H_{10}F_3NO_2^+$
17h 133
$C_8H_{10}F_9OS_3Si$
17b 302
$C_8H_{10}N$
9b 621
$C_8H_{10}NO$
1 55, *9b* 521ff, *9c1* 97, 100, 554, 612, *9c2* 43, *17c* 136ff, *17d1* 213ff, *17e* 69, 76
$C_8H_{10}NOS$
17d2 200
$C_8H_{10}NOS_2$
17c 543, 553
$C_8H_{10}NO_2$
9b 519, 520, 523, *9c1* 171, 550, 643, *17c* 133, 139ff, *17g* 88
$C_8H_{10}NO_2S$
17c 526
$C_8H_{10}NO_2Zn^{2+}$
17a 378, *17g* 88ff
$C_8H_{10}NO_3$
9c1 155, 947, *17c* 622, *17d1* 378
$C_8H_{10}NO_3P^-$
9c1 200
$C_8H_{10}NO_3S$
9c1 976, *17c* 494
$C_8H_{10}NO_3Zn^{2+}$
17a 378, *17g* 100
$C_8H_{10}NO_4$
9c1 950, *17d1* 381
$C_8H_{10}NO_5$
9b 300
$C_8H_{10}NO_6S$
17d1 385
$C_8H_{10}NS_2$
17c 554ff
$C_8H_{10}N_2OS_2^-$
17f 218
$C_8H_{10}N_2O_2^-$
1 97, *9d1* 201, *17g* 197
$C_8H_{10}N_2O_2^+$
9d2 30, *17h* 179
$C_8H_{10}N_3$
17b 307, *17c* 578
$C_8H_{10}N_4O_2^+$
17h 206
$C_8H_{10}N_4O_3^+$
9d2 71
$C_8H_{10}O^-$
9b 484, *17c* 100
$C_8H_{10}O^+$
17e 244
$C_8H_{10}O_2^-$
9d1 18, 31, 36, 40, 41, 44, 56, *17f* 186
$C_8H_{10}O_2^+$
9d1 353, 354, *9d2* 125ff, 129ff, *17h* 79
$C_8H_{10}O_2Tl$
17g 75
$C_8H_{10}O_3S^+$
17h 93
$C_8H_{10}O_4^+$
9d1 354, *9d2* 128
$C_8H_{10}O_4S^+$
17h 91
$C_8H_{10}O_5^-$
17g 237
$C_8H_{10}O_5S^+$
17h 92
$C_8H_{10}O_7$
9b 151
$C_8H_{10}S^-$
17c 100
$C_8H_{10}S_2^-$
17g 48
$C_8H_{10}S_2^+$
9d2 140, *17e* 383, *17h* 110
C_8H_{11}
9b 302, 454, *17b* 488, *17c* 52, 73ff, 78
$C_8H_{11}BrN_3O_3$
17d1 100
$C_8H_{11}BrP$
17e 279
$C_8H_{11}ClN$
17b 501
$C_8H_{11}ClP$
17e 280
$C_8H_{11}Cl_3NS$
17b 287
$C_8H_{11}D_3N_3O_4$
17d1 82
$C_8H_{11}D_4N_2O_3$
17b 244
$C_8H_{11}D_4N_4O_3$
17d1 83
$C_8H_{11}FN_3O_3$
17d1 101
$C_8H_{11}F_5NO_2$
17d1 204
$C_8H_{11}F_9O_4P$
9c2 237
$C_8H_{11}Ge$
9c2 310
$C_8H_{11}IP$
17e 280
$C_8H_{11}N^-$
17f 186

$C_8H_{11}N^+$
9d2 31, *17c* 183, *17h* 161
$C_8H_{11}NO^+$
9d2 31, *17c* 184
$C_8H_{11}NO_2^-$
9d1 47, *17g* 29
$C_8H_{11}NO_2^+$
9d2 31
$C_8H_{11}NO_3^-$
9d1 682
$C_8H_{11}N_2O_2$
17b 184
$C_8H_{11}N_2O_3$
17b 32, 134
$C_8H_{11}N_2S$
17b 286
$C_8H_{11}N_3OS_2M$
17d2 164
$C_8H_{11}N_3O_3^-$
17c 410
$C_8H_{11}N_4O_3$
9c1 31
$C_8H_{11}O$
9b 297
$C_8H_{11}O_3$
17b 516
$C_8H_{11}O_4^{2-}$
9d1 17, *17b* 205
$C_8H_{11}O_4$
1 37, *17b* 497, *17c* 18
$C_8H_{11}O_6$
9b 150
$C_8H_{11}P^+$
9c2 264
$C_8H_{11}Si$
9b 35, *17b* 93
$C_8H_{11}Sn$
17f 46
$C_8H_{12}^+$
17h 29
$C_8H_{12}^{2\cdot}$
17h 250
$C_8H_{12}Al_2N_2^-$
17a 396, *17f* 206
$C_8H_{12}B_2 \cdot (2DME)^-$
17f 226
$C_8H_{12}B_2F_2^-$
17f 227
$C_8H_{12}BrCl_3NO_2$
9c1 290
$C_8H_{12}BrD_2N_2O_2$
17d2 263
$C_8H_{12}Br_2DN_2O_2$
17d2 262
$C_8H_{12}Br_2NO$
17d2 233
$C_8H_{12}Br_3N_2O_2$
9c1 419
$C_8H_{12}Cl_5FNO$
9c1 385
$C_8H_{12}Cl_6NO$
9c1 377
$C_8H_{12}D_6^+$
17h 11
$C_8H_{12}F_3N_2O_2$
9c1 863
$C_8H_{12}F_3S$
9b 378
$C_8H_{12}F_4NO_2$
17d1 204
$C_8H_{12}GeO_8^-$
17f 43
$C_8H_{12}KO_2$
1 44
$C_8H_{12}LiO_2$
1 44
$C_8H_{12}N$
17c 135
$C_8H_{12}NO$
9c1 168, 169, 170, 172, 525, *17d2* 340
$C_8H_{12}NOS_2$
17c 553
$C_8H_{12}NO_2$
9c1 150, 525ff, *17d2* 341
$C_8H_{12}NO_3$
9b 126, *17b* 344
$C_8H_{12}NO_4$
9c1 146, 291
$C_8H_{12}NO_5M_2$
17d1 26
$C_8H_{12}NS_2$
17c 552
$C_8H_{12}N_2^+$
9d2 54, 99, *17h* 165ff
$C_8H_{12}N_2Na_2^+$
17a 208
$C_8H_{12}N_2O_2^{2\cdot}$
9d2 192, *17h* 271
$C_8H_{12}N_2O_2^+$
17h 143
$C_8H_{12}N_2O_2S^-$
9d1 585, 586
$C_8H_{12}N_2O_4^-$
9c1 864
$C_8H_{12}N_2S_2^-$
9d1 76, 77
$C_8H_{12}N_3O$
9c1 330, 698, 840, *17d2* 255
$C_8H_{12}N_3O_2$
9c1 863, *17d2* 262, 409
$C_8H_{12}N_3O_3$
9c1 896, *17d1* 100
$C_8H_{12}N_3O_4$
17d1 101
$C_8H_{12}N_4O_2^+$
17h 206
$C_8H_{12}N_4O_3^-$
17c 410
$C_8H_{12}N_5O_2$
9c1 31
$C_8H_{12}N_8O_2^{2\cdot}$
17h 262
$C_8H_{12}NaO_2$
1 44
$C_8H_{12}O^-$
9b 411, 412
$C_8H_{12}OS^-$
17b 333, *17g* 51
$C_8H_{12}O_2^-$
9d1 14, 45
$C_8H_{12}O_2^{2\cdot}$
17h 251
$C_8H_{12}O_3^-$
17g 28
$C_8H_{12}O_4^-$
9d1 68
$C_8H_{12}O_8S_2^+$
9c2 300
$C_8H_{12}O_8S_4^+$
17e 365
$C_8H_{12}O_8Sn^-$
17f 51
$C_8H_{12}S_2^-$
17b 333, 334, *17g* 51
$C_8H_{12}S_2^+$
9d2 144, *17e* 373, 381, 382, 383, *17h* 109, 110

$C_8H_{12}S_4^+$
17h 24, 110
$C_8H_{12}Si^-$
17f 11
$C_8H_{12}Si^+$
17f 33
C_8H_{13}
9b 302, 414, 432, *17b* 33, 96, 114, 485
$C_8H_{13}Ag^+$
9a 329
$C_8H_{13}BrNO$
17d2 233
$C_8H_{13}Br_2N_2OZn$
9a 333
$C_8H_{13}CdI_2N_2O$
9a 333
$C_8H_{13}ClHgNO$
17d2 233
$C_8H_{13}ClN$
17c 90
$C_8H_{13}ClNO$
17c 90
$C_8H_{13}Cl_2N_2O$
9a 334
$C_8H_{13}Cl_2N_2OZn$
9a 333
$C_8H_{13}Cl_4FNO$
9c1 385
$C_8H_{13}Cl_5NO$
9c1 328
$C_8H_{13}FN_3O_3$
17d1 94
$C_8H_{13}FN_3O_4$
9c1 301, *17d1* 94
$C_8H_{13}F_3NO$
17d1 209
$C_8H_{13}F_3O_5P$
9c2 237
$C_8H_{13}Ge$
17a 432, *17c* 84
$C_8H_{13}HgI_2N_2O$
9a 334
$C_8H_{13}I_2N_2OZn$
9a 333
$C_8H_{13}NO_2^+$
17h 133
$C_8H_{13}NO_3^+$
17h 133
$C_8H_{13}NO_3M$
17d2 191
$C_8H_{13}NO_3P$
17e 280
$C_8H_{13}NO_4PS$
17e 280
$C_8H_{13}N_2O^-$
9c1 223
$C_8H_{13}N_2O$
9c1 170, 526
$C_8H_{13}N_2OS$
17d1 165
$C_8H_{13}N_2O_2$
9c1 150, 783, *17d2* 255
$C_8H_{13}N_2O_3$
9c1 420, 513, 896, *17b* 342, 405
$C_8H_{13}N_2O_4$
9b 166, *17c* 452, *17d2* 262
$C_8H_{13}N_3O_3$
9b 78
$C_8H_{13}N_4O$
17d2 249
$C_8H_{13}N_4O_2$
17d1 80, 101, *17d2* 262
$C_8H_{13}N_4O_7Pd$
9a 327
$C_8H_{13}O$
9b 294, 298, 375, 434, *17b* 388, 389, 390, 391, 518
$C_8H_{13}OSi$
17c 64, 65
$C_8H_{13}O_2$
17b 30, 61, 249, 366, 370, 481, 515, 521, *17c* 37
$C_8H_{13}O_3$
9b 125, 190, 376, *17b* 25, 124, 170, 350
$C_8H_{13}O_3S$
17b 54, 179
$C_8H_{13}O_4$
9b 374, *17b* 23, 32, 41, 54, 64, 87, 119, 122, 132, 146, 151, 152, 153, 154, 161, 165, 172, 179, 368, *17c* 38
$C_8H_{13}O_4S$
9b 154
$C_8H_{13}O_5$
9b 148, 150, 198, *17b* 209
$C_8H_{13}O_9P^-$
17b 358
$C_8H_{13}S$
9b 417
$C_8H_{13}S_2Sn$
17a 452, *17c* 121
$C_8H_{13}Si$
17c 81ff
$C_8H_{13}Sn$
17a 453, *17c* 85
$C_8H_{14}^-$
9d1 845
$C_8H_{14}BrN_2O_2$
9c1 864, *17d2* 262
$C_8H_{14}ClN_2O_2$
9c1 864
$C_8H_{14}D$
17b 118
$C_8H_{14}DGe$
9b 409
$C_8H_{14}D_2O_2^+$
17b 59, 133, 248
$C_8H_{14}F_3OSe$
17b 321
$C_8H_{14}IN_2O_2$
9c1 865
$C_8H_{14}Mu$
17b 245
$C_8H_{14}N$
17b 507
$C_8H_{14}NO$
9b 201, *9c1* 152, 153, 162, 261, 526, *17b* 106, 257, 362, *17c* 399, 403, *17d2* 234, 341
$C_8H_{14}NO_2$
9c1 105, 389, 397, 777, 799, *17d2* 146, 191, 223, 224, 393, 399, 402, 403
$C_8H_{14}NO_3$
9c1 238, 259, 395, *17d1* 165, *17d2* 156
$C_8H_{14}NO_3S$
17d2 164
$C_8H_{14}NO_4$
9c1 291, 298, 783, 784, *17b* 207
$C_8H_{14}NO_4S$
9c1 993
$C_8H_{14}NO_5$
9c1 291, *17d1* 141

$C_8H_{14}NS_2$
17b 290, *17c* 548
$C_8H_{14}NSi$
9b 525, *17c* 148
$C_8H_{14}N_2^+$
9d2 71ff, 99, *17h* 180
$C_8H_{14}N_2O_3P$
17c 419
$C_8H_{14}N_2O_4$
17d1 26
$C_8H_{14}N_2O_4M$
17d1 8
$C_8H_{14}N_2O_4P$
17e 281
$C_8H_{14}N_2O_5M$
17d1 26
$C_8H_{14}N_3O_2$
9c1 420, 421, *17d2* 255
$C_8H_{14}N_3O_3$
9c1 223, 421, *17d1* 80ff, *17d2* 249, 263
$C_8H_{14}N_3O_4$
9c1 291, 301, *17d1* 82
$C_8H_{14}O^-$
9b 257
$C_8H_{14}O_2^-$
9d1 14, 17, 18, *17b* 118, 171
$C_8H_{14}O_2^+$
17h 51
$C_8H_{14}O_2S_2^-$
17g 46
$C_8H_{14}O_2S_2$
9c2 279
$C_8H_{14}O_2Si^-$
17g 30
$C_8H_{14}O_3PS$
17c 43
$C_8H_{14}O_4^-$
9d1 54
$C_8H_{14}O_4P$
17c 39
$C_8H_{14}S^-$
17b 333
$C_8H_{14}S_2^+$
17e 371
C_8H_{15}
9b 295, *17b* 112, 118, *17c* 11
$C_8H_{15}BD_3NOM$
17d2 183
$C_8H_{15}BF_3N_2O$
17d2 255
$C_8H_{15}BrNO_3$
17d1 115
$C_8H_{15}ClF_3OSi$
9b 244
$C_8H_{15}ClNO_3$
9c1 223
$C_8H_{15}Cl_2O_2Sn$
17a 453
$C_8H_{15}FNO$
17d2 184
$C_8H_{15}F_4OSi$
9b 248
$C_8H_{15}F_5O_3P$
9c2 237
$C_8H_{15}Ge$
9b 409
$C_8H_{15}INO$
9c1 397
$C_8H_{15}NO^+$
9d2 117, *17h* 133
$C_8H_{15}NO_2^-$
9d1 415
$C_8H_{15}NO_3M$
17d2 176
$C_8H_{15}NO_4M$
17d1 26
$C_8H_{15}NO_4P$
9c2 238, *17c* 608, *17d2* 169
$C_8H_{15}NO_5P$
17c 610
$C_8H_{15}NO_5SM$
17d2 176
$C_8H_{15}NO_6P$
17c 610
$C_8H_{15}N_2O$
9c1 151, 331, 840, 856, *17d1* 115, 157, *17d2* 256, 376
$C_8H_{15}N_2OS$
17d1 115, *17d2* 379
$C_8H_{15}N_2O_2$
9c1 398, 416, 421, 840, 853, 865, 890, *17d2* 148, 263, 336
$C_8H_{15}N_2O_3$
9c1 865, 896, *17b* 49, 250
$C_8H_{15}N_2O_4$
17d1 22
$C_8H_{15}N_2O_5$
9c1 908, *17d1* 26, *17d2* 405
$C_8H_{15}N_3O_2^-$
9d1 514
$C_8H_{15}N_3O_3$
9c1 302
$C_8H_{15}N_4O$
17c 468, *17d2* 175
$C_8H_{15}N_4O_2$
17d2 256
$C_8H_{15}N_4O_3$
17d1 82ff
$C_8H_{15}O$
9b 188, 295, 367, 368, *17b* 33, 252
$C_8H_{15}OS$
17b 199, 351
$C_8H_{15}OS_2$
17e 348
$C_8H_{15}OSi$
17b 222, *17c* 63
$C_8H_{15}O_2$
9b 16, 269, *9c2* 15, *17b* 118, 145, 171, 244, 253, 256, 350, 365, 440, 441, 467
$C_8H_{15}O_2S$
9b 146, *17b* 199, 470
$C_8H_{15}O_3$
9b 184, 212, *17b* 470
$C_8H_{15}S$
9b 377, *17b* 91, 198, 391
$C_8H_{15}Si$
9b 408
$C_8H_{16}Al_2N_2^+$
17a 397
$C_8H_{16}Br_2NO$
9c1 373
$C_8H_{16}Cl$
17b 17
$C_8H_{16}ClN_2PS^-$
17e 281
$C_8H_{16}Cl_2NO$
9c1 331, 378
$C_8H_{16}Cl_2O_4P$
17b 373
$C_8H_{16}FN_2O_2PS^-$
17e 281
$C_8H_{16}F_3O_2Si$
9b 249

$C_8H_{16}F_3S_2$
17e 345ff
$C_8H_{16}F_3Si$
9b 165
$C_8H_{16}Ga_2N_2^+$
17a 407
$(C_8H_{16}HgO_2)_n$
17b 55
$(C_8H_{16}HgO_3)_n$
17b 55
$C_8H_{16}In_2N_2^+$
17a 409
$C_8H_{16}KO$
1 44
$C_8H_{16}LiO$
1 44
$C_8H_{16}Mu$
17b 202, 244
$C_8H_{16}N$
17b 506, 507, *17c* 392, 393
$C_8H_{16}NO$
9b 132, *9c1* 125, 390, 398, 446, 524, *17c* 401, 402, *17d2* 128, 172, 224
$C_8H_{16}NOS$
17c 516, *17d2* 164
$C_8H_{16}NO_2^-$
9c1 223
$C_8H_{16}NO_2$
9c1 105, 219, 235, 236, 291, 297, 309, 390, 395, 399, 784, *17c* 402, *17d1* 115, 126, 141, 180, *17d2* 72, 75, 131ff, 156, 175, 393
$C_8H_{16}NO_2PbS$
17a 493, *17b* 360
$C_8H_{16}NO_2S$
9c1 371, *17d2* 164ff
$C_8H_{16}NO_2SSi$
17b 361
$C_8H_{16}NO_2SSn$
17a 453, *17b* 361
$C_8H_{16}NO_3$
9c1 292, 310, 368, 920, 925, 936, *17d1* 115, 141, 149, *17d2* 73, 75, 132, 156, 161, 176
$C_8H_{16}NO_3S$
9c1 326
$C_8H_{16}NO_4$
9c1 224, 936 , *17d2* 139, 393
$C_8H_{16}NO_5S$
9c1 993
$C_8H_{16}N_2^+$
9d2 99, *17h* 143, 230
$C_8H_{16}N_2O_2^-$
9d1 54
$C_8H_{16}N_2O_4S_2^-$
17e 386
$C_8H_{16}N_2O_5SM$
17d2 193
$C_8H_{16}N_2PS$
17e 264
$C_8H_{16}N_2P_2S_2^+$
17h 180
$C_8H_{16}N_2S^+$
17e 364, *17h* 138
$C_8H_{16}N_2S_2^-$
9d1 77
$C_8H_{16}N_3O_2$
9c1 866, *17d2* 149
$C_8H_{16}N_3O_3$
9c1 348, *17d1* 141, *17d2* 193
$C_8H_{16}N_3O_4$
9c1 302, *17d1* 22, 27
$C_8H_{16}N_3O_5$
9c1 331, 356, *17d2* 217
$C_8H_{16}N_4^+$
9d2 27
$C_8H_{16}N_4P$
17e 281
$C_8H_{16}N_5O_3$
17d2 149
$C_8H_{16}NaO$
1 44
$C_8H_{16}NaO_4S$
17b 118, 152
$C_8H_{16}O^-$
9b 231
$C_8H_{16}O^+$
17b 33
$C_8H_{16}O_2^+$
17b 59
$C_8H_{16}O_2PS_3$
17e 361
$C_8H_{16}O_4P$
9c2 238, *17b* 115
$C_8H_{16}O_5P$
9c2 239
$C_8H_{16}O_5PS_3$
17b 324
$C_8H_{16}S_2^+$
9c2 299, *17e* 377
C_8H_{17}
9b 15, 43, 82, 181, *17b* 244
$C_8H_{17}Cl_2Si$
17b 290
$C_8H_{17}Cl_3O_4P$
9c2 238
$C_8H_{17}F_2Si$
9b 225
$C_8H_{17}F_3O_4P$
9c2 239
$C_8H_{17}N^-$
17b 212
$C_8H_{17}N^+$
9b 262
$C_8H_{17}NO^+$
17b 102, 104, 107, 109
$C_8H_{17}NO_3^-$
9d1 395
$C_8H_{17}NO_5P$
17c 611, 612
$C_8H_{17}N_2O$
9c1 394, 399, *17c* 405, *17d2* 165, 187, 224, 249
$C_8H_{17}N_2O_2$
9c1 302, 821, 896, *17d1* 45, 116, 128, 129, 142, 157, *17d2* 193, 209
$C_8H_{17}N_2O_3$
17d1 16, 45, *17d2* 194
$C_8H_{17}N_2O_3S$
17d1 75
$C_8H_{17}N_2O_4$
17d1 16
$C_8H_{17}O$
9b 141, *17b* 191, 247
$C_8H_{17}OS$
17b 121
$C_8H_{17}OS_2$
17b 197
$C_8H_{17}OSi$
17b 265, 332

$C_8H_{17}O_2$
9b 100, *9c2* 12, 17, *17b* 56, 61, 130, 186, 191, 192, 221, 251, 273
$C_8H_{17}O_2S$
17b 21
$C_8H_{17}O_3$
17b 190
$C_8H_{17}O_4$
17b 191
$C_8H_{17}O_4P^-$
17g 54
$C_8H_{17}O_6S_2$
9c2 290
$C_8H_{17}S$
9b 144
$C_8H_{17}S_2$
9b 145
$C_8H_{17}Si$
17b 28, 332
$C_8H_{17}Sn$
17b 477
$C_8H_{18}^+$
17h 11, 14
$C_8H_{18}AlCl_3NO$
9a 295, *17a* 397
$C_8H_{18}AlO_2$
17a 397
$C_8H_{18}Al_2N$
17b 499
$C_8H_{18}BBr_3NO$
9a 294
$C_8H_{18}BCl_3NO$
9a 294
$C_8H_{18}BF_3NO$
9a 294
$C_8H_{18}Br_4NOSn$
17a 453
$C_8H_{18}ClN_2Si$
17c 453
$C_8H_{18}ClSi$
17b 181
$C_8H_{18}Cl_2NO_2Sn$
17d1 185
$C_8H_{18}Cl_3Si_3$
9b 309
$C_8H_{18}Cl_4GeNO$
17a 432
$C_8H_{18}Cl_4NOSn$
17a 453
$C_8H_{18}Cl_4NOTi$
17a 244
$C_8H_{18}DNO^+$
9d2 117
$C_8H_{18}FSi$
9b 130
$C_8H_{18}N$
9b 132, *9c1* 8, *17c* 393, 395
$C_8H_{18}NO$
1 54, *9c1* 214, 251, 331ff, *17c* 394, 482, *17d1* 116, 117, 132, 154ff, *17d2* 207
$C_8H_{18}NOS$
9c1 987, *17c* 524, *17d1* 191, 192
$C_8H_{18}NOSi$
17b 509, 510
$C_8H_{18}NO_2$
9c1 224, 243, 310, 937, *17d1* 117, 150, 185ff, *17d2* 74, 76
$C_8H_{18}NO_2S$
9c1 371, *17c* 524, *17d1* 63
$C_8H_{18}NO_3$
9c1 340, 369, 920, 982, *17d2* 209
$C_8H_{18}NO_3S$
17d1 196
$C_8H_{18}NO_4PS_2^-$
9d1 405
$C_8H_{18}NO_7$
9c1 340
$C_8H_{18}NS$
17c 499, 500
$C_8H_{18}NSi$
17b 511
$C_8H_{18}N_2^+$
9d2 100, *17b* 101, 108, *17h* 144, 154
$C_8H_{18}N_2O^-$
9d1 686
$C_8H_{18}N_2O^+$
9d2 100
$C_8H_{18}N_2OX$
17d1 45, *17d2* 224
$C_8H_{18}N_2O_2P$
9c2 240
$C_8H_{18}N_2S^-$
17e 387
$C_8H_{18}N_2Si_2^+$
17f 33, *17h* 180
$C_8H_{18}N_3$
17c 581
$C_8H_{18}N_3O_2$
17d1 46, 183
$C_8H_{18}N_3O_3$
17d1 46, 117
$C_8H_{18}O^+$
17b 247
$C_8H_{18}OP$
9c2 240
$C_8H_{18}OPS$
17e 282
$C_8H_{18}OSi^-$
17f 11
$C_8H_{18}O_2^+$
17h 51
$C_8H_{18}O_2P$
17b 195
$C_8H_{18}O_2PS_2$
17e 327
$C_8H_{18}O_3P$
9b 184, *9c2* 221, *17b* 74, 97, 195, 249, 252, *17e* 282
$C_8H_{18}O_3PS$
17e 282
$C_8H_{18}O_4P$
9c2 240, 241, *17b* 196, *17e* 283
$C_8H_{18}O_4PS$
17b 274, *17e* 283
$C_8H_{18}S^+$
17e 363
$C_8H_{18}S_2^+$
17e 366
$C_8H_{19}Al$
9b 95
$C_8H_{19}As$
9b 95
$C_8H_{19}As^+$
9b 28
$C_8H_{19}BN$
17b 252
$C_8H_{19}ClO_3P$
9c2 241
$C_8H_{19}FO_4P$
9c2 241
$C_8H_{19}Ge$
9b 29, 95

$C_8H_{19}NO^+$
9d2 117, *17d1* 156, *17h* 134
$C_8H_{19}NO_2P$
17c 489, 494
$C_8H_{19}NO_3P$
17b 42, *17c* 498, *17d2* 419
$C_8H_{19}NO_4P$
9c1 1007, *17c* 489, 494, *17d1* 200
$C_8H_{19}NO_4PSi$
17b 520, *17e* 283
$C_8H_{19}NO_5P$
17d1 186, *17d2* 419
$C_8H_{19}N_2$
9c1 42
$C_8H_{19}N_2^{2+}$
17b 103, 106
$C_8H_{19}N_2O$
9c1 224, 302, 897, *17d2* 72
$C_8H_{19}N_2OS$
17d1 46
$C_8H_{19}N_3^+$
9d2 61
$C_8H_{19}N_3OX$
17d1 46
$C_8H_{19}N_3O_2X$
17d1 13
$C_8H_{19}N_3O_3X$
17d1 14
$C_8H_{19}OSSn$
17b 239
$C_8H_{19}OSi$
17b 234, *17f* 6
$C_8H_{19}O_2SSi$
17b 18
$C_8H_{19}O_2Si$
9b 178
$C_8H_{19}O_4Si^-$
17b 147
$C_8H_{19}P^+$
9b 34, 103
$C_8H_{19}Si$
9b 35ff, 104
$C_8H_{19}Si_2$
9b 308
$C_8H_{19}Sn$
9b 36, 104, *17b* 46, 150
$C_8H_{20}BN_2$
17c 573
$C_8H_{20}Cl_2P_2S_2^+$
9c2 264
$C_8H_{20}FeNOP_2S_4$
17a 328
$C_8H_{20}FeN_2O_2P_2S_4$
17a 328
$C_8H_{20}NOSi$
17d1 131
$C_8H_{20}NO_6P_2$
9c1 1011
$C_8H_{20}N_2^+$
9d2 61, 100ff
$C_8H_{20}N_2OP$
17c 449
$C_8H_{20}N_2OSX$
17d1 46
$C_8H_{20}N_2OX$
17d1 47, 117, 132
$C_8H_{20}N_2O_2P$
9c2 241
$C_8H_{20}N_2O_3P$
17c 449
$C_8H_{20}N_2S^+$
17e 364, *17h* 138
$C_8H_{20}N_3Si$
17c 582
$C_8H_{20}O_2P$
9c2 242
$C_8H_{20}O_3P$
9c2 242
$C_8H_{20}O_4P$
9c2 242
$C_8H_{20}O_6P$
9c2 24
$C_8H_{20}O_6P_2S_2^-$
17e 284, 386
$C_8H_{20}PS_2$
17e 284
$C_8H_{20}P_2S_2^-$
17e 284, 308
$C_8H_{20}S_2^+$
9c2 299, *17e* 376
$C_8H_{20}Si_2^-$
17f 11
$C_8H_{21}ClN_2OP$
9c2 242
$C_8H_{21}NO_3P$
9c2 242, 243
$C_8H_{21}N_3O_3P$
17c 453
$C_8H_{21}OSSi$
17e 350, 351
$C_8H_{21}OSi_2$
17b 274
$C_8H_{21}SSi_2$
17b 327
$C_8H_{22}BNOP$
17d1 200
$C_8H_{22}NOSi_2$
17d2 207
$C_8H_{22}NSi_2$
17b 39
$C_8H_{22}N_2O_2P$
9c2 243
$C_8H_{22}N_3OX_2$
17d1 132, 146
$C_8H_{22}O_4PSi_2$
17e 284
$C_8H_{24}N_4Si_2^+$
9d2 101, *17f* 34, *17h* 144
$C_8H_{24}Si^-$
17f 26
$C_8N_4S^-$
9d1 891
$C_8N_4S_2^-$
9d1 890
$C_8N_5^{2-}$
1 79
$C_8Ni_2O_8^+$
17a 129
$C_8O_4S_4^+$
17h 114
C_9Cl_7
9b 616
$C_9Co_2FeO_9S^-$
17a 127
$C_9D_{15}N_2O_2$
17d2 236
C_9F_{11}
17c 191
$C_9F_{12}NO$
9c1 728
C_9F_{19}
17b 305
$C_9HN_5^-$
9d1 862
$C_9H_2F_{17}O$
9b 232
$C_9H_2O_6^{3-}$
9b 313, 314

$C_9H_3D_8NNaO_4S$
17d1 347
$C_9H_3D_8NNaO_5S$
17d1 347
$C_9H_3F_8$
17c 190
$C_9H_3O_7^{4-}$
17c 94
$C_9H_4ClF_6$
17c 190
$C_9H_4CdO_3^+$
17a 387
$C_9H_4F_5$
17c 26
$C_9H_4F_7$
17c 190
$C_9H_4FeO_8$
9a 312
$C_9H_4NO_3^{2-}$
9d1 48, *17g* 359
$C_9H_4NO_3$
9c1 163
$C_9H_4OS_2^-$
9b 711, 712
$C_9H_4O_3^-$
9d1 79, *17g* 57
$C_9H_4O_4^-$
9d1 159
$C_9H_4O_4S^{2-}$
9d1 48
$C_9H_4O_5^{3-}$
9d1 251, *17g* 254
$C_9H_4O_6^{3-}$
9b 460, *9d1* 90
$C_9H_4O_6^{2-}$
9b 306
$C_9H_4O_7^{3-}$
9b 469, 470
$C_9H_5Br_2O_4$
9c2 164
$C_9H_5Cl_2O_4$
9c2 165
$C_9H_5Cl_3N_3S$
17c 541
$C_9H_5D_2NO_3Zn^+$
17a 378, *17g* 353
$C_9H_5D_3MgNO_4$
17g 193
$C_9H_5D_3O_2^-$
17g 16
$C_9H_5D_6NS^-$
9b 585, *17c* 210
$C_9H_5F_6$
17c 101, 190
$C_9H_5F_7NO$
9c1 724
$C_9H_5I_2O_4$
9c2 165
$C_9H_5NO_2^-$
9d1 341, *17g* 358, 359
$C_9H_5NO_6^{3-}$
9d1 381
$C_9H_5OS_2$
17c 344
$C_9H_5O_2^-$
9d1 85
$C_9H_5O_3$
9c2 206
$C_9H_5O_4^{2-}$
9d1 126, 162, 177
$C_9H_5O_4$
9c2 206
$C_9H_6ClCoN_6O_2$
9c1 1016
$C_9H_6Cl_4NO_4$
9c1 649
$C_9H_6CoN_7O$
9c1 1016
$C_9H_6DNO_3Zn^+$
17a 379, *17g* 354
$C_9H_6D_2MgNO_4$
17g 193
$C_9H_6D_3NO_3^-$
9c2 40
$C_9H_6D_5O_2$
9c2 71
$C_9H_6F_3O$
17b 66
$C_9H_6F_6N_3O$
9c1 725
$C_9H_6F_9NO_2$
17d1 209
C_9H_6N
17b 505
C_9H_6NO
17b 67
$C_9H_6NO_2$
17g 348
$C_9H_6NO_3^{2-}$
9d1 49, *17b* 230
$C_9H_6N_3O_3^{2-}$
17f 290
$C_9H_6N_3O_4^{2-}$
17f 291
$C_9H_6O^-$
9b 616
$C_9H_6OS^-$
9b 708ff, 710, 711, *17c* 333ff
$C_9H_6OS_2^-$
17c 335, 336
$C_9H_6O_2^-$
9d1 36, *17g* 21
$C_9H_6O_2S^-$
9d1 48
$C_9H_6O_3^-$
9c2 39, *9d1* 49, 70, *17c* 328ff, 330ff, *17g* 38
$C_9H_6O_4^-$
17g 105
$C_9H_6O_5^{3-}$
17g 250, 251
C_9H_7
9b 434, 616, *17c* 242
$C_9H_7CdO_4$
17a 387, *17g* 86
$C_9H_7ClNaO_2$
9d1 158
$C_9H_7ClO_2^-$
9d1 22
$C_9H_7Cl_3NO_2$
17d1 238, 243
$C_9H_7CoN_6O$
9c1 1023
$C_9H_7DMgNO_4$
17g 193
$C_9H_7F_3NO$
9c1 135, 136
$C_9H_7F_3O^-$
9b 567
$C_9H_7F_5NO$
17d1 221
$C_9H_7F_6$
17c 101
$C_9H_7F_6O$
17c 103
$C_9H_7F_6OSi$
17b 312
$C_9H_7F_{12}N_2O_3$
17d2 221

$C_9H_7FeN_6^{3-}$
17f 203
$C_9H_7N^-$
9d1 862, 865ff, *17f* 186
$C_9H_7N^+$
17h 226, 227
$C_9H_7NO^-$
1 105, 106, *9d1* 863ff
$C_9H_7NO^+$
17e 212
$C_9H_7NO_2^-$
9d1 337, 553, *17g* 66
$C_9H_7NO_3Zn^+$
17a 379, *17g* 353, 354
$C_9H_7NO_4S^+$
17e 213
$C_9H_7NO_5^{3-}$
17g 253
$C_9H_7NS^-$
9d1 901, *17f* 217
$C_9H_7N_2O$
9c1 132
$C_9H_7N_2O_2$
9c1 595
$C_9H_7N_2O_3$
9c1 804
$C_9H_7OS_2$
17c 335
$C_9H_7O_2^{2-}$
9b 389
$C_9H_7O_2$
9b 461, *17c* 243
$C_9H_7O_3$
9c2 30, 35, 39
$C_9H_7O_4^{2-}$
9d1 93, 245, *17g* 86
$C_9H_7O_4$
9c2 58, *17e* 228
$C_9H_7O_4Zn$
17a 380, *17g* 86
$C_9H_7O_5^{2-}$
9d1 271, *17g* 251
$C_9H_7O_6$
9b 306
$C_9H_7O_7$
9b 469, 470
$C_9H_7O_8$
17c 60
$C_9H_7O_9$
17c 61
$C_9H_8BrN_2O_5$
17b 297
$C_9H_8Br_2N_2NaO_6S$
17d1 256
$C_9H_8ClCoN_6O$
9c1 1016
$C_9H_8ClN_2O_5$
17b 297
$C_9H_8Cl_2NO$
17c 591
$C_9H_8CoN_7O_3$
9c1 1016
$C_9H_8DNO_3^-$
9c2 39
$C_9H_8D_2^-$
9d1 835, 849
$C_9H_8FO_2$
17c 203
$C_9H_8F_3N_2O_6^{2-}$
9d1 428
$C_9H_8F_4NO$
17d1 216
$C_9H_8FeN_6^{3-}$
17a 328
$C_9H_8MgNO_4$
17g 193
C_9H_8N
9b 127, 622, 624, *17b* 167, *17c* 53, 180, 183, 212
C_9H_8NO
17c 213, *17d2* 361
$C_9H_8NO_2$
9c1 140, 141, 142, 768, *17c* 496
$C_9H_8NO_2S_2$
17c 545
$C_9H_8NO_3$
9c1 730, *17c* 487
$C_9H_8NO_4^{2-}$
9d1 381, *17g* 90, 196
$C_9H_8NO_4^-$
9c1 983, *9d1* 383
$C_9H_8NO_4$
17d2 413
$C_9H_8NO_4Zn^+$
17g 94
$C_9H_8NO_5^{2-}$
17g 104
$C_9H_8NO_5$
17d2 413
$C_9H_8NS_2$
17c 534
$C_9H_8N_2^-$
9d1 877
$C_9H_8N_2Na_2O_9P$
17b 234
$C_9H_8N_2O_2^-$
9d1 174, 558
$C_9H_8N_2O_3^-$
9d1 559
$C_9H_8N_2O_5^-$
9d1 543
$C_9H_8N_2O_6^-$
9d1 630
$C_9H_8N_2O_7^-$
9d1 631
$C_9H_8N_3$
9b 525
$C_9H_8N_3O_2$
9b 602, *17c* 456, 457
$C_9H_8N_3O_6^{2-}$
9d1 429
$C_9H_8NaO_2$
9d1 158
$C_9H_8O^-$
9b 429, 613ff
$C_9H_8O_2^-$
9b 615, *9d1* 22, 31, 60, 84, 177, *17b* 176, *17g* 16
$C_9H_8O_2S^-$
9d1 51
$C_9H_8O_3^-$
9d1 51, 71, 93
$C_9H_8O_3^+$
17h 83
$C_9H_8O_4^-$
17g 215
$C_9H_8O_4^+$
17e 246, *17h* 82
$C_9H_8S_2^+$
1 127, *17e* 371
C_9H_9
9b 546, 547, 616, *17c* 25
$C_9H_9AlNO_3^{2+}$
17a 397ff, *17g* 349
$C_9H_9AlNO_4^{2+}$
17a 398
$C_9H_9BrMgO_2^-$
17a 225, *17c* 204

$C_9H_9Br_2NNaO_4S$
17d1 258
$C_9H_9CaNO_3^+$
17a 231, *17g* 349
$C_9H_9CaNO_4^+$
17a 231, *17g* 92
$C_9H_9CdNO_3^+$
17a 388, *17g* 350
$C_9H_9CdNO_4^+$
17a 388, *17g* 93
$C_9H_9ClNO_4S$
9c1 994
$C_9H_9ClNaO_2$
9d1 153
$C_9H_9ClO^+$
17h 61
$C_9H_9Cl_2NiO_2$
17a 342
$C_9H_9Cl_3NO$
9c1 615
$C_9H_9Cl_4O_5P^+$
17e 190
$C_9H_9CoN_6O$
9c1 1016
$C_9H_9D_2NS^-$
9b 584
$C_9H_9F_3NO$
17d1 217
$C_9H_9F_4O_2Si$
17e 191
$C_9H_9InNO_4^{2+}$
17a 409
$C_9H_9KN_3O_8S$
9c1 105
$C_9H_9LaNO_2^{2+}$
17a 240
$C_9H_9MgNO_3^+$
17a 225, *17g* 350
$C_9H_9MgNO_4^+$
17a 225, *17g* 93
$C_9H_9MnNO_6$
9c1 1037, *17a* 282, *17d1* 202
$C_9H_9MnNO_7$
17d1 186
$C_9H_9N^+$
17h 61
$C_9H_9NO_2^-$
9d1 101, 463ff, 465, 466
$C_9H_9NO_2S_2^-$
17f 251
$C_9H_9NO_2Sr^+$
17a 234
$C_9H_9NO_3^-$
9c2 39, *17e* 41, *17g* 348
$C_9H_9NO_3P$
17c 575
$C_9H_9NO_3S^-$
17f 251
$C_9H_9NO_3Sc^{2+}$
17a 238, *17g* 351
$C_9H_9NO_3Y^{2+}$
17a 239, *17g* 352
$C_9H_9NO_3Zn^+$
17a 379ff, *17g* 352
$C_9H_9NO_4^-$
1 92, *9c2* 58, *9d1* 504, 531, 541, *17f* 232, *17g* 90, 91ff
$C_9H_9NO_4Sc^{2+}$
17a 238
$C_9H_9NO_4Sr^+$
17g 93
$C_9H_9NO_4Y^+$
17a 239, *17g* 94
$C_9H_9NO_4Zn^+$
17a 380, *17g* 94
$C_9H_9NO_5^-$
17f 283
$C_9H_9NO_6Re$
17d1 203
$C_9H_9N_2O_2$
9c1 892, *17c* 597
$C_9H_9N_2O_3$
9c1 179
$C_9H_9N_2O_4^{2-}$
9d1 423
$C_9H_9N_2O_5^{2-}$
9d1 426
$C_9H_9N_2O_8^{4-}$
9b 172
$C_9H_9N_3O_6^-$
1 103, *9d1* 664
$C_9H_9N_4O_2^{2-}$
17f 291
$C_9H_9N_5O_2^+$
9d2 72
C_9H_9O
9b 78, 364, 550, *17b* 526, *17c* 15, 175
C_9H_9OS
17c 177, 207, 241
$C_9H_9O_2^{2-}$
9b 572, *17c* 192, 194
$C_9H_9O_2$
9b 33, 118, *17b* 68, 176, *17c* 53, 175, 176, 203
$C_9H_9O_3^{2-}$
9b 574, *9d1* 250, 261, 273
$C_9H_9O_3$
1 59, 70, *9c2* 36
$C_9H_9O_4^{2-}$
9b 299, *9d1* 260, 271
$C_9H_9O_4$
9c2 170, *17e* 50, 154
$C_9H_9O_5$
1 71
C_9H_9S
9b 379
$C_9H_9S_2$
17c 208
$C_9H_9S_3$
17b 382
$C_9H_{10}^-$
1 85, 86, *9d1* 712, 716, 718ff, 720, 724, 834, 848ff, *17f* 71
$C_9H_{10}Br$
9b 559
$C_9H_{10}BrN_2O_5$
9c2 6, *17b* 216, *17e* 7, 8
$C_9H_{10}Cl$
17c 186
$C_9H_{10}ClN^+$ (?)
9d2 119
$C_9H_{10}ClNO_2^-$
9d1 384, 385
$C_9H_{10}ClN_2O_3$
9c1 981
$C_9H_{10}ClN_2O_5$
9c2 6, *17b* 216
$C_9H_{10}Cl_3NiO^-$
17a 342
$C_9H_{10}F$
17c 181, 186
$C_9H_{10}FNO^-$
9b 583
$C_9H_{10}FeO_5S_3$
17a 328
$C_9H_{10}IN_2O_5$
9b 156

$C_9H_{10}N^{2+}$
17h 233
$C_9H_{10}NO$
9b 98, *9c1* 124, 130, 132, 134, 757, *17b* 360, 525, *17c* 591
$C_9H_{10}NOS$
17c 614
$C_9H_{10}NO_2$
9c1 130, 132, 159, 766, 773, 791, *17c* 488, *17d2* 393, 394, 399
$C_9H_{10}NO_2S$
17c 542
$C_9H_{10}NO_3^{2-}$
17g 101
$C_9H_{10}NO_3$
9c1 207, 791, *9c2* 40
$C_9H_{10}NO_3S$
9c1 148, 149, *17d2* 416
$C_9H_{10}NO_3Zn^{2+}$
17a 380, *17g* 108
$C_9H_{10}NO_4$
17d1 386, *17d2* 413
$C_9H_{10}NO_5$
17d1 386
$C_9H_{10}NO_5S$
9c1 1000
$C_9H_{10}N_2Na_2O_9P$
17b 163, 222
$C_9H_{10}N_2Na_2O_{10}P$
17b 434
$C_9H_{10}N_2O_2S^-$
9d1 513, 545
$C_9H_{10}N_2O_4^-$
1 101, *9d1* 622
$C_9H_{10}N_2O_4^+$
17h 239
$C_9H_{10}N_2O_6^-$
9d1 696
$C_9H_{10}N_2O_9P^{2-}$
17e 8
$C_9H_{10}N_3O$
17d2 346
$C_9H_{10}N_3O_2$
17d1 365
$C_9H_{10}N_4O_5^-$
9d1 632
$C_9H_{10}N_4O_6^-$
17f 281
$C_9H_{10}NaO_2$
9d1 150
$C_9H_{10}O^-$
9b 541, 567, *17c* 193, *17f* 187
$C_9H_{10}OS^-$
9b 577, *17c* 169
$C_9H_{10}OS_2^-$
9b 579
$C_9H_{10}O_2^-$
1 81, *9b* 541, 568, 575, *9d1* 31, 32, 37, 60, 119, 123, 124, 175, 220, *17g* 231
$C_9H_{10}O_2^+$
17h 61, 62
$C_9H_{10}O_2S^-$
17c 169
$C_9H_{10}O_3^-$
9d1 221
$C_9H_{10}O_4^-$
17g 202, 231, 236
$C_9H_{10}O_4^+$
17h 62
$C_9H_{10}O_5^-$
17g 199, 202
$C_9H_{10}S_2^-$
9b 578, 579
C_9H_{11}
9b 19, 38, 83, 558, *17b* 23, *17c* 12, 52, 173, 185
$C_9H_{11}AlBr_3S$
17a 398, *17e* 320
$C_9H_{11}Al_2Br_6S$
17a 398
$C_9H_{11}BrN_2O_9P^-$
17b 418
$C_9H_{11}ClNS$
17c 500
$C_9H_{11}ClN_2O_9S^-$
17b 420
$C_9H_{11}Cl_2N_3O_3^-$
17f 282
$C_9H_{11}FNO$
17d1 369
$C_9H_{11}FN_2O_9S^-$
17b 420
$C_9H_{11}FO_2S^+$
17h 93
$C_9H_{11}FO_3S^+$
17h 92
$C_9H_{11}IN_2O_{10}S^-$
17b 421
$C_9H_{11}NNaO_4S$
17d1 351
$C_9H_{11}NNaO_5S$
17d1 351
$C_9H_{11}NO^-$
9b 554, 582, 611, *17c* 209, *17f* 306
$C_9H_{11}NO^+$
9d2 31, *17h* 62, 231
$C_9H_{11}NOM$
17d1 262
$C_9H_{11}NOX$
17d1 262
$C_9H_{11}NO_2^-$
1 90, 99, *9d1* 383, 385, 453ff, 458, 459ff
$C_9H_{11}NO_2^+$
9d2 32
$C_9H_{11}NO_3^-$
9d1 383, *17g* 100ff, 104
$C_9H_{11}NO_3Zn^+$
17a 381
$C_9H_{11}NO_4P^{2-}$
9d1 413
$C_9H_{11}NO_4S^-$
9d1 550, *17f* 235
$C_9H_{11}NO_6S_2^-$
9d1 693
$C_9H_{11}NO_7S_2^-$
9d1 552
$C_9H_{11}NS^-$
9b 584, *17c* 210
$C_9H_{11}NSe^-$
17c 216
$C_9H_{11}N_2$
9b 382, 524, *17c* 577
$C_9H_{11}N_2O$
17c 579, *17d2* 375
$C_9H_{11}N_2O_2$
17c 23
$C_9H_{11}N_2O_3$
9c1 969
$C_9H_{11}N_2O_4S$
9c1 773
$C_9H_{11}N_2O_5$
17b 215, 354
$C_9H_{11}N_2O_6$
9c1 951, *17b* 354

$C_9H_{11}N_2O_7^{2-}$
9d1 428
$C_9H_{11}N_3O_4^-$
9d1 628
$C_9H_{11}N_4O_2$
17c 433
$C_9H_{11}O$
1 58, 59, 69, *9b* 21, 194, 555, 568, *9c2* 30, 35, 38, 68, 174, *17b* 80, 186, *17c* 179, 182, 202, *17e* 75
$C_9H_{11}O_2$
1 57, 70, *9b* 427, *9c2* 17, 70, *17c* 203, *17e* 12, 77
$C_9H_{11}O_2S$
17e 340
$C_9H_{11}O_2S_2Sn$
17a 453
$C_9H_{11}O_3$
17b 86
$C_9H_{11}O_3Zn^+$
17g 102
$C_9H_{11}O_4$
9d2 166, *17b* 357, *17e* 154
$C_9H_{11}O_4P^-$
17g 54
$C_9H_{11}O_4S_2$
17c 19
$C_9H_{11}O_5^{2-}$
17b 100
$C_9H_{11}O_5$
9b 297, 299
$C_9H_{11}O_6S$
17e 154
$C_9H_{11}S_3$
17b 326
$C_9H_{12}^-$
1 85, *9d1* 713, 721, 722, *17f* 71
$C_9H_{12}Ag$
17a 375
$C_9H_{12}Au$
17a 376
$C_9H_{12}BrN_2O_5$
9b 286, 425, *17b* 417
$C_9H_{12}BrN_2O_6$
17b 417, 418
$C_9H_{12}ClN_2O_5$
17b 419
$C_9H_{12}ClN_2O_6$
17b 419, 420
$C_9H_{12}Cl_5FNO_2$
9c1 292
$C_9H_{12}DGe$
17f 39
$C_9H_{12}FN_2O_6$
17b 420
$C_9H_{12}IN_2O_5$
9b 286, 426, *17b* 421
$C_9H_{12}IN_2O_7$
17b 421
$C_9H_{12}N$
9b 551, *17b* 60, 245
$C_9H_{12}NO$
9c1 612, 624, *17d1* 217, *17d2* 209, *17e* 51
$C_9H_{12}NO_2$
9b 462, *9c1* 172, 173, 176, 177, 550, 643, *17c* 145
$C_9H_{12}NO_2S$
17c 527
$C_9H_{12}NO_2S_2$
17c 553
$C_9H_{12}NO_3$
9b 236, 462, *9c1* 207, 948
$C_9H_{12}NO_3P^-$
9c1 200
$C_9H_{12}NO_3S$
9c1 996, *17c* 495
$C_9H_{12}NO_3Zn^{2+}$
17a 381, *17g* 103, 104
$C_9H_{12}NO_4$
9b 298
$C_9H_{12}NO_5S_2$
9c1 995
$C_9H_{12}N_2Na_2O_8P$
17b 408
$C_9H_{12}N_2Na_2O_9P$
17b 408ff, 432
$C_9H_{12}N_2Na_2O_{10}P$
17b 434
$C_9H_{12}N_2O_2^-$
9d1 509, 513
$C_9H_{12}N_2O_3S$
9c1 980
$C_9H_{12}N_2O_5^-$
17c 47
$C_9H_{12}N_2O_6^-$
9b 423, *17c* 47
$C_9H_{12}N_2O_9S^-$
17b 410
$C_9H_{12}N_3O$
17d2 132
$C_9H_{12}N_3O_5$
17d1 101
$C_9H_{12}N_3O_6Sn^{2-}$
9d1 429
$C_9H_{12}N_3O_8P^-$
17b 216
$C_9H_{12}N_4OP$
17c 476
$C_9H_{12}N_4O_2^+$
9d2 72
$C_9H_{12}O^-$
9b 485
$C_9H_{12}OS^-$
9b 607
$C_9H_{12}O_2^-$
9d1 15, 37, 42, 44, 45
$C_9H_{12}O_2PS$
17e 285
$C_9H_{12}O_2Si^-$
9d1 191
$C_9H_{12}O_3^-$
9b 413, *9d1* 37
$C_9H_{12}O_3^+$
9d2 135, *17h* 81
$C_9H_{12}O_3S^+$
17h 93, 94
$C_9H_{12}O_4S^+$
17h 92
$C_9H_{12}S_2^-$
17c 119
$C_9H_{12}S_3^+$
17h 96
C_9H_{13}
9b 454, 463, *17b* 488, 489, *17c* 74, 80
$C_9H_{13}AlCl_3N_2O$
9a 296
$C_9H_{13}BD_3N^-$
17c 411
$C_9H_{13}BF_3N_2O$
9a 293
$C_9H_{13}BrNO_3$
17d2 234
$C_9H_{13}ClNO_2Sn$
17d1 377

$C_9H_{13}Cl_2N_2OZn$
9a 335
$C_9H_{13}D_3N_3O_4$
17d1 85, 96
$C_9H_{13}F_9O_3P$
9c2 243
$C_9H_{13}Ge$
9b 56, *17f* 39
$C_9H_{13}GeN_2$
9b 530
$C_9H_{13}N^-$
17f 187
$C_9H_{13}N^+$
9d2 32, 33
$C_9H_{13}NO^+$
9d2 33, *17h* 161
$C_9H_{13}NO_2^-$
17g 29
$C_9H_{13}NO_2^+$
9d2 33
$C_9H_{13}NO_2Si^-$
9d1 519, 520, 521
$C_9H_{13}NO_3^+$
9d2 34
$C_9H_{13}NO_3Si^-$
17f 255
$C_9H_{13}N_2O$
9c1 407, 624ff, 699ff, *17d2* 106, 234
$C_9H_{13}N_2O_2$
9d1 512, *17d1* 366
$C_9H_{13}N_2O_3$
17d1 366
$C_9H_{13}N_2O_5$
17b 407, 432
$C_9H_{13}N_2O_6$
9b 279, *17b* 408, 410
$C_9H_{13}N_2S$
17b 287
$C_9H_{13}N_2Si$
9b 527, 528, *17c* 149
$C_9H_{13}N_3O_2^-$
9d1 511
$C_9H_{13}N_3O_5^+$
17b 230
$C_9H_{13}N_3O_7P^-$
17e 8
$C_9H_{13}N_3O_7P$
17b 355
$C_9H_{13}N_3O_8P^-$
17e 9
$C_9H_{13}N_3O_8P$
9b 110, 156, *9c2* 8, *17b* 216
$C_9H_{13}N_4O_2$
17c 432
$C_9H_{13}N_4O_3$
9c1 31
$C_9H_{13}N_5O^+$
9d2 72, 73
$C_9H_{13}N_6O$
17d1 102
$C_9H_{13}O$
9b 297
$C_9H_{13}O_2^-$
9d1 58
$C_9H_{13}O_2Sn$
17g 74
$C_9H_{13}O_4$
9b 160, 374, 375
$C_9H_{13}O_5$
17b 100, 497
$C_9H_{13}O_7$
1 27
$C_9H_{13}S_3$
17c 66
$C_9H_{13}Si$
9b 75
$C_9H_{13}Sn$
9b 76
$C_9H_{14}BD_3Si^-$
17c 99
$C_9H_{14}BrNO_2$
9c1 451
$C_9H_{14}Cl_2NO_2Ti$
17d1 186
$C_9H_{14}Cl_2OP$
9c2 243
$C_9H_{14}Cl_3Si$
17c 34
$C_9H_{14}Cl_3Sn$
17c 34
$C_9H_{14}DGe$
17c 90
$C_9H_{14}Ge^-$
1 104
$C_9H_{14}GeNO_2$
17d1 231
$C_9H_{14}N$
17c 130, 135, 391
$C_9H_{14}NO$
9c1 168, 172, 173, *17d2* 341
$C_9H_{14}NOS$
9c1 225
$C_9H_{14}NOS_2$
17d2 200
$C_9H_{14}NO_2$
9c1 527, *17d2* 234, 342
$C_9H_{14}NO_2S$
17b 287, *17d2* 200
$C_9H_{14}NO_2Si$
17d1 231
$C_9H_{14}NO_2Sn$
17d1 231
$C_9H_{14}NO_3$
9c1 407, *17b* 286, *17d2* 224, 225, 234ff
$C_9H_{14}NO_3Sn$
17d1 381
$C_9H_{14}NO_4$
9c1 146, *17b* 466
$C_9H_{14}NO_5M_2$
17d1 27
$C_9H_{14}N_2^+$
9d2 101
$C_9H_{14}N_2O_2R$
17d2 854
$C_9H_{14}N_2O_2S^-$
9d1 586, *17g* 28
$C_9H_{14}N_3$
17b 284
$C_9H_{14}N_3O_3$
9c1 897, *17d1* 76, 94, 102
$C_9H_{14}N_3O_4$
17b 422, 436, *17c* 418
$C_9H_{14}N_3O_5$
17d1 83, 94
$C_9H_{14}N_3O_6$
9c1 302, 355, *17d1* 95
$C_9H_{14}O^-$
9b 407, 412, 436
$C_9H_{14}O_2^-$
9b 413, *9d1* 15, 45, *17b* 335, *17g* 25
$C_9H_{14}O_3Si^-$
17f 11
$C_9H_{14}O_5^-$
9d1 83
$C_9H_{14}O_5$
9b 124

$C_9H_{14}O_6^-$
17b 273
$C_9H_{14}S_2^+$
17e 372
$C_9H_{14}Si^-$
1 104, *17f* 11
$C_9H_{14}Si^+$
17f 34
C_9H_{15}
9b 19, *17b* 34, 115, 475, 485
$C_9H_{15}BN$
17c 156
$C_9H_{15}BrNO$
17d2 235
$C_9H_{15}ClCuN_5O_2$
17a 356
$C_9H_{15}ClF_5OSi$
9b 244
$C_9H_{15}Cl_5NO$
9c1 262
$C_9H_{15}F_3NOS$
17d2 178
$C_9H_{15}F_3NO_2$
17d2 176
$C_9H_{15}F_6OSi$
9b 243
$C_9H_{15}Ge$
17c 90
$C_9H_{15}GeN_2O$
17c 152
$C_9H_{15}INO$
17d2 235
$C_9H_{15}N^+$
9d2 27
$C_9H_{15}NNaO_3$
17d2 225
$C_9H_{15}NO_2^-$
9d1 417
$C_9H_{15}NO_2^+$
17h 134
$C_9H_{15}NO_3M$
17d2 174
$C_9H_{15}NO_4^-$
9d1 417
$C_9H_{15}NO_4^+$
17b 99
$C_9H_{15}N_2O$
9c1 399, 527, *17d2* 390
$C_9H_{15}N_2O_2$
9c1 408ff, 866, *17d2* 235ff, 390
$C_9H_{15}N_2O_3$
9c1 240, 840, 866
$C_9H_{15}N_2O_4$
9c1 300, 866, *17c* 452
$C_9H_{15}N_3O_5^+$
17b 422, 436
$C_9H_{15}N_4O_2$
17d1 77, 102
$C_9H_{15}N_4O_3$
17d2 194
$C_9H_{15}N_5O^+$
9d2 73
$C_9H_{15}NiO_2$
17c 46
$C_9H_{15}O$
9b 298, 408, 412, *17b* 389
$C_9H_{15}O_2$
17b 336, 372
$C_9H_{15}O_3S$
17b 62
$C_9H_{15}O_4$
9b 125, *17b* 19, 23, 32, 61, 85, 120, 122, 132, 151, 153, 154, 165, 172
$C_9H_{15}O_6$
17b 215, 354
$C_9H_{15}O_6S_2$
9c2 296
$C_9H_{15}Si$
17c 83, 97
$C_9H_{16}AlBr_3NO_2$
9a 295
$C_9H_{16}AlCl_3NO_2$
9a 295
$C_9H_{16}AlI_3NO_2$
9a 295
$C_9H_{16}BN_2O$
17d1 198
$C_9H_{16}BrN_2O_3$
17d2 281
$C_9H_{16}Br_3GaNO_2$
9a 295
$C_9H_{16}Br_3InNO_2$
9a 296
$C_9H_{16}ClN_2O_3$
17d2 281
$C_9H_{16}Cl_3GaNO_2$
9a 295
$C_9H_{16}CoF_3N_5O_6$
9c1 1022
$C_9H_{16}F_6O_3P$
9c2 244
$C_9H_{16}GaI_3NO_2$
9a 296
$C_9H_{16}I_3InNO_2$
9a 296
$C_9H_{16}Mu$
17b 121
$C_9H_{16}N$
17c 391
$C_9H_{16}NO$
9c1 24, 152, 225, 340, 448, 527, 528, *17b* 106, 359, 387, *17d2* 332
$C_9H_{16}NOS$
9c2 284, *17d2* 236
$C_9H_{16}NO_2$
1 54, *9c1* 451ff, *17d2* 129, 225, 236, 282ff, 399, 403
$C_9H_{16}NO_3$
9c1 225, 238, 298, 300, 400, *17d1* 123, 142, *17d2* 225
$C_9H_{16}NO_3S$
17d2 165
$C_9H_{16}NO_4$
9c1 426
$C_9H_{16}NO_5$
9c1 242, 320, *17d1* 142
$C_9H_{16}NO_6$
17d1 118
$C_9H_{16}NS$
9c1 1043, *17b* 287
$C_9H_{16}NSi$
9b 526
$C_9H_{16}N_2O_2^+$
17h 135, 154
$C_9H_{16}N_2O_4M$
17d1 8, 16, 28, 67, 146
$C_9H_{16}N_2O_5M$
17d1 9, 27
$C_9H_{16}N_3O$
9c1 400
$C_9H_{16}N_3O_2$
17d2 256

$C_9H_{16}N_3O_3$
9c1 226, 349, 867, *17d1* 77, 78, 83, 84ff, 95, *17d2* 264
$C_9H_{16}N_3O_4$
9c1 303, *17d1* 85, 86, 96
$C_9H_{16}N_3O_5$
9c1 303, *17d2* 286
$C_9H_{16}N_5OS$
17d2 257
$C_9H_{16}O^-$
9b 261, 295
$C_9H_{16}O_3^-$
17b 161
$C_9H_{16}S^-$
17b 340, 500
C_9H_{17}
17b 115
$C_9H_{17}BrNO$
17d2 226
$C_9H_{17}ClCuN_5O_2S$
17a 357
$C_9H_{17}ClNO$
9c1 461
$C_9H_{17}Cl_2NO_3P$
9c1 462
$C_9H_{17}Cl_3NO$
17d1 164
$C_9H_{17}NO^-$
9d1 687
$C_9H_{17}NO_2^-$
9d1 415
$C_9H_{17}NO_3M$
17d1 28, *17d2* 190
$C_9H_{17}NO_3P$
17e 285
$C_9H_{17}NO_3PS$
17b 287
$C_9H_{17}NO_3SM$
17d1 28, 29
$C_9H_{17}NO_5P$
17c 598, 613
$C_9H_{17}NO_6P$
17c 610
$C_9H_{17}NS^+$
17h 138
$C_9H_{17}N_2$
9c1 91, *17c* 480
$C_9H_{17}N_2O$
9c1 226, 901, *17c* 448, *17d2* 257, 335
$C_9H_{17}N_2O_2$
9c1 262, 390, 400ff, 462, 520, 867, *17d1* 123, 180, *17d2* 226, 286, 336, 338ff
$C_9H_{17}N_2O_2SSi$
9c1 980
$C_9H_{17}N_2O_3$
9c1 236, 401, 867, *17d1* 47, 63, *17d2* 129, 227,286, 336
$C_9H_{17}N_2O_3S$
17d2 165ff
$C_9H_{17}N_2O_4$
9c1 355, *17c* 445, *17d1* 9, 29, 47ff, 72
$C_9H_{17}N_2O_5$
9c1 292, *17d* 29, 64, 67
$C_9H_{17}N_4O$
17d2 257
$C_9H_{17}N_4O_2$
9c1 401, *17d1* 78, 86, 87, 97
$C_9H_{17}N_4O_3$
17d1 87, 97
$C_9H_{17}N_6$
17b 76
$C_9H_{17}O$
9b 112, *17b* 53, 203
$C_9H_{17}OS$
17b 249, 529, 530
$C_9H_{17}OS_2$
17e 349
$C_9H_{17}O_2$
9b 124, *17b* 244, 467
$C_9H_{17}O_2Si$
17b 517
$C_9H_{17}O_3$
9b 157, 184, *17b* 286
$C_9H_{17}S$
9b 291, *17b* 198
$C_9H_{18}AlCl_3NO$
9a 294, 295
$C_9H_{18}BBr_3NO$
9a 294
$C_9H_{18}BCl_3NO$
9a 294
$C_9H_{18}BF_3NO$
9a 294
$C_9H_{18}BrN^+$
17h 131
$C_9H_{18}Cl$
9b 238
$C_9H_{18}ClN^+$
17h 132
$C_9H_{18}ClN_2O_3$
17d1 30
$C_9H_{18}Cl_3O_3S_2$
9c2 296
$C_9H_{18}Cl_3SeSi$
17b 317
$C_9H_{18}Cl_3Si_2$
9b 308
$C_9H_{18}Cl_4GeNO$
17a 432
$C_9H_{18}Cl_4NOSn$
17a 454
$C_9H_{18}DNO^+$
9d2 117
$C_9H_{18}FN^+$
17h 132
$C_9H_{18}F_3N_2S$
9c1 108
$C_9H_{18}F_3N_2S_2$
9c1 108, *17c* 517
$C_9H_{18}F_3OSSi$
17b 321
$C_9H_{18}F_3OSi$
9b 213
$C_9H_{18}F_3OSi_2$
9b 308
$C_9H_{18}F_3O_2S$
17e 353
$C_9H_{18}F_3O_2Si$
9b 250
$C_9H_{18}F_3S_2$
17e 346
$C_9H_{18}F_3Si_2$
9b 308
$C_9H_{18}KO$
1 44
$C_9H_{18}LiO$
1 44
$C_9H_{18}N$
9b 310, *9c1* 24, 120, *17b* 506, 507, *17c* 410, 568
$C_9H_{18}NO$
1 57, *9c1* 107, 125, 251, 262, 341, 446ff, 462ff, *17b* 102, 360, 508, *17c* 404, 405, *17d2* 133, 172, 190, 286ff

$C_9H_{18}NOS$
17c 516, *17d2* 166, 178
$C_9H_{18}NO_2$
9c1 226, 237, 297, 390, 402, 471ff, 517, 784, *17d1* 180, *17d2* 73, 133, 139ff, 173, 290ff, 296, *17e* 16
$C_9H_{18}NO_2S$
9c1 777
$C_9H_{18}NO_3$
9c1 226, 227, 262, 349, 358, 369, 784, 920, 925, 937, *17d1* 118, 132, *17d2* 72, 73, 161
$C_9H_{18}NO_4$
17d2 217
$C_9H_{18}NO_4S$
9c1 237
$C_9H_{18}NO_5$
17d1 104, *17d2* 134, 140
$C_9H_{18}NS$
9c1 1044, *9c2* 284, *17b* 511, *17e* 325
$C_9H_{18}N_2^+$
9d2 101, 116
$C_9H_{18}N_2O^+$
17c 634, *17h* 154
$C_9H_{18}N_2OS^-$
9d1 77
$C_9H_{18}N_2OX$
17d2 257
$C_9H_{18}N_3O$
9c1 841, *17d2* 250
$C_9H_{18}N_3O_2$
9c1 867
$C_9H_{18}N_3O_3$
9c1 245, *17d2* 194
$C_9H_{18}N_3O_4$
9c1 303, 304, *17d1* 16, 17, 22, 30, 48, 64, 67, 68, *17d2* 195
$C_9H_{18}N_3O_5$
17d1 23, 64, 68
$C_9H_{18}NaO$
1 44
$C_9H_{18}O^-$
9b 241, *17b* 310
$C_9H_{18}O^+$
17b 332, *17h* 43
$C_9H_{18}O_4P$
17b 43, *17e* 285
$C_9H_{18}S^-$
17b 313
$C_9H_{18}Se^-$
17b 315
C_9H_{19}
1 26, *9b* 82, 159
$C_9H_{19}CoN_4O_5$
17a 334
$C_9H_{19}F_3O_3P$
9c2 244
$C_9H_{19}N^+$
9b 290, *9d2* 28
$C_9H_{19}NO^+$
9d2 117, *17b* 102, 105, 107, 402, *17h* 134
$C_9H_{19}NO_3P$
17b 510
$C_9H_{19}NO_4P$
9c1 1008, *17b* 520, *17e* 285, 286
$C_9H_{19}N_2$
9c1 17, 42, 92, 93
$C_9H_{19}N_2O$
9c1 304, 475, 520, 813, 890, 897, *17d1* 176, 183, *17d2* 187, 227, 250, 296ff
$C_9H_{19}N_2O_2$
9c1 416, 821, *17d1* 118, 129, 146, 157, 176, 180, *17d2* 195
$C_9H_{19}N_2O_3$
9c1 784, *17d1* 17, 30, 48, 73
$C_9H_{19}N_2O_3S$
17d1 64, 68
$C_9H_{19}N_2O_4$
17d1 31
$C_9H_{19}N_4Si$
17c 476
$C_9H_{19}O$
9b 138, *17b* 310
$C_9H_{19}OS$
17b 238
$C_9H_{19}OS_2$
9b 158, *17e* 347
$C_9H_{19}OSi$
17b 29, 50, 122, 337
$C_9H_{19}O_2$
17b 237
$C_9H_{19}O_2S$
17b 18
$C_9H_{19}O_2Si$
17b 181, *17c* 24
$C_9H_{19}O_3S$
17e 356
$C_9H_{19}O_4$
9b 41, 108, 230
$C_9H_{19}O_4P^-$
17g 54
$C_9H_{19}O_4S^-$
17b 320
$C_9H_{19}S_2$
17b 238
$C_9H_{19}Si$
9b 307, *17b* 245, 338, 474
$C_9H_{20}^+$
17h 11
$C_9H_{20}BO$
9b 194
$C_9H_{20}N$
17b 183, *17c* 394
$C_9H_{20}NO$
9c1 211, 227, 251, 341, *17c* 394, 482, *17d1* 118, 133, *17d2* 74
$C_9H_{20}NOSi$
17b 509
$C_9H_{20}NO_2$
9c1 227, 242, 243, 318, *17d1* 150, 187, *17d2* 74, 409
$C_9H_{20}NO_2Sn$
9c1 753
$C_9H_{20}NO_3$
9c1 920, 925
$C_9H_{20}NO_6$
17d1 104
$C_9H_{20}NO_7P_2$
17c 621
$C_9H_{20}N_2^+$
9d2 101, *17b* 101
$C_9H_{20}N_2OX$
17d2 297
$C_9H_{20}N_3O_2$
9c1 245, *17d1* 119, 146
$C_9H_{20}N_4O_3X$
17d1 14
$C_9H_{20}O_2P$
9c2 244

$C_9H_{20}O_3P$
17b 98, 196
$C_9H_{20}O_4P$
9b 39, 86, 195, *9c2* 244, *17b* 97
$C_9H_{20}O_4PS$
17b 239
$C_9H_{20}O_7P$
17b 190
$C_9H_{21}ClNO_2Sn$
17d2 409
$C_9H_{21}Cl_3N_3O_6S_2$
9c2 296ff
$C_9H_{21}Ge$
9c2 310
$C_9H_{21}GeO$
9b 195
$C_9H_{21}N^+$
17h 128
$C_9H_{21}NOPS$
9c1 247
$C_9H_{21}NOPS_2$
9c1 246
$C_9H_{21}NO_3P$
17b 185, *17e* 286
$C_9H_{21}NO_3PS_2$
9c1 246
$C_9H_{21}NO_3S$
9c1 247
$C_9H_{21}NO_4P$
9c1 247, *9c2* 244
$C_9H_{21}N_2^{2+}$
17b 103, 107, 108, 402
$C_9H_{21}N_2O$
9c1 240, *17d2* 75
$C_9H_{21}N_2O_2$
17d2 75, 76
$C_9H_{21}N_2Si$
9c1 95
$C_9H_{21}N_3O_2X$
17d1 48
$C_9H_{21}N_3O_3X$
17d1 14, 49
$C_9H_{21}OSSi$
17b 274, 321
$C_9H_{21}OSi$
9b 195, *17b* 27, 269
$C_9H_{21}O_2S$
17e 353
$C_9H_{21}O_3S$
17e 356
$C_9H_{21}PS_3^+$
9c2 264
$C_9H_{21}SSi$
17b 314
$C_9H_{21}Si$
9b 87, *17b* 165, *17f* 6, 7
$C_9H_{21}Sn$
9c2 314, *17f* 47
$C_9H_{22}BN_2O$
17d1 198
$C_9H_{22}NSi$
17c 560
$C_9H_{22}N_2OX$
17d1 49
$C_9H_{22}N_2O_2P$
9c2 245, *17c* 391
$C_9H_{22}N_2O_3P$
17e 286
$C_9H_{22}N_2O_3PSi_2$
17b 520
$C_9H_{22}N_3Si$
9c1 121, *17c* 582
$C_9H_{22}O_2P$
9c2 245
$C_9H_{22}O_4PSi$
17b 44, *17e* 286
$C_9H_{23}Ge_2$
17b 155
$C_9H_{23}OSSi$
17e 351
$C_9H_{23}OSiSn$
17b 22
$C_9H_{23}O_5P_2$
9c2 245
$C_9H_{23}Si_2$
17b 158ff
$C_9H_{23}Sn_2$
17b 160
$C_9H_{24}GeNOSi$
9c1 1006
$C_9H_{24}N_3OX_2$
17d1 49, 133, 146
$C_9H_{24}N_3Si_2$
17c 584
$C_9H_{24}N_5OX_2$
17d1 50, 133, 146
$C_9H_{24}O_4PSi$
17e 286
$C_9H_{25}GeSi_2$
17f 39
$C_9H_{25}NO_3PSi$
17e 287
$C_9H_{25}NPSi_2$
17e 255
$C_9H_{27}NO_3PSi_2$
17e 287
$C_9H_{27}N_2Si_3$
9c1 65, *17c* 451
$C_9H_{27}N_3O_3PSi_2$
17e 287
$C_9H_{27}Si_4$
9c2 309
$C_9H_{28}Si_5^-$
17f 26
$C_9Mn_2O_9^-$
17a 122
$C_{10}Br_4O_6Re$
17a 314, *17g* 203
$C_{10}Cl_4MnO_6$
17a 282ff, *17g* 207
$C_{10}Cl_4O_4^{3-}$
17g 282
$C_{10}Cl_4O_6Re$
17a 314, *17g* 209
$C_{10}Cl_8N_3$
17c 398
$C_{10}Co_3FO_9^-$
17a 127
$C_{10}D_8^-$
9d1 737
$C_{10}D_{24}N_4^+$
9d2 62
$C_{10}F_8^+$
9d2 20
$C_{10}F_{12}S_4^+$
17h 112
$C_{10}F_{21}$
17b 305
$C_{10}F_{21}O$
9b 242
$C_{10}HCo_3O_9^-$
17a 127
$C_{10}HD_8$
9b 618
$C_{10}HN_4O_2$
17e 189
$C_{10}HO_8^{4-}$
9b 314

$C_{10}H_2Cl_2O_6^{3-}$
9d1 301
$C_{10}H_2Cl_4O_4^-$
9d1 301, *17g* 282
$C_{10}H_2D_6^-$
9d1 737
$C_{10}H_2F_6^+$
9d2 20
$C_{10}H_2N_2O_6^{5-}$
9d1 337
$C_{10}H_2N_4^-$
1 106
$C_{10}H_2O_6^{2-}$
1 79
$C_{10}H_2O_6^-$
9d1 91
$C_{10}H_2O_8^{5-}$
9d1 90
$C_{10}H_2O_9^{5-}$
17c 95
$C_{10}H_3BrD_{15}N_2O_2$
17d2 228
$C_{10}H_3Cl_3O_4^-$
17g 281
$C_{10}H_3D_3O_4^-$
9d1 289
$C_{10}H_3D_5^-$
9d1 737
$C_{10}H_3D_{18}$
17b 296
$C_{10}H_3F_5^+$
9d2 20
$C_{10}H_3NO_5^-$
9d1 101
$C_{10}H_3O_5^{4-}$
17g 282
$C_{10}H_3O_7^{4-}$
9d1 251, 261
$C_{10}H_3O_8^{4-}$
9b 460, 461, *9d1* 90, *17c* 89
$C_{10}H_3O_9^{4-}$
9b 471ff, 472, 490, *17c* 95
$C_{10}H_4AsCl_2O_2S_2$
17g 367
$C_{10}H_4BiCl_2O_2S_2$
17a 501, *17g* 368
$C_{10}H_4ClO_3^{2-}$
9d1 295
$C_{10}H_4Cl_2O_2^-$
9d1 288
$C_{10}H_4Cl_2O_2S_2Sb$
17a 498, *17g* 375
$C_{10}H_4Cl_2O_4^-$
9d1 300, *17g* 280ff
$C_{10}H_4Cl_3O$
9c2 195
$C_{10}H_4Cl_3O_2S_2Sn$
17g 377
$C_{10}H_4CsO_2S_2$
9d1 349, 350, 351
$C_{10}H_4D_4^-$
9d1 736
$C_{10}H_4D_2O_4^-$
9d1 299
$C_{10}H_4D_4O^-$
17f 188
$C_{10}H_4F_2O_2^-$
9d1 288
$C_{10}H_4F_2O_2S_2Sb$
17a 498, *17g* 375
$C_{10}H_4F_4^+$
9d2 20
$C_{10}H_4KO_2S_2$
9d1 348, 349, 350, 351
$C_{10}H_4LiO_2S_2$
9d1 348, 349, 351
$C_{10}H_4MnO_6$
17a 283, *17g* 73
$C_{10}H_4NO_2S_2$
17c 549
$C_{10}H_4NO_5^{2-}$
9d1 296
$C_{10}H_4NO_8S_2^{2-}$
17c 630
$C_{10}H_4N_2O_2^-$
9d1 32
$C_{10}H_4N_2O_4^-$
9d1 101, *17g* 66
$C_{10}H_4N_2S_2^-$
17f 214
$C_{10}H_4N_3O_2^{2-}$
17g 358
$C_{10}H_4N_4O_8^-$
9d1 668
$C_{10}H_4NaO_2S_2$
9d1 348, 349, 350, 351
$C_{10}H_4O_2S_2^-$
9d1 349
$C_{10}H_4O_2S_3^-$
9d1 85
$C_{10}H_4O_4^{3-}$
9d1 303, 304
$C_{10}H_4O_4^-$
9d1 332
$C_{10}H_4O_8^{3-}$
9d1 90
$C_{10}H_4S_6^+$
17h 117
$C_{10}H_5Br_2D_4N_3O_7SM$
17d1 258
$C_{10}H_5ClNS_2$
17c 549
$C_{10}H_5ClO_3^-$
1 83
$C_{10}H_5ClO_4^-$
9d1 299, *17g* 280
$C_{10}H_5ClO_5^-$
9d1 300
$C_{10}H_5D_3^-$
9d1 736
$C_{10}H_5D_8NNaO_5S$
17d1 348
$C_{10}H_5F_4$
9b 619
$C_{10}H_5F_5N_2O_3^-$
9d1 387
$C_{10}H_5NO_5S^-$
17c 630
$C_{10}H_5O_2S_2$
17e 231ff, 233
$C_{10}H_5O_3^{2-}$
9d1 290, 291, 296, 297, 298
$C_{10}H_5O_3$
9c2 194
$C_{10}H_5O_4^{2-}$
9b 306, *9d1* 296
$C_{10}H_5O_7S_2^{2-}$
9c2 195
$C_{10}H_5O_8$
9c2 185
$C_{10}H_5O_8S_2^{2-}$
9c2 197
$C_{10}H_6AlBr_2O_2$
17g 257
$C_{10}H_6AlCl_2O_2$
17g 257
$C_{10}H_6AlI_2O_2$
17g 257
$C_{10}H_6Br_2D_4N_2O_7SM$
17d1 257

$C_{10}H_6Br_2GaO_2$
17g 258
$C_{10}H_6Br_2InO_2$
17g 260
$C_{10}H_6Br_2LaO_2$
17g 260
$C_{10}H_6Br_2O_2Y$
17g 264
$C_{10}H_6ClO$
17e 200
$C_{10}H_6ClInO_2$
17g 260
$C_{10}H_6Cl_2O_2Sc$
17g 263
$C_{10}H_6Cl_2O_2Y$
17g 264
$C_{10}H_6Cl_2O_3^+$ (?)
9d2 119
$C_{10}H_6CsO_2$
9d1 276, *17g* 266
$C_{10}H_6D_2^-$
9d1 736
$C_{10}H_6D_5N_4O$
17c 469
$C_{10}H_6D_5N_4S$
17c 472
$C_{10}H_6FO_4$
17b 498
$C_{10}H_6F_2^+$
9d2 20
$C_{10}H_6F_3$
9b 619, 622
$C_{10}H_6F_5N_2O$
9c1 709
$C_{10}H_6F_{15}N_2O_3$
17d2 221
$C_{10}H_6GaI_2O_2$
17g 259
$C_{10}H_6KO_2$
9d1 276, *17g* 260, 266
$C_{10}H_6LiO_2$
9d1 275, *17e* 195, *17g* 261
$C_{10}H_6NOS$
9c1 596
$C_{10}H_6NO_2$
9c1 165, 166, *17c* 629, 630
$C_{10}H_6NO_5S$
9c1 165, 166
$C_{10}H_6NO_8S_2$
9c1 165, 166
$C_{10}H_6NS_2$
17c 549
$C_{10}H_6NS_3$
17c 542
$C_{10}H_6N_2^-$
17c 169, *17f* 196, 197
$C_{10}H_6N_2O^-$
17f 196, 197
$C_{10}H_6N_2O_2^-$
9d1 609
$C_{10}H_6N_2O_4^-$
9d1 336, 653, 654, 655, *17f* 301
$C_{10}H_6N_2S^-$
9d1 900, 901
$C_{10}H_6N_2S^+$
17h 137
$C_{10}H_6N_4^-$
1 116
$C_{10}H_6NaO_2$
9d1 276, *17g* 261, 266
$C_{10}H_6O_2^-$
1 82, *9d1* 33, 275, 280ff, 290, 331, *17g* 257, 264ff
$C_{10}H_6O_2Rb$
9d1 276
$C_{10}H_6O_2S_2^-$
9d1 18ff, *17g* 12ff
$C_{10}H_6O_2Tl$
9d1 276
$C_{10}H_6O_3^-$
1 83, *9d1* 280, 285, 286, *17g* 277ff
$C_{10}H_6O_4^-$
1 83, *9d1* 19, 159, 288, 298, *17g* 13ff, 279, 280
$C_{10}H_6O_4^+$
9d1 358
$C_{10}H_6O_4S^-$
9c2 189, 191, 192, 193
$C_{10}H_6O_4X$
17c 38
$C_{10}H_6O_5^{3-}$
9d1 243, 244, 251
$C_{10}H_6O_5^-$
9d1 251, 299
$C_{10}H_6O_5P^{2-}$
9c2 189
$C_{10}H_6O_5S^-$
9c2 194
$C_{10}H_6O_6^-$
9d1 300, 301
$C_{10}H_6O_8^-$
9d1 302
$C_{10}H_6SSe^+$
17h 113
$C_{10}H_6STe^+$
17h 113
$C_{10}H_6S_2^-$
9d1 892
$C_{10}H_6S_2^+$
9d2 146, *17h* 113
$C_{10}H_6SeTe^+$
17h 119
$C_{10}H_6Se_2^+$
17h 119
$C_{10}H_7$
9b 317
$C_{10}H_7AlBr_3S$
17a 398, *17e* 322
$C_{10}H_7Al_2Br_6S$
17a 398
$C_{10}H_7BrNO_4$
9c1 588
$C_{10}H_7Br_2D_3N_2O_6SM$
17d1 257
$C_{10}H_7ClNO_2$
9b 590
$C_{10}H_7Cl_3GaS$
17a 407, *17e* 322
$C_{10}H_7Cl_3NO_2$
17d1 238
$C_{10}H_7Cl_6Ga_2S$
17a 407
$C_{10}H_7D^-$
9d1 735
$C_{10}H_7D_4NO_4^-$
17c 148
$C_{10}H_7FNO_2$
9b 590
$C_{10}H_7F_2$
9b 619, 622
$C_{10}H_7F_5NO_2$
9c1 651
$C_{10}H_7F_5NO_3$
9c1 969
$C_{10}H_7F_6O$
17c 191
$C_{10}H_7FeN_7^{2-}$
17a 329, *17f* 203, *17h* 180

$C_{10}H_7N^-$
1 117
$C_{10}H_7NO_2^-$
9d1 284, 571, 572, *17g* 269
$C_{10}H_7NO_2S^-$
9d1 582
$C_{10}H_7NO_3^-$
9d1 81, 571
$C_{10}H_7NO_3^+$
17e 213
$C_{10}H_7N_2$
9b 592, 623, *17c* 212, 216
$C_{10}H_7N_2O$
17b 479
$C_{10}H_7N_2O_2$
9c1 73, *17c* 464, *17e* 212
$C_{10}H_7N_2O_4$
9c1 139
$C_{10}H_7O$
9c2 187, *17e* 194, 199
$C_{10}H_7O_2$
9c2 188, 189, 190, 191, 192, 193, *17e* 194
$C_{10}H_7O_3^{2-}$
9b 389
$C_{10}H_7O_3$
9c2 206
$C_{10}H_7O_4^{2-}$
9b 152, 556, *9d1* 177, *17b* 206
$C_{10}H_7O_4$
9c2 207, *17b* 498, *17c* 243
$C_{10}H_7O_4S$
17e 194, 196
$C_{10}H_7O_4U^{2+}$
17a 241, *17g* 263
$C_{10}H_7O_6$
9c2 185
$C_{10}H_7O_7S_2$
17e 198
$C_{10}H_7O_9$
9b 472
$C_{10}H_7S$
1 77
$C_{10}H_8^-$
1 79, 111, *9d1* 731ff, 769ff, *17f* 71ff
$C_{10}H_8^+$
9d2 18
$C_{10}H_8Al_2Cl_6N_2^-$
17f 205
$C_{10}H_8Al_2D_6N_2^-$
17a 399
$C_{10}H_8Br_3N_2O$
9c1 707
$C_{10}H_8Cl_2NS$
17c 214
$C_{10}H_8Cl_2N_2O_2^-$
17g 245
$C_{10}H_8Cl_2Ti \cdot N_2$
9a 302
$C_{10}H_8Cl_3N_2O$
9c1 707, 708
$C_{10}H_8Cl_4NO_2$
9c1 649
$C_{10}H_8Cl_4NO_3$
9c1 618, 650, 967
$C_{10}H_8CoN_7O_3$
9c1 1017
$C_{10}H_8D_5N_4$
17c 464
$C_{10}H_8F$
9b 619
$C_{10}H_8FO_4$
9b 152
$C_{10}H_8F_2N_2O_2^-$
17g 245
$C_{10}H_8F_4NO_3$
9c1 968
$C_{10}H_8FeN_7^{2-}$
17h 180
$C_{10}H_8MnN_3OS_4$
17a 283
$C_{10}H_8N$
17b 505
$C_{10}H_8NO$
9c1 528, 551, *17d2* 367
$C_{10}H_8NOS_2$
17c 529
$C_{10}H_8NO_2$
9b 589, *9c1* 751, *17d2* 368
$C_{10}H_8NO_2S$
17c 633
$C_{10}H_8NO_3$
9c1 143, *17c* 599, 632
$C_{10}H_8NO_4$
9c1 588
$C_{10}H_8NO_5$
9c1 730
$C_{10}H_8NO_6^{2-}$
9c1 983, 984
$C_{10}H_8NO_8S_2$
17d2 368
$C_{10}H_8N_2^-$
1 114, 115, *9d1* 871ff, *17f* 205
$C_{10}H_8N_2O_2^-$
9d1 47, 342, *17g* 14
$C_{10}H_8N_2O_6^{3-}$
9d1 203
$C_{10}H_8N_3$
17b 307
$C_{10}H_8N_4O_2^+$
17h 207
$C_{10}H_8N_4O_5^-$
17f 283
$C_{10}H_8O^-$
9b 617, *9d1* 884, *17f* 187
$C_{10}H_8O_2^-$
9d1 42, 57, 236, *17g* 38
$C_{10}H_8O_2^+$
1 123, *9d1* 357, *9d2* 127
$C_{10}H_8O_3^-$
9d1 71
$C_{10}H_8O_4^-$
9d1 314
$C_{10}H_8O_4^+$
1 124, *9d1* 358, *9d2* 128, *17h* 82
$C_{10}H_8O_4S_6^+$
17h 112
$C_{10}H_8S^-$
9d1 891
$C_{10}H_8S_2^-$
9d1 893
$C_{10}H_8S_2^{2+}$
9d2 146
$C_{10}H_8S_4^+$
17h 116
$C_{10}H_8S_8^+$
17h 117
$C_{10}H_9$
9b 451, 540, 617ff, 621, *17c* 250
$C_{10}H_9BrCl_2NO$
17c 602
$C_{10}H_9BrNO_2$
9c1 801

$\mathrm{C_{10}H_9Br_2N_2O}$
9c1 705, 706

$\mathrm{C_{10}H_9ClNS}$
17c 214

$\mathrm{C_{10}H_9Cl_2N_2O}$
9c1 706, 707

$\mathrm{C_{10}H_9Cl_3NO}$
17c 603, *17d1* 243

$\mathrm{C_{10}H_9Cl_3NO_2}$
17d1 243, 249

$\mathrm{C_{10}H_9Cl_3NO_3}$
9c1 966

$\mathrm{C_{10}H_9CoN_6O}$
9c1 1013

$\mathrm{C_{10}H_9CoN_6O_2}$
9c1 1017

$\mathrm{C_{10}H_9CoN_6O_3}$
9c1 1017

$\mathrm{C_{10}H_9D_5NS}$
17c 501, 504

$\mathrm{C_{10}H_9F_4N_2O_4^{2-}}$
9d1 388

$\mathrm{C_{10}H_9F_5NO}$
9c1 692ff

$\mathrm{C_{10}H_9F_5NO_2}$
9c1 969

$\mathrm{C_{10}H_9F_6}$
17c 101

$\mathrm{C_{10}H_9F_6O}$
17c 103

$\mathrm{C_{10}H_9FeNO_2^-}$
9d1 570

$\mathrm{C_{10}H_9MgN_2}$
17a 226

$\mathrm{C_{10}H_9MgO_2}$
9b 421

$\mathrm{C_{10}H_9MgO_2^+}$
17a 226

$\mathrm{C_{10}H_9N^-}$
9d1 863, 866, 867, *17f* 188

$\mathrm{C_{10}H_9NO^-}$
9d1 864, 865

$\mathrm{C_{10}H_9NO_3^-}$
9d1 82

$\mathrm{C_{10}H_9NO_5^{3-}}$
17g 253

$\mathrm{C_{10}H_9N_2O}$
17d2 346

$\mathrm{C_{10}H_9N_2O_2S}$
17c 215

$\mathrm{C_{10}H_9N_2O_3}$
9c1 144, *17e* 213

$\mathrm{C_{10}H_9N_2O_4}$
9c1 141, 801

$\mathrm{C_{10}H_9N_2O_6^{2-}}$
9d1 427

$\mathrm{C_{10}H_9NiO_2}$
17c 46

$\mathrm{C_{10}H_9O}$
17b 521, *17c* 39

$\mathrm{C_{10}H_9O_4^-}$
9d1 93

$\mathrm{C_{10}H_9O_4}$
9b 152, *9c2* 59

$\mathrm{C_{10}H_9O_5^{2-}}$
9d1 272

$\mathrm{C_{10}H_9O_7}$
9b 476

$\mathrm{C_{10}H_9O_7S_2}$
9b 619, 620, 622

$\mathrm{C_{10}H_9O_8}$
17c 60

$\mathrm{C_{10}H_9O_9}$
17c 61

$\mathrm{C_{10}H_9S}$
17c 43

$\mathrm{C_{10}H_{10}^-}$
9d1 727, 763

$\mathrm{C_{10}H_{10}AlClD_3Ti}$
17a 244

$\mathrm{C_{10}H_{10}AlCl_2D_2Ti}$
17a 245

$\mathrm{C_{10}H_{10}AlCl_4N_2}$
9a 296

$\mathrm{C_{10}H_{10}AlCl_4O_2}$
17a 399, *17e* 192, *17g* 205

$\mathrm{C_{10}H_{10}AlCl_4Ti}$
9a 302

$\mathrm{C_{10}H_{10}AlD_4Ti}$
17a 244

$\mathrm{C_{10}H_{10}AlN_2}$
17a 399, *17f* 203

$\mathrm{C_{10}H_{10}BCl_4Ti}$
9a 302

$\mathrm{C_{10}H_{10}BF_4Fe}$
9a 314

$\mathrm{C_{10}H_{10}BrCoN_6O}$
9c1 1017

$\mathrm{C_{10}H_{10}BrN_2O}$
9c1 701, 702

$\mathrm{C_{10}H_{10}Br_2NNaO_5S}$
17d1 256

$\mathrm{C_{10}H_{10}ClCoN_6O}$
9c1 1017

$\mathrm{C_{10}H_{10}ClN_2O}$
9c1 702

$\mathrm{C_{10}H_{10}Cl_2NO_3}$
9c1 796, 961, 962

$\mathrm{C_{10}H_{10}Cl_3N_2O}$
17d1 238

$\mathrm{C_{10}H_{10}Cl_4GaTi}$
9a 302

$\mathrm{C_{10}H_{10}Cl_4InTi}$
9a 302

$\mathrm{C_{10}H_{10}Cl_4O_2Sb}$
17g 202

$\mathrm{C_{10}H_{10}Cl_4O_2Tl}$
9d1 156

$\mathrm{C_{10}H_{10}Cl_4Ti_2}$
9a 302

$\mathrm{C_{10}H_{10}CoN_7O_2}$
9c1 1018

$\mathrm{C_{10}H_{10}CoN_7O_3}$
9c1 1018

$\mathrm{C_{10}H_{10}D_2N_2O_6^{2-}}$
17b 414

$\mathrm{C_{10}H_{10}D_4N_2O^+}$
17b 78

$\mathrm{C_{10}H_{10}FN_2O}$
9c1 703

$\mathrm{C_{10}H_{10}FO_2}$
17b 370

$\mathrm{C_{10}H_{10}F_3}$
17c 186

$\mathrm{C_{10}H_{10}F_4NO}$
9c1 692

$\mathrm{C_{10}H_{10}F_6OP}$
17c 103

$\mathrm{C_{10}H_{10}FeI}$
9a 314

$\mathrm{C_{10}H_{10}IN_2O}$
9c1 703

$\mathrm{C_{10}H_{10}MgNO_4}$
17g 194

$\mathrm{C_{10}H_{10}MgO_2^+}$
17g 334, 335

$\mathrm{C_{10}H_{10}MnO_6S_3}$
17a 283

$C_{10}H_{10}N$
9b 622, 623, 624, *17c* 188

$C_{10}H_{10}NO$
9c1 164, 528, *17c* 202, 212, 214

$C_{10}H_{10}NOS$
17c 633

$C_{10}H_{10}NO_2$
9c1 139, 777, *17c* 631

$C_{10}H_{10}NO_3$
9b 209, *9c1* 804, 985

$C_{10}H_{10}NO_4^{2-}$
9d1 384, 385, *17g* 95, 96

$C_{10}H_{10}NO_4^-$
9c1 984

$C_{10}H_{10}NO_5S_2$
17c 555

$C_{10}H_{10}NS$
17c 214

$C_{10}H_{10}N_2^-$
9d1 874, 875

$C_{10}H_{10}N_2^+$
1 125, *9d2* 90

$C_{10}H_{10}N_2O_2^-$
17g 224

$C_{10}H_{10}N_2O_3^-$
17f 274

$C_{10}H_{10}N_3$
9b 525

$C_{10}H_{10}N_3O_6$
9c1 648

$C_{10}H_{10}N_3O_7$
9c1 648, 962

$C_{10}H_{10}N_3O_7S$
17d1 383

$C_{10}H_{10}N_3O_8$
9c1 963

$C_{10}H_{10}N_4^-$
9d1 881

$C_{10}H_{10}N_5O_3$
9c2 7, *17c* 469

$C_{10}H_{10}N_5O_5S$
9c1 351

$C_{10}H_{10}O^-$
9b 570

$C_{10}H_{10}O_2^-$
1 106, *9b* 116, 614, *9d1* 23, 33, 85

$C_{10}H_{10}O_2^+$
17h 82

$C_{10}H_{10}O_2S^-$
9d1 23

$C_{10}H_{10}O_2S_2^-$
9d1 63, 94, 104, *17g* 46

$C_{10}H_{10}O_2Se^-$
9d1 23

$C_{10}H_{10}O_2Tl$
17g 75

$C_{10}H_{10}O_3^-$
9b 615, *9d1* 23, 51, 71, 323, *17g* 34

$C_{10}H_{10}O_3S^-$
9d1 24, 105

$C_{10}H_{10}O_4^-$
9d1 94, 95, 140

$C_{10}H_{10}O_4^+$
17h 82

$C_{10}H_{10}O_6ReS_3$
17a 314

$C_{10}H_{10}S_2^+$
17e 371

$C_{10}H_{10}S_4^-$
9d1 105

$C_{10}H_{10}Si_2^-$
17f 23

$C_{10}H_{11}^{2-}$
9d1 723

$C_{10}H_{11}$
9b 79, 429, 451, 539, *17c* 240

$C_{10}H_{11}Br_2INO$
9c1 689

$C_{10}H_{11}Br_2NNaO_4S$
17d1 259

$C_{10}H_{11}Br_2NNaO_5S$
17d1 258

$C_{10}H_{11}Br_2N_2O_3$
9c1 689

$C_{10}H_{11}Br_3NO$
9c1 689, 690

$C_{10}H_{11}CdNO_4^+$
17a 389, *17g* 96

$C_{10}H_{11}ClNO$
17c 602

$C_{10}H_{11}ClNO_3$
9c1 794, 951

$C_{10}H_{11}ClNS$
9c1 107

$C_{10}H_{11}Cl_3NO$
9c1 690

$C_{10}H_{11}DN_2Na_2O_9P^{2-}$
17b 414

$C_{10}H_{11}D_3N$
17c 397

$C_{10}H_{11}D_4N_2O_6$
17b 413

$C_{10}H_{11}FNO_3$
9c1 952

$C_{10}H_{11}F_3NO_4$
9c1 725

$C_{10}H_{11}KN_3O_8S$
9c1 105

$C_{10}H_{11}N^+$
17h 62, 63

$C_{10}H_{11}NOS^-$
9d1 73

$C_{10}H_{11}NO_2^-$
9d1 464, 467

$C_{10}H_{11}NO_3^-$
9d1 682, 683, *17g* 89

$C_{10}H_{11}NO_4^-$
1 91, *9d1* 541, *17c* 147, *17g* 95, 96

$C_{10}H_{11}NO_4Zn^+$
17a 381, *17g* 96, 196

$C_{10}H_{11}NO_5$
9c1 794

$C_{10}H_{11}N_2O$
9c1 701, *17e* 215

$C_{10}H_{11}N_2O_2$
9c1 703, 812, 854, 861, 893, *17c* 589, *17d1* 231

$C_{10}H_{11}N_2O_5^{2-}$
9d1 426

$C_{10}H_{11}N_2O_5$
9c1 953

$C_{10}H_{11}N_2O_6^{2-}$
9d1 426, 427

$C_{10}H_{11}N_2O_8^{2-}$
9d1 429

$C_{10}H_{11}N_4O$
17c 468

$C_{10}H_{11}N_4O_4S$
17e 324

$C_{10}H_{11}N_4O_4S_2$
17e 359

$C_{10}H_{11}N_4O_6$
17b 216, *17c* 429

$C_{10}H_{11}N_4O_7$
9c1 690

$C_{10}H_{11}N_4O_7S$
17d1 383
$C_{10}H_{11}N_4S$
17c 472
$C_{10}H_{11}O$
9b 78, *17b* 254, *17c* 179, *17e* 200
$C_{10}H_{11}O_2^{2-}$
17c 192, 194, 197
$C_{10}H_{11}O_2$
9b 85, 559, *9c2* 14, 185, *17b* 369, *17c* 179, 180, 203, *17e* 12, 226
$C_{10}H_{11}O_2S$
17b 234
$C_{10}H_{11}O_3^{2-}$
9d1 250, 261, 272
$C_{10}H_{11}O_3$
9b 41, 109
$C_{10}H_{11}O_4$
9c2 170, *17e* 50
$C_{10}H_{11}O_7S$
9b 300
$C_{10}H_{11}S_2$
17c 208
$C_{10}H_{12}^-$
9d1 714, 719, 725, 726, 727, *17f* 73
$C_{10}H_{12}AlCl_2Ti$
9a 301, *17a* 245
$C_{10}H_{12}BrClNO$
9c1 682
$C_{10}H_{12}Br_2NO$
9c1 682, 683
$C_{10}H_{12}Cl_2NO$
9c1 683, *17d1* 236
$C_{10}H_{12}Cl_2NOSn$
17c 620
$C_{10}H_{12}Cl_2NO_2$
17d1 237
$C_{10}H_{12}Cl_2NS$
17c 502
$C_{10}H_{12}CsO_2$
9d1 223
$C_{10}H_{12}D_2O_2^+$
9d2 131
$C_{10}H_{12}D_2O_4^+$
9d2 136
$C_{10}H_{12}F$
17c 181
$C_{10}H_{12}KO_2$
9d1 223, 227ff, 229
$C_{10}H_{12}K_2O_2^+$
9d1 229
$C_{10}H_{12}LiO_2$
9d1 125, 223, 224, 225
$C_{10}H_{12}MoN_2O_4^-$
9a 309
$C_{10}H_{12}N^{2+}$
17h 234
$C_{10}H_{12}NO$
9c1 133, 134, 135, 178, 730, *17c* 585, 593
$C_{10}H_{12}NO_2$
9c1 134, 260, 643, 741, 777, 985, *17d2* 206, 394, 395, *17e* 76
$C_{10}H_{12}NO_2S$
9c1 791
$C_{10}H_{12}NO_3$
9c1 791, 948
$C_{10}H_{12}NO_3S$
9c1 149, 994, *17d2* 416
$C_{10}H_{12}NO_4$
9b 524, 954, *17d1* 231
$C_{10}H_{12}NO_4Zn^{2+}$
17a 381, *17g* 95
$C_{10}H_{12}NS$
9c1 1043
$C_{10}H_{12}NS_2$
17c 555
$C_{10}H_{12}N_2^+$
9d2 73, 90, *17h* 181
$C_{10}H_{12}N_2O_2S^+$
17h 139
$C_{10}H_{12}N_2O_2S_2^+$
17h 139
$C_{10}H_{12}N_2O_3S^+$
17h 139
$C_{10}H_{12}N_2O_4^-$
1 100, 102, *9d1* 514, 542, 622ff, 640
$C_{10}H_{12}N_2O_4^+$
17h 66
$C_{10}H_{12}N_2O_8^{3-}$
9b 172
$C_{10}H_{12}N_3O_5$
9c1 683
$C_{10}H_{12}N_3O_9$
9c1 356
$C_{10}H_{12}N_4Na_2O_8P$
17c 424, 428
$C_{10}H_{12}N_4Na_2O_9P$
17c 428
$C_{10}H_{12}N_5O$
17c 469
$C_{10}H_{12}N_5O_3$
9b 156
$C_{10}H_{12}N_5O_6$
9c1 44
$C_{10}H_{12}NaO_2$
9d1 125, 223, 225ff, 227, *17g* 82, 238
$C_{10}H_{12}Na_2O_2^+$
9d1 229
$C_{10}H_{12}O^-$
17c 193, 196
$C_{10}H_{12}O^+$
17h 62
$C_{10}H_{12}OS^-$
9b 577
$C_{10}H_{12}O_2^-$
1 81, *9b* 575, *9d1* 60, 124, 125, 176ff, 196, 221ff, *17g* 24, 34, 237
$C_{10}H_{12}O_2$ on Al_2O_3 surface
9d1 365
$C_{10}H_{12}O_2^+$
17h 64, 81
$C_{10}H_{12}O_2Rb$
9d1 223
$C_{10}H_{12}O_2S^-$
9d1 200
$C_{10}H_{12}O_2S_2^-$
9d1 205
$C_{10}H_{12}O_3^-$
9d1 221
$C_{10}H_{12}O_3Tl$
17g 106
$C_{10}H_{12}O_4^-$
9d1 204, 218ff, 221, 230, *17g* 224, 231, 240
$C_{10}H_{12}O_4^+$
17h 83
$C_{10}H_{12}O_5^+$
17h 83, 84, 86
$C_{10}H_{12}O_5P^{2-}$
9c2 178
$C_{10}H_{12}S_2^-$
9b 578, 579, 580

$C_{10}H_{12}S_2^+$
9d2 144, *17e* 384
$C_{10}H_{12}S_3^-$
17f 216
$C_{10}H_{12}S_4^+$
17h 112
$C_{10}H_{13}$
9b 17, 43, 182, 546, 558, *17b* 20, 27, *17c* 173, 179, 185
$C_{10}H_{13}AlClTi$
17a 245
$C_{10}H_{13}BrNO$
9c1 654, 655
$C_{10}H_{13}BrNOS$
17d1 192
$C_{10}H_{13}BrNO_3S$
17d1 196
$C_{10}H_{13}BrNS$
9c1 108, *17c* 501, 505
$C_{10}H_{13}ClN$
9c1 11
$C_{10}H_{13}ClNO$
9c1 101, 102, 655, 656, *17d1* 321
$C_{10}H_{13}ClNOS$
17d1 192
$C_{10}H_{13}ClNOSn$
17c 620
$C_{10}H_{13}ClNO_2$
9c1 954
$C_{10}H_{13}ClNO_3S$
17d1 196
$C_{10}H_{13}ClNS$
9c1 108, *17c* 501, 502, 505
$C_{10}H_{13}ClN_5O_2^-$
9c2 7
$C_{10}H_{13}FNO$
9c1 656ff
$C_{10}H_{13}FNS$
9c1 108
$C_{10}H_{13}FO_2S^+$
17h 94, 95
$C_{10}H_{13}F_3NO_6$
9c1 382
$C_{10}H_{13}INO$
9c1 658ff
$C_{10}H_{13}NNaO_5S$
17d1 351
$C_{10}H_{13}NO^-$
9b 582, *17f* 306
$C_{10}H_{13}NO^+$
17h 63, 231
$C_{10}H_{13}NOM$
17d1 268
$C_{10}H_{13}NOS^-$
9b 585
$C_{10}H_{13}NOX$
17d1 268
$C_{10}H_{13}NO_2^-$
1 87, 90, 100, *9d1* 416, 454, 458, 461
$C_{10}H_{13}NO_2S^-$
9d1 506, 507
$C_{10}H_{13}NO_3^-$
1 95, *9d1* 385, 391, 394, 502, *17f* 249
$C_{10}H_{13}NO_3S^-$
9d1 547
$C_{10}H_{13}NO_4S^-$
9d1 549, 550, *17f* 236
$C_{10}H_{13}NS_2^-$
9b 579
$C_{10}H_{13}N_2O$
9c1 179, 529, 810
$C_{10}H_{13}N_2OS$
17c 579
$C_{10}H_{13}N_2OS_2Sn$
17a 454, *17c* 346
$C_{10}H_{13}N_2O_2$
9c1 778
$C_{10}H_{13}N_2O_2S$
9c1 108, *17c* 506
$C_{10}H_{13}N_2O_2Sn$
17d1 382
$C_{10}H_{13}N_2O_3$
9c1 101, 659, *17d1* 168
$C_{10}H_{13}N_2O_3S$
17d1 192
$C_{10}H_{13}N_2O_4$
9c1 954
$C_{10}H_{13}N_2O_4S$
9c1 112, 773
$C_{10}H_{13}N_2O_5$
9c1 955, *9c2* 7, *17e* 9
$C_{10}H_{13}N_2O_5S$
17c 496, *17d1* 196
$C_{10}H_{13}N_3O^+$
17h 144
$C_{10}H_{13}N_3O_4^-$
9d1 627
$C_{10}H_{13}N_4O$
17d1 183
$C_{10}H_{13}N_4O_5$
17c 424, 428
$C_{10}H_{13}N_4O_6$
17c 425
$C_{10}H_{13}N_5Na_2O_7P$
17c 424
$C_{10}H_{13}O$
1 59, 60, 70, *9b* 17, 555, 569, *9c2* 39, 68, 176, *17b* 62, 84, *17c* 186
$C_{10}H_{13}O_2$
9c2 68, 176ff, *17c* 202, 203, *17e* 48, 52, 183, 187
$C_{10}H_{13}O_2S$
17e 338, 341
$C_{10}H_{13}O_4$
9c2 167, *17b* 358
$C_{10}H_{13}O_4U^{2+}$
17a 242, *17g* 83
$C_{10}H_{13}O_6S$
17c 19
$C_{10}H_{13}O_7$
9b 377, *17c* 17
$C_{10}H_{13}O_7S$
9b 474
$C_{10}H_{14}^-$
1 85, 86, *9d1* 714, 721
$C_{10}H_{14}^+$
9d2 6, *17h* 30
$C_{10}H_{14}AlTi$
9a 301, *17a* 245
$C_{10}H_{14}Al_2N_2^-$
17a 399, *17f* 205
$C_{10}H_{14}AsNO_2^-$
9d1 518, 519, *17f* 58
$C_{10}H_{14}AsNO_2S^-$
9d1 519
$C_{10}H_{14}AsNO_3^-$
9d1 519, *17f* 59
$C_{10}H_{14}BD_3^-$
17c 98
$C_{10}H_{14}BrClNO_2$
17d2 237
$C_{10}H_{14}ClGeN_2$
17c 452

$C_{10}H_{14}ClN_2Si$
17c 453
$C_{10}H_{14}CoN_2O_4$
17e 18
$C_{10}H_{14}F_3N_2O_2$
17d2 250
$C_{10}H_{14}N$
9c1 10
$C_{10}H_{14}NO$
1 55, *9c1* 100, 178, 551, 613, 651ff, *9c2* 43, *17d1* 217, 262, 269, *17d2* 211, *17e* 52
$C_{10}H_{14}NOS$
9c1 987, *17d1* 192ff, 232
$C_{10}H_{14}NO_2$
9c1 169, 529, 643, 660, 938, 948, *17c* 146, *17d1* 262, *17d2* 209
$C_{10}H_{14}NO_3$
9c1 948, *17d1* 143
$C_{10}H_{14}NO_3S$
9c1 998, *17c* 495, *17d1* 196
$C_{10}H_{14}NO_4$
9c1 975 [4])
$C_{10}H_{14}NO_4PS^-$
9d1 518
$C_{10}H_{14}NO_5P^-$
9d1 515
$C_{10}H_{14}NO_5S$
9c1 926
$C_{10}H_{14}NO_6P^-$
9d1 517
$C_{10}H_{14}NS$
9c1 108, *17c* 501, 504
$C_{10}H_{14}N_2^+$
17c 237, 249, *17h* 145, 155
$C_{10}H_{14}N_2Na_2O_8P$
17b 412
$C_{10}H_{14}N_2O^+$
9d2 34, *17b* 184
$C_{10}H_{14}N_2O_2^-$
1 99, *9d1* 201, 509, 511
$C_{10}H_{14}N_2O_2^+$
17h 145
$C_{10}H_{14}N_2O_5^-$
9b 425
$C_{10}H_{14}N_2O_9S^-$
17b 413
$C_{10}H_{14}N_3O$
17d2 227
$C_{10}H_{14}N_3O_3$
17d2 141
$C_{10}H_{14}N_3O_3M$
17d1 31
$C_{10}H_{14}N_3O_5P^-$
9d1 552
$C_{10}H_{14}N_4O$
17d2 93
$C_{10}H_{14}N_4O_3^+$
9d2 73
$C_{10}H_{14}N_5O_2$
17d2 86, 89
$C_{10}H_{14}N_5O_3$
9c1 30, *17c* 423, 427
$C_{10}H_{14}N_5O_4$
17c 423, 427
$C_{10}H_{14}N_5O_5$
17c 424
$C_{10}H_{14}O^-$
17c 70
$C_{10}H_{14}OS^-$
9b 606
$C_{10}H_{14}OSi^-$
17c 217
$C_{10}H_{14}O_2^-$
9d1 15, 19, 37, 42, 43, *17g* 21ff
$C_{10}H_{14}O_2^+$
1 121, *9d1* 355ff, *9d2* 127, 130, 132, *17h* 80
$C_{10}H_{14}O_2Tl$
17g 75
$C_{10}H_{14}O_3^+$
17h 84
$C_{10}H_{14}O_3S^+$
17h 94, 95
$C_{10}H_{14}O_4^-$
9d1 54
$C_{10}H_{14}O_4^+$
9d2 135, 136, *17h* 84, 85
$C_{10}H_{14}O_4S^+$
17h 94
$C_{10}H_{14}O_5P$
17e 46
$C_{10}H_{14}S^-$
17c 51
$C_{10}H_{14}S_2^-$
17g 48
$C_{10}H_{14}S_2^+$
9d2 141
$C_{10}H_{14}S_4^+$
9d2 141, *17h* 96, 97
$C_{10}H_{14}Si^-$
17f 11
$C_{10}H_{15}$
9b 302, 303, 427, 438, 454, *17b* 490, *17c* 74
$C_{10}H_{15}BrNO_2$
17d2 237
$C_{10}H_{15}BrNO_3$
17d2 237
$C_{10}H_{15}Cl_2O_2Zr$
17e 32
$C_{10}H_{15}DNO$
9c1 169
$C_{10}H_{15}D_2N_3^+$
17h 145
$C_{10}H_{15}D_9NSi$
17c 561, 562
$C_{10}H_{15}F_6OSi$
9b 258
$C_{10}H_{15}F_8OSi$
9b 248
$C_{10}H_{15}F_9O_4P$
9c2 246
$C_{10}H_{15}Ge$
17f 39
$C_{10}H_{15}HgO_6$
17a 390
$C_{10}H_{15}MnO_9P$
17e 17
$C_{10}H_{15}NOP$
9c1 1008
$C_{10}H_{15}NO_2^+$
9d2 34
$C_{10}H_{15}NO_2R$
(R unknown)
17d1 133

[4]) In II/9c1 erroneously listed as $[C_{10}H_{13}NO_4]^{\cdot}$.

$C_{10}H_{15}NO_3P$
9c1 1008
$C_{10}H_{15}N_2O$
9c1 509, 660
$C_{10}H_{15}N_2OS$
17d1 193, *17d2* 200, 407
$C_{10}H_{15}N_2O_2$
9d1 514, *17d2* 81, 93
$C_{10}H_{15}N_2O_3$
17d2 87, 89
$C_{10}H_{15}N_2O_3S$
17d1 168, 197
$C_{10}H_{15}N_2O_4$
17d2 88
$C_{10}H_{15}N_2O_5$
9b 284, *17b* 411
$C_{10}H_{15}N_2O_6$
17b 413
$C_{10}H_{15}N_2O_6S$
17d2 88
$C_{10}H_{15}N_2Si$
17c 149
$C_{10}H_{15}O$
9b 297, *17b* 515
$C_{10}H_{15}OSi$
9b 556, *17c* 166
$C_{10}H_{15}OSn$
9b 557
$C_{10}H_{15}O_2$
9b 125, 434, *17e* 12
$C_{10}H_{15}O_4$
1 29, *9b* 160
$C_{10}H_{15}Si$
9b 556
$C_{10}H_{15}Sn$
17f 47
$C_{10}H_{16}^-$
1 79, *9d1* 846
$C_{10}H_{16}^+$
17h 30
$C_{10}H_{16}BNNaO$
17d1 321
$C_{10}H_{16}ClN_2O_4^+$
1 122
$C_{10}H_{16}Cl_2NO$
9c1 378
$C_{10}H_{16}D_5N_4O_3$
17d1 31
$C_{10}H_{16}FO_2$
17b 369
$C_{10}H_{16}GeNO_2$
9b 529, *17c* 152
$C_{10}H_{16}N$
17c 136
$C_{10}H_{16}NO$
9c1 168, 178
$C_{10}H_{16}NOSi$
17d1 214
$C_{10}H_{16}NO_2$
9c1 154, 529ff, 801,
17d2 237
$C_{10}H_{16}NO_2Si$
9b 526, 527, *17c* 148,
17d1 378, 379, *17d2* 409
$C_{10}H_{16}NO_2Sn$
17d1 378
$C_{10}H_{16}NO_3$
9c1 298, 310, 409, 510,
17d1 123, *17d2* 238, 297,
342
$C_{10}H_{16}NO_4$
9c1 300, 784, 785, *17d2* 227
$C_{10}H_{16}NO_5$
17d1 165, *17d2* 228
$C_{10}H_{16}N_2^-$
17f 188
$C_{10}H_{16}N_2 \xrightarrow{CT} C_6Cl_4O_2$
9d1 363
$C_{10}H_{16}N_2^+$
9d2 55, 56, *17h* 145, 166ff
$C_{10}H_{16}N_2^+ \xleftarrow{CT} C_8Cl_2N_2O_2^-$
9d1 364
$C_{10}H_{16}N_2^+ \cdot Br$
9d2 55
$C_{10}H_{16}N_2^+ \cdot ClO_4$
9d2 55
$C_{10}H_{16}N_2^+ \cdot I$
9d2 55
$C_{10}H_{16}N_2O_2S^-$
9d1 586
$C_{10}H_{16}N_2O_4S_2^-$
9d1 77
$C_{10}H_{16}N_2S_2^-$
9d1 77
$C_{10}H_{16}N_3O$
9c1 341
$C_{10}H_{16}N_3O_2$
9c1 402
$C_{10}H_{16}N_3O_3$
17d1 78, 79, 103
$C_{10}H_{16}N_3O_4$
17b 42
$C_{10}H_{16}N_3O_5$
9c1 357
$C_{10}H_{16}N_3O_6S$
9c2 278
$C_{10}H_{16}N_3O_9$
9c1 228, 349
$C_{10}H_{16}N_8O_2^{2\cdot}$
9d2 289
$C_{10}H_{16}O^-$
9b 413
$C_{10}H_{16}OP$
9c2 246
$C_{10}H_{16}OSi^+$
17f 34
$C_{10}H_{16}O_2^-$
9b 413, *9d1* 45
$C_{10}H_{16}O_3^-$
17g 28
$C_{10}H_{16}O_3P$
17c 96
$C_{10}H_{16}S^-$
17b 486
$C_{10}H_{16}S_2^+$
17e 373
$C_{10}H_{16}Si^-$
17f 12
$C_{10}H_{17}$
9b 17, 296, *17b* 59, 245,
480
$C_{10}H_{17}BrNO$
17d2 332
$C_{10}H_{17}ClNO_2SSi$
9c1 976
$C_{10}H_{17}INO_2$
9c1 530
$C_{10}H_{17}NO_2Si^-$
9d1 521
$C_{10}H_{17}NO_3P$
17e 287
$C_{10}H_{17}N_2O$
9c1 391, *17d2* 140
$C_{10}H_{17}N_2O_2$
9c1 409, 416, 422, 476, 510,
868, *17d2* 250, 297, 342
$C_{10}H_{17}N_2O_4$
17d2 264
$C_{10}H_{17}N_2O_4SSi$
9c1 976

$C_{10}H_{17}N_2O_5$
17d1 133
$C_{10}H_{17}N_3O_5M$
17d1 9
$C_{10}H_{17}N_4O_2$
17d1 50, 103
$C_{10}H_{17}N_4O_3$
17d1 64, 74, *17d2* 134, 298
$C_{10}H_{17}N_5O^+$
17h 206
$C_{10}H_{17}N_8O_2$
9c1 856
$C_{10}H_{17}O$
9b 273, 299, *17b* 389, 390, 513, 518
$C_{10}H_{17}O_2$
9c2 14, *17e* 12
$C_{10}H_{17}O_2Sn$
17c 112
$C_{10}H_{17}O_3$
17b 439
$C_{10}H_{17}O_4$
9b 83, 116, *17b* 19, 23, 120, 122, 150, 151, 153, 154, 166, 172
$C_{10}H_{17}O_9P^-$
17b 359
$C_{10}H_{17}S_3Si$
17c 66
$C_{10}H_{18}^+$
17h 19
$C_{10}H_{18}Ag$
17b 499
$C_{10}H_{18}BN_3^-$
17c 581
$C_{10}H_{18}BrN_2O_2$
9c1 402
$C_{10}H_{18}ClSi_2$
17b 478
$C_{10}H_{18}Cl_3Se$
17b 316
$C_{10}H_{18}Cl_3Si$
9b 307
$C_{10}H_{18}Cu$
17b 499
$C_{10}H_{18}D_3OS_2$
17e 347
$C_{10}H_{18}D_6S_2^+$
17e 375
$C_{10}H_{18}F_3$
17b 304
$C_{10}H_{18}F_3N_2O$
9c1 17
$C_{10}H_{18}F_3O$
17b 311
$C_{10}H_{18}F_3S$
17b 314
$C_{10}H_{18}F_3SSe$
17b 317
$C_{10}H_{18}F_3Se$
17b 316
$C_{10}H_{18}F_5O_2Si$
9b 250
$C_{10}H_{18}F_6O_4P$
9c2 246
$C_{10}H_{18}FeNO_3S_4$
17a 329
$C_{10}H_{18}N$
17c 571
$C_{10}H_{18}NO$
9c1 152, 153, 215, 219, 251, 263, 341, 449, 530, 531, *17b* 336, 361
$C_{10}H_{18}NOS$
17b 287
$C_{10}H_{18}NO_2$
9c1 391, 426, 450, 531ff, 778, 938, *17d2* 270, 399
$C_{10}H_{18}NO_2S$
9c1 476
$C_{10}H_{18}NO_2SSi$
9c1 976ff
$C_{10}H_{18}NO_2Si$
9c1 1006
$C_{10}H_{18}NO_3$
9c1 215, 349, 369, 403, 476, *17d2* 210
$C_{10}H_{18}NO_3S$
17d2 166, 179
$C_{10}H_{18}NO_4$
9c1 426, 785, *17d2* 228
$C_{10}H_{18}NO_4S$
17d2 238
$C_{10}H_{18}NO_5$
9c1 341, 361, *17d1* 143
$C_{10}H_{18}N_2^+$
17h 181
$C_{10}H_{18}N_2O_3P$
17c 420
$C_{10}H_{18}N_2O_4^-$
17g 41
$C_{10}H_{18}N_2O_5M$
17d1 9, 31
$C_{10}H_{18}N_3O$
9c1 477
$C_{10}H_{18}N_3O_3$
9c1 841, *17d1* 79, 88, 89, 97, *17d2* 228
$C_{10}H_{18}N_3O_4$
9c1 304, 868, *17d1* 89, 98
$C_{10}H_{18}N_5O_2S$
17d2 264
$C_{10}H_{18}N_5O_3$
17d2 264
$C_{10}H_{18}O^-$
9b 275, 276
$C_{10}H_{18}OS^-$
17g 49
$C_{10}H_{18}O_2^-$
9d1 19, 20, *17g* 15
$C_{10}H_{18}O_2Si^-$
17g 31
$C_{10}H_{18}O_3P$
9b 78, *17b* 97
$C_{10}H_{18}O_4^-$
9d1 54
$C_{10}H_{18}S^-$
17b 394, 500
$C_{10}H_{18}S_2^-$
17g 48
$C_{10}H_{18}S_2^+$
17e 371
$C_{10}H_{18}Si_2^-$
17f 12ff
$C_{10}H_{19}$
9b 274
$C_{10}H_{19}BrNO$
17d2 298
$C_{10}H_{19}ClNO_2$
9c1 938
$C_{10}H_{19}Cl_6Si_2$
17b 296
$C_{10}H_{19}F_6Si_2$
9b 233
$C_{10}H_{19}NO^-$
9d1 687
$C_{10}H_{19}NOS^-$
9d1 73

$C_{10}H_{19}NO_2^-$
9d1 415, 416
$C_{10}H_{19}NO_3^+$
17b 105, 108
$C_{10}H_{19}NO_3M$
17d1 32, *17d2* 190
$C_{10}H_{19}NO_3P$
17c 607, *17e* 288
$C_{10}H_{19}NO_5P$
17c 611, 614, *17d2* 171
$C_{10}H_{19}N_2$
9c1 92
$C_{10}H_{19}N_2O$
9c1 842, 898, 902, *17d2* 134, 251, 257, 376
$C_{10}H_{19}N_2OS$
17d2 109
$C_{10}H_{19}N_2O_2$
9c1 263, 868, *17d1* 124, 180, *17d2* 264, 298, 342
$C_{10}H_{19}N_2O_3$
9c1 357, 477, *17d1* 181, *17d2* 149, 251, 298
$C_{10}H_{19}N_2O_3S$
17d2 166
$C_{10}H_{19}N_2O_4$
17c 446, *17d1* 32, 50
$C_{10}H_{19}N_4OS$
9c1 477
$C_{10}H_{19}N_4O_2$
17d1 90
$C_{10}H_{19}N_4O_3$
17d1 90
$C_{10}H_{19}N_4O_5$
17d1 23, 50
$C_{10}H_{19}O$
17b 296, 393, 514
$C_{10}H_{19}OSi$
17c 39
$C_{10}H_{19}O_2$
9b 123, 124, 242, *17b* 440
$C_{10}H_{19}O_5$
17b 443, 445
$C_{10}H_{19}S$
9b 378, *17b* 29, 198
$C_{10}H_{19}SSi$
17c 43
$C_{10}H_{19}Si_2$
17b 478, *17c* 84
$C_{10}H_{20}BrMgN_2$
9a 291
$C_{10}H_{20}BrN_2O_2$
9c1 358, *17d2* 80
$C_{10}H_{20}BrN_2Zn$
9a 331
$C_{10}H_{20}Br_2NO$
9c1 374
$C_{10}H_{20}Br_2N_2Zn^-$
9a 331
$C_{10}H_{20}CaN_2^+$
9a 291
$C_{10}H_{20}Cl$
9b 219
$C_{10}H_{20}ClMgN_2$
9a 291
$C_{10}H_{20}ClN_2O_2$
17d2 80
$C_{10}H_{20}ClN_2Zn$
9a 331
$C_{10}H_{20}Cl_2NO$
9c1 378
$C_{10}H_{20}Cl_2N_2Zn^-$
9a 331
$C_{10}H_{20}Cl_3Si$
9b 220, *17b* 283
$C_{10}H_{20}FN_2O_2$
17d2 81
$C_{10}H_{20}F_3O_2Si$
9b 250
$C_{10}H_{20}F_5S$
9b 220
$C_{10}H_{20}FeN_3OS_4$
17a 329
$(C_{10}H_{20}HgO_2)_n$
17b 55
$C_{10}H_{20}IN_2Zn$
9a 331
$C_{10}H_{20}N$
17b 506, 507
$C_{10}H_{20}NO$
9c1 228, 263, 391, *17b* 508, *17c* 401, 403, 405, *17d1* 119, 134, *17d2* 129, 192
$C_{10}H_{20}NOS$
17d2 167, 179
$C_{10}H_{20}NO_2$
9c1 228, 237, 391, 392, 396, 478, 785, 926, *17d1* 119, 126, 187, *17d2* 130, 135, 156ff, 190
$C_{10}H_{20}NO_2S$
9c1 372, *17c* 523, *17d2* 179
$C_{10}H_{20}NO_3$
9c1 320, 369, 938, 939, *17d1* 134, 156, 159, *17d2* 162
$C_{10}H_{20}NO_3S_2$
17d2 167
$C_{10}H_{20}NO_6$
17d1 104, 105, *17d2* 140
$C_{10}H_{20}NO_7$
17d1 105
$C_{10}H_{20}NS$
17c 500
$C_{10}H_{20}N_2^+$
9d2 62, 102, *17h* 146, 155
$C_{10}H_{20}N_2O_2^-$
9d1 55
$C_{10}H_{20}N_2O_4M$
17d2 79
$C_{10}H_{20}N_2S_2^-$
9d1 78
$C_{10}H_{20}N_2Zn^+$
9a 330
$C_{10}H_{20}N_3O$
9c1 842, *17d2* 299
$C_{10}H_{20}N_3O_2$
9c1 869
$C_{10}H_{20}N_3O_3$
17d1 50, *17d2* 195
$C_{10}H_{20}N_3O_4$
17d1 17, 23, 51, 68
$C_{10}H_{20}N_3O_5$
9c1 357, *17d1* 18, 65, 69
$C_{10}H_{20}N_4O_4PS$
17c 612
$C_{10}H_{20}N_5O_2$
17d2 79
$C_{10}H_{20}N_5Zn$
9a 332
$C_{10}H_{20}N_8Zn^-$
9a 332
$C_{10}H_{20}O^+$
17b 337
$C_{10}H_{20}O_2PS_2$
17e 288
$C_{10}H_{20}O_3P$
17b 29, 49, 50, 122

$C_{10}H_{20}O_4P$
17e 288
$C_{10}H_{20}O_5PS_3$
17b 324
$C_{10}H_{21}$
9b 15, 82, 206, *17b* 118, 295
$C_{10}H_{21}ClNOSi$
17d2 183
$C_{10}H_{21}Cl_2N_6OS_2Zn$
9a 333
$C_{10}H_{21}CoN_4O_5$
17a 335
$C_{10}H_{21}F_3O_4P$
9c2 246
$C_{10}H_{21}GeO_2$
9b 212
$C_{10}H_{21}N^+$
17b 108, 402
$C_{10}H_{21}NO^+$
17b 103, 105, 107
$C_{10}H_{21}NOPS$
17d2 169
$C_{10}H_{21}NO_2P$
17d2 170
$C_{10}H_{21}NO_3P$
17d2 170
$C_{10}H_{21}NO_3PS$
17d2 170
$C_{10}H_{21}NO_4P$
17d2 170
$C_{10}H_{21}N_2$
9c1 17, 43, 93, *17c* 479
$C_{10}H_{21}N_2O$
9c1 394, 521, 813, 898, *17d2* 135, 149, 175, 187, 299
$C_{10}H_{21}N_2OS$
17d2 179
$C_{10}H_{21}N_2O_2$
9c1 243, 416, 982, *17d1* 51, 119, 129, 143, 147, *17d2* 77
$C_{10}H_{21}N_2O_2^{2+}$
17b 103, 107, 403
$C_{10}H_{21}N_2O_2S$
17d1 51
$C_{10}H_{21}N_2O_3$
17d1 18ff, 33, 65, 69, 73, *17d2* 79
$C_{10}H_{21}N_2O_5S$
17d2 258
$C_{10}H_{21}N_3O_2X$
17d1 14
$C_{10}H_{21}N_3O_3P$
17d2 171
$C_{10}H_{21}N_3O_4X$
17d1 51
$C_{10}H_{21}N_4O_3$
17d1 51, 52
$C_{10}H_{21}O$
9b 134
$C_{10}H_{21}OS_2$
17e 347
$C_{10}H_{21}OSi$
17b 223
$C_{10}H_{21}O_2$
17b 56, 130, 191, 192
$C_{10}H_{21}O_2Pb$
17b 268
$C_{10}H_{21}O_2Si$
9b 213, *17b* 211
$C_{10}H_{21}O_3Si$
9b 216
$C_{10}H_{21}O_4P^-$
17g 55
$C_{10}H_{21}O_4S_2$
9c2 289
$C_{10}H_{21}O_7P_2$
9c2 247
$C_{10}H_{21}S$
17b 313
$C_{10}H_{21}SSe$
17b 314
$C_{10}H_{21}S_3$
17b 325
$C_{10}H_{21}Se$
17b 315
$C_{10}H_{21}Se_2$
17b 317
$C_{10}H_{21}Si$
9b 384, *17b* 329
$C_{10}H_{21}Sn$
9b 386
$C_{10}H_{22}^{\dot{+}}$
17h 14
$C_{10}H_{22}Al_2N_2^{\dot{-}}$
17a 400, *17f* 188, 200
$C_{10}H_{22}B_2P_2^{\dot{-}}$
17f 188
$C_{10}H_{22}NO$
9c1 215, 228, 229, 341, *17d1* 119, 156, *17d2* 213, 216, 217
$C_{10}H_{22}NOS$
9c1 370
$C_{10}H_{22}NO_2$
9c1 939
$C_{10}H_{22}NO_2Sn$
9c1 753
$C_{10}H_{22}NO_3$
17d1 159
$C_{10}H_{22}NO_4$
17d1 187
$C_{10}H_{22}NO_7$
17d1 105
$C_{10}H_{22}NSi$
17b 24
$C_{10}H_{22}N_2^{\dot{+}}$
9d2 62, *17b* 101, *17h* 21, 155, 160
$C_{10}H_{22}N_2O_4S_2^{\dot{+}}$
9c2 300, *17e* 374
$C_{10}H_{22}N_2Si_2^{\dot{+}}$
17f 34, *17h* 181
$C_{10}H_{22}N_3O_2$
9c1 243, 321, 899
$C_{10}H_{22}N_3O_3$
17d1 65
$C_{10}H_{22}O_2P$
17b 195
$C_{10}H_{22}O_2PS_2$
17e 327
$C_{10}H_{22}O_3P$
9b 105, *17b* 74, 195, 196, 252, *17e* 288
$C_{10}H_{22}O_4P$
9b 106, 111, *9c2* 247, *17b* 98
$C_{10}H_{22}O_4PSSi$
17e 289
$C_{10}H_{22}P$
17b 195
$C_{10}H_{22}S_2^{\dot{-}}$
9c2 302
$C_{10}H_{22}S_2^{\dot{+}}$
9c2 299, *17e* 366
$C_{10}H_{22}Si_2^{\dot{-}}$
17f 13
$C_{10}H_{23}NO_4P$
17d1 152

$C_{10}H_{23}N_2^{2+}$
17b 104, 107
$C_{10}H_{23}N_2O$
9c1 305
$C_{10}H_{23}N_3O_2P$
17d2 170
$C_{10}H_{23}N_3O_2SX$
17d1 52
$C_{10}H_{23}N_3O_2X$
17d1 52
$C_{10}H_{23}N_3O_3X$
17d1 52
$C_{10}H_{23}N_5O_3X$
17d1 52, 66
$C_{10}H_{23}OSeSn$
17b 321
$C_{10}H_{23}OSn$
17b 21
$C_{10}H_{23}O_2S$
17e 354
$C_{10}H_{23}O_2Si$
17b 46, 149
$C_{10}H_{23}O_2Si_2$
9b 385
$C_{10}H_{23}O_7P_2$
9c2 247
$C_{10}H_{23}SSi$
17b 21
$C_{10}H_{23}Si$
9b 185, *17b* 130, 250
$C_{10}H_{23}Si_2$
17b 159, 200, *17c* 23
$C_{10}H_{24}NOSi$
17d1 152, 201
$C_{10}H_{24}NO_2Si$
17d1 187
$C_{10}H_{24}NO_4P^+$
9c1 1008
$C_{10}H_{24}NSi$
17c 561
$C_{10}H_{24}N_2^+$
9d2 102
$C_{10}H_{24}N_2OP$
17b 74, *17e* 289
$C_{10}H_{24}N_2O_2P$
9c2 248
$C_{10}H_{24}N_2P$
17b 195
$C_{10}H_{24}N_3Si$
9c1 122, *17c* 582
$C_{10}H_{24}N_4^+$
9d2 62, 102
$C_{10}H_{24}OP$
9c2 248
$C_{10}H_{24}O_2P$
9c2 248
$C_{10}H_{24}O_3P$
9c2 248
$C_{10}H_{24}O_4P$
9c2 249
$C_{10}H_{24}O_6P$
9c2 25
$C_{10}H_{24}S_2^+$
17e 375
$C_{10}H_{24}Si_2^+$
17h 24
$C_{10}H_{25}NOP$
9c2 249
$C_{10}H_{25}NO_3P$
9c2 249
$C_{10}H_{25}N_3OP$
9c2 250
$C_{10}H_{25}OSSi$
17e 351
$C_{10}H_{25}OSiSn$
17b 19
$C_{10}H_{25}O_7P_2$
9c2 250
$C_{10}H_{25}Si_2$
17b 260
$C_{10}H_{26}BNOP$
17d1 199
$C_{10}H_{26}BN_2O$
17d1 199
$C_{10}H_{26}N_2O_2P$
9c2 250
$C_{10}H_{26}N_4O_2P$
17d1 129, 130
$C_{10}H_{26}N_7OX_2$
17d1 147
$C_{10}H_{27}Cl_3N_2PSi_3$
17e 264
$C_{10}H_{27}MnO_6P$
17e 17
$C_{10}H_{27}NO_3PSi$
17e 289
$C_{10}H_{27}NO_3PSi_2$
17b 43, *17e* 290
$C_{10}H_{27}N_4OP$
9c1 899
$C_{10}H_{27}N_4OP^+$
9c1 1009
$C_{10}H_{27}Si_3$
9b 235, *17b* 327
$C_{10}H_{28}N_2Si_4^+$
9d2 103, *17f* 34, *17h* 146
$C_{10}H_{29}Si_4$
17f 7
$C_{10}H_{30}Si_5^-$
17f 26
$C_{10}H_{40}Co$
9a 325
$C_{10}Mn_2O_{10}^-$
17a 122
$C_{10}Mo_2O_{10}^-$
17a 149
$C_{10}N_4O_2^-$
9d1 32, *17g* 242
$C_{10}N_4S_4^+$
17h 111
$C_{10}N_5^-$
9d1 862
$C_{10}O_{10}Re_2^-$
17a 172
$C_{11}Cl_4MnO_7$
17a 283, 284, *17e* 190, 193
$C_{11}F_{19}$
17b 480
$C_{11}Fe_3O_{11}^-$
17a 125
$C_{11}HCl_4CoN_6O^{3-}$
9a 319, 322, *9c1* 1036
$C_{11}HCl_4CoN_6O$
17d1 371
$C_{11}HFe_2O_{11}$
17a 125
$C_{11}HO_{11}^{6-}$
17c 95
$C_{11}H_2Cl_2CoFN_6O^{3-}$
9a 320
$C_{11}H_2Cl_2CoFN_6O$
9c1 1033
$C_{11}H_2Cl_2CoFN_7O_3$
9c1 1034
$C_{11}H_2Cl_3CoN_6O^{3-}$
9a 322, 324, *9c1* 1034, 1035
$C_{11}H_2D_8N_2NaO_4S$
17d1 349
$C_{11}H_2O_{10}^{5-}$
9b 461, *17c* 89

$C_{11}H_2O_{11}^{5-}$
9b 473
$C_{11}H_3BrD_{16}NO_3$
17d2 299
$C_{11}H_3ClCoN_7O_3^{3-}$
9c1 1032
$C_{11}H_3Cl_2CoN_6O^{3-}$
9a 322, *9c1* 1032, 1033
$C_{11}H_3CoN_8O_5^{3-}$
9a 323, *9c1* 1033
$C_{11}H_4BrCoN_6O$
9c1 1024
$C_{11}H_4ClCoN_6O^{3-}$
9a 321, *9c1* 1025
$C_{11}H_4CoFN_6O^{3-}$
9a 319, 320, *9c1* 1025, 1026
$C_{11}H_4CoIN_6O$
9c1 1026
$C_{11}H_4CoN_5O_2^{3-}$
17a 335, *17g* 214
$C_{11}H_4CoN_7O_3^{3-}$
9a 323
$C_{11}H_4CoN_7O_3$
9c1 1026, 1027
$C_{11}H_4CrN_2O_5^-$
17a 253
$C_{11}H_4D_4F_2^-$
17f 189
$C_{11}H_4D_8N_2NaO_4S$
17d1 348
$C_{11}H_4D_{12}NO_2$
17d1 270
$C_{11}H_4D_{12}NO_3$
17d2 112
$C_{11}H_4F_9O$
9b 239
$C_{11}H_4MoN_2O_5^-$
17a 267
$C_{11}H_4MoN_2O_5Se^-$
17a 268
$C_{11}H_4N_2O_5SW^-$
17a 275
$C_{11}H_4N_2O_5SeW^-$
17a 275
$C_{11}H_4N_2O_5W^-$
17a 275
$C_{11}H_4O_5^{3-}$
9d1 324
$C_{11}H_5CoN_6O^{3-}$
9a 322, *9c1* 1024
$C_{11}H_5CoN_6O$
17a 335
$C_{11}H_5D_4^{2-}$
17f 74
$C_{11}H_5D_4F^-$
17f 190
$C_{11}H_5D_9O^-$
17c 199
$C_{11}H_5F_9NO$
17c 191
$C_{11}H_5FeN_6O^{3-}$
9c1 1039
$C_{11}H_5Mn_2O_9S_2$
17a 284
$C_{11}H_5NO_6Re$
17d1 233
$C_{11}H_5O_5$
9c2 199
$C_{11}H_6D_4^-$
9d1 836, *17f* 75
$C_{11}H_6FeN_6O^{3-}$
9a 312, *17a* 329
$C_{11}H_6NO_3S_2$
17c 601
$C_{11}H_6N_2O^-$
9b 717, 718
$C_{11}H_6OS_3^-$
9b 713
$C_{11}H_7Br_2D_3N_2O_6SM$
17d1 257
$C_{11}H_7Br_2D_4N_3O_7SM$
17d1 257, 259
$C_{11}H_7ClO_4^-$
17g 279
$C_{11}H_7ClO_5^-$
9d1 300
$C_{11}H_7ClO_5P^{2-}$
9c2 196
$C_{11}H_7D_2^{2-}$
17f 74
$C_{11}H_7D_3MgNO_5$
17g 111
$C_{11}H_7D_8NNaO_5S$
17d1 349
$C_{11}H_7F_5NO_3$
9c1 621
$C_{11}H_7F_9NO$
9c1 250
$C_{11}H_7HgO_2S_2$
17g 369
$C_{11}H_7N_2O_3^{2-}$
9d1 690
$C_{11}H_7N_2O_4^{2-}$
9d1 425
$C_{11}H_7O$
17b 526, 527, 528
$C_{11}H_7OS_3$
17c 339, 340
$C_{11}H_7O_2$
9c2 198, 201, *17b* 528
$C_{11}H_7O_3^{2-}$
9d1 292
$C_{11}H_7O_3$
9c2 187, 189, 191, 193, *17e* 196, 199, 200
$C_{11}H_7O_4^{2-}$
17c 176
$C_{11}H_7O_5$
9c2 198, 199
$C_{11}H_7O_6^{2-}$
9d1 300
$C_{11}H_7O_8$
9c2 186
$C_{11}H_8Br_2D_2N_2O_6SM$
17d1 259
$C_{11}H_8D_2^-$
9d1 836
$C_{11}H_8D_2NO_2$
17b 177
$C_{11}H_8F_2^-$
17f 189
$C_{11}H_8FeO_3^-$
17a 329
$C_{11}H_8HgIO_2$
17a 390, *17g* 259
$C_{11}H_8N$
17c 565
$C_{11}H_8NO$
9c1 128, 129
$C_{11}H_8NO_2^{2-}$
9d1 559
$C_{11}H_8N_2O^-$
9b 717, *17f* 197
$C_{11}H_8N_2O_2^-$
17f 197
$C_{11}H_8N_2O_4^-$
9d1 613
$C_{11}H_8O^-$
17f 189

$C_{11}H_8O^{2\cdot}$
17h 261
$C_{11}H_8OS^-$
9b 706, 708, *17c* 332ff
$C_{11}H_8O_2^-$
9b 488, 626, *9d1* 38, 58, 237, 282ff, *17c* 230, *17g* 267ff
$C_{11}H_8O_2$ on Al_2O_3 surface
9d1 365
$C_{11}H_8O_3^-$
1 83, *9d1* 71, 285, 287, *17g* 278, 279
$C_{11}H_8O_5^{3-}$
9d1 251, 261
$C_{11}H_8O_5^-$
9d1 299
$C_{11}H_8O_5P^{2-}$
9c2 194
$C_{11}H_9^{2-}$
9d1 772, *17f* 73ff
$C_{11}H_9$
9b 301, *17b* 489, *17c* 169
$C_{11}H_9BrNS_2$
17c 558
$C_{11}H_9Br_2D_3N_2O_6SM$
17d1 260
$C_{11}H_9Cl_3NO_2$
17d1 243
$C_{11}H_9Cl_3NO_3$
17d1 243
$C_{11}H_9D^-$
9d1 836
$C_{11}H_9D_5NOS$
17c 516
$C_{11}H_9F^-$
17f 189
$C_{11}H_9F_{15}NO$
9c1 380
$C_{11}H_9HgO_2$
17a 390, *17g* 259
$C_{11}H_9K_2^-$
17a 216
$C_{11}H_9N^-$
17f 196
$C_{11}H_9N^{2\cdot}$
17h 264
$C_{11}H_9NO_2^-$
9d1 284, 468, 554
$C_{11}H_9NO_3^-$
9d1 469
$C_{11}H_9N_2$
9b 624
$C_{11}H_9N_3OX$
17d2 408
$C_{11}H_9N_3O_2S^+$
17h 191
$C_{11}H_9Na_2$
17a 209
$C_{11}H_9O_2$
9b 621, *9c2* 190, 191, 192, *17e* 194, 196
$C_{11}H_9O_2S$
17c 42
$C_{11}H_9O_3$
17c 38, 39
$C_{11}H_9O_4$
17c 176
$C_{11}H_{10}^-$
9d1 737, 740, 835, *17f* 74, 75, 76
$C_{11}H_{10}BrCoN_6O^{3-}$
9c1 1018
$C_{11}H_{10}ClCoN_6O^{3-}$
9c1 1018
$C_{11}H_{10}CoN_7O^{3-}$
9c1 1019
$C_{11}H_{10}CoN_7O_3^{3-}$
9c1 1019
$C_{11}H_{10}MgNO_5$
17g 111
$C_{11}H_{10}MnN_2O_2$
17a 284
$C_{11}H_{10}MnN_2O_5$
17a 284
$C_{11}H_{10}N$
17c 134
$C_{11}H_{10}NO$
17d2 347
$C_{11}H_{10}NO_2$
9b 589, 590, *17b* 176
$C_{11}H_{10}NO_2S_2$
17c 530
$C_{11}H_{10}NO_5$
9c1 589
$C_{11}H_{10}NS_2$
17c 556
$C_{11}H_{10}N_2O_3^-$
17e 215
$C_{11}H_{10}N_2O_4^-$
9d1 612
$C_{11}H_{10}N_2O_5^-$
9d1 612
$C_{11}H_{10}N_3O_6$
17d1 385
$C_{11}H_{10}N_4O_2^+$
9d2 74
$C_{11}H_{10}N_5O$
17c 469
$C_{11}H_{10}O_2^-$
9d1 57, 59, 236
$C_{11}H_{10}O_3^-$
9d1 72
$C_{11}H_{11}$
9b 301, 547, 618, 626, *17b* 489
$C_{11}H_{11}ClNO_5$
9c1 963, 964
$C_{11}H_{11}Cl_3NO_2$
17d1 239, 244, 250
$C_{11}H_{11}Cl_3NO_3$
17d1 239
$C_{11}H_{11}D^-$
17f 77
$C_{11}H_{11}D_5N_2O_2^+$
17b 141
$C_{11}H_{11}FNO_5$
9c1 964
$C_{11}H_{11}F_3NO_3$
9c1 955
$C_{11}H_{11}MgN_2O_5$
17g 194
$C_{11}H_{11}NO_2^-$
9d1 466, 468, 553
$C_{11}H_{11}NO_3Zn^+$
17a 381, *17g* 354
$C_{11}H_{11}NO_4^-$
9c1 985
$C_{11}H_{11}N_2$
17c 213
$C_{11}H_{11}N_2O$
9c1 891
$C_{11}H_{11}N_2O_2$
9c1 704, 812, *17c* 455
$C_{11}H_{11}N_2O_3$
9c1 69, 144, 812, 955
$C_{11}H_{11}N_2O_5$
9c1 141

$C_{11}H_{11}N_2O_7^{2-}$
9d1 427
$C_{11}H_{11}N_4O_6$
17d1 385
$C_{11}H_{11}O_2^{2-}$
9b 391
$C_{11}H_{11}O_4$
9b 575
$C_{11}H_{11}O_8$
17c 61
$C_{11}H_{11}O_9$
17c 61
$C_{11}H_{12}^-$
17f 76, 77
$C_{11}H_{12}BrCoN_6O^{3-}$
9c1 1019
$C_{11}H_{12}Br_2NO_3$
9c1 250
$C_{11}H_{12}ClN_2O_2$
17c 422
$C_{11}H_{12}Cl_2NO$
17c 595
$C_{11}H_{12}Cl_3N_2O$
17d1 239
$C_{11}H_{12}D_4N_2O^+$
17b 141
$C_{11}H_{12}FO_2$
17b 369, 370
$C_{11}H_{12}F_4NO_3$
9c1 621
$C_{11}H_{12}N$
9b 623
$C_{11}H_{12}NO$
17e 213
$C_{11}H_{12}NOS$
17c 215
$C_{11}H_{12}NO_2$
17c 632
$C_{11}H_{12}NO_3$
9c1 731, *17d2* 395
$C_{11}H_{12}NO_3S$
17d2 416
$C_{11}H_{12}NO_4$
9c1 956, *17b* 284
$C_{11}H_{12}NO_5$
9c1 956
$C_{11}H_{12}NS$
17c 214
$C_{11}H_{12}N_2O_2S^-$
17f 268
$C_{11}H_{12}N_3O_2$
17c 457
$C_{11}H_{12}N_3O_6$
17d1 181
$C_{11}H_{12}N_3O_7S$
17d1 383
$C_{11}H_{12}O^-$
9b 614
$C_{11}H_{12}OS^-$
17c 54, *17g* 52
$C_{11}H_{12}O_2^-$
9d1 24, 86, *17c* 53, *17g* 22
$C_{11}H_{12}O_2S_2^-$
17g 46
$C_{11}H_{12}O_3^-$
9d1 51
$C_{11}H_{12}O_3^+$
17h 85
$C_{11}H_{12}O_4^-$
9d1 24
$C_{11}H_{12}O_4Tl$
17g 105
$C_{11}H_{12}O_5^-$
9d1 9ff
$C_{11}H_{12}Si_2^-$
17f 13
$C_{11}H_{13}BrMgO_2^-$
17a 226, *17c* 204
$C_{11}H_{13}BrNO_3$
9c1 211
$C_{11}H_{13}ClNO$
17c 594
$C_{11}H_{13}ClNOS$
17c 516, 614
$C_{11}H_{13}ClNO_2$
17d1 315
$C_{11}H_{13}ClNS$
9c1 107
$C_{11}H_{13}Cl_3NO$
9c1 728, *17d1* 307
$C_{11}H_{13}CoN_6O^{3-}$
9c1 1013
$C_{11}H_{13}F_3NO$
9c1 661, 727
$C_{11}H_{13}F_3NO_5Pd$
17a 344
$C_{11}H_{13}F_3NO_5Pt$
17a 352
$C_{11}H_{13}N^+$
17h 65, 66
$C_{11}H_{13}NO_2^-$
9d1 464, 465, 466
$C_{11}H_{13}NO_2$
9c1 661
$C_{11}H_{13}NO_3^-$
9d1 505
$C_{11}H_{13}NO_4^-$
9d1 541
$C_{11}H_{13}N_2$
9c1 11
$C_{11}H_{13}N_2O$
9c1 661, 662, 704
$C_{11}H_{13}N_2OS$
9c1 822ff
$C_{11}H_{13}N_2O_2$
9b 620, *9c1* 704, 705, 823, 861, *17d2* 265
$C_{11}H_{13}N_2O_4$
9c1 106, 785, *17d1* 181
$C_{11}H_{13}N_2O_6$
9c1 964
$C_{11}H_{13}N_3^+$
17h 220
$C_{11}H_{13}N_3O_5^-$
17f 296
$C_{11}H_{13}N_4O$
17c 469
$C_{11}H_{13}N_4O_2$
17c 469
$C_{11}H_{13}N_4O_4$
17e 10
$C_{11}H_{13}N_4O_4S$
17b 92
$C_{11}H_{13}N_4O_7S$
17d1 384, 385
$C_{11}H_{13}N_4S$
17c 472
$C_{11}H_{13}O$
17c 187
$C_{11}H_{13}O_2^{2-}$
17c 192, 195, 197
$C_{11}H_{13}O_2$
17b 61, 249, *17c* 187, *17e* 223
$C_{11}H_{13}O_2SSn$
17a 454
$C_{11}H_{13}O_3$
9c2 169
$C_{11}H_{13}O_4$
9c2 170

$C_{11}H_{13}O_8$
17c 17

$C_{11}H_{14}^-$
9d1 714, 725

$C_{11}H_{14}ClN_2O_2$
17c 249

$C_{11}H_{14}Cl_2NO$
9c1 377, 639, *17d1* 284

$C_{11}H_{14}Cl_3N_2O_2$
17d2 85

$C_{11}H_{14}F_2NO$
9c1 381

$C_{11}H_{14}F_3OSi$
9b 595

$C_{11}H_{14}F_3OSn$
17a 454, *17c* 227

$C_{11}H_{14}Ge^-$
17f 43

$C_{11}H_{14}KO_2$
9d1 134

$C_{11}H_{14}LiO_2$
9d1 134

$C_{11}H_{14}N$
17b 508

$C_{11}H_{14}N^{2+}$
17h 235

$C_{11}H_{14}NO$
9c1 134, 135, 447, 731, *17c* 593, *17d1* 217

$C_{11}H_{14}NOS$
17c 516

$C_{11}H_{14}NOS_2$
17c 517

$C_{11}H_{14}NOS_2Sn$
17a 454, *17c* 345

$C_{11}H_{14}NOSi$
17c 221

$C_{11}H_{14}NO_2$
9c1 106, 662, 696, 741, 773, 785ff, *17c* 592, *17d1* 181, 267, 315, *17d2* 222, *17e* 212

$C_{11}H_{14}NO_2S$
9c1 987, *17c* 542

$C_{11}H_{14}NO_3^{2-}$
17g 103

$C_{11}H_{14}NO_3$
9c1 212, 552, 662, 663, 794, *17d1* 169

$C_{11}H_{14}NO_3S$
9c1 996, *17d2* 416

$C_{11}H_{14}NO_4$
9c1 794, *17d1* 379

$C_{11}H_{14}NO_6S$
9c1 921

$C_{11}H_{14}NS_2$
17c 558

$C_{11}H_{14}N_2O_2S^-$
9d1 513

$C_{11}H_{14}N_2O_2S^+$
17h 139

$C_{11}H_{14}N_2O_3^-$
9d1 514, *17g* 90

$C_{11}H_{14}N_2O_4M$
17d2 86

$C_{11}H_{14}N_3O_2$
17d1 366, *17d2* 86

$C_{11}H_{14}N_3O_3$
9c1 778, *17d2* 375

$C_{11}H_{14}N_4O_4SM$
17d2 95

$C_{11}H_{14}N_4O_5^-$
9d1 633

$C_{11}H_{14}NaO_2$
9d1 134

$C_{11}H_{14}O^-$
9b 488, 570, *17c* 194, 196, 199

$C_{11}H_{14}O^+$
17h 64

$C_{11}H_{14}O_2^-$
9b 542, *9d1* 43, 134, 212

$C_{11}H_{14}O_2^+$
17h 65, 85

$C_{11}H_{14}O_3^-$
9d1 142, 149, 199, 215ff, 221, *17g* 24

$C_{11}H_{14}O_3Tl$
17g 112, 113, 153

$C_{11}H_{14}O_5^+$
17h 85

$C_{11}H_{14}O_5P$
9c2 251

$C_{11}H_{14}S^-$
17c 206

$C_{11}H_{14}S_2^-$
9b 578, 580

$C_{11}H_{14}S_2^+$
17h 65

$C_{11}H_{14}Si^-$
17f 14

$C_{11}H_{15}$
9b 16, 437, 539, 559, *17b* 18, 31, *17c* 173, 185, 188

$C_{11}H_{15}AgBrN_3O^+$
9a 330

$C_{11}H_{15}AsNO_5M$
17d2 59

$C_{11}H_{15}BD_3NOM$
17d2 65

$C_{11}H_{15}BrNO$
17d2 61

$C_{11}H_{15}ClNO$
9c1 328, 786, *17d1* 269, *17d2* 61

$C_{11}H_{15}ClN_2NaO_3S$
17d1 124

$C_{11}H_{15}FNO$
9c1 330, 683, *17d2* 61

$C_{11}H_{15}FO_2S^+$
17h 95

$C_{11}H_{15}FO_3S^+$
17h 92

$C_{11}H_{15}F_3NO^-$
17c 200

$C_{11}H_{15}F_3NO$
9c1 381

$C_{11}H_{15}NO^-$
17c 209

$C_{11}H_{15}NO^+$
17h 231

$C_{11}H_{15}NO_2^-$
9d1 417, 455, 463, *17g* 30

$C_{11}H_{15}NO_2^+$
9d2 34, 35, *17h* 66

$C_{11}H_{15}NO_2S_2^-$
17f 251

$C_{11}H_{15}NO_3^-$
17g 103

$C_{11}H_{15}NO_3Tl$
17g 107

$C_{11}H_{15}NO_4PM_2$
17d2 69

$C_{11}H_{15}NO_4SM$
17d2 49, 93

$C_{11}H_{15}NO_5P$
17c 588

$C_{11}H_{15}NO_5PM_2$
17d2 59

$C_{11}H_{15}NO_5SM$
17d2 60, 96
$C_{11}H_{15}NO_6M$
17d1 352
$C_{11}H_{15}NS^-$
9b 585, *17c* 211
$C_{11}H_{15}NSe^-$
17c 216
$C_{11}H_{15}N_2$
17c 144
$C_{11}H_{15}N_2O$
9c1 813, 826, 833, *17d1* 176, 226, *17d2* 141
$C_{11}H_{15}N_2OS$
17c 580, *17d2* 407
$C_{11}H_{15}N_2O_2$
9c1 663, 786, 822, *17d1* 169
$C_{11}H_{15}N_2O_3$
9c1 229, *17d1* 124
$C_{11}H_{15}N_2O_4S$
9c1 774
$C_{11}H_{15}N_2O_5$
9c1 810
$C_{11}H_{15}N_4O$
9c1 305, *17d1* 152, *17d2* 46
$C_{11}H_{15}N_4O_4S$
17c 428
$C_{11}H_{15}O$
1 70, *9c2* 56, 60, *17b* 86, 145
$C_{11}H_{15}O_2$
1 60, 61, *9b* 324, 458, *9c2* 57, 67, 72, 159, 177, *17e* 187
$C_{11}H_{15}O_2S$
17e 332, 341
$C_{11}H_{15}O_2Si$
9b 556
$C_{11}H_{15}O_3$
9b 428, *9c2* 165
$C_{11}H_{15}O_3P^-$
17g 55
$C_{11}H_{15}O_4$
9c2 167
$C_{11}H_{15}O_4P^-$
17g 55
$C_{11}H_{15}O_6$
9b 372, *17b* 498
$C_{11}H_{15}O_7$
17b 442
$C_{11}H_{15}Si$
9b 308
$C_{11}H_{16}^-$
9d1 721
$C_{11}H_{16}^+$
9d2 6
$C_{11}H_{16}N$
17c 564
$C_{11}H_{16}NO$
9c1 11, 100, 101, 229, 263, 611, 664ff, *9c2* 163, *17d1* 124, 217, 263, 269ff, *17d2* 8ff, 135
$C_{11}H_{16}NOS$
9c1 246, *9c2* 665, 988, *17c* 506, *17d1* 193, *17d2* 201
$C_{11}H_{16}NO_2$
9b 381, *9c1* 101, 310, 623, 666ff, 921, 939, 949, *17c* 146, *17d1* 131, 169, 263, 318, *17d2* 49ff, 238, 409
$C_{11}H_{16}NO_2^{2+}$
17h 85
$C_{11}H_{16}NO_2S$
9c1 112, 667, *17d1* 193
$C_{11}H_{16}NO_3$
9c1 176, 311, 532, 956, 957, *17d2* 57ff, 238
$C_{11}H_{16}NO_3PM$
17d2 69
$C_{11}H_{16}NO_3S$
9c1 112, 369, 668, 998ff, *17c* 491, 495, *17d1* 197
$C_{11}H_{16}NO_3Zn^{2+}$
17a 382, *17g* 103
$C_{11}H_{16}NO_4S$
9c1 940, *17d1* 197
$C_{11}H_{16}NO_5P^-$
9d1 515ff
$C_{11}H_{16}NO_6$
17b 466
$C_{11}H_{16}NS$
17c 501, 504
$C_{11}H_{16}N_2^+$
17h 146
$C_{11}H_{16}N_3O$
17b 284, *17d1* 169, *17d2* 375
$C_{11}H_{16}N_3O_{10}U$
17a 242
$C_{11}H_{16}O_2^+$
9d2 128
$C_{11}H_{16}O_4S^+$
17h 95
$C_{11}H_{16}O_5^+$
9d2 136, *17h* 86
$C_{11}H_{17}$
17c 76
$C_{11}H_{17}BN_4^-$
17c 581
$C_{11}H_{17}ClN_2OX$
17d2 91
$C_{11}H_{17}F_6O_2$
17c 24
$C_{11}H_{17}F_6Si$
9b 256
$C_{11}H_{17}Ge$
17f 40
$C_{11}H_{17}NO_2Si^-$
9d1 522
$C_{11}H_{17}NO_3Tl$
17g 101
$C_{11}H_{17}NO_4P$
17c 493
$C_{11}H_{17}NO_5P$
17d2 419
$C_{11}H_{17}NO_5SMX$
17d2 91
$C_{11}H_{17}N_2$
9c1 92
$C_{11}H_{17}N_2O$
17d1 214, *17d2* 46, 75, 106, 107
$C_{11}H_{17}N_2O_2$
17d1 367, *17d2* 81ff
$C_{11}H_{17}N_2O_2SSi$
9c1 977
$C_{11}H_{17}N_2O_3$
17b 243, *17d1* 367, *17d2* 82
$C_{11}H_{17}N_2O_3MX$
17d2 91
$C_{11}H_{17}N_2O_4$
9c1 409, *17d2* 238
$C_{11}H_{17}N_2O_5$
9c1 514
$C_{11}H_{17}N_2O_5SMX$
17d2 92

$C_{11}H_{17}N_2O_6M_2$
17d1 9
$C_{11}H_{17}N_2S$
17b 287
$C_{11}H_{17}N_4O_3$
9c1 31
$C_{11}H_{17}O$
17b 487, 488
$C_{11}H_{17}OS_3$
17c 66
$C_{11}H_{17}OSi$
9b 592
$C_{11}H_{17}OSn$
9b 598
$C_{11}H_{17}O_2$
17c 50
$C_{11}H_{17}S$
9b 604, *17b* 487, 488
$C_{11}H_{18}^+$
17h 17, 30
$C_{11}H_{18}BD_2N_2$
17c 574
$C_{11}H_{18}BNO^-$
17d1 356
$C_{11}H_{18}BNO_6$
17d1 352
$C_{11}H_{18}F_3$
9b 307, 354, *17b* 330
$C_{11}H_{18}F_3O$
9b 307, *17b* 306
$C_{11}H_{18}F_5$
17b 304
$C_{11}H_{18}F_7O_2Si$
9b 250
$C_{11}H_{18}IN_2O_2$
17d2 251
$C_{11}H_{18}NO$
17d2 222
$C_{11}H_{18}NO_2$
9c1 427, 532, 533, *17b* 109, *17d1* 165, *17d2* 146, 239, 343
$C_{11}H_{18}NO_3$
9c1 299, 410, 510, *17d2* 135
$C_{11}H_{18}NO_3^{2+}$
17h 86
$C_{11}H_{18}NO_3SSi$
9c1 977
$C_{11}H_{18}NO_4$
9c1 146, *17d2* 299
$C_{11}H_{18}NO_5$
9c1 239, 300
$C_{11}H_{18}NO_7$
9c1 212
$C_{11}H_{18}N_2^+$
17h 146
$C_{11}H_{18}N_2OX$
17d1 52, *17d2* 89, 91
$C_{11}H_{18}N_2O_2S^-$
9d1 586
$C_{11}H_{18}N_2O_2X$
17d2 90, 92
$C_{11}H_{18}N_2O_4M$
17d1 9
$C_{11}H_{18}N_3O$
17d2 46
$C_{11}H_{18}N_3O_3$
9c1 478, *17d1* 79
$C_{11}H_{18}O_2^-$
9d1 15, *17b* 172, *17g* 25
$C_{11}H_{18}O_2Si^-$
17f 14
$C_{11}H_{18}S^-$
17b 500, *17c* 35
$C_{11}H_{18}S_2^-$
17g 48
$C_{11}H_{18}Si^-$
17f 14, 15
$C_{11}H_{19}$
9b 432
$C_{11}H_{19}Cl_2$
9b 357
$C_{11}H_{19}F_2$
17b 329
$C_{11}H_{19}GeN^+$
9d2 35
$C_{11}H_{19}IN_2O_2$
17d2 258
$C_{11}H_{19}MgO_2$
9b 421
$C_{11}H_{19}MgO_2^+$
17a 226
$C_{11}H_{19}NO_2$
17g 195
$C_{11}H_{19}NO_2^+$
17h 135
$C_{11}H_{19}NO_3^+$
17h 135
$C_{11}H_{19}NO_4PS$
17e 290
$C_{11}H_{19}NO_7P$
17d2 116
$C_{11}H_{19}NSi^+$
9d2 35
$C_{11}H_{19}N_2$
9c1 92
$C_{11}H_{19}N_2O$
9c1 902
$C_{11}H_{19}N_2O_2Si$
9c1 981, 982
$C_{11}H_{19}N_2O_4S$
17d2 167
$C_{11}H_{19}N_2O_5$
17d1 10
$C_{11}H_{19}N_2O_6$
17c 446
$C_{11}H_{19}N_3OS_2M$
17d2 110
$C_{11}H_{19}N_3O_5M$
17d1 10, 33
$C_{11}H_{19}N_3O_6M$
17d1 33, 69
$C_{11}H_{19}N_4O_3$
17d2 239
$C_{11}H_{19}NiO_2$
17c 46
$C_{11}H_{19}O_4$
9b 83, *17b* 20, 120, 153, 166
$C_{11}H_{19}S_2$
9b 418
$C_{11}H_{20}Cl_3$
9b 219, *17b* 279
$C_{11}H_{20}Cl_3Si$
9b 384
$C_{11}H_{20}F_3$
9b 219, *17b* 279
$C_{11}H_{20}F_3O$
9b 219, *17b* 279
$C_{11}H_{20}F_3S$
9b 220, *17b* 282
$C_{11}H_{20}NO$
9c1 403, *17d1* 134
$C_{11}H_{20}NOSSi$
17d2 201
$C_{11}H_{20}NO_2$
17d2 270
$C_{11}H_{20}NO_2SSi$
9c1 978

$C_{11}H_{20}NO_2S_2Si$
9c1 978
$C_{11}H_{20}NO_2Si$
17d2 201
$C_{11}H_{20}NO_3$
9c1 479, 940, *17d2* 146
$C_{11}H_{20}NO_3S$
17d2 180
$C_{11}H_{20}NO_3S_2$
17d2 333
$C_{11}H_{20}NO_4$
9c1 237, 427, *17d2* 196
$C_{11}H_{20}NO_4S_2Si$
9c1 978
$C_{11}H_{20}NO_5$
9c1 300, 311, *17d1* 143, *17d2* 117
$C_{11}H_{20}N_2O_4M$
17d1 10, 33
$C_{11}H_{20}N_2O_4SM$
17d1 10, 33
$C_{11}H_{20}N_2O_5M$
17d1 34
$C_{11}H_{20}N_3OZn$
9a 331
$C_{11}H_{20}N_3O_2S$
17d2 80
$C_{11}H_{20}N_3O_3$
17d1 103
$C_{11}H_{20}N_3O_4$
9c1 869, *17d1* 23, 90, 98
$C_{11}H_{20}N_3SZn$
9a 332
$C_{11}H_{20}O^-$
9b 294, 361
$C_{11}H_{20}O_2^+$
17h 40, 52
$C_{11}H_{20}O_4P$
17b 97
$C_{11}H_{20}S^-$
17b 395, 476
$C_{11}H_{20}Si^-$
17f 24
$C_{11}H_{20}Si_2^+$
17f 37
$C_{11}H_{21}$
9b 272, 351, *17b* 112
$C_{11}H_{21}BN$
17c 438
$C_{11}H_{21}Cl_2$
17b 279
$C_{11}H_{21}Cl_3NO_2$
9c1 377
$C_{11}H_{21}INO$
17d2 300
$C_{11}H_{21}NO^+$
17h 135
$C_{11}H_{21}NOPS$
17b 288
$C_{11}H_{21}NO_2^-$
9d1 416
$C_{11}H_{21}NO_2PS_2$
17b 290
$C_{11}H_{21}NO_3PS$
17b 288
$C_{11}H_{21}NO_4P$
17d2 300
$C_{11}H_{21}NO_8P$
17d2 117
$C_{11}H_{21}N_2O$
9c1 842, 843, *17d2* 136
$C_{11}H_{21}N_2O_2$
9c1 417, 521, 786, 869, 870, 902, *17d2* 174, 300, 335
$C_{11}H_{21}N_2O_3$
9c1 263, 869
$C_{11}H_{21}N_2O_3S$
17d2 167
$C_{11}H_{21}N_2O_4$
9c1 870, *17c* 445, *17d1* 70
$C_{11}H_{21}N_4O_3$
17d2 300
$C_{11}H_{21}N_4O_5$
17d1 19, 23, 24, 34, 70
$C_{11}H_{21}O$
9b 112, 113
$C_{11}H_{21}O_2$
9b 94, 124
$C_{11}H_{21}S$
17b 199
$C_{11}H_{21}Si_2$
17c 98
$C_{11}H_{22}ClN_2O_2$
17d2 77
$C_{11}H_{22}Cl_2CuN_3O_2$
17a 357
$C_{11}H_{22}GeNSi$
9b 528
$C_{11}H_{22}NO$
17b 58, *17c* 401, 403, *17d1* 119, 134, *17d2* 223, 301
$C_{11}H_{22}NO_2$
9c1 230, 238, 406, 427, 480ff, *17d1* 120, *17d2* 136
$C_{11}H_{22}NO_3$
9c1 292, 320, 940
$C_{11}H_{22}NO_4$
17d2 162
$C_{11}H_{22}NSiSn$
9b 529
$C_{11}H_{22}NSi_2$
9b 528
$C_{11}H_{22}N_2^+$
17h 129
$C_{11}H_{22}N_2O_4P$
17d2 301
$C_{11}H_{22}N_3O_2$
17d2 149, 301
$C_{11}H_{22}N_3O_3$
17d2 79
$C_{11}H_{22}N_3O_4$
17d1 24, 34, 70, 73
$C_{11}H_{22}N_3O_4S_2$
17d2 239
$C_{11}H_{22}N_3O_5$
17d1 35
$C_{11}H_{22}O_3P$
17b 98, 196
$C_{11}H_{23}$
9b 218, *17b* 296
$C_{11}H_{23}CoN_4O_5$
17a 335
$C_{11}H_{23}NO^+$
17b 106, 108, 109, 131, 156, 402, 403
$C_{11}H_{23}NO_3PSi$
17e 290
$C_{11}H_{23}NO_5P$
17d2 271, 301
$C_{11}H_{23}NO_6P$
9c1 324
$C_{11}H_{23}N_2$
9c1 93
$C_{11}H_{23}N_2O$
17d2 187

$C_{11}H_{23}N_2O_2$
9c1 264, *17d1* 53, 120, 147, *17d2* 77
$C_{11}H_{23}N_2O_3$
17d2 80
$C_{11}H_{23}N_2Zn$
17a 382
$C_{11}H_{23}N_3O_2X$
17d1 53
$C_{11}H_{23}N_3O_3P$
17d2 302
$C_{11}H_{23}N_3O_4M$
17d1 71
$C_{11}H_{23}N_4Si$
17c 476
$C_{11}H_{23}O$
9b 218, *17b* 279
$C_{11}H_{23}OS_2$
9b 158, *17e* 347, 348
$C_{11}H_{23}O_2$
17b 221
$C_{11}H_{23}O_2Si$
9b 204
$C_{11}H_{23}O_3S$
17e 356
$C_{11}H_{23}O_3Si$
17b 270
$C_{11}H_{23}O_4P^-$
17g 55
$C_{11}H_{23}SSe$
17b 313
$C_{11}H_{23}S_2$
17b 198
$C_{11}H_{23}Si$
17b 332
$C_{11}H_{24}BO$
9b 194, 195, 205
$C_{11}H_{24}GeNO$
17d2 183
$C_{11}H_{24}IN_2O$
9c1 521
$C_{11}H_{24}NO$
9c1 98, 264, 342, *17d1* 134, 156
$C_{11}H_{24}NOSi$
17b 509
$C_{11}H_{24}NO_2$
17d2 74
$C_{11}H_{24}NO_4P$
9d1 684
$C_{11}H_{24}NSi$
17b 511, 512, *17c* 561
$C_{11}H_{24}N_2^+$
17b 130, 155
$C_{11}H_{24}N_3O_2$
17d1 54
$C_{11}H_{24}N_3O_2Si$
9c1 122, *17c* 583
$C_{11}H_{24}N_3Si$
17c 583
$C_{11}H_{24}N_4O_3X$
17d1 54
$C_{11}H_{24}OP$
17b 311
$C_{11}H_{24}O_3P$
9b 106
$C_{11}H_{24}O_4P$
17b 97, *17e* 291
$C_{11}H_{24}O_4PSe$
17b 321
$C_{11}H_{24}O_5P$
17b 373
$C_{11}H_{24}O_5PS$
17b 374
$C_{11}H_{24}O_6P$
17b 374
$C_{11}H_{25}Ge$
17f 40
$C_{11}H_{25}NO_3PS_2$
9c1 246
$C_{11}H_{25}NO_4P$
9c1 372, *17c* 493, *17d1* 152
$C_{11}H_{25}NO_5P$
17d2 419
$C_{11}H_{25}N_2^{2+}$
17b 104, 107, 109, 402, 403
$C_{11}H_{25}N_3O_2SX$
17d1 55, 56
$C_{11}H_{25}N_3O_2X$
17d1 14, 54, 55
$C_{11}H_{25}N_3O_3SX$
17d1 56
$C_{11}H_{25}N_4O$
17b 184
$C_{11}H_{25}N_4O_2$
9c1 245
$C_{11}H_{25}O_2S$
17e 354
$C_{11}H_{26}NO_5S_2$
9c1 480
$C_{11}H_{26}NO_7P_2$
9c1 1008
$C_{11}H_{26}NSi$
17c 395, 560
$C_{11}H_{26}N_2O_2P$
17c 391
$C_{11}H_{26}N_2Si^+$
17h 128
$C_{11}H_{27}AlNO$
17a 400
$C_{11}H_{27}NOTl$
17a 410
$C_{11}H_{27}N_2SSi$
9c1 108
$C_{11}H_{27}N_2Si_2$
17b 508
$C_{11}H_{27}N_3OPS$
17e 291
$C_{11}H_{27}N_3O_3P$
17c 453
$C_{11}H_{27}OSiSn$
17b 18
$C_{11}H_{27}Si_2$
17b 57, 200
$C_{11}H_{27}Si_3$
9b 309
$C_{11}H_{28}NO_6P_2$
17e 291
$C_{11}H_{28}N_2O_3PSi_2$
17b 519, *17e* 291
$C_{11}H_{28}N_3OX_2$
17d1 135
$C_{11}H_{28}O_4PSi$
17e 291
$C_{11}H_{28}Si_3^-$
17f 15
$C_{11}H_{29}BNSi$
17b 200
$C_{11}H_{30}N_2OPSi_3$
17e 264
$C_{11}H_{30}N_4OX_3$
17d1 134, 147
$C_{11}H_{31}NO_3PSi_3$
17b 43, *17e* 292
$C_{11}H_{31}Si_4$
17f 7
$C_{11}H_{32}N_2OPSi_3$
17e 264
$C_{11}H_{34}Si_6^-$
17f 26

$C_{11}MnN_4O_5$
9b 388
$C_{11}N_4O_3^-$
17g 57
$C_{11}O_{10}^{5-}$
9b 315
$C_{12}Br_4D_4K_2N_3O_8S_2$
17d1 261
$C_{12}Cl_2F_{24}NO$
17d2 221
$C_{12}Cl_8NS$
9c1 37
$C_{12}Cl_{10}N$
9c1 16
$C_{12}Cl_{10}NO$
9c1 578
$C_{12}Cl_{10}O_4Tl$
9d1 157
$C_{12}CrD_{11}$
17a 254
$C_{12}CrD$
17a 254
$C_{12}CrD_{12}^+$
17a 254
$C_{12}D_{10}^-$
9d1 803
$C_{12}D_{10}NS_2$
17c 518
$C_{12}F_{26}NO_7S_2$
17d2 222
$C_{12}Fe_3O_{12}^-$
17a 125
$C_{12}H_2D_8^-$
9d1 803
$C_{12}H_2D_{18}F_3O$
9b 363
$C_{12}H_2F_{24}NO$
17d2 221
$C_{12}H_2N_4O_{11}^-$
9d1 668
$C_{12}H_2O_2S^-$
9d1 189
$C_{12}H_3BrCo_2N_{11}O$
9c1 1019
$C_{12}H_3CoN_7O_5^{4-}$
9a 323
$C_{12}H_3F_{12}O_2S$
17e 355
$C_{12}H_3Mn_3O_{12}^-$
17a 123
$C_{12}H_3N_3O_9^-$
9d1 667
$C_{12}H_3O_7^{4-}$
9d1 324
$C_{12}H_4Br_2N_2O_4^-$
17f 302
$C_{12}H_4Br_2N_4O_7^-$
9d1 334
$C_{12}H_4ClD_{18}S$
9b 378
$C_{12}H_4Cl_6NO$
9c1 574, *17d1* 251
$C_{12}H_4CoN_6O_3^{4-}$
9a 323, 324
$C_{12}H_4CoN_7O^{3-}$
9c1 1027
$C_{12}H_4CoN_7O_5^{3-}$
9c1 1033
$C_{12}H_4D_6^-$
9d1 803
$C_{12}H_4D_8N_2NaO_4S$
17d1 348
$C_{12}H_4F_6NO$
9c1 731
$C_{12}H_4MnO_6^{2-}$
9d1 331
$C_{12}H_4MnO_6S$
17g 365
$C_{12}H_4MnO_7$
17g 361
$C_{12}H_4N_2O_7^-$
9d1 656
$C_{12}H_4N_4^-$
1 84
$C_{12}H_4N_5O_9S^{2-}$
9d1 669
$C_{12}H_4N_7O_{13}$
9c1 575
$C_{12}H_4O_6^{3-}$
9d1 323
$C_{12}H_5Cl_2CoN_6O_2^{3-}$
9a 320
$C_{12}H_5Cl_2CoN_6O_2$
9c1 1035
$C_{12}H_5Cl_2O_3^{2-}$
9d1 258, 268
$C_{12}H_5Cl_4O_2$
17e 151, 189
$C_{12}H_5CoN_6O_2^{3-}$
9c1 1027
$C_{12}H_5CoN_6O_3^{3-}$
9c1 1028
$C_{12}H_5CoN_7O_2^{4-}$
9c1 1014
$C_{12}H_5D_8NNaO_6S$
17d1 350
$C_{12}H_5D_{18}$
9b 351
$C_{12}H_5D_{18}S$
9b 377
$C_{12}H_5F_5NO$
9c1 571
$C_{12}H_5F_6$
17c 104
$C_{12}H_5MnNO_5S_2$
17a 284
$C_{12}H_5MnNO_6$
17g 354
$C_{12}H_5O_4^{2-}$
9d1 343
$C_{12}H_5O_5^{2-}$
9d1 258, 269, *17g* 361
$C_{12}H_5O_6$
9c2 199
$C_{12}H_6^-$
17f 77
$C_{12}H_6BrO_3^{2-}$
9d1 253, 256, 262, 266
$C_{12}H_6Br_2S_2^+$
9d2 145
$C_{12}H_6ClO_3^{2-}$
9d1 253, 254, 256, 263, 264, 266
$C_{12}H_6Cl_2NS$
9c1 36, 37
$C_{12}H_6Cl_2OS^+$
9d2 143
$C_{12}H_6Cl_2O_2^-$
9d1 185
$C_{12}H_6Cl_2O_2^+$
9d2 142
$C_{12}H_6Cl_2S_2^+$
1 127, *9d2* 145
$C_{12}H_6Cl_3NO^-$
17e 152
$C_{12}H_6Cl_3O_2Sn$
17g 344
$C_{12}H_6Cl_4NO$
17d1 237

$C_{12}H_6Cl_4NS_2$
9c1 114, 115, *17c* 520, 521
$C_{12}H_6CoN_7O_3^{4-}$
9c1 1014
$C_{12}H_6CrN_4O_4^-$
17a 255, *17f* 208
$C_{12}H_6CsO_2$
9d1 328
$C_{12}H_6D_4^-$
9d1 802, 803
$C_{12}H_6D_4O_5^-$
9d1 300
$C_{12}H_6D_6N_4O_2^+$
17h 208
$C_{12}H_6D_8N_2NaO_4S$
17d1 350
$C_{12}H_6D_8N_2NaO_5S$
17d1 350
$C_{12}H_6D_{12}NO_2$
17d1 284
$C_{12}H_6D_{12}NO_3$
17d2 114
$C_{12}H_6FO_3^{2-}$
9d1 254, 256, 263, 264, 266
$C_{12}H_6F_2NO$
9c1 596
$C_{12}H_6F_2NS$
9c1 36
$C_{12}H_6F_2OS^+$
9d2 143
$C_{12}H_6F_2O_2^+$
9d2 142
$C_{12}H_6KO_2$
9d1 328
$C_{12}H_6LiO_2$
9d1 328
$C_{12}H_6MnO_8$
17a 285, *17b* 401
$C_{12}H_6NO_2^{2-}$
9d1 101
$C_{12}H_6NO_3$
9c1 807
$C_{12}H_6NO_5^{2-}$
9d1 254, 256, 266
$C_{12}H_6N_2O_2^-$
9d1 342
$C_{12}H_6N_2O_5Te^-$
9d1 661
$C_{12}H_6N_3OR_4$
(R = various substituents)
1 53
$C_{12}H_6N_3O_4S^{2-}$
9d1 661
$C_{12}H_6N_3O_4S$
9c1 36
$C_{12}H_6N_3O_5$
9c1 597
$C_{12}H_6N_3O_5S^{2-}$
9d1 661
$C_{12}H_6N_3R_4S$
(R = various substituents)
1 53
$C_{12}H_6N_4^-$
9d1 882
$C_{12}H_6N_4O_8^-$
9d1 668
$C_{12}H_6NaO_2$
9d1 328
$C_{12}H_6O_2^-$
1 84, *9d1* 327ff, *17g* 344
$C_{12}H_6O_2Rb$
9d1 328
$C_{12}H_6O_2S_2^-$
9d1 63
$C_{12}H_6O_3^-$
9d1 92, 343
$C_{12}H_6O_4^{3-}$
9d1 186, 253, 255, 257, 263, 264, 266
$C_{12}H_6O_4^-$
9d1 185, 347
$C_{12}H_7BrO_2^-$
9d1 180
$C_{12}H_7BrO_2S^-$
9d1 190
$C_{12}H_7Br_2D_3N_3O_6SM$
17d1 261
$C_{12}H_7Br_2NNaO_4S$
17d1 261
$C_{12}H_7Br_3NO$
17d1 255
$C_{12}H_7ClNOS$
9c1 605
$C_{12}H_7ClNS$
9c1 35, *17c* 435
$C_{12}H_7ClN_2O_5^-$
17f 302
$C_{12}H_7ClO_2^-$
9d1 180, 181
$C_{12}H_7ClO_2S^-$
9d1 190
$C_{12}H_7Cl_2O_2$
17e 152
$C_{12}H_7Cl_3NOS$
17d1 251
$C_{12}H_7CoN_6O^{3-}$
9c1 1012, 1028
$C_{12}H_7CoN_6O_2^{3-}$
9a 320, 321, *9c1* 1029
$C_{12}H_7D^-$
9d1 756
$C_{12}H_7D_8NNaO_4S$
17d1 349
$C_{12}H_7FNOS$
9c1 605
$C_{12}H_7FO_2^-$
9d1 181
$C_{12}H_7FO_2S^-$
9d1 190
$C_{12}H_7KN_3O_7S$
9c1 16
$C_{12}H_7NO_2^-$
9d1 102
$C_{12}H_7NO_3Te^-$
9d1 608
$C_{12}H_7NO_4^-$
9d1 181, 182, 608
$C_{12}H_7NO_4S^-$
9d1 191
$C_{12}H_7N_2O_3$
9c1 597
$C_{12}H_7N_2O_3S^{2-}$
9d1 608
$C_{12}H_7N_2O_3S$
9c1 605
$C_{12}H_7N_3O_7^-$
17f 305
$C_{12}H_7N_4O_5^{2-}$
9d1 553
$C_{12}H_7O_3^{2-}$
9d1 182, 183, 252, 262
$C_{12}H_7O_6$
9c2 199
$C_{12}H_7O_7$
9c2 200

$C_{12}H_8^-$
1 108, *9d1* 754ff, 774ff, *17f* 78
$C_{12}H_8^+$
1 121
$C_{12}H_8Al_2Br_6Cl_2S$
17e 321
$C_{12}H_8Al_2Br_6Cl_2S_2$
17e 321
$C_{12}H_8BrNO^-$
17e 43
$C_{12}H_8BrNO_2^-$
9d1 563
$C_{12}H_8Br_2NO$
9c1 561
$C_{12}H_8Br_2NS_2$
9c1 113, 114, *17c* 519
$C_{12}H_8ClNO_2^-$
9d1 562
$C_{12}H_8ClNS^+$
17h 193
$C_{12}H_8ClN_2O_2S$
17c 510
$C_{12}H_8ClO_4P^-$
17e 292
$C_{12}H_8Cl_2^+$
9d2 19
$C_{12}H_8Cl_2NO$
9c1 562
$C_{12}H_8Cl_2NO_5S_2$
17d2 417
$C_{12}H_8Cl_2NS$
9c1 111
$C_{12}H_8Cl_2NS_2$
9c1 113, 114, *17c* 519
$C_{12}H_8Cl_2O_6^-$
9d1 302
$C_{12}H_8Cl_4N_2O_2^+$
9d2 54
$C_{12}H_8Cl_8Ga_2S_2$
17e 322
$C_{12}H_8CrD_4^+$
17a 255
$C_{12}H_8D_2^-$
9d1 802
$C_{12}H_8D_3O_4$
9b 457
$C_{12}H_8D_4N_2^+$
9d2 90
$C_{12}H_8D_4N_4O_2^+$
17h 208
$C_{12}H_8FNO_2^-$
9d1 562
$C_{12}H_8F_2^-$
17f 190
$C_{12}H_8F_2^+$
9d2 19, *17h* 75
$C_{12}H_8F_2N$
9c1 13
$C_{12}H_8F_2NO$
9c1 562ff
$C_{12}H_8F_2NO_5S_2$
17d2 417
$C_{12}H_8F_2NS_2$
9c1 114, *17c* 519
$C_{12}H_8KN_2$
17a 217
$C_{12}H_8N$
9c1 28
$C_{12}H_8NO$
9c1 32, 598, *17c* 434
$C_{12}H_8NOS$
1 56, *9c1* 606, *17d2* 372
$C_{12}H_8NOSe$
9c1 608
$C_{12}H_8NO_2^{2-}$
9d1 334
$C_{12}H_8NO_2$
9c1 603ff
$C_{12}H_8NO_3$
17d1 223
$C_{12}H_8NS$
9c1 34, *17c* 435
$C_{12}H_8NS_2$
17c 544
$C_{12}H_8N_2^-$
1 116, *9d1* 877, 878
$C_{12}H_8N_2^+$
17h 230
$C_{12}H_8N_2O^-$
9d1 877
$C_{12}H_8N_2O_2^-$
9d1 877
$C_{12}H_8N_2O_2^+$
9d2 118
$C_{12}H_8N_2O_4^-$
1 107ff, *9d1* 641ff, 643, 655
$C_{12}H_8N_2O_4S^-$
1 109, *9d1* 644
$C_{12}H_8N_2O_5^-$
1 109, *9d1* 644
$C_{12}H_8N_2S^+$
17h 137
$C_{12}H_8N_3O_4$
9c1 13
$C_{12}H_8N_3O_4S_2$
9c1 114, *17c* 520
$C_{12}H_8N_3O_5$
9c1 563
$C_{12}H_8N_3S_2$
17c 545
$C_{12}H_8N_4O_2^-$
17f 293
$C_{12}H_8N_4O_4S^-$
17e 388
$C_{12}H_8O^-$
9b 627, *9d1* 887
$C_{12}H_8OS^-$
1 117
$C_{12}H_8OS^+$
1 127, *9d2* 143
$C_{12}H_8O_2^-$
9d1 178, 288, 324
$C_{12}H_8O_2^+$
1 127, *9d2* 142
$C_{12}H_8O_2S^-$
1 117, *17g* 346
$C_{12}H_8O_2S_2^-$
1 118
$C_{12}H_8O_2S_2^+$
9d2 145
$C_{12}H_8O_3^-$
9d1 72, 183, 283
$C_{12}H_8O_4^{3-}$
9d1 303
$C_{12}H_8O_4^-$
9d1 186
$C_{12}H_8O_4P$
17e 292
$C_{12}H_8O_4S_2^-$
1 118, *9d1* 889
$C_{12}H_8O_6^-$
9d1 300
$C_{12}H_8S^-$
1 117, *9d1* 888, 894
$C_{12}H_8S^+$
17h 114
$C_{12}H_8S_2^+$
1 127, *9d2* 145

$C_{12}H_8Se^-$
9d1 888ff

$C_{12}H_9Br^{2\cdot}$
17h 254

$C_{12}H_9BrNOS$
17d1 232

$C_{12}H_9BrNS$
9c1 111

$C_{12}H_9Br_2D_4N_3O_7SM$
17d1 258, 260

$C_{12}H_9Cl^{2\cdot}$
17h 254

$C_{12}H_9ClNO_3S$
17d2 414

$C_{12}H_9ClNS$
9c1 111

$C_{12}H_9ClO_2^-$
17c 230

$C_{12}H_9ClO_4^-$
17g 276, 277

$C_{12}H_9Cl_2N_2O_2$
9c1 970

$C_{12}H_9D^-$
9d1 753, 802

$C_{12}H_9F^+$
9d2 19

$C_{12}H_9F_5NO$
17d1 351

$C_{12}H_9HgNO_2^-$
9d1 523

$C_{12}H_9HgO_2S_2$
17g 369

$C_{12}H_9N^-$
9d1 868

$C_{12}H_9N^+$
9d2 90

$C_{12}H_9NO^-$
9b 715, 716, *9d1* 333, *17c* 346, 347, *17e* 43, *17f* 308

$C_{12}H_9NO^+$
1 127, *9d2* 74, *17h* 191

$C_{12}H_9NOS^+$
9d2 76

$C_{12}H_9NO_2^-$
1 91, 107, *9b* 716, *9d1* 560, 561ff, *17f* 276

$C_{12}H_9NO_2S^-$
9d1 190

$C_{12}H_9NO_3^-$
1 95, *9d1* 502

$C_{12}H_9NO_3P$
17d1 233, *17e* 293

$C_{12}H_9NO_4^-$
17f 276

$C_{12}H_9NS^+$
1 126, *17h* 192ff

$C_{12}H_9NSe^+$
9d2 76

$C_{12}H_9N_2$
17c 434

$C_{12}H_9N_2O_2S$
9c1 109, 110, *17c* 507

$C_{12}H_9N_2O_3$
9c1 558

$C_{12}H_9N_2O_5S$
9c1 1002

$C_{12}H_9N_3O_4^+$
9d2 40

$C_{12}H_9O$
17b 527, 529

$C_{12}H_9O_2$
1 59, *9c2* 34, *17b* 528

$C_{12}H_9O_2S$
17e 46

$C_{12}H_9O_3$
9c2 54

$C_{12}H_9S^+$
9d2 75

$C_{12}H_{10}^{2\cdot}$
1 107, *9d1* 751ff, 757, 772, 798ff, *17f* 78ff

$C_{12}H_{10}^+$
17h 71, 75

$C_{12}H_{10}^{2\cdot}$
17h 253

$C_{12}H_{10}Al_2Br_6S_2$
17e 320

$C_{12}H_{10}As$
9c2 317

$C_{12}H_{10}BrSe$
17e 353

$C_{12}H_{10}Br_2D_6N_3O_6SM$
17d1 259

$C_{12}H_{10}ClSe$
17e 353

$C_{12}H_{10}Cl_2P$
9c2 252

$C_{12}H_{10}Cl_3FeO_2$
9a 313

$C_{12}H_{10}Cl_6Ga_2S_2$
17e 321

$C_{12}H_{10}CoN_7O^{3-}$
9c1 1020

$C_{12}H_{10}Co_2O_2^-$
9a 326

$C_{12}H_{10}D_5N_3O^+$
17b 78

$C_{12}H_{10}MuN_2$
17c 441

$C_{12}H_{10}N$
1 52, *9b* 621, *9c1* 12

$C_{12}H_{10}NO$
1 55, *9c1* 136, 552, 556ff, *17d1* 224

$C_{12}H_{10}NOS$
17d1 232

$C_{12}H_{10}NO_2$
9c1 558

$C_{12}H_{10}NO_3$
9c1 154, *17c* 487

$C_{12}H_{10}NO_3S$
9c1 1002ff, *17d1* 232

$C_{12}H_{10}NO_4S$
17c 431

$C_{12}H_{10}NO_5$
9c1 589

$C_{12}H_{10}NO_5S$
9c1 1005

$C_{12}H_{10}NO_5S_2$
17d2 417

$C_{12}H_{10}NO_6$
9c1 964

$C_{12}H_{10}NS$
9c1 109, 1044, *17c* 506, *17e* 326

$C_{12}H_{10}NS_2$
9c1 113, *17c* 518

$C_{12}H_{10}N_2^+$
1 126, *9d2* 76, *17h* 182ff

$C_{12}H_{10}N_2O_2^-$
9d1 563, *17f* 252ff

$C_{12}H_{10}N_2O_2M$
17d1 217

$C_{12}H_{10}N_2O_2P$
17e 293

$C_{12}H_{10}N_2O_4^-$
9d1 336

$C_{12}H_{10}N_2S^-$
17e 387

$C_{12}H_{10}N_3O_2S$
17c 437

$C_{12}H_{10}N_3O_5$
9c1 563, 583

$C_{12}H_{10}N_3O_8$
17d1 386

$C_{12}H_{10}N_4O_2R$
17c 436

$C_{12}H_{10}N_4O_4^-$
17f 300

$C_{12}H_{10}O^-$
9b 626, *9d1* 883

$C_{12}H_{10}OP$
9c2 221

$C_{12}H_{10}O_2^-$
9d1 237, 239, 283, 287, *17f* 190, *17g* 268ff, 271

$C_{12}H_{10}O_2^+$
9d1 358, *9d2* 127, *17h* 88

$C_{12}H_{10}O_2S^-$
9b 708

$C_{12}H_{10}O_2S^+$
9d2 138, *17g* 386

$C_{12}H_{10}O_3^-$
9d1 72, 286

$C_{12}H_{10}O_3^+$
9d2 137

$C_{12}H_{10}O_3PS_2$
17e 360

$C_{12}H_{10}O_4^-$
9d1 288, 289, 299, *17g* 15ff, 276, 280

$C_{12}H_{10}O_5P^{2-}$
9c2 195

$C_{12}H_{10}O_5P^-$
1 83

$C_{12}H_{10}O_6^-$
9d1 289, 300, 301

$C_{12}H_{10}O_6S_2^{3-}$
9d1 205

$C_{12}H_{10}P$
9c2 217, *17e* 255

$C_{12}H_{10}PS$
17e 264

$C_{12}H_{10}S^-$
9d1 894

$C_{12}H_{10}S^+$
17h 114

$C_{12}H_{10}S_2^+$
1 128

$C_{12}H_{10}S_4^+$
17h 116

$C_{12}H_{10}Sb$
9c2 318

$C_{12}H_{11}$
9b 463, 540, *17b* 490, *17c* 90, 99, 251

$C_{12}H_{11}Br_3NO$
17d1 252

$C_{12}H_{11}Cl_3NO$
17d1 244

$C_{12}H_{11}Cl_3NO_2$
17d1 239, 244

$C_{12}H_{11}Cl_3NO_4$
17d1 244

$C_{12}H_{11}Cr$
17a 255

$C_{12}H_{11}D_2^{2-}$
17f 82ff

$C_{12}H_{11}D_3N_4O_2^+$
9d2 77

$C_{12}H_{11}F_5NO$
9c1 718

$C_{12}H_{11}F_6$
17c 101

$C_{12}H_{11}K^-$
17a 216

$C_{12}H_{11}MnNO_5$
17a 285

$C_{12}H_{11}N^+$
9d2 40

$C_{12}H_{11}NO_2^-$
9d1 284, 469

$C_{12}H_{11}NO_3^-$
9d1 470

$C_{12}H_{11}NO_3S^+$
17h 172

$C_{12}H_{11}N_2O_2$
9c1 970

$C_{12}H_{11}N_3O^-$
17f 308

$C_{12}H_{11}N_3O_2^-$
17f 276

$C_{12}H_{11}N_3O_2S^+$
17h 193

$C_{12}H_{11}N_3S^+$
1 126

$C_{12}H_{11}N_4O_2$
17c 458

$C_{12}H_{11}N_4O_2R^+$
(R=*D*-ribityl, *D*-ribitylphosphate, adenosyl-*D*-ribitylphosphate)
17h 217ff

$C_{12}H_{11}N_4O_8$
17d1 386

$C_{12}H_{11}Na^-$
17a 209

$C_{12}H_{11}O$
17c 242

$C_{12}H_{12}^-$
1 111, *9d1* 737ff, 741, 744, 747, 748, 749, 837, *17f* 80, 81

$C_{12}H_{12}^+$
9d2 6, 11, *17h* 70, 71, 72

$C_{12}H_{12}^{2\cdot}$
17h 250

$C_{12}H_{12}Cl_2N_2O_6^-$
17g 244

$C_{12}H_{12}Cr^-$
9a 305, *17a* 256

$C_{12}H_{12}Cr^+$
9a 306, *17a* 255ff

$C_{12}H_{12}CrI$
9a 303, 304

$C_{12}H_{12}D^{2-}$
17f 82

$C_{12}H_{12}KN_2$
17a 216

$C_{12}H_{12}N$
17c 145

$C_{12}H_{12}NO$
17d2 347

$C_{12}H_{12}NO_2$
9c1 731

$C_{12}H_{12}NO_3$
9c1 732

$C_{12}H_{12}NO_7$
9c1 965, 966

$C_{12}H_{12}N_2^-$
9d1 872

$C_{12}H_{12}N_2^+$
9d2 90, *17h* 168, 186

$C_{12}H_{12}N_2O_2^-$
17f 276, 277

$C_{12}H_{12}N_2O_4^-$
9d1 613
$C_{12}H_{12}N_3O_6^-$
17f 297
$C_{12}H_{12}N_3O_6$
17d1 386
$C_{12}H_{12}N_4^-$
9d1 882
$C_{12}H_{12}N_4O_2^+$
17h 207, 208, 209, 210
$C_{12}H_{12}NaO_2$
9d1 144
$C_{12}H_{12}O^-$
9b 621, *9d1* 884, *17f* 190
$C_{12}H_{12}O_2^-$
9d1 43, 198, 239
$C_{12}H_{12}O_2^+$
9d1 358, *9d2* 128, 137
$C_{12}H_{12}S_2^+$
17e 384, *17h* 98, 111
$C_{12}H_{12}S_4^+$
17h 112, 114
$C_{12}H_{12}V$
17a 121
$C_{12}H_{13}^{2-}$
17f 81ff
$C_{12}H_{13}BrF_3NOX$
(X unknown)
17d1 307
$C_{12}H_{13}ClF_3NOX$
(X unknown)
17d1 307
$C_{12}H_{13}Cl_3NO_2$
17d1 239
$C_{12}H_{13}Cl_5NO$
17d1 308
$C_{12}H_{13}CoN_6O_3^{3-}$
9c1 1020
$C_{12}H_{13}Cr$
9a 305
$C_{12}H_{13}D_3OS^-$
17c 207
$C_{12}H_{13}D_3OSe^-$
17c 216
$C_{12}H_{13}D_3S_2^-$
17c 208
$C_{12}H_{13}FNO$
9c1 742
$C_{12}H_{13}F_3NOSi$
17c 221
$C_{12}H_{13}F_5NO$
9c1 727
$C_{12}H_{13}F_6O$
17c 103
$C_{12}H_{13}NO_2^-$
9d1 467, 469
$C_{12}H_{13}N_2$
17c 568
$C_{12}H_{13}N_2OSi$
17c 222
$C_{12}H_{13}N_2O_2$
9c1 823, 859
$C_{12}H_{13}N_2O_3$
9c1 144, 802, 823, *17d1* 183
$C_{12}H_{13}N_2Zn^-$
17f 204
$C_{12}H_{13}N_4O_3$
17d1 226
$C_{12}H_{13}N_4O_4$
9c1 986
$C_{12}H_{13}O_4$
9b 576
$C_{12}H_{13}O_5$
9c2 64
$C_{12}H_{14}^-$
9d1 714
$C_{12}H_{14}^+$
9d2 8
$C_{12}H_{14}AlN_2^-$
17f 203
$C_{12}H_{14}AlN_2$
17a 400
$C_{12}H_{14}Al_2N_2^-$
17a 400, *17f* 207
$C_{12}H_{14}Cl_2NO$
9c1 716, *17d1* 235
$C_{12}H_{14}Cl_3NO_4SM$
17d2 94
$C_{12}H_{14}Cl_3N_2O$
9c1 813, *17d1* 240
$C_{12}H_{14}Cl_4NO$
17d1 308
$C_{12}H_{14}Cl_4NO_5P^+$
9c2 181
$C_{12}H_{14}MnN_2O_5$
17a 285
$C_{12}H_{14}NO$
9c1 533, 732, 742, *17d1* 218
$C_{12}H_{14}NO_3$
9c1 625, 957, *17d2* 362
$C_{12}H_{14}NO_4$
9c1 957
$C_{12}H_{14}N_2^+$
1 125, *17h* 186
$C_{12}H_{14}N_2O_2^-$
9d1 203
$C_{12}H_{14}N_2O_2S^-$
17f 268
$C_{12}H_{14}N_2O_3S^-$
17f 255
$C_{12}H_{14}N_2O_4SM$
17d2 95
$C_{12}H_{14}N_3O$
9c1 814
$C_{12}H_{14}N_3O_2$
9c1 68, *17c* 456, 457
$C_{12}H_{14}N_3O_5$
9c1 638
$C_{12}H_{14}N_4O_2^+$
9d2 77
$C_{12}H_{14}N_5O_7$
9c1 351
$C_{12}H_{14}OS^-$
17g 49
$C_{12}H_{14}O_2^-$
9b 615, *9d1* 10, 24, 25, *17g* 17, 22
$C_{12}H_{14}O_2S_2^-$
9d1 95, 105, 106, *17g* 47
$C_{12}H_{14}O_3^-$
9d1 51
$C_{12}H_{14}O_3S^-$
9d1 96, 106
$C_{12}H_{14}O_4^-$
9d1 96
$C_{12}H_{14}O_4^+$
17h 87
$C_{12}H_{14}S_4^-$
9d1 106
$C_{12}H_{14}S_4^+$
17h 115
$C_{12}H_{15}$
9b 18, 305, 437
$C_{12}H_{15}BrCdN_3O$
9a 334
$C_{12}H_{15}BrNOPb^{2+}$
9a 299
$C_{12}H_{15}BrN_2O_2$
9c1 870

$C_{12}H_{15}BrN_3O$
9a 335, *9c1* 843, *17d2* 379
$C_{12}H_{15}BrN_3OZn$
9a 334
$C_{12}H_{15}BrN_3O_5$
17d2 37
$C_{12}H_{15}Br_3NO$
17d2 37
$C_{12}H_{15}ClF_2NO$
9c1 264
$C_{12}H_{15}ClNO$
9c1 713, *17d2* 142
$C_{12}H_{15}ClNOS$
17d2 168
$C_{12}H_{15}ClNS$
17c 500
$C_{12}H_{15}ClN_3O_5$
17d2 38
$C_{12}H_{15}ClNaO_2$
9d1 150
$C_{12}H_{15}Cl_3NO$
9c1 264, *17d1* 152, 164,
17d2 38
$C_{12}H_{15}Cl_3NO_2$
9c1 723
$C_{12}H_{15}Cl_3NO_3$
17d2 58
$C_{12}H_{15}Cl_4O_5P^+$
9c2 181
$C_{12}H_{15}CrN_2O_4$
17d1 202 [5])
$C_{12}H_{15}D_2O_3$
9c2 166
$C_{12}H_{15}FNO$
9c1 742
$C_{12}H_{15}FN_2O_3$
17d2 39
$C_{12}H_{15}FN_3O_5$
17d2 39
$C_{12}H_{15}F_3NO$
9c1 265, 684, *17d2* 39
$C_{12}H_{15}F_6Si$
17c 103
$C_{12}H_{15}Mo$
17b 75
$C_{12}H_{15}N^+$
17h 66
$C_{12}H_{15}NO^-$
9d1 687
$C_{12}H_{15}NOS^-$
9d1 73
$C_{12}H_{15}NO_2^-$
9d1 418, 465, 526
$C_{12}H_{15}NO_3M$
17d1 35, *17d2* 43
$C_{12}H_{15}NO_4^-$
9d1 683, *17c* 148, *17f* 250
$C_{12}H_{15}NO_4SM$
17d2 95
$C_{12}H_{15}NO_5P$
17c 609
$C_{12}H_{15}N_2O$
9c1 265, 668, 705, 707, 899,
17d1 270ff, *17d2* 45, 147
$C_{12}H_{15}N_2OS$
17d2 62
$C_{12}H_{15}N_2O_2$
17d2 46
$C_{12}H_{15}N_2O_3$
9c1 742, 814, *17d2* 142,
269, 384
$C_{12}H_{15}N_2O_5$
9c1 854
$C_{12}H_{15}N_3O_5^-$
17f 297
$C_{12}H_{15}N_4O_3$
17c 433
$C_{12}H_{15}N_4O_7$
9c1 265, *17d2* 31
$C_{12}H_{15}O$
17b 110
$C_{12}H_{15}OS_2Sn$
17a 454, 455, *17c* 337, 338
$C_{12}H_{15}OS_3$
17b 382
$C_{12}H_{15}O_2^{2-}$
17c 193, 195, 197
$C_{12}H_{15}O_2$
1 71, *17e* 216, 227
$C_{12}H_{15}O_2Si$
17b 517
$C_{12}H_{15}O_2Sn$
17c 245
$C_{12}H_{15}W$
17b 94
$C_{12}H_{16}^-$
1 85, *9d1* 715, 725, 728,
17f 83
$C_{12}H_{16}AgN_3O^+$
9a 330
$C_{12}H_{16}CdN_3O^{2-}$
9a 334
$C_{12}H_{16}Cl_2GeNO$
17d2 171
$C_{12}H_{16}Cl_2NO$
9c1 265, *17d1* 308, *17d2* 18
$C_{12}H_{16}CrN_2O_4^-$
17a 256
$C_{12}H_{16}F$
17c 181
$C_{12}H_{16}F_2O_6S_2^+$
17h 67
$C_{12}H_{16}HgNO_3$
9c1 668
$C_{12}H_{16}Hg_{16}BrN_3O^{2-}$
9a 334
$C_{12}H_{16}LiO_2$
9d1 133
$C_{12}H_{16}MnN_2O_7$
17d1 188
$C_{12}H_{16}MoN_2O_4^-$
17a 268
$C_{12}H_{16}N$
17b 100, *17c* 568
$C_{12}H_{16}N^{2+}$
17h 235
$C_{12}H_{16}NO$
9c1 265, 392, 625, 626,
668, 669, 709ff, 743, 753,
17c 568, *17d1* 218,
17d2 142, 348
$C_{12}H_{16}NOS$
17c 505
$C_{12}H_{16}NOSi$
17c 221
$C_{12}H_{16}NO_2$
9c1 100, 101, 235, 238, 292,
669, 710, 718, 778, *17c* 405,
17d1 127, 315, *17d2* 348

[5]) In II/17d1 erroneously listed as $[C_{12}H_{15}CrN_2O_5]^{\cdot}$.

$C_{12}H_{16}NO_2Si$
17c 222
$C_{12}H_{16}NO_3$
9c1 106, 212, 235, 670, 957, *17d1* 181, 182, 271
$C_{12}H_{16}NO_3S$
9c1 753, 996, *17d1* 166
$C_{12}H_{16}NO_4$
9c1 787
$C_{12}H_{16}NO_4SZn^{2+}$
17g 196
$C_{12}H_{16}NO_5$
9c1 583ff
$C_{12}H_{16}NO_5S$
9c1 921, 999
$C_{12}H_{16}NO_6S$
9c1 926
$C_{12}H_{16}NO_9$
17c 624, 625
$C_{12}H_{16}NS_2$
17c 556
$C_{12}H_{16}NSi$
9b 624
$C_{12}H_{16}N_2^+$
17h 182, 236
$C_{12}H_{16}N_2O^+$
17h 236
$C_{12}H_{16}N_2O_2S^+$
17h 140
$C_{12}H_{16}N_2O_3^-$
9d1 543
$C_{12}H_{16}N_2O_4W^-$
17a 275
$C_{12}H_{16}N_2O_5^-$
9d1 626
$C_{12}H_{16}N_2O_5SM$
17d2 47
$C_{12}H_{16}N_3O$
9c1 824, 844, 845, *17d2* 379, 380
$C_{12}H_{16}N_3OPb^{2+}$
9a 299
$C_{12}H_{16}N_3OZn^{2+}$
9a 334
$C_{12}H_{16}N_3O_2$
9c1 814, 823, 871ff, *17d2* 384
$C_{12}H_{16}N_3O_3$
9c1 787
$C_{12}H_{16}NaO_2$
9d1 143
$C_{12}H_{16}O^-$
17c 196, *17f* 191
$C_{12}H_{16}O^+$
17h 66
$C_{12}H_{16}OS^-$
17c 207, *17g* 52
$C_{12}H_{16}O_2^-$
9d1 144, 160
$C_{12}H_{16}O_2^+$
17e 247
$C_{12}H_{16}O_2S_2^-$
9d1 205
$C_{12}H_{16}O_3^-$
9d1 142, 199, 216
$C_{12}H_{16}O_4^-$
9d1 204, 219, 230
$C_{12}H_{16}S_2^-$
9b 579, *17g* 52
$C_{12}H_{16}S_2^+$
9d2 144, *17e* 384, *17h* 110
$C_{12}H_{16}S_3^-$
17f 216
$C_{12}H_{17}$
9b 558, *17b* 23, *17c* 31, 186
$C_{12}H_{17}AsO_2^-$
9b 574
$C_{12}H_{17}AsO_2S^-$
9b 574
$C_{12}H_{17}AsO_3^-$
9b 574
$C_{12}H_{17}ClNO$
9c1 374ff, *17d1* 271, *17d2* 9ff
$C_{12}H_{17}D_{10}N_4O_3^{2+}$
17b 284
$C_{12}H_{17}FO_3S^+$
17h 67
$C_{12}H_{17}F_3NO$
17c 201
$C_{12}H_{17}N^+$
17b 100
$C_{12}H_{17}NO^+$
17b 103
$C_{12}H_{17}NO_2^-$
9d1 458, *17g* 30
$C_{12}H_{17}NO_2Tl$
17g 337, 348
$C_{12}H_{17}NO_3^-$
17f 249
$C_{12}H_{17}NO_4P$
9c2 251, *17c* 602
$C_{12}H_{17}NO_4SM$
17d2 93
$C_{12}H_{17}NO_5SM$
17d2 94
$C_{12}H_{17}N_2$
9c1 93
$C_{12}H_{17}N_2O$
9c1 814, 847, 848, *17d2* 142, 149, 375
$C_{12}H_{17}N_2O_2$
9c1 719, 899, *17d2* 44
$C_{12}H_{17}N_2O_3$
9c1 266, 899, *17b* 386, *17d2* 10
$C_{12}H_{17}N_2O_3S$
17d2 143
$C_{12}H_{17}N_2O_4S$
17d1 169
$C_{12}H_{17}N_4O_2$
17c 432
$C_{12}H_{17}O$
1 60, 70, *9c2* 60
$C_{12}H_{17}OSSn$
17c 242
$C_{12}H_{17}OS_3$
17b 326
$C_{12}H_{17}O_2$
9c2 62, 74, 160, 178, *17b* 193
$C_{12}H_{17}O_2Sn$
17c 241
$C_{12}H_{17}O_3$
9c2 166
$C_{12}H_{17}O_3P^-$
17g 56
$C_{12}H_{17}O_6$
17b 498
$C_{12}H_{17}S$
9b 576, *17b* 198
$C_{12}H_{18}^-$
9d1 721, 845, 846, *17f* 83
$C_{12}H_{18}^+$
1 121, *9d2* 7, *17h* 67
$C_{12}H_{18}ClCrN_2O_6$
17d1 202

$C_{12}H_{18}Cl_2N_5O$
9c1 481
$C_{12}H_{18}Cl_4O_2P^+$
17e 190
$C_{12}H_{18}CoN_2O_4$
17e 18
$C_{12}H_{18}CrN_2O_4S^-$
17a 256, *17e* 388
$C_{12}H_{18}F_7$
17b 304
$C_{12}H_{18}F_7N_2$
9c1 94
$C_{12}H_{18}MoN_2O_4S^-$
17a 268, *17e* 388
$C_{12}H_{18}NO$
9c1 230, 266, 611, 618ff, 670, 684, 685, *17d1* 125, 271ff, 356, *17d2* 10ff, 136, *17e* 70
$C_{12}H_{18}NOS$
9c1 671, *17d2* 62, 76, 201
$C_{12}H_{18}NOSe$
9c1 327
$C_{12}H_{18}NO_2$
9c1 230, 266, 293, 311ff, 634, 650, 686, 926, *17d1* 170, 263, 265, 272, *17d2* 11, 51, 399, 410
$C_{12}H_{18}NO_2S$
17d1 150, *17d2* 201
$C_{12}H_{18}NO_2S_2Si$
9c1 978, 979
$C_{12}H_{18}NO_3$
9c1 686, 958
$C_{12}H_{18}NO_3S$
17d1 120, *17d2* 76
$C_{12}H_{18}NO_4$
9c1 958
$C_{12}H_{18}NO_4S_2$
17c 530
$C_{12}H_{18}NO_4Si$
9b 527
$C_{12}H_{18}NO_5$
9c1 351, 927
$C_{12}H_{18}NO_5S$
9c1 921
$C_{12}H_{18}NO_6$
17b 466, *17d1* 353
$C_{12}H_{18}NO_9$
9c1 787
$C_{12}H_{18}N_2^+$
17b 101, *17h* 147
$C_{12}H_{18}N_2O_2^-$
1 99, *9d1* 202
$C_{12}H_{18}N_2O_4SW^-$
17a 275, *17e* 388
$C_{12}H_{18}N_3O$
9c1 815, *17d2* 228, 375
$C_{12}H_{18}N_3O_3S$
17d2 150
$C_{12}H_{18}N_3O_{10}U$
17a 242
$C_{12}H_{18}O_2^-$
9d1 20
$C_{12}H_{18}O_2^+$
9d1 354, *9d2* 126, 128, 132, 133, 134
$C_{12}H_{18}O_2S_2^+$
9d2 140
$C_{12}H_{18}O_3^-$
17g 29
$C_{12}H_{18}O_3P$
17e 293
$C_{12}H_{18}O_4^-$
9d1 55
$C_{12}H_{18}O_6^+$
9d2 136
$C_{12}H_{18}S_2^+$
17e 382, *17h* 96, 109
$C_{12}H_{18}S_6^+$
17h 97
$C_{12}H_{19}$
17b 490, *17c* 76ff
$C_{12}H_{19}AsNO_3^-$
17f 59
$C_{12}H_{19}ClOP$
9c2 252
$C_{12}H_{19}Ge$
17f 40
$C_{12}H_{19}GeNO_2^-$
9d1 522
$C_{12}H_{19}N^+$
9d2 36
$C_{12}H_{19}NO_2Tl$
17g 87
$C_{12}H_{19}N_2O$
9c1 240, 671, *17d1* 130, 317, *17d2* 107
$C_{12}H_{19}N_2O_2$
9c1 170, 171, *17d2* 82
$C_{12}H_{19}N_2O_3$
17d2 82, 83
$C_{12}H_{19}N_2O_4$
17d2 239
$C_{12}H_{19}N_2O_6M_2$
17d1 35
$C_{12}H_{19}N_3O_2$
17d2 112
$C_{12}H_{19}N_4O_2$
17c 432, *17d2* 112
$C_{12}H_{19}OSi$
9b 594
$C_{12}H_{19}O_6$
17b 354, 371, 373
$C_{12}H_{20}^+$
17h 30, 31
$C_{12}H_{20}Br_2NO$
9c1 374
$C_{12}H_{20}Cl_2NO$
9c1 378
$C_{12}H_{20}Cl_2O_2Sn_2$
17g 222, 229
$C_{12}H_{20}Cl_3N_2O_3$
9c1 522
$C_{12}H_{20}CoN_2O_5$
17e 18
$C_{12}H_{20}F_3$
9b 355
$C_{12}H_{20}NO$
9c1 584
$C_{12}H_{20}NOSi$
17c 148, *17d2* 66
$C_{12}H_{20}NO_2$
9c1 481, 482
$C_{12}H_{20}NO_2Si$
9c1 971, *17d1* 231, 267
$C_{12}H_{20}NO_2Sn$
17d1 267
$C_{12}H_{20}NO_3$
9c1 215, 299, *17d2* 228, 302
$C_{12}H_{20}NO_3SSi$
9c1 979
$C_{12}H_{20}NO_4$
17d1 143
$C_{12}H_{20}NO_4SSi$
9c1 979
$C_{12}H_{20}NO_5$
9c1 754, *17d1* 166, *17d2* 229

$C_{12}H_{20}NO_6$
17d2 202
$C_{12}H_{20}NO_7$
9c1 239
$C_{12}H_{20}NS$
9b 608
$C_{12}H_{20}NSi$
9b 596, *17c* 562
$C_{12}H_{20}N_2^+$
17h 147
$C_{12}H_{20}N_2OX$
17d1 56
$C_{12}H_{20}N_2O_3^{2\cdot}$
17h 281
$C_{12}H_{20}N_2O_4^+$
17h 147
$C_{12}H_{20}N_2O_4M$
17d1 35
$C_{12}H_{20}N_2O_5$
9c1 357
$C_{12}H_{20}N_2O_5M$
17d1 35
$C_{12}H_{20}N_2S_2^-$
9d1 78
$C_{12}H_{20}N_3O$
9c1 342
$C_{12}H_{20}N_3O_3$
17d2 302
$C_{12}H_{20}N_3O_4$
17d2 240
$C_{12}H_{20}N_3O_5$
9c1 873
$C_{12}H_{20}N_3O_7$
9c1 293, *17d2* 217
$C_{12}H_{20}N_3Si$
9c1 122, *17c* 583
$C_{12}H_{20}N_4O_2S_2M$
17d2 80
$C_{12}H_{20}N_4O_2Zn^-$
9a 332
$C_{12}H_{20}N_4S_2Zn^-$
9a 332
$C_{12}H_{20}N_4Zn^-$
9a 331
$C_{12}H_{20}N_5O_4$
17d1 24
$C_{12}H_{20}OPS_2$
9c2 252
$C_{12}H_{20}O_2^-$
9d1 46, 56, *17g* 25, 33
$C_{12}H_{20}O_2P$
9c2 252
$C_{12}H_{20}O_2Si_2^-$
9d1 206
$C_{12}H_{20}O_3P$
9c2 252
$C_{12}H_{20}O_{11}^-$
17b 218
$C_{12}H_{20}S_2^+$
9d2 144, *17e* 380, 381, 383
$C_{12}H_{20}Si^-$
17f 15, 16
$C_{12}H_{21}Cl_3NO_3$
9c1 297
$C_{12}H_{21}FO_2^-$
9b 366
$C_{12}H_{21}NO_2^-$
9d1 417ff
$C_{12}H_{21}NO_3^-$
9d1 682, 683
$C_{12}H_{21}N_2O$
17d2 173
$C_{12}H_{21}N_2OSi$
17d2 202
$C_{12}H_{21}N_2O_2$
9c1 163, *17d2* 136, 302
$C_{12}H_{21}N_2O_4$
17d2 302
$C_{12}H_{21}N_2O_6S$
17d2 90
$C_{12}H_{21}N_2O_7S$
17d2 91
$C_{12}H_{21}N_3O_2P$
17g 218
$C_{12}H_{21}N_3O_5M$
17d1 35
$C_{12}H_{21}N_4O_2$
9c1 856
$C_{12}H_{21}O_{10}$
9b 145
$C_{12}H_{21}O_{11}$
17b 218, 219, 231, 354, 446, 451, 452, 455, 459, 460, 463
$C_{12}H_{21}O_{12}^-$
17g 9
$C_{12}H_{21}Si$
17c 97
$C_{12}H_{22}Cl_6N_3O_3Rh_2$
17a 341
$C_{12}H_{22}FS$
9b 378
$C_{12}H_{22}GeO_2^-$
17g 32
$C_{12}H_{22}Ge_2^-$
17f 44, 191
$C_{12}H_{22}NO$
9c1 252, 392, *17b* 360, 387, *17d2* 213
$C_{12}H_{22}NOSn_2$
17d1 214
$C_{12}H_{22}NO_2$
9c1 428, 482, 483, 517, *17d2* 271
$C_{12}H_{22}NO_2SSi$
9c1 979
$C_{12}H_{22}NO_3$
9c1 215, 342, 359
$C_{12}H_{22}NO_3S$
17d2 180
$C_{12}H_{22}NO_4$
17d2 196
$C_{12}H_{22}NO_5$
9c1 343, 349, 361, *17d1* 120, *17d2* 303
$C_{12}H_{22}NSi$
9b 526
$C_{12}H_{22}N_2^+$
17h 183
$C_{12}H_{22}N_2O^+$
17c 634, *17h* 155
$C_{12}H_{22}N_2O_2$
17g 198
$C_{12}H_{22}N_2O_2^{2\cdot}$
9d2 192, *17h* 271
$C_{12}H_{22}N_2O_4M$
17d1 11, 36
$C_{12}H_{22}N_2O_4SM$
17d1 36
$C_{12}H_{22}N_2O_5SM$
17d1 36
$C_{12}H_{22}N_3O$
17d2 258
$C_{12}H_{22}N_3O_2$
17d2 265
$C_{12}H_{22}N_3O_4$
17d1 19, 70
$C_{12}H_{22}N_7O$
17d2 229

$C_{12}H_{22}O^-$
9b 361
$C_{12}H_{22}OSi_2^+$
17f 37
$C_{12}H_{22}O_2^-$
9b 366, *9d1* 20
$C_{12}H_{22}O_2Si^-$
17g 31
$C_{12}H_{22}O_4^-$
9d1 55
$C_{12}H_{22}O_4P$
17e 294
$C_{12}H_{22}O_4PSSi$
17c 232
$C_{12}H_{22}S^-$
17b 477
$C_{12}H_{22}S_2^+$
17e 371, 377
$C_{12}H_{22}Si_2^-$
17f 16, 191
$C_{12}H_{23}$
9b 272, 351, *17b* 115
$C_{12}H_{23}BN$
17c 438
$C_{12}H_{23}FNO_4$
9c1 380
$C_{12}H_{23}Ge_2$
17c 98
$C_{12}H_{23}N^+$
9d2 28
$C_{12}H_{23}NO_2^-$
9d1 416
$C_{12}H_{23}NO_3P$
17e 294
$C_{12}H_{23}NO_8P$
17d2 117
$C_{12}H_{23}N_2$
17c 440
$C_{12}H_{23}N_2O$
9c1 900, *17d2* 202, 251, 405
$C_{12}H_{23}N_2O_2$
9c1 342, *17d2* 136, 251, 303, 404
$C_{12}H_{23}N_2O_3S$
17d2 180
$C_{12}H_{23}N_2O_4$
9c1 873, *17d2* 150
$C_{12}H_{23}N_2Si$
17c 420
$C_{12}H_{23}N_5O_2X$
17d1 56
$C_{12}H_{23}N_6O_2$
9c1 321
$C_{12}H_{23}O_2$
9b 125, *9c2* 15
$C_{12}H_{23}S$
9b 377
$C_{12}H_{23}Si_2$
9b 477, 592, *17c* 97
$C_{12}H_{24}^-$
17f 83
$C_{12}H_{24}^+$
17h 19
$C_{12}H_{24}B_2^-$
17f 225
$C_{12}H_{24}Br_2NO$
9c1 374
$C_{12}H_{24}ClO_2Sn$
17a 455, *17b* 271
$C_{12}H_{24}N$
17b 506
$C_{12}H_{24}NO$
9c1 267
$C_{12}H_{24}NOS$
17d2 180, 181
$C_{12}H_{24}NO_2$
9c1 238, 343, 483, *17d2* 177
$C_{12}H_{24}NO_3$
17d1 143, 150, 159, *17d2* 229
$C_{12}H_{24}NO_4$
9c1 293, 311, 940
$C_{12}H_{24}NO_7$
17d2 137
$C_{12}H_{24}N_2^+$
9d2 116, *17h* 130
$C_{12}H_{24}N_2O_4P$
17d2 303
$C_{12}H_{24}N_2S_2^-$
9d1 78
$C_{12}H_{24}N_3O_2$
9c1 483, 873
$C_{12}H_{24}N_3O_4$
9c1 267, *17d1* 20, 24, 36, 70, 74
$C_{12}H_{24}N_3O_4X$
17d1 11
$C_{12}H_{24}N_5O_2X$
17d1 57
$C_{12}H_{24}N_5O_4$
17d1 11
$C_{12}H_{24}NaO_4S$
17b 152
$C_{12}H_{24}O_4P$
9c2 253, *17b* 98
$C_{12}H_{24}O_4PS$
17e 294
$C_{12}H_{24}Si_2^+$
17f 34, *17h* 31
$C_{12}H_{25}$
9b 230, *17b* 296
$C_{12}H_{25}ClNO_4P$
17d2 182
$C_{12}H_{25}GeO_4$
9b 249
$C_{12}H_{25}NO_3P$
9c2 253, *17b* 97
$C_{12}H_{25}NO_4P$
9c1 483
$C_{12}H_{25}N_2$
9c1 43, 94
$C_{12}H_{25}N_2O_2$
17d1 120, 135, 147, *17d2* 77, 109, 304
$C_{12}H_{25}N_2O_3$
17d1 143
$C_{12}H_{25}N_2O_4$
1 30
$C_{12}H_{25}N_2Zn$
17a 382
$C_{12}H_{25}N_3O_4M$
17d1 72
$C_{12}H_{25}N_4O_3$
17d1 57
$C_{12}H_{25}O$
9b 135
$C_{12}H_{25}OS_2$
17e 348
$C_{12}H_{25}O_3Si$
9b 204
$C_{12}H_{25}O_4Si$
9b 250
$C_{12}H_{25}S$
9b 144, *9c2* 272
$C_{12}H_{25}Se$
9c2 303
$C_{12}H_{25}Si$
17b 338

$C_{12}H_{25}Si_3$
17c 84
$C_{12}H_{25}Sn$
17b 385
$C_{12}H_{26}Al$
9b 155
$C_{12}H_{26}Al_2N$
17b 500
$C_{12}H_{26}Al_2N_2^-$
17f 202
$C_{12}H_{26}BrN_2O$
9c1 484
$C_{12}H_{26}FSi$
17b 283
$C_{12}H_{26}IN_2O$
17d2 229, 304
$C_{12}H_{26}N$
17b 183
$C_{12}H_{26}NO$
1 54, *9c1* 215, 343, *17c* 482, *17d1* 121, 157, *17d2* 73
$C_{12}H_{26}NOSi$
17d2 172
$C_{12}H_{26}NOSn$
9c1 99
$C_{12}H_{26}NO_5$
17d2 213
$C_{12}H_{26}N_3O$
17d2 229
$C_{12}H_{26}N_3O_2$
9c1 321
$C_{12}H_{26}N_3Si_3$
9b 80
$C_{12}H_{26}N_4^+$
9d2 63
$C_{12}H_{26}N_4O_3X$
17d1 57
$C_{12}H_{26}N_4O_4X$
17d1 71
$C_{12}H_{26}N_6O_4X$
17d1 57
$C_{12}H_{26}OP$
9b 143, *9c2* 221
$C_{12}H_{26}O_4P$
17b 98
$C_{12}H_{26}O_4PSi_2$
9b 308
$C_{12}H_{26}P$
9b 143
$C_{12}H_{27}Al^-$
17f 63
$C_{12}H_{27}AlO_3^-$
17f 63
$C_{12}H_{27}B^-$
17f 221
$C_{12}H_{27}ClNO_2Sn$
17d1 188
$C_{12}H_{27}Ge$
9c2 310
$C_{12}H_{27}GeSe$
17b 316
$C_{12}H_{27}N^+$
17h 128
$C_{12}H_{27}NOP$
9c1 1009
$C_{12}H_{27}NO_3P$
9c2 253
$C_{12}H_{27}NO_4P$
9c1 324, 1009
$C_{12}H_{27}N_2OS$
17c 525
$C_{12}H_{27}N_2Si$
9c1 95
$C_{12}H_{27}N_3O_2X$
17d1 57
$C_{12}H_{27}OSn$
17b 18
$C_{12}H_{27}O_2S$
17e 354
$C_{12}H_{27}O_2Si$
9c2 22
$C_{12}H_{27}O_2Sn$
9c2 24
$C_{12}H_{27}O_3S$
17e 356
$C_{12}H_{27}O_3Si$
17b 201
$C_{12}H_{27}O_4P_2S_4$
17b 259
$C_{12}H_{27}O_6S_5$
9c2 297
$C_{12}H_{27}P^+$
9b 105, *9c2* 264, *17e* 308
$C_{12}H_{27}SSi$
17b 314
$C_{12}H_{27}SSn$
17b 315
$C_{12}H_{27}SeSi$
17b 317
$C_{12}H_{27}SeSn$
17b 317
$C_{12}H_{27}Si$
17b 159, *17f* 7
$C_{12}H_{27}Si_2$
17b 159
$C_{12}H_{27}Sn$
9c2 314, *17b* 201, *17f* 47
$C_{12}H_{28}Cl_2P_2^+$
9c2 264
$C_{12}H_{28}F_3Si_3$
9b 233
$C_{12}H_{28}FeNO_5P_2S_4$
17a 330
$C_{12}H_{28}NOSi$
17d1 121, 152
$C_{12}H_{28}NSi$
9b 240
$C_{12}H_{28}N_2O_2P$
9c2 254
$C_{12}H_{28}N_2O_3PS$
9c1 108
$C_{12}H_{28}N_3O$
17d1 147
$C_{12}H_{28}N_4Zn^+$
9a 332
$C_{12}H_{28}O_2P$
9c2 254
$C_{12}H_{28}O_2S_2^+$
17e 376
$C_{12}H_{28}O_3PSi_2$
9b 308
$C_{12}H_{28}O_4P_2S_4^-$
17e 295, 386ff
$C_{12}H_{28}O_4S_4^+$
17e 366
$C_{12}H_{28}S_2^+$
17e 376
$C_{12}H_{28}Si_2^+$
17f 35, *17h* 24
$C_{12}H_{28}Si_4^-$
17f 17
$C_{12}H_{28}Si_4^+$
17f 35, *17h* 27
$C_{12}H_{29}BP$
17f 61
$C_{12}H_{29}BrN_3O$
17d1 183
$C_{12}H_{29}Ge_2O_2$
17b 140

$C_{12}H_{29}NO_2P$
9c2 254

$C_{12}H_{29}NO_3PSi_2$
17b 114, *17e* 295

$C_{12}H_{29}N_5O_3P$
9c1 245

$C_{12}H_{29}O_2Si_2$
17b 46, 149

$C_{12}H_{29}O_6P_2S$
9c2 254

$C_{12}H_{29}O_7P_2$
9c2 255

$C_{12}H_{29}Si_2$
9b 158

$C_{12}H_{30}As_2^+$
17f 59

$C_{12}H_{30}BN_2O$
17d1 199

$C_{12}H_{30}Ge_2N$
17c 439

$C_{12}H_{30}NOSi_2$
17c 496

$C_{12}H_{30}NSi_2$
17c 562

$C_{12}H_{30}N_2O_2P$
9c2 255

$C_{12}H_{30}N_4O_2P_2^+$
9c2 265

$C_{12}H_{30}O_2P_2^+$
9c2 265

$C_{12}H_{30}O_6P_2^+$
9c2 265, *17e* 309

$C_{12}H_{30}O_{10}P_2$
9c2 255

$C_{12}H_{30}P_2^+$
9c2 265

$C_{12}H_{31}Ge_3$
17b 162

$C_{12}H_{31}NO_3PSi_2$
17b 43, *17e* 295

$C_{12}H_{31}OSi_3$
17b 278

$C_{12}H_{32}N_2O_4P_2^+$
9c2 266

$C_{12}H_{32}N_2P$
17e 256

$C_{12}H_{32}Si_4^-$
17f 26

$C_{12}H_{33}GeSi_3$
17f 40

$C_{12}H_{33}NO_3PSi_2$
17e 296

$C_{12}H_{33}NSi_3^+$
9d2 28, *17h* 129

$C_{12}H_{33}OSi_4$
9b 195

$C_{12}H_{33}Si_4$
17f 7

$C_{12}H_{34}N_2PSi_3$
17e 265

$C_{12}H_{36}AsN_2Si_4$
9c2 317

$C_{12}H_{36}N_2PSi_4$
9c2 217

$C_{12}H_{36}N_2Si_4^+$
9d2 103, *17f* 35, *17h* 148

$C_{12}H_{36}N_3Si_4$
9c1 66

$C_{12}H_{36}N_4Si_4^+$
9d2 63, *17h* 160

$C_{12}H_{36}Si_6^-$
9d1 897, *17f* 27

$C_{12}H_{36}Si_6^+$
17f 37

$C_{12}O_{12}^{7-}$
9d1 90

$C_{13}ClD_8$
17c 256

$C_{13}Cl_2MnN_2O_7$
17a 285

$C_{13}Cl_9$
17c 300

$C_{13}Cl_{11}$
9b 642

$C_{13}D_8O^-$
17c 305

$C_{13}D_{10}O^-$
9b 651, *17c* 269

$C_{13}D_{36}NSi_4$
17b 327

$C_{13}F_{10}O^-$
9b 659, *17c* 273

$C_{13}Fe_4O_{13}^-$
17a 125

$C_{13}HCl_8$
17c 298

$C_{13}HCl_{10}$
9b 638

$C_{13}HD_8$
17c 252

$C_{13}HF_{10}$
9b 638

$C_{13}HF_{10}O$
9b 657ff, *17c* 273

$C_{13}HFe_4O_{13}$
17a 125

$C_{13}H_2D_{17}N_2O_3$
17d2 304

$C_{13}H_3D_{18}S$
9b 604

$C_{13}H_3D_{18}Se$
9b 608, *17c* 233

$C_{13}H_3NO_3^-$
17c 313

$C_{13}H_4Cl_7$
9b 641

$C_{13}H_4MnO_7$
17a 285, *17g* 378

$C_{13}H_5D_5NO$
9c1 137

$C_{13}H_5D_5NOS$
17c 517

$C_{13}H_5D_5N_3O$
17c 606

$C_{13}H_5F_5O^-$
9b 657

$C_{13}H_5FeN_4O_2$
9a 312, 388

$C_{13}H_5MnN_3O_7$
17c 458

$C_{13}H_5N_3O_7Re$
17c 459

$C_{13}H_6ClN_2S_2$
17c 435

$C_{13}H_6Cl_2O^-$
17c 311

$C_{13}H_6Cl_3N_2S$
17c 431

$C_{13}H_6Co_2N_{11}O^{3-}$
9c1 1020

$C_{13}H_6F_2O^-$
1 109, *9b* 686, *17c* 309

$C_{13}H_6F_4O^-$
9b 657

$C_{13}H_6F_5O_2$
9b 476

$C_{13}H_6N_2O_5^-$
17c 313

$C_{13}H_6N_2O_6^{3-}$
9c1 25

$C_{13}H_6OS_4^-$
9b 713, 714, 715
$C_{13}H_6O_5^{3-}$
9d1 256, 265
$C_{13}H_6O_6^{3-}$
9d1 323
$C_{13}H_7BF_2O_2^-$
17c 258
$C_{13}H_7BrNO$
9c1 164
$C_{13}H_7Br_2CoN_6O^{3-}$
9c1 1014
$C_{13}H_7ClD$
17c 255
$C_{13}H_7ClNO$
9c1 164
$C_{13}H_7ClO^-$
17c 309ff
$C_{13}H_7CoN_6O^{3-}$
9c1 1023
$C_{13}H_7D_2N_2O$
17c 430
$C_{13}H_7D_5NO_2S_2$
17c 521
$C_{13}H_7D_8NNaO_6S$
17d1 348
$C_{13}H_7FNO$
9c1 164
$C_{13}H_7FO^-$
17c 307, 308, 309
$C_{13}H_7F_3NOS$
9c1 606
$C_{13}H_7F_6O$
17c 102
$C_{13}H_7INO$
9c1 164
$C_{13}H_7NO_3^-$
17c 313
$C_{13}H_7N_2OS$
9c1 607
$C_{13}H_7N_2O_4^{2-}$
9d1 647
$C_{13}H_7N_2S$
17c 258, 550
$C_{13}H_7N_2S_2$
17c 435
$C_{13}H_7N_6O_{11}$
17d1 374
$C_{13}H_7OS_4$
17c 340, 341, 343
$C_{13}H_7O_2^{2-}$
17c 256
$C_{13}H_7O_4^{2-}$
9d1 180
$C_{13}H_7S_2$
17c 258
$C_{13}H_8BrCoN_6O^{3-}$
9c1 1014
$C_{13}H_8Cl$
17c 255
$C_{13}H_8ClN_2O$
17c 430
$C_{13}H_8ClN_2O_3S$
17c 517
$C_{13}H_8ClN_2S$
17c 431
$C_{13}H_8ClO$
9b 692
$C_{13}H_8ClO_3^{2-}$
9d1 258, 268
$C_{13}H_8Cl_2NO$
9c1 138
$C_{13}H_8Cl_2NOS$
17c 615
$C_{13}H_8Cl_2O^-$
1 110
$C_{13}H_8Cl_2S^-$
9b 664
$C_{13}H_8Cl_4MnO_6$
17a 286, *17g* 207
$C_{13}H_8CoN_7O_2$
9c1 1030
$C_{13}H_8CrO_6^-$
9a 306, *9b* 600
$C_{13}H_8CsO$
17a 223
$C_{13}H_8D$
17c 251
$C_{13}H_8DO$
9b 692
$C_{13}H_8D_{12}NOS$
17d2 115
$C_{13}H_8F$
17c 255
$C_{13}H_8F_2N$
9c1 120
$C_{13}H_8F_2N_2O_4^-$
9d1 644
$C_{13}H_8F_2O^-$
9b 656, 657
$C_{13}H_8KO$
17a 217
$C_{13}H_8LiO$
17a 203
$C_{13}H_8MnO_7$
17a 286, *17b* 395
$C_{13}H_8N$
17c 571
$C_{13}H_8NO$
9c1 32, 163, *17c* 629
$C_{13}H_8NOS_2$
17c 545
$C_{13}H_8NO_2$
9c1 602
$C_{13}H_8NO_3S$
17c 631
$C_{13}H_8N_2O_4S^-$
9b 664
$C_{13}H_8N_2O_5^-$
9b 661, *9d1* 648, 649, *17f* 300ff
$C_{13}H_8N_3O_2^{2-}$
17f 290
$C_{13}H_8N_3O_5$
9c1 137
$C_{13}H_8N_3O_6$
17d1 374
$C_{13}H_8N_3O_7$
17d1 382
$C_{13}H_8NaO$
17a 209
$C_{13}H_8O^-$
1 108, *9b* 631, 678ff, *17c* 301ff
$C_{13}H_8ORb$
17a 222
$C_{13}H_8OS^-$
1 118, *9b* 697, 699ff, *17c* 321
$C_{13}H_8OSe^-$
9b 703ff
$C_{13}H_8O_2^-$
9b 695ff, *17c* 256, 318
$C_{13}H_8O_2S^-$
9b 701, *17c* 323
$C_{13}H_8O_3S^-$
1 118, *9b* 701, *17c* 324
$C_{13}H_8O_4^-$
9d1 343

$C_{13}H_8O_6^{3-}$
9d1 303
$C_{13}H_8O_6W^-$
9a 306, *9b* 601
$C_{13}H_8S^-$
9b 687
$C_{13}H_8S_2^-$
9b 702ff
$C_{13}H_9^{2-}$
9d1 815
$C_{13}H_9$
1 42, *9b* 627ff, 674, *17c* 251
$C_{13}H_9Br_3NO$
17d1 252
$C_{13}H_9Br_3NO_2$
17d1 253
$C_{13}H_9Br_3NO_3$
17d1 253
$C_{13}H_9ClNOS$
17c 517, 615
$C_{13}H_9ClO_2^-$
9d1 185
$C_{13}H_9Cl_2NiO_2$
17a 343
$C_{13}H_9Cl_3NO$
17d1 240
$C_{13}H_9Cl_3NO_2$
17d1 244, 245
$C_{13}H_9DN$
9b 704
$C_{13}H_9D_5N_4O_2^+$
17h 211
$C_{13}H_9D_8NNaO_6S$
17d1 350
$C_{13}H_9FN$
9c1 120
$C_{13}H_9FO^-$
9b 655, 656
$C_{13}H_9F_2NO_2^-$
9d1 477
$C_{13}H_9KN_3O_7S$
9c1 16
$C_{13}H_9KN_3O_8S$
9c1 16
$C_{13}H_9N^-$
9d1 867
$C_{13}H_9NO^-$
9b 686, 687, *9d1* 867
$C_{13}H_9NO^+$
9d2 40
$C_{13}H_9NO_2^-$
9d1 693
$C_{13}H_9NO_3^-$
9d1 532, 533, *17f* 263ff
$C_{13}H_9N_2O$
9c1 859, *17c* 429, 497
$C_{13}H_9N_2O_2$
9c1 589, 893
$C_{13}H_9N_2O_4^{2-}$
9d1 646
$C_{13}H_9N_2O_4$
9b 639, *9c1* 599, *17d1* 374
$C_{13}H_9N_2O_5$
17d1 381
$C_{13}H_9N_3O_4S^-$
9d1 544
$C_{13}H_9N_4$
9c1 86
$C_{13}H_9N_4O_3$
9c1 148, *17c* 607
$C_{13}H_9N_4O_7$
9c1 15
$C_{13}H_9O$
9b 630, 676ff, 692, *17c* 300
$C_{13}H_9OS$
9b 699, *17c* 321
$C_{13}H_9O_2^{2-}$
17c 273
$C_{13}H_9O_2$
9b 631, 694ff, *17c* 91, 318
$C_{13}H_9O_3^{2-}$
9d1 252, 254, 255, 262, 264, 265
$C_{13}H_9O_3$
9c2 41
$C_{13}H_9O_4^{2-}$
9d1 253, 255, 257, 263, 265, 267
$C_{13}H_9O_6$
9c2 200
$C_{13}H_9S$
9b 699
$C_{13}H_9S_3$
9b 762ff, 764
$C_{13}H_{10}^-$
9d1 811ff, *17f* 84ff
$C_{13}H_{10}^{2\cdot}$
17h 253
$C_{13}H_{10}Ba_{1/2}O$
1 43
$C_{13}H_{10}BrO$
9b 659
$C_{13}H_{10}Ca_{1/2}O$
1 44
$C_{13}H_{10}ClO_3$
17e 229
$C_{13}H_{10}Cl_2NO$
17d1 233
$C_{13}H_{10}Cl_3NiO^-$
17a 343
$C_{13}H_{10}D_4N_4O_2^+$
17h 211
$C_{13}H_{10}FN_2O$
17d1 228
$C_{13}H_{10}F_3N_2O_2S$
9c1 67
$C_{13}H_{10}KO$
1 43
$C_{13}H_{10}Mg_{1/2}O$
1 44
$C_{13}H_{10}MnO_2Se$
17a 286
$C_{13}H_{10}N$
9c1 120, *17c* 327, 569
$C_{13}H_{10}NO$
9c1 107, 136ff, *17c* 595, *17d2* 372
$C_{13}H_{10}NOS$
9c1 607, *17c* 516, 615
$C_{13}H_{10}NO_2$
9c1 599, 791, *17d1* 227
$C_{13}H_{10}NO_2S$
9c1 607
$C_{13}H_{10}NO_3$
9c1 559, 792
$C_{13}H_{10}NS$
9c1 35, *9c2* 284, *17c* 634
$C_{13}H_{10}N_2^+$
17h 236
$C_{13}H_{10}N_2O^-$
9b 687
$C_{13}H_{10}N_2O_2S^-$
9d1 513, 544
$C_{13}H_{10}N_2O_4^-$
1 109, *9d1* 643
$C_{13}H_{10}N_2O_5^-$
17f 302
$C_{13}H_{10}N_3$
9c1 62

$C_{13}H_{10}N_3O$
9c1 147, *17c* 606
$C_{13}H_{10}N_3O_5$
9c1 15, 614
$C_{13}H_{10}N_4O_2^-$
17f 293
$C_{13}H_{10}N_5O_6$
9c1 45
$C_{13}H_{10}NaO$
1 43
$C_{13}H_{10}O^-$
1 109, *9b* 647ff, *9d1* 885, *17c* 268ff
$C_{13}H_{10}OS^-$
9b 577
$C_{13}H_{10}O_2^-$
9d1 160ff, 178, 179, 238, 288, *17g* 215
$C_{13}H_{10}O_2S^-$
9d1 189, 190, 886
$C_{13}H_{10}O_3^-$
9d1 183
$C_{13}H_{10}S^-$
9b 663
$C_{13}H_{10}S_2^-$
9b 580
$C_{13}H_{11}$
9b 494, 635, *17c* 260
$C_{13}H_{11}BrFeO_2^-$
9a 316
$C_{13}H_{11}BrO_4^-$
9d1 239
$C_{13}H_{11}ClNO_2S_2$
17c 521
$C_{13}H_{11}ClNS$
17c 507
$C_{13}H_{11}Cl_2N_2O$
9c1 902
$C_{13}H_{11}FeO_3^{2-}$
9d1 27
$C_{13}H_{11}HgO_2$
17g 78, 81
$C_{13}H_{11}NO^-$
9b 716
$C_{13}H_{11}NO_2^-$
9d1 562
$C_{13}H_{11}NO_2S^-$
9d1 508
$C_{13}H_{11}NO_2Zn^+$
17g 195
$C_{13}H_{11}NS^+$
1 126, *9d2* 77, *17h* 194
$C_{13}H_{11}N_2$
9c1 32, *17c* 257
$C_{13}H_{11}N_2O^-$
9c1 809
$C_{13}H_{11}N_2O$
9c1 826, *17d1* 228
$C_{13}H_{11}N_2O_2$
9c1 767, 792, 903, *17d1* 228
$C_{13}H_{11}N_2O_2S$
17c 507
$C_{13}H_{11}N_2O_3$
9c1 613
$C_{13}H_{11}N_2O_3S$
17c 510
$C_{13}H_{11}N_2O_4S_2$
17c 521
$C_{13}H_{11}O$
1 59, *9b* 643ff, *17b* 528, *17c* 175, 267
$C_{13}H_{11}O_2$
1 59, *9b* 659, *9c2* 62
$C_{13}H_{11}O_2S$
17c 177
$C_{13}H_{11}O_3$
17e 229
$C_{13}H_{11}O_3Ru^{2-}$
9d1 27
$C_{13}H_{12}^-$
1 107, *9d1* 753, 762, 804, 805, 822, 837
$C_{13}H_{12}D_4S^-$
17c 237
$C_{13}H_{12}D_5N_3O^+$
17b 142
$C_{13}H_{12}F_5N_2O$
9c1 846
$C_{13}H_{12}F_5N_2O_2$
9c1 874
$C_{13}H_{12}FeO_2^-$
9a 318, *9d1* 27
$C_{13}H_{12}MgN_3O_4$
17g 195
$C_{13}H_{12}MnN_2O_2$
17a 286
$C_{13}H_{12}NO$
9c1 559, 613, *17d1* 215, 225
$C_{13}H_{12}NOS$
9c1 110ff
$C_{13}H_{12}NO_2$
9c1 560
$C_{13}H_{12}NO_2S$
17d1 232
$C_{13}H_{12}NO_2S_2$
17c 521
$C_{13}H_{12}NO_3S$
9c1 995, 1003, *17c* 494, *17d1* 233
$C_{13}H_{12}NO_5P^-$
9d1 411
$C_{13}H_{12}NS$
9c1 110
$C_{13}H_{12}NS_2$
17c 559
$C_{13}H_{12}N_2^+$
9d2 78, *17h* 183
$C_{13}H_{12}N_2O_2Zn^+$
17g 195
$C_{13}H_{12}N_3O$
9c1 834, *17d1* 227
$C_{13}H_{12}O^-$
17c 167
$C_{13}H_{12}OP$
17b 88
$C_{13}H_{12}O_2^-$
9d1 237, 238, 240
$C_{13}H_{12}O_2Ru^-$
9a 317, *9d1* 28
$C_{13}H_{12}O_3^-$
9d1 72
$C_{13}H_{12}O_6^-$
9d1 289
$C_{13}H_{13}BrNO$
17d2 347
$C_{13}H_{13}Cl_3NO_4$
17d1 250
$C_{13}H_{13}Cl_3NO_5$
17d1 245
$C_{13}H_{13}Co^-$
17a 336
$C_{13}H_{13}CoN_6O_3^{3-}$
9c1 1020
$C_{13}H_{13}F_5NO$
17d1 352
$C_{13}H_{13}F_7NO$
9c1 672, 727, *17d1* 308
$C_{13}H_{13}Ge$
9c2 310

$C_{13}H_{13}NO^{-}$
9d1 284
$C_{13}H_{13}N_2O$
9c1 903
$C_{13}H_{13}N_2O_2$
9c1 73
$C_{13}H_{13}N_2O_2Zn^{2+}$
17g 195
$C_{13}H_{13}N_2S_2$
17c 23
$C_{13}H_{13}N_3O_2S^{+}$
17h 194ff
$C_{13}H_{13}O_2Si$
9c2 22
$C_{13}H_{13}O_3$
17e 196
$C_{13}H_{13}P^{+}$
9c2 266
$C_{13}H_{13}S_3$
17c 66
$C_{13}H_{13}Si$
17b 93
$C_{13}H_{13}Sn$
17f 47
$C_{13}H_{14}^{-}$
1 79, *9d1* 738, 741, *17f* 88ff
$C_{13}H_{14}^{+}$
9d2 18
$C_{13}H_{14}ClN_3OS_2M$
17d2 101
$C_{13}H_{14}D_5N_3O_2^{+}$
17b 141
$C_{13}H_{14}F_5N_2O_2$
9c1 417
$C_{13}H_{14}MnN_2O_5$
17a 286
$C_{13}H_{14}MnN_2O_6$
17d1 188
$C_{13}H_{14}NO$
17d2 347
$C_{13}H_{14}NO_2$
9b 590, *17d2* 395
$C_{13}H_{14}NO_2S_2$
17c 557
$C_{13}H_{14}NO_7$
9c1 104
$C_{13}H_{14}NS_2$
17c 552
$C_{13}H_{14}N_2O_3S^{-}$
17f 258
$C_{13}H_{14}N_3O_2$
9c1 824
$C_{13}H_{14}N_4O_2^{+}$
9d2 78, *17h* 210, 211ff
$C_{13}H_{14}O^{-}$
17c 168
$C_{13}H_{15}Br_3NO$
17d1 253
$C_{13}H_{15}Cl_3NO$
17d1 245
$C_{13}H_{15}Cl_3NO_2$
17d1 240
$C_{13}H_{15}Cl_5NO$
17d2 39
$C_{13}H_{15}CoN_6O_3^{3-}$
9c1 1020
$C_{13}H_{15}Cr^{-}$
9a 305
$C_{13}H_{15}Cr^{+}$
9a 305
$C_{13}H_{15}D_3NO_9$
17c 627, 628
$C_{13}H_{15}F_3NO$
17d2 143
$C_{13}H_{15}F_3NO_2$
9c1 293
$C_{13}H_{15}F_5NO$
9c1 267
$C_{13}H_{15}F_5NOSi$
17c 617
$C_{13}H_{15}MoO_4S_3$
17a 268
$C_{13}H_{15}NOS^{-}$
9d1 73
$C_{13}H_{15}NO_2^{-}$
9d1 467
$C_{13}H_{15}NO_5P$
9c2 255
$C_{13}H_{15}N_2O$
9c1 701, *17d1* 166, *17d2* 143
$C_{13}H_{15}N_2OS$
17d1 166
$C_{13}H_{15}N_2OSe$
17d1 166
$C_{13}H_{15}N_2O_2$
9c1 73, *17c* 455
$C_{13}H_{15}N_2O_3$
9c1 705, 958
$C_{13}H_{15}N_3^{+}$
17h 221
$C_{13}H_{15}N_3OS_2M$
17d2 62
$C_{13}H_{15}O_4$
17e 228
$C_{13}H_{15}O_5$
17b 444
$C_{13}H_{15}O_9$
17c 61
$C_{13}H_{15}Si$
17c 82
$C_{13}H_{16}AgBrN_2O^{+}$
9a 329
$C_{13}H_{16}BrHgN_2O^{2+}$
9a 335
$C_{13}H_{16}BrN_2O$
9c1 846, *17d2* 380
$C_{13}H_{16}BrN_2O_2$
9c1 874
$C_{13}H_{16}ClF_3NO$
17d2 18
$C_{13}H_{16}ClN_2O_2$
17d2 258, 265
$C_{13}H_{16}Cl_2N_2OTl$
17d2 259
$C_{13}H_{16}Cl_3N_2O$
17d1 245
$C_{13}H_{16}Cl_4NO$
17d1 284
$C_{13}H_{16}FN_2O_2$
9c1 874, 875, *17d2* 265
$C_{13}H_{16}Ge^{-}$
17f 44
$C_{13}H_{16}GeO_2^{-}$
9d1 25
$C_{13}H_{16}N^{2+}$
17h 236
$C_{13}H_{16}NO$
9c1 220, 619, 743
$C_{13}H_{16}NO_2$
9c1 392, 744, *17d2* 395
$C_{13}H_{16}NO_3$
9c1 396, *17d2* 143
$C_{13}H_{16}NO_4$
9c1 958
$C_{13}H_{16}NO_5$
9c1 644, 959
$C_{13}H_{16}NO_8$
9c1 975
$C_{13}H_{16}N_2O_3S^{-}$
17f 268

$C_{13}H_{16}N_3O_4$
9c1 875, *17d2* 265
$C_{13}H_{16}N_4O_6^-$
17f 282
$C_{13}H_{16}O^-$
9b 570
$C_{13}H_{16}O_2^-$
9d1 25
$C_{13}H_{16}O_2Si^-$
9d1 25
$C_{13}H_{16}S^-$
17b 479, *17c* 237
$C_{13}H_{16}Se^-$
17b 479
$C_{13}H_{16}Si^-$
17f 17
$C_{13}H_{16}Sn^-$
17f 51
$C_{13}H_{17}$
9b 16, 560
$C_{13}H_{17}AgN_2O^+$
9a 330
$C_{13}H_{17}Br_3NO$
17d1 253
$C_{13}H_{17}Cl_3NO$
17d1 272, 308
$C_{13}H_{17}Co^-$
17a 336
$C_{13}H_{17}D_2NS^-$
9b 586
$C_{13}H_{17}F_3NO_2$
9c1 267
$C_{13}H_{17}F_3NO_3$
17d2 304
$C_{13}H_{17}HgN_2O^{2+}$
9a 335
$C_{13}H_{17}MgN_2O_4$
17g 194
$C_{13}H_{17}N^+$
17h 64, 68
$C_{13}H_{17}NOS^-$
9d1 74
$C_{13}H_{17}NO_3M$
17d1 37
$C_{13}H_{17}NO_4M$
17d1 37
$C_{13}H_{17}N_2O$
9c1 267, 422, 708, 900, *17d1* 121, 125, *17d2* 12, 259, 376, 381
$C_{13}H_{17}N_2OS$
17d1 121
$C_{13}H_{17}N_2OSe$
17d1 121
$C_{13}H_{17}N_2O_2$
9c1 423, 779, 876ff, *17d2* 252, 266
$C_{13}H_{17}N_2O_2S$
9c1 268
$C_{13}H_{17}N_2O_3$
9c1 815, 877, *17d2* 130, 266, 384
$C_{13}H_{17}N_2O_4$
9c1 708, *17d2* 267
$C_{13}H_{17}N_2O_5^{2-}$
9d1 427
$C_{13}H_{17}N_2O_5$
9c1 854
$C_{13}H_{17}N_4O_2$
9c1 298, *17d2* 267
$C_{13}H_{17}O_2^{2-}$
9b 553, *17c* 195, 196, 198
$C_{13}H_{17}O_2$
9c2 184, *17e* 221, 222, 224, 228
$C_{13}H_{17}S$
17c 207
$C_{13}H_{18}^-$
9d1 715
$C_{13}H_{18}ClCuN_3O_4$
17a 357
$C_{13}H_{18}Cl_2GeNO$
17d2 189
$C_{13}H_{18}Cl_2NOSi$
17c 617
$C_{13}H_{18}DNS^-$
9b 586
$C_{13}H_{18}FN_2O_2$
9c1 417
$C_{13}H_{18}NO$
9c1 231, 268, 393, 619, 714, 743, *17d1* 124, 218, 272, *17d2* 144
$C_{13}H_{18}NOS$
17d2 168
$C_{13}H_{18}NO_2$
9c1 11, 293, 393, 714, 787, 798, *17d1* 121, 265, 267, 284, 316, 319, 359, *17d2* 44, 137, 144
$C_{13}H_{18}NO_2S$
17d2 168
$C_{13}H_{18}NO_3$
9c1 294, 312, 343, 672, *17d1* 170, 356, *17d2* 55
$C_{13}H_{18}NO_3S$
9c1 997
$C_{13}H_{18}NO_4$
9c1 312
$C_{13}H_{18}NO_5S$
9c1 927, 928
$C_{13}H_{18}NO_6S$
9c1 927, 941
$C_{13}H_{18}NO_9$
17c 624ff, 626, 627, 628
$C_{13}H_{18}N_2^+$
17h 148
$C_{13}H_{18}N_2NaO_3S$
17d2 150
$C_{13}H_{18}N_2OX$
17d2 260
$C_{13}H_{18}N_2O_2S^+$
17h 141
$C_{13}H_{18}N_2O_2X$
17d2 267
$C_{13}H_{18}N_2O_3^-$
17f 254
$C_{13}H_{18}N_3O$
9c1 824
$C_{13}H_{18}N_3O_5$
9c1 268, *17d2* 31, 60
$C_{13}H_{18}N_3O_6$
17d2 118
$C_{13}H_{18}O^-$
17c 194
$C_{13}H_{18}O_2^-$
9d1 212
$C_{13}H_{18}O_2^+$
17e 247, 248, *17h* 64, 68
$C_{13}H_{18}O_2Tl$
17g 336
$C_{13}H_{18}O_3^-$
9d1 142, 143, 199, 216
$C_{13}H_{18}O_5P$
9c2 256
$C_{13}H_{19}$
17b 19, *17c* 186
$C_{13}H_{19}BrNO$
17d2 12

$C_{13}H_{19}ClNO$
9c1 375, *17d1* 356
$C_{13}H_{19}ClNOSi$
17c 616
$C_{13}H_{19}ClNO_2$
9c1 941
$C_{13}H_{19}ClN_3O_7$
17d1 105
$C_{13}H_{19}FN_3O_7$
17d1 106
$C_{13}H_{19}IN_3O_5$
17d1 106
$C_{13}H_{19}IN_3O_7$
17d1 106
$C_{13}H_{19}NO^-$
9b 582, *17c* 209, 210
$C_{13}H_{19}NO^+$
17b 105
$C_{13}H_{19}NO_2^-$
9d1 199
$C_{13}H_{19}NO_4SM$
17d2 94
$C_{13}H_{19}NO_5SM$
17d2 94
$C_{13}H_{19}NS^-$
9b 585, 586, *17c* 211
$C_{13}H_{19}NSe^-$
17c 216
$C_{13}H_{19}N_2O$
9c1 815, *17d2* 188
$C_{13}H_{19}N_2O_2$
9c1 244, 322, *17d1* 58, *17d2* 44
$C_{13}H_{19}N_2O_3$
9c1 484, *17d1* 66, 147, *17d2* 60
$C_{13}H_{19}N_2O_4$
17d1 66
$C_{13}H_{19}N_2O_5$
9c1 811
$C_{13}H_{19}N_3O_{10}PM_2$
17d1 107
$C_{13}H_{19}N_4O_5$
17d1 80
$C_{13}H_{19}O_2$
9c2 74, *17b* 221
$C_{13}H_{19}O_2Sn$
17b 22
$C_{13}H_{19}O_8$
17b 447, 448, 457

$C_{13}H_{19}S$
9b 577, *17b* 197
$C_{13}H_{20}^-$
9d1 722
$C_{13}H_{20}ClNOSi$
17d2 12
$C_{13}H_{20}D_2N_3O_8$
17d1 90, 91
$C_{13}H_{20}D_3N_4O_7$
17d1 91
$C_{13}H_{20}N$
17b 60, 245
$C_{13}H_{20}NO$
9c1 268, 343, 639, 673, 691, *17d1* 125, 157, 266, 273, 285, 357, *17d2* 12ff
$C_{13}H_{20}NOS$
9c1 988, *17d2* 62
$C_{13}H_{20}NOSn$
17c 619, 620
$C_{13}H_{20}NO_2$
9c1 101, 102, 268ff, 312, 650, 691, 724, 941, 942, *17d1* 121, 263, 264, 266, 267, 285, 301, *17d2* 13, 18ff, 51, 410
$C_{13}H_{20}NO_2S$
17d1 160, 188, *17d2* 62
$C_{13}H_{20}NO_3$
17d1 166, 273, 285, *17d2* 13, 19
$C_{13}H_{20}NO_3S$
17c 494
$C_{13}H_{20}NO_3Si$
9c1 959
$C_{13}H_{20}NO_4$
9c1 691, *17d1* 151
$C_{13}H_{20}NO_5P^-$
9d1 516
$C_{13}H_{20}NO_5S$
9c1 928
$C_{13}H_{20}NO_6$
17d1 353, *17d2* 305
$C_{13}H_{20}NO_6S$
9c1 928, *17d1* 355
$C_{13}H_{20}NO_7$
17d1 354
$C_{13}H_{20}NS$
17c 501

$C_{13}H_{20}N_2^+$
17h 148
$C_{13}H_{20}N_2O_2M$
17d1 286
$C_{13}H_{20}N_2O_3^+$
17h 135
$C_{13}H_{20}N_3O$
9c1 815, *17d2* 230, 375
$C_{13}H_{20}N_3O_2$
17d2 13, 47
$C_{13}H_{20}N_3O_6$
17d1 100, 107
$C_{13}H_{20}N_3O_7$
17d1 108
$C_{13}H_{20}N_3O_8$
17d1 91
$C_{13}H_{20}N_3O_{10}U$
17a 242
$C_{13}H_{20}N_4Na_2O_8P$
17d1 107
$C_{13}H_{20}N_4Na_2O_9P$
17d1 108
$C_{13}H_{20}N_5O$
9c1 484
$C_{13}H_{20}O^-$
9b 362, 436, 485
$C_{13}H_{20}S_2^-$
17c 119
$C_{13}H_{21}$
17c 79
$C_{13}H_{21}BD_3N^-$
17c 411
$C_{13}H_{21}D_2N_2O_3$
17d1 121
$C_{13}H_{21}D_9NSi$
17c 562
$C_{13}H_{21}FN_3O_6$
17d1 98
$C_{13}H_{21}FN_3O_8$
17d1 98
$C_{13}H_{21}INO_2$
9c1 534
$C_{13}H_{21}NOPS$
9c1 324
$C_{13}H_{21}NO_3^+$
17h 136
$C_{13}H_{21}NO_3PS$
17d2 69
$C_{13}H_{21}NO_5PS$
17b 289

$C_{13}H_{21}N_2^{2+}$
17b 103
$C_{13}H_{21}N_2O$
9c1 687
$C_{13}H_{21}N_2O_2$
9c1 174, 175, 417, *17d2* 84
$C_{13}H_{21}N_2O_3$
17d1 66, *17d2* 84, 85
$C_{13}H_{21}N_2O_4$
17d1 66
$C_{13}H_{21}N_2O_6$
17d1 158
$C_{13}H_{21}N_3Na_2O_9P$
17d1 92
$C_{13}H_{21}N_3O_5M$
17d1 11
$C_{13}H_{21}N_4O_4X$
17d1 37
$C_{13}H_{21}N_4O_5$
17d1 99, 109
$C_{13}H_{21}N_4O_6$
17d1 109
$C_{13}H_{21}N_4O_8P$
17d1 109
$C_{13}H_{21}O$
9b 602, 603, *17c* 230
$C_{13}H_{21}OS$
17c 110
$C_{13}H_{21}OSi$
9b 593, 594
$C_{13}H_{21}OSn$
9b 598
$C_{13}H_{21}O_5$
17b 100
$C_{13}H_{21}S$
9b 604, 607, *17c* 118
$C_{13}H_{21}SSn$
17b 19
$C_{13}H_{21}Se$
9b 608
$C_{13}H_{22}Ge^-$
17f 44
$C_{13}H_{22}GeNO$
9c1 674
$C_{13}H_{22}GeOSi$
9b 597
$C_{13}H_{22}Ge_2$
9b 597
$C_{13}H_{22}IN_2O$
9c1 675
$C_{13}H_{22}I_2N_3O_3$
17d2 230
$C_{13}H_{22}NO$
9c1 252
$C_{13}H_{22}NOPb$
9c1 674
$C_{13}H_{22}NOSSi$
17d1 375, 376
$C_{13}H_{22}NOSi$
9c1 674, *17d2* 67, 202
$C_{13}H_{22}NOSn$
9c1 675, *17c* 493
$C_{13}H_{22}NO_2$
9c1 200, *9c1* 534, *17d2* 240
$C_{13}H_{22}NO_2SSi$
9c1 971
$C_{13}H_{22}NO_2Si$
9b 527, *9c1* 972, *17d1* 373, 378
$C_{13}H_{22}NO_2Sn$
17d1 379
$C_{13}H_{22}NO_3$
9c1 485, *17d2* 305
$C_{13}H_{22}NO_3SSi$
9c1 972
$C_{13}H_{22}NO_3Si$
9c1 959, 972, 973
$C_{13}H_{22}NO_4SSi$
9c1 973
$C_{13}H_{22}NO_7$
9c1 239
$C_{13}H_{22}NS$
9b 378
$C_{13}H_{22}NS_2$
17e 324
$C_{13}H_{22}N_2^+$
17h 148
$C_{13}H_{22}N_3O_3$
17d2 252
$C_{13}H_{22}N_3O_5$
17b 42
$C_{13}H_{22}N_3O_6$
17d1 92
$C_{13}H_{22}N_3O_7$
17d1 92, 109
$C_{13}H_{22}N_4Na_2O_8P$
17d1 93
$C_{13}H_{22}O^-$
9b 362
$C_{13}H_{22}O_2^-$
17g 26
$C_{13}H_{22}Si^-$
17f 17, 18
$C_{13}H_{22}Si^+$
17f 35
$C_{13}H_{23}GeN_2O$
17c 153
$C_{13}H_{23}Ge_2$
17a 432
$C_{13}H_{23}NO_2$
9c1 485
$C_{13}H_{23}NO_4^+$
17b 99
$C_{13}H_{23}NO_4PS$
17b 288
$C_{13}H_{23}NO_5^+$
17b 404
$C_{13}H_{23}NO_8P$
17d2 337
$C_{13}H_{23}N_2O$
17d1 157
$C_{13}H_{23}N_2OSi$
9b 527, *17c* 149
$C_{13}H_{23}N_2O_4$
17c 445, *17d2* 305
$C_{13}H_{23}N_2O_4S$
17d2 181
$C_{13}H_{23}N_2O_6S$
17d2 90
$C_{13}H_{23}N_4O_4S$
17d1 122
$C_{13}H_{23}N_4O_6$
17d1 93
$C_{13}H_{23}OSi$
17b 491
$C_{13}H_{23}OSi_2$
9b 596
$C_{13}H_{23}O_5$
17b 296
$C_{13}H_{23}Si$
17c 91
$C_{13}H_{24}BD_2N_2$
17c 412
$C_{13}H_{24}BNOP$
17d2 70
$C_{13}H_{24}BN_2O$
17d1 199
$C_{13}H_{24}D_2N_2^+$
17h 148

$C_{13}H_{24}NO$
17d2 333
$C_{13}H_{24}NO_2$
9c1 517, 518, *17d2* 271
$C_{13}H_{24}NO_4$
9c1 299, 428
$C_{13}H_{24}NO_5$
9c1 294
$C_{13}H_{24}N_2^+$
9d2 103
$C_{13}H_{24}N_2O_3^{2-\cdot}$
9d2 223, *17h* 281
$C_{13}H_{24}N_2O_4M$
17d1 37
$C_{13}H_{24}N_2O_5M$
17d1 37
$C_{13}H_{24}N_3O_4$
9c1 877, 878
$C_{13}H_{24}O_3P$
9b 79, 89, *9c2* 256, *17b* 111, 130
$C_{13}H_{24}O_4P$
9c2 256, *17b* 98
$C_{13}H_{24}S^-$
17b 395
$C_{13}H_{24}Si_2^-$
17f 18
$C_{13}H_{25}$
9b 352
$C_{13}H_{25}BN$
17c 438
$C_{13}H_{25}N^+$
17h 129
$C_{13}H_{25}NO_4P$
17b 510, 511
$C_{13}H_{25}NO_8P$
17d2 118, 119
$C_{13}H_{25}N_2O$
17d2 252
$C_{13}H_{25}N_2O_2$
17d2 305
$C_{13}H_{25}N_2O_3$
17d2 306
$C_{13}H_{25}N_2O_3S$
17d2 173
$C_{13}H_{25}N_2O_4$
17c 445
$C_{13}H_{25}N_2O_4^{2+}$
17b 404
$C_{13}H_{25}N_3O_2PS$
9c1 485, *17d2* 306
$C_{13}H_{25}N_3O_3P$
9c1 486
$C_{13}H_{25}N_4O_3$
17d1 176
$C_{13}H_{25}N_4O_5$
17d1 37, 38
$C_{13}H_{25}OSi$
9b 25
$C_{13}H_{25}O_5Si$
17b 312
$C_{13}H_{25}S$
9b 379
$C_{13}H_{25}SSn$
17a 455, *17b* 487
$C_{13}H_{26}ClN_2O$
17d2 306
$C_{13}H_{26}N$
17b 507
$C_{13}H_{26}NO$
9c1 99, 754, *17b* 58, *17d2* 130
$C_{13}H_{26}NO_2$
9c1 486, *17d1* 144, *17d2* 271, 306
$C_{13}H_{26}NO_3$
9c1 942, *17d1* 122
$C_{13}H_{26}NO_4$
9c1 350, 359
$C_{13}H_{26}NO_5S_2$
9c1 486
$C_{13}H_{26}N_2O_4P$
17d2 307
$C_{13}H_{26}N_3O_4$
17d1 20, 67
$C_{13}H_{26}N_3O_4S$
17d1 38
$C_{13}H_{26}N_3O_6S$
17d1 21
$C_{13}H_{27}$
9b 237
$C_{13}H_{27}Cl_2N_3O_2PS$
17d2 307
$C_{13}H_{27}NO^+$
17b 403
$C_{13}H_{27}NO_3P$
9c1 487
$C_{13}H_{27}NO_5P$
9c1 487, 1009
$C_{13}H_{27}N_2$
9c1 94
$C_{13}H_{27}N_2O$
17d2 188, 307
$C_{13}H_{27}N_2O_2$
9c1 318, *17d2* 77, 78
$C_{13}H_{27}N_2O_3$
17d1 144, 158, *17d2* 307, 308
$C_{13}H_{27}N_2Zn$
17a 382
$C_{13}H_{27}N_3O_2P$
17b 510
$C_{13}H_{27}N_3O_2SX$
17d1 58
$C_{13}H_{27}N_3O_4M$
17d1 72
$C_{13}H_{27}N_4O_5$
17d1 72
$C_{13}H_{27}OS_2$
17e 348
$C_{13}H_{27}OSe$
17b 316
$C_{13}H_{27}O_2S$
17e 354
$C_{13}H_{27}S$
17b 314
$C_{13}H_{27}SSe$
17b 313, 316, 317
$C_{13}H_{27}Se$
17b 315
$C_{13}H_{27}Si$
9b 307, *17b* 391
$C_{13}H_{27}Si_2$
17b 321
$C_{13}H_{27}Sn$
17b 474
$C_{13}H_{28}BO$
9b 231
$C_{13}H_{28}ClN_2O_4P$
17d2 308
$C_{13}H_{28}NO$
9c1 231
$C_{13}H_{28}NOSn$
9c1 99
$C_{13}H_{28}NO_2Sn$
9c1 754
$C_{13}H_{28}NO_3$
9c1 320

$C_{13}H_{28}N_2O_4P$
17d2 308
$C_{13}H_{28}N_3O_2$
9c1 322
$C_{13}H_{28}OP$
17b 311
$C_{13}H_{28}O_3PS$
17b 314
$C_{13}H_{28}O_3PSe$
17b 316
$C_{13}H_{29}B^-$
17f 221
$C_{13}H_{29}NO_3P$
9b 240
$C_{13}H_{29}NO_3PSSi_2$
17b 288
$C_{13}H_{29}N_2^{2+}$
17b 403
$C_{13}H_{29}N_3O_2S_2$
17d1 58
$C_{13}H_{29}N_7O_3X$
17d1 58
$C_{13}H_{29}OSn$
9b 71
$C_{13}H_{29}O_3P^+$
17b 281
$C_{13}H_{29}O_7P_2$
9c2 257
$C_{13}H_{29}P^+$
9b 72
$C_{13}H_{29}Si$
9b 220, *17b* 282
$C_{13}H_{30}NO_2Sn$
17d2 410
$C_{13}H_{30}NO_7P$
9c1 324
$C_{13}H_{30}NO_7P_2$
17d1 153
$C_{13}H_{30}NSi$
17c 562
$C_{13}H_{30}N_2O_2P$
17c 391
$C_{13}H_{30}O_4P$
9c2 257
$C_{13}H_{30}O_4PSi$
17b 374
$C_{13}H_{31}NO_3PSi_2$
17e 296
$C_{13}H_{32}BN_2O$
17d1 199
$C_{13}H_{32}NO_6P_2$
17e 296
$C_{13}H_{32}N_2PSi$
17e 256
$C_{13}H_{32}Si_4^-$
17f 18
$C_{13}H_{33}NPSi_2$
17e 256
$C_{13}H_{33}O_2Si_4$
9b 213
$C_{13}H_{33}Si_3$
17b 276
$C_{13}H_{35}Si_4$
17b 159
$C_{13}H_{36}NSi_4$
9b 252
$C_{13}H_{36}N_2OPSi_3$
17e 265
$C_{13}H_{36}N_2PSi_3$
17e 265
$C_{13}H_{36}OPSi_3$
17e 296
$C_{13}H_{37}NOSi_4$
17e 256
$C_{14}Br_6O_6Rh$
17g 362
$C_{14}Cl_8MnO_6$
17a 286, *17g* 208
$C_{14}D_{10}^-$
9d1 778, 782
$C_{14}F_{30}NO$
9c1 382, *17d2* 222
$C_{14}H_4Cl_2O_4^-$
9d1 332
$C_{14}H_4MnO_6S_2$
17a 287, *17g* 372
$C_{14}H_4MoN_2O_{10}W^-$
17f 203
$C_{14}H_4Mo_2N_2O_{10}^-$
17a 268, *17f* 201
$C_{14}H_4N_2O_4^-$
9d1 102
$C_{14}H_4N_2O_{10}W_2^-$
17a 276
$C_{14}H_4O_6^-$
9d1 93
$C_{14}H_5Br_4NO_5Re$
17a 314, *17g* 204
$C_{14}H_5CrN_4O_3$
9a 307, *9b* 388
$C_{14}H_5D_2O_5^{2-}$
17g 298
$C_{14}H_5D_3O_5^-$
17g 298
$C_{14}H_5D_8NNaO_4S_2$
17d1 350
$C_{14}H_5D_{15}F$
9b 561
$C_{14}H_5MoN_4O_3$
9a 307, *9b* 388
$C_{14}H_6Br_2KO_2$
9d1 321
$C_{14}H_6Cl_2O_2^-$
9d1 314, 316, *17g* 292
$C_{14}H_6Cl_2O_4^-$
9d1 317, *17g* 300
$C_{14}H_6Cl_2O_4^+$
9d1 359
$C_{14}H_6Cl_8NO$
17d1 245
$C_{14}H_6Co_2N_{11}O^{6-}$
9c1 1014
$C_{14}H_6D_2O_2^-$
17g 293
$C_{14}H_6D_2O_4^-$
9d1 314
$C_{14}H_6D_3O_3$
17e 204
$C_{14}H_6KN_2O_6$
9d1 322
$C_{14}H_6MnO_6$
17a 287, *17g* 261
$C_{14}H_6MnO_8$
17a 287
$C_{14}H_6NO_2S_2$
17c 550
$C_{14}H_6NO_9$
9c1 584
$C_{14}H_6N_2O_4^-$
9d1 102
$C_{14}H_6N_3S$
9c1 35
$C_{14}H_6NaO_4^{2-}$
17g 291
$C_{14}H_6Na_2O_8S_2^-$
9d1 315ff, 317, *17g* 296
$C_{14}H_6O_2^-$
9d1 329
$C_{14}H_6O_4^{3-}$
17g 295

$C_{14}H_6O_4^-$
9d1 332
$C_{14}H_6O_6^{3-}$
17g 300, 301
$C_{14}H_6O_6Re$
17a 314, *17g* 262
$C_{14}H_6O_8Re$
17a 315
$C_{14}H_6O_8S_2^{3-}$
17g 296
$C_{14}H_7ClO_2^-$
9d1 310, 312
$C_{14}H_7ClO_2$ on Al_2O_3 surface
9d1 365
$C_{14}H_7ClO_4^-$
17g 297
$C_{14}H_7CoN_6O_3^{3-}$
9a 321, *9c1* 1030
$C_{14}H_7D_2O_3$
17e 204
$C_{14}H_7D_2O_5S^+$
17g 386
$C_{14}H_7D_5^-$
9d1 725, *17f* 91
$C_{14}H_7D_5N_3O$
17c 606
$C_{14}H_7HgO_2S_3$
17g 371
$C_{14}H_7HgO_3S_2$
17g 370, 371
$C_{14}H_7KNO_4$
9d1 321
$C_{14}H_7NO_4^-$
9d1 311
$C_{14}H_7NO_6^{3-}$
9b 484
$C_{14}H_7N_2^+$
9d2 91
$C_{14}H_7N_3^-$
9d1 869
$C_{14}H_7NaO_5S^-$
9d1 311, 312, 321, *17g* 290
$C_{14}H_7O_4^{2-}$
17g 291
$C_{14}H_7O_5^{2-}$
17g 296
$C_{14}H_7O_5^-$
17g 298
$C_{14}H_7O_5S^{2-}$
17g 287, 290
$C_{14}H_7O_6^{2-}$
17g 300
$C_{14}H_7O_7S^{2-}$
17g 297
$C_{14}H_8^-$
9d1 757
$C_{14}H_8AlBr_2O_2$
17g 314
$C_{14}H_8AlCl_2O_2$
17a 400, *17g* 314
$C_{14}H_8AlI_2O_2$
17g 314
$C_{14}H_8BaO_2^+$
9d1 319
$C_{14}H_8Br_2GaO_2$
17g 317
$C_{14}H_8Br_2InO_2$
17g 319
$C_{14}H_8Br_2LaO_2$
17g 320
$C_{14}H_8Br_2O_2Y$
17g 328
$C_{14}H_8CaO_2^+$
9d1 319
$C_{14}H_8ClCoN_6O_2^{3-}$
9c1 1021
$C_{14}H_8ClN_2O_2$
9c1 802
$C_{14}H_8ClN_2O_4$
9c1 590
$C_{14}H_8Cl_2GaO_2$
17g 317
$C_{14}H_8Cl_2InO_2$
17g 319
$C_{14}H_8Cl_2LaO_2$
17g 320
$C_{14}H_8Cl_2O_2^-$
17g 17
$C_{14}H_8Cl_2O_2Sb$
17a 499, *17g* 324
$C_{14}H_8Cl_2O_2Sc$
17g 324
$C_{14}H_8Cl_2O_2Y$
17g 329
$C_{14}H_8Co_2N_{11}O^{3-}$
9c1 1021
$C_{14}H_8CrN_2NaO_4$
9a 30
$C_{14}H_8CsO_2$
9d1 309, 319
$C_{14}H_8D_2^-$
9d1 764
$C_{14}H_8D_2O^+$
17g 385
$C_{14}H_8D_3$
17c 253
$C_{14}H_8D_4^-$
9d1 780
$C_{14}H_8F_2O_2Sb$
17a 499, *17g* 324
$C_{14}H_8F_3$
9b 689
$C_{14}H_8GaI_2O_2$
17g 317
$C_{14}H_8I_2InO_2$
17g 319
$C_{14}H_8KO_2$
9d1 306, 309, 319, *17g* 284, 320
$C_{14}H_8K_2O_2^+$
9d1 310
$C_{14}H_8LiO_2$
9d1 306, 307, 319, *17g* 285
$C_{14}H_8Li_2O_2^+$
9d1 309
$C_{14}H_8MoN_2O_4^-$
9a 310
$C_{14}H_8NO_2^-$
9d1 103
$C_{14}H_8NO_2$
9c1 166, *17c* 631
$C_{14}H_8N_2^-$
1 108, *17c* 170
$C_{14}H_8N_2NaO_4W$
9a 311
$C_{14}H_8N_2O_4^-$
9d1 650, 656
$C_{14}H_8N_2O_4Re$
17a 315
$C_{14}H_8N_2O_6^-$
9d1 649
$C_{14}H_8N_3$
9c1 12
$C_{14}H_8N_3O$
9c1 564
$C_{14}H_8N_3OS_2$
9c1 564
$C_{14}H_8N_4^-$
9d1 882

$C_{14}H_8N_4O_6Re$
17a 315
$C_{14}H_8NaO_2$
9d1 306, 308, 319, *17g* 285, 323
$C_{14}H_8NaO_4$
17e 204
$C_{14}H_8Na_2O_2^+$
9d1 310
$C_{14}H_8OS_4^-$
9b 715
$C_{14}H_8O_2^{3-}$
9d1 318
$C_{14}H_8O_2^-$
1 83, 84, *9d1* 304ff, 318, 329, *17g* 283ff, 312ff
$C_{14}H_8O_2^-$ on metaloxide surface
9d1 366
$C_{14}H_8O_2Sr^+$
9d1 319
$C_{14}H_8O_2Tl$
9d1 319ff
$C_{14}H_8O_3^-$
9d1 92, 311, 312, *17g* 286
$C_{14}H_8O_4^{3-}$
9d1 91
$C_{14}H_8O_4^-$
1 83, *9d1* 313, 315, 316, *17g* 292ff, 294
$C_{14}H_8O_4^+$
9d1 359, *17g* 385
$C_{14}H_8O_4U^+$
17a 243, *17g* 327
$C_{14}H_8O_5^-$
9d1 317, *17g* 297, 298
$C_{14}H_8O_6^{5-}$
9d1 325
$C_{14}H_8O_6^{2-}$
9d1 323
$C_{14}H_8O_6^-$
9d1 317, *17g* 301
$C_{14}H_8O_8^-$
9d1 318
$C_{14}H_8O_8S_2^+$
17g 386
$C_{14}H_8S_4^+$
17h 117
$C_{14}H_9$
9b 317
$C_{14}H_9B_2O_5^{2-}$
17g 297
$C_{14}H_9Br_2CoN_6O$
9c1 1021
$C_{14}H_9ClNO_2$
17d2 351
$C_{14}H_9Cl_4N_4$
9c1 74
$C_{14}H_9CoN_6O$
9c1 1023
$C_{14}H_9CoN_6O_3$
9c1 1031
$C_{14}H_9D_2F_2N_4$
9c1 74
$C_{14}H_9D_3OS^-$
17f 212
$C_{14}H_9D_4N_4$
17c 465
$C_{14}H_9F_2$
9b 625, 689
$C_{14}H_9FeO_3^{2-}$
9d1 330
$C_{14}H_9FeO_6$
9a 312
$C_{14}H_9MnNO_5$
17a 287, *17g* 73
$C_{14}H_9N^+$
17h 221
$C_{14}H_9NO_2^-$
9d1 310, *17g* 34, 286
$C_{14}H_9NO_3^-$
9d1 313
$C_{14}H_9N_2O_2$
9c1 88, 167, 802, 834, 859
$C_{14}H_9N_2O_3$
9c1 802, 835
$C_{14}H_9N_2O_4^{2-}$
17g 358
$C_{14}H_9N_2O_4$
9c1 590, 591
$C_{14}H_9N_3^+$
9d2 40
$C_{14}H_9N_3O_2$
9c1 803
$C_{14}H_9N_3O_5$
9c1 638
$C_{14}H_9N_3O_6^-$
9d1 666
$C_{14}H_9N_6O_{11}$
17d1 375
$C_{14}H_9O$
9c2 201
$C_{14}H_9OS$
9b 578
$C_{14}H_9OS_4$
17c 341
$C_{14}H_9O_2$
9b 632, *9c2* 202, *17c* 296
$C_{14}H_9O_3$
17e 204
$C_{14}H_9O_4$
17e 205
$C_{14}H_9O_4U^{2+}$
17a 243, *17g* 327
$C_{14}H_9O_5S^+$
17g 385
$C_{14}H_{10}^-$
1 112, *9d1* 729, 759, 763, 776ff, 782, 826, *17f* 89, 90
$C_{14}H_{10}^+$
1 123, *9d2* 13, *17h* 76
$C_{14}H_{10}Ag$
17b 499
$C_{14}H_{10}Au$
17b 499
$C_{14}H_{10}BrNO_2^-$
9d1 527, 528
$C_{14}H_{10}BrO_3^{2-}$
9d1 259, 269
$C_{14}H_{10}ClNO_2^-$
9d1 527
$C_{14}H_{10}ClO_3^{2-}$
9d1 259, 269
$C_{14}H_{10}Cl_2NO_3S$
9c1 150•
$C_{14}H_{10}Cl_3N_2O$
17d1 240
$C_{14}H_{10}Cu$
17b 499
$C_{14}H_{10}D_2^-$
9d1 779, 824, *17f* 91, 92, 93
$C_{14}H_{10}D_{12}NOS$
17d2 116
$C_{14}H_{10}F$
9b 625, 689
$C_{14}H_{10}FeO_2^-$
9d1 330
$C_{14}H_{10}HgN_3O_2$
17c 458

$C_{14}H_{10}KN_2O_2$
9d1 322
$C_{14}H_{10}MoO_6^-$
9a 306, *9b* 600
$C_{14}H_{10}N$
9b 666, *17c* 285
$C_{14}H_{10}NO$
9c1 609
$C_{14}H_{10}NO_2$
9c1 142, 591, *17c* 599, *17d1* 376
$C_{14}H_{10}NO_2S$
9c1 608
$C_{14}H_{10}NO_3$
9c1 732, *17c* 488, 600
$C_{14}H_{10}NO_4Zn$
17g 195
$C_{14}H_{10}NS_2$
17c 534ff
$C_{14}H_{10}N_2O_2^-$
9d1 313
$C_{14}H_{10}N_2O_4^-$
9d1 420, 647, 650, *17f* 301
$C_{14}H_{10}N_2S^+$
17h 195
$C_{14}H_{10}N_3O_2$
17c 458
$C_{14}H_{10}N_3O_2S$
17c 459
$C_{14}H_{10}N_3S$
17c 541
$C_{14}H_{10}N_4^-$
9d1 880
$C_{14}H_{10}O^-$
9b 690, *9d1* 889, *17c* 306, 307
$C_{14}H_{10}OR$
(R unknown)
9b 662
$C_{14}H_{10}OS^-$
9d1 890, *17g* 49
$C_{14}H_{10}O_2^-$
1 110, *9d1* 25, 26, 240, 884ff, *17c* 311ff, *17g* 17ff, 345
$C_{14}H_{10}O_2^+$
1 124, *9d1* 359, *9d2* 128
$C_{14}H_{10}O_2S^-$
9d1 890
$C_{14}H_{10}O_2S_2^-$
9c2 303
$C_{14}H_{10}O_2S_2^+$
9c2 301
$C_{14}H_{10}O_3S^-$
9b 701, 702, *9d1* 191
$C_{14}H_{10}O_4^-$
9d1 179, 200
$C_{14}H_{10}O_6W^-$
9a 306, *9b* 601
$C_{14}H_{10}S^-$
9d1 889
$C_{14}H_{10}S_2^-$
17f 214
$C_{14}H_{10}S_2^+$
17e 372
$C_{14}H_{11}^{2-}$
9d1 815
$C_{14}H_{11}$
9b 305, 564, 625, 630, 689, *17c* 242, 243, 250, 252
$C_{14}H_{11}BrO_2^-$
9d1 187
$C_{14}H_{11}Br_3NO$
17d1 252, 254
$C_{14}H_{11}ClNO_2S$
17c 517
$C_{14}H_{11}ClNO_3S$
9c1 149, 150
$C_{14}H_{11}ClO_2^-$
9d1 187
$C_{14}H_{11}Cl_3NO$
17d1 241, 246, 370
$C_{14}H_{11}Cl_3NO_2$
17d1 246
$C_{14}H_{11}CoN_6O$
9c1 1031
$C_{14}H_{11}D_5N_4O_2^+$
17h 215, 216
$C_{14}H_{11}F_5NO_7$
9c1 725
$C_{14}H_{11}FeNO_2^-$
9a 317
$C_{14}H_{11}FeO_2$
9c2 186
$C_{14}H_{11}KN_3O_9S$
9c1 16
$C_{14}H_{11}MnN_2O_2^-$
17a 287
$C_{14}H_{11}N^+$
9d2 41
$C_{14}H_{11}NO^-$
9b 704
$C_{14}H_{11}NOS^+$
17h 195
$C_{14}H_{11}NO_2^-$
9d1 527, 528
$C_{14}H_{11}NO_2$
9c1 592
$C_{14}H_{11}NO_3^-$
9d1 504, 533, 563
$C_{14}H_{11}NO_4^-$
9d1 534, 563
$C_{14}H_{11}NO_4Zn^+$
17g 195
$C_{14}H_{11}NO_5S$
9c1 774
$C_{14}H_{11}NS^-$
9b 705
$C_{14}H_{11}N_2O$
9c1 626
$C_{14}H_{11}N_2O_2$
9c1 139, 854, *17c* 597, 598
$C_{14}H_{11}N_2O_3$
17d2 404
$C_{14}H_{11}N_2O_4$
17c 488, 489
$C_{14}H_{11}N_4$
9c1 87
$C_{14}H_{11}N_4O_4$
17c 478
$C_{14}H_{11}N_4O_6$
9c1 903
$C_{14}H_{11}N_4O_8$
9c1 15
$C_{14}H_{11}O$
9b 693, *17c* 175, 276, 317
$C_{14}H_{11}O_3^{2-}$
9d1 257, 267, 268
$C_{14}H_{11}O_6$
9c2 200
$C_{14}H_{12}^{3-}$
9d1 825
$C_{14}H_{12}^-$
1 110, 112, *9b* 540, *9d1* 748, 759, 761ff, 763, 765, 779, 812, 817ff, 823ff, 825, *17f* 90, 91, 92

$C_{14}H_{12}^{+}$
1 123, *9d2* 18, *17h* 73
$C_{14}H_{12}^{2\cdot}$
17h 254
$C_{14}H_{12}BD_3^-$
17c 295
$C_{14}H_{12}Br_2NO$
17d1 369
$C_{14}H_{12}CdCl_2N_4O_4^+$
9a 335
$C_{14}H_{12}Cl$
17c 266
$C_{14}H_{12}ClNS^+$
17h 195
$C_{14}H_{12}Cl_2NO$
17d1 234, 235, 370
$C_{14}H_{12}Cl_2NO_2$
17d1 237
$C_{14}H_{12}Cl_2NS_2$
9c1 114, *17c* 520
$C_{14}H_{12}Cl_2N_4O_4Zn^+$
9a 335
$C_{14}H_{12}D_2^+$
9d2 9
$C_{14}H_{12}D_4N_4O_2^+$
17h 216
$C_{14}H_{12}D_9NO_3S$
9c1 231
$C_{14}H_{12}D_{15}$
9b 352
$C_{14}H_{12}FO$
17c 202
$C_{14}H_{12}HgNO_2$
9c1 560
$C_{14}H_{12}KN_2$
17a 217
$C_{14}H_{12}KN_4O_7S$
9c1 16
$C_{14}H_{12}N$
17c 569
$C_{14}H_{12}NO$
9c1 167, 609, *17c* 570, 595
$C_{14}H_{12}NOS$
9c1 109, 110
$C_{14}H_{12}NOS_2$
9c1 216
$C_{14}H_{12}NO_2$
9c1 135, 767, 774, 795, *17c* 488, 596, *17d2* 365, 395
$C_{14}H_{12}NO_2S$
9c1 35, 36
$C_{14}H_{12}NO_3^-$
9c1 779, 793
$C_{14}H_{12}NO_3$
9c1 552, 792, *17d2* 395
$C_{14}H_{12}NO_3S$
9c1 148, 149, 150, 608
$C_{14}H_{12}NO_4S$
9c1 774
$C_{14}H_{12}NS$
9c1 35, 36
$C_{14}H_{12}N_2^+$
9d2 56, *17h* 167
$C_{14}H_{12}N_2O_2^-$
9d1 528
$C_{14}H_{12}N_2O_2S^-$
9d1 513, 545
$C_{14}H_{12}N_2O_4^-$
1 110, *9d1* 644
$C_{14}H_{12}N_3O$
17c 606
$C_{14}H_{12}N_3O_2$
9c1 148, *17c* 607
$C_{14}H_{12}N_3O_3$
9c1 792
$C_{14}H_{12}N_3O_6$
9c1 15
$C_{14}H_{12}N_3O_9S_2$
9c1 216
$C_{14}H_{12}N_4^+$
9d2 79, *17h* 184
$C_{14}H_{12}N_5O_6$
9c1 15, 45
$C_{14}H_{12}O^-$
9b 651, 652, *9d1* 886, 888, *17c* 269, 270
$C_{14}H_{12}O_2^-$
9d1 184, 240, 289, 324, *17g* 34, 274, 340, 382
$C_{14}H_{12}O_2S^-$
9d1 887
$C_{14}H_{12}O_4^-$
9d1 186, 302, *17g* 340
$C_{14}H_{12}O_4S_2^-$
1 118
$C_{14}H_{12}O_7^-$
9d1 290
$C_{14}H_{12}O_8S_4^+$
17h 112
$C_{14}H_{12}S^-$
9d1 888
$C_{14}H_{12}S_2^+$
1 127, *9d2* 145, *17e* 378
$C_{14}H_{12}S_4^+$
17h 115
$C_{14}H_{12}S_6^+$
17h 117
$C_{14}H_{13}$
9b 639ff, *17c* 180
$C_{14}H_{13}BrNO$
17d1 368
$C_{14}H_{13}ClNO$
17d1 368
$C_{14}H_{13}Cl_2N_2O$
9c1 904
$C_{14}H_{13}D^-$
9d1 810
$C_{14}H_{13}D_2N_3O_2S^+$
17h 197
$C_{14}H_{13}D_3N_4O_2^+$
17h 215
$C_{14}H_{13}D_6O_2$
17e 218
$C_{14}H_{13}FeO_3^{2-}$
9d1 28
$C_{14}H_{13}MnNO_2$
17a 287
$C_{14}H_{13}N^+$
.*9d2* 41
$C_{14}H_{13}NO^-$
9b 583, *17e* 76
$C_{14}H_{13}NO_2^-$
9d1 562
$C_{14}H_{13}NO_3Si^-$
17f 293ff
$C_{14}H_{13}NS^-$
9b 586
$C_{14}H_{13}NS^+$
1 126, *17h* 196
$C_{14}H_{13}N_2O$
9c1 826, *17d1* 228, 229
$C_{14}H_{13}N_2O_2$
9c1 793, 795, 904
$C_{14}H_{13}N_2O_3$
9c1 626
$C_{14}H_{13}N_2O_4$
17c 147
$C_{14}H_{13}N_3O_2S^+$
1 126

$C_{14}H_{13}N_3O_5^-$
17f 302
$C_{14}H_{13}N_4$
9c1 74, *17c* 145
$C_{14}H_{13}N_4O$
17c 470
$C_{14}H_{13}N_4O_4$
9c1 14
$C_{14}H_{13}O$
9b 555, 662, *17c* 183
$C_{14}H_{13}OS$
17c 177
$C_{14}H_{13}O_2$
9c2 17
$C_{14}H_{13}O_3P^-$
17g 56
$C_{14}H_{13}S$
9b 665, *17c* 176, 280
$C_{14}H_{13}S_3$
9b 252, *17b* 326
$C_{14}H_{14}^-$
1 107, *9d1* 743, 753, 767, 805, 806ff, 809, 823
$C_{14}H_{14}^+$
9d2 9, *17h* 72
$C_{14}H_{14}Al_2Br_6S_2$
17e 321
$C_{14}H_{14}Cl_6Ga_2S_2$
17e 321
$C_{14}H_{14}CrO_4^-$
17a 256
$C_{14}H_{14}D_2N_4O_2^+$
17h 215
$C_{14}H_{14}D_3O_2$
17e 217
$C_{14}H_{14}FeO_2^-$
9a 317, *9d1* 28
$C_{14}H_{14}MnO_6$
17a 288
$C_{14}H_{14}N$
9c1 12
$C_{14}H_{14}NO$
1 54, *9c1* 98, 216, 564, 565, 626, *17d1* 218, *17d2* 207
$C_{14}H_{14}NOS_2$
17c 553
$C_{14}H_{14}NO_2$
9b 525, *9c1* 13, 171, 172, 627, 644, *17c* 146
$C_{14}H_{14}NO_2S_2$
17c 521
$C_{14}H_{14}NO_3$
1 56, *9c1* 565ff, *17d2* 369
$C_{14}H_{14}NO_3S$
9c1 624, 995, 997, *17c* 495, *17d1* 215
$C_{14}H_{14}NO_3S_2$
17c 522
$C_{14}H_{14}NO_5P^-$
9d1 412
$C_{14}H_{14}NO_5S$
9c2 216
$C_{14}H_{14}NO_5S_2$
9c1 995, *17d2* 417
$C_{14}H_{14}NS_2$
9c1 113, 114, *17c* 518, 519
$C_{14}H_{14}N_2^+$
9d2 79, 105, *17h* 149, 168
$C_{14}H_{14}N_2O_2S^+$
17h 141
$C_{14}H_{14}N_2S^-$
17e 387
$C_{14}H_{14}N_2S^+$
1 126
$C_{14}H_{14}N_3O_2S$
17c 437
$C_{14}H_{14}N_3O_4S$
9c1 113
$C_{14}H_{14}N_4O_4^-$
17g 357
$C_{14}H_{14}O^-$
9b 618
$C_{14}H_{14}OS^-$
9b 706
$C_{14}H_{14}OS^+$
9d2 146
$C_{14}H_{14}O_2^-$
9d1 240
$C_{14}H_{14}O_2^+$
9d2 136, *17h* 88
$C_{14}H_{14}O_2PS_2$
17e 328
$C_{14}H_{14}O_2S^+$
9d2 139
$C_{14}H_{14}O_3^+$
9d2 137
$C_{14}H_{14}O_4^-$
17g 282
$C_{14}H_{14}O_6^-$
9d1 301
$C_{14}H_{14}O_6Re$
17a 315
$C_{14}H_{14}O_7^-$
9d1 302
$C_{14}H_{14}O_8^-$
9d1 302
$C_{14}H_{14}S^+$
9d2 141
$C_{14}H_{14}SSi^-$
17f 217
$C_{14}H_{15}$
17b 60, 246
$C_{14}H_{15}DN_4O_2^+$
17h 214
$C_{14}H_{15}F_{14}O_3Si$
9b 250
$C_{14}H_{15}HgO_2$
17a 390, *17g* 259
$C_{14}H_{15}N^+$
9d2 41
$C_{14}H_{15}NO$
17d1 170
$C_{14}H_{15}NO_2^-$
9d1 248, 285
$C_{14}H_{15}NO_2^+$
9d2 42, *17h* 172
$C_{14}H_{15}NO_3^+$
9d2 118
$C_{14}H_{15}N_2$
9c1 41, *17c* 248
$C_{14}H_{15}N_2O$
9c1 560, 904, *17d2* 346
$C_{14}H_{15}N_2O_3$
9c1 627
$C_{14}H_{15}N_2O_4$
9c1 970
$C_{14}H_{15}N_2O_5$
17c 452
$C_{14}H_{15}N_3O_2S^+$
17h 196, 197
$C_{14}H_{15}N_3O_2Zn^+$
17g 196
$C_{14}H_{15}N_3O_5^-$
17g 356
$C_{14}H_{15}N_4O_4$
17e 215
$C_{14}H_{15}OS_3Sn$
17a 455, *17c* 339, 340

$C_{14}H_{15}O_2P^-$
17g 30
$C_{14}H_{15}O_5$
17b 444
$C_{14}H_{15}Si$
9b 35
$C_{14}H_{15}Sn$
17f 47
$C_{14}H_{16}^-$
9d1 738, 741, 745, 748, 750, 751, *17f* 93, 94
$C_{14}H_{16}^+$
9d2 11, *17h* 73, 74
$C_{14}H_{16}Cl_3N_2O$
17d1 241
$C_{14}H_{16}Cr^+$
17a 257
$C_{14}H_{16}CrI$
9a 304
$C_{14}H_{16}D_3O_2$
17e 218
$C_{14}H_{16}Hg_2NO_5$
9c1 733
$C_{14}H_{16}NO$
9c1 269, 693
$C_{14}H_{16}NO_2$
9b 590, *17d1* 167, *17d2* 396
$C_{14}H_{16}NO_7$
9c1 104, 966
$C_{14}H_{16}NS_2$
17c 559, 560
$C_{14}H_{16}N_2^-$
9d1 872
$C_{14}H_{16}N_2^+$
9d2 105, *17h* 68
$C_{14}H_{16}N_2O_2S^-$
17f 257
$C_{14}H_{16}N_2O_4^-$
9d1 873
$C_{14}H_{16}N_3O$
9c1 269
$C_{14}H_{16}N_3O_2Zn^{2+}$
17g 196
$C_{14}H_{16}N_4O_2^+$
17h 213ff, 216ff
$C_{14}H_{16}O_2^-$
17g 339
$C_{14}H_{16}O_2P$
9c2 257
$C_{14}H_{16}O_2Tl$
9d1 277
$C_{14}H_{16}PS_2$
17e 297
$C_{14}H_{16}S_4^+$
17h 116
$C_{14}H_{16}Si^-$
17f 18
$C_{14}H_{17}ClNO_5$
17d1 218
$C_{14}H_{17}Cl_3NO$
17d1 241
$C_{14}H_{17}Cl_4CoN_5O_6$
9c1 1035
$C_{14}H_{17}NO_5^{2-}$
17g 197
$C_{14}H_{17}N_2O$
9c1 393, *17d2* 137
$C_{14}H_{17}N_2O_2$
17c 455
$C_{14}H_{17}O_4$
17e 225
$C_{14}H_{18}^-$
9d1 723
$C_{14}H_{18}^+$
9d2 8
$C_{14}H_{18}AlN_2$
17a 401, *17f* 204
$C_{14}H_{18}BD_2F_2O_2$
17g 157
$C_{14}H_{18}Cl_2N_5O_2P^+$
17e 189
$C_{14}H_{18}Cl_3CoN_5O_6$
9c1 1036
$C_{14}H_{18}Cl_3N_2O$
17d1 246
$C_{14}H_{18}F_4NO_3$
17d1 160
$C_{14}H_{18}F_6O_3P$
9c2 257
$C_{14}H_{18}FeNO^-$
9a 318
$C_{14}H_{18}FeNO$
9c1 696
$C_{14}H_{18}MnO_6$
17a 288
$C_{14}H_{18}NO$
9c1 410
$C_{14}H_{18}NO_2$
9c1 714, 744, 803, *17d1* 167
$C_{14}H_{18}NO_2S$
17d1 160, *17d2* 333
$C_{14}H_{18}NO_3$
9c1 299, 386, 407, 960, *17d2* 362
$C_{14}H_{18}NO_3Se$
17d1 167
$C_{14}H_{18}NO_5$
9c1 351, *17d2* 19
$C_{14}H_{18}NO_6$
17d1 222
$C_{14}H_{18}NSi$
17c 150
$C_{14}H_{18}N_2^+$
1 125, *9d2* 103
$C_{14}H_{18}N_2O_2S_2^-$
17g 248
$C_{14}H_{18}N_2O_4^-$
17g 248
$C_{14}H_{18}N_2O_4M$
17d1 71
$C_{14}H_{18}N_2O_5^-$
9d1 628
$C_{14}H_{18}N_2O_8Re$
17a 315
$C_{14}H_{18}N_2Zn^-$
17a 382
$C_{14}H_{18}N_3O_2$
17d2 259
$C_{14}H_{18}N_3O_3$
9c1 423, *17d2* 267
$C_{14}H_{18}N_4Ni^+$
9a 326
$C_{14}H_{18}N_5O$
17d2 150
$C_{14}H_{18}N_5O_7$
9c1 352
$C_{14}H_{18}O^-$
9b 571, *17b* 481
$C_{14}H_{18}O_2^-$
9d1 21
$C_{14}H_{18}O_4^-$
9d1 97, *17g* 240
$C_{14}H_{18}Si_2^-$
17f 24
$C_{14}H_{19}AlBr_2NO_4$
9d1 151
$C_{14}H_{19}AlCl_2NO_4$
9d1 151

$C_{14}H_{19}AlI_2NO_4$
9d1 151
$C_{14}H_{19}BrNO_2Si$
17d1 387
$C_{14}H_{19}Br_2GaNO_4$
9d1 152
$C_{14}H_{19}ClNO$
17d1 357
$C_{14}H_{19}ClNO_2Si$
17d1 387
$C_{14}H_{19}ClNaO_2$
9d1 150
$C_{14}H_{19}Cl_2GaNO_4$
9d1 151
$C_{14}H_{19}Cl_2InNO_4$
9a 298, *9d1* 152
$C_{14}H_{19}Cl_2LaNO_4$
9a 300, *9d1* 153
$C_{14}H_{19}Cl_2LuNO_4$
9a 301, *9d1* 152
$C_{14}H_{19}Cl_2NO_4Sc$
9d1 151
$C_{14}H_{19}Cl_2NO_4Y$
9a 300, *9d1* 152
$C_{14}H_{19}Cl_3NO_4Sn$
9d1 152
$C_{14}H_{19}Cl_4NO_4Sn$
9a 299
$C_{14}H_{19}Cl_5NO_4Sb$
9a 299
$C_{14}H_{19}CoN_5O_5$
9c1 1024
$C_{14}H_{19}CoO_7P$
17a 336, *17g* 72
$C_{14}H_{19}D_2O_3$
9c2 172
$C_{14}H_{19}GaI_2NO_4$
9d1 152
$C_{14}H_{19}NNaO_4$
9d1 150
$C_{14}H_{19}NOS^-$
9d1 74
$C_{14}H_{19}NO_5P$
17c 611, 613
$C_{14}H_{19}N_2O$
9c1 269, 270, 709, 824, *17d2* 260
$C_{14}H_{19}N_2O_2$
9c1 359, 812, 878, *17d2* 260, 267ff
$C_{14}H_{19}N_2O_3$
17d2 268
$C_{14}H_{19}N_2O_4$
9c1 879
$C_{14}H_{19}N_2O_{12}$
9b 150
$C_{14}H_{19}N_3O_4P$
17c 612, 613
$C_{14}H_{19}N_4O_2$
17d2 252, 343
$C_{14}H_{19}N_5O_2$
17g 232
$C_{14}H_{19}N_6O_4$
17d2 119
$C_{14}H_{19}OS$
17e 231
$C_{14}H_{19}O_2^{2-}$
9b 567, 569, *17c* 193, 198
$C_{14}H_{19}O_2$
9c2 184, *17e* 216ff, 227
$C_{14}H_{19}O_3^{2-}$
9d1 248, 271
$C_{14}H_{19}O_3$
9c2 183, *17e* 221
$C_{14}H_{19}O_4$
17e 225
$C_{14}H_{19}O_9$
17b 443ff, 445, 454
$C_{14}H_{20}AlBr_2O_2$
9d1 137, 148
$C_{14}H_{20}AlCl_2O_2$
9d1 137, 148, *17a* 401, *17e* 63, *17g* 154, 155
$C_{14}H_{20}AlI_2O_2$
9d1 138, 148
$C_{14}H_{20}AsO_2$
9c2 317
$C_{14}H_{20}BF_2O_2$
17g 157
$C_{14}H_{20}BaO_2$
9d1 137, *17a* 236
$C_{14}H_{20}BaO_2^+$
17g 226
$C_{14}H_{20}Br_2GaO_2$
9d1 138, 148
$C_{14}H_{20}Br_2InO_2$
9d1 147
$C_{14}H_{20}Br_2LaO_2$
9d1 147
$C_{14}H_{20}CaO_2$
9d1 137, *17g* 226
$C_{14}H_{20}CaO_2^+$
17a 231, *17g* 226
$C_{14}H_{20}Cl_2F_2O_2Sn^-$
17a 456, *17g* 190
$C_{14}H_{20}Cl_2GaO_2$
9d1 138
$C_{14}H_{20}Cl_2InO_2$
9a 297, *9d1* 138, 147
$C_{14}H_{20}Cl_2LaO_2$
9a 300, *9d1* 138, 147
$C_{14}H_{20}Cl_2LuO_2$
9a 301, *9d1* 138, 147
$C_{14}H_{20}Cl_2NO$
17d1 236
$C_{14}H_{20}Cl_2O_2Sb$
17g 137, 177
$C_{14}H_{20}Cl_2O_2Sc$
9d1 139, 147
$C_{14}H_{20}Cl_2O_2Y$
9a 300, *9d1* 139, 147
$C_{14}H_{20}Cl_3CoN_5O_6$
9c1 1031
$C_{14}H_{20}Cl_3FO_2Sn^-$
17a 456, *17g* 189
$C_{14}H_{20}Cl_3IO_2Sn^-$
17g 189
$C_{14}H_{20}Cl_3O_2Sn$
9d1 139, *17a* 456ff, *17g* 186ff
$C_{14}H_{20}Cl_4O_2Sn^-$
17a 457, *17g* 189
$C_{14}H_{20}Cl_4O_2Sn$
9a 299
$C_{14}H_{20}CrN_2O_4^-$
9a 307
$C_{14}H_{20}CsO_2$
9d1 214
$C_{14}H_{20}CsO_4$
9d1 148
$C_{14}H_{20}DO_2$
9c2 65
$C_{14}H_{20}FS$
9c2 282
$C_{14}H_{20}F_3OSi$
17c 223
$C_{14}H_{20}F_3O_2Si$
17g 137, 177

$(C_{14}H_{20}F_3O_2Si)_n$
17g 178
$C_{14}H_{20}F_4O_2Si^-$
17g 179
$C_{14}H_{20}GaI_2O_2$
9d1 138
$C_{14}H_{20}Ga_2N_2^+$
17a 408
$C_{14}H_{20}I_2InO_2$
9d1 148
$C_{14}H_{20}InO_2$
17g 164
$C_{14}H_{20}In_2N_2^+$
17a 409
$C_{14}H_{20}KO_2$
9d1 137, 197, 214, *17a* 217, *17g* 165, 227
$C_{14}H_{20}LiO_2$
9d1 137, 145, 197, 215
$C_{14}H_{20}MgO_2$
9d1 137
$C_{14}H_{20}MgO_2^+$
17a 226, *17g* 227
$C_{14}H_{20}MnN_2O_4$
17a 288
$C_{14}H_{20}MoN_2O_4^-$
9a 309
$C_{14}H_{20}NO$
9c1 269, 294, *17c* 591, 592, 594, *17d1* 273, 286, *17d2* 13, 173ff, *17e* 212
$C_{14}H_{20}NOS$
17d2 181
$C_{14}H_{20}NO_2$
9c1 162, 231, *17d1* 122, 302, 316, 360, *17d2* 13, 51
$C_{14}H_{20}NO_2S$
9c2 282
$C_{14}H_{20}NO_2Si$
17d1 387
$C_{14}H_{20}NO_3$
9c1 270, 294, 370, 960, *9c2* 130, *17d1* 273, 274, 286, 302, 357, *17d2* 271
$C_{14}H_{20}NO_3Se$
17d1 122
$C_{14}H_{20}NO_4$
9c1 270, 960, *17d2* 19, 96, 216
$C_{14}H_{20}NO_5S$
9c1 921, 929, 942
$C_{14}H_{20}NO_6$
17d1 353
$C_{14}H_{20}NO_7$
17d1 355
$C_{14}H_{20}NS_2$
17c 546
$C_{14}H_{20}N_2O_2^+$
17h 149
$C_{14}H_{20}N_2O_3^-$
9d1 531, *17f* 254
$C_{14}H_{20}N_2O_4^-$
1 102, *9d1* 640
$C_{14}H_{20}N_2O_4^+$
17h 136
$C_{14}H_{20}N_2O_4Re$
17a 315
$C_{14}H_{20}N_2O_4W^-$
9a 311
$C_{14}H_{20}N_2O_5SM$
17d2 95
$C_{14}H_{20}N_2O_7P$
17c 588, 589
$C_{14}H_{20}N_3O$
9c1 344
$C_{14}H_{20}N_3O_4S$
17d2 151
$C_{14}H_{20}N_4O_2^-$
17g 246
$C_{14}H_{20}N_5O_2P^+$
17e 55
$C_{14}H_{20}N_5O_4$
9c1 231
$C_{14}H_{20}N_7O$
9c1 487
$C_{14}H_{20}NaO_2$
9d1 137, 145, 197, 215, *17g* 125, 168, 220, 227
$C_{14}H_{20}O^-$
17c 197
$(C_{14}H_{20}O)_n$
17e 135
$C_{14}H_{20}OS^-$
17b 333
$C_{14}H_{20}O_2^-$
9d1 134, 196, 212ff, *17g* 114, 220, 225ff
$C_{14}H_{20}O_2^+$
17e 248
$C_{14}H_{20}O_2Rb$
9d1 137, 198, 215
$C_{14}H_{20}O_2S^-$
9d1 86
$C_{14}H_{20}O_2S_2^-$
9d1 205
$C_{14}H_{20}O_2Tl$
9d1 135, 146
$C_{14}H_{20}O_2Zn$
9d1 137
$C_{14}H_{20}S_2^-$
17b 334
$C_{14}H_{21}$
9b 559, 560, *17b* 59, 245
$C_{14}H_{21}CoN_4O_5$
17a 336
$C_{14}H_{21}F_2OSi$
17c 223
$C_{14}H_{21}F_3NO$
17c 201
$C_{14}H_{21}NO_2^-$
1 98, *9d1* 461ff, *17g* 232
$C_{14}H_{21}NO_2P$
17d2 338
$C_{14}H_{21}NO_2R_3$
(R = various substituents)
9d1 144
$C_{14}H_{21}NO_3^-$
17f 250
$C_{14}H_{21}NO_3Tl$
17g 89
$C_{14}H_{21}NO_4Tl$
17g 96
$C_{14}H_{21}NO_5P$
17c 588
$C_{14}H_{21}N_2O$
9c1 636, 811, 816, *17d1* 229, *17d2* 252
$C_{14}H_{21}N_2O_2$
9c1 318, 418, 779, 816, 910, *17d1* 58, *17d2* 44
$C_{14}H_{21}N_2O_3$
9c1 271, 514, 714, *17d1* 58ff, *17d2* 60
$C_{14}H_{21}N_2O_3S$
17d2 62
$C_{14}H_{21}N_2O_4$
9c1 848, *17d2* 381
$C_{14}H_{21}N_2O_5$
9c1 822, 879, *17d2* 385

$C_{14}H_{21}N_2O_6$
17d2 119
$C_{14}H_{21}N_6O$
17d2 308
$C_{14}H_{21}N_6O_4$
17d1 110
$C_{14}H_{21}O$
1 60, *9c2* 56, 60, 63, *17e* 51, 85
$C_{14}H_{21}O_2$
9c2 65, 66, 131, 160, *17e* 71, 134
$C_{14}H_{21}O_2S$
17e 332
$C_{14}H_{21}O_2Sn$
17b 19
$C_{14}H_{21}O_3$
9c2 172
$C_{14}H_{21}O_4$
9b 476
$C_{14}H_{21}O_4U^{2+}$
17a 243, *17g* 139
$C_{14}H_{21}S$
9c2 280
$C_{14}H_{22}^{-}$
9d1 722, *17c* 170, *17f* 94
$C_{14}H_{22}^{+}$
9d2 7
$C_{14}H_{22}Cl_3Si$
9b 478
$C_{14}H_{22}D_2N_3O_8$
17d1 93, 100
$C_{14}H_{22}FOSi$
17c 221, 222
$C_{14}H_{22}FSi$
17c 217
$C_{14}H_{22}N$
9b 609, *9c1* 11, *17c* 233, 234
$C_{14}H_{22}NO$
9c1 100, 232, 271, 313, 622, 642, 676, 692, *9c2* 119ff, 148, *17d1* 122, 222, 274, 301ff, *17d2* 14, 19ff, *17e* 71, 133, 148
$C_{14}H_{22}NOS$
17c 524, *17d1* 194, 320, *17d2* 63
$C_{14}H_{22}NO_2$
9c1 271, 272, 313, 428, 969, *17d1* 274, 286, 287, 302, 319, *17d2* 20, 31ff, 52, 272
$C_{14}H_{22}NO_2Si_2$
17c 222
$C_{14}H_{22}NO_3$
17d1 144, *17d2* 14, 20, 58, 137, 272
$C_{14}H_{22}NO_4$
17d2 32, 240
$C_{14}H_{22}NO_5$
9c1 352, *17d2* 97
$C_{14}H_{22}NO_5S$
9c1 922, 929, 943
$C_{14}H_{22}NO_6$
17d1 354
$C_{14}H_{22}NO_7$
17d2 393
$C_{14}H_{22}N_2^{+}$
17h 149
$C_{14}H_{22}N_2O_2^{-}$
1 99, 100, *9d1* 202, 230
$C_{14}H_{22}N_2O^{2\cdot}$
9d2 284
$C_{14}H_{22}N_2O_2S_2^{-}$
9d1 205
$C_{14}H_{22}N_2O_4$
17d2 106
$C_{14}H_{22}N_3O_2$
9c1 244, 245, *17d1* 317
$C_{14}H_{22}N_3O_3$
17d1 59
$C_{14}H_{22}N_3O_6$
17d1 80, 103
$C_{14}H_{22}N_3O_{10}U$
17a 243, 244
$C_{14}H_{22}N_4O_2S^{+}$
9c1 28
$C_{14}H_{22}N_5O_4$
9c1 848, *17d2* 381
$C_{14}H_{22}N_5O_5$
9c1 879, *17d2* 385
$C_{14}H_{22}OS^{-}$
9b 606, 608
$C_{14}H_{22}O_2^{-}$
9d1 21
$C_{14}H_{22}O_2^{+}$
9d1 354, 355, *9d2* 126, 131, 134, *17h* 69
$C_{14}H_{22}O_2S_2^{+}$
9d2 140
$C_{14}H_{22}O_2Tl$
17g 75
$C_{14}H_{22}O_4^{-}$
9d1 55
$C_{14}H_{22}O_4^{+}$
9d2 136
$C_{14}H_{23}ClNO$
9c1 375
$C_{14}H_{23}ClNO_2$
9c1 329, 943
$C_{14}H_{23}ClN_3O_6$
17d1 99
$C_{14}H_{23}CoO_4X$
(X = various solvents)
9c2 8
$C_{14}H_{23}CoO_6X$
(X = various solvents)
9c2 16
$C_{14}H_{23}CoO_7X$
(X = various solvents)
9c2 26
$C_{14}H_{23}DNOS$
9c1 987
$C_{14}H_{23}GaNO_3$
17a 408
$C_{14}H_{23}Ge$
17a 432
$C_{14}H_{23}N^{+}$
9d2 36, *17h* 162
$C_{14}H_{23}NO_3P$
17e 297
$C_{14}H_{23}N_2O$
9c1 305, 676, *17d2* 20, 32
$C_{14}H_{23}N_2O_2$
9c1 174, 175, 410, 534, 677, *17d1* 182
$C_{14}H_{23}N_2O_3$
9c1 358, *17d2* 85
$C_{14}H_{23}N_2O_4$
17d2 253
$C_{14}H_{23}N_2O_5S$
17d2 102
$C_{14}H_{23}N_2O_7$
9c1 355
$C_{14}H_{23}N_3O_2X$
17d1 15, 59
$C_{14}H_{23}N_4O_2$
17d2 112
$C_{14}H_{23}O$
17c 109

$C_{14}H_{23}OSi$
9b 593, 594
$C_{14}H_{23}OSn$
9b 598
$C_{14}H_{23}O_2S$
17e 354
$C_{14}H_{23}P^-$
9d1 895
$C_{14}H_{23}S$
9b 604
$C_{14}H_{23}Si$
9b 592, *17c* 217
$C_{14}H_{24}^{+}$
17h 31, 32
$C_{14}H_{24}AgN_4O_2S_2^{2\cdot}$
17h 286
$C_{14}H_{24}AlN_2^-$
17f 206
$C_{14}H_{24}Al_2N_2^-$
17a 401
$C_{14}H_{24}Cl_2GaO_2$
9d1 148
$C_{14}H_{24}MnNO_5P$
9c1 1037, *17a* 288
$C_{14}H_{24}MnNO_8P$
9c1 1037, *17a* 288
$C_{14}H_{24}NO$
9c1 271ff, 344, *17c* 482
$C_{14}H_{24}NOSi$
9c1 323, *17d1* 274, *17d2* 14, 67, 203
$C_{14}H_{24}NO_2$
9c1 154, 345, 429, *17b* 349
$C_{14}H_{24}NO_2SSi$
9c1 973
$C_{14}H_{24}NO_2Si$
17d2 203
$C_{14}H_{24}NO_3S_2$
17d2 168
$C_{14}H_{24}NO_3Si$
9c1 961, 973
$C_{14}H_{24}NO_4SSi$
9c1 974
$C_{14}H_{24}NO_6$
17d2 121
$C_{14}H_{24}NS$
17c 501
$C_{14}H_{24}N_2^{+}$
9d2 56, *17h* 149, 167
$C_{14}H_{24}N_2O_2X$
17d2 102
$C_{14}H_{24}N_2O_4^{2\cdot}$
9d2 213ff, 215, *17h* 282
$C_{14}H_{24}N_2O_4M$
17d1 38
$C_{14}H_{24}N_2S_2^-$
9d1 78
$C_{14}H_{24}N_3O_5$
9c1 880
$C_{14}H_{24}N_3O_6$
17d1 99
$C_{14}H_{24}N_3O_7$
17d1 99
$C_{14}H_{24}N_4^{+}$
17h 184
$C_{14}H_{24}N_4Na_2O_2^{+}$
17g 246
$C_{14}H_{24}N_4O_2$
9c1 857
$C_{14}H_{24}N_4O_2^{+}$
17g 384, *17h* 173
$C_{14}H_{24}N_4O_2^{2\cdot}$
9d2 287
$C_{14}H_{24}N_4O_2PdS_2^{2\cdot}$
17h 287
$C_{14}H_{24}N_4O_2PtS_2^{2\cdot}$
17h 287
$C_{14}H_{24}N_4O_3^{2\cdot}$
9d2 289
$C_{14}H_{24}N_4O_4^{2\cdot}$
9d2 290
$C_{14}H_{24}N_6O_4$
17d1 103
$C_{14}H_{24}O^+$
17h 32
$C_{14}H_{24}O_2^-$
17g 26
$C_{14}H_{24}O_3P$
9c2 258
$C_{14}H_{24}O_4PSi$
9b 595, *17c* 217, 222
$C_{14}H_{25}$
9b 402, *17b* 34, 477
$C_{14}H_{25}F_3NOSi_2$
17c 202
$C_{14}H_{25}NO_2Sn$
17d2 410
$C_{14}H_{25}N_2O_2$
17d1 124
$C_{14}H_{25}N_2O_3$
17d1 135, *17d2* 333
$C_{14}H_{25}N_2O_5$
17d2 309
$C_{14}H_{25}N_2O_6$
17d1 144
$C_{14}H_{25}N_2O_7$
9c1 305
$C_{14}H_{25}N_3O_2^{2\cdot}$
9d2 223
$C_{14}H_{25}N_3O_5M$
17d1 11, 38
$C_{14}H_{25}N_4O$
9c1 849
$C_{14}H_{25}N_4O_7S$
17d1 75
$C_{14}H_{25}O_{11}$
17e 10
$C_{14}H_{26}BNO_4P$
17d2 66
$C_{14}H_{26}BN_2O$
17d2 65
$C_{14}H_{26}Br_2N_4O_2PdS_2^{2\cdot}$
17h 287
$C_{14}H_{26}Cl_2N_4O_2PdS_2^{2\cdot}$
17h 288
$C_{14}H_{26}NO_2$
9c1 429, 430, *17b* 144, *17d2* 272
$C_{14}H_{26}NO_3$
9c1 216, *17d2* 309
$C_{14}H_{26}NO_5$
9c1 361
$C_{14}H_{26}N_2^{+}$
17h 130
$C_{14}H_{26}N_2O^+$
17b 103
$C_{14}H_{26}N_2O_4S_2M$
17d1 38
$C_{14}H_{26}N_2O_6S_2^{+}$
9c2 301, *17e* 375
$C_{14}H_{26}N_3O$
17d2 260
$C_{14}H_{26}N_3O_2$
9c1 272, 880, *17d2* 309
$C_{14}H_{26}N_3O_4$
17d1 38
$C_{14}H_{26}O_2Si^-$
17g 32

$C_{14}H_{26}O_4P$
9b 307, *9c2* 258, *17b* 111, 130
$C_{14}H_{26}O_{11}$
17e 11
$C_{14}H_{26}S^-$
17b 395
$C_{14}H_{26}S_2^+$
17e 378
$C_{14}H_{26}Si_2^+$
17h 60
$C_{14}H_{27}ClNO_3$
9c1 294, 375
$C_{14}H_{27}MnN_4O_6P$
17d1 202
$C_{14}H_{27}NO_8P$
17d2 120
$C_{14}H_{27}N_2O_3$
9c1 306
$C_{14}H_{27}N_2O_4$
17c 446
$C_{14}H_{27}N_2O_4S$
17d2 309
$C_{14}H_{27}N_2O_5$
9c1 908, *17d2* 405
$C_{14}H_{27}N_3^+$
17b 101
$C_{14}H_{27}N_4O_3$
17d1 176
$C_{14}H_{27}O$
9b 90
$C_{14}H_{27}O_2$
9b 125, 223
$C_{14}H_{27}S$
9b 380
$C_{14}H_{28}B_2^-$
17f 225
$C_{14}H_{28}BrN_2O$
9c1 522
$C_{14}H_{28}MnO_{10}P_2S_3$
17a 289
$C_{14}H_{28}NO$
9c1 99, 272, 754, *17d2* 184
$C_{14}H_{28}NOS_2$
17d1 160
$C_{14}H_{28}NO_2$
9c1 430, 755, *17d2* 185
$C_{14}H_{28}NO_3$
17d1 157
$C_{14}H_{28}NO_5$
17d2 339
$C_{14}H_{28}NO_7S_2$
17d2 230
$C_{14}H_{28}N_2O_2$
9c1 319
$C_{14}H_{28}N_3O_4$
17d1 39, 71
$C_{14}H_{28}N_3O_4S_2$
17d1 67
$C_{14}H_{28}N_4OPS$
17d2 310
$C_{14}H_{28}O^-$
9b 232
$C_{14}H_{28}O_3P$
9b 307
$C_{14}H_{28}O_4P$
9b 220, *17b* 99
$C_{14}H_{29}$
17b 279
$C_{14}H_{29}B^-$
9d1 699, *17f* 221
$C_{14}H_{29}NO_3P$
9c2 258, *17b* 97
$C_{14}H_{29}NO_4P$
9c1 488
$C_{14}H_{29}N_2O_2$
9c1 244, 443, *17d2* 77, 78, 109
$C_{14}H_{29}N_2O_3$
17d2 80
$C_{14}H_{29}OSi_2$
17b 474
$C_{14}H_{30}AlN_2$
17a 401
$C_{14}H_{30}Al_2N_2^-$
17f 202
$C_{14}H_{30}B_2N_2^-$
17f 225
$C_{14}H_{30}NOSi$
17d2 183
$C_{14}H_{30}NO_2P^+$
17b 280
$C_{14}H_{30}NO_2S$
9c1 327
$C_{14}H_{30}NO_2S_2$
17d1 188
$C_{14}H_{30}NO_2Sn$
9c1 755
$C_{14}H_{30}NO_3$
9c1 313
$C_{14}H_{30}N_2O_3P$
17c 397
$C_{14}H_{30}OP$
17b 280
$C_{14}H_{30}O_3P$
9b 220, *17b* 281
$C_{14}H_{31}B^-$
17f 221
$C_{14}H_{31}N_3O_2SX$
17d1 59
$C_{14}H_{31}N_3O_2X$
17d1 59
$C_{14}H_{31}N_3O_3P$
17c 454
$C_{14}H_{31}Sn$
9b 36, *17b* 46, 47
$C_{14}H_{32}NO_2Sn$
17d2 410
$C_{14}H_{32}NO_7P$
9c1 372
$C_{14}H_{32}NO_7P_2$
17d2 218
$C_{14}H_{32}N_2OP$
17b 280
$C_{14}H_{32}N_2O_2P$
17e 297
$C_{14}H_{32}O_2P$
9c2 258
$C_{14}H_{32}S_2^+$
17e 377
$C_{14}H_{33}N_3O_3P$
17c 454
$C_{14}H_{34}N_4P_2^+$
9c2 266
$C_{14}H_{35}Si_4$
17b 159
$C_{14}H_{36}N_2OPSi_3$
17e 265
$C_{14}H_{36}N_2PSi_2$
17e 256
$C_{14}H_{36}O_2PSi_2$
17e 298
$C_{14}H_{36}Si_4^-$
17f 18
$C_{14}H_{36}Si_4^+$
17f 35
$C_{14}H_{37}NO_3PSi_2$
17e 298

$C_{14}H_{37}Si_4$
9b 237, *17b* 261, 304
$C_{14}H_{38}AsSi_4$
9c2 317
$C_{14}H_{38}BrSi_4Sn$
9c2 314
$C_{14}H_{38}N_5OX_4$
17d1 135, 148
$C_{14}H_{38}PSi_4$
9c2 217
$C_{14}H_{42}Si_7^-$
17f 27
$C_{14}H_{42}Si_8^+$
17f 38
$C_{15}Fe_3O_{15}Pt_3^-$
17a 173
$C_{15}H_3D_{10}NO_2^-$
9d1 420
$C_{15}H_3D_{10}N_4$
17c 475
$C_{15}H_4D_5MnO_4^-$
17a 289
$C_{15}H_5Br_6O_4Pd$
17g 361
$C_{15}H_5Cl_6NiO_4$
17g 364
$C_{15}H_5Cl_6O_4Pd$
17g 364
$C_{15}H_5D_6^{2-}$
17f 95
$C_{15}H_5D_{18}$
9b 562
$C_{15}H_6Cl_9$
9b 642
$C_{15}H_6Cl_9O_2$
9b 642
$C_{15}H_6F_8O_3^-$
9b 661
$C_{15}H_6F_{11}OSi$
17c 289
$C_{15}H_6NO_8^{4-}$
9b 484
$C_{15}H_7Br_4NO_5Re$
17a 316, *17g* 204
$C_{15}H_7ClO_3^-$
9d1 311
$C_{15}H_7ClO_6^{3-}$
9b 480
$C_{15}H_7Cl_8O_2$
9b 639
$C_{15}H_7D_7^-$
9d1 725
$C_{15}H_7NO_8^{3-}$
9b 483
$C_{15}H_7O_7^{2-}$
9d1 246
$C_{15}H_7O_8^{2-}$
9d1 246
$C_{15}H_8ClMnO_4^-$
17a 289
$C_{15}H_8D_6OS^-$
17f 213
$C_{15}H_8D_{15}$
9b 561
$C_{15}H_8F_6NO_2$
17d1 376
$C_{15}H_8N_2O^-$
9b 662
$C_{15}H_8N_2O_2^-$
17g 34
$C_{15}H_8O^-$
1 113
$C_{15}H_8O_4^-$
9d1 312
$C_{15}H_8O_5^{3-}$
9d1 252
$C_{15}H_8O_6^{3-}$
9b 479, 481, 482, *17c* 99
$C_{15}H_8O_6^-$
9d1 130
$C_{15}H_8O_7^{3-}$
9b 481
$C_{15}H_8O_7^-$
9d1 131
$C_{15}H_9^{2-}$
9d1 783
$C_{15}H_9BrN_5O_4S$
9c1 57
$C_{15}H_9ClO_3^-$
17g 291
$C_{15}H_9D_2^{2-}$
17f 95
$C_{15}H_9F_5NO$
9c1 622
$C_{15}H_9MnO_4^-$
17a 289
$C_{15}H_9NO_3^-$
9d1 310
$C_{15}H_9N_6O_6S$
9c1 55
$C_{15}H_9O_4^{2-}$
9d1 178
$C_{15}H_9O_4$
1 43, *9c2* 207
$C_{15}H_9O_6^{2-}$
9d1 246
$C_{15}H_9O_7^{2-}$
17g 250, 253
$C_{15}H_{10}^-$
9d1 783
$C_{15}H_{10}Br_2NO_2$
17d2 351
$C_{15}H_{10}Co_2N_{11}O^{6-}$
9c1 1021
$C_{15}H_{10}D^{2-}$
9d1 774
$C_{15}H_{10}DO_2$
17c 205
$C_{15}H_{10}NO$
9c1 129
$C_{15}H_{10}NOS_2$
17c 529
$C_{15}H_{10}NO_2$
9c1 803
$C_{15}H_{10}NO_2S_2$
17c 559
$C_{15}H_{10}NO_3$
9c1 144, *17c* 600
$C_{15}H_{10}NO_8$
9b 483
$C_{15}H_{10}N_2O_2S^-$
17g 29
$C_{15}H_{10}N_2O_3^-$
9d1 82, *17f* 283
$C_{15}H_{10}N_2O_4^-$
9d1 645
$C_{15}H_{10}N_2S_3^-$
17g 49
$C_{15}H_{10}N_3O_2$
17c 570
$C_{15}H_{10}N_5O_4S$
9c1 47, 48
$C_{15}H_{10}O^-$
9b 405, 690ff
$C_{15}H_{10}O_2^-$
9d1 86, 310, 311, *17g* 35, 285, 287ff
$C_{15}H_{10}O_3^-$
9d1 79

$C_{15}H_{10}O_5^-$
9d1 130, *17g* 299

$C_{15}H_{10}O_6^-$
9d1 129, 143

$C_{15}H_{10}O_7S^-$
17g 299

$C_{15}H_{11}^{2-}$
9d1 774, *17f* 94

$C_{15}H_{11}$
17c 170

$C_{15}H_{11}BrNO_2$
17d2 351

$C_{15}H_{11}Cl_3NO_2$
17d1 241

$C_{15}H_{11}D^-$
17f 96

$C_{15}H_{11}DO_3^-$
17g 382

$C_{15}H_{11}D_{18}$
9b 353

$C_{15}H_{11}F_5NO$
9c1 622

$C_{15}H_{11}F_6OSi$
17b 312

$C_{15}H_{11}F_7NO_7$
9c1 726

$C_{15}H_{11}FeO_3^{2-}$
9d1 331

$C_{15}H_{11}HgO_2$
17a 391, *17g* 318

$C_{15}H_{11}K_2$
17a 217

$C_{15}H_{11}MgO_2$
9b 421

$C_{15}H_{11}MgO_2^+$
17a 227

$C_{15}H_{11}NO^-$
17f 198

$C_{15}H_{11}NO_2^-$
9d1 470, *17g* 35

$C_{15}H_{11}N_2$
9b 705, *17c* 327, 328

$C_{15}H_{11}N_2O$
17d1 223

$C_{15}H_{11}N_2O_2$
9c1 855

$C_{15}H_{11}N_2O_5$
9c1 592

$C_{15}H_{11}Na_2$
17a 210

$C_{15}H_{11}O_2^{2-}$
9b 391

$C_{15}H_{11}O_2$
9b 631, *9c2* 203, *17c* 205, 297, 314

$C_{15}H_{11}O_6$
9b 479

$C_{15}H_{11}O_7^{2-}$
17g 249, 253

$C_{15}H_{11}S_2$
9b 418, 762

$C_{15}H_{12}^-$
1 112, *9d1* 760, 773, 778, 818, *17f* 95, 96

$C_{15}H_{12}^+$
1 124

$C_{15}H_{12}BrN_8$
17c 460, 462

$C_{15}H_{12}BrO_3^{2-}$
9d1 270

$C_{15}H_{12}Cl_3$
17c 267

$C_{15}H_{12}Cl_4NO$
9c1 717

$C_{15}H_{12}Cl_4NO_2$
9c1 650

$C_{15}H_{12}D_2NO$
9c1 627

$C_{15}H_{12}D_{12}NOS$
17d2 116

$C_{15}H_{12}D_{12}NO_2$
17d1 276

$C_{15}H_{12}D_{12}NO_3$
17d1 275

$C_{15}H_{12}D_{18}O_3P$
9b 382

$C_{15}H_{12}F_3$
17c 265

$C_{15}H_{12}F_3S$
17c 284

$C_{15}H_{12}FeO_2^-$
9d1 330

$C_{15}H_{12}NO$
9c1 167, *17d2* 371

$C_{15}H_{12}NO_2$
9c1 768, 769, 796, *17d2* 352

$C_{15}H_{12}NO_3$
9c1 733, *17d2* 396

$C_{15}H_{12}N_2O$
9c1 733

$C_{15}H_{12}N_2S^+$
17h 197

$C_{15}H_{12}N_3O_2$
17c 457, *17d2* 368

$C_{15}H_{12}N_5O_5S^{2-}$
17c 437

$C_{15}H_{12}N_9O_2$
17c 460, 461, 462, 463

$C_{15}H_{12}O^-$
9b 691, *17c* 26, 317

$C_{15}H_{12}OS^-$
17g 50

$C_{15}H_{12}O_2+$
17g 385

$C_{15}H_{12}O_2S^-$
9d1 26

$C_{15}H_{12}O_2Se^-$
9d1 26

$C_{15}H_{12}O_3^-$
9d1 26, *17g* 35, 382

$C_{15}H_{12}O_3S^-$
9b 702

$C_{15}H_{12}O_4^-$
1 82

$C_{15}H_{12}O_6^-$
9d1 130

$C_{15}H_{12}S^-$
9d1 890

$C_{15}H_{13}^{2-}$
9d1 815

$C_{15}H_{13}$
17b 489, *17c* 253

$C_{15}H_{13}BrO_2^-$
9d1 187

$C_{15}H_{13}ClCrNO_6$
17d1 323

$C_{15}H_{13}ClNO_3$
9c1 644, 646

$C_{15}H_{13}Cl_2$
17c 266

$C_{15}H_{13}Cl_3NO_2$
17d1 241

$C_{15}H_{13}F_5NO_3$
17d2 240

$C_{15}H_{13}FeO_2$
9c2 186

$C_{15}H_{13}MnNO_6$
9a 311, *9c1* 1039, *17a* 289, *17d1* 325

$C_{15}H_{13}NO^-$
17c 237
$C_{15}H_{13}NOS^-$
9d1 74
$C_{15}H_{13}NO_2S^+$
17h 197
$C_{15}H_{13}NO_3^-$
9d1 504
$C_{15}H_{13}NO_3$
9c1 592
$C_{15}H_{13}NO_6Re$
9a 312, *9c1* 1041, *17a* 316, *17d1* 324
$C_{15}H_{13}N_2$
17c 215
$C_{15}H_{13}N_2O$
17d1 215
$C_{15}H_{13}N_8$
17c 460, 461
$C_{15}H_{13}O$
9b 693, *17c* 179, 276
$C_{15}H_{13}O_2$
1 70, *17c* 277
$C_{15}H_{13}O_3^{2-}$
9d1 258, 269
$C_{15}H_{13}S$
17c 281
$C_{15}H_{14}$
9d1 760, 813, 814, 818
$C_{15}H_{14}^{2\cdot}$
17h 253
$C_{15}H_{14}ClNS^+$
17h 198
$C_{15}H_{14}Cl_2NO$
9c1 716, *17d1* 234, 237, 370
$C_{15}H_{14}Cl_2NS$
17c 506
$C_{15}H_{14}FeNO_4$
9c1 627
$C_{15}H_{14}FeO_2^-$
9a 314
$C_{15}H_{14}FeO_3^-$
9a 316
$C_{15}H_{14}FeO_4^-$
9a 316
$C_{15}H_{14}KO_2S$
17a 217
$C_{15}H_{14}N$
9c1 31, *17c* 569
$C_{15}H_{14}NO$
9c1 124, 125, 137, 601ff, 613, *17c* 595, *17d2* 373
$C_{15}H_{14}NO_2$
9c1 627, 779, 803, *17d1* 215, 228, *17d2* 349, 365
$C_{15}H_{14}NO_2S$
9c1 774, *17d2* 396
$C_{15}H_{14}NO_3$
9c1 137, 138, 176, 628, 644
$C_{15}H_{14}NO_4S$
9c1 149, 150, 774
$C_{15}H_{14}NO_5S$
9c1 775
$C_{15}H_{14}N_2^+$
17h 237
$C_{15}H_{14}N_2O_3^-$
17f 254
$C_{15}H_{14}NaO_2S$
17a 210
$C_{15}H_{14}O^-$
1 110, *9b* 652ff, *9d1* 886
$C_{15}H_{14}OS^-$
17f 213
$C_{15}H_{14}OSi^-$
9b 706
$C_{15}H_{14}O_2^-$
9d1 186, 237, 238
$C_{15}H_{14}O_2S^-$
9b 664, *9d1* 887, *17c* 280
$C_{15}H_{14}O_3^-$
1 110, *9b* 660
$C_{15}H_{14}O_6^-$
9d1 885
$C_{15}H_{14}S^-$
9b 663
$C_{15}H_{15}$
9b 640, *17c* 261
$C_{15}H_{15}ClNO_2$
9c1 647
$C_{15}H_{15}ClNO_3S$
17c 510
$C_{15}H_{15}ClNS$
17c 507
$C_{15}H_{15}Cl_4MnO_8P$
17g 207
$C_{15}H_{15}FeO_3^{2-}$
9d1 28
$C_{15}H_{15}N^+$
9d2 42
$C_{15}H_{15}NOS^-$
9b 587
$C_{15}H_{15}NS^+$
17h 198
$C_{15}H_{15}N_2O$
17d1 215
$C_{15}H_{15}N_2O_2$
9c1 522, 797, 904
$C_{15}H_{15}N_2O_3$
9c1 710, 715
$C_{15}H_{15}N_2O_4S$
9c1 775
$C_{15}H_{15}N_3O_2P$
1 74
$C_{15}H_{15}N_3O_5^-$
17g 357
$C_{15}H_{15}N_4$
9c1 74
$C_{15}H_{15}N_4S$
9c1 86
$C_{15}H_{15}O$
1 70, *17b* 64, *17c* 187
$C_{15}H_{15}OS$
17b 295
$C_{15}H_{15}O_2S$
17c 187
$C_{15}H_{15}O_2Sn$
17g 342
$C_{15}H_{15}O_3$
9b 660
$C_{15}H_{15}O_4^{2-}$
17b 99
$C_{15}H_{15}S$
9b 665, *17c* 187, 280, 282
$C_{15}H_{16}^-$
9d1 768, 810, 822
$C_{15}H_{16}FeO_2^-$
9a 314, 315, *9d1* 29
$C_{15}H_{16}NO$
9c1 177, 178, 568, 628, 710, *17d1* 222, 371
$C_{15}H_{16}NO_2$
9c1 177, 644, *17c* 147, *17d1* 372, 373
$C_{15}H_{16}NO_3$
9c1 176
$C_{15}H_{16}NO_3S$
9c1 997

$C_{15}H_{16}NO_4$
9c1 568
$C_{15}H_{16}NO_4S$
9c1 929
$C_{15}H_{16}NO_5S$
9c1 922
$C_{15}H_{16}N_2^+$
9d2 106, *17h* 155
$C_{15}H_{16}N_2R_s$
(R_s = solvent radical)
9c1 44
$C_{15}H_{16}N_3O_5$
17e 215
$C_{15}H_{16}N_4OP$
17c 476
$C_{15}H_{16}OS^-$
9b 707
$C_{15}H_{16}O_2^-$
9d1 238, 240
$C_{15}H_{16}O_2S^-$
9b 708
$C_{15}H_{16}O_4^-$
9d1 239
$C_{15}H_{16}O_7^-$
9d1 302
$C_{15}H_{16}PS_2$
9b 668
$C_{15}H_{17}Cl_3NO_4$
17d2 310
$C_{15}H_{17}F_5NO$
17d1 352
$C_{15}H_{17}F_7NO_2$
9c1 272
$C_{15}H_{17}Ge$
17f 40
$C_{15}H_{17}MgN_4O_7$
17g 194
$C_{15}H_{17}N^+$
9d2 42
$C_{15}H_{17}NO_2^+$
9d2 42
$C_{15}H_{17}N_2O$
9c1 560, *17d1* 218,
17d2 108, 213
$C_{15}H_{17}N_2O_2$
9c1 880
$C_{15}H_{17}N_4O_3$
17d2 240
$C_{15}H_{18}F_5$
17c 188
$C_{15}H_{18}F_5Se$
17b 316
$C_{15}H_{18}F_{13}$
17b 304
$C_{15}H_{18}FeO^-$
9a 318
$C_{15}H_{18}Ge^-$
17f 44
$C_{15}H_{18}NO$
9c1 694, *17d1* 287, *17d2* 348
$C_{15}H_{18}NOSi$
17d1 216
$C_{15}H_{18}NO_2$
9c1 694, 745, *17b* 109,
17c 632, *17d2* 110
$C_{15}H_{18}NO_4$
9c1 535
$C_{15}H_{18}NO_7$
9c1 104
$C_{15}H_{18}N_2O$
17d2 130
$C_{15}H_{18}N_2OZn^-$
17a 383
$C_{15}H_{18}N_2O_3M$
17d1 39
$C_{15}H_{18}N_3O_2$
17c 458
$C_{15}H_{18}N_3O_5S$
17d2 241
$C_{15}H_{18}N_4O_5^-$
17g 354
$C_{15}H_{18}N_5O_5$
17d2 241
$C_{15}H_{18}OPS$
9c2 258
$C_{15}H_{18}O_4^-$
9d1 239
$C_{15}H_{18}Si^-$
17f 19
$C_{15}H_{19}Br_3NO$
17d1 252
$C_{15}H_{19}Cl_3NO$
17d1 242
$C_{15}H_{19}Cl_3NO_3$
17d2 103
$C_{15}H_{19}Cl_5NO$
17d1 287
$C_{15}H_{19}F_6$
17c 101
$C_{15}H_{19}MnO_8P$
17a 289, *17g* 73
$C_{15}H_{19}NO_4$
17d2 230
$C_{15}H_{19}N_2O$
9c1 411, 511, 849, *17d1* 135,
17d2 260, 269, 334
$C_{15}H_{19}N_2O_2$
9c1 411, 424, 881, *17d2* 253
$C_{15}H_{19}N_2O_2S$
9c1 345
$C_{15}H_{19}N_2O_3$
9c1 514, *17d1* 288
$C_{15}H_{19}N_3O_5^-$
17g 242
$C_{15}H_{19}N_4O_2$
17h 217
$C_{15}H_{19}N_4O_4$
17d2 82
$C_{15}H_{19}N_4O_{11}S$
17d1 384
$C_{15}H_{19}O_4$
17e 220
$C_{15}H_{20}ClO_3$
9c2 137
$C_{15}H_{20}Cl_3N_2O$
17d1 246
$C_{15}H_{20}CuO_3$
17g 117, 158
$C_{15}H_{20}D_3O$
9c2 80
$C_{15}H_{20}F_3O_2S$
17b 401
$C_{15}H_{20}F_6N$
9c1 25
$C_{15}H_{20}HgNO_3$
9c1 745
$C_{15}H_{20}NO$
1 66, *9c1* 411, *9c2* 117, 130,
17c 594, *17d2* 334
$C_{15}H_{20}NOS$
17d2 241
$C_{15}H_{20}NO_2$
17d2 174
$C_{15}H_{20}NO_3$
9c1 403, *17d2* 177
$C_{15}H_{20}NO_3S$
17d2 241
$C_{15}H_{20}NO_4$
9c1 677

$C_{15}H_{20}NO_4S$
9c1 294, 313, *17d2* 241
$C_{15}H_{20}NO_5$
17d1 151, *17d2* 104
$C_{15}H_{20}NO_8$
9c1 975
$C_{15}H_{20}NS$
9c2 282
$C_{15}H_{20}N_2O_4M$
17d1 12, 71
$C_{15}H_{20}N_2O_5M$
17d1 12, 39
$C_{15}H_{20}N_3O_2$
17c 456
$C_{15}H_{20}N_3O_5$
17e 188
$C_{15}H_{20}N_3O_6$
17d2 242
$C_{15}H_{20}N_3Si$
17b 308
$C_{15}H_{20}N_4O_{14}PS$
17d1 384
$C_{15}H_{20}N_5O_5$
9c1 488, 489, *17d2* 230, 310
$C_{15}H_{20}S^-$
17b 481
$C_{15}H_{20}Si_2^-$
17f 19
$C_{15}H_{21}BrMgO_2^-$
17a 227, *17c* 204
$C_{15}H_{21}BrNO_4S$
9c1 489
$C_{15}H_{21}ClNO$
9c1 375
$C_{15}H_{21}Cl_3NO$
17d1 275
$C_{15}H_{21}Cl_3NO_4$
17d2 97
$C_{15}H_{21}F_2$
17c 188
$C_{15}H_{21}NO_2PS$
17b 288
$C_{15}H_{21}NO_4P$
17d2 311
$C_{15}H_{21}NO_5P$
9c2 259
$C_{15}H_{21}NO_5S$
9c1 930
$C_{15}H_{21}N_2O$
9c1 273, 424, 709, *17d2* 32
$C_{15}H_{21}N_2O_2$
9c1 404, 824
$C_{15}H_{21}N_2O_3S$
17d1 144
$C_{15}H_{21}N_2O_4$
17d1 60, *17d2* 268
$C_{15}H_{21}N_2O_5$
17d2 106
$C_{15}H_{21}N_2O_6$
17d1 354
$C_{15}H_{21}N_3O_3X$
17d1 74
$C_{15}H_{21}N_4O_5$
17d2 84, 85
$C_{15}H_{21}O_2^{2-}$
9b 553, 568, *17c* 195
$C_{15}H_{21}O_2$
1 64, *9c2* 115, *17e* 224
$C_{15}H_{22}^-$
17c 170
$C_{15}H_{22}ClCoN_4O_5$
17a 337
$C_{15}H_{22}ClNO_4P$
9c1 489
$C_{15}H_{22}ClN_2S_2$
17c 511
$C_{15}H_{22}CoFN_4O_5$
17a 337
$C_{15}H_{22}CoN_5O_7$
17a 337
$C_{15}H_{22}F$
9b 562, *17c* 188
$C_{15}H_{22}F_3$
9b 464
$C_{15}H_{22}NO$
9c1 490, *17d1* 303, 357, 359, *17d2* 311
$C_{15}H_{22}NO_2$
9c1 273, *17d1* 358, *17d2* 20, 157, 311
$C_{15}H_{22}NO_3$
17d1 275, 288, *17d2* 21, 52, 102, 162
$C_{15}H_{22}NO_4$
17d2 137
$C_{15}H_{22}NO_6$
17d2 162
$C_{15}H_{22}NO_7$
17d1 355
$C_{15}H_{22}N_2O_2S^+$
17h 141
$C_{15}H_{22}N_2O_6P$
17c 612
$C_{15}H_{22}N_3O$
9c1 490
$C_{15}H_{22}N_3O_2$
9c1 816, 881, 882
$C_{15}H_{22}N_3O_4$
17d1 71
$C_{15}H_{22}N_3O_4S$
9c1 66, 882
$C_{15}H_{22}O^-$
9b 541, *17c* 167
$C_{15}H_{22}S^-$
17c 206
$C_{15}H_{23}$
9b 546, 561
$C_{15}H_{23}ClNO_2$
9c1 787
$C_{15}H_{23}Cl_2O_2Sn$
17a 457, *17g* 185
$C_{15}H_{23}CoN_4O_5$
17a 337
$C_{15}H_{23}HgO_2$
17g 162ff
$C_{15}H_{23}NO^-$
17c 209
$C_{15}H_{23}NO_2^-$
9d1 460
$C_{15}H_{23}NO_5SM$
17d2 96
$C_{15}H_{23}NS^-$
17c 211
$C_{15}H_{23}N_2$
9c1 94
$C_{15}H_{23}N_2O$
9c1 514, 816, 826, 907, *17d1* 170, 229, *17d2* 21
$C_{15}H_{23}N_2O_2$
9c1 359, 418, 788, *17d1* 170
$C_{15}H_{23}N_2O_2S$
9c1 66
$C_{15}H_{23}N_2O_3$
9c1 535, *17d2* 47, 203
$C_{15}H_{23}N_2O_4$
9c1 849, *17d2* 381
$C_{15}H_{23}N_2O_5$
9c1 883, *17d2* 385

$C_{15}H_{23}N_2O_5S$
17d2 261
$C_{15}H_{23}N_2S_2$
17c 510
$C_{15}H_{23}O$
1 61, 70, *9c2* 78ff, *17e* 86ff
$C_{15}H_{23}OS$
1 67, *9c2* 141, 283
$C_{15}H_{23}OSe$
1 68
$C_{15}H_{23}O_2$
1 63, 66, 72, *9c2* 66, 67, 131, 151, 160, *17b* 340, 398, *17e* 60, 71ff
$C_{15}H_{23}O_2S$
1 67, *17b* 401, *17e* 140
$C_{15}H_{23}O_2S_2$
9c2 283
$C_{15}H_{23}O_2Sn$
17b 18
$C_{15}H_{23}O_3S$
1 68
$C_{15}H_{23}O_4S$
9c2 140
$C_{15}H_{23}S$
9c2 281
$C_{15}H_{23}S_2$
9c2 283
$C_{15}H_{23}Se$
17b 315
$C_{15}H_{24}BrN_2O_3$
9c1 535
$C_{15}H_{24}IN_2O_3$
9c1 536
$C_{15}H_{24}NO$
9c1 126, 232, 273, 622, 693, *17d1* 275, 358, *17d2* 14ff, 32
$C_{15}H_{24}NOS$
17c 524, *17d2* 63
$C_{15}H_{24}NO_2$
9c1 274, 313, 314, 651, 687, 788, *17d1* 135, 373, *17d2* 21, 22, 32ff, 52ff, 54
$C_{15}H_{24}NO_2S$
17c 527
$C_{15}H_{24}NO_2Sn$
9c1 755
$C_{15}H_{24}NO_3$
9c1 314, 345, 584, 788, *17d2* 22, 58, *17e* 154
$C_{15}H_{24}NO_4$
17d2 210
$C_{15}H_{24}NO_5S$
9c1 930
$C_{15}H_{24}NO_6$
17d1 353, 354
$C_{15}H_{24}NO_8$
17d2 242
$C_{15}H_{24}N_2O_3X$
17d2 105
$C_{15}H_{24}N_3O$
9c1 817
$C_{15}H_{24}N_3O_4S$
17d1 170
$C_{15}H_{24}N_3O_5$
17d2 242
$C_{15}H_{24}O^+$
17e 246
$C_{15}H_{24}S_3^+$
17h 96
$C_{15}H_{25}CoN_3O_5$
17e 19
$C_{15}H_{25}F_3NOSi$
17c 202
$C_{15}H_{25}NO^+$
17h 162
$C_{15}H_{25}NOPS$
9c1 325
$C_{15}H_{25}NO_3PS$
9c1 325
$C_{15}H_{25}NO_3PS_2$
9c1 327
$C_{15}H_{25}NO_4P$
9c1 325, *17d2* 70
$C_{15}H_{25}N_2O$
9c1 536, *17d2* 22
$C_{15}H_{25}N_2O_2$
9c1 418
$C_{15}H_{25}N_2O_5$
9c1 850ff, *17d2* 382
$C_{15}H_{25}N_2O_6$
9c1 883, 884, *17d2* 385, 386
$C_{15}H_{25}N_3O_2X$
17d1 60
$C_{15}H_{25}N_3O_3X$
17d1 60
$C_{15}H_{25}N_4O_6$
17d2 138
$C_{15}H_{25}N_4O_7S$
17d2 181
$C_{15}H_{25}OSi$
9b 593, *17c* 220
$C_{15}H_{25}O_2$
17c 62
$C_{15}H_{25}O_4P^-$
17b 321, *17g* 56
$C_{15}H_{25}O_6P_2$
17c 216
$C_{15}H_{26}NOSi$
17d2 15
$C_{15}H_{26}NO_2Si$
17d1 380
$C_{15}H_{26}NO_2Sn$
17d1 380
$C_{15}H_{26}NO_3Si$
9c1 961
$C_{15}H_{26}NO_4$
9c1 146
$C_{15}H_{26}NO_4SSi$
9c1 974
$C_{15}H_{26}NO_7$
9c1 345
$C_{15}H_{26}N_2^+$
9d2 56
$C_{15}H_{26}N_2O_4M$
17d1 39
$C_{15}H_{26}N_2O_5^{2-\cdot}$
9d2 193
$C_{15}H_{26}N_3O_3$
17d2 253
$C_{15}H_{26}OS^-$
17g 50
$C_{15}H_{26}O_2^-$
9d1 46, *17g* 26
$C_{15}H_{26}O_4P$
9c2 259
$C_{15}H_{26}S^-$
17b 501
$C_{15}H_{27}$
9b 402, 404, 432
$C_{15}H_{27}F_6OSi$
9b 247
$C_{15}H_{27}NO_2Sn$
17d2 410
$C_{15}H_{27}N_2O$
9c1 536

$C_{15}H_{27}N_2O_6$
17c 446, *17d2* 253ff

$C_{15}H_{27}OSi$
17b 491

$C_{15}H_{27}Si$
17c 24

$C_{15}H_{28}BNOP$
17d2 70

$C_{15}H_{28}Cl_2O_3P$
9b 382

$C_{15}H_{28}NO$
17d2 334

$C_{15}H_{28}NO_2$
9c1 490, *17d2* 272

$C_{15}H_{28}NO_4$
17d2 196

$C_{15}H_{28}NO_7$
17d2 312

$C_{15}H_{28}N_2^+$
17h 156

$C_{15}H_{28}N_2O_3^{2\cdot}$
9d2 227

$C_{15}H_{28}N_2O_4SM$
17d1 39, 40

$C_{15}H_{28}N_3O_3$
9c1 900

$C_{15}H_{28}O^-$
9b 361

$C_{15}H_{28}O_4^-$
9b 374

$C_{15}H_{29}$
9b 353

$C_{15}H_{29}Cl_2N_3O_3P$
17d2 311

$C_{15}H_{29}NO_4P$
17d2 312

$C_{15}H_{29}N_2$
9c1 43

$C_{15}H_{29}N_2O_2$
17d2 78

$C_{15}H_{29}N_2O_4$
9c1 274

$C_{15}H_{29}N_3^+$
17b 102

$C_{15}H_{29}N_6$
17b 142

$C_{15}H_{29}OSi_3$
9c2 174

$C_{15}H_{29}O_3$
17b 286

$C_{15}H_{29}S_3Sn$
17b 385

$C_{15}H_{30}NO_2$
17d1 182, *17d2* 191, 312

$C_{15}H_{30}NO_3$
17d1 145

$C_{15}H_{30}N_3O_4$
17d1 40

$C_{15}H_{30}Si_3^-$
17f 19

$C_{15}H_{31}$
17b 276

$C_{15}H_{31}B^-$
9d1 699

$C_{15}H_{31}B_{10}$
17c 112

$C_{15}H_{31}NO_3P$
9c1 491

$C_{15}H_{31}NO_4P$
9c1 491

$C_{15}H_{31}NO_5P$
9c1 491

$C_{15}H_{31}N_2O$
17d2 188

$C_{15}H_{31}N_3O_4M$
17d1 72

$C_{15}H_{31}O_2Sn$
17b 164

$C_{15}H_{31}S_3Sn$
17b 383

$C_{15}H_{31}Si_2$
17b 321

$C_{15}H_{32}BO$
9b 195

$C_{15}H_{32}NO$
9c1 232

$C_{15}H_{32}N_2O_4P$
17e 298

$C_{15}H_{32}Si_4^-$
17f 24

$C_{15}H_{32}Si_5^-$
17f 27

$C_{15}H_{33}B^-$
9d1 699, *17f* 221

$C_{15}H_{33}Cl_3N_3O_6S_2$
9c2 297

$C_{15}H_{33}Ge$
17f 40

$C_{15}H_{33}N_3O_3P$
17c 454

$C_{15}H_{33}OSn$
9b 195, *17b* 194

$C_{15}H_{33}S_3Sn$
17b 327

$C_{15}H_{33}Sn$
17f 47

$C_{15}H_{34}NO_2Sn$
17d2 410

$C_{15}H_{34}N_4^+$
17h 150

$C_{15}H_{36}NO_6P_2$
17e 298

$C_{15}H_{37}O_4Si_4$
9b 250

$C_{15}H_{37}Si_3$
9b 232

$C_{15}H_{39}Ge_4$
17b 201

$C_{15}H_{39}IN$
1 27

$C_{15}H_{40}Si_5^-$
17f 27

$C_{15}H_{42}OPSi_5$
17e 298

$C_{15}H_{45}N_3PSi_5$
17e 266

$C_{15}H_{45}N_3Si_5^+$
17h 150

$C_{16}Cl_6MnO_8$
17a 290, *17g* 363

$C_{16}H_4Cr_2N_2O_{10}S^-$
17a 257

$C_{16}H_4Cr_2N_2O_{10}Se^-$
17a 257

$C_{16}H_4Cr_2N_2O_{11}^-$
17a 257

$C_{16}H_4Mo_2N_2O_{10}S^-$
17a 269

$C_{16}H_4Mo_2N_2O_{10}Se^-$
17a 269

$C_{16}H_4Mo_2N_2O_{11}^-$
17a 269

$C_{16}H_4N_2O_{10}SW_2^-$
17a 276

$C_{16}H_4N_2O_{10}SeW_2^-$
17a 276

$C_{16}H_4N_2O_{11}W_2^-$
17a 276

$C_{16}H_5Co_3O_9^-$
17a 127, 337

$C_{16}H_6Br_2O_2S_2^-$
1 119
$C_{16}H_6Cl_2O_2S_2^-$
1 118, 119
$C_{16}H_6Cr_2N_4O_8^-$
17a 258, *17f* 208
$C_{16}H_6D_4O_2^{2\cdot}$
17h 252
$C_{16}H_6D_5MnO_4^-$
17a 290
$C_{16}H_6D_6^-$
17f 103
$C_{16}H_6MnO_6$
17a 290, *17g* 342
$C_{16}H_6Mo_2N_4O_8^-$
17a 269, *17f* 208
$C_{16}H_6N_4O_8W_2^-$
17f 209, 277
$C_{16}H_6O_6Re$
17a 316, *17g* 342
$C_{16}H_7O_8^{4-}$
9b 481, 482
$C_{16}H_8^-$
9d1 834
$C_{16}H_8AsO_3S_3$
17g 367
$C_{16}H_8AsO_4S_2$
17g 367
$C_{16}H_8BrO_3^{2-}$
9d1 294, 295
$C_{16}H_8BrO_3^-$
9d1 295
$C_{16}H_8D_2^-$
9d1 833
$C_{16}H_8D_4^-$
9d1 833
$C_{16}H_8D_6N_4S_2^+$
17h 173
$C_{16}H_8F_8^-$
17f 191
$C_{16}H_8KO_2$
17a 218
$C_{16}H_8Mo_2N_2O_{10}^-$
17a 269
$C_{16}H_8N_2^-$
9b 633
$C_{16}H_8N_2O_2S_2^-$
17g 16
$C_{16}H_8N_2O_4^-$
9d1 103
$C_{16}H_8N_2O_6Re$
17a 316
$C_{16}H_8N_2O_8S_2^-$
1 117
$C_{16}H_8N_2O_{10}W_2^-$
17a 277
$C_{16}H_8O_2S_2^-$
1 118
$C_{16}H_8O_2S_4Sb$
17a 499
$C_{16}H_8O_2Se_2^-$
1 120
$C_{16}H_8O_3^-$
9d1 343
$C_{16}H_8O_3S_3Sb$
17a 499, *17g* 376
$C_{16}H_8O_4^{3-}$
9d1 293
$C_{16}H_8O_4^-$
1 120
$C_{16}H_8O_4S_2Sb$
17a 500, *17g* 376
$C_{16}H_9$
9b 317
$C_{16}H_9AsClO_2S_2$
17g 366
$C_{16}H_9ClO_2S_2Sb$
17a 500, *17g* 375
$C_{16}H_9Cl_2O_2S_2Sn$
17g 377
$C_{16}H_9D_5N_3O_3$
17e 214
$C_{16}H_9D_{12}N_2O_3$
17d2 113
$C_{16}H_9F_3O_2^+$
17h 89
$C_{16}H_9F_{10}NOP$
17d1 201
$C_{16}H_9F_{10}OSi$
17c 289
$C_{16}H_9HgO_2S_2$
17g 370
$C_{16}H_9N_2O_4$
9c1 855
$C_{16}H_9N_2O_5$
9c1 733
$C_{16}H_9O_3^{2-}$
9d1 259, 270, 293
$C_{16}H_9O_4^{2-}$
9d1 293
$C_{16}H_{10}^{3-}$
17f 97, 98ff, 100
$C_{16}H_{10}^-$
1 113, *9d1* 756, 765, 775, 783ff, 827, 833, *17f* 97, 98, 99, 100
$C_{16}H_{10}^+$
1 124, *9d2* 15, 18, *17h* 77, 99
$C_{16}H_{10}BD_9S_2^-$
17c 296
$C_{16}H_{10}ClMnO_4^-$
17a 290
$C_{16}H_{10}Co_2N_{11}O^{6-}$
9c1 1022
$C_{16}H_{10}CrN_2O_4S^-$
17a 258, *17e* 388
$C_{16}H_{10}D_2^-$
9d1 772
$C_{16}H_{10}D_2O_4^-$
17g 300
$C_{16}H_{10}D_8N_2^+$
17h 168
$C_{16}H_{10}NO$
9c1 599
$C_{16}H_{10}NS$
9c1 37
$C_{16}H_{10}NS_2$
17c 534
$C_{16}H_{10}N_2^-$
17f 198, 207
$C_{16}H_{10}N_2O^-$
17f 198
$C_{16}H_{10}N_2O_2^-$
1 117
$C_{16}H_{10}N_2O_4^-$
9d1 103
$C_{16}H_{10}N_4O_8^-$
9d1 669
$C_{16}H_{10}O_2^-$
9d1 33, 187, *17g* 20
$C_{16}H_{10}O_2^{2\cdot}$
17h 252
$C_{16}H_{10}O_2^{4\cdot}$
17h 261
$C_{16}H_{10}O_2S^-$
9d1 191, 286
$C_{16}H_{10}O_6^-$
9d1 332

$C_{16}H_{10}O_7^{3-}$
9b 481
$C_{16}H_{10}O_7^-$
9d1 131
$C_{16}H_{10}S_4^+$
17h 118
$C_{16}H_{11}$
17c 348
$C_{16}H_{11}ClN_5O_2$
17c 443
$C_{16}H_{11}Cl_3NO_3$
17d1 242
$C_{16}H_{11}Cl_4N_4O_2$
9c1 72
$C_{16}H_{11}F_{12}O_2S$
17e 355
$C_{16}H_{11}HgNO_2^-$
9d1 571
$C_{16}H_{11}HgO_2$
17a 391, *17g* 260
$C_{16}H_{11}MnO_4^-$
17a 290, 291
$C_{16}H_{11}MnO_5^-$
17a 291
$C_{16}H_{11}N^-$
9d1 869
$C_{16}H_{11}NO^-$
9d1 334
$C_{16}H_{11}NO_2^-$
9d1 566, 567
$C_{16}H_{11}O_5$
9c2 208
$C_{16}H_{11}O_8$
9b 482
$C_{16}H_{12}^-$
9d1 739ff, 742ff, 754, 764, 765, 771, 773, 786, 787, 832, *17f* 101, 102
$C_{16}H_{12}^+$
17h 100
$C_{16}H_{12}Co_2N_{11}O^{6-}$
9c1 1022
$C_{16}H_{12}CrN_2NaO_4$
9a 308
$C_{16}H_{12}FeN_3O_7$
9a 313
$C_{16}H_{12}NO$
9c1 579, *17d1* 225
$C_{16}H_{12}NO_2$
17d1 223
$C_{16}H_{12}NO_2S$
9c1 35
$C_{16}H_{12}NO_2S_2$
17c 560
$C_{16}H_{12}N_2^-$
9d1 879
$C_{16}H_{12}N_2O_3^-$
9d1 528, 529
$C_{16}H_{12}N_2O_4^-$
9d1 645
$C_{16}H_{12}N_3O_3$
9c1 68
$C_{16}H_{12}N_5O_2$
17c 443
$C_{16}H_{12}O_2^-$
9d1 56, 240, 311, 312, 313, 315, 316, *17g* 36, 289, 290, 295
$C_{16}H_{12}O_4^-$
17g 299, 301
$C_{16}H_{12}O_6^-$
17g 97
$C_{16}H_{12}S^-$
17f 212
$C_{16}H_{12}S_2^+$
17e 381
$C_{16}H_{13}$
9b 630, 634, *17c* 254
$C_{16}H_{13}Cl_3NO_3$
17d1 247
$C_{16}H_{13}D_{18}$
9b 353
$C_{16}H_{13}F_5NO$
9c1 578
$C_{16}H_{13}FeNO_2^-$
9a 313, *9d1* 559, 560
$C_{16}H_{13}FeO_3^{2-}$
9d1 331
$C_{16}H_{13}MnNO_7$
17d1 316
$C_{16}H_{13}N^+$
17h 221
$C_{16}H_{13}NO_2^-$
9d1 47
$C_{16}H_{13}N_2O$
9c1 757
$C_{16}H_{13}N_2O_2$
17c 597
$C_{16}H_{13}N_4O_2$
9c1 88
$C_{16}H_{13}N_6O_6$
9c1 72
$C_{16}H_{13}O_7$
9b 481
$C_{16}H_{14}^-$
1 112, *9d1* 760, 763, 764, 767, 778, 780, 788, 826, 827, 831, 838, *17f* 103, 104, 105
$C_{16}H_{14}^+$
1 124, *9d2* 10, 13, 14, *17h* 76
$C_{16}H_{14}Be_2N_2^+$
17a 224
$C_{16}H_{14}ClFN_3O_5$
17d2 337
$C_{16}H_{14}Cl_2N_3O_5$
9c1 252
$C_{16}H_{14}Cl_4CoO_6$
17g 206
$C_{16}H_{14}D_2^-$
9d1 839
$C_{16}H_{14}D_2O_4^-$
17g 340
$C_{16}H_{14}D_4N_2^+$
9d2 106
$C_{16}H_{14}D_{12}NO$
17d1 290, 303
$C_{16}H_{14}D_{12}NO_2$
17d1 290
$C_{16}H_{14}D_{12}NO_3$
17d2 114
$C_{16}H_{14}FO_2Si$
17e 203
$C_{16}H_{14}F_2N_3O_5$
9c1 252
$C_{16}H_{14}Fe^+$
9a 314
$C_{16}H_{14}FeNO_4$
9c1 628
$C_{16}H_{14}FeO_2^-$
9d1 330
$C_{16}H_{14}INO_2^-$
17g 36
$C_{16}H_{14}MgO_4$
17g 194
$C_{16}H_{14}Mg_2N_2^+$
17a 227
$C_{16}H_{14}NO$
9c1 745, *17c* 596, *17d2* 372

$C_{16}H_{14}NO_2$
9c1 734
$C_{16}H_{14}NO_3$
17d2 352
$C_{16}H_{14}NO_5$
9c1 566
$C_{16}H_{14}N_2^+$
9d2 63
$C_{16}H_{14}N_2O_4^-$
9d1 646
$C_{16}H_{14}N_2O_5^-$
9d1 647
$C_{16}H_{14}N_2S^+$
17h 198
$C_{16}H_{14}N_2Zn_2^+$
17a 383
$C_{16}H_{14}N_3O_3$
17e 214
$C_{16}H_{14}N_4S_2^+$
17h 173
$C_{16}H_{14}N_5O_9$
9c1 253
$C_{16}H_{14}O^-$
9b 690, 691
$C_{16}H_{14}O_2^-$
9d1 332
$C_{16}H_{14}O_2^+$
9d2 137
$C_{16}H_{14}O_2S_2^-$
17g 47
$C_{16}H_{14}O_3^-$
17g 382
$C_{16}H_{14}O_3S^-$
17g 50
$C_{16}H_{14}O_4^-$
9d1 97, 98, *17g* 20, 345
$C_{16}H_{14}O_8S_4^+$
17h 115
$C_{16}H_{14}S^-$
9d1 890
$C_{16}H_{14}S_2^-$
17f 214
$C_{16}H_{14}S_2^+$
17h 98
$C_{16}H_{15}^{2-}$
9d1 816
$C_{16}H_{15}$
17c 254
$C_{16}H_{15}Cl_3NO$
9c1 577
$C_{16}H_{15}FeO_2$
9c2 187
$C_{16}H_{15}NO^-$
17c 238
$C_{16}H_{15}NOS^-$
9d1 74
$C_{16}H_{15}NO_2^-$
9d1 159, 242, 467
$C_{16}H_{15}NO_2S^+$
17h 198
$C_{16}H_{15}NO_6Re$
17d2 71
$C_{16}H_{15}NS^+$
17h 221
$C_{16}H_{15}N_2$
17c 257
$C_{16}H_{15}N_2O$
9c1 835
$C_{16}H_{15}N_2O_2$
17c 597
$C_{16}H_{15}N_2O_3$
9c1 755
$C_{16}H_{15}N_2O_5$
9c1 893
$C_{16}H_{15}N_4O$
17c 470
$C_{16}H_{15}N_4O_2$
9c1 71
$C_{16}H_{15}N_4S$
17c 473
$C_{16}H_{15}N_8$
17c 462, 467
$C_{16}H_{15}N_8O$
17c 460, 462
$C_{16}H_{15}O$
9b 693, *17c* 187, 278
$C_{16}H_{15}OS_4Sn$
17a 457, 458, *17c* 341, 343
$C_{16}H_{15}O_2$
17c 187, 277
$C_{16}H_{15}O_5$
9c2 171
$C_{16}H_{16}^-$
9d1 730, 747, 754, 765, 788, 813, 825, 838, 839, *17f* 105, 106ff
$C_{16}H_{16}^+$
9d2 12, *17h* 74
$C_{16}H_{16}ClN_2O$
9c1 636
$C_{16}H_{16}Cl_2NO$
17d1 235
$C_{16}H_{16}Cr^+$
17a 258
$C_{16}H_{16}F_6NO_3$
17d2 22
$C_{16}H_{16}N$
17c 570
$C_{16}H_{16}NO$
9c1 600, 628
$C_{16}H_{16}NO_2$
9c1 604, 788, *17d2* 349, 365
$C_{16}H_{16}NO_3$
9c1 628, 647, *17d2* 396
$C_{16}H_{16}NO_4$
9c1 645
$C_{16}H_{16}N_2O_4S^+$
17h 141
$C_{16}H_{16}N_2S_2^-$
9d1 78
$C_{16}H_{16}N_3O_5$
9c1 253, *17d2* 214
$C_{16}H_{16}N_4^+$
9d2 79, *17h* 184
$C_{16}H_{16}NaO_2$
9d1 140
$C_{16}H_{16}O^-$
9b 653, *17c* 270
$C_{16}H_{16}O_2^-$
9d1 325, *17g* 344
$C_{16}H_{16}O_2S^-$
17g 346
$(C_{16}H_{16}O_2S)_n^-$
17e 78
$C_{16}H_{16}O_4^-$
17g 340
$C_{16}H_{16}O_4^+$
17h 89
$C_{16}H_{16}O_4S_2^+$
9d2 146, *17e* 378
$C_{16}H_{16}O_6^-$
17g 345
$C_{16}H_{16}O_9^-$
9d1 132
$C_{16}H_{16}O_{10}^-$
17g 106
$C_{16}H_{16}S_2^+$
17e 378
$C_{16}H_{16}Si^-$
17f 20

$C_{16}H_{17}BrNO$
9c1 571
$C_{16}H_{17}BrNO_2$
9c1 677
$C_{16}H_{17}ClNO$
17d1 310
$C_{16}H_{17}FNO$
17d1 310
$C_{16}H_{17}NO^-$
17c 346, 347
$C_{16}H_{17}NO_2S^-$
9b 587
$C_{16}H_{17}N_2O$
9c1 901, *17d1* 229
$C_{16}H_{17}N_2O_2$
9c1 901, 905, *17d1* 219
$C_{16}H_{17}N_2O_3$
9c1 572
$C_{16}H_{17}N_2O_3Se$
17d1 322
$C_{16}H_{17}N_2O_4$
9c1 905, *17d1* 229, *17d2* 386
$C_{16}H_{17}N_2O_4S$
9c1 775
$C_{16}H_{17}N_2O_5$
9c1 833
$C_{16}H_{17}N_2O_7$
9c1 708
$C_{16}H_{17}N_4$
9c1 75, 76
$C_{16}H_{17}O$
17c 276
$C_{16}H_{17}OSSn$
17a 458, *17c* 320, 322
$C_{16}H_{17}OSeSn$
17a 458, *17c* 325
$C_{16}H_{17}OSi$
9b 688, 698
$C_{16}H_{17}OSn$
9b 688, *17a* 458, *17c* 316
$C_{16}H_{17}O_2$
17e 75, 84
$C_{16}H_{17}O_2S$
17e 78, 84
$C_{16}H_{17}O_2Si$
9b 698
$C_{16}H_{17}O_2Sn$
9b 698, *17a* 459, *17c* 320
$C_{16}H_{17}S$
17c 281, 283
$C_{16}H_{17}SSeSn$
17a 459, *17c* 326
$C_{16}H_{17}S_2$
17b 302
$C_{16}H_{17}S_2Sn$
17a 459, *17c* 322
$C_{16}H_{17}Si$
9b 687
$C_{16}H_{18}^-$
9d1 758, 769, 806, 808, 810, 811, 823, *17f* 108
$C_{16}H_{18}^+$
9d2 9, 14, *17h* 102
$C_{16}H_{18}F_2N_2^+$
9d2 36
$C_{16}H_{18}F_5O$
17b 306
$C_{16}H_{18}FeNO_4$
17d1 323
$C_{16}H_{18}FeO_2^-$
9a 314, *9d1* 29
$C_{16}H_{18}MnN_2O_2$
17a 291
$C_{16}H_{18}N$
17c 285
$C_{16}H_{18}NO$
9c1 253, 567, 569, 629, 677, 710, 715, *17d1* 310ff, 371, 372, *17d2* 214
$C_{16}H_{18}NOS$
9c1 988, *17d1* 321
$C_{16}H_{18}NOSe$
17d1 322
$C_{16}H_{18}NOS_2$
9c1 217
$C_{16}H_{18}NO_2$
9c1 13, 572, 678
$C_{16}H_{18}NO_3$
1 56, *9c1* 569, 647
$C_{16}H_{18}NO_3S$
9c1 999
$C_{16}H_{18}NO_4$
9c1 14, 275
$C_{16}H_{18}NO_4S$
9c1 931
$C_{16}H_{18}NO_5$
1 56, *9c1* 537, 570, *17d2* 343
$C_{16}H_{18}NO_5S$
9c1 931
$C_{16}H_{18}NO_5S_2$
9c1 217, 997
$C_{16}H_{18}NO_6S_2$
9c1 931
$C_{16}H_{18}NO_7S_2$
9c1 217
$C_{16}H_{18}NS_2$
9c1 114, *17c* 520
$C_{16}H_{18}N_2^+$
9d2 106, 116, *17h* 150, 168
$C_{16}H_{18}N_2O_2^+$
9d2 79
$C_{16}H_{18}N_2O_4$
9c1 884
$C_{16}H_{18}N_2O_4^{2\cdot}$
9d2 158
$C_{16}H_{18}N_2S^+$
1 126
$C_{16}H_{18}N_2Zn^-$
17a 383
$C_{16}H_{18}N_3O_5$
17d2 122
$C_{16}H_{18}OS$
9b 707
$C_{16}H_{18}O_2^+$
9d2 137
$C_{16}H_{18}O_2S^+$
9d2 138, *17g* 386
$C_{16}H_{19}BrNS$
17c 503
$C_{16}H_{19}ClNS$
17c 503
$C_{16}H_{19}Cl_3NO$
17d1 242
$C_{16}H_{19}Cl_5NO$
17d1 288, 309
$C_{16}H_{19}Ge$
9b 671, *17a* 432
$C_{16}H_{19}GeO$
9b 671
$C_{16}H_{19}HgO_2$
17a 391
$C_{16}H_{19}NOP$
9c1 1009, *17d1* 200
$C_{16}H_{19}NO_2^+$
9d2 43
$C_{16}H_{19}NO_2P$
9c1 1009, *17c* 489, 494

$C_{16}H_{19}NO_3P$
17d2 419
$C_{16}H_{19}NO_4^+$
9d2 43
$C_{16}H_{19}NO_4P$
17c 489, 494
$C_{16}H_{19}NO_5P$
17d2 419
$C_{16}H_{19}NSi^-$
17f 217
$C_{16}H_{19}N_2O$
9c1 275, *17d1* 276, *17d2* 39
$C_{16}H_{19}N_2O_3$
17d1 219
$C_{16}H_{19}N_2Si$
9c1 96
$C_{16}H_{19}OSi$
9b 668, *17c* 288
$C_{16}H_{19}OSn$
9b 671, *17a* 459, *17c* 293
$C_{16}H_{19}O_6$
17b 448
$C_{16}H_{19}SSn$
9b 673, *17a* 459, *17c* 294
$C_{16}H_{19}Si$
9b 668
$C_{16}H_{19}Sn$
9b 671, *17a* 460
$C_{16}H_{20}^-$
9d1 746
$C_{16}H_{20}^+$
9d2 12
$C_{16}H_{20}AlF_3NO_7$
17a 401
$C_{16}H_{20}CoF_3NO_7$
17a 338
$C_{16}H_{20}Cr^+$
17a 258, 259
$C_{16}H_{20}CrI$
9a 304
$C_{16}H_{20}F_3NO_7Rh$
17a 342
$C_{16}H_{20}F_5$
9b 219
$C_{16}H_{20}Hg_2NO_5$
9c1 745
$C_{16}H_{20}MnNO_7P$
9c1 1037, *17a* 292
$C_{16}H_{20}NO$
9c1 694, 695, *17d1* 288, 303, *17d2* 23, 348
$C_{16}H_{20}NO_3$
17d2 242
$C_{16}H_{20}NO_4$
9c1 412, 537
$C_{16}H_{20}NO_7$
9c1 711
$C_{16}H_{20}NS$
17c 502
$C_{16}H_{20}N_2^+$
1 122, *9d2* 37, 104, *17h* 169
$C_{16}H_{20}N_2O^+$
9c1 306
$C_{16}H_{20}N_2O_2P$
1 74
$C_{16}H_{20}N_2O_3P$
1 52, *9c1* 66, *17c* 449, 451
$C_{16}H_{20}N_2P$
9c2 217
$C_{16}H_{20}N_3$
9c1 12
$C_{16}H_{20}N_3O_5$
17d2 33
$C_{16}H_{20}N_3O_7$
9c1 492
$C_{16}H_{20}N_3O_7^{2\cdot}$
17h 298
$C_{16}H_{20}N_4O_6^-$
17g 247
$C_{16}H_{20}N_5O_3S$
17d2 151
$C_{16}H_{20}O_2^-$
17g 27
$C_{16}H_{20}O_4Rh$
17g 174ff
$C_{16}H_{21}BrNO_4S$
17d2 243
$C_{16}H_{21}NO_6^-$
9d1 542
$C_{16}H_{21}N_2O$
9c1 511
$C_{16}H_{21}N_2O_2$
9c1 424, *17c* 455
$C_{16}H_{21}N_2O_3$
17d2 23, 268
$C_{16}H_{21}N_2O_4$
17d2 231, 386
$C_{16}H_{21}N_2O_5$
9c1 492
$C_{16}H_{21}N_2O_6$
17d1 219
$C_{16}H_{21}N_3S^{3+}$
17h 198
$C_{16}H_{21}N_4O_6$
17e 193
$C_{16}H_{21}O_4$
17e 220
$C_{16}H_{22}BrO_3$
17e 62
$C_{16}H_{22}NO$
9c1 711, *9c2* 83, *17c* 594, *17d1* 289
$C_{16}H_{22}NO_2$
17d1 319
$C_{16}H_{22}NO_2SSi$
9c1 979
$C_{16}H_{22}NO_3^-$
9c1 275
$C_{16}H_{22}NO_3$
9c1 298, 492, *17d2* 23, 312
$C_{16}H_{22}NO_3S$
9c1 360
$C_{16}H_{22}NO_4$
17d2 104, 105, 198
$C_{16}H_{22}N_2O_2^-$
9d1 203
$C_{16}H_{22}N_2O_4M$
17d1 40
$C_{16}H_{22}N_2O_5M$
17d1 40, 41
$C_{16}H_{22}N_3O$
17d2 313
$C_{16}H_{22}N_3O_3$
9c1 825, *17d2* 33, 47
$C_{16}H_{22}N_3O_4$
9c1 156
$C_{16}H_{22}N_3O_6$
17d2 243
$C_{16}H_{22}OS^-$
17g 50
$C_{16}H_{22}O_2^-$
9d1 21
$C_{16}H_{22}O_2^+$
17h 86, 87
$C_{16}H_{22}O_4^-$
9d1 98

$C_{16}H_{22}O_6^+$
17h 69
$C_{16}H_{22}Si_2^-$
17f 20
$C_{16}H_{23}F_3NO_2Sn$
17a 460
$C_{16}H_{23}NO_2^-$
9d1 419
$C_{16}H_{23}N_2O$
9c1 276
$C_{16}H_{23}N_2O_2$
9c1 493, 884ff
$C_{16}H_{23}N_2O_3$
9c1 412
$C_{16}H_{23}N_2O_4$
9c1 319
$C_{16}H_{23}N_3O_2P$
17g 270
$C_{16}H_{23}N_4O_2$
17d2 33, 48
$C_{16}H_{23}N_4O_3$
17d2 387
$C_{16}H_{23}N_4O_4$
17d2 243
$C_{16}H_{23}N_6O_4$
17d2 131
$(C_{16}H_{23}O)_n$
17e 101
$C_{16}H_{23}OS$
9c2 282
$C_{16}H_{23}O_2^{2-}$
9b 569, 570, *17c* 193, 198
$C_{16}H_{23}O_2$
9c2 116, *17e* 225
$C_{16}H_{23}O_3$
1 64, 66, *9c2* 83, 134, *17e* 61, 94
$C_{16}H_{24}ClNO_4P$
17d2 313
$C_{16}H_{24}ClO_2Pt$
17g 129
$C_{16}H_{24}Cl_2NO$
17d1 235, 289, 309
$C_{16}H_{24}Cl_2NO_2$
17d1 289, 309, 319
$C_{16}H_{24}CuN_4O_6$
9a 328, *17a* 357
$C_{16}H_{24}D_2O_2^+$
9d2 131
$C_{16}H_{24}F$
9b 562
$C_{16}H_{24}Ge_2^-$
17f 45, 192
$C_{16}H_{24}MnO_2S$
17a 292
$C_{16}H_{24}NO$
9c1 412, 640, 746, 756, 758, *9c2* 122, *17d1* 289, *17d2* 269
$C_{16}H_{24}NO_2$
1 65, *9c1* 315, 756, *9c2* 100, *17d2* 23
$C_{16}H_{24}NO_3$
9c1 276, *17d1* 289, 290, *17d2* 56, 273, *17e* 106
$C_{16}H_{24}NO_3P$
17d2 171
$C_{16}H_{24}NO_4$
17d2 103
$C_{16}H_{24}NO_4S$
9c1 493
$C_{16}H_{24}NO_5P$
9c1 1010 [6])
$C_{16}H_{24}NO_9$
17c 625, 626, 627, 628
$C_{16}H_{24}NS_2$
17c 558
$C_{16}H_{24}N_2O_2^+$
9d2 104
$C_{16}H_{24}N_2O_3$
9c1 789
$C_{16}H_{24}N_2O_4^-$
17g 39
$C_{16}H_{24}N_2O_4X$
17d2 106
$C_{16}H_{24}N_3O_2$
9c1 825, 885
$C_{16}H_{24}N_3O_3$
9c1 253
$C_{16}H_{24}N_4O_3^{2\cdot}$
9d2 290
$C_{16}H_{24}N_4O_6Pd$
17a 345
$C_{16}H_{24}N_4O_6Pd^{2\cdot}$
17h 288
$C_{16}H_{24}O^-$
17c 192
$C_{16}H_{24}OS^-$
17c 207
$C_{16}H_{24}Si_2^-$
17f 20, 21, 192
$C_{16}H_{24}Sn_2^-$
17f 192
$C_{16}H_{25}$
9b 219, *17b* 485
$C_{16}H_{25}Cl_2O_2Sn$
17e 185
$C_{16}H_{25}CoN_4O_5$
17a 338
$C_{16}H_{25}HgO_2$
17a 391
$C_{16}H_{25}NO^+$
17h 162
$C_{16}H_{25}NO_5P$
9c1 325
$C_{16}H_{25}N_2$
9c1 95
$C_{16}H_{25}N_2O$
9c1 413, *17d2* 24
$C_{16}H_{25}N_2O_2$
9c1 359, *17d1* 136, *17d2* 24, 79
$C_{16}H_{25}N_2O_2S$
9c1 96
$C_{16}H_{25}O$
1 61, 72, *9c2* 80, *17e* 87
$C_{16}H_{25}OS$
9c2 141
$C_{16}H_{25}OSi$
9b 595
$C_{16}H_{25}OSn$
9b 599
$C_{16}H_{25}O_2$
1 63, 66, *9c2* 74, 84, 131, 152, *9d1* 137, *17e* 94, 99, 109
$C_{16}H_{25}O_9$
1 36
$C_{16}H_{26}^+$
9d2 7

[6]) In II/9c1 erroneously listed as $[C_{16}H_{24}NO_5]$.

$C_{16}H_{26}Al_2N_2^-$
17a 402, *17f* 205
$C_{16}H_{26}Br_2N_4O_4^{2\cdot}$
17h 284
$C_{16}H_{26}ClO_2Pb$
17a 494
$C_{16}H_{26}ClO_2Sn$
17a 460, *17e* 151, *17g* 181
$C_{16}H_{26}Cl_2N_4O_4^{2\cdot}$
17h 284
$C_{16}H_{26}FO_2Si$
17e 141
$C_{16}H_{26}F_3O_3Sn$
17a 461
$C_{16}H_{26}HgN_2O_2^{2\cdot}$
17h 275
$C_{16}H_{26}NO$
9c1 254, *9c2* 120, *17d1* 125, 136, 216, *17d2* 15, 141
$C_{16}H_{26}NOSi$
17c 615, 616
$C_{16}H_{26}NO_2$
9c1 276, 277, 315, *17d2* 54
$C_{16}H_{26}NO_2Sn$
9c1 756
$C_{16}H_{26}NO_3$
9c1 315, *17d2* 114
$C_{16}H_{26}NO_3S$
17c 491, 495
$C_{16}H_{26}NO_4$
17d2 97
$C_{16}H_{26}NO_5$
9c1 350, 361, *17d1* 158
$C_{16}H_{26}NO_5S$
9c1 932, 943
$C_{16}H_{26}NO_9$
9c1 352
$C_{16}H_{26}N_2O_2^-$
1 100
$C_{16}H_{26}N_2Si_2^+$
17f 35
$C_{16}H_{26}N_3O_2$
9c1 322
$C_{16}H_{26}N_3O_5$
9c1 851, 885, *17d2* 382
$C_{16}H_{26}N_3O_6$
17d2 387
$C_{16}H_{26}N_4O_2^{2\cdot}$
9d2 287
$C_{16}H_{26}N_4O_3X$
17d1 15, 60, 61
$C_{16}H_{26}N_4O_4^{2\cdot}$
9d2 212, 291
$C_{16}H_{26}N_6O_2^{2\cdot}$
17h 289
$C_{16}H_{26}N_6O_3^{2\cdot}$
17h 297
$C_{16}H_{26}N_6O_4^{2\cdot}$
17h 284, 285
$C_{16}H_{26}N_6O_4Pd$
17a 345
$C_{16}H_{26}N_6O_4Pd^{2\cdot}$
17h 289
$C_{16}H_{26}N_7OS$
17d2 151
$C_{16}H_{26}O^-$
9b 485ff
$C_{16}H_{26}O^+$
17e 246
$C_{16}H_{26}O_2^-$
9d1 21
$C_{16}H_{26}O_2^+$
9d2 131, 133, 134
$C_{16}H_{26}O_2Tl$
17g 227, 228
$C_{16}H_{26}O_4^-$
9d1 55
$C_{16}H_{27}B^-$
9d1 701
$C_{16}H_{27}ClNOSi$
17d1 290
$C_{16}H_{27}F_8OSi$
9b 347
$C_{16}H_{27}N^+$
9d2 37, *17h* 162
$C_{16}H_{27}N_2O$
17d1 290
$C_{16}H_{27}N_2O_5$
17d2 382
$C_{16}H_{27}N_2O_6$
17d2 387
$C_{16}H_{27}N_3O_3^{2\cdot}$
9d2 222
$C_{16}H_{27}N_4O_7S$
17d2 168ff
$C_{16}H_{27}O$
9b 485
$C_{16}H_{28}^+$
17h 26, 32, 33
$C_{16}H_{28}BD_5^-$
17f 222
$C_{16}H_{28}NOSSi$
17d1 376
$C_{16}H_{28}NOSi$
17d1 322, 371
$C_{16}H_{28}NO_2$
9c1 518
$C_{16}H_{28}NO_2SSi$
9c1 974
$C_{16}H_{28}NO_2Si$
17d1 379
$C_{16}H_{28}NO_4$
17d2 313
$C_{16}H_{28}NO_4SSi$
9c1 974
$C_{16}H_{28}N_2^+$
9d2 104, *17h* 151, 184
$C_{16}H_{28}N_2O_3^{2\cdot}$
9d2 224, *17h* 282
$C_{16}H_{28}N_3O_2$
9c1 494, *17d2* 84
$C_{16}H_{28}N_4^+$
17h 160
$C_{16}H_{28}N_4O_2^{2\cdot}$
9d2 193ff, 288, *17h* 280
$C_{16}H_{28}N_4O_4^{2\cdot}$
9d2 291
$C_{16}H_{29}$
17c 11
$C_{16}H_{29}BD_4^-$
9d1 700
$C_{16}H_{29}N_2O_4$
17c 446, *17d2* 110
$C_{16}H_{29}N_2O_6$
9c1 295, *17d2* 110
$C_{16}H_{29}N_2O_7$
9c1 306
$C_{16}H_{29}N_4O$
9c1 404
$C_{16}H_{29}N_4O_2$
9c1 495
$C_{16}H_{29}OSi_2$
9c2 172
$C_{16}H_{29}S_4Sn$
17b 383
$C_{16}H_{30}F_3O_3Si_3$
9b 464
$C_{16}H_{30}NO$
9c1 100

$C_{16}H_{30}NOS_2Sn$
17a 461, *17c* 69
$C_{16}H_{30}NO_2$
9c1 519, *17d2* 273
$C_{16}H_{30}NO_2Si$
17d1 379
$C_{16}H_{30}NO_3$
9c1 345
$C_{16}H_{30}NO_4$
9c1 431, *17d2* 273
$C_{16}H_{30}NO_5$
9c1 362
$C_{16}H_{30}NSi$
9b 493, *9c1* 25, *17c* 412
$C_{16}H_{30}N_2O_4M$
17d1 41
$C_{16}H_{30}N_4O_2P$
17d1 276
$C_{16}H_{30}O_2^-$
9d1 22
$C_{16}H_{30}O_4P$
17b 130
$C_{16}H_{30}Si_2^+$
17h 64
$C_{16}H_{31}B^-$
9d1 700
$C_{16}H_{31}B_{10}Cl_2O_2Sn$
17a 461ff, *17g* 186
$C_{16}H_{31}B_{10}HgO_2$
17g 163
$C_{16}H_{31}Cl_2N_3O_3P$
17d2 313
$C_{16}H_{31}NNaO_6S$
17d2 274
$C_{16}H_{31}NO_3PSi$
17e 299
$C_{16}H_{31}N_2O_2$
17d2 78
$C_{16}H_{31}N_4O_2PX$
17d1 320, 322
$C_{16}H_{31}OSi$
17b 298
$C_{16}H_{31}O_2$
9b 115, 190, *17b* 171
$C_{16}H_{31}O_3P^+$
17b 281
$C_{16}H_{31}SSi$
17b 486
$C_{16}H_{32}ClN_2O_3$
17d2 314
$C_{16}H_{32}NO_2S$
17d1 160
$C_{16}H_{32}NO_3$
9c1 350
$C_{16}H_{32}NO_6$
17d2 339
$C_{16}H_{32}NO_7$
17d1 151
$C_{16}H_{32}N_3O_3$
17d2 231
$C_{16}H_{32}N_4O_4^{2\cdot}$
9d2 308
$C_{16}H_{32}O_2^-$
17b 291
$C_{16}H_{32}Si_2^+$
17f 36, *17h* 31, 33
$C_{16}H_{33}$
9b 82, 105, *17b* 275
$C_{16}H_{33}B^-$
9d1 699
$C_{16}H_{33}GeO_2$
17b 268
$C_{16}H_{33}NO_3$
9c1 779
$C_{16}H_{33}N_2O_2$
17d2 109
$C_{16}H_{33}N_2O_4Si$
17c 448
$C_{16}H_{33}O_2S$
9c2 289
$C_{16}H_{33}O_2Sn$
17a 462, *17b* 168, 202, 261, 270
$C_{16}H_{33}Si_4$
17f 7
$C_{16}H_{33}Sn$
9b 386
$C_{16}H_{34}Al_2N_2^-$
17f 201
$C_{16}H_{34}Al_2P_2^-$
17a 402
$C_{16}H_{34}Cl_2N_4O_2Pt^{2\cdot}$
9d2 194, *17h* 272
$C_{16}H_{34}NO$
9c1 217, 346
$C_{16}H_{34}NO_3$
9c1 254
$C_{16}H_{34}OSi_5^-$
17f 27, 28
$C_{16}H_{34}O_3P$
17b 275
$C_{16}H_{34}Si_6^-$
17f 28
$C_{16}H_{35}BNO$
17d1 152
$C_{16}H_{35}N_2O$
9c1 306, *17d1* 148
$C_{16}H_{35}OSi$
17b 283
$C_{16}H_{35}O_3P^+$
17b 281
$C_{16}H_{35}P^+$
9b 143, *17b* 280
$C_{16}H_{35}Si_2$
17b 159
$C_{16}H_{35}Sn$
9b 105, *17b* 201
$C_{16}H_{36}B_2^-$
17f 222
$C_{16}H_{36}B_2Si_2^-$
17f 226
$C_{16}H_{36}B_4^-$
17f 227
$C_{16}H_{36}NO_2P^+$
17b 280
$C_{16}H_{36}NO_2Sn$
17d1 188, *17d2* 411
$C_{16}H_{36}NO_3Sn$
17d2 411
$C_{16}H_{36}NO_7P_2$
9c1 326
$C_{16}H_{36}N_2O_4P_2^+$
9c2 266
$C_{16}H_{36}O_2P$
9c2 259
$C_{16}H_{36}O_4P$
9c2 259
$C_{16}H_{36}O_6P$
9c2 25
$C_{16}H_{36}P_4^-$
17e 309
$C_{16}H_{36}Si_2^-$
17f 24
$C_{16}H_{36}Si_4^+$
17h 28
$C_{16}H_{37}OSiSn$
17b 19
$C_{16}H_{37}Si_2$
9b 230, *17b* 296, *17f* 7

$C_{16}H_{38}BNOP$
17d1 199ff
$C_{16}H_{38}N_3P^+$
17b 280
$C_{16}H_{38}N_4O_2P_2^+$
9c2 266
$C_{16}H_{38}N_4P_2^+$
9c2 267
$C_{16}H_{39}BrN_4OP$
17d1 201
$C_{16}H_{40}O_{10}P_4^+$
9c2 267
$C_{16}H_{40}Si_4^-$
17f 28
$C_{17}H_2B_2D_{36}^-$
17f 223
$C_{17}H_4MnN_4O_5$
17a 292
$C_{17}H_5Cl_8MoO_4$
17a 270
$C_{17}H_6O_{10}^{5-}$
9b 483
$C_{17}H_8Cl_2O_2S_2^-$
1 119
$C_{17}H_9D_9N_3$
17c 478
$C_{17}H_9D_{12}N_2O_2$
17d2 113
$C_{17}H_9D_{16}N_2O_2$
17d2 314
$C_{17}H_{10}AsO_2S_4$
17g 368
$C_{17}H_{10}Cl_3D_6O_4S_2$
17c 267
$C_{17}H_{10}Cl_3D_6S_2$
17c 267
$C_{17}H_{10}D_6F_3O_2S$
17c 284
$C_{17}H_{10}D_6F_3O_4S_2$
17c 266
$C_{17}H_{10}D_6F_3O_4S_3$
17c 284
$C_{17}H_{10}D_6F_3S_2$
17c 266
$C_{17}H_{10}D_6F_3S_3$
17c 284
$C_{17}H_{10}D_{12}N_2O$
17d1 277
$C_{17}H_{10}N_7O_8$
9c1 59
$C_{17}H_{10}O_2S_4Sb$
17g 376
$C_{17}H_{11}$
9b 634, *17c* 348
$C_{17}H_{11}ClMnO_7$
17e 198
$C_{17}H_{11}ClO_7Re$
17e 199
$C_{17}H_{11}D_7N_3$
17c 477
$C_{17}H_{11}HgO_2S_2$
17g 370
$C_{17}H_{11}O_4^{2-}$
9d1 293, 294
$C_{17}H_{11}O_5^{2-}$
9d1 132
$C_{17}H_{12}^-$
9d1 837, *17f* 108, 111
$C_{17}H_{12}CrO_4^-$
17a 259
$C_{17}H_{12}F_6O_5^-$
9b 661
$C_{17}H_{12}MnO_7$
17e 197
$C_{17}H_{12}N_2O_3^-$
17f 283
$C_{17}H_{12}N_5O_4$
9c1 58
$C_{17}H_{12}O^-$
17c 275
$C_{17}H_{12}O^+$
17h 101
$C_{17}H_{12}O_3^-$
17g 278
$C_{17}H_{12}O_7Re$
17e 197
$C_{17}H_{12}S_2^-$
9b 506, *17c* 124
$C_{17}H_{12}S_3^-$
17f 216
$C_{17}H_{13}^{2-}$
17f 108
$C_{17}H_{13}Cl_3N_3O$
17d1 362
$C_{17}H_{13}Cl_4N_4O_2$
9c1 86
$C_{17}H_{13}D_6O_4S_3$
17c 282
$C_{17}H_{13}D_6S_2$
17c 261
$C_{17}H_{13}FFeO^-$
9a 317
$C_{17}H_{13}HgO_2$
17a 391, *17g* 259
$C_{17}H_{13}MnO_4^-$
17a 292
$C_{17}H_{13}MnO_5^-$
17a 293
$C_{17}H_{13}MoN_8OM_2$
17d1 324
$C_{17}H_{13}N_2O_2$
9c1 793
$C_{17}H_{13}N_4O_4^{2-\cdot}$
9d2 292
$C_{17}H_{13}S$
9b 762
$C_{17}H_{14}^-$
9d1 778, 837, 838, *17f* 109
$C_{17}H_{14}^+$
17h 100
$C_{17}H_{14}ClGe$
17a 433, *17c* 85
$C_{17}H_{14}ClSi$
17c 83
$C_{17}H_{14}D_2^-$
17f 110
$C_{17}H_{14}FeO^-$
9a 317, 318
$C_{17}H_{14}NO_2$
9b 576
$C_{17}H_{14}NO_4$
17d2 352ff
$C_{17}H_{14}N_2O_3S^-$
17f 263
$C_{17}H_{14}N_2O_3Se^-$
17f 263
$C_{17}H_{14}N_5O_3$
17c 443
$C_{17}H_{14}O^-$
9b 542
$C_{17}H_{14}ORu^-$
9a 317
$C_{17}H_{14}O_3^-$
9b 661
$C_{17}H_{15}^{2-}$
17f 109
$C_{17}H_{15}F_5NO$
9c1 277
$C_{17}H_{15}F_5NO_3$
17d2 243

$C_{17}H_{15}NO^-$
17f 198
$C_{17}H_{15}NO_2^-$
17f 198
$C_{17}H_{15}N_2O$
9c1 758
$C_{17}H_{15}N_2O_2$
9c1 88
$C_{17}H_{15}N_2O_3$
17d2 107
$C_{17}H_{16}^-$
17f 110
$C_{17}H_{16}ClN_2O$
9c1 636
$C_{17}H_{16}Cl_2NO$
17d1 236
$C_{17}H_{16}D_4N_2^+$
9d2 107
$C_{17}H_{16}FeO_2^-$
9d1 330
$C_{17}H_{16}NO_2$
9c1 746
$C_{17}H_{16}NO_3$
17d2 353
$C_{17}H_{16}N_2^+$
17h 130
$C_{17}H_{16}N_2S^+$
17h 198
$C_{17}H_{16}N_3O$
17d1 360
$C_{17}H_{16}N_3O_2$
17c 457
$C_{17}H_{16}N_3O_3$
9c1 629
$C_{17}H_{17}^{2-}$
9d1 816
$C_{17}H_{17}$
17c 254
$C_{17}H_{17}Br_3NO$
9c1 277
$C_{17}H_{17}Cl_3NO$
9c1 277
$C_{17}H_{17}Cl_3NO_2$
17d1 247
$C_{17}H_{17}NO^-$
17c 238
$C_{17}H_{17}NO_2^-$
9d1 868
$C_{17}H_{17}NO_2S^+$
17h 198
$C_{17}H_{17}N_2O$
9c1 629, 701, *17c* 430, 622
$C_{17}H_{17}N_2O_2$
9c1 523
$C_{17}H_{17}N_2O_3$
17d2 387
$C_{17}H_{17}N_3O_6S^-$
17f 253
$C_{17}H_{17}N_4$
9c1 87
$C_{17}H_{17}N_4O$
17c 470
$C_{17}H_{17}N_4O_2$
9c1 85
$C_{17}H_{17}N_4S$
17c 473
$C_{17}H_{17}N_6O_4$
9c1 70
$C_{17}H_{17}O$
9b 693
$C_{17}H_{17}OS$
9b 666, *17c* 281
$C_{17}H_{17}O_2^{2-}$
17c 274
$C_{17}H_{17}O_2$
17c 256, 277, *17e* 156
$C_{17}H_{17}O_2Pb$
17a 494, *17g* 323
$C_{17}H_{17}O_2Si$
17e 202
$C_{17}H_{17}O_2Sn$
17a 462, *17g* 325ff
$C_{17}H_{17}O_4$
17c 257
$C_{17}H_{17}O_6$
9c2 169, 171
$C_{17}H_{18}^-$
9d1 726, 813, 814
$C_{17}H_{18}CoN_7O_3^{3-}$
9c1 1031
$C_{17}H_{18}FN_2O$
9c1 817
$C_{17}H_{18}F_2NO$
17d2 40
$C_{17}H_{18}FeNO_3$
9a 312, *9c1* 1039, *17a* 330, *17d1* 323
$C_{17}H_{18}FeO_3^-$
9a 316
$C_{17}H_{18}MoNO_3$
17d1 323
$C_{17}H_{18}NO$
9c1 278, 798
$C_{17}H_{18}NO_2$
9c1 678, *17c* 482, *17d1* 316, *17d2* 366, 400
$C_{17}H_{18}NO_2SSi$
17b 360
$C_{17}H_{18}NO_3S$
9c1 779
$C_{17}H_{18}N_3$
9c1 89, *17c* 429, 477
$C_{17}H_{18}N_3O_2$
17d2 388
$C_{17}H_{18}N_3O_5$
9c1 278
$C_{17}H_{18}N_4O_4^-$
9c1 38
$C_{17}H_{18}N_4O_4^+$
9d2 80
$C_{17}H_{18}O^-$
9b 653ff, 655, *17c* 270, 271
$C_{17}H_{18}O_4^-$
9d1 239
$C_{17}H_{18}O_4^+$
17h 89
$C_{17}H_{18}O_5^-$
9b 660
$C_{17}H_{19}$
9b 635ff, 640, *17c* 262
$C_{17}H_{19}BrNOS$
17d2 63
$C_{17}H_{19}ClNO$
9c1 376
$C_{17}H_{19}ClNOS$
17d1 360, *17d2* 63
$C_{17}H_{19}ClNO_2$
9c1 789
$C_{17}H_{19}ClNO_3S$
17d2 64
$C_{17}H_{19}ClN_2S^+$
9d2 80, *17h* 199ff
$C_{17}H_{19}FNO$
17d2 40
$C_{17}H_{19}GeN_4$
17c 475
$C_{17}H_{19}NO_3$
17d1 311

$C_{17}H_{19}NO_4SM$
17d2 94
$C_{17}H_{19}N_2$
9c1 94
$C_{17}H_{19}N_2O$
9c1 818, 827, *17d1* 171, 229, 230
$C_{17}H_{19}N_2O_2$
17d1 171
$C_{17}H_{19}N_2O_3$
9c1 629, *17d1* 276, *17d2* 40
$C_{17}H_{19}N_2O_3S$
17d2 63
$C_{17}H_{19}N_3^+$
17h 185
$C_{17}H_{19}N_4$
9c1 69, 75, 77
$C_{17}H_{19}N_4O_2$
17d2 244
$C_{17}H_{19}N_4O_4$
9c1 38, *17c* 435
$C_{17}H_{19}N_4S$
9c1 86
$C_{17}H_{19}N_4Si$
17c 476
$C_{17}H_{19}OS$
9b 665
$C_{17}H_{19}OSn$
17c 317
$C_{17}H_{19}O_2$
17e 75
$C_{17}H_{19}O_2Si$
9b 595, *17c* 291
$C_{17}H_{19}O_2Sn$
17a 463
$C_{17}H_{19}O_6$
9c2 167, 168
$C_{17}H_{19}O_7$
17b 444
$C_{17}H_{19}S$
9b 665, *17c* 281
$C_{17}H_{20}^-$
9d1 740, 769
$C_{17}H_{20}Cl_2GeNO$
17d2 68
$C_{17}H_{20}Cl_2NOSi$
17d2 67
$C_{17}H_{20}Cl_3MnO_5P$
17g 167
$C_{17}H_{20}D_5$
9b 389, 390
$C_{17}H_{20}D_5O_2^{2-}$
9b 568
$C_{17}H_{20}FeO_2^-$
9a 315
$C_{17}H_{20}MgO_3^+$
17g 334, 338
$C_{17}H_{20}MgO_4^+$
17g 335, 339
$C_{17}H_{20}NO$
9c1 254, 278ff, 279, 572, 619, 630, 678, *17d1* 263, 264, 276ff, 311, 359, *17d2* 40ff, 216
$C_{17}H_{20}NOS$
9c1 573, *17d2* 63
$C_{17}H_{20}NO_2$
9c1 573, 623, 645, *17d1* 311, *17d2* 54, 214, 215
$C_{17}H_{20}NO_3$
9c1 716, *17d1* 219
$C_{17}H_{20}NO_3S$
17d1 321, *17d2* 65
$C_{17}H_{20}NO_4S$
9c1 932
$C_{17}H_{20}NO_5S$
9c1 944
$C_{17}H_{20}NO_6$
17d1 355
$C_{17}H_{20}NO_6S$
9c1 933
$C_{17}H_{20}N_2^+$
9d2 107
$C_{17}H_{20}N_2O^-$
9b 661
$C_{17}H_{20}N_2O^+$
9d2 37
$C_{17}H_{20}N_2O_2$
17d2 45
$C_{17}H_{20}N_2O_2S^+$
17h 142
$C_{17}H_{20}N_2O_4^{2\cdot}$
9d2 159
$C_{17}H_{20}N_2S^-$
9b 664
$C_{17}H_{20}N_2S^+$
9d2 80, *17h* 198ff
$C_{17}H_{20}N_3O$
9c1 818
$C_{17}H_{20}O_3^-$
17g 338
$C_{17}H_{20}O_3Zn^+$
17g 339
$C_{17}H_{20}O_4Zn^+$
17g 335, 339
$C_{17}H_{21}BrNO_3$
17d2 244
$C_{17}H_{21}ClNO$
9c1 250
$C_{17}H_{21}Cl_2OS$
9b 502
$C_{17}H_{21}D_2OS$
9b 501
$C_{17}H_{21}HgO_2$
17a 391
$C_{17}H_{21}NOPS$
9c1 247
$C_{17}H_{21}NOPS_2$
9c1 246
$C_{17}H_{21}NOX$
(X unknown group)
17d2 24
$C_{17}H_{21}NO_3PS_2$
9c1 247
$C_{17}H_{21}NO_3S$
9c1 248
$C_{17}H_{21}NO_4P$
9c1 248
$C_{17}H_{21}N_2O$
17d1 130
$C_{17}H_{21}N_2O_4$
17d2 244
$C_{17}H_{21}N_2S^{2+}$
17h 200
$C_{17}H_{21}N_3O_4M$
17d1 41
$C_{17}H_{22}AsO_2$
9c2 317
$C_{17}H_{22}BrOS$
9b 502
$C_{17}H_{22}BrS_2$
9b 508
$C_{17}H_{22}ClS_2$
9b 507, 508
$C_{17}H_{22}Cl_2NO$
17d2 24
$C_{17}H_{22}NO$
9c1 220, 279, 413, *17d2* 314

$C_{17}H_{22}NOSi$
9c1 323, *17d2* 67
$C_{17}H_{22}NO_2$
9c1 746
$C_{17}H_{22}NO_3$
17d2 244
$C_{17}H_{22}NO_5OsSi$
17d1 324
$C_{17}H_{22}NO_5RuSi$
17d1 324
$C_{17}H_{22}NO_5S_2$
9c1 996
$C_{17}H_{22}NO_7$
9c1 232
$C_{17}H_{22}NSi$
9b 671
$C_{17}H_{22}N_2OX$
17d2 90
$C_{17}H_{22}N_3O_3S$
17d2 86
$C_{17}H_{22}N_5O_7$
9c1 495
$C_{17}H_{23}AlCl_2NO_2$
17g 155
$C_{17}H_{23}Br_3NO_2$
9c1 295
$C_{17}H_{23}Cl_3NO$
17d2 24
$C_{17}H_{23}CoN_3O_4$
17e 19
$C_{17}H_{23}NO_3$
17d2 24
$C_{17}H_{23}N_2O$
17d2 25
$C_{17}H_{23}N_2O_3S$
9c1 307
$C_{17}H_{23}N_2O_3Tl$
17d2 261
$C_{17}H_{23}N_2O_5SM$
17d2 244
$C_{17}H_{23}N_3O_5M$
17d1 12, 41
$C_{17}H_{23}N_3O_6M$
17d1 41
$C_{17}H_{23}O$
9c2 114
$C_{17}H_{23}OS$
9b 500, 502
$C_{17}H_{23}S_2$
9b 508
$C_{17}H_{24}BrOPd$
17e 105
$C_{17}H_{24}ClOPd$
17e 105
$C_{17}H_{24}Cl_2N_3O_2$
17d2 315
$C_{17}H_{24}DO$
17e 101
$C_{17}H_{24}IOPd$
17e 106
$C_{17}H_{24}NO$
9b 502, 503, *17d1* 291
$C_{17}H_{24}NO_3$
17d2 196
$C_{17}H_{24}NO_4$
17d2 25, 197
$C_{17}H_{24}NO_5$
17d1 222, *17d2* 121
$C_{17}H_{24}NO_5S$
9c1 944
$C_{17}H_{24}N_3O_2$
9c1 494
$C_{17}H_{24}N_3O_3$
17d2 34
$C_{17}H_{25}Cl_3NO$
17d1 277, 291
$C_{17}H_{25}NO_2^-$
9d1 418
$C_{17}H_{25}N_2O_2$
17d2 131
$C_{17}H_{25}N_2O_4$
9c1 360, *17d2* 268
$C_{17}H_{25}N_2O_5$
9c1 515
$C_{17}H_{25}NiO_2$
17g 126
$C_{17}H_{25}O$
17c 189, *17e* 101
$C_{17}H_{25}O_2^{2-}$
9b 568, 570, *17c* 196, 199
$C_{17}H_{25}O_2$
17e 60, 224
$C_{17}H_{25}O_2Pd$
17g 127, 171
$C_{17}H_{25}O_3$
1 64, *9c2* 117, *17e* 61
$C_{17}H_{26}AlCl_2O_3$
17g 155
$C_{17}H_{26}Cl_2NO$
17d1 291, 309
$C_{17}H_{26}Cl_2NO_2$
17d1 320
$C_{17}H_{26}F_3O_3Sn$
17a 463
$C_{17}H_{26}NO$
9c1 279, 642, 756,
17d1 278, 291, 292, 358,
17d2 15, 25
$C_{17}H_{26}NO_2$
1 66, *9c1* 280, 315, *9c2* 100,
17d2 25, 34, 54, 177
$C_{17}H_{26}NO_3$
17d1 136, 278, 292, *17e* 106
$C_{17}H_{26}NO_3S$
9c1 315
$C_{17}H_{26}NO_6$
17d1 354
$C_{17}H_{26}N_2^+$
17h 156
$C_{17}H_{26}N_3O_3$
17d2 388
$C_{17}H_{26}N_3O_5$
17d1 61
$C_{17}H_{26}O^-$
17c 194
$C_{17}H_{27}$
9b 562, *17c* 186, 188
$C_{17}H_{27}F_{10}OSi$
9b 243
$C_{17}H_{27}NO^-$
17c 209, 210
$C_{17}H_{27}NO_4P$
17d2 315
$C_{17}H_{27}NS^-$
9b 585, *17c* 211, 212
$C_{17}H_{27}N_2O_2$
17d2 78
$C_{17}H_{27}N_3O_2^-$
17f 254
$C_{17}H_{27}N_4O_4S$
17d2 151
$C_{17}H_{27}O$
1 61, 72, *9c2* 75, 81,
17e 88, 100, 151
$C_{17}H_{27}OS$
1 67, *9c2* 141
$C_{17}H_{27}OSe$
1 68
$C_{17}H_{27}O_2$
1 66, *9c2* 131, 152, 161

$C_{17}H_{28}NO$
1 63, *9c2* 83, *17d1* 292
$C_{17}H_{28}NO_2$
9c1 280, 346, *17d2* 25, 34, 54
$C_{17}H_{28}NO_3$
17d2 26, 58
$C_{17}H_{28}NO_4$
17d2 96, 97, 192
$C_{17}H_{28}NO_5$
17d2 98
$C_{17}H_{28}NS$
17c 504
$C_{17}H_{28}N_6O_2^{2\cdot}$
17h 290
$C_{17}H_{28}N_6O_3^{2\cdot}$
17h 297
$C_{17}H_{28}O^-$
9b 486
$C_{17}H_{28}O^+$
17e 247
$C_{17}H_{29}ClNO_5$
9c2 126
$C_{17}H_{29}Cl_3NSi$
9b 529
$C_{17}H_{29}Cl_3PSi$
9b 534
$C_{17}H_{29}D_2S_3Sn$
17a 463, *17c* 67
$C_{17}H_{29}FOP$
17c 161
$C_{17}H_{29}GeO$
9c2 118, 152, *17a* 433
$C_{17}H_{29}GeO_2$
17a 433
$C_{17}H_{29}INO$
9c2 126
$C_{17}H_{29}N_2O$
17d2 26
$C_{17}H_{29}N_2O_5$
17d2 382
$C_{17}H_{29}N_2O_6$
9c1 851, *17d2* 388
$C_{17}H_{29}N_2O_7$
9c1 886
$C_{17}H_{29}N_4O$
9c1 857
$C_{17}H_{29}N_4O_7S$
17d2 182
$C_{17}H_{29}NaO_2P$
17c 162
$C_{17}H_{29}O$
9b 495
$C_{17}H_{29}OSi$
9b 593, 594, *9c2* 145, 152, *17c* 219, *17e* 143
$C_{17}H_{29}O_2Pb$
17a 494, *17e* 139, *17g* 169
$C_{17}H_{29}O_2Si$
17e 65, 141
$C_{17}H_{29}O_2Sn$
17a 464, *17b* 271, *17e* 141ff, *17g* 179
$C_{17}H_{29}P^-$
9d1 895
$C_{17}H_{29}S$
9b 605
$C_{17}H_{30}^{+}$
17h 26
$C_{17}H_{30}Cl_3Si$
9b 384
$C_{17}H_{30}GeNO$
17d2 68
$C_{17}H_{30}NOSi$
17c 492, *17d2* 67
$C_{17}H_{30}NOSn$
17c 493
$C_{17}H_{30}NO_2Si$
17d1 379
$C_{17}H_{30}NO_2Sn$
9c1 975, *17d1* 380
$C_{17}H_{30}NO_3$
17d2 274, 315
$C_{17}H_{30}NO_4$
9c1 146
$C_{17}H_{30}NO_6$
17d2 121
$C_{17}H_{30}N_2O_3^{2\cdot}$
9d2 224
$C_{17}H_{30}N_2O_3P$
17d2 70
$C_{17}H_{30}O^-$
17c 30
$C_{17}H_{30}O^+$
17h 33
$C_{17}H_{30}OP$
17c 158
$C_{17}H_{30}O_2P$
17c 162
$C_{17}H_{31}$
17c 34
$C_{17}H_{31}Ge$
17a 433, *17c* 84
$C_{17}H_{31}NO_2Sn$
17d1 189
$C_{17}H_{31}S_3Sn$
17a 464, *17c* 67
$C_{17}H_{31}Si$
9b 477
$C_{17}H_{32}BNOP$
17d2 66
$C_{17}H_{32}BN_2O$
17d2 66
$C_{17}H_{32}DSn$
9b 409
$C_{17}H_{32}NO_2$
17d2 274
$C_{17}H_{32}NO_3$
9c1 495, *17d2* 316
$C_{17}H_{32}NO_4$
17d2 274
$C_{17}H_{32}N_2O_6$
17d2 254
$C_{17}H_{32}N_3O_2$
17d2 245
$C_{17}H_{33}Sn$
9b 409
$C_{17}H_{34}NO_5$
17d2 316
$C_{17}H_{34}N_4O_3P$
17d2 316
$C_{17}H_{35}GeO_3$
17b 268
$C_{17}H_{35}NO_3P$
9c1 496
$C_{17}H_{35}NO_5P$
9c1 496
$C_{17}H_{35}N_2O_3$
17d2 281
$C_{17}H_{35}N_5O_3P$
17d2 316
$C_{17}H_{35}OSn$
17b 51
$C_{17}H_{35}O_3Sn$
17b 271
$C_{17}H_{35}Si_4$
17f 7
$C_{17}H_{37}N_2P^+$
17b 280

$C_{17}H_{37}N_3O_3P$
17d2 317

$C_{17}H_{38}NO_7P_2$
9c2 372

$C_{17}H_{38}Si_6^-$
17f 28

$C_{17}H_{39}N_2Si_2$
17b 508

$C_{17}H_{48}OPSi_6$
17e 299

$C_{18}Cl_{12}MnO_6$
17a 293, *17g* 208

$C_{18}Cl_{12}O_6 \overset{CT}{\leftarrow} C_{30}H_{40}N_{10}$
9d1 362

$C_{18}Cl_{12}O_6P$
17g 209

$C_{18}D_{12}^-$
9d1 793

$C_{18}H_2Cl_{10}O_2S_2^-$
9d1 206

$C_{18}H_2D_{10}N_5O_6$
17c 443

$C_{18}H_4B_2D_{36}^-$
17f 223ff

$C_{18}H_4Cl_6D_4NO$
17d1 247, 251

$C_{18}H_4F_{10}O_2Tl$
17g 77

$C_{18}H_6Cl_4O_4^+$
9d1 360

$C_{18}H_6Mo_2N_2O_{10}^-$
17a 270

$C_{18}H_6N_2O_{10}W_2^-$
17a 277

$C_{18}H_7Cl_5N_5O_6$
9c1 54

$C_{18}H_7F_5N_5O_6$
9c1 54

$C_{18}H_8CrO_6^-$
17a 259

$C_{18}H_8CrO_6$
17g 315

$C_{18}H_8D_4N_5O_6$
17c 442

$C_{18}H_8D_{12}NO_3$
17d1 317

$C_{18}H_8F_6O_4^+$
17h 89

$C_{18}H_8F_{13}$
17c 191

$C_{18}H_8MnO_6$
17a 293, *17g* 320ff

$C_{18}H_8MoO_6^-$
17a 270

$C_{18}H_8MoO_6$
17g 322

$C_{18}H_8O_2^-$
9d1 332

$C_{18}H_8O_4^-$
9d1 60, 333

$C_{18}H_8O_6Re$
17a 316, *17g* 323

$C_{18}H_8O_6W$
17g 327

$C_{18}H_8O_6W_2^-$
17a 277

$C_{18}H_9BD_6^-$
9d1 703

$C_{18}H_9BrN_5O_6$
1 49

$C_{18}H_9ClN_5O_6$
1 49, *9c1* 61

$C_{18}H_9CrMnO_7^-$
17a 294

$C_{18}H_9FO_3^-$
9d1 347

$C_{18}H_9MnNO_5S_2$
17a 294, *17g* 372

$C_{18}H_9MnN_5O_6S_2$
17g 373

$C_{18}H_{10}Br_2N_5O_6$
1 48, *9c1* 53

$C_{18}H_{10}Br_4O_2Tl$
9d1 158, *17a* 410

$C_{18}H_{10}Cl_2N_5O_6$
1 48, *9c1* 53

$C_{18}H_{10}Cl_2O_2^-$
9d1 199

$C_{18}H_{10}Cl_2O_2S_2^-$
1 119

$C_{18}H_{10}Cl_4O_2Tl$
9d1 156, *17a* 410

$C_{18}H_{10}Cl_6NO$
17d1 247

$C_{18}H_{10}D_2N_5O_6$
17c 442, 443

$C_{18}H_{10}D_{12}NOS$
17d2 116

$C_{18}H_{10}D_{12}N_2O_3$
17d1 279

$C_{18}H_{10}D_{17}N_2O_3$
17d2 315

$C_{18}H_{10}F_2N_5O_6$
1 48

$C_{18}H_{10}MnO_6$
17a 294

$C_{18}H_{10}N_5O_6$
1 48, *9c1* 60ff, *17c* 444

$C_{18}H_{10}N_7O_{10}$
1 47, *9c1* 53

$C_{18}H_{10}O_2^-$
9d1 333

$C_{18}H_{10}O_2S_2^-$
9d1 64

$C_{18}H_{10}O_4^-$
9d1 61, 333

$C_{18}H_{10}O_4^+$
9d1 360

$C_{18}H_{10}O_5W^-$
9a 307, *9b* 673

$C_{18}H_{10}O_6Re$
17a 316

$C_{18}H_{11}BrN_5O_6$
1 47

$C_{18}H_{11}ClN_5O_6$
1 47, *9c1* 52, 57

$C_{18}H_{11}ClO_3S_2^-$
1 119

$C_{18}H_{11}Cl_2O_2Sn$
17g 344

$C_{18}H_{11}D^-$
9d1 793

$C_{18}H_{11}D_{12}N_2O$
17d1 279

$C_{18}H_{11}FN_5O_6$
1 46

$C_{18}H_{11}FO_2^-$
9d1 241

$C_{18}H_{11}F_4O_2Si$
17e 191

$C_{18}H_{11}N^+$
17h 222

$C_{18}H_{11}NO_2^-$
17g 36

$C_{18}H_{11}NO_5Re$
17a 317, *17g* 262

$C_{18}H_{11}N_3O_6^-$
9d1 666

$C_{18}H_{11}N_4O_4$
9c1 60

$C_{18}H_{11}N_6O_8$
1 46, *9c1* 52
$C_{18}H_{11}O_3^{2-}$
9d1 252, 255, 262, 265
$C_{18}H_{12}^{3-}$
9d1 789
$C_{18}H_{12}^{-}$
1 113, *9d1* 761, 789, 791ff, *17f* 111ff
$C_{18}H_{12}^{+}$
1 124, *9d2* 15, *17h* 77
$C_{18}H_{12}BD_3^{-}$
9d1 702
$C_{18}H_{12}BrN_4O_4$
1 50
$C_{18}H_{12}Br_3N^{+}$
9d2 48
$C_{18}H_{12}Cl_2O_2Tl$
17g 77
$C_{18}H_{12}Cl_3N^{+}$
9d2 48
$C_{18}H_{12}Cl_4NO_4^{+}$
1 123
$C_{18}H_{12}F_3N^{+}$
9d2 47
$C_{18}H_{12}FeN_3OS_2$
17a 330
$C_{18}H_{12}KN_4O_7S$
1 49, 50
$C_{18}H_{12}MnN_3OS_2$
17a 294
$C_{18}H_{12}MoN_2O_4^{-}$
9a 309
$C_{18}H_{12}N_2O_4^{-}$
9d1 634, 651
$C_{18}H_{12}N_4NaO_7S$
1 49
$C_{18}H_{12}N_4O_2^{-}$
17f 293
$C_{18}H_{12}N_4O_6^{+}$
9d2 48
$C_{18}H_{12}N_5O_6$
1 45ff, *9c1* 49ff, *17c* 441ff
$C_{18}H_{12}O_2^{-}$
9d1 179, 198, *17g* 37, 274
$C_{18}H_{12}O_2S_2^{-}$
9d1 206, 220
$C_{18}H_{12}S_3^{-}$
9d1 894
$C_{18}H_{12}S_4^{+}$
17h 118
$C_{18}H_{13}D_5FeNO$
9c1 630
$C_{18}H_{13}F_3FeO^{-}$
9a 317
$C_{18}H_{13}NO^{+}$
9d2 80
$C_{18}H_{13}NO_2^{-}$
9d1 564, 565, 566
$C_{18}H_{13}NS^{+}$
9d2 81, *17h* 200
$C_{18}H_{13}N_2O$
9c1 905
$C_{18}H_{13}N_4O_4$
1 49, *9c1* 47, 48
$C_{18}H_{14}^{-}$
1 108, *9b* 761, *9d1* 766, 789ff, 822, 829, 830, *17c* 370, *17f* 126, 127ff, 133
$C_{18}H_{14}^{+}$
17h 100
$C_{18}H_{14}BrNO_2^{-}$
17g 272
$C_{18}H_{14}BrN_2O_2S$
9c1 67
$C_{18}H_{14}ClNO_2^{-}$
17g 273, 275ff
$C_{18}H_{14}Cl_2O_2^{+}$
17h 103
$C_{18}H_{14}Cl_5O_3$
9c2 162
$C_{18}H_{14}D_2^{-}$
9d1 787
$C_{18}H_{14}FeO_2^{-}$
9a 314, *9d1* 29
$C_{18}H_{14}NO$
9c1 560, *17d1* 226
$C_{18}H_{14}NO_2S_2$
17c 558
$C_{18}H_{14}NO_4P^{-}$
9d1 516
$C_{18}H_{14}NO_5PS^{-}$
9d1 518
$C_{18}H_{14}NO_6P^{-}$
9d1 517
$C_{18}H_{14}N_2^{+}$
9d2 81
$C_{18}H_{14}N_2O_2^{-}$
9d1 512, *17g* 39
$C_{18}H_{14}N_2O_4S^{2\cdot}$
9d2 244
$C_{18}H_{14}N_3O_2$
9c1 47
$C_{18}H_{14}N_7O_6$
1 47
$C_{18}H_{14}O^{-}$
9b 492, *17c* 104, 275
$C_{18}H_{14}O_2^{-}$
9b 614, *17c* 234ff
$C_{18}H_{14}O_2S^{2\cdot}$
17h 254
$C_{18}H_{14}O_2Tl$
9d1 118, *17a* 410ff, *17g* 76
$C_{18}H_{14}O_3Tl$
9d1 121, 132, *17a* 411, 412
$C_{18}H_{14}S^{2\cdot}$
17h 254
$C_{18}H_{15}^{2-}$
17f 134
$C_{18}H_{15}$
17c 347
$C_{18}H_{15}As^{+}$
17f 59
$C_{18}H_{15}AsBr$
17f 56
$C_{18}H_{15}AsCl$
17f 56
$C_{18}H_{15}AsF_2^{-}$
17f 59
$C_{18}H_{15}AsI$
17f 56
$C_{18}H_{15}AsO^{-}$
9c2 318, *17f* 56ff
$C_{18}H_{15}AsS^{-}$
17f 57
$C_{18}H_{15}AsSe^{-}$
17f 57
$C_{18}H_{15}B^{-}$
9d1 702
$C_{18}H_{15}BrGe^{-}$
17f 45
$C_{18}H_{15}BrP$
17e 299
$C_{18}H_{15}BrPb^{-}$
17f 54
$C_{18}H_{15}BrPb^{+}$
17f 55
$C_{18}H_{15}ClNO_4^{+}$
1 122

$C_{18}H_{15}ClP$
9c2 260, *17e* 299
$C_{18}H_{15}ClPb^-$
17f 54
$C_{18}H_{15}ClPb^+$
17f 55
$C_{18}H_{15}Cl_4N_4O_2$
9c1 86
$C_{18}H_{15}CrNO^-$
17a 259
$C_{18}H_{15}DPS$
17e 300
$C_{18}H_{15}D_6OS_2$
17c 276
$C_{18}H_{15}D_6S_3$
17c 283
$C_{18}H_{15}Ge$
9c2 311, *17f* 41
$C_{18}H_{15}GeO_2$
9c2 23
$C_{18}H_{15}IP$
17e 300
$C_{18}H_{15}K^-$
17a 218
$C_{18}H_{15}N^+$
9d2 48ff
$C_{18}H_{15}N^+ \cdot I_2$
9d2 49
$C_{18}H_{15}NO_2^-$
17f 199
$C_{18}H_{15}NO_3^-$
17f 199
$C_{18}H_{15}N_2$
9c1 46
$C_{18}H_{15}N_2O_2$
9c1 561, *17d1* 366
$C_{18}H_{15}N_2O_2S$
9c1 67
$C_{18}H_{15}N_5O^+$
9d2 81
$C_{18}H_{15}Na^-$
17a 210
$C_{18}H_{15}O_2Pb$
17e 33
$C_{18}H_{15}O_2Si$
9c2 22
$C_{18}H_{15}O_5P_2S_2$
17e 266
$C_{18}H_{15}P^-$
1 120
$C_{18}H_{15}P^+$
9c2 267, *17e* 309
$C_{18}H_{15}PS^-$
17e 300
$C_{18}H_{15}Si$
9c2 309, *17f* 8
$C_{18}H_{15}Sn$
9c2 314, *17f* 48
$C_{18}H_{16}^-$
9d1 786, 838, *17f* 134, 135
$C_{18}H_{16}^+$
17h 100, 101
$C_{18}H_{16}^{2\cdot}$
17h 251
$C_{18}H_{16}Cl_3O_4$
9c2 161
$C_{18}H_{16}D_4^-$
9d1 726
$C_{18}H_{16}N$
17c 408
$C_{18}H_{16}NO$
9c1 585
$C_{18}H_{16}NOS$
17c 318
$C_{18}H_{16}NO_4$
17d2 353
$C_{18}H_{16}N_2^+$
9d2 57, 65
$C_{18}H_{16}O_2^-$
9d1 317, *17c* 332
$C_{18}H_{16}O_3^+$
17h 104
$C_{18}H_{16}O_8S_4^+$
17h 116
$C_{18}H_{16}PS$
9c2 260, *17e* 300
$C_{18}H_{16}PSe$
9c2 260
$C_{18}H_{16}S_2^+$
17e 381
$C_{18}H_{17}Cl_3NO_3S$
17d2 144
$C_{18}H_{17}HgO_2$
17a 392, *17g* 318
$C_{18}H_{17}MgN_2$
17a 227
$C_{18}H_{17}N_2O$
9c1 758
$C_{18}H_{17}N_2O_2$
17d2 354
$C_{18}H_{17}Si$
17c 82
$C_{18}H_{18}^-$
9d1 773, 780, 788, 809, *17f* 135, 136, 137
$C_{18}H_{18}ClNO$
9c1 328
$C_{18}H_{18}ClN_2O$
9c1 637
$C_{18}H_{18}ClN_4O_4^+$
1 123
$C_{18}H_{18}Cl_3N_2O$
17d1 247
$C_{18}H_{18}Cl_3O_2$
17e 189
$C_{18}H_{18}Cl_4NO$
9c1 377, *17d1* 165
$C_{18}H_{18}F_2NO_3$
17d2 56
$C_{18}H_{18}MoNO_4$
9a 307, *9c1* 1042, *17a* 270
$C_{18}H_{18}N$
17c 285
$C_{18}H_{18}NO$
9c1 630
$C_{18}H_{18}NO_2$
9c1 747
$C_{18}H_{18}NO_4$
9c1 12
$C_{18}H_{18}NO_5$
9c1 567
$C_{18}H_{18}N_2S^+$
17h 200
$C_{18}H_{18}N_3O_3$
9c1 220
$C_{18}H_{18}N_4S_2^+$
17h 174
$C_{18}H_{18}O^-$
17c 26
$C_{18}H_{18}O_2^+$
17h 103
$C_{18}H_{18}O_2Tl$
9d1 320
$C_{18}H_{18}O_4^-$
9d1 98, 99
$C_{18}H_{18}Si^-$
9d1 896
$C_{18}H_{19}$
17c 298

$C_{18}H_{19}Cl_2O_2$
17e 55, 74, 152

$C_{18}H_{19}FNO_3$
17d2 56

$C_{18}H_{19}N^+$
17h 222

$C_{18}H_{19}NO^-$
17c 238

$C_{18}H_{19}NO_4^+$
9d2 43

$C_{18}H_{19}NO_5Re$
17a 317

$C_{18}H_{19}NO_5SM$
17d2 95

$C_{18}H_{19}N_2O$
9c1 95, 413, 637, *17c* 623, *17d1* 278, *17d2* 269

$C_{18}H_{19}N_2O_2$
9c1 630, 637, 719, *17c* 445

$C_{18}H_{19}N_2O_3$
17d2 268, 388

$C_{18}H_{19}N_4O_2$
9c1 72

$C_{18}H_{19}N_4O_4$
9c1 72

$C_{18}H_{19}N_6O_4$
9c1 78

$C_{18}H_{19}O_2$
9b 575, *17c* 278

$C_{18}H_{20}^-$
9d1 726, 768, 839, *17c* 171, *17f* 137

$C_{18}H_{20}ClN_2O_4P$
17d2 317

$C_{18}H_{20}ClO_2$
17e 48

$C_{18}H_{20}Cl_2NO_3S$
17c 510

$C_{18}H_{20}CoN_4O_2^{2-}$
17a 338, *17g* 116

$C_{18}H_{20}D_7O_3$
17b 398, 399

$C_{18}H_{20}FeO_3^-$
9a 316

$C_{18}H_{20}MnO_6$
17a 295, *17g* 123, 166

$C_{18}H_{20}NO$
9b 704, *9c1* 280, 679, *17d1* 167

$C_{18}H_{20}NO_2$
9c1 715, *17d1* 316, *17d2* 45

$C_{18}H_{20}NO_2S$
17d1 153

$C_{18}H_{20}NO_3$
9c1 316, *17d1* 317, *17d2* 56, 396

$C_{18}H_{20}NO_3S$
9c1 757, *17d1* 167

$C_{18}H_{20}NO_4$
17e 49

$C_{18}H_{20}NO_4S$
9c1 780

$C_{18}H_{20}N_2^+$
9d2 63, 108

$C_{18}H_{20}N_3$
9c1 89, 90

$C_{18}H_{20}N_3O_2$
17c 458

$C_{18}H_{20}N_3O_5$
9c1 254, 346, *17d1* 375, *17d2* 34, 60, 218

$C_{18}H_{20}N_4P$
9b 402

$C_{18}H_{20}O_2^-$
17c 275

$C_{18}H_{20}O_3PS_3$
17b 324

$C_{18}H_{20}O_4^+$
17h 89

$C_{18}H_{20}O_6Re$
17a 317, *17g* 130, 174

$C_{18}H_{20}S_2^+$
17e 381

$C_{18}H_{20}Si^-$
17f 21, 193

$C_{18}H_{21}$
17c 262

$C_{18}H_{21}MoO_6$
17a 270, *17g* 168

$C_{18}H_{21}NO_2X$
(X unknown)
17c 481

$C_{18}H_{21}NS^-$
9b 586

$C_{18}H_{21}N_2O$
9c1 307, 818, 819, *9c2* 93, *17d1* 136, *17e* 120

$C_{18}H_{21}N_2O_2$
1 55, *9c1* 719, *17d1* 219, *17d2* 48

$C_{18}H_{21}N_2O_3$
9c1 220, *17d2* 26, 60

$C_{18}H_{21}N_2O_4S$
9c1 651, *17d1* 219

$C_{18}H_{21}N_4$
9c1 75, 76, 77

$C_{18}H_{21}N_4O_4$
17c 436

$C_{18}H_{22}^-$
9d1 767, 806, 808, 809

$C_{18}H_{22}^+$
9d2 10, *17h* 103

$C_{18}H_{22}Al_2Br_6S_2$
17e 321

$C_{18}H_{22}CoN_4O_4$
17e 19

$C_{18}H_{22}N$
9c1 13

$C_{18}H_{22}NO$
9c1 281, 346, 575, 620, 640, *17d1* 266, 278, 357, *17d2* 16, 42, 218

$C_{18}H_{22}NOS$
9c1 281, *17d2* 64

$C_{18}H_{22}NO_2$
9c1 14, 282, *17d2* 16, 26

$C_{18}H_{22}NO_2S$
9c1 944, *17d1* 189

$C_{18}H_{22}NO_2S_2$
17d1 189

$C_{18}H_{22}NO_3$
1 56, *9c1* 575, 576, *17d2* 211

$C_{18}H_{22}NO_3S$
17d2 65

$C_{18}H_{22}NO_4$
9c1 14

$C_{18}H_{22}NO_5$
1 56, *9c1* 414, 576, *17d2* 121

$C_{18}H_{22}NO_6$
9c1 14, *17d1* 353

$C_{18}H_{22}NO_6S$
17d1 356

$C_{18}H_{22}NO_7$
1 56, *9c1* 577

$C_{18}H_{22}N_2O_2^+$
9d2 82
$C_{18}H_{22}N_2O_4^{2\cdot}$
9d2 159
$C_{18}H_{22}N_2O_{14}P_2W_2^-$
17a 278
$C_{18}H_{22}N_2S^+$
17h 201ff
$C_{18}H_{22}N_3O_2$
17d1 268
$C_{18}H_{22}O_2S^+$
9d2 139
$C_{18}H_{22}O_3^-$
9d1 92
$C_{18}H_{22}O_3P$
17c 184, 286
$C_{18}H_{22}O_4PS$
17b 295
$C_{18}H_{22}P$
17e 257
$C_{18}H_{23}N^+$
9d2 44
$C_{18}H_{23}NO_2^+$
9d2 44
$C_{18}H_{23}NO_4^+$
9d2 44
$C_{18}H_{23}NO_6^+$
9d2 45
$C_{18}H_{23}N_2O$
17d1 230, 311, *17d2* 107
$C_{18}H_{23}N_2O_2$
9c1 170, 171, *17d2* 83
$C_{18}H_{23}N_2O_3S$
17d2 48
$C_{18}H_{23}N_2O_5$
9c1 496
$C_{18}H_{23}N_2S^{2+}$
17h 203
$C_{18}H_{23}N_3O_4M$
17d1 42
$C_{18}H_{23}N_4O_9PM$
17d2 245
$C_{18}H_{23}O$
9c2 195
$C_{18}H_{23}OSi$
9b 669, *17b* 278, *17c* 288
$C_{18}H_{23}OSn$
9b 672
$C_{18}H_{23}O_2PbS$
17a 495, *17c* 295
$C_{18}H_{23}O_2SSi$
17c 290
$C_{18}H_{23}O_2SSn$
17a 464, *17c* 294
$C_{18}H_{23}O_3Si$
9b 669
$C_{18}H_{23}O_3Sn$
9b 672, *17a* 465, *17c* 293
$C_{18}H_{24}^-$
9d1 743, 747, 748, 750, 847, *17f* 137
$C_{18}H_{24}^+$
9d2 8, 12, *17h* 75
$C_{18}H_{24}AsO_2$
9c2 318
$C_{18}H_{24}Br_2N_2O_5^{2\cdot}$
17h 275
$C_{18}H_{24}Cr^+$
17a 260
$C_{18}H_{24}CrI$
9a 304
$C_{18}H_{24}LiO_2$
9d1 133
$C_{18}H_{24}MnNO_5P$
17a 295
$C_{18}H_{24}NO$
9b 504, *9c1* 255, 695, *17c* 116
$C_{18}H_{24}NO_2$
9b 505
$C_{18}H_{24}NO_3$
9c1 497
$C_{18}H_{24}NO_7$
9c1 282
$C_{18}H_{24}NS$
17c 124, 125
$C_{18}H_{24}N_2^+$
9d2 38
$C_{18}H_{24}N_2O_2^+$
9d2 38
$C_{18}H_{24}N_2O_6Pd^{2\cdot}$
17h 275
$C_{18}H_{24}N_2O_8^+$
9d2 57
$C_{18}H_{24}N_3O_4$
17d1 42
$C_{18}H_{24}N_3O_5$
9c1 851
$C_{18}H_{24}N_3O_6$
9c1 886
$C_{18}H_{24}N_4O_6^-$
17g 248
$C_{18}H_{24}N_5O_6$
9c1 497
$C_{18}H_{24}N_6^+$
17h 174
$C_{18}H_{24}O_2P$
9c2 260
$C_{18}H_{24}O_5$
9c2 135
$C_{18}H_{25}CoN_3O_4$
17e 20
$C_{18}H_{25}NO^+$
17c 116, 117
$C_{18}H_{25}NS^+$
17c 125
$C_{18}H_{25}N_2O$
17d2 346, *17e* 98
$C_{18}H_{25}N_2O_5$
17d2 389
$C_{18}H_{25}N_2O_6$
9c1 631
$C_{18}H_{25}N_3O_4P$
9c1 497
$C_{18}H_{25}N_4O_3$
17d2 245
$C_{18}H_{25}N_4O_7$
17d2 245
$C_{18}H_{25}O_3$
1 64, *17e* 129, 132
$C_{18}H_{25}O_5$
9c2 136
$C_{18}H_{26}AlN_2$
17a 402, *17f* 204
$C_{18}H_{26}Al_2N_2^-$
17a 402, *17f* 207
$C_{18}H_{26}Br_2N_2O_4^{2\cdot}$
9d2 221, *17h* 321
$C_{18}H_{26}CuN_2O_6$
17a 358
$C_{18}H_{26}D_3N_2O$
9c1 811
$C_{18}H_{26}F_3$
17c 191
$C_{18}H_{26}Ge_2^-$
17f 45
$C_{18}H_{26}NO$
17d1 292
$C_{18}H_{26}NOSi_2$
17d2 209

$C_{18}H_{26}NO_2$
9c1 803, *17d2* 404, *17g* 261
$C_{18}H_{26}NO_4$
17d2 56
$C_{18}H_{26}NO_5$
9c1 352, *17d2* 27
$C_{18}H_{26}N_2O_2^-$
9d1 202
$C_{18}H_{26}N_2O_2^+$
17h 162
$C_{18}H_{26}N_2O_4^{2\cdot}$
9d2 285, *17h* 328
$C_{18}H_{26}N_2O_5^{2\cdot}$
9d2 202
$C_{18}H_{26}N_5O_5$
17d2 246
$C_{18}H_{26}O_2^-$
9d1 22
$C_{18}H_{26}O_2^+$
17h 87
$C_{18}H_{26}Si_2^-$
17f 21
$C_{18}H_{26}Sn^-$
17f 51
$C_{18}H_{27}Cl_4O_2Sn$
17e 190
$C_{18}H_{27}D_2O_2$
9c2 133
$C_{18}H_{27}F_6Ge$
17c 104
$C_{18}H_{27}N_2O$
17d2 74, 377
$C_{18}H_{27}N_2O_2$
9c1 893
$C_{18}H_{27}N_2O_3$
9c1 498
$C_{18}H_{27}N_2O_4S$
17d2 138
$C_{18}H_{27}N_4Sn$
9a 298, *9b* 387
$C_{18}H_{27}O_2^{2-}$
9b 569, *17c* 198
$C_{18}H_{27}O_3$
1 65, *17b* 398, 399, *17e* 92
$C_{18}H_{27}O_4$
17b 398, *17b* 399
$C_{18}H_{28}^+$
17h 26
$C_{18}H_{28}AlCl_2O_3$
17g 155
$C_{18}H_{28}B^-$
9d1 705
$C_{18}H_{28}CrSi_2^-$
17a 261
$C_{18}H_{28}CrSi_2^+$
17a 260
$C_{18}H_{28}F_3$
9b 356
$C_{18}H_{28}F_3O_3Si$
17g 138, 178
$C_{18}H_{28}F_3O_3Sn$
17a 465
$C_{18}H_{28}NO$
9c1 729
$C_{18}H_{28}NO_2$
9c2 101, 123, *17d2* 54, 57, 397
$C_{18}H_{28}NO_3$
17d2 59
$C_{18}H_{28}NO_4$
17c 408, *17d2* 105
$C_{18}H_{28}NO_5$
9c1 585
$C_{18}H_{28}N_2O_4^{2\cdot}$
9d2 220, *17h* 282
$C_{18}H_{28}N_3O_2$
9c1 825
$C_{18}H_{28}N_3O_{11}$
9c1 356
$C_{18}H_{28}O^-$
17c 197, 199
$C_{18}H_{28}O_2^-$
9b 391
$C_{18}H_{29}$
9b 44, 313, *17b* 59, 275, 485, 502
$C_{18}H_{29}ClNO$
17d1 293
$C_{18}H_{29}ClOP$
17e 257
$C_{18}H_{29}ClP$
17e 257
$C_{18}H_{29}Cl_2O_2Sn$
17a 465, *17e* 74, 152, *17g* 185
$C_{18}H_{29}HgO_2$
17a 392, *17g* 121
$C_{18}H_{29}NO^-$
9d1 692, *17f* 307ff
$C_{18}H_{29}NO^+$
9d2 118, *17h* 232
$C_{18}H_{29}NOM$
17d1 328
$C_{18}H_{29}NOX$
17d1 328
$C_{18}H_{29}NO_2^-$
1 99, *9d1* 460ff, 462
$C_{18}H_{29}NO_3^-$
17f 250
$C_{18}H_{29}NO_4Tl$
9d1 151
$C_{18}H_{29}N_2O_6$
17d2 383
$C_{18}H_{29}N_2O_7$
17d2 389
$C_{18}H_{29}N_3O_6P$
17d2 317
$C_{18}H_{29}O$
1 62, *9c2* 75ff, *17e* 88, 103ff
$C_{18}H_{29}OS$
1 67, *9c2* 142
$C_{18}H_{29}OSe$
1 68, *9c2* 145
$C_{18}H_{29}O_2$
1 66, 71, *9c2* 132, 151, *17e* 99, 134
$C_{18}H_{29}O_2S_2Sn$
17a 465
$C_{18}H_{29}O_3$
17b 398
$C_{18}H_{29}O_4Sn$
17a 466, *17g* 182
$C_{18}H_{29}S$
9c2 281
$C_{18}H_{30}^+$
9d2 7
$C_{18}H_{30}AlCl_2O_3$
17a 402
$C_{18}H_{30}AlO_2$
17a 403, *17e* 63, *17g* 114, 154
$C_{18}H_{30}BD_2NO$
17d1 345
$C_{18}H_{30}Br_2N_6O_4Pd_2$
17a 345

$C_{18}H_{30}Br_2N_6O_4Pd_2^{2\cdot}$
17h 290
$C_{18}H_{30}ClN_2O_5$
17d2 317
$C_{18}H_{30}Cl_2N_6O_4Pd_2$
17a 346
$C_{18}H_{30}Cl_2N_6O_4Pd_2^{2\cdot}$
17h 291
$C_{18}H_{30}Cl_4N_6O_2Pd_2^{2\cdot}$
17h 291
$C_{18}H_{30}F_3$
9b 356
$C_{18}H_{30}F_3O$
9b 364
$C_{18}H_{30}F_3O_3Sn$
17a 466
$C_{18}H_{30}F_3S$
9b 378
$C_{18}H_{30}GaO_2$
17g 121, 161
$C_{18}H_{30}GeHgO_2$
9d1 139
$C_{18}H_{30}HgN_2O_2^{2\cdot}$
17h 299
$C_{18}H_{30}InO_2$
17g 122, 164
$C_{18}H_{30}N$
1 52, *9c1* 10, 11
$C_{18}H_{30}NO$
9b 496, *9c1* 102, 554ff, 687, 688, *9c2* 121, *17d1* 293, 328ff, *17d2* 16
$C_{18}H_{30}NO_2$
9c1 232
$C_{18}H_{30}NO_2S$
17d1 194
$C_{18}H_{30}NO_4$
17d2 97, 192
$C_{18}H_{30}NO_4S$
17d1 197
$C_{18}H_{30}NO_5$
9c1 350, 352, *17d2* 98
$C_{18}H_{30}NO_9$
17d2 123
$C_{18}H_{30}NS$
17c 504, 505
$C_{18}H_{30}N_2O_2Tl^{2\cdot}$
17h 299
$C_{18}H_{30}N_2O_4^{2\cdot}$
9d2 215, 216, *17h* 272, 283
$C_{18}H_{30}N_2O_5^{2\cdot}$
9d2 194
$C_{18}H_{30}N_3O_3$
17d1 61
$C_{18}H_{30}N_4^{+}$
17h 151
$C_{18}H_{30}N_4O_4X$
17d1 62
$C_{18}H_{30}N_6O_2^{2\cdot}$
17h 290
$C_{18}H_{30}N_6O_4^{2\cdot}$
17h 285
$C_{18}H_{30}N_6O_4Pd$
17a 346
$C_{18}H_{30}N_6O_4Pd^{2\cdot}$
17h 291
$C_{18}H_{30}O^{-}$
9b 487
$C_{18}H_{30}O_2^{+}$
9d2 133, 134
$C_{18}H_{30}O_2P$
1 73ff
$C_{18}H_{30}O_2PS$
17e 139
$C_{18}H_{30}O_2Tl$
9d1 136, 146, *17g* 139, 191
$C_{18}H_{30}O_3PS$
17e 138
$C_{18}H_{30}O_5P$
9c2 139, *17b* 400, *17e* 138
$C_{18}H_{30}S_4^{+}$
17h 97
$C_{18}H_{31}B^{-}$
9d1 701
$C_{18}H_{31}ClNO_2Sn$
17d1 377
$C_{18}H_{31}N^{+}$
9d2 38
$C_{18}H_{31}NO_3P$
17e 301
$C_{18}H_{31}N_2O_7Sn$
17d2 411
$C_{18}H_{31}N_2Sn$
17c 156
$C_{18}H_{31}N_3O_2^{2\cdot}$
17h 276, 321
$C_{18}H_{31}N_3O_2X$
17d1 62
$C_{18}H_{31}N_3O_3^{2\cdot}$
9d2 212, *17h* 321
$C_{18}H_{31}O$
9b 486
$C_{18}H_{31}OSi$
17c 184, 220
$C_{18}H_{31}O_2Sn$
17e 46, *17g* 74
$C_{18}H_{31}SSn$
17b 22
$C_{18}H_{31}Si$
17b 282, 283
$C_{18}H_{32}^{+}$
17h 26, 34
$C_{18}H_{32}CuN_2O_9P$
17a 358
$C_{18}H_{32}NOSn$
17c 493
$C_{18}H_{32}NO_2Si$
17d1 380
$C_{18}H_{32}NO_2Sn$
17d1 380
$C_{18}H_{32}NO_5S$
9c2 126
$C_{18}H_{32}N_2^{+}$
17h 151
$C_{18}H_{32}N_4O_2^{2\cdot}$
9d2 195, *17h* 272, 303
$C_{18}H_{32}N_4O_6^{2\cdot}$
17h 298
$C_{18}H_{32}OP$
17c 158
$C_{18}H_{32}O_2P$
9b 532, *17c* 163
$C_{18}H_{32}O_3P$
9b 479
$C_{18}H_{33}$
17c 11
$C_{18}H_{33}B^{-}$
9d1 700
$C_{18}H_{33}BN$
17c 156
$C_{18}H_{33}N_2O$
17d2 184
$C_{18}H_{33}N_4O$
9c1 498
$C_{18}H_{33}S_3Sn$
17a 466, *17c* 67
$C_{18}H_{34}ClN_2O_5P^{2\cdot}$
9d2 236
$C_{18}H_{34}Cl_4N_6O_4Pd_2^{2\cdot}$
17h 292

$C_{18}H_{34}FN_2O_5P^{2\cdot}$
9d2 236
$C_{18}H_{34}NOS_2Sn$
17a 466, *17c* 70
$C_{18}H_{34}NO_2$
17d2 275
$C_{18}H_{34}NO_4$
9c1 431
$C_{18}H_{34}NO_5$
9c1 362
$C_{18}H_{34}N_2O_2^{2\cdot}$
9d2 227
$C_{18}H_{34}N_2O_4^{2\cdot}$
17h 298
$C_{18}H_{34}N_2O_4S_2^{2\cdot}$
9d2 248
$C_{18}H_{34}N_2O_5S^{2\cdot}$
9d2 245
$C_{18}H_{34}N_2O_5Si^{2\cdot}$
9d2 234
$C_{18}H_{34}N_2O_6S^{2\cdot}$
9d2 245
$C_{18}H_{34}N_4O_2^{2\cdot}$
9d2 230
$C_{18}H_{34}NaO_2$
9b 115
$C_{18}H_{34}Si_2^{+}$
17h 69
$C_{18}H_{35}B_{10}Cl_2O_2Sn$
17a 466, *17g* 185
$C_{18}H_{35}NO_3PSi$
17e 301
$C_{18}H_{35}N_2O$
9c1 852
$C_{18}H_{35}N_2O_2$
9c1 886
$C_{18}H_{35}N_2O_3$
17d1 158
$C_{18}H_{35}N_3O_2^{2\cdot}$
17h 299
$C_{18}H_{35}N_4O_5$
17d1 24
$C_{18}H_{35}O_2$
9b 115, *17b* 118, 171
$C_{18}H_{36}MnN_3OS_4$
17a 295
$C_{18}H_{36}O_4P$
17b 282
$C_{18}H_{37}$
9b 230
$C_{18}H_{37}B^{-}$
9d1 699
$C_{18}H_{37}NO_5$
9c1 362
$C_{18}H_{37}O$
17b 296
$C_{18}H_{37}OSn$
17b 51
$C_{18}H_{37}S$
9c2 272
$C_{18}H_{37}Se$
9c2 303
$C_{18}H_{37}Se_2$
17b 317
$C_{18}H_{38}BNO_3P$
17e 301
$C_{18}H_{38}B_2Si^{-}$
17f 225
$C_{18}H_{38}Cl_2N_4O_2Pd^{2\cdot}$
17h 304
$C_{18}H_{38}Cl_2N_4O_2Pt^{2\cdot}$
9d2 230, *17h* 304
$C_{18}H_{38}NO_2Sn$
17d2 411
$C_{18}H_{38}N_4O_4^{2\cdot}$
17h 271
$C_{18}H_{39}Si$
17b 276
$C_{18}H_{40}B_2^{-}$
17f 223
$C_{18}H_{40}Si_4^{+}$
17f 36, *17h* 34
$C_{18}H_{40}Si_6^{-}$
17f 28
$C_{18}H_{41}OSi_2$
17b 126, 200
$C_{18}H_{42}B_{20}ClFO_2Sn^{-}$
17a 467
$C_{18}H_{42}B_{20}ClO_2Sn$
17a 467, *17e* 68, *17g* 184
$C_{18}H_{42}B_{20}Cl_2O_2Sn^{-}$
17g 188
$C_{18}H_{42}O_4PSi_2$
9b 308
$C_{18}H_{42}P_2S_6^{+}$
9c2 267
$C_{18}H_{44}Si_4^{+}$
17f 36, *17h* 24
$C_{18}H_{45}NO_3PSi_2$
17e 301, 302
$C_{18}H_{45}N_2Si_3$
9c1 66, *17c* 451
$C_{18}H_{48}Si_6^{-}$
17f 28
$C_{18}H_{48}Si_{10}^{-}$
17f 29
$C_{18}H_{54}GeN_3Si_6$
9c2 311
$C_{18}H_{54}N_3Si_6Sn$
9c2 315
$C_{19}BrCl_{14}$
9b 732
$C_{19}Br_3F_{12}$
9b 732
$C_{19}Cl_3F_{12}$
9b 729
$C_{19}Cl_{13}$
9b 751
$C_{19}Cl_{14}$
9b 642
$C_{19}Cl_{14}I$
9b 732
$C_{19}Cl_{14}NO_2$
9b 741
$C_{19}Cl_{15}$
9b 643, 731, *17c* 266, 352
$C_{19}D_{15}$
9b 721, *17c* 350
$C_{19}D_{16}CrO^{-}$
17a 261
$C_{19}F_{12}N_3O_6$
17c 364
$C_{19}F_{15}$
9a 40, *9b* 727
$C_{19}HCl_{14}$
9b 730
$C_{19}HCl_{14}NNa$
17c 364
$C_{19}HCl_{14}O$
9b 733, *17c* 356
$C_{19}H_2Cl_{14}N$
9b 740, *17c* 363
$C_{19}H_2D_{12}N_3$
17c 478
$C_{19}H_3Cl_{12}$
9b 729
$C_{19}H_3Cl_{12}O_3$
9b 733
$C_{19}H_5D_{10}N_2S_2$
17c 511

$C_{19}H_6Br_3D_5NO$ *9c1* 631

$C_{19}H_6D_5F_5NO$ *9c1* 631

$C_{19}H_7D_6S$ *9b* 759

$C_{19}H_6D_9$ *17c* 350

$C_{19}H_7F_{10}OSi$ *17c* 289

$C_{19}H_8D_5$ *17c* 257, 367

$C_{19}H_8D_5O$ *9b* 754

$C_{19}H_8D_5S$ *9b* 759

$C_{19}H_8D_5Se$ *9b* 760

$C_{19}H_8D_7N_2^+$ *17h* 186

$C_{19}H_9BrCo_2N_{11}O^{6-}$ *9c1* 1022

$C_{19}H_9D_4S$ *9b* 758

$C_{19}H_9D_5OS^-$ *17f* 213

$C_{19}H_9D_6$ *17c* 350

$C_{19}H_9F_6$ *9b* 727

$C_{19}H_9O_3$ *9b* 757

$C_{19}H_{10}BrD_5NO$ *9c1* 631, 632

$C_{19}H_{10}Cl_5N_2S_2$ *17c* 514

$C_{19}H_{10}D_5N_3^+$ *17h* 186

$C_{19}H_{10}O_3S^-$ *17c* 321

$C_{19}H_{11}$ *9b* 692, *17c* 349

$C_{19}H_{11}BO_4^-$ *17c* 259

$C_{19}H_{11}Cl_4N_2S_2$ *17c* 513, 514

$C_{19}H_{11}Cl_4N_4$ *9c1* 71

$C_{19}H_{11}CrMnO_7^-$ *17a* 295, 296

$C_{19}H_{11}CrMnO_8^-$ *17a* 296

$C_{19}H_{11}D_2O$ *9b* 754

$C_{19}H_{11}D_{13}NO_2$ *17d1* 279

$C_{19}H_{11}O_2$ *9b* 757

$C_{19}H_{12}Cl_3$ *9b* 729

$C_{19}H_{12}Cl_3N_2S_2$ *17c* 513

$C_{19}H_{12}DO$ *9b* 754

$C_{19}H_{12}DO_2$ *17c* 206

$C_{19}H_{12}DS$ *9b* 758

$C_{19}H_{12}D_3$ *17c* 349

$C_{19}H_{12}D_{12}NO$ *17d1* 303

$C_{19}H_{12}D_{12}NO_2$ *17d1* 320

$C_{19}H_{12}F_3$ *9b* 726

$C_{19}H_{12}NO_2$ *9c1* 582

$C_{19}H_{12}N_2O^-$ *9b* 717

$C_{19}H_{12}N_3O_6$ *1* 42, *17c* 364

$C_{19}H_{12}O_2^-$ *9d1* 330

$C_{19}H_{13}^{2-}$ *9d1* 817

$C_{19}H_{13}$ *9b* 751, *17c* 367

$C_{19}H_{13}Br_2N_2S_2$ *17c* 512

$C_{19}H_{13}Cl_2$ *9b* 728

$C_{19}H_{13}Cl_2N_2S_2$ *17c* 513

$C_{19}H_{13}D_2NS^-$ *9b* 588

$C_{19}H_{13}D_{12}N_2O$ *17d1* 280

$C_{19}H_{13}F_2$ *9b* 726

$C_{19}H_{13}MnNO_{10}^+$ *9c1* 1040

$C_{19}H_{13}N^-$ *9d1* 868

$C_{19}H_{13}NO_5Re$ *17a* 317, *17g* 262

$C_{19}H_{13}N_4$ *9c1* 88, *17c* 477

$C_{19}H_{13}N_4O_6$ *1* 49

$C_{19}H_{13}N_6O_4$ *9c1* 71

$C_{19}H_{13}O$ *9b* 753

$C_{19}H_{13}OS$ *9b* 698

$C_{19}H_{13}O_2$ *17c* 205, 206

$C_{19}H_{13}O_2S$ *9b* 759

$C_{19}H_{13}O_4Tl^-$ *17g* 109

$C_{19}H_{13}S$ *9b* 758

$C_{19}H_{13}Se$ *9b* 760

$C_{19}H_{14}^-$ *9d1* 814

$C_{19}H_{14}$ *9b* 633

$C_{19}H_{14}Cl$ *9b* 728

$C_{19}H_{14}ClN_2S_2$ *17c* 512

$C_{19}H_{14}Cl_2NO$ *17d1* 236, 370

$C_{19}H_{14}F$ *9b* 725

$C_{19}H_{14}I$ *9b* 732

$C_{19}H_{14}MnO_6$ *17a* 297, *17g* 342

$C_{19}H_{14}N$ *9b* 761

$C_{19}H_{14}NO$ *17d2* 373

$C_{19}H_{14}NOS$ *9c1* 632

$C_{19}H_{14}NO_2^-$ *17g* 37

$C_{19}H_{14}NO_2$
9c1 632, 793, 795
$C_{19}H_{14}N_2O$
9c1 593
$C_{19}H_{14}N_2O_2S^-$
9d1 545
$C_{19}H_{14}N_2O_3^-$
17f 253
$C_{19}H_{14}N_3$
9c1 91, *17c* 478
$C_{19}H_{14}N_3O$
17c 596
$C_{19}H_{14}N_3O_2S_2$
17c 515
$C_{19}H_{14}N_3O_3$
9c1 64
$C_{19}H_{14}N_5O_6$
1 46, *9c1* 56
$C_{19}H_{14}N_5O_7$
1 46, *9c1* 52, 56
$C_{19}H_{14}O^-$
9d1 886
$C_{19}H_{14}OR$
(R = unknown)
9b 662
$C_{19}H_{14}O_2S^-$
9d1 887
$C_{19}H_{14}O_3Tl$
9d1 126, *17a* 412
$C_{19}H_{14}O_4Tl$
17g 109
$C_{19}H_{15}$
1 41, *9b* 718ff, *17c* 349
$C_{19}H_{15}AsD_3$
17f 58
$C_{19}H_{15}ClNO_4$
9c1 734
$C_{19}H_{15}CrFO^-$
17a 261
$C_{19}H_{15}F_3PS$
17e 302
$C_{19}H_{15}F_4NO_4P$
9c1 1010
$C_{19}H_{15}FeO_2^-$
9a 315
$C_{19}H_{15}N^-$
9b 666
$C_{19}H_{15}NO^-$
9b 583
$C_{19}H_{15}NO_2S^-$
9d1 507
$C_{19}H_{15}NO_4Tl$
9d1 141, *17a* 413
$C_{19}H_{15}NS^-$
9b 587
$C_{19}H_{15}NS^+$
17h 203
$C_{19}H_{15}N_2O$
1 51, *9c1* 827, *17d1* 230, *17d2* 361
$C_{19}H_{15}N_2O_2$
9c1 796, 906
$C_{19}H_{15}N_2O_3$
9c1 637
$C_{19}H_{15}N_2S_2$
17c 511
$C_{19}H_{15}N_4$
1 53, *9c1* 70, *17c* 459ff
$C_{19}H_{15}N_6$
9c1 72, 87
$C_{19}H_{15}O_2$
9c2 17ff, *17e* 13
$C_{19}H_{15}O_3$
9b 496
$C_{19}H_{15}S$
17c 281, *17e* 317
$C_{19}H_{15}S_2$
17e 358
$C_{19}H_{15}S_3$
9b 253, *17b* 326
$C_{19}H_{15}Se$
17c 284
$C_{19}H_{16}^-$
9d1 767, 823, *17f* 138
$C_{19}H_{16}BrNO_2^-$
17g 273
$C_{19}H_{16}CrO^-$
17a 261
$C_{19}H_{16}NO$
9c1 632ff, *17d1* 220
$C_{19}H_{16}NO_2$
9c1 645
$C_{19}H_{16}NO_3S$
9c1 998
$C_{19}H_{16}NO_6$
17d2 354
$C_{19}H_{16}N_2O_2^-$
9d1 514
$C_{19}H_{16}N_3O_2$
17d1 220
$C_{19}H_{16}N_4O_2^+$
9d2 82
$C_{19}H_{16}O^+$
17h 101
$C_{19}H_{16}O_2Tl$
9d1 118ff, 123, *17a* 413ff, *17g* 79
$C_{19}H_{16}O_3Tl$
9d1 121ff, *17a* 414
$C_{19}H_{17}^{\cdot}ClPS_2$
17e 357
$C_{19}H_{17}NO_2^-$
17g 272, 273
$C_{19}H_{17}N_2O_2S$
9c1 67
$C_{19}H_{17}N_5O^+$
9d2 83
$C_{19}H_{17}N_6$
9c1 70
$C_{19}H_{17}P^+$
9b 72, *17b* 88
$C_{19}H_{18}^-$
17f 139
$C_{19}H_{18}^+$
17h 101
$C_{19}H_{18}As$
17f 57, 58
$C_{19}H_{18}Cl_3O_4$
9c2 162
$C_{19}H_{18}Cl_4NO_2$
9c1 295
$C_{19}H_{18}NO_2$
9b 576
$C_{19}H_{18}PS$
9c2 260, *17e* 302
$C_{19}H_{18}Si^-$
17f 21
$C_{19}H_{19}$
9b 675, *17c* 255
$C_{19}H_{19}MnO_8PS_2$
17a 297, *17g* 373
$C_{19}H_{19}NO_4PS$
17b 290
$C_{19}H_{19}N_2$
17c 328
$C_{19}H_{19}N_2O_3$
9c1 307

$C_{19}H_{19}N_3NaO_4S$
17d1 364

$C_{19}H_{19}N_3NaO_5S$
17d1 364

$C_{19}H_{19}N_5O^+$
9d2 82

$C_{19}H_{20}CoN_5O_2^{3-}$
17a 338, *17g* 116, 226

$C_{19}H_{20}CuF_{12}NO_6$
17a 358

$C_{19}H_{20}MnO_7$
17a 297, *17b* 399, *17e* 64

$C_{19}H_{20}NO$
17d1 376

$C_{19}H_{20}NO_2$
9c1 747, *17d1* 167

$C_{19}H_{20}NO_3$
17d2 354

$C_{19}H_{20}NO_4W$
17d2 71

$C_{19}H_{20}NO_5$
17d1 222

$C_{19}H_{20}N_3O$
17d1 362

$C_{19}H_{20}O_7Re$
17a 318, *17e* 69

$C_{19}H_{21}D_9O^-$
17c 199

$C_{19}H_{21}N^+$
17h 222

$C_{19}H_{21}NOPS$
17b 288

$C_{19}H_{21}NO_2Tl$
17g 355

$C_{19}H_{21}NO_5Re$
17a 318, *17g* 262

$C_{19}H_{21}N_2O$
9c1 282, 512, *17c* 623, *17d2* 334

$C_{19}H_{21}N_2O_2$
9c1 425, *17c* 597

$C_{19}H_{21}N_3O_2PS$
17b 289

$C_{19}H_{21}N_4$
9c1 87

$C_{19}H_{21}O_2^{2-}$
9b 660, *17c* 273

$C_{19}H_{21}O_3S$
9b 666

$C_{19}H_{22}MnN_3O_6PS_2$
17a 298, *17g* 373

$C_{19}H_{22}NO$
17d1 279, 293, 359, *17d2* 318

$C_{19}H_{22}NO_2$
17d1 122

$C_{19}H_{22}NO_2S$
9c1 282

$C_{19}H_{22}NO_3$
9c1 283

$C_{19}H_{22}NO_4$
17e 49

$C_{19}H_{22}N_3O_5$
9c1 255, *17d1* 375

$C_{19}H_{22}O^-$
9b 654, *17c* 270, 272

$(C_{19}H_{22}O_2)_n$
17e 75

$C_{19}H_{23}$
1 41, *9b* 636, *17c* 260

$C_{19}H_{23}MnNO_2$
17a 298

$C_{19}H_{23}MnNO_5$
17a 298, *17g* 124

$C_{19}H_{23}NO$
9c1 642

$C_{19}H_{23}N_2O$
9c1 828

$C_{19}H_{23}N_2O_2$
9c1 241, *17d1* 160, *17d2* 48, 57

$C_{19}H_{23}N_4$
9c1 70, 76, 78

$C_{19}H_{23}N_4O_2$
9c1 70

$C_{19}H_{23}OSn$
9b 688

$C_{19}H_{23}O_2$
17e 83

$C_{19}H_{23}O_2S$
9b 665, *17c* 282

$C_{19}H_{24}ClO$
17c 113

$C_{19}H_{24}ClS$
17c 123

$C_{19}H_{24}FO$
17c 112, 113

$C_{19}H_{24}FS$
17c 122

$C_{19}H_{24}NO$
9c1 640, 715, 718, *17d1* 358

$C_{19}H_{24}NO_2$
9c1 283, 316, *17d2* 27, 34ff

$C_{19}H_{24}NO_2S$
17c 123

$C_{19}H_{24}NO_3$
9c1 316, *17c* 114

$C_{19}H_{24}NO_3S$
17c 495

$C_{19}H_{24}N_2S^+$
17h 203

$C_{19}H_{24}N_3O$
17d2 318

$C_{19}H_{24}N_6O_9PM_2$
17d2 246

$C_{19}H_{24}OS^-$
9b 707

$C_{19}H_{24}O_3^-$
9d1 61

$C_{19}H_{25}ClNO$
9c1 376

$C_{19}H_{25}NO_3P$
17d2 70

$C_{19}H_{25}NO_4P$
9c1 373

$C_{19}H_{25}N_2O$
17d2 27

$C_{19}H_{25}N_2O_2$
9c1 174, 175

$C_{19}H_{25}N_2S^{2+}$
17h 203

$C_{19}H_{25}N_2Si$
9c1 96

$C_{19}H_{25}O$
9b 496

$C_{19}H_{25}OSn$
9b 672

$C_{19}H_{25}S$
9b 506

$C_{19}H_{26}BaClO_2$
17g 199

$C_{19}H_{26}BeClO_2$
17g 200

$C_{19}H_{26}CaClO_2$
17g 200

$C_{19}H_{26}CdClO_2$
17g 200

$C_{19}H_{26}ClMgO_2$
17g 200

$C_{19}H_{26}ClNO_2Sn$
17d1 377
$C_{19}H_{26}ClO_2Sr$
17g 201
$C_{19}H_{26}ClO_2Zn$
17g 201
$C_{19}H_{26}GeN$
9b 557
$C_{19}H_{26}NO$
9c1 283
$C_{19}H_{26}NOSn$
17d1 312
$C_{19}H_{26}NO_2$
17d2 246
$C_{19}H_{26}NSi$
9b 557
$C_{19}H_{26}O_5$
9c2 136
$C_{19}H_{27}BaO_4$
17a 236
$C_{19}H_{27}BeO_4$
17a 224
$C_{19}H_{27}CaO_4$
17a 231
$C_{19}H_{27}CdO_4$
17a 389
$C_{19}H_{27}F_{14}OSi$
9b 243
$C_{19}H_{27}MgO_4$
17a 228
$C_{19}H_{27}NO^-$
17c 212
$C_{19}H_{27}N_2O_2$
9c1 812
$C_{19}H_{27}N_3O_4P$
17d2 318
$C_{19}H_{27}O$
9c2 115
$C_{19}H_{27}O_2$
17b 482
$C_{19}H_{27}O_3$
1 64, *17e* 120
$C_{19}H_{27}O_4Sr$
17a 234
$C_{19}H_{27}O_4Zn$
17a 383
$C_{19}H_{27}O_5$
9c2 136
$C_{19}H_{28}AlCl_2O_4$
17a 403
$C_{19}H_{28}CoNO_7P$
9a 324, *9c1* 1036, *17a* 338
$C_{19}H_{28}D_3N_2O$
9c1 819
$C_{19}H_{28}F_2O_2Si$
17g 178
$C_{19}H_{28}NO_2$
9c2 102ff, *17d2* 231
$C_{19}H_{28}NO_5S$
9c1 944
$C_{19}H_{28}N_2O_4$
17d1 42
$C_{19}H_{28}N_2O_5M$
17d1 42
$C_{19}H_{28}N_3Si$
17b 308
$C_{19}H_{28}O_2^-$
17b 481
$C_{19}H_{29}Cl_3NO$
17d1 293
$C_{19}H_{29}F_3NO$
9c1 726
$C_{19}H_{29}NO_5Re$
17a 318
$C_{19}H_{29}N_2O$
9c1 859, 860, *17d2* 377
$C_{19}H_{29}N_5O_8S$
17d2 89
$C_{19}H_{29}O$
17b 526
$C_{19}H_{29}O_2$
9c2 116, 184, *17b* 482,
17c 55, 56, 66, *17e* 132
$C_{19}H_{29}O_3$
1 65, 66, *9c2* 135, *17b* 398,
399
$C_{19}H_{30}Cl_2NOSn$
17c 620
$C_{19}H_{30}D_2NO$
17d1 329
$C_{19}H_{30}NO$
9c2 123, *17d2* 349, *17e* 97
$C_{19}H_{30}NO_2$
9c2 100, 101
$C_{19}H_{30}NO_3$
17d1 280, 293
$C_{19}H_{30}NO_9$
17d2 123
$C_{19}H_{30}N_2O_6P$
9c1 499
$C_{19}H_{30}N_4O_2^{2\cdot}$
17h 276
$C_{19}H_{30}O^-$
17c 200
$C_{19}H_{31}$
9b 539, 546, 552, 559,
17c 180
$C_{19}H_{31}NO^-$
17c 210
$C_{19}H_{31}NOX$
(X unknown)
17c 484
$C_{19}H_{31}NS^-$
17c 211
$C_{19}H_{31}N_2O$
9c1 734, 819ff, 828,
17d1 325, *17d2* 231
$C_{19}H_{31}N_2O_4S$
17d2 319
$C_{19}H_{31}O$
17e 89
$C_{19}H_{31}O_2$
17c 64, *17e* 100
$C_{19}H_{32}ClN_7O_2^{2\cdot}$
9d2 196
$C_{19}H_{32}Cl_3O_2Sb$
9a 299
$C_{19}H_{32}KO_6$
17a 218
$C_{19}H_{32}NO$
9c1 158, 284, 448, 611ff,
17c 483, *17d1* 264, 329ff,
359
$C_{19}H_{32}NO_2$
9c1 233, 623, 967,
17d1 330, 342, *17d2* 27
$C_{19}H_{32}NO_2S$
17d2 76
$C_{19}H_{32}NO_2S_2$
17d1 161
$C_{19}H_{32}NO_3$
9c1 370
$C_{19}H_{32}N_2O_4^{2\cdot}$
9d2 228, *17h* 299
$C_{19}H_{32}N_2O_4P$
17d2 319
$C_{19}H_{32}N_4O_2S^{2\cdot}$
17h 276
$C_{19}H_{32}O^-$
9b 487

$C_{19}H_{32}O_5P$
17e 111
$C_{19}H_{33}ClN_3O_5$
17d2 319
$C_{19}H_{33}D_2S_3Sn$
17a 467, *17c* 68
$C_{19}H_{33}F_4NO_4P$
9c1 1010
$C_{19}H_{33}NO_3P$
17e 302
$C_{19}H_{33}N_2O$
17d1 330, *17d2* 28
$C_{19}H_{33}N_4O_8$
17d2 319
$C_{19}H_{33}OSi$
17c 220
$C_{19}H_{33}O_2$
17c 54, 55, 56, 62
$C_{19}H_{33}O_2Sn$
17e 48
$C_{19}H_{33}S_3$
17b 326
$C_{19}H_{34}NO_2$
17d1 343, *17d2* 320
$C_{19}H_{34}NO_4$
9c1 146
$C_{19}H_{34}N_2O_5^{2\cdot}$
9d2 232ff, *17h* 304
$C_{19}H_{34}N_4O_2^{2\cdot}$
9d2 232, *17h* 305
$C_{19}H_{34}N_4O_3^{2\cdot}$
9d2 228
$C_{19}H_{34}S^-$
17b 501
$C_{19}H_{35}N_2O_4$
17d1 123
$C_{19}H_{35}N_3O_4^{2\cdot}$
17h 305
$C_{19}H_{35}N_3O_3S_2^{2\cdot}$
17h 308
$C_{19}H_{35}O_2$
17c 18
$C_{19}H_{35}O_2P^+$
17c 160
$C_{19}H_{35}S_2Sn$
17a 467, *17c* 122
$C_{19}H_{35}S_3Sn$
17a 467, *17c* 68
$C_{19}H_{36}MnNO_5P$
17a 298
$C_{19}H_{36}NO$
17d2 335
$C_{19}H_{36}NO_2$
17d2 275
$C_{19}H_{36}N_2O_5P^{2\cdot}$
9d2 237
$C_{19}H_{36}N_4O_2S^{2\cdot}$
17h 305
$C_{19}H_{36}N_4O_3^{2\cdot}$
9d2 231
$C_{19}H_{37}$
17c 11
$C_{19}H_{37}NO_2$
9c1 431
$C_{19}H_{37}N_2O_4P^{2\cdot}$
9d2 237
$C_{19}H_{37}N_2O_5$
17d2 320
$C_{19}H_{37}N_2O_5P^{2\cdot}$
9d2 237
$C_{19}H_{37}O$
17b 392
$C_{19}H_{37}OSn$
17b 113
$C_{19}H_{37}O_2$
9b 123
$C_{19}H_{37}O_5Sn$
17b 313
$C_{19}H_{38}NO_2$
17d2 191, 321
$C_{19}H_{38}NSSn$
17b 290
$C_{19}H_{38}NSi_2$
9c1 25
$C_{19}H_{39}OSe$
17b 316
$C_{19}H_{39}O_2$
17e 13
$C_{19}H_{39}O_4Si_3$
9b 476, *17c* 105
$C_{19}H_{41}N_3O_3P$
9c1 498
$C_{19}H_{43}BNSi_4$
17c 157
$C_{20}HCl_{14}O_2$
9b 736
$C_{20}H_2BrCl_{14}$
17c 354
$C_{20}H_2Br_4Cl_4O_5^{3-}$
9d1 345
$C_{20}H_2Cl_4I_4O_5^{3-}$
9d1 346
$C_{20}H_2Cl_{15}$
17c 353
$C_{20}H_3Cl_{12}$
17c 368
$C_{20}H_3Cl_{12}O$
17c 368
$C_{20}H_3Cl_{14}$
9b 731
$C_{20}H_3Cl_{14}O$
9b 733, *17c* 360
$C_{20}H_4Cl_4I_4O_5^-$
9d1 347
$C_{20}H_4Cl_{14}NO_2$
9b 736
$C_{20}H_4D_{10}^-$
9d1 842
$C_{20}H_6BrO_5^{3-}$
9d1 344ff
$C_{20}H_6D_8^-$
9d1 842
$C_{20}H_8^-$
9d1 834
$C_{20}H_8B_2D_{36}^-$
17f 224
$C_{20}H_8Br_4O_5^-$
9d1 345
$C_{20}H_8Mo_2N_2O_{10}^-$
17a 271, *17f* 206
$C_{20}H_8Mo_2N_6O_8^-$
17a 271
$C_{20}H_8N_2O_{10}W_2^-$
17a 278, *17f* 206
$C_{20}H_8O_6^{5-}$
9d1 297
$C_{20}H_8O_6^-$
17g 218
$C_{20}H_9Br_2O_2$
9c2 197
$C_{20}H_9D_{12}N_2O_2$
17d1 303
$C_{20}H_{10}BrO_2$
9c2 197
$C_{20}H_{10}Cl_4^-$
9b 405
$C_{20}H_{10}D_2^-$
9d1 795
$C_{20}H_{10}D_4^-$
17f 139, 140

$C_{20}H_{10}D_8^-$
17f 144ff

$C_{20}H_{10}F_5N_2O$
9c1 718

$C_{20}H_{10}N_2^-$
9b 633

$C_{20}H_{10}O_4^{3-}$
9d1 280, *17g* 218

$C_{20}H_{10}O_4^-$
9d1 61, *17g* 218

$C_{20}H_{10}O_5^{3-}$
9b 756, *9d1* 344

$C_{20}H_{11}D_3^-$
17f 142

$C_{20}H_{11}FO_2^-$
9d1 241

$C_{20}H_{11}N_3O_6^-$
9d1 667

$C_{20}H_{11}O_2$
9c2 196

$C_{20}H_{11}O_5^{2-}$
9d1 344

$C_{20}H_{11}O_7^{3-}$
9d1 189

$C_{20}H_{12}^-$
1 112, 113, *9d1* 794, 795

$C_{20}H_{12}^+$
1 124, *9b* 635, *9d2* 8, 16, *17h* 77, 78

$C_{20}H_{12}Br_6O_4Rh$
17g 362

$C_{20}H_{12}D_2^-$
17f 141

$C_{20}H_{12}D_4^-$
17f 142

$C_{20}H_{12}HgN_2O_4^-$
9d1 656

$C_{20}H_{12}NO$
9c1 600

$C_{20}H_{12}NS$
9c1 37

$C_{20}H_{12}N_2Na_2O_{10}S_3^-$
9c1 44, 45

$C_{20}H_{12}N_2O_2^-$
9d1 342

$C_{20}H_{12}N_2O_4^-$
9d1 646

$C_{20}H_{12}O_2^-$
9d1 241

$C_{20}H_{12}O_5^-$
9d1 344

$C_{20}H_{13}Cl_4N_2OS_2$
17c 515

$C_{20}H_{13}Cl_4N_2S_2$
17c 514

$C_{20}H_{13}CrMnO_7^-$
17a 298, 299

$C_{20}H_{13}CrMnO_8^-$
17a 299, 300

$C_{20}H_{13}D_5NO$
9c1 633

$C_{20}H_{13}D_5NOS$
9c1 633

$C_{20}H_{13}D_5NO_2$
9c1 633

$C_{20}H_{13}HgO_2$
17a 392, *17g* 318ff

$C_{20}H_{13}NO_2^-$
9d1 386, 567, 568, 569

$C_{20}H_{13}NO_2^+$
9d2 8

$C_{20}H_{13}N_2NaO_7S_2^-$
9c1 44

$C_{20}H_{13}N_2O_2$
9c1 855

$C_{20}H_{13}N_4$
9c1 46

$C_{20}H_{13}N_6O_6S$
9c1 55

$C_{20}H_{13}O$
9c2 201, *17e* 200ff

$C_{20}H_{13}O_5$
17c 369

$C_{20}H_{14}^-$
9d1 778ff, 842, *17f* 139, 140, 141

$C_{20}H_{14}^+$
9d2 14

$C_{20}H_{14}Al_2Br_6S_2$
17e 322

$C_{20}H_{14}ClN_2O$
9c1 593

$C_{20}H_{14}Cl_2NO$
9c1 717

$C_{20}H_{14}Cl_2O_2S_2^-$
1 120

$C_{20}H_{14}Cl_6Ga_2S_2$
17e 322

$C_{20}H_{14}F_3$
17c 352

$C_{20}H_{14}N$
17c 365

$C_{20}H_{14}NO$
9c1 582, 600, 634, 751, *17d2* 372

$C_{20}H_{14}NO_2^-$
9d1 471

$C_{20}H_{14}NO_2$
9c1 634, 735, 797, *17d2* 355, 397

$C_{20}H_{14}NO_2S$
17d2 354

$C_{20}H_{14}NO_3$
17d2 355, 369

$C_{20}H_{14}NS_2$
17c 282

$C_{20}H_{14}N_2^-$
1 116, *9d1* 874, *17f* 207

$C_{20}H_{14}O_2^-$
9d1 86

$C_{20}H_{14}O_5W^-$
9b 673

$C_{20}H_{15}^{2-}$
9d1 817

$C_{20}H_{15}BNOS_2$
17c 345

$C_{20}H_{15}ClNO_3$
9c1 645, 647

$C_{20}H_{15}Cl_2N_2OS_2$
17c 515

$C_{20}H_{15}Cl_2N_2S_2$
17c 513

$C_{20}H_{15}CrF_3O^-$
17a 262

$C_{20}H_{15}N^+$
9d2 39

$C_{20}H_{15}NO^-$
17c 238

$C_{20}H_{15}NO_2^-$
9d1 420

$C_{20}H_{15}NO_4^-$
9d1 541

$C_{20}H_{15}N_2$
9c1 88

$C_{20}H_{15}N_2O^-$
9c1 593

$C_{20}H_{15}N_2O$
9c1 711

$C_{20}H_{15}N_2O_2$
9c1 735
$C_{20}H_{15}N_2O_5$
9c1 645
$C_{20}H_{15}O$
9b 694, 754, 755, *17c* 277, 367
$C_{20}H_{15}OS$
9b 759
$C_{20}H_{15}O_2$
9b 674, 755, *17c* 297
$C_{20}H_{15}S$
17c 314
$C_{20}H_{16}^-$
9d1 726, 770, 771, 814, *17f* 142
$C_{20}H_{16}^+$
9d2 11, *17h* 70
$C_{20}H_{16}ClMoO_5$
17e 198
$C_{20}H_{16}ClNO_3^-$
17g 274
$C_{20}H_{16}ClN_4$
17c 466
$C_{20}H_{16}ClO_5V$
17e 199
$C_{20}H_{16}Cr^{3-}$
9a 303
$C_{20}H_{16}Cr^-$
9a 303
$C_{20}H_{16}Cr^+$
9a 303
$C_{20}H_{16}D_4^-$
17f 148
$C_{20}H_{16}D_8NO$
17d1 171
$C_{20}H_{16}MoN_2O_4^-$
9a 309
$C_{20}H_{16}N$
9b 761
$C_{20}H_{16}NO$
9c1 634, 711, *17d1* 224, *17d2* 373
$C_{20}H_{16}NO_2$
17d2 350, 397
$C_{20}H_{16}NO_3$
9c1 646, *17d2* 397
$C_{20}H_{16}N_2^+$
9d2 64
$C_{20}H_{16}N_3O$
9c1 91, *17c* 478
$C_{20}H_{16}N_5O_2$
9c1 81, 82
$C_{20}H_{16}N_5O_6$
1 47
$C_{20}H_{16}N_5O_8$
1 47, *9c1* 53
$C_{20}H_{16}O_2^-$
17g 383
$C_{20}H_{16}O_2S^-$
9d1 887
$C_{20}H_{16}O_2S_2^-$
9d1 205, 220
$C_{20}H_{16}O_3Tl$
17g 106
$C_{20}H_{16}O_4S_2^-$
1 119
$C_{20}H_{16}O_4Tl$
9d1 120, 128, *17a* 414ff, *17g* 98, 110
$C_{20}H_{16}O_5^+$
17h 104
$C_{20}H_{16}O_6^-$
17g 302
$C_{20}H_{16}V$
9a 303
$C_{20}H_{17}$
9b 721, *17c* 264
$C_{20}H_{17}D_7NO$
17d1 172
$C_{20}H_{17}MgN_2$
17a 228
$C_{20}H_{17}MoO_5$
17e 197
$C_{20}H_{17}NOS^-$
9b 588
$C_{20}H_{17}NS^+$
17h 203
$C_{20}H_{17}N_2O$
17d2 361
$C_{20}H_{17}N_3O_2S^+$
9d2 88
$C_{20}H_{17}N_4$
9c1 80, *17c* 466
$C_{20}H_{17}N_4S$
9c1 86
$C_{20}H_{17}O$
17c 357
$C_{20}H_{17}OS$
17b 295
$C_{20}H_{17}O_2P^+$
9b 174
$C_{20}H_{17}O_5V$
17e 197
$C_{20}H_{17}S$
17c 183, 281, 283
$C_{20}H_{18}^-$
17f 143ff, 145, 146
$C_{20}H_{18}CrO^-$
17a 262
$C_{20}H_{18}F_{12}O_4P_2^+$
9c2 268
$C_{20}H_{18}NO$
9c1 212, 696
$C_{20}H_{18}NO_3$
9c1 720
$C_{20}H_{18}NO_4P^-$
9d1 516
$C_{20}H_{18}NO_6$
17d2 355
$C_{20}H_{18}NO_6P^-$
9d1 517ff
$C_{20}H_{18}N_2^+$
9d2 65, 108
$C_{20}H_{18}N_2O_2S^+$
17h 142
$C_{20}H_{18}N_2O_4^-$
17g 40
$C_{20}H_{18}N_4O_2^+$
9d2 83
$C_{20}H_{18}O_2Tl$
17g 76
$C_{20}H_{18}O_8S_4^+$
17h 117
$C_{20}H_{19}D_6O_2$
17c 262
$C_{20}H_{19}D_6O_4S_2$
17c 263
$C_{20}H_{19}D_6S_2$
17c 263
$C_{20}H_{19}NO_2Tl$
17g 87
$C_{20}H_{19}NO_3^-$
17g 273
$C_{20}H_{19}NO_3Tl$
17g 100
$C_{20}H_{19}N_2O_2$
17d1 368

$C_{20}H_{19}N_6$
9c1 78
$C_{20}H_{19}OS_2$
17c 319
$C_{20}H_{19}O_2S$
17c 319
$C_{20}H_{19}PS^+$
17b 238
$C_{20}H_{19}Si$
9b 36
$C_{20}H_{20}^-$
17f 147, 148
$C_{20}H_{20}Br_2NO_5$
9c1 362, 363
$C_{20}H_{20}Cl_2O_4^+$
17h 104
$C_{20}H_{20}Cl_2Ti_2$
9a 301
$C_{20}H_{20}Cl_5O_2$
9c2 134
$C_{20}H_{20}Cl_6NO$
17d1 248
$C_{20}H_{20}D_5O$
9c2 106, *17e* 122
$C_{20}H_{20}D_6O_3PS_2$
17c 287
$C_{20}H_{20}D_6O_5P$
17c 287
$C_{20}H_{20}D_6O_7PS_2$
17c 287
$C_{20}H_{20}NO$
9c1 579, *17d1* 226, *17d2* 369
$C_{20}H_{20}NO_2$
17c 408, *17d2* 347
$C_{20}H_{20}NO_3$
9c1 586
$C_{20}H_{20}N_2S^+$
17h 203
$C_{20}H_{20}N_4^+$
9d2 65
$C_{20}H_{20}O_4P$
9c2 261
$C_{20}H_{21}D_5NS$
17c 508
$C_{20}H_{21}NO_3$
9c1 317
$C_{20}H_{21}N_2O_6$
9c1 906
$C_{20}H_{21}N_3NaO_5S$
17d1 364
$C_{20}H_{21}N_3O_8S^-$
17g 270
$C_{20}H_{21}PS$
17c 96
$C_{20}H_{22}^-$
9d1 842
$C_{20}H_{22}N$
17c 184
$C_{20}H_{22}NO$
9c1 284
$C_{20}H_{22}NO_2$
17d2 138
$C_{20}H_{22}NO_3$
9c1 538, *17d1* 172ff, *17d2* 355
$C_{20}H_{22}NO_5$
9c1 363, *17d1* 173
$C_{20}H_{22}N_2^+$
17h 174
$C_{20}H_{22}N_2O_2P$
1 76
$C_{20}H_{22}N_3O$
17d1 362
$C_{20}H_{22}N_3O_2$
17d1 363
$C_{20}H_{22}N_4O_7^-$
9d1 334
$C_{20}H_{22}O_2PSSi$
17c 232
$C_{20}H_{22}O_4^-$
9d1 99, 198, 215
$C_{20}H_{22}O_4^+$
17h 104
$C_{20}H_{22}O_8^-$
9d1 26
$C_{20}H_{22}Si_2^-$
17f 22
$C_{20}H_{22}Ti_2$
9a 301
$C_{20}H_{23}BrNO$
9c1 697
$C_{20}H_{23}D_2O$
9c2 106
$C_{20}H_{23}NO_2^-$
9d1 104
$C_{20}H_{23}N_2O$
9c1 835, *17c* 623, *17d1* 216, *17d2* 48
$C_{20}H_{23}N_2O_2$
9c1 735
$C_{20}H_{23}N_2O_5$
9c1 894
$C_{20}H_{23}N_2O_5S$
1 67, *9c2* 143
$C_{20}H_{23}N_4O_6$
17c 398
$C_{20}H_{23}O_2^{2-}$
17c 273
$C_{20}H_{23}O_8$
9c2 168
$C_{20}H_{24}^-$
9d1 819, 839, *17f* 149
$C_{20}H_{24}^+$
9d2 8, *17h* 106
$C_{20}H_{24}BrO$
9c2 111
$C_{20}H_{24}ClNO^-$
17e 134
$C_{20}H_{24}ClO$
9c2 111
$C_{20}H_{24}Cl_2NS$
9c1 112, *17c* 509
$C_{20}H_{24}D_8NNa_2O_8S_2$
17d1 349
$C_{20}H_{24}FNO_3$
9c1 284
$C_{20}H_{24}FO$
9c2 107, 111
$C_{20}H_{24}N$
9c1 29
$C_{20}H_{24}NO$
9c1 32, 600, 697, *17c* 114, 434
$C_{20}H_{24}NO_3$
17d2 28, 35, 104, 191, 321
$C_{20}H_{24}NO_4$
17e 50
$C_{20}H_{24}NO_5S$
9c1 317
$C_{20}H_{24}NS$
9c1 36, *17c* 123
$C_{20}H_{24}N_2^+$
9d2 63
$C_{20}H_{24}N_3O_4$
17c 398
$C_{20}H_{24}N_3O_5$
9c1 255, 256

$C_{20}H_{24}N_3O_5S_2$
9c1 217
$C_{20}H_{24}N_4O_4^{2\cdot}$
9d2 308
$C_{20}H_{24}N_5O_4$
9c1 620
$C_{20}H_{24}NaO_2$
9d1 150
$C_{20}H_{24}O_2^-$
9d1 324
$C_{20}H_{24}S_2^+$
9d2 145
$C_{20}H_{25}BrNS$
9c1 112
$C_{20}H_{25}ClNO$
17d1 294
$C_{20}H_{25}ClNOS$
17c 525
$C_{20}H_{25}ClNS$
9c1 111
$C_{20}H_{25}ClO_2Tl$
9d1 139
$C_{20}H_{25}Cl_2O_2Pb$
17a 495
$C_{20}H_{25}Cl_2O_2Sn$
17a 468, *17g* 185
$C_{20}H_{25}FeO_3$
17a 330
$C_{20}H_{25}HgO_2$
17a 392, *17g* 122, 163
$C_{20}H_{25}NOP$
17d2 182
$C_{20}H_{25}NO_2X$
(X unknown)
17c 483
$C_{20}H_{25}NO_3P$
17d1 145
$C_{20}H_{25}NO_6^-$
17g 270
$C_{20}H_{25}N_2O$
9c1 828
$C_{20}H_{25}N_2O_2$
9c1 319, 499, 646, *17d2* 108
$C_{20}H_{25}N_2O_2S$
17c 509
$C_{20}H_{25}N_2O_2Si$
17c 448
$C_{20}H_{25}N_2O_4S$
17d2 111
$C_{20}H_{25}N_2O_5$
9c1 834
$C_{20}H_{25}N_3S^+$
17h 203
$C_{20}H_{25}N_4$
9c1 78, *17c* 465
$C_{20}H_{25}N_4O_2$
9c1 78
$C_{20}H_{25}O$
9c2 105, *17e* 121
$C_{20}H_{25}OS$
1 67
$C_{20}H_{25}O_2$
9c2 133
$C_{20}H_{25}O_2S$
1 67, *17c* 123, *17e* 140
$C_{20}H_{25}O_3S$
1 68
$C_{20}H_{25}O_4S$
9c2 140
$C_{20}H_{25}S$
9c2 281
$C_{20}H_{26}^-$
9d1 759, 809
$C_{20}H_{26}^+$
9d2 10
$C_{20}H_{26}ClNO_5P$
17d2 338
$C_{20}H_{26}ClN_2O$
17d1 312
$C_{20}H_{26}D_2^+$
17h 26
$C_{20}H_{26}D_5$
9b 390
$C_{20}H_{26}MnO_6$
17a 300, *17g* 123, 165
$C_{20}H_{26}N$
9c1 12, *17c* 285
$C_{20}H_{26}NO$
9c1 284, 285, 567,
17d1 136, 312, *17d2* 42,
218, *17e* 148
$C_{20}H_{26}NO_2$
17c 114, *17d1* 280
$C_{20}H_{26}NO_2S$
17c 528
$C_{20}H_{26}NO_3$
9c1 317, *9c2* 94, *17d2* 35
$C_{20}H_{26}NS$
17c 508
$C_{20}H_{26}NS_2$
9c1 114, *17c* 519
$C_{20}H_{26}N_2O_2S^{2\cdot}$
9d2 285
$C_{20}H_{26}N_2O_3^{2\cdot}$
9d2 285
$C_{20}H_{26}N_2O_3X$
17d2 105
$C_{20}H_{26}N_3O_2$
17d1 318
$C_{20}H_{26}N_4^+$
9d2 66
$C_{20}H_{26}N_4O_2^+$
9d2 39
$C_{20}H_{26}N_7O_5$
17d2 83
$C_{20}H_{26}O$
17c 26
$C_{20}H_{26}O_2S^+$
9d2 138
$C_{20}H_{27}Br_2N_4OZn$
9a 333
$C_{20}H_{27}Cl_2N_4OZn$
9a 333
$C_{20}H_{27}Cl_3NO_3$
17d1 294, 304
$C_{20}H_{27}D^+$
17h 27
$C_{20}H_{27}N^+$
17h 162
$C_{20}H_{27}N_2^+$
9d2 45
$C_{20}H_{27}N_2O_6$
9c1 500, *17d1* 161
$C_{20}H_{27}N_2S^{2+}$
17h 203
$C_{20}H_{27}N_4O$
17d1 317
$C_{20}H_{27}OSi$
17c 288
$C_{20}H_{27}O_2$
17c 113
$C_{20}H_{27}O_2Si$
17e 140
$C_{20}H_{27}O_5Si$
9b 669, *17c* 289
$C_{20}H_{28}^+$
17h 27
$C_{20}H_{28}AsO_2$
9c2 318

$C_{20}H_{28}Cr^+$
17a 262, 263
$C_{20}H_{28}CrI$
9a 304
$C_{20}H_{28}NO$
17d1 280, 294, 304
$C_{20}H_{28}N_2^+$
17h 169
$C_{20}H_{28}N_2O_2P$
1 75
$C_{20}H_{28}N_3O_3S$
9c1 405
$C_{20}H_{28}N_3O_5$
9c1 358
$C_{20}H_{28}N_4O_2^-$
17g 246
$C_{20}H_{28}N_4O_2^{2\cdot}$
9d2 288
$C_{20}H_{28}N_4O_3^{2\cdot}$
9d2 290
$C_{20}H_{28}N_4O_4^{2\cdot}$
9d2 292, 293
$C_{20}H_{28}N_4O_6^-$
17g 248
$C_{20}H_{28}N_4O_6P$
17d2 321
$C_{20}H_{28}O^+$
17h 34
$C_{20}H_{28}O_2^-$
17g 27
$C_{20}H_{28}O_4P$
9c2 261
$C_{20}H_{28}O_5$
9c2 136
$C_{20}H_{28}Si_2^+$
17h 103
$C_{20}H_{29}$
17b 502
$C_{20}H_{29}D_3S_2^-$
17c 208
$C_{20}H_{29}F_5NO$
9c1 726
$C_{20}H_{29}NO^+$
17h 163
$C_{20}H_{29}NS^-$
9b 587
$C_{20}H_{29}N_2O_2$
9c1 825
$C_{20}H_{29}O$
9c2 115
$C_{20}H_{29}OSi$
17c 221
$C_{20}H_{29}O_3$
1 64
$C_{20}H_{29}O_5$
9c2 137
$C_{20}H_{29}O_8PRe$
17a 318
$C_{20}H_{29}O_8Sn$
17a 468, *17g* 188
$C_{20}H_{30}^-$
17f 149ff
$C_{20}H_{30}^+$
17h 99
$C_{20}H_{30}Cl_3GaN_2O_2^{2\cdot}$
17h 303
$C_{20}H_{30}Cl_3Si$
17c 190
$C_{20}H_{30}Cl_4MnO_{10}P_2$
17g 208
$C_{20}H_{30}D_3S_2$
17c 208
$C_{20}H_{30}F_3$
9b 565
$C_{20}H_{30}NO$
9c1 256, 347, *9c2* 121,
17d2 218
$C_{20}H_{30}NO_2$
9c2 103
$C_{20}H_{30}NO_4$
17d2 198
$C_{20}H_{30}NO_5$
9c1 363, 364
$C_{20}H_{30}NO_{13}$
9c1 347, 353
$C_{20}H_{30}N_2O_2^{2\cdot}$
17h 303
$C_{20}H_{30}N_2O_6Pd^{2\cdot}$
17h 306
$C_{20}H_{30}N_4O_4M$
17d1 72
$C_{20}H_{30}N_4P_2^+$
9c2 268
$C_{20}H_{30}O_4P_2^+$
9c2 268
$C_{20}H_{30}Si_2^+$
17h 75
$C_{20}H_{31}MnN_4O_6$
17d1 325
$C_{20}H_{31}NO_2^-$
9d1 419
$C_{20}H_{31}N_2O$
17d2 321
$C_{20}H_{31}N_2O_3$
17d2 322
$C_{20}H_{31}O$
1 62, *9c2* 87, *17e* 101
$C_{20}H_{31}O_3Sn$
17a 468
$C_{20}H_{32}NO$
17d2 184, 350
$C_{20}H_{32}NO_2$
9c2 102, *17d1* 342, *17e* 97
$C_{20}H_{32}NO_3$
17d1 342
$C_{20}H_{32}N_2O_4^{2\cdot}$
9d2 229, *17h* 300
$C_{20}H_{32}N_3O_2Sn$
17c 459
$C_{20}H_{32}N_4O_4^{2\cdot}$
9d2 203
$C_{20}H_{32}N_6O_{12}S_2^-$
17e 386
$C_{20}H_{32}N_8O_2PdS_2^{2\cdot}$
17h 277
$C_{20}H_{32}NaO_7$
17g 239
$C_{20}H_{33}$
9b 560
$C_{20}H_{33}ClNO$
9c1 615
$C_{20}H_{33}F_2OSn$
17c 227
$C_{20}H_{33}NOX$
(X unknown)
17c 484
$C_{20}H_{33}N_2$
9c1 93
$C_{20}H_{33}N_2O$
9c1 820, *17d1* 325
$C_{20}H_{33}N_4O_2$
9c1 789
$C_{20}H_{33}N_4O_4^{2\cdot}$
9d2 229
$C_{20}H_{33}N_7O_4^{2\cdot}$
9d2 195
$C_{20}H_{33}O_2S_2Sn$
17a 468

$C_{20}H_{33}O_4Pb$
17g 127

$C_{20}H_{34}ClN_7O_2^{2\cdot}$
9d2 221

$C_{20}H_{34}ClO_2Si$
17e 184

$C_{20}H_{34}MnO_{10}P_2$
17a 300, *17g* 73

$C_{20}H_{34}NO$
9c1 160, 615, *17c* 481, 484, *17d1* 280, 330ff, *17e* 95

$C_{20}H_{34}NO_2$
9c1 285, 648, 967, *17d2* 343, 344

$C_{20}H_{34}NO_4$
9c1 616

$C_{20}H_{34}N_2^{+}$
17h 156

$C_{20}H_{34}N_2O_4^{2\cdot}$
17h 304

$C_{20}H_{34}N_2O_4S_4^{2\cdot}$
9d2 253, *17h* 311ff

$C_{20}H_{34}N_2O_6^{2\cdot}$
9d2 246

$C_{20}H_{34}N_6O_2^{2\cdot}$
17h 292

$C_{20}H_{34}O_2^{+}$
9d2 134, 135

$C_{20}H_{34}O_3PS$
17e 112

$C_{20}H_{34}O_5P$
17e 112

$C_{20}H_{35}$
17c 12

$C_{20}H_{35}CuO_2P$
17g 118

$C_{20}H_{35}GeHgO_2$
17a 392, 163, *17g* 163

$C_{20}H_{35}GeO_2$
17a 433, *17e* 64, 135, 148, *17g* 122, 162

$C_{20}H_{35}HgO_2Si$
17g 122, 164

$C_{20}H_{35}N^{+}$
17h 163

$C_{20}H_{35}NO_3P$
17e 303

$C_{20}H_{35}N_2O$
17d1 331

$C_{20}H_{35}OSi$
9b 593, *17c* 219

$C_{20}H_{35}O_2Pb$
17g 127, 170

$C_{20}H_{35}O_2Si$
17b 401, *17e* 65, 141, 151, *17g* 137

$C_{20}H_{35}O_2Sn$
17a 468, *17e* 77, *17g* 179

$C_{20}H_{35}O_4Sn$
17a 469, *17e* 155, 181

$C_{20}H_{35}S_2SeSn$
17a 469, *17c* 69

$C_{20}H_{35}S_3Sn$
17a 469, *17c* 69

$C_{20}H_{36}^{+}$
17h 34, 35

$C_{20}H_{36}Cl_2NO$
17d2 219

$C_{20}H_{36}CuN_2O_7$
17a 358

$C_{20}H_{36}MnNO_5P$
17d1 202

$C_{20}H_{36}MnNO_6P$
17d1 189

$C_{20}H_{36}NO_2$
9c1 519

$C_{20}H_{36}NO_6$
9c1 431

$C_{20}H_{36}N_3O_4S$
17d2 322

$C_{20}H_{36}N_4O_2^{2\cdot}$
9d2 248

$C_{20}H_{36}N_4O_4^{2\cdot}$
17h 272, 308

$C_{20}H_{36}O_4P$
17b 396

$C_{20}H_{36}S_2^{+}$
17e 381

$C_{20}H_{37}$
17c 11

$C_{20}H_{37}BD_4^{-}$
9d1 700

$C_{20}H_{37}N_2O_5P^{2\cdot}$
9d2 238

$C_{20}H_{37}N_3O_3S_2^{2\cdot}$
17h 308

$C_{20}H_{37}N_3O_4^{2\cdot}$
9d2 226

$C_{20}H_{37}Si_2$
9b 490

$C_{20}H_{38}BNOP$
17c 490

$C_{20}H_{38}NOS_2Sn$
17a 469, *17c* 69

$C_{20}H_{38}NO_3$
9c1 218, 500

$C_{20}H_{38}NO_4$
9c1 432

$C_{20}H_{38}N_3O_4PS^{2\cdot}$
9d2 238

$C_{20}H_{38}N_3O_5P^{2\cdot}$
9d2 238

$C_{20}H_{39}N_2O_2$
9c1 887

$C_{20}H_{39}N_2O_6P^{2\cdot}$
9d2 239

$C_{20}H_{39}N_3O_2^{2\cdot}$
9d2 232

$C_{20}H_{39}OSSiSn$
17a 469, *17c* 232

$C_{20}H_{39}Si$
9b 384

$C_{20}H_{39}Si_2$
9b 477, 490

$C_{20}H_{40}IN_3O_2^{2\cdot}$
9d2 227

$C_{20}H_{40}NPbSi$
9b 529

$C_{20}H_{40}N_2O_4Si^{2\cdot}$
9d2 234, *17h* 306

$C_{20}H_{40}N_4O_2^{2\cdot}$
17h 309

$C_{20}H_{40}N_4Zn^{+}$
9a 332

$C_{20}H_{40}Si_2^{-}$
17f 25

$C_{20}H_{40}Si_4^{-}$
17f 29

$C_{20}H_{40}Si_5^{-}$
17f 29

$C_{20}H_{41}$
9b 82

$C_{20}H_{41}B^{-}$
9d1 700

$C_{20}H_{42}Al_2N$
17b 500

$C_{20}H_{42}B_2N_2Si_2^{-}$
17f 226

$C_{20}H_{42}MoN_4O_4Si_4^-$
9a 310

$C_{20}H_{42}O_4P$
9b 307

$C_{20}H_{42}Si_4^+$
17h 60

$C_{20}H_{45}AlNO$
17a 403

$C_{20}H_{45}N_2SSi$
9c1 108

$C_{20}H_{48}NO_2PSi_2^+$
17b 281

$C_{20}H_{48}Si_4^-$
17f 29

$C_{20}H_{48}Si_4^+$
17f 38

$C_{20}H_{50}Si_5^-$
17f 30

$C_{20}H_{54}P_2Si_6^-$
17e 309, *17f* 22

$C_{20}H_{55}P_2$
17e 258

$C_{20}H_{56}N_2O_4P_2Si_4^+$
9c2 268

$C_{21}H_2BrCl_{14}O_2$
9b 738

$C_{21}H_3Cl_{14}O_2$
9b 737, 738

$C_{21}H_3Cl_{15}NO$
9b 741

$C_{21}H_6BrCl_{14}N_2O$
9b 741

$C_{21}H_6Cl_{13}$
9b 730

$C_{21}H_6Cl_{13}O_2$
9b 734

$C_{21}H_7D_{10}N_4$
17c 475

$C_{21}H_9D_{12}NO_3^+$
9d2 50

$C_{21}H_{10}Cl_2N_5$
9c1 46

$C_{21}H_{12}F_5O$
17c 279

$C_{21}H_{12}F_9N_2O_6S_3$
17c 443

$C_{21}H_{12}N_5$
9c1 46

$C_{21}H_{12}O_5^{3-}$
9d1 294

$C_{21}H_{13}BrN_5O_4S$
9c1 58

$C_{21}H_{13}Br_3NO_3$
17d1 254

$C_{21}H_{13}Cl_3NO$
17d1 204

$C_{21}H_{13}F_3NO_2$
9c1 735

$C_{21}H_{13}F_6$
17c 352

$C_{21}H_{13}N_2$
9c1 30, *17c* 365

$C_{21}H_{13}N_2O$
9c1 860

$C_{21}H_{13}N_2O_2$
9c1 894

$C_{21}H_{13}N_6O_6S$
9c1 55

$C_{21}H_{14}^+$
17h 78

$C_{21}H_{14}D_4O_8^-$
17g 307, 308

$C_{21}H_{14}N_2O$
9c1 593

$C_{21}H_{14}N_5O_4S$
9c1 47, 49

$C_{21}H_{14}O^-$
17c 168, 276

$C_{21}H_{14}O_2^-$
9d1 241

$C_{21}H_{15}ClNO$
17d2 362

$C_{21}H_{15}ClNO_4$
17d2 366

$C_{21}H_{15}Cl_3NO_2$
17d1 248

$C_{21}H_{15}D_3O_7^-$
17g 305, 306

$C_{21}H_{15}F_6GeO$
17a 434 ,*17b* 311

$C_{21}H_{15}F_6OPb$
17a 495, *17b* 312

$C_{21}H_{15}F_6OSi$
17b 312

$C_{21}H_{15}F_6OSn$
17a 469, *17b* 313

$C_{21}H_{15}HgO_2$
17a 392, *17g* 318

$C_{21}H_{15}N_2$
9c1 23

$C_{21}H_{15}N_2O$
9c1 107, 758, 759, 858, *17b* 309

$C_{21}H_{15}N_2O_2$
9c1 891

$C_{21}H_{15}N_2O_3$
17c 445

$C_{21}H_{16}BOS_2$
17c 337

$C_{21}H_{16}D_{12}NO_3$
17d1 295

$C_{21}H_{16}MnO_6$
17a 300, *17g* 321

$C_{21}H_{16}NO$
9c1 747, *17c* 365, *17d1* 205, *17d2* 362

$C_{21}H_{16}NO_2$
9c1 736, *17d1* 224

$C_{21}H_{16}NO_3$
9c1 736, *17d2* 356

$C_{21}H_{16}N_3O$
17d1 227

$C_{21}H_{16}N_3O_2$
17c 457

$C_{21}H_{16}O_4Tl$
17g 105

$C_{21}H_{17}Br_3NO$
17d1 254

$C_{21}H_{17}CoN_3O_4$
17e 20

$C_{21}H_{17}NO^-$
17c 239

$C_{21}H_{17}N_2O$
9c1 736

$C_{21}H_{17}N_2O_4S$
9c1 775

$C_{21}H_{17}N_2O_5$
9c1 221

$C_{21}H_{17}N_3^+$
17h 222

$C_{21}H_{17}N_4O_3$
17c 433

$C_{21}H_{17}O$
9b 755, *17c* 278, 360

$C_{21}H_{17}O_2$
17c 278, 367

$C_{21}H_{17}O_7$
17e 207

$C_{21}H_{17}O_8$
17e 208

$C_{21}H_{17}Se$
9b 760
$C_{21}H_{18}ClN_2S_2$
17c 512
$C_{21}H_{18}F_3OSi$
17b 270
$C_{21}H_{18}FeO_3^-$
9a 316
$C_{21}H_{18}NO$
9c1 759, *17d2* 373
$C_{21}H_{18}NO_2$
17d1 216
$C_{21}H_{18}NO_3$
9c1 221, 235, 647
$C_{21}H_{18}NO_4$
9c1 646, 648
$C_{21}H_{18}O^-$
9d1 886
$C_{21}H_{18}O_2Tl$
17a 415ff
$C_{21}H_{18}O_3Tl$
17g 112, 113, 153, 192
$C_{21}H_{18}O_4Tl$
9d1 120ff, 129, *17a* 416ff, *17g* 97
$C_{21}H_{18}O_5Tl$
9d1 141, *17a* 417
$C_{21}H_{18}O_7^-$
17g 304, 305ff
$C_{21}H_{18}O_8^-$
17g 307, 308
$C_{21}H_{19}$
17b 31
$C_{21}H_{19}NO_2S^-$
9b 588
$C_{21}H_{19}NO_3Tl$
17g 107, 351
$C_{21}H_{19}NS^+$
17h 204
$C_{21}H_{19}N_2S_2$
17c 511
$C_{21}H_{19}N_4$
9c1 76, 80, 81
$C_{21}H_{19}N_4O$
9c1 81, 82
$C_{21}H_{19}N_4O_2$
9c1 71, *17c* 433
$C_{21}H_{19}N_4S$
9c1 86
$C_{21}H_{19}O$
9b 735ff
$C_{21}H_{19}OS$
17b 302
$C_{21}H_{19}O_2$
17c 357, 358
$C_{21}H_{19}O_7$
17e 205
$C_{21}H_{19}S$
9b 739ff
$C_{21}H_{19}S_3$
17c 282
$C_{21}H_{20}^-$
17f 151
$C_{21}H_{20}NO$
9c1 177, 178, 233
$C_{21}H_{20}OPS$
17e 303
$C_{21}H_{20}O_2P$
9b 478
$C_{21}H_{20}O_2Tl$
9d1 119, *17a* 417ff
$C_{21}H_{20}O_4Tl$
9d1 128
$C_{21}H_{21}BO_3^-$
9d1 704
$C_{21}H_{21}ClNO_4^+$
1 123
$C_{21}H_{21}ClNO_7^+$
1 123
$C_{21}H_{21}D_9NO_3^+$
9d2 50
$C_{21}H_{21}Ge$
9c2 311, *17f* 41
$C_{21}H_{21}N^+$
9d2 49ff
$C_{21}H_{21}NO_2Tl$
17g 198
$C_{21}H_{21}NO_3^+$
9d2 50
$C_{21}H_{21}NO_3Tl$
17g 101
$C_{21}H_{21}N_4O_2$
17c 432
$C_{21}H_{21}N_4O_4$
9c1 87
$C_{21}H_{21}O$
17c 114ff
$C_{21}H_{21}O_2Pb$
17b 269
$C_{21}H_{21}P^+$
9b 39
$C_{21}H_{21}Sn$
17f 48
$C_{21}H_{22}D_4NO$
17d1 172
$C_{21}H_{22}GeNO_2$
17d2 411
$C_{21}H_{22}MnNO_5P$
9c1 1038, *17a* 300
$C_{21}H_{22}NO$
9c1 174
$C_{21}H_{22}NO_2$
17d2 370
$C_{21}H_{22}NO_2Si$
17d2 411
$C_{21}H_{22}NO_2Sn$
17d2 411
$C_{21}H_{22}N_3Si$
9c1 122, *17c* 582
$C_{21}H_{22}N_4O_2^+$
9d2 83
$C_{21}H_{23}N_3O_8S^-$
17g 275
$C_{21}H_{23}N_4O_4$
9c1 70
$C_{21}H_{24}CdNO_3$
17e 182, *17g* 233
$C_{21}H_{24}Cl_2NOS$
17d1 294, 309
$C_{21}H_{24}NO_2$
9c1 748
$C_{21}H_{24}NO_3^{2-}$
17g 232
$C_{21}H_{24}NO_3$
17e 230
$C_{21}H_{24}NO_3Pb$
17e 182, *17g* 233
$C_{21}H_{24}NO_3Zn$
17e 183, *17g* 233
$C_{21}H_{24}N_3O_6$
17d2 246
$C_{21}H_{24}N_3Si_2$
17c 584
$C_{21}H_{24}O^-$
17c 27
$C_{21}H_{25}$
9b 674
$C_{21}H_{25}ClNO$
9c2 127

$C_{21}H_{25}Cl_3NO_2$
17d1 242
$C_{21}H_{25}FeO_4$
17a 331
$C_{21}H_{25}MoO_4$
17a 271
$C_{21}H_{25}NO_2Tl$
17g 355
$C_{21}H_{25}NO_3^-$
17g 232
$C_{21}H_{25}N_2O$
9c1 860, *17c* 430, 623
$C_{21}H_{25}N_2O_3$
9c2 99, 127, *17d1* 173
$C_{21}H_{25}N_2O_3S$
17e 143
$C_{21}H_{25}N_4O$
17d2 28
$C_{21}H_{25}O_2^{2-}$
17c 274
$C_{21}H_{25}O_2$
1 64, *9c2* 116, *17e* 132
$C_{21}H_{25}O_3$
1 65, 67, *9c2* 135
$C_{21}H_{25}O_4$
17b 398
$C_{21}H_{26}^-$
9d1 819
$C_{21}H_{26}FO$
9c2 112, 113
$C_{21}H_{26}NO$
9c1 698, *9c2* 98, 126
$C_{21}H_{26}NO_2$
17d2 55
$C_{21}H_{26}NO_3$
17d2 59
$C_{21}H_{26}NO_4$
17e 153
$C_{21}H_{26}NO_{14}$
17c 625, 626
$C_{21}H_{26}O^-$
9b 654, *17c* 272
$C_{21}H_{26}S^-$
17c 280
$C_{21}H_{27}$
9b 636, 637
$C_{21}H_{27}ClNO$
17d2 28
$C_{21}H_{27}HgO_2$
17a 393, *17g* 122
$C_{21}H_{27}NO^-$
17e 134
$C_{21}H_{27}NO_2X$
(X unknown)
17c 483
$C_{21}H_{27}NO_3P$
9c1 500
$C_{21}H_{27}NO_5P$
9c1 501
$C_{21}H_{27}N_2O$
9c1 828, 829, *17d1* 174, 230, 312
$C_{21}H_{27}N_2O_2$
9c1 906
$C_{21}H_{27}N_2O_2S$
17c 508
$C_{21}H_{27}N_2O_3$
9c1 515
$C_{21}H_{27}N_4O_4$
17d2 247
$C_{21}H_{27}N_6O_4$
17d2 16
$C_{21}H_{27}O$
1 62, *9c2* 81, 106, 107, 108, *17e* 91
$C_{21}H_{27}O_2$
9c2 107, 108, 112, *17e* 109
$C_{21}H_{27}O_3$
17c 113
$C_{21}H_{27}O_4S$
9c2 141
$C_{21}H_{28}ClO_2Sn$
17a 470, *17e* 66
$C_{21}H_{28}FeO_2^-$
9a 314, 315
$C_{21}H_{28}MnO_6$
17a 301, *17g* 123, 165
$C_{21}H_{28}MnO_7$
17a 301, *17g* 123
$C_{21}H_{28}NO$
9c1 614, *9c2* 123, *17d2* 42
$C_{21}H_{28}NOS$
17c 525
$C_{21}H_{28}NO_2$
17d2 28
$C_{21}H_{28}NO_2S$
17c 528
$C_{21}H_{28}NO_3S$
9c1 1003
$C_{21}H_{28}NS$
17c 508
$C_{21}H_{28}N_2O_2^{2-\cdot}$
9d2 285
$C_{21}H_{28}N_3OS$
17d1 174
$C_{21}H_{28}N_3O_2$
9c1 241
$C_{21}H_{28}N_3O_2S$
17d1 174
$C_{21}H_{28}O_3P$
9c2 138
$C_{21}H_{28}O_3PS$
17b 400, *17e* 139
$C_{21}H_{28}O_4P$
17b 400, *17e* 111, 136
$C_{21}H_{28}O_6Re$
17a 318, *17g* 133
$C_{21}H_{29}F_7NO$
9c1 726
$C_{21}H_{29}GeO_2S$
17a 434, *17c* 292
$C_{21}H_{29}NO_5Re$
17a 318, *17g* 131
$C_{21}H_{29}NO_6Re$
17a 319, *17g* 132
$C_{21}H_{29}N_2O$
9c1 906, *17d1* 312
$C_{21}H_{29}N_2O_5$
9c1 852
$C_{21}H_{29}N_2O_6$
9c1 887
$C_{21}H_{29}N_3O_2X$
17d1 62
$C_{21}H_{29}N_3O_4M$
17d1 42
$C_{21}H_{29}N_4O_6$
9c1 887
$C_{21}H_{29}O_2$
17c 32
$C_{21}H_{30}Br$
17b 503
$C_{21}H_{30}Br_2NO$
9b 504
$C_{21}H_{30}Cl_2NO$
9b 504
$C_{21}H_{30}F_5O_3Si_3$
9b 480
$C_{21}H_{30}MnO_6$
17a 301, *17g* 123

$C_{21}H_{30}N$
9c1 120
$C_{21}H_{30}NO$
9c1 126, *17d2* 35, 36
$C_{21}H_{30}NO_2$
17d2 275
$C_{21}H_{30}NO_6$
17d1 355
$C_{21}H_{30}N_2O_4P$
17b 397, *17e* 114
$C_{21}H_{30}O^-$
17c 33
$C_{21}H_{30}O_3P$
17c 163
$C_{21}H_{31}$
17b 503
$C_{21}H_{31}CuN_4O_4$
17a 359
$C_{21}H_{31}DNO$
9b 503
$C_{21}H_{31}FNO_5$
17d1 327
$C_{21}H_{31}MnNO_5$
17g 166
$C_{21}H_{31}NO_5Tl$
9c1 748
$C_{21}H_{31}N_2O$
9c1 515
$C_{21}H_{31}N_2O_2$
9c1 515
$C_{21}H_{31}N_4O_4$
17d1 21, 62
$C_{21}H_{31}OS$
9b 501
$C_{21}H_{31}OS_2Sn$
17a 470, *17c* 344
$C_{21}H_{31}O_3$
1 65, *17e* 133
$C_{21}H_{31}O_3Sn$
17a 470
$C_{21}H_{31}S_2$
9b 507
$C_{21}H_{32}ClNO_5P$
17d2 322
$C_{21}H_{32}D_2NO_3$
17d1 331
$C_{21}H_{32}NO$
9b 504, *9c1* 616, *17d1* 281, 295
$C_{21}H_{32}NO_2$
9c2 104
$C_{21}H_{32}NO_{11}$
17d2 123
$C_{21}H_{32}N_2O_4^{2\cdot}$
17h 300
$C_{21}H_{32}N_3Sn$
17b 309
$C_{21}H_{33}$
9b 390, 560, *17c* 25
$C_{21}H_{33}D_2$
9b 560
$C_{17}H_{33}N_2O$
17d2 377
$C_{21}H_{33}OS_2Sn$
17a 470, *17c* 339
$C_{21}H_{33}O_2Sn$
17a 470, *17c* 245
$C_{21}H_{33}S_2Sn$
17a 471, *17c* 247
$C_{21}H_{34}FO_3Si_3$
9b 480
$C_{21}H_{34}NO$
9c1 616, *17d1* 332, *17e* 97
$C_{21}H_{34}NO_2$
9c1 616, *17d1* 332
$C_{21}H_{34}NO_3$
9c1 797, *17d1* 331
$C_{21}H_{34}NO_4$
17d2 98
$C_{21}H_{34}NO_7$
17d1 173
$C_{21}H_{34}NO_{10}$
17d2 123
$C_{21}H_{34}N_4O_4^{2\cdot}$
9d2 208
$C_{21}H_{35}$
9b 559, 560, *17b* 58
$C_{21}H_{35}NO^-$
17c 210
$C_{21}H_{35}NS^-$
17c 212
$C_{21}H_{35}N_2O$
9c1 820, *17d1* 326
$C_{21}H_{35}O_3Si_3$
9b 479
$C_{21}H_{36}ClN_5O_2^{2\cdot}$
9d2 249
$C_{21}H_{36}ClN_7O_2^{2\cdot}$
9d2 222
$C_{21}H_{36}KO_6$
17a 218
$C_{21}H_{36}NO$
9c1 103, 617, 620, 638, *17d2* 17
$C_{21}H_{36}NO_2$
9c1 649, 720, 967, *17c* 486, *17d1* 343
$C_{21}H_{36}NO_3$
17d1 332
$C_{21}H_{36}N_2O_3P$
17d2 70
$C_{21}H_{36}O_4P$
9c2 87
$C_{21}H_{36}N_2O_6^{2\cdot}$
17h 311
$C_{21}H_{36}PS$
17e 258
$C_{21}H_{37}$
17c 12
$C_{21}H_{37}BrOP$
9c2 84
$C_{21}H_{37}N^+$
17h 163
$C_{21}H_{37}N_2O$
17d1 332
$C_{21}H_{37}N_3O_4P$
17d2 323
$C_{21}H_{37}N_4O_5P^{2\cdot}$
9d2 239
$C_{21}H_{37}OSi$
17c 220
$C_{21}H_{37}O_3Si$
17e 184
$C_{21}H_{37}S_3Sn$
17a 471, *17c* 69
$C_{21}H_{38}BD_2N_2O$
17c 489, *17d1* 345
$C_{21}H_{38}B_{10}ClO_4Sn$
17a 471, *17g* 191
$C_{21}H_{38}GeNO_2$
17d1 343
$C_{21}H_{38}NOSi$
17c 492
$C_{21}H_{38}NO_2$
17d1 344, *17d2* 323
$C_{21}H_{38}NO_2Si$
17d1 343
$C_{21}H_{38}NO_2Sn$
17d1 343

$C_{21}H_{38}N_3O_2$
17d2 323

$C_{21}H_{38}N_4O_3P$
17d2 324

$C_{21}H_{38}O^-$
17c 60

$C_{21}H_{38}OP$
9b 531

$C_{21}H_{38}O_3P$
9b 383

$C_{21}H_{38}O_4P$
9b 383

$C_{21}H_{38}PSi$
9b 533

$C_{21}H_{39}Cl_2N_6O_3PdS_3^{3\cdot}$
17h 324

$C_{21}H_{39}NO_3PSi$
17e 303

$C_{21}H_{39}NO_5P$
9c1 502

$C_{21}H_{39}S_3Sn$
17a 471, *17c* 68

$C_{21}H_{39}Si_2$
9b 490

$C_{21}H_{40}NO$
17d2 335

$C_{21}H_{40}NO_3$
9c1 501, *17d2* 231, 232, 324

$C_{21}H_{40}NO_6P_2$
9c1 25

$C_{21}H_{40}N_2O_4Si^{2\cdot}$
9d2 235

$C_{21}H_{40}N_4O_2^{2\cdot}$
17h 309

$C_{21}H_{40}O_3P$
9b 383

$C_{21}H_{41}N_2O_2$
9c1 501

$C_{21}H_{41}N_2O_6$
17d2 324

$C_{21}H_{41}N_3O_3P$
9c1 502

$C_{21}H_{41}N_3O_5P^{2\cdot}$
9d2 239

$C_{21}H_{42}NO$
17d1 123

$C_{21}H_{42}Si_3^+$
17h 69

$C_{21}H_{43}B_{10}O$
17b 396

$C_{21}H_{43}OSn$
17b 53

$C_{21}H_{43}Si_3$
9b 478

$C_{21}H_{44}BrN_2O$
17d2 324

$C_{21}H_{45}AlNO$
17a 403

$C_{21}H_{45}O_4Si_4$
9b 478

$C_{21}H_{45}SSi$
17b 314

$C_{21}H_{45}SSn$
17b 315

$C_{21}H_{45}SeSi$
17b 317

$C_{21}H_{45}SeSn$
17b 318

$C_{21}H_{46}NSi$
9b 240

$C_{21}H_{54}GeN_2Si_3$
9c2 311

$C_{21}H_{54}N_3Si_3Sn$
9c2 315

$C_{21}H_{57}GeSi_6$
9c2 311

$C_{21}H_{57}Si_6Sn$
9c2 315

$C_{21}H_{57}Si_7$
9c2 309

$C_{22}Cl_{12}Na_3O_6$
9b 737

$C_{22}F_{21}$
9b 728

$C_{22}HD_{41}NO_4$
17d2 278

$C_{22}H_3Cl_{12}O_6$
9b 737

$C_{22}H_5Cl_{14}O_2$
17c 362

$C_{22}H_5Cl_{14}O_3$
9b 739, *17c* 362

$C_{22}H_6D_{10}^-$
17f 153

$C_{22}H_7BrCl_{14}NO_2$
9b 742

$C_{22}H_8F_{10}N_2^+$
17h 189

$C_{22}H_8Mo_2N_2O_{10}^-$
17a 271

$C_{22}H_8N_2O_{10}W_2^-$
17a 278

$C_{22}H_9Cl_{12}$
9b 730

$C_{22}H_9Cl_{12}O_3$
9b 735

$C_{22}H_9D_{24}$
17b 503

$C_{22}H_9F_{12}$
9b 727

$C_{22}H_9F_{12}O_3$
9b 734

$C_{22}H_{10}ClN_6$
9c1 46

$C_{22}H_{10}F_8N_2^+$
17h 188, 189

$C_{22}H_{10}Mo_2N_2O_{10}^-$
17a 271

$C_{22}H_{10}N_2O_{10}W_2^-$
17a 278

$C_{22}H_{11}F_3O_2^+$
17h 79

$C_{22}H_{12}^-$
9d1 757, 797

$C_{22}H_{12}^+$
9d2 16, 17, 19

$C_{22}H_{12}F_6N_2^+$
17h 188

$C_{22}H_{12}N_6O_4Re$
17a 319

$C_{22}H_{12}O_4^-$
9d1 61

$C_{22}H_{13}N_2O_4P^-$
9d1 651

$C_{22}H_{14}^-$
9d1 793

$C_{22}H_{14}^+$
9d2 16

$C_{22}H_{14}AsO_2S_2$
17g 366

$C_{22}H_{14}BiO_2S_2$
17a 501, *17g* 368

$C_{22}H_{14}ClO_2S_2Sn$
17g 377

$C_{22}H_{14}F_4N_2^+$
17h 188

$C_{22}H_{14}F_{10}NO$
17d1 304

$C_{22}H_{14}N_2O^-$
9b 633

$C_{22}H_{14}N_5O_6$
9c1 54
$C_{22}H_{14}O_2S_2Sb$
17a 500, *17g* 374
$C_{22}H_{14}O_4^-$
9d1 288
$C_{22}H_{14}O_{10}^{2-}$
9d1 247
$C_{22}H_{15}ClCrO_4P$
17a 263
$C_{22}H_{15}ClMoO_4P$
17a 271
$C_{22}H_{15}D_6O_2$
17c 264
$C_{22}H_{15}D_6O_2S$
17c 283
$C_{22}H_{15}D_6O_4S_2$
17c 264
$C_{22}H_{15}D_6S_2$
17c 264
$C_{22}H_{15}D_6S_3$
17c 283
$C_{22}H_{15}F_6O$
17b 311
$C_{22}H_{15}FeO_4P^+$
17a 331
$C_{22}H_{15}MnO_6P$
17e 17
$C_{22}H_{15}N_2O$
17d2 362
$C_{22}H_{15}NaO_5STl$
9d1 279, *17a* 418
$C_{22}H_{15}OS$
17c 318
$C_{22}H_{15}O_5^{2-}$
9d1 294
$C_{22}H_{16}^{3-}$
17f 151, 152
$C_{22}H_{16}^-$
17f 151, 152
$C_{22}H_{16}D_4O_7^-$
17g 303
$C_{22}H_{16}F_2N_2^+$
17h 187
$C_{22}H_{16}F_2N_3^+$
17h 187
$C_{22}H_{16}FeN_2O_4^-$
9a 313, *9d1* 652
$C_{22}H_{16}NO_4$
9c1 736, 737, *17d2* 356, 357
$C_{22}H_{16}N_5O_{10}$
1 47
$C_{22}H_{16}O_2Tl$
9d1 277, *17a* 418ff
$C_{22}H_{16}O_3Tl$
9d1 278, 279, *17a* 419
$C_{22}H_{16}O_4Tl$
17a 420
$C_{22}H_{16}S_4^+$
17h 115
$C_{22}H_{17}BrN_3O$
17d2 370
$C_{22}H_{17}ClNO_2$
9c1 749
$C_{22}H_{17}ClNO_4$
17d2 366
$C_{22}H_{17}Cl_2GeN_2$
17a 434, *17c* 421
$C_{22}H_{17}Cl_2N_2Si$
17c 421
$C_{22}H_{17}D_6N_2O_3$
9c1 640
$C_{22}H_{17}GeO_2S$
17b 377
$C_{22}H_{17}GeO_3$
17b 356
$C_{22}H_{17}N_4$
17c 248
$C_{22}H_{17}O_2SSi$
17b 377
$C_{22}H_{17}O_3Si$
17b 357
$C_{22}H_{17}O_3Sn$
17a 471
$C_{22}H_{18}AlN_2^-$
17f 204
$C_{22}H_{18}B_2D_{14}^-$
17f 224
$C_{22}H_{18}Be_2N_2^+$
17a 224
$C_{22}H_{18}ClGeN_2$
17a 434, *17c* 421
$C_{22}H_{18}ClN_2Si$
17c 421
$C_{22}H_{18}CuF_{12}N_3O_5$
17a 359
$C_{22}H_{18}F_{12}MnN_2O_5$
17a 301
$C_{22}H_{18}F_{12}MnN_3O_5$
17a 302
$C_{22}H_{18}F_{12}MnN_3O_6$
17a 302, 303
$C_{22}H_{18}F_{12}N_3O_6V$
17a 246
$C_{22}H_{18}Fe^+$
9a 314
$C_{22}H_{18}GeNO_2$
17b 344
$C_{22}H_{18}Mg_2N_2^+$
17a 228
$C_{22}H_{18}NO$
9c1 749, *17d1* 205, *17d2* 363
$C_{22}H_{18}NO_2$
9c1 749, *17d2* 363
$C_{22}H_{18}NO_2Si$
17b 345
$C_{22}H_{18}N_2^+$
1 125, *17h* 187
$C_{22}H_{18}N_2Zn_2^+$
17a 383
$C_{22}H_{18}N_3O$
17d1 363
$C_{22}H_{18}N_3O_2$
17d1 227
$C_{22}H_{18}O_2^+$
17h 90
$C_{22}H_{19}$
9b 751
$C_{22}H_{19}Cl_3NO$
9c1 578
$C_{22}H_{19}CoN_3O_4$
17e 20, 21
$C_{22}H_{19}CoN_3O_5$
17e 21
$C_{22}H_{19}F_5NO$
9c1 641
$C_{22}H_{19}GeN_2$
17a 434, *17c* 419
$C_{22}H_{19}Mg_2N_2$
17f 201
$C_{22}H_{19}NOS^-$
9d1 74
$C_{22}H_{19}NO_2^-$
9d1 338
$C_{22}H_{19}N_2O$
9c1 737
$C_{22}H_{19}N_2Si$
17c 420

$C_{22}H_{19}N_3^+$
17h 222
$C_{22}H_{19}N_4$
9c1 79, 82
$C_{22}H_{19}N_4O$
17c 470
$C_{22}H_{19}N_4S$
17c 473
$C_{22}H_{19}N_6O_7$
9c1 56
$C_{22}H_{19}O_2$
17c 361
$C_{22}H_{20}^+$
17h 70
$C_{22}H_{20}CoN_3O_4$
17e 21
$C_{22}H_{20}F_2NO$
17d1 313
$C_{22}H_{20}NO$
9c1 759
$C_{22}H_{20}NO_3$
9c1 235, 244, 296
$C_{22}H_{20}NO_4$
17c 408
$C_{22}H_{20}NO_5$
9c1 586
$C_{22}H_{20}N_2^+$
9d2 66
$C_{22}H_{20}O_2Tl$
17g 335
$C_{22}H_{20}O_3Tl$
17g 114
$C_{22}H_{20}O_4Tl$
9d1 129, *17a* 420
$C_{22}H_{20}O_7^-$
17g 302ff
$C_{22}H_{20}O_7^+$
17h 105
$C_{22}H_{21}$
9b 722
$C_{22}H_{21}BD_{18}^-$
9d1 701
$C_{22}H_{21}ClNO$
9c1 573
$C_{22}H_{21}Cl_3NO$
9c1 641
$C_{22}H_{21}NO_2^-$
9d1 564
$C_{22}H_{21}NO_2Tl$
17g 337
$C_{22}H_{21}NO_3Tl$
17g 89
$C_{22}H_{21}NO_3S^-$
9b 589
$C_{22}H_{21}NO_4Tl$
17g 95
$C_{22}H_{21}N_2O_2$
9c1 720
$C_{22}H_{21}N_2O_3$
9c1 573
$C_{22}H_{21}N_2S_2$
17c 511
$C_{22}H_{21}N_4$
9c1 77, 80, 82
$C_{22}H_{21}N_4O_2$
9c1 82, 83, 85
$C_{22}H_{21}N_4O_6$
9c1 88
$C_{22}H_{21}O_2Pb$
17b 269
$C_{22}H_{21}O_2Sn$
17b 271
$C_{22}H_{21}O_3$
9b 734, *17c* 358
$C_{22}H_{21}S_3$
17b 326
$C_{22}H_{22}^-$
17f 153
$C_{22}H_{22}D_{15}$
9b 561
$C_{22}H_{22}NO$
9c1 102, 233, 256, 285,
17d1 265
$C_{22}H_{22}NO_3$
17d1 305
$C_{22}H_{22}NO_4$
9c1 21
$C_{22}H_{22}NO_5$
9c1 353
$C_{22}H_{22}NO_8$
17d2 210
$C_{22}H_{22}N_2^+$
9d2 64
$C_{22}H_{22}N_4^+$
9d2 65
$C_{22}H_{22}O_2Tl$
9d1 125, *17a* 421, *17g* 76
$C_{22}H_{22}O_3PS$
17b 295
$C_{22}H_{23}ClN_4O_2$
17d2 247
$C_{22}H_{23}NOP$
9c1 679, 1010
$C_{22}H_{23}NOPS$
9c1 680, 681
$C_{22}H_{23}NO_2P$
9c1 680, 1011
$C_{22}H_{23}NS^-$
9b 588
$C_{22}H_{23}N_2O$
9c1 681
$C_{22}H_{23}N_2O_8$
17d2 210
$C_{22}H_{23}N_4O_6$
9c1 72
$C_{22}H_{23}Si$
9b 185
$C_{22}H_{24}AsO$
9c2 318
$C_{22}H_{24}AsO_3$
9c2 26
$C_{22}H_{24}D_5O_4$
17e 145
$C_{22}H_{24}F_6N_3O_7V$
17a 247
$C_{22}H_{24}MnNO_5$
9c1 1040
$C_{22}H_{24}MnNO_5P$
9a 311, *17a* 303
$C_{22}H_{24}NO$
17d1 296
$C_{22}H_{24}NO_2Pb$
17d1 189
$C_{22}H_{24}NO_2Sn$
17d1 189
$C_{22}H_{24}NO_8$
17d2 211
$C_{22}H_{24}N_2O_2Tl$
17g 198
$C_{22}H_{24}N_3O$
17d1 361, 362, 363
$C_{22}H_{24}N_3Si$
9c1 122, *17c* 583
$C_{22}H_{24}OP$
9c2 261, *17e* 304
$C_{22}H_{24}O_2PSi$
17c 217
$C_{22}H_{24}O_3P$
9c2 25

$C_{22}H_{24}O_4P$
9c2 261
$C_{22}H_{25}D_4O_4$
17e 145
$C_{22}H_{25}MnNO_5$
17g 125
$C_{22}H_{25}MoO_5$
17a 272
$C_{22}H_{25}N^+$
17h 223
$C_{22}H_{25}NO^-$
17c 285
$C_{22}H_{25}NO_4^-$
9d1 868
$C_{22}H_{25}NO_4^+$
9d2 91
$C_{22}H_{25}NO_5Re$
17a 319, *17g* 132
$C_{22}H_{25}N_2^{2+}$
9d2 39
$C_{22}H_{25}N_2O$
9c2 93
$C_{22}H_{25}N_3NaO_4S$
17d1 365
$C_{22}H_{25}N_4O_4$
9c1 78
$C_{22}H_{25}O_5W$
17a 278, 279
$C_{22}H_{26}^-$
9d1 761, 764, 809
$C_{22}H_{26}CdNO_4$
17g 235, 236
$C_{22}H_{26}Cl_3O_2$
9c2 133
$C_{22}H_{26}CuO_2$
17g 117
$C_{22}H_{26}DO_2S$
9c2 144
$C_{22}H_{26}NO_3$
17d2 198, 397
$C_{22}H_{26}NO_4^{2-}$
17g 234, 235
$C_{22}H_{26}NO_4Pb$
17g 234, 235, 236
$C_{22}H_{26}NO_4Zn$
17g 234, 235, 236
$C_{22}H_{26}NO_5$
9c1 364
$C_{22}H_{26}NO_5S$
9c1 260
$C_{22}H_{26}N_2O_4M$
17d1 43
$C_{22}H_{26}OS^-$
17g 50, 51
$C_{22}H_{26}O_2^-$
17g 20
$C_{22}H_{26}O_2Tl$
17a 421
$C_{22}H_{26}O_4^-$
9d1 100
$C_{22}H_{27}Cl_3NO_2$
17d1 248
$C_{22}H_{27}N_2O$
17e 98
$C_{22}H_{27}N_4O$
17d2 35
$C_{22}H_{27}N_4O_2$
17d2 325
$C_{22}H_{27}OS$
9c2 142
$C_{22}H_{27}O_2^{2-}$
17c 274
$C_{22}H_{27}O_2$
9c2 110
$C_{22}H_{27}O_2S$
9c2 144
$C_{22}H_{27}O_3$
17e 62
$C_{22}H_{27}O_3S$
9c2 145
$C_{22}H_{27}O_7$
9c2 168
$C_{22}H_{28}^-$
9d1 766, 819, 820
$C_{22}H_{28}^+$
17h 106
$C_{22}H_{28}DO_4$
17e 56
$C_{22}H_{28}NO$
9c2 99
$C_{22}H_{28}NO_2$
9c2 105, 127, *17d2* 115
$C_{22}H_{28}NO_3$
17d2 114
$C_{22}H_{28}NO_4$
17e 230
$C_{22}H_{28}NO_4S$
1 66
$C_{22}H_{28}N_2^+$
9d2 66
$C_{22}H_{28}N_3O_2$
9c1 888
$C_{22}H_{28}N_3O_3$
17d2 275
$C_{22}H_{28}N_3O_5$
9c1 256
$C_{22}H_{28}N_4^+$
9d2 66
$C_{22}H_{28}O_2^-$
9d1 325
$C_{22}H_{28}O_2S^-$
17g 346
$C_{22}H_{28}O_4^-$
17g 345
$C_{22}H_{29}$
17c 26
$C_{22}H_{29}BrO_4P$
17e 137
$C_{22}H_{29}ClO_4P$
17b 400, *17e* 137
$C_{22}H_{29}NO^-$
9b 583
$C_{22}H_{29}NO_5P$
17d2 325
$C_{22}H_{29}NO_5Re$
17a 319, *17g* 131
$C_{22}H_{29}N_2O$
9c1 829
$C_{22}H_{29}N_2O_2$
9c1 307, 907, *17d2* 108
$C_{22}H_{29}N_2O_4$
17d1 313
$C_{22}H_{29}O$
1 62, *9c2* 108, *17c* 270, *17e* 88, 102
$C_{22}H_{29}OPd$
17a 346, *17e* 105
$C_{22}H_{29}OS$
9c2 142
$C_{22}H_{29}O_2$
9c2 113, *17e* 81, 84, 144
$C_{22}H_{29}O_2S$
9c2 163, 164, *17e* 73, 81, 82, 84, 150
$C_{22}H_{29}O_2S_2$
17e 82, 150
$C_{22}H_{29}O_2S_3$
17e 82
$C_{22}H_{29}O_3$
17e 124

$C_{22}H_{29}O_3S$
17e 83
$C_{22}H_{29}O_4$
9c2 174, 175, *17e* 56, 145
$C_{22}H_{30}^-$
9d1 768
$C_{22}H_{30}AlCl_3N_2O_2^{2\cdot}$
17h 310
$C_{22}H_{30}ClO_2Sn$
17e 185
$C_{22}H_{30}D_9OSi$
17c 220
$C_{22}H_{30}HgN_2O_2^{2\cdot}$
17h 311
$C_{22}H_{30}NO$
9c2 121, 124
$C_{22}H_{30}NO_5$
9c1 364, *17d2* 215
$C_{22}H_{30}N_2O_2^{2\cdot}$
9d2 246, 285, *17h* 309
$C_{22}H_{30}N_3O_2$
17d1 360
$C_{22}H_{30}O_2^-$
9d1 22
$C_{22}H_{30}O_2^+$
9d1 358
$C_{22}H_{30}O_2S^+$
9d2 138, 139, *17g* 386
$C_{22}H_{30}O_3PS$
17e 139
$C_{22}H_{30}O_4P$
17b 400, *17e* 137
$C_{22}H_{30}Si_2^+$
17h 76
$C_{22}H_{31}NO_3PS$
17b 302
$C_{22}H_{31}N_2O$
9c1 907
$C_{22}H_{31}N_2O_3S$
17d1 313
$C_{22}H_{31}OSi$
17c 288
$C_{22}H_{31}O_2Sn$
17a 472, *17g* 180
$C_{22}H_{31}O_3Pd$
17a 346, *17e* 106, *17g* 171
$C_{22}H_{31}O_3Pt$
17a 352, *17g* 173
$C_{22}H_{32}^-$
9d1 718, 750

$C_{22}H_{32}Cr^+$
17a 263
$C_{22}H_{32}CuO_2$
17g 117, 158
$C_{22}H_{32}GeN$
9b 558
$C_{22}H_{32}NOSn$
17d1 313
$C_{22}H_{32}NO_{11}$
9c1 414
$C_{22}H_{32}NSi$
9b 557
$C_{22}H_{32}N_2^+$
9d2 57, *17h* 170
$C_{22}H_{32}N_2O_4^{2\cdot}$
17h 300
$C_{22}H_{32}N_2O_5^{2\cdot}$
17h 300
$C_{22}H_{32}N_6O_2S^{2\cdot}$
17h 280
$C_{22}H_{32}O_2^-$
17g 27
$C_{22}H_{32}O_2P$
1 75
$C_{22}H_{32}O_2Rh$
17g 175
$C_{22}H_{32}O_6$
9c2 137
$C_{22}H_{33}$
17b 503
$C_{22}H_{33}CuN_4O_4$
17a 360
$C_{22}H_{33}N_2O_6$
17d2 247
$C_{22}H_{33}N_4O_7S$
17d1 153
$C_{22}H_{33}O_2Sn$
17e 195
$C_{22}H_{34}MnO_{10}P_2S$
17g 366
$C_{22}H_{34}NOSe_2Sn$
17a 472, *17c* 70
$C_{22}H_{34}NO_2$
9c2 104
$C_{22}H_{34}NO_3$
17d1 281, 296
$C_{22}H_{34}NO_5$
9c1 365, 366
$C_{22}H_{34}NO_6$
17d1 327

$C_{22}H_{34}N_2O_2^{2\cdot}$
17h 273, 310
$C_{22}H_{34}N_2O_3^{2\cdot}$
9d2 228
$C_{22}H_{34}N_2O_4^{2\cdot}$
9d2 247, *17h* 310
$C_{22}H_{34}N_4O_2S^{2\cdot}$
9d2 197, *17h* 273
$C_{22}H_{34}N_4O_3P$
17d2 325
$C_{22}H_{34}N_4O_4^{2\cdot}$
9d2 203
$C_{22}H_{34}N_6O_4^-$
17g 247
$C_{22}H_{35}BrN_2O_4^{2\cdot}$
17h 301
$C_{22}H_{35}NO_5P$
17d2 325
$C_{22}H_{35}N_2O$
9c1 103, *17d1* 326,
17d2 377
$C_{22}H_{35}O_2$
17c 65
$C_{22}H_{35}O_3$
17b 262
$C_{22}H_{35}O_5Si_3$
17c 105
$C_{22}H_{35}O_6Sn$
17a 472, *17g* 185
$C_{22}H_{36}CuN_2O_9$
17a 360
$C_{22}H_{36}NO_2$
17c 486, *17d1* 337
$C_{22}H_{36}NO_3$
17d1 333
$C_{22}H_{36}NO_3X$
(X unknown)
17c 485
$C_{22}H_{36}NO_4$
17d2 98
$C_{22}H_{36}N_2O_4^{2\cdot}$
17h 301, 302
$C_{22}H_{36}N_2O_4P$ ··· *17d2* 326
$C_{22}H_{36}N_2O_6^{2\cdot}$
9d2 250
$C_{22}H_{36}N_4O_2^{2\cdot}$
9d2 198, *17h* 273
$C_{22}H_{36}N_4O_4^{2\cdot}$
9d2 203

$C_{22}H_{36}NaO_8$
17g 238
$C_{22}H_{36}O_7^-$
17g 60
$C_{22}H_{36}O_9U^+$
17a 244
$C_{22}H_{37}$
9b 561
$C_{22}H_{37}NO_5P$
17c 491
$C_{22}H_{38}AlCl_2O_3$
17a 403
$C_{22}H_{38}Al_2N_2^-$
17f 205
$C_{22}H_{38}ClN_7O_2^{2\cdot}$
9d2 252
$C_{22}H_{38}ClO_2Sn$
17a 472, *17g* 182
$C_{22}H_{38}FO_2Sn$
17a 472, *17e* 68
$C_{22}H_{38}NO$
9c1 103, 257, 617, 638, *9c2* 122, *17c* 484, 486, *17d1* 281, 332, 333
$C_{22}H_{38}NO_2$
9c1 103, 649, *17d1* 337, *17d2* 344
$C_{22}H_{38}NO_3$
9c1 285, *17d1* 136
$C_{22}H_{38}NO_5$
9c1 350
$C_{22}H_{38}N_2O_4^{2\cdot}$
17h 302
$C_{22}H_{38}N_2O_6^{2\cdot}$
9d2 251, *17h* 312
$C_{22}H_{38}OP$
17e 258
$C_{22}H_{38}PS$
17e 258
$C_{22}H_{38}O_2^+$
9d2 135
$C_{22}H_{38}O_3P$
9c2 138
$C_{22}H_{38}O_5P$
9c2 139
$C_{22}H_{38}O_{10}P_2Re$
17a 319, *17g* 137
$C_{22}H_{39}N^+$
17h 163
$C_{22}H_{39}NO_2P$
17c 490
$C_{22}H_{39}NO_3P$
17c 490, *17d1* 346, *17e* 304, 305
$C_{22}H_{39}NO_3PS$
17c 491, *17d1* 345
$C_{22}H_{39}NO_4P$
17d1 345
$C_{22}H_{39}N_2O$
17d1 296
$C_{22}H_{39}OSi$
9b 594, *17c* 220
$C_{22}H_{39}O_2Pb$
17b 269
$C_{22}H_{39}O_2Sn$
17a 473, *17e* 188
$C_{22}H_{40}^+$
17h 35
$C_{22}H_{40}AlO_3$
17a 404
$C_{22}H_{40}CuN_2O_7$
17a 360
$C_{22}H_{40}HgN_4O_2S_4^{2\cdot}$
17h 313
$C_{22}H_{40}NO_2Sn$
17d1 320
$C_{22}H_{40}NO_3$
17d1 137, *17d2* 326
$C_{22}H_{40}NO_4$
17d2 216
$C_{22}H_{40}N_2O_6^{2\cdot}$
9d2 240
$C_{22}H_{40}N_4O_2S_4^{2\cdot}$
17h 312
$C_{22}H_{40}N_4O_2S_4Zn^{2\cdot}$
17h 313
$C_{22}H_{40}N_4O_3P$
17d2 326
$C_{22}H_{40}N_4O_4^{2\cdot}$
9d2 251
$C_{22}H_{40}O_4P$
17b 396
$C_{22}H_{41}NNaO_4$
17d2 276
$C_{22}H_{41}N_4O_3$
9c1 888
$C_{22}H_{41}Si_2$
9b 490
$C_{22}H_{42}BNOP$
17c 490
$C_{22}H_{42}NO_2$
17d2 276
$C_{22}H_{42}NO_3$
17d2 232
$C_{22}H_{42}NO_4$
9c1 432, 433, 434, 435, 436, 437, *17d2* 275, 276, 277
$C_{22}H_{42}N_2O_4^{2\cdot}$
9d2 247
$C_{22}H_{42}N_3O_5P^{2\cdot}$
9d2 240, *17h* 306
$C_{22}H_{43}N_4O_4PS^{2\cdot}$
17h 306
$C_{22}H_{43}SSn$
17a 473, *17b* 487
$C_{22}H_{44}NO_2$
9c1 432
$C_{22}H_{44}N_2O_4^{2\cdot}$
9d2 247
$C_{22}H_{44}N_2O_4Si^{2\cdot}$
9d2 235
$C_{22}H_{45}N_2O_4Sn$
17c 448
$C_{22}H_{45}Si_3$
9b 491
$C_{22}H_{46}Al_2P_2^-$
17a 404
$C_{22}H_{46}N_2O_5Si_2^{2\cdot}$
9d2 249
$C_{22}H_{46}Si_4^+$
17h 65
$C_{22}H_{47}OSeSn$
17b 295
$C_{22}H_{47}Si$
9b 220, *17b* 282
$C_{22}H_{48}NO_7P$
9c1 373
$C_{22}H_{48}Si_4^+$
17f 36, *17h* 27
$C_{22}H_{52}B_{20}ClO_2Sn$
17a 473
$C_{22}H_{54}B_{20}ClO_2Sn$
17e 68, *17g* 184
$C_{22}H_{54}Si_6^+$
17h 28
$C_{23}H_3D_{15}O_2P$
17c 162

$C_{23}H_{13}D_{12}N_2O$
17d1 282
$C_{23}H_{14}BrO_3$
9b 757
$C_{23}H_{14}ClO_3$
9b 757
$C_{23}H_{15}D_6OS_2$
17c 279
$C_{23}H_{15}D_6O_2S_2$
17c 279
$C_{23}H_{15}D_6O_4$
17c 279
$C_{23}H_{15}N_4O_5$
9c1 65
$C_{23}H_{15}O$
9b 756
$C_{23}H_{15}O_3$
9b 756
$C_{23}H_{16}Cl_5O_3$
9c2 162
$C_{23}H_{16}N_3O_3$
9c1 64, 65
$C_{23}H_{16}N_5O_6$
9c1 57
$C_{23}H_{17}$
9b 750
$C_{23}H_{17}ClD_4NO_3$
9c1 574
$C_{23}H_{17}FOP$
17c 161
$C_{23}H_{17}F_2P^+$
17c 160
$C_{23}H_{17}N_2O$
9c1 64
$C_{23}H_{17}O$
9b 497, *17c* 115
$C_{23}H_{17}P^-$
9d1 896
$C_{23}H_{17}S$
9b 506, *17c* 124
$C_{23}H_{18}BN_2O$
17c 346
$C_{23}H_{18}NO_4$
9c1 737, *17d2* 357
$C_{23}H_{18}OP$
17c 158
$C_{23}H_{18}O_2P$
9b 532
$C_{23}H_{19}ClNO_4$
17d2 366
$C_{23}H_{19}GeOS$
17a 435, *17c* 111, 120
$C_{23}H_{19}GeO_2$
17a 435, *17c* 111
$C_{23}H_{19}GeS_2$
17a 435, *17c* 121
$C_{23}H_{19}OSSi$
17c 111, 119
$C_{23}H_{19}OSSn$
17a 473, *17c* 112, 121
$C_{23}H_{19}O_2$
17c 361
$C_{23}H_{19}O_2Si$
17c 111
$C_{23}H_{19}O_2Sn$
17a 474, *17c* 112
$C_{23}H_{19}O_3Sn$
17a 474
$C_{23}H_{19}O_5$
9b 756
$C_{23}H_{19}S_2Si$
17c 120
$C_{23}H_{19}S_2Sn$
17a 474, *17c* 122
$C_{23}H_{20}CaO_2^+$
17a 232
$C_{23}H_{20}FeNO$
9c1 635
$C_{23}H_{20}GeNOS$
17a 435, 436, *17d2* 203
$C_{23}H_{20}GeNO_2$
17a 435, *17b* 344, *17d2* 203
$C_{23}H_{20}MgO_2^+$
17a 228
$C_{23}H_{20}NO$
17d1 205, *17d2* 364
$C_{23}H_{20}NO_2$
9c1 749, 750, *17d2* 363
$C_{23}H_{20}NO_3$
17d2 363
$C_{23}H_{20}N_3O$
17d1 361
$C_{23}H_{20}O_2^-$
9d1 87, *17g* 61
$C_{23}H_{20}O_2Sr^+$
17a 234
$C_{23}H_{21}Cl_3NO_3S$
17d2 43
$C_{23}H_{21}CoN_3O_4$
17e 22
$C_{23}H_{21}F_5NO$
17d2 29
$C_{23}H_{21}F_{12}N_5O_4P_3$
17c 258
$C_{23}H_{21}N_2O$
9c1 737
$C_{23}H_{21}N_2O_2$
9c1 738
$C_{23}H_{21}N_2O_3$
17d2 360
$C_{23}H_{21}N_2O_4$
9c1 235
$C_{23}H_{21}N_2Si$
17c 151
$C_{23}H_{21}N_6O_6$
9c1 56
$C_{23}H_{22}Cl_5O_4$
9c2 163
$C_{23}H_{22}CoN_3O_4$
17e 22
$C_{23}H_{22}NO$
17d2 373
$C_{23}H_{22}NO_2SSi$
17b 361
$C_{23}H_{22}NO_3$
9c1 233, 296, 574
$C_{23}H_{22}NO_4$
9c1 296
$C_{23}H_{22}O_2Tl$
17g 336
$C_{23}H_{22}O_4Tl$
9d1 129
$C_{23}H_{23}$
1 42, *17c* 351
$C_{23}H_{23}Br_3NO$
9c1 286
$C_{23}H_{23}MoNO_5^-$
17a 272
$C_{23}H_{23}NO^-$
9b 583
$C_{23}H_{23}NO_3Tl$
17g 97, 110
$C_{23}H_{23}NO_5W^-$
17a 279
$C_{23}H_{23}NO_5W$
17g 328
$C_{23}H_{23}N_2O$
9c1 830
$C_{23}H_{23}N_2O_3$
9c1 221

$C_{23}H_{23}N_4$
9c1 76, 83
$C_{23}H_{23}N_4O$
9c1 84
$C_{23}H_{23}N_4O_3$
9c1 84
$C_{23}H_{23}O_4$
17c 359
$C_{23}H_{24}Cl_2NO$
9c2 92
$C_{23}H_{24}Cl_2NOSi$
17d1 161
$C_{23}H_{24}FeO_3^-$
9a 316
$C_{23}H_{24}NNiO$
17d1 268
$C_{23}H_{24}NO$
9c1 233, 234, 347, 641, 681, *17d1* 296
$C_{23}H_{24}NO_2$
9c1 104, *17d1* 281
$C_{23}H_{24}NO_2S$
17d1 161, 189
$C_{23}H_{24}NO_2S_2$
17d1 161, 190
$C_{23}H_{24}NO_3$
9c1 208
$C_{23}H_{24}NO_3S$
17d1 190
$C_{23}H_{24}NO_4$
9c1 22
$C_{23}H_{24}NO_4Sn$
17d1 381
$C_{23}H_{24}NS$
17c 503
$C_{23}H_{24}N_3Si$
17c 583
$C_{23}H_{24}OPS$
17e 305
$C_{23}H_{24}O_4^-$
9d1 343
$C_{23}H_{25}ClGeNO$
17d2 68
$C_{23}H_{25}ClNO$
9c1 286, *9c2* 89, 90
$C_{23}H_{25}ClNOSi$
17d2 67
$C_{23}H_{25}FNO$
9c2 89, 90, 91
$C_{23}H_{25}MnO_8P$
17a 303
$C_{23}H_{25}NO_3PS$
9c1 326
$C_{23}H_{25}NO_3Tl$
17g 103
$C_{23}H_{25}NO_4P$
9c1 326
$C_{23}H_{25}NO_6P$
17d2 59
$C_{23}H_{25}N_2O_3$
9c2 91
$C_{23}H_{26}Cl_3$
17c 299
$C_{23}H_{26}F_3$
17c 299
$C_{23}H_{26}LiO_7$
17g 287
$C_{23}H_{26}NO$
9c1 257, *9c2* 88
$C_{23}H_{26}NOSi$
9c1 323, *17d2* 68
$C_{23}H_{26}NO_3$
17e 94
$C_{23}H_{26}O_2^+$
17e 126
$C_{23}H_{26}O_7^-$
17g 286
$C_{23}H_{27}MnNO_5$
17g 166
$C_{23}H_{27}NO_5Re$
17a 320, *17g* 130
$C_{23}H_{27}N_2O_6$
17d2 107
$C_{23}H_{28}NO_3$
9c2 127
$C_{23}H_{28}NO_7$
17d2 278
$C_{23}H_{28}N_3O_3$
17d1 158
$C_{23}H_{29}$
17c 298
$C_{23}H_{29}MnNO_6$
9c1 1040, *17a* 303
$C_{23}H_{29}NO_6Re$
9c1 1041, *17a* 320
$C_{23}H_{29}N_2$
9b 591
$C_{23}H_{29}N_2O$
17d1 305, 318, *17e* 98
$C_{23}H_{29}N_2O_2$
17e 99
$C_{23}H_{29}O_2^{2-}$
17c 274
$C_{23}H_{29}O_2$
1 72, *9c2* 69, 72, *17e* 83, 156
$C_{23}H_{29}O_4$
9c2 169
$C_{23}H_{29}O_5$
9c2 171
$C_{23}H_{29}S$
17c 314
$C_{23}H_{30}^-$
9d1 820
$C_{23}H_{30}F_5$
9b 391
$C_{23}H_{30}N_3OS$
17d2 145
$C_{23}H_{30}N_4O_4^{2\cdot}$
9d2 308
$C_{23}H_{31}BrO_4P$
17e 110
$C_{23}H_{31}N_2O$
9c1 830
$C_{23}H_{31}O$
1 63, *9c2* 109, 113, *17e* 107
$C_{23}H_{31}O_2$
17e 79, 85
$C_{23}H_{31}O_4$
17e 73, 146
$C_{23}H_{32}NO$
17e 95, 96
$C_{23}H_{32}NO_3$
17d2 192
$C_{23}H_{32}NO_{12}$
9c1 414
$C_{23}H_{32}O_4P$
17b 400, *17e* 110, 137
$C_{23}H_{33}BrOP$
9c2 84
$C_{23}H_{33}N_3O_4^{2\cdot}$
17h 302
$C_{23}H_{34}O_2P$
1 76
$C_{23}H_{34}P$
9b 534
$C_{23}H_{35}MnNO_5$
17a 304, *17g* 124

$C_{23}H_{35}MnO_8P$
17a 304, *17g* 166

$C_{23}H_{35}NO_5Re$
17a 320, *17g* 131

$C_{23}H_{35}N_3O_4^{2\cdot}$
17h 322

$C_{23}H_{35}O_3Pd$
17a 347, *17g* 171

$C_{23}H_{35}O_3Pt$
17a 352, *17g* 173

$C_{23}H_{35}O_8PRe$
17g 174

$C_{23}H_{35}O_8Sn$
17a 474, *17g* 188

$C_{23}H_{36}HgNO_7S$
17a 393

$C_{23}H_{36}NO_2$
9c2 105

$C_{23}H_{36}NO_3$
17d2 278

$C_{23}H_{36}N_4O_3S^{2\cdot}$
9d2 197

$C_{23}H_{37}$
17c 189

$C_{23}H_{37}NOX$
(X unknown)
17c 485

$C_{23}H_{37}NO_2$
9c1 286

$C_{23}H_{37}NO_4P$
17b 397, *17e* 119

$C_{23}H_{38}D_2NO$
17c 484, *17d1* 333

$C_{23}H_{38}NO$
17c 485

$C_{23}H_{38}NO_2$
9c1 437, *17c* 486, 487

$C_{23}H_{38}NO_3$
9c1 437, *17d1* 333,
17d2 278, 279

$C_{23}H_{38}N_2O_6^{2\cdot}$
9d2 252

$C_{23}H_{38}N_3O_6PRe$
17a 320, *17g* 136

$C_{23}H_{39}N^+$
17h 163

$C_{23}H_{39}NNaO_5S$
17d2 29

$C_{23}H_{39}N_2O$
17d1 326

$C_{23}H_{39}O_4S_3Sn$
17a 474, *17c* 69

$C_{23}H_{39}Si_4$
9b 668

$C_{23}H_{39}Sn$
9b 564

$C_{23}H_{40}ClN_7O_2^{2\cdot}$
9d2 257

$C_{23}H_{40}NO$
17d1 337, *17d2* 17

$C_{23}H_{40}NO_3$
9c1 317, *17d1* 138, *17d2* 101

$C_{23}H_{40}N_2O_6^{2\cdot}$
9d2 255

$C_{23}H_{41}Cl_3O_6PSn$
17a 475, *17g* 190

$C_{23}H_{41}N_7O_4^{2\cdot}$
9d2 258

$C_{23}H_{41}OSi$
17c 219, 221

$C_{23}H_{41}O_3Pb$
17b 269

$C_{23}H_{42}NO_2$
17d2 61

$C_{23}H_{42}NO_3$
17d165 138

$C_{23}H_{42}O_4P$
17b 396

$C_{23}H_{43}NO_5P$
9c1 502

$C_{23}H_{43}O_7P_2$
17b 396

$C_{23}H_{44}BNOP$
17d2 66

$C_{23}H_{44}NO_3$
17d1 138, *17d2* 232, 326

$C_{23}H_{44}NO_4$
17d2 197, 279

$C_{23}H_{44}N_3O_4P^{2\cdot}$
17h 307

$C_{23}H_{45}N_3O_3P$
9c1 503

$C_{23}H_{45}N_4O_6P^{2\cdot}$
17h 307

$C_{23}H_{46}NO$
17d2 185

$C_{23}H_{46}NO_2$
17d2 279

$C_{23}H_{48}BrN_2O$
9c1 503

$C_{23}H_{51}N_3O_3P$
17c 454

$C_{24}B_2D_{20}^-$
9d1 705

$C_{24}H_4F_{15}O_2Sn$
17g 75

$C_{24}H_5D_{15}OP$
17c 161

$C_{24}H_5D_{15}O_2P$
17c 164

$C_{24}H_9BrCl_{14}NO_2$
9b 743

$C_{24}H_9Cl_{14}N_2O_3$
9b 741

$C_{24}H_{11}BrCl_{14}NO$
9b 742

$C_{24}H_{12}^-$
1 113, *9d1* 796, 828,
17f 154

$C_{24}H_{12}^+$
9d2 17

$C_{24}H_{12}BaO_4$
9d1 329

$C_{24}H_{13}D_{12}N_2O$
17d2 113

$C_{24}H_{13}O_2$
9b 627

$C_{24}H_{14}^-$
17f 154

$C_{24}H_{14}^+$
9d2 17

$C_{24}H_{14}Cl_4N_2^+$
9d2 84

$C_{24}H_{14}N_4O_4^+$
9d2 118

$C_{24}H_{14}N_5O_6$
1 48

$C_{24}H_{14}O_2$
9d1 289

$C_{24}H_{15}Br_4O_2P^+$
9c2 182

$C_{24}H_{15}Cl_3NO_2$
17d1 248

$C_{24}H_{15}Cl_4O_2P^+$
9c2 181

$C_{24}H_{15}Cl_4O_2Sn$
17a 475, *17g* 209

$C_{24}H_{15}D_{12}N_2O$
17d1 282

$C_{24}H_{15}F_4GeO_2$
17a 436, *17e* 191
$C_{24}H_{15}F_4O_2Si$
17e 191
$C_{24}H_{15}F_4O_2Sn$
17a 475, *17e* 192
$C_{24}H_{15}I_4O_2P^+$
9c2 182
$C_{24}H_{15}O_2$
9c2 196
$C_{24}H_{16}^-$
9d1 766, 794, 841, 843, *17f* 154, 155
$C_{24}H_{16}ClO_2Sn$
17g 343
$C_{24}H_{16}Cl_2N_4O_4^+$
9d2 108
$C_{24}H_{16}Cl_4N_2^+$
9d2 108
$C_{24}H_{16}KN_4O_7S$
9c1 16
$C_{24}H_{16}NO_3$
17d2 358
$C_{24}H_{16}N_2^+$
17h 151ff
$C_{24}H_{16}N_5O_6$
1 46, *9c1* 15, 52
$C_{24}H_{16}N_6O_8^+$
9d2 109
$C_{24}H_{16}O_2^-$
17g 37, 275, 383
$C_{24}H_{16}O_2Tl$
9d1 329, *17a* 421
$C_{24}H_{16}P$
9c2 262
$C_{24}H_{16}S_4^-$
17f 217
$C_{24}H_{16}Si$
1 120
$C_{24}H_{17}GeO_2S_2$
17a 436
$C_{24}H_{17}NO_2^-$
9d1 563ff
$C_{24}H_{17}N_3O_2^-$
9d1 338
$C_{24}H_{17}N_3O_6Se^-$
17f 301
$C_{24}H_{17}N_4O_4$
9c1 14
$C_{24}H_{17}O$
9c2 146ff
$C_{24}H_{17}O_2S_2Sn$
17a 475
$C_{24}H_{18}^-$
9d1 722, 828, 830
$C_{24}H_{18}^+$
9d2 10
$C_{24}H_{18}CrN_2O_2^-$
17a 263
$C_{24}H_{18}Mn_2N_2O_4^-$
17a 304
$C_{24}H_{18}NO$
9c1 635
$C_{24}H_{18}NO_6$
17d2 358
$C_{24}H_{18}N_2^+$
9d2 84, *17h* 152
$C_{24}H_{18}N_3O_4P^-$
9d1 651
$C_{24}H_{18}N_4O_4^+$
9d2 109
$C_{24}H_{18}O_2P$
1 73, *17g* 217
$C_{24}H_{18}P$
9c2 262
$C_{24}H_{19}GeN_2$
17c 154, 155
$C_{24}H_{19}GeO_2$
17e 46
$C_{24}H_{19}GeO_2S$
17a 436
$C_{24}H_{19}NOP$
9c2 33, 43
$C_{24}H_{19}NO_2P$
9d1 515
$C_{24}H_{19}N_2O_2$
9c1 561
$C_{24}H_{19}N_2Si$
17c 151
$C_{24}H_{19}N_4O_2$
17c 460
$C_{24}H_{19}O_2SSn$
17a 475
$C_{24}H_{19}O_2Sn$
9d1 117, *17a* 475, *17g* 74
$C_{24}H_{20}^-$
9d1 775, 840, 841, *17f* 155, 156, 157ff, 159
$C_{24}H_{20}^+$
17h 72
$C_{24}H_{20}As$
9c2 318
$C_{24}H_{20}Cl_2P_2^+$
9c2 268
$C_{24}H_{20}Cr^+$
17a 263
$C_{24}H_{20}CrI$
9a 305
$C_{24}H_{20}CuF_{12}N_3O_6$
17a 360, 361
$C_{24}H_{20}D_6NO_3$
9c1 641
$C_{24}H_{20}FOP^+$
17c 159
$C_{24}H_{20}FP^+$
17c 159
$C_{24}H_{20}F_{12}N_3O_7V$
17a 247
$C_{24}H_{20}NSSi$
17b 377
$C_{24}H_{20}N_2^+$
9d2 58, 109, *17h* 152
$C_{24}H_{20}N_2OP$
1 51
$C_{24}H_{20}N_2O_3P$
1 52
$C_{24}H_{20}N_2O_6S_2^+$
17h 152
$C_{24}H_{20}N_2PS$
17c 451
$C_{24}H_{20}N_3P_2S$
17c 541
$C_{24}H_{20}N_3Si$
9c1 122, *17c* 583
$C_{24}H_{20}N_4Zn^+$
9a 336
$C_{24}H_{20}O_2P$
9b 532
$C_{24}H_{20}O_6P_2S_2^-$
17e 306
$C_{24}H_{20}P_2S_2^-$
17e 306, 309
$C_{24}H_{20}Si^-$
1 120, *17f* 22
$C_{24}H_{21}D_{17}NO_3$
17d2 327
$C_{24}H_{21}GeN_2O$
17a 436, *17d2* 204

$C_{24}H_{21}N_2O_6$
9c1 46
$C_{24}H_{21}N_3NaO_4S$
17d1 365
$C_{24}H_{21}O_3Sn$
17a 476
$C_{24}H_{21}O_6Pb$
17a 495
$C_{24}H_{21}O_6Sn$
17a 476
$C_{24}H_{22}Cl_4NO$
17d1 310
$C_{24}H_{22}CuF_{12}N_3O_5$
17a 361, 362
$C_{24}H_{22}CuF_{12}N_3O_6$
17a 362
$C_{24}H_{22}F_3$
17c 352
$C_{24}H_{22}F_4NO_3$
17d2 29
$C_{24}H_{22}F_{12}N_3O_6V$
17a 248
$C_{24}H_{22}F_{12}N_3O_7V$
17a 248
$C_{24}H_{22}GeNO_2S$
17c 154
$C_{24}H_{22}GeNS$
17c 154
$C_{24}H_{22}N$
17c 365
$C_{24}H_{22}NO$
17d1 205
$C_{24}H_{22}NO_2$
17d1 205
$C_{24}H_{22}NO_2SSi$
17c 151
$C_{24}H_{22}NO_3$
17d2 357, 358
$C_{24}H_{22}NO_5$
9c1 321
$C_{24}H_{22}NSSi$
17c 150
$C_{24}H_{22}N_2^+$
17h 189, 190
$C_{24}H_{22}O_2^+$
17h 90
$C_{24}H_{22}O_4^+$
17h 90
$C_{24}H_{23}GeN_2$
17a 437, *17c* 419
$C_{24}H_{23}N^+$
9d2 51
$C_{24}H_{23}NO_2^-$
17c 238, 239
$C_{24}H_{23}N_2O$
9c1 715, 738
$C_{24}H_{23}N_2O_2$
17d1 296
$C_{24}H_{23}N_2O_4$
9c1 236
$C_{24}H_{23}N_2Si$
17c 420
$C_{24}H_{23}O$
17c 277
$C_{24}H_{24}^+$
9d2 11, *17h* 70, 71, 72
$C_{24}H_{24}NO$
9c1 287, 600, 642
$C_{24}H_{24}NO_2S$
17d1 158
$C_{24}H_{24}NO_6$
17c 409
$C_{24}H_{24}NO_7$
9c1 586
$C_{24}H_{24}N_2^+$
9d2 64
$C_{24}H_{24}O^-$
17c 272
$C_{24}H_{24}O_2S_2^-$
9d1 236
$C_{24}H_{24}O_2Tl$
17g 337
$C_{24}H_{24}O_5Tl$
17g 105
$C_{24}H_{25}GeN_2O_5$
17d2 406
$C_{24}H_{25}NO^+$
17h 164
$C_{24}H_{25}N_2O_2$
9c1 721
$C_{24}H_{25}N_2O_{10}$
17d1 222
$C_{24}H_{25}N_4$
9c1 83
$C_{24}H_{26}F_6N_3O_7V$
17a 248, 249
$C_{24}H_{26}NNiO$
17d1 324
$C_{24}H_{26}NO$
9c1 287, *17d1* 358
$C_{24}H_{26}NOS$
17d2 169
$C_{24}H_{26}NOS_2$
9c1 234, 371
$C_{24}H_{26}NO_2$
9c2 94, *17d1* 373, *17d2* 29, 36
$C_{24}H_{26}NO_2S$
17d1 161
$C_{24}H_{26}NO_5$
9c1 353, *17e* 107
$C_{24}H_{26}NO_7$
9c1 353
$C_{24}H_{26}N_2O_2^-$
9d1 236
$C_{24}H_{26}N_2O_{14}P_2W_2^-$
17a 279
$C_{24}H_{26}N_3O_2$
17d2 30
$C_{24}H_{26}N_4O_6Re$
17a 321, *17g* 132
$C_{24}H_{26}O_2Tl$
9d1 133, *17a* 422, *17g* 76ff
$C_{24}H_{26}O_4PS$
17b 303
$C_{24}H_{27}$
17c 264
$C_{24}H_{27}B^-$
17f 222
$C_{24}H_{27}Ge$
9c2 312, *17f* 41
$C_{24}H_{27}NO_3^+$
9d2 51
$C_{24}H_{27}N_4O_2$
17d2 112
$C_{24}H_{27}O$
9c2 114
$C_{24}H_{27}P^+$
17e 310
$C_{24}H_{27}Sn$
17f 48
$C_{24}H_{28}D_5O_4$
17e 146
$C_{24}H_{28}F_6N_3O_6V$
17a 249
$C_{24}H_{28}F_6N_3O_7V$
17a 250
$C_{24}H_{28}NO$
9c2 89, 90, *17d1* 226, *17d2* 370

$C_{24}H_{28}NO_2$
9c2 92
$C_{24}H_{28}N_2O_3P$
17b 397, *17e* 114
$C_{24}H_{29}D_4O_4$
17e 146
$C_{24}H_{29}MnNO_7$
9c1 797, *17a* 304
$C_{24}H_{30}^{-}$
9d1 809
$C_{24}H_{30}BN_3^{-}$
9d1 704
$C_{24}H_{30}Cl_6CuO_{10}P_2$
17g 363
$C_{24}H_{30}F_5$
9b 480
$C_{24}H_{30}GeP$
9b 534
$C_{24}H_{30}Ge_2^{-}$
17f 45
$C_{24}H_{30}MnNO_5$
17a 304, *17g* 124
$C_{24}H_{30}NO_5$
9c1 366, 367
$C_{24}H_{30}NO_7$
9c1 367
$C_{24}H_{30}N_2^{+}$
9d2 66
$C_{24}H_{30}N_2O_6^{2\cdot}$
9d2 204, 208
$C_{24}H_{30}N_4O_2^{2\cdot}$
9d2 205
$C_{24}H_{30}PSi$
9b 533
$C_{24}H_{30}Si_2^{-}$
17f 22
$C_{24}H_{31}Cl_3NO_2$
17d1 250
$C_{24}H_{31}N_2O$
17d2 36, 49, *17e* 99
$C_{24}H_{31}N_2O_4$
9c1 860, 861
$C_{24}H_{31}N_2O_5$
9c1 894
$C_{24}H_{31}O$
17c 314
$C_{24}H_{31}O_4$
17e 183, *17g* 346
$C_{24}H_{31}O_5$
9c2 117
$C_{24}H_{32}^{-}$
9d1 821
$C_{24}H_{32}^{+}$
17h 106
$C_{24}H_{32}AgBF_4N_6O_4$
17a 375
$C_{24}H_{32}Cl_2CuN_6O_4$
17a 363
$C_{24}H_{32}Cl_2NS$
17c 509
$C_{24}H_{32}Cl_2N_6O_4Pd$
17a 347, 348
$C_{24}H_{32}Cl_2N_6O_4Pt$
17a 353
$C_{24}H_{32}DO_4$
17e 56
$C_{24}H_{32}N$
9c1 29
$C_{24}H_{32}NO$
9c1 601
$C_{24}H_{32}NO_2$
9c2 73
$C_{24}H_{32}N_2O_2^{2\cdot}$
9d2 286
$C_{24}H_{32}N_4O_4^{2\cdot}$
9d2 209, *17h* 285
$C_{24}H_{32}N_8O_2^{2\cdot}$
17h 292ff
$C_{24}H_{32}N_8O_4^{2\cdot}$
17h 285
$C_{24}H_{32}O_2^{-}$
9d1 325
$C_{24}H_{33}BrO_4P$
17e 112
$C_{24}H_{33}ClCoO_6$
17g 200
$C_{24}H_{33}ClNS$
9c1 112, *17c* 509
$C_{24}H_{33}GeO_2$
17e 210, *17g* 341
$C_{24}H_{33}N^{+}$
9d2 45
$C_{24}H_{33}N_2O$
9c1 830
$C_{24}H_{33}N_2O_2$
17d1 265, *17d2* 110
$C_{24}H_{33}N_2O_2S$
17c 510
$C_{24}H_{33}N_2O_3$
17d1 341
$C_{24}H_{33}N_2O_3S$
17d1 344
$C_{24}H_{33}N_2O_4$
17d1 313
$C_{24}H_{33}N_2O_5$
9c1 834
$C_{24}H_{33}N_2O_6S_2$
17d2 111
$C_{24}H_{33}N_4O$
9c1 415, *17d2* 270
$C_{24}H_{33}O$
9c2 109
$C_{24}H_{33}O_2$
17e 79, 85
$C_{24}H_{33}O_2S$
17e 81
$C_{24}H_{33}O_2Sn$
17e 210ff, *17g* 343
$C_{24}H_{33}O_4$
17e 56, 146
$C_{24}H_{34}AsN_2O_8W_2^{-}$
17a 279
$C_{24}H_{34}CoO_6$
17g 116, 157
$C_{24}H_{34}MnO_{10}P_2S_2$
17a 305, *17g* 374
$C_{24}H_{34}NO$
9c1 568, *17d1* 341, 372, *17d2* 209, *17e* 95, 96
$C_{24}H_{34}NOS$
9c1 988, *17c* 525, *17d1* 344
$C_{24}H_{34}NOS_2$
17d2 215
$C_{24}H_{34}NO_2$
9c1 712, 949, *17d1* 344
$C_{24}H_{34}NO_4$
9c1 438
$C_{24}H_{34}N_2O_5Re$
17a 320, *17g* 132
$C_{24}H_{34}N_4O_3^{2\cdot}$
9d2 205
$C_{24}H_{34}N_4O_4^{2\cdot}$
9d2 213
$C_{24}H_{34}N_5O_5$
9c1 852
$C_{24}H_{34}N_5O_6$
9c1 888
$C_{24}H_{34}N_6O_2^{2\cdot}$
17h 280

$C_{24}H_{34}O_2S^+$
9d2 139
$C_{24}H_{34}O_4P$
17b 400, *17e* 112, 138
$C_{24}H_{34}O_7V$
17g 192
$C_{24}H_{34}P$
17e 259
$C_{24}H_{35}MnO_9P$
17a 305, *17e* 64
$C_{24}H_{35}N_2O_2$
9c1 681
$C_{24}H_{35}N_2O_4$
17d2 400
$C_{24}H_{35}OSi$
9b 669
$C_{24}H_{35}OSn$
9b 672
$C_{24}H_{36}^-$
9d1 847
$C_{24}H_{36}^+$
17h 67
$C_{24}H_{36}Cl_2N_6O_6Pt$
17a 352, 353
$C_{24}H_{36}Cr^+$
17a 264
$C_{24}H_{36}CrI$
9a 305
$C_{24}H_{36}DNOS_2$
9c1 371
$C_{24}H_{36}NO_8$
17d2 327
$C_{24}H_{36}N_2^+$
17h 169
$C_{24}H_{36}N_4O_2^{2\cdot}$
9d2 199, *17h* 273
$C_{24}H_{36}N_4O_8^{2\cdot}$
9d2 204
$C_{24}H_{36}N_6O_2^{2\cdot}$
17h 280
$C_{24}H_{37}N^-$
9b 626
$C_{24}H_{37}NO_4^{2\cdot}$
9d2 308, *17h* 265, 322
$C_{24}H_{38}Cl_2N_{12}O_2^{2\cdot}$
9d2 201
$C_{24}H_{38}CuN_2O_2$
17g 117
$C_{24}H_{38}NO_2$
9c1 503
$C_{24}H_{38}NO_4$
9c1 504
$C_{24}H_{38}NO_{11}$
17d2 123
$C_{24}H_{38}N_4O_2^{2\cdot}$
9d2 199, *17h* 273
$C_{24}H_{38}N_4O_2S^{2\cdot}$
9d2 258
$C_{24}H_{38}N_4O_2S_3^{2\cdot}$
17h 274
$C_{24}H_{38}N_4O_4^{2\cdot}$
9d2 209
$C_{24}H_{39}$
17c 189
$C_{24}H_{39}BD_4^-$
9d1 702
$C_{24}H_{39}NO_3$
17d2 99
$C_{24}H_{39}N_2O_3$
9c1 504, *17d2* 327
$C_{24}H_{39}N_2O_4P^{2\cdot}$
9d2 240
$C_{24}H_{39}N_2O_5P^{2\cdot}$
9d2 240
$C_{24}H_{39}N_2O_6P^{2\cdot}$
9d2 241
$C_{24}H_{40}Cl_2NO_3$
17d2 99
$C_{24}H_{40}NO$
9c1 103, *17d1* 334
$C_{24}H_{40}NO_3$
17d1 139, 333, *17d2* 138, 328
$C_{24}H_{40}NO_4$
17d1 139, *17d2* 157
$C_{24}H_{40}N_2O_4Si^{2\cdot}$
9d2 235
$C_{24}H_{40}N_2O_6^{2\cdot}$
9d2 196, 199, 255
$C_{24}H_{40}N_3O_5P^{2\cdot}$
9d2 241
$C_{24}H_{40}N_4O_2^{2\cdot}$
9d2 263
$C_{24}H_{40}N_4O_4Pd^{2\cdot}$
17h 293
$C_{24}H_{40}N_4O_8^{2\cdot}$
9d2 248
$C_{24}H_{40}O_3P$
9b 563
$C_{24}H_{41}$
9b 562
$C_{24}H_{41}NOX$
(X unknown)
17c 485
$C_{24}H_{41}O_4Sn$
17a 476, *17e* 69
$C_{24}H_{42}NO$
9c1 639, *17d1* 337
$C_{24}H_{42}NO_4$
17d2 99, 158
$C_{24}H_{42}N_2O_6^{2\cdot}$
9d2 260, *17h* 314
$C_{24}H_{42}N_2O_6S^{2\cdot}$
9d2 266
$C_{24}H_{42}N_3O_7$
17d2 345
$C_{24}H_{42}N_4O_4X$
17d1 62
$C_{24}H_{42}N_6O_4^{2\cdot}$
17h 286
$C_{24}H_{42}S_6^+$
17h 97
$C_{24}H_{43}AlNO$
17c 486
$C_{24}H_{43}NO_4P$
17d1 346
$C_{24}H_{43}OSn$
17c 227
$C_{24}H_{44}^+$
17h 35
$C_{24}H_{44}CrSi_4^-$
17a 264
$C_{24}H_{44}CrSi_4^+$
17a 264
$C_{24}H_{44}NOSi$
9c1 103, *17c* 492
$C_{24}H_{44}NOSn$
9c1 104
$C_{24}H_{44}NO_2$
9d1 212
$C_{24}H_{44}NO_4$
17d2 158
$C_{24}H_{44}NO_5$
17d2 163
$C_{24}H_{44}N_4O_4^{2\cdot}$
9d2 262
$C_{24}H_{44}O_2^-$
17g 26

$C_{24}H_{45}N_2O_3$
17d2 261
$C_{24}H_{45}N_2O_6P^{2\cdot}$
9d2 241
$C_{24}H_{46}NO$
9c1 257
$C_{24}H_{46}N_3O_5$
17d2 345
$C_{24}H_{47}B^-$
9d1 701
$C_{24}H_{47}NPSi_2$
17e 259
$C_{24}H_{47}N_6O_7$
17d2 400
$C_{24}H_{47}P_2Si_2$
17e 260
$C_{24}H_{48}Cl_2N_7O_2$
9d1 234
$C_{24}H_{48}MnO_{10}P_2$
17a 305
$C_{24}H_{49}O_2$
17e 13
$C_{24}H_{51}Al^-$
17f 63
$C_{24}H_{51}Si_4$
9b 491
$C_{24}H_{54}B_2^-$
17f 226
$C_{24}H_{54}P_2^+$
9c2 268
$C_{24}H_{56}N_4P_4^+$
17e 310
$C_{24}H_{56}Si_4^-$
17f 30
$C_{24}H_{72}N_4P_4Si_8^+$
17e 310
$C_{25}Cl_{17}$
9b 752, *17c* 368
$C_{25}Cl_{19}$
9b 643, 746
$C_{25}HCl_{18}$
9b 639
$C_{25}HD_{20}N_4$
9c1 62
$C_{25}H_8D_{15}NOP$
17c 164
$C_{25}H_8D_{15}O_2P^+$
17c 160
$C_{25}H_9Cl_{12}O_6$
9b 738, 739
$C_{25}H_{11}F_{10}OSi$
17c 290
$C_{25}H_{13}D_{12}N_2O_2$
9c1 721
$C_{25}H_{15}F_5GeNO$
17c 605
$C_{25}H_{15}F_5NOSi$
17c 619
$C_{25}H_{15}NO^+$
17h 223
$C_{25}H_{16}^-$
1 109, *9d1* 798
$C_{25}H_{16}N_5O_6$
1 48
$C_{25}H_{17}Cl_3O_2Sb$
17g 152
$C_{25}H_{17}N_4$
9c1 62
$C_{25}H_{18}^-$
9d1 815
$C_{25}H_{18}ClFGeNO$
17c 605
$C_{25}H_{18}ClFNOSi$
17c 619
$C_{25}H_{18}Cl_2GeNO$
17c 605
$C_{25}H_{18}Cl_2NOSi$
17c 618, 619
$C_{25}H_{18}Cl_2NS$
17c 504
$C_{25}H_{18}GeN_3$
17c 155
$C_{25}H_{18}NO$
17d2 364, 374
$C_{25}H_{18}N_5O_7$
1 46
$C_{25}H_{19}$
9b 744
$C_{25}H_{19}ClGeNO$
17c 604
$C_{25}H_{19}ClNOSi$
17c 618
$C_{25}H_{19}ClNS$
17c 503
$C_{25}H_{19}GeN_2OS_2$
17a 437, *17c* 345
$C_{25}H_{19}GeO_2S_2$
17a 437
$C_{25}H_{19}N_2O$
17d2 361
$C_{25}H_{19}N_2OS_2Si$
17c 345
$C_{25}H_{19}O$
17c 360
$C_{25}H_{19}O_2S_2Sn$
17a 476
$C_{25}H_{20}^-$
9d1 823
$C_{25}H_{20}BO$
17c 296
$C_{25}H_{20}NO$
9c1 712
$C_{25}H_{20}NS$
17c 503
$C_{25}H_{20}OSi^-$
17c 218
$C_{25}H_{21}CuF_{12}N_2O_7$
17a 364
$C_{25}H_{21}F_{12}N_2O_8V$
17a 250
$C_{25}H_{21}GeN_2$
17c 154
$C_{25}H_{21}GeO_2$
17e 47
$C_{25}H_{21}MnO_5P$
17a 305
$C_{25}H_{21}N_2Si$
9c1 96
$C_{25}H_{21}N_4O$
1 51
$C_{25}H_{21}OSi$
17c 218
$C_{25}H_{21}O_2Si$
17e 48
$C_{25}H_{21}Si$
9b 556
$C_{25}H_{22}CrNO_5$
17c 296
$C_{25}H_{22}CuF_{12}N_3O_4$
17a 364
$C_{25}H_{22}F_{12}N_3O_7V$
17a 250, 251
$C_{25}H_{22}GeNO$
17a 437, *17c* 153, *17d2* 204
$C_{25}H_{22}MnN_3O_9S_2$
17a 305
$C_{25}H_{22}NOSi$
17c 150, *17d2* 204

$C_{25}H_{22}N_4O_8^-$
17g 360
$C_{25}H_{22}O_3P$
17c 164
$C_{25}H_{23}$
17c 369
$C_{25}H_{23}CuF_{12}N_2O_7$
17a 364
$C_{25}H_{23}FNP^+$
17c 160
$C_{25}H_{23}F_{12}N_2O_8V$
17a 251
$C_{25}H_{23}GeOS$
17a 437, *17c* 120
$C_{25}H_{23}GeO_2$
17c 111
$C_{25}H_{23}GeS_2$
17a 438, *17c* 121
$C_{25}H_{23}N_2O$
9c1 759
$C_{25}H_{23}N_4O_8$
17e 214
$C_{25}H_{23}OSSi$
17c 120
$C_{25}H_{23}OSSn$
17a 476, *17c* 121
$C_{25}H_{23}O_2Si$
17c 111
$C_{25}H_{23}S_2Si$
17c 120
$C_{25}H_{24}ClCrNO_4P$
17a 264
$C_{25}H_{24}ClFeNO_4P$
17a 331, *17d1* 202
$C_{25}H_{24}CuF_{12}N_3O_5$
17a 365, 366
$C_{25}H_{24}F_{12}N_3O_6V$
17a 251
$C_{25}H_{24}NO$
17d1 205
$C_{25}H_{24}N_3O$
17d1 363
$C_{25}H_{24}N_3O_3$
17d1 223
$C_{25}H_{25}Cl_4MoO_4$
17a 272
$C_{25}H_{25}Cl_4O_4W$
17a 279
$C_{25}H_{25}NO$
9c1 602
$C_{25}H_{25}N_2O$
9c1 287, *17d1* 305
$C_{25}H_{25}N_2O_2$
9c1 722, *17d2* 30
$C_{25}H_{25}N_2O_2S$
9c1 308
$C_{25}H_{25}O_4Tl^-$
17g 109
$C_{25}H_{26}ClN_2S_2$
17c 512
$C_{25}H_{26}NO$
9c1 288
$C_{25}H_{26}NO_3S$
9c1 296
$C_{25}H_{26}N_5$
9c1 81
$C_{25}H_{26}O_4Tl$
17g 109
$C_{25}H_{27}F_6N_2O_8V$
17a 252
$C_{25}H_{27}NO_2X$
(X unknown)
17c 483
$C_{25}H_{27}N_2O_2$
17d1 174
$C_{25}H_{27}N_4O$
9c1 84
$C_{25}H_{27}O_6$
9b 735, *17c* 359
$C_{25}H_{28}F_6N_3O_7V$
17a 252
$C_{25}H_{28}NO$
9c1 288
$C_{25}H_{28}NOSi$
17d2 189
$C_{25}H_{28}N_3O_2$
9c1 322
$C_{25}H_{28}O_2Tl$
17g 82
$C_{25}H_{29}ClN_2O_3P$
17e 115
$C_{25}H_{29}D_{13}NO_3$
17d1 297
$C_{25}H_{29}F_6N_2O_8V$
17a 252
$C_{25}H_{29}NO_2Tl$
17g 99, 196
$C_{25}H_{29}NO_3Tl$
17g 101
$C_{25}H_{29}N_2O$
17d1 314
$C_{25}H_{30}F_6N_3O_6V$
17a 253
$C_{25}H_{31}NO_5Re$
17a 322, *17g* 131
$C_{25}H_{31}N_2O$
9c1 759
$C_{25}H_{32}NO_3S$
17d1 297, 305
$C_{25}H_{32}N_3O$
17d1 361
$C_{25}H_{33}NO_2Tl$
17g 355
$C_{25}H_{33}N_2O_6$
17d2 254
$C_{25}H_{33}OS_4Sn$
17a 477, *17c* 342
$C_{25}H_{33}O_2^{2-}$
17c 275
$C_{25}H_{34}^-$
9d1 821
$C_{25}H_{34}FeNO_3$
17d1 346
$C_{25}H_{34}NO_2$
9c1 103
$C_{25}H_{34}O^-$
17c 272
$C_{25}H_{34}O_2$
17e 131
$C_{25}H_{35}$
9b 637
$C_{25}H_{35}ClNO_3$
9c2 82
$C_{25}H_{35}MoN_2O_3P^-$
9a 310
$C_{25}H_{35}NO_3Tl$
17g 233
$C_{25}H_{35}N_2O$
9c1 821, 830, 831
$C_{25}H_{35}OSSn$
17a 477, *17c* 320
$C_{25}H_{35}OSn$
17c 316
$C_{25}H_{35}O_2$
17e 80, 83, 131, 143

$C_{25}H_{35}O_4$
17e 74, 147
$C_{25}H_{35}SSeSn$
17a 477, *17c* 326
$C_{25}H_{35}S_2Sn$
17a 477, *17c* 323
$C_{25}H_{36}BrO$
17e 89
$C_{25}H_{36}D_9OSi$
17c 222
$C_{25}H_{36}NO$
9c1 617, *17d1* 334, *17e* 96
$C_{25}H_{36}NO_2$
9c1 288, 624, *17d2* 100
$C_{25}H_{36}NO_2S$
17d2 64
$C_{25}H_{36}NO_4S$
17d2 65
$C_{25}H_{36}N_4O_3^{2\cdot}$
9d2 197
$C_{25}H_{37}BrOP$
9c2 85
$C_{25}H_{37}F_{12}N_2O_6P^{2\cdot}$
9d2 242
$C_{25}H_{37}IN_2O_3P$
17d2 328
$C_{25}H_{37}O$
17e 89
$C_{25}H_{37}OSn$
17c 293
$C_{25}H_{37}SSi$
9b 670
$C_{25}H_{37}SSn$
17a 477, *17c* 294
$C_{25}H_{38}F_7O_2Sn$
17a 478, *17e* 66
$C_{25}H_{38}NO$
17d1 282, 306
$C_{25}H_{38}NO_2$
17d2 328
$C_{25}H_{38}NSi$
9b 557
$C_{25}H_{38}N_4O_3^{2\cdot}$
9d2 198, *17h* 274
$C_{25}H_{39}$
9b 405
$C_{25}H_{39}NO_4$
17d2 99
$C_{25}H_{39}N_3O_6$
17d1 43
$C_{25}H_{40}MnO_5SSn$
17a 305, *17g* 124
$C_{25}H_{40}NO$
9b 503
$C_{25}H_{40}NO_2$
9c1 505, *17d2* 328, 329
$C_{25}H_{40}NO_5$
17d1 327
$C_{25}H_{40}N_2^{+}$
17h 190
$C_{25}H_{40}N_2O_5$
9c1 505
$C_{25}H_{40}N_3Sn$
17b 309
$C_{25}H_{41}$
9b 389, 390
$C_{25}H_{42}ClN_7O_2^{2\cdot}$
9d2 266
$C_{25}H_{42}NO$
17d1 334, 337
$C_{25}H_{42}NO_2$
17d2 329
$C_{25}H_{42}NO_3$
17d2 102, 103
$C_{25}H_{42}NO_4$
17d2 100
$C_{25}H_{42}N_3O_5P^{2\cdot}$
9d2 242
$C_{25}H_{43}$
9b 563
$C_{25}H_{43}O_5Pb$
17b 312
$C_{25}H_{44}NO$
17d1 334, 338
$C_{25}H_{44}NO_2$
9c1 505, *17d2* 329
$C_{25}H_{44}N_2O_6^{2\cdot}$
9d2 264, *17h* 314
$C_{25}H_{46}NO_4$
9c1 147
$C_{25}H_{46}N_4O_4^{2\cdot}$
9d2 265
$C_{25}H_{47}MoN_2O_3P^{-}$
9a 309
$C_{25}H_{48}NO_3$
9c1 506, *17d2* 330
$C_{25}H_{48}PSi_2$
17e 260
$C_{25}H_{49}N_2O_2$
9c1 506
$C_{25}H_{50}NO$
17d2 186
$C_{25}H_{50}Si_5^{-}$
17f 30
$C_{25}H_{60}Si_5^{-}$
17f 30, 31
$C_{26}CrD_{20}O_2^{-}$
17a 265
$C_{26}H_{12}S_4^{+}$
17h 118
$C_{26}H_{14}BO_4$
17c 259
$C_{26}H_{14}BeO_4^{-}$
17c 260
$C_{26}H_{16}^{-}$
9d1 795, 796, 832, *17f* 159ff
$C_{26}H_{16}^{+}$
9d2 9, 17
$C_{26}H_{16}Cl_6NO$
17d1 249
$C_{26}H_{16}N$
9b 401
$C_{26}H_{16}N_4O_2Re$
17a 321
$C_{26}H_{16}N_5O_6$
9c1 54
$C_{26}H_{17}$
9b 675
$C_{26}H_{17}Br_4N_4$
9c1 85
$C_{26}H_{17}N^{+}$
17h 223
$C_{26}H_{17}N_2O$
17d2 365
$C_{26}H_{18}^{-}$
9d1 780ff, *17f* 160
$C_{26}H_{18}^{+}$
9d2 14, 15, *17h* 76
$C_{26}H_{18}BiO_2$
17a 501, *17g* 315
$C_{26}H_{18}BrGeO_2S$
17a 438
$C_{26}H_{18}Br_2N_4^{+}$
9d2 84

$C_{26}H_{18}ClGeO_2S$
17a 438, 439

$C_{26}H_{18}ClO_2SSn$
17a 478

$C_{26}H_{18}Cl_2NO_6$
9c1 738

$C_{26}H_{18}CoN_2O_4$
9a 324

$C_{26}H_{18}CrF_2O_2^-$
17a 265

$C_{26}H_{18}FGeO_2S$
17a 439

$C_{26}H_{18}FO_2SSn$
17a 478, 479

$C_{26}H_{18}GeIO_2S$
17a 439, 440

$C_{26}H_{18}NO$
17d1 206

$C_{26}H_{18}N_5O_6$
9c1 51

$C_{26}H_{18}O^-$
9b 690, *17c* 366

$C_{26}H_{18}O_2Tl$
9d1 321, *17a* 422, *17g* 326

$C_{26}H_{19}$
9b 675

$C_{26}H_{19}BrGeNO_2$
17a 440, *17d1* 387

$C_{26}H_{19}BrNO_2Pb$
17d1 387

$C_{26}H_{19}BrNO_2Sn$
17a 479, *17d1* 387

$C_{26}H_{19}BrO_2Pb$
17a 496

$C_{26}H_{19}ClGeNO_2$
17a 440, *17d1* 388

$C_{26}H_{19}ClNO_2Pb$
17a 496, *17d1* 388

$C_{26}H_{19}ClNO_2Si$
17d1 388

$C_{26}H_{19}ClNO_2Sn$
17a 479, *17d1* 388

$C_{26}H_{19}GeO_2S$
17a 440

$C_{26}H_{19}GeO_3$
17a 440

$C_{26}H_{19}N_2O$
9c1 738

$C_{26}H_{19}N_4O_4$
9c1 48

$C_{26}H_{19}O$
17c 361

$C_{26}H_{19}O_2SSn$
17a 479

$C_{26}H_{19}O_2Si$
17e 202

$C_{26}H_{20}^-$
9d1 818, 824, 843

$C_{26}H_{20}^+$
9d2 9

$C_{26}H_{20}BNO_2^-$
17c 48

$C_{26}H_{20}BrN_4$
9c1 81

$C_{26}H_{20}CrO_2^-$
17a 265

$C_{26}H_{20}F_3OSi$
17c 223

$C_{26}H_{20}F_3OSn$
17a 480, *17c* 227

$C_{26}H_{20}GeNOS_2$
17a 441

$C_{26}H_{20}GeNO_2$
17a 441, *17d1* 388

$C_{26}H_{20}GeNS_2$
17c 345

$C_{26}H_{20}GeN_3O_2$
17c 458

$C_{26}H_{20}NO$
9c1 760

$C_{26}H_{20}NO_2$
17d2 398

$C_{26}H_{20}NO_2Pb$
17a 496, *17d1* 389

$C_{26}H_{20}NO_2Sn$
17a 480, *17d1* 388

$C_{26}H_{20}NO_3$
9c1 722

$C_{26}H_{20}NS_2Si$
17c 344

$C_{26}H_{20}N_2^+$
9d2 110, *17h* 230

$C_{26}H_{20}N_3O_2Pb$
17c 459

$C_{26}H_{20}N_3O_2Si$
17c 459

$C_{26}H_{20}N_4^+$
9d2 85

$C_{26}H_{20}O$
1 104

$C_{26}H_{20}O_2Tl$
17a 423

$C_{26}H_{20}S_4^+$
9d2 141

$C_{26}H_{21}$
9b 641

$C_{26}H_{21}GeO_2S_2$
17a 441

$C_{26}H_{21}N_2O$
9c1 712

$C_{26}H_{21}N_4$
9c1 81

$C_{26}H_{21}O$
9b 747

$C_{26}H_{21}OS$
17b 302

$C_{26}H_{21}O_2S_2Sn$
17a 480

$C_{26}H_{22}BrNOP$
9c1 760

$C_{26}H_{22}Mn_2N_2O_4^-$
17a 306

$C_{26}H_{22}NO$
9c1 234, 257, *17d2* 215

$C_{26}H_{22}N_2^+$
9d2 85, *17h* 170

$C_{26}H_{22}N_2O_2^+$
9d2 85

$C_{26}H_{22}N_3O_2$
17d2 107

$C_{26}H_{22}OP$
17c 286

$C_{26}H_{22}O_2PS$
17b 295

$C_{26}H_{22}O_3PS$
17b 303

$C_{26}H_{22}P$
17c 286

$C_{26}H_{23}GeO_2$
17a 441, *17e* 77

$C_{26}H_{23}GeO_4$
17a 441, *17e* 155

$C_{26}H_{23}NOP$
9c1 613, *9c2* 72, *17e* 70

$C_{26}H_{23}O_2Si$
17e 77
$C_{26}H_{23}O_2Sn$
17a 480, *17e* 78
$C_{26}H_{23}O_3$
9b 440
$C_{26}H_{23}O_4Pb$
17a 496, *17e* 155, 182
$C_{26}H_{23}O_4Si$
17e 155
$C_{26}H_{24}AsMnNO_5P$
9c1 1038
$C_{26}H_{24}Cl_7O_2Sn$
17a 480, *17e* 68
$C_{26}H_{24}GeNO$
17a 441, *17d2* 204
$C_{26}H_{24}GeNO_2$
17a 442, *17c* 153, *17d2* 204
$C_{26}H_{24}MnNO_5P$
17a 306
$C_{26}H_{24}NO$
17d1 205
$C_{26}H_{24}NO_2Si$
17c 150
$C_{26}H_{24}NO_3$
17d1 282, 297
$C_{26}H_{24}N_2^+$
9d2 110
$C_{26}H_{24}N_2O_2^+$
9d2 45, 110
$C_{26}H_{24}N_2O_6S_2^+$
17h 170
$C_{26}H_{26}^-$
17f 161
$C_{26}H_{26}^+$
17h 106
$C_{26}H_{26}BaO_2^+$
17a 236
$C_{26}H_{26}CaO_2^+$
17a 232
$C_{26}H_{26}GeP$
9b 534
$C_{26}H_{26}LiO_2$
17a 203
$C_{26}H_{26}MgO_2^+$
17a 228
$C_{26}H_{26}NOP^+$
17c 161
$C_{26}H_{26}NOSi$
17d1 322
$C_{26}H_{26}NO_3$
17d1 174
$C_{26}H_{26}NO_5$
9c1 297, 353
$C_{26}H_{26}NaO_2$
17a 210
$C_{26}H_{26}O_2^-$
9d1 87, *17g* 62
$C_{26}H_{26}O_2Sr^+$
17a 234
$C_{26}H_{26}PSi$
9b 533
$C_{26}H_{27}GeOSSi$
17a 442, *17c* 232
$C_{26}H_{27}NO_2X$
(X unknown)
17c 483
$C_{26}H_{27}N_2O$
17d1 326
$C_{26}H_{27}N_2Si$
17c 420
$C_{26}H_{27}N_4O$
17d2 36
$C_{26}H_{27}OSSi_2$
17c 231
$C_{26}H_{27}O_4$
17b 442
$C_{26}H_{28}ClO_2Sn$
17g 184
$C_{26}H_{28}Cl_3O_2Sn$
17a 481, *17e* 67
$C_{26}H_{28}N_4O_4M$
17d1 43
$C_{26}H_{28}O_3Tl$
17g 106
$C_{26}H_{28}O_4Tl$
17g 98
$C_{26}H_{28}O_5Tl$
17g 105
$C_{26}H_{29}ClO_2Tl$
17a 423
$C_{26}H_{29}CuN_4O_5$
17a 366
$C_{26}H_{29}N_2O_2Pd$
17g 127, 172
$C_{26}H_{29}N_2O_2Pt$
17g 129
$C_{26}H_{29}N_2O_3Pd$
17g 128
$C_{26}H_{29}N_2O_3Pt$
17g 129
$C_{26}H_{29}N_4$
9c1 83
$C_{26}H_{29}O$
9c2 110
$C_{26}H_{29}O_2$
9c2 110
$C_{26}H_{29}O_2Tl$
9d1 150
$C_{26}H_{30}AsO$
9c2 118
$C_{26}H_{30}AsOS$
9c2 118, 143
$C_{26}H_{30}AsO_2$
9c2 118
$C_{26}H_{30}BO_2$
17g 115
$C_{26}H_{30}ClO_2Sn$
17a 481, *17e* 67
$C_{26}H_{30}Cl_2O_2Sn^-$
17g 188
$C_{26}H_{30}Cl_2O_2Sn$
17a 481
$C_{26}H_{30}F_2O_2Sn^-$
17a 481, *17g* 189
$C_{26}H_{30}NO$
9c1 104, *9c2* 124
$C_{26}H_{30}NO_9$
9c1 354
$C_{26}H_{30}N_3O_4$
17c 409
$C_{26}H_{30}OP$
1 68
$C_{26}H_{30}OPS$
1 68
$C_{26}H_{30}O_2P$
1 69, *17e* 135ff
$C_{26}H_{30}O_2PS$
9c2 143
$C_{26}H_{30}O_2Tl$
9d1 126, 136, *17a* 423, 424,
17g 139, 192, 221, 228
$C_{26}H_{30}O_3P$
9c2 139, *17b* 400, *17e* 136
$C_{26}H_{30}O_4PS$
1 69, *9c2* 144

$C_{26}H_{30}O_5P$
9c2 140
$C_{26}H_{31}GeN_2$
17a 442, *17c* 396
$C_{26}H_{31}NO_2Tl$
17g 87ff
$C_{26}H_{31}NO_3Tl$
17g 100
$C_{26}H_{31}N_2Sn$
17a 482, *17c* 396
$C_{26}H_{31}N_4O_{10}$
17c 437
$C_{26}H_{31}O_2Si$
17e 140
$C_{26}H_{31}P^+$
17e 310
$C_{26}H_{32}Br_2N_4O_2Pd_2$
17a 348
$C_{26}H_{32}Br_2N_4O_2Pd_2^{2\cdot}$
17h 293
$C_{26}H_{32}Cl_2N_4O_2Pd$
17a 348
$C_{26}H_{32}Cl_2N_4O_2Pd_2^{2\cdot}$
17h 294
$C_{26}H_{32}D_5O_4$
17e 147
$C_{26}H_{32}I_2N_4O_2Pd$
17a 348
$C_{26}H_{32}I_2N_4O_2Pd_2^{2\cdot}$
17h 294
$C_{26}H_{32}N$
17c 315
$C_{26}H_{32}NO$
17d1 298
$C_{26}H_{32}NO_4$
9c2 93
$C_{26}H_{32}N_2^+$
9d2 67
$C_{26}H_{32}N_2O_3P$
17e 114
$C_{26}H_{32}N_3O_9$
17d1 175
$C_{26}H_{32}N_4O_2TlX^{2\cdot}$
17h 295
$C_{26}H_{33}D_4O_4$
17e 147
$C_{26}H_{34}F_{12}N_4O_2^{2\cdot}$
17h 307
$C_{26}H_{34}MoNO_4$
17d1 347
$C_{26}H_{34}NO_4W$
17d1 347
$C_{26}H_{34}NO_6$
17d2 344
$C_{26}H_{34}N_2O_7^{2\cdot}$
9d2 208, 210
$C_{26}H_{34}N_6O$
17d2 330
$C_{26}H_{35}NO_3P$
17e 118
$C_{26}H_{35}NO_4P$
17b 397, *17e* 113
$C_{26}H_{35}OS_4Sn$
17a 482, *17c* 343
$C_{26}H_{35}O_2$
17e 156
$C_{26}H_{35}O_2Sn$
17a 482, *17e* 203, *17g* 326
$C_{26}H_{36}^-$
9d1 821
$C_{26}H_{36}^+$
17h 35
$C_{26}H_{36}DO_4$
17e 57
$C_{26}H_{36}F_3O$
17e 89
$C_{26}H_{36}NO_3$
9c2 82
$C_{26}H_{36}NO_4$
17d1 175
$C_{26}H_{36}N_4O_2S_3^{2\cdot}$
9d2 200
$C_{26}H_{36}N_8O_2^{2\cdot}$
17h 295
$C_{26}H_{36}O_2^-$
9d1 325
$C_{26}H_{36}O_3P$
17c 315
$C_{26}H_{36}O_5P$
17e 113
$C_{26}H_{37}BrOP$
9c2 85
$C_{26}H_{37}NOX$
(X unknown)
17c 485
$C_{26}H_{37}N_2O$
9c1 832
$C_{26}H_{37}O_2$
17e 71, 79, 80, 130, 144
$C_{26}H_{38}HgN_2O_5S_2^{2\cdot}$
17h 316
$C_{26}H_{38}NO$
9c1 618, 639, *17d1* 335, 338
$C_{26}H_{38}NO_3$
17d1 335
$C_{26}H_{38}NO_{15}$
17d2 124
$C_{26}H_{38}N_2O_6^{2\cdot}$
9d2 253, 256, 261
$C_{26}H_{38}N_2O_{10}^{2\cdot}$
17h 317
$C_{26}H_{38}O_2^-$
17g 272
$C_{26}H_{39}NO_3P$
17c 490
$C_{26}H_{39}N_2O_2$
9c1 681
$C_{26}H_{39}O_2$
17e 198
$C_{26}H_{39}O_2Si$
17e 184
$C_{26}H_{40}^-$
9d1 718, 751
$C_{26}H_{40}ClNO_5P^{2\cdot}$
9d2 242
$C_{26}H_{40}ClSi_2$
17f 8
$C_{26}H_{40}Cl_2N_{12}O_2^{2\cdot}$
9d2 201
$C_{26}H_{40}MnNO_5P$
9c1 1040, *17a* 306
$C_{26}H_{40}NO$
9b 505
$C_{26}H_{40}NOS$
9b 508
$C_{26}H_{40}NO_2$
9b 505
$C_{26}H_{40}NS$
9b 508
$C_{26}H_{40}N_4O_2^{2\cdot}$
9d2 263
$C_{26}H_{40}N_4O_4^{2\cdot}$
9d2 254, 257, 262
$C_{26}H_{40}OP$
9b 531

$C_{26}H_{40}O_2^-$
17g 27
$C_{26}H_{40}O_2^+$
17c 117, *17h* 119
$C_{26}H_{40}STe^+$
17h 113
$C_{26}H_{40}S_2^+$
17c 126, *17h* 113
$C_{26}H_{40}Se_2^+$
17h 119
$C_{26}H_{40}Si_2^-$
17f 25
$C_{26}H_{40}Te_2^+$
17h 119
$C_{26}H_{41}MnO_8P$
17g 167
$C_{26}H_{41}NO^+$
17c 117
$C_{26}H_{41}NS^+$
17c 125
$C_{26}H_{41}N_2O_3$
17d1 338
$C_{26}H_{41}N_2O_5P^{2\cdot}$
9d2 243
$C_{26}H_{41}N_2O_6$
17d2 111
$C_{26}H_{41}N_4O_3$
9c1 506
$C_{26}H_{42}NO$
9c2 122
$C_{26}H_{42}N_2O_{10}^{2\cdot}$
17h 318
$C_{26}H_{42}N_4O_2^{2\cdot}$
9d2 200, 269, *17h* 274
$C_{26}H_{42}N_4O_2S^{2\cdot}$
9d2 266
$C_{26}H_{43}N_3O_5P^{2\cdot}$
9d2 242
$C_{26}H_{43}O_2$
9c2 183, *17e* 224
$C_{26}H_{45}Si$
9b 563
$C_{26}H_{46}NO$
17d1 283, 338
$C_{26}H_{46}NO_7$
9c1 354
$C_{26}H_{46}N_2O_6^{2\cdot}$
9d2 268, *17h* 316

$C_{26}H_{46}O_2P$
1 75
$C_{26}H_{47}Cl_3O_6PSn$
17a 482, *17g* 190
$C_{26}H_{47}NO_3P$
17e 306
$C_{26}H_{47}NO_4P$
17d1 346
$C_{26}H_{47}O_2Sn$
17a 483, *17e* 72, *17g* 179
$C_{26}H_{47}Si$
9b 592, *17b* 282
$C_{26}H_{48}NO_4$
17d2 178
$C_{26}H_{48}N_2O_2X$
17d2 92
$C_{26}H_{48}N_4O_4^{2\cdot}$
9d2 268
$C_{26}H_{48}N_4O_6^{2\cdot}$
9d2 275
$C_{26}H_{49}O_7P_2$
17b 396
$C_{26}H_{50}As_2CuO_2$
17g 118, 158
$C_{26}H_{50}CuO_2P_2$
17g 119
$C_{26}H_{50}CuO_8P_2$
17g 121, 161
$C_{26}H_{50}NO_4$
17d2 197
$C_{26}H_{57}Si_5$
9b 491
$C_{26}H_{58}Si_6^+$
17h 60
$C_{27}HD_{16}$
17c 27
$C_{27}H_9Cl_{14}O_3S$
17c 363
$C_{27}H_{11}^{2-}$
9d1 827
$C_{27}H_{13}Br_4$
9b 396
$C_{27}H_{13}Cl_{14}O_4$
17c 363
$C_{27}H_{14}Cl_{14}NO_3$
17c 366
$C_{27}H_{15}AsBr_4O_5Re$
17a 321, *17g* 204

$C_{27}H_{15}Br_4O_5PRe$
17a 321, *17g* 204
$C_{27}H_{15}Br_4O_5ReSb$
17a 321, *17g* 205
$C_{27}H_{15}Br_4O_8PRe$
17a 322, *17g* 204
$C_{27}H_{16}O_2^-$
17c 370
$C_{27}H_{17}$
1 43, *9b* 395ff, *17c* 27
$C_{27}H_{17}BrN_5O_4S$
9c1 58
$C_{27}H_{17}N_6O_6S$
9c1 55
$C_{27}H_{18}NO_2$
9b 576
$C_{27}H_{18}NO_4$
9c1 739, *17d2* 358, 359
$C_{27}H_{18}N_5O_4S$
9c1 47, 49
$C_{27}H_{18}O^-$
17c 366
$C_{27}H_{19}$
9b 395
$C_{27}H_{19}GeOS_2$
17a 442, *17c* 344
$C_{27}H_{19}GeO_3$
17a 442
$C_{27}H_{19}N_2ORh^{3+}$
9a 326
$C_{27}H_{19}N_2O_3$
17d2 360
$C_{27}H_{19}O$
9b 673
$C_{27}H_{19}OS_2Si$
17c 344
$C_{27}H_{20}GeN_3$
17b 308
$C_{27}H_{20}NO$
17d1 206
$C_{27}H_{20}N_3Si$
17b 308
$C_{27}H_{21}$
9b 392ff
$C_{27}H_{21}GeOS$
17a 443, *17c* 245, 246
$C_{27}H_{21}GeOS_2$
17a 443, *17c* 337, 338

$C_{27}H_{21}GeO_2$
17a 444, *17c* 244
$C_{27}H_{21}GeO_2S$
17a 444
$C_{27}H_{21}GeS_2$
17a 444, *17c* 247
$C_{27}H_{21}NO^+$
9d2 46
$C_{27}H_{21}N_2O$
9c1 739
$C_{27}H_{21}N_2O_2$
9c1 739, *17d2* 360
$C_{27}H_{21}OSSi$
17c 246
$C_{27}H_{21}OSSn$
17a 483, *17c* 247
$C_{27}H_{21}OS_2Si$
17c 337, 338
$C_{27}H_{21}OS_2Sn$
17a 483, *17c* 337
$C_{27}H_{21}O_2SSn$
17a 483
$C_{27}H_{21}O_2Si$
17c 243
$C_{27}H_{21}O_2Sn$
17a 483, *17c* 245
$C_{27}H_{21}S_2Si$
17c 246
$C_{27}H_{21}S_2Sn$
17a 484, *17c* 247
$C_{27}H_{22}Cl_3O_2Sn$
17e 186
$C_{27}H_{22}F_3O_3$
9b 440
$C_{27}H_{22}GeNS$
17c 156
$C_{27}H_{22}NO$
9c1 603
$C_{27}H_{22}NO_3$
9c1 244, 603
$C_{27}H_{22}NSSi$
17c 152
$C_{27}H_{22}N_4O_4^{2\cdot}$
9d2 308
$C_{27}H_{23}Cl_2O_2Sn$
17e 59
$C_{27}H_{23}GeOS$
17c 241
$C_{27}H_{23}GeO_2$
17c 241
$C_{27}H_{23}N_2O_2$
9c1 723
$C_{27}H_{23}OP^+$
9b 213
$C_{27}H_{23}OSSn$
17c 241
$C_{27}H_{24}ClO_2Sn$
17e 50
$C_{27}H_{24}D_6NO_{10}$
17g 310
$C_{27}H_{24}O_2PS_2$
17b 303
$C_{27}H_{24}O_3PS$
17b 303
$C_{27}H_{25}MnO_5PS_3$
17a 306
$C_{27}H_{25}MnO_8PS_3$
17a 306
$C_{27}H_{25}O_2Sn$
17g 75
$C_{27}H_{25}O_5PReS_3$
17a 322
$C_{27}H_{26}Fe_2NO$
9c1 635
$C_{27}H_{26}MnNO_5P$
17a 307
$C_{27}H_{26}NO$
17d2 330
$C_{27}H_{27}AsCl_6O_4Rh$
17g 203
$C_{27}H_{27}N^+$
9d2 51
$C_{27}H_{27}NO_6^+$
9d2 51
$C_{27}H_{27}NO_{11}^-$
17g 312
$C_{27}H_{28}GeN$
17c 153
$C_{27}H_{28}GeNO_2S$
17c 154
$C_{27}H_{28}NO_2SSi$
17c 151
$C_{27}H_{28}NO_{11}$
17e 208
$C_{27}H_{28}NSSi$
17c 151
$C_{27}H_{28}NSi$
17c 150
$C_{27}H_{28}N_3O$
17d1 314
$C_{27}H_{28}N_4NiO_4$
17a 343
$C_{27}H_{28}N_4O_4Pd$
17a 349
$C_{27}H_{29}ClNO_{10}$
17e 209
$C_{27}H_{29}NO_{10}^-$
17g 308ff
$C_{27}H_{29}NO_{11}^-$
17g 310ff
$C_{27}H_{29}O_5$
17c 359
$C_{27}H_{30}ClNO_{10}^-$
17g 310
$C_{27}H_{30}ClN_2S_2$
17c 512
$C_{27}H_{30}NO_{10}$
17e 205, 206, *17g* 309
$C_{27}H_{30}NO_{11}$
17e 206
$C_{27}H_{30}O^-$
17c 354
$C_{27}H_{30}O_3Tl$
17g 112, 113, 154, 193
$C_{27}H_{30}O_4Tl$
17g 85
$C_{27}H_{31}$
17c 351
$C_{27}H_{31}ClNO_{10}$
17e 206
$C_{27}H_{31}NO^-$
9b 584
$C_{27}H_{31}NO_2Tl$
17g 347
$C_{27}H_{31}NO_3Tl$
17g 107
$C_{27}H_{31}NO_4Tl$
17g 93
$C_{27}H_{31}NO_5Tl$
17g 104
$C_{27}H_{31}NS^-$
9b 588
$C_{27}H_{31}N_2$
9c1 95

$C_{27}H_{31}N_2O$
9c1 832
$C_{27}H_{31}N_4O$
9c1 84
$C_{27}H_{31}O$
1 63, *17e* 103
$C_{27}H_{32}NO$
9c1 288
$C_{27}H_{32}NO_3S$
17d1 314
$C_{27}H_{32}O_3P$
17e 110
$C_{27}H_{33}B^-$
1 120, *9d1* 703
$C_{27}H_{33}ClNO_2$
9c1 376
$C_{27}H_{33}ClO_2P$
1 69
$C_{27}H_{33}Cl_3O_5PSn$
17a 484, *17g* 190
$C_{27}H_{33}D_{13}NO_3$
17d1 298
$C_{27}H_{33}Ge$
9d2 312, *17f* 41
$C_{27}H_{33}Mo_2N_4O_7P^-$
17a 273, *17f* 209
$C_{27}H_{33}NOP$
9c2 129
$C_{27}H_{33}NO_3Tl$
17g 102
$C_{27}H_{33}O_2Sn$
17a 484, *17e* 65
$C_{27}H_{33}O_3P^+$
17e 311
$C_{27}H_{33}P^+$
17e 311
$C_{27}H_{33}Si$
17f 8
$C_{27}H_{33}Sn$
9c2 315, *17f* 48
$C_{27}H_{34}N_2O_3P$
17e 115
$C_{27}H_{35}N_2$
9b 380
$C_{27}H_{36}ClNO_3P$
17e 118
$C_{27}H_{36}CoN_4O_5$
17d2 163
$C_{27}H_{36}NO_2$
9c1 797
$C_{27}H_{36}NO_3$
17d2 247
$C_{27}H_{36}NO_5$
17d1 327
$C_{27}H_{36}N_3O$
17d1 361
$C_{27}H_{37}NO_2Tl$
17g 355
$C_{27}H_{37}NO_3P$
17e 118
$C_{27}H_{37}N_5O_7$
9c1 889
$C_{27}H_{38}^-$
9d1 821
$C_{27}H_{38}NO_4$
17d2 100
$C_{27}H_{38}NO_{10}$
17d2 124
$C_{27}H_{39}INO_3$
9c2 82
$C_{27}H_{39}O_4$
17e 74, 147
$C_{27}H_{40}MnN_2O_5$
17a 307
$C_{27}H_{40}N_2O_6^{2\cdot}$
9d2 256
$C_{27}H_{40}N_4O_3^{2\cdot}$
9d2 258
$C_{27}H_{40}O_2^-$
17g 381
$C_{27}H_{40}O_2^+$
17e 125
$C_{27}H_{40}P$
17e 260
$C_{27}H_{41}FOP$
17c 162
$C_{27}H_{41}NO_8P$
9c1 507
$C_{27}H_{41}NO_{10}P$
17d2 331
$C_{27}H_{41}NaO_2P$
17c 162
$C_{27}H_{41}O_2$
9b 498
$C_{27}H_{42}NO_3$
9c1 449
$C_{27}H_{42}N_2O_4^{2\cdot}$
9d2 254
$C_{27}H_{42}N_4O_2^{2\cdot}$
17h 314
$C_{27}H_{42}N_4O_3^{2\cdot}$
9d2 259
$C_{27}H_{42}OP$
17c 158
$C_{27}H_{42}O_2P$
17c 163
$C_{27}H_{43}N_2O_3$
17d1 338
$C_{27}H_{43}O$
17c 31
$C_{27}H_{44}Br$
17b 242
$C_{27}H_{44}N_2O_4^{2\cdot}$
9d2 217, 218
$C_{27}H_{44}N_4O_4^{2\cdot}$
9d2 265
$C_{27}H_{45}O$
17b 243, *17c* 35
$C_{27}H_{45}O_2$
9c2 183, *17e* 223
$C_{27}H_{46}NO_2$
17d1 339
$C_{27}H_{48}NO$
17d1 335, 339, *17d2* 17
$C_{27}H_{48}N_2O_6^{2\cdot}$
9d2 271
$C_{27}H_{49}BrOP$
9c2 85
$C_{27}H_{49}N_2O_7Sn$
17d2 412
$C_{27}H_{50}As_2O_3Rh$
17g 175
$C_{27}H_{51}N_2O$
17d2 185
$C_{27}H_{51}N_3O_6P^{3\cdot}$
9d2 295
$C_{27}H_{51}N_3O_6PS^{3\cdot}$
9d2 295
$C_{27}H_{52}NO_4$
9c1 438
$C_{27}H_{52}N_3O_5P^{2\cdot}$
9d2 243
$C_{27}H_{53}N_2O_2$
9c1 507, *17d2* 331

$C_{27}H_{56}BrN_2O$
17d2 331
$C_{27}H_{56}N_2PSi_3$
17e 266
$C_{28}H_{11}BrCl_{14}NO_2$
9b 743
$C_{28}H_{14}^-$
17f 162
$C_{28}H_{16}^-$
9d1 832, *17f* 162
$C_{28}H_{16}NS$
9c1 37
$C_{28}H_{16}O_2^-$
9d1 241, 304
$C_{28}H_{16}O_2^{2-\cdot}$
1 128
$C_{28}H_{16}O_4Sb$
17e 209
$C_{28}H_{17}O$
9c2 202
$C_{28}H_{17}O_2$
9c2 202, 203
$C_{28}H_{18}^-$
9d1 781, 843, *17f* 162
$C_{28}H_{18}Cl_2NO_6$
17d2 367
$C_{28}H_{18}CrF_6O_2^-$
17a 265
$C_{28}H_{18}F_9O_2Sn$
17g 78, 82
$C_{28}H_{18}N$
9c1 8
$C_{28}H_{18}O_6Tl$
17g 332, 333
$C_{28}H_{19}$
9b 396
$C_{28}H_{19}GeO_2S_2$
17a 444, 445, *17e* 232, 233, *17g* 369
$C_{28}H_{19}N_2O$
9c1 594
$C_{28}H_{19}N_2O_2$
9c1 594ff, 740
$C_{28}H_{19}N_2O_2Rh^{3+}$
9a 326
$C_{28}H_{19}N_2O_3$
9c1 740
$C_{28}H_{19}N_4O_3S$
9c1 65
$C_{28}H_{19}O_2S_2Si$
17e 232, 233, *17g* 376
$C_{28}H_{19}O_2S_2Sn$
17a 484, 485, *17e* 232, 233, *17g* 377
$C_{28}H_{20}^-$
9d1 782, 827, *17f* 163
$C_{28}H_{20}^+$
9d2 13, *17c* 265, *17h* 98, 99
$C_{28}H_{20}Br$
9b 406
$C_{28}H_{20}Cl$
9b 406
$C_{28}H_{20}D_6O_4PS_2$
17c 286
$C_{28}H_{20}D_6PS_2$
17c 286
$C_{28}H_{20}N$
1 53, *9c1* 21
$C_{28}H_{20}NO$
9c1 587
$C_{28}H_{20}N_3OS$
9c1 65
$C_{28}H_{20}O_2^-$
17f 211
$C_{28}H_{20}O_2P$
1 75
$C_{28}H_{20}O_4S^+$
9d2 146
$C_{28}H_{20}S_2^-$
17f 215
$C_{28}H_{20}S_2^+$
9d2 144, *17e* 382, 383
$C_{28}H_{21}D_{12}$
9b 723
$C_{28}H_{21}GeO_2$
17e 195
$C_{28}H_{21}N^+$
9d2 91
$C_{28}H_{21}N_2ORh^{3+}$
9a 326
$C_{28}H_{21}N_2O_2Rh^{3+}$
9a 326
$C_{28}H_{21}N_3^+$
17h 175
$C_{28}H_{21}O_2Pb$
17a 496, *17g* 261
$C_{28}H_{21}O_2Si$
17e 195
$C_{28}H_{21}O_2Sn$
17a 485, *17g* 263
$C_{28}H_{21}O_4Sn$
17a 485
$C_{28}H_{22}^-$
9d1 826, *17f* 163, 164
$C_{28}H_{22}N_2$
9b 132
$C_{28}H_{22}N_2^+$
9d2 91
$C_{28}H_{22}O_2Tl$
17g 329
$C_{28}H_{23}FeO_2P^-$
17a 331
$C_{28}H_{23}N_2O$
9b 705
$C_{28}H_{23}N_2O_2$
9c1 741
$C_{28}H_{24}^-$
17f 164
$C_{28}H_{24}^+$
17h 73
$C_{28}H_{24}Be_2N_2^-$
17f 200
$C_{28}H_{24}Cl_2N_2O_4^+$
9d2 86
$C_{28}H_{24}CrO_2^-$
17a 266
$C_{28}H_{24}Mg_2N_2^-$
17f 201
$C_{28}H_{24}N_2^+$
9d2 46, 58
$C_{28}H_{24}N_2Zn_2^-$
17f 201
$C_{28}H_{24}N_8^{2-\cdot}$
17h 262
$C_{28}H_{24}O_4P$
9c2 262
$C_{28}H_{25}ClO_2PS$
17b 303
$C_{28}H_{25}NO_6S_2$
17d1 220
$C_{28}H_{25}N_2O_2$
9c1 723, *17d1* 220
$C_{28}H_{26}GeNO$
17c 603, 604
$C_{28}H_{26}NO$
9c1 104, 691

$C_{28}H_{26}NOSi$
17c 617
$C_{28}H_{26}NS$
17c 503
$C_{28}H_{26}N_2^+$
9d2 86
$C_{28}H_{26}N_2O_4^+$
9d2 87
$C_{28}H_{26}O_3PS$
17b 303
$C_{28}H_{27}$
9b 723, *17c* 369
$C_{28}H_{27}GeN_2$
17c 155
$C_{28}H_{27}GeO_2$
17a 445, *17e* 187
$C_{28}H_{27}O_2Si$
17e 187
$C_{28}H_{27}O_2Sn$
17a 485, *17e* 188, *17g* 82, 83
$C_{28}H_{27}O_3$
9b 440
$C_{28}H_{28}^-$
17f 165
$C_{28}H_{28}BrN_3O_2P$
17d2 248
$C_{28}H_{28}N_2^+$
9d2 105, 111
$C_{28}H_{28}N_2O_4^+$
9d2 111
$C_{28}H_{28}N_4^+$
9d2 67
$C_{28}H_{28}N_4Zn^+$
9a 336
$C_{28}H_{29}GeOSi$
9b 597
$C_{28}H_{29}OSi$
9b 595
$C_{28}H_{29}OSiSn$
9b 597, 599, *17a* 485, *17c* 227
$C_{28}H_{29}OSi_2$
9b 596
$C_{28}H_{30}CuO_2$
17g 117, 158
$C_{28}H_{30}NO_4$
17d2 359
$C_{28}H_{30}N_2^+$
17h 190
$C_{28}H_{30}N_3Si$
17c 583
$C_{28}H_{31}$
17c 299
$C_{28}H_{31}GeS$
17a 445, *17b* 486
$C_{28}H_{31}KN_4O_5^+$
17a 219
$C_{28}H_{31}N_4$
9c1 79
$C_{28}H_{31}N_4O_5$
17c 467
$C_{28}H_{31}OSi$
17b 486
$C_{28}H_{31}O_3$
1 65, *17e* 62
$C_{28}H_{31}O_{15}$
9c2 208
$C_{28}H_{31}SSi$
17b 486
$C_{28}H_{31}SSn$
17a 486, *17b* 487
$C_{28}H_{32}^-$
9d2 11
$C_{28}H_{32}^+$
17h 73, 74
$C_{28}H_{32}O_2Tl$
17g 336
$C_{28}H_{33}GeN_2O_4$
17c 447
$C_{28}H_{33}GeN_2O_5$
17d2 406
$C_{28}H_{33}NO_2Tl$
17g 337, 347
$C_{28}H_{33}NO_3Tl$
17g 90
$C_{28}H_{33}NO_4Tl$
17g 95
$C_{28}H_{33}N_2O_4Si$
17c 448
$C_{28}H_{33}N_2O_4Sn$
17c 448
$C_{28}H_{33}N_2O_5Si$
17d2 406
$C_{28}H_{33}N_4O_2$
17d1 175
$C_{28}H_{33}O$
1 62, *17e* 88
$C_{28}H_{34}BO_2$
17g 115
$C_{28}H_{34}BO_4$
17g 115
$C_{28}H_{34}ClO_2Sn$
17a 486, *17e* 67
$C_{28}H_{34}NO_8$
17d2 248
$C_{28}H_{34}N_2^+$
9d2 67
$C_{28}H_{34}N_2O_2^{2\cdot}$
17h 315
$C_{28}H_{34}O_3P$
17e 111
$C_{28}H_{35}AsO_2Rh$
17g 324
$C_{28}H_{35}BF_4OP$
1 69
$C_{28}H_{35}BrOP$
9c2 86
$C_{28}H_{35}GeN_2$
17a 445, 446, *17c* 396ff
$C_{28}H_{35}GeN_2O_2$
17d2 81
$C_{28}H_{35}NO_9Re$
17a 322, *17g* 132
$C_{28}H_{35}N_2Si$
17c 397
$C_{28}H_{35}N_2Sn$
17a 486, *17c* 397
$C_{28}H_{36}CuN_4O_6$
17a 367
$C_{28}H_{36}NO$
17d2 370
$C_{28}H_{36}N_2O_2Tl$
17g 198
$C_{28}H_{37}$
17c 189
$C_{28}H_{37}N_4O_6$
17c 447
$C_{28}H_{37}N_5O_{11}^+$
17h 219
$C_{28}H_{38}AsCl_2O_4$
17e 186
$C_{28}H_{38}N_6O_2^{2\cdot}$
17h 281

$C_{28}H_{39}NO_2^-$
9c1 34
$C_{28}H_{39}NO_3P$
17e 118
$C_{28}H_{39}NO_7$
17d2 232
$C_{28}H_{39}O_6$
17e 124
$C_{28}H_{40}^-$
9d1 821
$C_{28}H_{40}AlO_4$
17g 155
$C_{28}H_{40}AsO_4$
17e 57, 62, 147, *17g* 156
$C_{28}H_{40}BO_4$
17g 157
$C_{28}H_{40}BiO_4$
17e 57, 64
$C_{28}H_{40}ClN_4O$
17d2 331
$C_{28}H_{40}MnO_{10}P_2$
17a 307
$C_{28}H_{40}N$
9c1 29
$C_{28}H_{40}NO$
9c1 33, 34, 601
$C_{28}H_{40}NO_2$
9c1 604, *9c2* 128, *17e* 181
$C_{28}H_{40}NO_4$
17d2 248
$C_{28}H_{40}N_4O_2^{2\cdot}$
9d2 200, *17h* 274
$C_{28}H_{40}N_8O_4Pd^{2\cdot}$
17h 296
$C_{28}H_{40}O_2^-$
9d1 325, *17g* 345
$C_{28}H_{40}O_2S^-$
17g 346
$C_{28}H_{40}O_3S^-$
17g 347
$C_{28}H_{40}O_4P$
17e 58, 64
$C_{28}H_{40}O_4Sb$
17e 65
$C_{28}H_{40}O_4Sn$
17a 486
$C_{28}H_{41}N^+$
9d2 46
$C_{28}H_{41}N_2O_6P^{2\cdot}$
9d2 243
$C_{28}H_{41}O_2$
9c2 112, *17e* 122, 135, 144
$C_{28}H_{41}O_2S$
1 67, *17e* 142, 150
$C_{28}H_{41}O_2S_2$
1 68, *17e* 143
$C_{28}H_{41}O_3S$
17e 142
$C_{28}H_{42}^-$
9d1 811
$C_{28}H_{42}N$
9c1 13
$C_{28}H_{42}NO$
9c1 567, 570, 571, 578, 621, *17d1* 372
$C_{28}H_{42}NO_2$
9c2 149
$C_{28}H_{42}N_2O_7ReSi_2$
17a 322, *17g* 133
$C_{28}H_{42}O_2^+$
9d1 359
$C_{28}H_{42}O_2S^+$
17g 387
$C_{28}H_{43}N^+$
9d2 46
$C_{28}H_{43}NO_3P$
17d1 346
$C_{28}H_{43}OSi$
17c 289
$C_{28}H_{43}O_2$
9b 498
$C_{28}H_{44}NOSn$
17d1 314
$C_{28}H_{44}NO_2^+$
17h 204
$C_{28}H_{44}N_4O_2^{2\cdot}$
9d2 198, 269
$C_{28}H_{44}O_2P$
9b 532, *17c* 163
$C_{28}H_{45}NO_4$
17d2 85
$C_{28}H_{45}N_2O_6$
17d2 88
$C_{28}H_{46}AsClO_2Rh$
17g 201
$C_{28}H_{46}AsO_3Pd$
17a 349, *17g* 171
$C_{28}H_{46}Cl_2N_{12}O_2^{2\cdot}$
9d2 278
$C_{28}H_{46}NO_3$
17d2 197
$C_{28}H_{46}N_2O_2^{2\cdot}$
17h 315
$C_{28}H_{46}N_2O_4^{2\cdot}$
9d2 219
$C_{28}H_{46}N_4O_2^{2\cdot}$
9d2 270, 275
$C_{28}H_{46}O_2^-$
17b 237
$C_{28}H_{47}AsO_2Rh$
17g 175, 176
$C_{28}H_{47}O_2$
9c2 183, *17e* 221ff, 222ff, 224
$C_{28}H_{48}N_2NaO_8$
17g 239
$C_{28}H_{48}N_8O_4Pd_2S_4^{4\cdot}$
17h 326
$C_{28}H_{50}MnO_{10}P_2$
17a 307, *17g* 167
$C_{28}H_{50}NO$
17d1 339
$C_{28}H_{50}N_2O_6^{2\cdot}$
9d2 274
$C_{28}H_{50}N_2O_6S^{2\cdot}$
9d2 277
$C_{28}H_{50}N_6O_8^{2\cdot}$
17h 316
$C_{28}H_{50}O_{10}P_2Re$
17g 174
$C_{28}H_{51}ClNO_3P_2Pt$
17a 353
$C_{28}H_{52}N_4O_4^{2\cdot}$
9d2 275
$C_{28}H_{54}NO$
17d2 186
$C_{28}H_{55}N_3O_5P^{2\cdot}$
9d2 244
$C_{28}H_{76}P_4Si_8^+$
17e 311
$C_{29}H_{11}Cl_{14}N_2O_3$
9b 742
$C_{29}H_{16}Br_2Cl_2O^-$
9b 445
$C_{29}H_{16}Br_2I_2O^-$
9b 447

$C_{29}H_{16}Br_4O^-$
9b 446
$C_{29}H_{16}Cl_4O^-$
9b 443
$C_{29}H_{17}Br_3O^-$
9b 445
$C_{29}H_{17}Cl_3O^-$
9b 443
$C_{29}H_{18}Br_2O^-$
9b 444
$C_{29}H_{18}Cl_2O^-$
9b 442
$C_{29}H_{18}I_2O^-$
9b 446
$C_{29}H_{19}$
9b 437, *17c* 63
$C_{29}H_{19}BrO^-$
9b 443
$C_{29}H_{19}ClO^-$
9b 441, 442
$C_{29}H_{20}^+$
17h 105
$C_{29}H_{20}NO$
17c 621
$C_{29}H_{20}NO_6$
17d2 359
$C_{29}H_{20}O^-$
9b 441
$C_{29}H_{21}GeOS_3$
17a 446, *17c* 339, 340
$C_{29}H_{21}O$
9b 440
$C_{29}H_{21}OS_3Si$
17c 339, 340
$C_{29}H_{22}^-$
9d1 829
$C_{29}H_{22}N_2O^-$
17c 239
$C_{29}H_{22}P$
9b 535
$C_{29}H_{23}$
9b 436
$C_{29}H_{23}NO_2P_2^-$
17g 44
$C_{29}H_{24}D_9N_4O$
17c 465
$C_{29}H_{24}GeN$
17c 155
$C_{29}H_{24}NO_2$
9c1 750
$C_{29}H_{24}NSi$
17c 152
$C_{29}H_{24}Si^-$
9d1 896
$C_{29}H_{25}NO_2Tl$
17g 356
$C_{29}H_{25}NO_6^+$
17h 164
$C_{29}H_{25}O_2$
9b 394
$C_{29}H_{26}DO_2$
17e 149
$C_{29}H_{26}F_3O_2Si$
17g 141
$C_{29}H_{26}KO_2$
9d1 140
$C_{29}H_{26}LiO_2$
9d1 140
$C_{29}H_{27}O_2$
17e 149
$C_{29}H_{28}NO$
9c2 153ff, *17d1* 298, *17e* 148
$C_{29}H_{28}N_2O_6^{2-\cdot}$
9d2 286
$C_{29}H_{28}O_3PS$
17b 303
$C_{29}H_{29}ClO_4PSi$
17c 219, 224
$C_{29}H_{29}D_8N_2$
9b 437
$C_{29}H_{29}NO_2PS$
17b 302
$C_{29}H_{29}N_2O_5$
17d2 122
$C_{29}H_{29}OSi$
9b 595
$C_{29}H_{30}GeNO$
17a 446, *17d2* 69
$C_{29}H_{30}MnNO_5P$
17a 307, 308
$C_{29}H_{30}NOPPt$
17d2 71
$C_{29}H_{30}NOSi$
9c1 323, *17d2* 68
$C_{29}H_{30}NO_7$
9c1 212, 347
$C_{29}H_{30}N_2^+$
9d2 58
$C_{29}H_{30}N_2O_2P$
17e 114
$C_{29}H_{30}O_4PSi$
17c 218, 224
$C_{29}H_{31}D_{13}NO_3$
17d1 298
$C_{29}H_{31}N_2O$
9c1 760
$C_{29}H_{31}O_2$
9b 752
$C_{29}H_{32}NO_5$
17d2 122
$C_{29}H_{33}NO_2PS$
17e 113
$C_{29}H_{33}NO_3P$
17e 119
$C_{29}H_{33}N_2O$
17d1 283, 298
$C_{29}H_{33}O_2$
17e 156
$C_{29}H_{34}D_{12}NO_3$
17d1 299
$C_{29}H_{34}NO$
9c1 761
$C_{29}H_{34}O_2Tl$
17g 336
$C_{29}H_{35}D_{13}NO_3$
17d1 299
$C_{29}H_{35}NO_3Tl$
17g 110
$C_{29}H_{35}N_4O$
9c1 85
$C_{29}H_{35}O_2$
17e 80
$C_{29}H_{35}O_2Sn$
17a 487, *17g* 181
$C_{29}H_{37}BrOP$
9c2 86
$C_{29}H_{37}D_4O_2$
17e 161
$C_{29}H_{37}D_{13}NO_3$
17d1 299
$C_{29}H_{38}NO_{10}$
17d2 124
$C_{29}H_{39}D_2O_2$
17e 160
$C_{29}H_{39}N_3NaO_4S$
17d1 364

$C_{29}H_{40}DO_2$
9c2 96, *17e* 159ff
$C_{29}H_{40}NO_5$
17d1 335, 339
$C_{29}H_{40}N_3O_6$
17d2 124
$C_{29}H_{41}D_2O_2$
17e 92
$C_{29}H_{41}N_2O_2$
17d2 100
$C_{29}H_{41}O_2^{2-}$
9b 638
$C_{29}H_{41}O_2$
1 64, *9c2* 94ff, *17e* 144, 157ff, 159
$C_{29}H_{41}O_4$
17e 60, 61
$C_{29}H_{41}O_5$
9c2 138
$C_{29}H_{41}O_8Zr$
17g 192
$C_{29}H_{42}N_3$
17c 444
$C_{29}H_{42}O^-$
17c 272
$C_{29}H_{42}O_4^-$
17g 240
$C_{29}H_{42}O_4^+$
17e 61
$C_{29}H_{43}$
9b 637, 638
$C_{29}H_{43}NO_5Re$
17a 323, *17g* 131
$C_{29}H_{43}N_2O$
9c1 832
$C_{29}H_{43}O_2$
9b 498ff, *9c2* 81, *17e* 86, 91ff
$C_{29}H_{44}NO_2$
9c1 614, *17d2* 101
$C_{29}H_{44}NO_3$
17d2 30ff
$C_{29}H_{44}NO_4$
17d2 55
$C_{29}H_{44}N_4O_3^{2\cdot}$
9d2 267$C_{29}H_{46}N_4O_3^{2\cdot}$
9d2 267
$C_{29}H_{47}MnO_8P$
17g 167
$C_{29}H_{47}N_2O_3^{2\cdot}$
9d2 225
$C_{29}H_{47}N_2O_6$
17d2 88
$C_{29}H_{47}O_2P^+$
17c 160
$C_{29}H_{47}O_5PRe$
17a 323, *17g* 134
$C_{29}H_{48}NO_2$
17d2 233
$C_{29}H_{48}N_2O_2^{2\cdot}$
9d2 219
$C_{29}H_{48}N_2O_4^{2\cdot}$
9d2 220
$C_{29}H_{49}AsO_3Rh$
17g 201
$C_{29}H_{49}O$
9b 575
$C_{29}H_{49}O_2$
9c2 184, *17e* 218, 219ff
$C_{29}H_{50}AsO_3Pd$
17a 349, *17g* 172
$C_{29}H_{50}AsO_3Pt$
17a 354, *17g* 173
$C_{29}H_{50}NO$
17d2 185
$C_{29}H_{50}NO_2$
9c1 612
$C_{29}H_{50}O_3^-$
17g 239ff
$C_{29}H_{51}ClNO_5P_2Pt$
9c1 289, *17a* 354
$C_{29}H_{51}N_2O_8P_2Pt$
17a 354
$C_{29}H_{52}NO$
9c1 618
$C_{29}H_{52}NO_2$
9c1 318
$C_{29}H_{52}NO_3$
9c1 318
$C_{29}H_{56}PSi$
9b 533
$C_{29}H_{59}O_3^-$
9d1 229
$C_{29}H_{61}MnNO_4P_2$
17a 308
$C_{30}H_{12}Cl_{14}NO_4$
9b 743
$C_{30}H_{16}CrO_6^-$
17a 266
$C_{30}H_{16}CrO_6$
17g 315
$C_{30}H_{16}MoO_6^-$
17a 273
$C_{30}H_{16}MoO_6$
17g 322
$C_{30}H_{16}O_2^-$
17e 202, *17g* 381
$C_{30}H_{16}O_6W^-$
17a 280
$C_{30}H_{16}O_6W$
17g 328
$C_{30}H_{16}S_4^+$
17h 118
$C_{30}H_{18}^{3-}$
9d1 797
$C_{30}H_{18}^-$
9d1 781, 796
$C_{30}H_{19}MnO_6P$
17a 308
$C_{30}H_{19}N_4O_6$
9c1 15
$C_{30}H_{20}^-$
9d1 828
$C_{30}H_{20}N_5O_6$
1 47, *9c1* 53
$C_{30}H_{20}S_2^-$
17f 215
$C_{30}H_{21}B^-$
9d1 704
$C_{30}H_{21}GeO_2$
17e 210, *17g* 341
$C_{30}H_{21}NO_2^-$
9d1 565
$C_{30}H_{21}N_7O_6^{2\cdot}$
9d2 172
$C_{30}H_{21}O_2Si$
17e 210
$C_{30}H_{21}O_2Sn$
17g 343
$C_{30}H_{22}^{3-}$
17f 166
$C_{30}H_{22}^{2-}$
17f 166
$C_{30}H_{22}^-$
9d1 830, *17f* 165

$C_{30}H_{22}^{+}$
17h 105
$C_{30}H_{22}Cl_2NO_6$
17d2 367
$C_{30}H_{22}FO_2P^{+}$
17c 159
$C_{30}H_{22}F_9O_2Sn$
17g 79
$C_{30}H_{22}NO_4P^{-}$
9d1 517
$C_{30}H_{22}NO_6$
17d2 359
$C_{30}H_{23}O_3Tl^{-}$
9d1 243, 249
$C_{30}H_{23}O_3Tl_2$
17a 424, 425
$C_{30}H_{24}NO_2$
17d2 398
$C_{30}H_{24}N_2O^{-}$
17c 239
$C_{30}H_{24}N_2O_2^{+}$
17h 171
$C_{30}H_{24}O_8S_4^{+}$
17h 118
$C_{30}H_{24}Si^{-}$
1 120
$C_{30}H_{25}N_2O_4$
17d1 220
$C_{30}H_{25}N_3O_2PS$
17c 450
$C_{30}H_{25}N_3O_3PSi$
17c 583
$C_{30}H_{26}^{-}$
17f 166
$C_{30}H_{26}NO$
9c1 761
$C_{30}H_{26}N_4OP$
17c 451
$C_{30}H_{26}N_4O_2P$
17c 451
$C_{30}H_{26}N_4O_4S_2$
17h 170
$C_{30}H_{26}O_2Tl$
17g 330, 331
$C_{30}H_{26}O_6Tl$
17g 332
$C_{30}H_{26}Si^{-}$
9d1 897
$C_{30}H_{27}NO_3$
9c1 750
$C_{30}H_{27}N_2O$
9c1 289
$C_{30}H_{28}N_4O_4^{+}$
9d2 87
$C_{30}H_{28}O_4^{+}$
9d2 137
$C_{30}H_{29}HgO_2$
17g 141
$C_{30}H_{29}N_2O_2$
17d1 221
$C_{30}H_{30}MnN_2O_4$
17a 308
$C_{30}H_{30}NO$
9c1 289, *17d2* 371
$C_{30}H_{30}NOS$
17d2 64
$C_{30}H_{30}NO_2S$
17d1 162
$C_{30}H_{30}O_3PS$
17b 303
$C_{30}H_{31}ClNOSn$
17d1 314
$C_{30}H_{32}NO_7$
9c1 347
$C_{30}H_{32}O_5PSi$
17c 219, 224
$C_{30}H_{34}BaO_2^{+}$
17a 237
$C_{30}H_{34}CaO_2^{+}$
17a 232
$C_{30}H_{34}Ge_2^{-}$
17f 45
$C_{30}H_{34}MgN_6^{-}$
17a 229
$C_{30}H_{34}MgO_2^{+}$
17a 229
$C_{30}H_{34}N_3O_3$
17d2 37
$C_{30}H_{34}N_6Zn^{-}$
17a 384
$C_{30}H_{34}O_2^{-}$
17g 62, 63ff
$C_{30}H_{34}O_2Sr^{+}$
17a 235
$C_{30}H_{34}Si_2^{-}$
17f 22
$C_{30}H_{35}N_2O$
17d1 283
$(C_{30}H_{36}AlTi_3)_n$
17a 245
$C_{30}H_{36}NO$
9c1 761, *17d2* 350
$C_{30}H_{36}N_2^{+}$
9d2 67
$C_{30}H_{36}N_3O_3^{3\cdot}$
9d2 293
$C_{30}H_{36}N_4O_4^{2\cdot}$
9d2 210
$C_{30}H_{36}N_4O_4Pd^{2\cdot}$
17h 296
$C_{30}H_{36}N_6O_{14}S_2^{-}$
17g 276
$C_{30}H_{36}NaO_8$
17g 238
$C_{30}H_{36}O_2Tl$
17g 338
$C_{30}H_{37}N^{+}$
17h 164
$C_{30}H_{37}N_2O_5$
17d2 37
$C_{30}H_{38}N_5O_7$
9c1 889
$C_{30}H_{38}O_2P$
1 75
$C_{30}H_{39}ClNO_3P$
17e 119
$C_{30}H_{39}Ge$
17f 42
$C_{30}H_{39}NO_2P$
17c 490
$C_{30}H_{39}P^{+}$
17e 311
$C_{30}H_{39}Sn$
9c2 315, *17f* 49
$C_{30}H_{40}Cl_2NO_2$
17c 118
$C_{30}H_{40}Cl_2NS_2$
17c 127
$C_{30}H_{40}FeN_2O_6^{2\cdot}$
17h 315
$C_{30}H_{40}MnN_2O_4$
17g 125
$C_{30}H_{40}N$
9c1 30

$C_{30}H_{40}N_3O$
9c2 124
$C_{30}H_{40}N_8O_6^{2\cdot}$
9d2 204
$C_{30}H_{40}N_{10} \xrightarrow{CT} C_{18}Cl_{12}O_6$
9d1 362
$C_{30}H_{40}O_2^{-}$
9d1 326
$C_{30}H_{41}Br_2NS_2^{+}$
17c 127
$C_{30}H_{41}Cl_2NO_2^{+}$
17c 118
$C_{30}H_{41}Cl_2NS_2^{+}$
17c 127
$C_{30}H_{42}^{-}$
9d1 730
$C_{30}H_{42}^{+}$
17h 102
$C_{30}H_{42}NO_2$
17c 118
$C_{30}H_{42}NO_7$
9c1 405, *17d2* 249
$C_{30}H_{42}NO_8$
17d2 345
$C_{30}H_{42}NS_2$
17c 126
$C_{30}H_{42}N_2^{+}$
9d2 112
$C_{30}H_{42}N_2O_4^{-}$
9d1 420
$C_{30}H_{42}N_2O_6^{2\cdot}$
9d2 260
$C_{30}H_{42}O_2S^{+}$
17c 117
$C_{30}H_{42}S_3^{+}$
17c 126
$C_{30}H_{43}Br_2O_2$
9c2 87
$C_{30}H_{43}NO_2^{-}$
9d1 419
$C_{30}H_{43}NO_2^{+}$
17c 118
$C_{30}H_{43}NS_2^{+}$
17c 126
$C_{30}H_{43}N_2O$
17d1 339
$C_{30}H_{43}N_8O_{15}P_2$
9c1 508
$C_{30}H_{43}O_2$
17e 161
$C_{30}H_{44}NO_4$
9c2 125
$C_{30}H_{44}N_2O_4Si^{2\cdot}$
9d2 236
$C_{30}H_{44}N_2O_9P_2^{2\cdot}$
9d2 250
$C_{30}H_{44}N_4O_2^{2\cdot}$
9d2 276
$C_{30}H_{45}$
9b 640
$C_{30}H_{45}N_2O$
9c1 833
$C_{30}H_{45}O_2$
17e 86, 102
$C_{30}H_{45}O_2S$
9c2 87, *17e* 100
$C_{30}H_{45}O_4$
9c2 175
$C_{30}H_{45}O_4S$
17e 100
$C_{30}H_{46}Mo_2N_2O_{14}P_2^{-}$
17a 273
$C_{30}H_{46}NO_5$
9c1 367
$C_{30}H_{46}NO_7$
9c1 368
$C_{30}H_{46}N_2O_6^{2\cdot}$
9d2 256, 268, 272, 274
$C_{30}H_{46}N_2O_8P_2W_2^{-}$
17a 280
$C_{30}H_{46}N_2O_{14}P_2W_2^{-}$
17a 280
$C_{30}H_{47}N_2O_3$
9c1 289, 757
$C_{30}H_{48}Cl_2N_{12}O_2^{2\cdot}$
9d2 278
$C_{30}H_{48}NO_3$
17d2 31
$C_{30}H_{48}NO_4$
17d2 55
$C_{30}H_{48}NO_5$
17d2 59
$C_{30}H_{48}N_3O_5$
9c1 348, *17d2* 219
$C_{30}H_{48}O_2^{-}$
17g 28
$C_{30}H_{50}B_{20}ClFO_2Sn^{-}$
17a 487, *17g* 191
$C_{30}H_{50}B_{20}ClO_2Sn$
17a 487, *17e* 68, *17g* 184
$C_{30}H_{50}B_{20}Cl_2O_2Sn^{-}$
17a 487, *17g* 188
$C_{30}H_{50}N_4O_2^{2\cdot}$
9d2 276
$C_{30}H_{51}AsO_4Rh$
17g 203
$C_{30}H_{51}O_2$
17b 256
$C_{30}H_{52}NO$
17d1 283, 299
$C_{30}H_{53}NO_5P$
17c 491
$C_{30}H_{54}Cl_4O_2Sn_2$
17g 244
$C_{30}H_{54}N_2O_5S^{2\cdot}$
9d2 279
$C_{30}H_{54}N_6O_8^{2\cdot}$
17h 317
$C_{30}H_{56}Cl_2O_2Sn_2$
17g 223, 229
$C_{30}H_{57}B$
9d1 701
$C_{30}H_{57}MnNO_4P_2S$
17a 308, 309
$C_{30}H_{57}N_6O_3^{3\cdot}$
17h 324
$C_{30}H_{58}O_2Sn_2$
17g 214
$C_{30}H_{60}CrSi_6^{-}$
17a 266
$C_{30}H_{60}CrSi_6^{+}$
17a 266
$C_{30}H_{61}MnNO_5P_2$
17d2 412
$C_{30}H_{63}Al^{-}$
17f 64
$C_{30}H_{63}MnNO_4P_2$
17a 309
$C_{30}H_{66}Si_6^{+}$
17h 70
$C_{30}H_{68}Si_4^{+}$
17f 37, *17h* 24
$C_{30}H_{70}Si_5^{-}$
17f 31

$C_{31}Cl_{23}$
9b 747
$C_{31}H_{15}F_{10}OSi$
17c 290
$C_{31}H_{15}F_{10}OSn$
17a 487, *17c* 294
$C_{31}H_{16}O_3^-$
17e 201, *17g* 380
$C_{31}H_{19}AsMnO_5S_2$
17a 309, *17g* 372
$C_{31}H_{19}MnNO_5S_2$
17a 309, *17g* 372
$C_{31}H_{19}MnO_5PS_2$
17a 309, *17g* 373
$C_{31}H_{19}MnO_5S_2Sb$
17a 310, *17g* 374
$C_{31}H_{21}AsO_5Re$
17a 323, *17g* 262
$C_{31}H_{21}GeOS_4$
17a 446, 447, *17c* 341, 342, 343
$C_{31}H_{21}OS_4Si$
17c 341, 342, 343
$C_{31}H_{21}O_5PRe$
17a 323, *17g* 263
$C_{31}H_{21}O_8PRe$
17a 323, *17g* 263
$C_{31}H_{22}N_5O_6$
1 51
$C_{31}H_{23}$
9b 747
$C_{31}H_{23}GeO$
17a 447, *17c* 316
$C_{31}H_{23}GeOS$
17a 447, *17c* 320, 322
$C_{31}H_{23}GeOSe$
17a 447, *17c* 325
$C_{31}H_{23}GeO_2$
17c 320
$C_{31}H_{23}GeSSe$
17a 448, *17c* 326
$C_{31}H_{23}GeS_2$
17a 448, *17c* 322
$C_{31}H_{23}N_2$
9c1 59
$C_{31}H_{23}OSSi$
17c 319, 321
$C_{31}H_{23}OSeSi$
17c 325
$C_{31}H_{23}OSeSn$
17a 488
$C_{31}H_{23}OSi$
9b 698, *17c* 315
$C_{31}H_{23}O_2$
17c 360, *17e* 151
$C_{31}H_{23}O_2Si$
17c 319
$C_{31}H_{23}O_4Tl^-$
9d1 127
$C_{31}H_{23}O_4Tl_2$
17a 425
$C_{31}H_{23}O_5Tl^-$
9d1 141
$C_{31}H_{23}O_5Tl_2$
17a 426
$C_{31}H_{23}SSeSi$
17c 326
$C_{31}H_{23}S_2Si$
17c 322
$C_{31}H_{23}Si$
9b 688
$C_{31}H_{24}^{3-}$
17f 167
$C_{31}H_{24}^{2-}$
17f 167
$C_{31}H_{24}^-$
17f 167
$C_{31}H_{24}^+$
17h 105
$C_{31}H_{24}MnNO_5P$
9c1 1038
$C_{31}H_{24}NO_4Tl_2^-$
9d1 248
$C_{31}H_{24}NO_4Tl_2$
17a 426
$C_{31}H_{24}O^-$
9b 441
$C_{31}H_{25}GeO$
17a 448, *17c* 292
$C_{31}H_{25}GeS$
17a 448, *17c* 292
$C_{31}H_{25}N_2$
1 50
$C_{31}H_{25}N_2Si$
9c1 96
$C_{31}H_{25}OSi$
9b 670, *17c* 290
$C_{31}H_{25}OSn$
9b 673
$C_{31}H_{25}P^+$
9b 667
$C_{31}H_{25}SSi$
17c 291
$C_{31}H_{25}SSn$
17a 488, *17c* 295
$C_{31}H_{25}Si$
9b 668
$C_{31}H_{25}Sn$
9b 671, *17a* 488
$C_{31}H_{26}NO_7$
9c1 105, 712
$C_{31}H_{27}NO_4^-$
9d1 338
$C_{31}H_{28}ClCrNO_4P$
17a 267
$C_{31}H_{28}ClMoNO_4P$
17a 273
$C_{31}H_{28}ClNO_4PW$
17a 280
$C_{31}H_{28}GeN_3$
17b 308
$C_{31}H_{28}N_2O_6^{2\cdot}$
17h 328
$C_{31}H_{28}N_3Si$
17b 308
$C_{31}H_{29}N_2O_4$
9c1 236
$C_{31}H_{29}O_4$
9b 394
$C_{31}H_{29}O_5PRe$
17a 323
$C_{31}H_{29}O_8PRe$
17a 324
$C_{31}H_{32}ClO_2Sn$
17e 58
$C_{31}H_{32}NO_5$
17e 107
$C_{31}H_{32}O_2^-$
17c 116
$C_{31}H_{32}O_2Tl$
17g 150
$C_{31}H_{33}ClNOSn$
17d1 300
$C_{31}H_{33}MnO_5P$
17a 310

$C_{31}H_{33}O_2$
17c 116
$C_{31}H_{35}GeOSi$
9b 597
$C_{31}H_{35}N_2O_7$
9c1 241
$C_{31}H_{35}N_5Zn^-$
17a 384
$C_{31}H_{35}OP^+$
17c 110
$C_{31}H_{35}OSiSn$
9b 598, 599
$C_{31}H_{35}OSi_2$
9b 596
$C_{31}H_{35}O_2Pb$
17a 497, *17g* 170
$C_{31}H_{36}NO$
17d1 206
$C_{31}H_{36}N_2O_7$
9c1 308
$C_{31}H_{37}NO_3P$
17e 120
$C_{31}H_{37}N_2O$
17d1 306
$C_{31}H_{39}$
9b 724, *17c* 351
$C_{31}H_{40}O_3^-$
9d1 326
$C_{31}H_{40}O_4P$
17b 397, *17e* 117
$C_{31}H_{41}Cl_3NO_4$
17d1 373
$C_{31}H_{41}D_{13}NO_3$
17d1 300
$C_{31}H_{42}N_2^+$
9d2 112
$C_{31}H_{42}N_4O_3^{2\cdot}$
9d2 202
$C_{31}H_{44}FO_2$
9b 499
$C_{31}H_{45}O_2$
9b 499, *17e* 161, 226
$C_{31}H_{45}O_3Pd$
17a 349, *17g* 171
$C_{31}H_{46}NO_4$
9c2 125
$C_{31}H_{46}NO_7$
17d2 332
$C_{31}H_{46}O_2^-$
9d1 287, *17g* 272
$C_{31}H_{47}O_2$
9c2 88, *17e* 85, 102, 107, 226
$C_{31}H_{50}NO_3$
17d2 113
$C_{31}H_{50}NO_4$
17d2 115
$C_{31}H_{50}NO_5$
17d2 115
$C_{31}H_{51}N_2OSi$
17c 447
$C_{31}H_{54}NO$
17d2 186
$C_{31}H_{54}NO_2$
9c1 438, 439, 440ff, *17d2* 279ff
$C_{31}H_{54}NO_3$
9c1 441
$C_{31}H_{57}MnNO_5P_2$
17d1 389
$C_{31}H_{57}MnNO_5P_2S$
17d1 389
$C_{31}H_{57}N_3O_8P^{3\cdot}$
9d2 296
$C_{31}H_{63}MnNO_5P_2$
17d1 190
$C_{32}HCl_{24}$
9b 731
$C_{32}H_8D_{15}$
17c 299
$C_{32}H_{14}Cl_{14}NO_4$
9b 744
$C_{32}H_{16}AlClN_8^{3-}$
9a 297
$C_{32}H_{16}AlClN_8^-$
9a 297
$C_{32}H_{16}Br_2CoN_8$
9a 325
$C_{32}H_{16}Cl_2CoN_8$
9a 325
$C_{32}H_{16}Cl_2FeN_8$
9a 318
$C_{32}H_{16}Cl_{14}NO_4$
9b 743
$C_{32}H_{16}Cl_{14}N_3O_4$
9b 742
$C_{32}H_{16}CoN_8^{4-}$
9a 325
$C_{32}H_{16}CoN_8^{2-}$
9a 325
$C_{32}H_{16}CuN_8^{2-}$
9a 328
$C_{32}H_{16}MgN_8^-$
9a 292
$C_{32}H_{16}NS$
9c1 37
$C_{32}H_{16}N_8Ni^{3-}$
9a 327
$C_{32}H_{16}N_8Ni^-$
9a 327
$C_{32}H_{16}N_8Zn^{3-}$
9a 338
$C_{32}H_{16}N_8Zn^-$
9a 338
$C_{32}H_{18}O_2^-$
9d1 289
$C_{32}H_{19}O_2$
9c2 196
$C_{32}H_{20}^-$
9d1 843, 844, *17f* 168
$C_{32}H_{20}^+$
9d2 15
$C_{32}H_{20}Cl_6NO$
17d1 250
$C_{32}H_{20}F_{15}O_2Sn$
17g 139
$C_{32}H_{20}O_2^+$
9d1 360
$C_{32}H_{20}S_4^+$
17h 114
$C_{32}H_{21}Cl_2N_2O$
9c1 580
$C_{32}H_{23}$
17c 299
$C_{32}H_{23}GeOS_4$
17a 448, *17c* 342
$C_{32}H_{23}GeO_2$
17e 202, *17g* 317
$C_{32}H_{23}NO_2^-$
9d1 338
$C_{32}H_{23}NO_2$
17d2 360
$C_{32}H_{23}N_2O$
9c1 580

$C_{32}H_{23}OS_4Si$
17c 342

$C_{32}H_{23}O_2Pb$
17g 323

$C_{32}H_{23}O_2Si$
17e 203, *17g* 285, 325

$C_{32}H_{23}O_2Sn$
17a 488, *17e* 203, *17g* 326

$C_{32}H_{24}^{-}$
9b 729, *9d1* 841, *17f* 168, 169, 170

$C_{32}H_{24}^{+}$
9d2 15

$C_{32}H_{24}^{2\cdot}$
9d2 151

$C_{32}H_{24}Cl_4NiO_2P_2^{+}$
17g 209

$C_{32}H_{24}N_2^{+}$
9d2 113

$C_{32}H_{24}N_3O_2$
17d2 359

$C_{32}H_{24}N_5O_7$
1 51

$C_{32}H_{24}N_8^{2\cdot}$
9d2 185

$C_{32}H_{24}O_2Tl$
17g 331

$C_{32}H_{25}$
9b 724, *17c* 265

$C_{32}H_{25}Ge$
17c 315

$C_{32}H_{25}GeO$
17c 317

$C_{32}H_{25}GeO_2$
17a 449

$C_{32}H_{25}N_8$
9c1 71

$C_{32}H_{25}O_2Sn$
17a 488

$C_{32}H_{25}O_4Tl_2^{-}$
9d1 124, 244

$C_{32}H_{25}O_4Tl_2$
17a 426ff

$C_{32}H_{25}O_5Tl_2^{-}$
9d1 245

$C_{32}H_{25}O_5Tl_2$
17a 427

$C_{32}H_{25}Si$
17c 315

$C_{32}H_{26}^{3-}$
17f 171

$C_{32}H_{26}^{2-}$
17f 170

$C_{32}H_{26}^{-}$
17f 170

$C_{32}H_{26}^{+}$
17h 105

$C_{32}H_{26}N_4^{+}$
17h 153

$C_{32}H_{26}O_2PS$
17b 303

$C_{32}H_{27}Ge$
17a 449, *17c* 185, 291

$C_{32}H_{27}N_2O$
1 50

$C_{32}H_{27}OSi$
9b 670

$C_{32}H_{27}O_2Si$
9b 670

$C_{32}H_{27}Si$
17c 184, 287

$C_{32}H_{27}Sn$
17a 489, *17c* 292

$C_{32}H_{28}^{-}$
9d2 13

$C_{32}H_{28}ClMoNO_5P$
17d1 323

$C_{32}H_{28}MnNO_5P$
9c1 1041, *17a* 310

$C_{32}H_{28}MnNO_8P$
9c1 1041, *17a* 310

$C_{32}H_{28}N$
9c1 21

$C_{32}H_{28}NO_4$
9c1 21

$C_{32}H_{28}N_3O_7$
17d2 219

$C_{32}H_{28}NiO_2P_2^{+}$
17g 73

$C_{32}H_{28}O_4S_2^{+}$
9d2 144, *17e* 384

$C_{32}H_{28}S_2^{+}$
9d2 144, *17e* 383

$C_{32}H_{29}N_3^{+}$
17h 175

$C_{32}H_{30}GeP$
9b 534

$C_{32}H_{30}O_2Tl$
17g 333

$C_{32}H_{30}PSi$
9b 533

$C_{32}H_{31}N_2O_9$
9c1 889

$C_{32}H_{32}F_3NOP$
9c2 150

$C_{32}H_{32}F_3O_2Sn$
17g 139

$C_{32}H_{32}N_4^{3+}$
17h 190

$C_{32}H_{33}O_2S$
9c2 164

$C_{32}H_{34}ClO_2Sn$
17e 185

$C_{32}H_{35}AsNO$
9c2 151

$C_{32}H_{35}ClO_2PPd$
17g 128

$C_{32}H_{35}Cl_3O_2PSn$
17a 489

$C_{32}H_{35}Cl_3O_3PSn$
17a 489, *17g* 189

$C_{32}H_{35}Cl_3O_6PSn$
17a 489, *17g* 190

$C_{32}H_{35}CuO_2P$
17g 118, 159

$C_{32}H_{35}GeHgO_2$
17g 122, 164

$C_{32}H_{35}GeO_2$
17e 72

$C_{32}H_{35}NOP$
9c2 130, 149, *17e* 149

$C_{32}H_{35}OSn$
9c2 145

$C_{32}H_{35}O_2Pb$
17a 497, *17e* 140, *17g* 170

$C_{32}H_{35}O_2Si$
17e 72, 141

$C_{32}H_{35}O_2Sn$
9d1 137, *17a* 489, 490, *17e* 58, 66, 142, *17g* 138, 142, 180

$C_{32}H_{36}BrN_4Zn$
9a 337

$C_{32}H_{36}ClN_4O_4Zn^{+}$
9a 336

$C_{32}H_{36}N_2O_8^-$
9d2 113

$C_{32}H_{36}N_4P$
9c2 262

$C_{32}H_{36}N_4Zn^+$
9a 336, *17a* 384

$C_{32}H_{36}P_2^+$
17e 311

$C_{32}H_{37}KNO_2$
17a 215

$C_{32}H_{37}N_2P^+$
9d2 59

$C_{32}H_{38}NO_4$
17d1 340

$C_{32}H_{38}NO_5$
17d2 122

$C_{32}H_{38}NO_5Ti$
17a 246

$C_{32}H_{39}NO_3P$
17e 120

$C_{32}H_{40}N_2O_4^{2\cdot}$
17h 283

$C_{32}H_{40}O_4^-$
9d1 326, *17e* 126, *17g* 381

$C_{32}H_{41}$
17c 298

$C_{32}H_{41}ClNO_3$
9c2 83

$C_{32}H_{41}N_4$
9c1 84

$C_{32}H_{41}O_4$
17e 126

$C_{32}H_{42}N_2O_6^{2\cdot}$
9d2 262, 264

$C_{32}H_{42}N_4O_6^{2\cdot}$
17h 313

$C_{32}H_{42}O_2Tl$
17g 222, 229

$C_{32}H_{42}O_4^-$
17e 103

$C_{32}H_{43}CoO_6S$
17a 339

$C_{32}H_{43}OSn$
17a 490, *17b* 147

$C_{32}H_{44}CuN_4O_6$
17a 367

$C_{32}H_{44}NO_3$
17d1 377

$C_{32}H_{44}N_2O_2Tl$
17g 199

$C_{32}H_{44}O_2^+$
17c 117

$C_{32}H_{44}S_2^+$
17c 126

$C_{32}H_{45}O_2$
17e 164ff

$C_{32}H_{45}O_7$
17e 80

$C_{32}H_{46}MnNO_5P$
17a 310

$C_{32}H_{46}N_4O_2^{2\cdot}$
9d2 206

$C_{32}H_{46}N_4O_3^{2\cdot}$
9d2 206

$C_{32}H_{47}O_2$
17e 162

$C_{32}H_{47}O_2Si$
17g 325

$C_{32}H_{48}NO_2$
9c2 146

$C_{32}H_{48}N_4O_2^{2\cdot}$
9d2 278

$C_{32}H_{49}O_2$
17c 261, *17e* 102, 108

$C_{32}H_{50}NO_2$
9c1 717

$C_{32}H_{50}NO_3$
17d2 371

$C_{32}H_{50}N_2^+$
9d2 113

$C_{32}H_{50}N_2O_5Si^{2\cdot}$
9d2 250

$C_{32}H_{50}O_5P$
17e 117

$C_{32}H_{51}OSi$
9b 669

$C_{32}H_{51}OSn$
9b 672

$C_{32}H_{54}Cl_2N_2O_2Sn_2$
17g 243

$C_{32}H_{54}CuN_2O_8$
17a 367

$C_{32}H_{54}CuN_2O_8^{2\cdot}$
9d2 260

$C_{32}H_{54}N_2O_6^{2\cdot}$
9d2 272

$C_{32}H_{54}N_2O_{12}^{2\cdot}$
9d2 212

$C_{32}H_{55}Si_2$
9b 491

$C_{32}H_{56}N_8O_6^{2\cdot}$
9d2 207

$C_{32}H_{56}O_4PSe$
17b 240

$C_{32}H_{57}Cl_2MnNO_4P_2$
17a 310

$C_{32}H_{58}N_2O_6^{2\cdot}$
9d2 279

$C_{32}H_{58}N_2O_6S^{2\cdot}$
9d2 280

$C_{32}H_{59}MnNO_4P_2$
17a 311

$C_{32}H_{72}Si_4^-$
17f 31

$C_{33}H_{16}Br_4Cl$
9b 399

$C_{33}H_{20}Br$
9b 399

$C_{33}H_{20}Cl$
9b 398ff, *17c* 28

$C_{33}H_{20}ClN_7^{2\cdot}$
9d2 172

$C_{33}H_{20}FN_7^{2\cdot}$
9d2 172

$C_{33}H_{21}^{2-}$
9d1 817

$C_{33}H_{21}$
9b 396ff, *17c* 28

$C_{33}H_{22}N_8^{2\cdot}$
9d2 172

$C_{33}H_{23}O_2$
17c 361

$C_{33}H_{25}GeOS$
17a 449, *17c* 247

$C_{33}H_{25}GeO_2$
17a 449, 450, *17c* 244

$C_{33}H_{25}OSSi$
17c 246

$C_{33}H_{25}O_2Si$
17c 244

$C_{33}H_{25}O_2Sn$
17a 490, *17c* 245

$C_{33}H_{26}O_2Tl$
17g 331

$C_{33}H_{27}O_5Tl_2^-$
9d1 245, 247
$C_{33}H_{27}O_5Tl_2$
17a 428
$C_{33}H_{28}ClO_2Sn$
17e 59
$C_{33}H_{28}O^-$
9b 441
$C_{33}H_{28}O_5^-$
9b 447
$C_{33}H_{29}OSi$
9b 670
$C_{33}H_{29}O_2P^+$
9b 667
$C_{33}H_{29}O_2PbS$
17a 497, *17c* 295
$C_{33}H_{29}O_2SSi$
17c 291
$C_{33}H_{29}O_2SSn$
17a 490, *17c* 295
$C_{33}H_{29}O_3Si$
17c 290
$C_{33}H_{29}O_3Sn$
17a 491, *17c* 294
$C_{33}H_{30}NiO_2P_2^+$
17g 81
$C_{33}H_{31}NO_2Si$
17d1 300
$C_{33}H_{32}O_2Tl$
17g 150
$C_{33}H_{33}D_{13}NO_3$
17d1 300
$C_{33}H_{33}N_2O_8$
9c1 853, *17d2* 383
$C_{33}H_{33}N_2O_9$
17d2 389
$C_{33}H_{34}F_3O_3Si$
17g 142
$C_{33}H_{35}AsO_3Rh$
17g 175
$C_{33}H_{35}O$
1 63
$C_{33}H_{35}O_2Sn$
17g 181
$C_{33}H_{35}O_3PRh$
17g 176
$C_{33}H_{35}O_6PRh$
17g 176
$C_{33}H_{36}O_2Tl$
17g 150
$C_{33}H_{37}BrOP$
9c2 86
$C_{33}H_{38}NO_7$
9c1 348
$C_{33}H_{40}N_3O_3$
17d1 327
$C_{33}H_{42}D_9$
9b 641
$C_{33}H_{42}NO_3$
17d1 340
$C_{33}H_{42}N_3O_9^{3\cdot}$
9d2 294
$C_{33}H_{45}Ge$
17f 42
$C_{33}H_{45}MoO_4$
17a 273, 274
$C_{33}H_{45}O_4W$
17a 280
$C_{33}H_{45}Sn$
17f 49
$C_{33}H_{46}N_4O_3^{2\cdot}$
9d2 273
$C_{33}H_{47}GeO_6$
17g 162
$C_{33}H_{47}O_2$
9b 500
$C_{33}H_{49}O_2$
17e 163
$C_{33}H_{50}NO_2$
17d1 221
$C_{33}H_{50}NO_3$
1 66, *9c2* 125
$C_{33}H_{51}$
9b 641
$C_{33}H_{51}F_{12}N_3O_8P^{3\cdot}$
9d2 296
$C_{33}H_{51}N_2$
9c1 94
$C_{33}H_{51}N_6O_3^{3\cdot}$
9d2 293
$C_{33}H_{51}O_2$
17c 262
$C_{33}H_{56}NO_4$
9c1 507
$C_{33}H_{56}N_3O_6Si^{3\cdot}$
9d2 296
$C_{33}H_{57}Cl_2MnNO_5P_2$
17d1 378
$C_{33}H_{58}N_2O_2^{2\cdot}$
9d2 192
$C_{33}H_{59}MnNO_5P_2$
17d1 231
$C_{33}H_{59}N_2O$
17d2 254
$C_{33}H_{61}MnNO_5P_2S$
17a 311
$C_{33}H_{63}N_3O_6^{3\cdot}$
9d2 297
$C_{33}H_{63}N_6O_3^{3\cdot}$
17h 324
$C_{33}H_{72}NSi_2$
9b 241
$C_{34}H_{16}O_2^-$
9d1 241
$C_{34}H_{18}^-$
17f 171
$C_{34}H_{20}Cl_2NO_6$
9c1 740
$C_{34}H_{20}S_4^+$
17h 116
$C_{34}H_{22}NO_6$
17d2 360
$C_{34}H_{22}O_2Tl$
17g 332
$C_{34}H_{25}D_6O_4S_2Si$
17c 288
$C_{34}H_{25}D_6S_2Si$
17c 287
$C_{34}H_{21}N_3O_2^-$
9d1 338
$C_{34}H_{22}^-$
17f 172
$C_{34}H_{22}^+$
9d2 14
$C_{34}H_{23}$
9b 398
$C_{34}H_{23}O$
9b 400
$C_{34}H_{24}^-$
17f 172
$C_{34}H_{24}N_{14}O_{12}S_2^{2\cdot}$
9d2 174
$C_{34}H_{25}D_6GeO_4S_2$
17c 292

$C_{34}H_{25}D_6GeS_2$
17c 291
$C_{34}H_{25}D_6O_4S_2Sn$
17c 293
$C_{34}H_{25}D_6S_2Sn$
17c 293
$C_{34}H_{25}O_3Tl_2$
9d1 292, *17a* 428
$C_{34}H_{26}NO_2$
9c1 750
$C_{34}H_{26}N_3S$
9c1 45
$C_{34}H_{26}Si^-$
9d1 897
$C_{34}H_{27}NO_2^-$
9d1 338
$C_{34}H_{27}N_2O$
9c1 581
$C_{34}H_{27}N_2O_3$
9c1 581, 582
$C_{34}H_{27}O_7Tl^-$
9d1 248
$C_{34}H_{27}O_7Tl_2$
17a 429
$C_{34}H_{28}N_8^{2\cdot}$
9d2 177ff
$C_{34}H_{30}MnNO_5P$
9c1 1038
$C_{34}H_{32}^-$
17f 172
$C_{34}H_{32}MnO_5P_2S_3$
17a 311
$C_{34}H_{32}NOSn$
17d1 315
$C_{34}H_{32}NO_7$
9c1 713
$C_{34}H_{32}O_5P_2ReS_3$
17a 324
$C_{34}H_{33}NO_4Tl$
17g 97
$C_{34}H_{34}NOSn$
17d1 300
$C_{34}H_{34}NO_2Si$
17d1 306
$C_{34}H_{34}N_4O_5^+$
9d2 92
$C_{34}H_{34}O_2Tl$
17g 334
$C_{34}H_{35}NO_2P$
17e 119
$C_{34}H_{36}NO$
9c1 761
$C_{34}H_{36}N_4O_5^+$
9d2 93
$C_{34}H_{37}O$
17e 89, 90
$C_{34}H_{38}NO_5$
9c1 368
$C_{34}H_{38}O_3P$
17e 115
$C_{34}H_{39}ClO_3P$
17e 116
$C_{34}H_{39}N_2O_2$
9c1 682
$C_{34}H_{40}Cl_4O_6P$
17e 149, *17g* 126, 169
$C_{34}H_{40}N_2O_2^-$
17e 127, *17g* 378
$C_{34}H_{40}N_3O_3$
9c1 442
$C_{34}H_{41}N_2O_7$
9c1 308
$C_{34}H_{44}NO_2$
17e 173
$C_{34}H_{44}N_3O$
17d2 369
$C_{34}H_{44}N_5O_6$
9c1 53, *17c* 443
$C_{34}H_{44}O_2^{2\cdot}$
17h 259
$C_{34}H_{45}CuN_4O_5$
17a 367
$C_{34}H_{45}N_2O_3Pd$
17g 128
$C_{34}H_{45}N_2O_3Pt$
17g 130
$C_{34}H_{45}O_2Sn$
17a 491
$C_{34}H_{46}^{3-}$
17f 173
$C_{34}H_{46}^-$
9d1 730, *17f* 173
$C_{34}H_{46}^+$
17h 102
$C_{34}H_{47}N_2O_2S$
9c1 67
$C_{34}H_{48}Br_2O_2Pd_2$
17a 350
$C_{34}H_{48}Cl_2O_2Pd_2$
17a 350
$C_{34}H_{48}CuN_4O_6$
17a 368
$C_{34}H_{48}I_2O_2Pd_2$
17a 350
$C_{34}H_{49}O_{10}Zr$
17g 192
$C_{34}H_{51}O_2$
17e 108, 162
$C_{34}H_{52}NO$
17d1 168
$C_{34}H_{52}NOSn$
17d2 189
$C_{34}H_{52}N_2OX$
17d2 93
$C_{34}H_{52}N_3O_5$
9c1 258
$C_{34}H_{53}O_2$
9c2 81, *17c* 261
$C_{34}H_{54}N_2^+$
9d2 114
$C_{34}H_{55}O_4Pb$
17g 127, 170
$C_{34}H_{56}KN_4O_{14}^-$
17a 219
$C_{34}H_{56}O_2^-$
17c 205
$C_{34}H_{58}N_4O_{14}^{2\cdot}$
17h 318, 319
$C_{34}H_{62}I_2N_8O_6^{2\cdot}$
9d2 207
$C_{34}H_{62}MnO_{10}P_2$
17g 168
$C_{34}H_{62}N_2O_6^{2\cdot}$
9d2 280
$C_{34}H_{62}N_6O_6S_2^{2\cdot}$
17h 317
$C_{34}H_{66}N_4O_4^{2\cdot}$
9d2 281
$C_{34}H_{73}Si_2$
17b 296
$C_{35}H_{23}AsMnO_5$
17a 311, *17g* 321, 322
$C_{35}H_{23}AsMoO_5^-$
17a 274

$C_{35}H_{23}AsO_5Re$
17a 324
$C_{35}H_{23}AsO_5W^-$
17a 281
$C_{35}H_{23}AsO_5W$
17g 327
$C_{35}H_{23}MnO_5Sb$
17a 311, *17g* 321
$C_{35}H_{23}O_5PRe$
17a 324
$C_{35}H_{23}O_5PW^-$
17a 281
$C_{35}H_{23}O_5PW$
17g 328
$C_{35}H_{24}F$
17c 87
$C_{35}H_{24}NO_2$
9c1 752
$C_{35}H_{24}N_8^{2\cdot}$
9d2 172
$C_{35}H_{25}^-$
9d1 829
$C_{35}H_{25}$
1 42, *9b* 448, *17c* 86
$C_{35}H_{25}MnO_5P$
17a 312
$C_{35}H_{25}O$
9b 410
$C_{35}H_{28}NO_3$
9c1 751
$C_{35}H_{28}Si^-$
9d1 897
$C_{35}H_{29}O_2Si$
17g 325
$C_{35}H_{30}NO$
9c1 33
$C_{35}H_{31}HgO_2$
17g 141
$C_{35}H_{31}NO_6PS$
17b 289
$C_{35}H_{31}N_3O^-$
9b 447
$C_{35}H_{32}F_9O_2Sn$
17g 138, 183
$C_{35}H_{33}MoNO_3P$
9c1 1042, *17a* 274
$C_{35}H_{34}Cl_2NOSi_2$
17d1 161
$C_{35}H_{34}NO_7$
9c1 234
$C_{35}H_{35}AsO_5Re$
17a 324, *17g* 133
$C_{35}H_{35}MnO_5P$
17a 312, *17g* 124
$C_{35}H_{35}O_5PRe$
17a 325, *17g* 134
$C_{35}H_{35}O_5ReSb$
17a 325, *17g* 136
$C_{35}H_{35}O_8PRe$
17a 325, *17g* 136
$C_{35}H_{37}NO_2Tl$
17g 356
$C_{35}H_{39}ClOPPd$
17a 350, *17e* 105
$C_{35}H_{40}^-$
9d1 798
$C_{35}H_{40}D_5O_2$
17e 168
$C_{35}H_{40}O_3P$
17e 113, 116
$C_{35}H_{41}D_4O_2$
17e 167ff
$C_{35}H_{41}NOP$
9c2 149, 158
$C_{35}H_{43}NO_{10}P$
17d2 125
$C_{35}H_{43}NO_{11}P$
17d2 125
$C_{35}H_{44}ClO_2$
17e 170, 171
$C_{35}H_{44}NO_2Si$
17d1 301, 306
$C_{35}H_{45}O_2$
1 64, *9c2* 98, *17e* 166ff
$C_{35}H_{46}NO_3$
9c2 125
$C_{35}H_{47}N_2O$
17c 444
$C_{35}H_{47}O_2$
17c 354
$C_{35}H_{48}O_3PS$
17b 398, *17e* 117
$C_{35}H_{50}N_4O_3^{2\cdot}$
9d2 273
$C_{35}H_{51}O_2$
17e 164
$C_{35}H_{53}$
9b 391
$C_{35}H_{53}O_2$
17c 263
$C_{35}H_{61}O_3$
17e 93
$C_{35}H_{69}ClNOP_2Pt$
17d2 71
$C_{36}H_{16}Cl_{14}NO_4$
9b 744
$C_{36}H_{18}^{3-}$
9d1 797
$C_{36}H_{20}N_8O_8^{2\cdot}$
17h 261
$C_{36}H_{20}N_8O_{12}S^{2\cdot}$
9d2 174
$C_{36}H_{20}O_2^-$
9d1 304
$C_{36}H_{24}^-$
9d1 831, *17f* 173
$C_{36}H_{24}^+$
17h 107, 108
$C_{36}H_{24}F_2^+$
9d2 19
$C_{36}H_{24}F_3$
17c 87
$C_{36}H_{24}N_2^+$
17h 175
$C_{36}H_{24}N_8O_{10}S^{2\cdot}$
9d2 174
$C_{36}H_{25}NO_2^-$
9d1 565
$C_{36}H_{25}N_4^{2+}$
9d2 88
$C_{36}H_{26}^-$
9d1 830, 831
$C_{36}H_{26}^{2\cdot}$
9d2 153
$C_{36}H_{27}$
9b 398, 449, *17c* 86
$C_{36}H_{27}ClNO_4^+$
1 123
$C_{36}H_{27}N^+$
9d2 52
$C_{36}H_{27}O$
17c 87
$C_{36}H_{28}N_2^+$
9d2 52

$C_{36}H_{28}Si^-$
1 120
$C_{36}H_{29}NO_4^-$
9d1 339
$C_{36}H_{30}Cl_6CuO_7P_2$
17g 363
$C_{36}H_{30}N_2^+$
9d2 59
$C_{36}H_{30}Pb_2^-$
17f 54
$C_{36}H_{30}Si_2^-$
17f 23, 25
$C_{36}H_{30}Sn^-$
17f 52
$C_{36}H_{31}O_4$
9c2 161
$C_{36}H_{32}NO$
9c1 33
$C_{36}H_{32}NO_2$
9c1 33
$C_{36}H_{36}^-$
17f 174ff
$C_{36}H_{36}^+$
17h 106
$C_{36}H_{36}NO_7$
9c1 290
$C_{36}H_{36}N_2O_2^+$
17h 171
$C_{36}H_{36}NiO_2P_2^+$
17g 82
$C_{36}H_{37}OS$
9b 501, 507
$C_{36}H_{37}O_2$
9b 497
$C_{36}H_{37}S_2$
9b 507
$C_{36}H_{40}Cl_2O_4^-$
17e 128, *17g* 379
$C_{36}H_{40}Cl_4O_2^-$
17e 127, *17g* 379
$C_{36}H_{40}N_4O_4^{2\cdot}$
9d2 211
$C_{36}H_{40}N_8^{2\cdot}$
9d2 179ff
$C_{36}H_{41}N_4O_4S$
9c1 57
$C_{36}H_{42}O_3P$
17e 116
$C_{36}H_{44}ClMgN_4O_4$
9a 292
$C_{36}H_{44}ClSi_2$
17f 8
$C_{36}H_{44}N_2^+$
9d2 114
$C_{36}H_{44}N_4Zn^+$
9a 336
$C_{36}H_{44}P_2^+$
17e 312
$C_{36}H_{44}Si_2^-$
17f 25
$C_{36}H_{45}Cl_2N_2$
9b 591
$C_{36}H_{46}ClN_2$
9b 591
$C_{36}H_{46}N_3O_5$
17d2 17
$C_{36}H_{46}O_4^-$
17e 127, *17g* 379
$C_{36}H_{47}$
17b 502
$C_{36}H_{47}N_2$
17c 215
$C_{36}H_{47}O_2$
17e 169, 170
$C_{36}H_{47}O_3$
17e 171
$C_{36}H_{47}S$
9c2 283
$C_{36}H_{50}O_4P$
17b 397, *17e* 117
$C_{36}H_{51}CoO_6$
17a 339
$C_{36}H_{51}Ge$
17f 42
$C_{36}H_{51}Sn$
9c2 316, *17f* 49
$C_{36}H_{54}N_3O_9^{3\cdot}$
9d2 296, *17h* 325
$C_{36}H_{56}N_2^+$
9d2 115
$C_{36}H_{57}N_2O_2$
17d1 336
$C_{36}H_{57}N_6O_3^{3\cdot}$
9d2 297
$C_{36}H_{58}Mo_2N_2O_8P_2^-$
17a 274
$C_{36}H_{58}NO$
17d1 341, *17d2* 219
$C_{36}H_{58}N_2O_8P_2W_2^-$
17a 281
$C_{36}H_{58}N_3O_2$
17d1 342
$C_{36}H_{58}OP$
17e 261, 266
$C_{36}H_{58}O_2P$
17e 261
$C_{36}H_{58}P$
17e 261
$C_{36}H_{58}P_2^-$
17e 312
$C_{36}H_{59}O_3Pb$
17g 170
$C_{36}H_{60}B_2^-$
17f 224
$C_{36}H_{62}MoN_2O_4P_2^-$
9a 310
$C_{36}H_{62}N_4O_{14}^{2\cdot}$
17h 319
$C_{36}H_{66}N_2O_{12}^{2\cdot}$
9d2 283
$C_{36}H_{68}N_4O_8Si^{4\cdot}$
9d2 299
$C_{36}H_{74}MoN_2O_2P_2^-$
9a 308
$C_{37}Cl_{25}$
9b 753
$C_{37}Cl_{27}$
9b 749
$C_{37}H_{15}Cl_{14}NP$
9b 744
$C_{37}H_{27}$
9b 748
$C_{37}H_{29}$
9b 449
$C_{37}H_{29}BClOSi$
17c 229
$C_{37}H_{29}ClO_2PSi$
17c 218
$C_{37}H_{30}BOSi$
17c 228
$C_{37}H_{30}O_2PSi$
17c 218
$C_{37}H_{32}^{2-}$
17f 176
$C_{37}H_{32}^-$
17f 175

$C_{37}H_{34}N_2O_2^-$
9d1 339
$C_{37}H_{36}D_9OSi$
17c 223
$C_{37}H_{37}NOP$
9c2 158
$C_{37}H_{39}OS$
9b 501
$C_{37}H_{39}S_2$
9b 507
$C_{37}H_{40}N_4NiO_6$
17a 344
$C_{37}H_{40}O_5Tl$
17g 333
$C_{37}H_{41}N_5Zn^+$
17a 385
$C_{37}H_{42}N_8^{2\cdot}$
9d2 181
$C_{37}H_{43}O$
17e 90
$C_{37}H_{44}N_2O_7Re$
17a 325
$C_{37}H_{44}O_2Tl$
17g 150
$C_{37}H_{44}O_3P$
17e 116
$C_{37}H_{45}O_2Sn$
17g 85
$C_{37}H_{46}ClOSiSn$
17a 491, *17c* 228
$C_{37}H_{47}N_2O$
17b 309
$C_{37}H_{47}OSiSn$
9b 600, *17a* 491, *17c* 228
$C_{37}H_{48}O_3^{2\cdot}$
17h 257
$C_{37}H_{49}O_2$
17e 171
$C_{37}H_{51}O_2$
17c 355
$C_{37}H_{53}NO_3^{2\cdot}$
17h 265, 277
$C_{37}H_{56}NO_2$
17d1 301
$C_{37}H_{57}NO_3^{2\cdot}$
17h 322
$C_{37}H_{64}N_2O_3^{2\cdot}$
9d2 225
$C_{37}H_{73}NO_8P$
17b 171
$C_{37}H_{74}NO_8P^-$
17b 294
$C_{38}Cl_{28}^-$
17c 353
$C_{38}Cl_{28}^+$
17c 353
$C_{38}H_6D_{20}F_2^{2\cdot}$
9d2 154
$C_{38}H_8D_{20}^{2\cdot}$
9d2 153
$C_{38}H_9D_{20}$
9b 745
$C_{38}H_{18}O_2^-$
9d1 242
$C_{38}H_{21}D_8$
9b 746
$C_{38}H_{24}Cl_4^{2\cdot}$
9d2 154
$C_{38}H_{28}^{2\cdot}$
1 128, *9d2* 152, 153, *17h* 252
$C_{38}H_{28}N_2^{2\cdot}$
9d2 155
$C_{38}H_{28}N_8O_{10}^{2\cdot}$
17h 262
$C_{38}H_{28}O^{2\cdot}$
1 128
$C_{38}H_{29}$
9b 745
$C_{38}H_{30}NOS$
17d2 413
$C_{38}H_{30}N_8^{2\cdot}$
9d2 173, 183
$C_{38}H_{31}$
9b 450
$C_{38}H_{31}NO_6^-$
9d1 339
$C_{38}H_{32}O_3PSi$
17c 218
$C_{38}H_{34}^{2-}$
17f 176
$C_{38}H_{34}^-$
17f 176
$C_{38}H_{38}HgN_2O_2^{2\cdot}$
17h 320
$C_{38}H_{38}NaO_6$
17a 211
$C_{38}H_{38}N_2O_2^{2\cdot}$
17h 320
$C_{38}H_{39}O_2Sn$
17e 184
$C_{38}H_{40}O_2P$
17e 115
$C_{38}H_{41}NO_4P$
9c2 151
$C_{38}H_{41}O_5PRe$
17a 326, *17g* 135
$C_{38}H_{44}N_4NiO_6$
17a 344
$C_{38}H_{44}N_8^{2\cdot}$
9d2 181, 182
$C_{38}H_{45}N_2O_7Re$
17g 133
$C_{38}H_{46}O_2PPd$
17a 350
$C_{38}H_{48}CoN_2O_4$
17g 116
$C_{38}H_{48}GeNO$
17d2 69
$C_{38}H_{48}NOSi$
17d2 68
$C_{38}H_{48}NOSn$
17d2 71
$C_{38}H_{50}^-$
9d1 731
$C_{38}H_{50}^+$
17h 102
$C_{38}H_{50}N_4OP$
9c2 150
$C_{38}H_{50}N_6O_{10}^{2\cdot}$
9d2 254
$C_{38}H_{53}O_2$
17c 355
$C_{38}H_{57}NO_3^{2\cdot}$
17h 265
$C_{38}H_{57}O_7$
17e 93
$C_{38}H_{60}Mo_2N_4O_6P_2^-$
17a 274, *17f* 209
$C_{38}H_{61}N_2O_2$
17d2 405
$C_{38}H_{63}MnNO_5P_2S$
17a 312
$C_{38}H_{68}N_4O_4^{2\cdot}$
9d2 207

$C_{38}H_{70}NO$
17d1 340
$C_{38}H_{94}N_2O_{13}Si_{10}^{2\cdot}$
9d2 282
$C_{39}H_{25}$
9b 400
$C_{39}H_{25}N_7S^{2\cdot}$
9d2 173
$C_{39}H_{28}^{2\cdot}$
9d2 154
$C_{39}H_{29}NO_2^-$
9d1 339
$C_{39}H_{30}FeO_3P_2^+$
17a 331
$C_{39}H_{30}N_8^{2\cdot}$
9d2 173
$C_{39}H_{31}$
9b 725
$C_{39}H_{32}NO$
9c1 290
$C_{39}H_{33}$
9b 450
$C_{39}H_{39}NOP$
9c2 158
$C_{39}H_{41}NOP$
9c2 158
$C_{39}H_{49}O$
9b 497
$C_{39}H_{55}O_2$
17e 164
$C_{39}H_{57}NO_3$
17d1 340
$C_{39}H_{57}N_6O_3^{3\cdot}$
9d2 294
$C_{39}H_{57}O_2$
17c 263
$C_{39}H_{58}N_2O_6Si_3^{2\cdot}$
9d2 259
$C_{39}H_{60}N_9O_3^{3\cdot}$
17h 325
$C_{39}H_{63}N_6O_3^{3\cdot}$
9d2 298
$C_{39}H_{65}P_2S$
17e 262
$C_{40}H_{20}D_{10}^{2\cdot}$
9d2 155
$C_{40}H_{24}^+$
9d2 16
$C_{40}H_{24}O_2S_2^-$
1 119
$C_{40}H_{26}NO$
17d1 206
$C_{40}H_{27}N_2O$
9c1 582
$C_{40}H_{28}N_2O_2^+$
17h 171
$C_{40}H_{30}^{2\cdot}$
9d2 155
$C_{40}H_{30}CrO_4P_2^+$
17a 267
$C_{40}H_{30}N_3S$
9c1 45
$C_{40}H_{30}N_{12}O_{12}^{2\cdot}$
9d2 174
$C_{40}H_{30}Si^-$
9d1 897
$C_{40}H_{32}^{2\cdot}$
9d2 154
$C_{40}H_{32}CuO_2P_2$
17g 316
$C_{40}H_{32}N_2O_2^+$
17h 171
$C_{40}H_{32}N_8^{2\cdot}$
9d2 183
$C_{40}H_{33}N_2O_6S_2$
17d1 223
$C_{40}H_{35}$
9b 451
$C_{40}H_{36}D_{10}N_8^{2\cdot}$
17h 263, 264
$C_{40}H_{40}^-$
17f 177
$C_{40}H_{40}N_3O_5$
9c1 354
$C_{40}H_{41}$
17c 265
$C_{40}H_{41}Sn$
17c 316
$C_{40}H_{42}MnO_7P_2S_3$
17a 312
$C_{40}H_{42}O_7P_2ReS_3$
17a 326
$C_{40}H_{44}^+$
17h 106
$C_{40}H_{44}CuO_2P_2$
17g 120, 159
$C_{40}H_{44}N_2O_5$
17d2 220
$C_{40}H_{44}O_4^-$
17e 128, *17g* 379
$C_{40}H_{46}O_3PPd$
17a 350, *17g* 172
$C_{40}H_{48}^+$
17h 106
$C_{40}H_{48}CuN_4O_4$
17a 368
$C_{40}H_{48}N_2^+$
9d2 47
$C_{40}H_{48}N_3$
9b 535
$C_{40}H_{48}N_8^{2\cdot}$
9d2 182
$C_{40}H_{48}O_2^{2\cdot}$
17h 260
$C_{40}H_{49}O$
17e 91
$C_{40}H_{49}O_2$
17e 122
$C_{40}H_{50}GeNO_2$
9a 298, *9c1* 38
$C_{40}H_{50}NO_2Pb$
9c1 39
$C_{40}H_{50}NO_2Sn$
9c1 38
$C_{40}H_{51}O_7Pb$
17a 498
$C_{40}H_{51}O_7Sn$
17a 492
$C_{40}H_{52}N_2^+$
9d2 115
$C_{40}H_{52}N_4O_2^{2\cdot}$
9d2 211
$C_{40}H_{52}P_2^+$
17e 312
$C_{40}H_{54}O_3^{2\cdot}$
17h 257
$C_{40}H_{55}$
17b 495
$C_{40}H_{60}O_3^+$
9b 498
$C_{40}H_{64}Cu_2N_4O_{14}$
17a 368
$C_{40}H_{67}OP_2$
17e 262

$C_{40}H_{74}MnO_{10}P_2$
17g 168
$C_{40}H_{74}N_2O_6S^{2\cdot}$
9d2 282
$C_{40}H_{74}N_2O_{14}^{2\cdot}$
9d2 283
$C_{40}H_{76}N_4O_8Si_2^{4\cdot}$
9d2 300
$C_{40}H_{78}N_4O_8Si_2^{4\cdot}$
9d2 300
$C_{40}H_{80}Cl_2N_7O_2$
9d1 234
$C_{40}H_{90}Si_5^-$
17f 31
$C_{41}H_{26}F_{10}O_2Tl$
17g 152
$C_{41}H_{29}$
17c 88
$C_{41}H_{30}O_2Tl$
9d1 278, *17a* 429
$C_{41}H_{31}NO_4^-$
9d1 339
$C_{41}H_{34}ClO_2Sn$
17g 149
$C_{41}H_{34}Cl_2O_2Tl$
17g 152
$C_{41}H_{34}F_2O_2Tl$
17g 152
$C_{41}H_{36}BO_2$
17g 140
$C_{41}H_{36}O_2Tl$
9d1 139, *17a* 429ff, *17g* 150
$C_{41}H_{41}CoN_5O_6$
17e 29
$C_{41}H_{41}NO_4P$
9c2 150
$C_{41}H_{44}O_2PPd$
17a 351
$C_{41}H_{46}CuO_2P_2$
17g 120, 159
$C_{41}H_{47}OSn$
17a 492, *17c* 85
$C_{41}H_{48}O_2^{2\cdot}$
9d2 160
$C_{41}H_{49}O_2$
9c2 88, *17e* 169ff
$C_{41}H_{50}N_3O_5PRe$
17a 326, *17g* 135
$C_{41}H_{50}O_3PPd$
17a 351, *17g* 172
$C_{41}H_{50}O_3PPt$
17a 354, *17g* 173
$C_{41}H_{51}O_2$
17c 356, *17e* 108
$C_{41}H_{52}O_3^-$
9d1 326
$C_{41}H_{53}NOP$
9c2 159
$C_{41}H_{53}O_2Sn$
17g 142
$C_{41}H_{53}O_3$
9c2 73
$C_{42}H_{20}O_2^-$
9d1 242
$C_{42}H_{24}O_6P$
17e 209
$C_{42}H_{26}^-$
9d1 787
$C_{42}H_{28}^-$
9d1 791
$C_{42}H_{28}^+$
9d2 15
$C_{42}H_{28}Cl_6CuO_2P_2$
17g 206
$C_{42}H_{30}^-$
9d1 832
$C_{42}H_{30}Br_4CuO_2P_2$
17g 203
$C_{42}H_{30}Cl_4CuO_2P_2$
17g 206
$C_{42}H_{30}Cl_4O_2P_2Pd^+$
9a 327
$C_{42}H_{30}Cl_4O_2P_2Pt^+$
9a 327
$C_{42}H_{30}Cl_4O_2Pd$
9d1 157
$C_{42}H_{30}Cl_4O_2Pt$
9d1 157
$C_{42}H_{32}N_2O_2^+$
17h 171
$C_{42}H_{33}N_3$
9b 132
$C_{42}H_{34}N_{10}O_2^{2\cdot}$
9d2 183
$C_{42}H_{34}N_{12}O_{14}^{2\cdot}$
9d2 175
$C_{42}H_{34}Si_2^-$
17f 23
$C_{42}H_{36}FeO_4P_2^+$
17a 332
$C_{42}H_{36}MnO_4P_2$
17a 312
$C_{42}H_{36}NO_3S$
9c1 297
$C_{42}H_{36}N_2O_2^+$
17h 171
$C_{42}H_{39}NOP$
9c2 157
$C_{42}H_{42}^-$
17f 177
$C_{42}H_{42}O_5P_2Re$
17a 326, *17g* 135
$C_{42}H_{48}B_2^-$
17f 227
$C_{42}H_{49}O_2$
17e 165
$C_{42}H_{51}N_2O_6Pd$
17a 351
$C_{42}H_{51}O_6Pd$
17e 133
$C_{42}H_{52}NO_{15}^-$
17g 304
$C_{42}H_{53}NO_{15}$
17e 207
$C_{42}H_{54}^-$
9d1 731
$C_{42}H_{54}GeNO_2$
9a 298, *9c1* 38
$C_{42}H_{54}NO_2Pb$
9c1 39
$C_{42}H_{54}NO_2Sn$
9c1 39
$C_{42}H_{54}NO_4Sn$
9c1 39
$C_{42}H_{57}AsCl_2O_4$
17e 186
$C_{42}H_{59}O_4$
17e 131
$C_{42}H_{60}AlO_6^{2-}$
17a 404, *17g* 156
$C_{42}H_{60}AlO_6$
17a 404

$C_{42}H_{60}AsO_6$
17e 63, *17g* 156
$C_{42}H_{60}GaO_6^{2-}$
17a 408, *17g* 162
$C_{42}H_{60}GaO_6$
17a 408
$C_{42}H_{60}InO_6^{2-}$
17a 409, *17g* 165
$C_{42}H_{60}InO_6$
17a 409
$C_{42}H_{60}KO_6Sn$
17a 492
$C_{42}H_{60}O_6P$
17e 58, 65, 150, *17g* 126, 169
$C_{42}H_{60}O_6Sn^-$
17g 191
$C_{42}H_{60}O_6Sn$
17a 492
$C_{42}H_{63}Ge$
17f 42
$C_{42}H_{63}N_6O_3^{3\cdot}$
9d2 298
$C_{42}H_{63}Si$
17f 9
$C_{42}H_{66}N_8O_8^{4\cdot}$
9d2 299
$C_{42}H_{67}NO_5P$
9c1 508
$C_{43}H_{27}$
9b 536, 750
$C_{43}H_{30}Cl_5IrO_3$
9d1 157
$C_{43}H_{30}Cl_5IrO_3P_2^+$
9a 326
$C_{43}H_{30}Cl_5O_3P_2Rh^+$
9a 326
$C_{43}H_{30}Cl_5O_3Rh$
9d1 157
$C_{43}H_{30}Cl_5O_3Ru$
9d1 158
$C_{43}H_{31}Cl_2O_2Sn$
17e 186
$C_{43}H_{32}ClO_2Sn$
17e 59
$C_{43}H_{32}O_5Tl_3^{2-}$
9d1 247, 260
$C_{43}H_{32}O_5Tl_3$
17a 430, 431
$C_{43}H_{33}O_2Sn$
17e 51, *17g* 83ff
$C_{43}H_{34}ClGeOSi$
17a 450, *17c* 226
$C_{43}H_{34}ClOSi_2$
17c 225
$C_{43}H_{35}GeOSi$
17a 450, *17c* 225, 226
$C_{43}H_{35}GeOSn$
17a 493, *17c* 226
$C_{43}H_{35}Ge_2O$
17a 451, *17c* 225
$C_{43}H_{35}N_2OP_2$
17e 153
$C_{43}H_{35}OSiSn$
9b 598, 600, *17a* 493, *17c* 226, 228
$C_{43}H_{35}OSi_2$
9b 596, *17c* 224
$C_{43}H_{37}NO_4^-$
9d1 339
$C_{43}H_{39}FeNO_4P_2$
17d1 202
$C_{43}H_{39}FeNO_4P_2^+$
17a 332
$C_{43}H_{40}BO_2$
17g 140
$C_{43}H_{40}BO_4$
17g 140
$C_{43}H_{40}O_2Tl$
17g 151
$C_{43}H_{41}NOP$
9c2 157
$C_{43}H_{42}D_{12}N_3$
9b 535
$C_{43}H_{44}AsO_5PRe$
17a 326, *17g* 135
$C_{43}H_{44}O_5P_2Re$
17a 327, *17g* 134, 135
$C_{43}H_{54}D_6O_3^{2\cdot}$
17h 259
$C_{43}H_{55}D_6O_3$
17e 173
$C_{43}H_{56}O_3$
9d1 327
$C_{43}H_{57}BN_2NaO_5$
17d2 320
$C_{43}H_{57}O_3$
9c2 97
$C_{43}H_{58}D_2O_3^{2\cdot}$
17h 258
$C_{43}H_{59}DO_3^{2\cdot}$
17h 258
$C_{43}H_{59}O_7$
17e 172
$C_{43}H_{60}^{2\cdot}$
9d2 161
$C_{43}H_{60}N_3O_5$
17d1 336
$C_{43}H_{60}O_3^-$
17e 172
$C_{43}H_{60}O_3^{2\cdot}$
17h 257
$C_{43}H_{61}D_2O$
17c 356
$C_{43}H_{61}O_3$
9c2 98, *17e* 172
$C_{43}H_{63}$
9b 724, *17c* 351
$C_{43}H_{63}O_3$
9b 500, *17c* 355
$C_{44}H_8D_{20}N_4Zn^+$
17a 385
$C_{44}H_{28}BClCoF_4N_4$
17a 339
$C_{44}H_{28}B_2CoF_8N_4$
17a 340
$C_{44}H_{28}BrCdN_4$
9a 337
$C_{44}H_{28}BrN_4Zn$
9a 337
$C_{44}H_{28}CdClN_4$
9a 337
$C_{44}H_{28}CdClN_4O_4$
9a 337
$C_{44}H_{28}ClCoN_4^+$
17a 340
$C_{44}H_{28}ClMgN_4$
9a 292
$C_{44}H_{28}ClMgN_4O_4$
9a 292
$C_{44}H_{28}ClN_4O_4Zn^+$
9a 336
$C_{44}H_{28}ClN_4Zn$
9a 336
$C_{44}H_{28}Cl_2CoN_4$
17a 340

$C_{44}H_{28}Cl_2CoN_4O_8$
17a 341
$C_{44}H_{28}CoN_4^{2+}$
9a 325
$C_{44}H_{28}FN_4Zn$
9a 336
$C_{44}H_{28}N_4Ni^{+}$
9a 327
$C_{44}H_{28}N_4Zn^{+}$
17a 385
$C_{44}H_{30}Br_4O_4P_2Ru$
9d1 158
$C_{44}H_{30}Br_4O_4P_2Ru^{+}$
9a 318
$C_{44}H_{30}Cl_4FeO_4P_2^{+}$
17a 332
$C_{44}H_{30}Cl_4O_4P_2Ru^{+}$
9a 318
$C_{44}H_{31}NO_2^{-}$
9d1 339
$C_{44}H_{32}^{2:}$
9d2 155
$C_{44}H_{32}N_4Zn^{-}$
9a 337
$C_{44}H_{34}N_4^{+}$
9d2 92
$C_{44}H_{39}N^{+}$
17h 175
$C_{44}H_{40}N_8^{2\cdot}$
9d2 184
$C_{44}H_{42}O_8P_2Rh_2^{+}$
17a 342
$C_{44}H_{44}CuO_2P_2$
17g 118
$C_{44}H_{44}N_2O_2PPd$
17g 128
$C_{44}H_{46}CuN_4O_4^{2+}$
9a 328
$C_{44}H_{46}NO_5$
9c1 368
$C_{44}H_{46}O_5P_2Re$
17a 327, *17g* 135
$C_{44}H_{50}N$
9c1 8
$C_{44}H_{52}CuO_2P_2$
17g 120, 159
$C_{44}H_{52}NO_3Tl_2$
17g 250, 252

$C_{44}H_{57}NO_5^{2\cdot}$
17h 266, 278, 279
$C_{44}H_{57}O_4S_2$
17e 183
$C_{44}H_{58}CuO_2P_2Si_2$
17g 121
$C_{44}H_{58}O_2Pd_2$
17a 351
$C_{44}H_{60}^{+}$
17h 36
$C_{44}H_{63}N_2$
9b 591
$C_{45}H_{24}O_3^{-}$
17g 380
$C_{45}H_{26}O_3^{-}$
17e 201
$C_{45}H_{30}N_9^{3\cdot}$
9d2 186
$C_{45}H_{33}^{3:}$
9d2 159
$C_{45}H_{35}O$
1 72
$C_{45}H_{35}Sn$
9b 395
$C_{45}H_{36}O_3^{-}$
17g 380
$C_{45}H_{36}O_3$
17e 125
$C_{45}H_{44}O_2Tl$
17g 151
$C_{45}H_{45}D_{12}N_3^{+}$
9b 536
$C_{45}H_{45}N_7S^{2\cdot}$
9d2 173
$C_{45}H_{45}Sn$
17f 49
$C_{45}H_{48}O_3^{-}$
17e 125, *17g* 380
$C_{45}H_{48}O_5P_2Re$
17a 327, *17g* 136
$C_{45}H_{51}$
17c 63
$C_{45}H_{60}O_3^{-}$
9d1 327
$C_{45}H_{61}NO_5^{2\cdot}$
17h 267, 323
$C_{45}H_{61}O_3$
9c2 97

$C_{45}H_{64}O_3^{2\cdot}$
17h 259
$C_{45}H_{65}$
9b 392
$C_{45}H_{69}N_6O_3^{3\cdot}$
9d2 298
$C_{45}H_{69}Sn$
17f 50
$C_{45}H_{80}N_4O_8Si_2^{4\cdot}$
9d2 300
$C_{45}H_{86}N_3O_{12}P^{2\cdot}$
17h 284
$C_{45}H_{87}N_2O_{10}P$
17d2 280
$C_{46}H_{32}N_6OZn^{+}$
17a 385
$C_{46}H_{32}N_{14}O_{12}S_2^{2\cdot}$
9d2 175
$C_{46}H_{34}Cl_2CuO_2P_2$
17g 258
$C_{46}H_{36}CuO_2P_2$
17g 258
$C_{46}H_{36}N_8^{2\cdot}$
9d2 185
$C_{46}H_{38}O_4^{-}$
9d1 88
$C_{46}H_{39}O_2Sn$
17g 84ff
$C_{46}H_{40}AlO_4$
17a 405
$C_{46}H_{40}CaO_4$
17a 232
$C_{46}H_{41}N_2OP_2$
17e 153
$C_{46}H_{43}N_4OP_2Pd$
17d1 325
$C_{46}H_{44}N_2O_2^{+}$
17h 153
$C_{46}H_{46}LiO_4$
17a 203
$C_{46}H_{46}Li_2O_4^{+}$
17a 204
$C_{46}H_{46}NaO_4$
17a 211
$C_{46}H_{46}Na_2O_4^{+}$
17a 211
$C_{46}H_{46}O_4^{-}$
17g 65

$C_{46}H_{49}O_{18}$
9c2 198
$C_{46}H_{54}NO_{11}$
17d2 125
$C_{46}H_{58}CuO_2P_2$
17g 119, 159
$C_{46}H_{66}O_3^{+}$
9b 500
$C_{46}H_{70}N_4O_{12}^{4\cdot}$
9d2 301, *17h* 326, 327
$C_{47}H_{26}F_{15}O_2Sn$
17g 148ff
$C_{47}H_{31}CrO_5P^-$
17a 267
$C_{47}H_{31}CrO_5P$
17g 315
$C_{47}H_{33}N_2$
9c1 59
$C_{47}H_{34}FOP^+$
17c 159
$C_{47}H_{35}GeO$
17a 451, *17c* 85
$C_{47}H_{35}N_2$
1 51, *9c1* 46
$C_{47}H_{35}OSn$
17a 493, *17c* 86
$C_{47}H_{38}F_3O_2Sn$
17g 148
$C_{47}H_{38}FeO_4P_2^{+}$
17a 332
$C_{47}H_{41}AsNO$
9c2 159
$C_{47}H_{41}NOP$
9c2 155
$C_{47}H_{41}O_2Sn$
9d1 140, *17e* 58, *17g* 143, 144
$C_{47}H_{42}NOSn$
17g 378
$C_{47}H_{43}N_2P^+$
9d2 59
$C_{47}H_{45}ClNOP_2Pt$
17d2 71
$C_{47}H_{45}N_4OP_2Pd$
17d1 360
$C_{47}H_{48}O_2Tl$
17g 151
$C_{47}H_{70}NO_3$
9c2 128
$C_{48}H_{24}^{+}$
9d2 18
$C_{48}H_{28}^{+}$
9d2 17
$C_{48}H_{28}N_8O_8S_2^{2\cdot}$
17h 262
$C_{48}H_{30}AgCl_6O_4P_2$
17g 362
$C_{48}H_{30}As_2Cl_6CuO_4$
17g 362
$C_{48}H_{30}Br_6CuO_4P_2$
17g 362
$C_{48}H_{30}Cl_6CuO_4P_2$
17g 363
$C_{48}H_{30}Cl_6O_4P_2Pt$
17g 364
$C_{48}H_{32}N_8O_8^{2\cdot}$
9d2 177
$C_{48}H_{34}MnO_4P_2S_2$
17a 313, *17g* 374
$C_{48}H_{34}N_2O_8P_2W_2^{-}$
17a 281
$C_{48}H_{34}N_{12}O_{12}^{2\cdot}$
9d2 175
$C_{48}H_{36}CuO_2P_2$
17g 341
$C_{48}H_{36}FeO_4P_2^{+}$
17a 332
$C_{48}H_{36}FeO_6P_2^{+}$
17a 333
$C_{48}H_{36}N_4^{+}$
9d2 88
$C_{48}H_{36}Si$
1 120
$C_{48}H_{37}O_2Sn$
17e 59
$C_{48}H_{39}N_{12}^{3\cdot}$
9d2 186
$C_{48}H_{40}Si_4^{-}$
17f 31
$C_{48}H_{41}O_2Sn$
17g 144
$C_{48}H_{43}N_2OP_2$
17e 153
$C_{48}H_{44}N_2O_4^{+}$
17h 168, 172
$C_{48}H_{46}N_{10}O_2^{2\cdot}$
9d2 184
$C_{48}H_{48}MnO_4P_2$
17a 313
$C_{48}H_{52}O_7P_2Re$
17a 327, *17g* 136
$C_{48}H_{63}O_3^{3\cdot}$
9d2 171
$C_{48}H_{64}N_2^{+}$
9d2 47
$C_{48}H_{64}O_3^{2\cdot}$
9d2 161
$C_{48}H_{65}O_3$
9c2 114, *17e* 123
$C_{48}H_{68}P_2^{+}$
17e 312
$C_{48}H_{74}N_2O_8^{2\cdot}$
9d2 270
$C_{48}H_{78}N_4O_9Si_2^{4\cdot}$
9d2 301
$C_{48}H_{84}N_2O_8P_4W_2^{-}$
17a 281
$C_{48}H_{90}N_2O_{18}^{2\cdot}$
9d2 284
$C_{48}H_{90}N_5O_{10}Si_2^{5\cdot}$
9d2 303
$C_{48}H_{92}N_5O_{10}Si_2^{5\cdot}$
9d2 304
$C_{48}H_{93}NO_9P$
9c1 405
$C_{49}H_{13}D_{20}N_5Zn^+$
17a 386
$C_{49}H_{29}Cl_4CoN_5O_2$
17e 22
$C_{49}H_{29}CoF_4N_5O_2$
17e 23
$C_{49}H_{29}CoN_9O_{10}$
17e 23
$C_{49}H_{33}ClCoN_5^{+}$
17a 341
$C_{49}H_{33}CoN_5O_2$
17e 24
$C_{49}H_{33}N_5Zn^+$
17a 386
$C_{49}H_{35}$
1 42
$C_{49}H_{35}N_{13}O_{12}S^{2\cdot}$
9d2 176
$C_{49}H_{53}N_7S^{2\cdot}$
9d2 173

$C_{49}H_{60}O_2PPd$
17a 351

$C_{49}H_{84}N_3O_4$
9c1 442

$C_{49}H_{86}NO_4$
17d2 280

$C_{49}H_{94}N_5O_{10}Si_2^{5\cdot}$
9d2 304

$C_{50}H_{31}Cl_4CoN_5O_2$
17e 24

$C_{50}H_{35}CoN_5O_2$
17e 25

$C_{50}H_{36}Cl_2CuO_2P_2$
17g 316

$C_{50}H_{38}CuO_2P_2$
17g 316

$C_{50}H_{38}F_9O_2Sn$
17g 146ff

$C_{50}H_{38}MoO_4P_2^-$
17a 275

$C_{50}H_{42}NO_{13}$
9c1 348

$C_{50}H_{47}NOP$
9c2 155

$C_{50}H_{47}NO_4P$
9c2 156

$C_{50}H_{47}O_2Sn$
17g 145ff

$C_{50}H_{47}O_5Sn$
17g 148

$C_{50}H_{48}Cl_2CuO_2P_2$
17g 120, 160

$C_{50}H_{50}AgO_2P_2$
17g 114, 154

$C_{50}H_{50}AsCuO_2P$
17g 119

$C_{50}H_{50}As_2CuO_2$
17g 118, 158

$C_{50}H_{50}ClOP_2Pd$
17a 352

$C_{50}H_{50}ClOP_2Pt$
17a 355

$C_{50}H_{50}CuO_2P_2$
17g 119, 160

$C_{50}H_{50}O_2P_2Pt$
17g 130, 174

$C_{50}H_{50}O_8P_2Rh$
17g 177

$C_{50}H_{60}NO_{13}Sn$
17d2 412

$C_{50}H_{61}NO_5^{2\cdot}$
17h 267, 279

$C_{50}H_{65}NO_5^{2\cdot}$
17h 323

$C_{50}H_{80}KO_{12}^+$
17a 219ff

$C_{50}H_{82}N_4O_8Si_2^{4\cdot}$
9d2 301, 302

$C_{51}H_{35}N_8P^{2\cdot}$
9d2 173

$C_{51}H_{50}O_3P_2Rh$
17g 176, 177

$C_{51}H_{64}O_3^-$
17e 123, *17g* 379

$C_{51}H_{65}NO_5^{2\cdot}$
17h 267

$C_{51}H_{78}NO_3$
9c2 129

$C_{51}H_{97}O_6$
17b 171

$C_{51}H_{98}O_6^-$
17b 294

$C_{52}H_{30}Br_4K_2O_2^{2\cdot}$
9d2 168

$C_{52}H_{30}Br_4Li_2O_2^{2\cdot}$
9d2 168

$C_{52}H_{30}Br_4Na_2O_2^{2\cdot}$
9d2 168

$C_{52}H_{34}K_2O_2^{2\cdot}$
9d2 166

$C_{52}H_{34}Li_2O_2^{2\cdot}$
9d2 166

$C_{52}H_{34}Na_2O_2^{2\cdot}$
9d2 167

$C_{52}H_{38}FeO_4P_2^+$
17a 333

$C_{52}H_{38}MoO_4P_2$
17g 322

$C_{52}H_{38}N_{12}O_{12}^{2\cdot}$
9d2 176

$C_{52}H_{40}FeO_4P_2^+$
17a 333

$C_{52}H_{40}MnO_4P_2$
17a 313

$C_{52}H_{40}N_8^{2\cdot}$
9d2 185

$C_{52}H_{40}N_{14}Ru_2^{2-}$
17a 333

$C_{52}H_{50}F_3O_4P_2Pd$
17g 129

$C_{52}H_{50}MnO_4P_2$
17a 313, *17g* 125

$C_{52}H_{50}MnO_{10}P_2$
17a 313, *17g* 125

$C_{52}H_{50}O_4P_2Re$
17a 327, *17g* 137

$C_{52}H_{52}AlO_4$
17a 405

$C_{52}H_{52}AlO_4^+$
17a 405

$C_{52}H_{52}BaO_4$
17a 237

$C_{52}H_{52}CaO_4$
17a 233

$C_{52}H_{52}LaO_4$
17a 241

$C_{52}H_{52}MgO_4$
17a 229

$C_{52}H_{52}O_4Sr$
17a 235

$C_{52}H_{52}O_4Y$
17a 240

$C_{52}H_{74}N_4O_{12}^{4\cdot}$
9d2 302, 303

$C_{52}H_{76}N_4Zn^+$
9a 336

$C_{52}H_{92}N_8O_{14}^{4\cdot}$
17h 327

$C_{53}H_{29}CoN_9O_2$
17e 25

$C_{53}H_{36}O^-$
9b 448

$C_{53}H_{41}CoN_5O_2$
17e 26

$C_{53}H_{41}CoN_5O_6$
17e 26

$C_{53}H_{47}NO_4P$
9c2 155, 156

$C_{53}H_{56}N_4OP$
9c2 156

$C_{53}H_{79}O_2^-$
9d1 221

$C_{53}H_{80}O_2^-$
17g 231

$C_{53}H_{81}O_2$
17e 182
$C_{53}H_{94}N_5O_{10}Si_2^{5\cdot}$
9d2 304
$C_{54}H_{31}CoN_9O_2$
17e 27
$C_{54}H_{36}N_{12}O_{13}P^{3\cdot}$
9d2 187
$C_{54}H_{43}CoN_5O_6$
17e 27
$C_{54}H_{46}NOS_2$
17d2 215
$C_{54}H_{60}^{2:}$
17h 252
$C_{54}H_{61}N_8NiO_2$
17e 128
$C_{54}H_{61}N_8O_2Pd$
17e 129
$C_{54}H_{77}O_3$
17e 92
$C_{54}H_{82}O_4^-$
17g 240
$C_{55}H_{45}CuN_6O_2$
17a 369
$C_{55}H_{53}NO_2P_4Ru^+$
17a 333
$C_{55}H_{53}NO_2P_4RuX$
17d1 203
$C_{55}H_{60}^{2:}$
17h 253
$C_{55}H_{92}N_3O_3$
9c1 509
$C_{55}H_{104}N_6O_{12}Si_2^{6\cdot}$
9d2 305
$C_{56}H_4D_{36}^{2:}$
9d2 151
$C_{56}H_{40}^{2:}$
9d2 151, 156
$C_{56}H_{42}CoN_7O_3$
17e 28
$C_{56}H_{42}K_2O_2^{2\cdot}$
9d2 169
$C_{56}H_{42}Li_2O_2^{2\cdot}$
9d2 169
$C_{56}H_{42}Na_2O_2^{2\cdot}$
9d2 169
$C_{56}H_{46}CuN_5O_3$
17a 369, 370, 371
$C_{56}H_{47}CuN_6O_2$
17a 371, 372
$C_{56}H_{50}CuN_5O_2$
17a 372
$C_{56}H_{50}NOS_2$
17d2 220
$C_{56}H_{52}NO_3P_4Ru^{2+}$
17a 334
$C_{56}H_{54}NO_3$
9c1 348
$C_{56}H_{58}NO_2Sn$
9c1 39
$C_{56}H_{62}N_8NaO_{10}^+$
17a 211
$C_{56}H_{80}BaO_{12}^{2+}$
17a 237
$C_{56}H_{80}CaO_{12}$
17a 233
$C_{56}H_{80}LiO_{12}^+$
17a 204
$C_{56}H_{80}MgO_{12}^{2+}$
17a 230
$C_{56}H_{80}NaO_{12}^+$
17a 212
$C_{56}H_{80}O_4^-$
9d1 327
$C_{56}H_{80}O_{12}Rb^+$
17a 222
$C_{56}H_{80}O_{12}Sr^{2+}$
17a 235
$C_{56}H_{84}N_2^+$
9d2 115, *17h* 153
$C_{56}H_{86}N_5O_{15}^{5:}$
17h 327
$C_{56}H_{104}N_6O_{12}Si_2^{6\cdot}$
9d2 305
$C_{56}H_{106}N_6O_{12}Si_2^{6\cdot}$
9d2 306
$C_{56}H_{112}Cl_2N_7O_2$
9d1 234
$C_{57}H_{41}CoN_5O_6$
17e 28
$C_{57}H_{41}CoN_5O_{10}$
17e 29
$C_{57}H_{51}CuN_6O_2$
17a 373
$C_{57}H_{57}NO_2OsP_4^+$
17a 334
$C_{57}H_{57}NO_2OsP_4X$
17d1 203
$C_{57}H_{57}NO_2P_4RuX$
17d1 203
$C_{57}H_{108}N_6O_{12}Si_2^{6\cdot}$
9d2 306
$C_{58}H_{40}N_{14}O_{12}S_2^{2\cdot}$
9d2 176
$C_{58}H_{56}NO_3P_4Ru^+$
17a 334
$C_{58}H_{80}O_4^{2\cdot}$
9d2 162
$C_{58}H_{81}O_4$
9c2 96, *17e* 179
$C_{59}H_{55}Ge_3OSi$
17a 451, *17b* 303
$C_{59}H_{55}OSi_4$
17b 304
$C_{59}H_{68}O_3^-$
17e 124
$C_{59}H_{90}O_4^-$
9d1 230, *17g* 241
$C_{59}H_{91}O_4$
17e 188
$C_{60}H_{42}^{2:}$
9d2 153
$C_{60}H_{48}N_4O_2^+$
17h 172
$C_{60}H_{50}Si_5^-$
17f 32
$C_{60}H_{51}Sn$
17f 50
$C_{60}H_{66}N_2O_4P_4Ru^{2+}$
17d1 153
$C_{60}H_{68}AlO_4$
17a 406
$C_{60}H_{68}AlO_4^+$
17a 406
$C_{60}H_{68}CaO_4$
17a 233
$C_{60}H_{68}LaO_4$
17a 241
$C_{60}H_{68}MgO_4$
17a 230
$C_{60}H_{68}O_4Sr$
17a 236
$C_{60}H_{68}O_4Y$
17a 240

$C_{60}H_{83}O_4$
17e 165
$C_{60}H_{85}O_4$
17e 121
$C_{60}H_{89}O_4$
17e 108
$C_{60}H_{106}N_6O_{12}Si_2^{6\cdot}$
9d2 307
$C_{60}H_{116}N_4O_4^{2\cdot}$
9d2 276
$C_{60}H_{116}N_6O_{12}Si_3^{6\cdot}$
9d2 307
$C_{61}H_{75}$
9b 749
$C_{62}H_{44}^{2\cdot}$
9d2 152, 154
$C_{62}H_{48}P_2^{+}$
9b 667
$C_{62}H_{122}I_2N_4O_2^{2\cdot}$
9d2 277
$C_{63}H_{46}^{2\cdot}$
9d2 153, 156
$C_{63}H_{83}NO_4^{2\cdot}$
9d2 162, 163
$C_{63}H_{84}NO_4$
17e 177
$C_{63}H_{100}N_5O_{12}Si_3^{5\cdot}$
9d2 304
$C_{64}H_4D_{40}^{2\cdot}$
9d2 157
$C_{64}H_{44}^{2\cdot}$
9d2 157
$C_{64}H_{48}^{2\cdot}$
9d2 156
$C_{64}H_{64}CuO_2P_4$
17g 160
$C_{64}H_{77}D_8O_4$
17e 176
$C_{64}H_{80}D_4O_4^{2\cdot}$
17h 258
$C_{64}H_{81}D_4O_4$
17e 176
$C_{64}H_{84}O_4^{2\cdot}$
9d2 163, 164
$C_{64}H_{85}O_4$
17e 173, 176
$C_{64}H_{87}O_4$
17e 174, 175
$C_{64}H_{122}N_6O_6^{2\cdot}$
9d2 281
$C_{65}H_{59}NO_4P$
9c2 157
$C_{66}H_{48}^{2\cdot}$
9d2 157
$C_{66}H_{52}^{2\cdot}$
9d2 155
$C_{66}H_{48}Cl_4CuO_2P_4$
17g 206
$C_{66}H_{102}N_6O_{18}^{6\cdot}$
17h 328
$C_{67}H_{39}$
9b 536
$C_{68}H_{48}^{2\cdot}$
9d2 156
$C_{68}H_{70}CoN_{10}O_6$
17e 30
$C_{68}H_{86}O_4^{2\cdot}$
9d2 164
$C_{68}H_{87}O_4$
17e 178, 179
$C_{70}H_{48}^{2\cdot}$
9d2 157, 158, *17h* 255
$C_{70}H_{54}CuO_2P_4$
17g 258
$C_{70}H_{64}P_2^{+}$
9b 667
$C_{70}H_{88}O_4^{2\cdot}$
9d2 165, 166
$C_{70}H_{89}O_4$
17e 174, 177
$C_{71}H_{52}^{2\cdot}$
9d2 158
$C_{71}H_{90}O_4^{2\cdot}$
9d2 166
$C_{71}H_{91}O_4$
17e 178
$C_{72}H_{52}^{2\cdot}$
9d2 158, *17h* 255
$C_{72}H_{60}^{2\cdot}$
17h 256
$C_{72}H_{144}Cl_2N_7O_2$
9d1 234
$C_{73}H_{80}N_{16}^{4\cdot}$
9d2 187
$C_{73}H_{107}O_{12}$
17e 93
$C_{74}H_{56}^{2\cdot}$
17h 255
$C_{74}H_{56}CuO_2P_4$
17g 316
$C_{74}H_{68}CuO_2P_4$
17g 120, 161
$C_{76}H_{50}K_2O_2^{2\cdot}$
9d2 170
$C_{76}H_{50}Li_2O_2^{2\cdot}$
9d2 170
$C_{76}H_{50}Na_2O_2^{2\cdot}$
9d2 170
$C_{76}H_{52}^{2\cdot}$
17h 256
$C_{76}H_{89}N_4O_4Zn$
17e 130
$C_{76}H_{91}N_4O_4$
17e 129
$C_{76}H_{93}O_4$
17e 175
$C_{76}H_{144}N_8O_{16}Si_3^{6\cdot}$
9d2 307
$C_{80}H_{56}^{2\cdot}$
9d2 151
$C_{81}H_{57}^{3\cdot}$
9d2 159
$C_{86}H_{76}CuO_2P_4$
17g 121, 161
$C_{86}H_{120}KO_{14}^{+}$
17a 221
$C_{86}H_{120}O_{14}Rb^{+}$
17a 223
$C_{93}H_{123}O_6^{3\cdot}$
9d2 171, *17h* 260
$C_{93}H_{124}O_6^{2\cdot}$
9d2 166
$C_{93}H_{125}O_6$
17e 180
$C_{99}H_{69}^{3\cdot}$
9d2 160
$C_{99}H_{71}$
9b 748
$C_{110}H_{214}N_2O_{49}^{2\cdot}$
9d2 284
$C_{129}H_{171}O_8$
17e 180
$C_{140}H_{178}O_8Si^{2\cdot}$
17h 260

$C_\infty AlN$
9a 78
$C_\infty N$
9a 76ff, 78, *17a* 47ff
$C_\infty NS$
9a 78
$C_\infty NSe$
9a 78
$C_\infty NSi$
9a 78
$C_\infty N_2$
9a 77, *17a* 50
C_nF_{2n-1}
9a 41
C_nO_2Si
17a 84
Ca^+
9a 200, *17a* 117
CaCl
17a 117
CaF
9a 200, *17a* 117
CaH
9a 200
CaI
17a 167
Cd
9a 243
Cd^+
9a 241ff, *17a* 159ff
CdCu
17a 162
CdF
17a 162
CdH
9a 243, *17a* 162
CdHO
17a 162
Cd_2^{3+}
9a 243ff
Cl^{2-}
17a 108
Cl
9a 178ff, *17a* 108
ClD_3Si^-
17a 83
ClF^-
9a 180ff
ClF^+
9a 180
$ClFO^+$
9a 120, 192
$ClFO_3^-$
17a 113
ClF_3OP
9a 154
ClF_3Si^-
17a 85
ClF_4
9a 195
ClF_6
9a 195, *17a* 113
ClH^-
17a 108
$ClHO^-$
9a 111
ClHNa
17a 12
ClH_3Si^-
17a 82
ClHg
9a 260
ClI^-
9a 250
ClO^{2-}
9a 179ff
ClO
1 15, *9a* 179, *17a* 108, *17e* 5
ClOS
9a 165
ClOSe
9a 218
ClO_2
1 16, *9a* 118ff, 187ff,
17a 68, 110ff, *17e* 5, 16
ClO_3^{2-}
9a 194
ClO_3
1 16, *9a* 192ff, *17a* 111ff
ClO_4^{2-}
9a 195, *17a* 112
ClO_4
1 16, *9a* 194ff, *17a* 112
ClO_5
17a 113
ClO_6
9a 195
ClS_2
9a 166
ClXe
9a 252
Cl_2^-
1 17, *9a* 181ff, *17a* 109ff
Cl_2^+
9a 180
$Cl_2Cr_2O_6^-$
17a 121
Cl_2F_2OP
9a 155
$Cl_2F_2Si^-$
17a 85
Cl_2Ga^{2-}
17a 134
Cl_2In^{2-}
17a 164
Cl_2N
9a 64
Cl_2O^+
9a 120
Cl_2OP
9a 147
$Cl_2O_2^-$
17a 111
Cl_2O_2P
9a 153
$Cl_2O_2S^-$
9a 175
$Cl_2O_4^-$
1 19, *9a* 196
$Cl_2O_8^-$
9a 196
Cl_2P
9a 142, 192, *17a* 88
Cl_2Sc
17a 118
Cl_2Sn^-
9a 246
Cl_3^{2-}
17a 111
Cl_3Cu^-
17a 131
Cl_3Ge
9a 208
Cl_3OP^-
9a 153ff

Cl_3Si
9a 139
Cl_4^+
9a 196
Cl_4P
9a 154
Cl_4Si^-
17a 86
$Cl_5N_3P_3$
17a 95
Cl_5P^-
9a 155
$Cl_6N_3P_3^-$
17a 95
Cl_6P^{2-}
9a 155
Cl_6P
9a 155
Cl_6Ru^{3-}
17a 149
$Cl_7N_4P_4$
17a 95
$Cl_8N_4P_4^-$
17a 96
$Co \cdots (CO)_n^{2+}$
17a 127
$CoH_{3n}N_nO_2^{2+}$
9a 119
Cs
1 13, *9a* 253ff, *17a* 169ff
CsF_2
9a 256
$CsNO_7S_2^-$
17a 223
CsO
9a 256
CsO_2
9a 120
CsO_4
17a 170
Cu
9a 202ff, *17a* 129ff
CuHg
17a 177
CuMg
17a 130
CuO
17a 130
CuO_2
17a 69, *17e* 32
CuO_2Si
17a 82
CuO_4
17e 32
CuZn
17a 132
Cu_3
17a 131
Cu_3O_2
17a 69, *17e* 16
Cu_5
17a 132
Cu_5O_2
17a 69, *17e* 16

D (^2H)
1 11, *9a* 13
DO (^2HO)
1 13, *9a* 94ff
DO_2 ($^2HO_2$)
1 13
$D_2H_2PS_2$
17e 267
D_2N
1 13
D_3ISi^-
17a 83

F^{2-}
9a 121
F
9a 120ff, *17a* 74
FHO^-
9a 110
FH_2Si
9a 138
FH_3P
9a 147ff
FH_3Sb
9a 247
FHg
17a 177
FI^-
9a 249
FIO_2^-
9a 252
FKr
9a 227
FKr_2
17a 76
FMg
9a 134, *17a* 79
FO^{2-}
9a 121ff, *17a* 75
FO
9a 121
FOS
17a 100
FO_2
9a 117
FO_2P^-
9a 147, *17a* 90
FO_2S
9a 171, *17a* 103
FO_3P^-
9a 151
FO_3S
9a 174
FO_5S_2
9a 170
FS
9a 158
FSe
9a 214
FSr
9a 231
FXe
1 16, *9a* 252
FYb
17a 171
FZn
17a 132
F_2^-
9a 122ff, *17a* 75ff
F_2HSi
9a 138
F_2H_2P
9a 148
F_2N
1 13, *9a* 64
F_2NO
9a 75, *9c1* 195, *17a* 46, *17d2* 420
$F_2NO_4S_2$
17c 528

F_2NS
17a 98, 103, *17c* 517
F_2O_2P
9a 151
F_2P
9a 141ff
F_2Sc
17a 118
F_2Si^-
17a 82
F_2Y
17a 148
F_3^{2-}
9a 131
$F_3FeN_3O_3P$
17a 123
F_3HP
9a 48
F_3ISi^-
17a 85
F_3N^+
9a 75, *17a* 47, *17h* 131
F_3NO^-
9a 78, *17a* 48
F_3OP^-
9a 151, *17a* 93
F_3OS
9a 174
F_3OSi
9a 140
F_3S
9a 171, *17a* 104
F_3Si
9a 138
F_3Ti
17a 118ff
F_4^{3-}
17a 77
F_4HOP
17a 95
F_4NS
9a 175
F_4NbO
9a 232
F_4OP^{2-}
17a 94
F_4OP
9a 154
F_4O_3PS
9a 152
F_4P
1 15, *9a* 152ff, *17a* 93
F_4Si^-
9a 140
F_5NbO^-
9a 232
$F_5O_3S_2$
9a 170, 176
F_5P^-
9a 155, *17a* 95
F_5S
9a 176, *17a* 106
F_5Sb^-
9a 247
F_5Se
9a 221
F_6Ge^{3-}
17a 136
F_6I
9a 252, *17a* 168
F_6Np
17a 185
F_6Pb^{3-}
17a 184
F_6S^-
9a 176, *17a* 106
F_6Sb^{2-}
17a 166
F_6Se^-
9a 222, *17a* 142
F_6Si^-
9a 140
F_6Sn^{3-}
17a 166
F_6Te^-
9a 248, *17a* 166
F_6Tl^{4-}
17a 182
F_7I^-
9a 252
F_9OS_2
9a 175
$F_{10}NS_2$
17c 521
$F_{11}O_2S_3$
9a 171
$F_{15}O_3S_4$
9a 171

Ga
9a 206ff, *17a* 133ff
Ga^{2+}
9a 206, *17a* 133
GaH_3^-
17a 134
Ge^{3+}
9a 207, *17a* 134
GeH
17a 135
$GeHO_2X_2$
17a 135
GeH_3
9a 208
GeH_3I^-
17a 135
GeO_3X_3
17a 135
GeO_5X_3
17a 69

H
1 9ff, *9a* 7ff, *17a* 7ff
^2H (D)
1 11, *9a* 13, *17a* 10ff
^3H (T)
1 11, *9a* 13ff, *17a* 11
$H \cdots Li^+$
17a 11, 12
$H \cdots Na^+$
9a 12
$H \cdots Pb^{2+}$
9a 13
HHg
9a 260
HHg^{2+}
9a 260
HHgO
9a 111
$HHgO^{2+}$
9a 111
HIO^-
9a 111
HMg
9a 134

$HMgO$
9a 110

HNO^+
9a 109

HNO_2^-
9a 67

HNO_3S^-
9a 55, *17a* 104

HNX_2^+
(X unknown)
1 20

HN_2O_2
17d2 404

HNa^+
9a 134

HO^{2-}
9a 95, *17a* 57

HO
1 13, *9a* 91ff, *17a* 54ff

^2HO (DO)
1 13, *9a* 94ff

HO_2
1 13, *9a* 109ff, *17a* 71, *17e* 16

$^2HO_2$ (DO_2)
1 13

HO_2P^-
1 19, *9a* 143, *17a* 88

HO_2S
9a 167

HO_2Se
9a 218

HO_3P^{3-}
9a 148

HO_3P^-
9a 146, *17a* 89

HO_3S
9a 170

HO_4P^-
9a 150, *17a* 92

HPb^{2+}
9a 267

HS
9a 157, *17a* 97

HSe
9a 214

HSn^{2+}
9a 246

HSr
9a 231

HTe
9a 248

HYb
17a 171

HZn
9a 206, *17a* 132

H_2^-
9a 14

H_2LiO
17a 13

H_2N^{2-}
9a 53, *17a* 48

H_2N
1 13, *9a* 51ff, *9c1* 6, *17a* 35ff, *17c* 388

^2H_2N (ND_2)
1 13

H_2NO
9a 67, *9c1* 195, *17a* 44, *17d2* 199

$H_2NO \cdots HF$
9c1 195

$H_2NO \cdots HOH$
9c1 195

H_2NO_2S
17a 44

H_2NO_3S
1 14, *9a* 67ff

H_2NO_4S
9c1 989

$H_2NO_5S_2$
9c1 1003

$H_2NO_7S_2$
9c1 1004ff

$H_2N_2^-$
17a 48

H_2O^-
17a 70

H_2O^+
9a 109, *17a* 70

H_2O_4P
17a 92

H_2P
9a 141, *17a* 87

H_2S^-
9a 159

$H_2S_2^-$
9a 166

H_2Sc
17a 117

H_2Y
17a 147

H_3ISi^-
17a 83

H_3LiN
17a 13

H_3N^+
1 20, *9a* 65ff, *9d1* 25, *17a* 42ff, *17h* 126

H_3N_2
9a 54, *9c1* 40, *17a* 48, *17c* 439

$H_3N_4S_4$
9a 64

H_3O
9a 120

$H_3O_3P^+$
9a 146

H_3P^+
9a 142

H_3S_2
9a 167

H_3Si
9a 137, *17a* 82

H_3Sn
9a 246

H_4LiO_2
17a 14

H_4N
1 14

$H_4N_2^+$
1 121, *9a* 80, *9d2* 93ff, *17a* 49, *17h* 142

H_4P
9a 147, *17a* 91

H_4PS_2
17e 267

H_4Sn^-
9a 246

H_4Sn^+
17a 165

H_5Si
17a 86

$H_6N_2^+$
17h 129

$H_{10}I_2O_{12}^+$
17a 74
He^+
17a 12
$He^{2+}\mu^-$
17a 7
$^4He^{2+}\mu^-e^-$
17a 7
He_2^+
17a 12
Hg
9a 260
Hg^+
9a 258ff, *17a* 176
Hg_2^{3+}
9a 261

I^{2-}
9a 249
I
9a 249, *17a* 167
IO
9a 249
IO_2
17a 167
IO_3^{2-}
9a 252
IO_4^{2-}
17a 168
IO_4
17a 167
I_2^-
9a 250ff
In
17a 163
In^{2+}
9a 244

K
1 12, *9a* 196ff, *17a* 114ff
KO_2
9a 119
KO_4
17a 116
$K_2NO_7S_2$
17d2 417ff
K_3
17a 116
K_5
17a 116
K_7
17a 116

LaO
9a 257
Li
1 12, *9a* 15, *17a* 12ff
Li_3
17a 14
Li_7
17a 14

Mg^+
9a 134, *17a* 79
MnO_3
17a 121
MnO_4
17a 121
Mu
17a 7

N^{2-}
17a 34
N
1 11ff, *17a* 33
NO
1 14, *9a* 95ff, *17a* 58
$NO \cdots Mg^{2+}$ (?)
17a 58
NO_2^{2-}
1 18, *9a* 63, *17a* 40ff
NO_2
1 14ff, *9a* 56ff, *17a* 37ff
NO_3^{2-}
1 18, *9a* 72ff, *17a* 45ff
NO_3
1 15, *9a* 70ff, *17a* 45
NO_3^{2+}
17a 44
NO_4^{2-}
9a 78
NO_4S^{2-}
9d1 688, *17a* 106
NO_6P_2
17a 41
$NO_6S_2^{2-}$
1 18, *9a* 64
NO_7RbS^-
17a 222
$NO_7S_2^{2-}$
1 19, *9a* 76, *17a* 41
NS
9a 158
NS_7
9a 64
NSi_∞
17a 47
N_2^-
9a 50ff, *17a* 35
N_2^+
17a 34, *17h* 232
N_2O^-
9a 56, *17a* 37
N_2O^+
9a 55ff
$N_2O_2^+$
9a 79, *9d2* 116
$N_2O_3^+$
17h 231
$N_2O_4^{3-}$
17a 49
$N_2O_4^-$
1 18, *9a* 80, *17a* 49
$N_2O_4^+$
17a 49
N_2S^-
17a 98
$N_2S_2^+$
9a 79
$N_2S_3^+$
9a 80, *17a* 107, *17h* 136
$N_2S_{14}^+$
9a 80
N_3^{2-}
9a 55, *17a* 37
N_4^-
1 17, *9a* 79
N_4^+
17h 232
$N_4S_4^-$
1 19, *9a* 177
$N_4S_4^+$
17h 136
$N_5S_4^{2-}$
17a 107

Na
1 12, *9a* 131ff, *17a* 77ff
NaO_2
9a 118
NaO_4
17a 78
Na_3
17a 78
Na_7
17a 78
NbO_4^{2-}
9a 232

O^-
1 17, *9a* 81ff, *17a* 50ff
$O^- \cdots Al$
17a 54
$O^- \cdots O_2$
17a 54
O
1 11
OPS_2^{2-}
17a 90
OPS_3^{4-}
17a 94
OPS_3^{2-}
17a 94
ORb
9a 231
OS^-
9a 158
OS_2^-
9a 164
OSc
9a 201
OSi^+
17a 81
OV^{2+}
17a 119
OY
9a 232
O_2^{3-}
9a 107ff
O_2^-
1 17, *9a* 99ff, *17a* 59ff, *17e* 15ff
$O_2^- \cdots Al^{3+}$
17a 68
$O_2^- \ldots Rh$
17a 70
O_2^+
9a 99, *17a* 59
O_2P^{2-}
9a 141, *17a* 87
O_2P
9a 141
O_2PS^{2-}
17a 90
$O_2PS_2^{4-}$
17a 94
$O_2PS_2^{2-}$
17a 93
O_2Rb
9a 119
O_2S^-
1 19, *9a* 160ff, *17a* 99ff
O_2S^+
9a 160, *17a* 98
$O_2S_2^-$
9a 171
O_2Se^-
9a 216ff, *17a* 140
O_2Si
9a 137
$O_2Sn_2^-$
17a 69
O_2Te^-
9a 248
O_2VL_n (L = ligand)
9a 119
O_3^{3-}
9a 116
O_3^-
1 17, *9a* 113ff, *17a* 71ff
$O_3^- \cdots ClO^-$
17a 73
O_3P^{2-}
1 18, *9a* 143ff, *17a* 88ff
O_3S^-
1 19, *9a* 167ff, *17a* 101ff
$O_3S_2^{3-}$
17a 106
$O_3S_2^-$
9a 174
O_3Se^-
9a 218ff, *17a* 141
O_3Si^{3-}
9a 138
O_3Si^-
9a 138
O_3Te^-
9a 248
O_4P^{4-}
9a 150, *17a* 93
O_4P^{2-}
9a 149ff, *17a* 91ff
O_4Rb
17a 147
O_4S^-
9a 172ff, *17a* 104ff
O_4Sb^{2-}
9a 247
O_4Se^{3-}
9a 220ff, *17a* 142
O_4Se^-
9a 219ff
O_4SiX_2
17a 84
O_4SiX_3
17a 84
O_4Si^{5-}
9a 140
O_4Si^{3-}
9a 139
$O_4SnX_4^{3+}$
17a 164
O_4V^{2-}
9a 201
O_4W^-
9a 257
O_5^-
17a 74
$O_5P_2^{3-}$
1 18
O_5S^-
9a 118, *17a* 68
$O_6S_2^{3-}$
17a 107
$O_6S_4^{3-}$
17a 107
$O_6Si_2^{5-}$
9a 140, *17a* 86
$O_7P_2^{5-}$
9a 150

$O_{10}P_4^-$
9a 155
Os^{5+}
17a 172
Os^{7+}
17a 172

P
1 12
PS_2^{2-}
17a 88
PS_3^{2-}
17a 90
PS_4^{4-}
17a 94
PS_4^{2-}
17a 94
P_4^-
9a 147
$P_4O_{10}^-$
17a 96
$P_4O_{10}^+$
17a 96
$P_4S_{10}^+$
17a 96
Pb^-
17a 184
Pb^+
9a 266, *17a* 183
Pb^{3+}
9a 263ff, *17a* 182ff
Pt^-
17a 173

Rb
1 12, *9a* 228ff, *17a* 146ff

S^-
9a 156ff, *17a* 97
S^+
9a 156
SSc
9a 201
SSe^-
9a 215
SY
9a 232
S_2^-
9a 158ff, *17a* 97ff
S_2^+
9a 158
S_3^-
9a 165ff, *17a* 100ff
$S_4^{\pm}$
9a 177
S_4^+
9a 177
S_5^+
9a 177
S_8^+
9a 178
S_n
1 16, *9a* 178
S_n^+
9a 178
Sb^{2+}
9a 247
Sc_3
17a 118
Sc_{13}
17a 118
Se^-
9a 213ff, *17a* 139
Se^+
17a 139
Se_2^{3-}
9a 215
Se_2^-
9a 215
Se_2^+
17a 143
Se_3^-
9a 218
$Se_4^{\pm}$
9a 222
Si^{3+}
9a 137, *17a* 81
$Si_2R_6^-$
9a 140
$Si_2R_6^+$
9a 140
Si_n
9a 139
Si_n^+
17a 86
$Si_n^{\pm}$
17a 86ff
Sn^+
9a 246, *17a* 165
Sn^{3+}
9a 245ff, *17a* 164
Sn_9^{3-}
17a 166
Sr^+
17a 147
T (^3H)
1 11, *9a* 13ff
Te^-
9a 248
Te_2^{3-}
9a 248
Ti_2^{7+}
17a 119
Tl
9a 262ff, *17a* 179ff
Tl^{2+}
9a 261ff, *17a* 177ff
Tl_2^+
17a 181ff
Tl_2^{3+}
17a 181
Xe_2^+
17a 168

Y_3
17a 148
Yb
9a 267

Zn
9a 206
Zn^+
9a 205, *17a* 132

Radicals derived from high polymers and compounds of biological interest

alimemazine
17h 201ff
caffeine
17h 206
chlorophyll α π
9a 292
chlorophyll α^-, $^{25}Mg^{\pm}$
9a 293
chlorophyll, bacterio-
9a 293
chlorpromazine
17h 199ff
cyclopentadiene, glassy
9b 407
dihydroalloxazine
17h 207
diquat
17h 186
DNA
9b 284
ethylene-trifluoromethyl-nitroxide copolymer
9c1 248, 383
flavin mononucleotide (FMN)
17h 218
glycine peptide
9b 61
guanine
17h 206
isocytosine
17h 204
isothiazine
17h 203
lumiflavin
17h 213
melanin-Al
17a 406
melanin-Cd
17a 389
melanin-In
17a 410
melanin-Mg
17a 230
melanin-Sc
17a 239
melanin-Zn
17a 386
methylviologen
17h 186
nitroxalkylcobanide
9a 326
norborn-5-en-2-yl(1-methyl-cyclopentadienyl) methane
9b 407
oxycobalt-myoglobin
17e 30ff
perazine
17h 203
phenoxazine
17h 191
phenothiazine
17h 192ff
polyacrylamide
9b 113, 217
polyacrylic acid
1 27, *9b* 117, 118
polyacrylonitrile
9b 126
polyalanine
9b 91, 207
polyamide
1 34
polyamide-6
9b 133, 319
polyarylene semiquinone
9d1 236
polyaspartic acid
9b 222
poly-*n*-butylmethacrylate
9b 201
poly-*p*-butylmethacrylate
9b 201
polycaproamide
9b 27, 105
polycaprolactam
9b 18, 50, 61, 133
poly-β-chloroethyl-acrylate
9b 105
polycyclopentadiene
9b 407
poly-2,6-diethylphenylene oxide
9c2 73
poly-2,6-diisopropylphenylene oxide
9c2 74
poly-2,6-dimethylphenylene oxide
9c2 71
poly-2,6-dimethoxy phenoxyl
9c2 173
polyester
9b 143
polyethylacrylate
9b 101, 125, 217
polyethylene
1 25, 39, *9b* 15, 105, 320, *9c2* 18
polyethylene glycol
9b 66
polyethylene terephthalate
9b 143, 312
polymethylmethacrylate
9b 200
polyglutamic acid
9b 222
polyhydroxyproline
9b 263
polyisobutylene
9b 159, 181
polyleucine
9b 181
polymethacrylate
9b 198
polymethacrylic acid
1 36, *9b* 198
polymethacrylnitrile
9b 201
poly-2-methoxyphenol
9c2 59
polymethylacrylate
9b 124, 200
poly-3-methyl-butene-1
9b 185
poly-2-methyl-6-*t*-butyl-phenylene oxide
9c2 160
polymethylene-bistriphenyl-methyl
1 128

polymethylmethacrylate
9b 69, 159
polymethylmethacrylate, deuterated
9b 200
poly-4-methylpent-1-ene
9b 181, 196
polyoxymethylene
9b 66
polypeptide
9b 166
poly-2-phenoxyphenylene oxide
9c2 173
polyphosphate
9a 146, 151
polyproline
9b 18
polypropylene
1 35, *9b* 40, 145, 196, *9c2* 18
polystyrene
1 50
polytetrafluoroethylene
1 39, *9a* 112, *9b* 245ff, *9c2* 20, 21
poly-2,3,5,6-tetramethylphenylene oxide
9c2 178
polyvaline
9b 182
polyvinylacetate
9b 215
promazine
17h 198ff
promethazine
17h 199
riboflavin
17h 217
saccharose
1 27
tetrafluoroethylene-hexafluoropropylene copolymer
9b 238, 239, 246
tetrafluoromethylene
9a 112
tricyclo[5.2.1.$0^{3,6}$]-deca-2,5-diene
9b 430
trimeprazine
17h 201ff
vitamin B_{12}, spin-labelled
9a 326
vitamin $B_{12r}O_2$
17e 31